ENZYME KINETICS

Behavior and Analysis of Rapid Equilibrium and Steady-State Enzyme Systems

IRWIN H. SEGEL

Department of Biochemistry and Biophysics
University of California
Davis, California

UNIVERSITY OF WOLVERHAMPTON
LIBRARY
Acc No. 2018960
CLASS 572.7 SEG
CONTROL 0471303097
DATE 16. SEP. 1996
SITE RS
FS32SAS
D22

Wiley Classics Library Edition Published 1993

A WILEY-INTERSCIENCE PUBLICATION

JOHN WILEY & SONS, INC.

New York • Chichester • Brisbane • Toronto • Singapore

This text is printed on acid-free paper.

Copyright © 1975, by John Wiley & Sons, Inc.

Wiley Classics Library Edition Published 1993

All rights reserved. Published simultaneously in Canada.

Reproduction or translation of any part of this work beyond that permitted by Sections 107 or 108 of the 1976 United States Copyright Act without the permission of the copyright owner is unlawful. Requests for permission or further information should be addressed to the Permissions Department, John Wiley & Sons, Inc.

Library of Congress Cataloging in Publication Data

Segel, Irwin H. 1935–
Enzyme kinetics.

"A Wiley-Interscience publication."
Includes bibliographies and index.
1. Enzymes. I. Title. [DNLM: 1. Enzymes.
2. Kinetics. QU135 S454e]

QP601.S45 574.1′925 74-26546
ISBN 0-471-30309-7 (pbk.)

Printed in the United States of America

10 9 8 7 6 5 4 3 2

To Leigh

For her tireless editorial assistance, expert artwork, keen professional advice, saintly patience, and excellent cooking, all of which made this book possible.

PREFACE

A thorough kinetic analysis is an essential part of the characterization of any enzyme. Yet, too often such studies are either omitted from research publications or are presented in a very superficial manner. This is unfortunate but somewhat understandable. Many biologists study enzymes but are intimidated by the apparent complexity of enzyme kinetics. I wrote *Enzyme Kinetics* for two purposes: to introduce the varieties of enzyme behavior to advanced undergraduates and graduate students in the biological sciences and to serve as a useful, accessible reference work on enzyme kinetics for professional researchers.

Enzyme Kinetics is a large book because my aim is to teach the subject and not just to present a multitude of equations. Thus the book is written in the same highly detailed, step-by-step fashion as a textbook and I have deliberately devoted as much attention to elementary and classical kinetics of unireactant enzymes as I have to modern steady-state kinetics of multireactant enzymes. I have tried to provide readers with a basic understanding of the procedures used to convert an idea for a model into a velocity equation (which can then be tested experimentally) and, conversely, the procedures for developing a model from experimental data. Simple equilibrium concepts are used to explain the basis of cooperative, concerted, and cumulative feedback inhibition, regulation by "energy charge", the major models of allosteric enzymes, and the effects of pH and metal ion activators. After seeing how multiple replot graphical analysis can be used to analyze simple rapid equilibrium systems, the reader should have no difficulty in following the same procedures as applied to more complicated steady-state and isotope exchange systems.

Since my objective is to describe kinetic behavior and analysis *in general*, specific enzymes are not discussed except where they represent a unique example of a mechanism. (Nevertheless, specific examples are given in the references.) Presteady-state transient kinetics is not covered because few biologists have access to the necessary rapid reaction instruments, and, also, books on this aspect of enzyme behavior are already available.

This book could not have been written without the cooperation, encouragement, and contributions of many people. Foremost are the scientists whose published works were the basis for the book. My colleagues in the

Department of Biochemistry and Biophysics and our students and post-doctorals contributed by providing me with an intellectual atmosphere conducive to the task and by helping me when I got stuck. I am grateful to Ms. Carolyn Sherwood, the world's greatest typist, who never missed a deadline. I thank my editor, Dr. Theodore P. Hoffman, of Wiley-Interscience for his confidence in me and for the guidelines that he set, and Ms. Joan Samuels of the Production Division, who calmly handled all the crises.

IRWIN H. SEGEL

Davis, California
1974

CONTENTS

THE AUTHOR

Irwin H. Segel was born on Staten Island, New York, in 1935. He attended Brooklyn Technical High School and Rensselaer Polytechnic Institute (B.S. in Chemistry, 1957). His undergraduate research on bacterial glucose metabolism was done under the direction of Professor Henry L. Ehrlich. Two summers were spent studying mammalian niacin metabolism with Dr. Lawrence V. Hankes at Brookhaven National Laboratory. His graduate research on sulfur metabolism of fungi was directed by Professor Marvin J. Johnson at the University of Wisconsin. After receiving his Ph.D. degree in 1962, Dr. Segel spent two years as a National Science Foundation and U.S. Public Health Service Fellow with Professor Jacques C. Senez at the Centre National de la Recherche Scientifique, Marseille, France. In 1964, he joined the faculty of the Department of Biochemistry and Biophysics at the University of California at Davis where he is now Professor of Biochemistry. His current research is on the enzymes of polysaccharide metabolism, inorganic sulfur and nitrogen metabolism, and on membrane transport systems of bacteria and fungi. Dr. Segel teaches introductory biochemistry and a course in enzyme kinetics. His first book *Biochemical Calculations* (John Wiley & Sons, 1968) is used as a supplementary text at many colleges and universities in this country and has been translated into Japanese and Spanish.

CHAPTER FOUR

RAPID EQUILIBRIUM PARTIAL AND MIXED-TYPE INHIBITION

CHAPTER FIVE

ENZYME ACTIVATION

CHAPTER SIX

RAPID EQUILIBRIUM BIREACTANT AND TERREACTANT SYSTEMS

CHAPTER SEVEN

MULTISITE AND ALLOSTERIC ENZYMES

CHAPTER EIGHT

MULTIPLE INHIBITION ANALYSIS

CHAPTER NINE

STEADY-STATE KINETICS OF MULTIREACTANT ENZYMES

CHAPTER TEN

ISOTOPE EXCHANGE

CHAPTER ELEVEN

EFFECTS OF pH AND TEMPERATURE

ENZYME KINETICS

CHAPTER ONE

INTRODUCTION—ENZYMES AS BIOLOGICAL CATALYSTS

A. THE DISCOVERY OF ENZYMES AND THE DEVELOPMENT OF ENZYMOLOGY

It is hard to pinpoint the exact discovery of enzymes. Cell-free activity was observed as early as 1783 when Spallanzani noted that meat was liquefied by gastric juice of hawks. In the following years, numerous similar observations were made. For example, in 1814 Kirchhoff observed that a "glutinous" (i.e., proteinaceous) component of wheat was capable of converting starch to sugar. Robiquet and Boutron and also Chaland discovered the hydrolysis of amygdalin by bitter almonds in 1830. Leuchs, in 1831, described the diastatic action of salivary ptyalin. The first discovery of an enzyme is usually credited to Payen and Persoz, who, in 1833, treated an aqueous extract of malt with ethanol and precipitated a heat-labile substance which promoted the hydrolysis of starch. They called their fraction "diastase," from the Greek word for *separation*, since their material separated soluble sugar from starch. Today we recognize that the diastase of Payen and Persoz was an impure preparation of amylase. The next enzyme to be partially purified was from an animal source. In 1834 Schwann described pepsin, and in 1836 he extracted the active agent with acid from the stomach wall. It is noteworthy that the first observations of enzyme activity preceded a clear notion of catalysis. Berzelius, in 1835 to 1837, described this unknown force, which by its mere presence could "exert its influence and arouse affinities and reactivities in the other complex bodies thereby causing a rearrangement of the constituents of the complex body." This concept of catalysis evolved from observations of the action of diastase and pepsin, and from the seemingly similar action produced by yeast during fermentation. In all cases, one substance was changed to another under the influence of an active agent—the catalyst. It was not yet recognized that yeast were living cells. In 1838, Cagniard de Latour showed that fermentation was caused by living organisms, an idea confirmed and extended by Pasteur between 1858 and 1871. Pasteur regarded the chemical changes occurring in fermentation as an essential part of the life processes of the microorganisms involved. The chemists of the day, most notably Stahl and Liebig, favored a purely chemical theory of fermentation. A distinction was made between the "organized ferments," such as those catalysts presumed to be present in or on the surface of yeast and lactic acid bacteria, and the "unorganized ferments," such as diastase and pepsin whose activities were clearly not associated with microorganisms. Liebig's theory visualized a ferment as a chemical substance produced by a decomposing organism. The atoms of the ferment were supposed to be in "ceaseless movement, constantly changing their position." This highly agitated state was somehow transmitted to the atoms of a sugar molecule "whose elements are held together by weak forces." As a result, the sugar breaks down to compounds (CO_2 and ethanol) whose atoms are held together more tightly. In 1860, Berthelot macerated yeast and obtained an alcohol-precipitable fraction which converted sucrose to glucose plus fructose. He concluded that this invertase (as the active agent was called) was one of many ferments present *in* yeast. In 1878, Kühne suggested the name *enzyme* (meaning "in yeast") for both organized and unorganized ferments. The suffix "*ase*" was proposed by Duclaux in 1898. The end of the Pasteur-Liebig controversy came in 1897 when Hans and Edouard Buchner were able to extract from yeast a cell-free juice which carried out the complete fermentation of sugar. The Buchner brothers were primarily interested in obtaining a yeast juice for therapeutic purposes. Since their preparation was intended for human consumption, it could not be preserved with the usual bacteriocidal agents. An assistant suggested that they add a large quantity of sucrose, since it was known that the growth of microorganisms was inhibited by high sugar concentration. Upon adding the sugar, the yeast juice bubbled vigorously as ethanol and CO_2 were produced. In the same year, Bertrand observed that some enzymes required dialyzable factors for catalytic activity. He named these substances *coenzymes*. By 1900, the catalysts of cellular oxidation were recognized as enzymes.

During the early part of the twentieth century, serious attempts were made to purify enzymes and describe their catalytic activity in precise mathematical terms. In 1902, Henri and Brown independently suggested that an enzyme-substrate complex was an obligate intermediate in the catalytic reaction. Their suggestion was based on the type of curve obtained when the initial velocity of the reaction was plotted against the substrate concentration and was in agreement with the lock-and-key concept proposed by Emil Fischer in 1894 to account for the high degree of specificity exhibited by enzymes. Henri derived a mathematical equation to account for the effect of substrate concentration on the velocity. The effect of pH on enzyme activity was pointed out by Sörensen in 1909. In 1913, Michaelis and Menten rediscovered the equation derived by Henri 11 years earlier. The Henri-Michaelis-Menten equation was based on simple chemical equilibrium principles. In 1925, Briggs and Haldane introduced the steady-state concept to enzyme kinetics. Today, both approaches are used to explain the kinetic properties of enzymes.

The fact that enzymes are proteins was not accepted until the late 1920's (although as early as 1877 Traube suggested that all cellular activities including fermentation, respiration, and putrefaction were catalyzed by substances allied to proteins). In 1926, Sumner crystallized the enzyme urease, but many argued that the enzyme was simply an impurity adsorbed onto or occluded within the protein crystals. However, during the 1930's, Northrop and co-workers crystallized pepsin, trypsin, and chymotrypsin, and demonstrated conclusively that the protein crystals were pure enzymes. By 1943, about 25 enzymes had been crystallized.

In the 1940's and 1950's hundreds of new enzymes were discovered, and many of them purified to homogeneity and crystallized. Dozens of key metabolic pathways were elucidated, and biochemists started focusing on the mechanisms of enzyme activity and regulation. Genetics and biochemistry joined to produce the field of molecular biology. New chemical and physical techniques were used to purify proteins and probe their structures. In 1955, Sanger reported the complete amino acid sequence of insulin, a small protein of molecular weight 6000. Five years later, the first enzyme (ribonuclease, molecular weight 13,700) was sequenced, and finally, in 1969, the first chemical synthesis of an enzyme (ribonuclease) was achieved. In 1957, Kendrew deduced the three-dimensional structure of myoglobin from X-ray diffraction studies.

Up until the 1950's most studies on the kinetics of enzyme activity were based on the Henri-Michaelis-Menten or Briggs-Haldane equations for unireactant enzymes. From the mid-1950's to the early 1960's attempts were made to analyze the kinetics of bireactant and terreactant enzymes. Equations based on the rapid equilibrium assumptions of Henri, Michaelis, and Menten could be derived quite easily, but many enzymes did not follow rapid equilibrium kinetics. Equations based on steady-state concepts were derived by Dalziel, Alberty, Hearon, and others, but, in general, these were rather complex and were not expressed in the familiar terms of K_m, K_i, and V_{max}. In 1963, Cleland presented a clear, uniform procedure for writing kinetic equations for multireactant steady-state enzyme systems together with a convenient shorthand nomenclature for describing the kinetic mechanisms (Chapter Nine). In 1965, Monod, Wyman, and Changeux presented a kinetic model for *allosteric enzymes* (regulatory enzymes which displayed sigmoidal rather than hyperbolic velocity curves). A year later, Koshland, Nemethy, and Filmer presented an alternate model based on the flexible enzyme-induced fit model of Koshland (1959). (These and other models are described in detail in Chapter Seven.)

Before embarking on our survey of enzyme kinetics, let us first examine exactly what an enzyme does.

B. LIFE, ENERGY, AND COUPLED REACTIONS

Chemical reactions can be classified as exergonic (energy-yielding) or endergonic (energy-requiring). A unique property of a living cell is its ability to couple exergonic and endergonic reactions and, thereby, grow and reproduce at the expense of its environment. It does not take a knowledge of thermodynamics to recognize that growth is an endergonic process. Living cells are composed of an organized assemblage of fragile macromolecules, each with a highly specific structure. It takes energy to build large molecules from small molecules, that is, work must be done to build the complex structures of proteins, nucleic acids, cell membranes, and such, from the basic building blocks. Indeed, the resulting structures are so fragile that work must be done continually just to maintain the integrity of the cell. The energy for this work is derived from exergonic reactions. Thus if we had to summarize "life" in a series of simple equations, we could write:

$$\text{A–B} \longrightarrow \text{A} + \text{B} + \text{energy} \qquad \Delta G' \text{ is negative} \qquad \text{(I-1)}$$

$$\text{E} + \text{F} + \text{energy} \longrightarrow \text{E–F} \qquad \Delta G' \text{ is positive} \qquad \text{(I-2)}$$

Reaction I-1 represents the catabolic reactions that occur in a living cell while reaction I-2 represents the anabolic (biosynthetic) reactions. The energy released by reaction I-1 must be made available for the endergonic reaction I-2; that is, the two reactions cannot take place randomly in different parts of the cell, but rather, they must somehow be coupled as in reaction I-3.

$$\begin{array}{ccc} \text{E + F} & \xrightarrow{\hspace{6em}} & \text{E—F} \\ & \text{A—B} \curvearrowright \text{A + B} & \end{array} \qquad \text{(I-3)}$$

The overall biosynthesis of E–F at the expense of A–B is more likely to proceed via a series of steps:

$$A-B + X \longrightarrow A\sim X + B \tag{I-4}$$

$$A\sim X + Y \longrightarrow A + X\sim Y \tag{I-5}$$

$$X\sim Y + E \longrightarrow E\sim X + Y \tag{I-6}$$

$$E\sim X + F \longrightarrow E-F + X \tag{I-7}$$

$$\text{Sum} \quad A-B + E + F \longrightarrow E-F + A + B \tag{I-8}$$

In reaction I-4, A–B is cleaved and a portion of the energy made available is used to condense A with X to yield a transient activated A. The potential energy of A$\sim$X is conserved when X is transferred to Y, producing a mobile, energy-rich X$\sim$Y (e.g., ATP). The condensation of E and F occurs in two steps: first E is activated to yield E$\sim$X, and then the potential energy of E$\sim$X is discharged by the formation of E–F.

C. ENZYMES AS CATALYSTS

The fact that a reaction has a negative ΔG does not mean that it will proceed at a detectable rate. A negative ΔG means that the existing [P]/[S] ratio is less than that at equilibrium. The rate at which the reaction approaches equilibrium cannot be deduced from the magnitude or sign of ΔG. For example, the oxidation of glucose to $CO_2 + H_2O$ has a $\Delta G°$ of $-686{,}000$ cal/mole; that is, glucose in the presence of oxygen is unstable in a *thermodynamic* sense. Yet, glucose does not immediately oxidize to $CO_2 + H_2O$ in the presence of oxygen. Thus glucose is quite stable in a *kinetic* sense. It is obvious that some barrier exists even for so-called spontaneous reactions. The barrier is the *activation energy* that is required. This is illustrated in Figure I-1 where we see that the reaction S→P has a negative ΔG, but before a molecule of S can become a molecule of P, it must possess a certain minimum energy to pass into an activated transition state, $S\cdot\cdot P^{\ddagger}$. The activated state represents a sort of halfway point where the bonds of S are distorted sufficiently so that conversion to P becomes possible. Molecules of S that attain less than the minimum energy simply fall back to the ground state. The rate at which S is converted to P depends on the number of molecules that make it to the transition state per unit time. Glucose is stable in air at room temperature because virtually none of the molecules are sufficiently activated. There are two ways of accelerating the reaction S→P. One is to raise the temperature until a significant number of S molecules attain the transition state. Another way is to lower the activation energy.

Living cells exist at relatively low temperatures—between 0°C and 100°C. At life temperatures few, if any, of the exergonic and endergonic reactions of intermediary metabolism would occur at a rate sufficient to permit cell maintenance and growth. Furthermore, even if a living cell could increase its temperature sufficiently, it would have no way of specifically increasing the temperature of one reaction relative to another. Living cells can operate under relatively mild environmental conditions because they possess *enzymes, which selectively lower the energies of activation* of the vital chemical reactions. In the presence of the appropriate enzyme, the ambient temperature provides a substantial fraction of the reactant molecules with the required activation energy. Enzymes, then, are *catalysts* which speed up the rate of a chemical reaction without themselves being consumed. In the process, enzymes act as mediators of the coupled reactions that constitute metabolism (e.g., reactions I-4 to I-7 would each be catalyzed by a specific enzyme). The equilibrium constant for a reaction is unaltered. Only the rate at which the reaction proceeds toward equilibrium is affected by an enzyme. For example, in the reaction $S \underset{k_{-1}}{\overset{k_1}{\rightleftharpoons}} P$, k_1 might be 10^{-3} min^{-1} while k_{-1} might be 10^{-5} min^{-1}. At equilibrium the forward and reverse velocities are equal. Therefore:

$$v_f = k_1[S]_{eq} = v_r = k_{-1}[P]_{eq}$$

The equilibrium constant for the reaction (defined as $[P]_{eq}/[S]_{eq}$) is:

$$K_{eq} = \frac{k_1}{k_{-1}} = \frac{10^{-3}}{10^{-5}} = 100$$

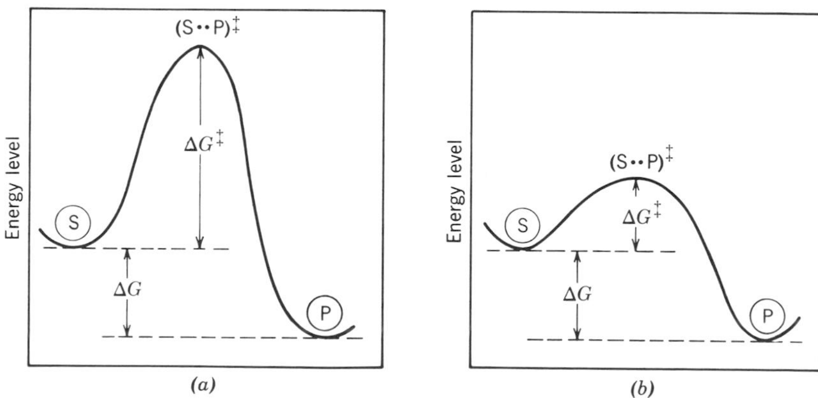

Fig. I-1. ΔG and $\Delta G^{\ddagger}$ of (*a*) nonenzymatic and (*b*) enzymatic reactions.

In the presence of an appropriate enzyme, both k_1 and k_{-1} are enhanced to the same degree. Thus k_1 might increase 10,000-fold to $10\,\text{min}^{-1}$; k_{-1} must also increase 10,000-fold to $10^{-1}\ \text{min}^{-1}$; and K_{eq} is unchanged.

$$K_{eq} = \frac{10}{10^{-1}} = 100$$

D. THE ACTIVE SITE

The enzyme-catalyzed production of P from S can be written:

$$S \xrightarrow{E} P$$

but it was recognized early that the enzyme and the reactant (hereafter called the substrate) must combine in some way during the course of the reaction. The overall catalytic sequence can be written:

$$S + E \longrightarrow ES \longrightarrow E + P$$

(with an arrow returning from E + P to E)

The existence of an enzyme-substrate complex, ES, was inferred from (*a*) the high degree of specificity exhibited by enzymes (Fischer, 1894), (*b*) the shape of the velocity versus substrate concentration curve (Brown, 1902; Henri, 1902), and (*c*) the fact that substrates frequently protected enzymes from inactivation (O'Sullivan and Tompson, 1890). The high degree of specificity of enzymes prompted Emil Fischer in 1894 to suggest the *template* or *lock-and-key* analogy of enzyme-substrate interaction. This relationship, shown schematically in Figure I-2, assumes that the enzyme possesses a region (called the substrate binding site, the active site, or the catalytic site) that is complimentary in size, shape, and chemical nature to the substrate molecule. Thus only a single substance, or, at most, a limited range of substances can bind to the enzyme and act as substrates. Only when the substrate is anchored in the site can it undergo the chemical change that converts it to the product. Today, we recognize that the active site need not be a rigid geometrical cavity or pocket, but rather a very specific and precise spatial arrangement of amino acid residue R-groups that can interact with complimentary groups on the substrate.

All enzymes are proteins with molecular weights in the tens of thousands or greater. Most substrates are low molecular weight substances. (The latter statement is true even for enzymes that accept high molecular weight polymers as substrates. The polymer itself is not recognized as the substrate, but rather, a specific region or bond of the polymer.) Thus only a small

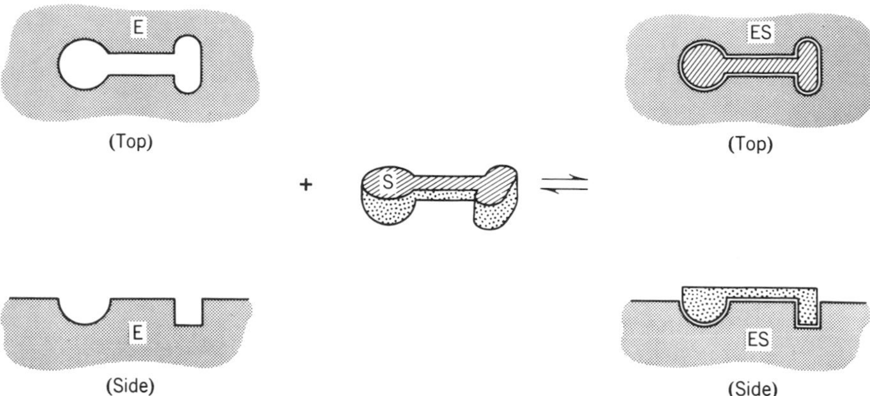

Fig. I-2. Lock and key (template) hypothesis of enzyme specificity.

fraction of the enzyme is actually involved in catalysis. To put it another way, the active site occupies only a very small fraction of the enzyme. In fact, there may be less than a dozen amino acid residues surrounding the absorption pocket of the active site, and, of these, only two or three may actually participate in substrate binding and/or catalysis. Why, then, are enzymes large proteins as opposed to small tripeptides or dodecapeptides? The answer is obvious when we consider that the two or three essential R-groups must be perfectly juxtaposed in three-dimensional space. A linear tripeptide might contain all the essential binding and catalytic groups, but the fixed bond distances and relatively fixed bond angles would not allow the essential R-groups to assume the required spatial relationship. With a large protein composed of a hundred or more amino acids the polypeptide chain could bend, twist, and fold back upon itself and in this way the positions of the three essential R-groups could be fixed exactly in space. Figure I-3 shows the tertiary structure of a hypothetical enzyme. The shaded area represents the absorption pocket of the active site while A, B, and C represent three essential R-groups that contribute to substrate binding and catalytic activity. The A and B might be only two or three residues apart but amino acid C might be 50 residues away from B. Even if the absorption pocket is lined with 12 amino acid residues, a dodecapeptide could never bend into the proper shape. Although only three amino acid residues are involved in the activity of our hypothetical enzyme, it is obvious that a great many of the other residues play an equally important role: that of maintaining the protein in its tertiary structure (via electrostatic interactions, hydrogen bonds, disulfide bonds, hydrophobic interactions, and dipole-dipole interactions).

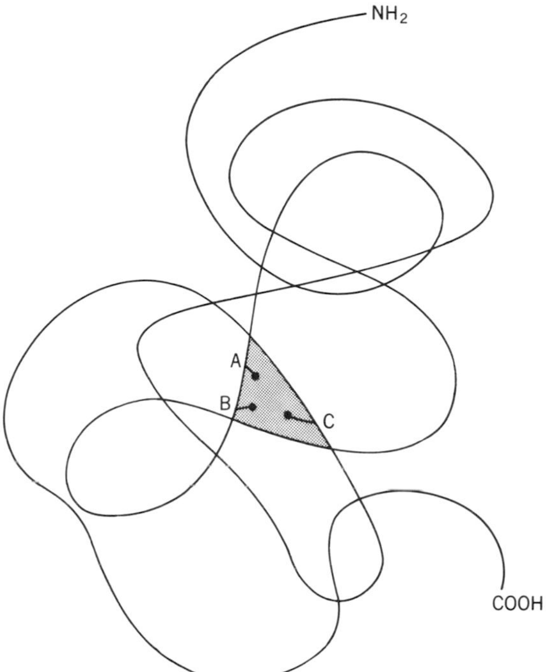

Fig. I-3. The active site (shaded area) occupies only a small region of the enzyme. A, B, and C are the amino acid R-groups responsible for substrate binding and catalytic activity.

E. THREE-POINT ATTACHMENT

The combination of enzyme and substrate can be even more specific than we might deduce from the lock-and-key concept. For example, alcohol dehydrogenase catalyzes the reaction:

$$CH_3-\overset{\overset{\text{(H)}}{|}}{\underset{\underset{H}{|}}{C}}-O\text{(H)} + NAD^+ \rightleftharpoons CH_3CHO + NADH + H^+$$

A given alcohol dehydrogenase always transfers the same methylene hydrogen to NAD^+ and vice versa. This high degree of stereospecificity can be explained if it is assumed that ethanol binds to the enzyme by a three-point attachment through the methyl group, the OH group, and one hydrogen of the methylene group (Fig. I-4*a*). Actually, only the methyl group and the OH group need be bound. When these two groups are bound by the complimentary sites on the enzyme only one of the two methylene hydrogens (always the same one) will be situated next to the NAD^+ site. Aconitase exhibits a similar specificity. In this case the substrate, citric acid, is a symmetrical molecule, yet the OH group is always transferred in the same direction.

$$\begin{matrix} H_2C-COOH \\ | \\ HO-C-COOH \\ | \\ H_2C-COOH \end{matrix} \rightleftharpoons \left[\begin{matrix} H_2C-COOH \\ | \\ C-COOH \\ \| \\ HC-COOH \end{matrix}\right]-E-H_2O \rightleftharpoons \begin{matrix} H_2C-COOH \\ | \\ H-C-COOH \\ | \\ H-C-COOH \\ | \\ OH \end{matrix}$$

The specificity can be explained if citric acid binds to the enzyme via a three-point attachment as shown in Figure I-4*b* and the catalytic group lies in the region of the A–B sites.

F. THE FLEXIBLE ENZYME-INDUCED FIT HYPOTHESIS

Although the lock-and-key hypothesis successfully explained the great majority of specificity patterns exhibited by enzymes, there were several phenomena that the hypothesis could not explain. For example, compounds that resembled the normal substrate chemically but possessed less bulky groups often failed to react, yet they certainly should have fit the template. Compounds with more bulky groups often failed to react (as expected), yet they were found to bind tightly to the enzyme. Many bireactant enzymes would not bind substrate B before substrate A, yet according to the lock-and-key hypothesis, the binding site for substrate B on the free enzyme should be accessible to B. These and other considerations led Koshland to propose the *flexible enzyme* or *induced fit* hypothesis. The hypothesis states that the substrate induces a conformational change in the enzyme that results in a precise alignment of the catalytic groups with the susceptible bonds on the substrate (Fig. I-5*a*, *b*). Substrate analogs with larger (Fig. I-5*c*) or smaller (Fig. I-5*d*) groups may bind to the enzyme, but may not induce the proper alignment of the catalytic groups. In ordered bireactant systems, substrate A is assumed to induce a conformational change that exposes the binding site for substrate B. There is considerable physical evidence supporting the induced fit hypothesis including (*a*) substrate-induced changes in the chemical reactivity of certain R-groups of the enzyme and (*b*) substrate-induced changes in the fluorescence and absorbance properties of certain R-groups. The concept of

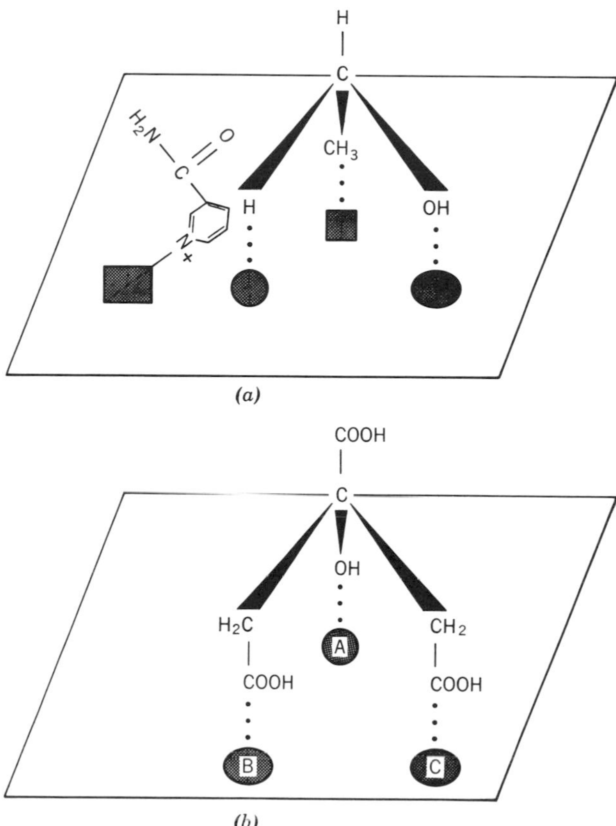

Fig. I-4 Three point attachment hypothesis to account for stereospecific catalysis. (*a*) Alcohol dehydrogenase. (*b*) Aconitase.

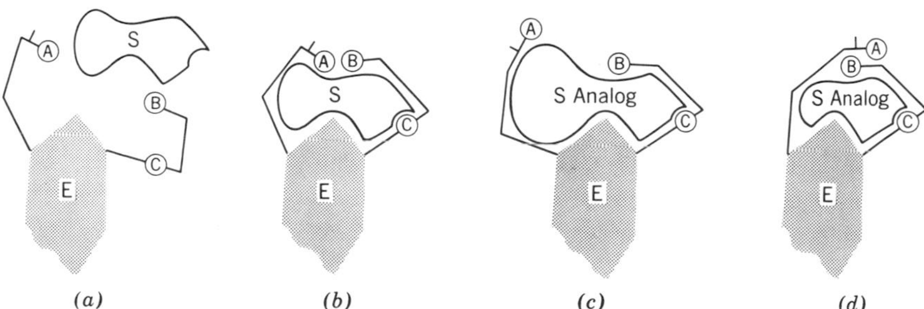

Fig. I-5. Induced-fit hypothesis of Koshland. (*a*) The substrate approaches the active site. (*b*) Substrate binding induces the proper alignment of the catalytic groups, A and B. (*c*) and (*d*) Substrate analogs (competitive inhibitors) bind to the enzyme (aided by group C) but the catalytic groups are not aligned properly. [Redrawn from Koshland, D. E. Jr., *Cold Spring Harbor Symposium on Quantatative Biology*, **28**, 473, (1963).]

a flexible enzyme is the basis for several theories of allosterism (Chapter Seven).

G. FACTORS RESPONSIBLE FOR THE CATALYTIC EFFICIENCY OF ENZYMES

Enzymes have a phenomenal ability to accelerate the rates of reactions. When it is possible to compare the nonenzymatic and enzymatic rates, we find that enzymes enhance the reaction rate by a factor of up to 10^{15}. By stating that "enzymes enhance reaction rates by lowering the activation energy required," we can discuss enhancement factors in a quantitative way, but the real question remains: How do enzymes lower the activation energy? A number of factors have been suggested which we can examine in a very qualitative way. First of all, it is generally agreed that most enzyme-catalyzed reactions proceed via recognized organic reaction mechanisms (e.g., general acid-base catalysis, nucleophilic and electrophilic displacements) in which the enzyme provides the catalytic groups. Certainly, some of the rate enhancement by an enzyme stems from *proximity and orientation* factors. In order for two substrate molecules (or a substrate molecule and a catalytic group) to react they must get close enough to each other and the approach must occur at the proper angle. In solution, the random motion of the two molecules would yield a low probability of an effective collision. When the two molecules are adsorbed onto the active site of the enzyme (or when one of the reactants is the substrate and the other reactant is an R-group of the active site), then both the intermolecular distance and orientation may be optimized. (The effective concentration of substrates in the volume of the active site is considerably greater than that in the solution from which they were adsorbed). A second-order intermolecular reaction between A and B in solution becomes a first-order intramolecular reaction when both A and B are enzyme-bound. Koshland and co-workers have proposed that the active sites of enzymes are so constructed that they align the orbitals of the substrate and catalytic groups optimally to enter the transition state. This concept of *orbital steering* is very similar to one proposed in 1960 by Bruice and Pandit. A qualitatively similar concept of *stereopopulation control* has been discussed by Milstien and Cohen. These workers point out that the combined effect of multipoint attachment and the precise fit of the substrate into the active site would tend to restrict the rotational freedom of the substrate and "freeze" it into a unique conformation. Also, substrates confined to the active site of an enzyme may have a relatively *long residence time* (compared to the time interval that the same substrates would be within striking distance of each other if they were in random motion in solution).

As a consequence of this *substrate anchoring* (as termed by Reuben), the number of substrate molecules attaining the activated transition state per unit time may be increased tremendously.

The idea that certain bonds of the substrate are distorted upon binding to the enzyme has been suggested by several workers. This so-called *rack mechanism* assumes that the substrate fits loosely into the active site, but the bonds that are formed between the enzyme and the substrate are so strong that a susceptible bond within the substrate is distorted producing an activated transition state (Fig. I-6). In this mechanism, a portion of the activation energy is provided directly by the binding forces between E and S. Although the mechanism is illustrated for a cleavage reaction, similar distortion models can easily be imagined for condensations and group transfer reactions in which the transition states of the substrates are more tightly bound to the enzyme than the unactivated substrates.

In the rack mechanism, the substrate molecule distorts to accommodate itself to a static enzyme. In 1947, Fano proposed that the substrate may be distorted by conformational changes in the enzyme. He visualized substrate binding as occurring in two steps (Fig. I-7). First, one part of the substrate binds to one site, then normal thermal motion brings the second site into a position that permits another part of the substrate to bind. As a result of the two-point binding of substrate, the enzyme is locked momentarily into a low entropy conformation. If the enzyme-substrate bonds are quite strong, then the substrate will be distorted to a transition state when the enzyme molecule swings back to its "open" conformation.

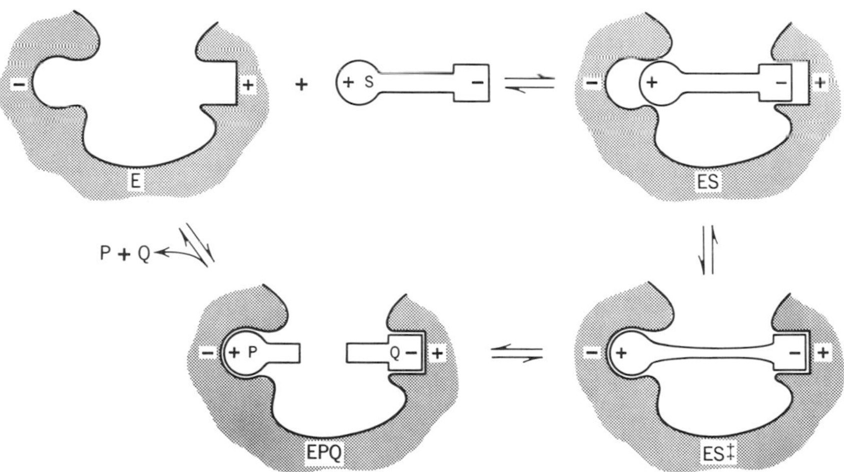

Fig. I-6. Distortion or rack mechanism.

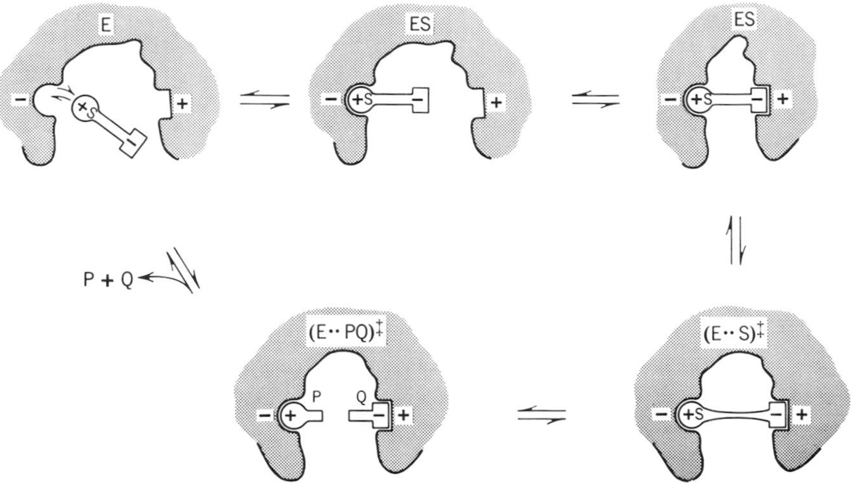

Fig. I-7. Two-step binding sequence leading to substrate distortion.

In spite of the many proposed mechanisms of rate enhancement and their success in accounting for the activity of some enzymes, we still understand very little about the factors responsible for the tremendous catalytic efficiency of the vast majority of enzymes.

H. ENZYME KINETICS

The sole function of an enzyme is to catalyze a reaction. *Enzyme kinetics* is that branch of enzymology that deals with the factors affecting the rates of enzyme-catalyzed reactions. The most important factors are: enzyme concentration, ligand concentrations (substrates, products, inhibitors, and activators), pH, ionic strength, and temperature. When all these factors are analyzed properly, it is possible to learn a great deal about the nature of the enzyme. For example, by varying the substrate and product concentrations, it is possible to deduce the *kinetic mechanism* of the reaction, that is, the order in which substrates add and products leave the enzyme. Such studies establish the kinds of enzyme-substrate and enzyme-product complexes that can form and thereby tell us something about the architecture of the active site. In some cases the kinetics of a reaction provide evidence for stable, covalently-bound intermediates that are undetectable by ordinary chemical analyses. Certain kinetic constants can be determined and from these we can make an educated guess concerning the usual intracellular concentrations of substrates and products and the physiological direction of the reaction. The kinetics of a reaction may indicate the way in which the activity of the enzyme is regulated *in vivo*. A study of the effect of varying pH and temperature on the kinetic constants can provide information concerning the identities of the amino acid R-groups of the active site. A kinetic analysis can lead to a model for an enzyme-catalyzed reaction and, conversely, the principles of enzyme kinetics can be used to write the kinetic equation for an attractive model. The kinetic equation tells us exactly how all the ligands of a system interact to affect the velocity of the reaction. Consequently, once we have a possible equation, the model can be tested experimentally. For many biologists, a thorough understanding of enzyme kinetics is indispensable to their research. A knowledge of the basic principles of enzyme kinetics is useful even to the average biologist who only occasionally assays an enzyme. With these principles, he will be able to design experiments and tell, for example, whether he is dealing with a single enzyme, or whether his preparation contains multiple enzymes that catalyze the same reaction. He will be able to tell whether his enzyme preparation contains inhibitors or activators. By comparing the kinetic constants of two seemingly-identical enzymes from different tissues, or from the same tissue at different stages of development, it is possible to decide whether the two enzymes are indeed the same gene product, or whether they are distinct proteins that catalyze the same reaction. For many biologists, the subject of enzyme kinetics provides no more than a satisfying intellectual experience while others are convinced that the maze of algebra is not worth going through. The following chapters are designed to act as a step-by-step guide through the maze. The basic principles and their application to specific types of enzyme systems are described in detail. An attempt is made to prove that the kinetic equations and graphical analyses are not as formidable as they seem at first glance. It is worth mentioning here that many of the velocity (kinetic) equations presented in Chapters Two through Eight and in Chapter Eleven are, in fact, equilibrium ligand binding equations. These equations, with only slight modification, are directly applicable to studies of drug and hormone binding to receptor sites, the interaction of inducers with repressor proteins, and the binding of substrates to membrane transport proteins.

REFERENCES

Early Studies of Enzymes

Brown, A. J., *Trans. Chem. Soc. (Lond.)* **81**, 373 (1902).

Fischer, E., *Berichte* **27**, 2985 (1894).

Henri, V., *Acad. Sci., Paris* **135**, 916 (1902); *Lois generales de l'action des diastases*, Herman, (1903).

Michaelis, L. and Menten, M. L., *Biochem. Z.* **49**, 333 (1913).

O'Sullivan, C. and Tomson, F. W., *J. Chem. Soc. (Lond.)* **57**, 834 (1890).

See also references at end of Chapter Two.

Bioenergetics and Coupled Reactions

Segel, I. H., *Biochemical Calculations*, John Wiley & Sons, 1968, Ch. IV and Appendix II.

Three-Point Attachment

Ogston, A. G., *Nature* **162**, 963 (1948).

The Flexible Enzyme-Induced Fit

Koshland, D. E., *Proc. Nat. Acad. Sci.* **44**, 98 (1958).

Koshland, D. E., *J. Cell. Comp. Physiol.* **54** (Suppl. 1), 245 (1959).

Koshland, D. E., *Cold Spring Harbor Symposium on Quantitative Biology* **28**, 473 (1963).

Factors Affecting the Catalytic Efficiency of Enzymes—Specific Articles

Bruice, T. C. and Pandit, U. K., *Proc. Nat. Acad. Sci.* **46**, 402 (1960).

Bruice, T. C., Brown, A. and Harris, D. O., *Proc. Nat. Acad. Sci.* **68**, 658 (1971).

Bruice, T. C., *Nature* **237**, 335 (1972).

Dafforn, A. and Koshland, D. E., *Proc. Nat. Acad. Sci.* **68**, 2463 (1971).

Fano, U., *J. Chem. Phys.* **15**, 845 (1947). Also see Reiner, J. M., *Behavior of Enzyme Systems*, Van Nostrand Reinhold, 1969, Ch. 11.

Jencks, W. P. and Page, M. I., *Biochem. Biophys. Res. Commun.* **57**, 887 (1974).

Koshland, D. E., Carraway, K. W., Dafforn, G. A., Gass, J. D. and Storm, D. R., *Cold Spring Harbor Symposium on Quantitative Biology* **36**, 13 (1971).

Lienhard, G. E., *Science* **180**, 149 (1973).

Marshall, J. J., *Eur. J. Biochem.* **33**, 494 (1973).

Milstien, S. and Cohen, L. A., *Proc. Nat. Acad. Sci.* **67**, 1143 (1970).

Reuben, J., *Proc. Nat. Acad. Sci.* **68**, 563 (1971).

Storm, D. R. and Koshland, D. E., *Proc. Nat. Acad. Sci.* **66**, 445 (1970).

Sykes, B. D. and Dolphin, D., *Nature* **233**, 421 (1971).

Tonnelat, J., *Biochimie* **56**, 21 (1974).

Covalent Enzyme-Substrate Intermediates

Bell, R. M. and Koshland, D. E., *Science* **172**, 1253 (1971).

General References

Dixon, M., "The History of Enzymes and of Biological Oxidations," in *The Chemistry of Life—Lectures on the History of Biochemistry*, Needham, J., ed., Cambridge, 1970, Ch. 2.

Gray, C. J., *Enzyme-Catalyzed Reactions*, Van Nostrand Reinhold, 1971.

Jencks, W. P., *Catalysis in Chemistry and Enzymology*, McGraw-Hill, 1969.

Segal, H. J., "The Development of Enzyme Kinetics," in *The Enzymes*, 2nd ed. Boyer, P. D., Lardy, J. and Myrbäck, K., eds., Vol. 1, Academic Press, 1959.

Medical Aspects of Enzymology

Bondy, P. K. and Rosenberg, L. E., *Duncan's Diseases of Metabolism*, Parts I and II, 7th ed., W. B. Saunders, 1974.

Goodley, E. L., *Diagnostic Enzymology*, Lea & Febiger, 1970.

Harris, H., *The Principles of Human Biochemical Genetics*, North Holland-American Elsevier, 1970.

Stanbury, B., Wyngaarden, J. B. and Fredrickson, D. S., *The Metabolic Basis of Inherited Disease*, 3rd ed., McGraw-Hill, 1972.

Wolf, P. W., Williams, D. and Von der Muehll, E., *Practical Clinical Enzymology and Biochemical Profiling*, Wiley-Interscience, 1973.

CHAPTER TWO

KINETICS OF UNIREACTANT ENZYMES

A. THE HENRI EQUATION AND THE MICHAELIS-MENTEN EQUATION

During the late nineteenth century, many workers sought to explain the progress of reactions involving enzymes in terms of the known principles of equilibrium and mass action. Most of the work at that time was concerned with measurements of substrate and product concentrations during the course of sucrose and starch hydrolyses, from which apparent rate constants were calculated. The first general rate equation for reactions involving enzymes was derived in 1903 by Henri. Henri's equation accounted for the observation that the initial rate of a reaction was directly proportional to the concentration of enzyme preparation, but increased in a nonlinear manner with increasing substrate concentration up to a limiting maximum rate. The derivation of Henri's equation was based on the assumptions that:

1. The enzyme is a catalyst (proposed in 1835 to 1837 by Berzelius).
2. The enzyme and substrate react rapidly to form an enzyme-substrate complex (proposed in 1902 by Brown).
3. Only a single substrate and a single enzyme-substrate complex are involved and the enzyme-substrate complex breaks down directly to form free enzyme and product.
4. Enzyme, substrate, and the enzyme-substrate complex are at equilibrium; that is, the rate at which ES dissociates to E + S is much faster than the rate at which ES breaks down to form E + P.
5. The substrate concentration is very much larger than the enzyme concentration so that the formation of an ES complex does not alter the substrate concentration.
6. The overall rate of the reaction is limited by the breakdown of the ES complex to form free enzyme and product.

7. The velocity is measured during the very early stages of the reaction so that the reverse reaction is insignificant.

The assumption that only the early components of the reaction are at equilibrium is called the quasi-equilibrium or rapid equilibrium assumption. The overall reaction was visualized as:

$$E + S \underset{k_{-1}}{\overset{k_1}{\rightleftharpoons}} ES \xrightarrow{k_p} E + P$$

The Henri equation is shown below.

$$\boxed{v = \frac{K[S]}{1 + \dfrac{[S]}{K_S}}} \qquad \text{(II-1)}$$

where $[S]$ = a fixed substrate concentration

v = initial velocity (the instantaneous velocity, $d[P]/dt$ or $-d[S]/dt$) at the given substrate concentration. In practice, v can be taken as $\Delta[P]/\Delta t$ or $-\Delta[S]/\Delta t$ provided the appearance of P is linear with time for the duration of the assay and no more than 5% of the original [S] is utilized.

k_p = rate constant for the breakdown of ES to E + P

K_S = the dissociation constant of the ES complex $= k_{-1}/k_1 = [E][S]/[ES]$

K = a constant, characteristic of the particular enzyme preparation $= k_p[E]_t/K_S$, where $[E]_t$ is the total concentration of enzyme, $[E] + [ES]$

Ten years later Michaelis and Menten confirmed Henri's experimental work and presented a slightly modified version of the rate equation.

$$v = \frac{k_p[E]_t[S]}{K_S + [S]} \qquad \text{(II-2)}$$

If $v = k_p[ES]$, then $k_p[E]_t$ can be taken as V_{max}, the limiting maximal velocity that would be observed when all the enzyme is present as ES. This gives the familiar "Michaelis-Menten" equation (II-3). As pointed out by a number of people, it would be appropriate to refer to the equation as the Henri-Michaelis-Menten equation.

$$\boxed{\frac{v}{V_{max}} = \frac{[S]}{K_S + [S]}} \qquad \text{(II-3)}$$

The derivation of the Henri-Michaelis-Menten equation from rapid equilibrium considerations is given below. The general procedure can be used to obtain velocity equations for all rapid equilibrium systems, including those involving multiple ligands.

1. Write the reactions involved in the overall conversion of S to P. For the simple reaction involving one ligand, one catalytic site, and one enzyme-substrate complex, the reactions are:

$$E + S \overset{K_S}{\rightleftharpoons} ES \overset{k_p}{\longrightarrow} E + P$$

2. Write the mass balance (conservation) equation expressing the distribution of the total enzyme, $[E]_t$, among the various species. For the simple reaction the equation is:

$$\boxed{[E]_t = [E] + [ES]} \tag{II-4}$$

3. Write the velocity-dependence equation. This equation states that v is equal to the concentrations of all product-forming species, each multiplied by its catalytic rate constant. When there is only one product-forming species, ES, the equation is:

$$\boxed{v = k_p[ES]} \tag{II-5}$$

4. Divide the left-hand term of the velocity-dependence equation by $[E]_t$, and the right-hand term by $[E]_t$ expressed as the sum of all the enzyme species:

$$\boxed{\frac{v}{[E]_t} = \frac{k_p[ES]}{[E] + [ES]}} \tag{II-6}$$

5. Express the concentration of each enzyme species in terms of free E. This is accomplished by rearranging the expressions for the various equilibria. For the simple reaction, there is only one equilibrium.

$$K_S = \frac{[E][S]}{[ES]}, \qquad \therefore \quad [ES] = \frac{[S]}{K_S}[E]$$

6. Substitute the expressions for each complex, in terms of [E], into equation II-6.

$$\frac{v}{[E]_t} = \frac{k_p \frac{[S]}{K_S}[E]}{[E] + \frac{[S]}{K_S}[E]} \qquad \text{(II-7)}$$

Or, canceling [E] and designating $k_p[E]_t$ as V_{max}:

$$\frac{v}{V_{max}} = \frac{\frac{[S]}{K_S}}{1 + \frac{[S]}{K_S}} \qquad \text{(II-8)}$$

The proportion of the total enzyme present as any one species is:

$$\frac{[E]}{[E]_t} = \frac{[E]}{[E]+[ES]} = \frac{[E]}{[E] + \frac{[S]}{K_S}[E]} = \frac{1}{1 + \frac{[S]}{K_S}}$$

$$\frac{[ES]}{[E]_t} = \frac{[ES]}{[E]+[ES]} = \frac{\frac{[S]}{K_S}[E]}{[E] + \frac{[S]}{K_S}[E]} = \frac{\frac{[S]}{K_S}}{1 + \frac{[S]}{K_S}} = Y_S = \frac{v}{V_{max}}$$

The ratio of occupied to total sites, Y_S, is equivalent to v/V_{max}.

Equation II-8 is a useful form of the velocity equation. The numerator contains one term expressing the fact that there is only one product-forming species. The denominator contains two terms expressing the fact that there are a total of two species (free E and one complex). Equation II-8 can be converted to the usual form of the Henri-Michaelis-Menten equation by multiplying the numerator and denominator of the right-hand part by K_S:

$$\frac{v}{V_{max}} = \frac{[S]}{K_S + [S]} \qquad \text{(II-9)}$$

The rapid equilibrium treatment described above permits us to express [ES] in terms of [E], [S], and K_S. The derivation gives nothing more than an *equilibrium* expression for the binding of the substrate to the enzyme. We obtain a *kinetic* expression (i.e., a velocity equation) only when we insert the expression for [ES] into the velocity dependence equation, or when we equate $[ES]/[E]_t$ to v/V_{max}. When k_p is of the same order of magnitude as k_{-1} (as is true for many, and very likely most, enzymes), the concentration of ES is no longer fixed solely by the concentrations of E and S and the equilibrium constant, K_S. In this case, the velocity equation must be derived by the more exact steady-state treatment (described below and in more detail in Chapter Nine). Nevertheless, we will make extensive use of the rapid equilibrium treatment for the following reasons. (*a*) The rapid equilibrium treatment is the simplest and most direct technique for deriving velocity equations in the absence of any prior knowledge of the relative magnitudes of the rate constants. The rapid equilibrium treatment permits us to write velocity equations for complex multiligand systems after simple inspection of the equilibria between the various enzyme species. If the experimental data fit the velocity equation, then we have the simplest kinetic mechanism for the system. If the data do not fit, then we can proceed to more complex models and velocity equations. (*b*) For many situations, the rapid equilibrium and steady-state treatments yield the same final velocity equation. That is, the *form* of the velocity equation is the same but the definitions of the constants are not the same. (*c*) The major theories of allosteric enzymes are based on rapid equilibrium assumptions. A steady-state treatment yields velocity equations that are, at present, too complex for practical consideration.

B. GENERAL RULES FOR WRITING VELOCITY EQUATIONS FOR RAPID EQUILIBRIUM SYSTEMS

Before proceeding further, it would be useful to point out the ease with which velocity equations for seemingly complex systems can be obtained if rapid equilibrium conditions prevail. No derivation is really necessary. In fact the velocity equation for any rapid equilibrium system can be written directly from an inspection of the equilibria between enzyme species. We start by writing the velocity dependence equation:

$$v = k_p[ES] + bk_p[ESA] + ak_p[ESI] \qquad \text{and so on}$$

where [ES], [ESA], [ESI], and so on are the concentrations of those species which yield product while k_p, bk_p, ak_p, and such are the respective catalytic rate constants. Next, both sides of the velocity dependence equation are divided by $[E]_t$, where $[E]_t$ on the right-hand side is expressed as the sum of the concentrations of all species: $[E]+[ES]+[EA]+[EI]+[ESA]+[ESI]$, and so on.

$$\frac{v}{[E]_t}=\frac{k_p[ES]+bk_p[ESA]+ak_p[ESI]+\cdots}{[E]+[ES]+[EA]+[EI]+[ESA]+[ESI]+\cdots} \qquad \text{(II-10)}$$

Now all we need do is express the concentration of each species in terms of [E]. The term for any given complex is composed of a numerator and a denominator. The numerator is the product of the concentrations of all ligands in the complex. The denominator is the product of all dissociation constants between the complex and free E. Suppose I is a mixed-type inhibitor and A is a nonessential activator. (Mixed-type inhibition and nonessential activation are described in Chapters Four and Five, respectively. For the moment, all we need to know is that the equilibria between enzyme species can be represented as shown below.)

$$\begin{array}{ccccccccc}
EA & + & S & \overset{\beta K_S}{\rightleftarrows} & ESA & \overset{bk_p}{\longrightarrow} & EA & + & P \\
K_A\downarrow\uparrow & & & & \uparrow\downarrow \beta K_A & & & & \\
A & & & & A & & & & \\
+ & & & & + & & & & \\
E & + & S & \overset{K_S}{\rightleftarrows} & ES & \overset{k_p}{\longrightarrow} & E & + & P \\
+ & & & & + & & & & \\
I & & & & I & & & & \\
K_i\uparrow\downarrow & & & & \downarrow\uparrow \alpha K_i & & & & \\
EI & + & S & \overset{\alpha K_S}{\rightleftarrows} & ESI & \overset{ak_p}{\longrightarrow} & EI & + & P
\end{array}$$

The term for [ESA] is $[S][A]/\beta K_S K_A$. The term for [ESI] is $[S][I]/\alpha K_S K_i$, and so on. Substituting into equation II-10, we obtain:

$$\frac{v}{[E]_t}=\frac{k_p\dfrac{[S]}{K_S}+bk_p\dfrac{[S][A]}{\beta K_S K_A}+ak_p\dfrac{[S][I]}{\alpha K_S K_i}}{1+\dfrac{[S]}{K_S}+\dfrac{[A]}{K_A}+\dfrac{[I]}{K_i}+\dfrac{[S][A]}{\beta K_S K_A}+\dfrac{[S][I]}{\alpha K_S K_i}}$$

where the 1 represents free [E]. If $k_p[\mathrm{E}]_t$ is taken as $V_{\max}$, the equation becomes:

$$\frac{v}{V_{\max}} = \frac{\dfrac{[\mathrm{S}]}{K_\mathrm{S}}\left(1+\dfrac{b[\mathrm{A}]}{\beta K_\mathrm{A}}+\dfrac{a[\mathrm{I}]}{\alpha K_i}\right)}{\text{same denominator}}$$

Multiplying numerator and denominator by K_S and factoring:

$$\frac{v}{V_{\max}} = \frac{[\mathrm{S}]\left(1+\dfrac{b[\mathrm{A}]}{\beta K_\mathrm{A}}+\dfrac{a[\mathrm{I}]}{\alpha K_i}\right)}{K_\mathrm{S}\left(1+\dfrac{[\mathrm{A}]}{K_\mathrm{A}}+\dfrac{[\mathrm{I}]}{K_i}\right)+[\mathrm{S}]\left(1+\dfrac{[\mathrm{A}]}{\beta K_\mathrm{A}}+\dfrac{[\mathrm{I}]}{\alpha K_i}\right)}$$

Or, in the usual Henri-Michaelis-Menten form:

$$\frac{v}{V_{\max}} = \frac{[\mathrm{S}]}{K_\mathrm{S}\dfrac{\left(1+\dfrac{[\mathrm{A}]}{K_\mathrm{A}}+\dfrac{[\mathrm{I}]}{K_i}\right)}{\left(1+\dfrac{b[\mathrm{A}]}{\beta K_\mathrm{A}}+\dfrac{a[\mathrm{I}]}{\alpha K_i}\right)}+[\mathrm{S}]\dfrac{\left(1+\dfrac{[\mathrm{A}]}{\beta K_\mathrm{A}}+\dfrac{[\mathrm{I}]}{\alpha K_i}\right)}{\left(1+\dfrac{b[\mathrm{A}]}{\beta K_\mathrm{A}}+\dfrac{a[\mathrm{I}]}{\alpha K_i}\right)}} \qquad \text{(II-11)}$$

Or, in general terms:

$$\frac{v}{V_{\max}} = \frac{[\mathrm{S}]}{K_\mathrm{S}(\text{slope factor})+[\mathrm{S}](\text{intercept factor})} \qquad \text{(II-12)}$$

As shown later, the slope and $1/v$-axis intercept of the reciprocal plot are given by:

$$slope_{1/\mathrm{S}} = \frac{K_\mathrm{S}}{V_{\max}}(\text{slope factor}), \qquad \frac{1}{V_{\max_{\mathrm{app}}}} = \frac{(\text{intercept factor})}{V_{\max}}$$

The $K_{\mathrm{S}_{\mathrm{app}}}$ (apparent Michaelis constant) is given by:

$$K_{\mathrm{S}_{\mathrm{app}}} = K_\mathrm{S}\frac{(\text{slope factor})}{(\text{intercept factor})}$$

C. THE VAN SLYKE EQUATION

In 1914, Van Slyke and co-workers independently derived a general rate equation based on their studies of the urease reaction. These workers assumed that the overall reaction occurred in two irreversible steps.

$$E + S \xrightarrow{k_1} ES \xrightarrow{k_p} E + P$$

The rate equation was obtained by assuming that the time required for the overall reaction was the sum of the times for each step. This may be expressed as:

$$t = \frac{1}{k_1[S]} + \frac{1}{k_p} = \frac{k_p + k_1[S]}{k_1 k_p [S]}$$

The units of t might be "minutes per mole of S converted to P per mole of enzyme." The reciprocal is:

$$\frac{1}{t} = \frac{k_1 k_p [S]}{k_p + k_1[S]} \qquad \text{moles } S \longrightarrow P/\text{min}/\text{mole of enzyme}$$

The overall velocity for a given amount of enzyme is $[E]_t(1/t)$:

$$v = \frac{[E]_t k_1 k_p [S]}{k_p + k_1[S]}$$

Dividing numerator and denominator by k_1:

$$\boxed{v = \frac{k_p[E]_t[S]}{\dfrac{k_p}{k_1} + [S]} = \frac{k_p[E]_t[S]}{K + [S]}} \qquad \text{(II-13)}$$

The equation is essentially identical to that derived by Henri and by Michaelis and Menten. Now, however, the constant K in the denominator is the ratio of forward rate constants instead of an equilibrium constant.

D. THE BRIGGS-HALDANE STEADY-STATE APPROACH

In 1925, Briggs and Haldane derived a general rate equation that did not require the restriction of equilibrium required by the methods of Henri, Michaelis, and Menten, nor the restriction of irreversible reactions required by the method of Van Slyke. Briggs and Haldane pointed out that the enzyme-substrate complex need not be in equilibrium with free enzyme and substrate, but within a very short time after starting the reaction ES would build up to a near-constant, or "steady-state," level (a concept introduced in 1913 by Bodenstein) (Fig. II-1). After the initial pre-steady-state period, ES would be formed at the same rate at which it decomposed. The steady-state level would be very close to the equilibrium level if the rate of equilibration is very rapid compared to the rate at which ES decomposes to E+P (i.e., if k_p is very small compared to k_{-1}). On the other hand, if k_p is comparable to k_{-1} or larger, then the steady-state level would be lower than the equilibrium level (i.e., ES decomposes to E+P so fast that it never can attain a level that would be in equilibrium with E and S). The rate at which P is formed will be proportional to the steady-state concentration of ES. The steps involved in deriving the general rate equation from steady-state considerations are similar to those described earlier for the derivation from equilibrium considerations. This time, however, the concentration of the product-forming species (ES) is obtained from steady-state equations instead of from equilibrium expressions. The derivation is outlined below.

ES is formed by one process: $\mathrm{E}+\mathrm{S} \xrightarrow{k_1} \mathrm{ES}$

(The reverse reaction E+P→ES is neglected because during the early stage of the reaction the concentration of product is essentially zero.)

ES decomposes by two processes: $\mathrm{ES} \xrightarrow{k_{-1}} \mathrm{E}+\mathrm{S}$

and $\mathrm{ES} \xrightarrow{k_p} \mathrm{E}+\mathrm{P}$

$$\therefore \quad \text{rate of ES formation} = \left(+\frac{d[\mathrm{ES}]}{dt}\right) = k_1[\mathrm{E}][\mathrm{S}]$$

$$\text{rate of ES decomposition} = \left(-\frac{d[\mathrm{ES}]}{dt}\right) = k_{-1}[\mathrm{ES}] + k_p[\mathrm{ES}]$$

$$= (k_{-1} + k_p)[\mathrm{ES}]$$

$$\text{at steady-state:} \quad \left(+\frac{d[\mathrm{ES}]}{dt}\right) = \left(-\frac{d[\mathrm{ES}]}{dt}\right) \quad \text{or} \quad \frac{d[\mathrm{ES}]}{dt} = 0$$

$$k_1[\mathrm{E}][\mathrm{S}] = (k_{-1} + k_p)[\mathrm{ES}]$$

Rearranging:

$$[\mathrm{ES}] = \frac{k_1[\mathrm{S}]}{(k_{-1} + k_p)}[\mathrm{E}] \qquad \text{(II-14)}$$

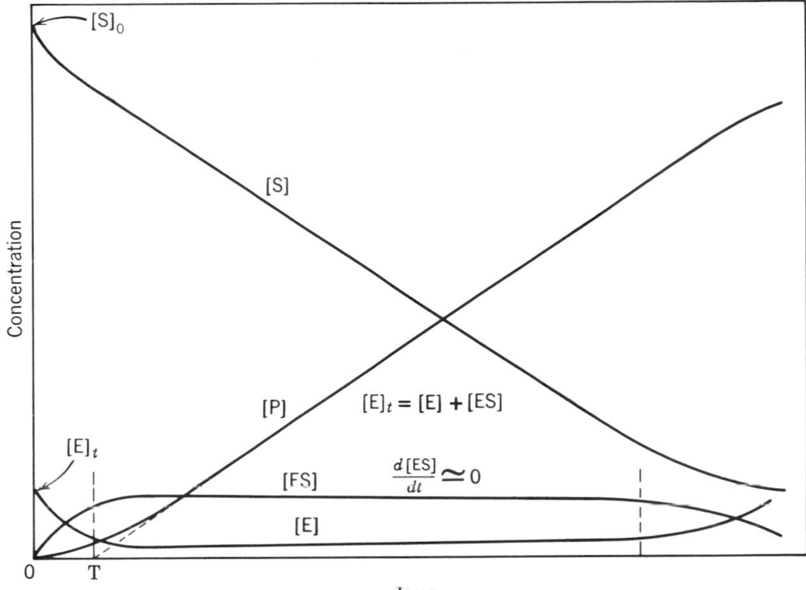

Fig. II-1. Progress curve for a catalyzed reaction where the initial substrate concentration, $[S]_0$, is significantly greater than the initial enzyme concentration, $[E]_t$. As the ratio of $[S]_0/[E]_t$ increases, the steady-state region accounts for an increasing fraction of the total reaction time.

Dividing both sides of the velocity-dependence equation by $[E]_t$:

$$\frac{v}{[E]_t} = \frac{k_p[ES]}{[E]+[ES]}$$

Rearranging and substituting for [ES]:

$$\frac{v}{k_p[E]_t} = \frac{\dfrac{k_1[S]}{(k_{-1}+k_p)}[E]}{[E]+\dfrac{k_1[S]}{(k_{-1}+k_p)}[E]}$$

Canceling [E] and substituting V_{max} for $k_p[E]_t$:

$$\frac{v}{V_{max}} = \frac{\dfrac{k_1[S]}{(k_{-1}+k_p)}}{1+\dfrac{k_1[S]}{(k_{-1}+k_p)}}$$

Grouping the rate constants as K_m, the Michaelis constant:

$$\boxed{\frac{v}{V_{\max}} = \frac{\dfrac{[S]}{K_m}}{1 + \dfrac{[S]}{K_m}}} \quad \text{or} \quad \boxed{\frac{v}{V_{\max}} = \frac{[S]}{K_m + [S]}} \tag{II-15}$$

where

$$K_m = \frac{k_{-1} + k_p}{k_1}$$

Equation II-14 can be rearranged to:

$$\frac{[E][S]}{[ES]} = \frac{k_{-1} + k_p}{k_1} = K_m \tag{II-15a}$$

Thus K_m is a dynamic or pseudoequilibrium constant expressing the relationship between the actual steady-state concentrations rather than equilibrium concentrations.

Note that it is possible to write four differential equations describing the overall reaction sequence. Each equation states that the rate at which the concentration of a given component changes equals the difference between the rate at which it is formed and the rate at which it is utilized.

$$\frac{d[E]}{dt} = (k_{-1} + k_p)[ES] - k_1[E][S]$$

$$\frac{d[ES]}{dt} = k_1[E][S] - (k_{-1} + k_p)[ES]$$

$$\frac{d[S]}{dt} = k_{-1}[ES] - k_1[E][S]$$

$$\frac{d[P]}{dt} = k_p[ES]$$

Together with the mass balance equation $[E]_t = [E] + [ES]$ we have five equations and four unknowns. Yet it is impossible to derive a general equation expressing v (i.e., $-d[S]/dt$ or $d[P]/dt$) as a function of [S] or an integrated equation expressing [S] or [P] as a function of time *unless* we assume steady-state conditions (i.e., $d[\mathrm{ES}]/dt=0$). The five equations can be solved numerically to yield [E], [S], [ES], and [P] at any time if $[\mathrm{S}]_0$, $[\mathrm{E}]_t$, k_1, k_{-1}, and k_p are known. (A computer would greatly facilitate the calculations.) The results would resemble those shown in Figure II-1. As the ratio of $[\mathrm{S}]_0/[\mathrm{E}]_t$ increases, the steady state assumption becomes more valid. That is, as $[\mathrm{S}]_0/[\mathrm{E}]_t$ increases, the time interval *before* $d[\mathrm{ES}]/dt \cong 0$ decreases and the extent of the reaction during which $d[\mathrm{ES}]/dt \cong 0$ increases. Because (a) most *in vitro* enzyme studies are conducted with "catalytic" concentrations of enzyme, and (b) v is taken as the velocity observed when only a small fraction of $[\mathrm{S}]_0$ is utilized, the steady-state assumption is quite valid.

Equation II-15 is identical to that derived by Henri, Michaelis and Menten, and Van Slyke, but now the constant has a different meaning. We see that the restrictive assumptions of Henri, Michaelis and Menten, and Van Slyke are special cases of the Briggs-Haldane steady-state treatment. When k_p is very small compared to k_{-1}, $K_m \cong k_{-1}/k_1 \cong K_\mathrm{S}$; that is, K_m is essentially the dissociation constant of the enzyme-substrate complex. When k_{-1} is $\ll k_p$, $K_m \cong k_p/k_1$; that is, K_m is a kinetic constant. The rate constants comprising K_m are not restricted to k_1, k_{-1}, and k_p. It may appear so because in deriving the velocity equation it was assumed that there was only a single central complex, ES, that decomposed directly to free E+P.

E. REVERSIBLE REACTIONS—EFFECT OF PRODUCT ON FORWARD VELOCITY

Strictly speaking, all enzyme-catalyzed reactions are reversible. The simplest representation of the overall reaction is:

$$\mathrm{E}+\mathrm{S} \underset{k_{-1}}{\overset{k_1}{\rightleftharpoons}} \mathrm{ES} \underset{k_{-2}}{\overset{k_2}{\rightleftharpoons}} \mathrm{E}+\mathrm{P}$$

A more realistic sequence involves two central complexes:

$$\mathrm{E}+\mathrm{S} \underset{k_{-1}}{\overset{k_1}{\rightleftharpoons}} \mathrm{ES} \underset{k_{-2}}{\overset{k_2}{\rightleftharpoons}} \mathrm{EP} \underset{k_{-3}}{\overset{k_3}{\rightleftharpoons}} \mathrm{E}+\mathrm{P}$$

Under the usual assay conditions, velocities are measured very early in the reaction before the product concentration has increased to a significant level. We can calculate the initial velocity for the reaction in each direction from the appropriate Henri-Michaelis-Menten equations:

$$v_f = \frac{V_{\max_f}[S]}{K_{m_S} + [S]} \qquad \text{in the absence of P}$$

$$v_r = \frac{V_{\max_r}[P]}{K_{m_P} + [P]} \qquad \text{in the absence of S}$$

It would be instructive to examine the effect of product on the initial forward velocity. For example, suppose we have a solution containing a certain concentration of S and a certain concentration of P. In the absence of an appropriate enzyme, the reaction does not occur at a measurable rate. Now we add an enzyme catalyzing the reversible reaction S⇌P. In which direction and at what rate will the reaction progress? The direction of the reaction will depend on the ratio of [P]/[S] relative to the equilibrium ratio. If $K_{eq} = 1$, [P]=5 mM, and [S]=1 mM, the reaction will proceed in the direction P→S and approach an equilibrium where [P]=3 mM and [S]=3 mM. The initial rate at which the reaction starts toward equilibrium *cannot* be calculated simply by taking the difference between v_f and v_r as given by their respective Henri-Michaelis-Menten expressions. These expressions were derived assuming [P]=0 or [S]=0. To calculate the initial rate when either [P] or [S] are not zero, we must derive a different general rate equation taking into account the reverse reactions. If we assume only one central complex the *net* velocity in the forward direction is given by:

$$v_{\text{net}} = k_2[ES] - k_{-2}[E][P]$$

The steady-state relationships are:

$$\left(+\frac{d[ES]}{dt}\right) = \left(-\frac{d[ES]}{dt}\right) \qquad \text{or} \qquad \frac{d[ES]}{dt} = 0$$

$$\left(+\frac{d[ES]}{dt}\right) = k_1[E][S] + k_{-2}[E][P]$$

$$\left(-\frac{d[ES]}{dt}\right) = k_2[ES] + k_{-1}[ES]$$

$$k_1[E][S] + k_{-2}[E][P] = (k_2 + k_{-1})[ES]$$

$$[ES] = \frac{k_1[S] + k_{-2}[P]}{k_2 + k_{-1}}[E]$$

Dividing the velocity-dependence equation by $[E]_t$, where $[E]_t=[E]+[ES]$:

$$\frac{v_{net}}{[E]_t}=\frac{k_2[ES]-k_{-2}[E][P]}{[E]+[ES]}$$

Substituting for [ES]:

$$\frac{v_{net}}{[E]_t}=\frac{k_2\left(\dfrac{k_1[S]+k_{-2}[P]}{k_2+k_{-1}}\right)[E]-k_{-2}[E][P]}{[E]+\left(\dfrac{k_1[S]+k_{-2}[P]}{k_2+k_{-1}}\right)[E]}$$

$$=\frac{\dfrac{k_2k_1[S]}{k_2+k_{-1}}+\dfrac{k_2k_{-2}[P]}{k_2+k_{-1}}-\dfrac{k_{-2}(k_2+k_{-1})[P]}{k_2+k_{-1}}}{1+\dfrac{k_1[S]}{k_2+k_{-1}}+\dfrac{k_{-2}[P]}{k_2+k_{-1}}}$$

$$v_{net}=\frac{\dfrac{k_2k_1[S][E]_t}{k_2+k_{-1}}-\dfrac{k_{-2}k_{-1}[P][E]_t}{k_2+k_{-1}}}{1+\dfrac{k_1[S]}{k_2+k_{-1}}+\dfrac{k_{-2}[P]}{k_2+k_{-1}}}$$

$$\boxed{v_{net}=\frac{V_{max_f}\dfrac{[S]}{K_{m_S}}-V_{max_r}\dfrac{[P]}{K_{m_P}}}{1+\dfrac{[S]}{K_{m_S}}+\dfrac{[P]}{K_{m_P}}}} \qquad \text{(II-16)}$$

where:

$$k_2[E]_t=V_{max_f}, \qquad k_{-1}[E]_t=V_{max_r}$$

$$\frac{k_2+k_{-1}}{k_1}=K_{m_S}, \qquad \frac{k_2+k_{-1}}{k_{-2}}=K_{m_P}$$

Solving for the individual rate constants:

$$k_1=\frac{V_{max_f}+V_{max_r}}{K_{m_S}[E]_t}, \qquad k_{-1}=\frac{V_{max_r}}{[E]_t}$$

$$k_2=\frac{V_{max_f}}{[E]_t}, \qquad k_{-2}=\frac{V_{max_f}+V_{max_r}}{K_{m_P}[E]_t}$$

The $[E]_t$ in the equations above represents the concentration of catalytic sites. If the enzyme contains one catalytic site per molecule, $[E]_t$ is the molar concentration of enzyme. As shown later, an enzyme with multiple identical and independent catalytic sites is indistinguishable kinetically from an enzyme with only one site. The constants k_1 and k_{-2} are second-order rate constants:

$$k_1 = \frac{V_{\max_f} + V_{\max_r}}{K_{m_S}[E]_t} = \frac{\dfrac{\text{moles}}{1 \times \text{min}} + \dfrac{\text{moles}}{1 \times \text{min}}}{\left(\dfrac{\text{moles}}{1}\right)\left(\dfrac{\text{moles}}{1}\right)} = M^{-1} \times \text{min}^{-1}$$

At a fixed enzyme concentration, the reaction $E + S \rightleftharpoons ES$ is pseudo-first-order with an observed first-order rate constant equal to $k_1[E]$. The constants k_{-1} and k_2 are true first-order rate constants:

$$k_2 = \frac{V_{\max_f}}{[E]_t} = \frac{M \times \text{min}^{-1}}{M} = \text{min}^{-1}$$

At a fixed enzyme concentration, $k_2[E]_t$ is a pseudo-zero-order constant:

$$k_2[E]_t = V_{\max_f} = M \times \text{min}^{-1}$$

The constant, k_2 (or k_p or k_{cat}), is called the *turnover number* (or *molecular activity* or *catalytic rate constant*) and represents the maximum velocity per mole of enzyme (or per mole of catalytic site if $[E]_t$ is expressed in concentration of catalytic sites).

$$k_2 = \frac{V_{\max_f}}{[E]_t} = \text{moles of product formed per minute per mole of enzyme}$$

The reciprocal of k_2 represents the time required to complete one catalytic cycle. The k_1 values are usually in the range 10^7 to 10^{10} $M^{-1} \times \text{min}^{-1}$. The maximum value is about $10^{11} M^{-1} \times \text{min}^{-1}$, limited by the rate of diffusion of a small molecule in aqueous solution to the active site of the enzyme. The k_{-1} values are usually 10^2 to 10^6 min^{-1}, while k_2 values vary from 50 to 10^7 min^{-1}. The K_m values are usually in the range 10^{-6} to 10^{-2} M.

The complete velocity equation for the more realistic reaction sequence involving two central complexes is easily derived for rapid equilibrium conditions, as shown below.

$$\frac{v_{\text{net}}}{[\mathrm{E}]_t} = \frac{k_2[\mathrm{ES}] - k_{-2}[\mathrm{EP}]}{[\mathrm{E}] + [\mathrm{ES}] + [\mathrm{EP}]}$$

$$v_{\text{net}} = \frac{k_2[\mathrm{E}]_t \dfrac{[\mathrm{S}]}{K_\mathrm{S}} - k_{-2}[\mathrm{E}]_t \dfrac{[\mathrm{P}]}{K_\mathrm{P}}}{1 + \dfrac{[\mathrm{S}]}{K_\mathrm{S}} + \dfrac{[\mathrm{P}]}{K_\mathrm{P}}}$$

$$\boxed{v_{\text{net}} = \frac{V_{\max_f} \dfrac{[\mathrm{S}]}{K_\mathrm{S}} - V_{\max_r} \dfrac{[\mathrm{P}]}{K_\mathrm{P}}}{1 + \dfrac{[\mathrm{S}]}{K_\mathrm{S}} + \dfrac{[\mathrm{P}]}{K_\mathrm{P}}}} \tag{II-17}$$

A steady-state treatment (Chapter Nine) yields:

$$v_{\text{net}} = \frac{[\mathrm{E}]_t(k_1k_2k_3[\mathrm{S}] - k_{-1}k_{-2}k_{-3}[\mathrm{P}])}{k_{-1}k_{-2} + k_{-1}k_3 + k_2k_3 + k_1(k_2 + k_{-2} + k_3)[\mathrm{S}] + k_{-3}(k_{-1} + k_2 + k_{-2})[\mathrm{P}]}$$

The equation above may be further modified by grouping rate constants into K_m and $V_{\max}$ values. The final equation has the same form as that derived for rapid equilibrium conditions and for steady-state conditions assuming only one central complex. Only the definitions of K_m and $V_{\max}$ in terms of rate constants change.

$$K_{m_\mathrm{S}} = \frac{k_{-1}k_3 + k_{-1}k_{-2} + k_2k_3}{k_1(k_2 + k_{-2} + k_3)}, \qquad K_{m_\mathrm{P}} = \frac{k_{-1}k_3 + k_{-1}k_{-2} + k_2k_3}{k_{-3}(k_{-1} + k_2 + k_{-2})}$$

$$V_{\max_f} = \frac{k_2k_3[\mathrm{E}]_t}{k_2 + k_{-2} + k_3}, \qquad V_{\max_r} = \frac{k_{-1}k_{-2}[\mathrm{E}]_t}{k_{-1} + k_2 + k_{-2}}$$

It is obvious that the physical significance of K_m cannot be stated with any certainty in the absence of other data concerning the relative magnitudes of the various rate constants. Nevertheless, K_m represents a valuable constant that relates the velocity of an enzyme-catalyzed reaction to the substrate concentration. Inspection of the Henri-Michaelis-Menten equation shows that K_m is numerically equivalent to the substrate concentration that yields half-maximal velocity:

$$v = \frac{[S]}{K_m + [S]} V_{max} \quad \therefore \quad \text{when} \quad [S] = K_m:$$

$$v = \frac{K_m}{K_m + K_m} V_{max} = \tfrac{1}{2} V_{max}$$

The numerical value of K_m is of interest for several reasons. (*a*) The K_m establishes an approximate value for the intracellular level of the substrate. It is unlikely that this level would be significantly greater or significantly lower than K_m. If $[S]_{intracell} \ll K_m$, v would be very sensitive to changes in [S], but most of the catalytic potential of the enzyme would be wasted, since v would be $\ll V_{max}$. There is also no physiological sense in maintaining $[S] \gg K_m$, since v cannot exceed V_{max}, and the difference between v at $[S] = K_m$ and $[S] = 1000K_m$ is only twofold. Also at $[S] \gg K_m$, v becomes insensitive to small changes in [S]. (*b*) Since K_m is a constant for a given enzyme, its numerical value provides a means of comparing enzymes from different organisms or from different tissues of the same organism, or from the same tissue at different stages of development. In this way, we might determine whether enzyme A is identical to enzyme B, or whether they are different proteins that catalyze the same reaction. (*c*) A ligand-induced change in the effective value of K_m is one mode of regulating the activity of an enzyme. If K_m determined *in vitro* seems "unphysiologically" high then we might search for activators that function *in vivo* to lower the effective K_m. By measuring the effects of different compounds on K_m we might identify physiologically important inhibitors as well. (*d*) If we know K_m, we can adjust the assay conditions so that $[S] \gg K_m$, and thereby determine V_{max}, which is a measure of $[E]_t$.

F. HALDANE RELATIONSHIP BETWEEN KINETIC CONSTANTS AND EQUILIBRIUM CONSTANT

The constants K_m and V_{max} were derived in terms of the various rate constants of the overall reaction. The equilibrium constant for the overall reaction is composed of the same rate constants. Consequently, it should be possible to express K_{eq} in terms of K_m and V_{max}. For example, consider the simple two-step reaction shown below.

$$E + S \underset{k_{-1}}{\overset{k_1}{\rightleftharpoons}} ES \underset{k_{-2}}{\overset{k_2}{\rightleftharpoons}} E + P$$

The overall equilibrium constant for the reaction reading left to right is the product of the equilibrium constants for the individual steps, which may be expressed in terms of the rate constants:

$$K_{eq} = \frac{[P]_{eq}}{[S]_{eq}} = K_1 K_2 = \frac{k_1 k_2}{k_{-1} k_{-2}}$$

We can express this grouping of the rate constants in terms of K_m and V_{max} values as shown below.

$$\frac{V_{max_f}}{K_{m_S}} = \frac{k_1 k_2 [E]_t}{k_2 + k_{-1}}, \qquad \frac{V_{max_r}}{K_{m_P}} = \frac{k_{-2} k_{-1} [E]_t}{k_2 + k_{-1}}$$

Now dividing one ratio by the other:

$$\frac{V_{max_f}/K_{m_S}}{V_{max_r}/K_{m_P}} = \frac{(k_1 k_2 [E]_t)(k_2 + k_{-1})}{(k_2 + k_{-1})(k_{-2} k_{-1} [E]_t)}$$

$$\boxed{\frac{V_{max_f} K_{m_P}}{V_{max_r} K_{m_S}} = \frac{k_1 k_2}{k_{-1} k_{-2}} = \frac{[P]_{eq}}{[S]_{eq}} = K_{eq}} \qquad \text{(II-18)}$$

The relationship between K_{eq} and the K_m and V_{max} values is known as the Haldane equation. The Haldane equation can be obtained directly from the equation for v_{net}.

$$v_{net} = \frac{V_{max_f} \dfrac{[S]}{K_{m_S}} - V_{max_r} \dfrac{[P]}{K_{m_P}}}{1 + \dfrac{[S]}{K_{m_S}} + \dfrac{[P]}{K_{m_P}}} \qquad \text{(II-19)}$$

At equilibrium, $v_{net} = 0$. Thus the numerator of equation II-19 must equal zero, or:

$$\frac{V_{max_f}[S]_{eq}}{K_{m_S}} = \frac{V_{max_r}[P]_{eq}}{K_{m_P}} \qquad \text{or} \qquad \frac{V_{max_f} K_{m_P}}{V_{max_r} K_{m_S}} = \frac{[P]_{eq}}{[S]_{eq}} = K_{eq}$$

Substituting $V_{\max_f}/K_{m_S}K_{eq}$ for $V_{\max_r}/K_{m_P}$ in equation II-19 yields equation II-20 for v_{net}:

$$\boxed{v_{net} = \frac{V_{\max_f}\left([S] - \frac{[P]}{K_{eq}}\right)}{K_{m_S}\left(1 + \frac{[P]}{K_{m_P}}\right) + [S]}} \quad \text{where} \quad \frac{[P]}{K_{eq}} = [S]_{eq} \qquad \text{(II-20)}$$

Equation II-20 can be written as:

$$v_{net} = \frac{V_{\max_f}\Delta S}{K_{m_S}\left(1 + \frac{[P]}{K_{m_P}}\right) + \frac{[P]}{K_{eq}} + \Delta S} \quad \text{where} \quad \Delta S = [S] - \frac{[P]}{K_{eq}} \qquad \text{(II-21)}$$

In place of the usual [S] in the numerator, we have the *difference* between [S] and the equilibrium value of [S]. The K_{m_S} term in the denominator is modified in a manner consistent with the product acting as a competitive inhibitor with respect to the substrate. In other words, the initial net velocity depends on the displacement of the system from equilibrium, (that is, the thermodynamic driving force) and the amount of enzyme tied up with product. A more detailed account of competitive inhibition is given in a later section. The effect of a fixed [P] on the [S]-dependence of initial net velocity is shown in Figure II-2. We see that at any substrate concentration below the equilibrium value of 5 mM the net velocity is negative, that is, in the direction of P→S. When [S] = 5 mM the reaction is at equilibrium and $v = 0$.

The net velocity of S→P in the presence of preexisting P is always less than the initial velocity at the same [S] in the absence of preexisting P. There are two reasons for the decreased rate: (*a*) at any time, some of the P is being converted back to S, and (*b*) at any time, some of the enzyme is combined with P so that less enzyme is available for combination with S. Reason *a* is expressed by the $([S] - [P]/K_{eq})$ factor in the numerator of the velocity equation, which makes it seem as if less substrate is available. Reason *b* is expressed by the $(1 + [P]/K_{m_P})$ factor in the denominator, which increases the apparent K_m value for S. The net rate (equation II-19) can be expressed as:

$$v_{net} = \frac{V_{\max_f}[S]}{K_{m_S}\left(1 + \frac{[P]}{K_{m_P}}\right) + [S]} - \frac{V_{\max_r}[P]}{K_{m_P}\left(1 + \frac{[S]}{K_{m_S}}\right) + [P]}$$

$$= v_f - v_r \qquad \text{(II-22)}$$

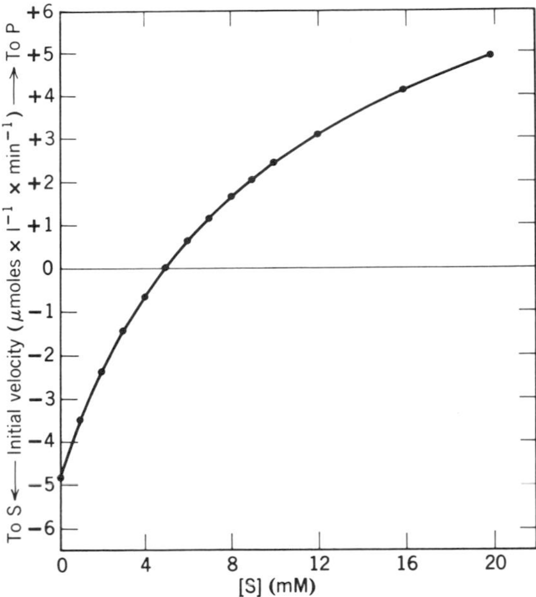

Fig. II-2. Velocity curve in the presence of product. [P]=5 mM, $K_{eq}=1.0$, $K_{m_S}=0.4$ mM, $K_{m_P}=0.2$ mM, $V_{max_f}=10$ μmoles$\times$l$^{-1}\times$min^{-1}, $V_{max_r}=5$ μmoles$\times$l$^{-1}\times$min^{-1}.

The absolute rates in either direction (v_f or v_r) can be measured by using radioactively labeled S or P. For example, suppose we have a solution containing 3 mM S, 3 mM P, and an enzyme that catalyzes the reversible reaction S$\rightleftharpoons$P. Assume that $K_{eq}=1$ so that the reaction is at equilibrium. If the concentrations of S and P are determined by chemical means, no change in either will be detected with time (i.e., $v_{net}=0$). However, the equilibrium is a dynamic one. The v_{net} equals zero because $v_f=v_r$. If S is made radioactive by introducing a small amount of high specific activity S (such that the concentration of S is unchanged) then v_f can be measured as the initial rate at which radioactivity appears in P (see Chapter Ten).

G. SPECIFIC (OR RELATIVE OR REDUCED) SUBSTRATE CONCENTRATION AND VELOCITY

Strictly speaking, the initial velocity of an enzyme-catalyzed reaction depends not on the substrate concentration, but rather on the ratio of [S] to K_m. This ratio, $[S]/K_m$, has been called "specific substrate concentration," "reduced substrate concentration," "relative substrate concentration," or "normalized substrate concentration" and denoted [S′], α, or σ. The Henri-

Michaelis-Menten equation may be expressed in terms of [S′].

$$\frac{v}{V_{max}} = \frac{\frac{[S]}{K_m}}{1+\frac{[S]}{K_m}} = \frac{[S']}{1+[S']}$$

When $[S']=1$, $v=\frac{1}{2}V_{max}$. The initial velocity at any given [S] may also be expressed as a fraction of V_{max}, that is, in terms of "specific velocity" or "relative velocity," v' or ϕ, giving:

$$\boxed{v' = \frac{[S']}{1+[S']}} \qquad \text{(II-23)}$$

The use of "specific" or "relative" values provides a way of normalizing data for a variety of enzymes. For example, when $[S']=6$, $v'=0.857$, or in other words, when the substrate concentration is six times the K_m value, the initial velocity is always $\frac{6}{7}$ or 85.7% of V_{max}, regardless of the actual values of K_m or V_{max}. Equation II-23 may be rearranged to solve for the specific substrate concentration required for any fraction of V_{max}:

$$\boxed{[S'] = \frac{v'}{1-v'}} \qquad \text{(II-24)}$$

H. VELOCITY VERSUS SUBSTRATE CONCENTRATION CURVE

The velocity equation can be written as $(V_{max}-v)(K_m+[S])=K_m V_{max}$ or $(a-y)(b+x)=$ a constant. This equation describes a right rectangular hyperbola with limits of V_{max} and $-K_m$. The curvature is fixed regardless of the values of K_m and V_{max}. Consequently, the ratio of substrate concentrations for any two fractions of V_{max} is constant for all enzymes that obey Henri-Michaelis-Menten kinetics. For example, the ratio of substrate required for 90% of V_{max}, $[S]_{0.9}$, to the substrate required for 10% of V_{max}, $[S]_{0.1}$, is always 81 as shown below and illustrated in Figure II-3.

When $v=0.9\,V_{max}$:

$$[S']_{0.9} = \frac{v'}{1-v'} = \frac{0.9}{1-0.9} = \frac{0.9}{0.1} = 9$$

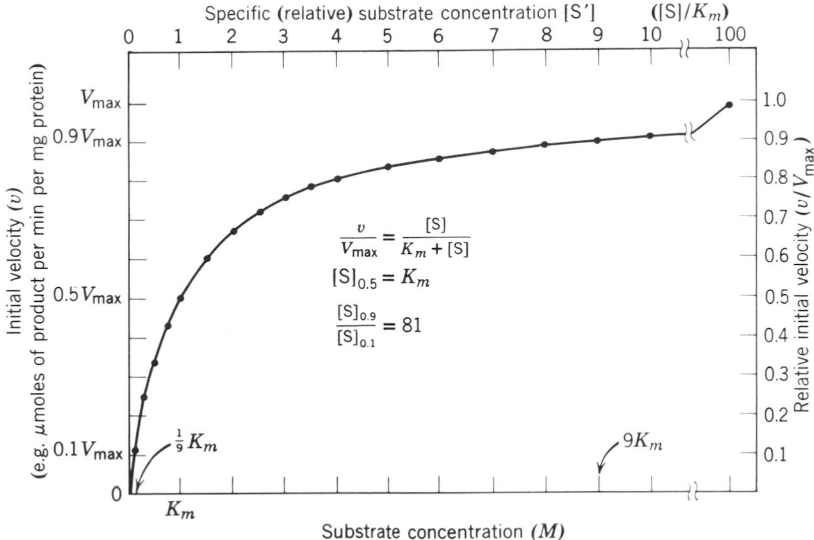

Fig. II-3. The curvature of the v versus [S] plot is constant; $[S]_{0.9}/[S]_{0.1}$ always equals 81 regardless of the absolute values of K_m and V_{max}.

When $v = 0.1\, V_{max}$:

$$[S']_{0.1} = \frac{v'}{1-v'} = \frac{0.1}{1-0.1} = \frac{0.1}{0.9} = 0.111$$

$$\therefore \quad \frac{[S]_{0.9}}{[S]_{0.1}} = \frac{9}{0.111} \quad \text{or} \quad \boxed{\frac{[S]_{0.9}}{[S]_{0.1}} = 81} \qquad \text{(II-25)}$$

I. REACTION ORDER

If we examine the v versus [S] curve, we find three distinct regions where the velocity responds in a characteristic way to increasing [S] (Fig. II-4*a*). At very low substrate concentrations (e.g., $[S] < 0.01 K_m$), the v versus [S] curve is essentially linear; that is, the velocity (for all practical purposes) is directly proportional to the substrate concentration (Fig. II-4*b*). This is the region of *first-order kinetics*. At very high substrate concentrations (e.g., $[S] > 100 K_m$), the velocity is essentially independent of the substrate concentration. This is the region of *zero-order kinetics* (Fig. II-4*c*). At intermediate substrate concentrations, the relationship between v and [S] follows neither first-order nor

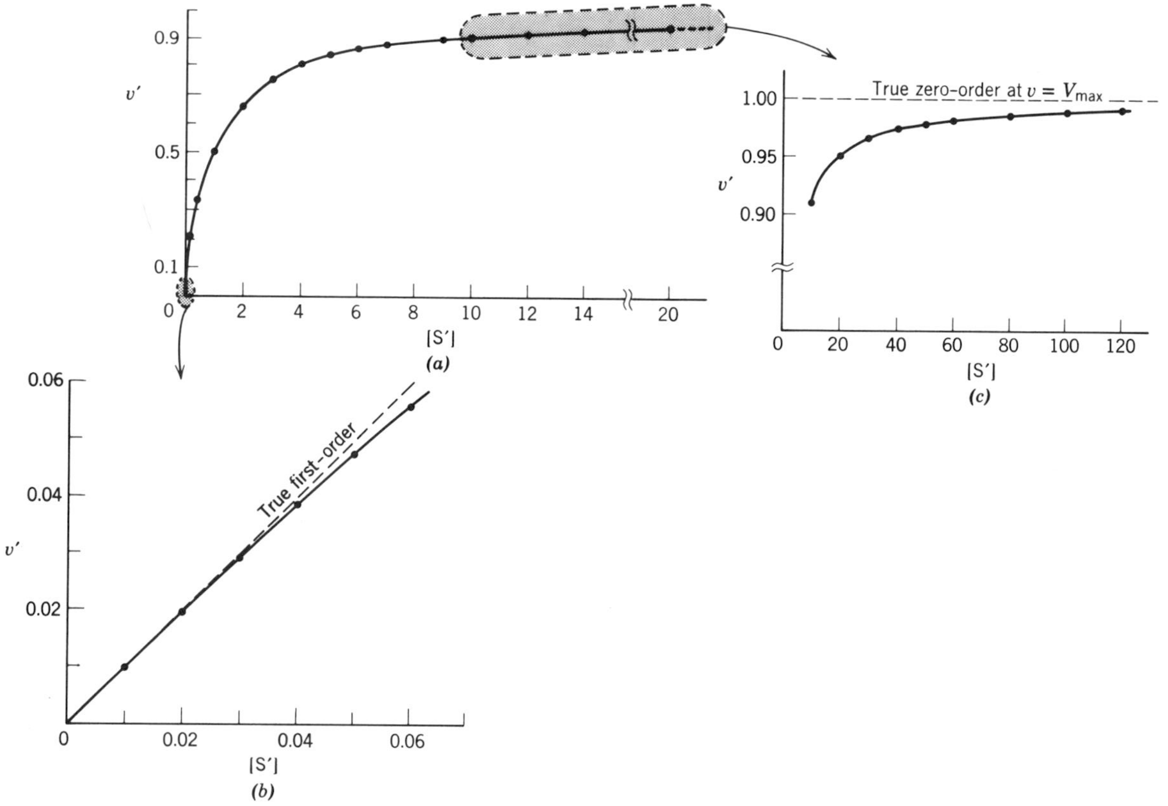

Fig. II-4. (*a*) v versus [S] plot over a wide range of [S]. (*b*) v versus [S] plot in a narrow range where $[S] \ll K_m$. (*c*) v versus [S] plot in a range where $[S] > K_m$.

zero-order kinetics. The characteristics of the first-order and zero-order regions are described below.

First-Order Kinetics

The linear relationship between v and [S] when $[S] \ll K_m$ can be derived from the Henri-Michaelis-Menten equation.

$$v = \frac{V_{max}[S]}{K_m + [S]}$$

When $[S] \ll K_m$, the [S] in the denominator may be ignored and the equation reduces to:

$$v = \frac{V_{\max}}{K_m}[S] \qquad \text{or} \qquad v = k[S] \tag{II-26}$$

where k is a first-order rate constant equivalent to $V_{\max}/K_m$. The units of k are $\min^{-1}$ if v is expressed as $\text{moles} \times l^{-1} \times \min^{-1}$ and K_m is expressed as $\text{moles} \times l^{-1}$:

$$k = \frac{V_{\max}}{K_m} = \frac{\text{moles} \times l^{-1} \times \min^{-1}}{\text{moles} \times l^{-1}} = \min^{-1}$$

In terms of the rate constants of the individual steps, $k = k_1 k_p [E]_t / (k_p + k_{-1})$. Equation II-26 expresses the fact that when [S] is very small, the absolute velocity decreases from moment to moment as [S] decreases (Fig. II-5*a*). However, at any given moment, a constant *fraction* of the substrate present undergoes conversion to product:

$-\dfrac{d[S]}{dt} =$	$v =$	k	$[S]$
The amount of S used up per small increment of time...	that is, the velocity...	is some constant fraction...	of the substrate present at that time

Thus the physical significance of the first-order rate constant is that it approximates the fraction of the substrate present that is converted to product per unit of time. For example, if $k = 0.02 \min^{-1}$, then approximately 2% of the substrate present at any time is converted to product in a minute. A $k > 1 \min^{-1}$ means that more than 100% of the substrate present at

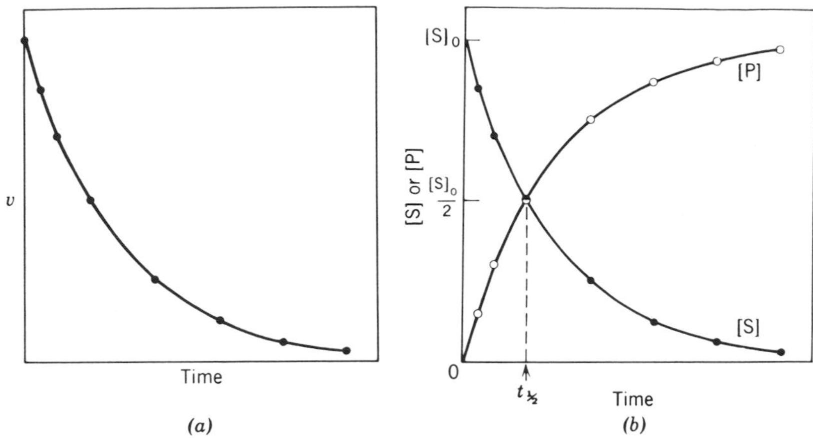

Fig. II-5. First-order region of the velocity curve. (*a*) v decreases continually with time. (*b*) The appearance of P and the disappearance of S are not linear with time.

zero-time could be utilized in a minute *if* v remained constant for a minute. It may be more meaningful to express k in units that yield numerical values less than unity. For example, if $k = 2.3\ \text{min}^{-1}$:

$$k = \frac{2.3\ \text{min}^{-1}}{60\ \text{sec} \times \text{min}^{-1}} = 0.0383\ \text{sec}^{-1}$$

Thus $k = 2.3\ \text{min}^{-1}$ means that approximately 3.83% of [S] is utilized per second.

Because v decreases with time in the first-order region, the plots of [S] versus time and [P] versus time are curved (Fig. II-5*b*). We can determine the amount of substrate utilized or product formed during any given time interval by using the integrated first-order rate equation:

$$v = -\frac{d[\text{S}]}{dt} = k[\text{S}] \qquad \text{or} \qquad -\frac{d[\text{S}]}{[\text{S}]} = k\,dt$$

Integrating between $[\text{S}]_0$ at $t = 0$ and [S] at any other time, t, we obtain:

$$-\int_{[\text{S}]_0}^{[\text{S}]} \frac{d[\text{S}]}{[\text{S}]} = k\int_0^t dt$$

$$\boxed{2.3 \log \frac{[\text{S}]_0}{[\text{S}]} = kt} \qquad \text{or} \qquad \boxed{[\text{S}] = [\text{S}]_0 e^{-kt}} \qquad \text{(II-27)}$$

Equation II-27 may be rearranged to:

$$\boxed{\log[S] = -\frac{k}{2.3}t + \log[S]_0} \qquad \text{(II-28)}$$

Thus a plot of log [S] versus t is linear with a slope of $-k/2.3$ and an intercept of log $[S]_0$ on the log [S]-axis (Fig. II-6). When $[S] = \frac{1}{2}[S]_0$, $t=$ the "halflife," $t_{1/2}$, the time required to convert half the substrate originally present to product. The $t_{1/2}$ is constant for first-order reactions and is related to k as shown below.

$$2.3\log\frac{1}{0.5} = kt_{1/2} \qquad \therefore \frac{0.693}{k} = t_{1/2} \qquad \text{(II-29)}$$

Zero-Order Kinetics

When $[S] \gg K_m$, the K_m in the denominator of the Henri-Michaelis-Menten equation may be ignored and the equation simplifies as shown below.

$$v = \frac{V_{max}[S]}{K_m + [S]} \xrightarrow{[S]\gg K_m} \frac{V_{max}[S]}{[S]} \qquad \text{or} \qquad v = V_{max}$$

$$\therefore \boxed{[P] = [S]_0 - [S] = V_{max}t} \qquad \text{(II-30)}$$

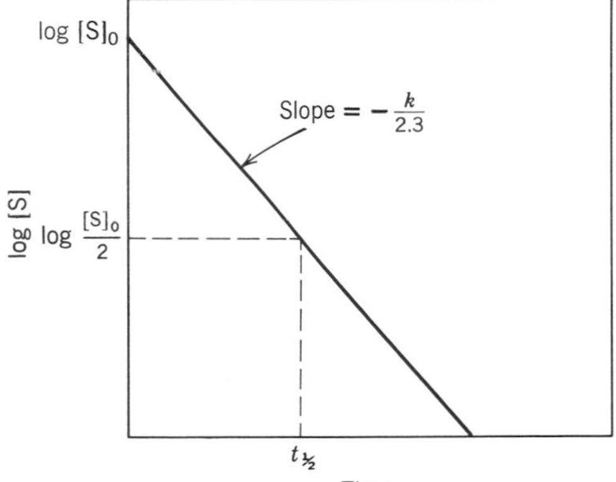

Fig. II-6. Semilog plot of the integrated first-order velocity equation.

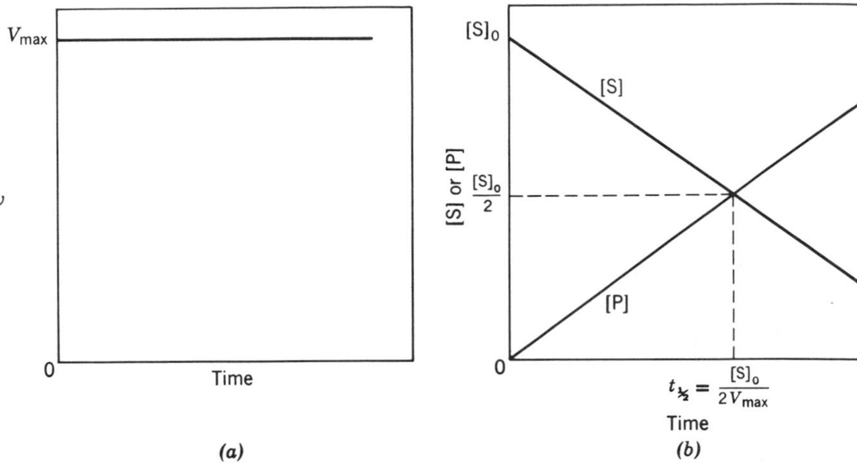

Fig. II-7. Zero-order region of the velocity curve. (*a*) The velocity is constant over time. (*b*) P appears and S disappears linearly with time.

For all practical purposes, the velocity is constant and independent of [S] (Fig. II-7*a*). Plots of [S] versus time and [P] versus time are linear (Fig. II-7*b*).

J. GRAPHICAL DETERMINATION OF K_m AND V_{max}

Because the v versus [S] curve is a hyperbola, it is extremely difficult to determine V_{max} and K_m. An early attempt to remedy the situation involved plotting v versus log [S] as shown in Figure II-8. This plot is based on the rearrangement of the Henri-Michaelis-Menten equation as shown below:

$$\frac{v}{V_{max}} = \frac{[S]}{K_m + [S]}$$

Inverting:

$$\frac{V_{max}}{v} = \frac{K_m + [S]}{[S]} = \frac{K_m}{[S]} + 1$$

$$\frac{V_{max}}{v} - 1 = \frac{K_m}{[S]}$$

$$\boxed{\log\left(\frac{V_{max}}{v} - 1\right) = \log K_m - \log [S]} \qquad \text{(II-31)}$$

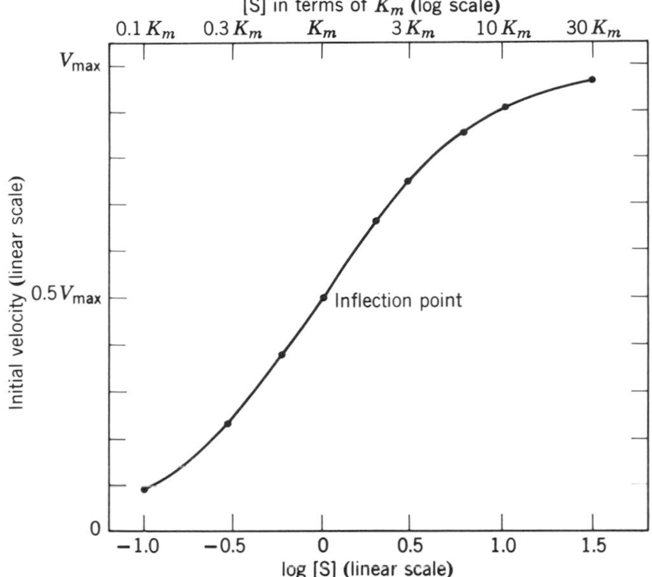

Fig. II-8. Plot of v versus log [S]; [S] at the inflection point equals K_m.

When $v = 0.5\,V_{max}$, $(V_{max}/v - 1)$ equals 1 and log $(V_{max}/v - 1)$ equals zero. At this point log [S] = log K_m, or [S] = K_m. To determine K_m, it is only necessary to identify the midpoint (inflection point) of the curve.

A slightly different rearrangement will yield a familiar equation:

$$\boxed{\mathrm{p[S]} = \mathrm{p}K_m + \log \frac{V_{max} - v}{v}} \qquad \text{(II-32)}$$

The equation is similar to the Henderson-Hasselbalch equation relating pH to pK_a and the ratio of conjugate base to conjugate acid; $(V_{max} - v)/v$ is analogous to [original HA − amount titrated to base]/[base] or [HA]/[A^-].

In 1934, the more useful Lineweaver-Burk double reciprocal plot was introduced. (The double reciprocal plot was first proposed in 1932 by Haldane and Stern, as a result of a suggestion by Woolf.) This plot is based on the rearrangement of the Henri-Michaelis-Menten equation into a linear $(y = mx + b)$ form.

Lineweaver-Burk Reciprocal Plot: $1/v$ versus $1/[S]$

$$\frac{v}{V_{max}} = \frac{[S]}{K_m + [S]}$$

Inverting:

$$\frac{V_{max}}{v} = \frac{K_m + [S]}{[S]} = \frac{K_m}{[S]} + 1$$

Cross multiplying V_{max}:

$$\boxed{\frac{1}{v} = \frac{K_m}{V_{max}} \frac{1}{[S]} + \frac{1}{V_{max}}} \qquad \text{(II-33)}$$

Thus if we plot $1/v$ versus $1/[S]$, the slope $= K_m/V_{max}$ and the intercept on the $1/v$ axis $= 1/V_{max}$. We can also see that when $1/v = 0$, $1/[S] = -1/K_m$. As we see later, any factor that multiplies the K_m term of the original Henri-Michaelis-Menten equation will turn out to be a factor of the slope (i.e., of K_m/V_{max}) in the reciprocal equation. Any factor that multiplies the denominator [S] term of the original equation will turn out to be a factor of the $1/v$-axis intercept (i.e., of $1/V_{max}$) in the reciprocal equation.

Substrate Concentration Range

The concentrations of substrate chosen to generate the reciprocal plot should be in the neighborhood of K_m (Fig. II-9). If the concentrations chosen are very high relative to K_m, the curve will be essentially horizontal (Fig. II-10). This will allow V_{max} to be determined, but the slope of the line will be near zero. Consequently, it will be difficult to determine K_m accurately. If the substrate concentrations chosen are very low relative to K_m, the curve will intercept both axes too close to the origin to allow either V_{max} or K_m to be determined accurately (Fig. II-11). (At very low substrate concentrations, the reaction is essentially first-order. There is no hint of saturation. V_{max} and K_m appear to be infinite.)

Generally, substrate concentrations are chosen that give evenly spaced reciprocals (e.g., 1.0, 1.11, 1.25, 1.43, 1.67, 2.0, 2.5, 3.33, 5.0, 10). If a constant increment of substrate concentration is used (e.g., 1.0, 2.0, 3.0, 4.0, etc.) the points will cluster close to the $1/v$-axis.

Labeling the Axes of Reciprocal Plots

The beginning student sometimes is uncertain about the units used in labeling axes or columns of data. The uncertainty arises because there are two ways of interpreting units containing factors (see Table II-1). For

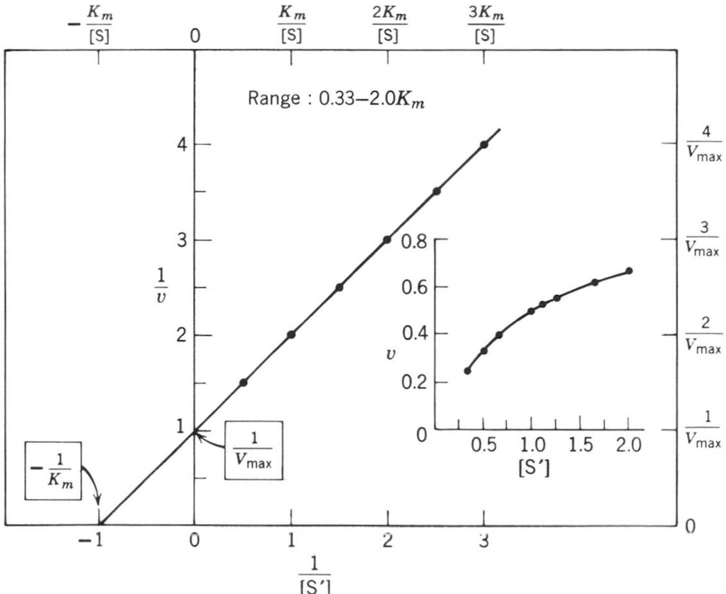

Fig. II-9. Double reciprocal ($1/v$ versus 1/[S]) Lineweaver-Burk plot. The [S] range chosen is optimal for the determination of K_m and V_{max}.

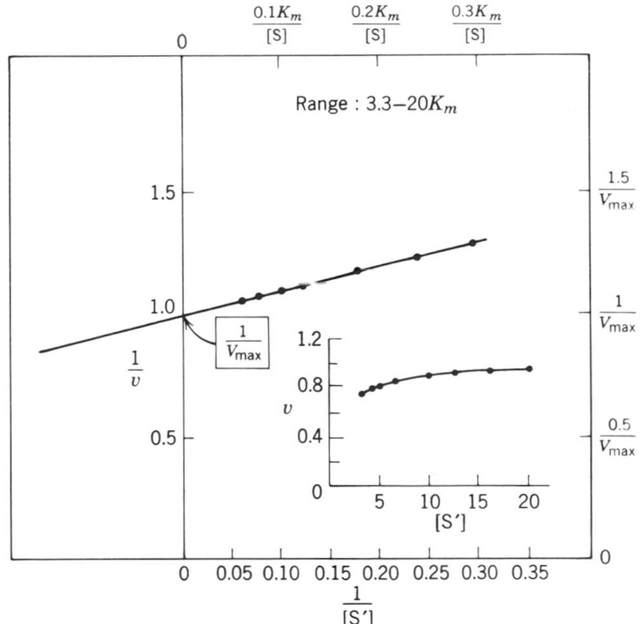

Fig. II-10. $1/v$ versus 1/[S] plot. The [S] range chosen is higher than optimal; v is relatively insensitive to changes in [S].

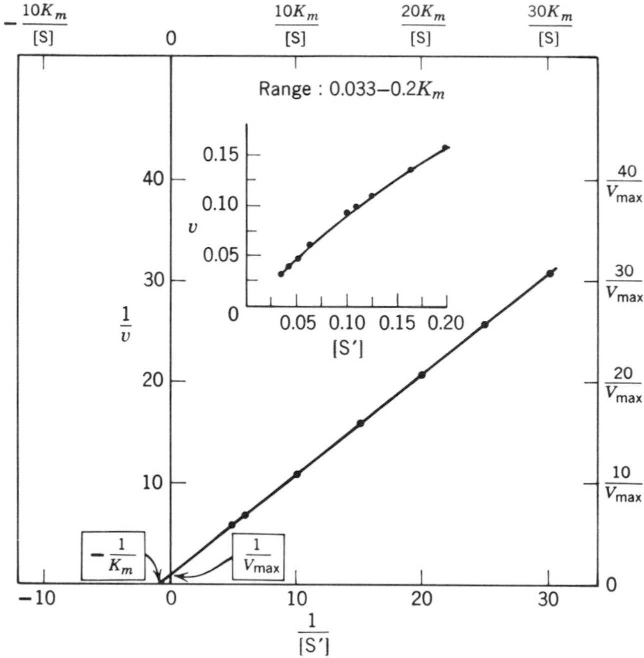

Fig. II-11. $1/v$ versus 1/[S] plot. The [S] range chosen is lower than optimal. The reaction is almost first-order with respect to [S].

example, we might find a column headed "substrate concentration" with units of $mM \times 10^2$. Below the heading we might find the figure 0.1. Some people interpret the heading as "the units of the data;" hence the 0.1 really represents 0.1×10^2 mM or 10 mM. Others interpret the heading as "the numbers shown below are 100 times the mM concentrations." The figure 0.1 then represents 0.1×10^{-2} mM or 0.001 mM. Most biochemists use the latter convention. To avoid confusion, it is desirable, whenever possible, to reduce the data to units that do not include factors. In the example above, the column may have been headed "substrate concentration, μM." Then 0.001 mM could have been entered as 1.0.

In the reciprocal plot shown in Figure II-9, the substrate concentration range was $0.33K_m$ to $2.0K_m$. If $K_m = 1 \times 10^{-5}$ M, the range becomes 0.33×10^{-5} M to 2.0×10^{-5} M. The various ways of expressing the 1/[S]-axis are shown below. Obviously, the most convenient way would be to use units of μM^{-1} and label the axis at 0.1, 0.2, and 0.3.

Table II-1 Different Ways of Labeling the 1/[S]-Axis of Reciprocal Plots

Values	Low Substrate (a)	Low Substrate (b)	Low Substrate (c)	High Substrate (a)	High Substrate (b)	High Substrate (c)
[S]	$0.33 \times 10^{-5}\,M$	$3.3 \times 10^{-6}\,M$	$3.3\ \mu M$	$2 \times 10^{-5}\,M$	$20 \times 10^{-6}\,M$	$20\ \mu M$
$\frac{1}{[S]}$	$3 \times 10^{5}\,M^{-1}$	$0.3 \times 10^{6}\,M^{-1}$	$0.3\ \mu M^{-1}$	$0.5 \times 10^{5}\,M^{-1}$	$0.05 \times 10^{6}\,M^{-1}$	$0.05\ \mu M^{-1}$
Number on axis	3	0.3	0.3	0.5	0.05	0.05
Units	$M^{-1} \times 10^{-5}$ or $10^{-5}\,M^{-1}$ or $10^{-5}/M$ or $(M \times 10^{5})^{-1}$ or $(10^{5}\,M)^{-1}$	$M^{-1} \times 10^{-6}$ or $10^{-6}\,M^{-1}$ or $10^{-6}/M$ or $(M \times 10^{6})^{-1}$ or $(10^{6}\,M)^{-1}$	μM^{-1}		(same as low [S])	

Graphical Analysis as a Method of Solving Simultaneous Equations

If we know the velocity, v_1, at one substrate concentration, $[S]_1$, and we also know the velocity, v_2, at a different higher substrate concentration, $[S]_2$, it is a simple matter to solve the two simultaneous equations for the two unknowns, K_m and V_{max}. All we need do is take the ratio of v_2/v_1:

$$\frac{v_2}{v_1} = \frac{\dfrac{[S]_2 V_{max}}{K_m + [S]_2}}{\dfrac{[S]_1 V_{max}}{K_m + [S]_1}} = \frac{[S]_2(K_m + [S]_1)}{[S]_1(K_m + [S]_2)}$$

Solving for K_m:

$$\boxed{K_m = \frac{[S]_2[S]_1(v_1 - v_2)}{v_2[S]_1 - v_1[S]_2}} \tag{II-34}$$

The V_{max} can be obtained by substituting the value of K_m into the original Henri-Michaelis-Menten expression for v_2 or v_1. In effect, the Lineweaver-Burk reciprocal plot has solved two simultaneous equations. We need at least two $1/v - 1/[S]$ points to draw a straight line. Once the line is drawn, we automatically obtain the two unknowns as intercepts. In Chapter Three we see that in the presence of an inhibitor, the velocity equation contains an additional constant, K_i. Again, we could determine all the constants by solving three simultaneous equations (or four equations if we do not know the type of inhibition), but it is far simpler to plot the data as two (or more) straight lines and extract the constants from intercepts and slopes (or replots thereof). Other methods of plotting enzyme kinetics data are described in Chapter Four.

Effect of Impure Substrate on K_m and V_{max}

Suppose that the substrate is quite impure but the contaminant is not inhibitory. The velocity is given by:

$$\boxed{\frac{v}{V_{max}} = \frac{y[S]_{add}}{K_m + y[S]_{add}} = \frac{[S]_{add}}{\dfrac{K_m}{y} + [S]_{add}}} \tag{II-35}$$

where $[S]_{add}$ = the concentration of added substrate (i.e., the assumed concentration of S which will be plotted)

y = the fractional purity of added S (as a decimal)

$\therefore\ y[S]_{add}$ = the true concentration of S.

The intercept on the $1/v$-axis gives the true $1/V_{max}$ but the observed K_m will be higher than the true K_m because the assumed (plotted) values of $1/[S]$ will be lower than the true values. If, for example, $y=0.50$ (S is only 50% pure), the K_m determined from the reciprocal plot will be high by a factor of 2. If the impurity is inhibitory, then both the K_m and V_{max} will change, as described in Chapter Three.

Eisenthal, Cornish-Bowden Plot and New Dixon Plot

While we might assume that all the properties of the hyperbolic velocity curve have been discovered by now, every few years something new turns up. For example, in 1974, Eisenthal and Cornish-Bowden showed that if the experimental [S] values are plotted on a negative horizontal axis, and the observed v values are plotted on a vertical axis, then straight lines drawn through the corresponding $-[S]$ and v points intersect at $[S]=K_m$ and $v=V_{max}$ (Fig. II-12a) (lines drawn through positive [S] and v points intersect at $-K_m$).

Another property of the hyperbolic velocity curve described by Dixon in 1972 is illustrated in Figure II-12b. If V_{max} can be determined easily (as v observed with a large excess of substrate), then straight lines can be drawn from the origin through various $[(n-1)/n]\ V_{max}$ points on the velocity curve to intersect the horizontal V_{max} line at some value of [S], called $[S]_n$. The n represents various whole numbers such that the lines are drawn through $\frac{1}{2}V_{max}$ $(n=2)$, $\frac{2}{3}V_{max}$ $(n=3)$, $\frac{3}{4}V_{max}$ $(n=4)$, and so on. The intercept, $[S]_n$, is related to [S], the actual substrate concentration needed for $[(n-1)/n]\ V_{max}$, as shown in the insert to Figure II-12b and equation II-36.

$$\boxed{[S]_n=\left(\frac{n}{n-1}\right)[S]} \tag{II-36}$$

The distance between one intercept and the next is always equal to K_m. For example, when $n=4$, $v=\frac{3}{4}V_{max}$. Therefore, $[S]_1=3K_m$ and $[S]_{n_1}=\frac{4}{3}[S]_1$; when $n=5$, $v=\frac{4}{5}V_{max}$. Therefore, $[S]_2=4K_m$ and $[S]_{n_2}=\frac{5}{4}[S]_2$. Then $\Delta=[S]_{n_2}-[S]_{n_1}=(1.25)(4K_m)-(1.33)(3K_m)=5K_m-4K_m=K_m$. If the V_{max} line is drawn too low, the increments will decrease in size as n increases. If the

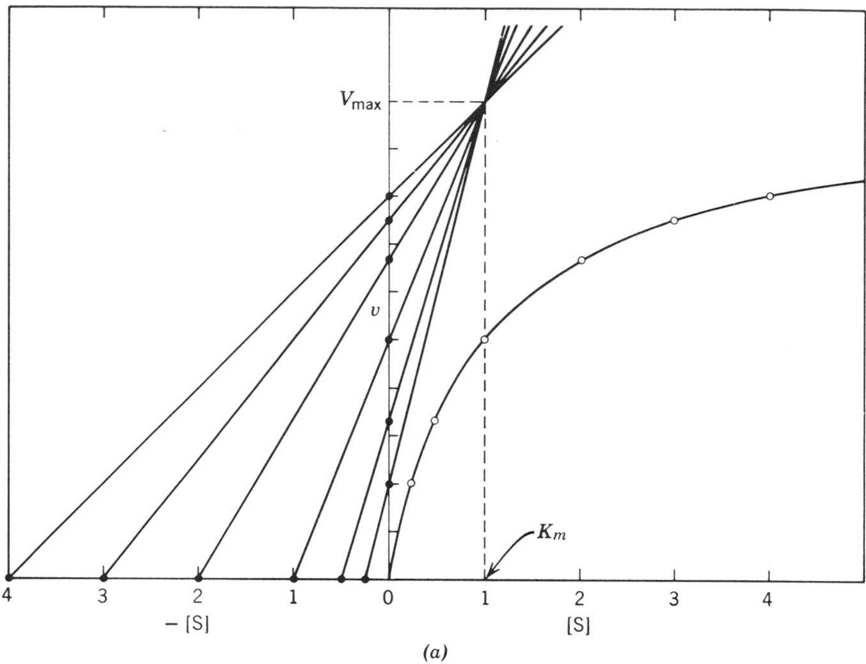

(*a*)

Fig. II-12. (*a*) Direct linear plot of $v-[S]$ data described by Eisenthal and Cornish-Bowden [*Biochem. J.* **139**, 715 (1974)] and by Merino [*Biochem. J.* **143**, 93 (1974)].

V_{max} line is drawn too high, the increments will increase in size as n increases. Thus the method provides a check on the assumed value of V_{max}. The first line is drawn through $\frac{1}{2}V_{max}$. The distance between the v-axis and the intercept of this line on the V_{max} line equal $2K_m$. The hypothetical line for $n=1$ divides this distance into two equal lengths of K_m each. The $n=1$ line is tangent to the velocity curve at the origin. If, after setting off an increment of K_m to the left of the $n=2$ line (i.e., after drawing the $n=1$ line) the remaining increment (between the v-axis and the $n=1$ line) is greater than K_m, then we may conclude that a substantial portion of the total substrate added is bound by the enzyme. In fact, the major advantage of this new Dixon plot is that it provides a way of determining K_m and $[E]_t$ for an enzyme that has a very high affinity for its substrate such that under assay conditions $[S]_{free} \neq [S]_{total}$. A similar plot can be used for inhibition systems where I is very strongly bound by the enzyme such that $[I]_{free} \neq [I]_{total}$. The equations and the plots are described in more detail in a following section (p. 74) and Chapter Three.

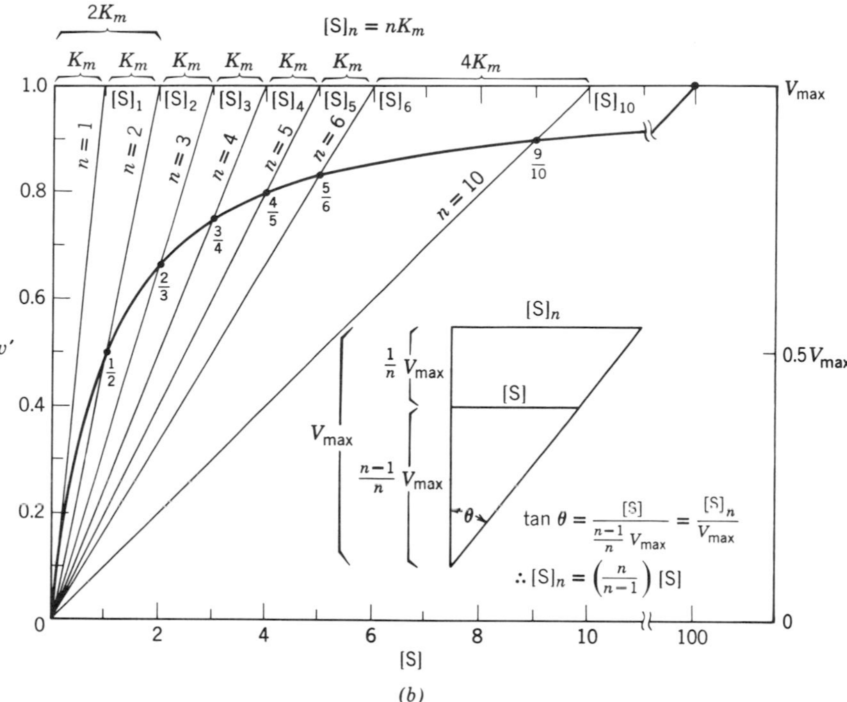

(b)

Fig. II-12. (b) An interesting geometric property of the velocity curve described by Dixon [*Biochem. J.* **129**, 197 (1972)]. Lines drawn from the origin through various $V_{max}(n-1)/n$ points intersect a horizontal line at V_{max} at increments of K_m.

Log v Versus Log [S] Plot

The log v versus log [S] plot is the most convenient way of expressing initial velocity data obtained over a very wide range of substrate concentrations. The plot provides a preliminary estimate of K_m and V_{max} and in some cases will disclose the presence of multiple enzymes catalyzing the same reaction (as described in a following section). Figure II-13 shows the log-log plot obtained in the author's laboratory for methylamine transport by *Pencillium chrysogenum*. The plot shows that methylamine uptake by nitrogen-sufficient cells is apparently first-order throughout the entire concentration range studied. Nitrogen deficiency derepresses (or deinhibits) a saturable transport system with an apparent K_m in the region of 10^{-5} M. At high methylamine concentrations (ca. 10^{-3} M), there is little difference between the two types of cells in their ability to transport methylamine. However, at low substrate concentrations ($<10^{-5}$ M), the nitrogen deficient cells can transport methylamine about 1000-times faster than the nitrogen sufficient cells.

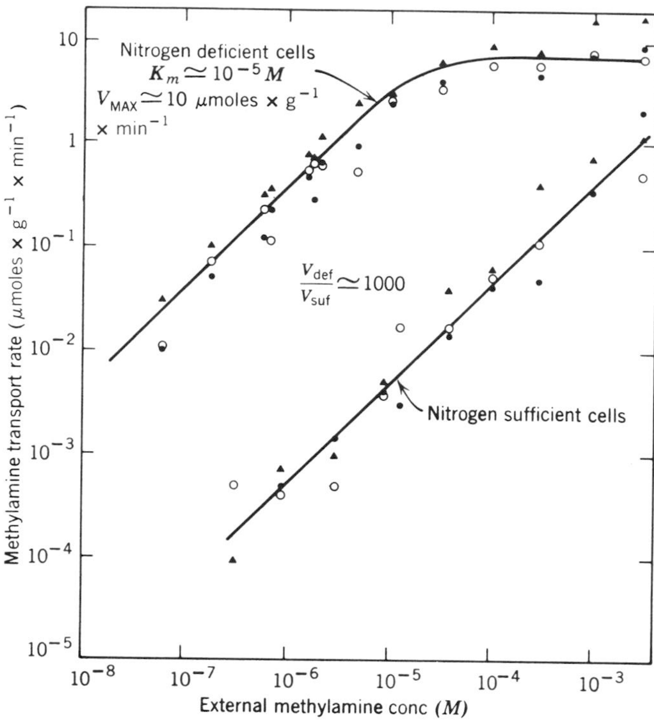

Fig. II-13. Log v versus log [S] plot for methylamine transport by *Penicillum chrysogenum*. [Redrawn from Hackette, S. L., Skye, G. E., Burton, C. and Segel, I. H., *J. Biol. Chem.* **245**, 4241, (1970).]

K. INTEGRATED FORM OF THE HENRI-MICHAELIS-MENTEN EQUATION

Integrated Rate Equation Assuming No Product Inhibition ($K_{m_S} \ll K_{m_P}$) and that K_{eq} Is Very Large

The Henri-Michaelis-Menten equation is a differential velocity equation where v represents $d[\text{P}]/dt$ or $-d[\text{S}]/dt$. Thus, to determine K_m and $\text{V}_{\max}$ by one of the usual linear plots based on the Henri-Michaelis-Menten equation (e.g., $1/v$ versus $1/[\text{S}]$), the reaction must proceed to a negligible extent during the course of the assay (e.g., less than 5% of $[\text{S}]_0$ converted to P). Under some experimental conditions it may not be feasible to restrict the reaction extent to 5% or less. For example, it may be difficult to determine very low product concentrations and, consequently, it becomes necessary to allow the reaction to proceed until a substantial fraction of the initial $[\text{S}]_0$ is converted to P. In some assays, P is not measured directly, but rather is determined from the decrease in substrate concentration: $[P]=[S]_0-[S]$. Again, a reliable estimate of [P] may require that [S] be significantly less than $[S]_0$. These situations can be handled by using the integrated rate equation and incorporating as many points as desired between, for example, 10 and 90% conversion of S to P. The integrated equation is valid over the entire course of the reaction. The simplest integrated rate equation is derived below. The derivation assumes that the decrease in velocity with time results only from decreasing saturation of the enzyme, and not from product inhibition or approach to equilibrium.

Let

$$v=-\frac{d[S]}{dt}=\frac{V_{max}[S]}{K_m+[S]}$$

Rearranging and inverting:

$$V_{max}\,dt=-\frac{K_m+[S]}{[S]}\,d[S]$$

Integrating between any two times (e.g., zero-time, t_0, and any other time, t) and the corresponding two substrate concentrations ($[S]_0$ and [S]):

$$V_{max}\int_{t_0}^{t}dt=-\int_{[S]_0}^{[S]}\frac{K_m+[S]}{[S]}\,d[S]$$

Separating the terms in the right-hand expression:

$$V_{max}\int_{t_0}^{t}dt=-K_m\int_{[S]_0}^{[S]}\frac{d[S]}{[S]}-\int_{[S]_0}^{[S]}d[S]$$

$$V_{max}t=-K_m\ln\frac{[S]}{[S]_0}-([S]-[S]_0)$$

or

$$\boxed{V_{max}t=2.3K_m\log\frac{[S]_0}{[S]}+([S]_0-[S])} \qquad \text{(II-37)}$$

where

$$([\mathrm{S}]_0 - [\mathrm{S}]) = \text{concentration of substrate utilized by time } t$$

$$= [\mathrm{P}], \text{ the concentration of product produced by time } t$$

Unlike the integrated first-order rate equation (II-27) or integrated zero-order rate equatiion (II-30), equation II-37 can not be solved for [S] or [P] at any time t even if K_m and $V_{\max}$ are known. On the other hand, if $[\mathrm{S}]_0$ and [S] at two or more times are known, K_m and $V_{\max}$ can be calculated. Equation II-37 can be rearranged to a linear form:

$$\frac{2.3}{t}\log\frac{[\mathrm{S}]_0}{[\mathrm{S}]} = -\frac{1}{K_m}\frac{([\mathrm{S}]_0 - [\mathrm{S}])}{t} + \frac{V_{\max}}{K_m} \qquad \text{(II-38)}$$

Thus K_m and $V_{\max}$ may be determined by measuring the concentration of substrate utilized (or product produced) several times during the reaction and then plotting the appropriate values as shown in Figure II-14.

There are several other linear forms of the integrated rate equation:

$$\boxed{\frac{t}{[\mathrm{P}]} = \frac{K_m}{V_{\max}}\left(\frac{2.3\log\dfrac{[\mathrm{S}]_0}{[\mathrm{S}]}}{[\mathrm{P}]}\right) + \frac{1}{V_{\max}}} \qquad \text{(II-39)}$$

$$\boxed{\frac{t}{2.3\log\dfrac{[\mathrm{S}]_0}{[\mathrm{S}]}} = \frac{1}{V_{\max}}\left(\frac{[\mathrm{P}]}{2.3\log\dfrac{[\mathrm{S}]_0}{[\mathrm{S}]}}\right) + \frac{K_m}{V_{\max}}} \qquad \text{(II-40)}$$

$$\boxed{\frac{[\mathrm{P}]}{t} = -K_m\left(\frac{2.3}{t}\log\frac{[\mathrm{S}]_0}{[\mathrm{S}]}\right) + V_{\max}} \qquad \text{(II-41)}$$

where $[\mathrm{P}] = [\mathrm{S}]_0 - [\mathrm{S}]$. Equation II-39 has the same slope and vertical axis intercept as the plot of $1/v$ versus $1/[\mathrm{S}]$. Equation II-40 is analogous to the plot of $[\mathrm{S}]/v$ versus [S], while equation II-41 is analogous to the plot of v versus $v/[\mathrm{S}]$. These alternate linear plots for initial velocity studies are described in Chapter Four. When plotting data according to one of the integrated rate equations a good spread of points can be obtained by starting with an $[S]_0$ that is several-fold greater than K_m (e.g., $10K_m$) and allowing the reaction to proceed until [S] is significantly below K_m (e.g., $0.1K_m$).

Determination of K_m and V_{max} from $[\bar{S}]$ and $\bar{v}$

If a substantial fraction of the substrate is utilized during the assay, the K_m and V_{max} values determined from the reciprocal plot (or any other linear plot described in Chapter Four) will be in error. First of all, if $[S]_0 - [S] = \Delta[S]$ is an appreciable fraction of $[S]_0$, then neither $[S]_0$ nor [S] can be taken as the [S] variable. Secondly, $[P]/t$ represents an overall velocity which is neither the initial velocity $(-d[S]/dt)$ at $[S]=[S]_0$ nor the final velocity at [S]. The Lineweaver-Burk plot could be used if we knew exactly which substrate concentration between $[S]_0$ and [S] yields an instantaneous velocity the same as the observed overall velocity. Certainly, there is *some* value of [S] that gives a $-d[S]/dt = \Delta S/\Delta t$. Lee and Wilson (1971) have shown that the arithmetic mean substrate concentration, $\frac{1}{2}([S]_0 + [S])$ is an excellent approximation and with this value the modified Lineweaver-Burk equation can yield reliable estimates of the kinetic constants even though the extent of the reaction is significant. The modified plot is based on equation II-42:

$$\boxed{\frac{1}{\bar{v}} = \frac{K_m}{V_{max}} \frac{1}{[\bar{S}]} + \frac{1}{V_{max}}} \qquad \text{(II-42)}$$

where

$$\bar{v} = \frac{[P]}{t} = \frac{[S]_0 - [S]}{t} = \text{the overall velocity}$$

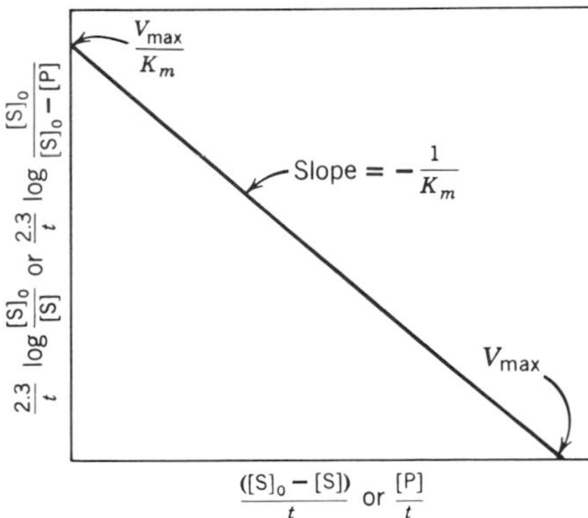

Fig. II-14. Plot of the integrated Henri-Michaelis-Menten equation.

and

$$[\bar{S}] = \frac{[S]_0 + [S]}{2} = \text{the arithmetical average substrate concentration over the course of the assay}$$

The modified equation can be compared to the integrated rate equation II-39, which can be written as:

$$\boxed{\frac{1}{\bar{v}} = \frac{K_m}{V_{max}}\left(\frac{2.3\log\dfrac{[S]_0}{[S]}}{[S]_0 - [S]}\right) + \frac{1}{V_{max}}} \tag{II-43}$$

The validity of equation II-42 depends on how closely $1/[\bar{S}]$ approximates the parenthetical term in equation II-43. (Remember that the integrated rate equation is valid for any extent of the reaction.) We can easily check the approximation. For example, suppose that the reaction proceeds until 30% of $[S]_0$ is utilized. Let $[S]_0 = 1.0$:

$$\frac{[S]_0 + [S]}{2} = [\bar{S}] = \frac{1.7}{2} = 0.85 \qquad \frac{2.3\log\dfrac{[S]_0}{[S]}}{[S]_0 - [S]} = \frac{2.3\log 1.43}{0.3}$$

$$\frac{1}{[\bar{S}]} = 1.18 \qquad = 1.19$$

Thus the modified Lineweaver-Burk plot introduces only about a 1% error in the determination of K_m even when the reaction has proceeded to the extent of 30%. A similar calculation will show an error of only about 4% when the reaction has proceeded to 50% utilization of initial substrate. Figure II-15 shows the relative error in K_m as a function of the extent of the reaction when the modified equation is used. For comparison, the error obtained by using $[S]_0$ instead of $[\bar{S}]$ is also shown. It is apparent that the Lineweaver-Burk plot (and the other linear plots described in Chapter Four) can be used to determine K_m and V_{max} even when the reaction extent is large, *provided* that (*a*) $[\bar{S}]$ rather than $[S]_0$ is used as the horizontal axis variable, (*b*) all cosubstrates are saturating, (*c*) the reaction is not significantly reversible, and (*d*) there is no product inhibition. Factors *c* and *d* can be checked easily. If necessary, the reaction can be made irreversible and product inhibition minimized by continually removing the products with trapping agents or coupling enzymes.

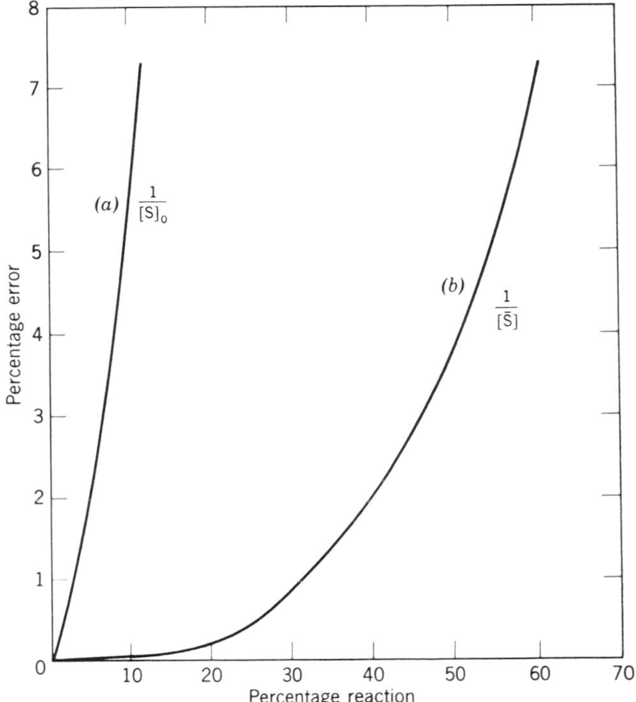

Fig. II-15. Percentage error in the value of K_m as a function of the extent of the reaction. Curve *a* plotting $1/[S]_0$. Curve *b* plotting $1/[\bar{S}]$. [Redrawn from Lee, H-J. and Wilson I. B., *Biochim. Biophys. Acta* **242**, 519, (1971).]

Integrated Rate Equation Where $K_{m_P} \cong K_{m_S}$ and K_{eq} Is Very Large

If the enzyme has an appreciable affinity for the product, but K_{eq} is still very large, then the decrease in velocity with time will result from a combination of decreasing saturation of the enzyme with substrate and increasing product inhibition. The velocity at any time t is given by:

$$v = -\frac{d[S]}{dt} = \frac{V_{max}[S]}{K_{m_S}\left(1 + \frac{[S]_0 - [S]}{K_{m_P}}\right) + [S]}$$

where

$$[S]_0 - [S] = [P] = \text{the product concentration at time } t.$$

Rearranging and inverting:

$$V_{\max}dt = \left(-\frac{K_{m_S}}{[S]} - \frac{K_{m_S}[S]_0}{K_{m_P}[S]} + \frac{K_{m_S}[S]}{K_{m_P}[S]} - \frac{[S]}{[S]} \right) d[S]$$

Grouping the constants K_{m_S}, K_{m_P}, and $[S]_0$ and integrating between $[S]_0$ and $[S]$ and time zero and time t:

$$V_{\max}\int_{t_0}^{t} dt = -K_{m_S}\left(1 + \frac{[S]_0}{K_{m_P}}\right)\int_{[S]_0}^{[S]} \frac{d[S]}{[S]} + \left(\frac{K_{m_S}}{K_{m_P}} - 1\right)\int_{[S]_0}^{[S]} d[S]$$

$$\boxed{V_{\max}t = 2.3K_{m_S}\left(1 + \frac{[S]_0}{K_{m_P}}\right)\log\frac{[S]_0}{[S]} + \left(1 - \frac{K_{m_S}}{K_{m_P}}\right)([S]_0 - [S])} \qquad \text{(II-44)}$$

The equation can be rearranged to several linear forms analogous to those shown above for $K_{m_P} \gg K_{m_S}$:

$$\boxed{\frac{2.3}{t}\log\frac{[S]_0}{[S]} = -\frac{1}{K_{m_S}}\left(\frac{K_{m_P} - K_{m_S}}{K_{m_P} + [S]_0}\right)\frac{[P]}{t} + \frac{V_{\max}}{K_{m_S}\left(1 + \dfrac{[S]_0}{K_{m_P}}\right)}} \qquad \text{(II-45)}$$

$$\boxed{\frac{t}{[P]} = \frac{K_{m_S}}{V_{\max}}\left(1 + \frac{[S]_0}{K_{m_P}}\right)\left(\frac{2.3\log\dfrac{[S]_0}{[S]}}{[P]}\right) + \frac{1}{V_{\max}}\left(1 - \frac{K_{m_S}}{K_{m_P}}\right)} \qquad \text{(II-46)}$$

The series of plots based on equation II-46 will have a common intersection point on the vertical axis for all values of $[S]_0$.

$$\boxed{\frac{t}{2.3\log\dfrac{[S]_0}{[S]}} = \frac{1}{V_{\max}}\left(1 - \frac{K_{m_S}}{K_{m_P}}\right)\left(\frac{[P]}{2.3\log\dfrac{[S]_0}{[S]}}\right) + \frac{K_{m_S}}{V_{\max}}\left(1 + \frac{[S]_0}{K_{m_P}}\right)} \qquad \text{(II-47)}$$

The family of curves for different values of $[S]_0$ will be parallel, since the slope of equation II-47 is independent of $[S]_0$.

$$\frac{[P]}{t}=-K_{m_S}\left(\frac{K_{m_P}+[S]_0}{K_{m_P}-K_{m_S}}\right)\left(\frac{2.3}{t}\log\frac{[S]_0}{[S]}\right)+\frac{V_{max}}{\left(1-\frac{K_{m_S}}{K_{m_P}}\right)} \qquad \text{(II-48)}$$

Plots of equation II-48 will intersect on the vertical axis for all values of $[S]_0$. Plots based on equations II-45, II-47, and II-48 can have positive or negative slopes, depending on the relative values of K_{m_S} and K_{m_P}. Similarly, vertical axis intercepts of plots based on equations II-46 and II-48 can be positive or negative. When $K_{m_S}=K_{m_P}$, equations II-45 and II-47 yield plots with zero slope; equation II-48 will yield vertical lines (infinite slope and vertical axis intercept). Equation II-46 will yield plots that intersect at the origin. All three unknowns, K_{m_S}, K_{m_P}, and V_{max} can be determined from a series of plots obtained for different initial substrate concentrations, $[S]_0$. For example, the slope of the plot described by equation II-48 is given by:

$$slope=-K_{m_S}\left(\frac{K_{m_P}+[S]_0}{K_{m_P}-K_{m_S}}\right)=-\frac{K_{m_S}}{(K_{m_P}-K_{m_S})}[S]_0-\frac{K_{m_S}K_{m_P}}{(K_{m_P}-K_{m_S})}$$

Thus the replot of *slope* versus $[S]_0$ is linear. When $slope=0$, the intercept on the horizontal axis is $-K_{m_P}$. With K_{m_P} known, K_{m_S} can be calculated from the slope or vertical axis intercept of the replot. Then V_{max} can be calculated from the vertical axis intercept of the original plot. The plot of equation II-48 and the *slope* replot are shown in Figure II-16 for a system where $K_{m_P}>K_{m_S}$. For convenience, $-$*slope* is plotted versus $[S]_0$ in the replot.

Integrated Rate Equation Where $K_{m_P}\simeq K_{m_S}$ and K_{eq} Is Not Very Large

The complete integrated rate equation, taking into account both product inhibition and the reverse reaction is:

$$\left(\frac{V_{max_f}}{K_{m_S}}+\frac{V_{max_r}}{K_{m_P}}\right)t=\left(\frac{1}{K_{m_S}}-\frac{1}{K_{m_P}}\right)[P]$$
$$-2.3\log\left(1-\frac{[P]}{[P]_{eq}}\right)\left(1+\frac{[S]_0}{K_{m_P}}+\frac{\left(\frac{1}{K_{m_S}}-\frac{1}{K_{m_P}}\right)}{\left(\frac{V_{max_f}}{K_{m_S}}+\frac{V_{max_r}}{K_{m_P}}\right)}\frac{V_{max_r}}{K_{m_P}}[S]_0\right) \qquad \text{(II-49)}$$

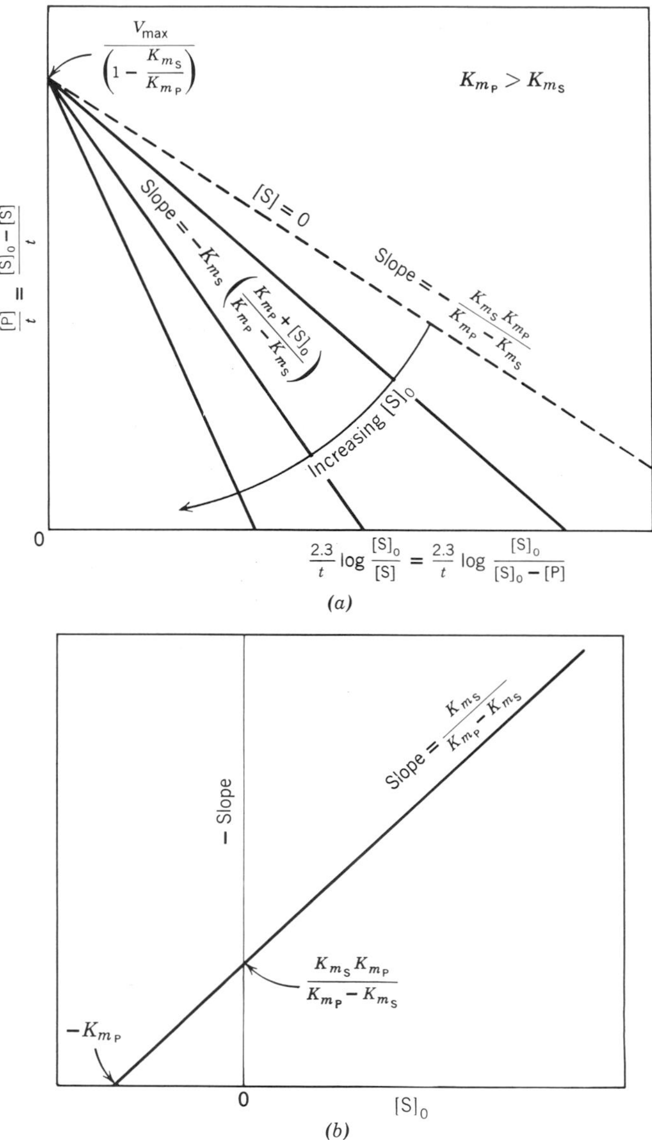

Fig. II-16. (*a*) Plot of the integrated velocity equation II-48 where $K_{m_P} > K_{m_S}$ and $K_{eq} \gg 1$. (*b*) Slope replot.

The four kinetic constants are not independent, but rather are related through the Haldane equation:

$$K_{eq} = \frac{[P]_{eq}}{[S]_{eq}} = \frac{V_{max_f} K_{m_P}}{V_{max_r} K_{m_S}}$$

Substituting for V_{max_r} from the Haldane equation:

$$\frac{V_{max_f}}{K_{m_S}}\left(1 + \frac{1}{K_{eq}}\right)t = \left(\frac{1}{K_{m_S}} - \frac{1}{K_{m_P}}\right)[P] - 2.3 \log\left(1 - \frac{[P]}{[P]_{eq}}\right)\left(1 + \frac{[S]_0}{K_{m_P}} + \frac{\left(\frac{1}{K_{m_S}} - \frac{1}{K_{m_P}}\right)}{1 + K_{eq}}[S]_0\right) \quad \text{(II-50)}$$

The equation can be rearranged to a linear form:

$$\frac{2.3}{t} \log\left(1 - \frac{[P]}{[P]_{eq}}\right) = \left(\frac{(K_{m_P} - K_{m_S})(1 + K_{eq})}{K_{m_P}K_{m_S}(1 + K_{eq}) + (K_{m_P} + K_{m_S}K_{eq})[S]_0}\right)\frac{[P]}{t} - \frac{V_{max}\left(1 + \frac{1}{K_{eq}}\right)}{K_{m_S} + \left(\frac{K_{m_S}K_{eq} + K_{m_P}}{K_{m_P}(1 + K_{eq})}\right)[S]_0} \quad \text{(II-51)}$$

The equations are complex, yet all that is required is a knowledge of K_{eq} and the product concentration during the course of the reaction at two or more initial substrate concentrations, $[S]_0$, and all four kinetic constants can be determined.

The integrated rate equations were derived for a one substrate-one product (Uni Uni) system. The corresponding equations for multisubstrate-multiproduct systems are exceedingly complex and are rarely employed, although sometimes it may be possible to treat a complex reaction as a Uni Uni reaction. For example, consider the one substrate-two product (Uni Bi) reaction: $E + S \rightleftharpoons ES \rightleftharpoons E + P + Q$. Fructose diphosphate aldolase catalyzes a

Uni Bi reaction, while many hydrolytic reactions yield apparent Uni Bi kinetics because the concentration of the second substrate (water) is constant (Chapter Nine). One or both of the products, P and Q, may compete with the substrate for free enzyme, depending on whether the release of P and Q is random or ordered. In a random release, P and Q both can combine with free enzyme, and thus act as competitive inhibitors. In an ordered reaction, only the last product released (Q) is a competitive inhibitor with respect to S, but P may act as a noncompetitive inhibitor if it binds appreciably to the EQ complex. If P is trapped chemically, or removed as it is formed by a coupled enzyme reaction, then the reaction can be treated as a Uni Uni reaction with Q as the sole accumulating competitive product inhibitor. The removal of P will also eliminate the reverse reaction. (In the ordered reaction, it is necessary to know beforehand which product is P and which is Q.) The calculated K_Q constant is an inhibition constant and may not equal the K_m value for Q in the reverse direction. In general, however, integrated rate studies for multireactant enzymes are far less profitable than initial velocity studies (Chapter Nine).

L. MULTIPLE ENZYMES CATALYZING THE SAME REACTION

Occasionally, a preparation may contain two or more enzymes (or multiple forms of the same enzyme) that catalyze the same reaction. The velocity at any substrate concentration is the sum of the velocities contributed by each enzyme. For two enzymes the velocity is given by:

$$v = \frac{[S]\,V_{max_1}}{K_{m_1} + [S]} + \frac{[S]\,V_{max_2}}{K_{m_2} + [S]} \qquad \text{(II-52)}$$

The multiplicity may go undetected even if the K_m and V_{max} values of the enzymes are quite different. Figure II-17 shows the reciprocal plots for five different combinations of two enzymes with K_m values differing by a factor of 10. All the plots for two enzymes are curved although the points over any limited range of $1/[S]$ may appear to fall on a straight line. In all cases the greatest curvature is in the region close to the $1/v$-axis, that is, at very high substrate concentrations. Unfortunately, this is also the region where the limitations of the assay frequently make it difficult to detect changes in initial velocity as the substrate concentration is increased. The relative contributions of two enzymes to the observed total initial velocity are shown in Figures II-18 to II-22. In the system where the enzyme with the higher

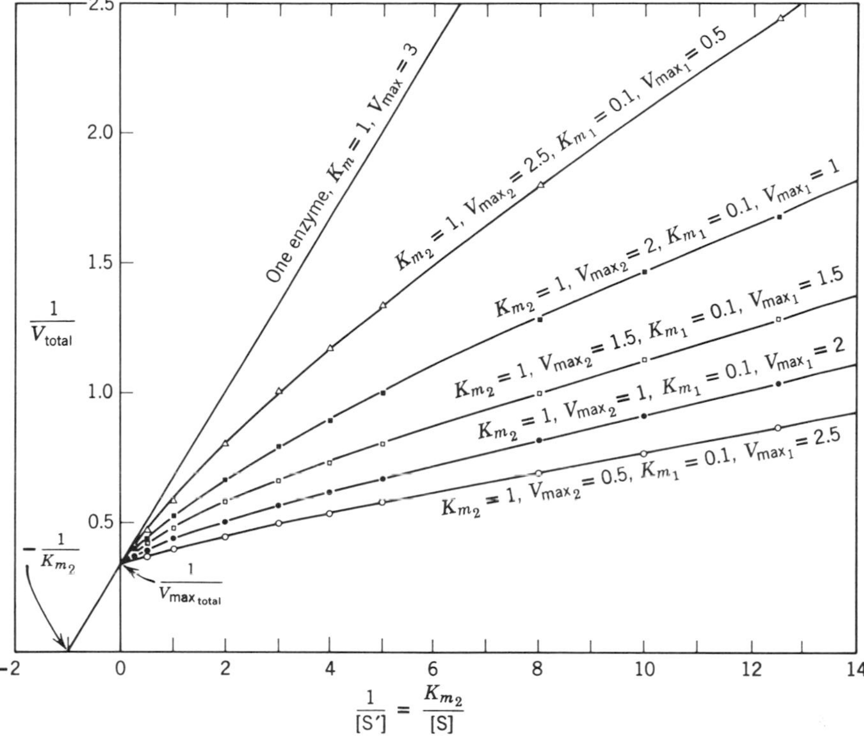

Fig. II-17. The $1/v$ versus $1/[S]$ plots when two enzymes are present: $V_{max_1} + V_{max_2} = 3$, and $K_{m_1} = 0.1 K_{m_2}$.

V_{max} has the lower K_m, the contribution of the lower V_{max}–higher K_m enzyme will be completely obscured (Fig. II-18). A reciprocal plot will appear linear over a wide range of $1/[S]$ and extrapolate close to the $1/V_{max}$ and $-1/K_m$ values for the lower K_m–higher V_{max} enzyme. If the two enzymes have the same K_m but different V_{max} values (Fig. II-19), the mixture behaves as a single enzyme with $V_{max} = V_{max_1} + V_{max_2}$. When the enzymes have the same V_{max} but different K_m values (Fig. II-20), the reciprocal plot bends downward close to the vertical axis as the contribution of the higher K_m enzyme to the total velocity increases at high substrate concentrations. The greatest curvature of the reciprocal plot occurs when the V_{max} and K_m of one enzyme are significantly higher than that of the other (Fig. II-22 and Fig. II-23). At low substrate concentrations, the relative contributions of the two enzymes will depend on the V_{max}/K_m ratios (i.e., the first-order rate constants) of the two enzymes. For example, in the system described in Figure II-21, both enzymes have V_{max}/K_m ratios of 100, and both contribute about equally to

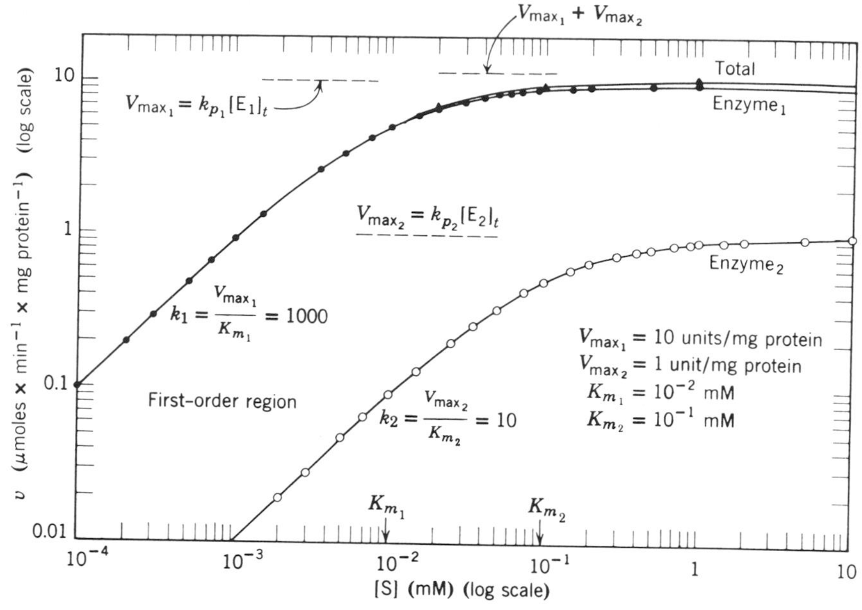

Fig. II-18. Log v versus log [S] plot in the presence of two enzymes: $V_{max_1} > V_{max_2}$, $K_{m_1} < K_{m_2}$.

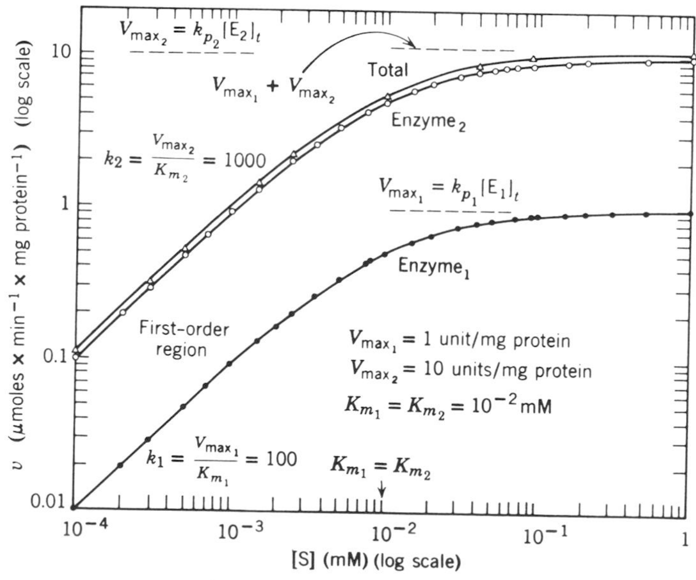

Fig. II-19. Log v versus log [S] plot in the presence of two enzymes: $V_{max_1} < V_{max_2}$, $K_{m_1} = K_{m_2}$.

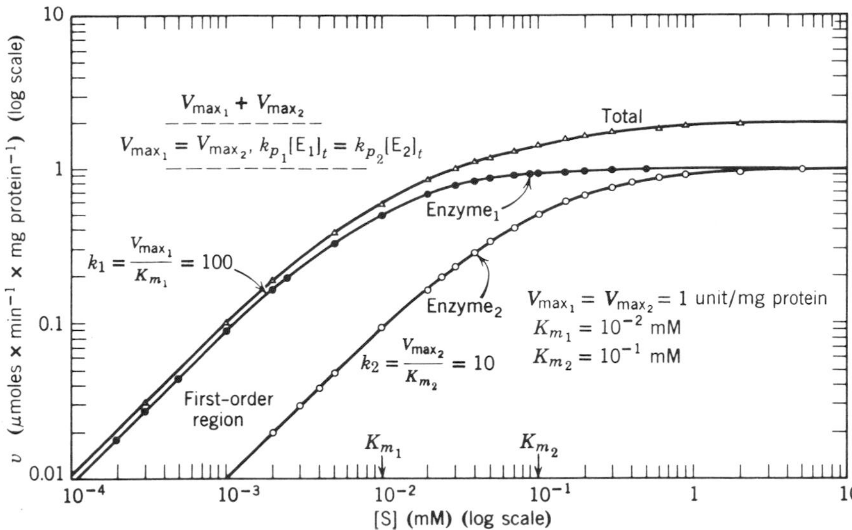

Fig. II-20. Log v versus log [S] plot in the presence of two enzymes: $V_{max_1} = V_{max_2}$, $K_{m_1} < K_{m_2}$.

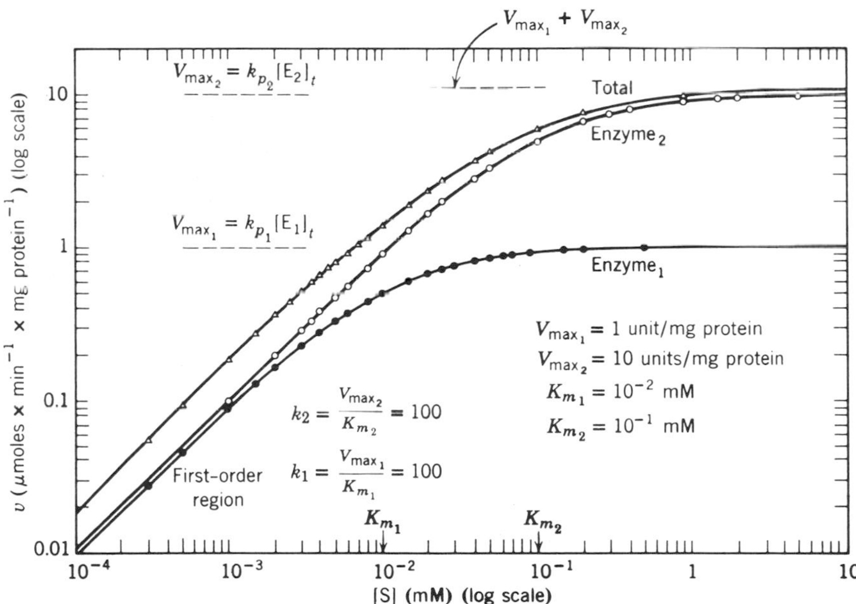

Fig. II-21. Log v versus log [S] plot in the presence of two enzymes: $V_{max_1} < V_{max_2}$, $K_{m_1} < K_{m_2}$.

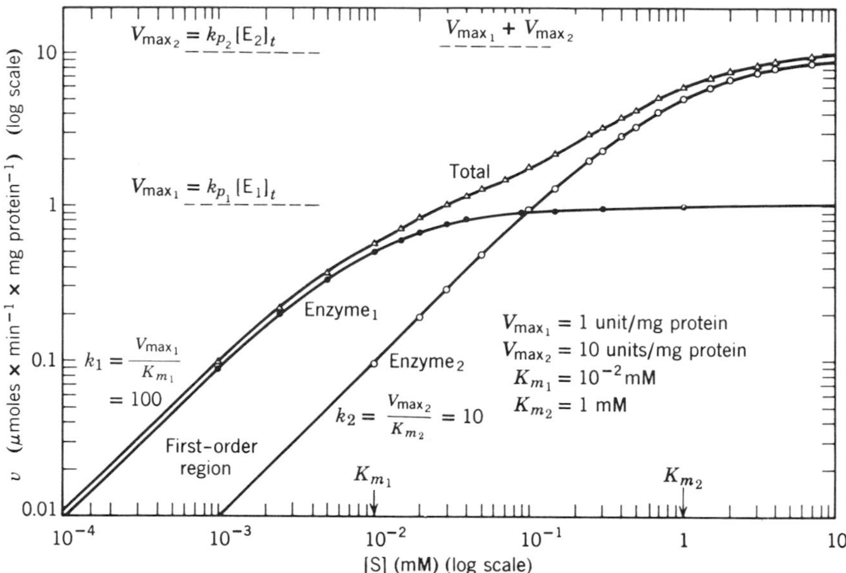

Fig. II-22. Log v versus log [S] plot in the presence of two enzymes: $V_{max_1} < V_{max_2}$, $K_{m_1} < K_{m_2}$.

the observed initial velocity in the first-order region. In the system described in Figure II-22, the V_{max}/K_m ratio of the low K_m–low V_{max} enzyme is 100, while the ratio for the high K_m–high V_{max} enzyme is 10. Accordingly, the low K_m–low V_{max} enzyme is the major contributor to the observed initial velocity at low [S]. In this system, the presence of two different enzymes is clearly indicated by the obvious inflection in the (total) velocity curve. Estimates of the K_m and V_{max} values can be obtained from reciprocal plots taking points from the region below the first inflection for the estimation of K_{m_1} and V_{max_1}, and points well within the second phase of the velocity curve for the estimation of K_{m_2} and $V_{max_1} + V_{max_2}$. At very low substrate concentrations, only the lower K_m enzyme contributes significantly to the observed velocity. Consequently, the reciprocal plot for the low substrate region approximates a straight line as 1/[S] increases (Fig. II-23). Extrapolation of the highest 1/[S]–1/v points to the 1/[S]-axis gives $-1/K_{m_1}$. Under experimental conditions, it may be difficult to obtain accurate measurements of the initial velocity at sufficiently low substrate concentrations, and the highest 1/[S] points may not be high enough to approximate a good straight line.

Occasionally, linear plots are obtained which seem to have two distinct regions that can be fitted with different straight lines. Barring experimental error, the plot suggests the presence of two different enzymes. This is illustrated in Figures II-24, II-25, and II-26. The curves were calculated

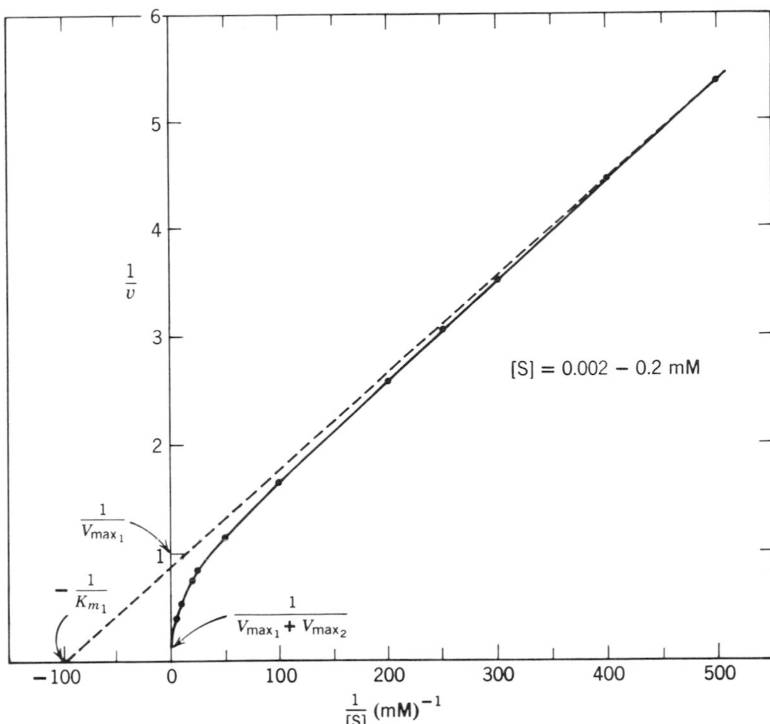

Fig. II-23. The $1/v$ versus $1/[S]$ plot in the presence of two enzymes: $V_{max_1} < V_{max_2}$, $K_{m_1} < K_{m_2}$ as in Fig. II-22. The [S] plotted is in the region of K_{m_1}.

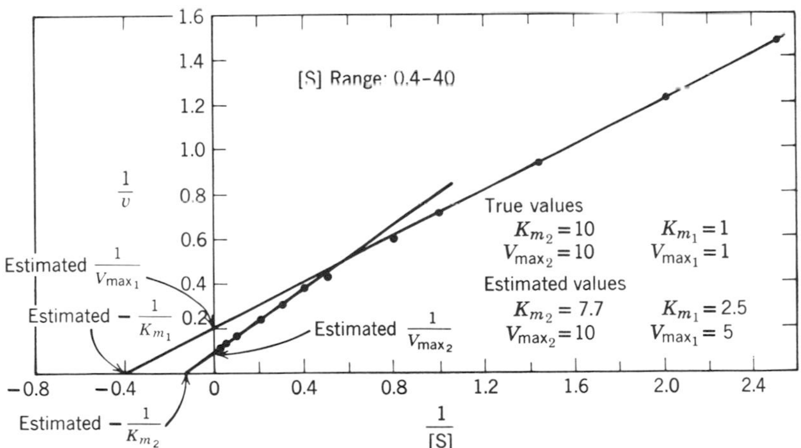

Fig. II-24. The $1/v$ versus $1/[S]$ plot in the presence of two enzymes.

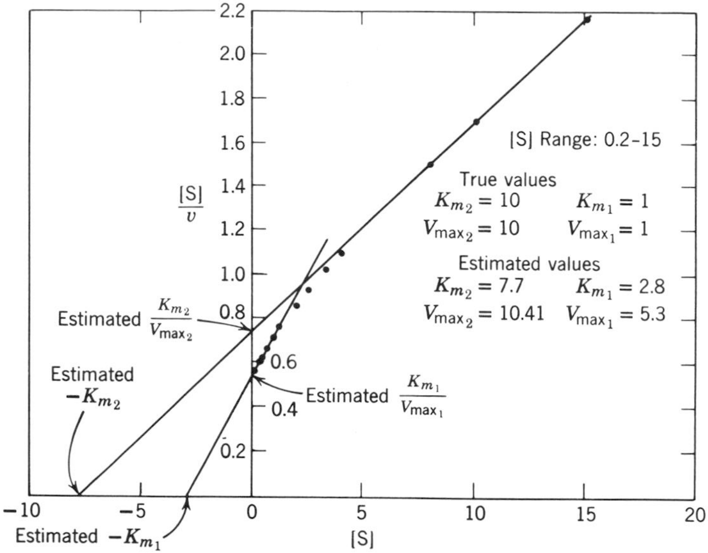

Fig. II-25. The [S]/v versus [S] plot in the presence of two enzymes.

assuming the additive velocities of two enzymes where $K_{m_1} = 1$, $V_{max_1} = 1$, and $K_{m_2} = 10$, $V_{max_2} = 10$. (The log-log plot would resemble that shown in Figure II-21.) The [S]/v versus [S] and v/[S] versus v plots are described in detail in Chapter Four. For the moment, however, we can see that these plots weight points at high and low v or high and low [S] more equally than the

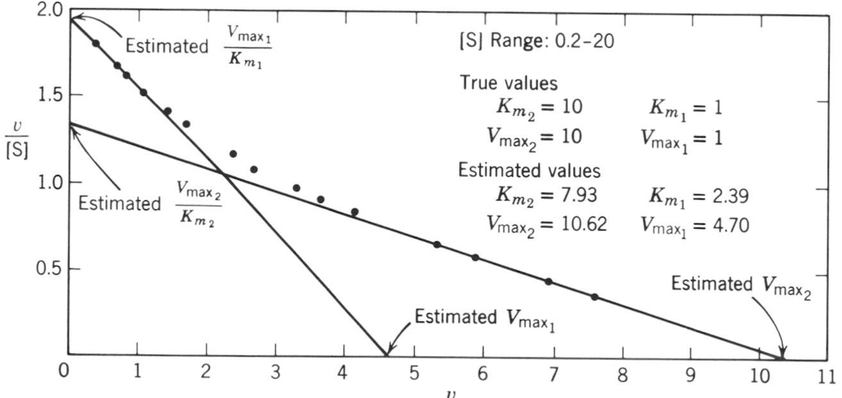

Fig. II-26. v/[S] versus v plot in the presence of two enzymes. Of the three linear plots, this one gives the best indication that more than one enzyme is present.

$1/v$ versus $1/[S]$ plot, and consequently, are more useful in detecting the presence of multiple enzymes. In spite of the apparent fit of the two straight lines on all three plots, the extrapolated intercepts do not yield the correct values for the constants. Spears, Sneyd, and Loten (1971) have described a procedure whereby the constants for two enzymes can be obtained by successive approximations. First, a linear plot is constructed from velocity measurements obtained at substrate concentrations above the approximate K_m of the high K_m enzyme (enzyme 2) (e.g., $[S]=5$ to 100 for the example shown in Figures II-24 to II-26). The plot will give a first estimate of V_{max_2} and K_{m_2} (of about 10.8 and 8.3, respectively). The estimated constants are then used to calculate the velocity of enzyme 2 at substrate concentrations below the approximate K_m of the low K_m enzyme (enzyme 1) (e.g., $[S]=0.1$ to 2 for the system shown in Figures II-24 to II-26). The contribution of enzyme 2 in this [S] region is subtracted from the observed velocity, and a linear plot is constructed from the corrected v values. This plot gives a first estimate of V_{max_1} and K_{m_1} (of about 0.46 and 0.67, respectively). These values in turn are then used to calculate the contribution of enzyme 1 in the high [S] region that is used to determine V_{max_2} and K_{m_2}. The contribution of enzyme 1 is subtracted from the observed velocities, and the corrected values plotted to obtain a second estimate of V_{max_2} and K_{m_2} (of about 10.4 and 9.1, respectively). The better estimates of V_{max_2} and K_{m_2} are used to correct the observed velocity at low [S] and the corrected values are plotted to obtain a second (and closer) estimate of V_{max_1} and K_{m_1} (0.71 and 0.80, respectively). The entire procedure can be repeated if closer estimates are desired. In general, the presence of two enzymes will be obvious, and reliable estimates of the four constants can be obtained only if $K_{m_2} > K_{m_1}$ and $V_{max_2} \geqslant V_{max_1}$. In addition, V_{max_2}/K_{m_2} (i.e., the first order rate constant, k_2) must be about equal to or less than V_{max_1}/K_{m_1} (i.e., k_1) so that a substrate concentration region exists where the low K_m enzyme contributes significantly to the observed velocity. If k_2 is much greater than k_1, enzyme 2 will obscure enzyme 1 even at low [S] concentrations. (If $k_2 > k_1$, the *estimated* contribution of enzyme 2 to the velocity at low [S] may be equal to or greater than the actual observed velocity because of experimental error and the fact that the first estimate of K_{m_2} will be slightly low.) In this case, it will be necessary to fractionate the preparation to obtain separate fractions significantly enriched with only one of the enzymes. Alternatively, it may be possible to increase the apparent K_{m_2}, or decrease the apparent V_{max_2}, or both, with an inhibitor that is specific for enzyme 2. Finally, keep in mind that curved reciprocal plots and multiphasic velocity curves do not necessarily indicate the presence of multiple enzymes. For example, multisite enzymes that have substrate binding sites of different affinities or display negative cooperativity also yield curved reciprocal plots.

M. KINETIC BEHAVIOR AT HIGH ENZYME CONCENTRATIONS

The [S] term in the Henri-Michaelis-Menten equation refers to *free* substrate, $[S]_f$. Generally, we do not bother to discriminate between $[S]_f$ and total substrate, $[S]_t$. That is, the derivation of the Henri-Michaelis-Menten equation assumes that $[S]_t$ is much greater than the total enzyme concentration, $[E]_t$, so that any decrease in the concentration of S by formation of ES is negligible. Thus we include the mass balance $[E]_t = [E] + [ES]$ in the derivation, but we neglect the mass balance $[S]_t = [S]_f + [ES]$. Under most *in vitro* assay conditions the enzyme concentration ranges from about 10^{-12} to $10^{-7}M$, while the substrate concentration ranges from about 10^{-6} to $10^{-2}M$. Under these conditions, the assumption that $[S]_f = [S]_t$ is quite valid.

Intracellular Concentrations of Enzymes

We might ask ourselves how valid is the assumption that enzymes are present at "catalytic" concentrations *in vivo*? We can estimate the intracellular enzyme concentration in several ways. For example, the protein content of most cells represents about 15 to 20% of the fresh weight. The average molecular weight of soluble proteins is about 100,000 to 150,000. If we assume that the entire fresh weight of cells is water and that all the soluble proteins are enzymes, then the total molarity of enzymes is:

$$\frac{150\,\text{g/l}}{150{,}000\,\text{g/mole}} = 10^{-3}M$$

If we assume that there are about 1000 different enzymes in the average cell, then the average concentration of an enzyme is about $10^{-6}M$ and we might expect a range between 10^{-8} and $10^{-4}M$ for individual enzymes.

Another estimate can be obtained from the enzymic activity of a given tissue and the known molecular weight and turnover number of the enzyme. For example, suppose we observe 40 units of enzyme activity per gram of fresh tissue for an enzyme with a specific activity of 3×10^9 units/mole. The molarity of the enzyme in the tissue is:

$$\frac{40 \times 10^3\,\text{units/l tissue water}}{3 \times 10^9\,\text{units/mole enzyme}} = 1.33 \times 10^{-5}M$$

The estimates above assume that the entire fresh weight of tissue is water and that there is no compartmentation. For mitochondrial enzymes, the apparent concentrations would be about five times higher because in most cells of higher organisms mitochondria occupy 20% of the cell volume.

Most enzymes are composed of subunits which range in molecular weight from about 15,000 to 60,000. The average subunit molecular weight is about 40,000. If each subunit carries a substrate binding site, then the actual molar concentration of the catalytic unit is increased by another factor of 2 to 4. The K_m values for different enzymes range from about 10^{-6} to $10^{-2} M$. It seems reasonable to expect that the intracellular concentrations of substrates are in the neighborhood of their K_m values (otherwise the full potential of the enzymes would not be realized). Thus it seems likely that enzyme concentrations within cells are of the same order of magnitude as their substrates and, consequently, a significant fraction of $[\mathrm{S}]_t$ may be bound as ES complexes. Substrate depletion by binding will also occur when $[\mathrm{E}]_t$ is very low if the enzyme has a very high affinity for S. For example, if $[\mathrm{E}]_t = 2 \times 10^{-8} M$ and $K_\mathrm{S} = 10^{-8} M$, then 50% of the total substrate will be bound as ES when $[\mathrm{S}]_t = 2 \times 10^{-8} M$. A general velocity equation, taking substrate depletion into account, is derived below. The equation expresses $v/V_{\max}$ in terms of $[\mathrm{E}]_t$ and $[\mathrm{S}]_t$.

$$\frac{[\mathrm{E}][\mathrm{S}]}{[\mathrm{ES}]} = K_\mathrm{S}$$

$$[\mathrm{E}] = [\mathrm{E}]_t - [\mathrm{ES}]$$

$$[\mathrm{S}] = [\mathrm{S}]_f = [\mathrm{S}]_t - [\mathrm{ES}]$$

$$\frac{([\mathrm{E}]_t - [\mathrm{ES}])([\mathrm{S}]_t - [\mathrm{ES}])}{[\mathrm{ES}]} = K_\mathrm{S} \qquad \text{(II-53)}$$

$$[\mathrm{E}]_t[\mathrm{S}]_t - [\mathrm{E}]_t[\mathrm{ES}] - [\mathrm{S}]_t[\mathrm{ES}] + [\mathrm{ES}]^2 = K_\mathrm{S}[\mathrm{ES}] \qquad \text{(II-53a)}$$

$$[\mathrm{ES}]^2 - ([\mathrm{E}]_t + [\mathrm{S}]_t + K_\mathrm{S})[\mathrm{ES}] + [\mathrm{E}]_t[\mathrm{S}]_t = 0$$

$$[\mathrm{ES}] = \frac{([\mathrm{E}]_t + [\mathrm{S}]_t + K_\mathrm{S}) - \sqrt{([\mathrm{E}]_t + [\mathrm{S}]_t + K_\mathrm{S})^2 - 4[\mathrm{E}]_t[\mathrm{S}]_t}}{2} \qquad \text{(II-53b)}$$

The positive root solution of the quadratic equation was neglected because the situation requires that $[\mathrm{ES}] = 0$ when either $[\mathrm{S}_t] = 0$ or $[\mathrm{E}]_t = 0$. This can only be true when the negative root solution is taken. A binding or velocity equation can be obtained by dividing by $[E]_t$:

$$\frac{[ES]}{[E]_t} = Y_S = \text{fraction of total sites occupied} = \frac{v}{V_{max}}$$

$$\therefore \boxed{\frac{v}{V_{max}} = \frac{([E]_t + [S]_t + K_S) - \sqrt{([E]_t + [S]_t + K_S)^2 - 4[E]_t[S]_t}}{2[E]_t}} \tag{II-54}$$

Equation II-54 is valid for all values of $[S]_t$ and $[E]_t$. When $[E]_t \ll K_S$, equation II-54 yields the same value for v/V_{max} or $[ES]/[E]_t$ as the Henri-Michaelis-Menten equation. As $[E]_t$ increases, the ratio at any fixed $[S]_t$ decreases because $[S]_f$ decreases.

The quadratic equation II-53a can be rearranged to a linear form which can be plotted to obtain K_S and $[E]_t$ if V_{max} is known (Henderson, 1973):

$$\boxed{\frac{[S]_t}{v} = K_S \frac{1}{V_{max} - v} + \frac{[E]_t}{V_{max}}} \tag{II-54a}$$

For binding studies, v and V_{max} can be replaced with $k_p[ES]$ and $k_p[E]_t$, respectively. The equation becomes:

$$\boxed{\frac{[S]_t}{[ES]} = K_S \frac{1}{[E]_t - [ES]} + 1} \tag{II-54b}$$

(See Chapter Four for a further discussion of equilibrium binding studies.)

Dixon Plot for Determining K_m and $[E]_t$

Dixon (1972) has described an elegant and simple direct plot for determining K_m and $[E]_t$ when a substantial fraction of the added $[S]_t$ is bound. Consider some point on the velocity curve where:

$$v = V_{max}\left(\frac{n-1}{n}\right) \tag{II-55}$$

where n is a whole number. The velocity at any [S] is given by:

$$v = k_p[\mathrm{ES}] \qquad \text{(II-56)}$$

Therefore, at the given point on the velocity curve:

$$k_p[\mathrm{ES}] = V_{\max}\left(\frac{n-1}{n}\right) \qquad \text{(II-57)}$$

As usual:

$$V_{\max} = k_p[\mathrm{E}]_t \qquad \text{(II-58)}$$

$$\therefore \quad k_p[\mathrm{ES}] = k_p[\mathrm{E}]_t\left(\frac{n-1}{n}\right) \quad \text{or} \quad [\mathrm{ES}] = [\mathrm{E}]_t\left(\frac{n-1}{n}\right) \qquad \text{(II-59)}$$

From equations II-15a and II-53, we can write:

$$K_m = \frac{([\mathrm{E}]_t - [\mathrm{ES}])([\mathrm{S}]_t - [\mathrm{ES}])}{[\mathrm{ES}]} = ([\mathrm{E}]_t - [\mathrm{ES}])\left(\frac{[\mathrm{S}]_t}{[\mathrm{ES}]} - 1\right) \qquad \text{(II-60)}$$

Substituting for [ES] from equation II-59 and simplifying:

$$K_m = \frac{[\mathrm{S}]_t}{n-1} - \frac{[\mathrm{E}]_t}{n} \qquad \text{(II-61)}$$

A series of lines are drawn from the origin through points on the velocity curve where $v = ((n-1)/n)V_{\max}$ (Fig. II-27*a*). Since n is a whole number (2, 3, 4, etc.), the points correspond to $\frac{1}{2}V_{\max}$, $\frac{2}{3}V_{\max}$, $\frac{3}{4}V_{\max}$, and so on. Each line intersects a horizontal line of height $V_{\max}$ at different $[\mathrm{S}]_n$ values, called $[\mathrm{S}]_2$ (for the line drawn through $\frac{1}{2}V_{\max}$), $[\mathrm{S}]_3$ (for the line drawn through $\frac{2}{3}V_{\max}$), and so on. The value of each $[\mathrm{S}]_n$ (as shown earlier in Fig. II-12*b*) is given by:

$$[\mathrm{S}]_n = \left(\frac{n}{n-1}\right)[\mathrm{S}]_t \qquad \text{(II-62)}$$

where $[\mathrm{S}]_t$ is the total substrate concentration required for a given n. Consequently,

$$[\mathrm{S}]_t = \left(\frac{n-1}{n}\right)[\mathrm{S}]_n \qquad \text{(II-63)}$$

and from equation II-61:

$$K_m = \left(\frac{n-1}{n}\right)\frac{[\mathrm{S}]_n}{n-1} - \frac{[\mathrm{E}]_t}{n} = \frac{[\mathrm{S}]_n}{n} - \frac{[\mathrm{E}]_t}{n}$$

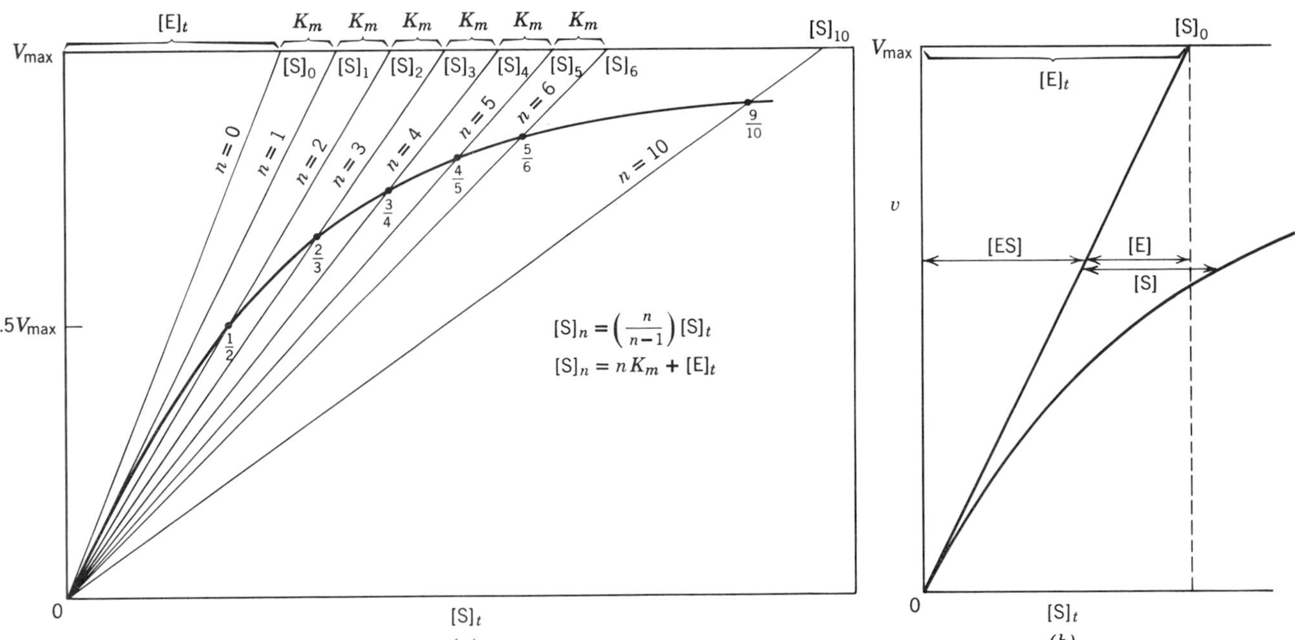

Fig. II-27. (*a*) Plot suggested by Dixon for determining $[E]_t$ when a significant fraction of the added substrate is bound to the enzyme. (*b*) Left-hand portion of (*a*).

or

$$[S]_n = nK_m + [E]_t \tag{II-64}$$

The intercepts on the V_{max} line occur at increments of K_m; that is,

$$[S]_n - [S]_{n-1} = K_m \tag{II-65}$$

A line drawn from the origin to a point on the horizontal V_{max} line one K_m distance to the left of the $n=2$ intercept represents the $n=1$ line. This line is tangent to the velocity curve at the origin. A line drawn similarly for $n=0$ intercepts the horizontal V_{max} line at $[E]_t$. Thus if $[E]_t \ll [S]_t$, this line is vertical and corresponds to the v-axis (as shown earlier in Fig. II-12*b*). The procedure requires a knowledge of V_{max}. As noted earlier, if the V_{max} value chosen is too low, the distances between intercepts (i.e., K_m) will decrease toward the right. If the V_{max} chosen is too high, the intervals will increase toward the right. Thus the method gives a check on the assumed value of V_{max}. The method also allows the concentration of all components of the system to be determined, as shown in Figure II-27*b*, which is just the left hand part of Figure II-27*a*. All that is required is a vertical line from $[S]_0$ to the $[S]_t$-axis and a diagonal line from $[S]_0$ to the origin. The components [ES], [E], and [S] (i.e., free S) as well as $[E]_t$ can be read off directly as appropriate distances between lines.

N. ENZYME ASSAYS

Initial Velocity as a Function of $[E]_t$

Under the usual *in vitro* assay conditions, the enzyme is present in limiting or "catalytic" amounts. The $[E]_t$ is generally 10^{-12} to $10^{-7} M$ while $[S]_t$ is generally 10^{-6} to $10^{-2} M$. At any substrate concentration the instantaneous or initial velocity is given by:

$$v = \frac{[S]\,V_{max}}{K_m + [S]} = \frac{[S]k_p[E]_t}{K_m + [S]} = \frac{k_p}{\left(1 + \dfrac{K_m}{[S]}\right)}[E]_t \tag{II-66}$$

Thus the initial velocity is directly proportional to $[E]_t$ at all substrate concentrations, and this fact can be used to quantitate the concentration of enzyme in any preparation, at any stage of purification. It should be stressed that the relationship between v and $[E]_t$ is linear only if true initial velocities are measured; that is, the rate of product formation must be constant over the entire time interval of the assay. Since v varies with [S], the assay period must be short enough to ensure that only a small fraction of the substrate is utilized (about 5% or less). Figure II-28*a* shows the appearance of product at different concentrations of enzyme and a fixed substrate concentration. The rate of product formation, $d[\mathrm{P}]/dt$, is constant for $[\mathrm{E}]_{t_1}$ to $[\mathrm{E}]_{t_4}$ up until time $= t_1$ (Fig. II-28*b*). If a longer assay time is chosen (e.g., t_2), the response would not be linear over the entire range of $[\mathrm{E}]_t$. Similarly, if an enzyme concentration greater than $[\mathrm{E}]_{t_4}$ is used, the response would not be linear for an assay time of t_1. Thus the first thing to do in any kinetic study of an enzyme is to establish the limits of linearity, that is, establish the maximum concentration of product that can accumulate before the [P] versus time and v versus $[\mathrm{E}]_t$ responses become nonlinear. This applies to all kinetic studies where v means initial velocity. In initial velocity studies with multireactant enzymes, a preliminary check for linearity should be made at several combinations of [A], [B], [C], and so on, especially at (*a*) the lowest concentrations of all substrates that will be used together and (*b*) the lowest concentration of each substrate that will be used in the presence of the highest concentrations of the others. If a linear response of [P] versus time cannot be obtained with any $[\mathrm{E}]_t$, because the concentration of one of the substrates changes markedly during the assay, then the equation of Lee and Wilson (equation II-42) can be used when that substrate is varied provided the concentrations of the other substrates do not change significantly.

Enzyme Units and Specific Activities—Quantitating $[\mathrm{E}]_t$

In most preparations the actual molar concentration of enzyme is unknown. Consequently, the amount of enzyme present can be expressed only in terms of its activity. To standardize the reporting of enzyme activities, the Commission on Enzymes of the International Union of Biochemistry has defined a standard unit:

> *One International Unit (1 U) of enzyme is that amount which catalyzes the formation of 1 μmole of product per minute under defined conditions.*

The concentration of enzyme in an impure preparation can be expressed in terms of units/ml (i.e., mM/min), and the specific activity of the preparation as units/mg protein. As the enzyme is purified, the specific activity will increase to a limit (that of the pure enzyme). Since v varies with [S], pH, ionic strength, temperature, and so on, a given preparation can have an infinite number of specific activities. Consequently, specific activities are usually reported for optimal assay conditions at a fixed temperature (usually 25°C, 30°C, or 37°C), with all substrates present at saturating concentrations. Thus the specific activity of a preparation represents V_{max}/mg protein.

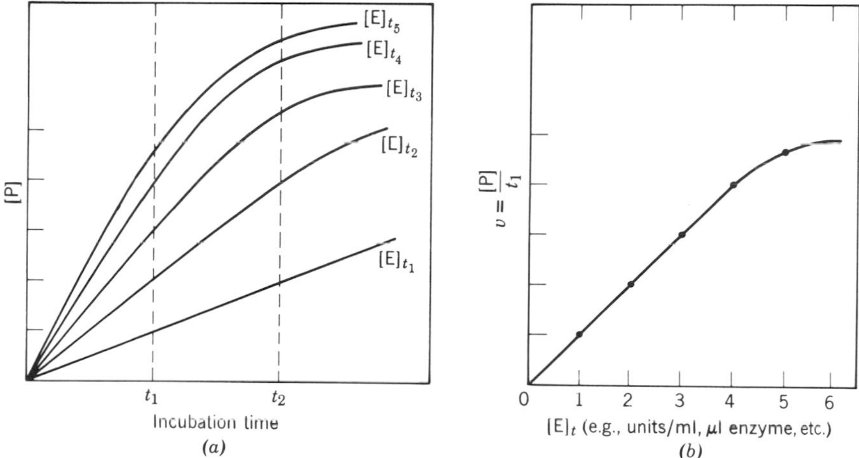

Fig. II-28. Enzyme assays: (*a*) Product formation as a function of time at different concentrations of enzyme. (*b*) Initial velocity (calculated as $[P]/t_1$) as a function of enzyme concentration.

Turnover Number

The term "turnover number" can be used in two ways. One way, which has been recently redefined as "molecular activity," is *the number of moles of substrate transformed per minute per mole of enzyme (units per micromole of enzyme) under optimum conditions*. Since many enzymes are oligomers containing n subunits, another possible "turnover number" is *the number of moles of substrate transformed per minute per mole of active subunit or catalytic center under optimum conditions*. This latter definition of catalytic power is called "catalytic center activity." Both values are sometimes given simply as a number with dimensions of min^{-1}.

$$k_p = \frac{V_{max}}{[E]_t} = \frac{\mu\text{moles (of S} \longrightarrow \text{P)} \times \text{min}^{-1} \times \text{ml}^{-1}}{\mu\text{moles (of E)} \times \text{ml}^{-1}} = \text{min}^{-1}$$

Values of k_p range from about 50 to about 10^7min^{-1}. Carbonic anhydrase has one of the highest turnover numbers known ($36 \times 10^6 \text{min}^{-1}$). The reciprocal of k_p gives the time required for a single catalytic cycle. Thus, for carbonic anhydrase, $1/k_p = 1/(36 \times 10^6) = 0.028 \times 10^{-6} \text{min} = 1.7\ \mu\text{sec}$.

Quantitation of $[E]_t$ Using the Integrated Velocity Equation

One form of the integrated velocity equation is:

$$\boxed{V_{max}t = 2.3K_m \log \frac{[S]_0}{[S]_0 - [P]} + [P]} \qquad \text{(II-67)}$$

Thus for any fixed $[S]_0$, the time required for the formation of a given [P] is inversely proportional to V_{max}; that is, $V_{max}t$ is constant for a given $[S]_0$ and [P]. Keep in mind that V_{max} is a measure of $[E]_t$. Thus the rule is usually stated as:

$$\boxed{[E]_t t = \text{constant}} \qquad \text{(II-68)}$$

If n units of enzyme produce 1 mM product in 5 min, then $2n$ units will yield 1 mM product in 2.5 min while $0.5n$ units will require 10 min. This relationship holds for all regions of $[S]_0$. Although $2n$ units yields 1 mM product in 2.5 min, it is not necessarily true that $2n$ units will yield 2 mM product in 5 min (unless $[S]_0 \gg K_m$ so that the reaction is zero order).

Reporting Data

Published data should always be reported in meaningful terms. Velocities and specific activities should be given in terms of units rather than as "cpm incorporated" or "ΔOD/min," although it is a good idea to include an extra vertical axis on at least one of the published figures where the actual raw data are shown. This serves to inform readers of the magnitudes of the actual observations. Raw data should not be reported in misleading terms. For example, do not report "cpm/mg protein," if only 5 μg of protein are used. Also, nonstandard "units" should be clearly defined (e.g., μmoles/hr or *n*moles/min).

Enzyme Purification

Enzymes are purified by employing successive chemical or physical fractionation procedures. The object of each step is to retain as much of the desired enzyme as possible while getting rid of as much of the other proteins, nucleic acids, and such, as possible. The efficiency of each step is given by the "yield" or "recovery" (the percentage of the total enzyme activity originally present that is retained) and the "purification" or "purification factor" (the factor by which the specific activity of the preparation has increased). The object is to optimize both factors. Somtimes a good yield is sacrificed for the sake of an excellent purification step; sometimes a good purification step is not used because the yield is too low. If the crude cell-free extract contains inhibitors, yields greater than 100% may be observed in the early stages of purification. Table II-2 shows a hypothetical purification scheme. The crude cell-free extract may be prepared by a number of means depending on the nature of the starting tissue or cells and the size of the preparation. Some common cell breakage methods include autolysis, freeze-thaw, sonic oscillation, mechanical grinding (with or without an abrasive), ballistic homogenization, or disruption in any one of a number of pressure cells (X-press, French press). The resulting homogenate is usually centrifuged to remove unbroken cells and large debris. There are no general rules concerning the order of the purification steps, although heat treatment (where possible) and ammonium sulfate precipitations are usually done early in the purification sequence. Gel filtration can follow ammonium sulfate precipitation and, thereby, desalt the preparation as well as fractionate the proteins according to size. If ion-exchange chromatography is to follow the ammonium sulfate step, then it is a good idea to dialyze the preparation first, or pass the preparation through a rapid gel filtration column (e.g., Sephadex G-25). The removal of the ammonium sulfate will facilitate the binding of the proteins to the ion-exchange column. Other steps not shown in Table II-2 that may be highly effective for certain enzymes include differential centrifugation (for mitochondria, chloroplasts, nuclei, microsomes, ribosomes), pH precipitation, organic solvent precipitation (e.g., ethanol, acetone), protamine sulfate or streptomycin sulfate pre-

Table II-2 Hypothetical Enzyme Purification Scheme

Step	Protein			Enzyme				
	Volume of Fraction (ml)	Concentration (mg/ml)	Total Amount (mg)	Concentration (units/ml)	Specific Activity (units/mg protein)	Total Amount (units)	Yield (%)	Purification Factor Fold
Crude cell-free extract	1000	12	12,000	5	0.416	5000	"100"	"1.00"
Heat step: 50°C for 5 min, then remove denatured protein	1000	8	8,000	4.8	0.60	4800	96	1.44
Ammonium sulfate precipitation: 30 to 50% saturation fraction	250	3	750	11.0	3.67	2730	55	8.83
Ion-exchange chromatography: *DEAE-sephadex*: elution via pH gradient. Fractions 50 to 60, 5 ml each, pooled, dialyzed and concentrated	25	9	225	88	9.8	2200	44	23.6
Ion exchange chromatography: *DEAE-sephadex*: elution via KCl gradient. Fractions 21 to 31, 2 ml each, pooled and concentrated	5	7	35	364	52	1820	36.4	125
Gel filtration: BioGel P-100. Fractions 30 to 40, 1 ml each, pooled.	10	0.92	9.2	170	185	1700	34	444
Hydroxyl apatite chromatography: Elution via phosphate buffer gradient. Fractions 15 to 18, 1 ml each, pooled.	4	0.75	3	375	500	1500	30	1200

cipitation (to precipitate nucleic acids and acidic proteins), affinity chromatography, and preparative gel electrophoresis. The purity of the final preparation should be checked by several methods before concluding that the preparation is homogeneous. Suitable methods include analytical disc gel electrophoresis at several pH values and gel concentrations, and ultracentrifugation. A homogeneous enzyme preparation should elute from an ion-exchange or gel filtration column as a single symmetrical activity and protein peak with a constant specific activity throughout. A homogeneous preparation is by no means necessary for kinetic analyses, but the purer the enzyme, the less the complications from competing reactions that may use up the substrate or the product.

Determination of v

An assay requires some method of measuring the rate of appearance of a product, $+d[\mathrm{P}]/dt$, or the rate of disappearance of a substrate, $-d[\mathrm{S}]/dt$. Theoretically, either measurement is sufficient but in practice the measurement of product appearance is usually more precise because the value is based on the difference between zero and a finite concentration. To measure v as $-d[\mathrm{S}]/dt$, the difference between $[\mathrm{S}]_0$ and $[\mathrm{S}]$ at time t must be large enough to determine accurately, but, at the same time, the difference must not be so large as to cause a significant change in the substrate concentration. The same problem arises in product inhibition studies. In this case, it is desirable to arrange conditions so that the product formed during the reaction can be distinguished from the product added to the reaction mixture. This can be accomplished by using radioactively labeled S and measuring the rate of labeled P appearance (assuming P* can be separated from S*). When two different products are formed, P and Q, it would be desirable to measure v as $d[\mathrm{Q}]/dt$ when P is the product inhibitor, and vice versa.

Assays with Auxiliary Enzymes

Frequently, the product of a reaction cannot be detected and quantitated directly, but it is possible to add an auxiliary enzyme which converts the product quantitatively to another substance which can be measured. The overall reaction sequence is:

$$\mathrm{A} \xrightarrow{\mathrm{E}_1} \mathrm{S} \xrightarrow{\mathrm{E}_2} \mathrm{P}$$

where E_1 is the enzyme being assayed and E_2 is the auxiliary enzyme. The conditions for the conversion of S to P by E_2 may not be compatible with those for the conversion of A to S. (One of the cosubstrates for the E_2

reaction may be an inhibitor of E_1; the pH optimum for the E_2 reaction may be quite different from that of the E_1 reaction; etc.) In this case, the assay is run in two stages. First A is incubated with E_1 (plus any cosubstrates) for a time sufficient to accumulate a detectable concentration of S. The reaction is then stopped (by boiling, or changing the pH, etc.). Then E_2 and all necessary cosubstrates are added, and the reaction is allowed to proceed until all the S accumulated in stage 1 is converted to P, which is then measured. The question is: how long should the second stage incubation time be to convert the maximum level of S to P for a fixed amount of E_2 and saturating cosubstrates of E_2? The incubation time for any level of S can be calculated from the integrated velocity equation:

$$\boxed{t=\frac{2.3K_{m_S}}{V_{\max_{E2}}}\log\frac{[S]_0}{[S]_0-[P]}+\frac{[P]}{V_{\max_{E2}}}} \qquad \text{(II-69)}$$

Since part of the reaction will be first-order, a 100% conversion of S to P will take an infinite time. Suppose we settle for 98% conversion, which would not introduce any significant error. If $K_{m_S}=2\times10^{-4}M$, $V_{\max_{E2}}=5\times10^{-5}M/\text{min}$ for the amount of auxillary enzyme used (i.e., $0.05\,\text{mM/min}=0.05\,\text{units/ml}$) and the maximum concentration of S that is allowed to accumulate is $1\times10^{-4}M$, the incubation time required to yield $0.98\times10^{-4}M$ P is:

$$t_{98\%}=\frac{(2.3)(2\times10^{-4})}{(5\times10^{-5})}\log\frac{10^{-4}}{2\times10^{-6}}+\frac{0.98\times10^{-4}}{5\times10^{-5}}$$

$$=(9.2)(1.7)+1.96=18\,\text{min}$$

If the auxillary enzyme is very expensive, then we may wish to calculate the minimum amount needed to "complete" the reaction in a reasonable time. For example, to obtain 98% conversion of $10^{-4}M$ S in 30 min, we require:

$$V_{\max_{E2}}=\frac{(2.3)(2\times10^{-4})}{30}\log\frac{10^{-4}}{2\times10^{-6}}+\frac{0.98\times10^{-4}}{30}$$

$$=2.94\times10^{-5}M/\text{min}=0.0294\,\text{mM/min}\cong0.03\text{ units of }E_2/\text{ml}$$

If the K_{m_S} is 100 times or more greater than the maximum concentration of S that is allowed to accumulate, then the reaction S→P will always be first-order with respect to [S]. In this case, the concentration of S can be determined by measuring the *initial velocity* of the second stage reaction.

$$v=k[S]=\frac{V_{\max_{E2}}}{K_{m_S}}[S] \qquad \text{or} \qquad [S]=\frac{vK_{m_S}}{V_{\max_{E2}}}$$

Kinetics of Coupled Assays

If none of the conditions required for the $S \xrightarrow{E_2} P$ reaction are detrimental to the reaction $A \xrightarrow{E_1} S$ (and vice versa), then both stages of the assay can be carried out simultaneously. The amount of E_1 present can be determined by measuring the velocity of P formation. The following conditions are necessary for a valid coupled assay: (*a*) the primary reaction must be zero-order with respect to [A] over the assay time and irreversible, and (*b*) the second-stage reaction must be first order with respect to [S] and irreversible. Condition *a* is easily met if only a small fraction of $[A]_0$ is utilized during the assay period or if $[A]_0 \gg K_{m_A}$ for E_1 and all cosubstrates are saturating. Irreversibility is assumed by the removal of S in the second-stage reaction. Condition *b* is met if $[S]_{ss}$, the steady-state concentration of S, is $\ll K_{m_S}$ for E_2. The $[S]_{ss}$ can be maintained $\ll K_{m_S}$ by using a sufficient excess of E_2. Irreversibility can be assumed if the equilibrium of the E_2 reaction lies far to the right, or one of the coproducts of the reaction is continuously removed, or if the reaction proceeds to only a small extent. Under these conditions, A will yield a certain $[S]_{ss}$ *after a short lag*; thereafter, the rate of P formation will be constant and proportional to $[E_1]$ (Fig. II-29). If $[E]_1$ is doubled, $[S]_{ss}$ will double, and the rate of P formation will double. On the other hand, doubling $[E_2]$ halves $[S]_{ss}$ but because the velocity of the reaction S→P is given by $v_2 = k_2[S]_{ss}[E_2]$, v_2 is unchanged. Thus once a sufficient excess of E_2 is present, the rate of P formation will be independent of $[E_2]$. The amount of E_2 required to make the overall velocity dependent only on $[E_1]$ can be determined by trial and error, or calculated as shown by McClure (1969) and outlined below.

The differential rate equation for the conditions of the coupled assay is:

$$\frac{d[S]}{dt} = k_1 - k_2[S] \qquad \text{(II-70)}$$

where $k_1 = v_1 =$ the pseudo-zero-order velocity of the A→S reaction
$k_2 =$ the first-order rate constant of the S→P reaction

Equation II-70 integrates to:

$$[S] = \frac{k_1}{k_2}(1 - e^{-k_2 t}) \qquad \text{or} \qquad [S] = \frac{k_1}{k_2}(1 - 0.5^{t/t_{1/2}}) \qquad \text{(II-71)}$$

where $[S] =$ the concentration of S at any time, t
$t_{1/2} =$ the half-life of the second stage reaction $= 0.693/k_2$

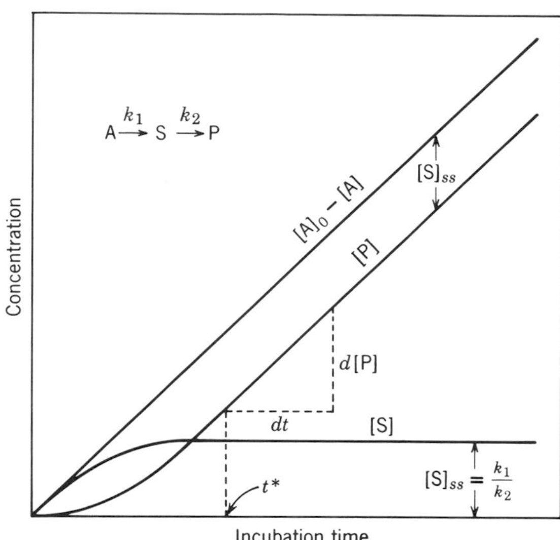

Fig. II-29. Kinetics of coupled assays: Substrate A disappears linearly from time zero, but the appearance of P becomes linear with time only after a lag period.

As $t \to \infty$, $[\mathrm{S}] \to [\mathrm{S}]_{ss}$, which is given by:

$$[\mathrm{S}]_{ss} = \frac{k_1}{k_2} = 1.44 t_{1/2} k_1 \qquad \text{(II-72)}$$

Thus the steady-state level of S represents the amount of A converted to S in 1.44 half-lives of the second stage reaction. Equation II-72 can be obtained directly by recalling that $v_1 = v_2$ when the steady-state is attained (i.e., $d[\mathrm{S}]/dt = 0$).

$$v_1 = k_1, \qquad v_2 = k_2[\mathrm{S}]_{ss} \qquad \therefore \quad [\mathrm{S}]_{ss} = \frac{k_1}{k_2} \qquad \text{(II-73)}$$

To obtain an expression for the time required to attain a practical fraction of $[\mathrm{S}]_{ss}$, equation II-71 is rearranged and put into logarithmic form:

$$1 - \frac{k_2}{k_1}[\mathrm{S}] = e^{-k_2 t}$$

$$2.3 \log\left(1 - \frac{k_2}{k_1}[\mathrm{S}]\right) = -k_2 t \qquad \text{or} \qquad 2.3 \log\left(1 - \frac{[\mathrm{S}]}{[\mathrm{S}]_{ss}}\right) = -k_2 t \qquad \text{(II-74)}$$

Letting F_S = the fraction of $[S]_{ss}$ desired at time = t^*, we obtain:

$$t^* = -\frac{2.3\log(1-F_S)}{k_2} \qquad \text{(II-75)}$$

Substituting $V_{\max_{E2}}/K_{m_S}$ for k_2 and rearranging:

$$V_{\max_{E2}} = -\frac{K_{m_S} 2.3\log(1-F_S)}{t^*} \qquad \text{(II-76)}$$

Equation II-76 gives the amount of auxillary E_2 required to obtain some practical fraction of $[S]_{ss}$ in time = t^*. Note that the time required to achieve a practical fraction of $[S]_{ss}$ is independent of k_1. As an example, suppose K_{m_S} for $E_2 = 2 \times 10^{-4} M$ at the pH and temperature used to assay E_1 and we accept $0.99[S]_{ss}$ as being experimentally equivalent to $[S]_{ss}$. How much E_2 must be added so that $[S]_t = 0.99[S]_{ss}$ in 6 sec?

$$V_{\max_{E2}} = \frac{-2\times 10^{-4} M(2.3\log 0.01)}{0.10\,\text{min}} = \frac{-(-9.2)\times 10^{-4}}{10^{-1}} = 9.2\,\text{mM/min} = 9.2\,\text{units/ml}$$

The 9.2 units/ml is the amount of auxillary enzyme required under the specific conditions used to assay E_1. If the stock solution of E_2 is given as 4600 units/ml at pH 8.5 and 37°C (optimum for E_2), but E_1 is being assayed at pH 6.8 and 25°C, then 2 μl of stock E_2 (i.e., "9.2 units") will not provide 9.2 units of activity. Obviously, the K_m and $V_{\max}$ of the auxillary enzyme should be determined under the conditions of use.

When two auxillary enzymes are used, the overall sequence is:

$$A \underset{E_1}{\overset{k_1}{\longrightarrow}} B \underset{E_2}{\overset{k_2}{\longrightarrow}} S \underset{E_3}{\overset{k_3}{\longrightarrow}} P$$

The object is to arrange conditions so that k_1 is a pseudo-zero-order rate constant while k_2 and k_3 are first-order rate constants; that is, $[B]_{ss} \ll K_{m_B}$ of E_2 and $[S]_{ss} \ll K_{m_S}$ of E_3, and only a small fraction of $[A]_0$ is utilized over the assay period. Under these conditions, the rate of P appearance will be constant (after a short lag) and proportional to $[E_1]$. The differential rate equation is:

$$\frac{d[S]}{dt} = k_2[B] - k_3[S]$$

Substituting for [B] from equation II-71:

$$\frac{d[\mathrm{S}]}{dt} = k_1(1 - e^{-k_2 t}) - k_3[\mathrm{S}]$$

or

$$\frac{d[\mathrm{S}]}{dt} = k_1 - k_1 e^{-k_2 t} - k_3[\mathrm{S}] \tag{II-77}$$

As $t \to \infty$, $d[\mathrm{S}]/dt \to 0$ and $[\mathrm{S}] \to [\mathrm{S}]_{ss}$ given by:

$$[\mathrm{S}]_{ss} = \frac{k_1}{k_3} \tag{II-78}$$

At any time, t, the concentration of S is given by:

$$[\mathrm{S}] = \frac{k_1}{k_3} - \frac{k_1}{k_3 - k_2}\left(e^{-k_2 t} - \frac{k_2}{k_3} e^{-k_3 t}\right) \tag{II-79}$$

When $t = \infty$, the parenthetical term reduces to zero and $[\mathrm{S}] = k_1/k_3 = [\mathrm{S}]_{ss}$. Equation II-79 can be rearranged to:

$$(k_3 - k_2)(1 - F_{\mathrm{S}}) = k_3 e^{-k_2 t^*} - k_2 e^{-k_3 t^*} \tag{II-80}$$

where F_{S} = some fraction of $[\mathrm{S}]_{ss}$ attained at time $= t^*$. Equations II-79 and II-80 are symmetrical with respect to k_2 and k_3; that is, a pair of values for k_2 and k_3 can be interchanged without affecting the time required to attain F_{S}. When $k_3 \gg k_2$, equation II-80 becomes:

$$\boxed{(1 - F_{\mathrm{S}}) = e^{-k_2 t^*}} \quad \text{or} \quad \boxed{t^* = -\frac{2.3 \log(1 - F_{\mathrm{S}})}{k_2}} \tag{II-81}$$

In other words, as k_3 becomes very large compared to k_2, [S] approaches $[\mathrm{S}]_{ss}$ as fast as [B] approaches $[\mathrm{B}]_{ss}$. When $k_2 \gg k_3$ the equations are:

$$\boxed{(1 - F_{\mathrm{S}}) = e^{-k_3 t^*}} \quad \text{or} \quad \boxed{t^* = -\frac{2.3 \log(1 - F_{\mathrm{S}})}{k_3}} \tag{II-82}$$

Thus if a large excess of one of the auxillary enzymes is used, the amount of the other auxillary enzyme required for a given lag time, t^*, can be calculated quite easily. In fact, if one constant is only 4 to 5 times greater than the other, equation II-81 or II-82 will predict t^* satisfactorily (a small difference between k_2 and k_3 causes a large difference in the exponential terms of equation II-80). When neither auxillary enzyme is in significant excess (i.e., $k_2 \cong k_3$), then there is no single solution for t^* (any number of combinations of $[E_1]$ and $[E_2]$ can produce a given lag). McClure (1969) has solved equation II-80 numerically and presented the solutions as a nomogram. Figures II-30*a* and *b* show the times required to reach 99% of $[S]_{ss}$ for different values of k_2 and k_3.

Although we have been concerned with using coupled assays to quantitate E_1, the same procedures can be used for initial velocity and product inhibition studies. For the latter, the inhibitory product of the E_1 reaction must be one other than that utilized by E_2 and must not be inhibitory to E_2 or E_3.

O. EFFECTS OF ENDOGENOUS SUBSTRATES

A crude cell-free extract may contain endogenous substrates of the enzyme under investigation. This extra substrate will be introduced into the assay mixture together with the enzyme and may contribute to the observed reaction velocity. A plot of v versus $[E]_t$ at some subsaturating substrate concentration will not be linear since increasing $[E]_t$ also increases the total [S]. For example, suppose the substrate concentration added to the assay mixture is equivalent to $0.1K_m$. Suppose further that the enzyme preparation contains 100 units of activity per 10 μl and is contaminated with endogenous substrate such that each 10 μl of extract contributes additional substrate equivalent to $0.1K_m$. The observed velocity for any amount of preparation is given by:

$$v=\frac{k_p[E]_t\dfrac{[S]_t}{K_m}}{1+\dfrac{[S]_t}{K_m}}=\frac{k_p[E]_t\dfrac{[S]_{add}+[S]_{endog}}{K_m}}{1+\dfrac{[S]_{add}+[S]_{endog}}{K_m}} \qquad \text{(II-83)}$$

where $k_p[E]_t$ = the units of enzyme activity
= V_{max} for the given enzyme concentration at saturating [S]
$[S]_{add}$ = the substrate concentration added to the assay mixture
$[S]_{endog}$ = the endogenous substrate concentration introduced along with the enzyme preparation
$[S]_t$ = the total substrate concentration in the assay mixture

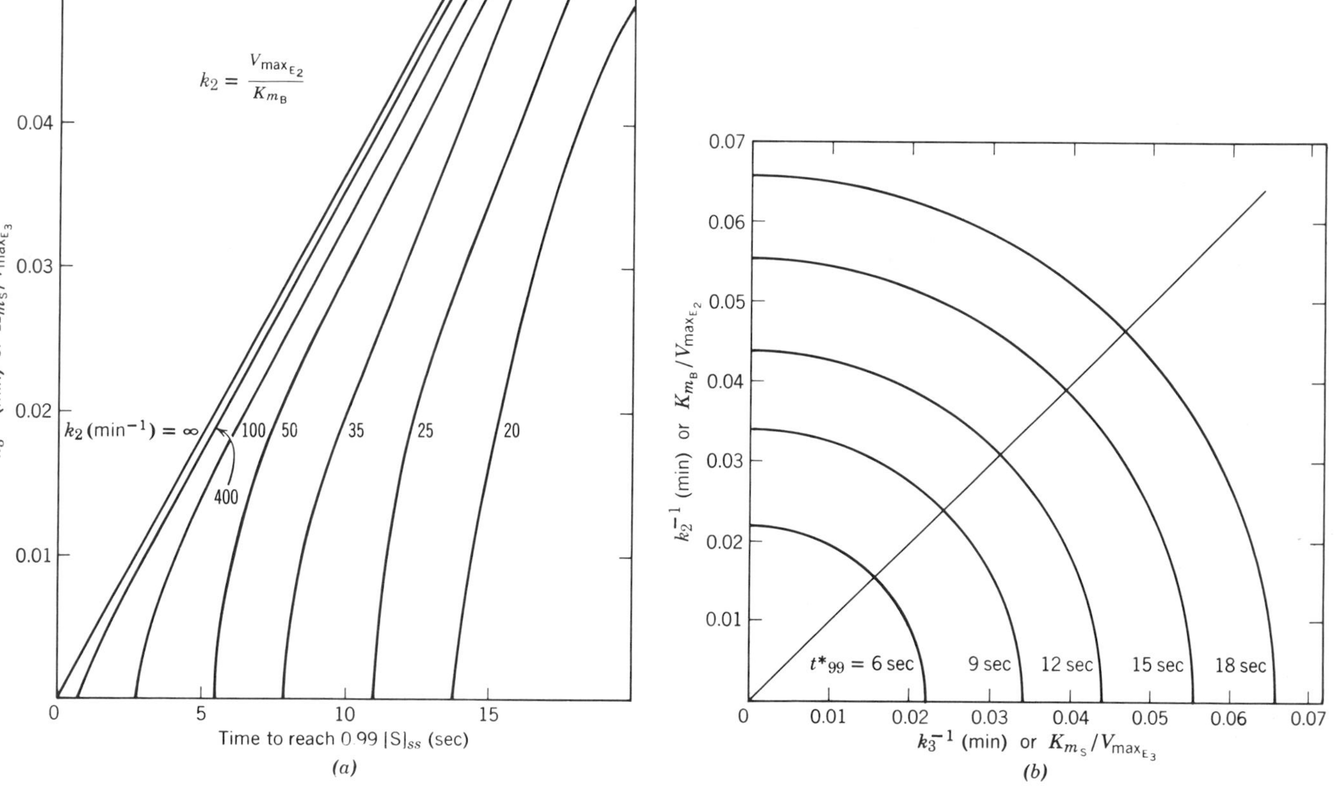

Fig. II-30. Nomograms of McClure showing the time required for [S] to attain 99% of its steady-state level in a coupled assay; k_2 and k_3 represent the effective first-order rate constants (V_{max}/K_m) of the two coupling enzymes. [Redrawn with permission from McClure, W. R., *Biochemistry* **8**, 2782 (1969) Copyright by the American Chemical Society.]

At 10 μl of preparation:

$$v = \frac{(100)(0.1+0.1)}{1+(0.1+0.1)} = 16.65 \text{ units}$$

At 20 μl of preparation:

$$v = \frac{(200)(0.1+0.2)}{1+(0.1+0.2)} = 46.2 \text{ units}$$

Thus doubling the enzyme concentration more than doubles the observed velocity; halving the enzyme concentration yields a velocity less than half the original velocity (Fig. II-31). As the amount of enzyme preparation increases, $[S]_t$ increases. Eventually, the enzyme becomes saturated, and the v versus $[E]_t$ curve approaches a straight line with a limiting positive slope. If the assays were conducted with $[S]_{add} \gg K_m$, then the enzyme would always be saturated with S and the endogenous substrate would have no significant effect on the observed velocity. The observed v would always equal V_{max} for the amount of enzyme added, and the V_{max} versus $[E]_t$ curve would be linear. Nonlinear v versus $[E]_t$ curves may also be observed if the enzyme preparation contains endogenous activators. If the activator functions by decreasing the apparent K_m, then the nonlinearity will disappear when the assay is run at saturating [S]. However, if the activator increases the apparent V_{max}, then nonlinear v versus $[E]_t$ plots will be seen even when $[S] \gg K_m$.

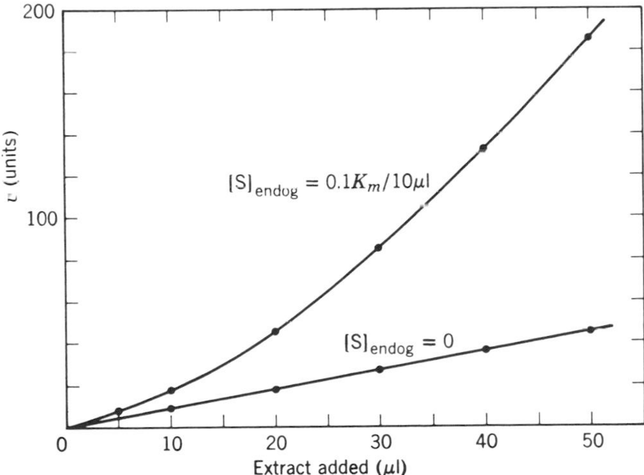

Fig. II-31. Effect of endogenous substrate on the plot of v versus amount of enzyme preparation added. $[E]_t = 100$ Units/10 μl of extract; $[S]_{endog} = 0.1\ K_m/10\ \mu$l of extract. The assays are run at $[S]_{add} = 0.1\ K_m$.

We can also consider the effect of endogenous substrates on the determination of K_m and V_{max}. The raw data would consist of a series of initial velocities observed at various concentrations of substrate and a fixed concentration of enzyme. If the enzyme preparation contains endogenous substrates, then we would observe a reaction velocity at zero added substrate. (The endogenous substrate would not have been used up prior to the assay if the cells were broken and centrifuged at 4°C, or if the reaction requires additional substrates or cofactors that are not present in the crude extract.) The v versus $[S]_{add}$ data might be as shown below in columns 1 and 2. The third column represents an attempt to "correct" the observed initial velocity for the contribution of the endogenous substrate. The numbers in the table were chosen for an enzyme with a K_m of 1.0, an $[E]_t$ per assay equivalent to a V_{max} of 1.0, and a level of contaminating endogenous substrate in the preparation sufficient to add $[S]_{endog} = K_m$ to each assay mixture. Figure II-32 shows the reciprocal plots of $1/v_{observed}$ versus $1/[S]_{add}$, $1/v_{\text{"corrected"}}$ versus $1/[S]_{add}$, and the true curve that would be observed if no endogenous substrate were present. The plot of $1/v_{observed}$ versus $1/[S]_{add}$ is curved and intercepts the $1/v$-axis at $1/V_{max}$. This is not surprising since the contribution of the endogenous substrate disappears as the concentration of added substrate approaches infinity. At high values of $1/[S]_{add}$, the plot approaches a horizontal limit. In other words, as the concentration of added substrate approaches zero, the velocity plateaus at the value dictated by the fixed concentration of endogenous substrate. The "corrected" data yield a linear reciprocal plot, but the intercepts do not give the correct values for K_m and V_{max}. The fact that the plot is linear can easily deceive an investigator into believing that the "correction" was valid. The correction would be valid if the "zero added substrate" value represented endogenous noninhibitory product or a compound that behaves like the true product in the final

$[S]_{add}$	Observed v	"Corrected" v
0	0.500	0.000
0.5	0.600	0.100
1	0.667	0.167
2	0.750	0.250
3	0.800	0.300
5	0.858	0.358
8	0.900	0.400
20	0.955	0.455
100	0.990	0.490

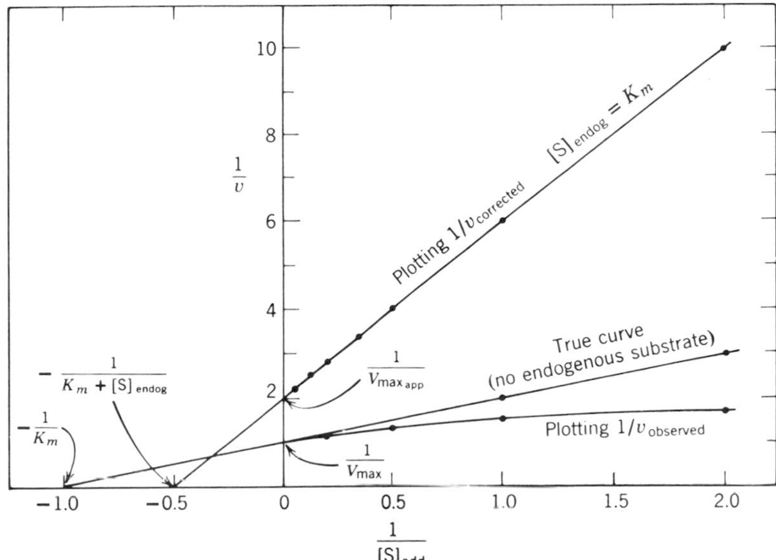

Fig. II-32. $1/v$ versus $1/[S]$ plot in the presence of an endogenous substrate.

analysis. (Non-enzymatic formation of P from endogenous or added S will also complicate the determination of K_m and V_{max}. In this case, a "minus enzyme" control should be run at each added substrate concentration.) Because valid corrections for endogenous substrates or activators cannot be made easily, it is good practce to dialyze or pass a crude extract through a gel filtration column before assaying the enzyme.

If an assay is based on the rate of production of radioactively labeled P from labeled S, endogenous unlabeled substrate will not only affect the assay substrate concentration, but also the specific radioactivity of the substrate. The effect on the calculated reaction rate is illustrated in Table II-3. Column 1 indicates the concentration of labeled substrate in the assay mixture (i.e., the assumed concentration). Column 2 indicates the true substrate concentration assuming that the amount of crude cell-free extract contributes $[S] = K_m$. Column 3 indicates the actual specific radioactivity of the substrate after dilution of the labeled added substrate (S.A. $= 10^5$ cpm/μmole) by endogenous unlabeled substrate. Column 4 indicates the true reaction rate. Column 5 indicates the observed rate in terms of cpm/min. If this rate is converted to μmoles/min using the assumed specific activity of the substrate (10^5 cpm/μmole), we obtain the calculated rate shown in column 6. Surprisingly, a reciprocal plot of $1/v_{calc}$ versus $1/[S]_{add}$ is linear (Fig. II-33). The curve extrapolates to the true $1/V_{max}$, which is not

able II-3 Effect of Unlabeled Endogenous Substrate on the Calculated Reaction Rates in an Assay Employing a Radioactive Substrate[a]

Assumed Substrate Concentration $[S^*]_{add}$(mM)	Actual Substrate Concentration $[S^*]_{add}+[S]_{endog}$	True Specific Activity (cpm/μmole)	True Rate (μmoles/min)	Observed Rate (cpm/min)	Calculated Rate (μmoles/min)
0.1	1.1	0.91×10^4	5.24	4.76×10^4	0.476
0.2	1.2	1.66×10^4	5.46	9.06×10^4	0.906
0.25	1.25	2.0×10^4	5.56	11.1×10^4	1.11
0.33	1.33	2.5×10^4	5.72	14.3×10^4	1.43
0.50	1.5	3.33×10^4	6.0	20×10^4	2.0
1.0	2.0	5.0×10^4	6.67	33.4×10^4	3.34
2.0	3.0	6.66×10^4	7.5	50×10^4	5.0
3.33	4.33	7.67×10^4	8.13	62.4×10^4	6.24
4.0	5.0	8.0×10^4	8.33	66.7×10^4	6.67
5	6.0	8.33×10^4	8.57	71.5×10^4	7.15
10	11.0	9.11×10^4	9.16	83.6×10^4	8.36

[a] $V_{max}=10$ μmoles/min, $K_m=1$ mM, S.A. of $S^*=10^5$ cpm/μmole, $[S]_{endog}=1$ mM. Vol = 10 ml.

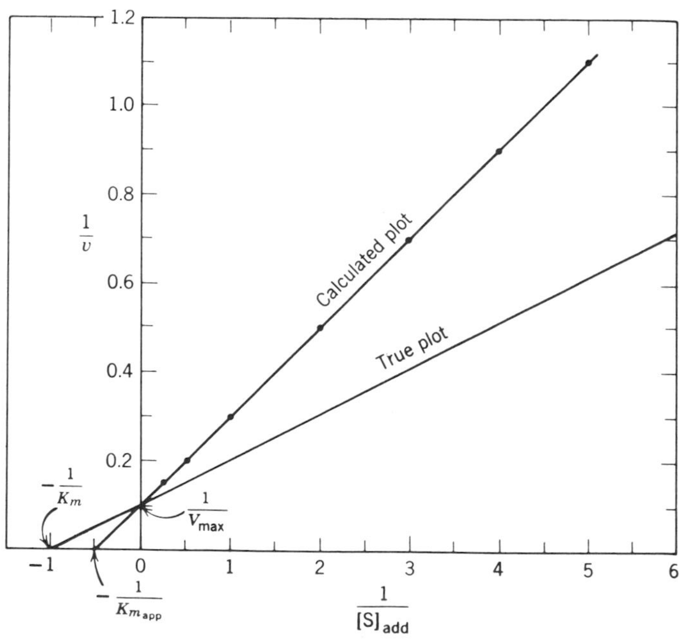

Fig. II-33. Effect of endogenous substrate ($[S]=K_m$) on the $1/v$ versus $1/[S]$ plot when the assay depends on the rate of radioactive P appearance (see data in Table II-3).

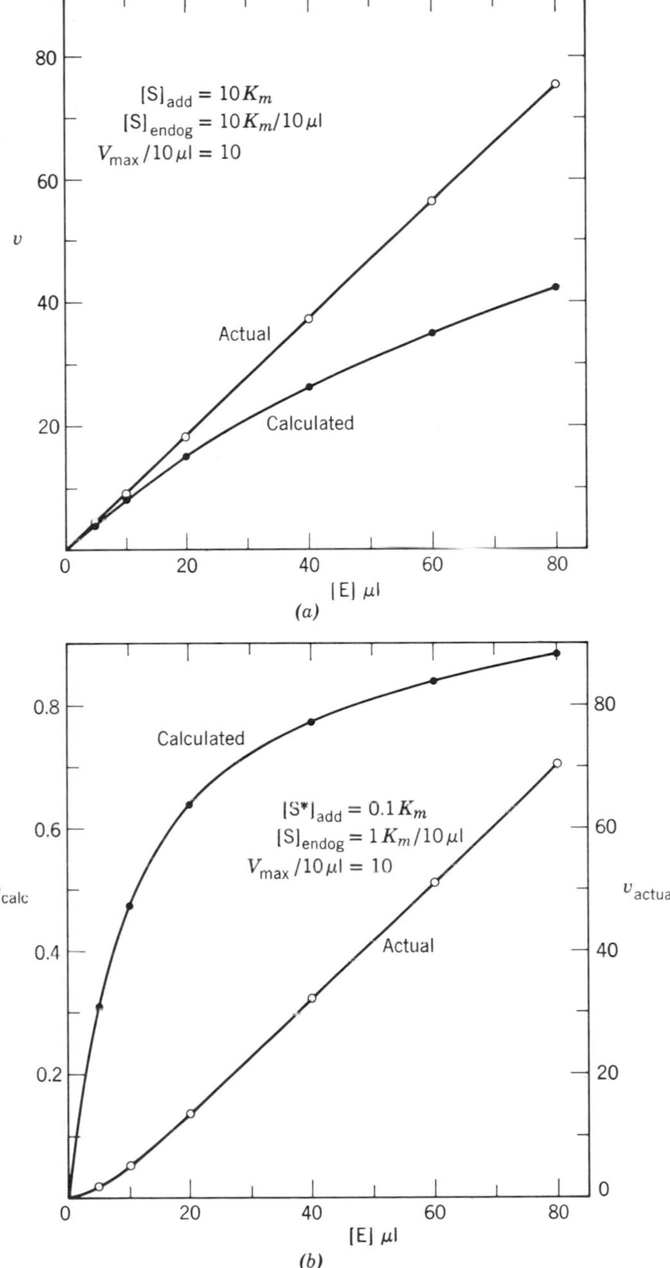

Fig. II-34. Effect of endogenous substrate on the v versus [E] plot when the assay depends on the rate of radioactive P appearance.

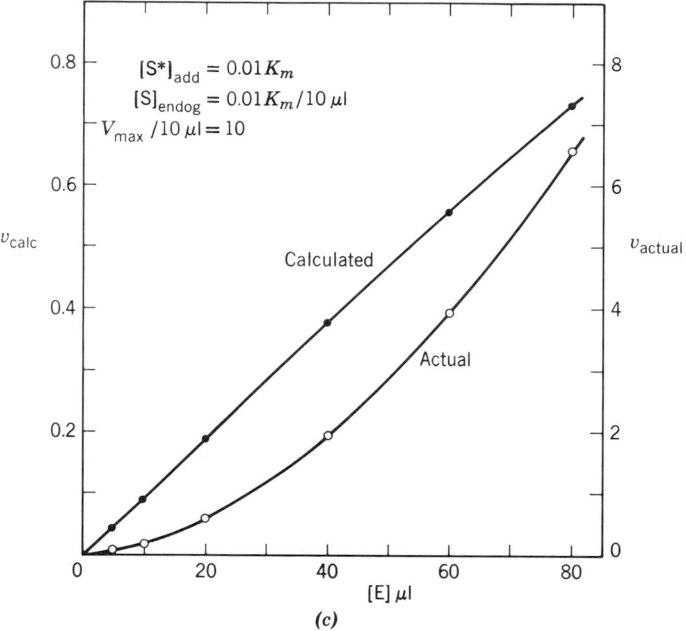

Fig. II-34. (Cont.)

unexpected. (As $[S^*]_{add}$ increases, the difference between the assumed and true substrate concentrations decreases and at the same time the specific activity of the substrate approaches the assumed value.) The extrapolated K_m, however, is incorrect. In fact, the unlabeled endogenous substrate has exactly the same effect on K_m as a competitive inhibitor (Chapter Three).

$$K_{m_{app}} = K_m\left(1 + \frac{[S]_{endog}}{K_m}\right) \tag{II-84}$$

The v versus [E] plots will be nonlinear. The curvature is opposite to that observed when a nonradioactive assay is used. Three examples are shown in Figure II-34*a*, *b*, and *c*. At low $[S^*]_{add}$ and low $[S]_{endog}$, the calculated velocities may yield a near-linear v versus [E] plot (Fig. II-34*c*), yet the calculated velocities (hence the calculation of $[E_t]$) can be significantly in error. The effects of contaminating inhibitors in the enzyme preparation and in the substrate are discussed in Chapter Three.

REFERENCES

For Historical Background

Haldane, J. B. S., *Enzymes*, M.I.T. Press, 1965. (Reprint of 1930 book originally published by Longmans, Green and Company.)

Haldane, J. B. S. and Stern, K. G., *Allegemeine Chemie der Enzyme*, Verlag von Steinkopff, 1932.

Segal, H. L., "The Development of Enzyme Kinetics," in *The Enzymes*, 2nd ed., P. D. Boyer, J. Lardy, and K. Myrback, eds., Vol. 1, Academic Press, 1959, p. 1.

Summer, J. B. and Somers, G. G., *Chemistry and Methods of Enzymes*, Academic Press, 1943.

Reference Books

Barman, T. F., *Enzyme Handbook*, Springer Verlag, 1969.

Bender, M. L. and Brubacher, L. J., *Catalysis and Enzyme Action*, McGraw-Hill, 1973.

Bernhard, S., *The Structure and Function of Enzymes*, W. A. Benjamin, 1968.

Dixon, M. and Webb, E. C., *Enzymes*, 2nd ed, Academic Press, 1964.

Gray, C. J., *Enzyme Catalyzed Reactions*, Van Nostrand Reinhold, 1971.

Gutfreund, H., *Enzymes: Physical Principles*, John Wiley & Sons, 1972.

Jencks, W. P., *Catalysis in Chemistry and Enzymology*, McGraw-Hill, 1969.

Laidler, K. J., *The Chemical Kinetics of Enzyme Action*, Oxford-Clarendon Press, 1958.

Nielands, J. B. and Stumpf, P. K., *Outlines of Enzyme Chemistry*, 2nd ed, John Wiley & Sons, 1958.

Plowman, K. M., *Enzyme Kinetics*, McGraw-Hill, 1972.

Reiner, J. M., *Behavior of Enzyme Systems*, 2nd ed., Von Nostrand Reinhold, 1969.

Walter, C., *Steady State Applications in Enzyme Kinetics*, The Ronald Press, 1965.

Webb, J. L., *Enzyme and Metabolic Inhibitors*, Vol. I, Academic Press, 1963.

Westley, J., *Enzymic Catalysis*, Harper and Row, 1969.

Whitaker, J. R., *Principles of Enzymology for the Food Sciences*, Marcel Dekker, 1972.

Zeffren, E. and Hall, P. L., *The Study of Enzyme Mechanisms*, Wiley-Interscience, 1973.

Chemical Kinetics

Capellos, C. and Bielski, H. J., *Kinetic Systems—The Mathematical Description of Chemical Kinetics in Solutions*, Wiley-Interscience, 1972.

Daniels, F. and Alberty, R. A., *Physical Chemistry*, 3rd ed., John Wiley & Sons, 1966, Ch. 10.

Glasstone, S., *Textbook of Physical Chemistry*, 2nd ed., Van Nostrand Company, 1946, Ch. 13.

Gucker, F. T. and Siefert, R. L., *Physical Chemistry*, W. W. Norton and Company, 1966, Ch. 23.

Laidler, K. J., *Chemical Kinetics*, 2nd ed., McGraw-Hill, 1965, Ch. 1.

Laidler, K. J., *Can. J. Chem.* **33**, 1614 (1955).

Laidler, K. J., *The Chemical Kinetics of Enzyme Action*, Oxford-Clarendon Press, 1958, Ch. 2.

Mahler, H. R. and Cordes, E. H., *Biological Chemistry*, 2nd ed., Harper and Row, 1971, Ch. 6.

Maron, S. H. and Prutton, C. F., *Principles of Physical Chemistry*, 4th ed., Macmillan, 1965, Ch. 13.

Moore, W. J., *Physical Chemistry*, 3rd ed., Prentice-Hall, 1962, Ch. 8.

Simple Enzyme Kinetics Calculations

Christensen, H. N. and Palmer, G. A., *Enzyme Kinetics* (a programmed text), 2nd ed., W. B. Saunders, 1974.

Dawes, E. A., *Quantitative Problems in Biochemistry*, 4th ed., Williams and Wilkins, 1967, Ch. 4.

Finlayson, J. S., *Basic Biochemical Calculations*, Addison-Wesley, 1969, Ch. 9.

Montgomery, R. and Swenson, C. A., *Quantitative Problems in the Biochemical Sciences*, Freeman, 1969, Ch. 11.

Segel, I. H., *Biochemical Calculations*, 2nd ed., John Wiley & Sons, 1976. Ch. 4.

Integrated Velocity Equations

Alberty, R. A., "The Rate Equation for an Enzymic Reaction," in *The Enzymes*, Vol. 1, 2nd ed., P. D. Boyer, A. Lardy, and K. Myrback, eds., Academic Press, 1954, p. 143.

Atkins, G. L. and Nimmo, I. A., *Biochem. J.* **135**, 779 (1973).

Bizzozero, S. A., Kaiser, A. W. and Dutler, H., *Eur. J. Biochem.* **33**, 292 (1973).

Darvey, I. G. and Williams, J. F., *Biochim. Biophys. Acta* **85**, 1 (1964).

Huang, H. T. and Niemann, C., *J. Am. Chem. Soc.* **73**, 1541 (1951).

Jennings, R. R. and Niemann, C., *J. Am. Chem. Soc.* **77**, 5432 (1955).

Lee, H.-J. and Wilson, I. B., *Biochim. Biophys. Acta* **242**, 519 (1971).

Philo, R. D. and Selwyn, M. J., *Biochem. J.* **135**, 525 (1973).

Walker, A. C. and Schmidt, C. L. A., *Arch. Biochem. Biophys.* **5**, 445 (1944).

Kinetic Behavior at High Enzyme Concentrations—Kinetics of Enzymes with Tightly Bound Ligands

Cha, S., *J. Biol. Chem.* **245**, 4814 (1970).

Dixon, M., *Biochem. J.* **129**, 197 (1972).

Henderson, P. J. F., *Biochem. J.* **135**, 101 (1973).

Sols, A. and Marco, R., "Concentrations of Metabolites and Binding Sites. Implications in Metabolic Regulation," in *Current Topics in Cellular Regulation*, Vol. 2, B. L. Horecker and E. R. Stadtman, eds., Academic Press, 1970, p. 227.

Srere, P. A., *Science* **158**, 936 (1967).

See also references on Tightly Bound Inhibitors at the end of Chapter Three.

Multiple Enzymes Acting on the Same Substrate

Gross, W., Geck, P., Burkhardt, K.-L. and Ring, K., *Biophys.* **8**, 271 (1972).

Neal, J. L., *J. Theoret. Biol.* **35**, 113 (1972).

Spears, G., Sneyd, J. G. T. and Loten, E. G., *Biochem. J.* **125**, 1149 (1971).

Thompson, W. J., and Appleman, M. M., *Biochemistry* **10**, 311 (1971) (Cyclic nucleotide phosphodiesterases).

Diffusion Controlled Enzyme Reactions

Alberty, R. A. and Hammes, G. G., *J. Phys. Chem.* **62**, 154 (1958).

Enzyme Assays

Easterby, J. S., *Biochim. Biophys. Acta* **293**, 552 (1972). (Kinetics of coupled assays.)

Layne, E., "Spectrophotometric and Turbidometric Methods for Measuring Proteins," in *Methods in Enzymology*, Vol. 3, S. P. Colowick and N. O. Kaplan, eds., Academic Press, 1957, p. 447.

Levitzki, A., *Anal. Biochem.* **33**, 335 (1970). (Measuring NH_3 by the first-order response of glutamic dehydrogenase.)

Lowry, O. H. and Passonneau, J. V., *A Flexible System of Enzymatic Analysis*. Academic Press, 1972.

McClure, W. R. *Biochemistry* **8**, 2782 (1969). (Kinetics of coupled assays.)

Storer, A. C. and Cornish-Bowden, A., *Biochem. J.* **141**, 205 (1974). (Kinetics of coupled assays.)

See also references on Methods of Plotting Enzyme Kinetics Data at the end of Chapter Four.

CHAPTER THREE

SIMPLE INHIBITION SYSTEMS

Any substance that reduces the velocity of an enzyme-catalyzed reaction can be considered to be an "inhibitor." The inhibition of enzyme activity is one of the major regulatory devices of living cells, and one of the most important diagnostic procedures of the enzymologist. Inhibition studies often tell us something about the specificity of an enzyme, the physical and chemical architecture of the active site, and the kinetic mechanism of the reaction. In our every day life, enzyme inhibitors can be found masquerading as drugs, antibiotics, preservatives, poisons, and toxins. In this chapter we examine three simple types of enzyme inhibitors. We assume that only a single substrate is involved in the reaction, and that only one type of inhibitor is present at any time. The effects of inhibitors on multisubstrate enzymes are discussed in Chapters Six and Nine. The effects of multiple inhibitors are discussed in Chapter Eight.

A. COMPETITIVE INHIBITION (SIMPLE INTERSECTING LINEAR COMPETITIVE INHIBITION)

A competitive inhibitor is a substance that combines with free enzyme in a manner that prevents substrate binding. That is, the inhibitor and the substrate are *mutually exclusive*, often because of true competition for the same site. A competitive inhibitor might be a nonmetabolizable analog or derivative of the true substrate, or an alternate substrate of the enzyme, or a product of the reaction.

Malonic acid is a classical example of a true competitive inhibitor. Malonic acid inhibits succinic dehydrogenase, which catalyzes the oxidation of succinic acid to fumaric acid, as shown below.

$$\begin{matrix} CH_2\text{—}COOH \\ | \\ CH_2\text{—}COOH \end{matrix} + FAD \rightleftharpoons \begin{matrix} H\text{—}C\text{—}COOH \\ \| \\ HOOC\text{—}CH \end{matrix} + FADH_2$$

Malonic acid resembles succinic acid sufficiently to combine with the enzyme at the active site.

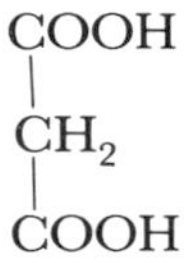

$$\begin{array}{c} COOH \\ | \\ CH_2 \\ | \\ COOH \end{array}$$

malonic acid

However, because malonic acid has only one methylene group, no oxidation-reduction can take place.

Another classical example of a competitive inhibitor is the sulfa drug sulfanilamide, which interferes with the biosynthesis of folic acid from the precursor *p*-amino-benzoic acid (PABA).

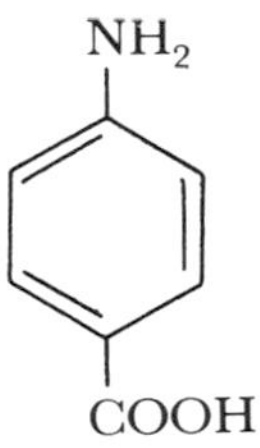

NH_2 / $COOH$ (benzene ring, para-substituted)

PABA

NH_2 / SO_2NH_2 (benzene ring, para-substituted)

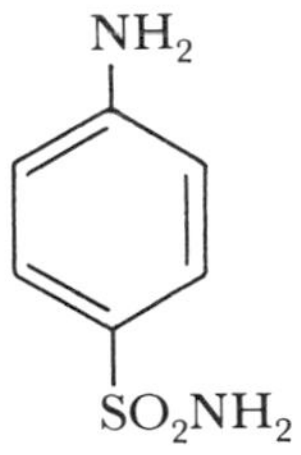

sulfanilamide

Figure III-1 illustrates several situations that lead to the mutually exclusive binding of S and I, Model 1 represents classical competitive inhibition. Models 2 to 4 yield the same results.

There are many examples of "competitive" inhibition by compounds that bear no structural relationship to the substrate. The inhibitor might be an end-product or near end-product of a metabolic pathway; the enzyme is one that catalyzes an early reaction (or a branch-point reaction) in the pathway. The phenomenon is called feedback inhibition. The inhibitor ("effector," "modulator," or "regulator") combines with the enzyme at a position other than the active (substrate) site. The combination of the inhibitor with the enzyme causes a change in the conformation (tertiary or quaternary structure) of the enzyme that distorts the substrate site and thereby prevents the substrate from binding (Model 5).

The inhibition of the hexokinase-catalyzed reaction between glucose and ATP by fructose or mannose is an example of competitive inhibition by alternate substrates. Glucose, fructose, and mannose are all substrates of hexokinase and can be converted to product (hexose-6-phosphate). All three hexoses combine with the enzyme at the same active site. Consequently, the

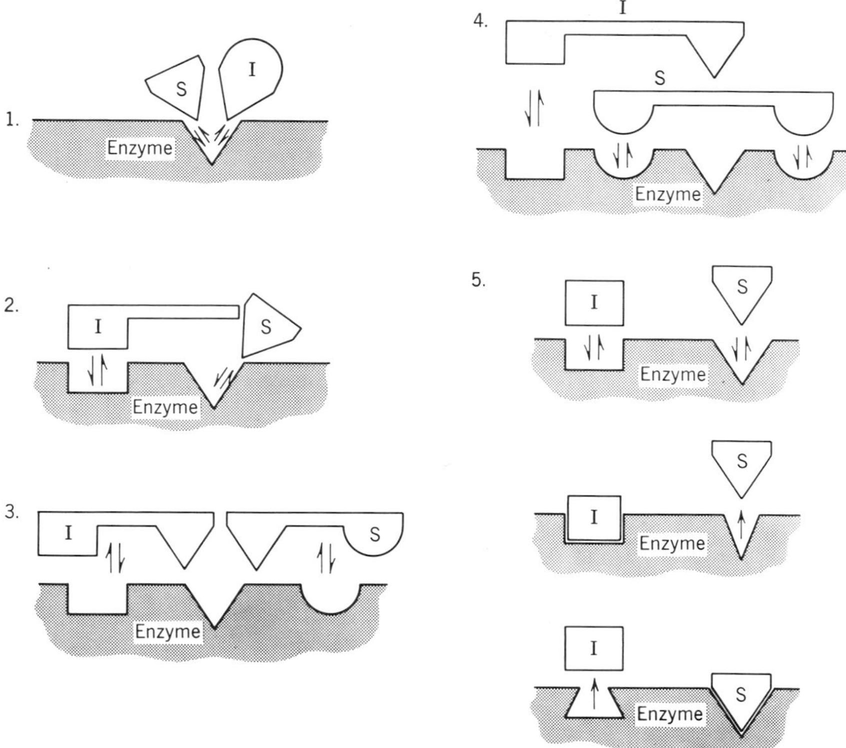

Fig. III-1. Models of competitive inhibition: S and I are mutually exclusive. (1) Classical model. S and I compete for the same binding site. I must resemble S structurally. (2) I and S are mutually exclusive because of steric hinderance. (3) I and S share a common binding group on the enzyme. (4) The binding sites for I and S are distinct, but overlapping. (5) The binding of I to a distinct inhibitor site causes a conformational change in the enzyme that distorts or masks the substrate binding site (and vice versa).

utilization of any one of the hexoses is inhibited in the presence of either of the other two.

The equilibria describing competitive inhibition are shown below. The competition and mutual exclusion of S and I are clearly seen.

$$\begin{array}{l} \mathrm{E} + \mathrm{S} \overset{K_S}{\rightleftharpoons} \mathrm{ES} \overset{k_p}{\longrightarrow} \mathrm{E} + \mathrm{P} \\ + \\ \mathrm{I} \\ K_i \updownarrow \\ \mathrm{EI} \end{array}$$

where $K_i = [E][I]/[EI]$, $K_S = [E][S]/[ES]$, and k_p = rate constant for the breakdown of ES to E+P.

The initial velocity of the reaction is proportional to the steady-state concentration of the enzyme-substrate complex, ES. All the enzyme species are reversibly connected. Consequently, we can predict that at any fixed unsaturating concentration of inhibitor (*a*) v_i (the velocity in the presence of a competitive inhibitor) can be made to equal v (the velocity in the absence of the inhibitor), but that a higher substrate concentration will be required (to obtain the same ES concentration), and (*b*) in the presence of an infinitely high (saturating) substrate concentration all the enzyme can be driven to the ES form. Consequently, the maximal initial velocity in the presence of the competitive inhibitor equals V_{max} (the maximum initial velocity in the absence of inhibitor). However, the apparent K_m (measured as [S] required for $\frac{1}{2}V_{max}$) will increase in the presence of a competitive inhibitor because at any inhibitor concentration, a portion of the enzyme exists in the EI form which has no affinity for S.

An expression relating v, V_{max}, [S], K_m, [I], and K_i in the presence of a competitive inhibitor can be derived from either rapid equilibrium or steady-state assumptions. This time we must recognize that the total enzyme $[E]_t$ is present in three forms: free enzyme, [E]; enzyme-substrate complex [ES]; and enzyme-inhibitor complex, [EI].

$$v = k_p[ES]$$

$$\frac{v}{[E]_t} = \frac{k_p[ES]}{[E]+[ES]+[EI]}$$

$$\frac{v}{k_p[E]_t} = \frac{\dfrac{[S]}{K_S}}{1+\dfrac{[S]}{K_S}+\dfrac{[I]}{K_i}}$$

$$\boxed{\frac{v}{V_{max}} = \frac{\dfrac{[S]}{K_S}}{1+\dfrac{[S]}{K_S}+\dfrac{[I]}{K_i}}} \qquad \text{(III-1)}$$

If we compare equation III-1 to the usual velocity equation II-8, we see that the denominator has gained an additional $[I]/K_i$ term representing the

EI complex. The numerator still has one term indicating that there is still only one product-forming complex (ES). To obtain a more familiar form, the numerator and denominator of the right-hand part of equation III-1 can be multiplied by K_S and factored:

$$\frac{v}{V_{max}} = \frac{[S]}{K_S\left(1+\frac{[I]}{K_i}\right)+[S]} \tag{III-2}$$

We obtain the same final velocity equation for steady-state conditions; however, K_m replaces K_S. This is not surprising, since the steady-state assumption does not change the form of the velocity equation for the uninhibited reaction while the reaction between E and I to yield EI must be at equilibrium. (There is nowhere for EI to go but back to E + I).

The velocity equation differs from the usual Henri-Michaelis-Menten equation in that the K_m term is multiplied by the factor $(1+[I]/K_i)$. This confirms our original prediction that V_{max} is unaffected by a competitive inhibitor, but that the *apparent* K_m value is increased. The increase in the K_m value does *not* mean that the EI complex has a lower affinity for the substrate. The EI has no affinity at all for the substrate, while the affinity of E (the only form that can bind substrate) is unchanged. The apparent increase in K_m results from a distribution of available enzyme between the "full affinity" and "no affinity" forms. The factor $(1+[I]/K_i)$ may be considered as an [I]-dependent statistical factor describing the distribution of enzyme between the E and EI forms. There are systems in which EI has an altered affinity for S. This type of system, called partial competitive inhibition, is discussed in Chapter Four.

The effect of a competitive inhibitor on the kinetics of an enzyme-catalyzed reaction is illustrated in Figure III-2. The [I] was arbitrarily chosen as $3K_i$; $K_{m_{app}}$ then is $(1+3)K_m = 4K_m$. In the presence of the inhibitor it takes four times as much substrate to attain $0.5\,V_{max}$. In general:

$$\frac{[S]_i}{[S]} = \left(1+\frac{[I]}{K_i}\right) \tag{III-3}$$

where $[S]_i/[S]$ represents the ratio of substrate concentration required in the presence of inhibitor to substrate concentration required in the absence of inhibitor for any given velocity. A competitive inhibitor will increase $[S]_{0.9}$ and $[S]_{0.1}$. However, since both concentrations are increased by the same factor, the $[S]_{0.9}/[S]_{0.1}$ ratio is still 81 at all inhibitor concentrations. The

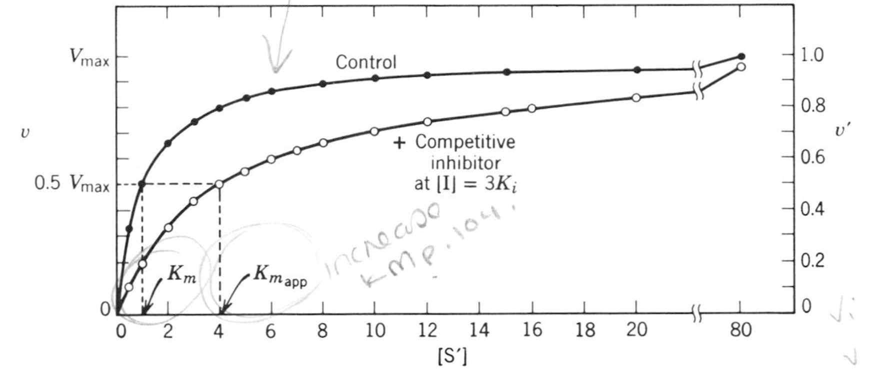

Fig. III-2. The v versus [S] plot in the presence and in the absence of a fixed concentration of a competitive inhibitor.

constant ratio is expected, since the form of the equation is unchanged; only the numerical value of K_m is changed.

An expression for the relative velocity or fractional activity in the presence and absence of a competitive inhibitor can be derived readily:

Let v_i = the initial velocity at a given [S] in the presence of inhibitor

v_0 = the initial velocity at the same [S] in the absence of inhibitor

$\frac{v_i}{v_0} = a =$ the relative activity

$$\frac{v_i}{v_0} = a = \frac{\dfrac{V_{max}[S]}{K_m\left(1+\dfrac{[I]}{K_i}\right)+[S]}}{\dfrac{V_{max}[S]}{K_m+[S]}} \quad \text{or} \quad \boxed{a = \frac{K_m+[S]}{K_m\left(1+\dfrac{[I]}{K_i}\right)+[S]}} \qquad \text{(III-4)}$$

Relative velocity data are frequently expressed in terms of fractional inhibition (i) or "percent inhibition" ($i_{\%}$) where:

$$i = 1 - \frac{v_i}{v_0} = 1 - a \qquad \text{and "percent inhibition"} \qquad = 100i$$

$$\boxed{i = \frac{[I]}{[I]+K_i\left(1+\dfrac{[S]}{K_m}\right)}} \qquad \text{(III-5)}$$

Effect of Concentration Range on Degree of Inhibition

A point not always appreciated is that the degree of inhibition caused by an *n-fold excess* of competitive inhibitor is maximum when both [I] and [S] are very high compared to K_i and K_m, respectively. To put it another way, if [S] is very low compared to K_m, an excess of competitive inhibitor will not cause much inhibition even though K_i is of the same order of magnitude as K_m. A simple example illustrates the point. Suppose $[S]=0.01K_m$ and $K_i=K_m$. What is the degree of inhibition caused by a tenfold excess of inhibitor (i.e., by an $[I]=10[S]$)?

$$i=\frac{0.1K_m}{0.1K_m+K_m(1+0.01)}=\frac{0.1K_m}{1.11K_m}=0.09$$

$$i_{\%}=9\%$$

In other words, the inhibited velocity is 91% of the control velocity; we observe only 9% inhibition. On the other hand, when $[S]=10K_m$, a tenfold excess of inhibitor ($[I]=100K_m$) will inhibit 90% as shown below.

$$i=\frac{100K_m}{100K_m+K_m+10K_m}=\frac{100K_m}{111K_m}$$

$$\therefore \quad i=0.90 \qquad \text{or} \qquad i_{\%}=90$$

Figure III-3 shows the effect of increasing competitive inhibitor concentration on the initial velocity at three different substrate concentrations. The degree of inhibition depends on the substrate concentration, decreasing as [S] increases, as predicted by equation III-5. To obtain 50% inhibition:

$$\boxed{[I]_{0.5}=\left(1+\frac{[S]}{K_m}\right)K_i} \qquad \text{(III-6)}$$

Similarly, we can show that $[I]_{0.9}$ and $[I]_{0.1}$ (the inhibitor concentrations required for 90% and 10% inhibition, respectively) are:

$$[I]_{0.9}=9\left(1+\frac{[S]}{K_m}\right)K_i, \qquad [I]_{0.1}=\frac{1}{9}\left(1+\frac{[S]}{K_m}\right)K_i$$

Thus the $[I]_{0.9}/[I]_{0.1}$ ratio is always 81, regardless of the substrate concentration or the values of K_m and K_i.

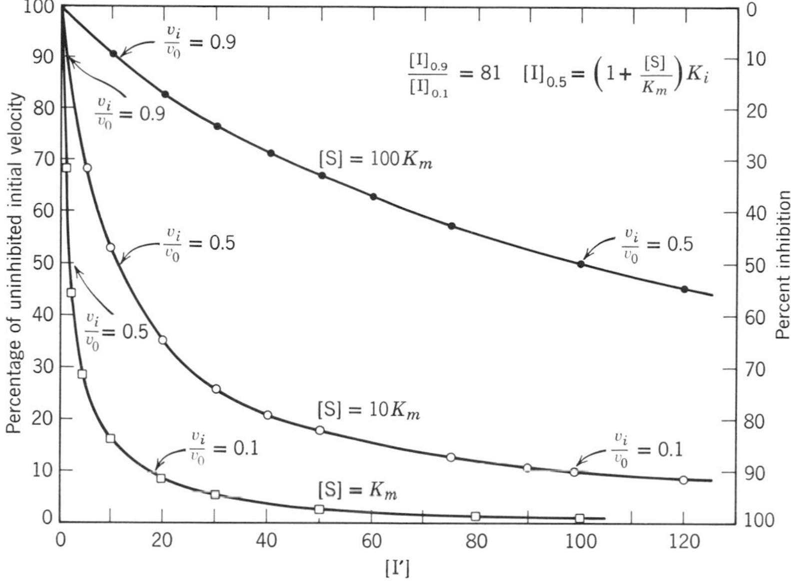

Fig. III-3. Relative activity as a function of competitive inhibitor concentration in the presence of different fixed concentrations of substrate.

Reciprocal Plot for Competitive Inhibition Systems

The velocity equation for competitive inhibition in reciprocal form is:

$$\frac{1}{v} = \frac{K_m}{V_{\max}}\left(1 + \frac{[I]}{K_i}\right)\frac{1}{[S]} + \frac{1}{V_{\max}} \tag{III-7}$$

Thus the slope of the plot increases by the factor $(1 + [I]/K_i)$ (which multiplied K_m in the original equation), but the $1/v$-axis intercept remains $1/V_{\max}$. For each inhibitor concentration, a new reciprocal plot can be drawn. As [I] increases, the "plus inhibitor" curves increase in slope (Fig. III-4) pivoting counterclockwise about the point of intersection with the control curve (at $1/V_{\max}$ on the $1/v$-axis). Because the initial velocity can be driven to zero by a saturating inhibitor concentration, the limiting plot will be a vertical line on the $1/v$ axis. As [I] increases, the intercept on the $1/[S]$ axis moves closer to the origin; that is, $K_{m_{\text{app}}}$ continually increases. The K_i can be calculated from the slope of any reciprocal plot or from any $K_{m_{\text{app}}}$. However, a replot as described below is better.

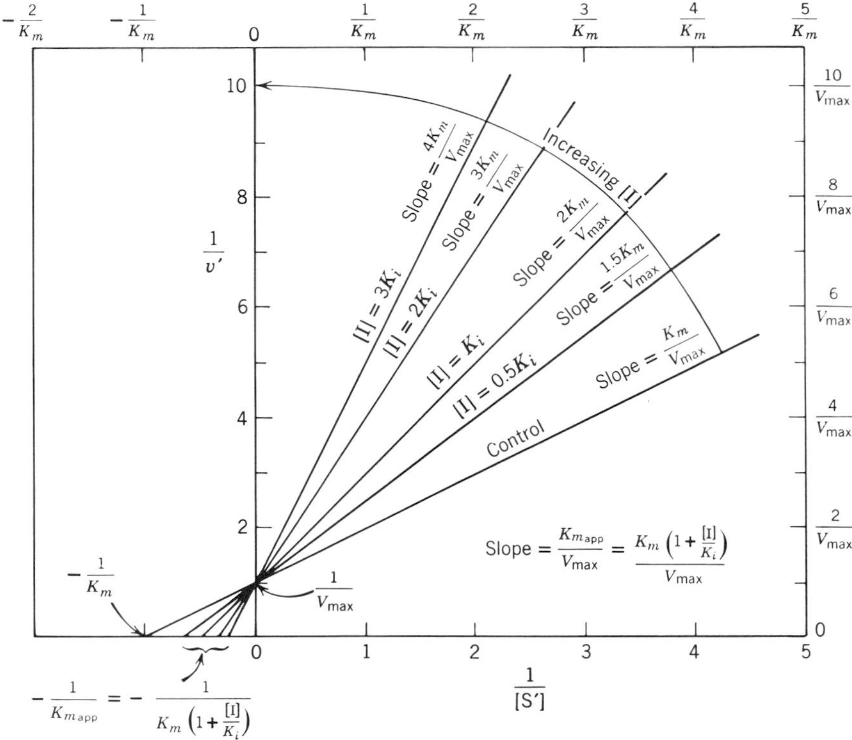

Fig. III-4. The $1/v$ versus $1/[S]$ plot in the presence of different fixed concentrations of a competitive inhibitor.

Replots of Slope and $K_{m_{app}}$ Versus [I]

The slope of the reciprocal plot in the presence of a competitive inhibitor is given by:

$$slope_{1/S} = \frac{K_m}{V_{max}}\left(1 + \frac{[I]}{K_i}\right) \quad \text{or} \quad \boxed{slope_{1/S} = \frac{K_m}{V_{max}K_i}[I] + \frac{K_m}{V_{max}}} \tag{III-8}$$

A replot of the slope of each reciprocal plot versus the corresponding inhibitor concentration will be a straight line with a slope of $K_m/V_{max}K_i$ and an intercept on the $slope_{1/S}$-axis of K_m/V_{max} (i.e., control slope at $[I]=0$) (Fig. III-5a). When $slope_{1/S}=0$, the intercept on the [I]-axis gives $-K_i$. For

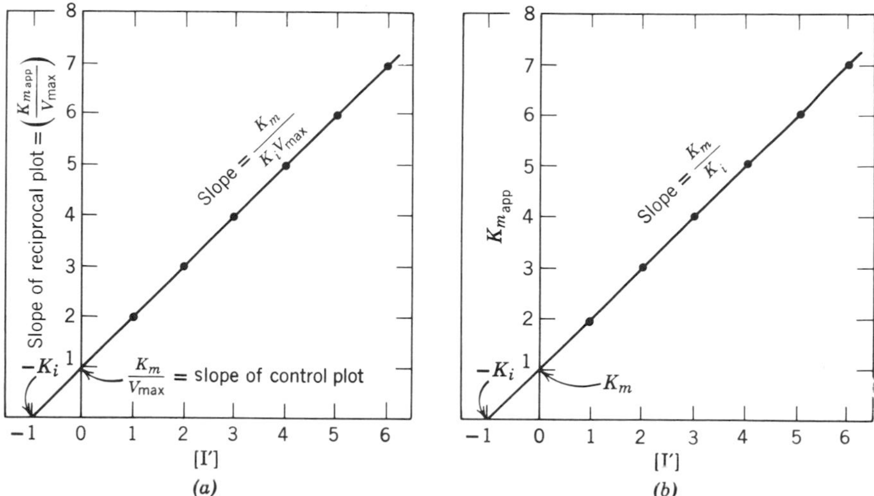

Fig. III-5. Replots of data taken from the reciprocal plot. (*a*) $Slope_{1/S}$ versus [I]. (*b*) $K_{m_{app}}$ versus [I].

convenience the slope of the reciprocal plots can be read off directly as the ratio (absolute values) of the vertical axis intercept to horizontal axis intercept. A linear $slope_{1/S}$ versus [I] replot distinguishes pure competitive inhibition from partial competitive inhibition. The latter gives hyperbolic $slope_{1/S}$ versus [I] replots (Chapter Four). The $K_{m_{app}}$ is also a linear function of the inhibitor concentration in pure competitive systems:

$$\boxed{K_{m_{app}} = \frac{K_m}{K_i}[I] + K_m} \qquad \text{(III-9)}$$

A replot of $K_{m_{app}}$ versus [I] has intercepts of K_m (on the $K_{m_{app}}$-axis) and $-K_i$ (on the [I]-axis) (Fig. III-5*b*).

Dixon Plot for Competitive Inhibition: $1/v$ Versus [I]

The Dixon plot is used frequently to identify the type of inhibition and to determine the K_i value. The velocity equation for competitive inhibition may be converted to a linear form in which the varied ligand is [I]. Starting with the reciprocal equation:

$$\frac{1}{v}=\frac{K_m}{V_{\max}[S]}\left(1+\frac{[I]}{K_i}\right)+\frac{1}{V_{\max}}=\frac{K_m}{V_{\max}[S]}+\frac{K_m[I]}{V_{\max}[S]K_i}+\frac{1}{V_{\max}}$$

$$\boxed{\frac{1}{v}=\frac{K_m}{V_{\max}[S]K_i}[I]+\frac{1}{V_{\max}}\left(1+\frac{K_m}{[S]}\right)} \qquad \text{(III-10)}$$

A plot of $1/v$ versus [I] at some unsaturating [S] will yield a straight line with a positive slope as shown in Figure III-6*a*. If the inhibition is known to be competitive and $V_{\max}$ is known, a horizontal line at a height of $1/V_{\max}$ can be drawn. The $-[I]$ value at the intersection of two lines gives K_i as shown below. When $1/v=1/V_{\max}$:

$$\frac{1}{V_{\max}}=\frac{K_m[I]}{V_{\max}[S]K_i}+\frac{1}{V_{\max}}\left(1+\frac{K_m}{[S]}\right)$$

$$1=\frac{K_m[I]}{[S]K_i}+1+\frac{K_m}{[S]}$$

$$-\frac{K_m[I]}{[S]K_i}=\frac{K_m}{[S]} \qquad \text{or} \qquad [I]=-K_i$$

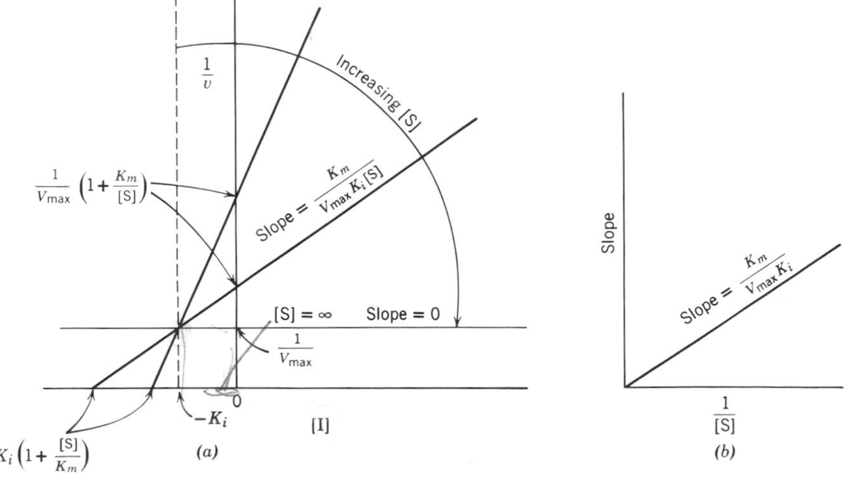

Fig. III-6. (*a*) Dixon plot for a competitive inhibitor: $1/v$ versus [I] in the presence of different fixed concentrations of substrate. If $V_{\max}$ is known, a horizontal line at a height of $1/V_{\max}$ can be drawn directly. (*b*) Replot of the slopes of the Dixon plot.

The horizontal line at $1/V_{max}$ also signifies that at an infinitely high [S], increasing the inhibitor concentration will have no effect on v. If the inhibition is not known for sure to be competitive, or if V_{max} is unknown, another series of experiments at a different unsaturating [S] will yield a second line with a different positive slope. The intersection of the $[S]_1$ and $[S]_2$ lines, where $1/v_1 = 1/v_2$, gives K_i as shown below. When $1/v_1 = 1/v_2$:

$$\frac{K_m[I]}{V_{max}[S]_1 K_i} + \frac{[S]_1 + K_m}{V_{max}[S]_1} = \frac{K_m[I]}{V_{max}[S]_2 K_i} + \frac{[S]_2 + K_m}{V_{max}[S]_2} \quad \text{or} \quad \frac{[I]+K_i}{[S]_1} = \frac{[I]+K_i}{[S]_2}$$

The equation above holds true only if $[S]_1 = [S]_2$, which is not the case, or when both sides equal zero; that is, when $[I] = -K_i$.

The slope of the Dixon plot is given by:

$$slope = \frac{K_m}{V_{max}K_i}\frac{1}{[S]} \qquad \text{(III-11)}$$

Thus a replot of *slope* versus the corresponding $1/[S]$ (Fig. III-6*b*) will be a straight line through the origin with a slope of $K_m/V_{max}K_i$.

The family of Dixon plots for pure competitive inhibition intersects above the [I]-axis at $[I] = -K_i$ and $1/v = 1/V_{max}$. Certain types of mixed inhibition systems also yield lines that intersect above the [I]-axis. Consequently, a Dixon plot such as that shown in Figure III-6*a* establishes only that the inhibition is neither noncompetitive nor uncompetitive. Other plots for competitive inhibition systems are described in Chapter Four.

General Principles

A competitive inhibitor acts only to increase the apparent K_m for the substrate. As [I] increases, $K_{m_{app}}$ increases. The V_{max} remains unchanged, but in the presence of a competitive inhibitor a much greater substrate concentration is required to attain any given fraction of V_{max}. The v_i may be considered equal to V_{max} when $[S] \geqslant 100K_{m_{app}}$.

The degree of inhibition caused by a competitive inhibitor depends on [S], [I], K_m, and K_i. An increase in [S] at constant [I] decreases the degree of inhibition. An increase in [I] at constant [S] increases the degree of inhibition. The lower the value of K_i, the greater is the degree of inhibition at any given [S] and [I]. The K_i is equivalent to the concentration of I that doubles the slope of the $1/v$ versus $1/[S]$ plot. (K_i is not equivalent to the [I] that yields 50% inhibition).

Integrated Rate Equation in the Presence of a Competitive Inhibitor

If the reaction has a very large K_{eq} and none of the products have an appreciable affinity for the enzyme, then the integrated Henri-Michaelis-Menten equation in the presence of a competitive inhibitor can be written as:

$$\frac{2.3}{t} \log \frac{[S]_0}{[S]} = -\frac{1}{K_m\left(1+\frac{[I]}{K_i}\right)} \frac{[P]}{t} + \frac{V_{max}}{K_m\left(1+\frac{[I]}{K_i}\right)} \qquad \text{(III-12)}$$

where $[P]=[S]_0-[S]$. The equation assumes that [I] remains constant as [S] decreases. Consequently, I cannot be an alternate substrate. The determination of [P] at various times during the course of the reaction will permit $K_{m_{app}}$ and V_{max} to be determined. A family of curves can be obtained for different inhibitor concentrations (Fig. III-7). The values of K_m and K_i can be determined from appropriate replots of the slopes or vertical axis intercepts.

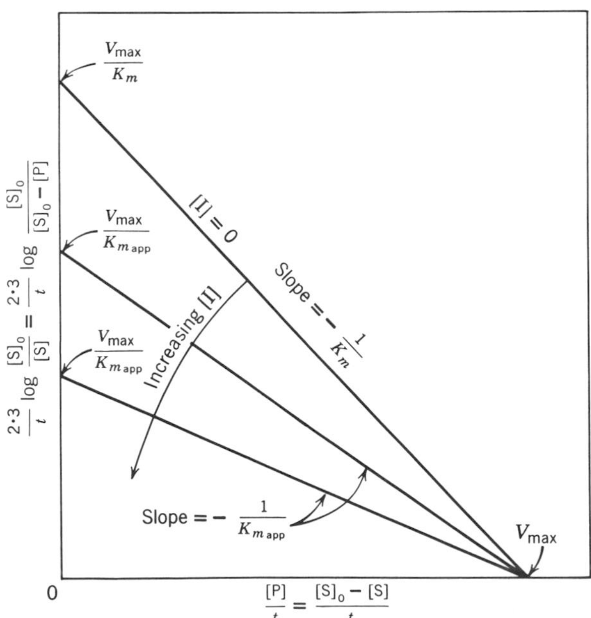

Fig. III-7. Plot of the integrated velocity equation in the presence of a competitive inhibitor.

Competitive Inhibition and Total Velocity with Mixed Alternative Substrates

When a single enzyme acts on two different substrates, and both are present simultaneously, each will act as a competitive inhibitor with respect to the other. If the products of the two substrates can be distinguished from each other, the system may be treated by the usual competitive inhibition relationships. If, on the other hand, the products are identical, or appear so by the assay method used, then the situation becomes more complex. For example, suppose the enzyme is a nonspecific phosphatase that catalyzes the reactions $A \rightarrow Q + P$ and $B \rightarrow R + P$ where A and B are two phosphate esters, Q and R are two distinct alcohols, and P is inorganic phosphate. In the presence of A and B the equilibria are:

$$
\begin{array}{lllll}
E + A & \overset{K_A}{\rightleftharpoons} & EA & \overset{k_{p_A}}{\longrightarrow} & E + Q + P \\
+ & & & & \\
B & & & & \\
{\scriptstyle K_B}\updownarrow & & & & \\
EB & \overset{k_{p_B}}{\longrightarrow} & E + R + P & &
\end{array}
$$

If the rate of P formation is measured, the observed velocity, v_t, is the sum of two reactions:

$$v_t = k_{p_A}[EA] + k_{p_B}[EB] \qquad \text{and} \qquad \frac{v_t}{[E]_t} = \frac{k_{p_A}[EA] + k_{p_B}[EB]}{[E] + [EA] + [EB]}$$

Under rapid equilibrium or steady-state conditions:

$$\boxed{v_t = \frac{V_{\max_A}\dfrac{[A]}{K_{m_A}} + V_{\max_B}\dfrac{[B]}{K_{m_B}}}{1 + \dfrac{[A]}{K_{m_A}} + \dfrac{[B]}{K_{m_B}}}} \qquad \text{(III-13)}$$

where $V_{\max_A} = k_{p_A}[E]_t$ = the maximal velocity with A as a substrate
$V_{\max_B} = k_{p_B}[E]_t$ = the maximal velocity with B as the substrate

Equation III-13 can be rearranged to:

$$\boxed{v_t = \frac{V_{\max_A}[A]}{K_{m_A}\left(1+\frac{[B]}{K_{m_B}}\right)+[A]} + \frac{V_{\max_B}[B]}{K_{m_B}\left(1+\frac{[A]}{K_{m_A}}\right)+[B]}} \qquad \text{(III-14)}$$

Thus one enzyme that accepts two substrates will yield results identical to two enzymes, each of which is specific for one substrate but is competitively inhibited by the other's substrate.

The v_t can be expressed in terms of v_A (the velocity observed at the given [A] in the absence of B) and v_B (the velocity observed at the given [B] in the absence of A).

Substituting:

$$V_{\max_A} = v_A\left(1+\frac{K_{m_A}}{[A]}\right) \qquad \text{and} \qquad V_{\max_B} = v_B\left(1+\frac{K_{m_B}}{[B]}\right)$$

into equation III-13 and simplifying:

$$\boxed{v_t = \frac{v_A\left(1+\frac{[A]}{K_{m_A}}\right)+v_B\left(1+\frac{[B]}{K_{m_B}}\right)}{1+\frac{[A]}{K_{m_A}}+\frac{[B]}{K_{m_B}}}} \qquad \text{(III-15)}$$

If A and B are equally acceptable substrates (i.e., same K_m and $V_{\max}$), then the v_t observed in the presence of any given mixture of A and B will be the same as the v observed with either A or B alone at the same total specific concentration. For example, if $V_{\max_A}=1$ and $V_{\max_B}=1$, then at $[A]=0.2K_{m_A}$ plus $[B]=0.6K_{m_B}=0.6K_{m_A}$, v_t equals 0.444. This is the same v observed at $[A]=0.8K_{m_A}$ and $[B]=0$, or $[B]=0.8K_{m_B}$ and $[A]=0$. On the other hand, v_t is *always less than* the sum of the velocities observed with each substrate alone at a given concentration. In the example above, $v_A=0.167$ at $[A]=0.2K_{m_A}$ and $[B]=0$. The $v_B=0.375$ at $[B]=0.6K_{m_B}$ and $[A]=0$. The sum, $v_A+v_B=$ 0.542, is greater than the observed v_t of 0.444 at $[A]=0.2K_{m_A}$ plus $[B]=0.6K_{m_B}$. The fact that v_t is always less than v_A+v_B may seem odd at first. But suppose A and B are really the same compound and $V_{\max}=1$. At $[A]=K_{m_A}$, $v_A=0.5$. At $[B]=K_{m_B}$, $v_B=0.5$. $v_A+v_B=1$, yet we know that at $[A]=2K_{m_A}$ or $[B]=2K_{m_B}$, v is only 0.667. Note that $v_t<v_A+v_B$ holds regardless of the relative values of K_{m_A}, K_{m_B}, $V_{\max_A}$ and $V_{\max_B}$. As with ordinary competitive inhibition, the degree of inhibition (in this case, the difference between v_t and v_A+v_B)

decreases as the concentration of either A or B becomes small compared to the respective K_m values. The maximum difference between v_t and $v_A + v_B$ is observed when both A and B are present at concentrations that are very high compared to their respective K_m values.

If the specific concentrations of A and B are equal:

$$v_t = \frac{v_A\left(1 + \frac{[A]}{K_{m_A}}\right) + v_B\left(1 + \frac{[A]}{K_{m_A}}\right)}{1 + \frac{[A]}{K_{m_A}} + \frac{[A]}{K_{m_A}}} = (v_A + v_B)\frac{1 + \frac{[A]}{K_{m_A}}}{1 + \frac{2[A]}{K_{m_A}}} \qquad \text{(III-16)}$$

Thus when $[A]/K_{m_A}$ and $[B]/K_{m_B}$ are very small, $v_t \cong v_A + v_B$. As $[A]/K_{m_A}$ and $[B]/K_{m_B}$ increase (but remain equal), v_t increases to a limit of $(v_A + v_B)/2$.

An interesting relationship can be derived for the special case where A and B are present at equimolar concentrations (not equal specific concentrations). If $[A] = [B]$:

$$v_t = \frac{V_{\max_A}\frac{[A]}{K_{m_A}} + V_{\max_B}\frac{[A]}{K_{m_B}}}{1 + \frac{[A]}{K_{m_A}} + \frac{[A]}{K_{m_B}}}$$

Dividing numerator and denominator by [A]:

$$v_t = \frac{\frac{V_{\max_A}}{K_{m_A}} + \frac{V_{\max_B}}{K_{m_B}}}{\frac{1}{[A]} + \frac{1}{K_{m_A}} + \frac{1}{K_{m_B}}}$$

When [A] and [B] are very high compared to their respective K_m values, the $1/[A]$ term becomes negligible and the observed combined velocity is maximal.

$$v_t \cong \frac{\frac{V_{\max_A}}{K_{m_A}} + \frac{V_{\max_B}}{K_{m_B}}}{\frac{1}{K_{m_A}} + \frac{1}{K_{m_B}}} = V_{\max_t}$$

or

$$\boxed{\frac{K_{m_B}}{K_{m_A}} = \frac{V_{\max_B} - V_{\max_t}}{V_{\max_t} - V_{\max_A}}} \qquad \text{(III-17)}$$

Thus the relative K_m values can be determined from three measurements, namely, V_{max_A}, V_{max_B}, and V_{max_t} (maximal mixed velocity with an equimolar mixture of A and B).

If two specific enzymes are present and each is unaffected by the other's substrate, then v_t will equal $v_A + v_B$. However, there are at least five conditions where a mixture of two enzymes yields $v_t < v_A + v_B$: (*a*) each enzyme is catalytically active with only one of the substrates, but is competitively inhibited by the other substrate (equation III-14); (*b*) each enzyme is catalytically active with only one of the substrates but one of the enzymes (e.g., the A-specific enzyme) is competitively inhibited by the other's substrate (B); (*c*) one enzyme is catalytically active with only one of the substrates (e.g., A) and is unaffected by the other substrate (B), and the second enzyme is nonspecific; (*d*) one enzyme is catalytically active with only one of the substrates (e.g., A), but is competitively inhibited by the other substrate, and the second enzyme is nonspecific; (*e*) two nonspecific enzymes. Under conditions *c* and *d*, nonlinear reciprocal and Eadie-Scatchard plots may be seen when the varied substrate is the one acted on by both enzymes. Under condition *e*, nonlinear plots may be seen for both substrates. Conditions *a* and *b* yield linear plots for both substrates. (In both cases, there is only one enzyme active on a given substrate.) Additional distinctions may be made if v is measured as the rate of unique product (Q and R) formation. Under conditions *a*, *d*, and *e*, a saturating concentration of either substrate will inhibit completely the formation of the unique product of the other substrate. Under condition *b*, a saturating concentration of one substrate (e.g., B) will inhibit completely the formation of the unique product of the other substrate (e.g., Q from A), but not vice versa (i.e., [A] will have no effect on R formation from B). Under condition *c*, a saturating concentration of one substrate (e.g., A) will inhibit completely the formation of the unique product of the other substrate (e.g., R from B). The reverse experiment yields partial inhibition. A saturating concentration of B will inhibit only the activity of the nonspecific enzyme on A. The activity of the A-specific enzyme is unaffected.

The mixed substrate phenomenon was applied in an interesting way to a study of NH_4^+ transport by *Penicillium chrysogenum*. In this study methylammonium-^{14}C was used as an NH_4^+ analog. The K_m for methylammonium-^{14}C transport was 10^{-5} M, and the V_{max} was 10 μmoles$\times$g dry weight cells$^{-1}\times$min^{-1}. The physiologically important substrate of the transport system, NH_4^+ was a potent inhibitor of methylammonium-^{14}C transport ($K_{i_{NH_4^+}} = 2.5\times 10^{-7}$ M). The K_i value for NH_4^+ as an inhibitor was assumed to be equivalent to the K_m value for NH_4^+ as a substrate. The mixed substrate method was used to estimate V_{max} for NH_4^+ transport. Varying concentrations of NH_4^+ (10^{-5} to 10^{-4} M) were mixed with a constant concentration(10^{-4} M) of methylammonium-^{14}C and the uptake of the methyl ammonium-^{14}C from each mixture was determined over a period of time. The results are shown in Figure III-8. During the early stages of the assay, NH_4^+ almost completely displaced methylammonium-^{14}C from the transport system. During the same time period, NH_4^+ was transported into the cells thereby reducing its external concentration. This resulted in a progressive decrease in the inhibition of methylammonium-^{14}C transport with time. When the NH_4^+ had been depleted, the methylammonium-^{14}C transport rate attained the control rate. The lag period (between zero-time and the time when the methylammonium-^{14}C transport rate attained the control rate) was taken as the time required to transport the NH_4^+ present. The length of the lag period was proportional to the initial NH_4^+ concentration.

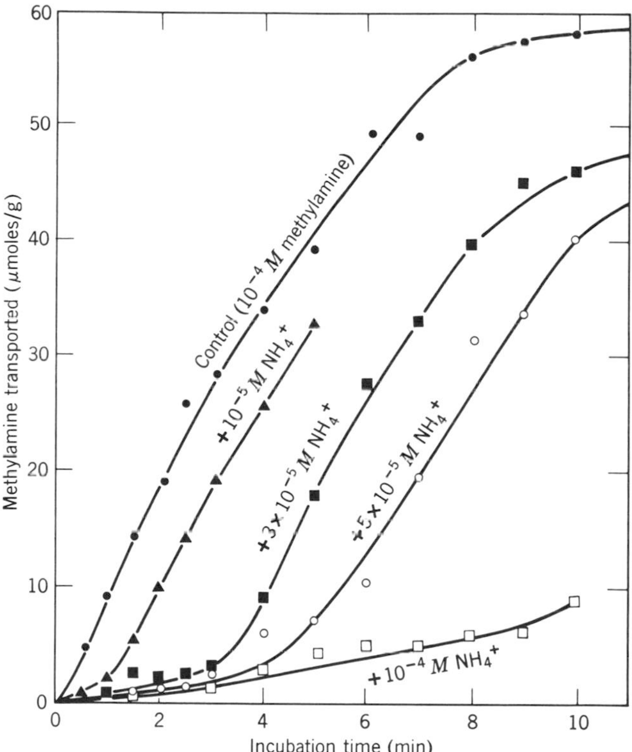

Fig. III-8. Effect of NH_4^+ on methylamine–^{14}C transport by *Penicillium chrysogenum*. Methylamine–^{14}C is excluded from the transport system as long as NH_4^+ is present in the medium. [Redrawn from Hackette, S. L., Skye, G. E., Burton, C. and Segel, I. H., *J. Biol. Chem.* **245**, 4241, (1970).] (See Dixon, M. and Webb, E. C., *Enzymes*, 2nd ed., Ch. 4, p. 88 for another example observed by Willstätter *et al.*, in 1927.)

The rate thus calculated was estimated as V_{max} for NH_4^+ transport. The estimate assumes that (*a*) 10^{-4} *M* methylammonium had little effect on the rate of NH_4^+ transport from solutions containing 10^{-5} to 10^{-4} *M* NH_4^+, and (*b*) the rate of NH_4^+ transport was essentially constant over the lag period. Assumption *a* is valid because the affinity of the transport system for NH_4^+ is 40-fold greater than the affinity for methylammonium. Assumption *b* is also valid because more than 97% of the NH_4^+ would have been transported before its concentration decreased to the K_m value.

Apparent Competitive Inhibition by Carrier Dilution (Isotope Competition)

In assays employing radioactive substrates, the addition of unlabeled substrate will produce the same *apparent* degree of inhibition as an equivalent amount of an alternative substrate with the same K_m value, or a nonsubstrate competitive inhibitor whose K_i value equals the K_m of the radioactive substrate. This method can be used to obtain a rapid comparison of the relative affinities of a variety of alternative substrates or nonsubstrate inhibitors. The carrier dilution method is illustrated below.

Suppose an enzyme catalyzes a reaction with a certain substrate, S, where $K_m = 2 \times 10^{-5}$ *M* and $V_{max} = 25$ μmoles/min. If radioactive S is used with a specific activity of 300,000 cpm/μmole, then at $[S] = K_m$, for example, the initial velocity will be $0.5\,V_{max}$, or 12.5 μmoles/min. The experimental raw data value of v would be (12.5 μmoles/min)$\times$(3×10^5 cpm/μmole)$= 37.5 \times 10^5$ cpm/min. If a tenfold excess of unlabeled substrate (2×10^{-4} *M*) is added to the assay mixture together with the radioactive substrate, the specific activity of the substrate will be reduced to 1/11 of the original specific activity. The new specific activity will be 27,272 cpm/μmole. The new [S] will be $11K_m$ and the new velocity will be $11/12\,V_{max}$, or 22.9 μmoles/min. The raw data value of v will be (22.9 μmoles/min)$\times$(27,272 cpm/μmole)$= 6.25 \times 10^5$ cpm/min. Compared to the original rate of 37.5×10^5 cpm/min, we observe an *apparent* 83.3% inhibition.

$$a = \frac{6.25 \times 10^5}{37.5 \times 10^5} = 0.167 \qquad i = 1 - 0.167 = 0.833$$

This corresponds to an *apparent* v of 2.08 μmoles/min. The true velocity, of course, has not decreased. It has increased on adding the additional substrate. However, v will appear to decrease if the raw data rate in terms of cpm/min are compared to the original raw data rate, or if the rate in terms of μmoles/min is calculated using the original, undiluted specific activity.

Now, let us calculate the degree of inhibition caused either by a tenfold excess of an alternate substrate where the K_m value equals the K_m of the radioactive substrate, or by a tenfold excess of a nonsubstrate competitive inhibitor where K_i equals the K_m of the substrate.

$$v = \frac{(25)(K_m)}{K_m\left(1 + \frac{10K_i}{K_i}\right) + K_m} = \frac{25K_m}{12K_m} = 2.08\ \mu\text{moles/min}$$

The specific activity of the substrate is unchanged, so the observed raw data value of v_0 would be (2.08 μmoles/min)×(3×10^5 cpm/μmole)=6.25×10^5 cpm/min. Thus whether we compare the raw data velocities in terms of cpm/min, or velocities in terms of μmoles/min, the inhibitor and the unlabeled substrate produced the same degree of inhibition—real in the presence of the inhibitor, but only apparent in the presence of the excess unlabeled substrate. If the inhibitor produced a lower degree of inhibition than an equivalent amount of excess substrate, then we could conclude that $K_i > K_m$. If the inhibitor produced a greater degree of inhibition, then K_i must be less than K_m. It is not necessary that K_m and V_{max} be known to compare affinities by this method. However, the degree of inhibition by an n-fold excess of unlabeled substrate or inhibitor will be maximum when the concentration of radioactive substrate is high compared to K_m.

Isotope competition can be used to determine unknown concentrations of unlabeled substrate in solutions known to be free of real inhibitors. If we substitute $[S]/K_m$ for $[I]/K_i$ in equation III-4 we obtain:

$$\boxed{\frac{v_i}{v_0} = a = \frac{K_m + [S^*]}{K_m + [S] + [S^*]}} \qquad \text{(III-17a)}$$

where $[S^*]$ = the known concentration of radioactive substrate in the assay mixture

$[S]$ = the added unknown concentration of unlabeled substrate

$v_i/v_0 = a$ = the relative activity in terms of cpm/min or velocities (μmoles/min) if the original, undiluted specific activity of the radioactive substrate is used to calculate v_i

Equation III-17a can be solved for the concentration of unlabeled substrate present:

$$\boxed{[S] = (K_m + [S^*])\frac{1-a}{a}} \qquad \text{(III-17b)}$$

(See also Chapter Two, Section O. The concentration of unlabeled S can also be calculated from $K_{m_{app}}$ as given by equation II-84, provided the true K_m is known).

Competitive Product Inhibition Where [S] + [P] is Constant (Regulation Via "Energy Charge")

Consider a system in which the substrate and product are interconvertible, but the total pool of [S] + [P] remains essentially constant. The rate of the S→P reaction will depend on the relative concentrations of the substrate and the product which competes with the substrate for the enzyme. Because [S] + [P] is constant, an increase in [S] automatically means that [P] must decrease. Consequently, any increase in [S] is accompanied by a decrease in the degree of competitive product inhibition. The velocity curve can be concave (decreasing slope), convex (increasing slope), or linear, depending on the relative affinities of the enzyme for S and P. To simplify matters, we will assume that K_{eq} is very large (because V_{max_r} is very small) so that even at the lowest [S]/[P] ratio the observed initial velocity is the true forward velocity of the S→P reaction, uncomplicated by the P→S reaction. This assumption eliminates the $[P]/K_{eq}$ term from the numerator of equation II-20. Alternately, we can assume that the reaction yields two products, one of which is removed in a subsequent reaction. Under either of these conditions the velocity equation is:

$$\frac{v}{V_{max}} = \frac{[S]}{K_S\left(1 + \frac{[P]}{K_P}\right) + [S]} \qquad \text{(III-18)}$$

Figure III-9 shows the velocity curves for a system where $[S] + [P] = 10K_S$. Note that it is possible to obtain very steep (convex) curves if $K_P < K_S$. Steep velocity curves are usually associated with multisite enzymes that display cooperative binding and possess specific effector sites. In the present system, the [S]/[P] ratio exerts a very sensitive control over the velocity when $K_P < K_S$, yet only a single binding site for S and P is involved. Atkinson and co-workers (1970) have shown that a number of ATP-utilizing enzymes are strongly inhibited by their product, ADP or AMP. The initial velocities of these reactions are markedly influenced by the ATP-ADP-AMP

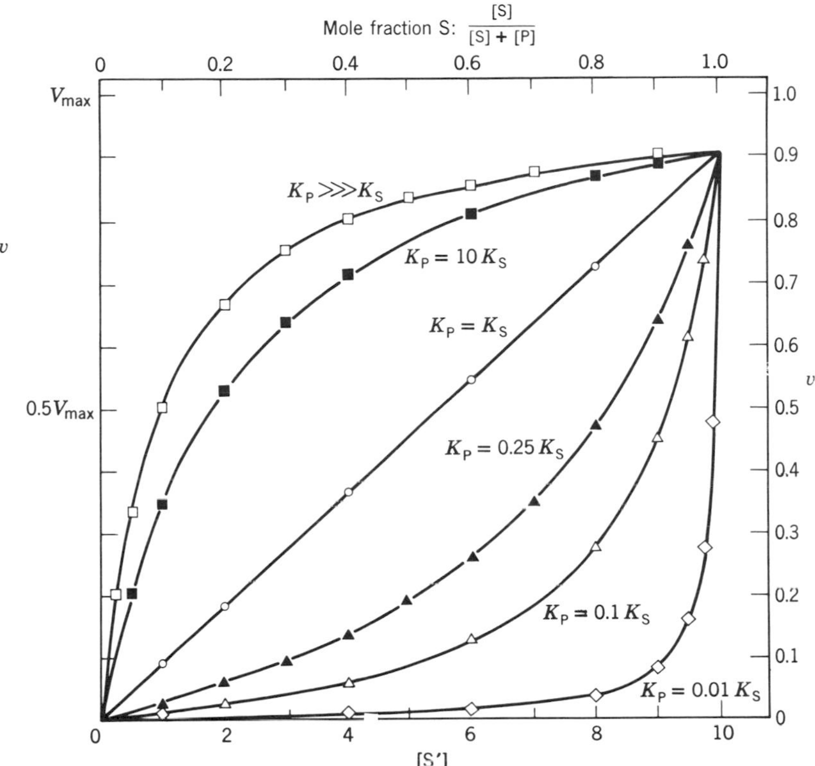

Fig. III-9. The v versus [S] plots where the total pool of [S]+[P] is constant at $10K_S$ and P has some affinity for E.

balance, called the "energy charge" of the system, where:

$$\text{"energy charge"} = \frac{[\text{ATP}] + \frac{1}{2}[\text{ADP}]}{[\text{ATP}] + [\text{ADP}] + [\text{AMP}]} \qquad \text{(III-19)}$$

The use of "energy charge" in place of the [ATP]/[ADP] or [ATP]/[AMP] ratios represents an attempt to simulate a given total adenine nucleotide pool under conditions that would exist *in vivo* where (presumably) adenylate kinase maintains the three nucleotides at equilibrium:

$$\text{ATP} + \text{AMP} \xrightleftharpoons{\text{adenylate kinase}} 2\text{ADP}$$

Thus "energy charge" represents the mole fraction of adenine nucleotides represented by ATP or its equivalent ($\frac{1}{2}$ADP). The adenylate system is analogous to an electrochemical storage battery. The system is fully charged when all adenylate is present as ATP ("energy charge" = 1.0), and completely discharged when all adenylate is present as AMP ("energy charge" = 0). A system in which all the ATP had been converted to ADP would have an "energy charge" of 0.5. After equilibration via adenylate kinase, the "energy charge" would still be 0.5, since the total concentration of phosphate anhydride bonds would be unchanged. (Energy charge can also be defined as one half the phosphate anhydride bonds per adenosine.) The distribution of adenylates as a function of the "energy charge" is shown in Figure III-10. Many biosynthetic (i.e., energy utilizing) reactions are promoted by a high energy charge and inhibited by a low energy charge while the converse is true for energy-producing reactions. For example, Figure III-11 shows the velocity response to the [S]/[P] ratio for two enzymes, one of which catalyzes an essentially irreversible ATP-utilizing reaction (indicated as S→P) while the other catalyzes an essentially irreversible ATP-generating reaction (indicated as P→S). In both cases, $K_P = 0.1K_S$. The pool of [S] + [P] is fixed at $10K_S$. The velocities are given by the usual equations taking into account the product inhibition by P in the S→P reaction and the product inhibition by S in the P→S reaction:

$$\frac{v}{V_{max}} = \frac{[S]}{K_S\left(1 + \frac{[P]}{K_P}\right) + [S]} \qquad \text{(for S} \longrightarrow \text{P)}$$

$$\frac{v}{V_{max}} = \frac{[P]}{K_P\left(1 + \frac{[S]}{K_S}\right) + [P]} \qquad \text{(for P} \longrightarrow \text{S)}$$

If, *in vivo*, the mole fraction of [S] is poised at about 0.9 (for the particular parameters chosen), then the velocities of the S-utilizing and S-generating reactions will proceed at about $0.5V_{max}$ and small changes in the [S]/[P] ratio will tend to reestablish the original [S]/[P] ratio. (When [S] decreases, the velocity of the S→P reaction slows up while that of the P→S reaction increases.) If we consider only the S→P reaction, we see that the "energy charge" model can provide an effective "off-on" switch; that is, for a wide range of S concentrations, the velocity of the reaction can be relatively low and insensitive to increasing [S]. Then, for a relatively small increase in [S], v can increase markedly. To be an effective control system, K_P must be significantly less than K_S and the total concentration of [S] + [P] must be

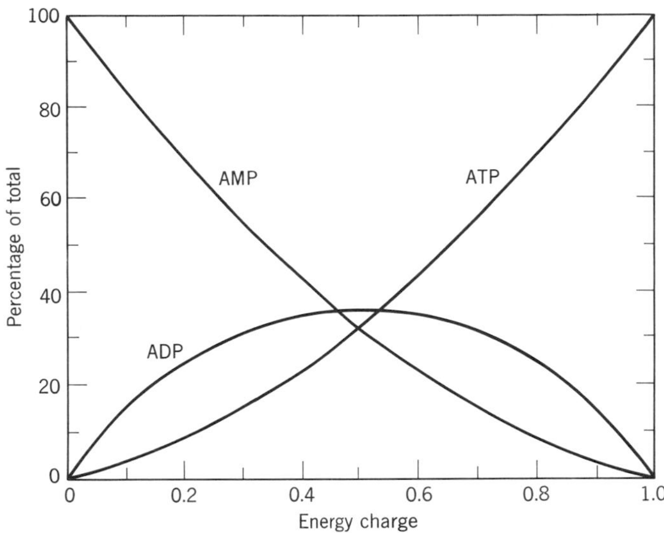

Fig. III-10. Relative concentrations of adenine nucleotides as a funtion of energy charge when the adenylate kinase reaction is at equilibrium ($K'_{eq}=0.8$). [Redrawn with permission from Atkinson, D.E., *Biochemistry* **7** 4030 (1968). Copyright by the American Chemical Society.]

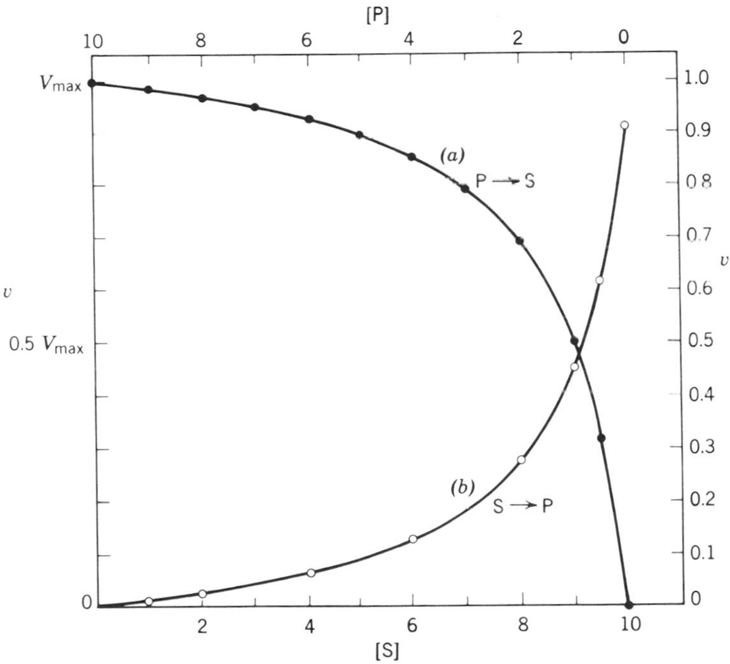

Fig. III-11. Velocity response to changing [S]/[P] ratio where [S] + [P] is constant. (*a*) A P-utilizing enzyme. (*b*) An S-utilizing enzyme. The product of one enzyme is the substrate of the other and vice versa; $K_P=0.1$, $K_S=1.0$, [S] + [P] = 10.

large compared to K_S. (If $[S]+[P]=100K_S$, the velocity curves would be essentially the same as those shown in Figure III-11 for the same K_S and K_P values.) In this respect, it is noteworthy that the intracellular levels of adenine nucleotides are quite high compared to their K_m values. It seems likely that many biological oxidation-reduction reactions will be regulated by an analogous "reduction charge," that is, the [NADH]/[NADH + NAD^+] or [NADPH]/[NADPH + $NADP^+$] ratio. For most dehydrogenases involved in energy metabolism, K_{NADH} is less than K_{NAD^+}, which is exactly the condition necessary for a steep velocity response to the "reduction charge." Furthermore, the intracellular concentration of pyridine nucleotides is high compared to their K_m values (another required condition). The pyruvic dehydrogenase of *E. coli* responds to both the adenylate "energy charge" and the oxidation level of the NADH + NAD^+ pool. The response of pyruvic dehydrogenase to the adenylate "energy charge" can be treated as described in activation system A-5 (Chapter Five) where the enzyme activity is regulated by the [I]/[A] ratio and neither effector is a substrate or product of the reaction. Other factors, in addition to the [S] + [P] concentration and the relative K_S and K_P values, can influence the velocity response to "energy charge" or "reduction charge." These factors include (*a*) the concentrations of effectors which alter K_S or K_P, (*b*) the concentration of Mg^{2+} (when the true S and P species binding to the enzyme are the Mg complexes), and, for reactions involving two or more substrates and/or products, (*c*) the kinetic mechanism, and (*d*) the number and nature of dead-end complexes that can form. Regulation via "energy charge" may have been one of the first control devices evolved by living cells. In its simplest form, the model requires only effective competition between P and S for a single binding site. Examples of enzymes regulated by the energy charge are (*a*) citrate cleavage enzyme from rat liver, which produces extramitochondrial acetyl-*S*-CoA for fatty acid biosynthesis via the reaction:

$$\text{ATP} + \text{citrate} + \text{CoASH} \rightleftharpoons \text{acetyl-}S\text{-CoA} + \text{oxalacetate} + \text{ADP} + \text{P}_i$$

and (*b*) phosphoribosylpyrophosphate (PRPP) synthetase of *E. coli*, which furnishes PRPP for histidine, trytophan, and purine and pyrimidine nucleotides via the reaction:

$$\text{ATP} + \text{ribose-5-phosphate} \rightleftharpoons \text{PRPP} + \text{AMP}$$

It is noteworthy that ADP is much better than AMP as an inhibitor of PRPP synthetase. Thus ADP must have a higher affinity than AMP for the AMP-ATP site, or ADP binds to a distinct regulatory site. In either case, the "energy charge" of the system controls the velocity of the reaction, since a high ADP or high AMP level signifies a low ATP level, and vice versa. Another example of a biosynthetic enzyme regulated by the "energy charge" is the adenosinediphosphoglucose (ADPG) synthetase (pyrophosphorylase) of bacteria. This enzyme catalyzes the production of ADPG from ATP and glucose-1-phosphate.

$$\text{ATP} + \text{glucose-1-phosphate} \rightleftharpoons \text{ADPG} + \text{PP}_i$$

In this case neither ADP nor AMP are products of the reaction, yet both are inhibitors, with AMP being the more potent. ADP is produced in the subsequent reaction where ADPG serves as the glucosyl donor for the bacterial glycogen synthetase:

$$\text{ADPG} + (\text{glycogen})_n \longrightarrow \text{ADP} + (\text{glycogen})_{n+1}$$

With ADPG synthetase, we do not have a simple case of competitive product inhibition, yet the response of the system to the "energy charge" still serves to insure that glycogen synthesis will proceed only when the cell is energy-sufficient.

B. NONCOMPETITIVE INHIBITION (SIMPLE INTERSECTING LINEAR NONCOMPETITIVE INHIBITION)

A classical noncompetitive inhibitor has no effect on substrate binding, and vice versa. The inhibitor and the substrate bind reversibly, randomly, and independently at different sites. That is, I binds to E and to ES; S binds to E and to EI. The binding of one ligand has no effect on the dissociation constant of the other. However, the resulting ESI complex is inactive. Noncompetitive inhibition is common in steady-state multireactant systems (Chapter Nine), but for reasons somewhat different than those presented here. A model for classical noncompetitive inhibition is shown in Figure III-12. It is assumed that I distorts the enzyme sufficiently to prevent the proper positioning of the catalytic center and thus ESI is nonproductive. A similar situation is shown in Figure III-13. Here, there is no direct path from ES to ESI, but the same four enzyme species are at equilibrium. If ES is assumed to exist in two forms, one "open" (and able to bind I) and one "closed" (as shown), the properties of the system would be unchanged since multiple central complexes (e.g., ES, ES′, EP, EP′) do not affect the velocity

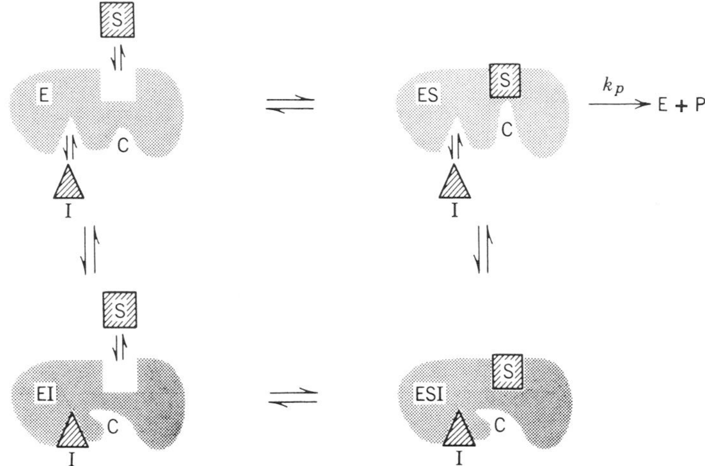

Fig. III-12. Noncompetitive inhibition; S and I are not mutually exclusive but ESI is catalytically inactive. When S binds, the enzyme undergoes a conformational change which aligns the catalytic center, C, with the susceptible bonds of S; I interferes with the conformational change, but has no effect on S binding.

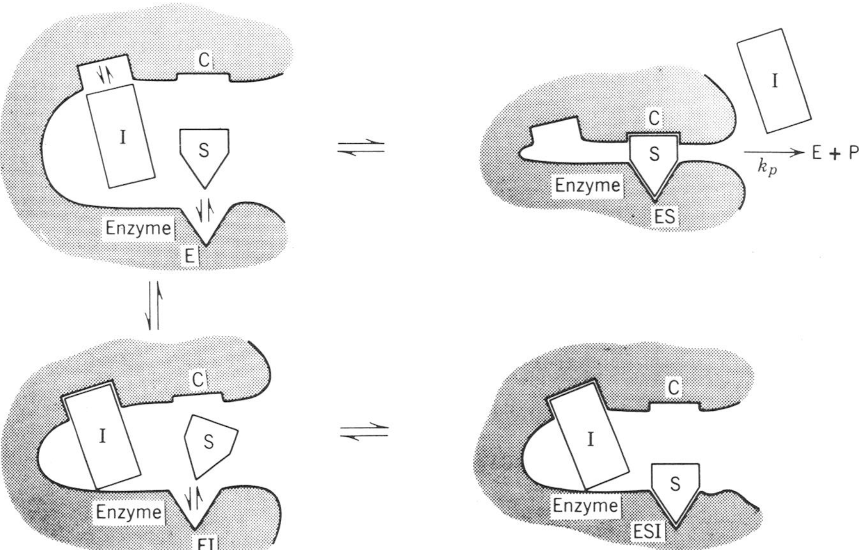

Fig. III-13. Noncompetitive inhibition. In this model, I cannot bind to ES, but the properties of the system are identical to that shown in Fig. III-12 because the same four enzyme species are at equilibrium. In steady-state conditions substrate inhibition is observed.

equations. The equilibria are:

$$
\begin{array}{ccccc}
E + S & \underset{}{\overset{K_S}{\rightleftharpoons}} & ES & \overset{k_p}{\longrightarrow} & E + P \\
+ & & + & & \\
I & & I & & \\
K_i \updownarrow & & K_i \updownarrow & & \\
EI + S & \overset{K_S}{\rightleftharpoons} & ESI & &
\end{array}
$$

$$K_S = \frac{[E][S]}{[ES]} = \frac{[EI][S]}{[ESI]}$$

$$K_i = \frac{[E][I]}{[EI]} = \frac{[ES][I]}{[ESI]}$$

We can see from the equilibria that at any inhibitor concentration, an infinitely high substrate concentration cannot drive all the enzyme to the productive ES form. At any [I] a portion of the enzyme will remain as the nonproductive ESI complex. Consequently, we can predict that the V_{max} in the presence of a noncompetitive inhibitor (V_{max_i}) will be less than the V_{max} observed in the absence of inhibitor. The K_m value (measured as the [S] required for $\frac{1}{2} V_{max_i}$) will be unchanged by a noncompetitive inhibitor because at any inhibitor concentration the enzyme forms which can combine with S (E and EI) have equal affinities for S. The net effect of a reversibly bound noncompetitive inhibitor is to make it appear as if less total enzyme is present.

A substance that *irreversibly* inactivates an enzyme is sometimes (incorrectly) called a noncompetitive inhibitior because V_{max} is decreased. Irreversible inhibition and reversible noncompetitive inhibition may be distinguished by plotting V_{max} versus $[E]_t$, where $[E]_t$ represents total units of enzyme activity added to the assay (Fig. III-14). For a reversible noncompetitive inhibitor, the "plus inhibitor" curve will have a smaller slope than the control curve and will go through the origin. If an irreversible inhibitor is present, the "plus inhibitor" curve will have the same slope as the control curve, but will intersect the horizontal axis at a position equivalent to the amount of enzyme that is irreversibly inactivated.

An expression relating v, V_{max}, [S], K_S, [I], and K_i in the presence of a noncompetitive inhibitor can be derived easily from rapid equilibrium assumptions. This time we must recognize that the total enzyme, $[E]_t$, is present in four forms: free enzyme, [E]; enzyme-substrate complex, [ES];

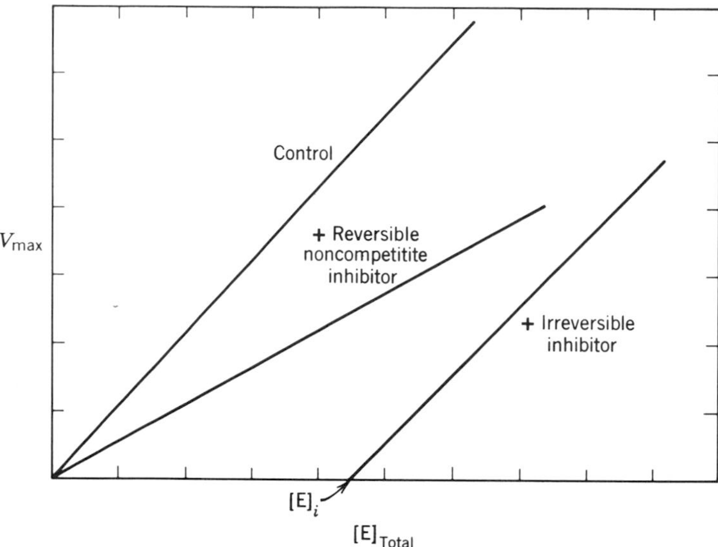

Fig. III-14. A plot of $V_{\max}$ versus amount of enzyme added will distinguish between a reversible and an irreversible noncompetitive inhibitor. $[E]_i$ represents the amount of enzyme titrated by the irreversible inhibitor.

enzyme-inhibitor complex, [EI]; and enzyme-substrate-inhibitor complex, [ESI].

$$v = k_p[\mathrm{ES}]$$

$$\frac{v}{[\mathrm{E}]_t} = \frac{k_p[\mathrm{ES}]}{[\mathrm{E}]+[\mathrm{ES}]+[\mathrm{EI}]+[\mathrm{ESI}]}$$

$$\boxed{\frac{v}{V_{\max}} = \frac{\dfrac{[\mathrm{S}]}{K_\mathrm{S}}}{1+\dfrac{[\mathrm{S}]}{K_\mathrm{S}}+\dfrac{[\mathrm{I}]}{K_i}+\dfrac{[\mathrm{I}][\mathrm{S}]}{K_\mathrm{S}K_i}}} \qquad \text{(III-20)}$$

where $V_{\max} = k_p[\mathrm{E}]_t$. In this form, we see that the denominator has two additional terms compared to the normal velocity equation. The $[\mathrm{I}]/K_i$ term is contributed by the EI complex, while the $[\mathrm{I}][\mathrm{S}]/K_\mathrm{S}K_i$ term results from the ESI complex. The latter term does not appear in the velocity equation for competitive inhibition because there is no ESI complex. The numerator still contains only one term as there is only one product-forming complex (ES). To obtain a more familiar form, we can multiply the numerator and denominator of equation III-20 by K_S and factor:

$$\boxed{\frac{v}{V_{max}} = \frac{[S]}{K_S\left(1+\frac{[I]}{K_i}\right)+[S]\left(1+\frac{[I]}{K_i}\right)}} \qquad \text{(III-21)}$$

The expression above differs from the usual Henri-Michaelis-Menten equation and from that derived earlier for competitive inhibition in that the K_S and [S] terms in the denominator are both multiplied by the factor $(1+[I]/K_i)$. We can better appreciate the effect of a noncompetitive inhibitor by dividing the denominators of both sides of the equation by $(1+[I]/K_i)$:

$$\boxed{\frac{v}{\dfrac{V_{max}}{\left(1+\frac{[I]}{K_i}\right)}} = \frac{[S]}{K_S+[S]}} \qquad \text{or} \qquad \frac{v}{V_{max_i}} = \frac{[S]}{K_S+[S]} \qquad \text{(III-22)}$$

where:

$$V_{max_i} = \frac{V_{max}}{\left(1+\frac{[I]}{K_i}\right)} = \text{the apparent } V_{max} \text{ at the given [I]}$$

As predicted, the only effect of a noncompetitive inhibitor is to decrease V_{max}. The K_S value remains unchanged. (The reader should not be confused by the fact that the K_S term in the original equation III-21 was multiplied by $(1+[I]/K_i)$. Before deciding whether K_S is affected, we must first modify the equation by removing any factor of the variable, [S].) The decrease in V_{max} does not mean that the inhibitor has decreased the rate constant for the breakdown of ES to E+P. This constant, k_p, is unchanged. It is the steady-state level of ES that is decreased. At any [S] and [I], the enzyme-substrate complex is present as a mixture of productive ES and nonproductive ESI forms ($k_p=0$ for the ESI complex). The factor $(1+[I]/K_i)$ may be considered to be an [I]-dependent statistical factor describing the distribution of the enzyme-substrate complexes between the ES and ESI forms.

Systems are known in which the inhibitor does affect k_p (i.e., in which ESI forms product slower than ES). This type of system, called partial noncompetitive inhibition, is described in Chapter Four.

If k_p contributes significantly to the $K_m = (k_{-1} + k_p)/k_1$ relationship, then an inhibitor that affects the apparent k_p value will also affect the K_m value. In pure and partial noncompetitive systems, the apparent k_p changes without K_m changing. Consequently, k_p must be very small compared to k_{-1}, and K_m must be equivalent to K_S. In other words, classical noncompetitive inhibition in unireactant systems is obtained only under rapid equilibrium conditions. In fact, a steady-state treatment does not yield an equation of the Henri-Michaelis-Menten form, but rather a complex expression containing $[S]^2$ and $[I]^2$ terms. The reciprocal plots are theoretically nonlinear (although they may appear so because the nonlinear region may occur close to the $1/v$-axis). Figure III-15 shows a situation that could be at steady-state and still yield a simple velocity equation without $[S]^2$ terms (because there is only one reaction in which S adds).

An expression for the relative velocity in the presence of a noncompetitive inhibitor can be derived readily:

$$a = \frac{v_i}{v_0} = \frac{\dfrac{V_{max}[S]}{(K_m + [S])\left(1 + \dfrac{[I]}{K_i}\right)}}{\dfrac{V_{max}[S]}{K_m + [S]}} \quad \text{or} \quad \boxed{a = \frac{K_i}{K_i + [I]}} \tag{III-23}$$

Thus a is independent of [S], or, in other words, a given concentration of I reduces the velocity by exactly the same factor at all substrate concentrations.

$$\therefore \quad \boxed{\frac{V_{max_i}}{V_{max}} = \frac{K_i}{K_i + [I]}} \tag{III-24}$$

Although V_{max} is reduced to V_{max_i}, the specific substrate concentration required for any fraction of V_{max_i} is unchanged. For example, $[S]_{0.5} = K_m$, $[S]_{0.9} = 9K_m$, and $[S]_{0.1} = \frac{1}{9}K_m$. Thus the $[S]_{0.9}/[S]_{0.1}$ ratio is 81 at all inhibitor concentrations. As before, the constant ratio is expected, since the form of the equation is unchanged; only the absolute value of V_{max} is changed. The effect of a noncompetitive inhibitor is shown in Figure III-16.

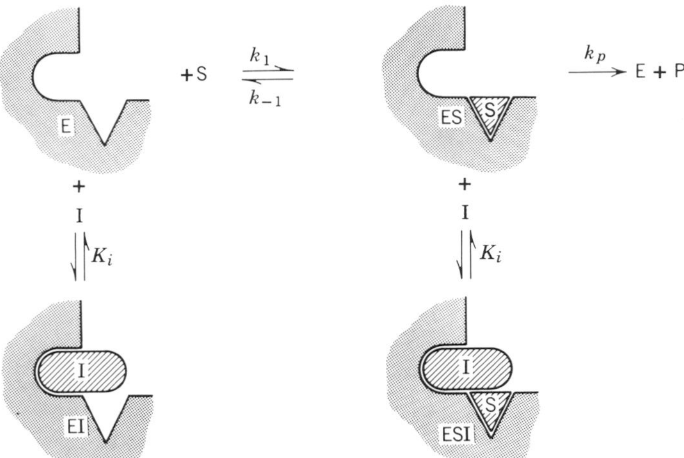

Fig. III-15. A third model for noncompetitive inhibition; I sterically hinders S binding. The velocity equation derived from steady-state assumptions would be the same as that derived from rapid equilibrium assumptions.

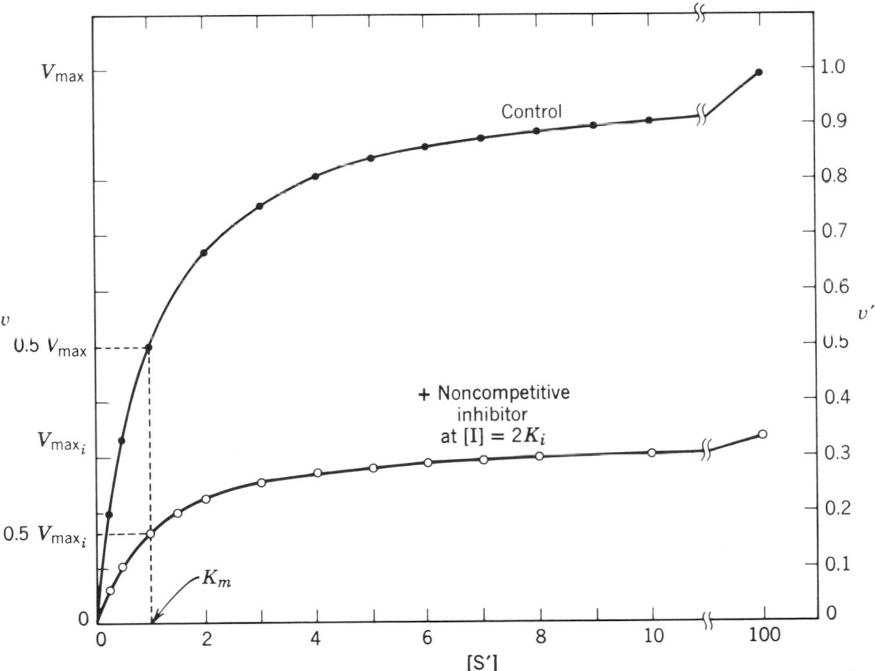

Fig. III-16. The v versus [S] plot in the presence of a noncompetitive inhibitor.

An expression for the fractional inhibition can be derived from equation III-23:

$$i = 1 - \frac{v_i}{v_0} = 1 - a = 1 - \frac{K_i}{K_i + [\mathrm{I}]}$$

$$\boxed{i = \frac{[\mathrm{I}]}{K_i + [\mathrm{I}]}} \qquad \boxed{i_{\%} = 100\left(\frac{[\mathrm{I}]}{K_i + [\mathrm{I}]}\right)} \qquad \text{(III-25)}$$

At any [S], $[\mathrm{I}]_{0.5} = K_i$, $[\mathrm{I}]_{0.9} = 9K_i$, and $[\mathrm{I}]_{0.1} = \frac{1}{9}K_i$. Thus the $[\mathrm{I}]_{0.9}/[\mathrm{I}]_{0.1}$ ratio is always 81.

General Principles

A classical noncompetitive inhibitor decreases V_{max}, but has no effect on the K_m value. The degree of inhibition in the presence of a noncompetitive inhibitor depends only upon [I] and K_i. The inhibited velocity (v_i) is always a constant fraction of v_0, regardless of the substrate concentration or the value of K_m. An increase in [S] causes both v_0 and v_i to increase by the same factor. The net effect of a noncompetitive inhibitor is to make it seem as if less enzyme were present. When $[\mathrm{I}] = K_i$, we observe 50% inhibition at all substrate concentrations.

Reciprocal Plot for Noncompetitive Inhibition Systems

In the reciprocal form, the velocity equation for noncompetitive inhibition is:

$$\boxed{\frac{1}{v} = \frac{K_m}{V_{max}}\left(1 + \frac{[\mathrm{I}]}{K_i}\right)\frac{1}{[\mathrm{S}]} + \frac{1}{V_{max}}\left(1 + \frac{[\mathrm{I}]}{K_i}\right)} \qquad \text{(III-26)}$$

The equation indicates that both the slope and the $1/v$-axis intercept of the reciprocal plot are increased by the factor $(1 + [\mathrm{I}]/K_i)$ compared to the "control" plot. If the slope *and* the $1/v$-axis intercept increase by the same factor, then the $1/[\mathrm{S}]$-axis intercept will remain the same (equal to $-1/K_m$). The K_i can be calculated from the slope or the $1/v$-axis intercept. When $1/[\mathrm{S}] = 0$, the $1/v$ intercept gives $1/V_{max_i}$, where $1/V_{max_i} = 1/V_{max}(1 + [\mathrm{I}]/K_i)$. For each fixed inhibitor concentration, a new reciprocal plot can be drawn (Fig. III-17). As [I] increases, the "plus inhibitor" curves increase in slope and $1/v$-axis intercept, pivoting counterclockwise about the point of intersection with the control curve (at $-1/K_m$ on the $1/[S]$-axis). Because the initial velocity can be driven to zero at a saturating inhibitor concentration, the limiting slope will be a vertical line through $-1/K_m$ and parallel to the $1/v$-axis.

Replots of $Slope_{1/S}$ and $1/V_{max_i}$ Versus [I]

The slope of the reciprocal plot in the presence of a pure noncompetitive inhibitor is a linear function of [I] (Fig. III-18*a*) as shown earlier for pure competitive inhibition (Fig. III-5). The $1/v$-axis intercept ($1/V_{max_i}$) is also a linear function of [I] (Fig. III-18*b*):

$$\boxed{\frac{1}{V_{max_i}} = \frac{1}{V_{max}K_i}[I] + \frac{1}{V_{max}}} \qquad \text{(III-27)}$$

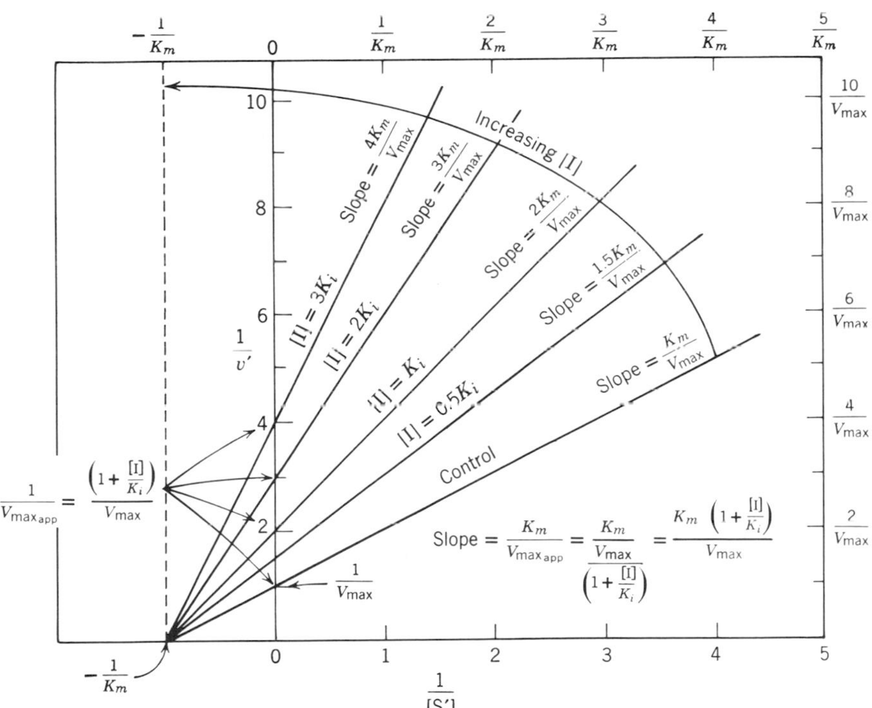

Fig. III-17. The $1/v$ versus $1/[S]$ plot in the presence of different fixed concentrations of a noncompetitive inhibitor.

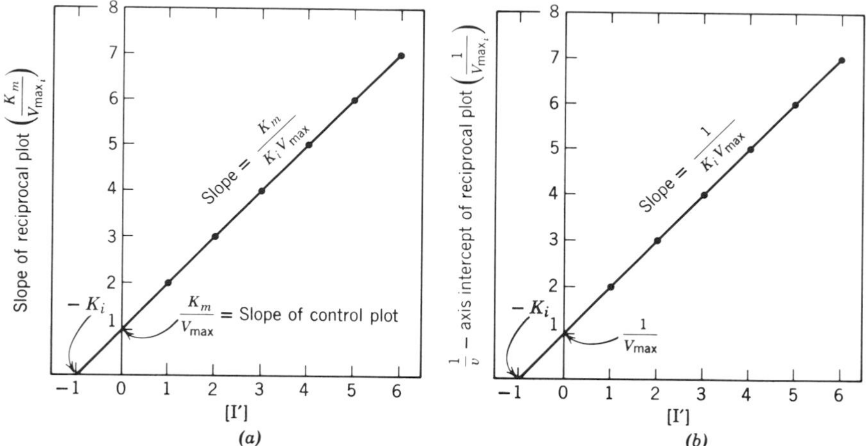

Fig. III-18. Replots of data taken from the reciprocal plot. (*a*) $Slope_{1/S}$ versus [I]. (*b*) $1/v$-axis intercept (i.e., $1/V_{max_i}$) versus [I].

Linear replots of $slope_{1/S}$ versus [I] and $1/V_{max_i}$ versus [I] distinguish pure noncompetitive inhibition from partial noncompetitive inhibition. The latter yields hyperbolic replots (Chapter Four).

Dixon Plot for Noncompetitive Inhibition: $1/v$ Versus [I]

The reciprocal equation for noncompetitive inhibition may be rearranged to:

$$\frac{1}{v} = \frac{\left(1 + \frac{K_m}{[S]}\right)}{V_{max} K_i}[I] + \frac{1}{V_{max}}\left(1 + \frac{K_m}{[S]}\right) \tag{III-28}$$

A plot of $1/v$ versus [I] will be a straight line (Fig. III-19*a*). The slope of the Dixon plot is given by:

$$slope = \frac{K_m}{V_{max} K_i}\frac{1}{[S]} + \frac{1}{V_{max} K_i} \tag{III-29}$$

A replot of *slope* versus the corresponding $1/[S]$ (Fig. III-19*b*) is a straight line with a slope of $K_m/V_{max}K_i$ and an intercept of $1/V_{max}K_i$ on the *slope*-axis. When $slope = 0$, the intercept on the $1/[S]$-axis gives $-1/K_m$. (In contrast, the slope replot for competitive inhibition passes through the origin.) The replot of $1/v$-axis intercept versus the corresponding $1/[S]$ is

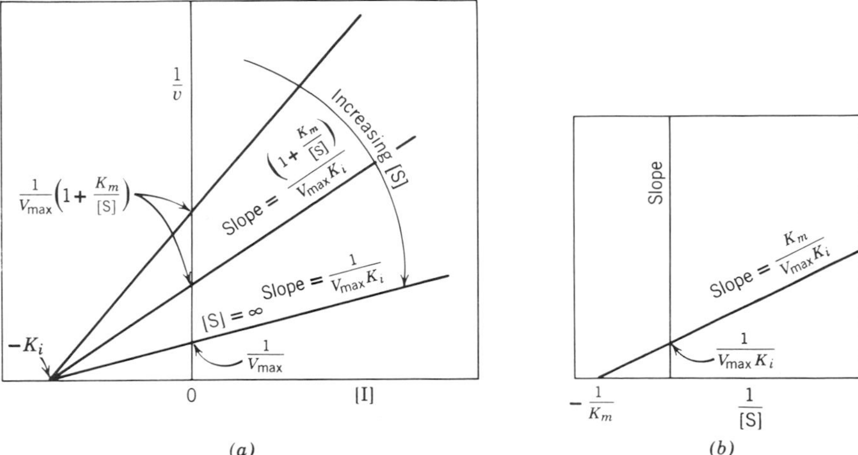

Fig. III-19. (*a*) Dixon plot for a noncompetitive inhibitor: $1/v$ versus [I] in the presence of different fixed concentrations of substrate. (*b*) *Slope* replot.

identical to the usual reciprocal plot in the absence of I. Other plots for noncompetitive inhibition are described in Chapter Four.

Integrated Rate Equation in the Presence of a Noncompetitive Inhibitor

If the reaction has a large K_{eq} and none of the products have an appreciable affinity for the enzyme, then the time course of the reaction in the presence of a noncompetitive inhibitor can be expressed as:

$$\frac{2.3}{t}\log\frac{[S]_0}{[S]} = -\frac{1}{K_m}\frac{[P]}{t} + \frac{V_{max}}{K_m\left(1+\frac{[I]}{K_i}\right)} \qquad \text{(III-30)}$$

where $[P]=[S]_0-[S]$.

The equation assumes a constant inhibitor concentration during the time course of the reaction. A determination of [P] at various times during the course of the reaction will allow K_m and $V_{max_{app}}$ to be determined. A family of plots can be obtained for different inhibitor concentrations (Fig. III-20). The lines will be parallel since the slope is independent of [I]. The values of V_{max} and K_i can be obtained from appropriate replots of vertical or horizontal axes intercepts.

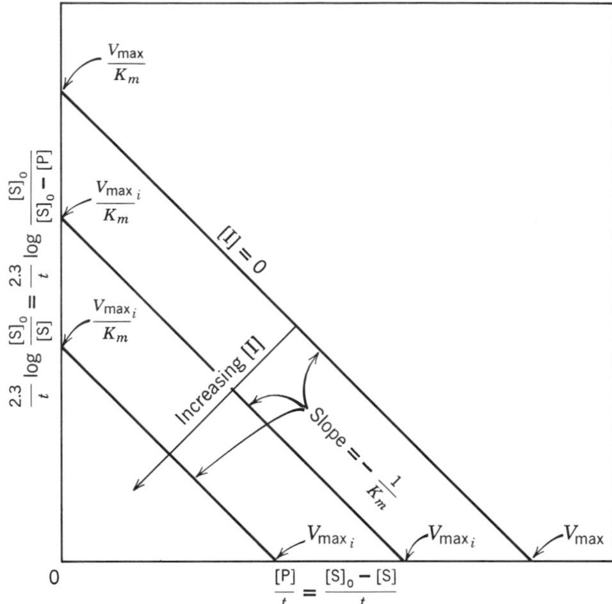

Fig. III-20. Plot of the integrated velocity equation in the presence of a noncompetitive inhibitor.

C. UNCOMPETITIVE INHIBITION (SIMPLE LINEAR UNCOMPETITIVE INHIBITION)

A classical uncompetitive inhibitor is a compound that binds reversibly to the enzyme-substrate complex yielding an inactive ESI complex (Fig. III-21). The inhibitor does not bind to the free enzyme. Pure uncompetitive inhibition (also called anticompetitive inhibition and coupling inhibition) may be rare in unireactant systems. Nevertheless, it is worth considering because it is a simple example of the sequential addition of two enzyme ligands in an obligate order. Uncompetitive inhibition is common in steady-state multireactant systems (Chapter Nine) for reasons similar to those described here. That is, I will be uncompetitive with respect to a given substrate if I binds to the enzyme only after the substrate binds (although I rarely binds to a central complex where all the substrate binding sites are filled). Classical uncompetitive inhibition is described by the following

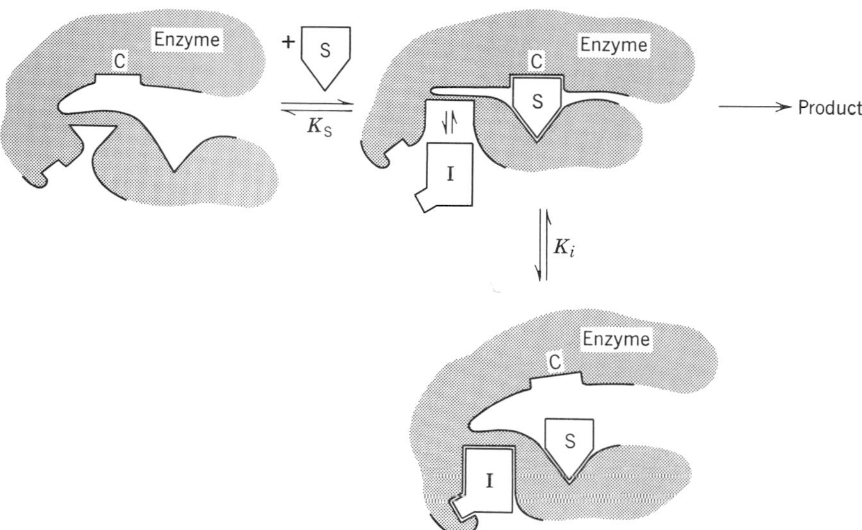

Fig. III-21. Uncompetitive inhibition; I binds only to the ES complex. When S binds, a conformational change occurs in the enzyme which forms or unmasks the I site. The resulting ESI complex is catalytically inactive; C represents the catalytic center of the enzyme.

equilibria:

$$\begin{array}{ccccc} E + S & \overset{K_S}{\rightleftharpoons} & ES & \overset{k_p}{\longrightarrow} & E + P \\ & & + & & \\ & & I & & \\ & & K_i \upharpoonleft\downharpoonright & & \\ & & ESI & & \end{array}$$

The equilibria show that at any [I] an infinitely high substrate concentration will not drive all of the enzyme to the ES form; some nonproductive ESI complex will always be present. Consequently, we can predict that V_{max} in the presence of an uncompetitive inhibitor (V_{max_i}) will be lower than the V_{max} in the absence of inhibitor. Unlike noncompetitive inhibition, however, the apparent K_m value will decrease. The decrease occurs because the reaction $ES + I \rightarrow ESI$ removes some ES causing the reaction $E + S \rightarrow ES$ to proceed to the right. Under certain conditions, a mixed-type inhibitor can produce the same effects as an uncompetitive inhibitor. The specific conditions are discussed in Chapter Four.

An expression relating v, V_{max}, [S], K_S, [I], and K_i in the presence of an uncompetitive inhibitor is derived below:

$$v = k_p[\mathrm{ES}]$$

$$\frac{v}{[\mathrm{E}]_t} = \frac{k_p[\mathrm{ES}]}{[\mathrm{E}]+[\mathrm{ES}]+[\mathrm{ESI}]}$$

$$\boxed{\frac{v}{V_{max}} = \frac{\dfrac{[\mathrm{S}]}{K_S}}{1+\dfrac{[\mathrm{S}]}{K_S}+\dfrac{[\mathrm{S}][\mathrm{I}]}{K_S K_i}}} \qquad \text{(III-31)}$$

The equation does not contain an $[\mathrm{I}]/K_i$ term because no EI complex forms. Multiplying numerator and denominator of equation III-31 by K_S and factoring:

$$\boxed{\frac{v}{V_{max}} = \frac{[\mathrm{S}]}{K_S+[\mathrm{S}]\left(1+\dfrac{[\mathrm{I}]}{K_i}\right)}} \qquad \text{(III-32)}$$

A steady-state treatment yields the same equation with K_m replacing K_S.

The velocity equation differs from the usual Henri-Michaelis-Menten expression in that the [S] term in the denominator is multiplied by the factor $(1+[\mathrm{I}]/K_i)$. We can better appreciate the effect of an uncompetitive inhibitor on the kinetic constants by dividing the denominators of both sides of the equation by $(1+[\mathrm{I}]/K_i)$.

$$\boxed{\frac{v}{\dfrac{V_{max}}{\left(1+\dfrac{[\mathrm{I}]}{K_i}\right)}} = \frac{[\mathrm{S}]}{\dfrac{K_m}{\left(1+\dfrac{[\mathrm{I}]}{K_i}\right)}+[\mathrm{S}]}} \qquad \text{(III-33)}$$

or

$$\boxed{\frac{v}{V_{\max_i}} = \frac{[\mathrm{S}]}{K_{m_{\mathrm{app}}} + [\mathrm{S}]}} \tag{III-34}$$

where

$$V_{\max_i} = \frac{V_{\max}}{\left(1 + \frac{[\mathrm{I}]}{K_i}\right)} \qquad \text{and} \qquad K_{m_{\mathrm{app}}} = \frac{K_m}{\left(1 + \frac{[\mathrm{I}]}{K_i}\right)}$$

In other words, an uncompetitive inhibitor decreases $V_{\max}$ and K_m to the same extent (Fig. III-22).

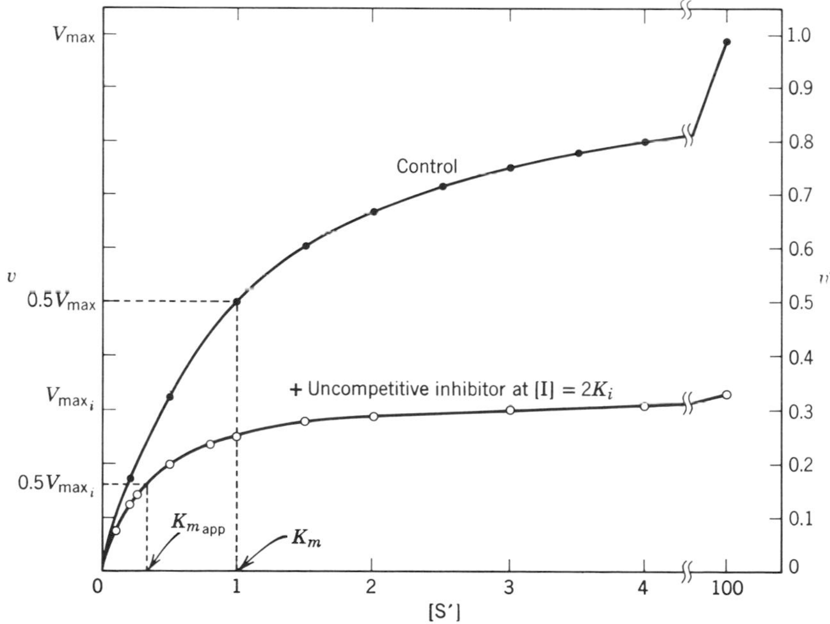

Fig. III-22. The v versus [S] plot in the presence of an uncompetitive inhibitor.

An expression for the relative velocity in the presence of an uncompetitive inhibitor can be derived:

$$a = \frac{v_i}{v_0} = \frac{\dfrac{V_{\max}[S]}{K_m + [S]\left(1 + \dfrac{[I]}{K_i}\right)}}{\dfrac{V_{\max}[S]}{K_m + [S]}}$$

$$\boxed{a = \frac{K_m + [S]}{K_m + [S]\left(1 + \dfrac{[I]}{K_i}\right)}} \qquad \text{(III-35)}$$

The fractional inhibition caused by an uncompetitive inhibitor is given by:

$$i = 1 - a \qquad \text{or} \qquad \boxed{i = \frac{[I]}{K_i\left(1 + \dfrac{K_m}{[S]}\right) + [I]}} \qquad \text{(III-36)}$$

When $i = 0.5$:

$$\boxed{[I]_{0.5} = \left(1 + \frac{K_m}{[S]}\right)K_i} \qquad \text{(III-37)}$$

Note that the relationship between K_m and [S] is opposite to that for competitive inhibition. Similarly, we can show that:

$$[I]_{0.9} = 9\left(1 + \frac{K_m}{[S]}\right)K_i, \qquad [I]_{0.1} = \frac{1}{9}\left(1 + \frac{K_m}{[S]}\right)K_i$$

Thus the $[I]_{0.9}/[I]_{0.1}$ ratio is 81 for all substrate concentrations regardless of the absolute values of K_m and K_i. The degree of inhibition depends on the substrate concentration, but, unlike competitive inhibition, the degree of inhibition *increases* as [S] increases. This is to be expected, because an uncompetitive inhibitor combines only with the ES complex, and the concentration of ES increases as [S] increases. An uncompetitive inhibitor inhibits because of its effect on V_{max}. The inhibitor is actually an activator with respect to K_m ($K_{m_{app}} < K_m$). If the substrate concentration is low enough so that the reaction is essentially first-order, the effect of an uncompetitive inhibitor on V_{max} will be almost completely canceled by its opposite effect on K_m and little or no inhibition will be observed.

Reciprocal Plot for Uncompetitive Inhibition

The reciprocal form of the velocity equation for uncompetitive inhibition is:

$$\frac{1}{v} = \frac{K_m}{V_{max}} \frac{1}{[S]} + \frac{1}{V_{max}} \left(1 + \frac{[I]}{K_i}\right) \qquad \text{(III-38)}$$

The slope of the plot is still K_m/V_{max}, but the $1/v$-axis intercept is increased by the factor $(1+[I]/K_i)$ which multiplied the [S] term in the original equation. Consequently, the "plus inhibitor" and control curves will be parallel. As [I] increases, the $1/v$-axis intercepts increase, yielding a series of parallel plots (Fig. III-23). A saturating inhibitor concentration will drive the velocity to zero. Consequently, the displacement of the "plus inhibitor" plots from the control plot increases without limit.

Replots of $1/V_{max_i}$ and $1/K_{m_{app}}$ Versus [I]

A replot of $1/V_{max_i}$ versus [I] will be linear (Fig. III-24*a*) with intercepts of $1/V_{max}$ and $-K_i$ as shown for pure noncompetitive inhibition. The $K_{m_{app}}$ varies inversely with [I] (Fig. III-24*b*):

$$\frac{1}{K_{m_{app}}} = \frac{1}{K_i K_m}[I] + \frac{1}{K_m} \qquad \text{(III-39)}$$

The linear replots will distinguish pure uncompetitive inhibition from a mixed-type system in which K_m and V_{max} change by the same factor (see mixed-type inhibition System C4 where $\alpha = \beta$, in Chapter Four).

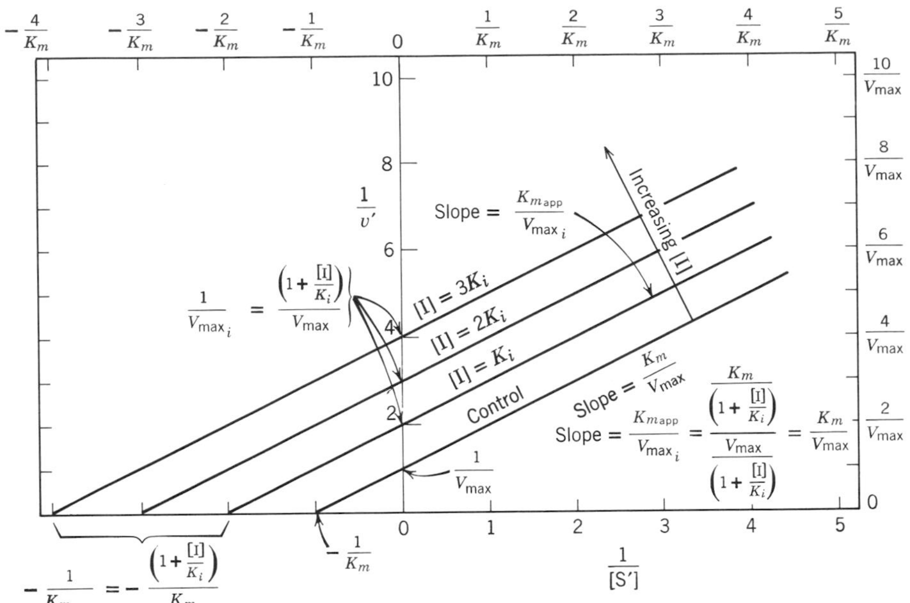

Fig. III-23. The $1/v$ versus $1/[S]$ plot in the presence of different fixed concentrations of an uncompetitive inhibitor.

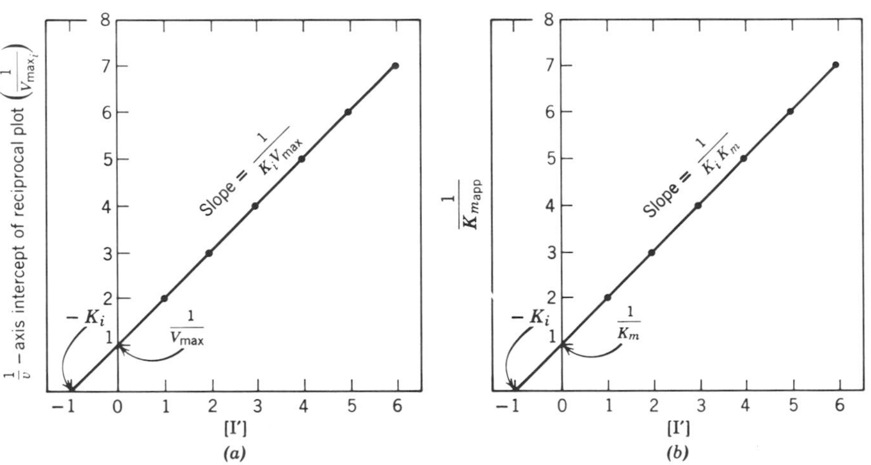

Fig. III-24. Replots of data taken from the reciprocal plot. (*a*) $1/v$-axis intercept (i.e., $1/V_{max_i}$) versus [I]. (*b*) $1/K_{m_{app}}$ versus [I].

Dixon Plot for Uncompetitive Inhibition: $1/v$ Versus [I]

The equation for the Dixon plot is:

$$\frac{1}{v} = \frac{1}{V_{max}K_i}[\mathrm{I}] + \frac{1}{V_{max}}\left(1 + \frac{K_m}{[\mathrm{S}]}\right) \qquad \text{(III-40)}$$

The slope expression does not contain an [S] term. Hence we can expect the plots to be parallel at all substrate concentrations (Fig. III-25).

Integrated Rate Equation in the Presence of an Uncompetitive Inhibitor

The integrated rate equation in the presence of an uncompetitive inhibitor is:

$$\frac{2.3}{t}\log\frac{[\mathrm{S}]_0}{[\mathrm{S}]} = -\frac{\left(1 + \frac{[\mathrm{I}]}{K_i}\right)}{K_m}\frac{[\mathrm{P}]}{t} + \frac{V_{max}}{K_m} \qquad \text{(III-41)}$$

where $[\mathrm{P}] = [\mathrm{S}]_0 - [\mathrm{S}]$. Figure III-26 shows the time course of the reaction at various fixed inhibitor concentrations plotted according to equation III-41.

D. EFFECTS OF CONTAMINATING INHIBITORS ON THE INITIAL VELOCITY VERSUS ENZYME CONCENTRATION PLOT

Enzyme preparations frequently contain inhibitors. These may be endogenous metabolites or substances introduced during the cell breakage or fractionation procedures. If the presence of inhibitors is not recognized, errors may be introduced into the determination of K_m or K_i values, or total enzyme units. The problems caused by contaminating endogenous inhibitors can be avoided usually if the enzyme is partially purified. Indeed, recoveries greater than 100% of the original activity after a preliminary purification step are not uncommon. However, many times a large number of cell free extracts must be assayed under conditions where a preliminary purification of each one is not feasible. Under such circumstances, it would be desirable

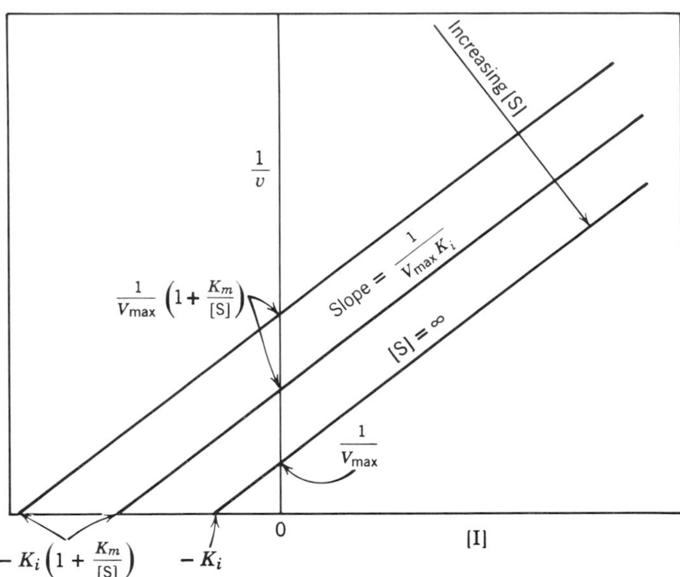

Fig. III-25. Dixon plot for an uncompetitive inhibitor: $1/v$ versus [I] at different fixed concentrations of S.

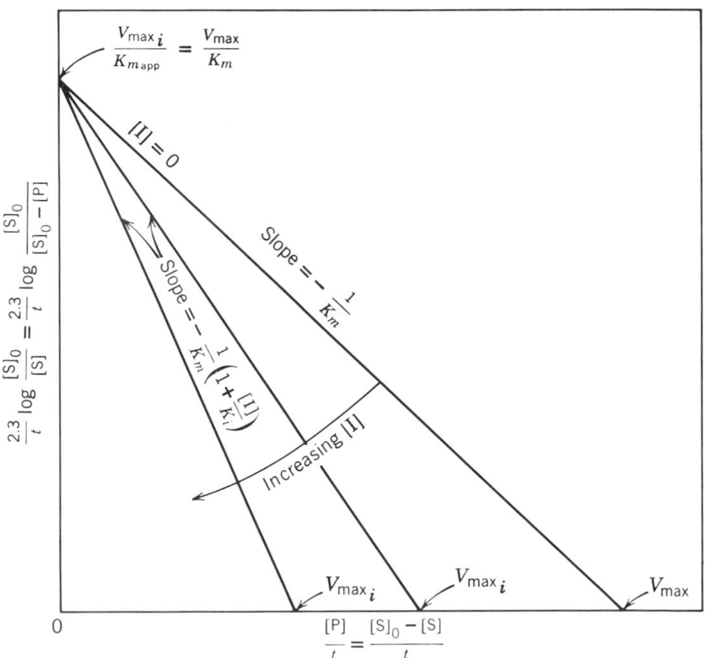

Fig. III-26. Plot of the integrated velocity equation in the presence of an uncompetitive inhibitor.

to know whether the enzyme preparation contains inhibitors. This can be checked by determining the response of the assay to increasing amounts of enzyme preparation. A departure of the v versus $[E]_t$ plot from linearity indicates the presence of inhibitors in the enzyme preparation. Suppose an extract contains a competitive inhibitor such that 10 μl gives a final concentration of $3K_i$ in the assay mixture. The assay is conducted at $[S]=10K_m$. The 10 μl of extract contains an amount of enzyme sufficient to yield a V_{max} under the assay conditions of 100 arbitrary units in the absence of any inhibitors. At the assay substrate concentration ($10K_m$), the theoretical (uninhibited) v is 91 units. (Of course, the amount of enzyme present is not known—the object of the assay is to determine how much enzyme is present.) The observed activity in the presence of $[I]=3K_i$ will be:

$$v_{10\,\mu l}=\frac{k_p[E]_t[S]}{K_{m_{app}}+[S]}=\frac{V_{max}[S]}{K_m\left(1+\dfrac{[I]}{K_i}\right)+[S]}$$

$$=\frac{(100)(10K_m)}{K(1+3)+10K_m}=\frac{1000K_m}{14K_m}$$

$$v_{10\,\mu l}=71.4 \text{ units}$$

If the presence of the inhibitor is unrecognized, the value of 71.4 units would be taken as a measure of the amount of enzyme present. The true velocity should have been 91 units; thus the assay will give a value 20% lower than the true value. Now, let us double the amount of enzyme preparation used. If 10 μl produced a v of 71.4 units and this represents an uninhibited velocity, then 20 μl should yield a v of 142.8 units (doubling $[E]_t$ doubles V_{max}, since $V_{max}=k_p[E]_t$). If the preparation contains an inhibitor, then doubling $[E]_t$ will also double [I].

$$v_{20\,\mu l}=\frac{2V_{max}[S]}{K_m\left(1+\dfrac{2[I]}{K_i}\right)+[S]}=\frac{(2)(100)(10K_m)}{K_m(1+6)+10K_m}$$

$$v_{20\,\mu l}=\frac{2000K_m}{17K_m}=118 \text{ units}$$

Similarly, if only 5 μl of preparation are used and the v obtained at 10 μl represents an uninhibited v, then $v_{5\ \mu l}$ should be 35.7 units.

$$v_{5\ \mu l} = \frac{0.5\,V_{max}[\mathrm{S}]}{K_m\left(1+\dfrac{0.5[\mathrm{I}]}{K_i}\right)+[\mathrm{S}]} = \frac{(0.5)(100)(10K_m)}{K_m(1+1.5)+10K_m}$$

$$= \frac{500K_m}{12.5K_m} = 40 \text{ units}$$

We see that doubling the enzyme concentration produced less than a doubling of the velocity, and halving the enzyme concentration produced more than half the velocity. It is obvious that v is not directly proportional to $[\mathrm{E}]_t$.

Figure III-27 shows the effect of a contaminating competitive inhibitor and a contaminating noncompetitive inhibitor on the v versus $[\mathrm{E}]_t$ curve at $[\mathrm{S}]=10K_m$ and $[\mathrm{I}]=3K_i$ in the 10 μl assay. The curve for a contaminating uncompetitive inhibitor would be slightly higher than that shown for the noncompetitive inhibitor.

If the enzyme preparation contains an activator then the v versus $[\mathrm{E}]_t$ plot

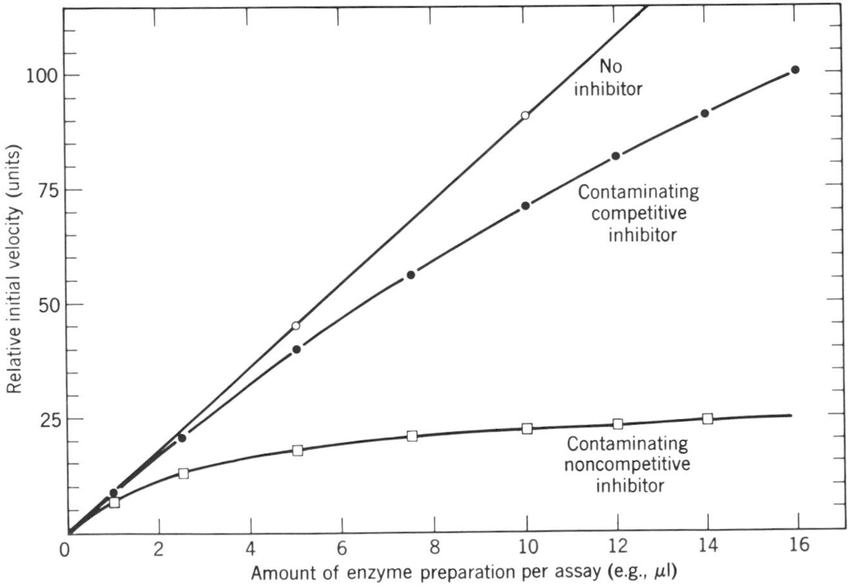

Fig. III-27. Effect of endogenous inhibitors on the plot of v versus amount of enzyme preparation added. Each 10 μl enzyme preparation contains 100 units of activity at saturating [S], and $[\mathrm{I}]=3K_i$; $[\mathrm{S}]=10K_m$ in all assays.

will curve upward, giving a greater than proportionate increase in v for a given increment of $[E]_t$.

Other Factors Producing Nonlinear v Versus $[E]_t$ Plots

Nonlinear v versus $[E]_t$ plots can result from a number of factors besides the presence of endogenous inhibitors in the enzyme preparation. Some of these are as follows. (*a*) The enzyme preparation may be stored at a pH or ionic strength significantly different from the optimum of the reaction. As the amount of enzyme preparation in the assay is increased, the "carry-over" will cause the assay pH or ionic strength to depart more and more from the optimum. Similarly, the enzyme may be stored in the presence of certain stabilizing agents (e.g., EDTA, thiols, and specific cations) which may inhibit the reaction. (*b*) The measured velocity may not be a true initial velocity. This could occur if the substrate concentration decreased significantly (i.e., out of the zero-order range) before the first product measurement. Similarly, if the pH of the assay mixture changes or a product inhibitor accumulates to a significant level before the first measurement, the calculated velocity will be lower than the true initial velocity. (*c*) The enzyme preparation may contain enzymes that convert the product to another compound that escapes detection by the assay method. (*d*) The enzyme may be unstable at the assay temperature, pH, ionic strength, and so on. If a significant amount of denaturation occurs before the first product measurement, the calculated velocity will be lower than the true initial velocity. (*e*) The enzyme preparation may contain proteolytic enzymes that are inactive under the storage conditions but degrade the enzyme rapidly in the assay mixture. (*f*) The assay method may be inaccurate at high product concentrations. For example, in spectrophotometric assays, high optical densities may be impossible to read accurately with the spectrophotometer used. In assays employing radioactive substrates, the radioactivity in the product may exceed the resolution time of the detector. These problems can be avoided by diluting the product before final measurement. For radioactive assays, a "coincidence" correction can be applied. (*g*) The reagents used to convert the product into a measureable form may be limiting. Similarly, "coupling" enzymes included in the original assay mixture may be the rate limiting factor, rather than the enzyme being assayed. (See Chapter Two, Section H.)

Contaminating Inhibitors in the Substrate

Inhibitors may be present as contaminants in substrates. These inhibitors may be structural or stereoisomers of the true substrate, or unrelated compounds (e.g., the buffer used to prepare the substrate stock solution, or a preservative). A contaminant may go undetected if it acts as a competitive inhibitor. For example, suppose [I], the concentration of the contaminant in the stock solution of substrate, is x[S]. As [S] is varied, [I] will also vary. However, [I] in the assay mixture will always be a constant fraction or multiple of [S]. In other words, [S] and [I] will be varied together at a constant ratio. The velocity equation is:

$$\frac{v}{V_{\max}} = \frac{\dfrac{[\mathrm{S}]}{K_m}}{1 + \dfrac{[\mathrm{S}]}{K_m} + \dfrac{[\mathrm{I}]}{K_i}} = \frac{\dfrac{[\mathrm{S}]}{K_m}}{1 + \dfrac{[\mathrm{S}]}{K_m} + \dfrac{x[\mathrm{S}]}{K_i}} = \frac{[\mathrm{S}]}{K_m + [\mathrm{S}]\left(1 + \dfrac{xK_m}{K_i}\right)}$$

or

$$\boxed{\frac{\dfrac{v}{V_{\max}}}{\left(1 + \dfrac{xK_m}{K_i}\right)} = \frac{[\mathrm{S}]}{\dfrac{K_m}{\left(1 + \dfrac{xK_m}{K_i}\right)} + [\mathrm{S}]}} \qquad \text{(III-42)}$$

The velocity curve and reciprocal plot will appear normal, but the kinetic constants are only apparent constants. The observed reciprocal plot will be parallel to and above the true plot.

The situation is different if I acts noncompetitively. In this case the velocity equation is:

$$\frac{v}{V_{\max}} = \frac{\dfrac{[\mathrm{S}]}{K_m}}{1 + \dfrac{[\mathrm{S}]}{K_m} + \dfrac{[\mathrm{I}]}{K_i} + \dfrac{[\mathrm{S}][\mathrm{I}]}{K_m K_i}} = \frac{\dfrac{[\mathrm{S}]}{K_m}}{1 + \dfrac{[\mathrm{S}]}{K_m} + \dfrac{x[\mathrm{S}]}{K_i} + \dfrac{x[\mathrm{S}]^2}{K_m K_i}}$$

$$\boxed{\frac{v}{V_{\max}} = \frac{[\mathrm{S}]}{K_m + [\mathrm{S}]\left(1 + \dfrac{xK_m}{K_i} + \dfrac{x[\mathrm{S}]}{K_i}\right)}} \qquad \text{(III-43)}$$

The equation does not have the usual form of the Henri-Michaelis-Menten equation. In effect, the denominator contains an $[\mathrm{S}]^2$ term resulting in apparent substrate inhibition. The equation for the reciprocal plot can be written as:

$$\frac{1}{v} = \frac{K_m}{V_{\max}}\left(\frac{1}{[\mathrm{S}]}\right) + \frac{1}{V_{\max}}\left(1 + \frac{xK_m}{K_i}\right) + \frac{x}{V_{\max}K_i}\left(\frac{1}{[\mathrm{S}]}\right)^{-1} \quad \text{(III-44)}$$

If we take the first derivative of equation III-44 and set it equal to zero, we find that the plot has a minimum.

$$\frac{d(1/v)}{d(1/[\mathrm{S}])} = \frac{K_m}{V_{\max}} - \frac{x}{V_{\max}K_i}\left(\frac{1}{[\mathrm{S}]}\right)^{-2} = 0$$

$$\therefore (1/[\mathrm{S}])^{-2} = \frac{K_m K_i}{x}, \qquad (1/[\mathrm{S}])^2 = \frac{x}{K_m K_i}$$

The minimum occurs at:

$$\frac{1}{[\mathrm{S}]} = \sqrt{\frac{x}{K_m K_i}} \qquad \text{or} \qquad [\mathrm{S}] = \sqrt{\frac{K_m K_i}{x}}$$

The plot of v versus [S] will increase, pass through a maximum, and then decrease to zero. The plot of $1/v$ versus $1/[\mathrm{S}]$ will bend upward as it approaches the $1/v$-axis (Fig. III-28). At very low [S] (i.e., high $1/[\mathrm{S}]$), the

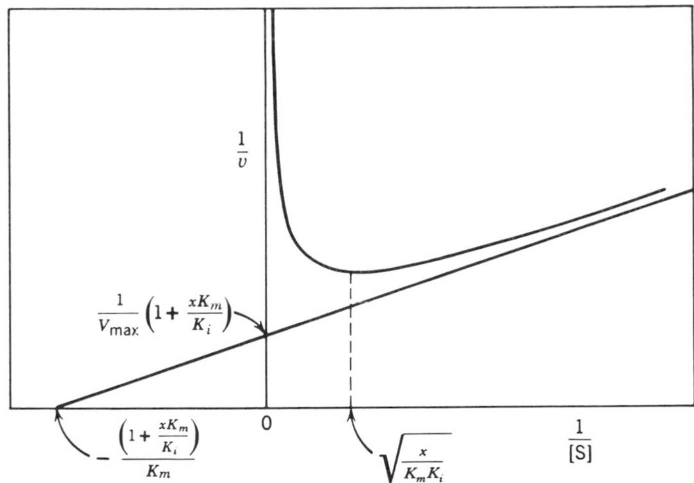

Fig. III-28. Effect of a contaminating noncompetitive inhibitor in the substrate on the $1/v$ versus $1/[\mathrm{S}]$ plot.

reciprocal plot approaches a straight line asymptotically. In this region of $1/[S]$, the effect of I is minimal ([I] is very low compared to K_i). Extrapolation of the asymptote yields apparent values for K_m and V_{max}. An uncompetitive inhibitor will behave similarly since the velocity equation will contain the $x[S]^2/K_mK_i$ term. However, the extrapolated K_m and V_{max} values will be the real values, since the velocity equation will not contain $x[S]/K_i$ (i.e., an $[I]/K_i$) term.

The treatment above assumes that I represents a very small fraction of the added substrate so that $[S] \cong [S]_{added}$. If, however, x is not $\ll 1$, then the true substrate concentration will be significantly less than $[S]_{added}$ and the $1/[S]$ coordinate of a given velocity point will be in error. Suppose that I is a competitive inhibitor. Let $[I] = x[S]_{added}$, then $[S] = (1-x)[S]_{added}$, where [S] is the true substrate concentration. The velocity is given by:

$$\frac{v}{V_{max}} = \frac{\dfrac{(1-x)[S]_{added}}{K_m}}{1 + \dfrac{(1-x)[S]_{added}}{K_m} + \dfrac{x[S]_{added}}{K_i}} = \frac{[S]_{added}}{\dfrac{K_m}{(1-x)} + [S]_{added} + \dfrac{xK_m[S]_{added}}{(1-x)K_i}}$$

or

$$\boxed{\frac{v}{\dfrac{V_{max}}{\left(1 + \dfrac{xK_m}{(1-x)K_i}\right)}} = \frac{[S]_{added}}{\left[\dfrac{K_m}{(1-x) + \dfrac{xK_m}{K_i}}\right] + [S]_{added}}} \qquad \text{(III-45)}$$

The reciprocal plot is still linear but now the observed K_m and V_{max} are altered by different factors. When $x \ll 1$, equation III-45 reduces to equation III-42, and the effects of the incorrect assumed values of [S] disappear (only the effects of the competitive inhibition by I remain). When $K_i \gg K_m$ (i.e., the impurity is not inhibitory), equation III-45 reduces to equation II-35 [with $(1-x)$ representing the fractional purity of S, called y in deriving equation II-35].

E. TIGHTLY BOUND INHIBITORS

The equations derived in the previous sections assume that there is no depletion of the inhibitor by the enzyme. That is, that the formation of EI and ESI does not significantly change the concentration of free inhibitor.

Thus the [I] of the equations (which represents the concentration of free inhibitor) is assumed to be identical to $[I]_t$, the concentration of added I. If, however, the enzyme has a very high affinity for the inhibitor, it will be necessary to use very low inhibitor concentrations in initial velocity studies. Consequently, a significant proportion of the total inhibitor present may be enzyme-bound and the usual graphical methods of determining K_i will not be valid. Dixon has devised a simple graphical method that permits K_i to be determined and at the same time tells us whether a significant fraction of $[I]_t$ is enzyme-bound. The method is based on the same procedures outlined in Chapter Two for systems involving tightly bound substrates (or systems in which $[E]_t \cong [S]_t$). In the derivations given below, it is assumed that only the inhibitor is depleted by the enzyme and $[S]=[S]_t$.

Competitive Inhibitors

In the presence of a competitive inhibitor, the enzyme is distributed among three forms:

$$[E]_t=[E]+[ES]+[EI] \tag{III-46}$$

From the velocity-dependence equation, $v=k_p[ES]$, we obtain:

$$[ES]=\frac{v}{k_p} \tag{III-47}$$

From the definition of $K_m=[E][S]/[ES]$, we obtain, after substituting [ES] as given by equation III-47:

$$K_m=\frac{[E][S]k_p}{v} \qquad \therefore \quad [E]=\frac{K_m v}{k_p[S]} \tag{III-48}$$

From the definition of $V_{max}=k_p[E]_t$, we obtain:

$$[E]_t=\frac{V_{max}}{k_p} \tag{III-49}$$

Now substituting for [E], [ES], and $[E]_t$ in the mass balance equation III-46, we obtain:

$$\frac{V_{max}}{k_p}=\frac{K_m v}{k_p[S]}+\frac{v}{k_p}+[EI] \tag{III-50}$$

$$V_{max}-\frac{K_m}{[S]}v-v=k_p[EI]$$

$$V_{max}-v\left(1+\frac{K_m}{[S]}\right)=k_p[EI] \tag{III-51}$$

The K_i is given by:

$$K_i = \frac{[E][I]}{[EI]} = \frac{[E]([I]_t - [EI])}{[EI]} = [E]\left(\frac{[I]_t}{[EI]} - 1\right) \qquad \text{(III-52)}$$

Substituting for [E] from equation III-48:

$$K_i = \frac{K_m v}{k_p[S]}\left(\frac{[I]_t}{[EI]} - 1\right) = \frac{K_m v [I]_t}{k_p[S][EI]} - \frac{K_m v}{k_p[S]} \qquad \text{(III-53)}$$

Rearranging:

$$K_i + \frac{K_m v}{k_p[S]} = \frac{K_m v [I]_t}{k_p[S][EI]}$$

$$\frac{[I]_t}{k_p[EI]} = \frac{K_i[S]}{K_m v} + \frac{1}{k_p} \qquad \text{(III-54)}$$

From equation III-53:

$$[EI] = \frac{V_{max}}{k_p} - \frac{K_m v}{k_p[S]} - \frac{v}{k_p} = \frac{1}{k_p}\left[V_{max} - v\left(1 + \frac{K_m}{[S]}\right)\right] \qquad \text{(III-55)}$$

Substituting the solution above for [EI] into equation III-54:

$$\frac{[I]_t}{V_{max} - v\left(1 + \frac{K_m}{[S]}\right)} = \frac{K_i[S]}{K_m v} + \frac{1}{k_p} \qquad \text{(III-56)}$$

At any substrate concentration in the absence of I, the velocity, v_0, is given by:

$$v_0 = \frac{[S]\, V_{max}}{K_m + [S]} = \frac{V_{max}}{\left(1 + \frac{K_m}{[S]}\right)} \qquad \text{(III-57)}$$

Now consider some point on the v versus [I] plot where the velocity, v_i, is some fraction of v_0; that is,

$$v_i = \frac{v_0}{n} = \frac{V_{max}}{n\left(1 + \frac{K_m}{[S]}\right)} \qquad \text{(III-58)}$$

Substituting v_i given by equation III-58 for v in the left-hand part of equation III-56, we obtain:

$$\frac{[\mathrm{I}]_t}{V_{\max} - \dfrac{V_{\max}}{n}} = \frac{[\mathrm{I}]_t}{V_{\max}\left(1 - \dfrac{1}{n}\right)} = \frac{n[\mathrm{I}]_t}{(n-1)\,V_{\max}} \qquad \text{(III-59)}$$

And substituting v_0/n for v in the right-hand part of equation III-56, we obtain:

$$\frac{n[\mathrm{I}]_t}{(n-1)\,V_{\max}} = \frac{nK_i[\mathrm{S}]}{K_m v_0} + \frac{1}{k_p} \qquad \text{(III-60)}$$

A line drawn from v_0 through any v_i intersects the $[\mathrm{I}]_t$-axis at some value, $[\mathrm{I}]_n$ (e.g., $[\mathrm{I}]_2$ for $v_i = v_0/2$, $[\mathrm{I}]_3$ for $v_i = v_0/3$, etc.). The geometry of the situation is shown below.

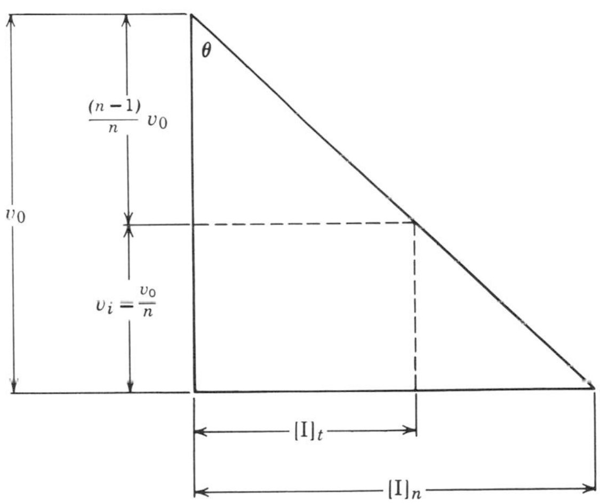

$$\tan\theta = \frac{[\mathrm{I}]_n}{v_0} = \frac{n[\mathrm{I}]_t}{(n-1)v_0}$$

or

$$[\mathrm{I}]_n = \frac{n}{(n-1)}[\mathrm{I}]_t \quad \text{and} \quad [\mathrm{I}]_t = \frac{(n-1)}{n}[\mathrm{I}]_n \qquad \text{(III-61)}$$

Substituting for $[\mathrm{I}]_t$ in equation III-60:

$$\frac{[\mathrm{I}]_n}{V_{\max}} = \frac{nK_i[\mathrm{S}]}{K_m v_0} + \frac{1}{k_p}$$

$$[\mathrm{I}]_n = nK_i \frac{[\mathrm{S}]\,V_{\max}}{K_m v_0} + \frac{V_{\max}}{k_p} \qquad \text{(III-62)}$$

but

$$\frac{V_{\max}}{v_0} = \left(1 + \frac{K_m}{[\mathrm{S}]}\right) \qquad \text{and} \qquad \frac{V_{\max}}{k_p} = [\mathrm{E}]_t$$

$$\therefore \quad \boxed{[\mathrm{I}]_n = nK_i\left(1 + \frac{[\mathrm{S}]}{K_m}\right) + [\mathrm{E}]_t} \qquad \text{(III-63)}$$

The plot is shown in Figure III-29. The distance between intercepts for successive values of n is $K_{i_{\mathrm{app}}}$, where:

$$K_{i_{\mathrm{app}}} = K_i\left(1 + \frac{[\mathrm{S}]}{K_m}\right)$$

If several plots are made at different fixed substrate concentrations, $K_{i_{\mathrm{app}}}$ can be replotted against [S] (Fig. III-29*b*). The intercepts of the replot give K_m and K_i.

A line drawn from v_0 to the $[\mathrm{I}]_t$-axis that intercepts the $[\mathrm{I}]_t$-axis one $K_{i_{\mathrm{app}}}$ distance to the left of $[\mathrm{I}]_2$, represents the line for $n = 1$. This is a line tangent to the v versus $[\mathrm{I}]_t$ curve at v_0. Another line drawn from v_0 that intercepts the $[\mathrm{I}]_t$-axis one $K_{i_{\mathrm{app}}}$ distance to the left of $[\mathrm{I}]_1$ gives $[\mathrm{E}]_t$. If only a very small fraction of the inhibitor is tied up as EI, this $n = 0$ line will coincide with the vertical axis. Figure III-29*c*, which represents the left hand portion of the v versus $[\mathrm{I}]_t$ plot, shows how the distribution of enzyme forms can be calculated.

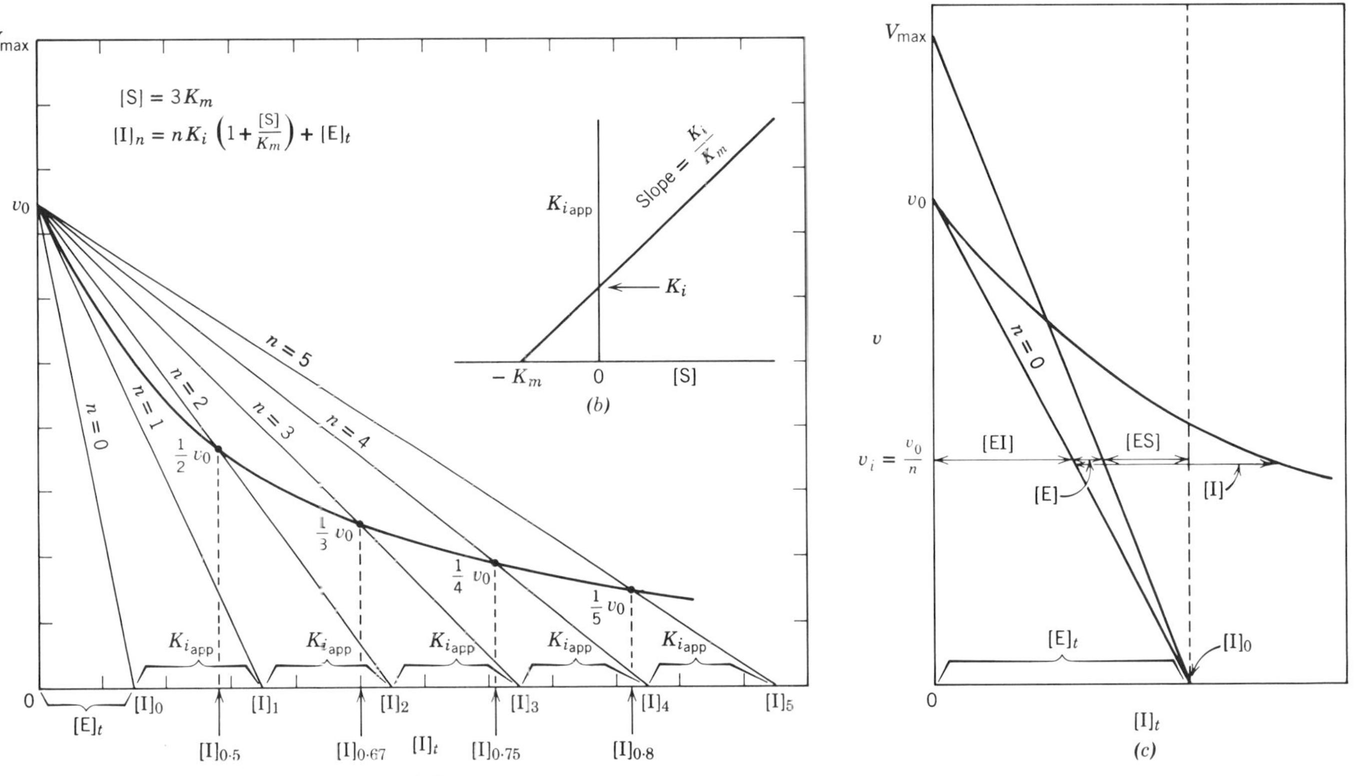

Fig. III-29. (*a*) Plot suggested by D xon to determine $[E]_t$ when a significant fraction of competitive inhibitor is bound by the enzyme; [S] is assumed to be equal to $3K_m$ ($\therefore v_0 = 0.75\,V_{max}$). (*b*) Replot of $K_{i_{app}}$ versus [S]. (*c*) Left-hand portion of Fig. III-29*a* showing distribution of enzyme species at a given v_0/n.

Noncompetitive Inhibitors

In the presence of a noncompetitive inhibitor, the enzyme is distributed among four forms:

$$[E]_t = [E] + [ES] + [EI] + [ESI]$$

$$[EI] + [ESI] = [E]_t - [E] - [ES]$$

$$= \frac{V_{max}}{k_p} - \frac{K_m v}{k_p[S]} - \frac{v}{k_p}$$

$$= \frac{V_{max} - v\left(1 + \frac{K_m}{[S]}\right)}{k_p} = \frac{U}{k_p} \qquad \text{(III-64)}$$

where, for convenience, the numerator of equation III-64 is written as U. The mass balance equation for I is:

$$[I]_t = [I] + [EI] + [ESI]$$

$$[I] = [I]_t - ([EI] + [ESI]) = [I]_t - \frac{U}{k_p} \qquad \text{(III-65)}$$

The K_i is defined as:

$$K_i = \frac{[E][I]}{[EI]} = \frac{[ES][I]}{[ESI]}$$

$$\therefore \quad [EI] = \frac{[E][I]}{K_i} \quad \text{and} \quad [ESI] = \frac{[ES][I]}{K_i}$$

Since $[ES] = v/k_p$,

$$[ESI] = \frac{v[I]}{k_p K_i} = \frac{v}{k_p K_i}\left([I]_t - \frac{U}{k_p}\right) \qquad \text{(III-66)}$$

From equation III-64:

$$[EI] + [ESI] = \frac{U}{k_p}$$

$$[EI] = \frac{U}{k_p} - [ESI] = \frac{U}{k_p} - \frac{v}{k_p K_i}\left([I]_t - \frac{U}{k_p}\right) \qquad \text{(III-67)}$$

Now, substituting into the expression for K_i where:

$$K_i = \frac{[\mathrm{E}][\mathrm{I}]}{[\mathrm{EI}]} \qquad \text{and} \qquad [\mathrm{E}] = \frac{K_m v}{k_p[\mathrm{S}]}$$

$$K_i = \frac{K_m v}{k_p[\mathrm{S}]}\left[\frac{[\mathrm{I}]_t - \dfrac{U}{k_p}}{\dfrac{U}{k_p} - \dfrac{v}{k_p K_i}\left([\mathrm{I}]_t - \dfrac{U}{k_p}\right)}\right] \tag{III-68}$$

which simplifies to:

$$\frac{[\mathrm{I}]_t}{V_{\max} - v\left(1 + \dfrac{K_m}{[\mathrm{S}]}\right)} = \frac{K_i}{v}\left(\frac{[\mathrm{S}]}{K_m + [\mathrm{S}]}\right) + \frac{1}{k_p} \tag{III-69}$$

Now considering some point on the v versus $[\mathrm{I}]_t$ curve where $v_i = v_0/n$ and proceeding as shown earlier for a competitive inhibitor, we obtain:

$$[\mathrm{I}]_n = nK_i\left(\frac{[\mathrm{S}]}{K_m + [\mathrm{S}]}\right)\frac{V_{\max}}{v_0} + [\mathrm{E}]_t \tag{III-70}$$

But

$$\frac{V_{\max}}{v_0} = \left(1 + \frac{K_m}{[\mathrm{S}]}\right) = \frac{[\mathrm{S}] + K_m}{[\mathrm{S}]}$$

so that

$$\boxed{[\mathrm{I}]_n = nK_i + [\mathrm{E}]_t} \tag{III-71}$$

The plot is identical to that shown in Figure III-29 except now the distance between successive intercepts on the $[\mathrm{I}]_t$-axis gives K_i directly. Figure III-30 shows how the concentrations of the four enzyme forms can be determined.

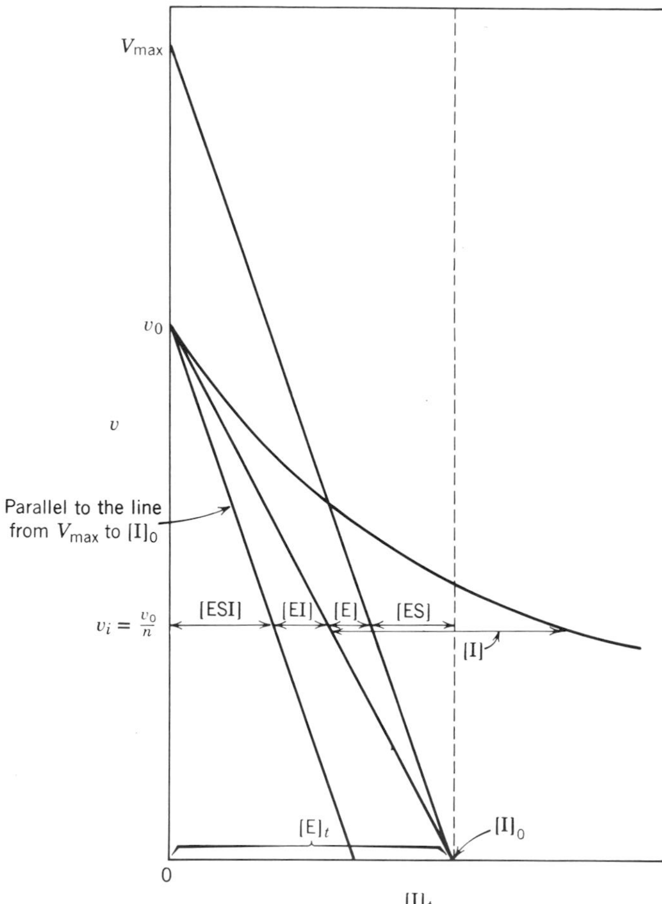

Fig. III-30. Distribution of enzyme species at a given v_0/n in the presence of a noncompetitive inhibitor.

Henderson (1972) has suggested a linear plot of $[I]_t/(1 - v_i/v_0)$ versus v_0/v_i to determine K_i and $[E]_t$ in systems where a substantial fraction of the added inhibitor is bound. The equations for the plots are shown below.

Competitive $$\boxed{\frac{[I]_t}{\left(1 - \frac{v_i}{v_0}\right)} = K_i\left(1 + \frac{[S]}{K_m}\right)\frac{v_0}{v_i} + [E]_t} \qquad \text{(III-72)}$$

$$\text{Noncompetitive} \quad \boxed{\frac{[\mathrm{I}]_t}{\left(1-\frac{v_i}{v_0}\right)} = K_i\frac{v_0}{v_i} + [\mathrm{E}]_t} \tag{III-73}$$

$$\text{Uncompetitive} \quad \boxed{\frac{[\mathrm{I}]_t}{\left(1-\frac{v_i}{v_0}\right)} = K_i\left(1+\frac{K_m}{[\mathrm{S}]}\right)\frac{v_0}{v_i} + [\mathrm{E}]_t} \tag{III-74}$$

The concentration of inhibitor is varied at different fixed substrate concentrations and a constant enzyme concentration. For each fixed [S] and $[\mathrm{E}]_t$, the values of v_i/v_0 and v_0/v_i are calculated and plotted as described above. The family of plots for different fixed [S] intersect on the vertical-axis at $[\mathrm{E}]_t$. The value of K_i can be calculated from the slopes of the plots or from a *slope* versus [S] (competitive) or *slope* versus 1/[S] (uncompetitive) replot. The value of [S] in equations III-72 to III-74 can be taken as $[\mathrm{S}]_t$ as long as there is no depletion because of tight substrate binding. Otherwise, $[\mathrm{S}]_t -$ [ES] can be substituted for [S]. The concentration of ES can be calculated from equation II-53*b*. The plot remains linear with a vertical-axis intercept of $[\mathrm{E}]_t$, but slope replots will be nonlinear. (See equations IV-76 to IV-80 for a plot useful when both S and I are tightly bound to the enzyme [Best-Belpomme and Dessen, 1973]).

REFERENCES

See General References at the end of Chapter Two and also references on plotting inhibition data at the end of Chapter Four.

Tightly Bound Inhibitors

Best-Belpomme, M. and Dessen, P., *Biochimie* 55, 11 (1973).

Dixon, M., *Biochem. J.* **129**, 197 (1972).

Henderson, P. J. F., *Biochem. J.* **127**, 321 (1972).

Henderson, P. J. F., *Biochem. J.* **135**, 101 (1973).

Morrison, J. F., *Biochim. Biophys. Acta* **185**, 269 (1969).

Kinetics of Enzyme-Catalyzed Reactions with Competing Alternate Substrates

Cha, S., *Molecular Pharmacol.* **4**, 621 (1968).

Pocklington, T. and Jeffery, J., *Biochem. J.* **112**, 331 (1969).

See also references on the effects of alternate substrates in multireactant systems at the end of Chapter Nine.

Regulation by "Energy Charge"

Atkinson, D. E., "Enzymes as Control Elements in Metabolic Regulation" in *The Enzymes*, Vol. 1, 3rd ed., P. Boyer, ed., Academic Press, 1970, p. 461.

See also additional references on Regulation by "Energy Charge" at the end of Chapter Five.

Contaminating Inhibitors in the Substrate (Varying [I] and [S] Together at Constant Ratio)

Cleland, W. W., Gross, M. and Folk, J. F., *J. Biol. Chem.* **248**, 6541 (1973).

CHAPTER FOUR

RAPID EQUILIBRIUM PARTIAL AND MIXED-TYPE INHIBITION

In the preceding sections we examined the properties of three simple types of inhibition systems. All three types have in common a dead-end EI, or nonproductive ESI complex, or both. In the following sections, a few more complicated systems are described. Several of these systems involve the formation of an ESI complex that can yield product with equal or less facility than can the ES complex.

A. PARTIAL COMPETITIVE INHIBITION (SIMPLE INTERSECTING HYPERBOLIC COMPETITIVE INHIBITION)

Consider a situation where (*a*) the substrate and the inhibitor bind to the enzyme at different sites to yield ES, EI, and ESI complexes, (*b*) the substrate binds to free enzyme with a greater affinity than to the EI complex, and (*c*) the ES and ESI complexes both yield product with equal facility. The situation above is called *partial competitive inhibition* or *hyperbolic competitive inhibition*, the latter because of the shape of the replots of the primary reciprocal plots. The equilibria describing this situation are shown below.

$$\begin{array}{ccccccc}
E + S & \overset{K_S}{\rightleftharpoons} & ES & \overset{k_p}{\longrightarrow} & E + P \\
+ & & + & & \\
I & & I & & \\
K_i \updownarrow & & \alpha K_i \updownarrow & & \\
EI + S & \overset{\alpha K_S}{\rightleftharpoons} & ESI & \overset{k_p}{\longrightarrow} & EI + P
\end{array}$$

The various equilibrium constants are:

$$K_S = \frac{[E][S]}{[ES]}, \qquad \alpha K_S = \frac{[EI][S]}{[ESI]}, \qquad \alpha > 1$$

$$K_i = \frac{[E][I]}{[EI]}, \qquad \alpha K_i = \frac{[ES][I]}{[ESI]}$$

k_p = rate constant for the breakdown of ES and ESI to P

The factor α is the factor by which K_S changes when I occupies the enzyme. For the four enzyme species to be at equilibrium, the dissociation constant for I must change to αK_i when S occupies the enzyme. In other words, the overall equilibrium constant for the formation of ESI must be the same regardless of the path (E→ES→ESI or E→EI→ESI). At any inhibitor concentration, an infinitely high substrate concentration will drive all the enzyme into a mixture of ES and ESI forms. Because both ES and ESI produce product with equal facility, the V_{max} will be unchanged. The $K_{S_{app}}$ value (measured as [S] required for $\frac{1}{2} V_{max}$) will increase, because at any inhibitor concentration a portion of the available enzyme exists in a form (EI) that has a decreased affinity for S ($\alpha K_S > K_S$).

Partial competitive inhibition cannot be distinguished from pure competitive inhibition simply by plotting v versus [S] in the presence and absence of inhibitor, or by the corresponding reciprocal plots. It can be distinguished, however, by plotting v versus [I] at a fixed [S]. If we examine the equilibria for pure competitive inhibition, we can see that at a fixed [S], an infinitely high [I] will drive all the enzyme to the EI form; that is, the velocity can be reduced to zero. From the equilibria describing partial competitive inhibition, we can see that at a fixed [S], an infinitely high [I] will drive all the enzyme to the EI and ESI forms. The ESI can produce product. Therefore, the velocity of the reaction can never be driven to zero. The net effect of an infinitely high [I] is to produce a modified, but functional, enzyme, EI, with the same V_{max}, but with a different K_S value. As [I] increases from zero to infinity, $K_{S_{app}}$ increases from K_S to a limit of αK_S. This is in contrast to pure competitive inhibition where the $K_{S_{app}}$ increases without limit as [I] increases. The velocity equation for partial competitive inhibition can be derived easily assuming rapid equilibrium conditions as shown below.

$$v = k_p[\mathrm{ES}] + k_p[\mathrm{ESI}]$$

$$\frac{v}{[\mathrm{E}]_t} = \frac{k_p[\mathrm{ES}] + k_p[\mathrm{ESI}]}{[\mathrm{E}] + [\mathrm{ES}] + [\mathrm{EI}] + [\mathrm{ESI}]}$$

$$\frac{v}{k_p[\mathrm{E}]_t} = \frac{\dfrac{[\mathrm{S}]}{K_\mathrm{S}} + \dfrac{[\mathrm{S}][\mathrm{I}]}{\alpha K_\mathrm{S} K_i}}{1 + \dfrac{[\mathrm{S}]}{K_\mathrm{S}} + \dfrac{[\mathrm{I}]}{K_i} + \dfrac{[\mathrm{S}][\mathrm{I}]}{\alpha K_\mathrm{S} K_i}}$$

$$\boxed{\frac{v}{V_{\max}} = \frac{\dfrac{[\mathrm{S}]}{K_\mathrm{S}} + \dfrac{[\mathrm{S}][\mathrm{I}]}{\alpha K_\mathrm{S} K_i}}{1 + \dfrac{[\mathrm{S}]}{K_\mathrm{S}} + \dfrac{[\mathrm{I}]}{K_i} + \dfrac{[\mathrm{S}][\mathrm{I}]}{\alpha K_\mathrm{S} K_i}}} \qquad \text{(IV-1)}$$

The numerator of equation IV-1 contains two terms corresponding to the two product-forming species, ES and ESI. The denominator contains four terms, reflecting the total of four enzyme species, E, ES, EI, and ESI. We can see that equation IV-1 reduces to the velocity equation for pure competitive inhibition when $\alpha \gg 1$. In this case, the EI complex has very little affinity for S (and the ES complex has very little affinity for I). Equation IV-1 can be rearranged to the usual form of the Henri-Michaelis-Menten equation:

$$\boxed{\frac{v}{V_{\max}} = \frac{[\mathrm{S}]}{K_\mathrm{S}\dfrac{\left(1 + \dfrac{[\mathrm{I}]}{K_i}\right)}{\left(1 + \dfrac{[\mathrm{I}]}{\alpha K_i}\right)} + [\mathrm{S}]}} \qquad \text{(IV-2)}$$

or

$$\frac{v}{V_{\max}} = \frac{[\mathrm{S}]}{K_{\mathrm{S}_{\mathrm{app}}} + [\mathrm{S}]}$$

where $K_{S_{app}}$ = the apparent K_S for a given inhibitor concentration

$$= K_S \frac{\left(1+\frac{[I]}{K_i}\right)}{\left(1+\frac{[I]}{\alpha K_i}\right)}$$

The term multiplying K_S is the factor by which the slope of the reciprocal plot is increased in the presence of I.

At an infinitely high (saturating) inhibitor concentration, the velocity equation reduces to:

$$\frac{v}{V_{max}} = \frac{[S]}{\alpha K_S + [S]} \tag{IV-3}$$

Figure IV-1 shows the effect of increasing partial competitive inhibitor concentration at a fixed substrate concentration of $[S] = 10K_S$. For comparison, a curve for a pure competitive inhibitor is shown. As predicted from the equilibria and the velocity equations, the curves for the partial competitive inhibitor plateau at limiting velocities which are dictated by the value of α. Although the curves plateau at some finite initial velocity, the $[I]_{0.9}/[I]_{0.1}$ ratio is still 81 if we define $[I]_{0.9}$ and $[I]_{0.1}$ as the inhibitor concentrations required to decrease the velocity by 90% and 10% of the *difference*, respectively, between the original and plateau values. If we define $v_{\frac{1}{2}}$ as the specific velocity half way between the specific velocity at the given [S] in the absence of inhibitor, and the specific velocity at the same [S] in the presence of a saturating concentration of inhibitor, then $v_{\frac{1}{2}}$ occurs at:

$$\boxed{[I]_{\frac{1}{2}} = \frac{\alpha K_i\left(1+\frac{K_S}{[S]}\right)}{\left(1+\frac{\alpha K_S}{[S]}\right)}} \tag{IV-4}$$

As [S] decreases, $[I]_{\frac{1}{2}}$ approaches K_i; as [S] increases, $[I]_{\frac{1}{2}}$ approaches αK_i.

The reciprocal form of the velocity equation for partial competitive

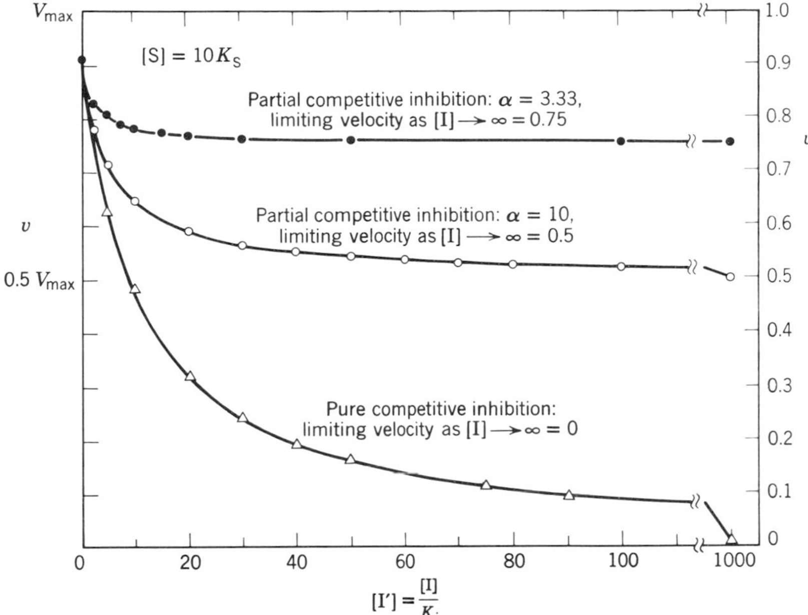

Fig. IV-1. Plot of v versus [I] at $[S]=10K_S$ for a partial competitive inhibitor ($\alpha=3.33$ and $\alpha=10$) and for a pure competitive inhibitor ($\alpha=\infty$).

inhibition is:

$$\frac{1}{v} = \frac{K_S}{V_{max}} \frac{\left(1 + \frac{[I]}{K_i}\right)}{\left(1 + \frac{[I]}{\alpha K_i}\right)} \frac{1}{[S]} + \frac{1}{V_{max}} \qquad \text{(IV-5)}$$

As [I] increases, the slope of the $1/v$ versus $1/[S]$ plot increases by the factor multiplying K_S/V_{max}. The curves pivot counterclockwise about the point of intersection with the control curve (at $1/V_{max}$ on the $1/v$-axis) (Fig. IV-2). Unlike pure competitive inhibition, the slope of the reciprocal plot approaches a finite limit as [I] becomes infinitely high. The $1/[S]$-axis intercept of the limiting plot gives $-1/\alpha K_S$. The kinetic constants can be obtained from appropriate $1/\Delta$ *slope* replots as described later for hyperbolic mixed-type inhibition systems.

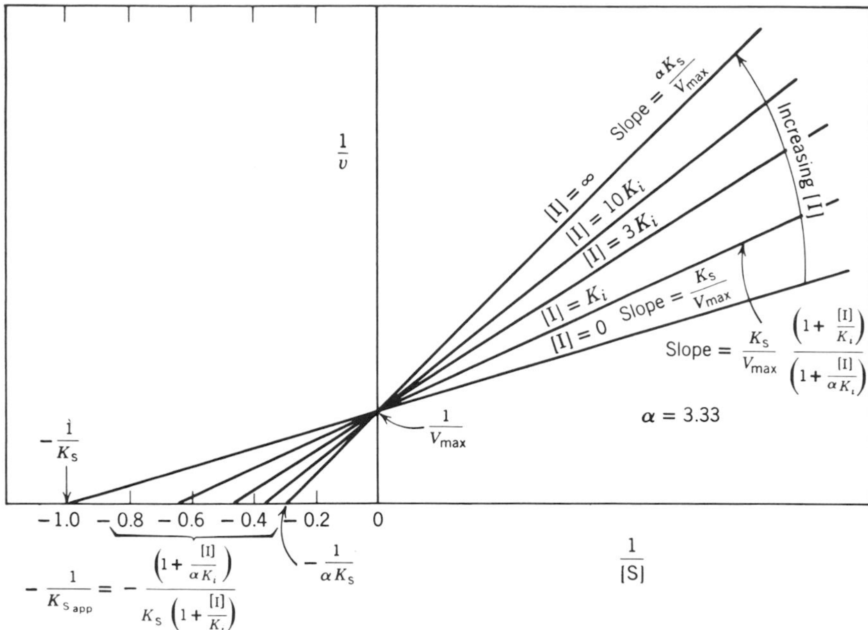

Fig. IV-2. The $1/v$ versus $1/[S]$ plot in the presence of different fixed concentrations of a partial competitive inhibitor.

B. PARTIAL NONCOMPETITIVE INHIBITION (SIMPLE INTERSECTING HYPERBOLIC NONCOMPETITIVE INHIBITION)

Consider a system in which the substrate and inhibitor combine independently and reversibly to the enzyme at different sites to produce ES, EI, and ESI complexes. The ESI can produce product, but not as effectively as ES. It is in this latter respect that this system, called partial noncompetitive inhibition, differs from pure noncompetitive inhibition. The equilibria describing a partial noncompetitive inhibition system are shown below.

$$\begin{array}{ccccc}
E + S & \overset{K_S}{\rightleftharpoons} & ES & \overset{k_p}{\longrightarrow} & E + P \\
+ & & + & & \\
I & & I & & \\
K_i \upharpoonleft\downharpoonright & & K_i \upharpoonleft\downharpoonright & & \\
EI + S & \overset{K_S}{\rightleftharpoons} & ESI & \overset{\beta k_p}{\longrightarrow} & EI + P
\end{array}$$

At any [I], the overall velocity at which P is produced is $k_p[\mathrm{ES}] + \beta k_p[\mathrm{ESI}]$, where $\beta < 1$. From the equilibria we can see that an infinitely high substrate concentration will not drive all the enzyme to the ES form. At any given inhibitor concentration, a portion of the enzyme will exist as the less productive ESI form. Consequently, $V_{\max}$ will decrease in the presence of a partial noncompetitive inhibitor. The K_S value remains the same because the two forms of enzyme available for combination with S (E and EI) have equal affinities for S. Partial noncompetitive inhibition cannot be distinguished from pure noncompetitive inhibition simply by plotting v versus [S] or $1/v$ versus $1/[\mathrm{S}]$ in the presence and absence of inhibitor. It can be distinguished, however, by plotting v versus [I] at a fixed [S]. In pure noncompetitive systems, an infinitely high [I] drives the velocity to zero. In a partial noncompetitive system, an infinitely high [I] will, in effect, convert the enzyme to a modified (but functional) enzyme (EI) with a decreased rate of product formation from the ESI form. Thus the v versus [I] plot of the partial system is hyperbolic and plateaus at some limiting velocity as shown earlier for partial competitive inhibition in Figure IV-1.

The velocity equation can be derived in the usual manner if rapid equilibrium conditions prevail.

$$v = k_p[\mathrm{ES}] + \beta k_p[\mathrm{ESI}]$$

$$\frac{v}{[\mathrm{E}]_t} = \frac{k_p \dfrac{[\mathrm{S}]}{K_S} + \beta k_p \dfrac{[\mathrm{S}][\mathrm{I}]}{K_S K_i}}{1 + \dfrac{[\mathrm{S}]}{K_S} + \dfrac{[\mathrm{I}]}{K_i} + \dfrac{[\mathrm{S}][\mathrm{I}]}{K_S K_i}}$$

$$\boxed{v = \frac{V_{\max} \dfrac{[\mathrm{S}]}{K_S} + \beta V_{\max} \dfrac{[\mathrm{S}][\mathrm{I}]}{K_S K_i}}{1 + \dfrac{[\mathrm{S}]}{K_S} + \dfrac{[\mathrm{I}]}{K_i} + \dfrac{[\mathrm{S}][\mathrm{I}]}{K_S K_i}}} \qquad \text{(IV-6)}$$

where $k_p[\mathrm{E}]_t = V_{\max}$. In the form above, there are two numerator terms reflecting the two product-forming complexes and four denominator terms corresponding to the four different enzyme species (E, ES, EI, and ESI). The equation can be rearranged to the usual Henri-Michaelis-Menten form:

$$\boxed{\frac{v}{V_{\max}} = \frac{[\mathrm{S}]}{K_{\mathrm{S}}\dfrac{\left(1+\dfrac{[\mathrm{I}]}{K_i}\right)}{\left(1+\dfrac{\beta[\mathrm{I}]}{K_i}\right)} + [\mathrm{S}]\dfrac{\left(1+\dfrac{[\mathrm{I}]}{K_i}\right)}{\left(1+\dfrac{\beta[\mathrm{I}]}{K_i}\right)}}} \tag{IV-7}$$

As in pure noncompetitive inhibition, both denominator terms are multiplied by a factor. At any fixed [I], the velocity equation can be expressed as:

$$\frac{v}{V_{\max_i}} = \frac{[\mathrm{S}]}{K_{\mathrm{S}}+[\mathrm{S}]} \tag{IV-8}$$

$$V_{\max_i} = \text{the apparent } V_{\max} \text{ at the given } [\mathrm{I}]$$

$$= \frac{V_{\max}\left(1+\dfrac{\beta[\mathrm{I}]}{K_i}\right)}{\left(1+\dfrac{[\mathrm{I}]}{K_i}\right)}$$

We can see from equation IV-7 that (*a*) when $[\mathrm{I}]=0$, the equation reduces to the usual Henri-Michaelis-Menten equation, (*b*) when $\beta \ll 1$ (i.e., the ESI complex is extremely poor in producing product), the equation reduces to the usual modified Henri-Michaelis-Menten equation for pure noncompetitive inhibition, and (*c*) when $[\mathrm{I}] \gg K_i$, the velocity equation reduces to:

$$\frac{v}{\beta V_{\max}} = \frac{[\mathrm{S}]}{K_{\mathrm{S}}+[\mathrm{S}]} \tag{IV-9}$$

The reciprocal form of the velocity equation for partial noncompetitive inhibition is:

$$\frac{1}{v} = \frac{K_S}{V_{max}} \frac{\left(1 + \frac{[I]}{K_i}\right)}{\left(1 + \frac{\beta[I]}{K_i}\right)} \frac{1}{[S]} + \frac{1}{V_{max}} \frac{\left(1 + \frac{[I]}{K_i}\right)}{\left(1 + \frac{\beta[I]}{K_i}\right)} \qquad \text{(IV-10)}$$

As [I] increases, both the slope and the $1/v$-axis intercept of the reciprocal plot increase by the factors in parentheses. The curves pivot counterclockwise about the point of intersection with the control curve (at $-1/K_S$ on the $1/[S]$-axis) (Fig. IV-3). Unlike pure noncompetitive inhibition, the slope of the reciprocal plot reaches a limit as [I] becomes infinitely high. The $1/v$-axis intercept of the limiting plot gives $1/\beta V_{max}$. The kinetic constants can be obtained from appropriate $1/\Delta$ *slope* and $1/\Delta$ *intercept* replots as shown later for hyperbolic mixed-type inhibition.

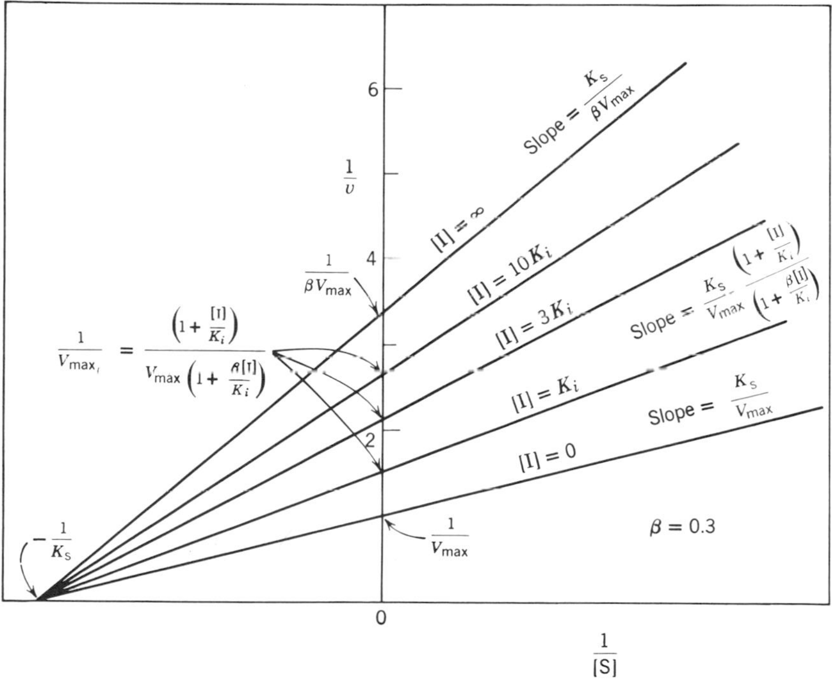

Fig. IV-3. The $1/v$ versus $1/[S]$ plot in the presence of different fixed concentrations of a partial noncompetitive inhibitor.

C. MIXED-TYPE INHIBITION

A mixed-type inhibitor, as the name suggests, affects both the V_{max} and K_m values of an enzyme-catalyzed reaction. Mixed inhibition can arise from several situations. Some are described below.

Linear Mixed-Type Inhibition

System C1. ($\alpha > 1, \beta = 0$, "Intersecting, Linear Noncompetitive Inhibition")

The simplest mixed system is one in which EI has a lower affinity than E for S, and the ESI complex is nonproductive. The system may be considered a mixture of partial competitive inhibition and pure noncompetitive inhibition. The equilibria describing this system are shown below.

$$\begin{array}{ccccc} E + S & \overset{K_S}{\rightleftharpoons} & ES & \overset{k_p}{\longrightarrow} & E + P \\ + & & + & & \\ I & & I & & \\ K_i \upharpoonleft\downharpoonright & & \alpha K_i \upharpoonleft\downharpoonright & & \\ EI + S & \underset{}{\overset{\alpha K_S}{\rightleftharpoons}} & ESI & & \end{array}$$

As long as I is present, some of the enzyme will always be in the nonproductive ESI form, even at an infinitely high [S]. Consequently, $V_{max_i} < V_{max}$. Also, at any [I], a portion of the enzyme available for combination with S will exist in the lower affinity EI form. Consequently, $K_{S_{app}} > K_S$. At an infinitely high [I], all the enzyme can be driven to the EI and ESI forms. Because ESI is nonproductive, the velocity can be driven to zero by increasing [I].

The velocity equation for this type of mixed inhibition may be derived easily from rapid equilibrium assumptions.

$$v = k_p[ES], \qquad \frac{v}{[E]_t} = \frac{k_p[ES]}{[E]+[ES]+[EI]+[ESI]}$$

$$\boxed{\frac{v}{V_{max}} = \frac{\dfrac{[S]}{K_S}}{1 + \dfrac{[S]}{K_S} + \dfrac{[I]}{K_i} + \dfrac{[S][I]}{\alpha K_S K_i}}} \qquad \text{(IV-11)}$$

The numerator $[S]/K_S$ term reflects the single product-forming complex (ES), while the denominator reflects the total of four species, E, ES, EI, and ESI.

Rearranging to the usual Henri-Michaelis-Menten form:

$$\boxed{\frac{v}{V_{max}} = \frac{[S]}{K_S\left(1+\frac{[I]}{K_i}\right)+[S]\left(1+\frac{[I]}{\alpha K_i}\right)}} \qquad \text{(IV-12)}$$

The noncompetitive aspect is seen from the fact that both the K_S and denominator [S] terms are multiplied by factors containing [I]. As we might expect, there is no β term because the ESI complex is nonproductive (i.e., $\beta = 0$). The effect of the inhibitor on the kinetic constants can be seen by rearranging the equation.

$$\boxed{\frac{v}{\dfrac{V_{max}}{\left(1+\dfrac{[I]}{\alpha K_i}\right)}} = \frac{[S]}{K_S\dfrac{\left(1+\dfrac{[I]}{K_i}\right)}{\left(1+\dfrac{[I]}{\alpha K_i}\right)}+[S]}} \qquad \text{(IV-13)}$$

Now we can see that the expression is similar to that for partial competitive inhibition with an additional term affecting V_{max}. The equation can be written:

$$\frac{v}{V_{max_i}} = \frac{[S]}{K_{S_{app}}+[S]} \qquad \text{(IV-14)}$$

where:

$$V_{max_i} = \frac{V_{max}}{\left(1+\dfrac{[I]}{\alpha K_i}\right)} \qquad \text{and} \qquad K_{S_{app}} = K_S\frac{\left(1+\dfrac{[I]}{K_i}\right)}{\left(1+\dfrac{[I]}{\alpha K_i}\right)}$$

The velocity curve for a mixed inhibitor where $\alpha = 2$ and $\beta = 0$ is shown in Figure IV-4. At $[I] = 3K_i$, V_{max_i} is $0.4V_{max}$ and $K_{S_{app}}$ is $1.6K_S$. The $[S]_{0.9}/[S]_{0.1}$ ratio remains 81 in the presence of the inhibitor.

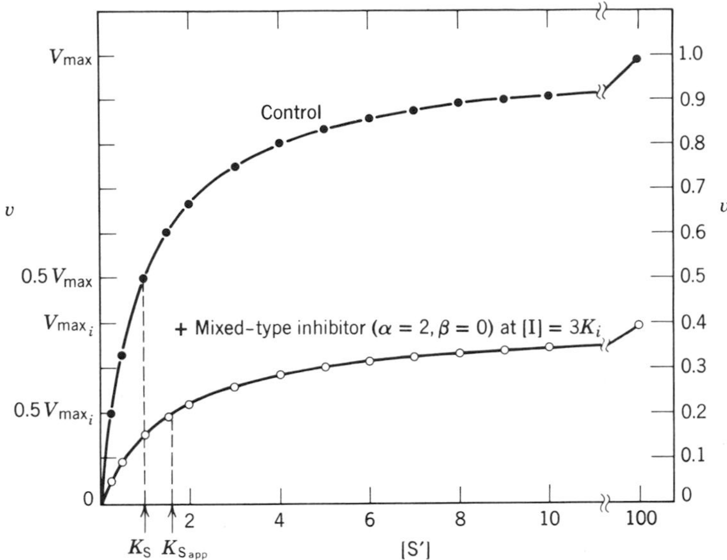

Fig. IV-4. The v versus [S] plot in the presence of a linear mixed-type inhibitor where $\alpha=2$ and $\beta=0$.

The reciprocal form of the velocity equation for mixed inhibition System C1 is:

$$\frac{1}{v} = \frac{K_S}{V_{max}}\left(1 + \frac{[I]}{K_i}\right)\frac{1}{[S]} + \frac{1}{V_{max}}\left(1 + \frac{[I]}{\alpha K_i}\right) \quad \text{(IV-15)}$$

Figure IV-5 shows the reciprocal plot. The control and "plus inhibitor" curves intersect above the 1/[S]-axis. As shown later, this intersection point is at a height of $(1/V_{max})(1 - K_{i_{slope}}/K_{i_{int}})$ or $(\alpha - 1)/\alpha V_{max}$ above the 1/[S]-axis for a mixed system where $\beta = 0$. The corresponding 1/[S]-coordinate equals $-1/\alpha K_S$. As [I] increases, the slopes and $1/v$-axis intercepts of the reciprocal plots increase, pivoting counterclockwise about the point of intersection with the control curve. Because the ESI complex is nonproductive, the velocity can be driven to zero by an infinitely high inhibitor concentration. Thus the limiting slope is a vertical line through the intersection point, parallel to the $1/v$-axis. If $\alpha < 1$ and $\beta = 0$, the reciprocal plots will intersect below the 1/[S]-axis.

Replots of Primary Reciprocal Plot Data: Slope Versus [I] *and* $1/v$*-Axis Intercept Versus* [I] The constants K_i and αK_i can be calculated from the values of

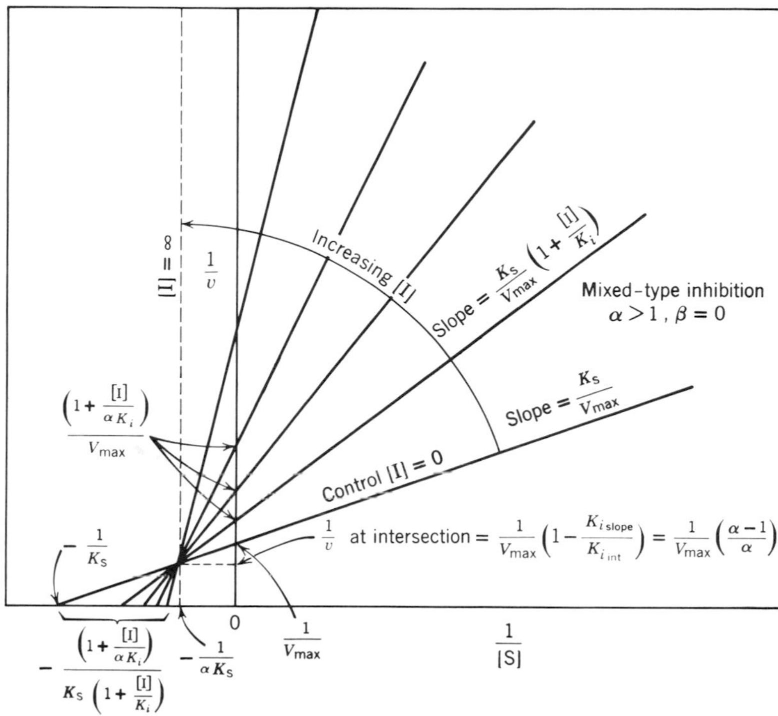

Fig. IV-5. The $1/v$ versus $1/[S]$ plot in the presence of different fixed concentrations of a linear mixed-type inhibitor ($\alpha > 1, \beta = 0$). Because $\beta = 0$, the velocity can be driven to zero at $[I] = \infty$. Thus the limiting plot is a vertical line through the intersection point.

$K_{S_{app}}$ and V_{max_i}. However, a more accurate determination can be made if a series of reciprocal plots are constructed for a wide range of inhibitor concentrations. Then, the slope and the $1/v$-axis intercept of each can be replotted versus the corresponding inhibitor concentration. When $\beta = 0$, the slope versus [I] and the $1/v$-axis intercept versus [I] replots are linear (which we could have predicted, since there is no finite asymptotic limit to v). Unlike pure noncompetitive systems, the two replots do not superimpose or extrapolate to the same point on the [I]-axis. (We also could have predicted this because the slope factor and the intercept factor are not identical in mixed-type inhibition systems.)

$$slope_{1/S} = \frac{K_S}{K_i V_{max}}[I] + \frac{K_S}{V_{max}} \qquad \text{(IV-16)}$$

$$\frac{1}{V_{max_i}} = \frac{1}{\alpha K_i V_{max}}[I] + \frac{1}{V_{max}} \qquad \text{(IV-17)}$$

Figure IV-6 shows the replots for the mixed-type system where $\alpha = 2$ and

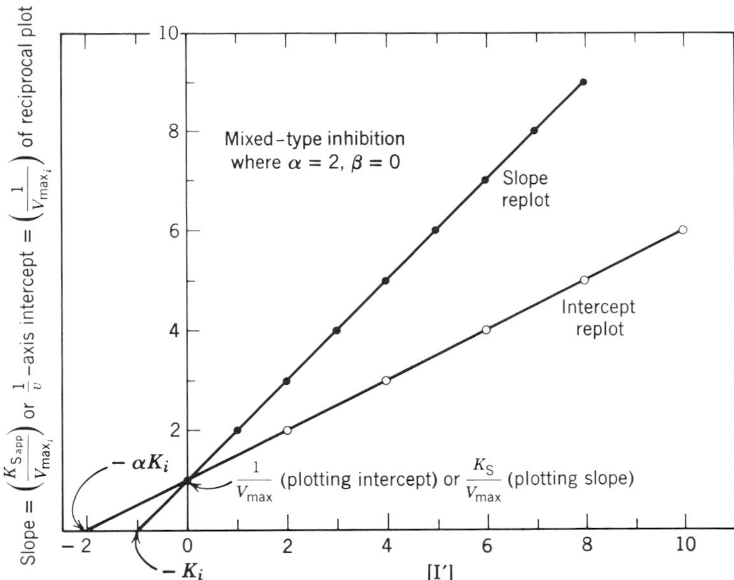

Fig. IV-6. *Slope*$_{1/S}$ and $1/v$-axis intercept replots of reciprocal plot data for a linear mixed-type inhibitor where $\alpha > 1$ and $\beta = 0$.

$\beta = 0$. Because the replots are linear, this type of mixed system has been called linear noncompetitive inhibition or intersecting, linear noncompetitive inhibition (because all of the plots intersect at one point). (Pure noncompetitive inhibition would be "simple, intersecting, linear noncompetitive inhibition.") With K_i determined from the slope versus [I] replot, α can be determined from the intercept versus [I] replot, and checked against the value determined from the intersection point of the primary reciprocal plot.

Dixon Plot Dixon plots for partial and most mixed-type inhibition systems are curved. However, when the ESI complex is not catalytically active ($\beta = 0$), the plot is linear. The equation for the Dixon plot is:

$$\frac{1}{v} = \frac{\left(1 + \frac{\alpha K_S}{[S]}\right)}{\alpha K_i V_{max}}[I] + \frac{1}{V_{max}}\left(1 + \frac{K_S}{[S]}\right) \qquad \text{(IV-18)}$$

The plot is shown in Figure IV-7. Setting $1/v$ for $[S]_1$ equal to $1/v$ for $[S]_2$,

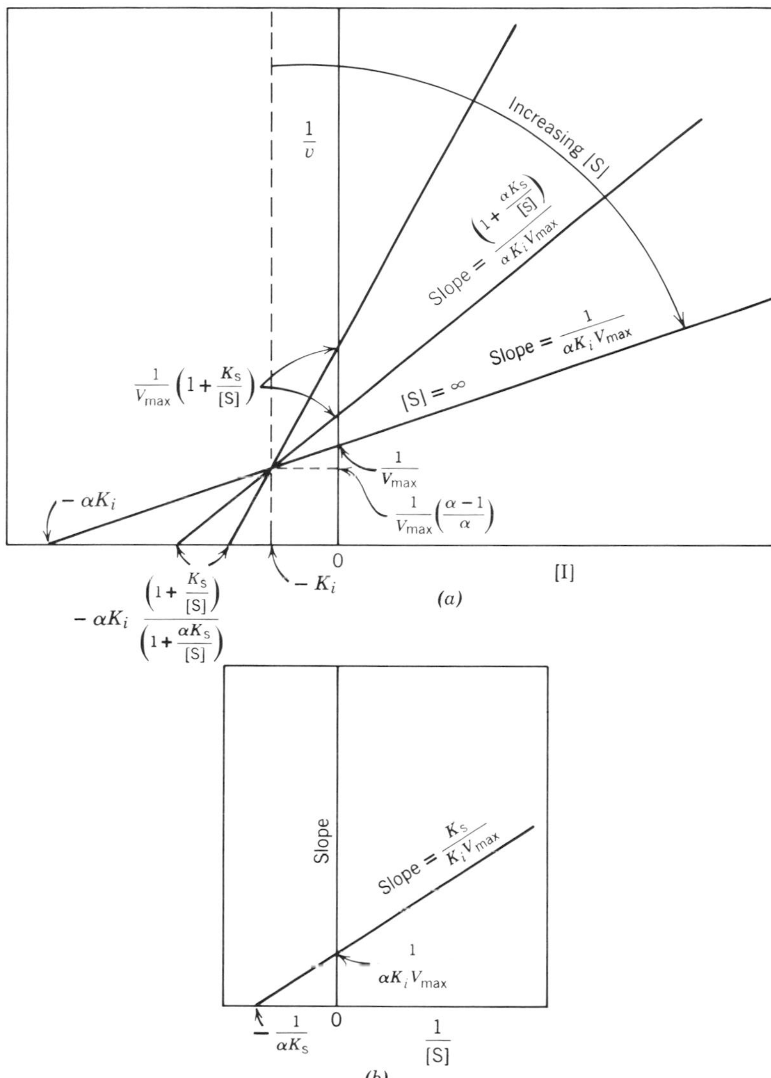

Fig. IV-7. Dixon plot: (*a*) $1/v$ versus [I] at different fixed concentrations of substrate. The characteristics of the plot are almost identical to that obtained for a pure competitive inhibitor (except the intersection point is lower than $1/V_{max}$). (*b*) *Slope* replot (which does not go through the origin as does the replot for pure competitive inhibition).

we find that the family of curves intersect to the left of the $1/v$-axis at $[\mathrm{I}] = -K_i$ and $1/v = (\alpha - 1)/\alpha V_{max}$. The slope of the Dixon plot is given by:

$$\boxed{slope = \frac{K_S}{K_i V_{max}} \frac{1}{[S]} + \frac{1}{\alpha K_i V_{max}}} \qquad \text{(IV-19)}$$

If $\alpha > 1$, the family of Dixon plots intersect above the horizontal axis, and the plot resembles that obtained for pure competitive inhibition. If $\alpha = 1$, the curves intersect on the horizontal axis, and the system is pure noncompetitive inhibition. If $\alpha < 1$, the family of curves intersect below the horizontal axis. We see then that the Dixon plot *by itself* is not sufficient to diagnose inhibition systems. Only pure uncompetitive systems yield unique plots (parallel lines). The Dixon plot together with the Cornish-Bowden plot (1974) of $[\mathrm{S}]/v$ versus [I] will distinguish between all four types of simple linear inhibition.

Linear Mixed Inhibition When I Binds at Two Sites

Linear mixed-type inhibition can also result from an inhibitor binding at two different mutually exclusive sites. Binding at one site completely excludes S. Binding of I at the second site has no effect on the binding of S, but the resulting ESI complex is catalytically inactive. The equilibria are:

$$\begin{array}{ccccccccc}
\mathrm{IE} & & & & & & & & \\
\gamma K_i \updownarrow & & & & & & & & \\
\mathrm{I} & & & & & & & & \\
+ & & & & & & & & \\
\mathrm{E} & + & \mathrm{S} & \overset{K_S}{\rightleftharpoons} & \mathrm{ES} & \overset{k_p}{\longrightarrow} & \mathrm{E} & + & \mathrm{P} \\
+ & & & & + & & & & \\
\mathrm{I} & & & & \mathrm{I} & & & & \\
K_i \updownarrow & & & & K_i \updownarrow & & & & \\
\mathrm{EI} & + & \mathrm{S} & \overset{K_S}{\rightleftharpoons} & \mathrm{ESI} & & & &
\end{array}$$

The system is a mixture of pure competitive and pure noncompetitive inhibition. The velocity equation is:

$$\frac{v}{V_{\max}} = \frac{\frac{[S]}{K_S}}{1+\frac{[S]}{K_S}+\frac{[I]}{K_i}+\frac{[I]}{\gamma K_i}+\frac{[S][I]}{K_S K_i}} = \frac{[S]}{K_S\left(1+\frac{[I]}{K_i}+\frac{[I]}{\gamma K_i}\right)+[S]\left(1+\frac{[I]}{K_i}\right)} \tag{IV-20}$$

or

$$\boxed{\frac{v}{V_{\max}} = \frac{[S]}{K_S\left(1+\frac{[I]}{\frac{\gamma}{1+\gamma}K_i}\right)+[S]\left(1+\frac{[I]}{K_i}\right)}} \tag{IV-21}$$

The system has all the properties of a mixed-type system where $\beta=0$ (mixed-type System C1) even if $\gamma=1$. If I at the second site changes the dissociation constant for S by a factor α, then the system will appear to be pure noncompetitive if coincidentally $\alpha=(1+\gamma)/\gamma$ (i.e., $K_{i_{\text{slope}}}=K_{i_{\text{int}}}$). If the two inhibitor sites are not mutually exclusive, then an EI_2 complex could form. This introduces an $[I]^2/\delta\gamma K_i^2$ term into the slope factor (δ is the interaction factor of the two I sites). The reciprocal plot patterns are essentially unchanged, but now all the plots will not intersect at a common point and slope replots will be parabolic.

If I can bind to free E and exclude S, but a second I site becomes accessible after S binds (to yield a dead end ESI), then we have a mixture of pure competitive and pure uncompetitive inhibition.

$$\begin{array}{ccccc}
E + S & \underset{}{\overset{K_S}{\rightleftharpoons}} & ES & \overset{k_p}{\longrightarrow} & E + P \\
+ & & + & & \\
I & & I & & \\
\gamma K_i \upharpoonleft\downharpoonright & & K_i \upharpoonleft\downharpoonright & & \\
IE & & ESI & &
\end{array}$$

The final velocity equation is:

$$\frac{v}{V_{\max}} = \frac{[\mathrm{S}]}{K_\mathrm{S}\left(1+\frac{[\mathrm{I}]}{\gamma K_i}\right)+[\mathrm{S}]\left(1+\frac{[\mathrm{I}]}{K_i}\right)} \qquad \text{(IV-22)}$$

Again, the kinetics are identical to a mixed-type inhibition system with one site for I if $\gamma \neq 1$. If $\gamma = 1$, the results are identical to those for pure noncompetitive inhibition yet the I on IE does not occupy the same site as the I on ESI. If ESI is catalytically active with a rate constant βk_p, we observe hyperbolic mixed-type inhibition indistinguishable from the single I site model described in the following section. The velocity equation would be identical to equation IV-22 with the slope and intercept factors divided by $(1+\beta[\mathrm{I}]/K_i)$.

The more complex mixed-type inhibition systems involve changes in both K_S and k_p where $\beta k_p \neq 0$. These systems are conveniently discussed in terms of the following general scheme.

Hyperbolic Mixed-Type Inhibition

Inhibition may be expressed in a general way:

$$\begin{array}{ccccccccc}
\mathrm{E} & + & \mathrm{S} & \underset{}{\overset{K_\mathrm{S}}{\rightleftharpoons}} & \mathrm{ES} & \overset{k_p}{\longrightarrow} & \mathrm{E} & + & \mathrm{P} \\
+ & & & & + & & & & \\
\mathrm{I} & & & & \mathrm{I} & & & & \\
K_i \updownarrow & & & & \alpha K_i \updownarrow & & & & \\
\mathrm{EI} & + & \mathrm{S} & \overset{\alpha K_\mathrm{S}}{\rightleftharpoons} & \mathrm{ESI} & \overset{\beta k_p}{\longrightarrow} & \mathrm{EI} & + & \mathrm{P}
\end{array}$$

For example, $\alpha = 1$ and $\beta = 0$ describes pure noncompetitive inhibition. When $\alpha = 1$ and $0 < \beta < 1$, we have partial noncompetitive inhibition. When $1 < \alpha < \infty$ and $\beta = 1$, we have partial competitive inhibition. When $\alpha = \infty$, we have pure competitive inhibition. Mixed-type System C1 would be described as $1 < \alpha < \infty$ and $\beta = 0$. The general velocity equation may be derived from rapid equilibrium assumptions as shown below.

$$v = k_p[\mathrm{ES}] + \beta k_p[\mathrm{ESI}]$$

$$\frac{v}{[\mathrm{E}]_t} = \frac{k_p[\mathrm{ES}] + \beta k_p[\mathrm{ESI}]}{[\mathrm{E}] + [\mathrm{ES}] + [\mathrm{EI}] + [\mathrm{ESI}]}$$

$$\frac{v}{[\mathrm{E}]_t} = \frac{k_p \dfrac{[\mathrm{S}]}{K_\mathrm{S}}[\mathrm{E}] + \beta k_p \dfrac{[\mathrm{S}][\mathrm{I}]}{\alpha K_\mathrm{S} K_i}[\mathrm{E}]}{[\mathrm{E}] + \dfrac{[\mathrm{S}]}{K_\mathrm{S}}[\mathrm{E}] + \dfrac{[\mathrm{I}]}{K_i}[\mathrm{E}] + \dfrac{[\mathrm{S}][\mathrm{I}]}{\alpha K_\mathrm{S} K_i}[\mathrm{E}]}$$

$$\boxed{\frac{v}{V_{\max}} = \frac{\dfrac{[\mathrm{S}]}{K_\mathrm{S}} + \dfrac{\beta[\mathrm{S}][\mathrm{I}]}{\alpha K_\mathrm{S} K_i}}{1 + \dfrac{[\mathrm{S}]}{K_\mathrm{S}} + \dfrac{[\mathrm{I}]}{K_i} + \dfrac{[\mathrm{S}][\mathrm{I}]}{\alpha K_\mathrm{S} K_i}}} \qquad \text{(IV-23)}$$

As usual, the number of different terms in the numerator of equation IV-23 corresponds to the number of product-forming complexes. The four terms in the denominator correspond to E, ES, EI, and ESI, in that order. The $V_{\max}$ was taken as $k_p[\mathrm{E}]_t$. Equation IV-23 may be rearranged into a number of different forms:

$$\boxed{\frac{v}{V_{\max}} = \frac{[\mathrm{S}]}{K_\mathrm{S}\dfrac{\left(1 + \dfrac{[\mathrm{I}]}{K_i}\right)}{\left(1 + \dfrac{\beta[\mathrm{I}]}{\alpha K_i}\right)} + [\mathrm{S}]\dfrac{\left(1 + \dfrac{[\mathrm{I}]}{\alpha K_i}\right)}{\left(1 + \dfrac{\beta[\mathrm{I}]}{\alpha K_i}\right)}}} \qquad \text{(IV-24)}$$

or

$$\boxed{\frac{v}{V_{\max}} = \frac{[\mathrm{S}]}{\alpha K_\mathrm{S}\left(\dfrac{[\mathrm{I}] + K_i}{\beta[\mathrm{I}] + \alpha K_i}\right) + [\mathrm{S}]\left(\dfrac{[\mathrm{I}] + \alpha K_i}{\beta[\mathrm{I}] + \alpha K_i}\right)}} \qquad \text{(IV-25)}$$

The parenthetical phrases can be expressed in terms of apparent constants that are obtained from the reciprocal plots.

$$\boxed{\frac{v}{V_{\max}} = \frac{[S]}{K_S\left(1 + \frac{[I]}{K_{i_{slope}}}\right) + [S]\left(1 + \frac{[I]}{K_{i_{int}}}\right)}} \qquad \text{(IV-26)}$$

where $K_{i_{slope}}$ = an apparent K_i value that accounts for the change in slope (i.e., the K_i value calculated from the slope of the reciprocal plot)

$K_{i_{int}}$ = an apparent K_i value that accounts for the change in intercept (i.e., the K_i value calculated from the $1/v$-axis intercept)

Note that $K_{i_{slope}}$ and $K_{i_{int}}$ as they are defined here are not true constants since they vary with [I].

$$K_{i_{slope}} = \frac{\beta[I] + \alpha K_i}{(\alpha - \beta)}, \qquad K_{i_{int}} = \frac{\beta[I] + \alpha K_i}{(1 - \beta)}$$

and

$$\frac{K_{i_{slope}}}{K_{i_{int}}} = \frac{1 - \beta}{\alpha - \beta}$$

Another form of the equation then is:

$$\boxed{\frac{v}{V_{\max}} = \frac{[S]}{K_S\left(1 + \frac{[I](\alpha - \beta)}{\beta[I] + \alpha K_i}\right) + [S]\left(1 + \frac{[I](1 - \beta)}{\beta[I] + \alpha K_i}\right)}} \qquad \text{(IV-27)}$$

At any inhibitor concentration:

$$\boxed{\frac{v}{V_{\max_i}} = \frac{[S]}{K_{S_{app}} + [S]}} \qquad \text{(IV-28)}$$

where

$$V_{\max_i} = \frac{V_{\max}}{\left(\dfrac{[\mathrm{I}]+\alpha K_i}{\beta[\mathrm{I}]+\alpha K_i}\right)} = \frac{V_{\max}}{\left(1+\dfrac{[\mathrm{I}](1-\beta)}{\beta[\mathrm{I}]+\alpha K_i}\right)} = \frac{V_{\max}}{\dfrac{\left(1+\dfrac{[\mathrm{I}]}{\alpha K_i}\right)}{\left(1+\dfrac{\beta[\mathrm{I}]}{\alpha K_i}\right)}}$$

$$= V_{\max}\frac{\left(1+\dfrac{\beta[\mathrm{I}]}{\alpha K_i}\right)}{\left(1+\dfrac{[\mathrm{I}]}{\alpha K_i}\right)} = \frac{V_{\max}}{\left(1+\dfrac{[\mathrm{I}]}{K_{i_{\mathrm{int}}}}\right)}$$

$$K_{\mathrm{S_{app}}} = \alpha K_{\mathrm{S}}\left(\frac{[\mathrm{I}]+K_i}{[\mathrm{I}]+\alpha K_i}\right) = K_{\mathrm{S}}\frac{\left(1+\dfrac{[\mathrm{I}]}{K_i}\right)}{\left(1+\dfrac{[\mathrm{I}]}{\alpha K_i}\right)} = K_{\mathrm{S}}\frac{\left(1+\dfrac{[\mathrm{I}]}{K_{i_{\mathrm{slope}}}}\right)}{\left(1+\dfrac{[\mathrm{I}]}{K_{i_{\mathrm{int}}}}\right)}$$

At an infinitely high [I], the equation reduces to:

$$\frac{v}{\beta V_{\max}} = \frac{[\mathrm{S}]}{\alpha K_{\mathrm{S}}+[\mathrm{S}]} \qquad \text{(IV-29)}$$

The reciprocal form of the general velocity equation is:

$$\boxed{\frac{1}{v} = \frac{K_{\mathrm{S}}}{V_{\max}}\frac{\left(1+\dfrac{[\mathrm{I}]}{K_i}\right)}{\left(1+\dfrac{\beta[\mathrm{I}]}{\alpha K_i}\right)}\frac{1}{[\mathrm{S}]} + \frac{1}{V_{\max}}\frac{\left(1+\dfrac{[\mathrm{I}]}{\alpha K_i}\right)}{\left(1+\dfrac{\beta[\mathrm{I}]}{\alpha K_i}\right)}} \qquad \text{(IV-30)}$$

or

$$\boxed{\frac{1}{v} = \frac{\alpha K_{\mathrm{S}}}{V_{\max}}\left(\frac{[\mathrm{I}]+K_i}{\beta[\mathrm{I}]+\alpha K_i}\right)\frac{1}{[\mathrm{S}]} + \frac{1}{V_{\max}}\left(\frac{[\mathrm{I}]+\alpha K_i}{\beta[\mathrm{I}]+\alpha K_i}\right)} \qquad \text{(IV-31)}$$

or

$$\frac{1}{v} = \frac{K_S}{V_{max}}\left(1 + \frac{[I](\alpha - \beta)}{\beta[I] + \alpha K_i}\right)\frac{1}{[S]} + \frac{1}{V_{max}}\left(1 + \frac{[I](1-\beta)}{\beta[I] + \alpha K_i}\right) \quad \text{(IV-32)}$$

or

$$\frac{1}{v} = \frac{K_S}{V_{max}}\left(1 + \frac{[I]}{K_{i_{slope}}}\right)\frac{1}{[S]} + \frac{1}{V_{max}}\left(1 + \frac{[I]}{K_{i_{int}}}\right) \quad \text{(IV-33)}$$

Intersection Points in Mixed Inhibition Systems

In mixed-type systems both K_S and k_p change. The slope factor and the intercept factor will not be equal, and $K_{i_{slope}}$ will not equal $K_{i_{int}}$. These conditions result in the intersection of the "plus inhibitor" and control curves above or below the $1/[S]$-axis. The $1/[S]$-axis coordinate of the intersection between the "plus inhibitor" and control lines can be determined by setting the two velocity equations equal to each other and solving for the value of $1/[S]$ that yields the same $1/v$. The coordinate is:

$$\frac{1}{[S]} = -\frac{K_{i_{slope}}}{K_{i_{int}}K_S} = -\frac{(1-\beta)}{K_S(\alpha - \beta)} \quad \text{(IV-34)}$$

The point of intersection contains no [I] term; that is, all "plus inhibitor" lines, as well as the control line, intersect at a common point to the left of the vertical axis. The $1/v$-axis coordinate of the point of intersection can be calculated by substituting $-K_{i_{slope}}/K_{i_{int}}K_S$ or $(\beta - 1)/K_S(\alpha - \beta)$ for $1/[S]$.

$$\frac{1}{v} = \frac{1}{V_{max}}\left(1 - \frac{K_{i_{slope}}}{K_{i_{int}}}\right) = \frac{1}{V_{max}}\left(\frac{\alpha - 1}{\alpha - \beta}\right) \quad \text{(IV-35)}$$

Let us examine some specific examples of hyperbolic mixed-type inhibition.

System C2. $(1 < \alpha < \infty, 0 < \beta < 1)$

Consider an inhibition system in which E and EI bind S, but with different affinities. Both ES and ESI form product, but at different rates. This mixed-type system may be considered to be a mixture of partial competitive and partial noncompetitive inhibition. Because of the shape of the replots of the primary reciprocal plots, mixed System C2 can be called intersecting, slope-hyperbolic, intercept-hyperbolic, noncompetitive inhibition.

At any inhibitor concentration, an infinitely high [S] will drive all the enzyme to a mixture of ES and ESI forms. Because ESI is less productive than ES, $V_{\max_i}$ will be less than $V_{\max}$. At any inhibitor concentration, a portion of the enzyme available for combination with S will exist in the lower affinity EI form. Consequently, $K_{S_{app}}$ will be greater than K_S. At an infinitely high [I] all the enzyme will be converted to EI and ESI. Because ESI can form product, the velocity cannot be driven to zero by increasing [I].

Mixed System C2 cannot be distinguished from mixed System C1 by a single v versus [S] plot, or by a reciprocal plot at a single inhibitor concentration. Both systems yield reciprocal plots that intersect the control plot *above* the 1/[S]-axis. However, mixed System C2 approaches a limiting finite velocity at any [S] as the inhibitor concentration is increased. As [I] increases, the reciprocal plot pivots counterclockwise about the point of intersection with the control curve at $1/[S] = (\beta - 1)/K_S(\alpha - \beta)$ and $1/v = (\alpha - 1)/V_{\max}(\alpha - \beta)$ (Fig. IV-8). As [I] becomes infinitely high, all the enzyme is converted to EI and ESI and the reciprocal plot attains a limiting slope. The limiting 1/[S]-axis intercept gives $-1/\alpha K_S$. The limiting $1/v$-axis intercept gives $1/\beta V_{\max}$. Mixed System C2 can be distinguished easily from mixed System C1 by the replots of slope versus [I] and $1/v$-axis intercept versus [I] (Fig. IV-9). Both replots are hyperbolic; hence the designation "slope-hyperbolic, intercept-hyperbolic noncompetitive inhibition". Consequently, the limits (which would allow α, β, and then K_i to be determined) are difficult to identify. A plot of $1/\Delta$ *slope* versus 1/[I] and $1/\Delta$ *intercept* versus 1/[I] converts the relationship to a straight line (Fig. IV-10). By using Δ *slope* and Δ *intercept*, we are, in effect, raising the horizontal-axis of Figure IV-9 so that the curves become hyperbolae that start at the origin.

$$\Delta\,(1/v\text{-axis } intercept) = \frac{1}{V_{\max}}\left(\frac{[I] + \alpha K_i}{\beta[I] + \alpha K_i}\right) - \frac{1}{V_{\max}}$$

$$\boxed{\frac{1}{\Delta\ intercept} = \frac{\alpha K_i V_{\max}}{(1-\beta)}\frac{1}{[I]} + \frac{\beta V_{\max}}{(1-\beta)}} \qquad \text{(IV-36)}$$

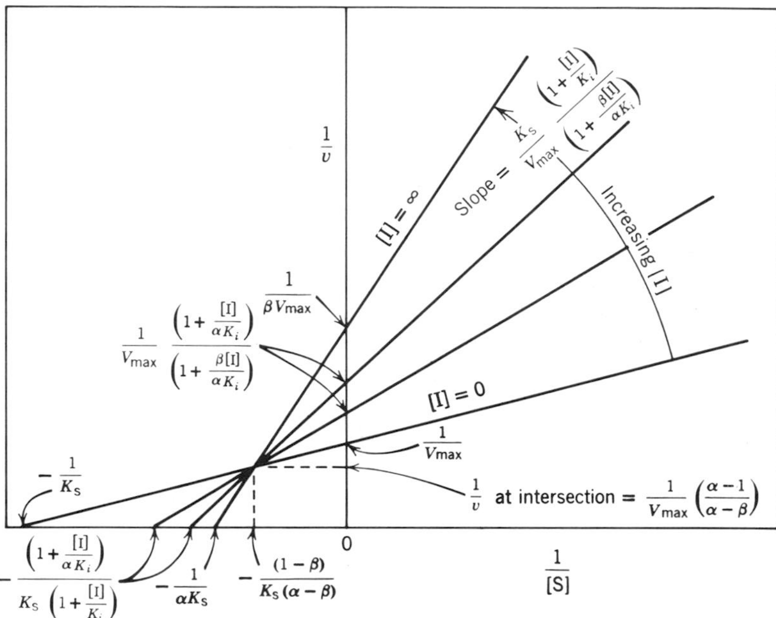

Fig. IV-8. The $1/v$ versus $1/[S]$ plot in the presence of different fixed concentrations of a hyperbolic (partial) mixed-type inhibitor ($1<\alpha<\infty, 0<\beta<1$); ESI is catalytically active. Consequently, the limiting plot at $[I]=\infty$ has a finite positive slope.

Thus the replot has a slope of $\alpha K_i V_{max}/(1-\beta)$ and an intercept on the $1/\Delta$ *intercept*-axis of $\beta V_{max}/(1-\beta)$. When $1/\Delta$ *intercept* $=0$, the intercept on the $1/[I]$-axis gives $-\beta/\alpha K_i$.

$$\Delta\, slope = \frac{\alpha K_S}{V_{max}}\left(\frac{[I]+K_i}{\beta[I]+\alpha K_i}\right) - \frac{K_S}{V_{max}}$$

$$\boxed{\frac{1}{\Delta\, slope} = \frac{\alpha K_i V_{max}}{K_S(\alpha-\beta)}\frac{1}{[I]} + \frac{\beta V_{max}}{K_S(\alpha-\beta)}} \tag{IV-37}$$

The plot is linear with a slope of $\alpha K_i V_{max}/K_S(\alpha-\beta)$ and an intercept on the $1/\Delta$ *slope* axis of $\beta V_{max}/K_S(\alpha-\beta)$. When $1/\Delta$ *slope* is zero, the intercept on the $1/[I]$-axis gives $-\beta/\alpha K_i$.

With β determined from the $1/\Delta$ *intercept* replot, α can be determined from the $1/\Delta$ *slope* replot. With both α and β known (and K_S and V_{max} from the control plots), K_i may be calculated. The equations apply equally well to partial competitive inhibition ($\beta=1$) and partial noncompetitive inhibition ($\alpha=1$).

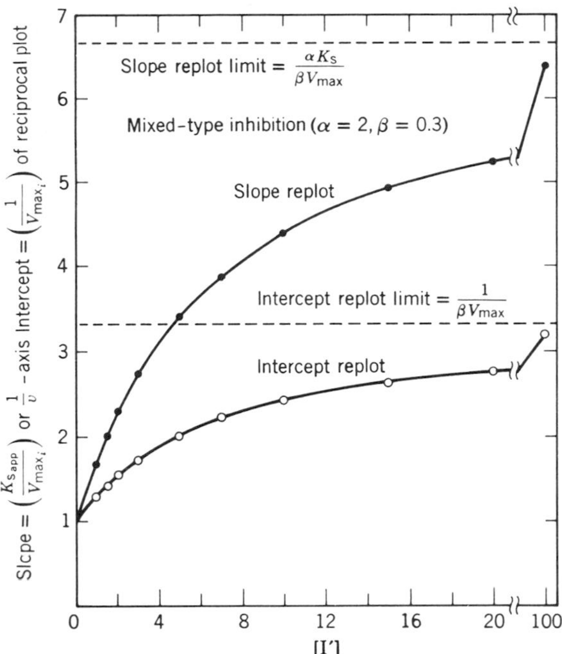

Fig. IV-9. *Slope*$_{1/S}$ and $1/v$-axis intercept replots for a hyperbolic (partial) mixed-type inhibitor ($\alpha = 2, \beta = 0.3$).

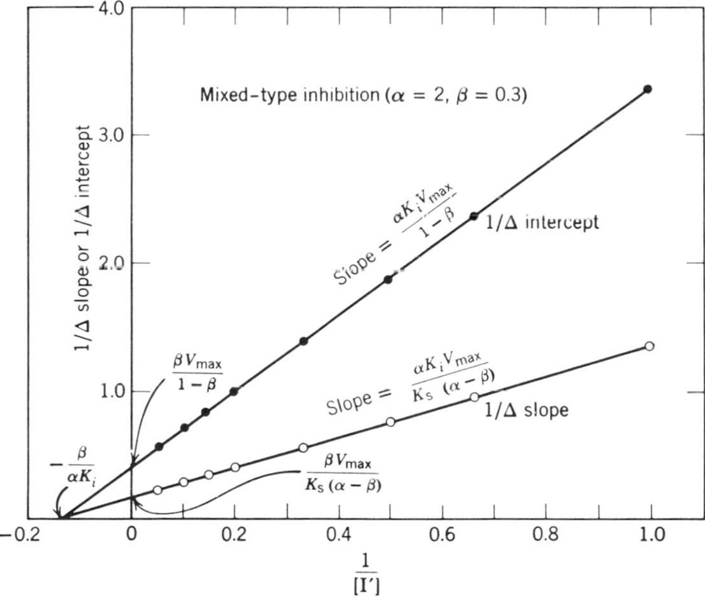

Fig. IV-10. Secondary replot of 1/Δ *slope* and 1/Δ *intercept* for a hyperbolic (partial) mixed-type inhibitor ($\alpha = 2, \beta = 0.3$).

Variation of $K_{i_{slope}}$ *and* $K_{i_{int}}$ *with* [I] We can use $K_{i_{slope}}$ and $K_{i_{int}}$ in an alternate procedure for determining α, β, and K_i. As pointed out earlier, the quantities $K_{i_{slope}}$ and $K_{i_{int}}$ vary with [I] for partial and mixed systems and thus are only apparent constants for a single inhibitor concentration.

$$K_{i_{slope}} = \frac{\beta[\mathrm{I}] + \alpha K_i}{(\alpha - \beta)}, \qquad K_{i_{int}} = \frac{\beta[\mathrm{I}] + \alpha K_i}{(1 - \beta)}$$

The expressions can be written in the forms:

$$\boxed{K_{i_{int}} = \frac{\beta}{(1-\beta)}[\mathrm{I}] + \frac{\alpha K_i}{(1-\beta)}} \tag{IV-38}$$

$$\boxed{K_{i_{slope}} = \frac{\beta}{(\alpha-\beta)}[\mathrm{I}] + \frac{\alpha K_i}{(\alpha-\beta)}} \tag{IV-39}$$

From the slope of the $K_{i_{int}}$ versus [I] curve, β can be determined. With this value and the value of the slope of the $K_{i_{slope}}$ versus [I] curve, α can be determined. With both α and β known, K_i can be determined from the value of any intercept. This is illustrated in Figure IV-11 for the mixed-type inhibition system where $\alpha = 2$ and $\beta = 0.3$.

When $\beta = 0$, $K_{i_{slope}}$ equals K_i and $K_{i_{int}}$ equals αK_i. The plots of $K_{i_{slope}}$ and $K_{i_{int}}$ versus [I] will be horizontal. When $\beta > \alpha$. The plot of $K_{i_{slope}}$ versus [I] will have a negative slope and a negative intercept on the $K_{i_{slope}}$-axis.

Alternate Plots of ΔK_S *and* ΔV_{max} A plot of $K_{S_{app}}$ versus [I] is hyperbolic with a limit of αK_S. A general reciprocal equation relating ΔK_S and [I] is:

$$\boxed{\frac{1}{\Delta K_S} = \frac{\alpha K_i}{K_S(\alpha - 1)} \frac{1}{[\mathrm{I}]} + \frac{1}{K_S(\alpha - 1)}} \tag{IV-40}$$

Thus a plot of $1/\Delta K_S$ versus $1/[\mathrm{I}]$ is a straight line with a slope of $\alpha K_i / K_S(\alpha - 1)$ and a vertical-axis intercept of $1/K_S(\alpha - 1)$. When $1/\Delta K_S = 0$, the intercept on the $1/[\mathrm{I}]$-axis gives $-1/\alpha K_i$.

A plot of V_{max_i} versus [I] is also hyperbolic with a limit of βV_{max}. The reciprocal equation is:

$$\boxed{\frac{1}{\Delta V_{max}} = \frac{\alpha K_i}{V_{max}(1-\beta)} \frac{1}{[\mathrm{I}]} + \frac{1}{V_{max}(1-\beta)}} \tag{IV-41}$$

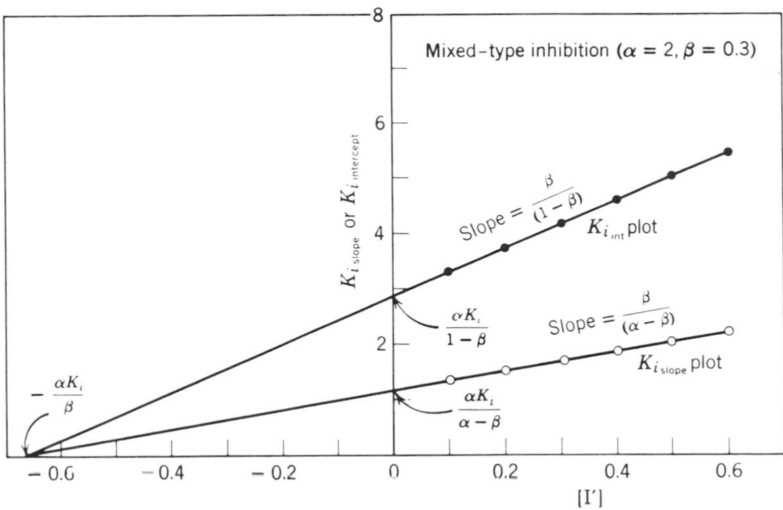

Fig. IV-11. Replot of $K_{i_{slope}}$ and $K_{i_{int}}$ for a hyperbolic (partial) mixed-type inhibitor ($\alpha = 2, \beta = 0.3$).

The plot is linear with a slope of $\alpha K_i / V_{max}(1-\beta)$ and an intercept on the $1/\Delta V_{max}$-axis of $1/V_{max}(1-\beta)$. When $1/\Delta V_{max} = 0$, the intercept on the $1/[\mathrm{I}]$-axis gives $-1/\alpha K_i$. This is essentially a variation of the $1/\Delta$ *intercept* versus $1/[\mathrm{I}]$ plot. From the $1/\Delta K_S$ versus $1/[\mathrm{I}]$ plot, α can be determined (from the intercept on the $1/\Delta K_S$-axis). The quantity β can be determined from the $1/\Delta V_{max}$-axis intercept of the $1/\Delta V_{max}$ versus $1/[\mathrm{I}]$ plot. With α and β known, K_i can be calculated from the slopes or intercepts on the $1/[\mathrm{I}]$-axis.

In the preceding examples, we considered inhibitors which decreased both the apparent affinity of the enzyme for the substrate ($\alpha > 1$) *and* the V_{max} of the reaction ($\beta < 1$, giving $V_{max_i} < V_{max}$). It is also possible for an inhibitor to convert the enzyme to an EI form which has an increased affinity for the substrate ($\alpha < 1$), but the resulting ESI forms product slower than ES. If, by chance, the inhibitor decreases K_S by exactly the same factor that it decreases V_{max}, the system will resemble that described as uncompetitive inhibition (System C4). (We might also consider the remaining possibility where $\alpha < 1$, but $\beta > 1$. A little thought will reveal that in this situation the effector is actually an activator.) The positions of the intercepts, intersection points, and limiting slope (if any) of the reciprocal plots for these systems will depend on which effect of the inhibitor is dominant. The various curves arise because an increase in affinity ($\alpha < 1$) contributes to a decrease in the slope of the reciprocal plot, while a decrease in k_p ($\beta < 1$) contributes to an increase in the slope. (The $1/v$-axis intercept is controlled solely by k_p; thus the intercept increases if $\beta < 1$ regardless of the value of α).

System C3. $(\alpha<1, \beta<1)$

Consider first System C3, where k_p decreases and the affinity increases, but to a lesser degree. This can be stated as $0<\beta<1$ and $0<\alpha<1$, but also $\alpha>\beta$, or $\alpha/\beta>1$. In other words, the increase in slope caused by a decrease in k_p will be *partially* compensated for by a decrease in slope caused by a decrease in K_S. The "plus inhibitor" reciprocal plot will still have a greater slope than the control plot, but the curves will now intersect below the $1/[S]$-axis and at a $-1/[S]$ value more negative than $-1/K_S$ (Fig. IV-12). In this system, the intercept factor is greater than the slope factor so that $K_{i_{slope}}$ will be greater than $K_{i_{int}}$. The plot of $K_{S_{app}}$ versus [I] is hyperbolic, but *decreases* to a limit as [I] increases. The plot of $1/\Delta K_S$ versus $1/[I]$ will have a negative slope. Alternatively, if ΔK_S is defined as $K_S - K_{S_{app}}$, the reciprocal equation becomes:

$$\boxed{\frac{1}{\Delta K_S} = \frac{\alpha K_i}{K_S(1-\alpha)}\frac{1}{[I]} + \frac{1}{K_S(1-\alpha)}} \qquad \text{(IV-42)}$$

In the form above, the slope of the plot is positive.

System C4. $(\alpha=\beta)$

Now consider the situation where k_p decreases and the affinity increases to the same degree. This can be expressed as: $0<\alpha<1$, $0<\beta<1$, and $\alpha=\beta$. The increase in slope caused by the decrease in k_p will be exactly canceled by the decrease in slope caused by the decrease in K_S. Thus the "plus inhibitor" reciprocal plot will be parallel to the control plot. In other words, in the presence of inhibitor, the slope remains constant and the $1/v$-axis and $1/[S]$-axis intercepts increase by the same amount relative to the control. Increasing [I] will cause the "plus inhibitor" curve to move further and further away from the control curve but all curves will be parallel (Fig. IV-13). This system yields the same reciprocal plot pattern as uncompetitive inhibition. In pure uncompetitive inhibition, however, the curves continue to move apart as [I] increases. In the mixed system, the displacement reaches a limit as [I] becomes infinite. A replot of $1/v$-axis intercept versus [I] will be hyperbolic for the mixed system and linear for the pure uncompetitive system. This mixed system is sometimes called hyperbolic or partial uncompetitive inhibition.

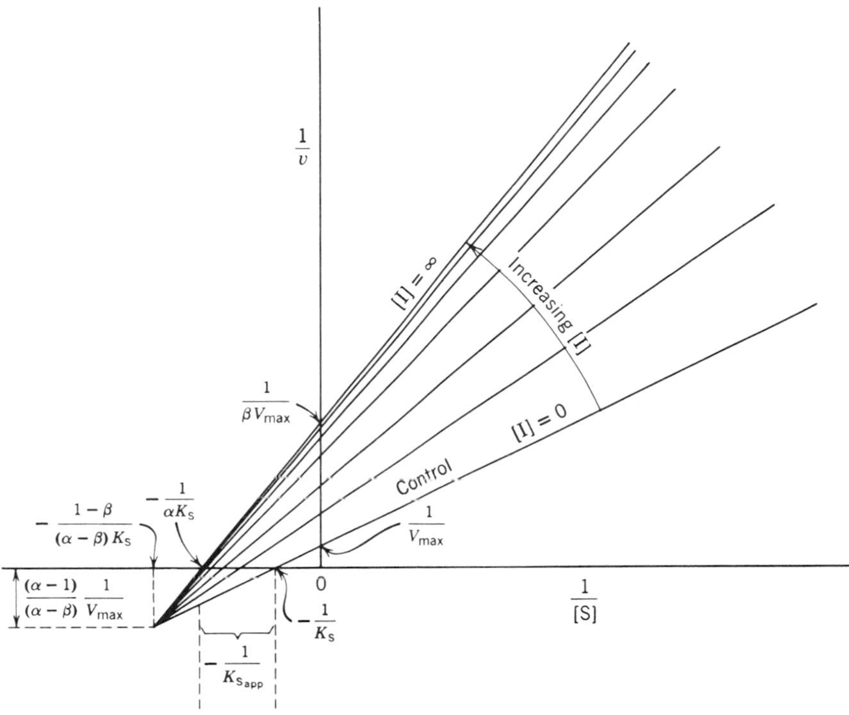

Fig. IV-12. The $1/v$ versus $1/[S]$ plot in the presence of different fixed concentrations of a hyperbolic (partial) mixed-type inhibitor ($0<\alpha<1, 0<\beta<1$). When $\alpha<1$, the family of curves intersect below the $1/[S]$-axis.

System C5. $(\alpha<1, \beta<1, \beta>\alpha)$

The final possibility arises when the inhibitor decreases the rate constant for product formation, but also markedly increases the affinity of the enzyme for the substrate. This can be described as $\alpha<1$, $\beta<1$, and also $\beta>\alpha$. In other words, k_p has decreased, but K_S has decreased much more. Figure IV-14 shows the velocity curve for such a system, where $\alpha=0.2$, $\beta=0.5$, and $[I]=10K_i$. There may be some doubt whether to call I an inhibitor or an activator. Clearly, at low substrate concentrations (up to the intersection point at $[S]=0.6K_S$) I activates, but at higher substrate concentrations, I inhibits. This is not unexpected given the values of α and β. In the presence of I, there will be a higher concentration of ESI than of ES at low [S] (because EI has a higher affinity than E for S). The greater concentration of ESI more than compensates for the fact that ESI is not as effective as ES in

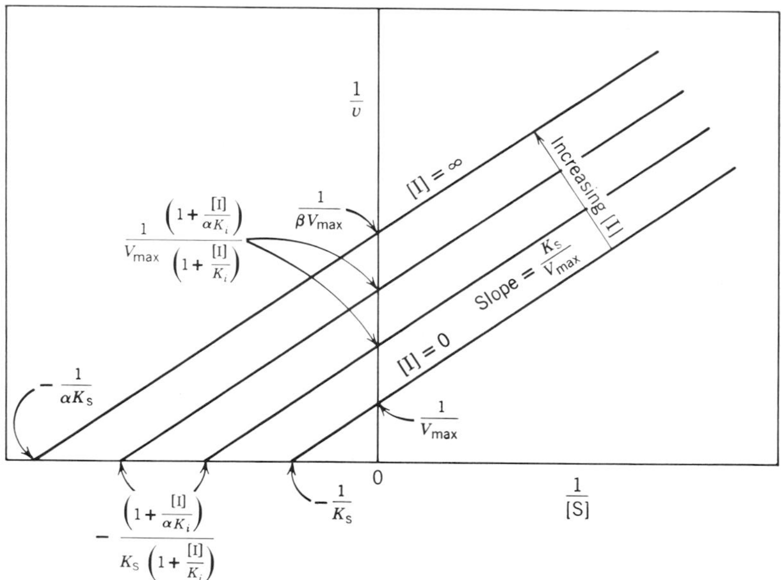

Fig. IV-13. Hyperbolic uncompetitive inhibition: $1/v$ versus 1/[S] plot in the presence of different fixed concentrations of a mixed-type inhibitor where $\alpha<1$ and $0<\beta<1$, but $\alpha=\beta$.

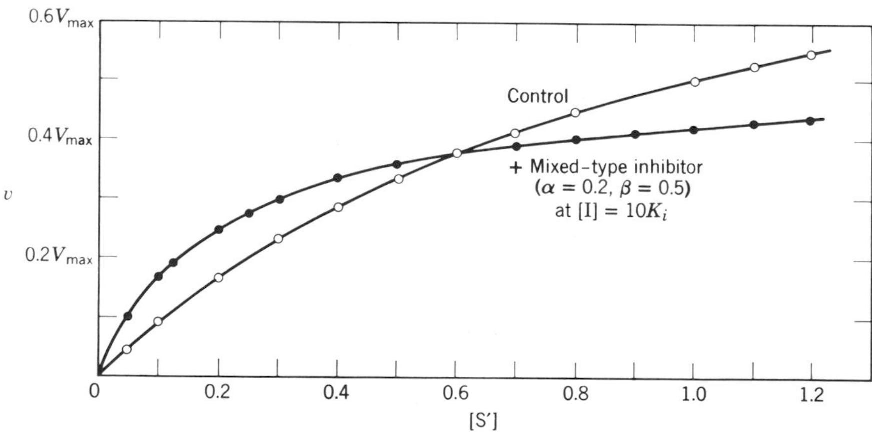

Fig. IV-14. The v versus [S] plot in the presence of a mixed-type inhibitor where $\alpha<1$ and $\beta<1$, but $\beta>\alpha$; I is an activator at low substrate concentrations and an inhibitor at high substrate concentrations.

forming product. However, as [S] increases, the advantage is lost because V_{max_i} is only 0.51 V_{max} at [I] = $10K_i$.

The reciprocal plot is shown in Figure IV-15. The increase in slope caused by the decrease in k_p is now more than compensated for by a larger decrease in slope caused by the decrease in K_S. The slope of the "plus inhibitor" curve is now less than the control curve. The "plus inhibitor" and control curves intersect above the 1/[S]-axis at a point greater than $1/V_{max_i}$ and at a positive 1/[S] value. In this system, the slope factor is less than unity and $K_{i_{slope}}$ turns out to be a negative number. As [I] increases, the reciprocal plot pivots *clockwise* about the intersection point and approaches a limiting positive slope at an infinitely high [I]. The replots will be hyperbolic, but this time the replot of slope versus [I] will be a descending curve. For example, if $\alpha = 0.2$ and $\beta = 0.5$, the limiting slope will be $(\alpha/\beta) \times$ original slope, or $0.4 \times$ original slope, but the limiting $1/v$-axis intercept will be $1/\beta \times$ original intercept, or $2.0 \times$ original intercept.

A plot of v versus [I] provides some interesting results. At a fixed substrate concentration below the intersection point, increasing [I] results in activation. At a fixed substrate concentration above the intersection point, increasing [I] results in inhibition. When [S] = $0.6K_S$, I has no effect on the initial velocity at any concentration.

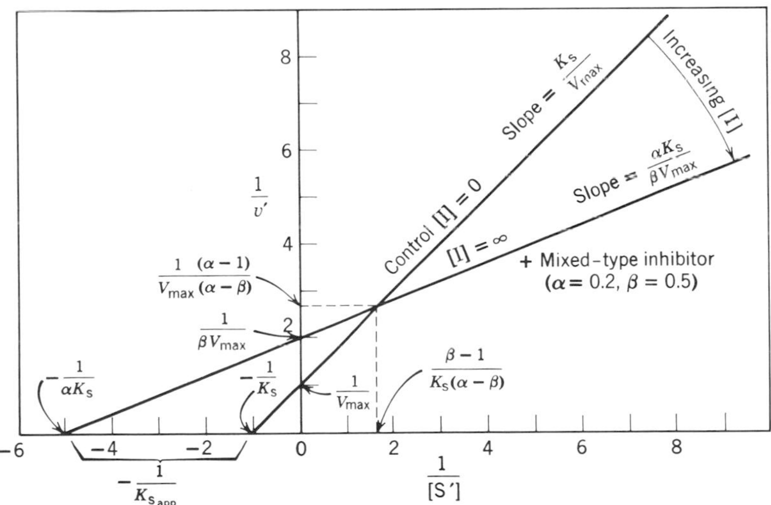

Fig. IV-15. The $1/v$ versus 1/[S] plot in the presence of the mixed-type inhibitor described in Fig. IV-14 ($\alpha < 1, \beta < 1, \beta > \alpha$).

As a limiting case, consider the system shown below in which I binds only to ES and $\beta < 1$.

$$
\begin{array}{ccccc}
E + S & \overset{K_S}{\rightleftharpoons} & ES & \overset{k_p}{\longrightarrow} & E + P \\
 & & + & & \\
 & & I & & \\
 & & K_i \updownarrow & & \\
 & & ESI & \overset{\beta k_p}{\longrightarrow} & E + P + I
\end{array}
$$

The velocity equation is:

$$
\boxed{\frac{v}{V_{max}} = \frac{[S]}{\dfrac{K_S}{\left(1+\dfrac{\beta[I]}{K_i}\right)} + [S]\dfrac{\left(1+\dfrac{[I]}{K_i}\right)}{\left(1+\dfrac{\beta[I]}{K_i}\right)}}} \qquad \text{(IV-43)}
$$

The family of reciprocal plots is similar to that shown in Figure IV-15 except now the limiting plot at $[I]=\infty$ is a horizontal line (i.e., at saturating [I], the K_m for S appears to be zero). The plots intersect at $1/[S]=(1-\beta)/\beta K_S$ and $1/v=1/\beta V_{max}$. If $\beta=1$, the family of plots intersect on the $1/v$-axis at $1/V_{max}$. (The equilibria suggest that this system should be called partial uncompetitive inhibition. However, the reciprocal plots are not parallel.)

Two-Site Model for Partial Inhibition

The equilibria shown below describe an enzyme with two distinct and mutually exclusive inhibitor sites. When the inhibitor binds to one of the sites (the "inhibitory" site), it induces an electronic or conformational change in the enzyme that either prevents the substrate from binding to the catalytic site (pure competitive action) (Fig. IV-16) or prevents catalytic activity even though S is bound (pure noncompetitive action). The binding of the inhibitor to the other inhibitor site (the "noninhibitory" site) is also prevented. When the inhibitor binds to the noninhibitory site, binding to the

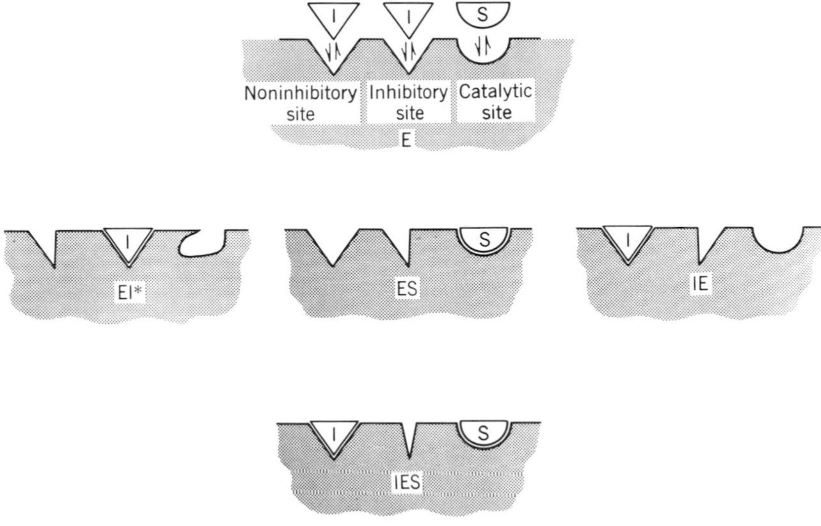

Fig. IV-16. Partial competitive inhibition resulting from two distinct types of sites for I. The binding of a ligand to any site distorts the adjacent sites.

inhibitory site is prevented, but there is no effect on substrate binding or the catalytic activity.

System C6. Competitive

$$
\begin{array}{ccccccccc}
\mathrm{IE} & + & \mathrm{S} & \overset{K_S}{\rightleftharpoons} & \mathrm{IES} & \overset{k_p}{\longrightarrow} & \mathrm{IE} & + & \mathrm{P} \\
{\scriptstyle K_i}\downarrow\uparrow & & & & {\scriptstyle K_i}\downarrow\uparrow & & & & \\
\mathrm{I} & & & & \mathrm{I} & & & & \\
+ & & & & + & & & & \\
\mathrm{E} & + & \mathrm{S} & \overset{K_S}{\rightleftharpoons} & \mathrm{ES} & \overset{k_p}{\longrightarrow} & \mathrm{E} & + & \mathrm{P} \\
+ & & & & & & & & \\
\mathrm{I} & & & & & & & & \\
{\scriptstyle K_i^*}\uparrow\downarrow & & & & & & & & \\
\mathrm{EI^*} & & & & & & & &
\end{array}
$$

System C7. Noncompetitive

$$
\begin{array}{ccccccc}
IE & + S & \underset{\longrightarrow}{\overset{K_S}{\longleftarrow}} & IES & \overset{k_p}{\longrightarrow} & IE + P \\
K_i \Downarrow\Uparrow & & & K_i \Downarrow\Uparrow & & \\
I & & & I & & \\
+ & & & + & & \\
E & + S & \underset{\longrightarrow}{\overset{K_S}{\longleftarrow}} & ES & \overset{k_p}{\longrightarrow} & E + P \\
+ & & & + & & \\
I & & & I & & \\
K_i^* \Uparrow\Downarrow & & & K_i^* \Uparrow\Downarrow & & \\
EI^* & + S & \underset{\longrightarrow}{\overset{K_S}{\longleftarrow}} & EI^*S & &
\end{array}
$$

The systems above are simplified versions of the model proposed by Stadtman and co-workers (1968) to explain both the partial and the cumulative nature of the feedback inhibition of glutamine synthetase. We can see from the equilibria that an infinitely high inhibitor concentration will drive all the enzyme to a mixture of EI* and IE forms. Because IE is as effective as free E in substrate binding and catalytic activity, the velocity can never be driven to zero. The distribution of enzyme between EI* and IE will depend on the relative values of K_i and K_i*. The velocity equations for the systems above can be written in the usual manner.

Competitive

$$\frac{v}{V_{\max}} = \frac{\dfrac{[S]}{K_S} + \dfrac{[I][S]}{K_i K_S}}{1 + \dfrac{[S]}{K_S} + \dfrac{[I]}{K_i} + \dfrac{[I]}{K_i^*} + \dfrac{[I][S]}{K_i K_S}} \tag{IV-44}$$

or

$$\frac{v}{V_{\max}} = \frac{[S]}{K_S \dfrac{\left(1 + \dfrac{[I]}{K_i} + \dfrac{[I]}{K_i^*}\right)}{\left(1 + \dfrac{[I]}{K_i}\right)} + [S]} \tag{IV-45}$$

If K_i^* is expressed in terms of K_i where $K_i^* = \gamma K_i$, the velocity equation can be written as:

$$\frac{v}{V_{\max}} = \frac{[\mathrm{S}]}{K_{\mathrm{S}}\left(1 + \frac{[\mathrm{I}]}{\gamma(K_i + [\mathrm{I}])}\right) + [\mathrm{S}]} \tag{IV-46}$$

As [I] increases, $K_{\mathrm{S}_{app}}$ approaches $K_{\mathrm{S}}(1 + 1/\gamma)$ as a limit.

Noncompetitive

$$\frac{v}{V_{\max}} = \frac{\frac{[\mathrm{S}]}{K_{\mathrm{S}}} + \frac{[\mathrm{I}][\mathrm{S}]}{K_i K_{\mathrm{S}}}}{1 + \frac{[\mathrm{S}]}{K_{\mathrm{S}}} + \frac{[\mathrm{I}]}{K_i} + \frac{[\mathrm{I}]}{K_i^*} + \frac{[\mathrm{I}][\mathrm{S}]}{K_i K_{\mathrm{S}}} + \frac{[\mathrm{I}][\mathrm{S}]}{K_i^* K_{\mathrm{S}}}} \tag{IV-47}$$

or

$$\frac{v}{V_{\max}} = \frac{[\mathrm{S}]}{K_{\mathrm{S}}\left(1 + \frac{[\mathrm{I}]}{\gamma(K_i + [\mathrm{I}])}\right) + [\mathrm{S}]\left(1 + \frac{[\mathrm{I}]}{\gamma(K_i + [\mathrm{I}])}\right)} \tag{IV-48}$$

As [I] increases, $V_{\max_i}$ approaches $V_{\max}/(1 + 1/\gamma)$ as a limit.

The partial nature of the systems above is evident from the hyberbolic replots of slope (System C6) or slope and intercept (System C7) versus [I]. The secondary reciprocal replots of $1/\Delta$ *slope* versus 1/[I] (for Systems C6 and C7) or $1/\Delta$ *intercept* versus 1/[I] (for System C7) are linear. For example, for System C6:

$$\Delta\, slope = \frac{K_{\mathrm{S}}}{V_{\max}}\left(1 + \frac{[\mathrm{I}]}{\gamma(K_i + [\mathrm{I}])}\right) - \frac{K_{\mathrm{S}}}{V_{\max}}$$

$$\frac{1}{\Delta\, slope} = \frac{\gamma K_i V_{\max}}{K_{\mathrm{S}}}\frac{1}{[\mathrm{I}]} + \frac{\gamma V_{\max}}{K_{\mathrm{S}}} \tag{IV-49}$$

Thus the intercept on the $1/\Delta$ *slope*-axis is $\gamma V_{max}/K_S$; the intercept on the $1/[I]$-axis is $-1/K_i$, and the slope of the plot is $\gamma K_i V_{max}/K_S$. The two-site models can be made formally equivalent to the analogous single site partial inhibition models if $\gamma = 1/(\alpha - 1)$ and the K_i of the noninhibitory site on the two-site model is numerically equal to $(\alpha) \times (K_i$ of one-site model). Thus the single-site and two-site models cannot be distinguished by the kinetics. The number of inhibitor sites for either model would have to be established by other means, for example, differential inactivation of one of the two sites combined with kinetic and binding studies.

Apparent Partial or Mixed-Type Inhibition Resulting from Multiple Enzymes

A preparation may contain multiple enzymes (or multiple forms of the same enzyme) that catalyze the same reaction but are unequally sensitive to a given inhibitor. The presence of multiple enzymes can easily be overlooked if their kinetic constants are similar (Chapter Two), and, consequently, an inhibition study leads to erroneous conclusions. For example, consider a system containing two enzymes that catalyze the same $S \rightarrow P$ reaction with the same K_m and V_{max}. However, only one of the enzymes is competitively inhibited by I. The velocity at any substrate and inhibitor concentration is given by:

$$v_t = v_1 + v_2$$

$$v_t = \frac{\frac{[S]}{K_m} V_{max}}{1 + \frac{[S]}{K_m}} + \frac{\frac{[S]}{K_m} V_{max}}{1 + \frac{[S]}{K_m} + \frac{[I]}{K_i}} \tag{IV-50}$$

where v_1 = the velocity of the uninhibited enzyme
v_2 = the velocity of the inhibited enzyme
v_t = the observed total velocity.

A v versus [I] plot at various fixed substrate concentrations resembles the plot for a partial or mixed-type inhibition system where $\beta \neq 0$ (Fig. IV-17). That is, the velocity cannot be driven to zero by an infinitely high inhibitor concentration. The reciprocal plot in the absence of inhibitor is linear with intercepts of $1/V_{max_t} = 1/(V_{max_1} + V_{max_2})$ and $-1/K_m$ (Fig. IV-18). The reciprocal plot in the presence of inhibitor will appear linear unless points at

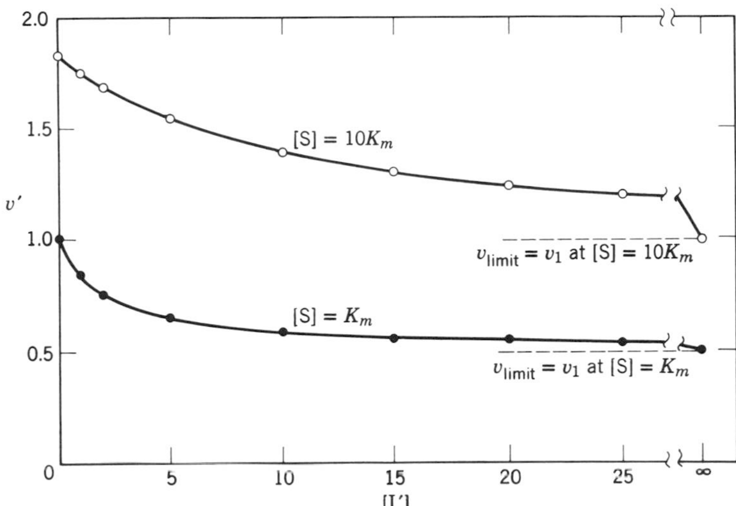

Fig. IV-17. Apparent partial inhibition when two enzymes are present and only one is competitively inhibited by I; $K_{m_1} = K_{m_2} = 1$; $V_{max_1} = V_{max_2} = 1$.

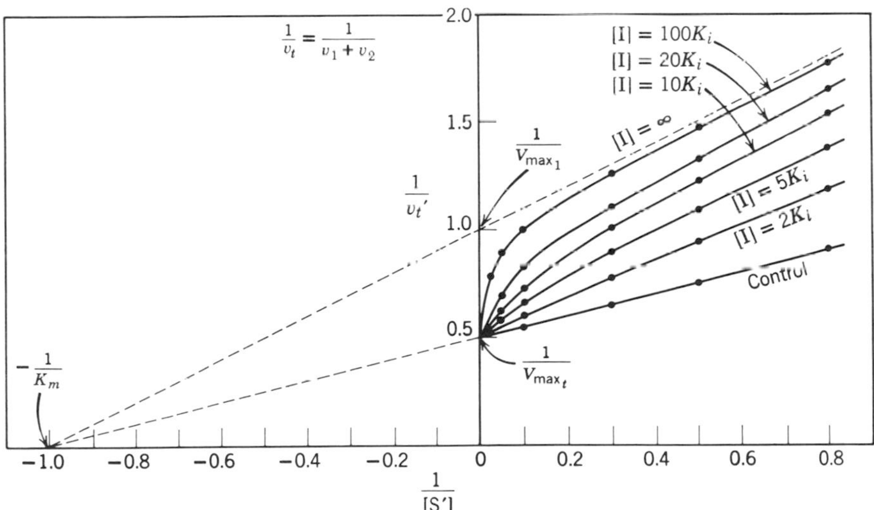

Fig. IV-18. The $1/v$ versus $1/[S]$ plots at different fixed concentrations of I. Two enzymes are present but only one is competitively inhibited by I; $K_{m_1} = K_{m_2} = 1$; $V_{max_1} = V_{max_2} = 1$.

low $1/[S]$ are included. An experiment consisting of initial velocity measurements at $[S]=1-10K_m$ for a control and one or two low fixed inhibitor concentrations will yield what appears to be straight lines that intersect above the $1/[S]$-axis and to the left of the $1/v$-axis. The inhibition will appear to be mixed-type with $\alpha>1$ and $0<\beta<1$. The plots are actually curved but the slight departure from linearity within the [S] range studied can be masked by (or blamed on) experimental error. The situation becomes clearer as the substrate and inhibitor concentration ranges are increased. At high substrate concentrations, the curvature of the reciprocal plots in the region close to the $1/v$-axis becomes more obvious. Also, as the fixed inhibitor concentration is increased, it will become more obvious that the apparent linear portions of the plots do not extrapolate to a common point. The same type of kinetics would be observed for an enzyme with two identical substrate binding sites and one inhibitor site where the inhibitor excludes the substrate from only one of the substrate sites.

Reduction of Steady-State Velocity Equation to Rapid Equilibrium Form

The general scheme for mixed and partial inhibition systems is shown below in terms of individual rate constants.

$$
\begin{array}{ccccccc}
E + S & \underset{k_{-1}}{\overset{k_1}{\rightleftharpoons}} & ES & \underset{k_{-5}}{\overset{k_5}{\rightleftharpoons}} & E + P \\
+ & & + & & \\
I & & I & & \\
k_2 \downarrow\uparrow k_{-2} & & k_{-3} \uparrow\downarrow k_3 & & \\
EI + S & \underset{k_{-4}}{\overset{k_4}{\rightleftharpoons}} & ESI & \underset{k_{-6}}{\overset{k_6}{\rightleftharpoons}} & EI + P
\end{array}
$$

The velocity equation in the absence of products can be expressed as:

$$\frac{v}{[E]_t} = \frac{k_5[ES]+k_6[ESI]}{[E]+[ES]+[EI]+[ESI]} \qquad \text{(IV-51)}$$

In the rapid equilibrium treatment, the concentrations of the various complexes are expressed in terms of dissociation constants and the concentrations of S and I:

$$[\mathrm{ES}]=\frac{[\mathrm{S}]}{K_{\mathrm{S}}}[\mathrm{E}], \qquad [\mathrm{EI}]=\frac{[\mathrm{I}]}{K_i}[\mathrm{E}], \qquad [\mathrm{ESI}]=\frac{[\mathrm{S}][\mathrm{I}]}{\alpha K_{\mathrm{S}} K_i}[\mathrm{E}]$$

The steady-state treatment of the general scheme yields very complex expressions for the concentrations of the various species. These expressions contain $[\mathrm{S}]^2$, $[\mathrm{S}]^2[\mathrm{I}]$, $[\mathrm{I}]^2$, and $[\mathrm{I}]^2[\mathrm{S}]$ terms in addition to the usual [S] and [I] terms. As a result, nonlinear $1/v$ versus $1/[\mathrm{S}]$ plots are expected in the presence of I. (The reciprocal plots are always linear in the absence of I, since neither the rapid equilibrium nor the steady-state treatment yield $[\mathrm{S}]^2$ or greater power terms.) The expressions for the relative concentrations of each species under steady-state conditions can be obtained by the King-Altman schematic method as described in Chapter Nine. Frieden (1964) has shown that these expressions can be rearranged and factored to the forms shown below when $[\mathrm{P}]=0$.

$$\frac{[\mathrm{E}]}{[\mathrm{E}]_t}=\frac{k_{-1}k_{-3}k_4\left(1+\dfrac{k_5}{k_{-1}}\right)\left[[\mathrm{S}]+\dfrac{k_2}{k_1}\dfrac{\left(1+\dfrac{k_6}{k_{-4}}\right)}{\left(1+\dfrac{k_5}{k_{-1}}\right)}[\mathrm{I}]+\dfrac{k_{-2}(k_{-3}+k_{-4}+k_6)}{k_{-3}k_4}\right]}{\text{denominator}}$$

$$\frac{[\mathrm{ES}]}{[\mathrm{E}]_t}=\frac{k_1k_{-3}k_4[\mathrm{S}]\left[[\mathrm{S}]+\dfrac{k_2}{k_1}[\mathrm{I}]+\dfrac{k_{-2}(k_{-3}+k_{-4}+k_6)}{k_{-3}k_4}\right]}{\text{denominator}}$$

$$\frac{[\mathrm{EI}]}{[\mathrm{E}]_t}=\frac{k_1k_3k_{-4}\left(1+\dfrac{k_6}{k_{-4}}\right)[\mathrm{I}]\left[[\mathrm{S}]+\dfrac{k_2}{k_1}[\mathrm{I}]+\dfrac{k_{-2}}{k_{-3}k_4}\dfrac{\left(1+\dfrac{k_5}{k_{-1}}\right)}{\left(1+\dfrac{k_6}{k_{-4}}\right)}(k_{-3}+k_{-4}+k_6)\right]}{\text{denominator}}$$

$$\frac{[\mathrm{ESI}]}{[\mathrm{E}]_t}=\frac{k_1k_3k_4[\mathrm{S}][\mathrm{I}]\left[[\mathrm{S}]+\dfrac{k_2}{k_1}[\mathrm{I}]+\dfrac{k_{-2}}{k_{-3}k_4}\left(k_{-3}+k_{-4}+\dfrac{k_{-4}k_5}{k_{-1}}\right)\right]}{\text{denominator}}$$

where the denominator of the right-hand expressions also equals $[E]_t$ and is the sum of all the numerators. It can be shown that the steady-state velocity equation simplifies to the same *form* as the rapid equilibrium velocity equation when the numerical value of the bracketed part of each equation is the same. (The bracketed terms then cancel out when the expressions for the various species are substituted into equation IV-51.) The values of the square bracketed terms are identical when:

$$\left(1+\frac{k_5}{k_{-1}}\right)=\left(1+\frac{k_6}{k_{-4}}\right)$$

This can occur under the conditions described below.

(*a*) $k_6/k_{-4}=k_5/k_{-1}$: In this situation, the inhibitor has no effect on the *ratio* of the forward to reverse rate constants for the breakdown of the ES and ESI complexes, although the absolute values of k_6 and k_5 (or k_{-4} and k_{-1}) may be quite different. Under this condition, it is also true that $k_{-6}/k_4 = k_{-5}/k_1$.

(*b*) $k_5=k_{-1}$, $k_6=k_{-4}$: In this situation, the rate constant for the breakdown of ES to E+P equals the rate constant for the breakdown of ES back to E+S. Similarly, the rate constant for the reactions ESI→EI+S and ESI→EI+P are equal. The absolute values of k_5 and k_6 may be quite different.

(*c*) $k_5=k_6$, $k_{-4}=k_{-1}$: In this situation, the inhibitor has no effect on the catalytic activity of the ES complex. Consequently, V_{max} is unaffected and I behaves as a partial competitive inhibitor. However, as shown below, K_m and αK_m are not equal to K_S and αK_S unless k_5 and k_6 are very small compared to k_{-1} and k_{-4}, respectively. Although we are considering I as an inhibitor, the same reasoning can be applied if I is an activator. We can also see that if $k_3=k_4=0$, then no ESI complex forms and I behaves as a pure competitive inhibitor. As shown in Chapter Three, the steady-state and rapid equilibrium forms of the velocity equation for pure competitive inhibition are the same (although $K_m \neq K_S$).

Now let us examine the velocity equation under the conditions described above. When the terms in the square brackets are canceled, we obtain for $v=k_5[ES]+k_6[ESI]$:

$$v=\frac{k_5k_1k_{-3}k_4[S][E]_t+k_6k_1k_3k_4[S][I][E]_t}{k_{-1}k_{-3}k_4\left(1+\dfrac{k_5}{k_{-1}}\right)+k_1k_{-3}k_4[S]+k_1k_3k_{-4}\left(1+\dfrac{k_5}{k_{-1}}\right)[I]+k_1k_3k_4[S][I]}$$

Dividing by the constant term in the denominator and cancelling terms:

$$v = \frac{\dfrac{k_5 k_1[\mathrm{S}][\mathrm{E}]_t}{k_{-1}\left(1+\dfrac{k_5}{k_{-1}}\right)} + \dfrac{k_6 k_{-1} k_3[\mathrm{S}][\mathrm{I}][\mathrm{E}]_t}{k_{-1}k_{-3}\left(1+\dfrac{k_5}{k_{-1}}\right)}}{1 + \dfrac{k_1[\mathrm{S}]}{k_{-1}\left(1+\dfrac{k_5}{k_{-1}}\right)} + \dfrac{k_1 k_3 k_{-4}[\mathrm{I}]}{k_{-1}k_{-3}k_4} + \dfrac{k_1 k_3[\mathrm{S}][\mathrm{I}]}{k_{-1}k_{-3}\left(1+\dfrac{k_5}{k_{-1}}\right)}}$$

If we make the substitutions $k_6 = \beta k_5$, $V_{\max} = k_5[\mathrm{E}]_t$, $K_\mathrm{S} = k_{-1}/k_1$, $K_i = k_{-2}/k_2$, and $\alpha K_i = k_{-3}/k_3$, and rearrange, we obtain:

$$\boxed{\frac{v}{V_{\max}} = \frac{[\mathrm{S}]}{\alpha K_\mathrm{S}\left(1+\dfrac{k_5}{k_{-1}}\right)\left(\dfrac{[\mathrm{I}]+K_i}{\beta[\mathrm{I}]+\alpha K_i}\right) + [\mathrm{S}]\left(\dfrac{[\mathrm{I}]+\alpha K_i}{\beta[\mathrm{I}]+\alpha K_i}\right)}} \qquad \text{(IV-52)}$$

An additional substitution of $k_{-2}/k_2 = K_i = (k_1 k_3 k_{-4})/(k_{-1}k_{-3}k_4)$ was made in the coefficient of the [I] term in the denominator because of the rule of microscopic reversibility:

$$k_1 k_3 k_{-4} k_{-2} = k_2 k_4 k_{-3} k_{-1}$$

or

$$\alpha K_\mathrm{S} K_i = K_\mathrm{S} \alpha K_i$$

where

$$k_{-4}/k_4 = \alpha K_\mathrm{S}.$$

Note that the realtionship between the equilibrium constants must hold whether or not the system is at equilibrium. The velocity equations are identical to those derived earlier for rapid equilibrium conditions, except now the K_m value for S is no longer equivalent to K_S, but rather:

$$K_m = K_\mathrm{S}\left(1+\frac{k_5}{k_{-1}}\right) = \frac{k_{-1}}{k_1}\left(1+\frac{k_5}{k_{-1}}\right) = \frac{k_{-1}+k_5}{k_1}$$

Alberty and Bloomfield (1963) studied the effect of pH on the velocity of an enzyme-catalyzed reaction involving a single substrate. They concluded that the steady-state velocity could be described by the usual Henri-

Michaelis-Menten equation if the reactions of the enzyme involving H^+ were very rapid compared to the reactions involving substrate. By analogy, if the vertical reactions $E + I \rightleftharpoons EI$ and $ES + I \rightleftharpoons ESI$ are very rapid compared to the horizontal reactions $E + S \rightleftharpoons ES \rightleftharpoons E + P$ and $EI + S \rightleftharpoons ESI \rightleftharpoons EI + P$, then two of the four binding reactions shown in the general scheme will be at equilibrium, and the velocity equation will have the usual Henri-Michaelis-Menten form. However, the slope and $1/v$-axis intercept will be complex functions of I. The final equation can be written:

$$\boxed{\frac{v}{[E]_t} = \frac{[S]}{\left(\dfrac{d + e[I] + f[I]^2}{a + b[I] + c[I]^2}\right) + [S]\left(\dfrac{g + h[I] + i[I]^2}{a + b[I] + c[I]^2}\right)}} \qquad \text{(IV-53)}$$

where a through i represent combinations of rate constants. The equation can be derived using Cha's modification of the King-Altman method, described in detail in Chapter Nine. (See the derivation of equation IX-419.)

D. RECIPROCAL PLOT NOMENCLATURE

The description of reciprocal plots used in this book is based on the classical nomenclature of enzyme inhibitors. Thus the term competitive is used to describe systems yielding reciprocal plots that intersect on the $1/v$-axis. The term noncompetitive refers to systems yielding reciprocal plots that intersect on the $1/[S]$-axis. "Uncompetitive" refers to systems yielding parallel reciprocal plots, while a "mixed-type" system yields reciprocal plots that do not intersect on either axis. Most of the "mixed-type" systems described in this chapter yield reciprocal plots that intersect to the left of the $1/v$-axis, above or below the $1/[S]$-axis depending on the relative values of the dissociation constants (or interaction factors). One unusual "mixed-type" system yields plots that intersect to the right of the $1/v$-axis, while another yields an "uncompetitive" pattern. In Chapter Nine we observe "mixed-type" patterns that do not intersect at a common point, although any two lines within a family of reciprocal plots intersect to the left of the $1/v$-axis, above or below the $1/[S]$-axis. The classical terminology is not perfect. An inhibitor may influence only the slope of a plot (hence yield a "competitive" pattern) without actually competing with a substrate for a common binding site. Some workers have employed a modified nomenclature to avoid implications about the mode of action of an inhibitor or ligand. For example, uncompetitive inhibition has been called anticompetitive inhibition and coupling inhibition. Reiner (1969) uses "exclusive E inhibition" to describe what has classically been called competitive inhibition. Cleland classifies as "noncompetitive" all plots that are not competitive or uncompetitive. Noncompetitive and mixed-type patterns observed for steady-state ordered systems (Chapter Nine) usually result from the combination of the inhibitor and varied substrate with different enzyme species. In fact, most noncompetitive and mixed-type patterns have nothing to do with the classical unireactant models described in Chapters Three and Four. A comparison of nomenclature is given in Table IV-1, which describes inhibition systems, but the same classical and Cleland nomenclature can be used for many other multiple ligand systems (e.g., substrate plus activator or multiple substrates). The complete Cleland nomenclature conveys the most information. For example, one of the terreactant systems described in Chapter Nine yields "mixed-type" reciprocal plots that do not intersect at a common point. The replot of *slope* versus 1/[nonvaried ligand] is parabolic, while the replot of $1/v$-axis intercept versus 1/[nonvaried ligand] is linear. In the Cleland nomenclature this system can be described as nonintersecting, slope-parabolic, intercept-linear, noncompetitive.

E. INTERACTION BETWEEN INHIBITOR AND SUBSTRATE

System E1. Inhibition by Substrate Depletion

A compound will act as an inhibitor if it reacts with the substrate to form an inactive SI complex, thereby decreasing the effective substrate concentration. No EI or ESI complexes need form. The equilibria are shown below.

$$\begin{array}{l} \mathrm{S} + \mathrm{E} \underset{}{\overset{K_S}{\rightleftharpoons}} \mathrm{ES} \overset{k_p}{\longrightarrow} \mathrm{E} + \mathrm{P} \\ + \\ \mathrm{I} \\ K_0 \updownarrow \\ \mathrm{SI} \end{array}$$

The velocity equation is still given by:

$$\frac{v}{V_{\max}} = \frac{[\mathrm{S}]}{K_S + [\mathrm{S}]} \qquad \text{(IV-54)}$$

Table IV-1 Reciprocal Plot Nomenclature for Inhibitors Described in Chapter IV

Classical	Cleland	Reiner	Other	Characteristics
Pure competitive	Simple intersecting linear competitive	Complete exclusive E		Family of reciprocal plots intersect on the $1/v$-axis; slope replot is linear
Partial competitive	Simple intersecting hyperbolic competitive	Partial exclusive E		Family of reciprocal plots intersect on the $1/v$-axis; slope replot is hyperbolic
Pure noncompetitive	Simple intersecting (slope and intercept linear) noncompetitive	Complete inclusive		Family of reciprocal plots intersect on the 1/[S]-axis; slope and $1/v$-axis intercept replots are linear
Partial noncompetitive	Simple intersecting (slope and intercept hyperbolic) noncompetitive	Partial inclusive		Family of reciprocal plots intersect on the 1/[S]-axis; slope and $1/v$-axis intercept replots are hyperbolic
Uncompetitive	Linear uncompetitive	Exclusive C	Coupling; anti-competitive	Family of reciprocal plots are parallel; $1/v$-axis intercept replot is linear
Mixed-type ($\beta=0$)	Intersecting (slope and intercept linear) noncompetitive			Family of reciprocal plots intersect above or below the 1/[S]-axis, to the left of the $1/v$-axis; slope and $1/v$-axis intercept replots are linear
Mixed-type ($\beta\neq0$)	Intersecting (slope and intercept hyperbolic) noncompetitive			Family of reciprocal plots intersect above or below the 1/[S]-axis, usually to the left of the $1/v$-axis; slope and $1/v$-axis intercept replots are hyperbolic
Mixed-type ($\alpha=\beta$)	Hyperbolic uncompetitive			Family of reciprocal plots are parallel; $1/v$-axis intercept replot is hyperbolic

where [S] represents the concentration of *free* substrate available for combination with the enzyme. The [S] can be calculated for any given *total* inhibitor and *total* substrate concentrations, $[S]_t$ and $[I]_t$, as shown below.

$$K_0 = \frac{[S][I]}{[SI]}$$

The mass balance equations for S and I are:

$$[S]_t = [S] + [SI] \qquad \text{or} \qquad [SI] = [S]_t - [S]$$

$$[I]_t = [I] + [SI] \qquad \text{or} \qquad [I] = [I]_t - [SI]$$

The mass balance equations assume that the concentration of S complexed with the enzyme is negligible.

Substituting for [I] and [SI] in the expression for K_0:

$$K_0 = \frac{[S]([I]_t - [SI])}{[S]_t - [S]} = \frac{[S][I]_t - [S]([S]_t - [S])}{[S]_t - [S]}$$

$$= \frac{[S][I]_t - [S][S]_t + [S]^2}{[S]_t - [S]}$$

$$K_0[S]_t - K_0[S] = [S][I]_t - [S][S]_t + [S]^2$$

$$[S]^2 + ([I]_t - [S]_t + K_0)[S] - K_0[S]_t = 0$$

$$[S] = \frac{-([I]_t - [S]_t + K_0) \pm \sqrt{([I]_t - [S]_t + K_0)^2 + 4K_0[S]_t}}{2}$$

The negative solution of the square root can be ignored because [S] must be a positive number.

$$\boxed{[S] = \frac{\sqrt{([I]_t - [S]_t + K_0)^2 + 4K_0[S]_t} - ([I]_t - [S]_t + K_0)}{2}} \qquad \text{(IV-55)}$$

The expression for [S] may be substituted into the velocity equation IV-54.

The velocity curve is sigmoidal as shown in Figure IV-19 for a system where $K_0 = 0.1$ mM, $K_S = 1$ mM, and $[I]_t = 1$ mM. Because K_0 is relatively small in this case, almost all the inhibitor is complexed as SI when $[S]_t > [I]_t$. For example, when $[S]_t = 2$ mM, [SI] is about 0.92 mM. Therefore, [S] is about 1.08 mM, and since $K_S = 1$ mM, v is approximately 0.5 V_{max}. The v versus $[I]_t$ curve (Fig. IV-20) is also sigmoidal.

System E2. An SI Complex Is the True Inhibitor

The true inhibitor might not be free I, but rather an SI complex that combines with the enzyme in competition with free S.

$$
\begin{array}{ccccl}
E & + & S & \overset{K_S}{\rightleftharpoons} & ES \xrightarrow{k_p} E + P \\
 & & + & & \\
 & & I & & \\
 & & K_0 \updownarrows & & \\
E & + & SI & \overset{K_{SI}}{\rightleftharpoons} & ESI
\end{array}
$$

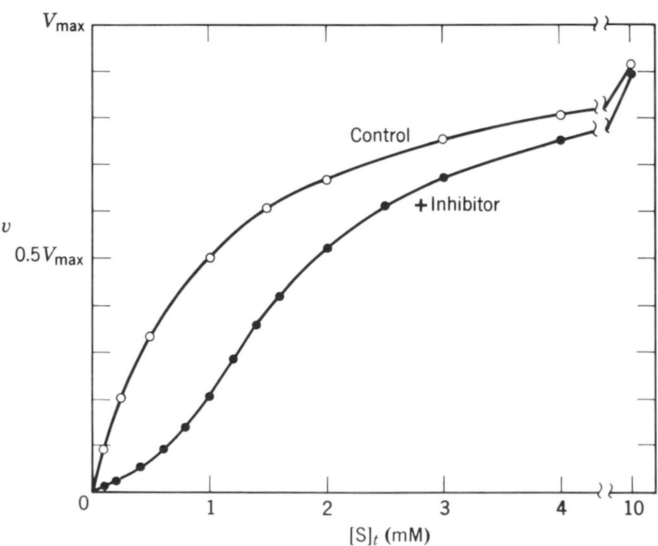

Fig. IV-19. The v versus $[S]_t$ plot in the presence of a fixed concentration of a compound, I, that combines with the substrate to produce an inactive SI complex; $K_0 = 0.1$mM, $K_S = 1$mM, $[I]_t = 1$mM.

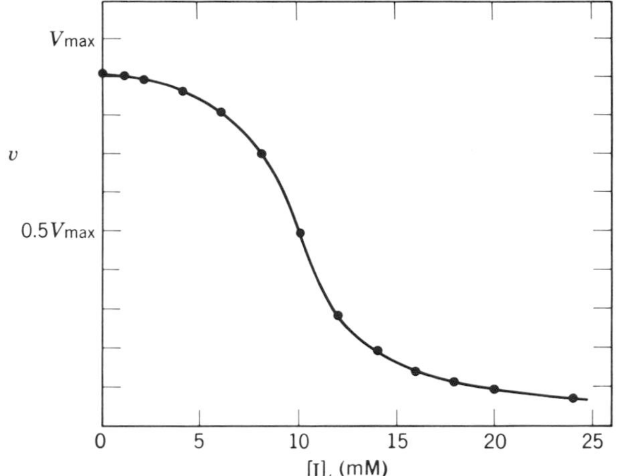

Fig. IV-20. The v versus $[I]_t$ plot at a fixed substrate concentration; I combines with S forming an inactive SI complex. $K_0 = 0.1\text{mM}$, $K_S = 1\text{mM}$, $[S]_t = 10\text{mM}$.

If the enzyme-bound substrate is accessible to the inhibitor, ESI might also form by the reaction $ES + I \rightleftharpoons ESI$.

The velocity equation in terms of *free* [S] and [SI] is:

$$\frac{v}{V_{max}} = \frac{\frac{[S]}{K_S}}{1 + \frac{[S]}{K_S} + \frac{[SI]}{K_{SI}}} \qquad \text{(IV-56)}$$

or

$$\frac{v}{V_{max}} = \frac{[S]}{K_S\left(1 + \frac{[SI]}{K_{SI}}\right) + [S]} \qquad \text{(IV-57)}$$

In terms of *free* [S] and *free* [I], the velocity equation becomes:

$$\frac{v}{V_{max}} = \frac{[S]}{K_S\left(1 + \frac{[S][I]}{K_0 K_{SI}}\right) + [S]} \qquad \text{(IV-58)}$$

where $[S][I]/K_0 = [SI]$.

Equations IV-56 through IV-58 are not too useful, since the concentrations of *free* SI or *free* I must be recalculated for each value of [S]; that is, [SI] and [I] do not remain constant as [S] varies. If the substrate concentration is not significantly reduced by SI formation and the concentration of enzyme-bound S and SI is very small compared to *free* S and *free* SI, then the concentration of the true inhibitor, SI, can be expressed in terms of *total* inhibitor and substrate concentrations, $[\mathrm{I}]_t$ and $[\mathrm{S}]_t$, as shown below.

$$K_0 = \frac{[\mathrm{I}][\mathrm{S}]}{[\mathrm{SI}]}, \qquad [\mathrm{S}] \cong [\mathrm{S}]_t, \qquad [\mathrm{I}] = [\mathrm{I}]_t - [\mathrm{SI}]$$

$$\therefore \quad K_0 = \frac{([\mathrm{I}]_t - [\mathrm{SI}])[\mathrm{S}]_t}{[\mathrm{SI}]}$$

$$[\mathrm{SI}] = \frac{[\mathrm{I}]_t[\mathrm{S}]_t}{K_0 + [\mathrm{S}]_t}$$

The more useful velocity equation in terms of $[\mathrm{S}]_t$ and $[\mathrm{I}]_t$ is obtained from equation IV-58 by substituting the expression for [SI] as given above:

$$\frac{v}{V_{\max}} = \frac{[\mathrm{S}]_t}{K_{\mathrm{S}}\left(1 + \dfrac{[\mathrm{I}]_t[\mathrm{S}]_t}{K_{\mathrm{SI}}(K_0 + [\mathrm{S}]_t)}\right) + [\mathrm{S}]_t} \qquad \text{(IV-59)}$$

The equations assume that $[\mathrm{S}] \cong [\mathrm{S}]_t$, but unless $[\mathrm{S}]_t$ is large compared to $[\mathrm{I}]_t$, some substrate depletion may occur, especially if K_0 is relatively small. As long as substrate depletion is negligible, hyperbolic v versus [S] and v versus [I] curves are obtained and the system resembles pure competitive inhibition. (See Reiner [1969] p. 196 for a system that combines Systems E1 and E2.)

F. OTHER METHODS OF PLOTTING ENZYME KINETICS DATA

The Lineweaver-Burk reciprocal plot is the most widely used primary diagnostic plot. However, the use of the plot has been criticized on two grounds: first, equal increments of [S] that yield equally spaced points on the basic v versus [S] plot do not yield equally spaced points on the reciprocal plot. For example, relative values of [S] equal to 1, 2, 3, and so on, will yield reciprocal values that tend to cluster close to the $1/v$-axis. Thus there will be relatively few points at the high end of the $1/[S]$ scale, and it is these points that are most heavily weighted in the subjective visual fitting of the line. The second, and more important, criticism is that small errors in the determination of v are magnified when reciprocals are taken. Errors in the determination of v are most likely to be significant at low substrate concentrations (and low values of v). One or two "bad" points at high $1/v$-$1/[S]$ values can introduce a marked error to the slope of the plot.

The first criticism is dealt with simply by including assay points that yield equal reciprocal increments. This means that relative substrate concentrations of 1.00, 1.11, 1.25, 1.43, 1.67, 2.0, 2.5, 3.33, 5, and 10, and so on, must be used. The second criticism cannot be dismissed. At best, errors in the determination of initial velocities at low substrate concentrations can be minimized. For example, if the assay involves quantitation of a radioactive product, the specific radioactivity of the substrate can be increased for the low substrate concentrations. The reliability of the assay depends on the observed count-rate of the product above background and the blank not on the absolute amount of product formed.

The Lineweaver-Burk reciprocal plot is not the only linear transformation of the basic velocity (or ligand binding) curve. Indeed, under some circumstances one of the other linear plots described below may be more suitable or may yield more reliable estimates of the kinetic constants. For example, the Hanes-Woolf plot of $[S]/v$ versus $[S]$ may be more convenient for data obtained at equally spaced increments of $[S]$. The Woolf-Augustinsson-Hofstee plot of v versus $v/[S]$ and the Eadie-Scatchard plot of $v/[S]$ versus v do not involve reciprocals of v and, consequently, may be more reliable when the error in v is significant. These latter two plots have a further advantage of calling attention to points that deviate significantly from the theoretical relationship because both plotted variables are influenced in the same direction by an error in v. Dowd and Riggs (1965) have investigated the reliability of various linear plots by programming a computer to calculate the kinetic constants from each of 500 replicate "experiments" which differed from each other only by the normally distributed error in v. The $1/v$ versus $1/[S]$ plot was found to be the least reliable whatever the assumed error in v. The $1/v$ versus $1/[S]$ plot tended to give a deceptively "good" visual fit even with unreliable points. In spite of the advantages of the alternate linear plots, they are not as popular as the $1/v$ versus $1/[S]$ plot because they cannot be easily evaluated visually (either v or $[S]$ appears on both axes). For all practical purposes any one of the linear plots will suffice if the data are good or properly weighted. If the data are scattered, it is best to revise the assay procedure (if possible) to minimize errors rather than rely on a particular plot that seemingly hides the scatter. The linear plots described below have been variously known by the names of the investigators who originally proposed them, or rediscovered or strongly advocated them. In fact, the Lineweaver-Burk plot was actually suggested by Woolf two years before Lineweaver and Burk's publication.

The Hanes-Woolf Plot: $[S]/v$ Versus $[S]$

The Lineweaver-Burk equation may be rearranged to yield the linear equation for the Hanes-Woolf plot:

$$\frac{1}{v} = \frac{K_S}{V_{max}} \frac{1}{[S]} + \frac{1}{V_{max}}$$

Multiplying both sides of the equation by [S]:

$$\frac{[S]}{v} = \frac{[S]K_S}{V_{max}} \frac{1}{[S]} + \frac{[S]}{V_{max}}$$

or

$$\boxed{\frac{[S]}{v} = \frac{1}{V_{max}}[S] + \frac{K_S}{V_{max}}} \qquad \text{(IV-60)}$$

Thus a plot of $[S]/v$ versus $[S]$ is linear with a slope of $1/V_{max}$. The intercept on the $[S]/v$-axis gives K_S/V_{max}. When $[S]/v=0$, the intercept on the [S]-axis gives $-K_S$. As usual, care should be exercised in choosing the range of substrate concentration. If the substrate concentration range is very low compared to K_S, the plot will be near horizontal. If the substrate range is very high compared to K_S, the plot will intersect the axes very close to the origin. Figure IV-21 shows the effect of different types of inhibitors on the characteristics of the plot. The nature of the inhibition (linear or hyperbolic) and the values of K_i, α, and β can be determined from replots of the appropriate functions. For example, the replot of the $[S]/v$-axis intercepts versus the corresponding [I] provides the same information as the replot of the slopes of the Lineweaver-Burk reciprocal plot versus [I]. Similarly, the replot of the slopes of the Hanes-Woolf plot versus the corresponding [I] is analogous to the replot of $1/v$-axis intercepts (of the Lineweaver-Burk plot) versus [I].

The Woolf-Augustinsson-Hofstee Plot: v Versus $v/[S]$

Another linear form is obtained by rearranging the basic velocity equation as shown below.

$$v = \frac{V_{max}[S]}{K_S + [S]}$$

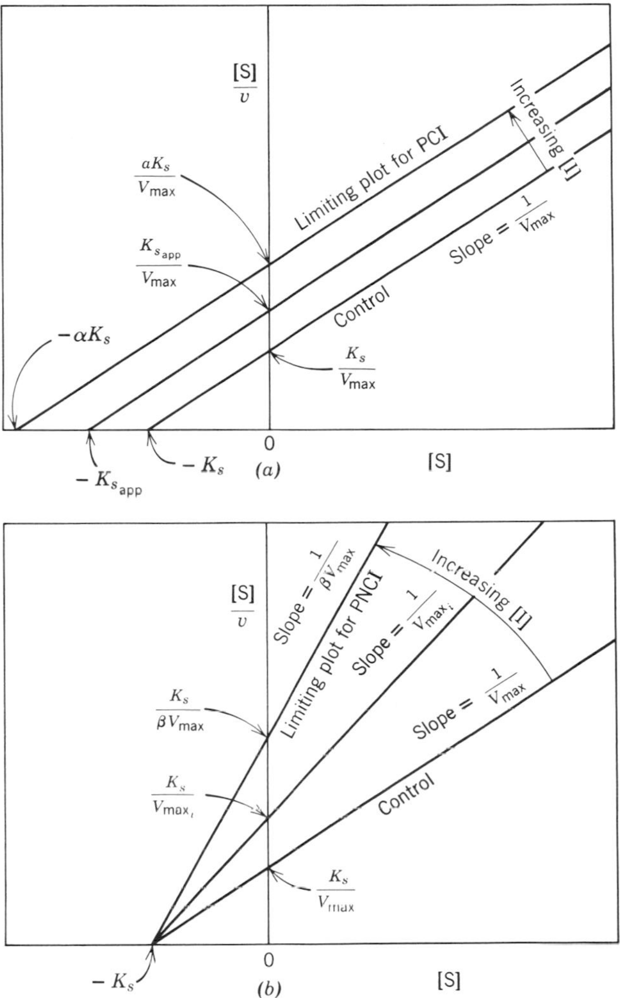

Fig. IV-21. Hanes-Woolf Plot: [S]/v versus [S]. (*a*) Competitive inhibition. (*b*) Noncompetitive inhibition.

$$\frac{[S]}{v} = \frac{1}{V_{\mathbf{max}_i}}[S] + \frac{K_{S_{app}}}{V_{\mathbf{max}_i}}$$

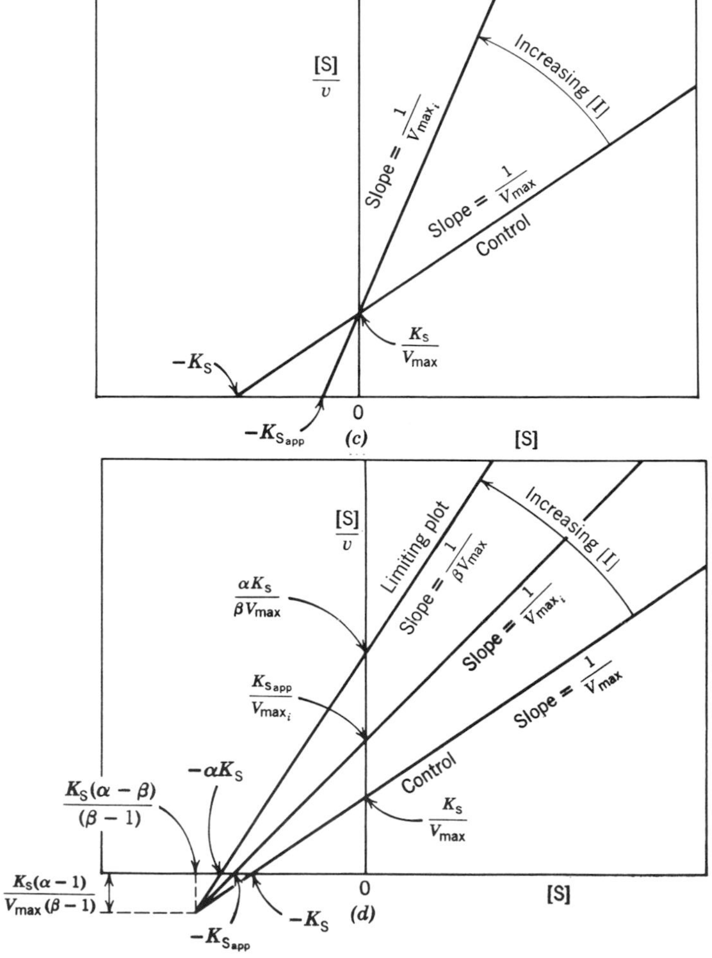

Fig. IV-21. Hanes-Woolf Plot: $[S]/v$ versus $[S]$. (*Cont.*) (*c*) Uncompetitive inhibition. (*d*) Mixed-type inhibition ($\alpha > 1, 0 < \beta < 1$).

Dividing numerator and denominator by [S]:

$$v = \frac{V_{max}}{\dfrac{K_S}{[S]} + 1}, \qquad V_{max} = \frac{vK_S}{[S]} + v, \qquad v = V_{max} - \frac{vK_S}{[S]}$$

or

$$\boxed{v = -K_S \frac{v}{[S]} + V_{max}} \qquad \text{(IV-61)}$$

The plot of v versus $v/[S]$ is linear with a slope of $-K_S$. The intercept on the v-axis gives V_{max}. When $v=0$, the intercept on the $v/[S]$-axis gives V_{max}/K_S. If the substrate concentration range is very low compared to K_S, the plot will have an extremely steep slope, approaching a vertical line that intersects the horizontal axis at V_{max}/K_S (i.e., the first-order rate constant for the reaction). If the substrate concentration range is very high compared to K_S, the plot will be nearly horizontal at a height of V_{max} above the $v/[S]$-axis. Figure IV-22 shows the effects of different types of inhibitors.

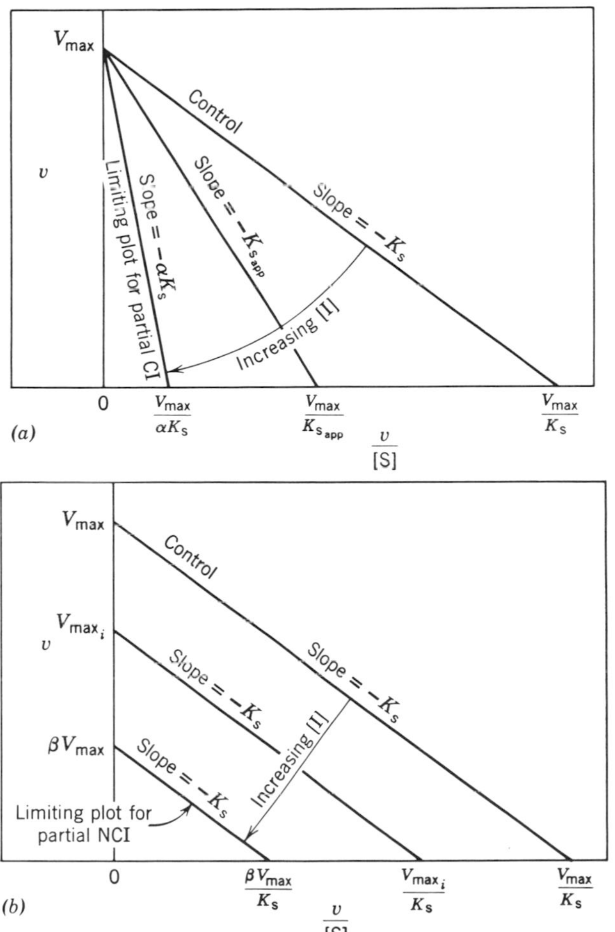

Fig. IV-22. Woolf-Augustinsson-Hofstee Plot: v versus $v/[S]$. (*a*) Competitive inhibition. (*b*) Noncompetitive inhibition.

$$v = -K_{S_{app}} \frac{v}{[S]} + V_{max_i}$$

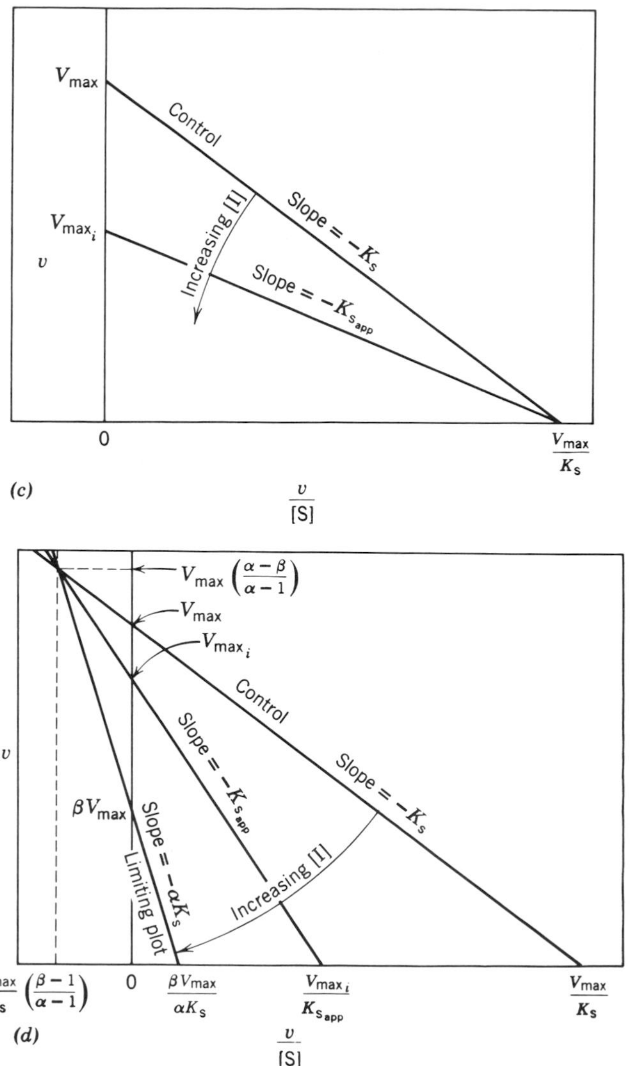

Fig. IV-22. Woolf-Augustinsson-Hofstee Plot: v versus $v/[S]$. (*Cont.*) (*c*) Uncompetitive inhibition. (*d*) Mixed-type inhibition ($\alpha > 1, 0 < \beta < 1$).

The Eadie-Scatchard Plot: $v/[S]$ Versus v

If the Henri-Michaelis-Menten equation is rearranged we obtain:

$$V_{max} = \frac{vK_S}{[S]} + v$$

Dividing both sides of the equation by K_S:

$$\frac{V_{max}}{K_S} = \frac{v}{[S]} + \frac{v}{K_S}$$

or

$$\boxed{\frac{v}{[S]} = -\frac{1}{K_S}v + \frac{V_{max}}{K_S}} \tag{IV-62}$$

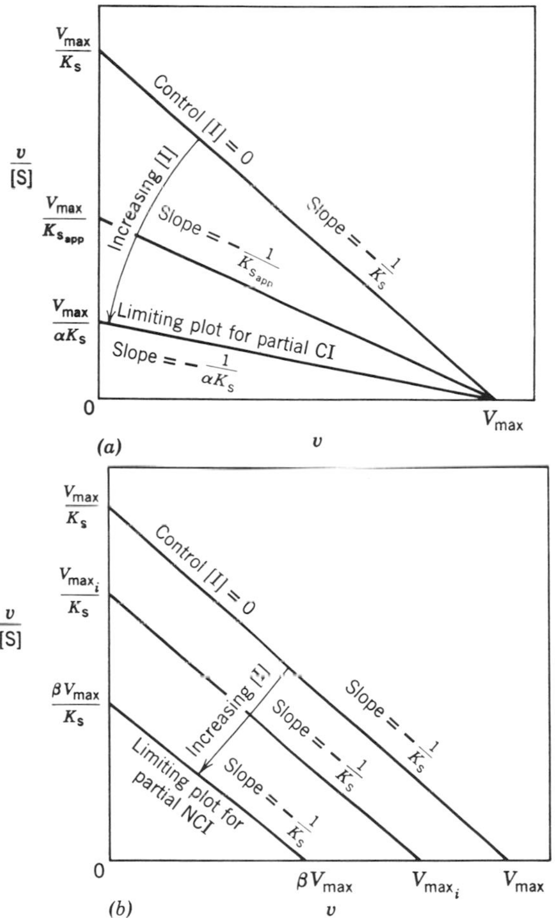

Fig. IV-23. Eadie-Scatchard plot: $v/[S]$ versus v. (*a*) Competitive inhibition. (*b*) Noncompetitive inhibition.

$$\boxed{\frac{v}{[S]} = -\frac{1}{K_{S_{app}}}v + \frac{V_{max_i}}{K_S}}$$

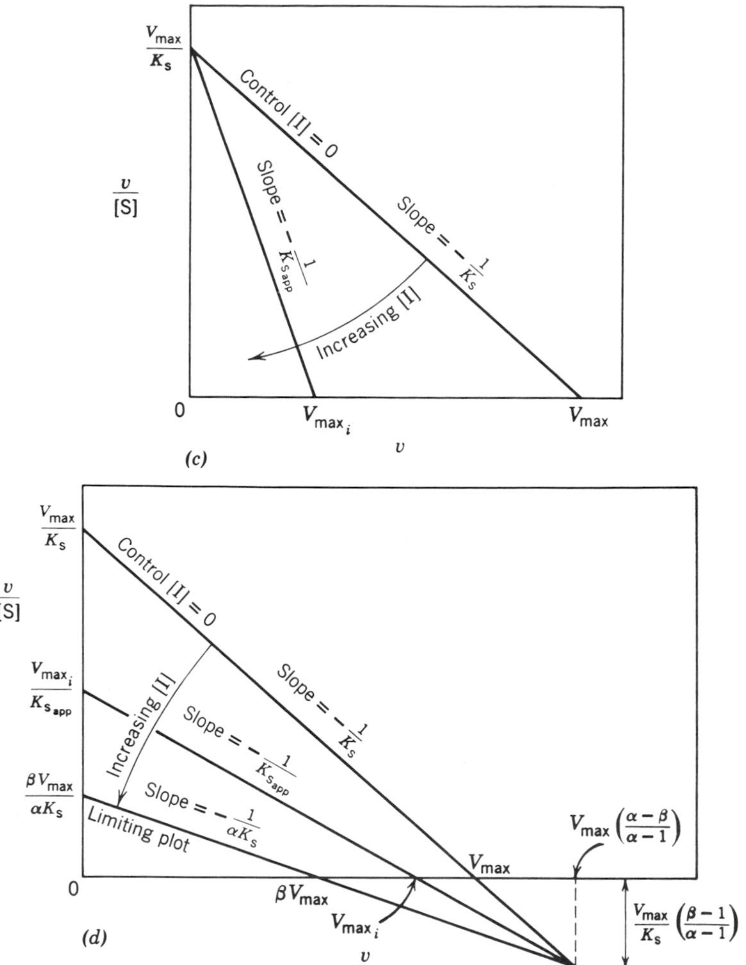

Fig. IV-23. Eadie-Scatchard plot: v/[S] versus v. (*Cont.*) (*c*) Uncompetitive inhibition. (*d*) Mixed-type inhibition ($\alpha > 1, 0 < \beta < 1$).

Thus a plot of $v/[S]$ versus v is linear with a slope of $-1/K_S$ and an intercept of V_{max}/K_S on the $v/[S]$-axis. When $v/[S]=0$, the intercept on the v-axis gives V_{max}. The effects of different types of inhibitors are shown in Figure IV-23.

The $v/[S]$ versus v and the v versus $v/[S]$ plot are more useful than the $1/v$ versus $1/[S]$ plot in detecting the presence of multiple enzymes that catalyze the same reaction. Figure IV-24 shows the $1/v$ versus $1/[S]$ plot for a mixture of two enzymes with the same V_{max} values but with K_S values that differ by a factor of 10. Figure IV-25 shows the $v/[S]$ versus v plot for the same mixture over the same range of substrate concentrations. The $1/v$

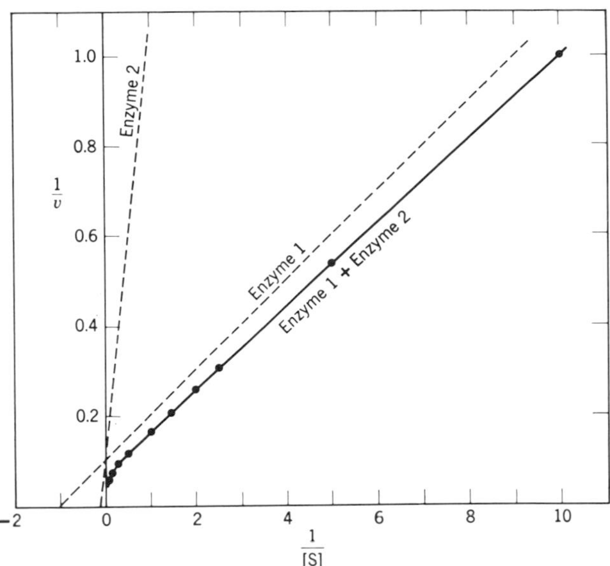

Fig. IV-24. Lineweaver-Burk Plot ($1/v$ versus $1/[S]$) in the presence of two enzymes catalyzing the same reaction; $V_{\max_1}=10$, $K_{S_1}=1$, $V_{\max_2}=10$, $K_{S_2}=10$. The [S] range plotted is 0.1 to 25.

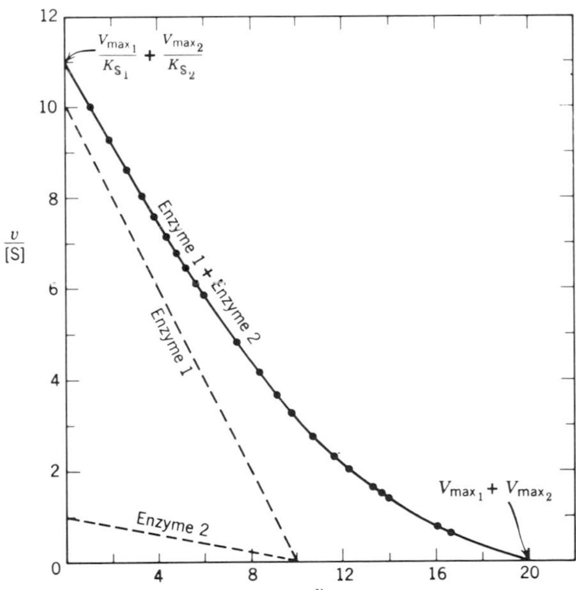

Fig. IV-25. Eadie-Scatchard Plot ($v/[S]$ versus v) in the presence of two enzymes catalyzing the same reaction; $V_{\max_1}=10$, $K_{S_1}=1$, $V_{\max_2}=10$, $K_{S_2}=10$. The [S] range plotted is 0.1 to 25.

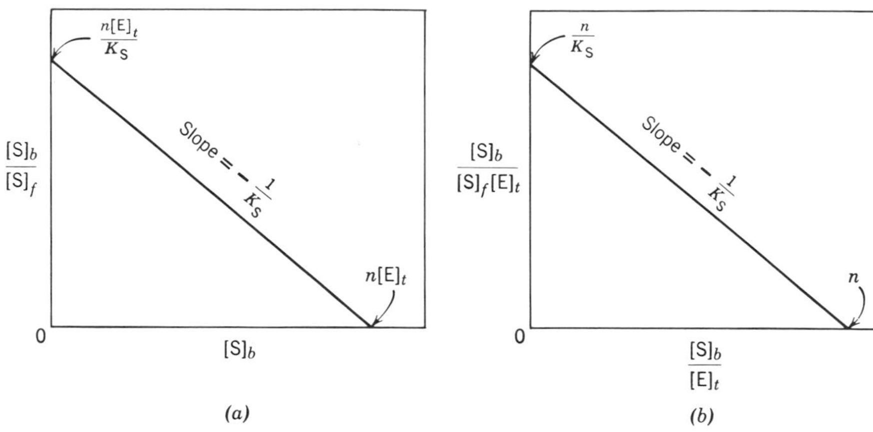

Fig. IV-26. Scatchard Plot of equilibrium substrate binding data. (*a*) Plot of $[S]_b/[S]_f$ versus $[S]_b$. (*b*) Plot of $[S]_b/[S]_f[E]_t$ versus $[S]_b/[E]_t$.

versus $1/[S]$ curve is quite linear over most of the range studied. Only at very low $1/[S]$ values is there a marked deviation from linearity. With actual experimental data, the deviation might be missed, and we would conclude that only a single enzyme is present. The $v/[S]$ versus v plot weights the points more evenly and the deviation from linearity is seen throughout most of the v range plotted (see also Figures II-25 and II-26).

The Scatchard Plot for Equilibrium Binding Data: $[S]_b/[S]_f$ Versus $[S]_b$ or $[S]_b/[S]_f[E]_t$ Versus $[S]_b/[E]_t$

A plot very similar to that shown in Figure IV-25 has been used to study ligand binding by equilibrium dialysis or other methods. For this purpose, equation IV-62 can be modified by substituting [ES] or $[S]_b$ for v and $n[E]_t$ for V_{max} since:

$$v \propto [S]_b \qquad \text{and} \qquad V_{max} \propto n[E]_t$$

where $[S]_b = [ES] =$ the concentration of bound ligand
$=$ the concentration of occupied sites
$[E]_t =$ the total enzyme concentration
$n =$ the number of identical and independent ligand binding sites per molecule of enzyme
$n[E]_t =$ the total concentration of ligand binding sites.

The equation becomes:

$$\boxed{\frac{[S]_b}{[S]_f} = -\frac{1}{K_S}[S]_b + \frac{n[E]_t}{K_S}} \qquad \text{(IV-63)}$$

where K_S is the intrinsic substrate dissociation constant of a site. The $[S]_f$ term in the equation above stands for the concentration of *free* substrate. The $[S]_b/[S]_f$ represents the ratio of *bound* to *free* substrate. In most *in vitro* initial velocity studies, the concentration of enzyme is many orders of magnitude lower than the concentration of substrate. Consequently, the formation of ES does not significantly decrease the concentration of S, and it is safe to assume that the concentration of *free* substrate, $[S]_f$, is the same as the total added substrate concentration, $[S]_t$. This is not true for equilibrium binding studies, where relatively high enzyme concentrations are used, and a relatively large proportion of the added substrate is bound (see equation II-54 and the accompanying discussion).

A plot of the ratio of bound to free ligand versus the concentration of bound ligand is linear with a slope of $-1/K_S$ (Fig. IV-26*a*). The vertical axis intercept gives $n[E]_t/K_S$. The horizontal axis intercept gives $n[E]_t$ (i.e., the total concentration of ligand binding sites). If the molar concentration of enzyme is known, the data can be plotted according to equation IV-64 (obtained by dividing equation IV-63 by $[E]_t$).

$$\boxed{\frac{[S]_b}{[S]_f[E]_t} = -\frac{1}{K_S}\frac{[S]_b}{[E]_t} + \frac{n}{K_S}} \qquad \text{(IV-64)}$$

A plot of $[S]_b/[S]_f[E]_t$ (i.e., the moles of ligand bound per mole of enzyme divided by the concentration of *free* substrate) versus $[S]_b/[E]_t$ (i.e., moles of ligand bound per mole of enzyme) is linear with a slope of $-1/K_S$ (Fig. IV-26*b*). The intercept on the vertical axis gives n/K_S. The intercept on the horizontal axis gives n, the number of ligand binding sites per molecule of enzyme. If the enzyme possesses multiple independent binding sites with different intrinsic affinities for the ligand, the plot will be curved as shown in Figure IV-25. Of course, the data could just as easily be plotted in reciprocal form analogous to $1/v$ versus $1/[S]$ (Klotz plot):

$$\boxed{\frac{1}{[S]_b} = \frac{K_S}{n[E]_t}\frac{1}{[S]_f} + \frac{1}{n[E]_t}} \qquad \text{(IV-65)}$$

Or, multiplying by $[E]_t$:

$$\boxed{\frac{[E]_t}{[S]_b} = \frac{K_S}{n}\frac{1}{[S]_f} + \frac{1}{n}} \qquad \text{(IV-66)}$$

To measure ligand binding by equilibrium dialysis, $[E]_t$ should be in the region of K_S. If $[E]_t$ is much lower than K_S, it may be impossible to measure $[S]_b$ accurately, since $[S]_b$ is calculated as the difference between $[S]_b + [S]_f$ (in the dialysis compartment containing the enzyme) and $[S]_f$ (in the dialysis compartment without the enzyme).

Just as it is possible to saturate an enzyme with substrate, it is possible to saturate a substrate with enzyme. The reverse technique is impractical for initial velocity studies, but it can be done when enzyme-substrate binding is measured by equilibrium dialysis or some other suitable technique provided no catalytic reaction takes place. This condition is realized when (*a*) S is only one of several cosubstrates, all of which must be present before any reaction occurs and (*b*) S can bind independently to the enzyme (e.g., S binds first in an obligatory ordered sequence, or S and the other substrates bind in a random order). The binding equations are symmetrical to those derived earlier. For example, the Eadie-Scatchard equation is:

$$\boxed{\frac{[E]_b}{n[E]_f} = -\frac{1}{K_S}[E]_b + \frac{[S]_t}{K_S}} \qquad \text{(IV-67)}$$

where $[E]_b$ = the concentration of occupied enzyme sites
= the concentration of bound substrate
$n[E]_f$ = the concentration of free binding sites
$[S]_t$ = the total concentration of substrate

Isotope Competition in Equilibrium Ligand Binding

Unlabeled substrate will appear to act as a competitive inhibitor with respect to the binding of labeled substrate provided that the concentration of bound substrate, [ES], is measured in terms of cpm/ml (or M, if the specific activity of the labeled S is not corrected for isotope dilution). This apparent competition can be used to determine unknown concentrations of unlabeled S in solutions free of real inhibitors. All we need know is K_S, $[E]_t$, and $[S^*]_0$, the initial concentration of labeled substrate. In binding measurements, equations III-17a and III-17b cannot be used to calculate the concentration of unlabeled S. These equations give *free* [S] after equilibration, while we are interested in the initial concentration of unlabeled S added to the dialysis chamber. An appropriate equation can be derived if we consider that the specific activity of the free substrate must equal the specific activity of the bound substrate at all times (even after adding an unknown concentration of unlabeled S). From now on, the free S concentration will be designated simply as [S]; the concentration of bound S will be designated as [ES]. We can write:

$$\frac{(\text{cpm})_S}{[S]} = \frac{(\text{cpm})_{ES}}{[ES]} \tag{IV-68}$$

The $(\text{cpm})_S$ might be cpm in 10 μl of solution from the compartment without the protein, and [S] = μmoles of S in 10 μl of this solution; $(\text{cpm})_{ES}$ would then mean cpm of bound S in 10 μl of solution from the compartment containing the protein. The $(\text{cpm})_{ES}$ is calculated as $(\text{cpm})_t - (\text{cpm})_S$, where $(\text{cpm})_t$ is the total cpm in 10 μl of solution from the compartment containing the protein. Rearranging equation IV-68:

$$\frac{(\text{cpm})_S}{(\text{cpm})_{ES}}[ES] = [S] \qquad \text{or} \qquad r[ES] = [S]$$

where r is the ratio $(\text{cpm})_S/(\text{cpm})_{ES}$. Substituting into the equilibrium equation:

$$\frac{[S][E]}{[ES]} = K_S = \frac{r[ES]([E]_t - [ES])}{[ES]} = r([E]_t - [ES])$$

Solving for [ES]:

$$\therefore \quad [ES] = [E]_t - \frac{K_S}{r} \tag{IV-69}$$

Since $[E] = [E]_t - [ES]$, we can write:

$$[E] = [E]_t - [E]_t + \frac{K_S}{r} \qquad \text{or} \qquad [E] = \frac{K_S}{r} \tag{IV-70}$$

Substituting for [E] and [ES] in the equilibrium expression:

$$\frac{[S]\left(\dfrac{K_S}{r}\right)}{[E]_t - \dfrac{K_S}{r}} = K_S$$

Solving for [S]:

$$[S] = r[E]_t - K_S \tag{IV-71}$$

We now have expressions for [S], [E], and [ES] in terms of $[E]_t$, K_S, and r.

If the volumes of the two compartments of the dialysis chamber are equal, then:

$$[\mathrm{S}]_0 - [\mathrm{S}] = 0.5[\mathrm{ES}] \qquad \text{or} \qquad [\mathrm{S}]_0 = [\mathrm{S}] + 0.5[\mathrm{ES}] \tag{IV-72}$$

That is, the decrease in the concentration of free S equals one-half the concentration of bound S (the binding takes place in a volume of n ml but S is removed from a volume of $2n$ ml). (The symbol $[\mathrm{S}]_0$ is used here, rather than $[\mathrm{S}]_t$, because in equilibrium ligand binding studies where the bound and free substrate do not occupy the same volume, $[\mathrm{S}]+[\mathrm{ES}]$ does not equal $[\mathrm{S}]_0$.) Substituting for [S] and [ES]:

$$[\mathrm{S}]_0 = r[\mathrm{E}]_t - K_\mathrm{S} + 0.5[\mathrm{E}]_t - 0.5\frac{K_\mathrm{S}}{r}$$

or

$$\boxed{[\mathrm{S}]_0 = \left([\mathrm{E}]_t - \frac{K_\mathrm{S}}{r}\right)(r + 0.5)} \tag{IV-73}$$

Equation IV-72 gives the initial concentration of substrate present in the dialysis chamber (the same in both compartments at zero time, or the equilibrium concentration if no protein is present). All we need to know is K_S, $[\mathrm{E}]_t$, and the ratio, r. The $[\mathrm{S}]_0$ calculated from equation IV-73 will be greater than the known, experimental $[\mathrm{S}^*]_0$ by an amount equal to the concentration of added, unlabeled S.

$$\therefore \quad [\mathrm{S}]_{\mathrm{added}} = [\mathrm{S}]_{0_{\mathrm{calc}}} - [\mathrm{S}^*]_0 \tag{IV-74}$$

If no unlabeled S is added, then $[\mathrm{S}]_0$ calculated from $[\mathrm{E}]_t$, K_S, and r by equation IV-73 should be identical to the known $[\mathrm{S}^*]_0$.

An expression for $[\mathrm{S}]_0$ can be derived in terms of a different ratio, R, where $\mathrm{R} = (\mathrm{cpm})_t/(\mathrm{cpm})_\mathrm{S}$, that is, the equilibrium ratio of cpm in 10 μl of solution from the compartment containing the protein to cpm in 10 μl of solution from the compartment without the protein:

$$r = \frac{(\mathrm{cpm})_\mathrm{S}}{(\mathrm{cpm})_\mathrm{ES}} = \frac{(\mathrm{cpm})_\mathrm{S}}{(\mathrm{cpm})_t - (\mathrm{cpm})_\mathrm{S}}$$

$$\frac{1}{r} = \frac{(\mathrm{cpm})_t - (\mathrm{cpm})_\mathrm{S}}{(\mathrm{cpm})_\mathrm{S}} = \frac{(\mathrm{cpm})_t}{(\mathrm{cpm})_\mathrm{S}} - 1 = \mathrm{R} - 1$$

Substituting for $1/r$ in equation IV-73:

$$\boxed{[\mathrm{S}]_0=0.5\Big([\mathrm{E}]_t-K_\mathrm{S}(\mathrm{R}-1)\Big)\frac{\mathrm{R}+1}{\mathrm{R}-1}} \tag{IV-75}$$

If the protein and the substrate are in a single compartment, the 0.5 in equations IV-73 and IV-75 becomes 1.0.

Best-Belpomme and Dessen (1973) have derived several equations which can be plotted to obtain K_i and $[\mathrm{I}]_t$ when both the substrate and a competitive inhibitor are tightly bound to the protein. One of the equations is derived below in terms of dissociation constants.

$$Y_{\mathrm{S}_i}=\frac{[\mathrm{PS}]}{[\mathrm{P}]_t}=\frac{[\mathrm{PS}]}{[\mathrm{P}]+[\mathrm{PS}]+[\mathrm{PI}]}=\left(1+\frac{K_\mathrm{S}}{[\mathrm{S}]}+\frac{K_\mathrm{S}[\mathrm{I}]}{K_i[\mathrm{S}]}\right)^{-1} \tag{IV-76}$$

where Y_{S_i} is the fraction of total sites occupied by S in the presence of I.

$$[\mathrm{I}]_t=[\mathrm{I}]+[\mathrm{PI}]=[\mathrm{I}]+\frac{[\mathrm{I}]}{K_i}[\mathrm{P}]$$

$$[\mathrm{P}]=\frac{K_\mathrm{S}[\mathrm{PS}]}{[\mathrm{S}]}\quad\therefore[\mathrm{I}]_t=[\mathrm{I}]+\frac{[\mathrm{I}]K_\mathrm{S}[\mathrm{PS}]}{K_i[\mathrm{S}]}=\frac{[\mathrm{I}](K_i[\mathrm{S}]+K_\mathrm{S}[\mathrm{PS}])}{[\mathrm{S}]K_i}$$

$$\therefore\frac{[\mathrm{I}]}{[\mathrm{S}]}=\frac{[\mathrm{I}]_tK_i}{K_i[\mathrm{S}]+K_\mathrm{S}[\mathrm{PS}]}$$

Substituting for [I]/[S] in equation IV-76 and inverting:

$$\frac{1}{Y_{\mathrm{S}_i}}=1+\frac{K_\mathrm{S}}{[\mathrm{S}]}+\frac{K_\mathrm{S}[\mathrm{I}]_t}{K_i[\mathrm{S}]+K_\mathrm{S}[\mathrm{PS}]}$$

$$1+\frac{K_\mathrm{S}}{[\mathrm{S}]}=\frac{1}{Y_{\mathrm{S}_0}}$$

where $1/Y_{\mathrm{S}_0}$ is the reciprocal of the fraction of total sites occupied in the absence of I.

$$\therefore\left(\frac{1}{Y_{\mathrm{S}_i}}-\frac{1}{Y_{\mathrm{S}_0}}\right)=\frac{K_\mathrm{S}[\mathrm{I}]_t}{K_i[\mathrm{S}]+K_\mathrm{S}[\mathrm{PS}]} \tag{IV-77}$$

Multiplying the right-hand term of equation IV-77 by [PS] and the left-hand term by $[PS]=Y_{S_i}[P]_t$:

$$\left(\frac{Y_{S_i}[P]_t}{Y_{S_i}}-\frac{Y_{S_i}[P]_t}{Y_{S_0}}\right)=\frac{[PS]K_S[I]_t}{K_i[S]+K_S[PS]} \qquad \text{(IV-78)}$$

Inverting, separating terms, and cancelling:

$$\left[[P]_t\left(1-\frac{Y_{S_i}}{Y_{S_0}}\right)\right]^{-1}=\frac{K_i[S]}{K_S[I]_t[PS]}+\frac{1}{[I]_t} \qquad \text{(IV-79)}$$

or

$$\boxed{f=\frac{K_i}{K_S[I]_t}r+\frac{1}{[I]_t}} \qquad \text{(IV-80)}$$

The procedure is to vary $[S]_t$ in the presence of a fixed $[I]_t$. The ratio of free/bound substrate, r, is measured. The reciprocal function, f, is calculated for each $[S]_t$ and is plotted against the corresponding r. The plot is linear with a slope of $K_i/K_S[I]_t$, a vertical-axis intercept of $1/[I]_t$, and a horizontal-axis intercept of $-K_S/K_i$ for all $[I]_t$. If the inhibitor is, in fact, unlabeled S and r is expressed as cpm_S/cpm_{PS}, then the vertical-axis intercept gives $1/[S]_{t_{added}}$ and the horizontal-axis intercept is -1. The equations assume that the protein, substrate, and inhibitor are all in one compartment. Thus equation IV-80 can be used directly when binding is determined by membrane filtration. In equilibrium dialysis experiments with equal volume compartments, $[I]_0=[I]+0.5[PI]$. In this case, the intercepts become $0.5/[I]_t$ and $0.5K_S/K_i$.

REFERENCES ON PARTIAL AND MIXED-TYPE INHIBITION

General

Alberty, R. A. and Bloomfield, V., *J. Biol. Chem.* **238**, 2804 (1963).

Dixon, M. and Webb, E. C., *Enzymes*, 2nd ed., Academic Press, 1964, Ch. 8.

Freiden, C., *J. Biol. Chem.* **239**, 3522 (1964).

Reiner, J., *Behavior of Enzyme Systems*, 2nd ed., Van Nostrand-Reinhold, 1969, Ch. 8.

Segal, H. L., "The Development of Enzyme Kinetics," in *The Enzymes*, Vol. I, 2nd ed., P. D. Boyer, H. Lardy and K. Myrback, eds., Academic Press, 1959, p. 1.

Segal, H. L., Kachmar, J. F. and Boyer, P. D., "Kinetic Analysis of Enzyme Reactions," in *Enzymologia* **15**, 187 (1952).

Webb, J. L., *Enzyme and Metabolic Inhibitors*, Vol. I, Academic Press, 1963, Ch. 3 and 5.

Two-Site (Glutamine Synthetase) Model for Partial Inhibition

Stadtman, E. R., Shapiro, B. M., Kingdon, H. S., Woolfolk, C. A. and Hubbard, J. S., "Cellular Regulation of Glutamine Synthetase Activity in *Escherichia coli*," in *Advances in Enzyme Regulation*, Vol. 6, G. Weber, ed., Pergamon Press, 1968, p. 257.

Examples of Partial Inhibition

Gold, M. H., Farrand, R. J., Livoni, J. P. and Segel, I. H., *Arch. Biochem. Biophys.* **161**, 515 (1974) (Glycogen phosphorylase).

Wang, J. H., Tu, J. I. and Lo, F. M., *J. Biol. Chem.* **245**, 3115 (1970) (Glycogen phosphorylase).

Weinrib, I. and Michel, I. M., *Biochem. Biophys. Acta* **334**, 218 (1974) (Adenylate cyclase).

Dixon and Cornish-Bowden Plots for Linear Mixed-Typed Inhibition

Butterworth, P. J., *Biochim. Biophys. Acta* **289**, 251 (1972).

Cornish-Bowden, A., *Biochem. J.* **137**, 143 (1974).

Purich, D. L. and Fromm, H. J., *Biochim. Biophys. Acta* **268**, 1 (1972).

Formation of an SI Complex

Sluyterman, L. A. and Wijdenes, J., *Biochim. Biophys. Acta* **321**, 697 (1973).

REFERENCES ON METHODS OF PLOTTING ENZYME KINETICS DATA

General

Dixon, M. and Webb, E. C., *Enzymes*, 2nd ed., Academic Press, 1964, Ch. 8.

Dowd, J. E. and Riggs, D. S., *J. Biol. Chem.* **240**, 863 (1965).

Walter, C., *Steady State Applications in Enzyme Kinetics*, The Ronald Press, 1965, Ch. 5.

Webb, J. L., *Enzyme and Metabolic Inhibitors*, Vol. I, Academic Press, 1963, Ch. 5.

References for Specific Methods

Augustinsson, K. -B., *Acta Physiol. Scand.* **15**, Suppl. 52 (1948).

Best-Belpomme, M. and Dessen, P., *Biochimie* **55**, 11 (1973).

Dixon, M., *Biochem. J.* **55**, 170 (1953).

Eadie, G. S., *J. Biol. Chem.* **146**, 85 (1942).

Eisenthal, R. and Cornish-Bowden, A., *Biochem. J.* **139**, 715 (1974).

Fajszi, Cs. and Endrenyi, L., *FEBS Letters* **44**, 240 (1974).

Hanes, C. S., *Biochem. J.* **26**, 1406 (1932).

Hofstee, B. H. J., *Science* **116**, 329 (1952).

Hofstee, B. H. J., *Nature* **184**, 1296 (1959).

Hofstee, B. H. J., *Science* **131**, 39 (1960).

Hunter, A. and Downs, C. E., *J. Biol. Chem.* **157**, 427 (1945).

Klotz, I. M., Walker, F. M. and Pivan, R. B., *J. Am. Chem. Soc.* **68**, 1486 (1946).

Lineweaver, H. and Burk, D., *J. Am. Chem. Soc.* **56**, 658 (1934).

Reiner, J., *Behavior of Enzyme Systems*, 2nd ed., Van Nostrand-Reinhold, 1969, Ch. 8.

Scatchard, G., *Ann. N.Y. Acad. Sci.* **51**, 660 (1949).

Weigard, J. H., *Science* **131**, 1068 (1960).

Wilkinson, G. N., *Biochem. J.* **80**, 324 (1961).

Woolf, B., quoted by Haldane J. B. S. and K. G. Stern, in *Allegemeine Chemie der Enzyme*, Verlag von Steinkopff, 1932, p. 119.

See also references on Tightly Bound Inhibitors at the end of Chapter Three.

Plots of the Integrated Henri-Michaelis-Menten Equation

Alberty, R. A., "The Rate Equation for an Enzymic Reaction," in *The Enzymes*, Vol. 1, 2nd ed., P. D. Boyer, H. Lardy and K. Myrback, eds., Academic Press, 1959, p. 143.

Huang, H. T. and Niemann, C., *J. Am. Chem. Soc.* **73**, 1541 (1951).

Jennings, R. R. and Niemann, C., *J. Am. Chem. Soc.* **77**, 5432 (1955).

Lee, H. -J. and Wilson, I. B., *Biochim. Biophys. Acta* **242**, 519 (1971).

Walker, A. C. and Schmidt, C. L. A., *Arch. Biochem. Biophys.* **5**, 445 (1944).

Determination of Initial Velocities from Non-Linear Reaction Curves

Algranati, I., *Biochim. Biophys. Acta* **73**, 152 (1963).

Plots for Multiple Inhibitors (See Chapter VIII)

Yagi, K. and Ozawa, T., *Biochim. Biophys. Acta* **39**, 304 (1960).

Yagi, K. and Ozawa, T., *Biochim. Biophys. Acta* **42**, 381 (1960).

Yonetani, T. and Theorell, H., *Arch. Biochem. Biophys.* **106**, 243 (1964).

Two Enzymes Acting on the Same Substrate

Spears, G., Sneyd, J. G. T. and Loten, E. G., *Biochem. J.* **125**, 1149 (1971).

Specific Methods for Multisite (Allosteric) Enzymes (See Chapter VII)

Blangy, D., Buc, H. and Monod, J., *J. Mol. Biol.* **31**, 13 (1968).

Horn, A. and Börnig, H., *FEBS Letters* **3**, 325 (1969).

Endrenyi, L., Chan, M. -S. and Wong, J. T. -F., *Can. J. Biochem.* **49**, 581 (1971).

Frieden, C., *J. Biol. Chem.* **242**, 4045 (1967).

Rubin, M. M. and Changeux, J. -P., *J. Mol. Biol.* **21**, 265 (1966).

CHAPTER FIVE

ENZYME ACTIVATION

Activators are modifiers that increase the velocity of an enzyme-catalyzed reaction. In this chapter, we consider two major types of activation: (*a*) nonessential activation, in which the reaction can occur in the absence of the activator; and (*b*) systems in which the true substrate is an SA complex where A is a metal ion.

A. NONESSENTIAL ACTIVATION

System A1. General Scheme for Nonessential Activation

Nonessential activation can be treated in the same manner as partial and mixed-type inhibition; however, now the changes are in the opposite direction. In general, α will be less than unity, and β will be greater than unity. Mixed-type systems, where $\alpha > 1$ and $\beta > 1$, are also possible, as are systems where $\alpha < 1$ and $\beta < 1$. In the latter case, the activator might activate at low [S] but inhibit at high [S] (see mixed-type inhibition System C5 in Chapter Four). The example described below is analogous to mixed-type inhibition System C2.

$$\begin{array}{ccccccccc}
E & + & S & \overset{K_S}{\rightleftharpoons} & ES & \overset{k_p}{\longrightarrow} & E & + & P \\
+ & & & & + & & & & \\
A & & & & A & & & & \\
{\scriptstyle K_A}\upharpoonleft\downharpoonright & & & & {\scriptstyle \alpha K_A}\upharpoonleft\downharpoonright & & & & \\
EA & + & S & \overset{\alpha K_S}{\rightleftharpoons} & ESA & \overset{\beta k_p}{\longrightarrow} & EA & + & P
\end{array}$$

The velocity equation is:

$$v = k_p[\mathrm{ES}] + \beta k_p[\mathrm{ESA}]$$

$$v = \frac{V_{max}\dfrac{[\mathrm{S}]}{K_\mathrm{S}} + \beta V_{max}\dfrac{[\mathrm{A}][\mathrm{S}]}{\alpha K_\mathrm{A} K_\mathrm{S}}}{1 + \dfrac{[\mathrm{S}]}{K_\mathrm{S}} + \dfrac{[\mathrm{A}]}{K_\mathrm{A}} + \dfrac{[\mathrm{A}][\mathrm{S}]}{\alpha K_\mathrm{A} K_\mathrm{S}}} \qquad \text{(V-1)}$$

where V_{max} represents the maximal velocity of the unactivated reaction. The maximal velocity in the presence of a saturating concentration of activator is βV_{max}. In Henri-Michaelis-Menten form the velocity equation is:

$$\frac{v}{V_{max}} = \frac{[\mathrm{S}]}{K_\mathrm{S}\dfrac{\left(1 + \dfrac{[\mathrm{A}]}{K_\mathrm{A}}\right)}{\left(1 + \dfrac{\beta[\mathrm{A}]}{\alpha K_\mathrm{A}}\right)} + [\mathrm{S}]\dfrac{\left(1 + \dfrac{[\mathrm{A}]}{\alpha K_\mathrm{A}}\right)}{\left(1 + \dfrac{\beta[\mathrm{A}]}{\alpha K_\mathrm{A}}\right)}} \qquad \text{(V-2)}$$

At any activator concentration:

$$V_{max_{app}} = \frac{V_{max}\left(1 + \dfrac{\beta[\mathrm{A}]}{\alpha K_\mathrm{A}}\right)}{\left(1 + \dfrac{[\mathrm{A}]}{\alpha K_\mathrm{A}}\right)} \qquad \text{and} \qquad K_{\mathrm{S}_{app}} = K_\mathrm{S}\frac{\left(1 + \dfrac{[\mathrm{A}]}{K_\mathrm{A}}\right)}{\left(1 + \dfrac{[\mathrm{A}]}{\alpha K_\mathrm{A}}\right)}$$

At an infinitely high activator concentration $K_{\mathrm{S}_{app}} = \alpha K_\mathrm{S}$, $V_{max_{app}} = \beta V_{max}$.

Figure V-1 shows the velocity curve for a mixed-type nonessential activator at $[\mathrm{A}] = 3K_\mathrm{A}$ where $\alpha = 0.3$, $\beta = 2.5$. The family of reciprocal plots for this type of activator is shown in Figure V-2. As [A] increases, the reciprocal plot pivots clockwise about the point of intersection with the control plot at $(\beta - 1)/K_\mathrm{S}(\alpha - \beta)$. The $1/v$-axis and $1/[\mathrm{S}]$-axis intercepts of the limiting plot (at a saturating concentration of activator) give $1/\beta V_{max}$ and $-1/\alpha K_\mathrm{S}$, respectively. At intermediate activator concentrations, the intercepts give $-1/K_{\mathrm{S}_{app}}$ and $1/V_{max_{app}}$. The replots of $slope_{1/\mathrm{S}}$ versus [A] and $1/v$-axis

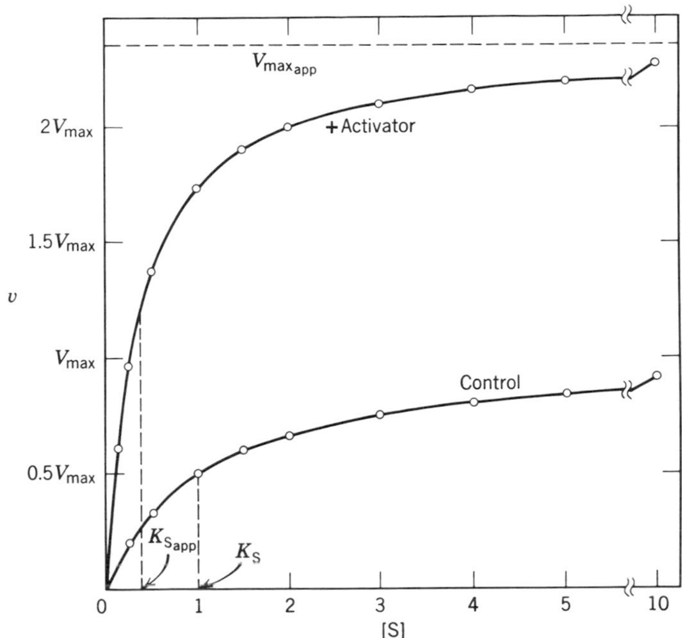

Fig. V-1. The v versus [S] plot in the absence and in the presence of a mixed-type, nonessential activator at $[A]=3K_A(\alpha=0.3, \beta=2.5)$.

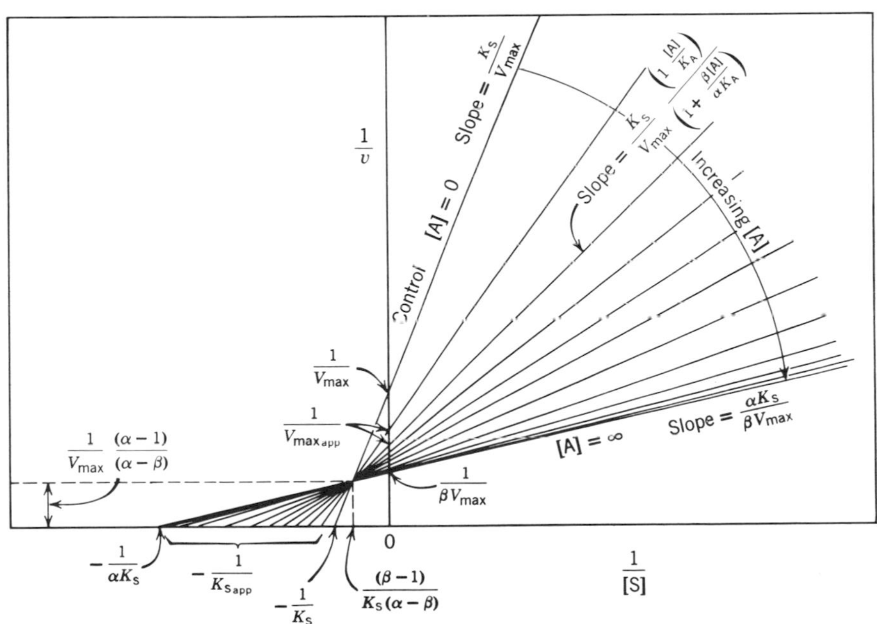

Fig. V-2. The $1/v$ versus $1/[S]$ plot in the presence of different fixed concentrations of a mixed-type, nonessential activator.

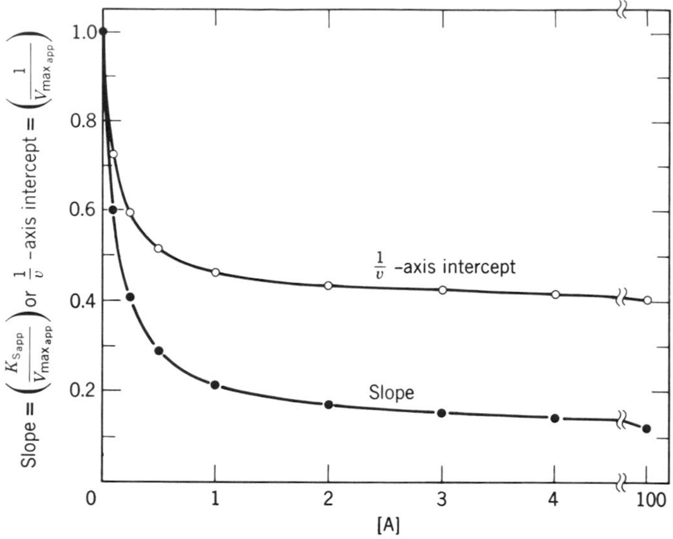

Fig. V-3. *Slope*$_{1/S}$ and $1/v$-axis intercept replots for a mixed-type, nonessential activator ($\alpha = 0.3, \beta = 2.5$).

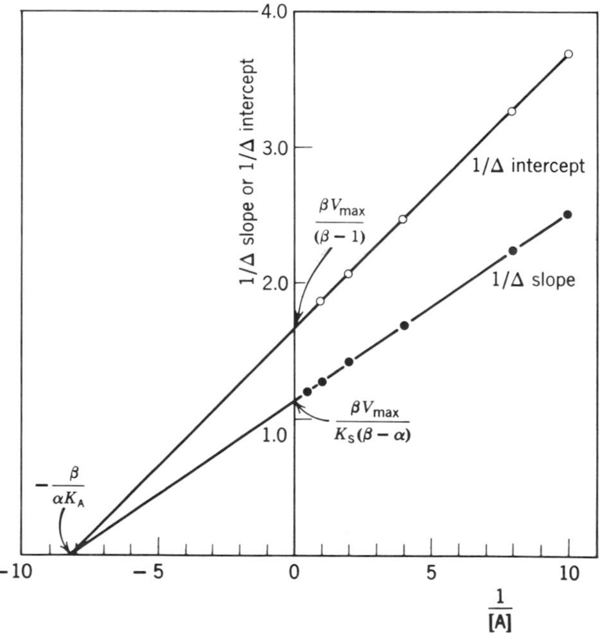

Fig. V-4. Secondary replot of $1/\Delta$*slope* and $1/\Delta$*intercept* versus 1/[A] for a mixed-type, nonessential activator ($\alpha = 0.3, \beta = 2.5$).

intercept versus [A] are hyperbolic (Fig. V-3). In contrast to mixed-type inhibition, these replots are descending curves.

The secondary reciprocal plots of $1/\Delta$ *slope* and $1/\Delta$ *intercept* versus $1/[A]$ are shown in Figure V-4. The Δ *slope* and Δ *intercept* were taken as control value minus "plus activator" value to obtain curves with positive slopes. This changes the $1/\Delta$ *slope*-axis intercept to $\beta V_{max}/K_S(\beta-\alpha)$ and the $1/\Delta$ *intercept*-axis intercept to $\beta V_{max}/(\beta-1.)$ The $1/[A]$-axis intercept gives $-\beta/\alpha K_A$ as in mixed-type inhibition.

System A2. Inhibitor Competitive with Nonessential Activator

$$
\begin{array}{ccccccc}
EA & + & S & \underset{\longrightarrow}{\overset{\beta K_S}{\longleftarrow}} & ESA & \xrightarrow{bk_p} & EA + P \\
K_A \Downarrow\Uparrow & & & & \beta K_A \Downarrow\Uparrow & & \\
A & & & & A & & \\
+ & & & & + & & \\
E & + & S & \underset{\longrightarrow}{\overset{K_S}{\longleftarrow}} & ES & \xrightarrow{k_p} & E + P \\
+ & & & & + & & \\
I & & & & I & & \\
K_i \Uparrow\Downarrow & & & & \alpha K_i \Uparrow\Downarrow & & \\
EI & + & S & \underset{\longrightarrow}{\overset{\alpha K_S}{\longleftarrow}} & ESI & \xrightarrow{ak_p} & EI + P
\end{array}
$$

In System A2, the enzyme posesses binding sites for substrate, inhibitor, and activator. When the inhibitor binds, the affinity of the enzyme for the substrate is reduced (partial competitive action), or the k_p value for product formation is reduced (partial noncompetitive action), or both occur (mixed-type inhibition). The activator has the same, but opposite, effects (i.e., the affinity is increased, or k_p is increased, or both). As the system is shown, the inhibitor and the activator are competitive with respect to each other; that is, the binding of one excludes the binding of the other. Because ES is catalytically active, A is a nonessential activator. There are many variations on the general theme of System A2. For example, A may simply compete with I, but have no effect on K_S or k_p ($\beta=1, b=1$). In this case, A is a "deinhibitor" and "activates" only in the presence of I. Conversely I may have no effect on K_S or k_p ($\alpha=1, a=1$), but if I competes with A (which does have an effect on K_S and k_p), then I is a "deactivator" and "inhibits" only in the presence of A.

The velocity equation for System A2 is:

$$v = k_p[\text{ES}] + ak_p[\text{ESI}] + bk_p[\text{ESA}]$$

$$\frac{v}{V_{max}} = \frac{\dfrac{[\text{S}]}{K_S} + \dfrac{a[\text{S}][\text{I}]}{\alpha K_S K_i} + \dfrac{b[\text{S}][\text{A}]}{\beta K_S K_A}}{1 + \dfrac{[\text{S}]}{K_S} + \dfrac{[\text{I}]}{K_i} + \dfrac{[\text{A}]}{K_A} + \dfrac{[\text{S}][\text{I}]}{\alpha K_S K_i} + \dfrac{[\text{S}][\text{A}]}{\beta K_S K_A}} \qquad \text{(V-3)}$$

or

$$\frac{v}{V_{max}} = \frac{[\text{S}]}{K_S \dfrac{\left(1 + \dfrac{[\text{I}]}{K_i} + \dfrac{[\text{A}]}{K_A}\right)}{\left(1 + \dfrac{a[\text{I}]}{\alpha K_i} + \dfrac{b[\text{A}]}{\beta K_A}\right)} + [\text{S}] \dfrac{\left(1 + \dfrac{[\text{I}]}{\alpha K_i} + \dfrac{[\text{A}]}{\beta K_A}\right)}{\left(1 + \dfrac{a[\text{I}]}{\alpha K_i} + \dfrac{b[\text{A}]}{\beta K_A}\right)}} \qquad \text{(V-4)}$$

where $V_{max} = k_p[\text{E}]_t$ = the maximal velocity in the absence of either effector.

Figure V-5 shows the effect of varying [A] on the initial velocity in the presence of a constant [S] and several fixed concentrations of I. Some calculations will confirm that the initial velocity at any fixed [S] depends more on the *ratio* of [I]/[A] than on the actual concentrations of [I] or [A].

System A3. Nonessential Activation by Two Competing Activators that Alter only K_S

Frequently, more than one compound will serve as an activator for a given enzyme. Alternate activators will compete for the activator site on the enzyme, or for free S in systems where the true substrate is the SA complex (Section B). The equilibria shown for System A2 could just as well describe a situation where A and I are two competing nonessential activators. In general, α and β will be <1 (partial competitive activation), or a and b will

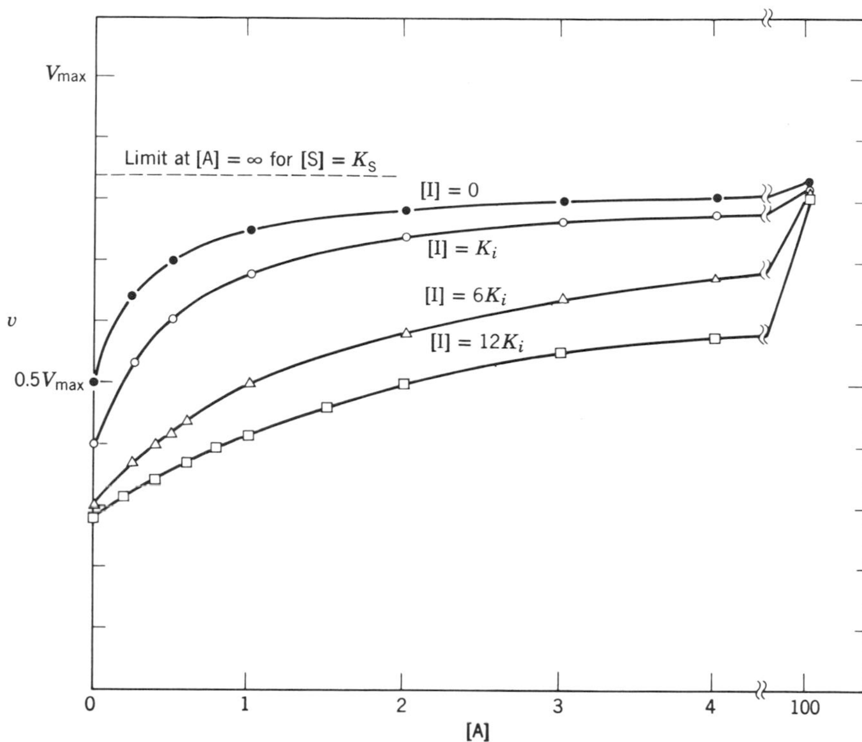

Fig. V-5. The v versus [A] plot at $[S]=K_S$ at different fixed concentrations of I; I is a partial competitive inhibitor with respect to S ($\alpha=3, a=1$); A is a partial competitive, nonessential activator with respect to S ($\beta=0.2, b=1$); A is competitive with I.

be >1 (partial noncompetitive activation), or both (mixed-type activation).

Consider a situation where the activators affect only the substrate dissociation constant and one activator is more effective than the other (e.g., $\alpha<\beta$, but both <1). At low (unsaturating) concentrations of S, the addition of A will increase the velocity, but not as much as will an equal specific concentration of I. Mixtures of A and I will activate more than the same total specific concentration of A alone, but less than the same total specific concentration of I alone. The addition of A to mixtures of S and I will increase v at very low [I] and decrease v at high [I]. A or I (individually or together) will have no effect on the velocity at a saturating concentration of S. The constants K_i, K_A, α, and β can be determined from separate experiments using only one activator at a time, as described for System A1.

System A4. Nonessential Activator Acts as Deinhibitor (Anti-inhibitor)

System A4a. I Is a Pure Competitive Inhibitor with Respect to S and A

In this system the inhibitor prevents the substrate from binding although the inhibitor itself does not bind to the substrate site. The "activator" competes with the inhibitor for the inhibitor site. When the activator occupies the inhibitor site, the substrate site is not distorted and S binds normally. Thus A is a deinhibitor and we can have ES, EA, ESA, and EI complexes, but not ESI or EIA. The equilibria are shown below. It is assumed that the binding of A has no effect on either the dissociation constant for S or on the catalytic activity of the complex. System A4a then is a special case of System A2 ($\beta=1, b=1, \alpha=\infty$).

$$
\begin{array}{ccccccccc}
\mathrm{EI} & & & & & & & & \\
K_i \Big\downarrow\!\Big\uparrow & & & & & & & & \\
\mathrm{I} & & & & & & & & \\
+ & & & & & & & & \\
\mathrm{E} & + & \mathrm{S} & \overset{K_S}{\rightleftharpoons} & \mathrm{ES} & \overset{k_p}{\longrightarrow} & \mathrm{E} & + & \mathrm{P} \\
+ & & & & + & & & & \\
\mathrm{A} & & & & \mathrm{A} & & & & \\
K_A \Big\uparrow\!\Big\downarrow & & & & K_A \Big\uparrow\!\Big\downarrow & & & & \\
\mathrm{EA} & + & \mathrm{S} & \overset{K_S}{\rightleftharpoons} & \mathrm{ESA} & \overset{k_p}{\longrightarrow} & \mathrm{EA} & + & \mathrm{P}
\end{array}
$$

The velocity equation is:

$$
\boxed{\frac{v}{V_{\max}} = \frac{\dfrac{[\mathrm{S}]}{K_S} + \dfrac{[\mathrm{S}][\mathrm{A}]}{K_S K_A}}{1 + \dfrac{[\mathrm{S}]}{K_S} + \dfrac{[\mathrm{A}]}{K_A} + \dfrac{[\mathrm{I}]}{K_i} + \dfrac{[\mathrm{S}][\mathrm{A}]}{K_S K_A}}} \qquad \text{(V-5)}
$$

The equation can be rearranged to the usual form of the Henri-Michaelis-

Menten equation:

$$\boxed{\frac{v}{V_{\max}} = \frac{[S]}{K_S \dfrac{\left(1 + \dfrac{[A]}{K_A} + \dfrac{[I]}{K_i}\right)}{\left(1 + \dfrac{[A]}{K_A}\right)} + [S]}} \tag{V-6}$$

or

$$\boxed{\frac{v}{V_{\max}} = \frac{[S]}{K_S\left[1 + \dfrac{[I]}{K_i\left(1 + \dfrac{[A]}{K_A}\right)}\right] + [S]}} \tag{V-7}$$

We see that when $[A]=0$, the equation reduces to the usual velocity equation for pure competitive inhibition. When $[I]=0$, the parenthetical term reduces to unity, and we obtain the usual Henri-Michaelis-Menten equation. At any unsaturating S and I concentrations, A partially restores the velocity to the uninhibited value. Saturating [A] will completely overcome the inhibition. This is evident from the equilibria where we see that at any unsaturating [S] and [I], an infinitely high [A] will drive all the enzyme to the EA and ESA complexes because A and I are mutually exclusive. The same conclusion is reached by examining the velocity equation where we see that an infinitely high [A] reduces the slope factor (i.e., the factor multiplying K_S) to unity. The velocity in the presence of I and A depends more on the ratio of I : A than on the actual specific concentrations. In the presence of A at a constant [S], $[I]_{0.5}$ is increased.

In systems where one ligand (e.g., A) antagonizes the effect of another (e.g., I) but has no effect on K_S or k_p by itself, the K_A value can be determined by measuring the effect of different concentrations of A on the inhibition by I. This is most conveniently accomplished by Dixon plots of $1/v$ versus [I] at a constant [S] and different fixed concentrations of A. The equation for the Dixon plot is:

$$\frac{1}{v}=\frac{K_S}{V_{max}K_i\left(1+\frac{[A]}{K_A}\right)[S]}[I]+\frac{1}{V_{max}}\left(1+\frac{K_S}{[S]}\right) \tag{V-8}$$

The $1/v$-axis intercept is independent of [A] at a constant [S], but the slope of the plot decreases as the fixed concentration of A increases. The plot is shown in Figure V-6*a*. The K_A can be calculated from either the slopes or the [I]-axis intercepts provided the other kinetic constants are known. Alternately, the reciprocal slopes can be replotted as a function of the corresponding [A] (Fig. V-6*b*).

$$\frac{1}{slope}=\frac{V_{max}K_i[S]}{K_SK_A}[A]+\frac{V_{max}K_i[S]}{K_S} \tag{V-9}$$

Similarly, the negative [I]-axis intercepts may be replotted:

$$-\,intercept=\frac{K_i}{K_A}\left(1+\frac{[S]}{K_S}\right)[A]+K_i\left(1+\frac{[S]}{K_S}\right) \tag{V-10}$$

The replot is shown in Figure V-6*c*.

System A4b. I Is a Noncompetitive Inhibitor with Respect to S; I and A are Competitive

If the inhibitor acts noncompetitively or in a mixed fashion with respect to the substrate but the activator and inhibitor are mutually exclusive, then the equilibria are:

$$\begin{array}{ccccccccc}
EI & + & S & \overset{\alpha K_S}{\rightleftharpoons} & ESI & & & & \\
{\scriptstyle K_i}\updownarrow & & & & {\scriptstyle \alpha K_i}\updownarrow & & & & \\
I & & & & I & & & & \\
+ & & & & + & & & & \\
E & + & S & \overset{K_S}{\rightleftharpoons} & ES & \overset{k_p}{\longrightarrow} & E & + & P \\
+ & & & & + & & & & \\
A & & & & A & & & & \\
{\scriptstyle K_A}\updownarrow & & & & {\scriptstyle K_A}\updownarrow & & & & \\
EA & + & S & \overset{K_S}{\rightleftharpoons} & ESA & \overset{k_p}{\longrightarrow} & EA & + & P
\end{array}$$

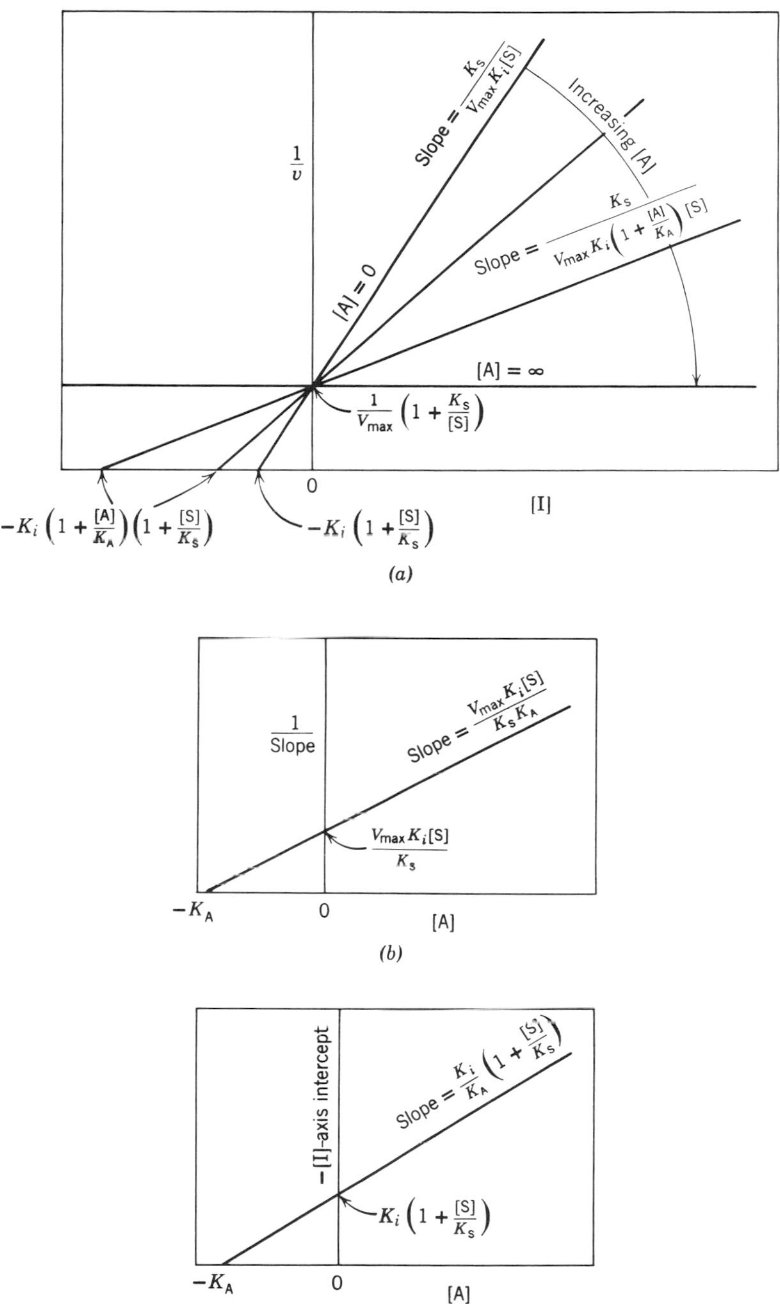

Fig. V-6. (*a*) Dixon plot for System A4a: $1/v$ versus [I] at different fixed concentrations of A (a deinhibitor) and a constant concentration of S. (*b*) Replot of 1/*slope* of the Dixon plot versus [A]. (*c*) Replot of −[I]-axis intercept of Dixon plot versus [A].

The velocity equation is:

$$\frac{v}{V_{\max}} = \frac{[\mathrm{S}]}{K_{\mathrm{S}}\left[1 + \dfrac{[\mathrm{I}]}{K_i\left(1 + \dfrac{[\mathrm{A}]}{K_{\mathrm{A}}}\right)}\right] + [\mathrm{S}]\left[1 + \dfrac{[\mathrm{I}]}{\alpha K_i\left(1 + \dfrac{[\mathrm{A}]}{K_{\mathrm{A}}}\right)}\right]} \tag{V-11}$$

The equation for the Dixon plot is:

$$\frac{1}{v} = \frac{\left(1 + \dfrac{\alpha K_{\mathrm{S}}}{[\mathrm{S}]}\right)}{\alpha K_i V_{\max}\left(1 + \dfrac{[\mathrm{A}]}{K_{\mathrm{A}}}\right)}[\mathrm{I}] + \frac{1}{V_{\max}}\left(1 + \frac{K_{\mathrm{S}}}{[\mathrm{S}]}\right) \tag{V-12}$$

The intercept on the [I]-axis gives $-\alpha K_i(1 + K_{\mathrm{S}}/[\mathrm{S}])(1 + [\mathrm{A}]/K_{\mathrm{A}})/(1 + \alpha K_{\mathrm{S}}/[\mathrm{S}])$. The slope replot is given by:

$$\frac{1}{slope} = \frac{\alpha K_i V_{\max}}{K_{\mathrm{A}}\left(1 + \dfrac{\alpha K_{\mathrm{S}}}{[\mathrm{S}]}\right)}[\mathrm{A}] + \frac{\alpha K_i V_{\max}}{\left(1 + \dfrac{\alpha K_{\mathrm{S}}}{[\mathrm{S}]}\right)} \tag{V-13}$$

The horizontal-axis intercept gives $-K_{\mathrm{A}}$. The [I]-axis intercept of the original Dixon plot can also be replotted to give $-K_{\mathrm{A}}$ as shown for the competitive situation (Fig. V-6*c*).

System A4c. I and A Are Not Mutually Exclusive

Figure V-7 shows the equilibria for a system in which the binding of A reverses the inhibitory action of a pure noncompetitive inhibitor, I. Both I and A can occupy the enzyme at the same time, since they do not compete for a common site. This introduces two new species, EIA and ESIA. If we assume that A simply restores the catalytic activity to k_p, the velocity

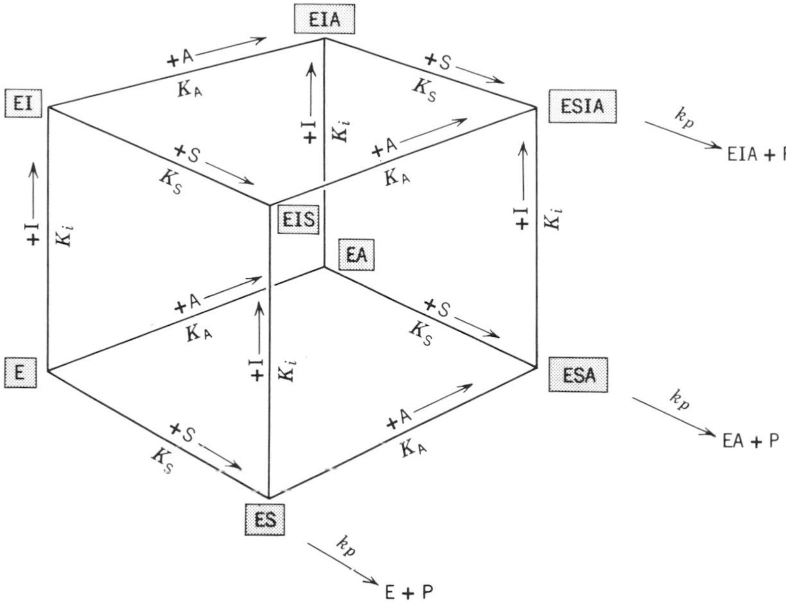

Fig. V-7. Equilibria for a system in which I is a pure noncompetitive inhibitor and A is a deinhibitor which does not exclude I.

equation is:

$$\frac{v}{V_{max}} = \frac{\dfrac{[S]}{K_S} + \dfrac{[S][A]}{K_S K_A} + \dfrac{[S][A][I]}{K_S K_A K_i}}{1 + \dfrac{[S]}{K_S} + \dfrac{[A]}{K_A} + \dfrac{[I]}{K_i} + \dfrac{[S][A]}{K_S K_A} + \dfrac{[S][I]}{K_S K_i} + \dfrac{[A][I]}{K_A K_i} + \dfrac{[S][A][I]}{K_S K_A K_i}} \tag{V-14}$$

or

$$\frac{v}{V_{max}} = \frac{[S]}{K_S \dfrac{\left(1 + \dfrac{[I]}{K_i}\right)}{\left[1 + \dfrac{[I]}{K_i\left(1 + \dfrac{K_A}{[A]}\right)}\right]} + [S] \dfrac{\left(1 + \dfrac{[I]}{K_i}\right)}{\left[1 + \dfrac{[I]}{K_i\left(1 + \dfrac{K_A}{[A]}\right)}\right]}} \tag{V-15}$$

where $V_{max} = k_p[E]_t$. In the absence of A, the system is described by the front face of the equilibrium cube. I is a pure noncompetitive inhibitor, and an infinitely high [I] will drive the velocity to zero. This can also be seen from the velocity equation where setting $[A]=0$, we obtain the usual equation for pure noncompetitive inhibition. Thus K_i can be determined in the usual manner. In the presence of a finite concentration of A, the velocity can no longer be driven to zero by saturating [I]. At any [A], an infinitely high inhibitor concentration will drive all the enzyme to the species shown on the top surface of the equilibrium cube which includes the catalytically active ESIA complex. Thus in the presence of A, I behaves as a partial noncompetitive inhibitor. However, the limiting maximal velocity, "βV_{max}", (as defined in Chapter Four) is not constant, but rather depends on the concentration of A.

$$\beta V_{max} = \frac{V_{max}}{\left(1 + \frac{K_A}{[A]}\right)} \qquad \therefore \quad \beta = \frac{1}{\left(1 + \frac{K_A}{[A]}\right)}$$

System A4c can be distinguished from System A4b by replots of the primary reciprocal plot data. The primary reciprocal plots for both systems are linear at a fixed [I] and fixed [A], but in System A4c, the replots of *slope* versus [I] and $1/v$-axis intercept versus [I] are hyperbolic in the presence of A. The replots for System A4b are linear in the absence or presence of A. System A4c also yields nonlinear Dixon plots in the presence of A. The K_A can be calculated from the slopes or $1/v$-axis intercepts of the primary reciprocal plots if the other kinetic constants are known. If a series of reciprocal plots are constructed for various fixed concentrations of I at a constant A concentration, the $1/v$-axis intercepts can be replotted according to equation V-16, which is identical to equation IV-36, with $\beta = 1/(1 + K_A/[A])$.

$$\boxed{\frac{1}{\Delta\ intercept} = K_i V_{max}\left(1 + \frac{[A]}{K_A}\right)\frac{1}{[I]} + \frac{V_{max}[A]}{K_A}} \qquad \text{(V-16)}$$

System A5. "Energy Charge" Regulation. [I] + [A] Pool Is Constant

Consider a system identical to System A4a with the restriction that the total inhibitor plus activator pool is constant. I and A might be closely related compounds, one a metabolite of the other. *In vivo*, the ratio of [I] : [A] may vary greatly according to the nutritional status of the organism, but the total concentration, [I]+[A], remains relatively constant. The A and I could represent AMP (or ADP) and ATP, respectively, or NAD^+ and NADH, respectively. The rate of the S→P reaction would then be regulated according to the energy sufficiency of the cell. A high ATP/AMP or $NADH/NAD^+$ ratio would retard the S→P reaction, while a low ratio would promote the reaction. The velocity curve would be the usual hyperbola, with $[S]_{0.5}$ dependent on the [I]/[A] ratio. The v versus [I] and v versus [A] curves, at a given [S], are not necessarily hyperbolic because of the restriction that [I]+[A] remain constant.

The v versus [I] curve for System A5 at $[S]=10K_S$ is shown in Figure V-8 where [I]+[A] is assumed to be equal to $100K_A$. If we start with all the modifier in the inhibitor form and plot v versus [A], the resulting curves will be the ascending mirror images of those shown in Figure V-8. System A5 is similar to the "energy-charge" system described in Chapter Two, where P is

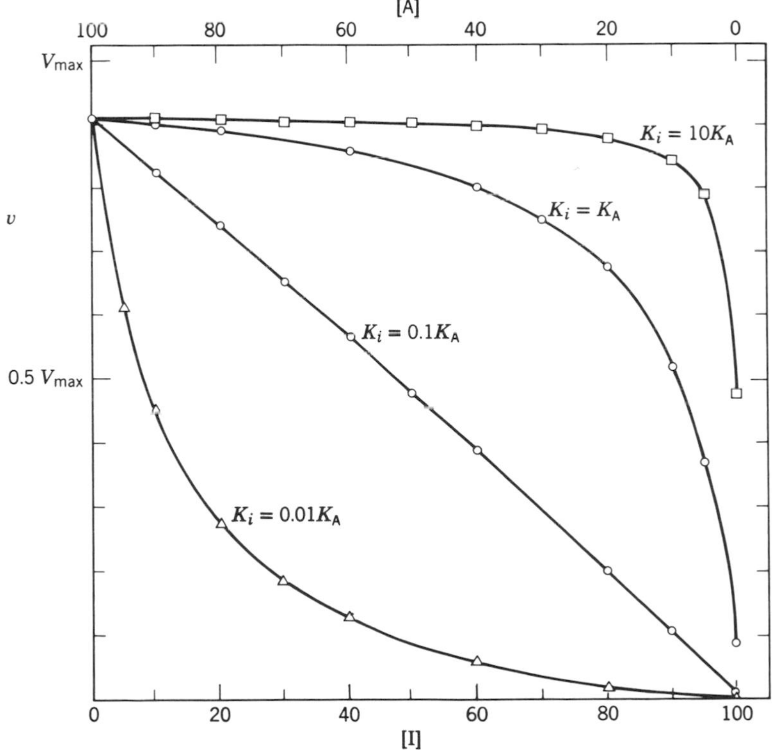

Fig. V-8. Effect of the [I]/[A] ratio on v where $[S]=10K_S$ and $[I]+[A]=100K_A$.

a competitive product inhibitor and [S] + [P] is constant. Other special cases would have A as a nonessential activator that either increases the effective k_p or decreases the apparent K_S, or both, while I excludes A but has no effect on K_S or k_p; that is, I is a "deactivator." Results similar to those shown for System A5 would be obtained, except now the limiting velocity at 100% I would be the unactivated control velocity. The shape and limits of the velocity curves depend on a number of factors including (*a*) the substrate concentration, (*b*) the [I] + [A] concentration, (*c*) the relative values of K_i and K_A, and (*d*) the nature of the inhibition by I (competitive or noncompetitive).

B. SUBSTRATE-ACTIVATOR COMPLEX IS THE TRUE SUBSTRATE

The true substrate of an enzyme may be a substrate-activator (SA) complex where, usually, A is a metal ion. This phenomenon is common in reactions involving phosphorylated intermediates, especially nucleotides. For example, almost all reactions involving ATP require Mg^{2+}. The true substrate is $MgATP^{2-}$. In addition to forming an active complex with S, metal ions may also combine with the enzyme at a specific activator site. The additional activation may be essential or nonessential. The analysis of initial velocity studies involving an SA complex can be quite complicated because of the additional equilibrium $SA \overset{K_0}{\rightleftharpoons} S + A$. Thus an assay mixture containing 1 mM S and 1 mM A may not yield 1 mM SA unless $K_0 \ll 1$ mM. Further complications arise when more than one metal-complexed substrate is involved in the reaction and each substrate has a different affinity for the metal. In such cases, the exact concentrations of S_1A, S_2A, S_1, S_2, and A must be recalculated whenever the concentration of any ligand is changed. A computer would help matters. Sometimes, complications can be avoided by including a very large excess of A in the assay mixture. While this procedure may ensure that all substrates are present as SA complexes, potential information may be lost (e.g., can free S bind to the enzyme?). There is also the possibility that free A may inhibit the enzyme.

In this section the properties of some unireactant systems are explored. A general rapid equilibrium model for a unireactant enzyme that combines with S, A, and SA has been presented by London and Steck (1969). The model (with slightly different notation) is shown in Figure V-9. The bottom face of the cube represents the equilibria for an enzyme with no additional activator sites. The top face represents a system in which the enzyme

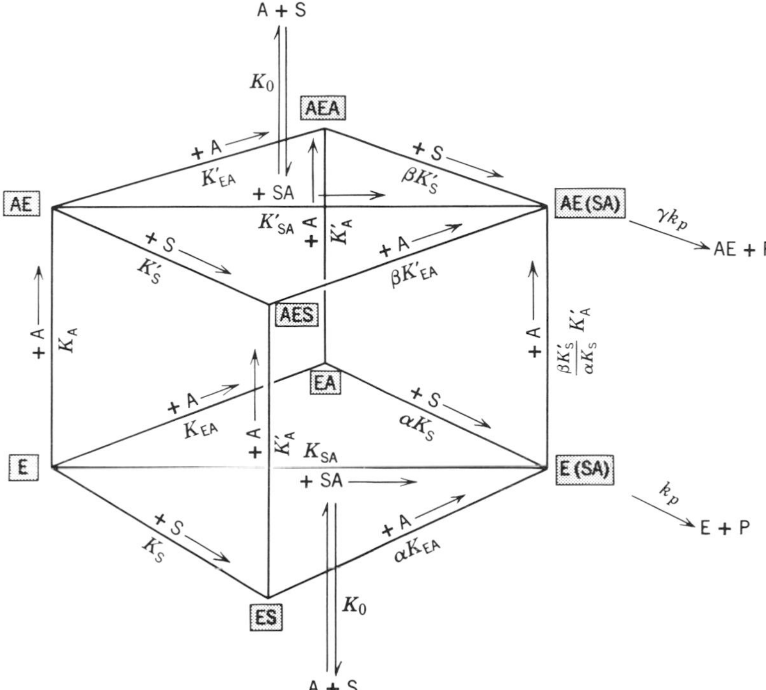

Fig. V-9. General model (equilibria) for metal ion activation. The substrate-activator complex, SA, is the reactive species. E and A can bind randomly (bottom square) or SA can bind directly (diagonal of bottom square). In addition, A may bind to a distinct activator site (top square).

possesses a specific activator site in addition to requiring SA as the substrate. If the activation is nonessential, both E(SA) and AE(SA) are catalytically active. The general velocity equation is:

$$v = k_p[\text{ESA}] + \gamma k_p[\text{AESA}]$$

$$\frac{v}{[\text{E}]_t} = \frac{k_p \dfrac{[\text{SA}]}{K_{\text{SA}}} + \gamma k_p \dfrac{[\text{A}][\text{SA}]}{K_{\text{A}} K'_{\text{SA}}}}{1 + \dfrac{[\text{S}]}{K_{\text{S}}} + \dfrac{[\text{A}]}{K_{\text{EA}}} + \dfrac{[\text{SA}]}{K_{\text{SA}}} + \dfrac{[\text{A}]}{K_{\text{A}}} + \dfrac{[\text{A}][\text{S}]}{K_{\text{A}} K'_{\text{S}}} + \dfrac{[\text{A}]^2}{K_{\text{A}} K_{\text{EA}}} + \dfrac{[\text{A}][\text{SA}]}{K_{\text{A}} K'_{\text{SA}}}}$$

or

$$v = \frac{\frac{[SA]}{K_{SA}} V_{max} + \frac{[A][SA]}{K_A K'_{SA}} V'_{max}}{\left(1 + \frac{[S]}{K_S} + \frac{[A]}{K_{EA}} + \frac{[SA]}{K_{SA}}\right) + \frac{[A]}{K_A}\left(1 + \frac{[S]}{K'_S} + \frac{[A]}{K_{EA}} + \frac{[SA]}{K'_{SA}}\right)}$$

(V-17)

where $V_{max} = k_p[E]_t$ = the maximal velocity of the unactivated enzyme and $V'_{max} = \gamma k_p[E]_t$ = the maximal velocity of the activated enzyme. If the activation is essential, the numerator $[SA]/K_{SA}$ term is omitted.

In addition to the usual equalities around the edges of the cube (e.g., $K_A K'_S = K_S K'_A$, $K_{EA} \alpha K_S = K_S \alpha K_{EA}$), the reaction $S + A \rightleftharpoons SA$ coupled with the diagonal reactions in which SA adds to E (bottom face) or AE (top face) provide alternate routes from free E to the E(SA) and AE(SA) complexes. Consequently:

$$\frac{[S][A]}{K_0 K_{SA}} = \frac{[S][A]}{K_S \alpha K_{EA}} = \frac{[SA]}{K_{SA}}$$

and

$$\frac{[S][A]^2}{K_A K_0 K'_{SA}} = \frac{[S][A]^2}{K_A K'_S \beta K'_{EA}} = \frac{[A][SA]}{K_A K'_{SA}}$$

Thus the general velocity equation can be expressed entirely in terms of *free* A and *free* S or entirely in terms of *total* A and *total* S, where $[S]_t = [S] + [SA]$, and $[A]_t = [A] + [SA]$, as shown in the following sections.

The models described below represent special cases of the general model. For example, System B1 represents just the diagonal of the bottom face of the equilibrium cube. System B2 represents the forward triangle of the bottom face; System B3 represents the rear triangle. System B4 represents the entire bottom face (i.e., the enzyme combines with free A and free S, as well as with SA).

System B1. Only SA Binds to the Enzyme

$$
\begin{array}{l}
S + A \overset{K_0}{\rightleftharpoons} SA \\
\qquad\qquad + \\
\qquad\qquad E \\
\qquad K_{SA} \upharpoonleft\downharpoonright \\
\qquad\qquad ESA \xrightarrow{k_p} E + P + A \\
\qquad\qquad\qquad (E + PA)
\end{array}
$$

The velocity equation can be written in the usual terms.

$$\boxed{\frac{v}{V_{max}} = \frac{[SA]}{K_{SA} + [SA]}} \quad \text{or} \quad \boxed{\frac{v}{V_{max}} = \frac{\frac{[SA]}{K_{SA}}}{1 + \frac{[SA]}{K_{SA}}}} \tag{V-18}$$

where $v = k_p[ESA]$, $V_{max} = k_p[E]_t$. The concentration of SA can be expressed in terms of *free* A, *free* S, and the dissociation constant for the SA complex. However, the resulting equations have limited use because, in general, S and A cannot be independently varied.

$$K_0 = \frac{[A][S]}{[SA]} \qquad \therefore \quad [SA] = \frac{[A][S]}{K_0}$$

$$\therefore \quad \frac{v}{V_{max}} = \frac{\frac{[A][S]}{K_0}}{K_{SA} + \frac{[A][S]}{K_0}}$$

$$\boxed{\frac{v}{V_{max}} = \frac{[A][S]}{K_0 K_{SA} + [A][S]}} \quad \text{or} \quad \boxed{\frac{v}{V_{max}} = \frac{\frac{[A][S]}{K_{SA}K_0}}{1 + \frac{[A][S]}{K_{SA}K_0}}} \tag{V-19}$$

At a fixed activator concentration, increasing the concentration of S results in an increasing velocity. If the activator is present in a large excess, SA can build up to concentration that is very high compared to K_{SA}, and the initial velocity will approach the true V_{max} as [S] increases. If the activator is not present in a large excess, the initial velocity will plateau at an apparent V_{max} as all the available activator is converted to SA. Above this point, the velocity is independent of [S]. The $V_{max_{app}}$ will not equal the true V_{max} if the final concentration of SA is not $\gg K_{SA}$.

The velocity equation can be rearranged to express either free S or free A as the varied ligand:

$$\boxed{\frac{v}{V_{max}} = \frac{[S]}{\dfrac{K_0 K_{SA}}{[A]} + [S]}} \quad \text{and} \quad \boxed{\frac{v}{V_{max}} = \frac{[A]}{\dfrac{K_0 K_{SA}}{[S]} + [A]}} \qquad \text{(V-20)}$$

The equations show that the velocity-dependence of the reaction is symmetrical with respect to *free* A and *free* S. Consequently, the velocity-dependence will also be symmetrical with respect to *total* A and *total* S. (The equilibrium concentration of *free* A at some fixed concentration of *total* S equals the equilibrium concentration of *free* S at the same fixed concentration of *total* A.) In other words, the initial velocities determined at $[S]_t = a$, $[S]_t = b$, and $[S]_t = c$ in the presence of a fixed concentration of $[A]_t = d$ will be identical to the initial velocities determined at $[A]_t = a$, $[A]_t = b$, and $[A]_t = c$ in the presence of a fixed concentration of $[S]_t = d$. If the reciprocal plots are linear for the concentration range above, then the apparent K_S at $[A]_t = d$ will equal the apparent K_A at $[S]_t = d$.

The velocity equation can be expressed in terms of total concentrations as shown below.

$$[A] = [A]_t - [SA] \qquad \text{and} \qquad [S] = [S]_t - [SA]$$

$$[A][S] = ([S]_t - [SA])([A]_t - [SA])$$

$$= [A]_t[S]_t - [SA]([A]_t + [S]_t) + [SA]^2$$

$$[SA] = \frac{[A][S]}{K_0}$$

$$[A][S] = [A]_t[S]_t - \frac{[A][S]}{K_0}([A]_t + [S]_t) + \frac{[A]^2[S]^2}{K_0^2}$$

Rearranging to the usual form of a quadratic equation and solving for [A][S]:

$$0=\frac{1}{K_0^2}[A]^2[S]^2-\left(\frac{[A]_t+[S]_t+K_0}{K_0}\right)[A][S]+[A]_t[S]_t$$

$$[A][S]=\frac{\left(\dfrac{[A]_t+[S]_t+K_0}{K_0}\right)\pm\sqrt{\dfrac{([A]_t+[S]_t+K_0)^2-4[A]_t[S]_t}{K_0^2}}}{\dfrac{2}{K_0^2}}$$

$$=\frac{([A]_t+[S]_t+K_0)-\sqrt{([A]_t+[S]_t+K_0)^2-4[A]_t[S]_t}}{\dfrac{2}{K_0}}$$

The concentration of SA cannot exceed $[S]_t$ or $[A]_t$ (whichever is less). Consequently the larger solution for [A][S] can be ignored, since it can lead to $[A][S]/K_0$ values greater than $[A]_t$ or $[S]_t$. Substituting the expression for [A][S] into equation V-19:

$$v=\frac{V_{max}}{1+\dfrac{2K_{SA}}{([A]_t+[S]_t+K_0)-\sqrt{([A]_t+[S]_t+K_0)^2-4[A]_t[S]_t}}} \qquad \text{(V-21)}$$

The symmetry of the velocity equation with respect to total concentrations is easily seen in the equation above. Interchanging the values of $[A]_t$ and $[S]_t$ yields the same value for v/V_{max}. If the concentration of S or A is large compared to K_0, it may be safely assumed that $[SA]=[S]_t$ (in the presence of excess A) or $[SA]=[A]_t$ (in the presence of excess S). The system can then be analyzed in terms of equation V-18 and K_{SA} determined from the reciprocal plot of $1/v$ versus $1/[S]_t$ or $1/v$ versus $1/[A]_t$. We can examine the approximation by considering a system involving $MgATP^{2-}$ as the true substrate. The dissociation constant for $MgATP^{2-}$ is approximately 10^{-5} *M*.

$$MgATP^{2-}\overset{K_0}{\rightleftharpoons}Mg^{2+}+ATP^{4-} \qquad K_0=10^{-5}\,M$$

Suppose we wish to measure the initial velocity in solutions containing 10^{-3} M *total* Mg^{2+} and varying concentrations of *total* ATP (e.g., 10^{-5} to 10^{-3} M). Can we assume that the concentration of the true substrate, $MgATP^{2-}$, varies from 10^{-5} to 10^{-3} M? In the solution containing 10^{-5} M *total* ATP, let:

$$[MgATP^{2-}]=X, \qquad [Mg^{2+}]=10^{-3}-X, \qquad \text{and} \qquad [ATP^{4-}]=10^{-5}-X$$

$$K_0=\frac{[Mg^{2+}][ATP^{4-}]}{[MgATP^{2-}]}=\frac{(10^{-3}-X)(10^{-5}-X)}{(X)}=10^{-5}$$

$$X^2-(1.02\times10^{-3})X+10^{-8}=0$$

$$X=\frac{(1.02\times10^{-3})\pm\sqrt{(1.02\times10^{-3})^2-4\times10^{-8}}}{2}$$

$$=1.01\times10^{-3}\,M \qquad \text{and} \qquad 0.99\times10^{-5}\,M$$

The larger solution can be ignored because $[MgATP^{2-}]$ cannot exceed 10^{-5} M (ATP is limiting). We see that essentially all the ATP is present as $MgATP^{2-}$ in the solution containing the lowest concentration of total ATP, so certainly the assumption that $[ATP]_t=[MgATP^{2-}]$ will hold throughout the range of $[ATP]_t$ concentrations used. As the Mg^{2+} concentration decreases, the approximation becomes less valid. For example, at 10^{-4} M total Mg^{2+} and 10^{-5} M total ATP, $[MgATP^{2-}]$ is only about 0.9×10^{-5} M.

Figure V-10 shows the velocity curves for a system where $[S]_t$ and $[A]_t$ are large compared to K_0 throughout the concentration range studied. The velocity is relatively independent of $[A]_t$ at low values of $[S]_t$. For example, at $[S]_t=0.1$ mM, v is about 0.16 V_{max} at $[A]_t$ ranging from 0.2 to 2 mM. This odd result makes sense if we remember that $[S]_t$ and $[A]_t$ are large compared to K_0, so that the concentration of the reactive SA complex will equal the concentration of the limiting ligand. When $[S]_t=0.1$ mM, $[SA] \cong 0.1$ mM at all four concentrations of $[A]_t$. As long as $[A]_t>[S]_t$, the velocity curve rises in a normal manner. When all the A has been converted to SA, increasing $[S]_t$ has no effect on the velocity, and an apparent V_{max} is attained where:

$$V_{max_{app}}=\frac{[A]_t}{K_{SA}+[A]_t}V_{max}$$

The reciprocal plot in the region where $[A]_t \gg [S]_t$ is reasonably linear and

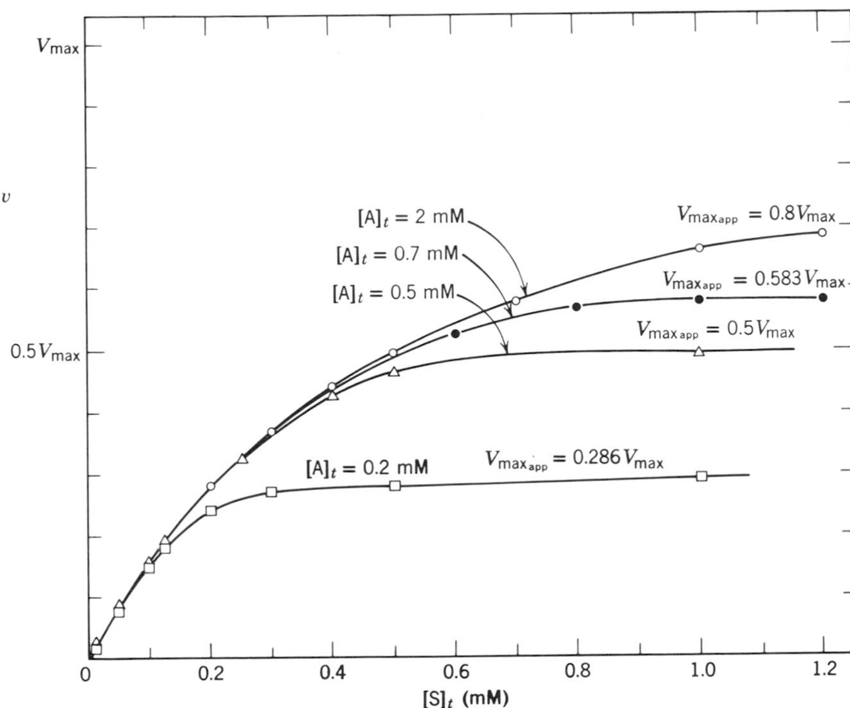

Fig. V-10. The v versus $[S]_t$ plot at different fixed concentrations of $[A]_t$. The true substrate is the SA complex; $K_{SA}=0.5$mM, $K_0=0.01$mM.

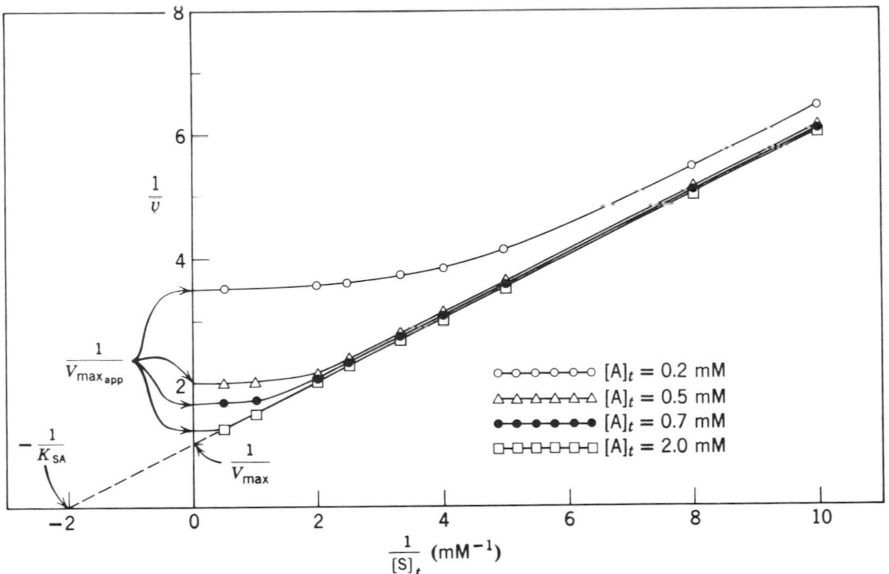

Fig. V-11. Reciprocal plot of the data shown in Fig. V-10.

extrapolates to $1/V_{max}$ and $-1/K_{SA}$ (Fig. V-11). Note that the reciprocal plot for $[A]_t = 0.2$ mM approaches linearity only at high values of $1/[S]$ and cannot be extrapolated to obtain $1/K_{SA}$. In the region where $[S]_t \cong [A]_t$, the reciprocal plots are curved and become horizontal when $[S]_t > [A]_t$. The horizontal portion intersects the $1/v$-axis at $1/V_{max_{app}}$.

For preliminary experiments, it may be convenient to pre-mix A and S in some desired ratio (e.g., 1 : 1 or, more usually, 10 : 1) and then determine the initial velocity at varying concentrations of A + S. This will allow V_{max} to be determined, but the apparent K_m will equal K_{SA} only if K_0 is very small compared to the concentration range studied. When [A + S] is low compared to K_0, the velocity curve will be sigmoidal because increasing [A + S] by a given factor will increase [SA] by a greater factor. For example, at 10^{-6} M total Mg^{2+} and 10^{-6} M total ATP^{4-}, the equilibrium concentration of $MgATP^{2-}$ is 8.4×10^{-8} M. At 2×10^{-6} M Mg^{2+} and 2×10^{-6} M ATP^{4-}, the equilibrium concentration of $MgATP^{2-}$ is 29.2×10^{-8} M.

System B2. SA and S Bind to the Enzyme

An enzyme may combine with both SA and free S with different or equal affinities. The equilibria are:

$$
\begin{array}{ccccccccc}
S & + & A & & & & & & \\
+ & & K_0 \updownarrow & & & & & & \\
E & + & SA & \underset{}{\overset{K_{SA}}{\rightleftharpoons}} & ESA & \overset{k_p}{\longrightarrow} & E + P + A & & \\
K_S \updownarrow & & & & & & (E + PA) & & \\
ES & & & & & & & &
\end{array}
$$

The A might also combine with ES to yield ESA without changing the characteristics of the system. The velocity equation for System B2 in terms of *free* S and *free* A is:

$$
\boxed{\frac{v}{V_{max}} = \frac{\dfrac{[A][S]}{K_0 K_{SA}}}{1 + \dfrac{[S]}{K_S} + \dfrac{[A][S]}{K_0 K_{SA}}}} \qquad \text{(V-22)}
$$

where $v = k_p[\mathrm{ESA}]$ and $V_{max} = k_p[\mathrm{E}]_t$. Alternate forms of the velocity equation are:

$$\frac{v}{V_{max}} = \frac{[\mathrm{A}][\mathrm{S}]}{K_0 K_{SA}\left(1+\frac{[\mathrm{S}]}{K_S}\right)+[\mathrm{A}][\mathrm{S}]} \tag{V-23}$$

or

$$v = \frac{V_{max}}{\frac{K_0 K_{SA}}{[\mathrm{A}][\mathrm{S}]}\left(1+\frac{[\mathrm{S}]}{K_S}\right)+1} \tag{V-24}$$

The velocity equation for System B2 is not symmetrical with respect to S and A. When S is the varied ligand:

$$\frac{v}{V_{max}} = \frac{[\mathrm{S}]}{\frac{K_0 K_{SA}}{[\mathrm{A}]}+[\mathrm{S}]\left(1+\frac{K_0 K_{SA}}{[\mathrm{A}]K_S}\right)} \tag{V-25}$$

When A is the varied ligand:

$$\frac{v}{V_{max}} = \frac{[\mathrm{A}]}{\frac{K_0 K_{SA}}{K_S}\left(1+\frac{K_S}{[\mathrm{S}]}\right)+[\mathrm{A}]} \tag{V-26}$$

Thus the initial velocity determined at $[\mathrm{S}]_t = a$ and $[\mathrm{A}]_t = b$ will not equal the initial velocity obtained at $[\mathrm{A}]_t = a$ and $[\mathrm{S}]_t = b$.

The velocity equation can be expressed in terms of [SA] and [S], the two species that combine with free enzyme.

$$\frac{v}{V_{max}} = \frac{\dfrac{[SA]}{K_{SA}}}{1 + \dfrac{[S]}{K_S} + \dfrac{[SA]}{K_{SA}}} \qquad \text{(V-27)}$$

or

$$\frac{v}{V_{max}} = \frac{[SA]}{K_{SA}\left(1 + \dfrac{[S]}{K_S}\right) + [SA]} \qquad \text{(V-28)}$$

In the form above, free S is seen to act as a competitive inhibitor with respect to SA.

The velocity equation for this system can also be expressed in terms of *total* [S] and *total* [A]:

$$[S] = [S]_t - [SA] = [S]_t - \frac{[A][S]}{K_0}$$

Substituting into equation V-24:

$$v = \frac{V_{max}}{\dfrac{K_0 K_{SA}}{[A][S]}\left(1 + \dfrac{[S]_t}{K_S} - \dfrac{[A][S]}{K_0 K_S}\right) + 1} = \frac{V_{max}}{\dfrac{K_0 K_{SA}}{[A][S]} + \dfrac{K_0 K_{SA}[S]_t}{[A][S]K_S} - \dfrac{K_{SA}}{K_S} + 1}$$

$$= \frac{V_{max}}{\left(1 - \dfrac{K_{SA}}{K_S}\right) + \left(1 + \dfrac{[S]_t}{K_S}\right)\dfrac{K_0 K_{SA}}{[A][S]}}$$

$$[A][S] = \frac{[A]_t + [S]_t + K_0 - \sqrt{([A]_t + [S]_t + K_0)^2 - 4[A]_t[S]_t}}{\dfrac{2}{K_0}}$$

$$v = \frac{V_{max}}{\left(1 - \dfrac{K_{SA}}{K_S}\right) + \left(1 + \dfrac{[S]_t}{K_S}\right)\dfrac{2K_{SA}}{([A]_t + [S]_t + K_0) - \sqrt{([A]_t + [S]_t + K_0)^2 - 4[A]_t[S]_t}}}$$

(V-29)

If the enzyme combines with SA and free S with the same affinity, then $K_{SA} = K_S$ and the velocity equations reduce to:

$$v = \frac{V_{max}}{\frac{K_0}{[A]}\left(1 + \frac{K_S}{[S]}\right) + 1} \tag{V-30}$$

or

$$\frac{v}{V_{max}} = \frac{[SA]}{K_S + [S] + [SA]} \tag{V-31}$$

and

$$\frac{v}{V_{max}} = \frac{([A]_t + [S]_t + K_0) - \sqrt{([A]_t + [S]_t + K_0)^2 - 4[A]_t[S]_t}}{2(K_{SA} + [S]_t)} \tag{V-32}$$

In System B2, SA and free S compete for the enzyme. If S is present in excess over A, some of the enzyme will be tied up as the nonproductive ES complex yielding an apparent case of substrate inhibition. This may be easier to see if we write the equilibria describing the system in the presence of excess S.

$$\begin{array}{l} E + SA \overset{K_{SA}}{\rightleftharpoons} ESA \overset{k_p}{\longrightarrow} E + P + A \\ + \qquad\qquad\qquad\qquad\quad (E + PA) \\ S \\ K_S \updownarrow \\ ES \end{array}$$

At a constant $[A]_t$, increasing $[S]_t$ will result in an increasing velocity up to the point where all the A is converted to SA. (This will occur when $[S]_t = [A]_t$ if K_0 is very small compared to $[A]_t$.) At higher S concentrations the velocity will decrease as S acts as a competitive inhibitor (Fig. V-12). System B2 may be distinguished from System B1 by this apparent substrate inhibition and also by the lack of symmetry between S and A.

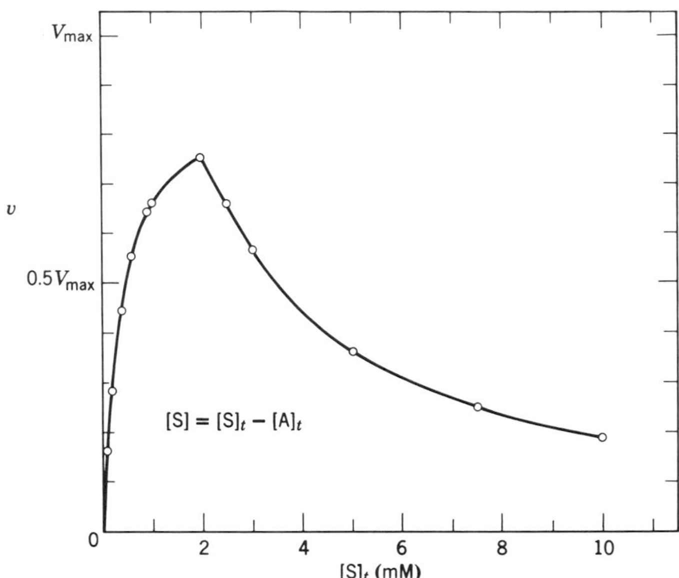

Fig. V-12. Inhibition by free S. SA is the true substrate but S also binds to the enzyme. For the constants chosen (K_0=0.01mM, K_{SA}=0.5mM, K_S=0.5mM, and $[A]_t$=2mM), $[SA] \cong [S]_t$ when $[S]_t \leqslant [A]_t$ and the peak velocity occurs at $[S]_t = [A]_t$. When $[S]_t > [A]_t$, $[S] \cong [S]_t - [A]_t$. The peak can occur at $[S]_t > [A]_t$, $[S]_t = [A]_t$, or $[S]_t < [A]_t$, depending on the relative values of K_0, K_{SA}, K_S and $[A]_t$.

Determination of K_{SA}

If the fixed $[A]_t$ and the varied $[S]_t$ range studied are very high compared to K_0, it may be assumed that $[SA] \cong [S]_t$; that is, virtually all the S is complexed as SA as long as $[A]_t > [S]_t$. The velocity equation becomes:

$$\boxed{\frac{v}{V_{max}} = \frac{[SA]}{K_{SA} + [SA]} = \frac{[S]_t}{K_{SA} + [S]_t}} \tag{V-33}$$

The v versus $[S]_t$ curve is essentially hyperbolic with a limit of V_{max}. The K_{SA} may be determined from a reciprocal plot of $1/v$ versus $1/[S]_t$. As an alternate procedure, A and S can be varied together at a constant ratio where $[A]_t = x[S]_t$ and $x > 1$.

Determination of K_S

When $[S]_t$ is present in excess over $[A]_t$, and $[A]_t \gg K_0$, it may be assumed that $[SA] \cong [A]_t$; that is, all the A is complexed as SA. Free [S] is then given by:

$$[S] = [S]_t - [SA] = [S]_t - [A]_t$$

The K_S may be determined by treating *free* S as a competitive inhibitor. For example, a series of velocity curves and reciprocal plots can be constructed for varied $[SA] \cong [A]_t$ at several fixed concentrations of *free* S. To maintain a constant [S] for each [SA], $[S]_t$ must be increased along with $[A]_t$ so that $[S]_t - [A]_t$ equals the desired fixed concentration of free S. For example, if we want to construct a velocity curve for increasing $[A]_t$ in the presence of a fixed excess of 1 mM S, then at $[A]_t \cong [SA] = 0.2$ mM, $[S]_t$ must be 1.2 mM. At $[A]_t \cong [SA] = 0.75$ mM, $[S]_t$ must be 1.75 mM, and so on. The K_S may also be determined from a Dixon plot of $1/v$ versus *free* [S] at several different fixed concentrations of SA (i.e., $[A]_t$). The primary data would be obtained from the inhibition phase of the v versus $[S]_t$ plots. Figure V-12 would represent the plot for one fixed $[A]_t$.

The effect of varying $[A]_t$ in the presence of a fixed $[S]_t$ is shown in Figure V-13. The shape of the velocity curve depends on the relative values of K_{SA} and K_S. When $K_S < K_{SA}$ (i.e., the affinity of the enzyme for S is greater than for SA), the curve is quite sigmoidal resembling a titration curve. The sigmoidicity results from a combination of two effects: as $[A]_t$ increases, the concentration of the true substrate, SA, increases, and, simultaneously, the concentration of the inhibitor, free S, decreases.

System B3. SA and A Bind to the Enzyme

An enzyme may combine with both SA and free A with different or equal affinities.

$$\begin{array}{ccccccc}
 & A + S & & & & & \\
 & + \quad K_0 \updownarrow & & & & & \\
 & E \quad + SA & \overset{K_{SA}}{\rightleftarrows} & ESA & \overset{k_p}{\longrightarrow} & E + P + A & \\
K_{EA} \updownarrow & & & & & (E + PA) & \\
 & EA & & & & &
\end{array}$$

The S might also combine with EA to yield ESA. This system is analogous to System B2 except now free A is competitive with SA. In terms of *free* A and

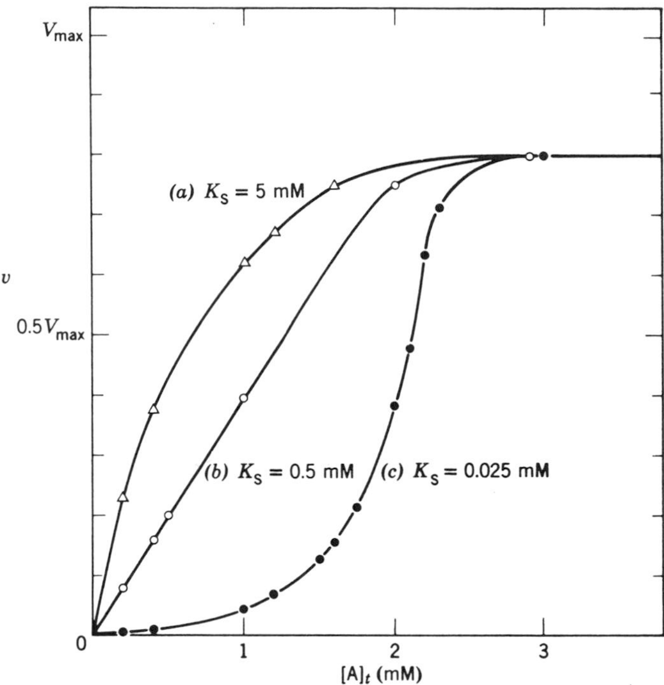

Fig. V-13. The v versus $[A]_t$ in the presence of a fixed concentration of $[S]_t$. The true substrate is SA but free S also binds to the enzyme; K_0=0.01mM, K_{SA}=0.5mM, $[S]_t$ =2mM.

free S, the velocity equation is:

$$\frac{v}{V_{max}} = \frac{\dfrac{[A][S]}{K_0 K_{SA}}}{1 + \dfrac{[A]}{K_{EA}} + \dfrac{[A][S]}{K_0 K_{SA}}} \tag{V-34}$$

Alternate forms are:

$$\frac{v}{V_{max}} = \frac{[A][S]}{K_0 K_{SA}\left(1 + \dfrac{[A]}{K_{EA}}\right) + [A][S]} \tag{V-35}$$

$$v = \frac{V_{max}}{\frac{K_0 K_{SA}}{[A][S]}\left(1 + \frac{[A]}{K_{EA}}\right) + 1} \tag{V-36}$$

In terms of [SA] and free (excess) [A], the two species that can combine with E:

$$\frac{v}{V_{max}} = \frac{\frac{[SA]}{K_{SA}}}{1 + \frac{[A]}{K_{EA}} + \frac{[SA]}{K_{SA}}} \quad \text{or} \quad \frac{v}{V_{max}} = \frac{[SA]}{K_{SA}\left(1 + \frac{[A]}{K_{EA}}\right) + [SA]} \tag{V-37}$$

In the form above the role of A as a competitive inhibitor with respect to SA is clearly seen.

The velocity equation in terms of *total* concentrations of S and A is similar to that shown for System B2.

$$v = \frac{V_{max}}{\left(1 - \frac{K_{SA}}{K_{EA}}\right) + \left(1 + \frac{[A]_t}{K_{EA}}\right)\frac{2K_{SA}}{([A]_t + [S]_t + K_0) - \sqrt{([A]_t + [S]_t + K_0)^2 - 4[A]_t[S]_t}}} \tag{V-38}$$

If both SA and A combine with E with equal affinity, then $K_{SA} = K_{EA}$ and the velocity equations reduce to:

$$\frac{v}{V_{max}} = \frac{[A][S]}{K_0(K_{EA} + [A]) + [A][S]} \tag{V-39}$$

$$v = \frac{V_{max}}{\frac{K_0}{[S]}\left(1 + \frac{K_{EA}}{[A]}\right) + 1} \tag{V-40}$$

$$\boxed{\frac{v}{V_{\max}} = \frac{[\mathrm{SA}]}{K_{\mathrm{EA}} + [\mathrm{A}] + [\mathrm{SA}]}} \qquad \text{(V-41)}$$

$$\boxed{\frac{v}{V_{\max}} = \frac{([\mathrm{A}]_t + [\mathrm{S}]_t + K_0) - \sqrt{([\mathrm{A}]_t + [\mathrm{S}]_t + K_0)^2 - 4[\mathrm{A}]_t[\mathrm{S}]_t}}{2(K_{\mathrm{SA}} + [\mathrm{A}]_t)}} \qquad \text{(V-42)}$$

System B3 may be distinguished from System B1 and System B2 by the lack of symmetry between S and A and by the inhibition by excess A. The K_{SA} can be determined from reciprocal plots of velocity data obtained in the presence of a concentration of S that is very large compared to K_0. The K_{EA} can be determined by treating excess A as an inhibitor, as described for excess S in System B2.

System B4. SA, S, and A Bind to the Enzyme

An enzyme may combine with SA, free S, and free A with varying affinities. The equilibria corresponding to the bottom square of Figure V-9 are shown below.

$$\mathrm{E} + \mathrm{S} \overset{K_S}{\rightleftharpoons} \mathrm{ES}$$
$$\mathrm{A} + \mathrm{S} \overset{K_0}{\rightleftharpoons} \mathrm{SA}$$
$$\mathrm{E} + \mathrm{A} \overset{K_{EA}}{\rightleftharpoons} \mathrm{EA}$$
$$\mathrm{E} + \mathrm{SA} \overset{K_{SA}}{\rightleftharpoons} \mathrm{ESA} \overset{k_p}{\longrightarrow} \mathrm{E} + \mathrm{A} + \mathrm{P} \quad (\mathrm{E} + \mathrm{PA})$$

Or, if S binds to EA, and A binds to ES:

$$\mathrm{E} + \mathrm{S} \overset{K_S}{\rightleftharpoons} \mathrm{ES}$$
$$\mathrm{A} + \mathrm{S} \overset{K_0}{\rightleftharpoons} \mathrm{SA}$$
$$\mathrm{E} + \mathrm{A} \overset{K_{EA}}{\rightleftharpoons} \mathrm{EA}$$
$$\mathrm{E} + \mathrm{SA} \overset{K_{SA}}{\rightleftharpoons} \mathrm{ESA}$$
$$\mathrm{ES} + \mathrm{A} \overset{\alpha K_{EA}}{\rightleftharpoons} \mathrm{ESA}$$
$$\mathrm{EA} + \mathrm{S} \underset{\alpha K_S}{\rightleftharpoons} \mathrm{ESA} \overset{k_p}{\longrightarrow} \mathrm{E} + \mathrm{A} + \mathrm{P} \quad (\mathrm{E} + \mathrm{PA})$$

The velocity equation in terms of *free* A and S is:

$$\frac{v}{V_{\max}} = \frac{\dfrac{[S][A]}{K_0 K_{SA}}}{1 + \dfrac{[S]}{K_S} + \dfrac{[A]}{K_{EA}} + \dfrac{[S][A]}{K_0 K_{SA}}} \qquad \text{(V-43)}$$

or

$$\frac{v}{V_{\max}} = \frac{\dfrac{[S][A]}{K_S \alpha K_{EA}}}{1 + \dfrac{[S]}{K_S} + \dfrac{[A]}{K_{EA}} + \dfrac{[S][A]}{K_S \alpha K_{EA}}} \qquad \text{(V-44)}$$

or

$$v = \frac{V_{\max}}{\dfrac{K_0 K_{SA}}{[A][S]}\left(1 + \dfrac{[S]}{K_S} + \dfrac{[A]}{K_{EA}}\right) + 1} \qquad \text{(V-45)}$$

Alternately, the equation can be expressed in terms of SA, *free* S, and *free* A.

$$\frac{v}{V_{\max}} = \frac{\dfrac{[SA]}{K_{SA}}}{1 + \dfrac{[S]}{K_S} + \dfrac{[A]}{K_{EA}} + \dfrac{[SA]}{K_{SA}}} \qquad \text{(V-46)}$$

or

$$\frac{v}{V_{\max}} = \frac{[SA]}{K_{SA}\left(1 + \dfrac{[S]}{K_S} + \dfrac{[A]}{K_{EA}}\right) + [SA]} \qquad \text{(V-47)}$$

Equation V-47 clearly shows that both free S and free A are competitive inhibitors with respect to SA. The equation can also be written in terms of SA, *total* S, and *total* A:

$$\boxed{\frac{v}{V_{\max}} = \frac{\dfrac{[\mathrm{SA}]}{K_{\mathrm{SA}}}}{1 + \dfrac{[\mathrm{S}]_t - [\mathrm{SA}]}{K_{\mathrm{S}}} + \dfrac{[\mathrm{A}]_t - [\mathrm{SA}]}{K_{\mathrm{EA}}} + \dfrac{[\mathrm{SA}]}{K_{\mathrm{SA}}}}} \qquad \text{(V-48)}$$

or

$$\boxed{\frac{v}{V_{\max}} = \frac{\dfrac{[\mathrm{SA}]}{K_{\mathrm{SA}}}}{1 + \dfrac{[\mathrm{S}]_t}{K_{\mathrm{S}}} + \dfrac{[\mathrm{A}]_t}{K_{\mathrm{EA}}} + [\mathrm{SA}]\left(\dfrac{1}{K_{\mathrm{SA}}} - \dfrac{1}{K_{\mathrm{EA}}} - \dfrac{1}{K_{\mathrm{S}}}\right)}} \qquad \text{(V-49)}$$

where $[\mathrm{S}] = [\mathrm{S}]_t - [\mathrm{SA}]$
$[\mathrm{A}] = [\mathrm{A}]_t - [\mathrm{SA}]$
$[\mathrm{S}][\mathrm{A}] = K_0[\mathrm{SA}]$

A velocity equation expressed completely in terms of *total* concentrations of S and A can be derived. The relationships shown below are substituted into equation V-45.

$$[\mathrm{S}] = [\mathrm{S}]_t - \frac{[\mathrm{A}][\mathrm{S}]}{K_0}$$

$$[\mathrm{A}] = [\mathrm{A}]_t - \frac{[\mathrm{A}][\mathrm{S}]}{K_0}$$

$$\therefore \quad v = \frac{V_{\max}}{\dfrac{K_0 K_{\mathrm{SA}}}{[\mathrm{A}][\mathrm{S}]}\left(1 + \dfrac{[\mathrm{S}]_t}{K_{\mathrm{S}}} - \dfrac{[\mathrm{A}][\mathrm{S}]}{K_{\mathrm{S}} K_0} + \dfrac{[\mathrm{A}]_t}{K_{\mathrm{EA}}} - \dfrac{[\mathrm{A}][\mathrm{S}]}{K_{\mathrm{EA}} K_0}\right) + 1}$$

$$= \frac{V_{\max}}{\dfrac{K_0 K_{\mathrm{SA}}}{[\mathrm{A}][\mathrm{S}]} + \dfrac{K_0 K_{\mathrm{SA}}[\mathrm{S}]_t}{[\mathrm{A}][\mathrm{S}]K_{\mathrm{S}}} - \dfrac{K_{\mathrm{SA}}}{K_{\mathrm{S}}} + \dfrac{K_0 K_{\mathrm{SA}}[\mathrm{A}]_t}{[\mathrm{A}][\mathrm{S}]K_{\mathrm{A}}} - \dfrac{K_{\mathrm{SA}}}{K_{\mathrm{EA}}} + 1}$$

$$v = \frac{V_{\max}}{\left(1 - \frac{K_{SA}}{K_S} - \frac{K_{SA}}{K_{EA}}\right) + \left(1 + \frac{[S]_t}{K_S} + \frac{[A]_t}{K_{EA}}\right) \frac{K_0 K_{SA}}{[A][S]}}$$

$$[A][S] = \frac{([A]_t + [S]_t + K_0) - \sqrt{([A]_t + [S]_t + K_0)^2 - 4[A]_t[S]_t}}{\frac{2}{K_0}}$$

$$\boxed{v = \frac{V_{\max}}{\left(1 - \frac{K_{SA}}{K_S} - \frac{K_{SA}}{K_{EA}}\right) + \left(1 + \frac{[S]_t}{K_S} + \frac{[A]_t}{K_{EA}}\right) \frac{2K_{SA}}{([A]_t + [S]_t + K_0) - \sqrt{([A]_t + [S]_t + K_0)^2 - 4[A]_t[S]_t}}}}$$

(V-50)

In System B4, both free S and free A compete with the SA complex for the free enzyme. The shape of the velocity curve will depend on the relative values of the three dissociation constants and the concentration of the fixed ligand. Figure V-14 shows the velocity-dependence on $[S]_t$ in the presence of a fixed $[A]_t$. Curves *B* and *D* illustrate the inhibition by free S in systems where E has a reasonable affinity for S. Curves *C* and *D* are sigmoidal illustrating the "deinhibition" effect. This effect is quite noticeable in systems where the enzyme has a higher affinity for free A than for SA. The "deinhibition" effect is not observed in curves *A* and *B* where $K_{SA} < K_{EA}$. Figure V-14 shows the velocity curves for a single fixed excess of $[A]_t$. If $[A]_t$ is varied, a family of curves can be generated. The peak velocities of curves *B* and *D* move to higher v' and $[S]_t$ values as $[A]_t$ is increased. An analogous family of curves can be obtained by determining the initial velocity as a function $[A]_t$ at several fixed values of $[S]_t$. Figure V-15 shows the curves generated by London and Steck (1969) for the lombricine kinase reaction. The peaks do not necessarily occur at $[S]_t = [A]_t$, but rather depend on the relative values of K_{EA} and K_S. If the affinity of the enzyme for free substrate is greater than the affinity for free activator ($K_S < K_{EA}$), the peak velocity on the v versus $[A]_t$ curves occur where $[A]_t$ is greater than $[S]_t$. The smaller the value of K_0 relative to K_S, the smaller the excess of $[A]_t$ over $[S]_t$ at a peak velocity. The peak velocities on the v versus $[S]_t$ curves occur where $[S]_t > [A]_t$ at very low fixed $[A]_t$ values, but as $[A]_t$ increases, the peak velocities occur where $[A]_t > [S]_t$. Opposite results are obtained if $K_{EA} < K_S$.

If the peak velocities of the v versus $[S]_t$ plot are plotted against the corresponding $[S]_t$, a nonlinear curve that intersects the $[S]_t$-axis at $\sqrt{K_0 K_S}$ is obtained. Similarly, the analogous replot of peak velocity versus the corresponding $[A]_t$ from the v versus $[A]_t$ family of curves yields $\sqrt{K_0 K_{EA}}$ (London and Steck, 1969).

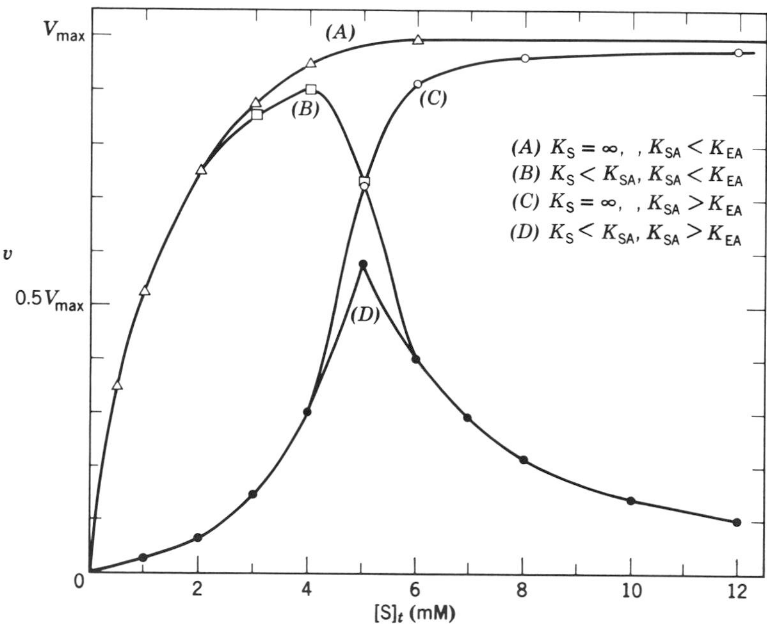

Fig. V-14. System B4; v versus $[S]_t$ curves for $[A]_t = 5$mM, $K_0 = 10^{-5}M$, and $K_{SA} = 2 \times 10^{-5}M$. Curve A: $K_S = \infty$, $K_{EA} = 10^{-4}M$. Curve B: $K_S = 3.33 \times 10^{-6}M$, $K_{EA} = 10^{-4}M$. Curve C: $K_S = \infty$, $K_{EA} = 2 \times 10^{-6}M$. Curve D: $K_S = 3.33 \times 10^{-6}M$, $K_{EA} = 2 \times 10^{-6}M$. [Redrawn with permission from London, W. P. and Steck, T. L., *Biochemistry* **8**, 1767 (1969). Copyright by the American Chemical Society.]

The glutamine synthetase of *Bacillus licheniformis* (Hubbard and Stadtman, 1967) is an example of an enzyme inhibited by free A and free S. The enzyme utilizes the $ATPMn^{2-}$ complex as the substrate, while an excess of either free ATP^{4-} or free Mn^{2+} is inhibitory. Other nucleoside triphosphates are unable to replace ATP. However, because their binding constants for Mn^{2+} are comparable to that of ATP, they will stimulate the enzymatic activity when Mn^{2+} is in excess, or inhibit when the concentration of Mn^{2+} is equal to or less than the ATP concentration.

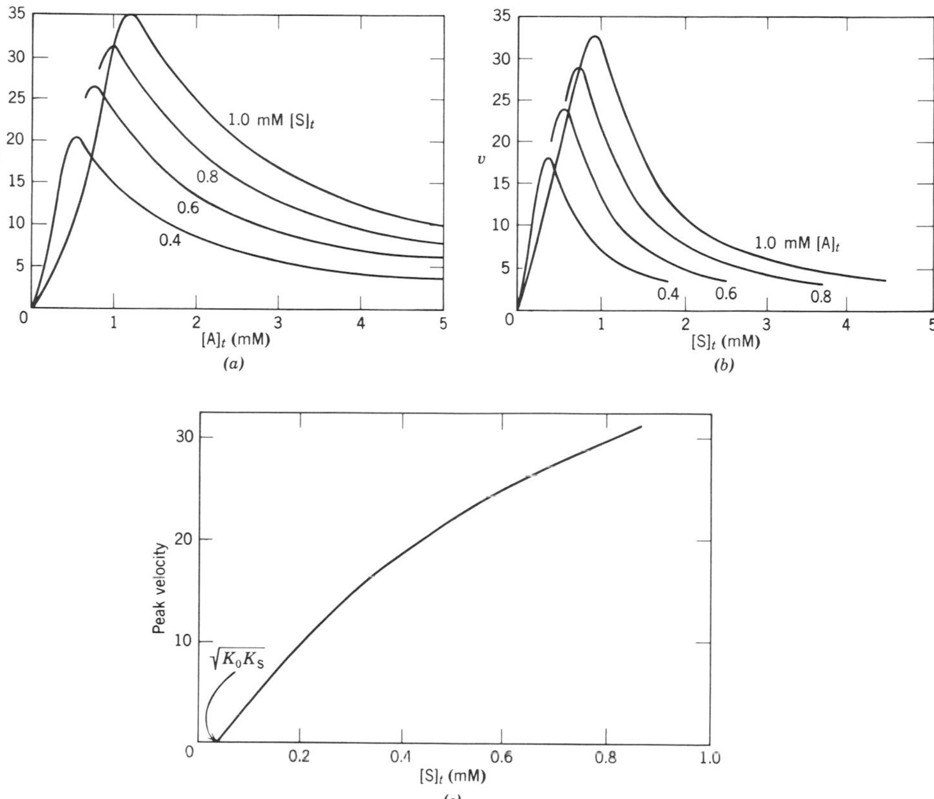

Fig. V-15. System B4. (*a*) v versus $[A]_t$ at different fixed concentrations of $[S]_t$. (*b*) v versus $[S]_t$ at different fixed concentrations of $[A]_t$. (*c*) Replot of the peak velocities of *b* against the corresponding $[S]_t$. The kinetic constants are: $K_0 = 1.4 \times 10^{-5} M$, $K_{SA} = 9 \times 10^{-4} M$, $K_S = 1.3 \times 10^{-4} M$, and $K_{EA} = 4.5 \times 10^{-4} M$. [Redrawn with permission from London, W. P. and Steck, T. L., *Biochemistry* **8**, 1767 (1969). Copyright by the American Chemical Society.]

System B5. A Is an Essential Activator

System B5a. A Is Essential for the Binding of S, A, and SA to the Catalytic Site

The equilibria shown on the top face of the equilibrium cube (Fig. V-9) plus the reaction $E + A \rightleftharpoons AE$ represent an essential activation system where A binds to a specific activator site and promotes the binding of S, A, and SA to the catalytic site. The velocity equation for this system is:

$$\boxed{\frac{v}{V_{\max}} = \frac{\dfrac{[\mathrm{A}][\mathrm{SA}]}{K_{\mathrm{A}}K'_{\mathrm{SA}}}}{1+\dfrac{[\mathrm{A}]}{K_{\mathrm{A}}}\left(1+\dfrac{[\mathrm{S}]}{K'_{\mathrm{S}}}+\dfrac{[\mathrm{A}]}{K'_{\mathrm{EA}}}+\dfrac{[\mathrm{SA}]}{K'_{\mathrm{SA}}}\right)}} \qquad \text{(V-51)}$$

where $V_{\max} = \gamma k_p[\mathrm{E}]_t$.

The v versus $[\mathrm{S}]_t$ curves at fixed concentrations of $[\mathrm{A}]_t$ and v versus $[\mathrm{A}]_t$ curves at fixed concentrations of $[\mathrm{S}]_t$ are similar to those of System B4. The ascending portion of the v versus $[\mathrm{S}]_t$ curves may be concave or sigmoidal. The inhibition after the peak velocity results from the "deactivation" of the enzyme as the $\mathrm{S}+\mathrm{A}\rightleftharpoons\mathrm{SA}$ reaction ties up A, as well as from the formation of the inactive AES complex. The ascending portion of the v versus $[\mathrm{A}]_t$ curves is always sigmoidal because increasing $[\mathrm{A}]_t$ both adds activator and "de-inhibits" the enzyme by complexing S. The velocity curve obtained with equimolar $[\mathrm{S}]_t$ and $[\mathrm{A}]_t$ is always sigmoidal, more so than for System B1. Increasing $[\mathrm{S}]_t+[\mathrm{A}]_t$ by a given factor not only increases the concentration of the true substrate, SA, by a greater factor, but also increases the concentration of the activator, A. In fact, the velocity depends on $[\mathrm{A}]^2$ as we can see by writing the velocity equation in terms of *free* A and *free* S.

$$\boxed{\frac{v}{V_{\max}} = \frac{\dfrac{[\mathrm{A}]^2[\mathrm{S}]}{K_{\mathrm{A}}K_0K'_{\mathrm{SA}}}}{1+\dfrac{[\mathrm{A}]}{K_{\mathrm{A}}}\left(1+\dfrac{[\mathrm{S}]}{K'_{\mathrm{S}}}+\dfrac{[\mathrm{A}]}{K'_{\mathrm{EA}}}+\dfrac{[\mathrm{A}][\mathrm{S}]}{K_0K'_{\mathrm{SA}}}\right)}} \qquad \text{(V-52)}$$

System B5b. A Is an Essential Activator and Only SA Binds to the Catalytic Site

System B5b is a simplified version of System B5a chosen by London and Steck to illustrate the effect of varying the fixed excess of $[\mathrm{S}]_t$ and $[\mathrm{A}]_t$ on the shape of the velocity curve. The equilibria (assuming A is essential for SA binding) are:

$$\begin{array}{ccccc} E & + & A & \overset{K_A}{\rightleftharpoons} & AE \\ & & + & & + \\ & & S & \overset{K_0}{\rightleftharpoons} & SA \\ & & & & \downarrow\uparrow K'_{SA} \\ & & & & AE(SA) \xrightarrow{k_p} AE + PA \end{array}$$

The velocity equation is:

$$\boxed{\frac{v}{V_{max}} = \frac{\dfrac{[A][SA]}{K_A K'_{SA}}}{1 + \dfrac{[A]}{K_A} + \dfrac{[A][SA]}{K_A K'_{SA}}}} \tag{V-53}$$

or

$$\boxed{\frac{v}{V_{max}} = \frac{\dfrac{[A]^2[S]}{K_A K_0 K'_{SA}}}{1 + \dfrac{[A]}{K_A} + \dfrac{[A]^2[S]}{K_A K_0 K'_{SA}}}} \tag{V-54}$$

The velocity curves where $[A]_t$ and $[S]_t$ are increased together are shown in Figure V-16. The curve is sigmoidal when $[A]_t = [S]_t$ and increases in sigmoidicity when $[S]_t$ is present at a fixed excess over $[A]_t$. In contrast to System B1, the sigmoidicity is apparent at concentrations greater than K_0. In this system free S does not combine with the enzyme, and, consequently, the sigmoidicity results solely from the reaction $S + A \rightleftharpoons SA$, which decreases the concentration of activator available for combination at the activator site. Reciprocal plots of $1/v$ versus $1/([A]_t = [S]_t)^2$ are approximately linear. When $[A]_t$ is maintained at a fixed excess over $[S]_t$, the velocity curve becomes less sigmoidal. At a sufficient fixed excess $[A]_t$, the curve is hyperbolic.

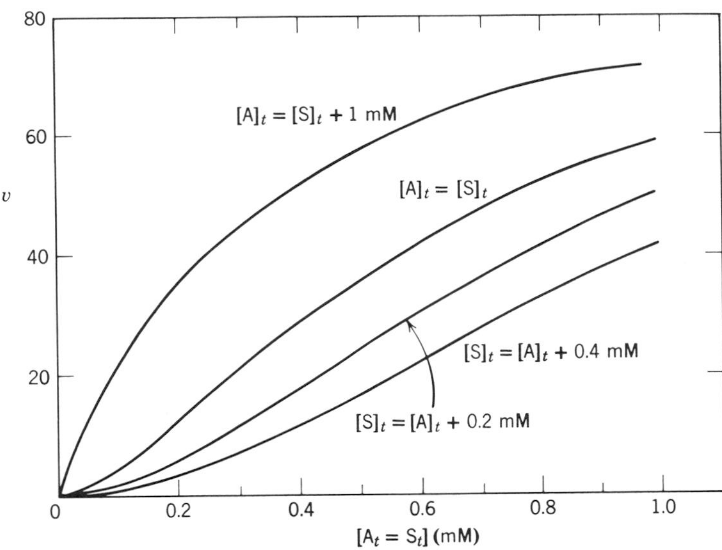

Fig. V-16. System B5b. Plot of v versus $[A]_t=[S]_t$; $[A]_t$ and $[S]_t$ are increased together. In addition, a fixed excess of $[A]_t$ or $[S]_t$ may be present. It is assumed that A is an essential activator required for SA binding and that free A and free S do not bind to the catalytic site. $K'_{SA}=3.2\times10^{-4}M$, $K_A=10^{-4}M$, $K_0=1.37\times10^{-5}M$. [Redrawn with permission from London, W. P. and Steck, T. L., *Biochemistry* **8**, 1767 (1969). Copyright by the American Chemical Society.]

System B5c. A Is an Essential Activator; A, S, and SA Bind Randomly to E and AE

In this case, A, S, and SA bind to free E and also to AE. The A is essential for catalytic activity as well as for formation of SA, but not for ligand binding to the catalytic site. The velocity equation is:

$$\frac{v}{V_{max}} = \frac{\dfrac{[A][SA]}{K_A K'_{SA}}}{\left(1+\dfrac{[S]}{K_S}+\dfrac{[A]}{K_{EA}}+\dfrac{[SA]}{K_{SA}}\right)+\dfrac{[A]}{K_A}\left(1+\dfrac{[S]}{K'_S}+\dfrac{[A]}{K'_{EA}}+\dfrac{[SA]}{K'_{SA}}\right)} \qquad \text{(V-55)}$$

The equation expresses the fact that all ligands bind in a rapid equilibrium random manner, but only the AE(SA) complex is catalytically active. The enzyme PRPP synthetase may have a mechanism similar to System B5a or System B5c. Free ATP is a competitive inhibitor with respect to $MgATP^{2-}$, while Ca^{2+} is a partial competitive inhibitor with respect to Mg^{2+}. Apparently, Ca^{2+} binds to the Mg^{2+} activator site and decreases (but does not abolish) the catalytic activity of the enzyme. The possibility that Ca^{2+} inhibited by forming an inactive $CaATP^{2-}$ complex could be discarded, since Ca^{2+} has only half the affinity of Mg^{2+} for ATP and yet the inhibition was maximal (ca. 60%) at 0.3 mM Ca^{2+} in the presence of 2 mM ATP and 10 mM Mg^{2+}(Switzer, 1971).

System B6. A Is a Nonessential Activator; Only SA Binds to the Catalytic Site

System B6 is described by the diagonal plane of the equilibrium cube shown in Figure V-9, plus the reaction $S+A \rightleftharpoons SA$. In effect, A binds to an activator site on E or E(SA), while SA binds to either free E or to AE. Both ESA and AESA are catalytically active. The velocity equation is obtained from:

$$v = k_p[E(SA)] + \gamma k_p[AE(SA)]$$

$$\boxed{v = \frac{\dfrac{[SA]}{K_{SA}} V_{max} + \dfrac{[A][SA]}{K_A K'_{SA}} V'_{max}}{1 + \dfrac{[A]}{K_A} + \dfrac{[SA]}{K_{SA}} + \dfrac{[A][SA]}{K_A K'_{SA}}}} \tag{V-56}$$

where $V_{max} = k_p[E]_t =$ the theoretical V_{max} of the unactivated enzyme
$V'_{max} = \gamma k_p[E]_t =$ the theoretical V_{max} of the activated enzyme
$= \gamma V_{max}$

Other forms of the velocity equation may be obtained by substituting:

$$[A]_t - [SA] = [A] \qquad \text{and} \qquad [SA] = \frac{[S][A]}{K_0}$$

The velocity curve with $[S]_t = [A]_t$ is slightly sigmoidal in the region of K_0 and approaches V'_{max} as a limit. The v versus $[S]_t$ curves are never sigmoidal, but show a peak velocity for each fixed $[A]_t$ (Fig. V-17). A peak occurs because as $[S]_t$ is increased [SA] increases, but the reaction $S+A \rightleftharpoons SA$ also decreases the A available for the activator site. When $[S]_t \gg [A]_t$, all the A is complexed as SA and the enzyme is completely unactivated. Thus for each fixed $[A]_t$, the velocity curve rises to a peak, then falls to a lower limiting

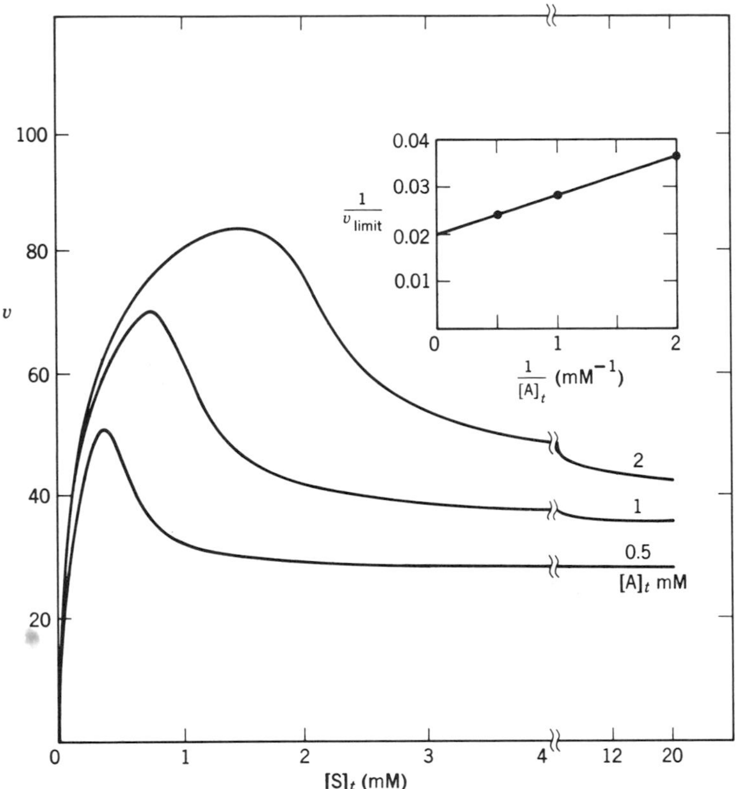

Fig. V-17. System B6. Plot of v versus $[S]_t$ at different fixed concentrations of $[A]_t$. It is assumed that A is a mixed-type, nonessential activator and SA is the true substrate. Free A and free S do not bind to the catalytic site. Insert: Replot of the limiting velocities to determine V_{max} and K_{SA}. The kinetic constants are: $K_0 = 10^{-5}M$, $K_{SA} = 4\times10^{-4}M$, $K'_{SA} = 2\times10^{-4}M$, $K_A = 10^{-4}M$, $V_{max} = 50$, and $V'_{max} = 100$ (i.e., $\gamma = 2$). [Redrawn with permission from London, W. P. and Steck, T. L., *Biochemistry* **8**, 1767 (1969). Copyright by the American Chemical Society.]

velocity given by:

$$\boxed{\frac{v_{limit}}{V_{max}} = \frac{[SA]}{K_{SA} + [SA]} = \frac{[A]_t}{K_{SA} + [A]_t}} \tag{V-57}$$

The V_{max} and K_{SA} can be determined from the intercepts of a reciprocal plot of $1/v_{limit}$ versus $1/[A]_t$. The v versus $[A]_t$ curves at several fixed concentrations of $[S]_t$ are shown in Figure V-18. At low fixed $[S]_t$, the curve is

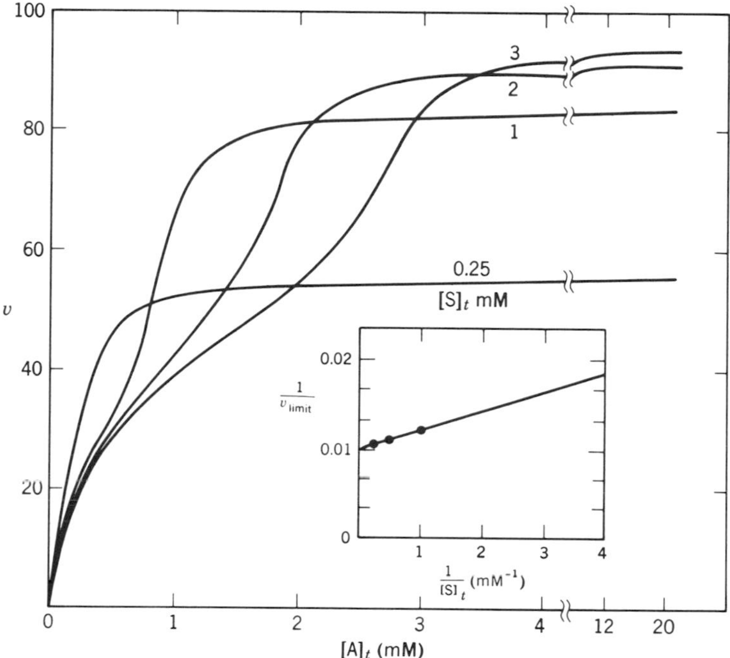

Fig. V-18. System B6. Plot of v versus $[A]_t$ at different fixed concentrations of $[S]_t$. Insert: Replot of limiting velocities to determine K'_{SA} and V'_{max}. [Redrawn with permission from London, W. P. and Steck, T. L., *Biochemistry* **8**, 1767 (1969). Copyright by the American Chemical Society.]

concave. At higher fixed $[S]_t$ the curves are initially concave, then sigmoidal. At low $[A]_t$ (and $[S]_t > [A]_t$) the observed velocity results primarily from the activity of the unactivated enzyme. As $[A]_t$ increases, the unactivated velocity approaches V_{max}, and, simultaneously, the activator site binds A resulting in an increase in the velocity as the relative proportion of the activated enzyme increases. When $[A]_t \gg [S]_t$, the enzyme is completely in the activated form and v approaches a limiting velocity for each $[S]_t$ given by:

$$\boxed{\frac{v_{limit}}{V'_{max}} = \frac{[SA]}{K'_{SA} + [SA]} = \frac{[S]_t}{K'_{SA} + [S]_t}} \qquad \text{(V-58)}$$

The V'_{max} and K'_{SA} can be determined from the intercepts of a reciprocal plot of $1/v_{limit}$ versus $1/[S]_t$. The curves shown in Figures V-17 and V-18 were drawn for a particular set of constants. With different constants, the shapes will differ.

It is noteworthy that the systems described above can yield sigmoidal velocity curves and curves suggesting negative cooperativity yet there are no multiple interacting sites (Chapter Seven). Sigmoidal v versus [I] plots are also possible where I is a metal ion that competes with A for S, or a compound that competes with S for the metal ion, A. Partial inhibition will be observed if the resulting SI (or IA) is active as a substrate. For a more complete discussion of metal ion modifier systems the reader is referred to the fine articles by London and Steck (1969), Switzer (1971), and DeWeer and Lowe (1973) listed in the references.

System B7. Both S and SA Are Substrates (ES and ESA Are Catalytically Active)

When both S and SA are substrates and free A does not bind to E, the equilibria are:

$$\begin{array}{ccccc} E & + & S & \overset{K_S}{\rightleftharpoons} & ES \overset{k_p}{\longrightarrow} E+P \\ + & & + & & \\ SA & \underset{K_0}{\rightleftharpoons} & A & & \\ K_{SA}\Big\updownarrow & & & & \\ ESA & \overset{\beta k_p}{\longrightarrow} & E & + & PA \end{array}$$

A may also combine with ES to form ESA. The velocity is given by:

$$v = k_p[ES] + \beta k_p[ESA]$$

and

$$\frac{v}{[E]_t} = \frac{k_p \dfrac{[S]}{K_S} + \beta k_p \dfrac{[SA]}{K_{SA}}}{1 + \dfrac{[S]}{K_S} + \dfrac{[SA]}{K_{SA}}} \qquad \text{(V-59)}$$

where β will usually be >1. The best way to study the system is to use S and SA separately as substrates (i.e., determine v as a function of [S] in the absence of A and in the presence of a large excess of A). This will permit K_S, V_{max}, K_{SA}, and βV_{max} to be determined. If the A concentration is not maintained at a saturating concentration, then the relative proportions of S and SA will vary as [S] varies and a complicated velocity curve may result. The complications can be avoided and a normal velocity curve obtained even at unsaturating [A] provided the ratio of [S]/[SA] is maintained constant. This can be accomplished if the concentration of free A is maintained constant, since:

$$K_0 = \frac{[\mathrm{S}][\mathrm{A}]}{[\mathrm{SA}]} \qquad \text{or} \qquad \frac{[\mathrm{SA}]}{[\mathrm{S}]} = \frac{[\mathrm{A}]}{K_0} = x, \quad \text{a constant}$$

$$\therefore \quad [\mathrm{SA}] = x[\mathrm{S}]$$

The velocity equation becomes:

$$\frac{v}{[\mathrm{E}]_t} = \frac{k_p \dfrac{[\mathrm{S}]}{K_\mathrm{S}} + \beta k_p \dfrac{x[\mathrm{S}]}{K_\mathrm{SA}}}{1 + \dfrac{[\mathrm{S}]}{K_\mathrm{S}} + \dfrac{x[\mathrm{S}]}{K_\mathrm{SA}}} = \frac{\beta k_p \dfrac{[\mathrm{S}]}{K_\mathrm{S}}\left(\dfrac{1}{\beta} + \dfrac{xK_\mathrm{S}}{K_\mathrm{SA}}\right)}{1 + \dfrac{[\mathrm{S}]}{K_\mathrm{S}}\left(1 + \dfrac{xK_\mathrm{S}}{K_\mathrm{SA}}\right)} \tag{V-60}$$

Designating $\beta k_p[\mathrm{E}]_t$ as $V'_{\max}$ and dividing numerator and denominator by the numerator parenthetical term:

$$\frac{v}{V'_{\max}} = \frac{\dfrac{[\mathrm{S}]}{K_\mathrm{S}}}{1 + \dfrac{[\mathrm{S}]}{K_\mathrm{S}} \dfrac{\left(1 + \dfrac{xK_\mathrm{S}}{K_\mathrm{SA}}\right)}{\left(\dfrac{1}{\beta} + \dfrac{xK_\mathrm{S}}{K_\mathrm{SA}}\right)}} = \frac{[\mathrm{S}]}{K_\mathrm{S} + [\mathrm{S}] \dfrac{\left(1 + \dfrac{xK_\mathrm{S}}{K_\mathrm{SA}}\right)}{\left(\dfrac{1}{\beta} + \dfrac{xK_\mathrm{S}}{K_\mathrm{SA}}\right)}} \tag{V-61}$$

The parenthetical terms are composed solely of constants. Thus the equation has the usual form:

$$\frac{v}{V_{\max_\mathrm{app}}} = \frac{[\mathrm{S}]}{K_{\mathrm{S}_\mathrm{app}} + [\mathrm{S}]} \tag{V-62}$$

where

$$V_{\max_\mathrm{app}} = V'_{\max} \frac{\left(\dfrac{1}{\beta} + \dfrac{xK_\mathrm{S}}{K_\mathrm{SA}}\right)}{\left(1 + \dfrac{xK_\mathrm{S}}{K_\mathrm{SA}}\right)}, \qquad K_{\mathrm{S}_\mathrm{app}} = K_\mathrm{S} \frac{\left(\dfrac{1}{\beta} + \dfrac{xK_\mathrm{S}}{K_\mathrm{SA}}\right)}{\left(1 + \dfrac{xK_\mathrm{S}}{K_\mathrm{SA}}\right)}$$

The apparent kinetic constants vary with x. When x is very large (i.e., all the substrate is present as SA) $V_{\max_\mathrm{app}} = V'_{\max} = \beta k_p[\mathrm{E}]_t$. When x is very small (i.e., all the substrate is present as free S), $V_{\max_\mathrm{app}} = V'_{\max}/\beta = k_p[\mathrm{E}]_t = V_{\max}$.

REFERENCES

Regulation by "Energy Charge"

Atkinson, D. E. and Walton, G. M., *J. Biol. Chem.* **242**, 3239 (1967).

Atkinson, D. E. and Fall, L., *J. Biol. Chem.* **242**, 3241 (1967).

Atkinson, D. E., "Enzymes as Control Elements in Metabolic Regulation," in *The Enzymes*, Vol. 1, 3rd ed., P. Boyer, ed., Academic Press, 1970, p. 461.

Purich, D. L. and Fromm, H. J., *J. Biol. Chem.* **247**, 249 (1972).

Purich, D. L. and Fromm, H. J., *J. Biol. Chem.* **248**, 461 (1973).

Shen, L. C. and Atkinson, D. E., *J. Biol. Chem.* **245**, 3996 (1970).

Shen, L. C. and Atkinson, D. E., *J. Biol. Chem.* **245**, 5974 (1970).

Metal Ion Activation (Substrate-Activator Complexes)

Dixon, M. and Webb, E. C., *Enzymes*, 2nd ed., Academic Press, 1964, Ch. 9.

Janson, C. A. and Cleland, W. W., *J. Biol. Chem.* **249**, 2567, 2572 (1974) (Chromium ATP as a dead-end inhibitor competitive with MgATP).

London, W. P. and Steck, T. L., *Biochemistry* **8**, 1767 (1969).

Reiner J., *Behavior of Enzyme Systems*, 2nd ed., Van Nostrand-Reinhold, 1969, Ch. 8.

Specific Examples

Ainsworth, S. and Macfarlane, N., *Biochem. J.* **131**, 223 (1973) (Pyruvate kinase).

Baykov, A. A. and Avaeva, S. M., *Eur. J. Biochem.* **32**, 136 (1973) (Inorganic pyrophosphatase).

Colomb, M. G., Cheruy, A. and Vignais, V., *Biochemistry* **11**, 3370 (1972) (Nucleoside diphosphokinase).

DeWeer, P. and Lowe, A. G., *J. Biol. Chem.* **248**, 2829 (1973) (Myokinase).

Gerber, G., Preissler, H., Heinrich, R. and Rapoport, S. M., *Eur. J. Biochem.* **45**, 39 (1974) (Hexokinase).

Gold, M. H. and Segel, I. H., *J. Biol. Chem.* **249**, 2417 (1974) (Protein kinase).

Hammes, G. G. and Kochavi, D., *J. Am. Chem. Soc.* **84**, 2069, 2073, 2076 (1962) (Hexokinase).

Hubbard, J. S. and Stadtman, E. R., *J. Bacteriol.* **94**, 1007 (1967) (Glutamine synthetase).

McGregor, W. G., Phillips, J. and Suelter, C. H., *J. Biol. Chem.* **249**, 3132 (1974) (Glycerol dehyrdogenase).

Miziorko, H. M., Nowak, T. and Mildvan, A. S., *Arch. Biochem. Biophys.* **163**, 378 (1974) (PEP carboxylase).

Purich, D. L. and Fromm, H. J., *Biochem. J.* **130**, 63 (1972) (Hexokinase).

Switzer, R. L., *J. Biol. Chem.* **246**, 2447 (1971) (PRPP synthetase).

Tatibana, M. and Shigesada, K., *J. Biochemistry* **72**, 537 (1972) (Carbamylphosphate synthetase).

Tenn, J. P., Viratelle, O. M. and Yon, J., *Eur. J. Biochem.* **26**, 112 (1972) (β-galactosidase).

Tweedie, J. W. and Segel, I. H., *J. Biol. Chem.* **246**, 3438 (1971) (ATP sulfurylase).

CHAPTER SIX

RAPID EQUILIBRIUM BIREACTANT AND TERREACTANT SYSTEMS

Most enzymes catalyze reactions between two or more substrates to yield two or more products. Yet, most introductory biochemistry texts restrict their discussion to unireactant systems. As a result, it is not always appreciated that the K_m for a particular substrate at one fixed set of cosubstrate concentrations may not be the "real" K_m, but, rather, an apparent value which changes as the cosubstrate concentrations vary. Similarly, the observed $V_{\max}$ of a preparation at a saturating concentration of one substrate may not be the same $V_{\max}$ observed when another substrate is saturating. The true K_m for a particular substrate is that observed when all other substrates are saturating. The true $V_{\max}$ is observed when all substrates are present at saturating concentrations. Inhibition constants are also affected by the fixed concentrations of substrates. Thus an observed K_i may not represent a true inhibitor dissociation constant.

In this chapter the subject of multireactant enzymes is introduced by examining some rapid equilibrium bireactant and terreactant systems. We indicate the ligands as A and B, where B is a substrate and A can be a cosubstrate or coenzyme or an essential activator. Since we are assuming that all binding and dissociation steps are very rapid compared to the catalytic step, the rate limiting step will be indicated simply as $EAB \xrightarrow{k_p} E + P$. This rapid equilibrium approach is useful and reasonably valid for the random binding systems described in this chapter. However, the reader should keep in mind that for most enzymes that catalyze ordered reaction sequences, the steady-state approach (described in Chapter Nine) is preferred.

A. RANDOM BIREACTANT SYSTEMS

System A1. Initial Velocity Studies

If A and B bind randomly, and if the binding of one substrate changes the dissociation constant for the other substrate by a factor α, the system can be described by the equilibria shown below.

$$\begin{array}{ccccc} E & + & B & \overset{K_B}{\rightleftharpoons} & EB \\ + & & & & + \\ A & & & & A \\ K_A \updownarrow & & & & \updownarrow \alpha K_A \\ EA & + & B & \overset{\alpha K_B}{\rightleftharpoons} & EAB \xrightarrow{k_p} E + P \end{array}$$

The complexes EA, EB, and EAB are called *transitory complexes*. A transitory complex is defined as any enzyme species capable of a unimolecular reaction releasing a substrate or product, or capable of isomerizing to such an intermediate. A transitory complex which cannot participate in a bimolecular reaction with another ligand (because all binding sites are occupied), but can only undergo a unimolecular reaction releasing a substrate or product, is called a *central complex*. Thus EAB is a central complex. If EAB isomerizes to EPQ which then releases P or Q, EPQ would be another central complex, while EP and EQ are ordinary transitory complexes. Enzyme forms which cannot undergo a unimolecular dissociation reaction (or isomerize to a form which can) are called *stable enzyme forms*. In this system, free E is the only stable enzyme form.

The term *sequential* is used to denote systems in which all substrates must bind to the enzyme before any product is released. Thus the random system is *sequential* even though A and B bind randomly. An ordered system in which A must bind before B is also *sequential* because no product is released before the formation of the ternary EAB complex. Ping Pong systems, in which a product is released between substrate additions (Chapter Nine), are *nonsequential*.

The velocity equation for the random system can be derived from rapid equilibrium assumptions in the usual way:

$$v = k_p[\mathrm{EAB}] \qquad \text{and} \qquad \frac{v}{[\mathrm{E}]_t} = \frac{k_p[\mathrm{EAB}]}{[\mathrm{E}]+[\mathrm{EA}]+[\mathrm{EB}]+[\mathrm{EAB}]}$$

Substituting and designating $k_p[E]_t$ as V_{max}, we obtain:

$$\frac{v}{V_{max}} = \frac{\dfrac{[A][B]}{\alpha K_A K_B}}{1 + \dfrac{[A]}{K_A} + \dfrac{[B]}{K_B} + \dfrac{[A][B]}{\alpha K_A K_B}} \tag{VI-1}$$

or

$$\frac{v}{V_{max}} = \frac{[A][B]}{\alpha K_A K_B + \alpha K_A[B] + \alpha K_B[A] + [A][B]} \tag{VI-2}$$

When rearranged to reflect the dependence of the initial velocity on the concentration of the varied ligand, the equations are:

$$\frac{v}{V_{max}} = \frac{[A]}{\alpha K_A\left(1 + \dfrac{K_B}{[B]}\right) + [A]\left(1 + \dfrac{\alpha K_B}{[B]}\right)} \tag{VI-3}$$

$$\frac{v}{V_{max}} = \frac{[B]}{\alpha K_B\left(1 + \dfrac{K_A}{[A]}\right) + [B]\left(1 + \dfrac{\alpha K_A}{[A]}\right)} \tag{VI-4}$$

The velocity equation can be expressed as:

$$\frac{v}{V_{max_{app}}} = \frac{[A]}{K_{A_{app}} + [A]} \qquad \text{(for varied [A])}$$

$$\frac{v}{V_{max_{app}}} = \frac{[B]}{K_{B_{app}} + [B]} \qquad \text{(for varied [B])}$$

where

$$V_{\max_{app}} = \frac{V_{\max}}{\left(1 + \frac{\alpha K_B}{[B]}\right)}, \qquad V_{\max_{app}} = \frac{V_{\max}}{\left(1 + \frac{\alpha K_A}{[A]}\right)}$$

$$K_{A_{app}} = \alpha K_A \frac{\left(1 + \frac{K_B}{[B]}\right)}{\left(1 + \frac{\alpha K_B}{[B]}\right)}, \qquad K_{B_{app}} = \alpha K_B \frac{\left(1 + \frac{K_A}{[A]}\right)}{\left(1 + \frac{\alpha K_A}{[A]}\right)}$$

In reciprocal form, the equations become:

$$\boxed{\frac{1}{v} = \frac{\alpha K_A}{V_{\max}}\left(1 + \frac{K_B}{[B]}\right)\frac{1}{[A]} + \frac{1}{V_{\max}}\left(1 + \frac{\alpha K_B}{[B]}\right)} \qquad \text{(VI-5)}$$

$$\boxed{\frac{1}{v} = \frac{\alpha K_B}{V_{\max}}\left(1 + \frac{K_A}{[A]}\right)\frac{1}{[B]} + \frac{1}{V_{\max}}\left(1 + \frac{\alpha K_A}{[A]}\right)} \qquad \text{(VI-6)}$$

If α is less than unity (i.e., the binding of one ligand increases the affinity of the enzyme for the other ligand), the K_{app} for the varied ligand decreases as the concentration of the fixed ligand increases. The family of reciprocal plots, shown in Figure VI-1 for varied [B], intersect at a point above the 1/[B]-axis and pivot clockwise about this point. The family of plots for varied [A] is symmetrical to that shown for varied [B]. The limiting plot (at a saturating concentration of fixed ligand) will intersect the $1/v$ and 1/[varied ligand]-axes at $1/V_{\max}$ and $1/\alpha K_{var}$, respectively. This is expected because at a saturating concentration of A, for example, all the enzyme available for combination with B exists as EA, and the reciprocal plot gives the $-1/K_{B_{app}}$ for the reaction EA + B$\rightleftharpoons$EAB, which equals $-1/\alpha K_B$. As [A] decreases to near zero concentration, the major reaction of enzyme with B becomes E + B$\rightleftharpoons$EB. Thus $K_{B_{app}}$ approaches K_B as [A] decreases. Because A is essential for the reaction, a reciprocal plot whose $1/K_{B_{app}}$ equals $1/K_B$ can never be obtained experimentally. However, we can see from Figure VI-1 that the limiting plot is a vertical line through the intersection point which intercepts the 1/[B]-axis at $-1/K_B$. The position of the intersection point

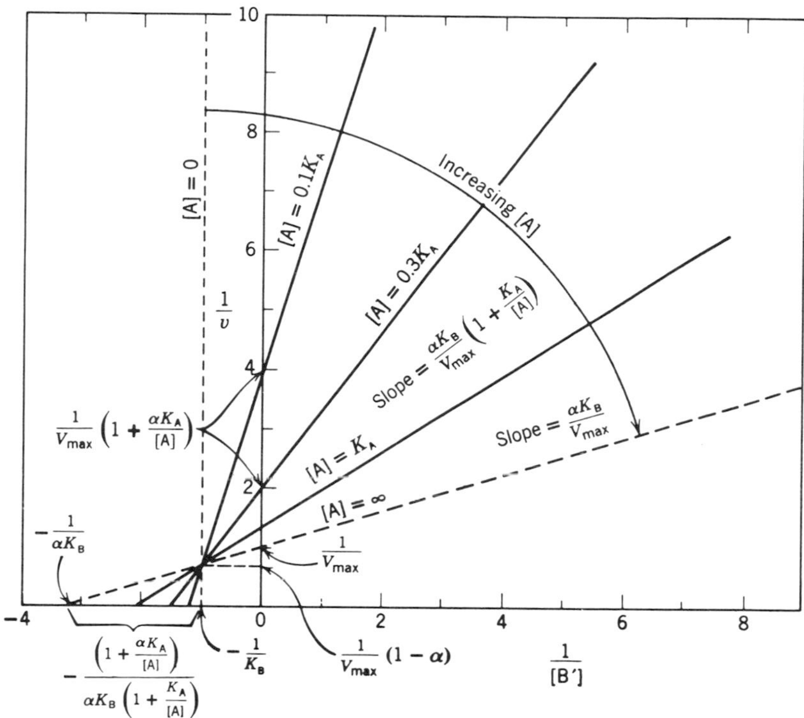

Fig. VI-1. The $1/v$ versus $1/[B]$ plot at different fixed concentrations of A for a Rapid Equilibrium Random system where $\alpha = 0.3$. The family of $1/v$ versus $1/[A]$ plots are symmetrical to those shown for varied [B].

can be verified by solving for the value of $1/[B]$ that makes the value of $1/v$ equal for two different concentrations of A. Similarly, the family of $1/v$ versus $1/[A]$ plots intersect at $-1/K_A$. The $1/v$ coordinate of the intersection point can be determined by substituting $-1/K_B$ for $1/[B]$ or $-1/K_A$ for $1/[A]$ in the reciprocal equation. The coordinate is:

$$\frac{1}{v} = \frac{1}{V_{max}}(1-\alpha)$$

Thus if $\alpha = 1$ (i.e., one ligand has no effect on the binding of the other), the plots intersect on the horizontal axis at $-1/K_B$ or $-1/K_A$. A more general solution to the intersection coordinates is given by:

$$\frac{1}{[B]} = -\frac{1}{K_B^*}\left(\frac{K_{int}}{K_{slope}}\right) \quad \text{and} \quad \frac{1}{v} = \frac{1}{V_{max}^*}\left(1-\frac{K_{int}}{K_{slope}}\right) \qquad \text{(VI-7)}$$

where K_B^* and V_{max}^* are the limiting constants at the concentrations of substrates and activators which are held constant while two others are varied. The K_{int} and K_{slope} are apparent intercept and slope constants for the substrate which is maintained at a different fixed concentration for each plot. In the present system, there are only two ligands so that $K_B^* = \alpha K_B$, $V_{max}^* = V_{max}$, $K_{int} = \alpha K_A$, and $K_{slope} = K_A$. If a third ligand (e.g., C) were present, K_B^*, V_{max}^*, K_{int}, and K_{slope} would be functions of $K_C/[C]$.

The value of V_{max} is conveniently obtained from replots of $1/V_{max_{app}}$ versus $1/[A]$ or $1/V_{max_{app}}$ versus $1/[B]$. From the $1/v$ versus $1/[A]$ plot:

$$\frac{1}{V_{max_{app}}} = \frac{\alpha K_B}{V_{max}} \frac{1}{[B]} + \frac{1}{V_{max}} \tag{VI-8}$$

From the $1/v$ versus $1/[B]$ plot:

$$\frac{1}{V_{max_{app}}} = \frac{\alpha K_A}{V_{max}} \frac{1}{[A]} + \frac{1}{V_{max}} \tag{VI-9}$$

The replots also give αK_A and αK_B (Fig. VI-2*a*, *d*). The slope replots give K_A and K_B (Fig. VI-2*b*, *e*). From the $1/v$ versus $1/[A]$ plot:

$$slope_{1/A} = \frac{\alpha K_A K_B}{V_{max}} \frac{1}{[B]} + \frac{\alpha K_A}{V_{max}} \tag{VI-10}$$

From the $1/v$ versus $1/[B]$ plot:

$$slope_{1/B} = \frac{\alpha K_B K_A}{V_{max}} \frac{1}{[A]} + \frac{\alpha K_B}{V_{max}} \tag{VI-11}$$

With αK_A and K_A (or αK_B and K_B) known, α can be calculated. Replots of $1/\Delta K_B$ versus $1/[A]$ and $1/\Delta K_A$ versus $1/[B]$ (Fig. VI-2*c*, *f*) can also be constructed in order to determine α. The equations are:

$$\frac{1}{\Delta K_B} = \frac{\alpha K_A}{K_B(1-\alpha)} \frac{1}{[A]} + \frac{1}{K_B(1-\alpha)} \tag{VI-12}$$

$$\frac{1}{\Delta K_A} = \frac{\alpha K_B}{K_A(1-\alpha)} \frac{1}{[B]} + \frac{1}{K_A(1-\alpha)} \tag{VI-13}$$

Figure VI-3 shows the alternate linear plots that can be used to determine all the kinetic constants.

If the binding of one ligand decreases the affinity for the second ligand ($\alpha > 1$), the family of reciprocal plots will intersect below the $1/[B]$-and $1/[A]$-axes at $-1/K_B$ and $-1/K_A$, respectively (Fig. VI-4). In this case, the apparent K value for the varied ligand increases as the concentration of

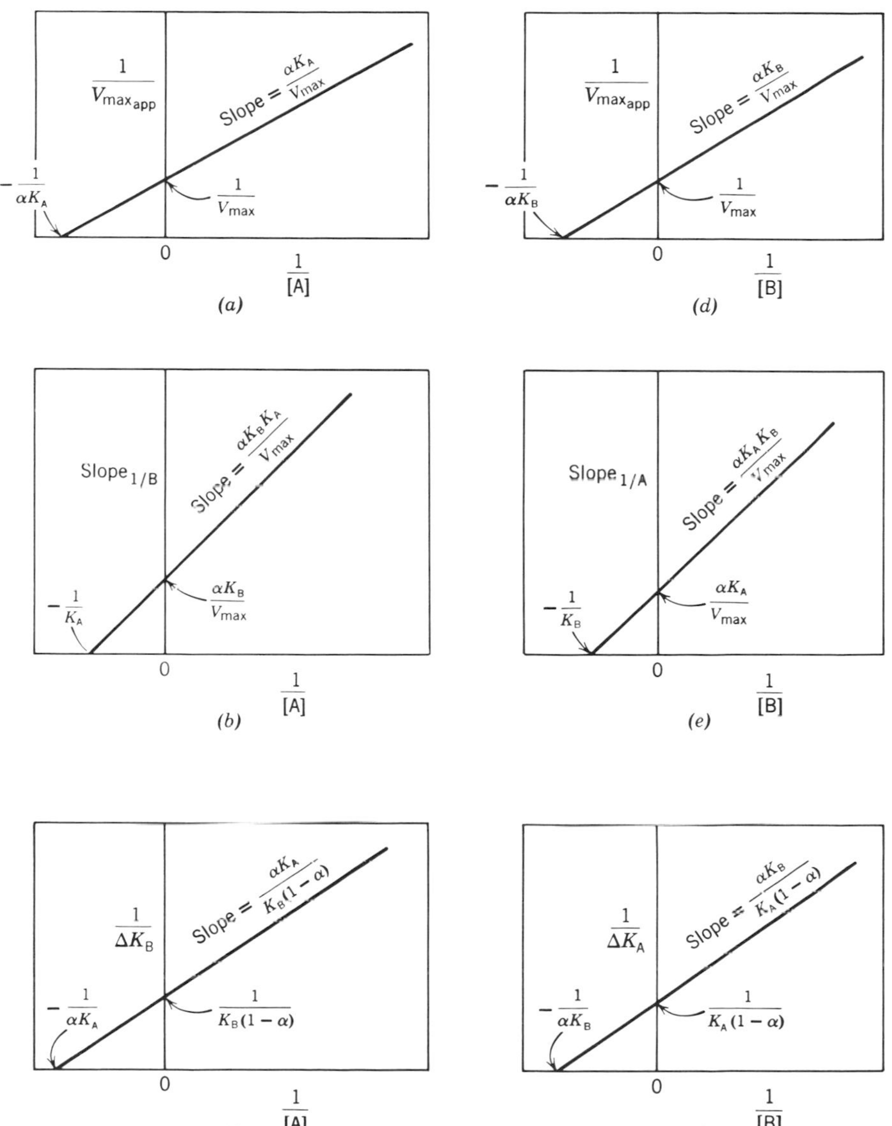

Fig. VI-2. (*a*)–(*c*) Replots of data taken from plots of $1/v$ versus $1/[B]$ at different fixed concentrations of A. The functions are replotted against the corresponding $1/[A]$. (*d*)–(*f*) Replots of data taken from plots of $1/v$ versus $1/[A]$ at different fixed concentrations of B. The functions are replotted against the corresponding $1/[B]$.

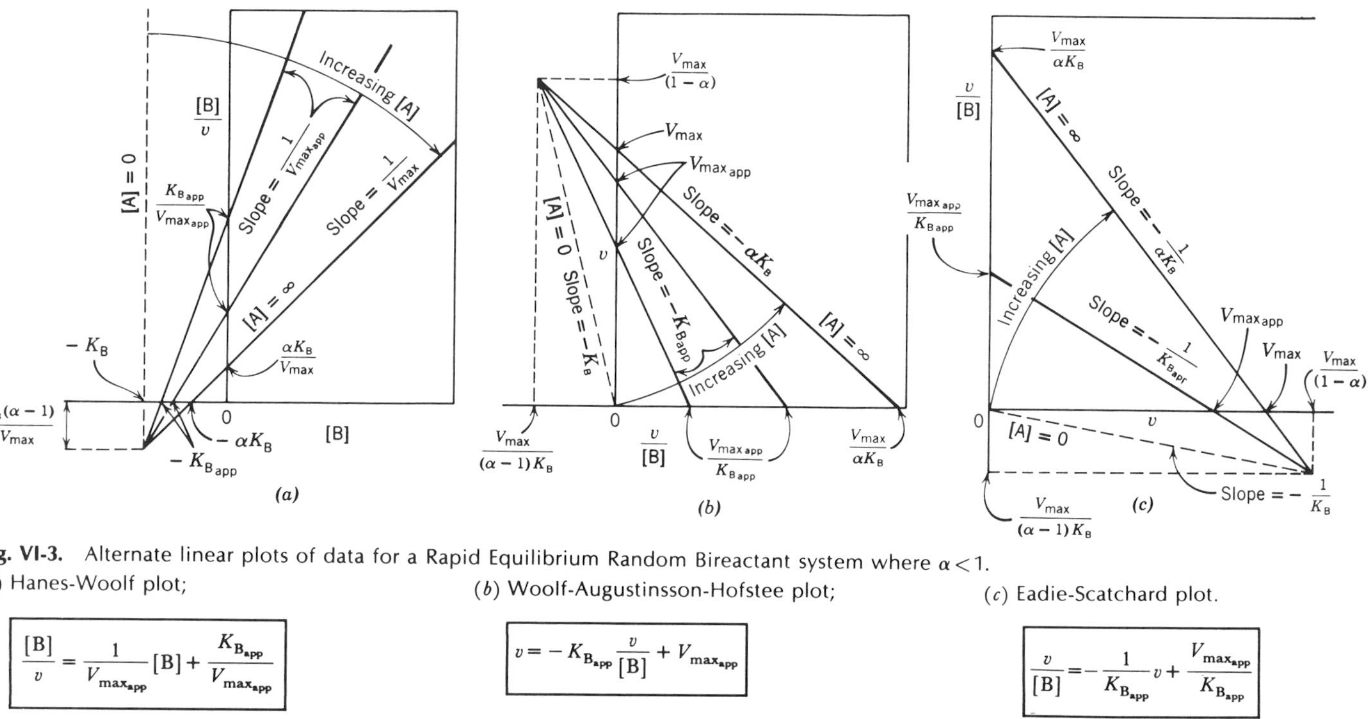

Fig. VI-3. Alternate linear plots of data for a Rapid Equilibrium Random Bireactant system where $\alpha < 1$. (*a*) Hanes-Woolf plot; (*b*) Woolf-Augustinsson-Hofstee plot; (*c*) Eadie-Scatchard plot.

$$\frac{[B]}{v} = \frac{1}{V_{max_{app}}}[B] + \frac{K_{B_{app}}}{V_{max_{app}}}$$

$$v = -K_{B_{app}}\frac{v}{[B]} + V_{max_{app}}$$

$$\frac{v}{[B]} = -\frac{1}{K_{B_{app}}}v + \frac{V_{max_{app}}}{K_{B_{app}}}$$

fixed ligand increases. The limiting plots give $-1/\alpha K_B$, $-1/\alpha K_A$, and $1/V_{max}$, as usual. If a replot of $1/\Delta K_B$ versus $1/[A]$ or $1/\Delta K_A$ versus $1/[B]$ is constructed, positive slopes can be obtained by designating ΔK as $K_{app} - K$ instead of $K - K_{app}$. The intercept on the $1/\Delta K_B$-axis then becomes $1/K_B(\alpha - 1)$ and the slope becomes $\alpha K_A / K_B(\alpha - 1)$. The alternate linear plots are shown in Figure VI-5.

Replotting $1/v$ Versus $1/[B]$ Data to Obtain $1/v$ Versus $1/[A]$ Plots

It is not always necessary to perform two separate experiments to construct reciprocal plots for two varied ligands. One family of plots (e.g., $1/v$ versus $1/[B]$) may contain all the information necessary for the other family of plots (e.g., $1/v$ versus $1/[A]$). This is illustrated in Figure VI-6 for $\alpha = 1$. The experiment was designed to provide data for six different concentrations of B at five different fixed concentrations of A (Fig. VI-6*a*). The data of Figure VI-6*a* are replotted in Figure VI-6*b*.

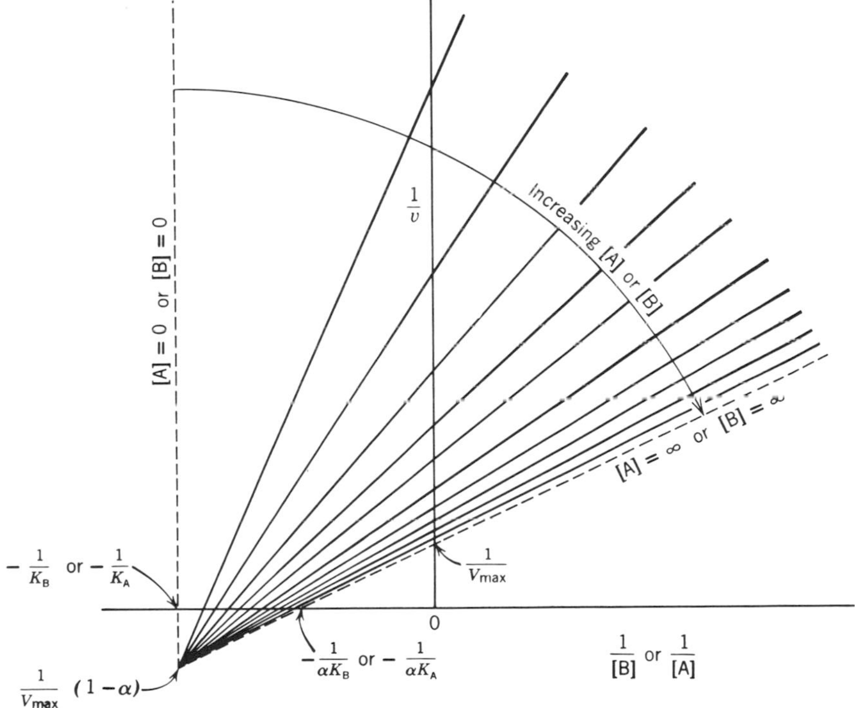

Fig. VI-4. Plot of $1/v$ versus $1/[B]$ or $1/[A]$ at different fixed concentrations of the nonvaried ligand for a Rapid Equilibrium Random Bireactant system where $\alpha > 1$.

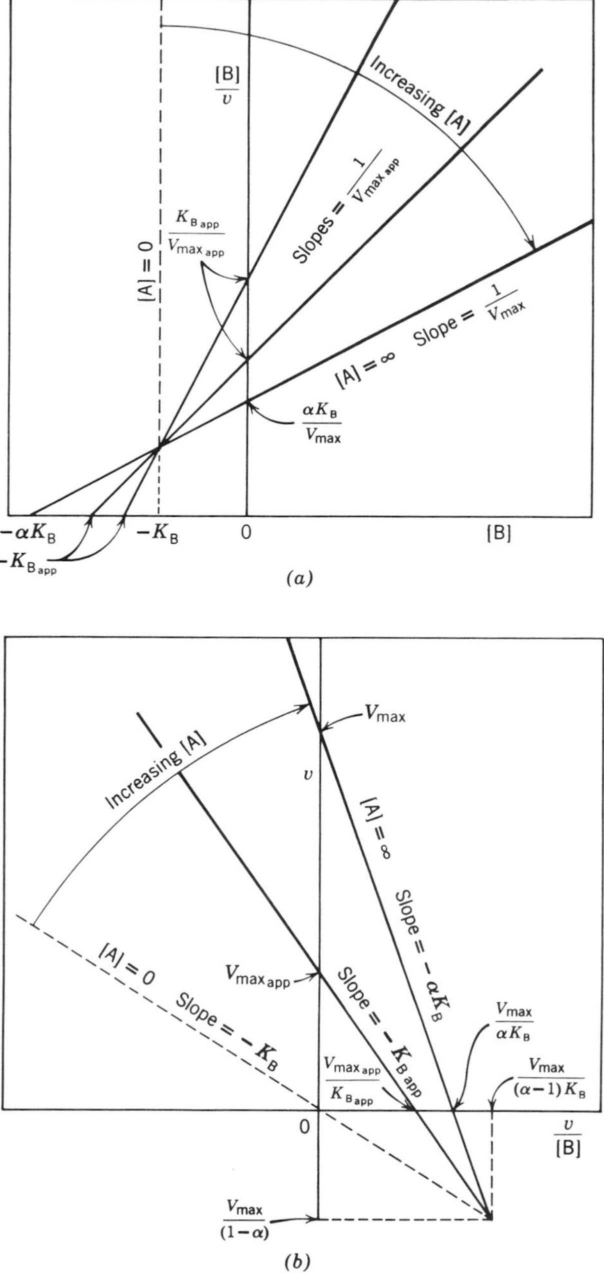

Fig. VI-5. Alternate linear plots for a Rapid Equilibrium Random Bireactant system where $\alpha > 1$. (*a*) Hanes-Woolf plot; (*b*) Woolf-Augustinsson-Hofstee plot.

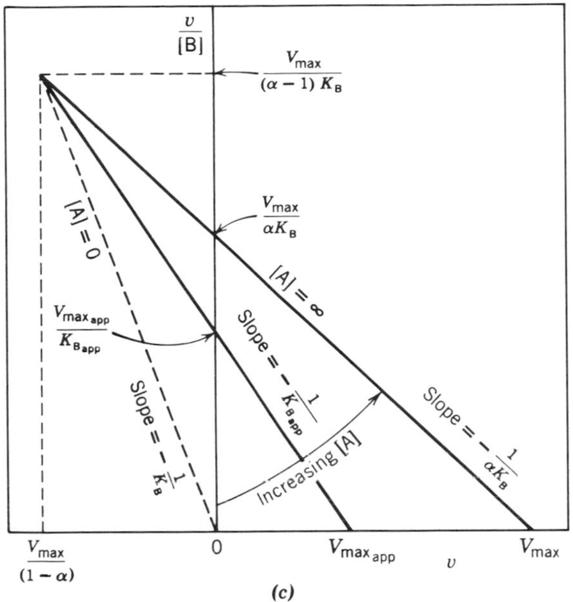

(c)

Fig. VI-5. (*Cont.*) (c) Eadie-Scatchard plot.

System A2. Inhibitor Competes with One Substrate

System A2a. EIB Is Catalytically Inactive

The equilibria shown below describe a rapid equilibrium random system in which an inhibitor competes with A but allows B to bind.

$$
\begin{array}{ccccc}
EI & + & B & \overset{\beta K_B}{\rightleftharpoons} & EIB \\
{\scriptstyle K_i}\downarrow\uparrow & & & & \uparrow\downarrow{\scriptstyle \beta K_i} \\
I & & & & I \\
+ & & & & + \\
E & + & B & \overset{K_B}{\rightleftharpoons} & EB \\
+ & & & & + \\
A & & & & A \\
{\scriptstyle K_A}\uparrow\downarrow & & & & \downarrow\uparrow{\scriptstyle \alpha K_A} \\
EA & + & B & \overset{\alpha K_B}{\rightleftharpoons} & EAB \overset{k_p}{\longrightarrow} E+P
\end{array}
$$

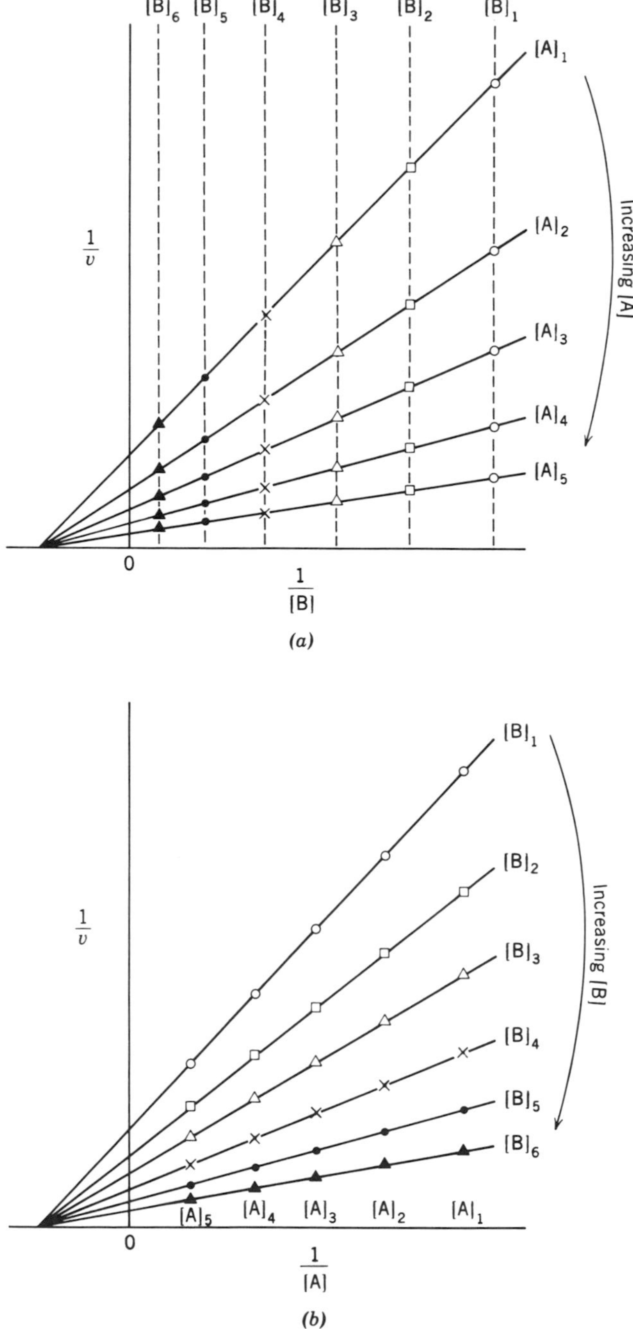

Fig. VI-6. Experiment designed so that the same thirty assays provide data which can be plotted as $1/v$ versus $1/[B]$ or as $1/v$ versus $1/[A]$.

The β represents the factor by which the dissociation constant for B is changed by the binding of I (and the factor by which the inhibitor dissociation constant is changed by the binding of B). The general velocity equation is:

$$\frac{v}{V_{max}} = \frac{\dfrac{[A][B]}{\alpha K_A K_B}}{1 + \dfrac{[A]}{K_A} + \dfrac{[B]}{K_B} + \dfrac{[I]}{K_i} + \dfrac{[A][B]}{\alpha K_A K_B} + \dfrac{[I][B]}{\beta K_i K_B}} \tag{VI-14}$$

The velocity equation can be rearranged to show either A or B as the varied ligand.

$$\frac{v}{V_{max}} = \frac{[A]}{\alpha K_A\left(1 + \dfrac{K_B}{[B]} + \dfrac{[I]K_B}{K_i[B]} + \dfrac{[I]}{\beta K_i}\right) + [A]\left(1 + \dfrac{\alpha K_B}{[B]}\right)} \tag{VI-15}$$

$$\frac{v}{V_{max}} = \frac{[B]}{\alpha K_B\left(1 + \dfrac{K_A}{[A]} + \dfrac{[I]K_A}{K_i[A]}\right) + [B]\left(1 + \dfrac{\alpha K_A}{[A]} + \dfrac{\alpha K_A[I]}{\beta K_i[A]}\right)} \tag{VI-16}$$

The reciprocal plots of $1/v$ versus $1/[A]$ will intersect on the $1/v$-axis at any constant [B], regardless of the value of β (Fig. VI-7). The intercept, which is independent of [I], gives $1/V_{max_{app}}$, where:

$$V_{max_{app}} = \frac{V_{max}}{\left(1 + \dfrac{\alpha K_B}{[B]}\right)} = \text{an apparent } V_{max} \text{ at the constant [B]}$$

The $V_{max_{app}}$ will equal V_{max} when $[B] \gg \alpha K_B$.

The $1/v$ versus $1/[B]$ plots at a constant [A] will intersect on, above, or

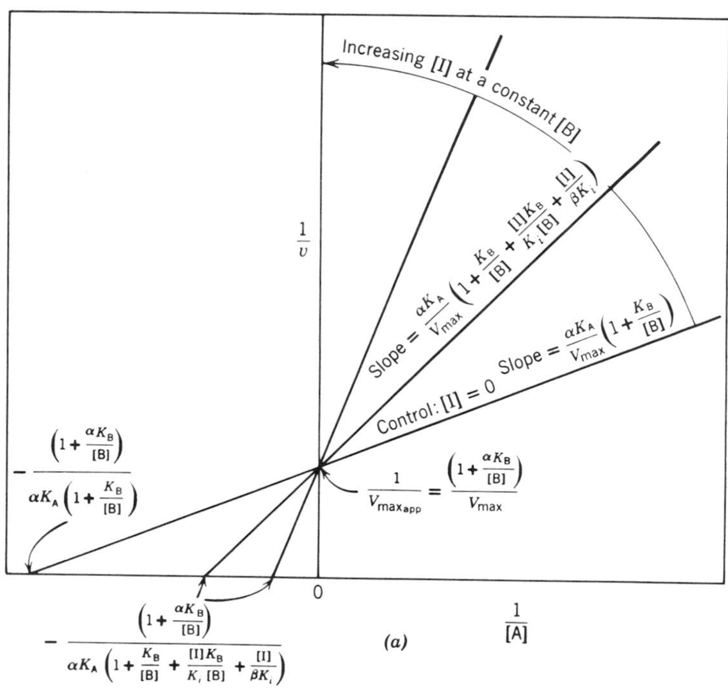

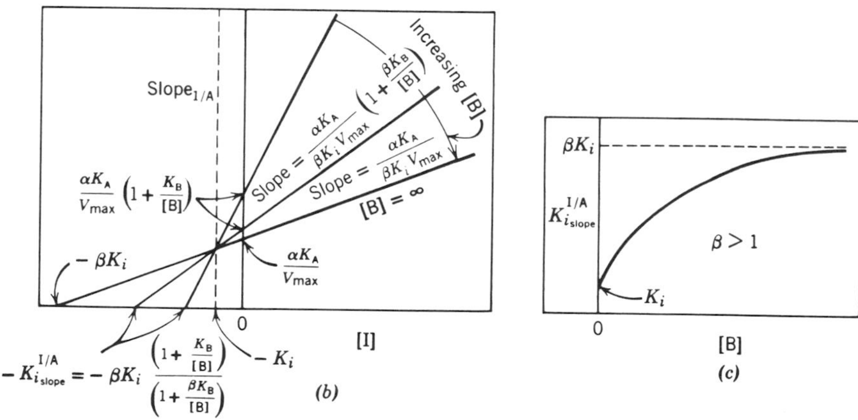

Fig. VI-7. (*a*) Plot of $1/v$ versus $1/[A]$ at different fixed concentrations of I and a constant [B]. A and B bind in a Rapid Equilibrium Random fashion; I competes with A but does not exclude B. EIB is catalytically inactive. (*b*) $Slope_{1/A}$ replot. Each line represents data from a family of reciprocal plots obtained at a different concentration of the constant ligand, B. (*c*) Secondary replot of $K_{i_{slope}}^{I/A}$ versus [B] (data taken from *b*).

below the $1/[B]$-axis depending on the values of α and β (Fig. VI-8*a*). For example, if $\alpha = \beta = 1$, the slope factor (i.e., the factor multiplying the K_B term) and the $1/v$-axis intercept factor (i.e., the factor multiplying the denominator [B] term) are equal. In this case, I behaves as a pure noncompetitive inhibitor with respect to B. If, coincidentally, $[A] = \alpha K_A(1-\beta)/(\beta - \alpha)$, the family of plots will intersect on the horizontal axis even though $\alpha \neq 1$ and $\beta \neq 1$. On the other hand, I will behave as a mixed-type inhibitor with respect to B if $\beta \neq 1$, even if $\alpha = 1$ or $\alpha = \beta \neq 1$ or $\beta = 1 \neq \alpha$.

The effect of a saturating concentration of one ligand on the inhibition observed at a fixed unsaturating concentration of the other ligand can be predicted easily from the equilibria and the velocity equations. For example, we can see from the equilibria that at a fixed unsaturating [A] and [I], saturating [B] cannot drive all the enzyme to a catalytically active form. Some enzyme will remain tied up as EIB. Thus saturating [B] will not overcome the inhibition. The same conclusion can be drawn by seeing what happens to the velocity equation VI-15 when [B] is infinitely high. The $K_B/[B]$ and $[I]K_B/K_i[B]$ terms in the slope factor (i.e., the factor multiplying K_B) reduce to zero, but the $[I]/\beta K_i$ term remains. Thus I is still competitive with A even at saturating [B]. Saturating [B] reduces the $1/v$-axis intercept factor (i.e., the factor multiplying [A]) to unity. Thus the $1/v$-axis intercept gives $1/V_{max}$ when [B] is saturating. Similarly, we can see from the equilibria that at a fixed unsaturating [B] and [I], saturating [A] will drive all the enzyme to EA and EAB; none of the enzyme will be tied up as EI or EIB and, consequently, no inhibition is observed. Inspection of the velocity equation VI-16 shows that when [A] is infinitely high the slope and $1/v$-axis intercept factors reduce to unity. Thus $K_{B_{app}} = \alpha K_B$ and $V_{max_{app}} = V_{max}$, that is, the same values that would be obtained at saturating [A] in the absence of I.

The value of K_i can be calculated from the slope of the $1/v$ versus $1/[B]$ plot if α, K_B, and K_A are known. The latter constants can be obtained from control plots in the absence of I. With K_i known, β can be calculated from the value of the $1/v$-axis intercepts. Alternately, the slopes and intercepts can be replotted as a function of the corresponding [I] (Fig. VI-8*b–e*):

$$slope_{1/B} = \frac{\alpha K_B K_A}{V_{max} K_i [A]}[I] + \frac{\alpha K_B}{V_{max}}\left(1 + \frac{K_A}{[A]}\right) \qquad \text{(VI-17)}$$

The [I]-axis intercept (when *slope* = 0) gives a constant that we can call $K_{i_{slope}}^{I/B}$ (i.e., a K_i calculated from the slopes of the $1/v$ versus $1/[B]$ plots):

$$K_{i_{slope}}^{I/B} = K_i\left(1 + \frac{[A]}{K_A}\right)$$

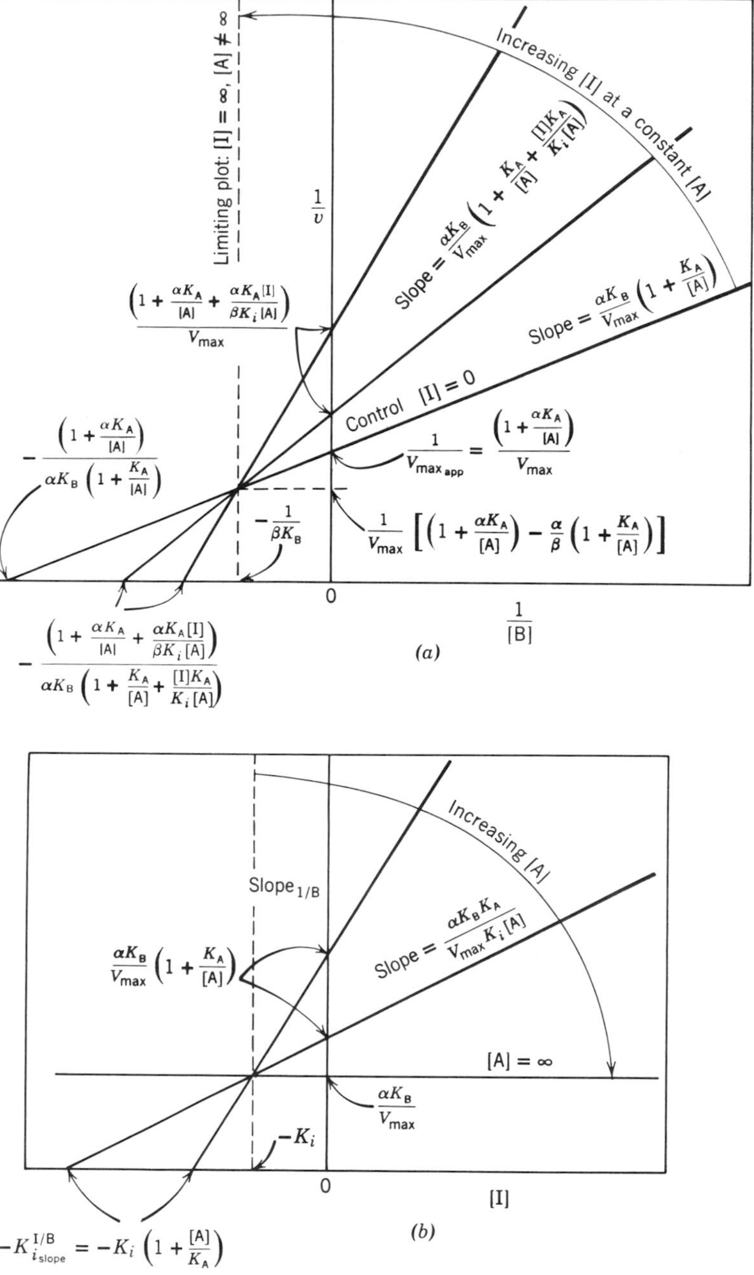

Fig. VI-8. (*a*) Plot of $1/v$ versus $1/[B]$ at different fixed concentrations of I and a constant [A] for the system described in Fig. VI-7. The intersection coordinates depend on the values of α, β, and [A]. (*b*) $Slope_{1/B}$ replot. Each line represents data from a family of reciprocal plots obtained at a different concentration of the constant ligand, A.

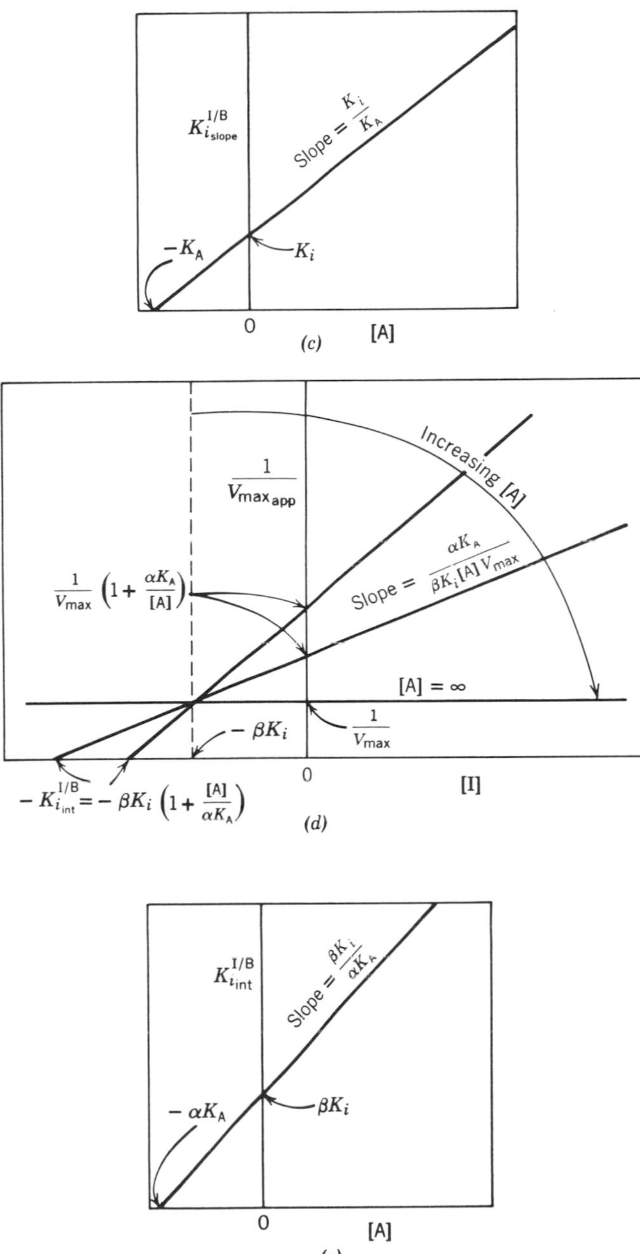

Fig. VI-8. (*Cont.*) (*c*) Secondary replot of $K_{i_{slope}}^{I/B}$ versus [A] (data taken from *b*). (*d*) Replot of $1/v$-axis intercept of *a*. Each line represents data from a family of reciprocal plots obtained at a different concentration of the constant ligand, A. (*e*) Secondary replot of $K_{i_{int}}^{I/B}$ versus [A] (data taken from *d*).

Similarly:

$$\frac{1}{V_{\max_{app}}} = \frac{1}{V_{\max}}\left(\frac{\alpha K_A}{\beta K_i[A]}\right)[I] + \frac{1}{V_{\max}}\left(1 + \frac{\alpha K_A}{[A]}\right) \qquad \text{(VI-18)}$$

In effect, the replot represents a Dixon plot of $1/v$ versus [I] at saturating [B]. The [I]-axis intercept (when $1/V_{\max_{app}} = 0$) gives $-K_{i_{int}}^{I/B}$:

$$K_{i_{int}}^{I/B} = \beta K_i\left(1 + \frac{[A]}{\alpha K_A}\right)$$

Replotting the slopes of the $1/v$ versus $1/[A]$ plots gives K_i directly if $\beta = 1$ (Fig. VI-7*b*). If $\beta \neq 1$, the equation is:

$$slope_{1/A} = \frac{\alpha K_A}{\beta K_i V_{\max}}\left(1 + \frac{\beta K_B}{[B]}\right)[I] + \frac{\alpha K_A}{V_{\max}}\left(1 + \frac{K_B}{[B]}\right) \qquad \text{(VI-19)}$$

The intercept on the [I]-axis gives $-K_{i_{slope}}^{I/A}$:

$$K_{i_{slope}}^{I/A} = \beta K_i \frac{\left(1 + \dfrac{K_B}{[B]}\right)}{\left(1 + \dfrac{\beta K_B}{[B]}\right)} \qquad \text{(VI-19a)}$$

The interaction factor β can also be obtained from the $1/[B]$ coordinate of the intersection point of the reciprocal plot, which equals $-1/\beta K_B$. (As [I] becomes infinitely high at a constant unsaturating [A], all the enzyme is driven to the EI form. The only B binding reaction becomes $EI + B \rightleftharpoons EIB$ with a $K_{B_{app}}$ of βK_B.) The $1/v$ coordinate of the intersection point can be solved for by setting the expression for $1/v$ at one value of [I] equal to $1/v$ at another [I] (same constant [A]) and solving for the value of $1/[B]$. Substitution of the $1/[B]$ coordinate back into the reciprocal equation allows the $1/v$ coordinate to be solved. The intersection coordinates can also be found from the general relationships:

$$\frac{1}{[B]} = -\frac{1}{K_B^*}\left(\frac{K_{i_{slope}}}{K_{i_{int}}}\right) \quad \text{and} \quad \frac{1}{v} = \frac{1}{V_{\max}^*}\left(1 - \frac{K_{i_{slope}}}{K_{i_{int}}}\right) \qquad \text{(VI-20)}$$

where K_B^* and $V_{\max}^*$ are the apparent K_B and $V_{\max}$ values at the given concentrations of nonvaried substrates and activators in the absence of I. The $K_{i_{slope}}$ and $K_{i_{int}}$ are apparent slope and intercept inhibitor constants. They can be derived directly by rearranging the velocity equation to obtain

$(1+[\mathrm{I}]/K_i)$ factors in the slope and intercept terms. This is illustrated below for the present system where [A] is constant and B is the varied substrate. First, the terms not containing [I] are factored out (making it necessary to divide the [I]-containing term by the same factor):

$$\frac{v}{V_{max}} = \frac{[\mathrm{B}]}{\alpha K_{\mathrm{B}}\left(1+\frac{K_{\mathrm{A}}}{[\mathrm{A}]}\right)\left[1+\frac{\left(\frac{[\mathrm{I}]K_{\mathrm{A}}}{K_i[\mathrm{A}]}\right)}{\left(1+\frac{K_{\mathrm{A}}}{[\mathrm{A}]}\right)}\right]+[\mathrm{B}]\left(1+\frac{\alpha K_{\mathrm{A}}}{[\mathrm{A}]}\right)\left[1+\frac{\left(\frac{\alpha K_{\mathrm{A}}[\mathrm{I}]}{\beta K_i[\mathrm{A}]}\right)}{\left(1+\frac{\alpha K_{\mathrm{A}}}{[\mathrm{A}]}\right)}\right]}$$

$$= \frac{[\mathrm{B}]}{\alpha K_{\mathrm{B}}\left(1+\frac{K_{\mathrm{A}}}{[\mathrm{A}]}\right)\left[1+\frac{[\mathrm{I}]}{K_i\left(1+\frac{[\mathrm{A}]}{K_{\mathrm{A}}}\right)}\right]+[\mathrm{B}]\left(1+\frac{\alpha K_{\mathrm{A}}}{[\mathrm{A}]}\right)\left[1+\frac{[\mathrm{I}]}{\beta K_i\left(1+\frac{[\mathrm{A}]}{\alpha K_{\mathrm{A}}}\right)}\right]}$$

Thus

$$K_{i_{slope}}^{\mathrm{I/B}} = K_i\left(1+\frac{[\mathrm{A}]}{K_{\mathrm{A}}}\right), \qquad K_{i_{int}}^{\mathrm{I/B}} = \beta K_i\left(1+\frac{[\mathrm{A}]}{\alpha K_i}\right)$$

$$K_{\mathrm{B}}^{*} = \alpha K_{\mathrm{B}}\frac{\left(1+\frac{K_{\mathrm{A}}}{[\mathrm{A}]}\right)}{\left(1+\frac{\alpha K_{\mathrm{A}}}{[\mathrm{A}]}\right)}, \qquad V_{max}^{*} = \frac{V_{max}}{\left(1+\frac{\alpha K_{\mathrm{A}}}{[\mathrm{A}]}\right)}$$

If the inhibitor competes with B, the denominator of the velocity equation will contain an $[\mathrm{A}][\mathrm{I}]/\beta K_{\mathrm{A}}K_i$ term (representing the EAI complex) in place of the $[\mathrm{B}][\mathrm{I}]/\beta K_{\mathrm{B}}K_i$ term. The reciprocal plots will be symmetrical but opposite to those of System A2a; that is, I will behave as a competitive inhibitor with respect to B and as a noncompetitive or mixed-type inhibitor with respect to A.

System A2b. I Is an Alternative Substrate for A

If I is an alternative substrate for A, and EIB forms product with a rate constant, γk_p, then the kinetic constants associated with I can be obtained simply by substituting I for A in the assay. If, however, the assay depends on measuring the radioactive product of A, and I cannot be obtained in labeled form, then I can be treated as an inhibitor competitive with A as described above as long as v is taken as the rate of product formation via the EAB complex; that is, as long as the rate of appearance of the unique product of A is measured. For example, if A = ATP, I = ITP, and B = sugar, v would be measured as the rate of ADP formation or ATP utilization. The $K_{i_{slope}}$ of the competitive pattern (equation VI-19a) will give the apparent K_m for I as a substrate (in place of A) at the fixed [B]. If, on the other hand, we measure v as the rate of sugar-phosphate appearance, or if I and A are alternate activators (e.g., Mg^{2+} and Mn^{2+}) and the product of EIB and EAB are the same, then the situation is somewhat different. The velocity equation would be:

$$\frac{v}{[\mathrm{E}]_t} = \frac{k_p[\mathrm{EAB}] + \gamma k_p[\mathrm{EIB}]}{[\mathrm{E}]+[\mathrm{EA}]+[\mathrm{EB}]+[\mathrm{EAB}]+[\mathrm{EI}]+[\mathrm{EIB}]}$$

$$= \frac{k_p\dfrac{[\mathrm{A}][\mathrm{B}]}{\alpha K_\mathrm{A} K_\mathrm{B}} + \gamma k_p \dfrac{[\mathrm{I}][\mathrm{B}]}{\beta K_i K_\mathrm{B}}}{1 + \dfrac{[\mathrm{A}]}{K_\mathrm{A}} + \dfrac{[\mathrm{B}]}{K_\mathrm{B}} + \dfrac{[\mathrm{A}][\mathrm{B}]}{\alpha K_\mathrm{A} K_\mathrm{B}} + \dfrac{[\mathrm{I}]}{K_i} + \dfrac{[\mathrm{I}][\mathrm{B}]}{\beta K_i K_\mathrm{B}}}$$

We can designate $k_p[\mathrm{E}]_t$ as V_{max}, although if $\gamma > 1$, then the true maximal velocity will be γV_{max}. When [A] is varied:

$$\boxed{\frac{v}{V_{max}} = \frac{[\mathrm{A}]}{\alpha K_\mathrm{A}\dfrac{\left(1 + \dfrac{K_\mathrm{B}}{[\mathrm{B}]} + \dfrac{K_\mathrm{B}[\mathrm{I}]}{K_i[\mathrm{B}]} + \dfrac{[\mathrm{I}]}{\beta K_i}\right)}{\left(1 + \dfrac{\alpha\gamma K_\mathrm{A}[\mathrm{I}]}{\beta K_i[\mathrm{A}]}\right)} + [\mathrm{A}]\dfrac{\left(1 + \dfrac{\alpha K_\mathrm{B}}{[\mathrm{B}]}\right)}{\left(1 + \dfrac{\alpha\gamma K_\mathrm{A}[\mathrm{I}]}{\beta K_i[\mathrm{A}]}\right)}}} \tag{VI-21}$$

The slope and intercept factors contain an [A] term. Consequently, the reciprocal plots will be nonlinear. As [A] decreases (1/[A] increases), the plot approaches a horizontal asymptote representing the reciprocal velocity of product formation from EIB at the fixed B and I concentrations. As [A] increases, an increasing fraction of the total product formed occurs via EAB. Thus as the plot approaches the $1/v$-axis, the line curves up or down and intersects the $1/v$-axis at the $1/V_{\max_{\text{app}}}$ for the fixed [B]. When [B] is varied, the velocity equation becomes:

$$\frac{v}{V_{\max}} = \frac{[\mathrm{B}]}{\alpha K_{\mathrm{B}}\dfrac{\left(1+\dfrac{K_{\mathrm{A}}}{[\mathrm{A}]}+\dfrac{K_{\mathrm{A}}[\mathrm{I}]}{K_i[\mathrm{A}]}\right)}{\left(1+\dfrac{\alpha\gamma K_{\mathrm{A}}[\mathrm{I}]}{\beta K_i[\mathrm{A}]}\right)} + [\mathrm{B}]\dfrac{\left(1+\dfrac{\alpha K_{\mathrm{A}}}{[\mathrm{A}]}+\dfrac{\alpha K_{\mathrm{A}}[\mathrm{I}]}{\beta K_i[\mathrm{A}]}\right)}{\left(1+\dfrac{\alpha\gamma K_{\mathrm{A}}[\mathrm{I}]}{\beta K_i[\mathrm{A}]}\right)}} \tag{VI-22}$$

The reciprocal plots are linear and the family of plots obtained at different fixed I concentrations intersect above, below, or, coincidentally, on the $1/[\mathrm{B}]$-axis. $Slope_{1/\mathrm{B}}$ and intercept (i.e., $1/V_{\max_{\text{app}}}$) versus [I] replots are hyperbolic. As [I] increases the $slope_{1/\mathrm{B}}$ and intercept approach limiting values of $\beta K_{\mathrm{B}}/\gamma V_{\max}$ and $1/\gamma V_{\max}$, respectively. The interaction factors, β and γ, can be determined from a replot of $1/\Delta$ *intercept* versus $1/[\mathrm{I}]$ as described earlier for mixed-type hyperbolic systems. The intercepts of the replot are:

$$\frac{1}{[\mathrm{I}]} = -\frac{\alpha\gamma K_{\mathrm{A}}}{\beta K_i[\mathrm{A}]}, \quad \text{and} \quad \frac{1}{\Delta\, intercept} = \frac{\gamma V_{\max}}{1-\gamma\left(1+\dfrac{\alpha K_{\mathrm{A}}}{[\mathrm{A}]}\right)}$$

System A3. I Is a Nonexclusive Inhibitor

In the equilibrium scheme shown in Figure VI-9, A and B are cosubstrates while I is a nonexclusive inhibitor. The interaction factors α, β, and γ represent, respectively, the effect of A on the binding of B (and vice versa), the effect of I on the binding of B (and vice versa), and the effect of I on the binding of A (and vice versa). The factor δ represents the effect of I on the catalytic activity of the EAB complex. The general velocity equation is

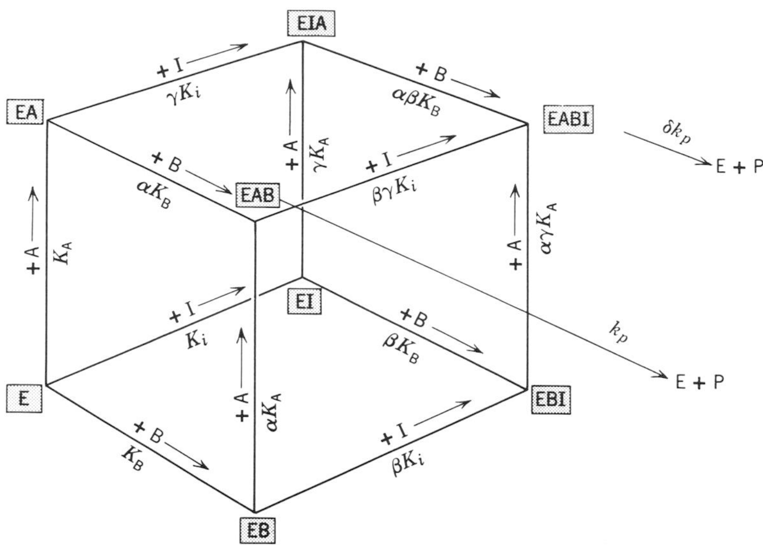

Fig. VI-9. General model (equilibria) for a nonexclusive inhibitor; A and B are cosubstrates of a Rapid Equilibrium Random Bireactant system.

obtained from:

$$\frac{v}{[\mathrm{E}]_t} = \frac{k_p[\mathrm{EAB}] + \delta k_p[\mathrm{EABI}]}{[\mathrm{E}]+[\mathrm{EA}]+[\mathrm{EB}]+[\mathrm{EI}]+[\mathrm{EAB}]+[\mathrm{EAI}]+[\mathrm{EBI}]+[\mathrm{EABI}]}$$

$$\boxed{\frac{v}{V_{\max}} = \frac{\dfrac{[\mathrm{A}][\mathrm{B}]}{\alpha K_{\mathrm{A}}K_{\mathrm{B}}} + \dfrac{\delta[\mathrm{A}][\mathrm{B}][\mathrm{I}]}{\alpha\beta\gamma K_{\mathrm{A}}K_{\mathrm{B}}K_i}}{1+\dfrac{[\mathrm{A}]}{K_{\mathrm{A}}}+\dfrac{[\mathrm{B}]}{K_{\mathrm{B}}}+\dfrac{[\mathrm{I}]}{K_i}+\dfrac{[\mathrm{A}][\mathrm{B}]}{\alpha K_{\mathrm{A}}K_{\mathrm{B}}}+\dfrac{[\mathrm{A}][\mathrm{I}]}{\gamma K_{\mathrm{A}}K_i}+\dfrac{[\mathrm{B}][\mathrm{I}]}{\beta K_{\mathrm{B}}K_i}+\dfrac{[\mathrm{A}][\mathrm{B}][\mathrm{I}]}{\alpha\beta\gamma K_{\mathrm{A}}K_{\mathrm{B}}K_i}}}$$

(VI-23)

where $V_{\max} = k_p[\mathrm{E}]_t$. The equations for varied [A] and varied [B] are:

$$\boxed{\frac{v}{V_{\max}} = \frac{[\mathrm{A}]}{\alpha K_{\mathrm{A}}\dfrac{\left(1+\dfrac{K_{\mathrm{B}}}{[\mathrm{B}]}+\dfrac{K_{\mathrm{B}}[\mathrm{I}]}{[\mathrm{B}]K_i}+\dfrac{[\mathrm{I}]}{\beta K_i}\right)}{\left(1+\dfrac{\delta[\mathrm{I}]}{\beta\gamma K_i}\right)} + [\mathrm{A}]\dfrac{\left(1+\dfrac{\alpha K_{\mathrm{B}}}{[\mathrm{B}]}+\dfrac{\alpha K_{\mathrm{B}}[\mathrm{I}]}{\gamma K_i[\mathrm{B}]}+\dfrac{[\mathrm{I}]}{\beta\gamma K_i}\right)}{\left(1+\dfrac{\delta[\mathrm{I}]}{\beta\gamma K_i}\right)}}}$$

(VI-24)

$$\frac{v}{V_{\max}}=\frac{[\mathrm{B}]}{\alpha K_{\mathrm{B}}\dfrac{\left(1+\dfrac{K_{\mathrm{A}}}{[\mathrm{A}]}+\dfrac{K_{\mathrm{A}}[\mathrm{I}]}{[\mathrm{A}]K_i}+\dfrac{[\mathrm{I}]}{\gamma K_i}\right)}{\left(1+\dfrac{\delta[\mathrm{I}]}{\beta\gamma K_i}\right)}+[\mathrm{B}]\dfrac{\left(1+\dfrac{\alpha K_{\mathrm{A}}}{[\mathrm{A}]}+\dfrac{\alpha K_{\mathrm{A}}[\mathrm{I}]}{\beta K_i[\mathrm{A}]}+\dfrac{[\mathrm{I}]}{\beta\gamma K_i}\right)}{\left(1+\dfrac{\delta[\mathrm{I}]}{\beta\gamma K_i}\right)}} \tag{VI-25}$$

The equation can be simplified for special cases. For example, in the simplest case where I has no effect on the binding of A or B and EABI is catalytically inactive ($\beta=1, \gamma=1, \delta=0$), the velocity equation for B as the varied ligand reduces to:

$$\frac{v}{V_{\max}}=\frac{[\mathrm{B}]}{\alpha K_{\mathrm{B}}\left(1+\dfrac{K_{\mathrm{A}}}{[\mathrm{A}]}\right)\left(1+\dfrac{[\mathrm{I}]}{K_i}\right)+[\mathrm{B}]\left(1+\dfrac{\alpha K_{\mathrm{A}}}{[\mathrm{A}]}\right)\left(1+\dfrac{[\mathrm{I}]}{K_i}\right)} \tag{VI-26}$$

A symmetrical equation is obtained when A is the varied ligand. Thus I behaves as a pure noncompetitive inhibitor with respect to A and B. The K_i can be determined in the usual way from reciprocal plots of $1/v$ versus $1/[\mathrm{B}]$ at different fixed concentrations of I and a constant concentration of A:

$$\frac{1}{v}=\frac{K_{\mathrm{B}_{\mathrm{app}}}}{V_{\max_{\mathrm{app}}}}\left(1+\frac{[\mathrm{I}]}{K_i}\right)\frac{1}{[\mathrm{B}]}+\frac{1}{V_{\max_{\mathrm{app}}}}\left(1+\frac{[\mathrm{I}]}{K_i}\right) \tag{VI-27}$$

The $K_{\mathrm{B}_{\mathrm{app}}}$ and $V_{\max_{\mathrm{app}}}$ are the observed constants for the given concentration of A in the absence of I. If the constant A concentration is not saturating:

$$K_{\mathrm{B}_{\mathrm{app}}}=\alpha K_{\mathrm{B}}\frac{\left(1+\dfrac{K_{\mathrm{A}}}{[\mathrm{A}]}\right)}{\left(1+\dfrac{\alpha K_{\mathrm{A}}}{[\mathrm{A}]}\right)}, \qquad V_{\max_{\mathrm{app}}}=\frac{V_{\max}}{\left(1+\dfrac{\alpha K_{\mathrm{A}}}{[\mathrm{A}]}\right)}$$

If the constant A concentration is saturating, $K_{\mathrm{B}_{\mathrm{app}}}=\alpha K_{\mathrm{B}}$. Similar reciprocal plots of $1/v$ versus $1/[\mathrm{A}]$ can be constructed. Replots of *slope* or $1/v$-axis intercept versus [I] are linear. For example, replotting $slope_{1/B}$ versus [I] at a constant unsaturating [A]:

$$slope_{1/B} = \frac{\alpha K_B}{K_i V_{max}}\left(1 + \frac{K_A}{[A]}\right)[I] + \frac{\alpha K_B}{V_{max}}\left(1 + \frac{K_A}{[A]}\right) \quad \text{(VI-28)}$$

The intercept on the [I]-axis gives $-K_i$.

When EABI is catalytically inactive but $\beta \neq 1$ and $\gamma \neq 1$, the velocity equations are:

$$\frac{v}{V_{max}} = \frac{[A]}{\alpha K_A\left(1 + \frac{K_B}{[B]} + \frac{K_B[I]}{[B]K_i} + \frac{[I]}{\beta K_i}\right) + [A]\left(1 + \frac{\alpha K_B}{[B]} + \frac{\alpha K_B[I]}{\gamma K_i[B]} + \frac{[I]}{\beta\gamma K_i}\right)} \quad \text{(VI-29)}$$

$$\frac{v}{V_{max}} = \frac{[B]}{\alpha K_B\left(1 + \frac{K_A}{[A]} + \frac{K_A[I]}{[A]K_i} + \frac{[I]}{\gamma K_i}\right) + [B]\left(1 + \frac{\alpha K_A}{[A]} + \frac{\alpha K_A[I]}{\beta K_i[A]} + \frac{[I]}{\beta\gamma K_i}\right)} \quad \text{(VI-30)}$$

I behaves as a linear mixed-type inhibitor with respect to A and B. The inhibition with respect to the varied ligand is not overcome by saturation with the nonvaried ligand. The constants β, γ, and K_i can be determined from slope and intercept replots obtained from primary $1/v$ versus $1/[A]$ and $1/v$ versus $1/[B]$ data at different I concentrations and saturating nonvaried ligand. When A is saturating, the slope of the $1/v$ versus $1/[B]$ plot is given by:

$$slope_{1/B} = \frac{\alpha K_B}{\gamma K_i V_{max}}[I] + \frac{\alpha K_B}{V_{max}} \quad \text{(VI-31)}$$

The horizontal axis intercept of the replot gives $-\gamma K_i$. The $1/v$-axis intercept is given by:

$$\frac{1}{V_{max_{app}}} = \frac{1}{\beta\gamma K_i V_{max}}[I] + \frac{1}{V_{max}} \quad \text{(VI-32)}$$

The horizontal axis intercept of the replot gives $-\beta\gamma K_i$. Replots of the $1/v$ versus $1/[A]$ data at saturating [B] give βK_i and $\beta\gamma K_i$. The same mixed-type patterns are obtained if either β or γ is unity. In this case we need only replot the data from one family of reciprocal plots. For example, if $\gamma=1$ and $\beta\neq 1$, the replots of the $1/v$ versus $1/[\mathrm{B}]$ data give K_i and βK_i.

If it is impractical to maintain either A or B at a saturating concentration, the various constants can still be determined, although additional experiments are necessary. For example, we can determine γ and K_i from several families of reciprocal plots of $1/v$ versus $1/[\mathrm{B}]$. Any one family consists of plots obtained at different I concentrations and a constant, unsaturating A concentration. A different constant [A] is used for each family. The slope of the plots within a family is given by:

$$slope_{1/\mathrm{B}} = \frac{\alpha K_\mathrm{B}}{\gamma K_i V_{\max}}\left(1+\frac{\gamma K_\mathrm{A}}{[\mathrm{A}]}\right)[\mathrm{I}] + \frac{\alpha K_\mathrm{B}}{V_{\max}}\left(1+\frac{K_\mathrm{A}}{[\mathrm{A}]}\right) \quad \text{(VI-33)}$$

The replot is linear for each family. The slope of the replot depends on the constant [A] for that family.

$$slope_{\text{family}} = \frac{\alpha K_\mathrm{B} K_\mathrm{A}}{K_i V_{\max}}\frac{1}{[\mathrm{A}]} + \frac{\alpha K_\mathrm{B}}{\gamma K_i V_{\max}} \quad \text{(VI-34)}$$

Thus a secondary replot of the slopes of each family replot against the reciprocal of the corresponding A concentration yields a straight line. The intercept on the $1/[\mathrm{A}]$-axis gives $-1/\gamma K_\mathrm{A}$. The intercept on the vertical axis gives $\alpha K_\mathrm{B}/\gamma K_i V_{\max}$. If αK_B, K_A, and $V_{\max}$ are known, γ may be calculated. Then K_i can be calculated from the slope of the secondary replot. A similar series of $1/v$ versus $1/[\mathrm{A}]$ reciprocal plot families can be constructed for different constant B concentrations. The replot procedure described above allows β and K_i to be determined.

In the general case, I is a hyperbolic mixed-type inhibitor with respect to B and A. The inhibition with respect to a varied ligand cannot be overcome by saturation with the nonvaried ligand. Slope and intercept replots are hyperbolic. At a saturating concentration of A, the family of $1/v$ versus $1/[\mathrm{B}]$ plots obtained at different fixed inhibitor concentrations intersect at:

$$\frac{1}{[\mathrm{B}]} = \frac{(\delta-1)}{\alpha K_\mathrm{B}(\beta-\delta)}$$

When [A] is varied at saturating [B], the $1/[\mathrm{A}]$ intersection coordinate is given by:

$$\frac{1}{[\mathrm{A}]} = \frac{(\delta-1)}{\alpha K_\mathrm{A}(\gamma-\delta)}$$

The inhibitor constants are most conveniently obtained from replots of $1/\Delta\, slope$ and $1/\Delta\, intercept$ versus $1/[\mathrm{I}]$ obtained at a saturating concentration of the constant ligand. For example, when [A] is saturating, the slope and intercept of the $1/v$ versus $1/[\mathrm{B}]$ plots are given by:

$$slope_{1/\mathrm{B}} = \frac{\alpha K_\mathrm{B}}{V_{\max}} \frac{\left(1+\dfrac{[\mathrm{I}]}{\gamma K_i}\right)}{\left(1+\dfrac{\delta[\mathrm{I}]}{\beta\gamma K_i}\right)}, \qquad 1/v\text{-axis intercept} = \frac{1}{V_{\max}} \frac{\left(1+\dfrac{[\mathrm{I}]}{\beta\gamma K_i}\right)}{\left(1+\dfrac{\delta[\mathrm{I}]}{\beta\gamma K_i}\right)}$$

(VI-35)

The system behaves as a single substrate, hyperbolic, mixed-type inhibition system where "K_B" $=\alpha K_\mathrm{B}$, "K_i" $=\gamma K_i$, and the interaction factors affecting K_B and k_p are β and δ, respectively (Chapter Four). Similar replots of $1/v$ versus $1/[\mathrm{A}]$ data obtained at saturating [B] allows βK_i, γ, and δ to be determined.

If it is impractical to maintain A or B at saturation, the inhibitor constants can still be determined, although the procedure involves many more experiments. Several families of $1/v$ versus $1/[\mathrm{B}]$ plots are needed. Any one family consists of the plots obtained at different fixed concentrations of I and a constant unsaturating A concentration. A different constant unsaturating [A] is used for each family. The data within each family are plotted as $1/\Delta\, slope$ versus $1/[\mathrm{I}]$ and $1/\Delta\, intercept$ versus $1/[\mathrm{I}]$. For the general system, the equations are:

$$\frac{1}{\Delta\, slope} = \frac{\beta\gamma K_i V_{\max}[\mathrm{A}]}{\alpha K_\mathrm{B}[K_\mathrm{A}(\beta\gamma-\delta)+[\mathrm{A}](\beta-\delta)]} \frac{1}{[\mathrm{I}]} + \frac{\delta V_{\max}[\mathrm{A}]}{\alpha K_\mathrm{B}[K_\mathrm{A}(\beta\gamma-\delta)+[\mathrm{A}](\beta-\delta)]}$$

(VI-36)

$$\frac{1}{\Delta\, intercept} = \frac{\beta\gamma K_i V_{\max}[\mathrm{A}]}{\alpha K_\mathrm{A}(\gamma-\delta)+[\mathrm{A}](1-\delta)} \frac{1}{[\mathrm{I}]} + \frac{\delta V_{\max}[\mathrm{A}]}{\alpha K_\mathrm{A}(\gamma-\delta)+[\mathrm{A}](1-\delta)}$$

(VI-37)

The replots for each family are linear. The horizontal axis intercept gives $-\delta/\beta\gamma K_i$. The slopes and vertical axis intercepts depend on the constant A concentration. The reciprocals of the vertical axis intercepts of the replots are replotted against the corresponding $1/[\mathrm{A}]$. (The reciprocals of the vertical axis intercepts represent $\Delta slope$ and $\Delta intercept$ at saturating [I] for a given A concentration.) For the $1/\Delta intercept$ replots:

$$\frac{1}{intercept} = \frac{K_A(\gamma - \delta)}{\delta V_{max}} \frac{1}{[A]} + \frac{1-\delta}{\delta V_{max}} \qquad \text{(VI-38)}$$

The value of δ can be calculated from the vertical axis intercept of the replot. With δ known, γ can be calculated from the slope or horizontal axis intercept. With δ and γ known, βK_i can be calculated from the slope of the original $1/\Delta intercept$ versus $1/[I]$ replot. (It is assumed that V_{max}, α, K_A, etc., have already been determined from earlier experiments performed in the absence of inhibitor.) The β and K_i still are unknown. The vertical axis intercepts of the $1/\Delta slope$ replots can be replotted:

$$\frac{1}{intercept} = \frac{\alpha K_B K_A(\beta\gamma - \delta)}{V_{max}} \frac{1}{[A]} + \frac{\alpha K_B(\beta - \delta)}{V_{max}} \qquad \text{(VI-39)}$$

Then β can be calculated from the vertical or horizontal axis intercepts. The replotting procedure is outlined in Figure VI-10. If I is a partial noncompetitive inhibitor, $\beta = 1$, $\gamma = 1$, and $\delta < 1$. In this case, δ can be obtained directly from the $1/\Delta intercept$ or $1/\Delta slope$ replots. For example, the $1/\Delta intercept$ replot has intercepts of $-\delta/K_i$ and $\delta V_{max}/(1+\alpha K_A/[A])(1-\delta) = \delta V_{max_{app}}/(1-\delta)$. If I is a partial competitive inhibitor $\beta \neq 1$, $\gamma \neq 1$, and $\delta = 1$. In this case β and γ can be obtained from the $1/\Delta slope$ replots II and III shown in Figure VI-10.

System A4. Product Inhibition in Rapid Equilibrium Random Bireactant Systems

The products of a reaction are formed at the catalytic site on the surface of the enzyme and are the substrates for the reverse reaction. Consequently, it is not unexpected that a product may act as an inhibitor by occupying the same site as the substrate from which it is derived. The type of inhibition observed depends on the number and type of enzyme-product complexes that can form. Consequently, product inhibition studies can be quite useful in diagnosing kinetic mechanisms. In the examples described below, it is assumed that the two ligands, A and B, are cosubstrates which yield two products, P and Q; Q is the "product" of A, and P is the "product" of B. In the reverse direction with P and Q as substrates, A would be the product of Q and B the product of P. The A–Q, B–P relationship can be arbitrary and often meaningless in random systems. For example, consider a random kinase reaction: ATP + sugar$\rightleftharpoons$sugar-phosphate + ADP. The ATP and ADP can be considered an A–Q pair, while sugar and sugar-phosphate represent a B–P pair. Most of A is retained in Q. If the adenine serves as the binding

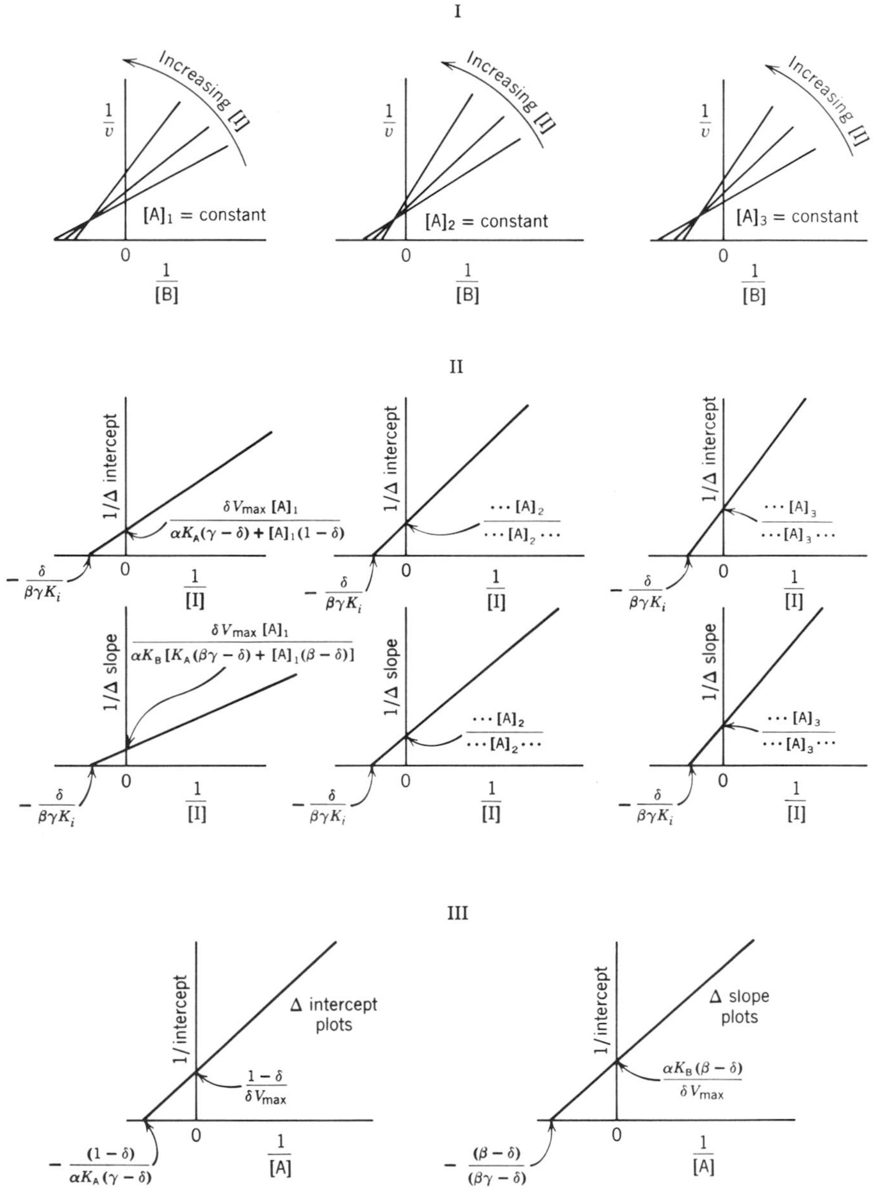

I

II

III

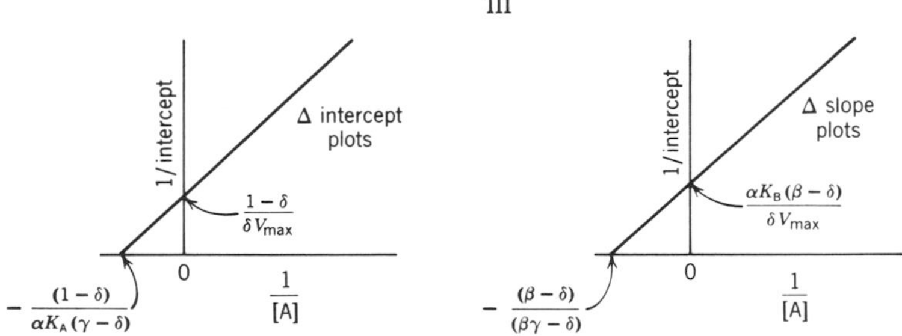

Fig. VI-10. Plots and replot procedure for determining the kinetic constants and interaction factors of a Rapid Equilibrium Random system in which I is a hyperbolic mixed-type or partial inhibitor. I. Plot $1/v$ versus $1/[B]$ at different fixed [I] and constant [A]. II. Replot $1/\Delta$ *intercept* versus $1/[I]$ and $1/\Delta$ *slope* versus $1/[I]$ for each family. III. Replot reciprocal of vertical-axis intercepts of II against the corresponding $1/[A]$.

site recognition group, then there is no doubt that A and Q occupy the same site on the enzyme. Similarly, most of B is retained in P; B and P fit the same binding site. (Presumably the site recognizes the sugar moiety.) Now consider a random AMP transferase reaction: ATP + B $\rightleftharpoons$ B-AMP + PPi. Here an A–Q, B–P designation is impossible. In ordered reactions, the symbols A, B, P, and Q are used to indicate the order of substrate addition and product release; A adds before B, and P is released before Q. The overall Rapid Equilibrium Random sequence is shown below.

$$\mathrm{E} + \mathrm{A} \overset{K_A}{\rightleftharpoons} \mathrm{EA}, \quad \mathrm{EA} + \mathrm{B} \overset{\alpha K_B}{\rightleftharpoons} \mathrm{EAB}$$

$$\mathrm{E} + \mathrm{B} \overset{K_B}{\rightleftharpoons} \mathrm{EB}, \quad \mathrm{EB} + \mathrm{A} \overset{\alpha K_A}{\rightleftharpoons} \mathrm{EAB}$$

$$\mathrm{EAB} \underset{k_{-p}}{\overset{k_p}{\rightleftharpoons}} \mathrm{EPQ}$$

$$\mathrm{EPQ} \overset{\beta K_Q}{\rightleftharpoons} \mathrm{EP} + \mathrm{Q}, \quad \mathrm{EP} \overset{K_P}{\rightleftharpoons} \mathrm{P} + \mathrm{E}$$

$$\mathrm{EPQ} \overset{\beta K_P}{\rightleftharpoons} \mathrm{EQ} + \mathrm{P}, \quad \mathrm{EQ} \overset{K_Q}{\rightleftharpoons} \mathrm{Q} + \mathrm{E}$$

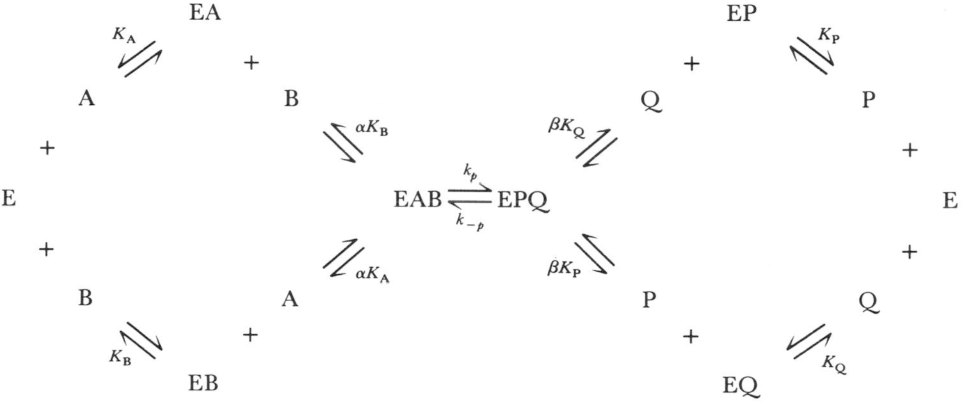

The overall sequence is shown below using the notation of Cleland:

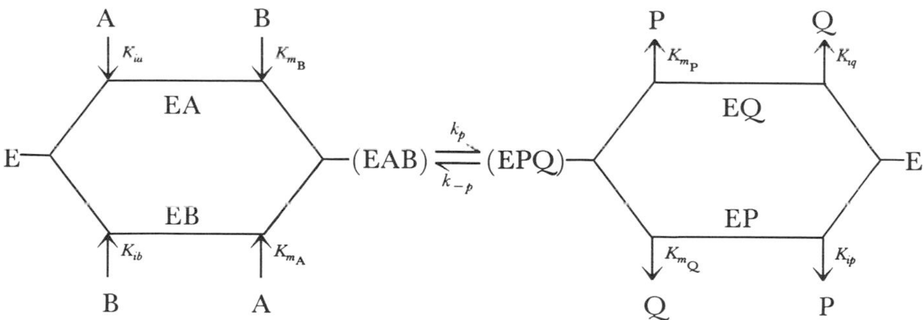

K_A, K_B, K_P, and K_Q (or K_{ia}, K_{ib}, K_{ip}, and K_{iq}) are dissociation constants for A, B, P, and Q from the EA, EB, EP, and EQ complexes, respectively, while αK_A, αK_B, βK_P, and βK_Q (or K_{m_A}, K_{m_B}, K_{m_P}, and K_{m_Q}) are dissociation constants for the respective ligands from the EAB or EPQ complexes. These latter constants also represent the Michaelis constants for the respective ligand in the presence of a saturating concentration of the coligand for the reaction in the given direction. Since we are assuming rapid equilibrium conditions, all ligand dissociations are extremely rapid compared to the interconversion of EAB and EPQ. Thus the levels of EP and EQ are essentially zero in the absence of added P and Q. In the presence of a finite concentration of only one of the products, the reverse reaction can be neglected (the concentration of the other product is essentially zero during the early part of the reaction). Nevertheless, the forward velocity will be inhibited because the finite P (or Q) ties up some of the enzyme.

In the following sections, the shorthand nomenclature of Cleland will be used to describe the number of substrates and products involved in a reaction sequence. A reaction between two substrates yielding two products is called Bi Bi and said to be bireactant in both directions. A Bi Uni reaction is one in which two substrates yield one product (i.e., bireactant in the forward direction and unireactant in the reverse direction), while a reaction in which one substrate yields two products is called Uni Bi (i.e., unireactant in the forward direction and bireactant in the reverse direction). The schematic "lock-and-key" models shown in the following sections are used to illustrate the possible steric interactions between products and substrates and do not imply any specific reaction mechanism.

System A4a. Rapid Equilibrium Random Bi Bi with Dead-End EBQ Complex

Figure VI-11 shows a random bireactant scheme in which a portion of substrate A is transferred to substrate B. In the presence of a finite concentration of P, an EP complex can form. P excludes both A and B. Hence there are no EAP or EBP complexes. This situation is observed with muscle pyruvate kinase where $MgATP^{2-}$ excludes both PEP and ADP (Ainsworth and MacFarlane, 1973). In the presence of a finite concentration of Q, an EQ complex can form. Q will not interfere with the binding of B. Consequently, a dead-end EBQ complex can form. The equilibria are as follows. In the presence of P:

$$
\begin{array}{ccccc}
\text{EP} & & & & \\
K_P \updownarrow & & & & \\
\text{P} & & & & \\
+ & & & & \\
\text{E} + \text{A} & \overset{K_A}{\rightleftharpoons} & \text{EA} & & \\
+ & & + & & \\
\text{B} & & \text{B} & & \\
K_B \updownarrow & & \updownarrow \alpha K_B & & \\
\text{EB} + \text{A} & \overset{\alpha K_A}{\rightleftharpoons} & \text{EAB} & \xrightarrow{k_p} & \text{E} + \text{P} + \text{Q}
\end{array}
$$

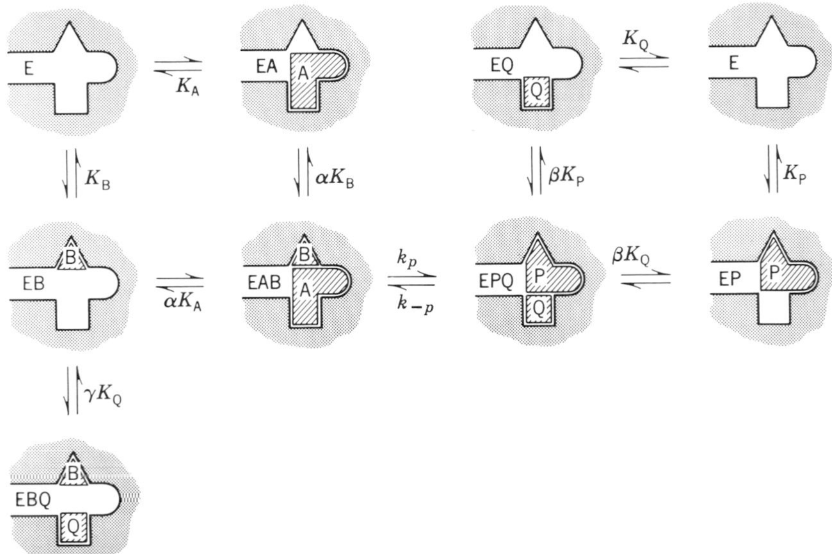

Fig. VI-11. Lock-and-key model of a Rapid Equilibrium Random Bireactant system showing the formation of a dead-end EBQ complex in the presence of added Q.

The velocity equation is obtained from:

$$\frac{v}{[E]_t} = \frac{k_p[EAB]}{[E]+[EA]+[EB]+[EP]+[EAB]}$$

When the concentration of each species is expressed in terms of [E], we obtain:

$$\boxed{\frac{v}{V_{max}} = \frac{\dfrac{[A][B]}{\alpha K_A K_B}}{1+\dfrac{[A]}{K_A}+\dfrac{[B]}{K_B}+\dfrac{[P]}{K_P}+\dfrac{[A][B]}{\alpha K_A K_B}}} \qquad \text{(VI-40)}$$

Rearranging to express each of the substrates as the varied ligand:

$$\boxed{\frac{v}{V_{max}} = \frac{[A]}{\alpha K_A\left(1+\dfrac{K_B}{[B]}+\dfrac{K_B[P]}{[B]K_P}\right)+[A]\left(1+\dfrac{\alpha K_B}{[B]}\right)}} \qquad \text{(VI-41)}$$

$$\frac{v}{V_{max}} = \frac{[B]}{\alpha K_B\left(1 + \frac{K_A}{[A]} + \frac{K_A[P]}{[A]K_P}\right) + [B]\left(1 + \frac{\alpha K_A}{[A]}\right)} \qquad \text{(VI-42)}$$

We see that only the slope factors are affected by P (and only $K_{A_{app}}$ and $K_{B_{app}}$ change in the presence of P). The $1/v$-axis intercept factor (i.e., the factor multiplying the [A] or [B]) is independent of P. Consequently, $V_{max_{app}}$ is unchanged and remains that fixed by the specific concentration of the cosubstrate. Thus P acts as a competitive inhibitor with respect to both A and B. We could have predicted this from the equilibria or Figure VI-11, where we see that the binding of P excludes both A and B. When A is the varied ligand and $[B] \gg K_B$ and $\gg \alpha K_B$, all the enzyme will be driven to EB and EAB. In this case, P will have no effect on the velocity (B and P are mutually exclusive. Also note that the slope factor approaches unity as [B] becomes very large.) Similarly when B is the varied ligand and $[A] \gg K_A$ and $\gg \alpha K_A$, all the enzyme will be driven to the EA and EAB complexes. Again, P will not affect the velocity (A and P are mutually exclusive). The primary reciprocal plot and replots are shown in Figure VI-12 for A as the varied substrate. The plots for B as the varied substrate are symmetrical to those shown for varied [A].

In the presence of Q:

$$\begin{array}{ccccccccc}
EQ & \overset{K_Q}{\rightleftharpoons} & Q + & E & + A & \overset{K_A}{\rightleftharpoons} & EA & & \\
+ & & & + & & & + & & \\
B & & & B & & & B & & \\
\gamma K_B \updownarrow & & & K_B \updownarrow & & & \updownarrow \alpha K_B & & \\
EBQ & \overset{\gamma K_Q}{\rightleftharpoons} & Q + & EB & + A & \overset{\alpha K_A}{\rightleftharpoons} & EAB & \overset{k_p}{\longrightarrow} & E+P+Q
\end{array}$$

The velocity equation is obtained from:

$$\frac{v}{[E]_t} = \frac{k_p[EAB]}{[E] + [EA] + [EB] + [EQ] + [EAB] + [EBQ]}$$

$$\frac{v}{V_{max}} = \frac{\dfrac{[A][B]}{\alpha K_A K_B}}{1 + \dfrac{[A]}{K_A} + \dfrac{[B]}{K_B} + \dfrac{[Q]}{K_Q} + \dfrac{[A][B]}{\alpha K_A K_B} + \dfrac{[B][Q]}{\gamma K_B K_Q}} \qquad \text{(VI-43)}$$

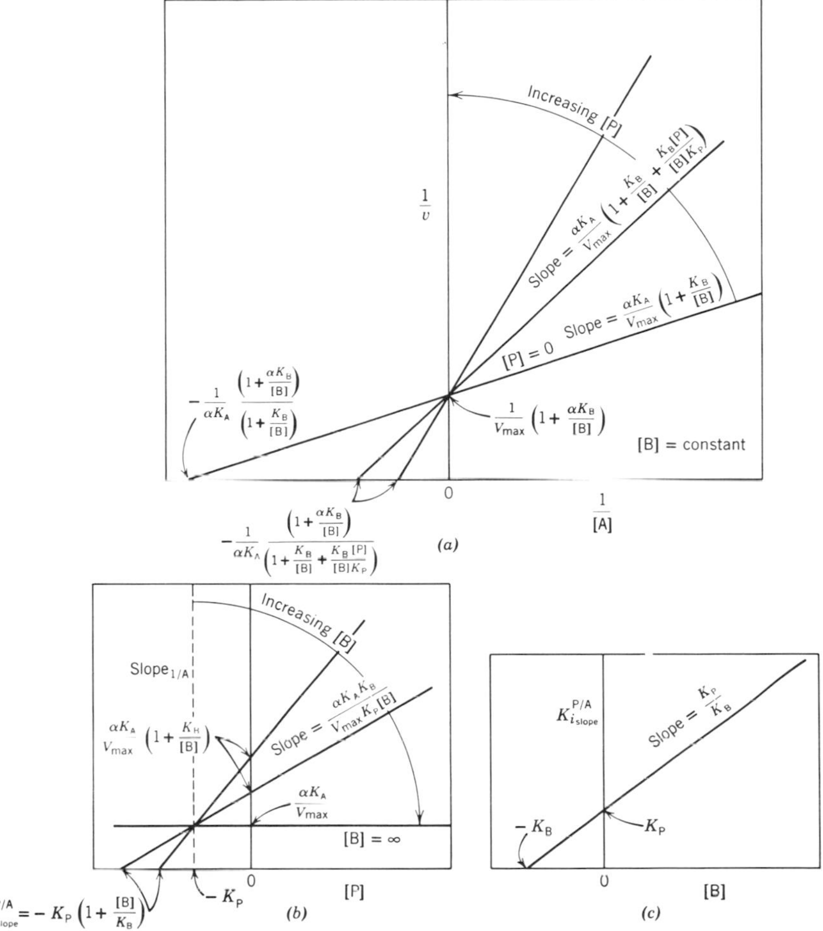

Fig. VI-12. (*a*) $1/v$ versus $1/[A]$ plot at different fixed concentrations of product P and a constant [B]. (*b*) $Slope_{1/A}$ replot. Each line represents data from a family of reciprocal plots at a different concentration of the constant ligand, B. (*c*) Replot of $K_{i_{slope}}^{P/A}$ versus [B] (data taken from *b*). The plots for B as the varied substrate are symmetrical to those for A.

We can predict the effect of Q by inspecting the equilibria and Figure VI-11. We see that Q excludes A and that increasing B to a saturating level will not prevent Q from binding. Consequently, Q will act as a competitive inhibitor with respect to A at all B concentrations. Ligands B and Q are not mutually exclusive, but the EBQ complex is catalytically inactive. Consequently, Q acts as a noncompetitive or mixed-type inhibitor with respect to B. Saturation with A will drive all the enzyme to EA and EAB. In this case, Q has no effect (Q and A are mutually exclusive). To verify our predictions, we can rearrange the velocity equation to show either A or B as the varied substrate.

$$\frac{v}{V_{max}} = \frac{[A]}{\alpha K_A\left(1 + \frac{K_B}{[B]} + \frac{K_B[Q]}{[B]K_Q} + \frac{[Q]}{\gamma K_Q}\right) + [A]\left(1 + \frac{\alpha K_B}{[B]}\right)} \qquad \text{(VI-44)}$$

$$\frac{v}{V_{max}} = \frac{[B]}{\alpha K_B\left(1 + \frac{K_A}{[A]} + \frac{K_A[Q]}{[A]K_Q}\right) + [B]\left(1 + \frac{\alpha K_A}{[A]} + \frac{\alpha K_A[Q]}{\gamma K_Q[A]}\right)} \qquad \text{(VI-45)}$$

As predicted, Q affects only the slope of the $1/v$ versus $1/[A]$ plot (i.e., only the factor multiplying αK_A contains a Q term). Saturation with B will reduce the $K_B/[B]$ and $K_B[Q]/[B]K_Q$ terms in the slope factor to zero, but the $[Q]/\gamma K_Q$ term remains. Thus as predicted, Q is a competitive inhibitor with respect to A at all B concentrations. When B is the varied substrate, Q affects both the slope and $1/v$-axis intercept (i.e., both the αK_B and [B] terms are multiplied by factors containing Q). If $\alpha = \gamma = 1$, the slope factor and the $1/v$-axis intercept factor are identical and Q behaves as a noncompetitive inhibitor. If $\alpha \neq 1$ and $\gamma \neq 1$, Q behaves as a mixed-type inhibitor (even when $\alpha = \gamma$). Saturation with A reduces both the slope factor and the $1/v$-axis intercept factor to unity. Thus as predicted, Q is not an inhibitor with respect to B when A is saturating. The equilibria, equations, and plots are identical to those given in System A2a for an inhibitor that competes with one substrate but allows the other substrate to bind.

System A4b. Rapid Equilibrium Random Bi Bi with Dead-End EAP and EBQ Complexes

When neither substrate hinders the binding of the product of the cosubstrate, then both EAP and EBQ dead-end complexes can form (Fig. VI-13). In the presence of Q, the equilibria and velocity equations are those given for System A2a. The equations for P as a product inhibitor are symmetrical to those shown for Q.

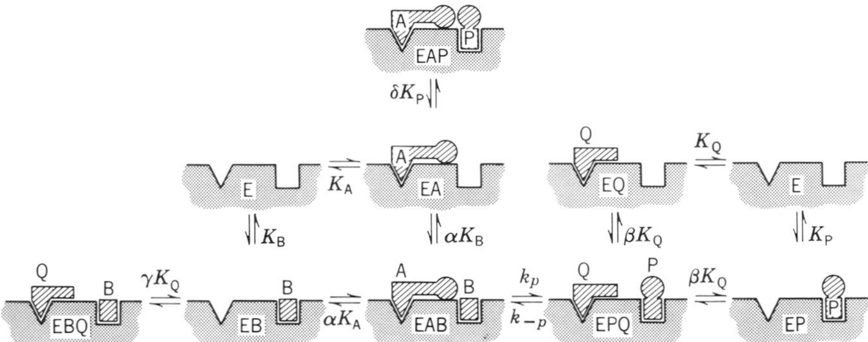

Fig. VI-13. Lock-and-key model of a Rapid Equilibrium Random Bireactant system showing the formation of a dead-end EAP complex (in the presence of added P) and a dead-end EBQ complex (in the presence of added Q).

System A4c. Rapid Equilibrium Random Bi Uni Reaction

When A and B condense to yield a single product, P, then it is likely that P will exclude both A and B. Saturation with either A or B will prevent EP formation, since P requires both sites for binding. As we might expect, P acts as a competitive inhibitor with respect to A and B at unsaturating concentrations of the nonvaried substrate. To measure the initial forward velocity without regard to the reverse reaction, radioactive A can be used and v taken as the rate of label appearance in P. If the assay times are short enough, the specific radioactivity of the substrate will not be diluted significantly by the reverse reaction. The equilibria and velocity equations in the presence of P are the same as those shown for System A4a.

Written backwards, System A4c becomes a Rapid Equilibrium Random Uni Bi reaction. Both products, A and B, act as competitive inhibitors with respect to the substrate, P.

System A4d. Equilibrium Random Bi Bi Reaction with No Dead-End Complexes

Figure VI-14 shows a hypothetical double displacement reaction. As shown, A and B or P and Q can occupy the enzyme simultaneously and all four ligands can bind by themselves, but neither product can bind when either substrate is bound, and vice versa. In this case both products act as competitive inhibitors with respect to the varied substrate at unsaturating concentrations of the nonvaried substrate. Saturation with either substrate excludes both products. The equilibria and velocity equations in the presence of P are the same as that given for System A4a. The equilibria and velocity equations in the presence of Q are symmetrical to those for P. Random systems with no dead end complexes are unlikely.

Complete Velocity Equation. When both substrates and both products are present and no "unnatural" dead-end complexes form the enzyme will be distributed among the nine different species shown below.

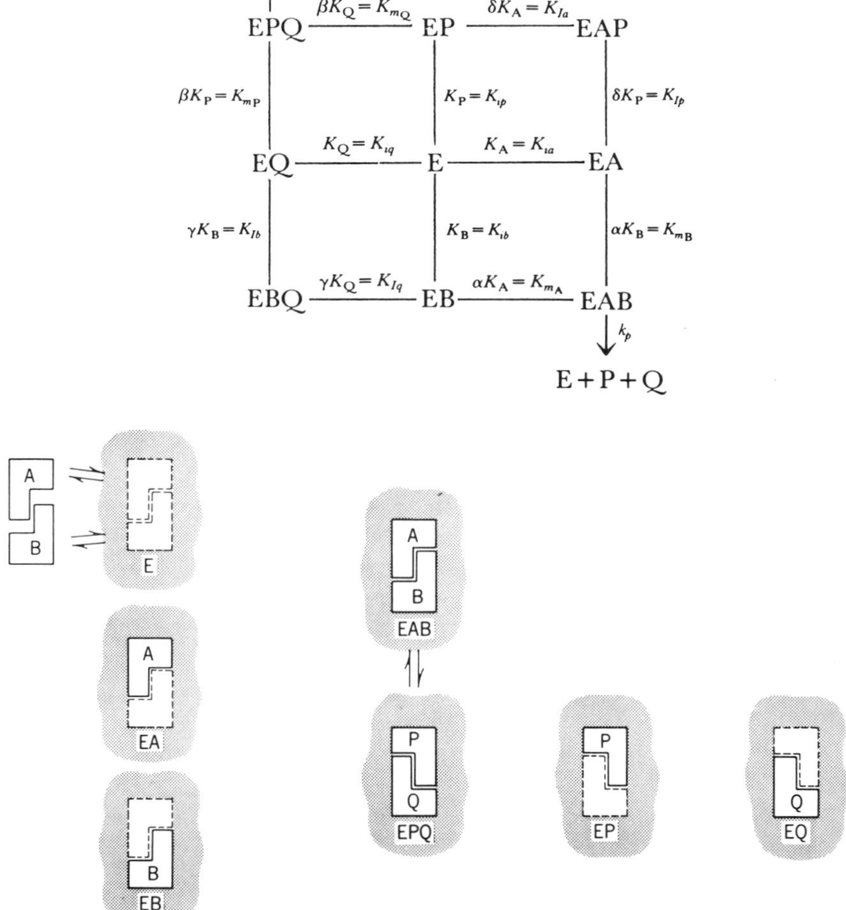

Fig. VI-14. Complexes that can form in a hypothetical double displacement reaction in which A and B add randomly and P and Q leave randomly. No dead-end complexes can form. Each product is competitive with each substrate.

The reaction will proceed in both directions when all four ligands are present. The net velocity in the forward direction (A + B→P + Q) is:

$$v_{\text{net}_f} = v_f - v_r$$

$$= \frac{\dfrac{[\mathrm{A}][\mathrm{B}]}{\alpha K_\mathrm{A} K_\mathrm{B}} V_{\max_f}}{\text{denominator}} - \frac{\dfrac{[\mathrm{P}][\mathrm{Q}]}{\beta K_\mathrm{P} K_\mathrm{Q}} V_{\max_r}}{\text{denominator}}$$

where $V_{\max_f} = k_p[\mathrm{E}]_t$, $V_{\max_r} = k_{-p}[\mathrm{E}]_t$ and the denominator is the sum of the terms for all nine enzyme species (free E plus eight complexes). The general equation is:

$$v_{\text{net}_f} = \frac{\dfrac{[\mathrm{A}][\mathrm{B}]}{\alpha K_\mathrm{A} K_\mathrm{B}} V_{\max_f} - \dfrac{[\mathrm{P}][\mathrm{Q}]}{\beta K_\mathrm{P} K_\mathrm{Q}} V_{\max_r}}{1 + \dfrac{[\mathrm{A}]}{K_\mathrm{A}} + \dfrac{[\mathrm{B}]}{K_\mathrm{B}} + \dfrac{[\mathrm{P}]}{K_\mathrm{P}} + \dfrac{[\mathrm{Q}]}{K_\mathrm{Q}} + \dfrac{[\mathrm{A}][\mathrm{B}]}{\alpha K_\mathrm{A} K_\mathrm{B}} + \dfrac{[\mathrm{P}][\mathrm{Q}]}{\beta K_\mathrm{P} K_\mathrm{Q}} + \dfrac{[\mathrm{B}][\mathrm{Q}]}{\gamma K_\mathrm{B} K_\mathrm{Q}} + \dfrac{[\mathrm{A}][\mathrm{P}]}{\delta K_\mathrm{A} K_\mathrm{P}}} \quad \text{(VI-46)}$$

System A5. Substrate Inhibition in Rapid Equilibrium Random Systems

System A5a. Dead-End BE and BEB Complexes

One substrate may have an appreciable affinity for the other substrate's binding site, particularly when the two substrates are chemically similar and the reaction is being studied in the nonphysiological direction. Two Rapid Equilibrium Random situations are shown in Figure VI-15. In Figure VI-15*a*, it is assumed that the binding of B to the A site does not prevent B from binding to its usual site so that two dead-end complexes form, BE and BEB. The equilibria are:

$$\begin{array}{ccccccc}
\mathrm{BE} & \overset{K_i}{\rightleftharpoons} & \mathrm{B} + \mathrm{E} + \mathrm{A} & \overset{K_\mathrm{A}}{\rightleftharpoons} & \mathrm{EA} & & \\
+ & & + & & + & & \\
\mathrm{B} & & \mathrm{B} & & \mathrm{B} & & \\
\beta K_\mathrm{B} \upharpoonleft\downharpoonright & & K_\mathrm{B} \upharpoonleft\downharpoonright & & \downharpoonleft\upharpoonright \alpha K_\mathrm{B} & & \\
\mathrm{BEB} & \overset{\beta K_i}{\rightleftharpoons} & \mathrm{B} + \mathrm{EB} + \mathrm{A} & \overset{\alpha K_\mathrm{A}}{\rightleftharpoons} & \mathrm{EAB} & \overset{k_p}{\longrightarrow} & \mathrm{E} + \mathrm{P} + \mathrm{Q}
\end{array}$$

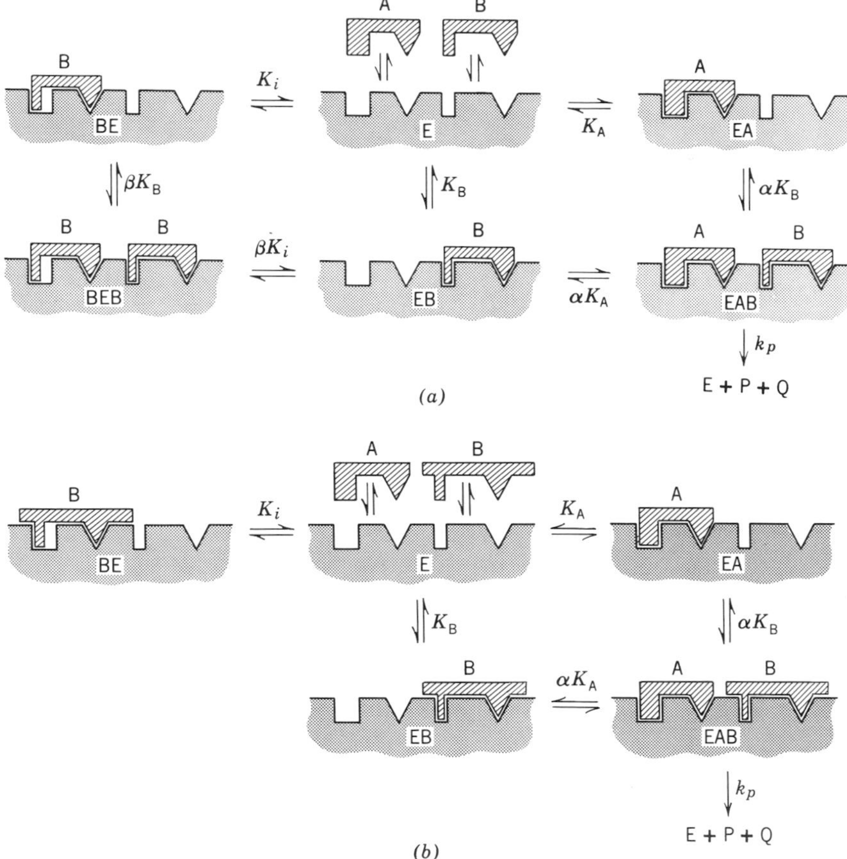

Fig. VI-15. Two models of competitive substrate inhibition in a Rapid Equilibrium Random system. (*a*) B binds to the A site as well as to the normal B site. (*b*) B binds to the A site and prevents another molecule of B from binding normally.

The velocity equation is:

$$\frac{v}{[\mathrm{E}]_t} = \frac{k_p[\mathrm{EAB}]}{[\mathrm{E}]+[\mathrm{EA}]+[\mathrm{EB}]+[\mathrm{EAB}]+[\mathrm{BE}]+[\mathrm{BEB}]}$$

$$\boxed{\frac{v}{V_{\max}} = \frac{\dfrac{[\mathrm{A}][\mathrm{B}]}{\alpha K_\mathrm{A} K_\mathrm{B}}}{1+\dfrac{[\mathrm{A}]}{K_\mathrm{A}}+\dfrac{[\mathrm{B}]}{K_\mathrm{B}}+\dfrac{[\mathrm{A}][\mathrm{B}]}{\alpha K_\mathrm{A} K_\mathrm{B}}+\dfrac{[\mathrm{B}]}{K_i}+\dfrac{[\mathrm{B}]^2}{\beta K_i K_\mathrm{B}}}} \qquad \text{(VI-47)}$$

If the "incorrectly" bound B excludes B from the normal B site (Fig. VI-15*b*), the $[B]^2/\beta K_i K_B$ denominator term would be missing. When [A] is varied (and both BE and BEB complexes form):

$$\frac{v}{V_{max}} = \frac{[A]}{\alpha K_A\left(1 + \frac{K_B}{[B]} + \frac{K_B}{K_i} + \frac{[B]}{\beta K_i}\right) + [A]\left(1 + \frac{\alpha K_B}{[B]}\right)} \tag{VI-48}$$

The $1/v$ versus $1/[A]$ plot at any fixed [B] is linear, as usual. However, because of the inhibitory $[B]/\beta K_i$ term in the slope factor, the family of reciprocal plots may look quite unusual. Figure VI-16*a* shows a family of plots for $K_A = 1$, $K_B = 1$, $V_{max} = 1$, $\alpha = 1$, and $\beta K_i = 10$ (i.e., all plotted values are specific or relative values and $\beta K_i = 10K_B$). The assumption that $\beta K_i > K_B$ is reasonable, since the affinity of the A site for B will probably be much lower than the affinity of the B site for B. At low B concentrations the curves appear normal. For example, between $[B] = 0.1$ and $[B] = 1$, the slope decreases with increasing [B] and the lines almost intersect at a common point $(-1/K_A)$. However, between $[B] = 1$ and $[B] = 10$, the lines appear almost parallel. Between $[B] = 10$ and $[B] = 100$, the slopes clearly increase with increasing [B] and the lines intersect on the $1/v$-axis at $1/V_{max}$. In this range of B concentrations, B acts predominantly as a competitive inhibitor. The effect of changing B concentrations is understandable. When [B] is low compared to K_B (thus very low compared to βK_i), the $K_B/[B]$ term is the major variable contributor to the total value of the slope factor. Any increase in [B] causes a marked decrease in $K_B/[B]$, hence in the $slope_{1/A}$. At the same time, the $[B]/\beta K_i$ term contributes little to the total value of the slope factor and an increase in [B] has little effect on the contribution of the term. When [B] is large compared to K_B, the $K_B/[B]$ term is small and has little influence on the total value of the slope factor. Now the $[B]/\beta K_i$ term becomes the major contributor and its value increases with increasing [B]. Thus the $slope_{1/A}$ decreases, passes through a minimum, and then increases again. The B concentration at the minimum can be calculated by setting the first derivative of the $slope_{1/A}$ equal to zero:

$$slope_{1/A} = \frac{\alpha K_A}{V_{max}}\left(1 + \frac{K_B}{K_i}\right) + \frac{\alpha K_A K_B}{V_{max}}[B]^{-1} + \frac{\alpha K_A}{\beta K_i V_{max}}[B] \tag{VI-49}$$

$$\frac{d\ slope_{1/A}}{d[B]} = \frac{\alpha K_A}{\beta K_i V_{max}} - \frac{\alpha K_A K_B}{V_{max}}[B]^{-2} = 0$$

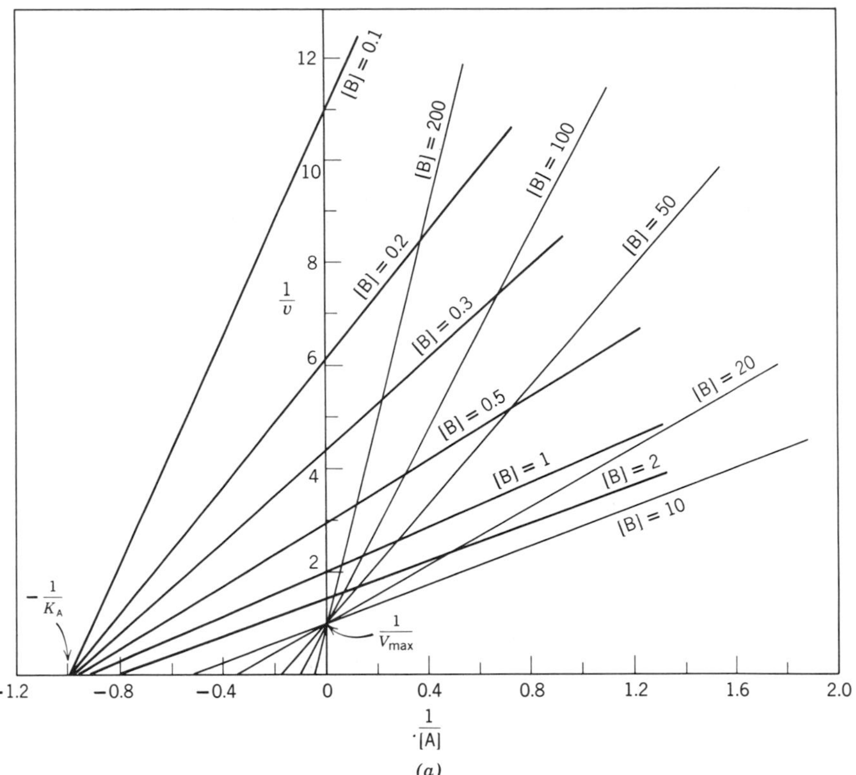

Fig. VI-16. (*a*) $1/v$ versus $1/[A]$ plots for a Rapid Equilibrium Random Bireactant system in which B binds to the A site yielding BE and BEB dead-end complexes.

Solving for [B]:

$$[B]_{min} = \sqrt{\beta K_i K_B}$$

The slope replot is shown in Figure VI-16*b*. As [B] decreases (i.e., as $1/[B]$ increases), the replot approaches a straight line given by the first two terms of equation VI-49. The extrapolated $1/[B]$-axis intercept gives:

$$\frac{1}{[B]} = -\left(\frac{1}{K_B} + \frac{1}{K_i}\right) = -\frac{1}{K_B}\left(1 + \frac{1}{\gamma}\right)$$

where $\gamma = K_i/K_B$.

If the $slope_{1/A}$ is replotted against [B] (as for an inhibitor) the shape of the replot is the same as that shown in Figure VI-16*b*, except now as [B]

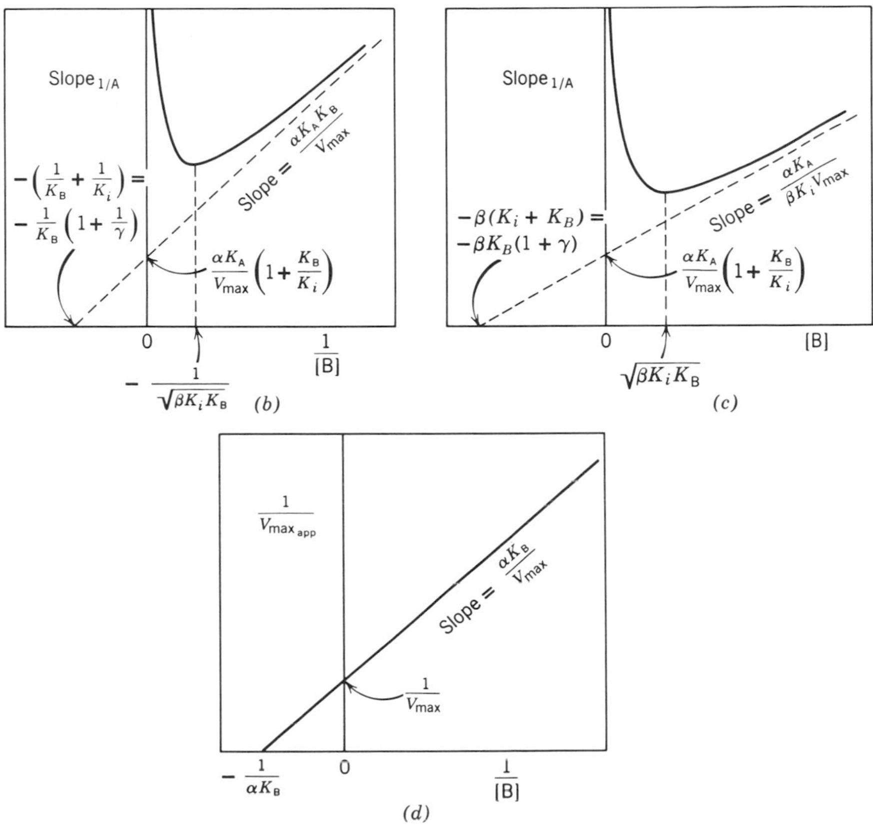

Fig. VI-16. (*Cont*). (*b*) $Slope_{1/A}$ replot in a low range of B concentrations. (*c*) $Slope_{1/A}$ replot in a high range of B concentrations. (*d*) $1/v$-axis intercept replot.

increases, the replot approaches a straight line given by the first and third terms of equation VI-49 (Fig. VI-16*c*). The extrapolated [B]-axis intercept now gives:

$$[B] = -\beta(K_i + K_B) = -\beta K_B(1+\gamma)$$

Obviously, a large range of B concentrations are necessary to construct both types of $slope_{1/A}$ replots.

The replot of $1/V_{max_{app}}$ versus 1/[B] is normal and permits αK_B and V_{max} to be determined (Fig. VI-16*d*). With these values, K_A can be calculated from the asymptotic slope of the $slope_{1/A}$ versus 1/[B] replot or from the ratio of asymptotic slopes of the $slope_{1/A}$ replot and $1/V_{max_{app}}$ replot.

When [B] is varied, the velocity equation is:

$$\frac{v}{V_{max}} = \frac{[B]}{\alpha K_B\left(1+\frac{K_A}{[A]}\right)+[B]\left(1+\frac{\alpha K_A}{[A]}+\frac{\alpha K_A K_B}{K_i[A]}+\frac{\alpha K_A[B]}{\beta K_i[A]}\right)} \quad \text{(VI-50)}$$

The family of reciprocal plots approach a straight line at high values of 1/[B] (low [B]). As the lines approach the $1/v$-axis, they pass through a minimum and then bend upward. The B concentration at the minimum is given by:

$$[B]_{min} = \sqrt{\beta K_i K_B\left(1+\frac{[A]}{K_A}\right)}$$

Thus as the fixed [A] increases, the minimum moves closer to the $1/v$-axis, and at high [A] the plot may appear to be completely linear over the entire 1/[B] range plotted. (A is competitive with B at the A site, and, consequently, saturating [A] eliminates the substrate inhibition.) The replot of the asymptotic $slope_{1/B}$ versus 1/[A] yields $-1/K_A$ as the 1/[A]-axis intercept. A replot of the extrapolated $1/v$-axis intercepts versus 1/[A] yields:

$$-\frac{1}{\alpha K_{A_{app}}} = -\frac{1}{\alpha K_A\left(1+\frac{1}{\gamma}\right)}$$

If the fixed [A] is small, the minimum will occur well within the 1/[B] range plotted. In this case it may be difficult to esitmate the aymptote. The usual mistake is to draw the asymptote too close to the original plot. A procedure outlined by Cleland (1970) can be used here and for all substrate inhibitions resulting from the dead-end combination of the varied substrate with the wrong enzyme form. In such situations, the velocity equation can be written as:

$$\frac{v}{V^*_{max}} = \frac{[B]}{K^*_B+[B]+\frac{[B]^2}{K^*_i}} \quad \text{(VI-51)}$$

where V^*_{max}, K^*_B, and K_i^* are apparent constants at the constant cosubstrate concentrations. For the present system, equation VI-51 is obtained by factoring all the non-[B]-containing terms from the intercept factor of equation VI-50 and then dividing both sides of the equation by the factored quantity. We obtain:

$$V^*_{max} = \frac{V_{max}}{1+\dfrac{\alpha K_A}{[A]}\left(1+\dfrac{1}{\gamma}\right)}, \qquad K^*_B = \frac{\alpha K_B\left(1+\dfrac{K_A}{[A]}\right)}{1+\dfrac{\alpha K_A}{[A]}\left(1+\dfrac{1}{\gamma}\right)}$$

and

$$K_i^* = \beta K_i\left(1+\frac{1}{\gamma}+\frac{[A]}{\alpha K_A}\right)$$

where $\gamma = K_i/K_B$. When $1/v$ is plotted against $1/[B]$, the coordinates of the minimum point are:

$$\frac{1}{[B]_{min}} = \frac{1}{\sqrt{K^*_B K_i^*}} \qquad \text{and} \qquad \frac{1}{v} = \frac{1}{V^*_{max}}\left(1+2\sqrt{\frac{K^*_B}{K_i^*}}\right)$$

The line connecting the $1/v$-axis intercept of the asymptote with the minimum point has a slope of $2K^*_B/V^*_{max}$, or double that of the asymptote itself. Thus the correct position of the asymptote is determined by adjusting it until its slope is half that of the line connecting the minimum point with the $1/v$-axis intercept. The value of $1/K_i^*$ can be determined by finding the $1/[B]$ position where the vertical distance between the asymptote and the original curve equals $1/V^*_{max}$. At this $1/[B]$ value, the curve has a vertical coordinate of $(1/V^*_{max})\ (2+K^*_B/K_i^*)$ while the asymptote coordinate is $(1/V^*_{max})\ (1+K^*_B/K_i^*)$. The procedure is shown in Figure VI-17.

System A5b. No BEB Complex

The preceding discussion assumes that the binding of B to the A site does not exclude the normal binding of B to the B site. If normal B binding is excluded (Fig. VI-15*b*), then the denominator $[B]/\beta K_i$ term will be missing from equations VI-48 and VI-50. As a result, there will be no obvious substrate inhibition. The system will behave as an ordinary Rapid Equilibrium Random Bireactant system with apparent dissociation constants

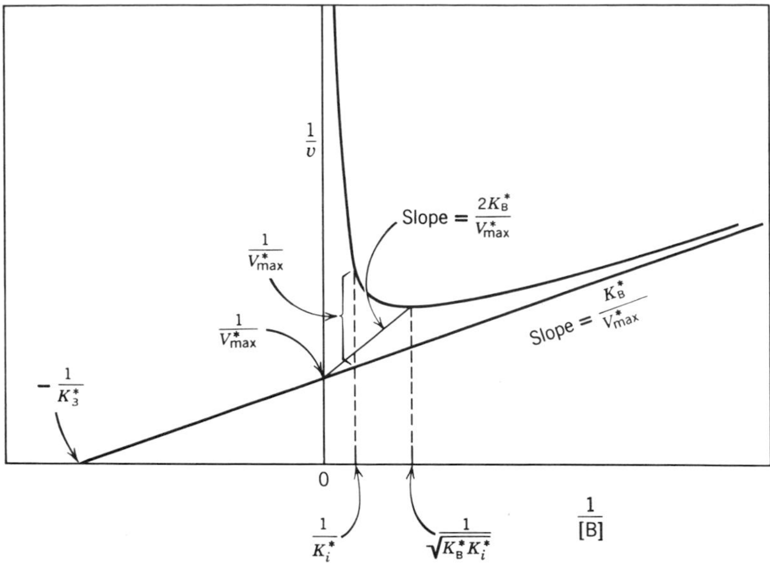

Fig. VI-17. Procedure described by Cleland for determining the correct position of the asymptote. The position of the asymptote is adjusted until its slope is half that of a line connecting the minimum point of the original plot with the $1/v$-axis intercept of the asymptote; $1/K_i^*$ is obtained by finding the value of 1/[B] where the distance between the original plot and the asymptote equals $1/V_{max}^*$.

shown below.

$$\begin{array}{ccccc}
E + A & \overset{K_A}{\rightleftharpoons} & EA & & \\
+ & & + & & \\
B & & B & & \\
\frac{K_B}{\left(1+\frac{1}{\gamma}\right)} \Big\updownarrow & & \Big\updownarrow \alpha K_B & & \\
EB + A & \overset{\alpha K_A\left(1+\frac{1}{\gamma}\right)}{\rightleftharpoons} & EAB & \xrightarrow{k_p} & E+P+Q
\end{array}$$

The apparent interaction factor is $\alpha(1+1/\gamma)$.

System A5c. Glycogen Phosphorylase Model (B Competes with B)

Glycogen phosphorylase catalyzes the reaction:

$$(\text{glucose})_n + P_i \rightleftharpoons (\text{glucose})_{n-1} + \text{glucose-1-phosphate}$$

The reaction is freely reversible and is often assayed in the direction of glycogen synthesis. It is reasonable to assume that there are two ways that glycogen can bind to the enzyme: the phosphorolysis mode and the synthesis mode (Fig. VI-18). The kinetic mechanism is believed to be Rapid Equilibrium Random. The reaction sequence showing the expected complexes is:

$$
\begin{array}{ccccccccc}
\mathrm{BE} & + & \mathrm{A} & \overset{\beta K_A}{\rightleftharpoons} & \mathrm{BEA} & & & & \mathrm{EB} \\
\gamma K_B \updownarrow & & & & \updownarrow \beta\gamma K_B & & & & \updownarrow K_B \\
\mathrm{B} & & & & \mathrm{B} & & & & \mathrm{B} \\
+ & & & & + & & & & + \\
\mathrm{E} & + & \mathrm{A} & \overset{K_A}{\rightleftharpoons} & \mathrm{EA} & & \mathrm{BE} & \overset{\gamma K_B}{\rightleftharpoons} & \mathrm{B} + \mathrm{E} \\
+ & & & & + & & + & & + \\
\mathrm{B}\ (\text{phosphorolysis}) & & & & \mathrm{B} & & \mathrm{Q}\ (\text{synthesis}) & & \mathrm{Q} \\
K_B \updownarrow & & & & \updownarrow \alpha K_B & & \delta K_Q \updownarrow & & \updownarrow K_Q \\
\mathrm{EB} & + & \mathrm{A} & \overset{\alpha K_A}{\rightleftharpoons} & \mathrm{EAB} & \underset{k_{-p}}{\overset{k_p}{\rightleftharpoons}} & \mathrm{BEQ} & \overset{\gamma\delta K_B}{\rightleftharpoons} & \mathrm{EQ}
\end{array}
$$

where EB = glycogen bound in the phosphorolysis mode
BE = glycogen bound in the synthesis mode
A = inorganic phosphate
Q = glucose-1-phosphate

It is clear from Figure VI-18 that glycogen bound in the synthesis mode will be competitive with glycogen bound in the phosphorolysis mode, but will

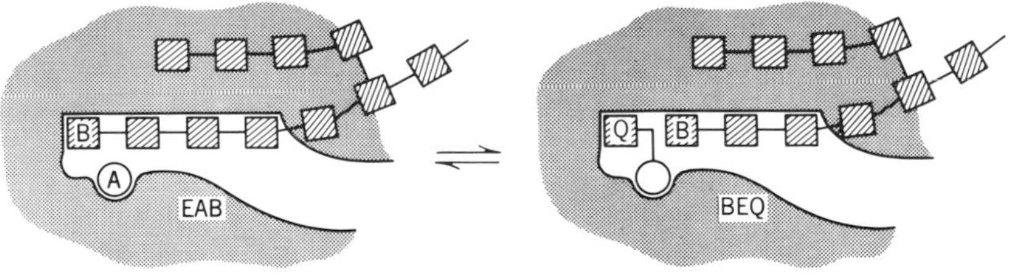

Phosphorolysis mode Synthesis mode

Fig. VI-18. Central complexes of glycogen phosphorylase showing two mutually exclusive modes of glycogen binding. A = P_i, B = glycogen, Q = glucose-1-phosphate.

allow P_i to bind. The velocity equation for the phosphorolysis direction is:

$$\frac{v}{V_{max}} = \frac{\dfrac{[A][B]}{\alpha K_A K_B}}{1 + \dfrac{[A]}{K_A} + \dfrac{[B]}{K_B} + \dfrac{[B]}{\gamma K_B} + \dfrac{[A][B]}{\alpha K_A K_B} + \dfrac{[A][B]}{\beta\gamma K_A K_B}} \qquad \text{(VI-52)}$$

When $[P_i]$ is varied:

$$\frac{v}{V_{max}} = \frac{[A]}{\alpha K_A\left(1 + \dfrac{1}{\gamma} + \dfrac{K_B}{[B]}\right) + [A]\left(1 + \dfrac{\alpha}{\beta\gamma} + \dfrac{\alpha K_B}{[B]}\right)} \qquad \text{(VI-53)}$$

When [glycogen] is varied:

$$\frac{v}{V_{max}} = \frac{[B]}{\alpha K_B\left(1 + \dfrac{K_A}{[A]}\right) + [B]\left(1 + \dfrac{\alpha}{\beta\gamma} + \dfrac{\alpha K_A}{[A]} + \dfrac{\alpha K_A}{\gamma[A]}\right)} \qquad \text{(VI-54)}$$

The $1/v$ versus $1/[A]$ plots are linear and intersect at $-1/K_A$. The $slope_{1/A}$ and $1/V_{max_{app}}$ versus $1/[B]$ replots yield:

$$K_{slope}^{B/A} = K_{B_{app}} = \frac{K_B}{\left(1 + \dfrac{1}{\gamma}\right)}, \qquad K_{int}^{B/A} = K_{m_B} = \frac{\alpha K_B}{\left(1 + \dfrac{\alpha}{\beta\gamma}\right)}$$

If $\alpha = \beta$ (i.e., both glycogen binding modes affect the binding of P_i equally) then $K_{int}^{B/A} = \alpha K_B/(1 + 1/\gamma)$. The $1/v$ versus $1/[B]$ plots are also linear, intersecting at $-(1 + 1/\gamma)/K_B$. The $slope_{1/B}$ and $1/V_{max_{app}}$ versus $1/[A]$ replots yield:

$$K_{slope}^{A/B} = K_A, \qquad K_{int}^{A/B} = K_{m_A} = \alpha K_A \frac{\left(1 + \dfrac{1}{\gamma}\right)}{\left(1 + \dfrac{\alpha}{\beta\gamma}\right)}$$

If $\alpha = \beta$, then $K_{int}^{A/B} = \alpha K_A$.

In the synthesis direction, glycogen bound in the phosphorolysis mode excludes glycogen bound in the synthesis mode as well as glucose-1-phosphate. The velocity equation is:

$$\frac{v}{V_{\max}} = \frac{\dfrac{[B][Q]}{\gamma\delta K_B K_Q}}{1 + \dfrac{[B]}{\gamma K_B} + \dfrac{[B]}{K_B} + \dfrac{[Q]}{K_Q} + \dfrac{[B][Q]}{\gamma\delta K_B K_Q}} \quad \text{(VI-55)}$$

When [glycogen] is varied:

$$\frac{v}{V_{\max}} = \frac{[B]}{\gamma\delta K_B\left(1 + \dfrac{K_Q}{[Q]}\right) + [B]\left(1 + \dfrac{\gamma\delta K_Q}{[Q]} + \dfrac{\delta K_Q}{[Q]}\right)} \quad \text{(VI-56)}$$

When [glucose-1-phosphate] is varied:

$$\frac{v}{V_{\max}} = \frac{[Q]}{\gamma\delta K_Q\left(1 + \dfrac{1}{\gamma} + \dfrac{K_B}{[B]}\right) + [Q]\left(1 + \dfrac{\delta\gamma K_B}{[B]}\right)} \quad \text{(VI-57)}$$

Again, both families of reciprocal plots are linear intersecting at $-(1+1/\gamma)/K_B$ (plotting $1/v$ versus $1/[B]$) and $-1/K_Q$ (plotting $1/v$ versus $1/[Q]$). The kinetic constants are:

$$K_{\text{slope}}^{B/Q} = K_{B_{\text{app}}} = \frac{K_B}{\left(1 + \dfrac{1}{\gamma}\right)}, \qquad K_{\text{int}}^{B/Q} = K_{m_B} = \delta\gamma K_B$$

$$K_{\text{slope}}^{Q/B} = K_Q, \qquad K_{\text{int}}^{Q/B} = K_{m_Q} = \delta(1+\gamma)K_Q$$

The system behaves in an ordinary Rapid Equilibrium Random fashion in both directions.

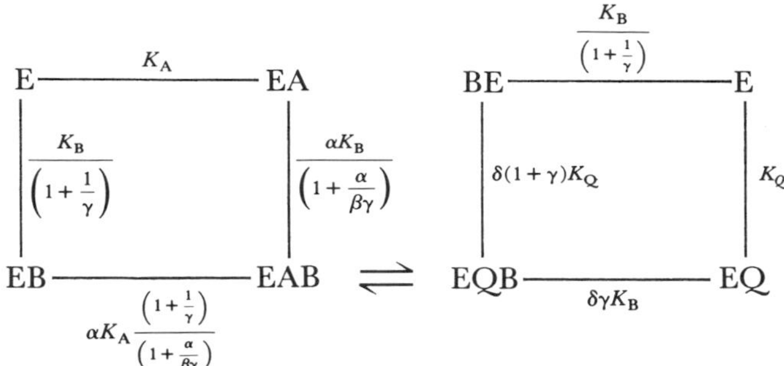

$$\text{Phosphorolysis: apparent interaction factor} = \frac{\alpha\left(1+\dfrac{1}{\gamma}\right)}{\left(1+\dfrac{\alpha}{\beta\gamma}\right)}$$

$$\text{Synthesis: apparent interaction factor} = \delta(1+\gamma)$$

If all the actual interaction factors are unity, the K_m for glycogen in the phosphorolysis direction will be $K_B/2$, while in the synthesis direction $K_{m_B} = K_B$. The difference in K_{m_B} values for the two directions is not hard to understand. The K_{m_B} value represents the dissociation constant for glycogen at saturating cosubstrate. In the phosphorolysis direction, saturating $[P_i]$ will still permit both binding modes for glycogen. The observed K_{m_B} is really a composite of K_B and γK_B (which equals $K_B/2$ if $\gamma = 1$). In the synthesis direction, saturating [glucose-1-phosphate] excludes glycogen binding in the phosphorolysis mode. Glycogen can bind only in the synthesis mode with a dissociation constant K_B (if $\delta = 1$ and $\gamma = 1$).

B. ORDERED BIREACTANT SYSTEMS

The equilibria shown below describe a rapid equilibrium system in which an activator, A, and a substrate, B, (or two substrates) combine with the enzyme in an obligate order: B can add only to the EA complex.

$$\begin{array}{ccc} E + A & \overset{K_A}{\rightleftharpoons} & EA \\ & & + \\ & & B \\ & & \Big\downarrow\!\uparrow K_B \\ & & EAB \xrightarrow{k_p} E + P \end{array}$$

The activator might induce a conformational change in the enzyme so that substrate binding groups become available. Another possibility is that the activator is modified by the enzyme to a form that then participates in positioning the substrate properly in relation to the catalytic groups. The system resembles uncompetitive inhibition in that the two ligands add to the enzyme in a definite order. By analogy, we could reason that the reaction EA + B→EAB pulls the reaction E + A→EA to the right. Consequently, we would expect the apparent affinity of E for A to increase as B increases (i.e., $K_{A_{app}}$ decreases as B increases). The analogy with uncompetitive inhibition is not perfect because now EAB is catalytically active, while ESI is not. We can see from the equilibria that an infinitely high concentration of B can pull all the enzyme to the reactive EAB form *as long as some A is present*, even if [A] is far below K_A. (Keep in mind that a "very low" concentration of A is still far in excess over the concentration of enzyme.) Thus the system also resembles competitive inhibition in that V_{max} is unaffected by the activator concentration, or, in other words, the requirement for the activator appears to disappear when [B] is very high. The apparent K_B should vary with varying [A] because the concentration of A affects the distribution of enzyme between a form that combines with B and a form that does not. We can check these predictions by examining the velocity equations, which can be derived in the usual manner from rapid equilibrium assumptions.

$$v = k_p[\mathrm{EAB}], \qquad [\mathrm{E}]_t = [\mathrm{E}] + [\mathrm{EA}] + [\mathrm{EAB}]$$

$$\therefore \quad \frac{v}{[\mathrm{E}]_t} = \frac{k_p[\mathrm{EAB}]}{[\mathrm{E}] + [\mathrm{EA}] + [\mathrm{EAB}]}$$

When the concentrations of EA and EAB are expressed in terms of [E], we obtain:

$$\boxed{\frac{v}{V_{max}} = \frac{\dfrac{[\mathrm{A}][\mathrm{B}]}{K_A K_B}}{1 + \dfrac{[\mathrm{A}]}{K_A} + \dfrac{[\mathrm{A}][\mathrm{B}]}{K_A K_B}}} \qquad \text{(VI-58)}$$

or

$$\boxed{\frac{v}{V_{max}} = \frac{[\mathrm{A}][\mathrm{B}]}{K_A K_B + K_B[\mathrm{A}] + [\mathrm{A}][\mathrm{B}]}} \qquad \text{(VI-59)}$$

where $V_{max} = k_p[E]_t$. Rearranging the velocity equation to express B as the varied ligand.

$$\frac{v}{V_{max}} = \frac{[B]}{K_B\left(1+\frac{K_A}{[A]}\right)+[B]} \tag{VI-60}$$

In the above form, the competitive activation by A is clear. We can also see that at any [A]:

$$K_{B_{app}} = K_B\left(1+\frac{K_A}{[A]}\right) \tag{VI-61}$$

As [A] increases, $K_{B_{app}}$ decreases and approaches K_B as a limit. The reciprocal form of the velocity equation for B as the varied ligand is:

$$\frac{1}{v} = \frac{K_B}{V_{max}}\left(1+\frac{K_A}{[A]}\right)\frac{1}{[B]} + \frac{1}{V_{max}} \tag{VI-62}$$

The family of reciprocal plots are shown in Figure VI-19*a*. Although the limiting plot gives $-1/K_B$ as the 1/[B]-axis intercept, it is more practical to determine K_B from a replot. The variation of K_B with [A] is given by:

$$K_{B_{app}} = K_B K_A \frac{1}{[A]} + K_B \tag{VI-63}$$

Thus a plot of $K_{B_{app}}$ versus 1/[A] is linear with a slope of $K_B K_A$ and an intercept on the $K_{B_{app}}$-axis of K_B.The intercept on the 1/[A]-axis gives $-1/K_A$ (Fig. VI-19*b*). The slope of the $1/v$ versus 1/[B] plot is given by:

$$slope_{1/B} = \frac{K_B K_A}{V_{max}}\frac{1}{[A]} + \frac{K_B}{V_{max}} \tag{VI-64}$$

A replot of $slope_{1/B}$ versus 1/[A] (Fig. VI-19*c*) is linear with a slope of $K_B K_A/V_{max}$ and an intercept on the slope-axis of K_B/V_{max} (i.e., when [A] is

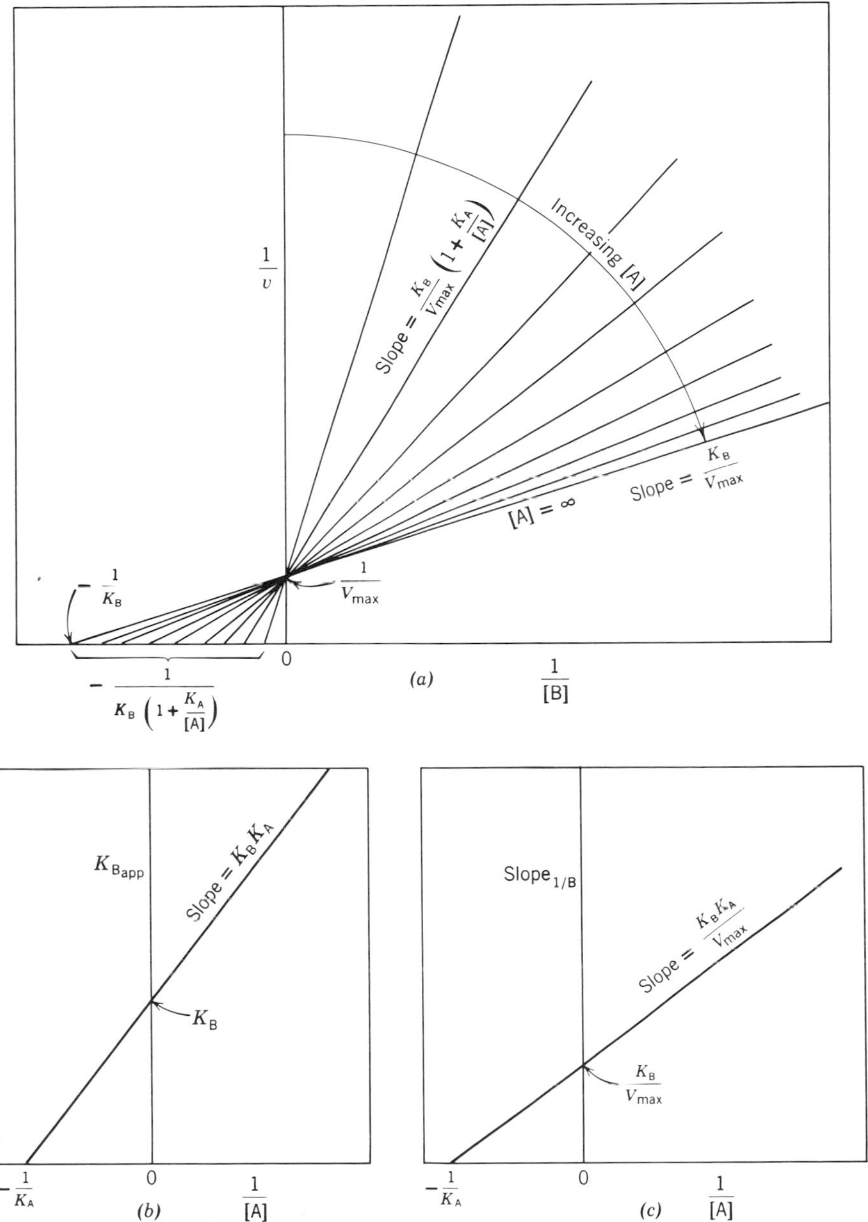

Fig. VI-19. (*a*) $1/v$ versus $1/[B]$ plot for a Rapid Equilibrium Ordered Bireactant system where A adds before B. (*b*) Replot of $K_{B_{app}}$ versus $1/[A]$. (*c*) Replot of $Slope_{1/B}$ versus $1/[A]$.

infinitely high $K_{B_{app}} = K_B$ and the slope, which equals $K_{B_{app}}/V_{max}$ at all values of [A], now gives K_B/V_{max}). When $slope_{1/B} = 0$, the intercept on the 1/[A]-axis gives $-1/K_A$. The competitive activation by A is one way of distinguishing the Rapid Equilibrium Ordered reaction from the Rapid Equilibrium Random and Steady-State Ordered reactions.

The velocity equation can be rearranged to express A as the varied ligand:

$$\frac{v}{V_{max}} = \frac{[A]}{K_A\left(\frac{K_B}{[B]}\right) + [A]\left(1 + \frac{K_B}{[B]}\right)} \qquad \text{(VI-65)}$$

To see the effect of [B] on the constants, the denominators of both sides of the equation can be divided by $(1 + K_B/[B])$.

$$\frac{v}{\frac{V_{max}}{\left(1 + \frac{K_B}{[B]}\right)}} = \frac{[A]}{\frac{K_A}{\left(1 + \frac{[B]}{K_B}\right)} + [A]} \qquad \text{(VI-66)}$$

Both the apparent V_{max} and the apparent K_A vary with [B], but in opposite directions.

$$V_{max_{app}} = \frac{V_{max}}{\left(1 + \frac{K_B}{[B]}\right)}, \qquad K_{A_{app}} = \frac{K_A}{\left(1 + \frac{[B]}{K_B}\right)}$$

As [B] increases, $V_{max_{app}}$ increases and approaches V_{max} as a limit. The $K_{A_{app}}$ decreases as [B] increases and approaches a limit of zero. As [B] decreases, $K_{A_{app}}$ approaches K_A as a limit. The reciprocal form of the velocity equation for A as the varied ligand is:

$$\frac{1}{v} = \frac{K_A}{V_{max}}\left(\frac{K_B}{[B]}\right)\frac{1}{[A]} + \frac{1}{V_{max}}\left(1 + \frac{K_B}{[B]}\right) \qquad \text{(VI-67)}$$

The reciprocal plot is shown in Figure VI-20*a*. The family of curves intersect at $1/[\mathrm{A}] = -1/K_{\mathrm{A}}$ and $1/v = 1/V_{\max}$. As [B] increases, the plots pivot clockwise about the intersection point. When [B] is infinitely high, the slope of the plot is zero and the $1/v$-axis intercept gives $1/V_{\max}$. Again, we see that the dependence of the velocity on [A] seems to disappear as [B] becomes very large. The variation of $V_{\max_{\mathrm{app}}}$ with [B] is given by:

$$\frac{1}{V_{\max_{\mathrm{app}}}} = \frac{K_{\mathrm{B}}}{V_{\max}} \frac{1}{[\mathrm{B}]} + \frac{1}{V_{\max}} \qquad \text{(VI-68)}$$

The replot of $1/V_{\max_{\mathrm{app}}}$ versus $1/[\mathrm{B}]$ (Fig. VI-20*b*) is linear with a slope of $K_{\mathrm{B}}/V_{\max}$ and a $1/V_{\max_{\mathrm{app}}}$-axis intercept of $1/V_{\max}$. The intercept on the $1/[\mathrm{B}]$-axis gives $-1/K_{\mathrm{B}}$. The variation of $K_{\mathrm{A}_{\mathrm{app}}}$ with [B] is given by:

$$\frac{1}{K_{\mathrm{A}_{\mathrm{app}}}} = \frac{1}{K_{\mathrm{A}}K_{\mathrm{B}}}[\mathrm{B}] + \frac{1}{K_{\mathrm{A}}} \qquad \text{(VI 69)}$$

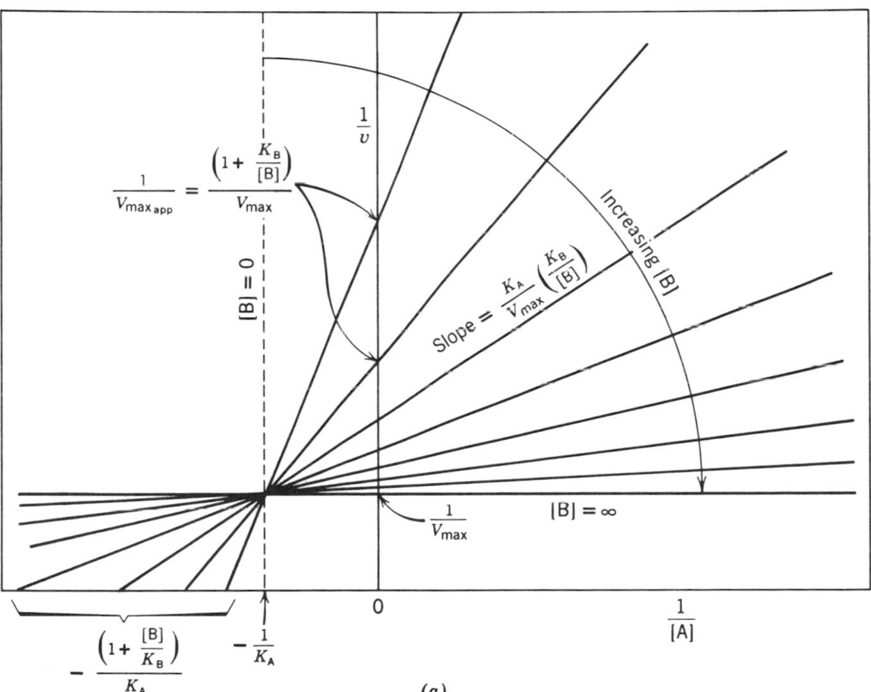

(*a*)

Fig. VI-20. (*a*) $1/v$ versus $1/[\mathrm{A}]$ plot for a Rapid Equilibrium Ordered Bireactant system where A adds before B.

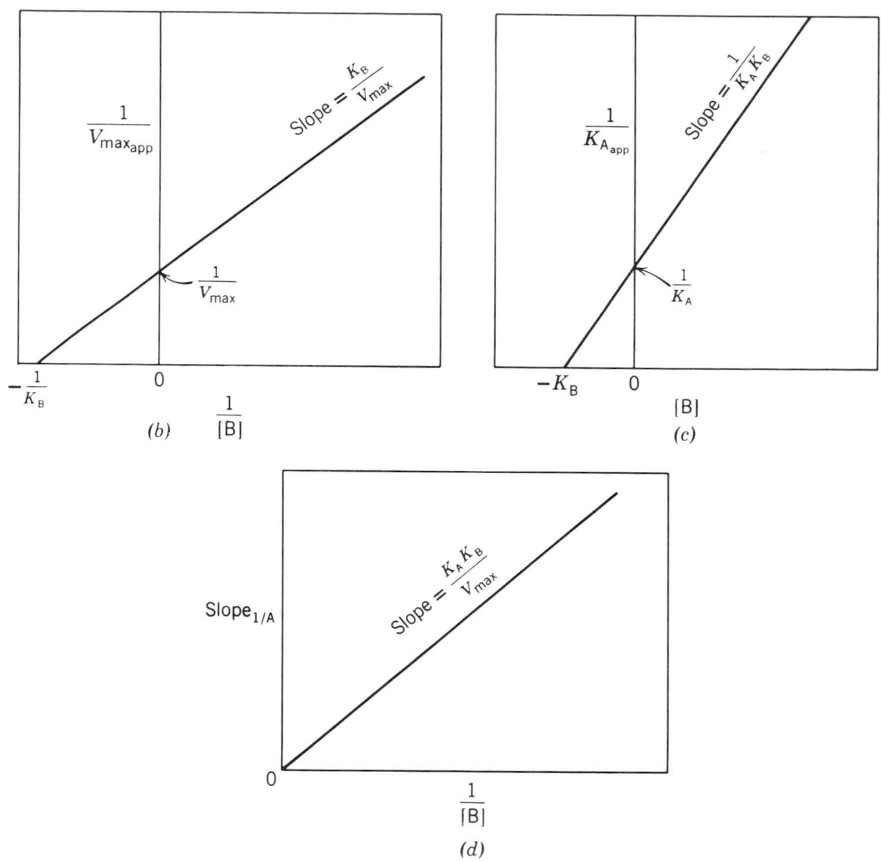

Fig. VI-20. (*Cont.*) (*b*) Replot of $1/v$-axis intercept versus $1/[B]$. (*c*) Replot of $1/K_{A_{app}}$ versus [B]. (*d*) Replot of $Slope_{1/A}$ versus $1/[B]$.

A plot of $1/K_{A_{app}}$ versus [B] (Fig. VI-20*c*) is linear with a slope of $1/K_AK_B$ and an intercept on the $1/K_{A_{app}}$-axis of $1/K_A$ (i.e., when $[B]=0$, $K_{A_{app}} = K_A$). When $1/K_{A_{app}}=0$, the intercept on the [B]-axis gives $-K_B$.

The slope of the $1/v$ versus $1/[A]$ plot is given by:

$$slope_{1/A} = \frac{K_AK_B}{V_{max}} \frac{1}{[B]} \tag{VI-70}$$

A plot of $slope_{1/A}$ versus $1/[B]$ is linear with a slope of K_AK_B/V_{max} and an intercept at the origin [i.e., when [B] is infinitely high, the slope of the reciprocal plot is zero (Fig. VI-20*d*)]. The plot does not give K_A, K_B, or V_{max} as an intercept, but it verifies that the limiting slope is indeed zero and that

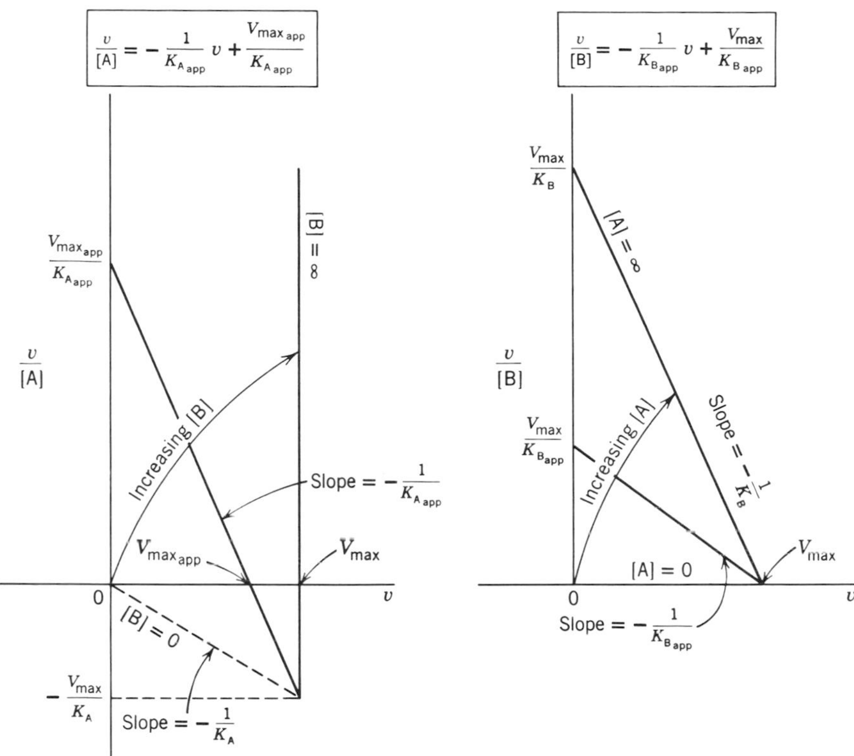

Fig. VI-21. Hanes-Woolf plots for a Rapid Equilibrium Ordered Bireactant system.

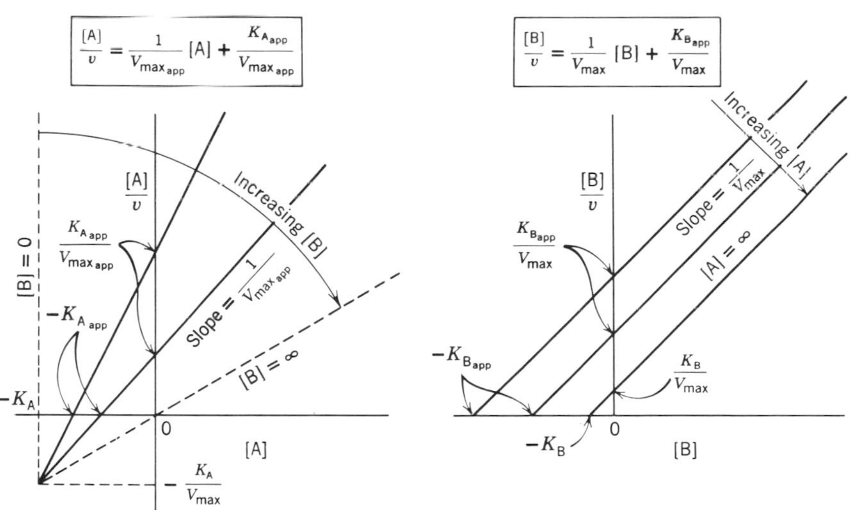

Fig. VI-22. Woolf-Augustinsson-Hofstee plot for a Rapid Equilibrium Ordered Bireactant system.

the reaction is ordered with A adding to the enzyme before B. The ratio of the slopes of the $slope_{1/A}$ replot and $1/v$-axis intercept replot gives K_A.

$$\text{slope of } slope_{1/A} \text{ replot} = \frac{K_A K_B}{V_{max}}$$

$$\text{slope of } 1/v\text{-axis intercept replot} = \frac{K_B}{V_{max}}$$

$$\text{ratio} = \frac{K_A K_B / V_{max}}{K_B / V_{max}} = K_A$$

The alternate linear plots are shown in Figures VI-21, VI-22, and VI-23.

Some of the properties of the Rapid Equilibrium Random and Rapid Equilibrium Ordered systems are summarized in Table VI-1. The reader should keep in mind that the Rapid Equilibrium Ordered system is a limiting case of the more realistic steady-state mechanism. In unireactant systems, rapid equilibrium and steady-state assumptions yield the same final velocity equation. In multireactant systems, the two approaches yield different velocity equations. As shown in Chapter Nine, the steady-state velocity equation for an ordered bireactant system contains both a K_{m_A} and a K_A (i.e., K_{ia}) term. When $K_{m_A} \ll K_{ia}$, the system will appear to be rapid equilibrium and can be analyzed as such.

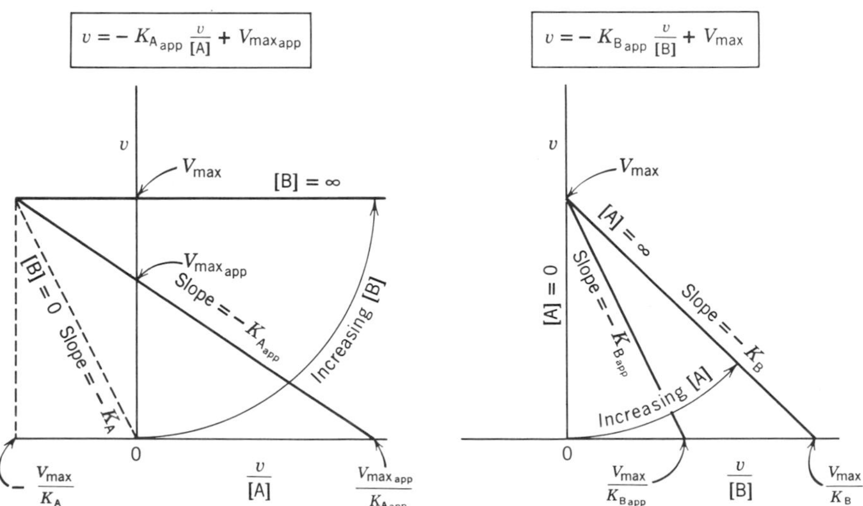

Fig. VI-23. Eadie-Scatchard plot for a Rapid Equilibrium Ordered Bireactant system.

Table VI-1 Characteristics and Differentiating Features of Rapid Equilibrium Bireactant Systems

System	Control Reciprocal Plots		Reciprocal Plots With Inhibitor Competitive With A		Reciptocal Plots With Inhibitor Competitive With B		Reciprocal Plots With Alternative Substrate for A		Reciprocal Plots With Alternative Substrate for B	
	$1/v$ vs. 1/[A] at Various Fixed [B]	$1/v$ vs. 1/[B] at Various Fixed [A]	$1/v$ vs. 1/[A] at Fixed [B]	$1/v$ vs. 1/[B] at Fixed [A]	$1/v$ vs. 1/[A] at Fixed [B]	$1/v$ vs. 1/[B] at Fixed [A]	$1/v$ vs. 1/[A] at Fixed [B]	$1/v$ vs. 1/[B] at Fixed [A]	$1/v$ vs. 1/[A] at Fixed [B]	$1/v$ vs. 1/[B] at Fixed [A]
Random[a]	NC($\alpha=1$) MT($\alpha\neq1$)	NC($\alpha=1$) MT($\alpha\neq1$)	C	NC($\beta=1$) MT($\beta\neq1$) MT	NC($\beta=1$) MT($\beta\neq1$) MT	C	Same as competitive inhibitor			
Ordered (A before B)[b]	$\frac{\text{MT}}{(slope\to0)}$	C	C	C	U	C	C	MT	U	C

C = competitive; that is, the family of reciprocal plots intersect on the $1/v$-axis and above the horizontal-axis.
NC = noncompetitive; that is, the family of reciprocal plots intersect on the horizontal-axis and to the left of the $1/v$-axis.
U = uncompetitive; that is, the family of reciprocal plots are parallel with positive slopes.
MT = mixed-type; that is, the family of reciprocal plots intersect above or below the horizontal-axis and to the left of the $1/v$-axis.

When an alternative substrate is used as an inhibitor, v is measured as the rate of appearance of the product of the normal substrate. The $K_{i_{slope}}$ of the competitive pattern is the same as the apparent K_m for I as a substrate (in place of the normal substrate) at the fixed concentration of cosubstrate. Note that an alternative substrate for A in a Rapid Equilibrium Ordered system does not yield the same inhibition patterns as a dead-end inhibitor competitive with A.

[a]It is assumed that competitive inhibitors do not prevent cosubstrate binding in random systems.

[b]It is assumed that an inhibitor competitive with A in the ordered system does not promote the binding of B, whereas an alternative substrate for A will promote B binding.

C. RANDOM TERREACTANT SYSTEMS

The equilibria shown in Figure VI-24 describe an enzyme that requires the simultaneous presence of all three substrates for catalytic activity. The equilibria could just as well describe a random terreactant system with two cosubstrates and one activator, or one substrate and two coactivators that add in a rapid equilibrium fashion. If A, B, and C are three substrates that yield three products (e.g., P, Q, and R), the system can be designated Rapid Equilibrium Random Ter Ter. If two products are formed, the system can be called Rapid Equilibrium Ter Bi. The velocity equation is:

$$\frac{v}{V_{\max}} = \frac{\dfrac{[A][B][C]}{\alpha\beta\gamma K_A K_B K_C}}{1 + \dfrac{[A]}{K_A} + \dfrac{[B]}{K_B} + \dfrac{[C]}{K_C} + \dfrac{[A][B]}{\gamma K_A K_B} + \dfrac{[A][C]}{\beta K_A K_C} + \dfrac{[B][C]}{\alpha K_B K_C} + \dfrac{[A][B][C]}{\alpha\beta\gamma K_A K_B K_C}} \qquad \text{(VI-71)}$$

The numerator reflects the existence of a single product-forming complex, while the eight terms in the denominator reflect free E (the 1) and seven

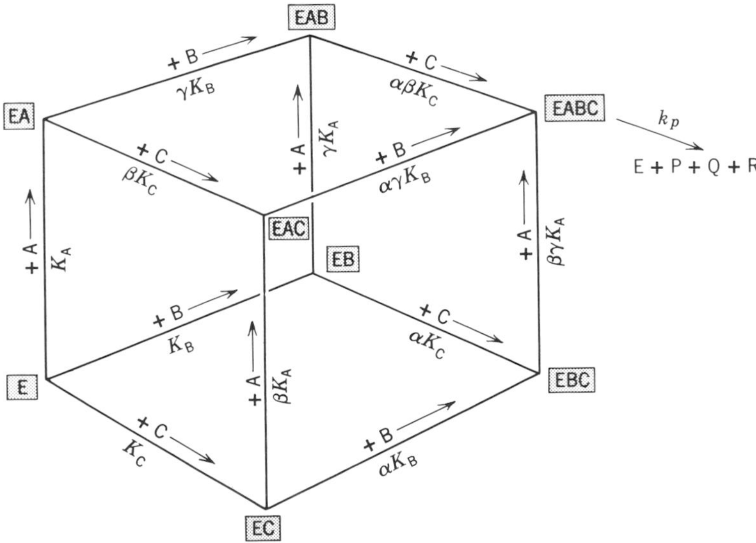

Fig. VI-24. Equilibria among enzyme species in a Rapid Equilibrium Random Terreactant system.

different complexes. The interaction factors, α, β, and γ, are factors by which the dissociation constant for a given ligand changes when other ligands are bound. The velocity equation can be rearranged to show any of the three ligands as the varied ligand at fixed concentrations of the other two ligands. For example, when C is the varied ligand, we obtain:

$$\frac{v}{V_{\max}} = \frac{[\mathrm{C}]}{\alpha\beta K_{\mathrm{C}}\left(1 + \frac{\gamma K_{\mathrm{A}}}{[\mathrm{A}]} + \frac{\gamma K_{\mathrm{B}}}{[\mathrm{B}]} + \frac{\gamma K_{\mathrm{A}} K_{\mathrm{B}}}{[\mathrm{A}][\mathrm{B}]}\right) + [\mathrm{C}]\left(1 + \frac{\beta\gamma K_{\mathrm{A}}}{[\mathrm{A}]} + \frac{\alpha\gamma K_{\mathrm{B}}}{[\mathrm{B}]} + \frac{\alpha\beta\gamma K_{\mathrm{A}} K_{\mathrm{B}}}{[\mathrm{A}][\mathrm{B}]}\right)} \tag{VI-72}$$

The family of reciprocal plots at any constant concentration of A and different fixed concentrations of B (or any constant [B] and different fixed [A]) will intersect to the left of the $1/v$-axis on, above, or below the $1/[\mathrm{C}]$-axis, depending on the values of the interaction factors, α, β, and γ. If $\alpha = \beta = \gamma = 1$, the reciprocal plots will intersect on the horizontal axis. Equation VI-72 can be rearranged as shown below to express C as the varied ligand, B as the changing fixed ligand, and A as the constant ligand.

$$\frac{v}{V_{\max}} = \frac{[\mathrm{C}]}{\alpha\beta K_{\mathrm{C}}\left(1 + \frac{\gamma K_{\mathrm{A}}}{[\mathrm{A}]}\right)\left[1 + \frac{\gamma K_{\mathrm{B}}\left(1 + \frac{K_{\mathrm{A}}}{[\mathrm{A}]}\right)}{[\mathrm{B}]\left(1 + \frac{\gamma K_{\mathrm{A}}}{[\mathrm{A}]}\right)}\right] + [\mathrm{C}]\left(1 + \frac{\beta\gamma K_{\mathrm{A}}}{[\mathrm{A}]}\right)\left[1 + \frac{\alpha\gamma K_{\mathrm{B}}\left(1 + \frac{\beta K_{\mathrm{A}}}{[\mathrm{A}]}\right)}{[\mathrm{B}]\left(1 + \frac{\beta\gamma K_{\mathrm{A}}}{[\mathrm{A}]}\right)}\right]}$$

When [B] is saturating, the terms in the large brackets reduce to unity. But the limiting K_{C} and $V_{\max}$ values are still apparent values for the constant [A]:

$$K_{\mathrm{C}}^{*} = \alpha\beta K_{\mathrm{C}} \frac{\left(1 + \frac{\gamma K_{\mathrm{A}}}{[\mathrm{A}]}\right)}{\left(1 + \frac{\beta\gamma K_{\mathrm{A}}}{[\mathrm{A}]}\right)} \quad \text{and} \quad V_{\max}^{*} = \frac{V_{\max}}{\left(1 + \frac{\beta\gamma K_{\mathrm{A}}}{[\mathrm{A}]}\right)}$$

The apparent $K_{\text{slope}}^{B/C}$ and $K_{\text{int}}^{B/C}$ are given by:

$$K_{\text{slope}}^{B/C} = \frac{\gamma K_B\left(1 + \frac{K_A}{[A]}\right)}{\left(1 + \frac{\gamma K_A}{[A]}\right)} \quad \text{and} \quad K_{\text{int}}^{B/C} = \frac{\alpha\gamma K_B\left(1 + \frac{\beta K_A}{[A]}\right)}{\left(1 + \frac{\beta\gamma K_A}{[A]}\right)}$$

where $K_{\text{slope}}^{B/C}$ and $K_{\text{int}}^{B/C}$ are constants relating the effect of changing fixed [B] on the slope and $1/v$-axis intercept of the $1/v$ versus $1/[C]$ plot. The intersection coordinates can be calculated in the usual way, or from the relationship:

$$\frac{1}{[C]} = -\frac{1}{K_C^*}\left(\frac{K_{\text{int}}^{B/C}}{K_{\text{slope}}^{B/C}}\right) \quad \text{and} \quad \frac{1}{v} = \frac{1}{V_{\max}^*}\left(1 - \frac{K_{\text{int}}^{B/C}}{K_{\text{slope}}^{B/C}}\right) \qquad \text{(VI-73)}$$

The $1/[C]$ coordinate is given by:

$$\frac{1}{[C]} = -\frac{1}{\beta K_C}\frac{\left(1 + \frac{\beta K_A}{[A]}\right)}{\left(1 + \frac{K_A}{[A]}\right)}$$

When [A] is saturating, the $1/[C]$ coordinate gives $-1/\beta K_C$.

The random terreactant system can be analyzed in the manner described earilier for the random bireactant system, if, in any one series of experiments, one of the ligands is maintained at a saturating concentration. Thus the terreactant system can be completely analyzed by six families of reciprocal plots and the corresponding replots:

Reciprocal Plot Family	Varied ligand	Changing Fixed Ligand	Constant Saturating Ligand
1	A	C	B
2	C	A	B
3	B	C	A
4	C	B	A
5	A	B	C
6	B	A	C

For example, when B is the saturating ligand, all the enzyme will be driven to the complexes shown on the rear face of the equilibrium cube. The velocity equation becomes:

$$\frac{v}{V_{\max}} = \frac{\dfrac{[A][B][C]}{\alpha\beta\gamma K_A K_B K_C}}{\dfrac{[B]}{K_B} + \dfrac{[A][B]}{\gamma K_A K_B} + \dfrac{[B][C]}{\alpha K_B K_C} + \dfrac{[A][B][C]}{\alpha\beta\gamma K_A K_B K_C}} \qquad \text{(VI-74)}$$

The denominator now contains only the terms corresponding to the EB, EAB, EBC, and EABC complexes. Since B is present at a saturating concentration, the concentrations of E, EA, EC, and EAC are negligible. Dividing by $[B]/K_B$ the velocity is shown to be independent of [B]:

$$\frac{v}{V_{\max}} = \frac{\dfrac{[A][C]}{\alpha\beta\gamma K_A K_C}}{1 + \dfrac{[A]}{\gamma K_A} + \dfrac{[C]}{\alpha K_C} + \dfrac{[A][C]}{\alpha\beta\gamma K_A K_C}} \qquad \text{(VI-75)}$$

The system behaves as a two-ligand system where the dissociation constant of C is αK_C, the dissociation constant of A is γK_A, and the interaction factor is β. The constants αK_C, γK_A, and β can be determined from reciprocal plot families 1 and 2 and the corresponding replots as described earlier for bireactant systems. Similarly, in the presence of a saturating concentration of A, all the enzyme will be driven to the complexes shown on the top face of the equilibrium cube. The velocity is given by:

$$\frac{v}{V_{\max}} = \frac{\dfrac{[B][C]}{\alpha\beta\gamma K_B K_C}}{1 + \dfrac{[B]}{\gamma K_B} + \dfrac{[C]}{\beta K_C} + \dfrac{[B][C]}{\alpha\beta\gamma K_B K_C}} \qquad \text{(VI-76)}$$

Reciprocal plot families 3 and 4 and the corresponding replots will allow βK_C, γK_B, and the interaction factor, α, to be determined. With α and β known, K_C can be calculated. When C is the saturating ligand all the enzyme will be driven to the complexes shown on the right-hand face of the equilibrium cube. The denominator of the velocity equation will contain only terms corresponding to the EC, EAC, EBC, and EABC complexes. The concentrations of free E, EA, EB, and EAB are negligible. Dividing by

$[\mathrm{C}]/K_{\mathrm{C}}$, the velocity equation reduces to:

$$\frac{v}{V_{\max}} = \frac{\dfrac{[\mathrm{A}][\mathrm{B}]}{\alpha\beta\gamma K_{\mathrm{A}} K_{\mathrm{B}}}}{1 + \dfrac{[\mathrm{A}]}{\beta K_{\mathrm{A}}} + \dfrac{[\mathrm{B}]}{\alpha K_{\mathrm{B}}} + \dfrac{[\mathrm{A}][\mathrm{B}]}{\alpha\beta\gamma K_{\mathrm{A}} K_{\mathrm{B}}}} \tag{VI-77}$$

The system behaves as a two-ligand system where the dissociation constant for A is βK_{A}, the dissociation constant for B is αK_{B}, and the interaction factor is γ. Thus all three dissociation constants and all three interaction factors can be determined.

If it is impractical to maintain one of the ligands at a saturating concentration, the six constants can still be determined, but now the procedure becomes far more tedious. In place of any one family of reciprocal plots, several families are required, each obtained at a different constant unsaturating concentration of the third ligand. The procedure is illustrated in Figure VI-25. First we obtain several families of reciprocal plots of, for example, $1/v$ versus $1/[\mathrm{C}]$. The plots within each family represent different fixed concentrations of, for example, B. Each family represents a different constant concentration of A. The slope of each plot is given by:

$$slope_{1/\mathrm{C}} = \frac{\alpha\beta\gamma K_{\mathrm{C}} K_{\mathrm{B}}}{V_{\max}} \left(1 + \frac{K_{\mathrm{A}}}{[\mathrm{A}]}\right) \frac{1}{[\mathrm{B}]} + \frac{\alpha\beta K_{\mathrm{C}}}{V_{\max}} \left(1 + \frac{\gamma K_{\mathrm{A}}}{[\mathrm{A}]}\right) \tag{VI-78}$$

The slopes of each family are replotted against the corresponding $1/[\mathrm{B}]$. The intercept on the $slope_{1/\mathrm{C}}$-axis is given by the last term in equation VI-78, which can be rewritten as:

$$slope_{1/\mathrm{C}}\text{-axis intercept} = \frac{\alpha\beta\gamma K_{\mathrm{C}} K_{\mathrm{A}}}{V_{\max}} \frac{1}{[\mathrm{A}]} + \frac{\alpha\beta K_{\mathrm{C}}}{V_{\max}} \tag{VI-79}$$

The $slope_{1/\mathrm{C}}$-axis intercepts are then replotted against the corresponding $1/[\mathrm{A}]$ for each family. This final replot gives $\alpha\beta K_{\mathrm{C}}/V_{\max}$ and $-1/\gamma K_{\mathrm{A}}$ as intercepts. Now going back to the original $1/v$ versus $1/[\mathrm{C}]$ plots, we can rearrange the $1/v$-axis intercept (i.e., $1/V_{\max_{app}}$) term to:

$$\frac{1}{V_{\max_{app}}} = \frac{\alpha\gamma K_{\mathrm{B}}}{V_{\max}} \left(1 + \frac{\beta K_{\mathrm{A}}}{[\mathrm{A}]}\right) \frac{1}{[\mathrm{B}]} + \frac{1}{V_{\max}} \left(1 + \frac{\beta\gamma K_{\mathrm{A}}}{[\mathrm{A}]}\right) \tag{VI-80}$$

The $1/V_{\max_{app}}$ of each family is replotted against the corresponding $1/[\mathrm{B}]$. The intercept on the $1/V_{\max_{app}}$-axis is given by the last term in equation

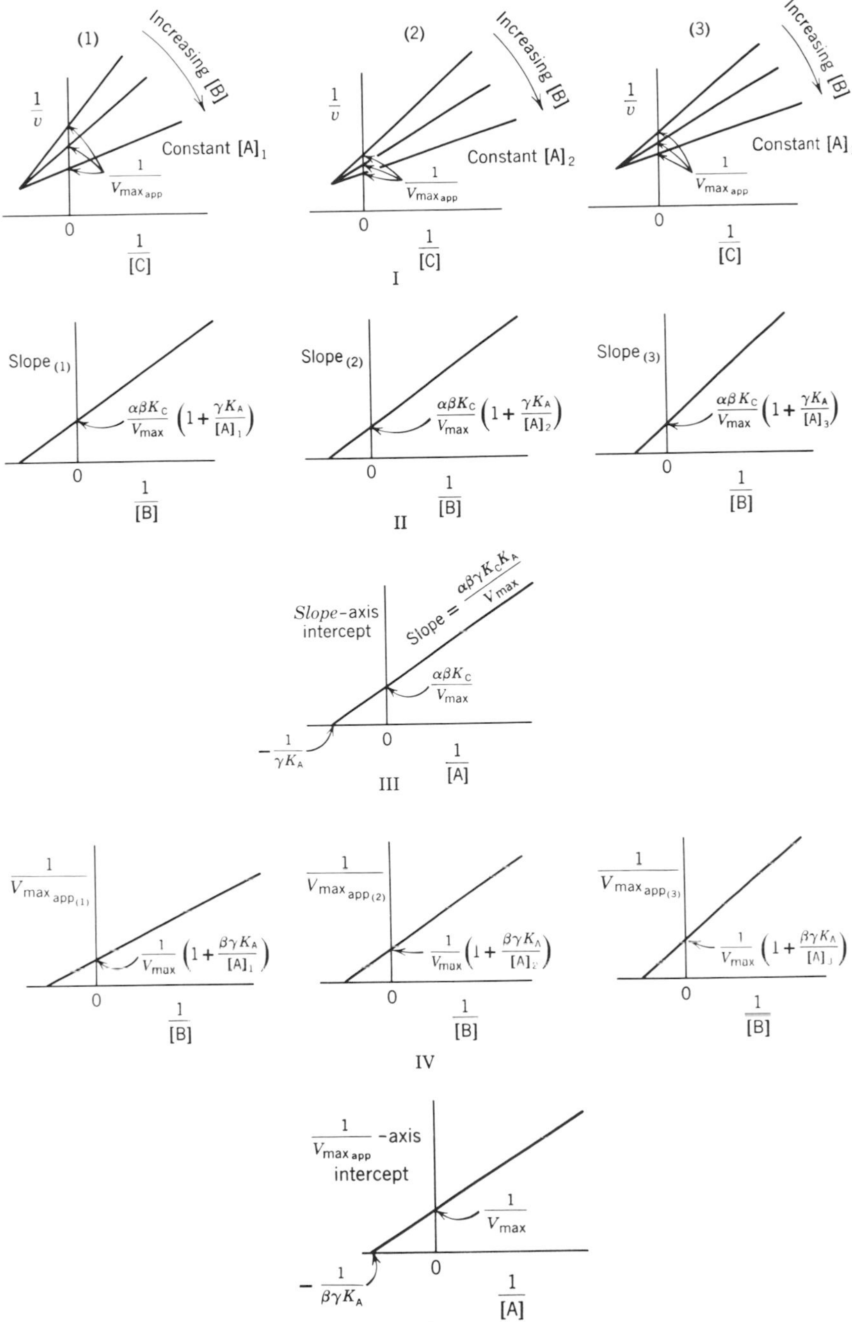

Fig. VI-25. Procedure for obtaining all the dissociation constants and interaction factors of a Rapid Equilibrium Random Terreactant system. The procedure is repeated three times, each time one of the ligands is treated as the "constant" ligand.

VI-80, which can be rewritten as:

$$1/V_{\max_{app}}\text{-axis intercept} = \frac{\beta\gamma K_A}{V_{\max}}\frac{1}{[A]} + \frac{1}{V_{\max}} \tag{VI-81}$$

The final replot of $1/V_{\max_{app}}$-axis intercept against the corresponding $1/[A]$ gives $1/V_{\max}$ and $-1/\beta\gamma K_A$ as intercepts. With γK_A and $\beta\gamma K_A$ known, β can be calculated. The process is repeated rearranging the plotted variables. For example, if A is the varied ligand, C is the changing fixed ligand, and B is the constant ligand, the equations are:

$$\boxed{\frac{v}{V_{\max}} = \frac{[A]}{\beta\gamma K_A\left(1 + \frac{\alpha K_B}{[B]} + \frac{\alpha K_C}{[C]} + \frac{\alpha K_B K_C}{[B][C]}\right) + [A]\left(1 + \frac{\alpha\gamma K_B}{[B]} + \frac{\alpha\beta K_C}{[C]} + \frac{\alpha\beta\gamma K_B K_C}{[B][C]}\right)}} \tag{VI-82}$$

$$slope_{1/A} = \frac{\alpha\beta\gamma K_A K_C}{V_{\max}}\left(1 + \frac{K_B}{[B]}\right)\frac{1}{[C]} + \frac{\beta\gamma K_A}{V_{\max}}\left(1 + \frac{\alpha K_B}{[B]}\right) \tag{VI-83}$$

$$slope_{1/A}\text{-axis intercept} = \frac{\alpha\beta\gamma K_A K_B}{V_{\max}}\frac{1}{[B]} + \frac{\beta\gamma K_A}{V_{\max}} \tag{VI-84}$$

$$\frac{1}{V_{\max_{app}}} = \frac{\alpha\beta K_C}{V_{\max}}\left(1 + \frac{\gamma K_B}{[B]}\right)\frac{1}{[C]} + \frac{1}{V_{\max}}\left(1 + \frac{\alpha\gamma K_B}{[B]}\right) \tag{VI-85}$$

$$1/V_{\max_{app}}\text{-axis intercept} = \frac{\alpha\gamma K_B}{V_{\max}}\frac{1}{[B]} + \frac{1}{V_{\max}} \tag{VI-86}$$

The replots give αK_B, $\beta\gamma K_A$, $\alpha\gamma K_B$, and $V_{\max}$, some of which were already obtained from the $1/[C]$ series of replots. We can now calculate γ from the values of αK_B and $\alpha\gamma K_B$. The K_A can be calculated from the values of $\beta\gamma K_A$, β, and γ.

The final set of plots treat B as the varied ligand, A as the changing fixed ligand, and C as the constant ligand. The equations are:

$$\frac{v}{V_{max}} = \frac{[B]}{\alpha\gamma K_B\left(1 + \frac{\beta K_C}{[C]} + \frac{\beta K_A}{[A]} + \frac{\beta K_A K_C}{[A][C]}\right) + [B]\left(1 + \frac{\alpha\beta K_C}{[C]} + \frac{\beta\gamma K_A}{[A]} + \frac{\alpha\beta\gamma K_A K_C}{[A][C]}\right)} \tag{VI-87}$$

$$slope_{1/B} = \frac{\alpha\beta\gamma K_B K_A}{V_{max}}\left(1 + \frac{K_C}{[C]}\right)\frac{1}{[A]} + \frac{\alpha\gamma K_B}{V_{max}}\left(1 + \frac{\beta K_C}{[C]}\right) \tag{VI-88}$$

$$slope_{1/B}\text{-axis intercept} = \frac{\alpha\beta\gamma K_B K_C}{V_{max}}\frac{1}{[C]} + \frac{\alpha\gamma K_B}{V_{max}} \tag{VI-89}$$

$$\frac{1}{V_{max_{app}}} = \frac{\beta\gamma K_A}{V_{max}}\left(1 + \frac{\gamma K_C}{[C]}\right)\frac{1}{[A]} + \frac{1}{V_{max}}\left(1 + \frac{\alpha\beta K_C}{[C]}\right) \tag{VI-90}$$

$$1/V_{max_{app}}\text{-axis intercept} = \frac{\alpha\beta K_C}{V_{max}}\frac{1}{[C]} + \frac{1}{V_{max}} \tag{VI-91}$$

Only the $slope_{1/B}$-axis intercept replot gives unique information (βK_C).With βK_C known, the values of the remaining constants, α, K_C, and K_B, can be calculated. As noted earlier, it may not be necessary to run three separate experiments to obtain the three sets of reciprocal plot families. The information in any one family frequently can be replotted to obtain another family. For example, the $1/v$ versus $1/[A]$ plots at different fixed B concentrations and a constant [C] can be replotted to obtain the $1/v$ versus $1/[B]$ family at different fixed A concentrations and constant [C]. The usefulness of the replot depends on the range and spacing of the B concentrations used to generate the original plot.

D. ORDERED AND HYBRID RANDOM-ORDERED TERREACTANT SYSTEMS

Tables VI-2 to VI-4 summarize the properties of several rapid equilibrium terreactant systems. (See also Table IX-7 for the properties of some steady-state systems with rapid equilibrium segments.)

Table VI-2 Kinetic Mechanism and Velocity Equation for Several Rapid Equilibrium Ordered and Hybrid Random-Ordered Terreactant Systems

System	Kinetic Mechanism	Velocity Equation
Ordered	A (K_A↓), B (K_B↓), C (K_C↓); E EA EAB $\left(\mathrm{EABC} \overset{k_p}{\rightleftharpoons} \mathrm{EPQR}\right)$ EQR ER E; P (K_P↑), Q (K_Q↑), R (K_R↑)	$\dfrac{v}{V_{max}} = \dfrac{\dfrac{[A][B][C]}{K_A K_B K_C}}{1 + \dfrac{[A]}{K_A} + \dfrac{[A][B]}{K_A K_B} + \dfrac{[A][B][C]}{K_A K_B K_C}}$
Random A–B, ordered C	E —K_A— EA —αK_B— EAB; E —K_B— EB —αK_A— EAB; EAB + C (K_C) — EABC $\xrightarrow{k_p}$	$\dfrac{v}{V_{max}} = \dfrac{\dfrac{[A][B][C]}{\alpha K_A K_B K_C}}{1 + \dfrac{[A]}{K_A} + \dfrac{[B]}{K_B} + \dfrac{[A][B]}{\alpha K_A K_B} + \dfrac{[A][B][C]}{\alpha K_A K_B K_C}}$
Ordered A, random B–C	E + A (K_A) — EA; EA —K_B— EAB —αK_C— EABC; EA —K_C— EAC —αK_B— EABC; EABC $\xrightarrow{k_p}$	$\dfrac{v}{V_{max}} = \dfrac{\dfrac{[A][B][C]}{\alpha K_A K_B K_C}}{1 + \dfrac{[A]}{K_A} + \dfrac{[A][B]}{K_A K_B} + \dfrac{[A][C]}{K_A K_C} + \dfrac{[A][B][C]}{\alpha K_A K_B K_C}}$
Random A–C, ordered B	E —K_A— EA —K_B— EAB —βK_C— EABC; E —K_C— EC —αK_B— EBC —$\frac{\beta}{\alpha} K_A$— EABC; EABC $\xrightarrow{k_p}$	$\dfrac{v}{V_{max}} = \dfrac{\dfrac{[A][B][C]}{\beta K_A K_B K_C}}{1 + \dfrac{[A]}{K_A} + \dfrac{[C]}{K_C} + \dfrac{[A][B]}{K_A K_B} + \dfrac{[B][C]}{\alpha K_B K_C} + \dfrac{[A][B][C]}{\beta K_A K_B K_C}}$

Table VI-3. Slope and Intercept Replots of Rapid Equilibrium Terreactant Systems

System	Varied Ligand[a]	Slope Replot[b]	$\frac{1}{V_{max_{app}}}$ Replot[c]
Completely Random	A	P[d](int = $\alpha\beta K_A/V_{max}$)	P
	B	P(int = $\beta\gamma K_B/V_{max}$)	P
	C	P(int = $\alpha\gamma K_C/V_{max}$)	P
Completely Ordered	A	P(int = 0)[e]	P
	B	P(int = 0)	L[f]
	C	P(int = K_C/V_{max})	—[g]
Random A-B, ordered C	A	P(int = 0)	P
	B	P(int = 0)	P
	C	P(int = K_C/V_{max})	—[g]
Random B-C, ordered A	A	P(int = 0)	P
	B	P(int = $\alpha K_B/V_{max}$)	L
	C	P(int = $\alpha K_C/V_{max}$)	L
Random A-C, ordered B	A	P(int = $\beta K_A/\alpha V_{max}$)	P
	B	P(int = 0)[h]	L
	C	P(int = $\beta K_C/V_{max}$)	P

[a] The initial plots are $1/v$ versus 1/[varied ligand] at different fixed concentrations of the other two nonvaried ligands which are maintained at a constant ratio. All plots show a mixed-type pattern except the $1/v$ versus 1/[C] plot in the Completely Ordered and the Random A-B, ordered C systems (see footnote g).

[b] The slopes of the reciprocal plots are replotted against the reciprocal concentration of one of the changing fixed ligands.

[c] The $1/v$-axis intercepts of the reciprocal plots are replotted against the reciprocal concentration of one of the changing fixed ligands. The intercept on the vertical axis gives $1/V_{max}$.

[d] P = parabolic replot with a finite intercept on the vertical axis (given in parentheses).

[e] P (int = 0) = parabolic replot that goes through the origin.

[f] L = linear replot with a finite intercept on the vertical axis.

[g] The $1/v$-axis intercepts are the same for all concentrations of nonvaried ligands.

[h] The intercept of the replot is $\beta K_B/\gamma V_{max}$ if a dead-end EAC complex forms.

Table VI-4. Characteristics and Differentiating Features of Rapid Equilibrium Terreactant Systems

Mechanism	Control Plots			Effect of Strict Inhibitor Competitive					
				With A			With B		
	1/[A]	1/[B]	1/[C]	1/[A]	1/[B]	1/[C]	1/[A]	1/[B]	1/[C]
Completely Random	MT[a]	MT[a]	MT[a]	C	MT[a]	MT[a]	MT[a]	C	MT[a]
Completely Ordered	MT	MT C at Constant [C]	C	C	C	C	U	C	C
Random A-B, Ordered C	MT	MT MT at Constant [C]	C	C	MT[b]	C	MT[b]	C	C
Random B-C, Ordered A	MT	MT C at Constant [C]	MT C at Constant [B]	C	C	C	U	C	MT[b]
Random A-C, Ordered B	MT	MT (Always)	MT	C	MT[b]	MT[b]	MT	C	MT

[a] The reciprocal plots will show a noncompetitive pattern (i.e., intersection on the 1/[varied ligand]-axis) if the interaction factors $\alpha=\beta=\gamma=1$.
[b] I will act as a competitive inhibitor if steric hindrance prevents the varied ligand from binding.

The control plots are $1/v$ versus 1/[varied ligand] at different fixed concentrations of the other two ligands. The ratio of the two coligands is maintained constant as their concentrations are changed. Many of these control plots yield a mixed-type (MT) pattern. That is, any two reciprocal plots within a family will intersect to the left of the $1/v$-axis. However, all the plots within the family will not necessarily intersect at a common point (as in Fig. IX-56). A second series of control plots in which one of the two fixed coligands is maintained constant at various fixed concentrations of the other coligand will permit further distinctions between the above systems. This second series

With C			Effect of Alternative Substrate For A			For B			For C		
1/[A]	1/[B]	1/[C]	1/[A]	1/[B]	1/[C]	1/[A]	1/[B]	1/[C]	1/[A]	1/[B]	1/[C]
MT[a]	MT[a]	C	Same as Strict Competitive Inhibitor								
U	U	C	C	MT	MT	U	C	MT			
U	U	C	C	MT	MT	MT	C	MT			
U	MT[b]	C	C	MT	MT						
MT[b]	MT[b]	C									

of control plots will have a common intersection point. The effects of strict competitive inhibitors are measured by plots of $1/v$ versus 1/[varied ligand] at different fixed concentrations of inhibitor and constant unsaturating concentrations of the other two coligands. The effects of alternative substrates for a given varied ligand are determined in the same way. It is assumed that competitive inhibitors in random sequences do not prevent cosubstrate binding, but in ordered sequences, a competitive inhibitor will not promote the binding of the next cosubstrate. Alternative substrates, on the other hand, do promote the ordered binding of cosubstrates and the $K_{i_{slope}}$ of the competitive pattern is equal to the apparent K_m of the alternative substrate at the fixed concentrations of the cosubstrates. The shaded area indicates that the alternative substrate has the same effect as a strict competitive inhibitor.

E. RULES FOR PREDICTING INHIBITION PATTERNS IN RAPID EQUILIBRIUM SYSTEMS

The velocity equations for rapid equilibrium systems are quite easily derived (or written directly from inspection of the equilibria, as described in Chapter Two). Consequently, we can easily determine the nature of the inhibition by rearranging the velocity equation to the usual Henri-Michaelis-Menten form and seeing whether the K_S or the [S] term in the denominator (or both) is multiplied by a factor containing I. The factor multiplying K_S becomes the slope factor in the reciprocal form, while the factor multiplying [S] becomes the $1/v$-axis intercept factor. If only K_S is multiplied by a factor containing I, the inhibition will be competitive. If only [S] is multiplied by an I-containing factor, the inhibition will be uncompetitive. If both K_S and [S] are multiplied by factors containing I, the inhibition will be noncompetitive (if the factors are identical), or mixed-type (if the factors are different). The nature of the inhibition can also be predicted directly from inspection of the equilibria. The following rules apply to rapid equilibrium systems.

Competitive Inhibition (I Affects Slope Only)

A compound (I) will act as a competitive inhibitor with respect to a varied substrate (S) only *if I and S are mutually exclusive.* Either I and S compete for the same enzyme form, or I adds before S in an ordered reaction sequence to yield an enzyme form that does not bind S. Increasing [S] decreases the concentration of the enzyme form that combines with I. Thus the inhibition can be completely overcome by saturating [S], provided [I] is not also saturating.

Uncompetitive Inhibition (I Affects $1/v$-Axis Intercept Only)

I will act as an uncompetitive inhibitior with respect to S only *if S must add before I* in an ordered reaction sequence. I and S combine with different enzyme forms and are not mutually exclusive. Increasing [S] increases the concentration of the enzyme form that combines with I. Thus the inhibition cannot be overcome by saturating [S].

Noncompetitive and Mixed-Type Inhibition (I Affects Slope and $1/v$-Axis Intercept)

I will act as a noncompetitive or mixed-type inhibitor with respect to S *if I and S are not mutually exclusive and I can add before S*. I and S can combine with the same and/or different enzyme forms. I will be a noncompetitive or mixed-type inhibitor if it promotes the binding of S by mimicing a substrate that adds before S in an ordered reaction sequence.

Effect of Saturation by a Nonvaried Substrate

The inhibition with respect to a varied substrate can be overcome by saturation with a nonvaried substrate which is mutually exclusive (i.e., competitive) with the inhibitor, provided the inhibitor is not also saturating.

Velocity Equation

A dead-end inhibitor introduces a $(1+[\mathrm{I}]/K_i)$ factor to that term in the denominator of the velocity equation which represents the enzyme form combining with the inhibitor. For example, the velocity equation for the Rapid Equilibrium Ordered Terreactant system is shown below with the denominator terms identified.

$$\frac{v}{V_{\max}} = \frac{\dfrac{[\mathrm{A}][\mathrm{B}][\mathrm{C}]}{K_\mathrm{A}K_\mathrm{B}K_\mathrm{C}}}{\underset{\mathrm{E}}{\underset{\uparrow}{1}} + \underset{\mathrm{EA}}{\underset{\uparrow}{\dfrac{[\mathrm{A}]}{K_\mathrm{A}}}} + \underset{\mathrm{EAB}}{\underset{\uparrow}{\dfrac{[\mathrm{A}][\mathrm{B}]}{K_\mathrm{A}K_\mathrm{B}}}} + \underset{\mathrm{EABC}}{\underset{\uparrow}{\dfrac{[\mathrm{A}][\mathrm{B}][\mathrm{C}]}{K_\mathrm{A}K_\mathrm{B}K_\mathrm{C}}}}} \tag{VI-93}$$

In the presence of a dead-end inhibitor that combines with EAB, the $[\mathrm{A}][\mathrm{B}]/K_\mathrm{A}K_\mathrm{B}$ term in the denominator is multiplied by $(1+[\mathrm{I}]/K_i)$. After multiplying out the equation becomes:

$$\frac{v}{V_{\max}} = \frac{\dfrac{[\mathrm{A}][\mathrm{B}][\mathrm{C}]}{K_\mathrm{A}K_\mathrm{B}K_\mathrm{C}}}{1 + \dfrac{[\mathrm{A}]}{K_\mathrm{A}} + \dfrac{[\mathrm{A}][\mathrm{B}]}{K_\mathrm{A}K_\mathrm{B}} + \dfrac{[\mathrm{A}][\mathrm{B}][\mathrm{I}]}{K_\mathrm{A}K_\mathrm{B}K_i} + \dfrac{[\mathrm{A}][\mathrm{B}][\mathrm{C}]}{K_\mathrm{A}K_\mathrm{B}K_\mathrm{C}}} \tag{VI-94}$$

If I combines with more than one enzyme form, then the terms for each enzyme form adding I is multiplied by the corresponding factor, that is $(1+[\mathrm{I}]/K_{i_1})$, $(1+[\mathrm{I}]/K_{i_2})$, $(1+[\mathrm{I}]/K_{i_3})$, and so on.

The rules are summarized in Figure VI-26 for a number of different inhibitors acting on a Rapid Equilibrium Ordered Terreactant system. For example, consider inhibitor Y. Y adds only after A. Thus Y is an uncompetitive inhibitor with respect to A. Y and B are not mutually exclusive and Y can add before B. Thus Y is a mixed-type inhibitor with respect to B. The

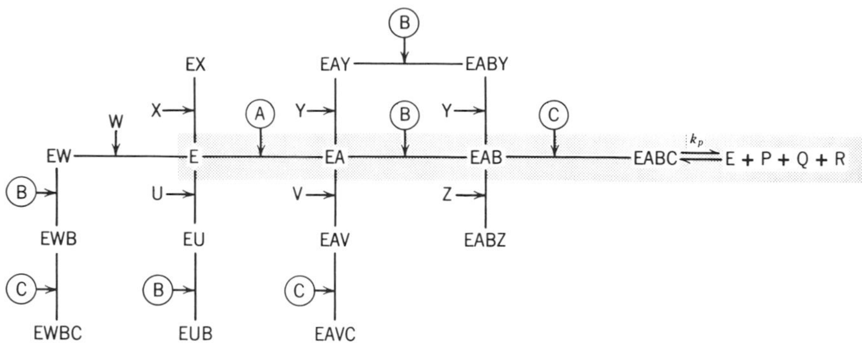

Fig. VI-26. Inhibition patterns in a Rapid Equilibrium Ordered Terreactant system. A, B, and C are the three substrates. U–Z represent inhibitors that either bind in a dead-end fashion or promote the binding of other substrates.

Inhibition patterns

Inhibitor	Inhibition relative to varied substrate A	B	C
U	C	MT	C
V	UC	C	MT
W	C	MT	MT
X	C	C	C
Y	UC	MT	C
Z	UC	UC	C

EABY complex cannot bind C (i.e., Y excludes C). Consequently, Y is a competitive inhibitor with respect to C. Saturation with C will drive all the enzyme to the productive EABC and thus overcome the inhibition with respect to A (at constant [B]) or B (at constant [A]). Similarly, we can predict that V will act uncompetitively with respect to A, competitively with respect to B, and as a mixed-type inhibitor with respect to C. Saturation with B will exclude V and overcome the inhibition with respect to A and C.

The rules above apply to all rapid equilibrium systems with one product-forming species. Cleland has stated a more extensive series of rules for steady-state systems. These are described in Chapter Nine.

REFERENCES

General

Cleland, W. W., "Steady State Kinetics," in *The Enzymes*, Vol. 2, 3rd ed., P. Boyer, ed., Academic Press, 1970 (Substrate inhibition), p. 1.

Dalziel, K., *Biochem. J.* **114**, 547 (1969) (Criticism of partially random systems).

Fromm, H. J., *Biochim. Biophys. Acta* **139**, 221 (1967).

Rudolph, F. B. and Fromm, H. J., *J. Biol. Chem.* **244**, 3832 (1969).

Rudolph, F. B. and Fromm, H. J., *Arch. Biochem. Biophys.* **147**, 515 (1971).

Rudolph, F. B. and Fromm, H. J., *J. Theor. Biol.* **39**, 363 (1973) (Support of partially random systems).

Webb, J. L., *Enzyme and Metabolic Inhibitors*, Vol. 1, Academic Press, 1963, p. 45.

Specific Examples

Ainsworth, S. and Macfarlane, N., *Biochem. J.* **131**, 223 (1973) (Pyruvate kinase: Rapid Equilibrium Random Bi Bi; P excludes both A and B).

Chao, J., Johnson, G. F. and Graves, D. T., *Biochemistry* **8**, 1459 (1969) (Maltodextrin phosphorylase: Rapid Equilibrium Random Bi Bi).

Gulbinsky, J. S. and Cleland, W. W., *Biochemistry* **7**, 566 (1968) (Galactokinase: Rapid Equilibrium Random Bi Bi).

Gold, M. H., Livoni, J., Farrand, R. and Segel, I. H., *Arch. Biochem. Biophys.* **161**, 515 (1974) (Glycogen phosphorylase: Rapid Equilibrium Random Bi Bi).

Gold, M. H., and Segel, I. H., *J. Biol. Chem.* **249**, 2417 (1974) (Protein kinase: Random Bi Bi).

Hanson, R. L., Rudolph, F. B. and Lardy, H. A., *J. Biol. Chem.* **248**, 7852 (1973) (Muscle phosphofructokinase: Rapid Equilibrium Random Bi Bi).

Katiyar, S. S. and Porter, J. W., *Arch. Biochem. Biophys.* **163**, 324 (1974) (Competitive substrate inhibition with fatty acid synthetase).

Lueck, J. D. and Fromm, H. J., *J. Biol. Chem.* **249**, 1341 (1974) (Skeletal muscle hexokinase: Random Bi Bi).

Matsuoka, Y. and Srere, P A., *J. Biol. Chem.* **248**, 8022 (1973) (Citrate synthetase of rat kidney and brain: Random Bi Bi with two dead-end complexes).

Moffet, F. J. and Bridger, W. A., *Can. J. Biochem.* **51**, 44 (1973) (Succinyl-CoA synthetase: Random B-C, Ordered A).

Morrison, J. F. and James, E., *Biochem. J.* **97**, 37 (1965) (Creatine kinase: Rapid Equilibrium Random Bi Bi).

Plesner, L., Plesner, I. W. and Esmann, V., *J. Biol. Chem.* **249**, 1119 (1974) (Glycogen synthase D from leukocytes: Rapid Equilibrium Ordered A, Random B-C, where A = glucose-6-phosphate).

Rudolph, F. B. and Fromm, H. J., *J. Biol. Chem.* **244**, 3832 (1969)) (Adenylosuccinate synthetase: Rapid Equilibrium Random Ter Ter).

CHAPTER SEVEN

MULTISITE AND ALLOSTERIC ENZYMES

A. ENZYMES WITH MULTIPLE CATALYTIC SITES

Many enzymes are oligomers composed of distinct subunits or monomers. Often, the subunits are identical, each bearing a catalytic site. If the sites are identical and completely independent of each other, then the presence of substrate at one site will have no effect on the binding properties of the vacant sites or on the catalytic activities of other occupied sites. If the enzyme is a tetramer, then at any fixed substrate concentration $[E]_t$ will be distributed among five different species: E, ES_1, ES_2, ES_3, and ES_4. Yet, as we shall soon see, the S binding or velocity curve will be the usual hyperbola. In other words, n molecules of a one-site enzyme behave identically to one molecule of an n-site enzyme. Although there may be no obvious interactions between the sites, the isolated monomers are often completely inactive. Association to a tetramer may cause small changes in the tertiary structure of each monomer, resulting in the formation of the substrate binding site or the proper juxtaposition of the substrate binding site and the catalytic groups. Oligomerization may also contribute to the stability of enzymes *in vivo*.

If the presence of substrate on one site does influence the binding of substrate to vacant sites, or the rate of product formation at other occupied sites, then we have a situation where the substrate itself acts as a modifier or effector yielding substrate activation (including sigmoidal v versus [S] responses) or substrate inhibition.

Noncooperative Sites

Let us first examine a dimer (two-site) model in which both sites are identical and independent. The substrate binding sequence can be shown in two ways. First, showing only the different kinds of complexes that form:

$$\begin{array}{ccccc} E + S & \overset{K_{S_1}}{\rightleftharpoons} & ES_1 & \overset{k_{p_1}}{\longrightarrow} & E + P \\ & & + & & \\ & & S & & \\ & & K_{S_2} \updownarrow & & \\ & & ES_2 & \overset{k_{p_2}}{\longrightarrow} & ES_1 + P \end{array}$$

The K_{S_1} and K_{S_2} are called *effective* dissociation constants. A better representation considers each site separately, as shown below:

$$\begin{array}{ccccccc} & & E + S & \overset{K_S}{\rightleftharpoons} & ES & \overset{k_p}{\longrightarrow} & E + P \\ & & + & & + & & \\ & & S & & S & & \\ & & K_S \updownarrow & & \updownarrow K_S & & \\ E + P & \overset{k_p}{\longleftarrow} & SE + S & \overset{K_S}{\rightleftharpoons} & SES & \overset{k_p}{\longrightarrow} & SE + P \\ & & & & \downarrow k_p & & \\ & & & & ES + P & & \end{array}$$

The K_S is called an *intrinsic* dissociation constant. The velocity equation is obtained in the usual way:

$$v = k_p[\text{ES}] + k_p[\text{SE}] + 2k_p[\text{SES}]$$

$$\frac{v}{[\text{E}]_t} = \frac{k_p \dfrac{[\text{S}]}{K_S} + k_p \dfrac{[\text{S}]}{K_S} + 2k_p \dfrac{[\text{S}]^2}{K_S^2}}{1 + \dfrac{[\text{S}]}{K_S} + \dfrac{[\text{S}]}{K_S} + \dfrac{[\text{S}]^2}{K_S^2}}$$

or

$$\boxed{\frac{v}{V_{\max}} = \frac{\dfrac{[\text{S}]}{K_S} + \dfrac{[\text{S}]^2}{K_S^2}}{1 + \dfrac{2[\text{S}]}{K_S} + \dfrac{[\text{S}]^2}{K_S^2}}} \qquad \text{(VII-1)}$$

where $V_{max}=2k_p[E]_t$. When the two sites are considered separately, the existence of two different singly occupied species (i.e., two ways of arranging S as ES and SE) is automatically taken care of (the K_S and k_p values shown are for a given *site*). Equation VII-1 reduces to the Henri-Michaelis-Menten equation, as shown below. Factoring $[S]/K_S$:

$$\frac{v}{V_{max}}=\frac{\dfrac{[S]}{K_S}\left(1+\dfrac{[S]}{K_S}\right)}{\left(1+\dfrac{[S]}{K_S}\right)\left(1+\dfrac{[S]}{K_S}\right)}=\frac{\dfrac{[S]}{K_S}}{1+\dfrac{[S]}{K_S}}=\frac{[S]}{K_S+[S]}$$

When an enzyme contains more than two sites, it becomes too complicated to write the equilibria in terms of intrinsic constants showing each site separately, but it is relatively easy to show the equilibria between the various complexes and then convert the effective K_S constants to intrinsic K_S constants. An intrinsic constant (also called microscopic constant) is a constant for the site in isolation; that is, a constant for the site without regard to its association with other sites on the same molecule. An intrinsic dissociation constant describes the equilibrium between free substrate, free *site*, and the site-substrate complex. The effective dissociation constant describes the equilibrium between free substrate, available *enzyme*, and enzyme-substrate complex. Consider an enzyme with two identical substrate binding sites per molecule. The rate at which the first molecule binds to form a singly occupied complex, ES_1, from free E and S is proportional to the concentration of S and to the *concentration of free sites*. The concentration of free sites is *twice* the concentration of enzyme molecules.

$$v_{E+S\rightarrow ES_1} \quad \propto[S] \quad \text{and} \quad \propto[\text{free sites}]$$

$$\text{or} \qquad \propto[S] \quad \text{and} \quad \propto 2[E]$$

$$\therefore \quad v_{E+S\rightarrow ES_1}=2k_1[E][S]$$

where k_1 is the intrinsic rate constant for the binding of S to a free site. The rate at which ES_1 dissociates back to E + S is proportional to the concentration of *occupied sites*. There is one occupied site per ES_1 complex.

$$v_{ES_1\rightarrow E+S}\propto[ES_1] \qquad \text{or} \qquad v_{ES_1\rightarrow E+S}=k_{-1}[ES_1]$$

In this case, the effective rate constant is identical to the intrinsic rate constant. When E, S, and ES_1 are at equilibrium, the rate of association equals the rate of dissociation.

$$v_{E+S \rightarrow ES_1} = v_{ES_1 \rightarrow E+S}$$

$$2k_1[E][S] = k_{-1}[ES_1]$$

$$\frac{[E][S]}{[ES_1]} = \frac{k_{-1}}{2k_1} = K_{S_1} = \frac{K_S}{2}$$

where K_S represents the intrinsic dissociation constant k_{-1}/k_1, and K_{S_1} is the effective dissociation constant for the reaction $ES_1 \rightleftharpoons E+S$.

The rate of ES_2 formation from ES_1 and S is proportional to the concentration of S and to the concentration of *free sites*. There is one free site per ES_1 molecule.

$$v_{ES_1+S \rightarrow ES_2} \propto [S] \qquad \text{and} \qquad \propto [\text{free sites}]$$

$$\therefore \quad v_{ES_1+S \rightarrow ES_2} = k_2[S][ES_1]$$

where k_2 is the intrinsic (and, in this case, also the effective) rate constant for the binding of S to the vacant site. The rate at which ES_2 dissociates to yield ES_1 and S is proportional to the concentration of *occupied sites*. There are two occupied sites per ES_2 molecule.

$$v_{ES_2 \rightarrow ES_1+S} \propto [\text{occupied sites}]$$

$$\propto 2[ES_2]$$

$$\therefore \quad v_{ES_2 \rightarrow ES_1+S} = 2k_{-2}[ES_2]$$

where k_{-2} is the intrinsic rate constant for the dissociation of S from either occupied site on ES_2.

At equilibrium:

$$v_{ES_1+S \rightarrow ES_2} = v_{ES_2 \rightarrow ES_1+S}$$

$$k_2[ES_1][S] = 2k_{-2}[ES_2]$$

$$\frac{[ES_1][S]}{[ES_2]} = \frac{2k_{-2}}{k_2} = K_{S_2} = 2K_S^*$$

where K_S^* is the intrinsic dissociation constant of the second site, and K_{S_2} is the effective dissociation constant for the reaction $ES_2 \rightleftharpoons ES_1 + S$. When we state that the two sites are "identical and independent," we mean that the chemical and electronic environments of the two sites are the same, and that the binding of a substrate molecule to one site has no effect on the intrinsic properties of the other site; that is, $k_2 = k_1$, $k_{-2} = k_{-1}$, and $K_S = K_S^*$.

$$\therefore \quad \text{if} \quad K_{S_1} = \frac{k_{-1}}{2k_1} = \frac{K_S}{2}, \qquad \text{and} \qquad K_{S_2} = \frac{2k_{-1}}{k_1} = 2K_S$$

then $K_{S_2} = 4K_{S_1}$. Also, the rate at which ES_2 forms product must be twice the rate of ES_1, or, in other words, $k_{p_2} = 2k_{p_1} = 2k_p$. With these ideas in mind, let us examine substrate binding to an enzyme with four identical sites. The reaction sequence in terms of effective constants is shown below.

$$\begin{array}{ccccccc}
 & & E + P & & & & \\
 & & \uparrow k_{p_1} & & & & \\
E + S & \overset{K_{S_1}}{\rightleftharpoons} & ES_1 & & ES_4 & \xrightarrow{k_{p_4}} & ES_3 + P \\
 & & + & & \updownarrow K_{S_4} & & \\
 & & S & & S & & \\
 & & \updownarrow K_{S_2} & & + & & \\
P + ES_1 \xleftarrow{k_{p_2}} & & ES_2 + S & \overset{K_{S_3}}{\rightleftharpoons} & ES_3 & \xrightarrow{k_{p_3}} & ES_2 + P
\end{array}$$

The effective K_S and k_p values in terms of intrinsic constants are:

$$K_{S_1} = \frac{K_S}{4}, \qquad K_{S_2} = \frac{2K_S}{3}, \qquad K_{S_3} = \frac{3K_S}{2}, \qquad K_{S_4} = 4K_S$$

$$k_{p_1} = k_p, \qquad k_{p_2} = 2k_p, \qquad k_{p_3} = 3k_p, \qquad k_{p_4} = 4k_p$$

The values for K_{S_1}, K_{S_2}, and so on, in terms of the intrinsic constant were obtained in the usual manner. For example, for K_{S_2}, the fraction $\frac{2}{3}$ indicates that there are two ways of dissociating S from ES_2 to yield ES_1 and three ways of binding S to ES_1 to yield ES_2. From the original velocity equation in terms of effective constants:

$$v = k_{p_1}[ES_1] + k_{p_2}[ES_2] + k_{p_3}[ES_3] + k_{p_4}[ES_4]$$

$$\frac{v}{[E]_t} = \frac{k_{p_1}\dfrac{[S]}{K_{S_1}} + k_{p_2}\dfrac{[S]^2}{K_{S_1}K_{S_2}} + k_{p_3}\dfrac{[S]^3}{K_{S_1}K_{S_2}K_{S_3}} + k_{p_4}\dfrac{[S]^4}{K_{S_1}K_{S_2}K_{S_3}K_{S_4}}}{1 + \dfrac{[S]}{K_{S_1}} + \dfrac{[S]^2}{K_{S_1}K_{S_2}} + \dfrac{[S]^3}{K_{S_1}K_{S_2}K_{S_3}} + \dfrac{[S]^4}{K_{S_1}K_{S_2}K_{S_3}K_{S_4}}}$$

we obtain the velocity equation in terms of intrinsic constants:

$$\frac{v}{[\mathrm{E}]_t} = \frac{k_p\left(\dfrac{4[\mathrm{S}]}{K_\mathrm{S}} + \dfrac{(6)(2)[\mathrm{S}]^2}{K_\mathrm{S}^2} + \dfrac{(4)(3)[\mathrm{S}]^3}{K_\mathrm{S}^3} + \dfrac{(4)[\mathrm{S}]^4}{K_\mathrm{S}^4}\right)}{1 + \dfrac{4[\mathrm{S}]}{K_\mathrm{S}} + \dfrac{6[\mathrm{S}]^2}{K_\mathrm{S}^2} + \dfrac{4[\mathrm{S}]^3}{K_\mathrm{S}^3} + \dfrac{[\mathrm{S}]^4}{K_\mathrm{S}^4}} \qquad \text{(VII-2)}$$

The numerator of equation VII-2 reflects the existence of four singly occupied complexes each with a given catalytic rate constant, k_p; six doubly occupied complexes, each twice as effective as a singly occupied complex in forming product; four triply occupied complexes, each with a catalytic rate constant three times that of the singly occupied complex; and one complex with all four sites filled with a catalytic rate constant four times that of the singly occupied complex (Fig. VII-1). The conversion of effective dissociation and catalytic rate constants to intrinsic constants simply introduces the necessary statistical factors. The velocity equation can be written directly in

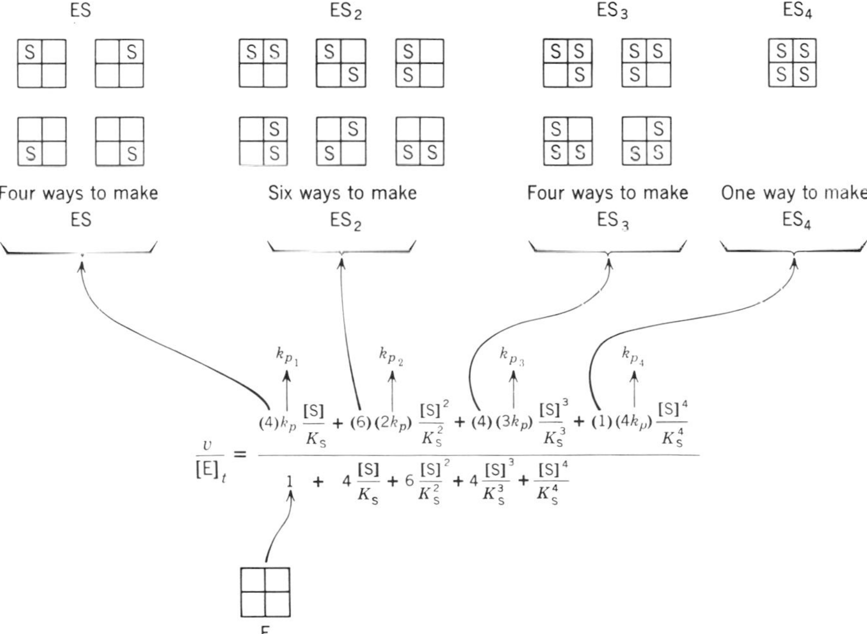

Fig. VII-1. Kinds of species for an enzyme with four identical sites and relationship to terms in the velocity equation.

terms of intrinsic constants and the appropriate statistical factors:

$$\frac{v}{[\mathrm{E}]_t} = \frac{k_p m_1 \dfrac{[\mathrm{S}]}{K_\mathrm{S}} + 2k_p m_2 \dfrac{[\mathrm{S}]^2}{K_\mathrm{S}^2} + \cdots + nk_p m_n \dfrac{[\mathrm{S}]^n}{K_\mathrm{S}^n}}{1 + \dfrac{m_1[\mathrm{S}]}{K_\mathrm{S}} + \dfrac{m_2[\mathrm{S}]^2}{K_\mathrm{S}^2} + \cdots + \dfrac{m_n[\mathrm{S}]^n}{K_\mathrm{S}^n}} \qquad \text{(VII-3)}$$

where n is the total number of identical substrate binding sites and m represents the number of ways of arranging S in the given complex. If $4k_p[\mathrm{E}]_t$ is factored out as $V_{\max}$, we obtain:

$$\boxed{\frac{v}{V_{\max}} = \frac{\dfrac{[\mathrm{S}]}{K_\mathrm{S}} + \dfrac{3[\mathrm{S}]^2}{K_\mathrm{S}^2} + \dfrac{3[\mathrm{S}]^3}{K_\mathrm{S}^3} + \dfrac{[\mathrm{S}]^4}{K_\mathrm{S}^4}}{1 + \dfrac{4[\mathrm{S}]}{K_\mathrm{S}} + \dfrac{6[\mathrm{S}]^2}{K_\mathrm{S}^2} + \dfrac{4[\mathrm{S}]^3}{K_\mathrm{S}^3} + \dfrac{[\mathrm{S}]^4}{K_\mathrm{S}^4}}} \qquad \text{(VII-4)}$$

The velocity curve for the four-site enzyme will be the usual hyperbola with $v = 0.5\,V_{\max}$ when $[\mathrm{S}] = K_\mathrm{S}$, $v = 0.667\,V_{\max}$ when $[\mathrm{S}] = 2K_\mathrm{S}$, and so on. Indeed, the equation reduces to the usual equation for a single-site enzyme: Factoring $[\mathrm{S}]/K_\mathrm{S}$:

$$\frac{v}{V_{\max}} = \frac{\dfrac{[\mathrm{S}]}{K_\mathrm{S}}\left(1 + \dfrac{[\mathrm{S}]}{K_\mathrm{S}}\right)^3}{\left(1 + \dfrac{[\mathrm{S}]}{K_\mathrm{S}}\right)^4} = \frac{\dfrac{[\mathrm{S}]}{K_\mathrm{S}}}{1 + \dfrac{[\mathrm{S}]}{K_\mathrm{S}}} = \frac{[\mathrm{S}]}{K_\mathrm{S} + [\mathrm{S}]}$$

Note that the experimentally measured K is the intrinsic dissociation constant, not any of the effective constants. Also, keep in mind that rapid equilibrium velocity equations are really equilibrium ligand binding equations. They become velocity equations when Y_S, the fraction of occupied sites, is equated to $v/V_{\max}$. For example, consider the velocity dependence equation $v = k_{p_1}[\mathrm{ES}_1] + k_{p_2}[\mathrm{ES}_2] \ldots$, and so on. This could just as easily be written as $[\text{bound S}] = k_{p_1}[\mathrm{ES}_1] + k_{p_2}[\mathrm{ES}_2] \ldots$, and so on, where k_{p_1}, k_{p_2}, and so on, represent the number of occupied sites on ES_1, ES_2, and so on. Thus $4k_p[\mathrm{E}]_t$ represents the total concentration of available sites, so that $[\text{bound S}]/4k_p[\mathrm{E}]_t = Y_\mathrm{S}$.

While the analysis above is instructional, it really is not necessary to go through the entire procedure outlined to obtain the final velocity equation. If we examine the velocity equations for two-site, four-site, and six-site enzymes, we see that the coefficients of each term in the numerator and denominator follow a symmetrical distribution (e.g., 1-3-3-1 and 1-4-6-4-1 for a four-site enzyme; 1-5-10-10-5-1 and 1-6-15-20-15-6-1 for a six-site enzyme). This distribution is characteristic of the expansion of $(1+X)^n$. Close examination will show that the velocity equations for enzymes containing n identical sites can be expressed as:

$$\frac{v}{V_{\max}} = \frac{\dfrac{[\mathrm{S}]}{K_{\mathrm{S}}}\left(1+\dfrac{[\mathrm{S}]}{K_{\mathrm{S}}}\right)^{n-1}}{\left(1+\dfrac{[\mathrm{S}]}{K_{\mathrm{S}}}\right)^{n}} \tag{VII-5}$$

where n is the number of identical sites and K_S is the intrinsic dissociation constant of a site. Equation VII-5 reduces to the simple Henri-Michaelis-Menten equation, again demonstrating that if all the sites are equivalent (as assumed in the derivation), n molecules of a single-site enzyme yield the same velocity curve as one molecule of an n-site enzyme.

Allosteric Enzymes—Cooperative Binding

So far, we have considered enzymes that possess multiple, but independent, substrate binding sites; that is, the binding of one molecule of substrate has no effect on the intrinsic dissociation constants of the vacant sites. Such enzymes yield normal hyperbolic velocity curves. However, if the binding of one substrate molecule induces structural or electronic changes that result in altered affinities for the vacant sites, the velocity curve will no longer follow Henri-Michaelis-Menten kinetics and the enzyme will be classified as an "allosteric" enzyme. Generally, allosteric enzymes yield sigmoidal velocity curves. The binding of one substrate molecule facilitates the binding of the next substrate molecule by increasing the affinities of the vacant binding sites. The phenomenon has been called cooperative binding, or positive cooperativity with respect to substrate binding, or a positive homotropic response. Interactions between unlike ligands (e.g., substrate and inhibitor or substrate and activator) are termed heterotropic responses and may be positive (activator) or negative (inhibitor).

The potential advantages of a sigmoidal response to varying substrate are illustrated in Figure VII-2. For comparison, a normal hyperbolic velocity

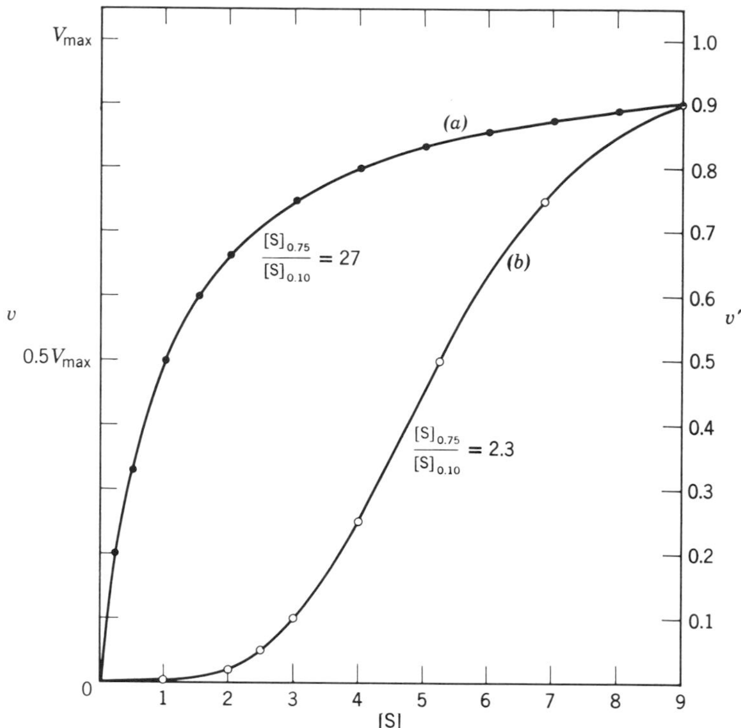

Fig. VII-2. Comparison of velocity curves for two different enzymes that coincidentally have the same v at [S] = 9. (*a*) Hyperbolic response. (*b*) Sigmoidal response.

with the same $[S]_{0.9}$ is shown. Between [S] = 0 and [S] = 3, the hyperbolic response curve decelerates, but still rises to $0.75\,V_{max}$. The sigmoidal curve accelerates exponentially, but only attains $0.10\,V_{max}$ between the same limits of [S]. However, the sigmoidal curve increases from $0.10\,V_{max}$ to $0.75\,V_{max}$ with only an additional 2.3-fold increase in [S]. To cover the same specific velocity range, the hyperbolic curve requires a 27-fold increase in [S]. Thus the sigmoidal response acts, in a sense, as an "off-on switch." Also, at intermediate specific velocities, the sigmoidal response provides a much more sensitive control of the reaction rate by variations in the substrate concentration.

The term allosteric was originally applied by Monod, Changeux, and Jacob (1963) to enzymes that display altered kinetic properties (usually a change in $[S]_{0.5}$) in the presence of effectors that have no structural resemblance to the substrate. The allosteric response is usually quite understandable in terms of metabolic control and cellular economy (e.g., the feedback inhibition of the first reaction of a sequence by the ultimate product). Most allosteric enzymes display sigmoidal ligand saturation curves. Consequently, allosterism has become synonomous with sigmoidal responses.

Two types of models for allosteric enzymes have been proposed. These are the sequential interaction models and the concerted-symmetry model. As the name suggests, the sequential models assume sequential or progressive changes in the affinities of vacant sites as sites are occupied. The concerted-symmetry model assumes that the enzyme preexists as an equilibrium mixture of a high affinity oligomer and a low affinity oligomer. Ligands, including the substrate itself, act by displacing the equilibrium in favor of one state or the other. During the transition, the conformation of all subunits change at the same time and the oligomer retains its symmetry. The details of the various models are described below. It is notewrothy that the sequential interaction and concerted-symmetry models are based on rapid equilibrium assumptions.

Adair-Pauling Simple Sequential Interaction Model

The simple interaction model evolved from attempts to explain the sigmoidal oxygen saturation curve of hemoglobin, a phenomenon first reported by Bohr in 1903. In 1910, Hill proposed the empirical equation $Y = KP^n_{O_2}/(1 + KP^n_{O_2})$ which gave a reasonably good fit to the data when $n = 2.6$. In 1925, Adair obtained a closer fit using an equation containing four different binding constants for the four different heme groups. Pauling, in 1935, made the first attempt to relate the different binding constants to the geometry of the protein by assuming a single binding constant and a series of interaction factors. More recently, several workers, notably Koshland and Atkinson, have extended the Adair-Pauling model and equations to account for the behavior of allosteric enzymes.

Interaction Factors

First, let us consider an enzyme with two identical substrate binding sites. If the binding of one substrate molecule changes the intrinsic dissociation constant by a factor, a, the reaction sequence is:

$$
\begin{array}{ccccccccc}
 & & & \mathrm{E} + \mathrm{S} & \overset{K_S}{\rightleftharpoons} & \mathrm{ES} & \xrightarrow{k_p} & \mathrm{E} + \mathrm{P} & \\
 & & & + & & + & & & \\
 & & & \mathrm{S} & & \mathrm{S} & & & \\
 & & & K_S \updownarrow & & \updownarrow aK_S & & & \\
\mathrm{P} + \mathrm{E} & \xleftarrow{k_p} & & \mathrm{SE} + \mathrm{S} & \overset{aK_S}{\rightleftharpoons} & \mathrm{SES} & \xrightarrow{k_p} & \mathrm{SE} + \mathrm{P} & \\
 & & & & & \downarrow k_p & & & \\
 & & & & & \mathrm{ES} + \mathrm{P} & & &
\end{array}
$$

The distribution of enzyme between free E, ES, SE, and SES is:

$$\frac{[ES]}{[E]_t} = \frac{\dfrac{[S]}{K_S}}{1 + \dfrac{2[S]}{K_S} + \dfrac{[S]^2}{aK_S^2}}, \qquad \frac{[SE]}{[E]_t} = \frac{\dfrac{[S]}{K_S}}{1 + \dfrac{2[S]}{K_S} + \dfrac{[S]^2}{aK_S^2}}$$

$$\frac{[SES]}{[E]_t} = \frac{\dfrac{[S]^2}{aK_S^2}}{1 + \dfrac{2[S]}{K_S} + \dfrac{[S]^2}{aK_S^2}}, \qquad \frac{[E]}{[E]_t} = \frac{1}{1 + \dfrac{2[S]}{K_S} + \dfrac{[S]^2}{aK_S^2}}$$

The velocity equation is:

$$\boxed{\frac{v}{V_{max}} = \frac{\dfrac{[S]}{K_S} + \dfrac{[S]^2}{aK_S^2}}{1 + \dfrac{2[S]}{K_S} + \dfrac{[S]^2}{aK_S^2}}} \tag{VII-6}$$

where $V_{max} = 2k_p[E]_t$. An enzyme with four identical sites may be treated similarly, as described below.

If the binding of the first substrate molecule changes the intrinsic dissociation constants of the vacant sites by a factor, a, then, in terms of effective constants:

$$K_{S_1} = \frac{K_S}{4} \qquad \text{(all four sites are equivalent before any substrate binds)}$$

$$K_{S_2} = \frac{a2K_S}{3}$$

If the binding of the second substrate molecule changes the intrinsic dissociation constants for the remaining vacant sites by a factor, b, then:

$$K_{S_3} = \frac{ab3K_S}{2}$$

Note that the change induced by the first molecule of substrate is retained;

that is, the interactions are cumulative. Similarly, if the third substrate molecule introduces an interaction factor, c, then the effective dissociation constant for the fourth substrate molecule is:

$$K_{S_4} = abc4K_S$$

The concentration of each species relative to free E is:

$$[ES_1] = \frac{4[S]}{K_S}[E], \qquad [ES_2] = \frac{6[S]^2}{aK_S^2}[E]$$

$$[ES_3] = \frac{4[S]^3}{a^2bK_S^3}[E], \qquad [ES_4] = \frac{[S]^4}{a^3b^2cK_S^4}[E]$$

The fraction of the total enzyme represented by each species is:

$$\frac{[ES_1]}{[E]_t} = \frac{\dfrac{4[S]}{K_S}}{1 + \dfrac{4[S]}{K_S} + \dfrac{6[S]^2}{aK_S^2} + \dfrac{4[S]^3}{a^2bK_S^3} + \dfrac{[S]^4}{a^3b^2cK_S^4}}$$

$$\frac{[ES_2]}{[E]_t} = \frac{\dfrac{6[S]^2}{aK_S^2}}{\text{same denominator}}$$

$$\frac{[ES_3]}{[E]_t} = \frac{\dfrac{4[S]^3}{a^2bK_S^3}}{\text{same denominator}}$$

$$\frac{[ES_4]}{[E]_t} = \frac{\dfrac{[S]^4}{a^3b^2cK_S^4}}{\text{same denominator}}$$

$$\frac{[E]}{[E]_t} = \frac{1}{\text{same denominator}}$$

The velocity equation is obtained from:

$$\frac{v}{[E]_t} = \frac{k_p[ES_1] + 2k_p[ES_2] + 3k_p[ES_3] + 4k_p[ES_4]}{[E] + [ES_1] + [ES_2] + [ES_3] + [ES_4]}$$

$$\boxed{\frac{v}{V_{max}} = \frac{\dfrac{[S]}{K_S} + \dfrac{3[S]^2}{aK_S^2} + \dfrac{3[S]^3}{a^2bK_S^3} + \dfrac{[S]^4}{a^3b^2cK_S^4}}{1 + \dfrac{4[S]}{K_S} + \dfrac{6[S]^2}{aK_S^2} + \dfrac{4[S]^3}{a^2bK_S^3} + \dfrac{[S]^4}{a^3b^2cK_S^4}}} \qquad \text{(VII-7)}$$

where $V_{max} = 4k_p[E]_t$. Figure VII-3*a* shows the distribution of complexes as a function of [S] for an enzyme with four identical catalytic sites and no cooperativity. Figure VII-3*b* shows the distribution if the binding of each substrate molecule increases the binding constants of the vacant sites by a factor of about 25; that is $a = b = c = 0.04$. The curves were adapted from those presented by Atkinson, Hathaway, and Smith (1965) for the yeast NAD-linked isocitrate dehydrogenase. We see that if the cooperativity is strong, the enzyme exists predominantly as free E and ES_4 over a wide range of [S].

A Note on Terminology Regarding "Interaction Factors"

An equation could have been derived calling *a*, *b*, and *c* "factors by which the intrinsic binding constants have been increased." Indeed, we generally discuss enzyme-substrate interactions in terms of "affinities," and "changes in affinities or binding constants." However, velocity equations are usually derived in terms of dissociation constants, and it seems preferable to indicate changes in affinities as factors of K_S. Thus an increase in the intrinsic affinity by a factor of 10 [or an "interaction (or cooperative) factor of 10" as it might be described] appears as a factor of 0.1 with respect to K_S. If the alternate definition of *a*, *b*, *c*, and so on, is used, then these factors will appear in the numerator of each fraction term in the velocity equation. Of course, the equation remains the same—it makes no difference whether we divide by 0.1 or multiply by 10. For example, equation VII-7 could be written as:

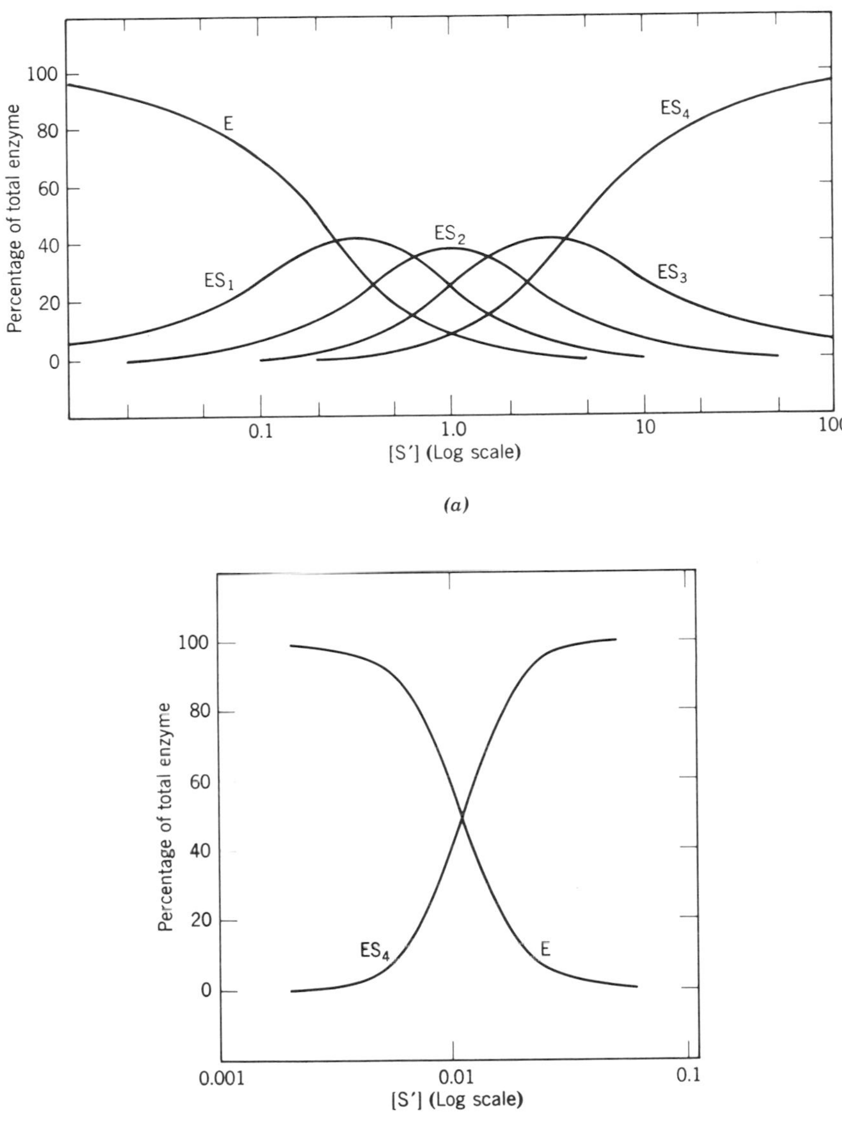

Fig. VII-3. Distribution of enzyme species for a four-site enzyme as a function of the substrate concentration. (*a*) No cooperative binding ($a = b = c = 1$). (*b*) Cooperative binding ($a \cong b \cong c \cong 0.04$). The ES_1, ES_2, and ES_3 complexes which represent small fractions of the total enzyme, are not shown. [Redrawn from Atkinson, D. E., Hathaway, J. A. and Smith, E. C., *J. Biol. Chem.* **240**, 2682 (1965).]

$$\frac{v}{V_{max}} = \frac{\frac{[S]}{K_S} + 3i\left(\frac{[S]}{K_S}\right)^2 + 3i^2j\left(\frac{[S]}{K_S}\right)^3 + i^3j^2k\left(\frac{[S]}{K_S}\right)^4}{1 + 4\left(\frac{[S]}{K_S}\right) + 6i\left(\frac{[S]}{K_S}\right)^2 + 4i^2j\left(\frac{[S]}{K_S}\right)^3 + i^3j^2k\left(\frac{[S]}{K_S}\right)^4} \qquad \text{(VII-8)}$$

where $i = 1/a$, $j = 1/b$, and $k = 1/c$; that is, i, j, and k can be considered as factors by which the substrate binding constants change as sites are filled.

A Simplified Velocity Equation for Allosteric Enzymes—The Hill Equation

Consider an enzyme with n equivalent substrate binding sites. If the cooperativity in substrate binding is very marked (i.e., the factors a, b, c, and so on, are very small numbers), then the concentrations of all enzyme-substrate complexes containing less than n molecules of substrate will be negligible at any [S] that is appreciable compared to K_S. Under this condition, the velocity equation will be dominated by the $[S]^n$ terms. For example, the equation for the four-site enzyme reduces to:

$$\frac{v}{V_{max}} = \frac{\frac{[S]^4}{a^3b^2cK_S^4}}{1 + \frac{[S]^4}{a^3b^2cK_S^4}} = \frac{\frac{[S]^4}{K'}}{1 + \frac{[S]^4}{K'}} = \frac{[S]^4}{K' + [S]^4}$$

where $K' = a^3b^2cK_S^4$. In general:

$$\frac{v}{V_{max}} = \frac{[S]^n}{K' + [S]^n} \qquad \text{(VII-9)}$$

Equation VII-9 is known as the Hill equation.

n = the number of substrate binding sites per molecule of enzyme

K' = a constant comprising the interaction factors a, b, c, and so on, and the intrinsic dissociation constant, K_S

$= K_S^n(a^{n-1}b^{n-2}c^{n-3}\ldots z^1)$

The constant K' in the equation above no longer equals the substrate concentration that yields half-maximal velocity (except when $n=1$, when the equation reduces to the Henri-Michaelis-Menten equation).

When $v=0.5\,V_{max}$, $[S]_{0.5}^{n}=K'$

$$\therefore\quad [S]_{0.5}=\sqrt[n]{K'} \qquad \text{or} \qquad n\log[S]_{0.5}=\log K' \qquad \text{(VII-10)}$$

A new kind of specific (reduced, normalized, relative) substrate concentration can be introduced: $[S]^*$, which means [S] relative to $[S]_{0.5}$ and can be expressed as:

$$[S]^*=\frac{[S]}{K_S^*} \qquad \text{or} \qquad [S]^{*n}=\frac{[S]^n}{K_S^{*n}}=\frac{[S]^n}{K'}$$

where $K_S^*=\sqrt[n]{K'}$ or $K_S^{*n}=K'$. The Hill equation can then be written as:

$$\boxed{\frac{v}{V_{max}}=\frac{[S]^n}{K_S^{*n}+[S]^n}} \qquad \text{or} \qquad \boxed{v'=\frac{[S]^{*n}}{1+[S]^{*n}}} \qquad \text{(VII-11)}$$

The v' is the usual specific velocity, v/V_{max}.

If the cooperativity of substrate binding is not very high, then the velocity equation will not reduce to the Hill equation. Nevertheless, velocity curves can be expressed in terms of the Hill equation although n will no longer equal the actual number of substrate binding sites. In this case, the n in the equation should be designated n_H or n_{app}. For example, if the cooperativity of substrate binding to a four-site enzyme is such that the major species present throughout a major portion of the velocity curve (e.g., between 10% and 90% of V_{max}) are ES_3 and ES_4, then the velocity data can be made to fit the Hill equation if n is taken as some nonintegral value between 3 and 4 (e.g., $n=3.6$). To put it another way, if experimental velocity data are analyzed in terms of the Hill equation, the calculated value of n will almost always be less than the actual number of sites. The next highest integer above this *apparent* n value represents the minimum number of actual sites. Therefore, if the experimental data yield an n_{app} value of 1.8 based on the Hill equation, we are in effect saying that the enzyme behaves as if it possesses exactly 1.8 substrate binding sites with very strong cooperativity. We know that there are at least two sites with relatively strong cooperativity but there could just as well be four sites with poor cooperativity, or many sites that act in highly cooperative pairs.

Sigmoidicity of the Velocity Curve

The shape of the v versus [S] curve can be expressed in terms of the ratio of substrate concentration required for any two fractions of V_{max}, for example, $0.9\,V_{max}$ and $0.1\,V_{max}$. This ratio depends on the value of n as shown below.

$$\frac{v}{V_{max}}=\frac{[S]^n}{K'+[S]^n}$$

When $v=0.9\,V_{max}$:

$$0.9=\frac{[S]_{0.9}^n}{K'+[S]_{0.9}^n}, \qquad [S]_{0.9}=\sqrt[n]{9K'}$$

When $v=0.1\,V_{max}$:

$$0.1=\frac{[S]_{0.1}^n}{K'+[S]_{0.1}^n}, \qquad [S]_{0.1}=\sqrt[n]{\frac{K'}{9}}$$

$$\therefore \quad \frac{[S]_{0.9}}{[S]_{0.1}}=\frac{\sqrt[n]{9K'}}{\sqrt[n]{\dfrac{K'}{9}}}=\sqrt[n]{\frac{9}{1/9}}$$

$$\boxed{\frac{[S]_{0.9}}{[S]_{0.1}}=\sqrt[n]{81}} \quad \text{or} \quad \boxed{n=\frac{\log 81}{\log\dfrac{[S]_{0.9}}{[S]_{0.1}}}} \qquad \text{(VII-12)}$$

The relative sigmoidicity of two curves is not always obvious from simple inspection. For example, all the curves shown in Figure VII-4 have n values of 4.0 ($[S]_{0.9}/[S]_{0.1}=3.0$). The displacement along the [S]-axis results from different values of the intrinsic K_S.

Inflection Point of the Velocity Curve

The first derivative of the velocity equation, $dv/d[S]$, gives the slope of the velocity curve.

$$v=\frac{V_{max}[S]^n}{K'+[S]^n}=\frac{V_{max}[S]^n}{K_S^{*n}+[S]^n}$$

$$\frac{dv}{d[S]}=\textit{slope}$$

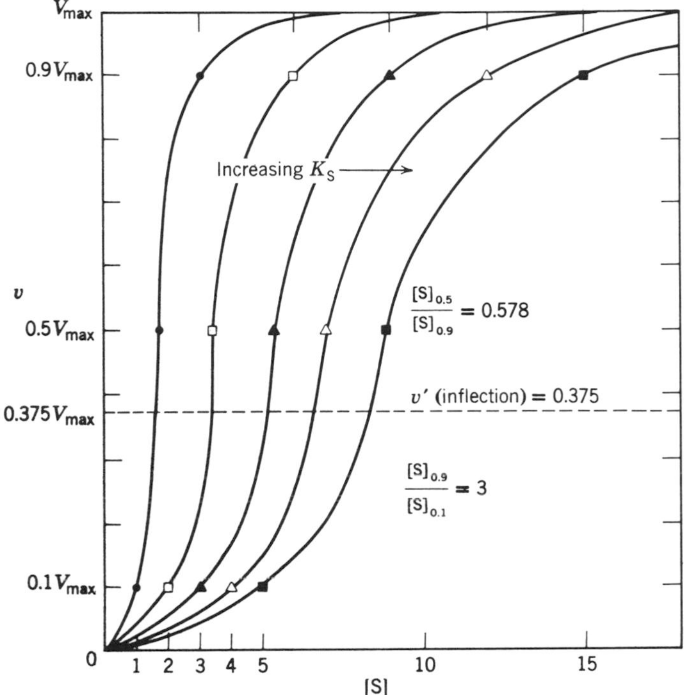

Fig. VII-4. Sigmoidal curves; $n=4$ in all cases, but the numerical value of K_S differs.

For a fractional equation such as the simplified velocity equation:

$$\frac{dv}{d[S]} = \frac{f'g - g'f}{g^2}$$

where f = the numerator
g = the denominator
f' = the derivative of the numerator
g' = the derivative of the denominator

$$\boxed{\frac{dv}{d[S]} = \frac{nK'V_{max}[S]^{n-1}}{(K' + [S]^n)^2}} \qquad \text{(VII-13)}$$

In terms of specific values, where $v' = v/V_{max}$ and $[S]^* = [S]/K_S^*$, the derivative can be expressed as:

$$\boxed{\frac{dv'}{d[S]^*} = \frac{n[S]^{*n-1}}{(1+[S]^{*n})^2}} \tag{VII-14}$$

Setting $dv/d[S]$ equal to zero, we find that [S] must be infinitely high in order to obtain a zero slope. This is obvious from the shape of the velocity curve, which is asymptotic to V_{max}. The second derivative, $d^2v/d[S]^2$, gives the rate of change of the slope. The early part of the velocity curve is concave upward (i.e., we observe an acceleration). The curve passes through an inflection point and then decelerates as v approaches V_{max}. At the inflection point, the acceleration is zero; that is, $d^2v/d[S]^2 = 0$. If we determine $d^2v/d[S]^2$, and solve for the value of [S] that makes $d^2v/d[S]^2$ equal zero, we will have the substrate concentration (in terms of K') at the inflection point.

The second derivative is obtained as shown above. To obtain the derivative of the parenthetical term $(K' + [S]^n)^2$, we apply the rule:

$$\frac{d(f(x))^n}{dx} = n(f(x))^{n-1}\frac{df}{dx}$$

$$\therefore \quad \frac{d(K' + [S]^n)^2}{d[S]^n} = 2(K' + [S]^n)n[S]^{n-1}$$

and

$$\frac{d^2v}{d[S]^2} = \frac{n(n-1)K'V_{max}[S]^{n-2}(K' + [S]^n)^2 - 2(K' + [S]^n)n[S]^{n-1}nK'V_{max}[S]^{n-1}}{(K' + [S]^n)^4}$$

When $d^2v/d[S]^2 = 0$:

$$\frac{n(n-1)K'V_{max}[S]^{n-2}(K' + [S]^n)^2}{(K' + [S]^n)^4} = \frac{2(K' + [S]^n)n[S]^{n-1}nK'V_{max}[S]^{n-1}}{(K' + [S]^n)^4}$$

Solving for $[S]^n$ and [S]:

$$[S]^n = \frac{(n-1)}{(n+1)}K' \qquad \text{or} \qquad [S] = \sqrt[n]{\frac{(n-1)K'}{(n+1)}} \qquad \text{(VII-15)}$$

If we substitute the value of $[S]^n$ into the Hill equation, we obtain the specific velocity (v/V_{max}) at the inflection point:

$$\frac{v}{V_{max}} = \frac{(n-1)}{2n} \qquad \text{or} \qquad n = \frac{1}{1-2v'} \qquad \text{(VII-16)}$$

Thus when $n=2$, the inflection point occurs at $v=0.25\,V_{max}$. When $n=4$, the inflection occurs at $0.375\,V_{max}$. As n increases, the specific velocity at the inflection point approaches $n/2n$, or $0.5\,V_{max}$.

A departure of the velocity curve from the usual hyperbolic response may be overlooked if the substrate range examined is high compared to K_S and the cooperativity is low. Figure VII-5, curve A, shows the velocity curve for a two-site enzyme with an interaction factor, a, of 0.25. The curve appears normal. Furthermore, small experimental errors in the determination of initial velocity at low substrate concentrations might completely obscure the inflection in the curve. However, compared to curve B (the velocity curve for a two-site enzyme with no cooperativity), it is apparent that curve A rises more rapidly. A check of the $[S]_{0.9}/[S]_{0.1}$ ratio immediately establishes the departure of curve A from Henri-Michaelis-Menten kinetics. Figure VII-6 shows several other velocity curves for a two-site enzyme with different interaction factors. The substrate concentration axis has been expanded to magnify the sigmoidicities of the curves. It is interesting that curve B rises faster than curve A and has an $[S]_{0.9}/[S]_{0.1}$ of 50, yet curve B has no inflection point. (The reason for this is described in a following section and illustrated in Figure VII-16).

Lineweaver-Burk Plot for Allosteric Enzymes

The plot of $1/v$ versus $1/[S]$ for an allosteric enzyme is curved and approaches a horizontal line which intersects the $1/v$-axis at $1/V_{max}$. An example is shown in Figure VII-7a for an enzyme where $n=2$ and the cooperativity is very high. Although the plot of $1/v$ versus $1/[S]$ is curved, the plot of $1/v$ versus $1/[S]^2$ is linear (Fig. VII-7b). The $1/v$-axis intercept

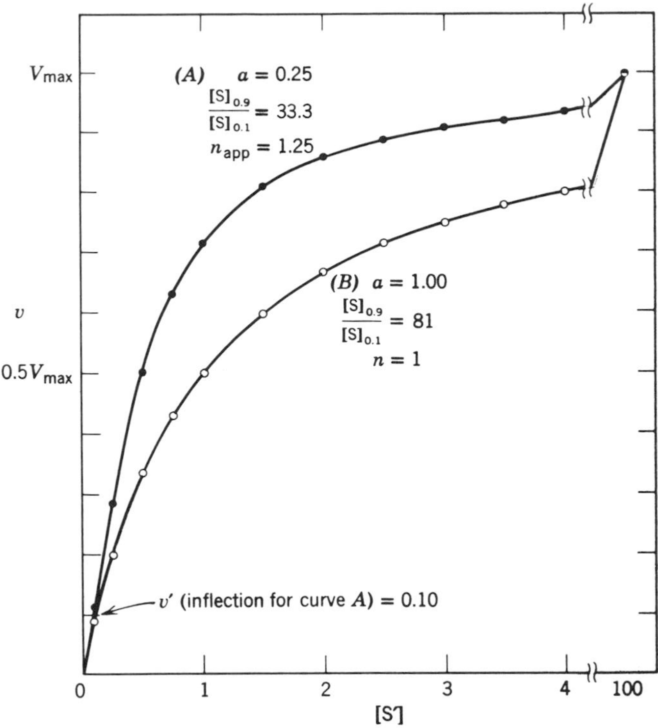

Fig. VII-5. Velocity curves for a two-site enzyme. Curve A: Cooperative binding ($a=0.25$). Curve B: Noncooperative binding ($a=1.0$). Curve A shows an inflection point, but it occurs at $v=0.10\ V_{max}$ and may be missed. The $[S]_{0.9}/[S]_{0.1}$ ratio, however, is considerably less than 81. $[S']=[S]/K_S$.

gives $1/V_{max}$, while the $1/[S]^2$-axis intercept gives $-1/K'$. Theoretically, it is possible to determine n by constructing plots of $1/v$ versus the reciprocal of [S] raised to various powers. The n value would be the power that yields a straight line. However, keep in mind that the simplified velocity equation VII-9 assumes that all complexes containing less than n molecules of substrate are negligible (i.e., the cooperativity of substrate binding is quite high). If this is not the case, then the n value that linearizes the reciprocal plot will be less than the actual number of substrate sites and could, in fact, be a nonintegral value. Thus the curve-fitting procedure yields an *apparent n* value (n_{app}). In the example shown in Figure VII-7*b*, the cooperativity was assumed to be high. Hence $n_{app}=n=2$. The n_{app} value of 2 indicates there are *at least* two substrate sites, but an enzyme with four sites and low cooperativity could also yield an n_{app} of 2.

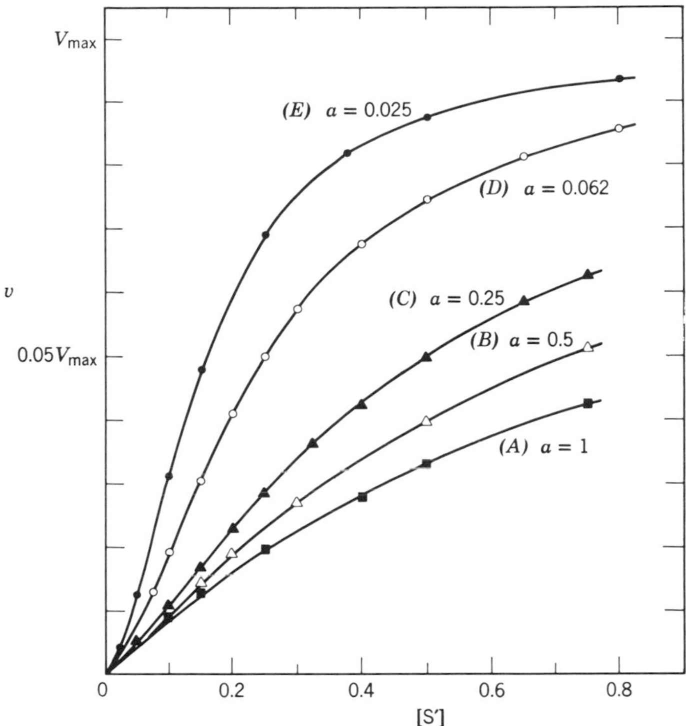

Fig. VII-6. Velocity curves for a variety of two-site enzymes. Curve A: $a=1$ (no cooperative binding). Curve B: $a=0.5$ (cooperative binding, but no inflection point). Curve C: $a=0.25$. Curve D: $a=0.062$. Curve E: $a=0.025$. $[S']=[S]/K_S$.

Eadie-Scatchard Plot for Allosteric Enzymes

The Hill equation can be rearranged into a form analogous to the Eadie-Scatchard equation IV-62:

$$\frac{v}{[S]^n} = -\frac{1}{K'}v + \frac{V_{max}}{K'} \qquad \text{or} \qquad v = -\frac{[S]^n}{K'}v + \frac{V_{max}[S]^n}{K'}$$

In terms of specific values ($v/V_{max} = v'$; $[S]^n/K' = [S]^{*n}$):

$$v' = -[S]^{*n}v' + [S]^{*n}$$

Dividing both sides of the equation by $[S]^*$:

$$\frac{v'}{[S]^*} = -v[S]^{*n-1} + [S]^{*n-1} \tag{VII-17}$$

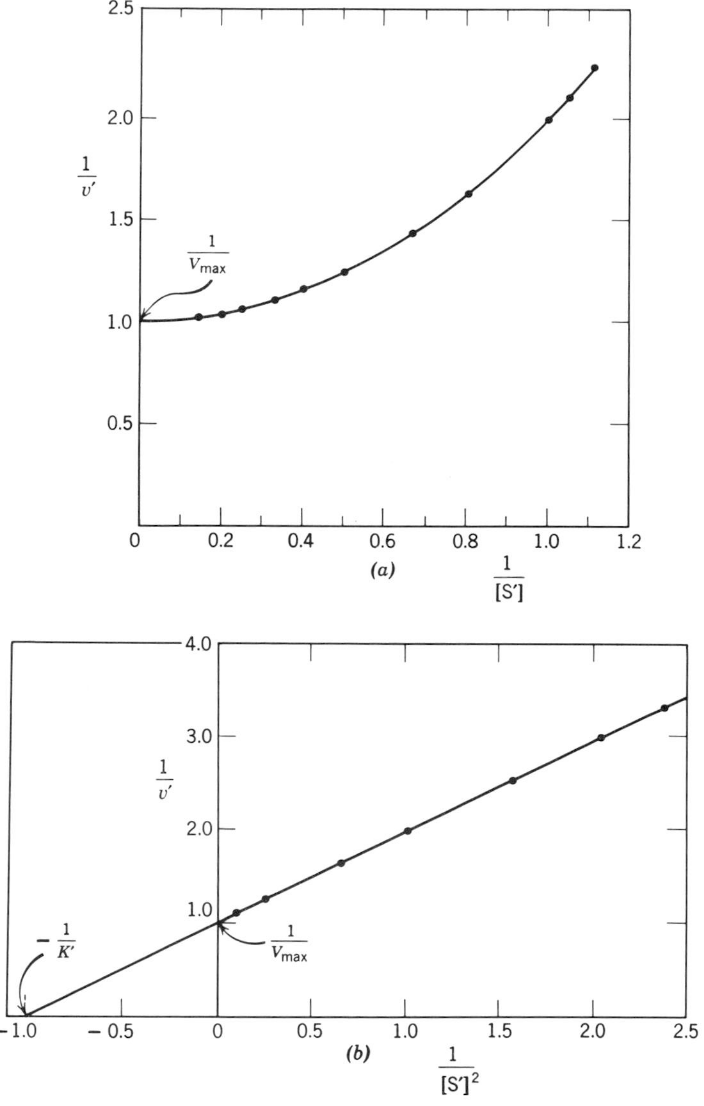

Fig. VII-7. (*a*) $1/v$ versus $1/[S]$ plot for an enzyme exhibiting strongly cooperative substrate binding ($n_{app} = 2$). (*b*) Plot of $1/v$ versus $1/[S]^2$ for the same enzyme.

When $n=1$, the equation reduces to the usual Eadie-Scatchard equation and the plot of $v/[S]$ versus v is linear with a slope of $-1/K_S$. When $n>1$, the plot is curved passing through a maximum (Fig. VII-8). The height of the maximum (i.e., the $v/[S]$ coordinate) depends on the value of K' but the v coordinate depends only on n. The maximum occurs where the slope of the plot (i.e., the first derivative, $d(v/[S])/dv$) equals zero. We can determine the coordinates of the maximum by solving for the values of v and [S] that make the first derivative equal zero.

$$slope = \frac{d\left(\dfrac{v'}{[S]^*}\right)}{dv'} = 0$$

$$0 = -[S]^{*n-1} - v'(n-1)[S]^{*n-2}\frac{d[S]'}{dv'} + (n-1)[S]^{*n-2}\frac{d[S]^*}{dv'}$$

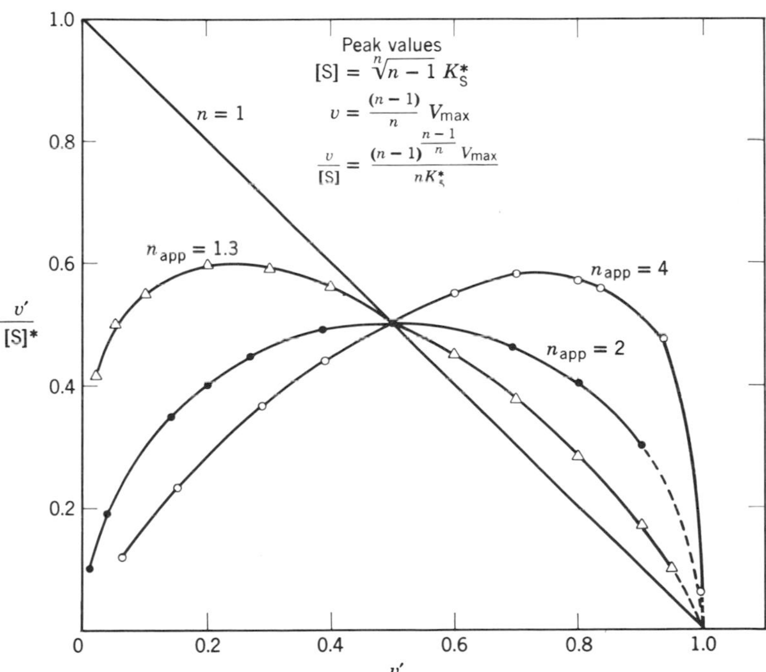

Fig. VII-8. The $v'/[S]^*$ versus v' plot for enzymes with different n_{app} values. Normalized values are plotted: $v'=v/V_{max}$, $[S]^*=[S]/[S]_{0.5}=[S]/K_S^*$. In practice $v/[S]$ versus v is plotted.

Solving for [S]*:

$$[S]^* = \frac{(1-v')(n-1)}{\dfrac{dv'}{d[S]^*}} \qquad \text{(VII-18)}$$

The $dv'/d[S]^*$ is the slope of the v versus [S] plot (equation VII-14):

$$\frac{dv'}{d[S]^*} = \frac{n[S]^{*n-1}}{(1+[S]^{*n})^2}$$

Substituting into equation VII-18:

$$[S]^* = \frac{(1-v')(n-1)(1+[S]^{*n})^2}{n[S]^{*n-1}}$$

$$v' = \frac{[S]^{*n}}{1+[S]^{*n}}$$

$$\therefore \quad [S]^* = \frac{\left(1 - \dfrac{[S]^{*n}}{1+[S]^{*n}}\right)(n-1)(1+[S]^{*n})^2}{n[S]^{*n-1}}$$

We now have an equation that gives [S]* at the peak of the Eadie-Scatchard plot in terms of n. The equation simplifies to:

$$\boxed{[S]^* = \sqrt[n]{n-1}} \quad \text{or} \quad \boxed{[S] = \sqrt[n]{n-1}\, K_S^*} \qquad \text{(VII-19)}$$

Substituting the expression above for [S]* into the Hill equation, we can find the corresponding v' coordinate:

$$v' = \frac{[S]^{*n}}{1+[S]^{*n}} = \frac{(\sqrt[n]{n-1})^n}{1+(\sqrt[n]{n-1})^n} = \frac{n-1}{1+n-1}$$

$$\boxed{v' = \frac{(n-1)}{n}} \quad \text{or} \quad \boxed{n = \frac{1}{1-v'}} \qquad \text{(VII-20)}$$

Thus the v' coordinate of the peak shifts to the right as n increases. The $v'/[S]^*$ coordinate is:

$$\frac{v'}{[S]^*} = \frac{\frac{(n-1)}{n} V_{max}}{(n-1)^{1/n} K_S^*} = \frac{(n-1)^{n-1/n} V_{max}}{n K_S^*} \qquad \text{(VII-21)}$$

The height of the peak decreases in the presence of inhibitors that increase $[S]_{0.5}$ (i.e., increase $K^*_{S_{app}}$) and increases in the presence of activators that decrease $[S]_{0.5}$. The most useful parameter is the v' coordinate, from which the apparent n value can be estimated. Again, it should be stressed that the n value obtained in this manner is only an apparent value (n_{app}). The derivation assumes that the enzyme obeys the Hill equation. Consequently, n_{app} does not represent the actual number of sites unless the cooperativity is extremely high.

So far we have seen several methods of estimating n_{app}: (a) from the $[S]_{0.9}/[S]_{0.1}$ ratio, (b) from the v' value of the inflection point of the velocity curve, (c) by fitting the data to plots of $1/v$ versus $1/[S]^n$, and (d) from the v' value of the peak on the Eadie-Scatchard plot. A fifth, and the most widely used, method is the logarithmic Hill plot described below.

The Hill Plot—Logarithmic Form of the Hill Equation

The simplified velocity equation can be converted to a useful linear form as shown below.

$$\frac{v}{V_{max}} = \frac{[S]^n}{K' + [S]^n}, \qquad V_{max}[S]^n = vK' + v[S]^n$$

$$\frac{[S]^n (V_{max} - v)}{v} = K', \qquad n \log[S] + \log \frac{V_{max} - v}{v} = \log K'$$

$$\boxed{\log \frac{v}{V_{max} - v} = n \log[S] - \log K'} \qquad \text{(VII-22)}$$

A plot of $\log v/(V_{max} - v)$ versus $\log[S]$ is a straight line with a slope of n. When $\log v/(V_{max} - v) = 0$, $v/(V_{max} - v) = 1$ and the corresponding position on the $\log[S]$ axis gives $\log[S]_{0.5}$. The K' may be calculated from the relationship $K' = [S]^n_{0.5}$.

Theoretically, the Hill plot is linear over the entire range of substrate concentration (by virtue of the derivation which assumes no intermediates between E and ES_n). With experimental data, the Hill plot usually deviates from linearity at low specific velocities, where complexes containing less than n molecules of substrate contribute significantly to the initial velocity. The limiting slope at very low substrate concentrations (which may never be observed experimentally) is 1.0. On the other hand, if the enzyme contains noncatalytic regulatory sites that must be occupied before the substrate can bind to the catalytic site, the slope of the Hill plot will increase as the substrate concentration decreases. At very low [S], the slope will approach the number of sites that must be occupied before any reaction occurs. Deviations from linearity at high substrate concentrations are also common, mainly because of the difficulty in detecting small changes in the velocity as [S] is increased, but also if catalytic sites of different affinities contribute to the observed velocity.

Hill plots can be constructed by plotting $\log v/(V_{max} - v)$ versus $\log[S]$ on a linear scale. The slope (n) can then be read directly from the plot. However, it is usually more convenient to plot $v/(V_{max} - v)$ versus [S] directly on a log-log scale. When $v/(V_{max} - v) = 1.0$, then $v = 0.5\,V_{max}$ and the corresponding [S] gives $[S]_{0.5}$. If the decades of the log-log scale are the same size on both axes, the slope can be determined by measuring suitable vertical and horizontal distances with a ruler. If the slope is determined using the points corresponding to $[S]_{0.9}$ and $[S]_{0.1}$, n will be the same as that calculated from equation VII-12. However, because the plot may be nonlinear over the entire range of $v = 0.1\,V_{max}$ to $0.9\,V_{max}$, n is usually determined as the slope of the Hill plot at $v = 0.5\,V_{max}$. This n value can be designated n_H or n_{app}.

Figure VII-9 shows the Hill plots for three different four-site enzymes, each with the same interaction factors ($a = b = c = 0.1$), but with different intrinsic K_S values. As K_S increases, the curves move to the right. The $[S]_{0.5}$ and K' values increase, but the slopes remain the same. Figure VII-10 shows the Hill plots for three different enzymes where $n = 1$, $n = 2$, and $n = 4$. The intrinsic K_S is the same for all three enzymes. The interaction factors (a, for the $n = 2$ enzyme, and a, b, and c for the $n = 4$ enzyme) are all 0.1. As n increases, the curves move to the left; $[S]_{0.5}$ and K' decrease.

If equilibrium substrate binding is measured instead of initial velocities, $\log Y_S/(1 - Y_S)$ is plotted, where Y_S is the fraction of total sites occupied. In both initial velocity and binding studies the [S] variable plotted must be *free* [S]. In fact, a Hill plot using the *total* substrate concentration can have an abnormally high slope at low $[S]_t$ if a significant fraction of the total substrate is bound.

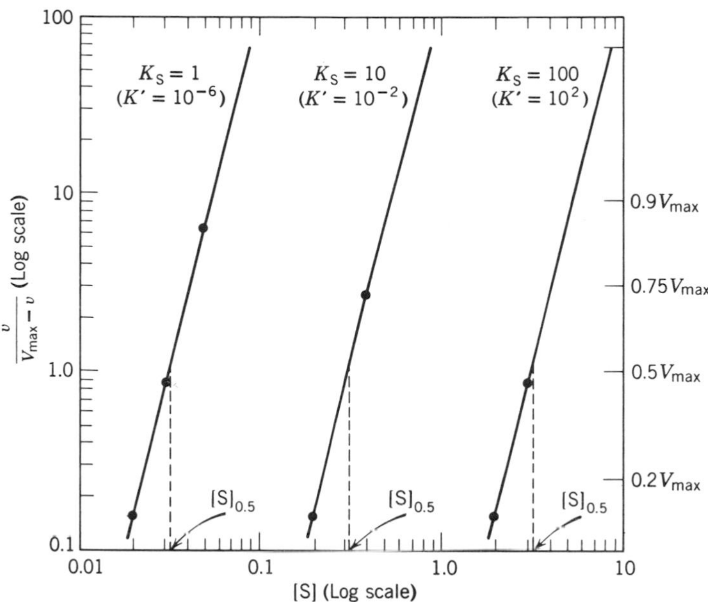

Fig. VII-9. Hill plots for enzymes with $n_{app} = 4$. The slope is 4.0 for all curves. The position of the curve depends on the value of K_S; $a = b = c = 0.1$. $[S]_{0.5} = \sqrt[4]{K'}$, $K' = a^3b^2cK_S^4$.

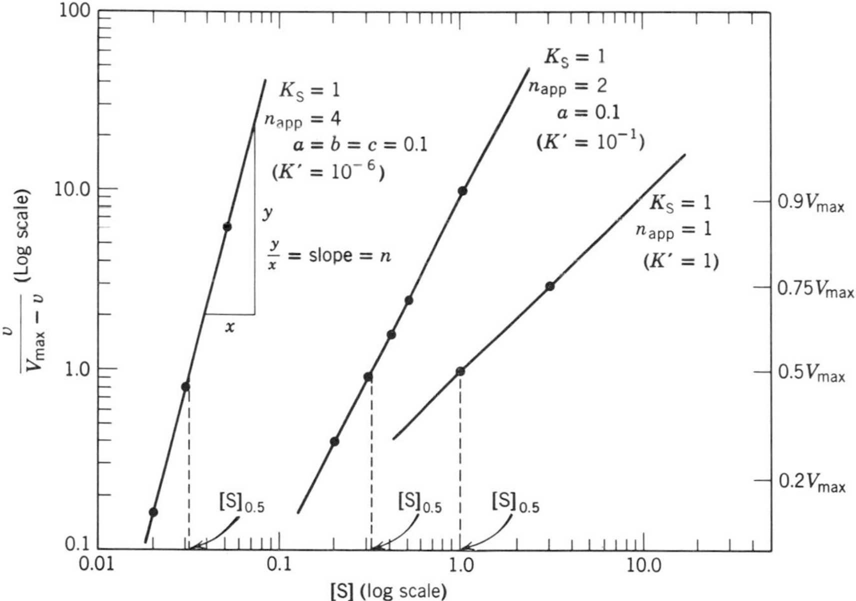

Fig. VII-10. Hill plots for enzymes with different n_{app} values and the same intrinsic K_S.

Summary of Different Uses of the Symbol n

The symbol n is used in different ways with respect to allosteric enzymes:

n — n refers to the actual number of ligand binding sites on a molecule of enzyme. Usually, n can be determined by equilibrium ligand binding studies. The data are plotted as $[S]_b/[S]_f[E]_t$ versus $[S]_b/[E]_t$ (Scatchard plot, Chapter Four) and n is obtained as the horizontal-axis intercept, that is, "moles ligand bound per mole of enzyme at saturating ligand concentration." Scatchard plots for allosteric enzymes exhibiting positive cooperativity are curved, as shown in Figure VII-8.

n_H — n_H is the slope of the Hill plot usually measured at the point where $\log v/(V_{max} - v) = 0$; that is, where $v = 0.5\,V_{max}$. Effectors that alter $[S]_{0.5}$ may change n_H.

n_{app} — n_{app} is a value calculated from any of the following:

1. The $[S]_{0.9}/[S]_{0.1}$ ratio where

$$\frac{[S]_{0.9}}{[S]_{0.1}} = \sqrt[n]{81}$$

2. The slope of the Hill plot measured between the points corresponding to $[S]_{0.9}$ and $[S]_{0.1}$, that is, between $v/(V_{max} - v) = 9$ and $v/(V_{max} - v) = 0.111$ on a log-log scale
3. The v' value of the inflection point of the v versus [S] curve, where

$$v'_{\text{inflection}} = \frac{(n-1)}{2n}$$

4. The v' value of the peak of the Eadie-Scatchard plot of $v/[S]$ versus v, where

$$v'_{\text{peak}} = \frac{(n-1)}{n}$$

5. The trial-and-error value of n that linearizes the $1/v$ versus $1/[S]^n$ plot.

Generally, n_{app} and n_H will be identical or very close.

$n_{H_{max}}$ or $n_{app_{max}}$

$n_{H_{max}}$ or $n_{app_{max}}$ is the theoretical maximum slope of the Hill plot, which may occur in the region of 0.5 V_{max} or at an extreme value of v, depending on the system.

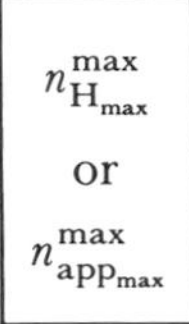

The maximum value of $n_{H_{max}}$ or $n_{app_{max}}$ among the family of Hill plots obtained in the presence of different fixed concentrations of effectors.

Effect of Interaction Strengths on the Velocity Curve

Figure VII-11 shows a plot of v versus [S] for three different enzymes, each with four identical sites (with identical intrinsic dissociation constants). For enzyme A, the interaction factors are taken to be $a=b=c=0.05$; for enzyme B, $a=b=c=0.1$, and for enzyme C, $a=b=c=0.2$. Curves A, B, and C were calculated using the complete velocity equation. Curves A', B', and C' were calculated using the Hill equation (i.e., the terms containing powers of $[S]/K_S$ less than 4 were ignored). We see that as the strength of the interactions decreases (i.e., as a, b, and c increase), the complete equation and the Hill equation yield less comparable results. This is to be expected because the Hill equation deals only with the E and ES_4 forms, while the complete equation accounts for all enzyme-substrate complexes. As the strength of the interactions decreases, a smaller fraction of the enzyme will exist as the ES_4 form at any given substrate concentration. We also see that as the strength of the interactions decreases, the substrate concentration required for any given velocity increases. For example, when $a=b=c=0.05$, $[S]_{0.5}=0.011K_S$. When $a=b=c=0.1$, $[S]_{0.5}$ increases to $0.032K_S$, and when $a=b=c=0.20$, $[S]_{0.5}$ is $0.09K_S$. Again, the results are expected. The initial velocity at any substrate concentration is proportional to the concentration of occupied sites. As the strength of the interactions decreases, a smaller fraction of the total enzyme will have multiple sites occupied at any given substrate concentration. To state it another way: as the interaction strength increases a lower substrate concentration is needed to obtain any given concentration of a multiply occupied species.

The sigmoidicity of the v versus [S] plot (i.e., the cooperativity of substrate binding) also changes as the interaction factors change. When $a=b=c=$

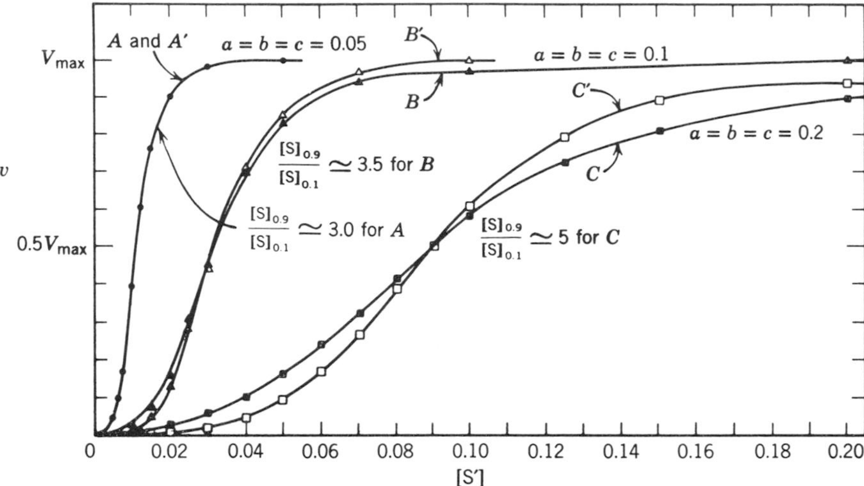

Fig. VII-11. Effect of the interaction factors on the sigmoidicity and $[S]_{0.5}$ of a four-site enzyme. Curve *A*: $a=b=c=0.05$. Curve *B*: $a=b=c=0.1$. Curve *C*: $a=b=c=0.2$. Curves *A*, *B*, and *C* were calculated using the complete velocity equation (VII-7). Curves *A'*, *B'*, and *C'* were calculated from the corresponding Hill equation (VII-9), i.e., only the terms corresponding to E and ES_4 are taken into account.

0.05, the $[S]_{0.9}/[S]_{0.1}$ ratio is close to 3.0 (the theoretical minimum ratio for a four-site enzyme), and the n_{app} or n_H value is very close to 4. When $a=b=c=0.2$, the $[S]_{0.9}/[S]_{0.1}$ ratio is about 5 and n_H is about 2.8. If we continue to decrease the interaction strengths we would eventually obtain a curve with an n_{app} of about 2.0. In fact, with the proper combination of interaction factors (not necessarily $a=b=c$) a four-site enzyme could yield a velocity curve indistinguishable from that for an enzyme with two sites having the same intrinsic K_S. To recapitulate, the sigmodicity and the $[S]_{0.5}$ depend on two factors: (*a*) the actual number of ligand binding sites, and (*b*) the strength of the interactions between them. The velocity curve, Hill plot, or Eadie-Scatchard plot, and so on, do not allow us to determine each of these two factors separately, but they can provide an indication of the *combined effect* of the two factors. Thus *if we fit the experimental velocity data to the Hill equation*, we can calculate an *apparent n* value and a K' value. Keep in mind that the Hill equation was derived assuming that only a single multiply occupied species is present besides free E. The equation ignores the contributions toward the observed velocity of species containing less than *n* molecules of substrate. This contribution may be significant if the interactions are weak, thus leading to nonintegral values of n_{app} or n_H. If *n*, the actual number of sites, is known, then curve-fitting procedures can be used to estimate the interaction factors. Although we have assumed that $a = b = c$ in plotting Figure VII-11, the interaction factors need not be identical. For that matter, the intrinsic K_S values for the different sites might also vary.

Negative Cooperativity

So far, we have considered allosteric enzymes that exhibit positive cooperativity with respect to substrate binding. Negative cooperativity is also possible. In this case, the binding of each substrate molecule *decreases* the intrinsic affinities of the vacant sites. For example, consider an enzyme with four identical active sites with intrinsic dissociation constants K_S. Assume that the binding of each substrate molecule increases the intrinsic K_S of the vacant sites by a factor of 2.0. As we might predict, the velocity curve (curve B, Fig. VII-12) rises rapidly at low substrate concentrations, but then slopes off markedly as the affinities of the vacant sites decrease. The curve is not sigmoidal, so we cannot describe its "sigmoidicity." Nevertheless, we can describe its shape by the ratio of substrate required for any two fractions of

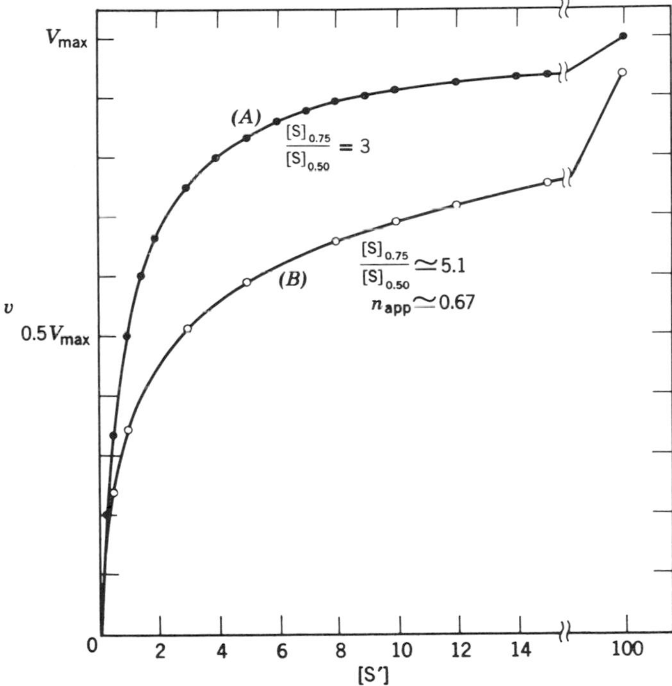

Fig. VII-12. The v versus [S] plots for a four-site enzyme. Curve A: no cooperativity. Curve B: negative cooperativity, four sites: $a = b = c = 2$.

V_{max}. For example, the $[S]_{0.75}/[S]_{0.5}$ ratio is approximately 5.1, compared to 3.0 for a normal hyperbolic velocity curve (curve A, Fig. VII-12). We can fit the data to the Hill equation and calculate an apparent value of n.

$$\frac{[S]_{0.75}}{[S]_{0.5}} = \sqrt[n]{3} \qquad \text{or} \qquad \log \frac{[S]_{0.75}}{[S]_{0.5}} = \frac{\log 3}{n}$$

For the negatively cooperative enzyme:

$$\frac{[S]_{0.75}}{[S]_{0.5}} = 5.17 \qquad n = \frac{\log 3}{\log 5.17} = \frac{0.477}{0.714} = 0.67$$

The reciprocal plot is shown in Figure VII-13a. The linear region at high $1/[S]$ values corresponds to the normal hyperbolic saturation of the first (normal affinity) site at low substrate concentrations. As the substrate concentration increases, the second, third, and fourth sites fill (with lower affinities). The Eadie-Scatchard plot is shown in Figure VII-13b. The reciprocal plot and the Eadie-Scatchard plot are indistinguishable from the plots obtained for a mixture of enzymes with different substrate affinities, or

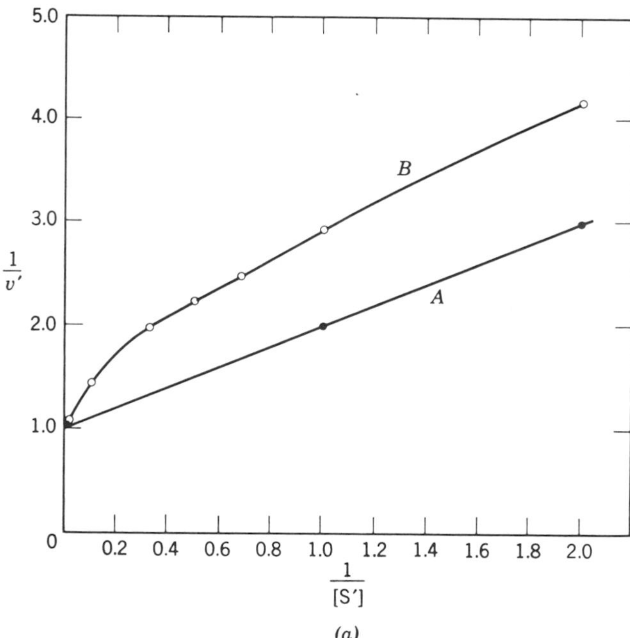

(a)

Fig. VII-13. Negative cooperativity in a four-site enzyme: $a = b = c = 2$. (a) $1/v$ versus $1/[S]$; Curve A: no cooperativity; Curve B: negative cooperativity.

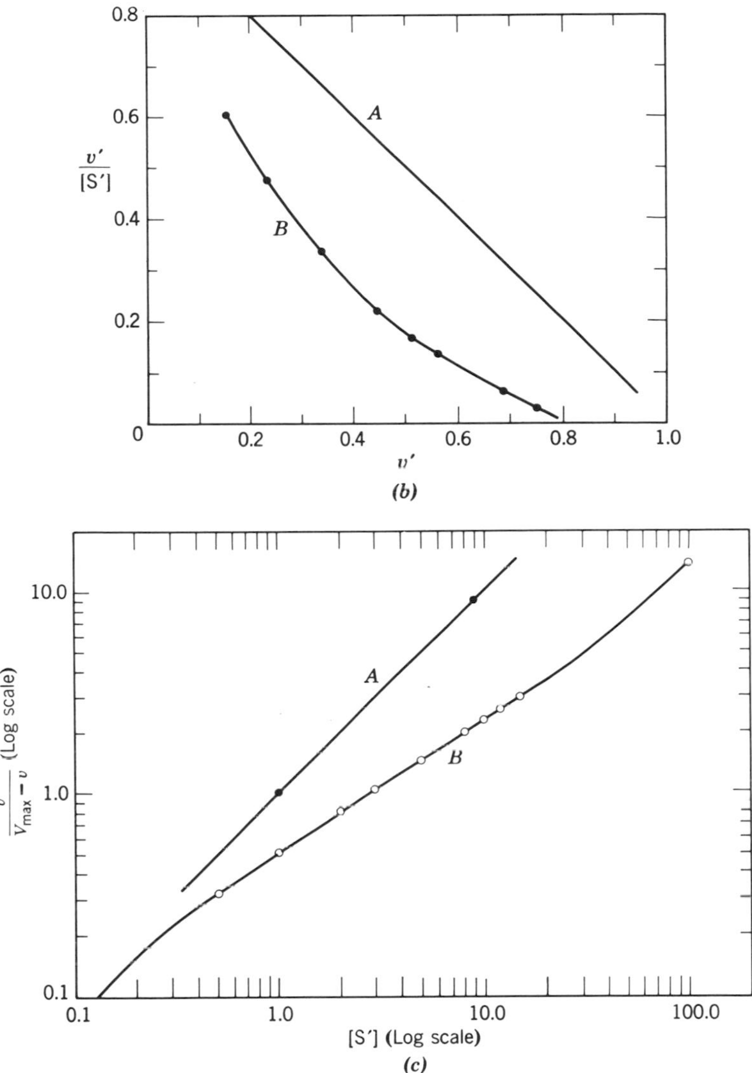

Fig. VII-13. (*Cont.*) (*b*) $v/[S]$ versus v; Curve *A*: no cooperativity; Curve *B*: negative cooperativity. (*c*) Hill plot; Curve *A*: no cooperativity; Curve *B*: negative cooperativity.

one enzyme with multiple sites of different affinities. The Hill plot (Fig. VII-13*c*) has a slope of 0.67 measured between the points corresponding to $0.5\,V_{max}$ and $0.75\,V_{max}$, but approaches a slope of 1.0 at very high and very low substrate concentrations.

A physiological role for a negatively cooperative enzyme would not be hard to imagine. With the proper combination of interaction factors (e.g., $a=1$, b and $c>1$) we would have an enzyme that responds almost normally to variations in [S] at low substrate ranges, but exhibits marked substrate inhibition at high substrate concentrations. This would tend to limit the initial velocity to a narrower than normal range.

Figure VII-14*a* shows the velocity curve for a four-site enzyme which shows positive cooperativity for the second substrate molecule bound, negative cooperativity for the third, and positive cooperativity for the fourth ($a=0.1$, $b=100$, $c=0.1$). At substrate concentrations from $0.1K_S$ to $10K_S$, the predominant enzyme form is the ES_2 complex. [Between $[S]=0.1K_S$ and $[S]=10K_S$, the $([S]/K_S)^2$ term is the largest term in the numerator of the velocity equation.] The presistence of the ES_2 form results in the unusual velocity curve. The reciprocal plot (Fig. VII-14*b*), and the Hill plot (Fig. VII-14*c*) reflect the intermediate negatively cooperative step.

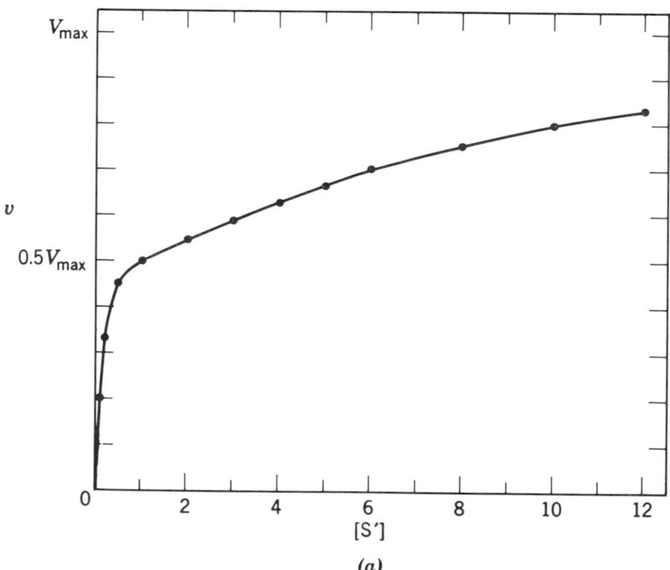

(a)

Fig. VII-14. Positive-negative-positive cooperativity: $a=0.1$, $b=100$, $c=0.1$. (*a*) v versus [S].

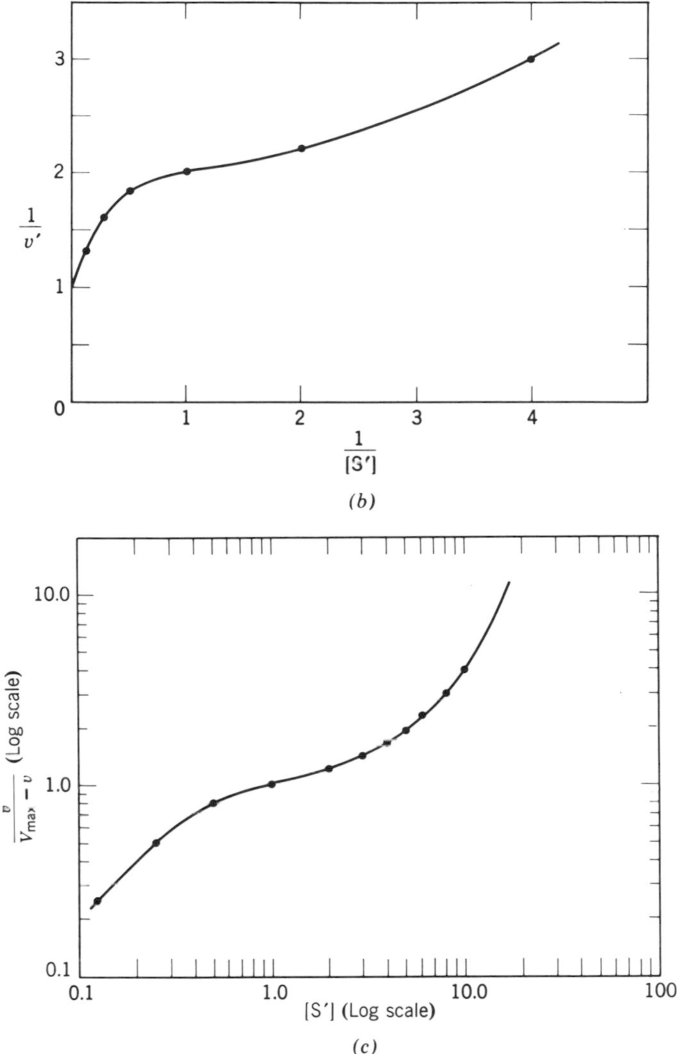

Fig. VII-14. (*Cont.*) (*b*) $1/v$ versus $1/[S]$. (*c*) Hill plot.

Negative cooperativity for one or more binding steps has been observed for a number of allosteric enzymes, including deoxythymidine kinase, homoserine dehydrogenase, isocitrate dehydrogenase, glutamic dehydrogenase, human heart lactic dehydrogenase, phosphoenolpyruvate carboxylase, and triose phosphate dehydrogenase.

Interaction that Affects V_{max}

Two occupied substrate sites may interact to increase their respective catalytic activities without affecting the intrinsic substrate dissociation constants. The equilibria shown below describe a system in which the SES complex is ten-times more active than either ES or SE (i.e., each site in the doubly occupied complex is five-times more active than the same site in a singly occupied complex).

$$\begin{array}{ccccccc}
E & + & S & \overset{K_S}{\rightleftharpoons} & ES & \xrightarrow{k_p} & E + P \\
+ & & & & + & & \\
S & & & & S & & \\
{\scriptstyle K_S}\updownarrow & & & & \updownarrow{\scriptstyle K_S} & & \\
SE & + & S & \overset{K_S}{\rightleftharpoons} & SES & \xrightarrow{10k_p} & SE + P \\
\downarrow{\scriptstyle k_p} & & & & & & (ES + P) \\
E + P & & & & & &
\end{array}$$

The velocity equation is:

$$v = \frac{\dfrac{2[S]}{K_S} k_p[E]_t + \dfrac{[S]^2}{K_S^2} 10k_p[E]_t}{1 + \dfrac{2[S]}{K_S} + \dfrac{[S]^2}{K_S^2}}$$

or

$$\boxed{\frac{v}{V_{max}} = \frac{\dfrac{0.2[S]}{K_S} + \dfrac{[S]^2}{K_S^2}}{1 + \dfrac{2[S]}{K_S} + \dfrac{[S]^2}{K_S^2}}} \qquad \text{(VII-23)}$$

where $V_{max} = 10k_p[E]_t$.

At low substrate concentrations the observed velocity results mainly from the activities of the singly occupied SE and ES complexes. As [S] increases, the far more productive SES complex builds up. As a result, the velocity curve (curve *A*, Fig. VII-15) is sigmoidal. Curve *B* would be obtained if it were possible to dissociate the cooperative enzyme into subunits, each carrying one site that retained its intrinsic catalytic activity. For comparison, curve *C* shows the hyperbolic response of a two-site enzyme with no cooperativity, but with the same intrinsic K_S and V_{max} of the cooperative

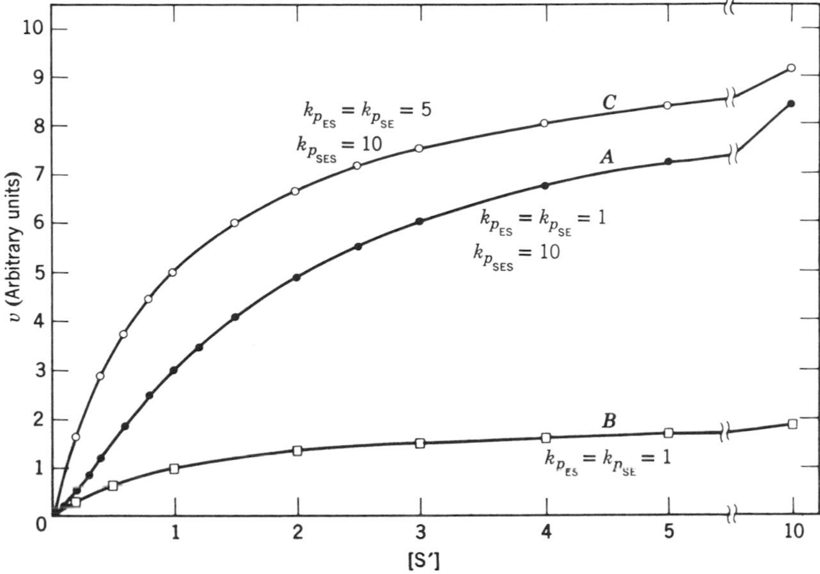

Fig. VII-15. Curve A: v versus [S] plot for a two-site enzyme where the interaction results in a fivefold increase in catalytic activity of each site ($b=5$ or $\beta=10$). Curve B: v versus [S] plot if the subunits were dissociated preventing interaction, or if no interaction occurred in the original dimer. Curve C: v versus [S] plot for a noncooperative dimer with a V_{max} equal to that of the cooperative dimer.

enzyme (each site has a catalytic activity of five arbitrary units). The Hill plot for curve A is curved with limiting slopes of 2.0 at very low substrate concentrations, and 1.0 at very high substrate concentrations. Interactions that change K_S as well as k_p are, of course, also possible. The general equilibria and velocity equation for a two-site enzyme are shown below.

$$
\begin{array}{ccccccc}
\mathrm{E} + \mathrm{S} & \overset{K_S}{\rightleftharpoons} & \mathrm{ES} & \xrightarrow{k_p} & \mathrm{E} + \mathrm{P} \\
+ & & + & & \\
\mathrm{S} & & \mathrm{S} & & \\
K_S \upharpoonleft\downharpoonright & & \downharpoonleft\upharpoonright aK_S & & \\
\mathrm{SE} + \mathrm{S} & \overset{aK_S}{\rightleftharpoons} & \mathrm{SES} & \xrightarrow{bk_p} & \mathrm{SE} + \mathrm{P} \\
k_p \downarrow & & \downarrow bk_p & & \\
\mathrm{E} + \mathrm{P} & & \mathrm{ES} + \mathrm{P} & & \\
\end{array}
$$

$$
\text{or} \quad \mathrm{SES} \xrightarrow{\beta k_p} \mathrm{ES}\ (\text{or SE}) + \mathrm{P}
$$

where b represents the factor by which the intrinsic catalytic rate constant of each site is changed in the SES complex, or β represents the factor by which the effective catalytic rate constant of the SES complex is changed relative to that of the singly occupied ES or SE. If there is no change in the intrinsic catalytic rate constant, $b=1$ (or $\beta=2$). The general velocity equation is:

$$\frac{v}{[\mathrm{E}]_t} = \frac{k_p \dfrac{2[\mathrm{S}]}{K_\mathrm{S}} + 2bk_p \dfrac{[\mathrm{S}]^2}{aK_\mathrm{S}^2}}{1 + \dfrac{2[\mathrm{S}]}{K_\mathrm{S}} + \dfrac{[\mathrm{S}]^2}{aK_\mathrm{S}^2}} \qquad \text{(VII-24)}$$

where $2bk_p[\mathrm{E}]_t = \beta k_p[\mathrm{E}]_t$ = the velocity observed at saturating [S] when all the enzyme is driven to the SES complex. The shape of the curve depends on the values of a and b and on the b/a ratio (Fig. VII-16). If we take the first derivative of the equation we will find that the curve will pass through a maximum when $b<0.5$ (i.e., $\beta<1$). (See the analysis based on equations IX-182 and IX-187 to IX-189 for random bireactant systems.) In other words, if the SES complex is less active catalytically than the singly occupied species, the velocity will increase at first as ES and SE build up but then decrease as a significant fraction of the enzyme is converted to the less active

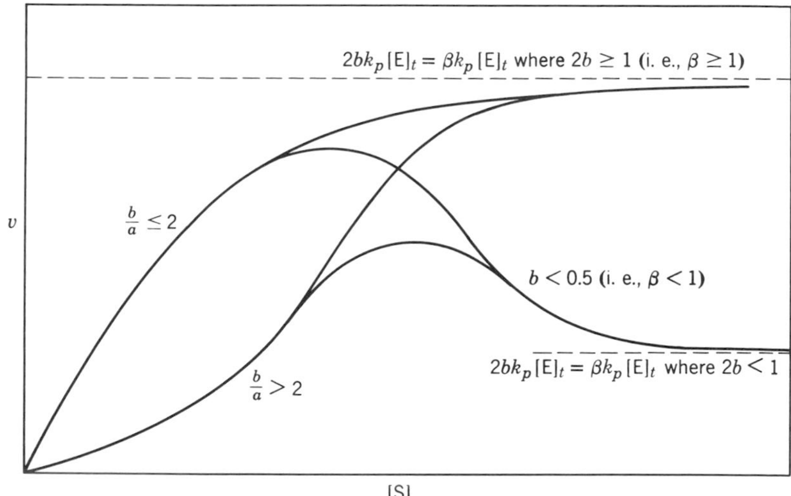

Fig. VII-16. Shapes of v versus [S] plot of a two-site enzyme as affected by the ratio of interaction factors; a represents the substrate binding interaction factor; b represents the catalytic activity interaction factor of a site; β represents the factor by which the catalytic activity of the SES complex increases relative to the catalytic activity of an ES_1 complex ($\beta=2b$).

SES. The limiting velocity is $\beta k_p[\mathrm{E}]_t$ (which could be zero if SES is catalytically inactive). The second derivative of the equation shows that the curve will be sigmoidal when the ratio of b/a is >2. Thus a could be <1 leading to cooperative *substrate binding*, but a sigmoidal *velocity response* will not be observed if at the same time the intrinsic catalytic rate constants are decreased so that b/a is not >2. On the other hand, the velocity curve will be sigmoidal even if there is no cooperative substrate binding ($a=1$) if b is >2 (as in Fig. VII-15). When $b/a>2$ but $b<0.5$, the curve will have two inflection points and a maximum; that is, we will observe substrate activation at low [S] and substrate inhibition at high [S]. The curves shown in Figure VII-16 were sketched to illustrate the four basic types of velocity curves and do not correspond to any particular values of a and b. Inflection points and maxima on experimental curves may not be as obvious. For example, when $a=1$ and $b=5$ (Fig. VII-15), the inflection occurs at about $v=0.1\,V_{\max}$. Experimental error at low v could obscure the sigmoidicity of the curve.

B. INHIBITION IN MULTISITE SYSTEMS

System B1. Pure Competitive Inhibition, Exclusive at Both Substrate Sites ("Ligand Exclusion")

Consider a pure competitive situation in which the binding of an inhibitor to a single site prevents the substrate from binding to either of two identical sites. The competition can result from a distortion in the substrate binding sites caused by the binding of the inhibitor, or mutual steric hinderance, or, alternately, the inhibitor binding site may overlap or utilize part of each substrate binding site. The reaction sequence is:

$$
\begin{array}{ccccccccc}
 & & & & & \mathrm{SE} + \mathrm{P} & & & \\
 & & & & & \uparrow{\scriptstyle k_p} & & & \\
\mathrm{E} + \mathrm{P} & \xleftarrow{k_p} & \mathrm{SE} & + \mathrm{S} & \underset{}{\overset{K_\mathrm{S}}{\rightleftharpoons}} & \mathrm{SES} & \xrightarrow{k_p} & \mathrm{ES} + \mathrm{P} \\
 & & \upharpoonleft\!\downharpoonright{\scriptstyle K_\mathrm{S}} & & & {\scriptstyle K_\mathrm{S}}\downharpoonleft\!\upharpoonright & & \\
 & & \mathrm{S} & & & \mathrm{S} & & \\
 & & + & & & + & & \\
 & & \mathrm{E} & + \mathrm{S} & \overset{K_\mathrm{S}}{\rightleftharpoons} & \mathrm{ES} & \xrightarrow{k_p} & \mathrm{E} + \mathrm{P} \\
 & & + & & & & & \\
 & & \mathrm{I} & & & & & \\
 & & {\scriptstyle K_i}\upharpoonleft\!\downharpoonright & & & & & \\
 & & \mathrm{EI} & & & & &
\end{array}
$$

The velocity equation is:

$$\frac{v}{V_{max}} = \frac{\dfrac{[S]}{K_S} + \dfrac{[S]^2}{K_S^2}}{1 + \dfrac{2[S]}{K_S} + \dfrac{[S]^2}{K_S^2} + \dfrac{[I]}{K_i}} \qquad \text{(VII-25)}$$

In the absence of inhibitor, the velocity equation reduces to the Henri-Michaelis-Menten equation, and the velocity curve is the usual hyperbola. In the presence of inhibitor $[S]_{0.5}$ increases, and the velocity curve becomes sigmoidal (Fig. VII-17). As [I] increases, $[S]_{0.5}$ increases and the $[S]_{0.9}/[S]_{0.1}$ ratio decreases to a limit of 9 (i.e., n_{app} approaches a limit of 2). The v versus [I] curve at a fixed [S] remains hyperbolic with an $[I]_{0.9}/[I]_{0.1}$ ratio of 81. The same results are obtained if only one of the substrate binding sites is catalytically active but both must be occupied before catalysis occurs. The velocity equation would be identical to equation VII-25 without the $[S]/K_S$ numerator term. In this case, $V_{max} = k_p[E]_t$ instead of $2k_p[E]_t$. The "ligand exclusion" theory of inhibitor-induced sigmoidal behavior was first proposed by Fisher, Gates, and Cross (1970). The model has some interesting features.

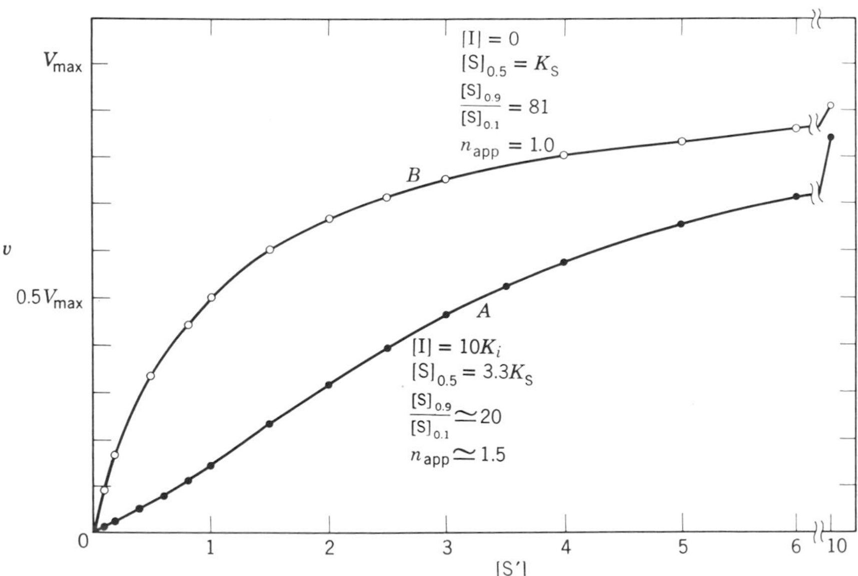

Fig. VII-17. Curve A: sigmoidal v versus [S] plot caused by substrate exclusion. In the absence of I, the plot is hyperbolic (Curve B).

For example, if the two substrate binding sites are on different subunits, and only one is catalytically active, then dissociation of the enzyme by physical or chemical reagents would yield a catalytic subunit that is insensitive to the inhibitor, and a "regulatory" subunit that binds substrate but is not catalytically active. If the inhibitor site on the native enzyme overlaps the two substrate sites, or utilizes part of each substrate site, then neither dissociated subunit would bind the inhibitor, although reassociation might occur faster in the presence of inhibitor.

If the enzyme displays cooperative substrate binding in the absence of inhibitor, then the inhibitor will increase the sigmoidicity of the velocity curve; $[I]_{0.9}/[I]_{0.1}$ will still be 81. However, if the cooperativity of substrate binding is relatively high to begin with (so that the apparent n value is close to the number of substrate binding sites), then a competitive inhibitor that increases $[S]_{0.5}$ may have little or no detectable effect on the $[S]_{0.9}/[S]_{0.1}$ ratio (or the slope of the Hill plot).

System B2. Inhibition Competitive at Two Sites

System B2a. Cooperative Substrate Binding; Inhibitor Mimics Substrate

If the substrate binds cooperatively with an interaction factor a, and a competitive inhibitor induces the same conformational change (i.e., inhibitor binding changes the dissociation constant of the vacant site for I or S by a), the equilibria are those shown in abbreviated form below.

$$
\begin{array}{ccccc}
\text{IEI} & \xrightarrow[aK_i]{} & \text{IE} & \xleftarrow[aK_S]{} & \text{IES} \\
\downarrow{\scriptstyle aK_i} & & \downarrow{\scriptstyle K_i} & & {\scriptstyle aK_i}\downarrow \\
\text{EI} & \xrightarrow{K_i} & \text{E} & \xleftarrow{K_S} & \text{ES} \\
\uparrow{\scriptstyle aK_S} & & \uparrow{\scriptstyle K_S} & & {\scriptstyle aK_S}\uparrow \\
\text{SEI} & \xrightarrow{aK_i} & \text{SE} & \xleftarrow{aK_S} & \text{SES}
\end{array}
$$

The velocity equation is:

$$
\frac{v}{V_{\max}} = \frac{\dfrac{[S]}{K_S} + \dfrac{[S]^2}{aK_S^2} + \dfrac{[S][I]}{aK_SK_i}}{1 + \dfrac{2[S]}{K_S} + \dfrac{[S]^2}{aK_S^2} + \dfrac{2[S][I]}{aK_SK_i} + \dfrac{2[I]}{K_i} + \dfrac{[I]^2}{aK_i^2}} \qquad \text{(VII-26)}
$$

where $V_{max} = 2k_p[\mathrm{E}]_t$. The interaction factor appears only in those terms corresponding to the doubly occupied complexes. As [I] increases, $[\mathrm{S}]_{0.5}$ increases (as expected) and the $[\mathrm{S}]_{0.9}/[\mathrm{S}]_{0.1}$ ratio approaches 81. At very low specific substrate and inhibitor concentrations, the increased (inhibitor-induced) substrate binding may more than compensate for the small degree of competitive inhibition, and the inhibitor, in effect, may activate the reaction.

System B2b. Cooperative Substrate Binding; Inhibitor Does Not Mimic Substrate

If the substrate binds cooperatively but the competitive inhibitor does not, the equilibria are those shown below.

$$
\begin{array}{ccccc}
\mathrm{IEI} & \xrightarrow[K_i]{} & \mathrm{IE} & \xleftarrow[K_\mathrm{S}]{} & \mathrm{IES} \\
\downarrow K_i & & \downarrow K_i & & K_i \downarrow \\
\mathrm{EI} & \xrightarrow{K_i} & \mathrm{E} & \xleftarrow{K_\mathrm{S}} & \mathrm{ES} \\
\uparrow K_\mathrm{S} & & \uparrow K_\mathrm{S} & & aK_\mathrm{S} \uparrow \\
\mathrm{SEI} & \xrightarrow{K_i} & \mathrm{SE} & \xleftarrow{aK_\mathrm{S}} & \mathrm{SES}
\end{array}
$$

To fulfill the equilibrium requirements of the model, the binding of the first substrate molecule must not affect the affinity of the vacant site for I, although the affinity of the site for S is altered to $1/aK_\mathrm{S}$. The velocity equation is:

$$
\frac{v}{V_{max}} = \frac{\dfrac{[\mathrm{S}]}{K_\mathrm{S}} + \dfrac{[\mathrm{S}]^2}{aK_\mathrm{S}^2} + \dfrac{[\mathrm{S}][\mathrm{I}]}{K_\mathrm{S}K_i}}{1 + \dfrac{2[\mathrm{S}]}{K_\mathrm{S}} + \dfrac{[\mathrm{S}]^2}{aK_\mathrm{S}^2} + \dfrac{2[\mathrm{S}][\mathrm{I}]}{K_\mathrm{S}K_i} + \dfrac{2[\mathrm{I}]}{K_i} + \dfrac{[\mathrm{I}]^2}{K_i^2}} \qquad \text{(VII-27)}
$$

where $V_{max} = 2k_p[\mathrm{E}]_t$. The interaction factor appears only in the term corresponding to the SES complex. The inhibitor increases $[\mathrm{S}]_{0.5}$, but the $[\mathrm{S}]_{0.9}/[\mathrm{S}]_{0.1}$ values remain constant. The v versus [I] curves become more cooperative as [S] increases.

System B2c. Inhibitor Binds Cooperatively, But Substrate Does Not

System B2c is the reverse of the situation described in System B2b. Now the competitive inhibitor induces structural changes that facilitate further inhibitor binding, but the affinity of the vacant site for S remains $1/K_S$. The equilibria are:

$$
\begin{array}{ccccc}
\text{IEI} & \underset{cK_i}{\longrightarrow} & \text{IE} & \underset{K_S}{\longleftarrow} & \text{IES} \\
\downarrow{\scriptstyle cK_i} & & \downarrow{\scriptstyle K_i} & & {\scriptstyle K_i}\downarrow \\
\text{EI} & \overset{K_i}{\longrightarrow} & \text{E} & \overset{K_S}{\longleftarrow} & \text{ES} \\
\uparrow{\scriptstyle K_S} & & \uparrow{\scriptstyle K_S} & & {\scriptstyle K_S}\uparrow \\
\text{SEI} & \overset{K_i}{\longrightarrow} & \text{SE} & \overset{K_S}{\longleftarrow} & \text{SES}
\end{array}
$$

The velocity equation is:

$$
\boxed{\frac{v}{V_{\max}} = \frac{\dfrac{[S]}{K_S} + \dfrac{[S]^2}{K_S^2} + \dfrac{[S][I]}{K_S K_i}}{1 + \dfrac{2[S]}{K_S} + \dfrac{[S]^2}{K_S^2} + \dfrac{2[S][I]}{K_S K_i} + \dfrac{2[I]}{K_i} + \dfrac{[I]^2}{cK_i^2}}} \qquad \text{(VII-28)}
$$

where $2k_p[E]_t = V_{\max}$. The interaction factor, c, appears only in the term corresponding to IEI. In the absence of inhibitor, the velocity curve is hyperbolic. The inhibitor increases $[S]_{0.5}$ and the cooperativity of the v versus [S] curve. However, the cooperativity of the v versus [I] curves do not change as [S] increases.

General Equation for the Two-Site Pure Competitive System

The pure competitive System B2 can be expressed in general terms as shown below:

$$
\begin{array}{ccccc}
\text{IEI} & \underset{cK_i}{\longrightarrow} & \text{IE} & \underset{bK_S}{\longleftarrow} & \text{IES} \\
\downarrow{\scriptstyle cK_i} & & \downarrow{\scriptstyle K_i} & & {\scriptstyle bK_i}\downarrow \\
\text{EI} & \overset{K_i}{\longrightarrow} & \text{E} & \overset{K_S}{\longleftarrow} & \text{ES} \\
\uparrow{\scriptstyle bK_S} & & \uparrow{\scriptstyle K_S} & & {\scriptstyle aK_S}\uparrow \\
\text{SEI} & \overset{bK_i}{\longrightarrow} & \text{SE} & \overset{aK_S}{\longleftarrow} & \text{SES}
\end{array}
$$

where a is the factor by which the substrate dissociation constant of the vacant site changes when the first molecule of substrate binds; b is the factor by which the substrate dissociation constant of the vacant site changes when the first molecule of inhibitor is bound (and vice versa); and c is the factor by which the inhibitor dissociation constant of the vacant site changes when the first molecule of inhibitor is bound. The general velocity equation would be:

$$\frac{v}{V_{max}} = \frac{\frac{[S]}{K_S} + \frac{[S]^2}{aK_S^2} + \frac{[S][I]}{bK_S K_i}}{1 + \frac{2[S]}{K_S} + \frac{[S]^2}{aK_S^2} + \frac{2[S][I]}{bK_S K_i} + \frac{2[I]}{K_i} + \frac{[I]^2}{cK_i^2}} \quad \text{(VII-29)}$$

In System B2a, all three interaction factors are <1 and $a=b=c$. In System B2b, $a<1$, and $b=c=1$. In System B2c, $a=b=1$ and $c<1$. Obviously, a variety of pure competitive two-site systems can be devised by assuming various combinations of values for the interaction factors. The relative values of the interaction factors will determine whether the cooperativity of the v versus [S] and v versus [I] curves increase, decrease, or remain constant as the concentration of the alternate ligand is varied. System B2 is described in terms of the general sequential interaction model of Koshland, Nemethy and Filmer in Section C.

System B3. Partial Competitive Inhibition

Systems B3a through System B3e assume two substrate binding sites and one effector binding site. This combination of sites represents the minimal unit required for the variety of allosteric effects that are described below. Also, by restricting ourselves to a minimal three-site unit, we are able to visualize the equilibria between the various complexes by means of a simple equilibrium cube. The different dissociation constants can then be written easily. Physically, the minimal unit could be a monomer with two substrate sites and one effector site. A more likely minimal unit model would place the two substrate sites on two different catalytic subunits and the effector site on a distinct regulatory subunit. Since trimers are not fashionable in quaternary protein structure, the association of two independent minimal trimers could yield a more stable hexameric enzyme. As long as the interactions were restricted to the minimal trimer, the equations and curves developed for Systems B3a through System B3e would still hold (i.e., n_{app} values would not exceed 2.0 in spite of the fact that the oligomer package contains four catalytic subunits).

System B3a. Cooperative Substrate Binding in the Presence or Absence of Inhibitor

Figure VII-18 shows the equilibria for an enzyme with cooperative substrate binding in the absence of inhibitor. The inhibitor increases K_S by a factor, α, but has no effect on the interaction factor. Consequently, when the first substrate molecule binds to EI, the dissociation constant of the vacant site is changed (from αK_S) by the factor a (to $a\alpha K_S$). The velocity equation is:

$$\frac{v}{V_{\max}} = \frac{\dfrac{[S]}{K_S} + \dfrac{[S]^2}{aK_S^2} + \dfrac{[I][S]}{\alpha K_i K_S} + \dfrac{[I][S]^2}{a\alpha^2 K_i K_S^2}}{1 + \dfrac{2[S]}{K_S} + \dfrac{[S]^2}{aK_S^2} + \dfrac{[I]}{K_i} + \dfrac{2[I][S]}{\alpha K_i K_S} + \dfrac{[I][S]^2}{a\alpha^2 K_i K_S^2}} \qquad \text{(VII-30)}$$

The velocity curve is shown in Figure VII-19. The slope of the Hill plot at $v = 0.5\,V_{\max}$ is about 1.3 for all three curves; that is, the cooperativity is

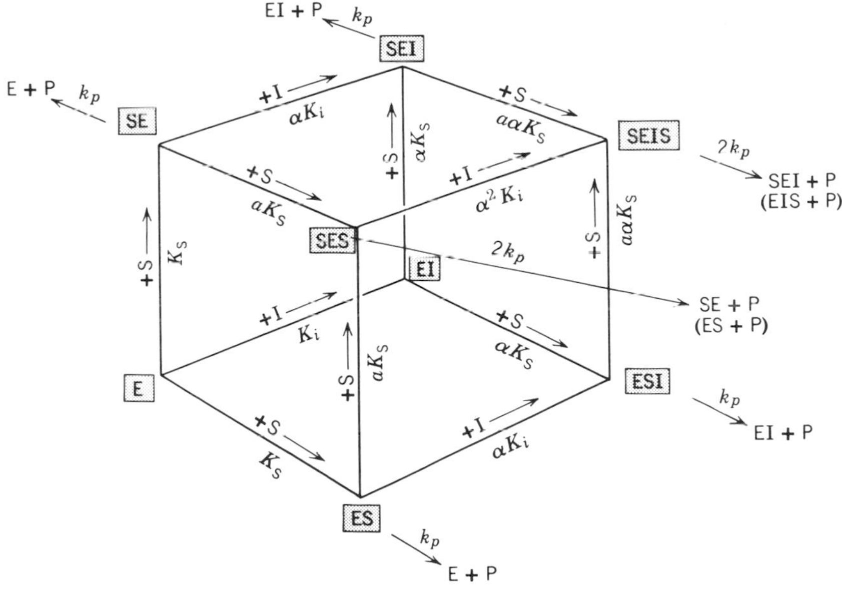

Fig. VII-18. Equilibria among enzyme species of a trimer unit consisting of two substrate sites and one effector site; S binds cooperatively with an interaction factor a; I increases the dissociation constant of the S sites by a factor α, but has no effect on the interaction between substrate sites.

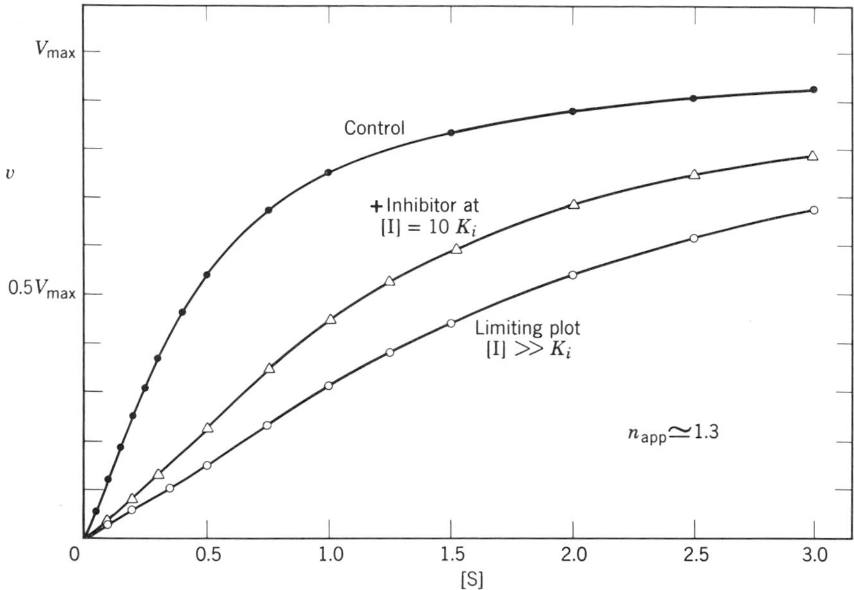

Fig. VII-19. System B3a; S binds cooperatively to a two-site enzyme with an interaction factor of a; I is a partial competitive inhibitor; I changes the dissociation constant for S to αK_S but has no effect on the interaction between the two S sites.

unchanged. At an infinitely high inhibitor concentration, all the enzyme will exist as the complexes shown on the rear face of the cube in Figure VII-18 and the velocity equation reduces to:

$$\boxed{\frac{v}{V_{max}} = \frac{\dfrac{[S]}{\alpha K_S} + \dfrac{[S]^2}{a\alpha^2 K_S^2}}{1 + \dfrac{2[S]}{\alpha K_S} + \dfrac{[S]^2}{a\alpha^2 K_S^2}}} \qquad \text{(VII-31)}$$

(All terms not containing I were ignored, then the factor $[I]/K_i$ was canceled in the numerator and denominator.) The enzyme behaves as a new enzyme with the same interaction factor a, and a new dissociation constant, αK_S.

System B3b. Substrate Binding Is Noncooperative in the Absence of Inhibitor; Substrate Reverses Effect of Inhibitor

Consider an enzyme with two identical noncooperative substrate sites and a single inhibitor site. When the inhibitor binds, it induces a conformational change that increases the dissociation constants of both sites to αK_S, where $\alpha > 1$. However, when the first molecule of substrate binds (with the decreased affinity), the effect of the inhibitor is partially or completely abolished. The dissociation constant of the vacant substrate site is changed to $a\alpha K_S$, where $a < 1$. This model is essentially the partial competitive version of the ligand exclusion model (System B1). Thus in the presence of inhibitor, the previously noncooperative enzyme now displays cooperative substrate binding. The equilibria are shown in Figure VII-20. The velocity equation for the situation where $a = 1/\alpha$ is shown below. In this case, the first molecule of substrate bound restores the intrinsic K_S of the vacant site to its original value.

$$\frac{v}{V_{max}} = \frac{\dfrac{[S]}{K_S} + \dfrac{[S]^2}{K_S^2} + \dfrac{[S][I]}{\alpha K_S K_i} + \dfrac{[S]^2[I]}{\alpha K_S^2 K_i}}{1 + \dfrac{2[S]}{K_S} + \dfrac{[S]^2}{K_S^2} + \dfrac{[I]}{K_i} + \dfrac{2[S][I]}{\alpha K_S K_i} + \dfrac{[S]^2[I]}{\alpha K_S^2 K_i}} \qquad \text{(VII-32)}$$

The result is similar to that seen in System B1; namely, $[S]_{0.5}$ increases and the velocity curve becomes sigmoidal. However, the inhibition in System B3 is partial in that an infinitely high inhibitor concentration will drive the enzyme to the catalytically active SEI, EIS, and SEIS complexes shown on the rear face of the cube. The enzyme behaves as a new enzyme with an intrinsic substrate dissociation complex of αK_S and an interaction factor of a (which we have assumed equals $1/\alpha$). The velocity equation at infinite [I] reduces to:

$$\frac{v}{V_{max}} = \frac{\dfrac{[S]}{\alpha K_S} + \dfrac{[S]^2}{\alpha K_S^2}}{1 + \dfrac{2[S]}{\alpha K_S} + \dfrac{[S]^2}{\alpha K_S^2}} \qquad \text{(VII-33)}$$

In contrast, in System B1 and System B2 an infinitely high [I] will drive the velocity to zero.

System B3c. Cooperative Substrate Binding in the Absence of Inhibitor; Substrate Reverses Effect of Inhibitor

If S binds cooperatively and the first molecule of substrate bound changes the conformation of the enzyme to the higher affinity state (i.e., changes the

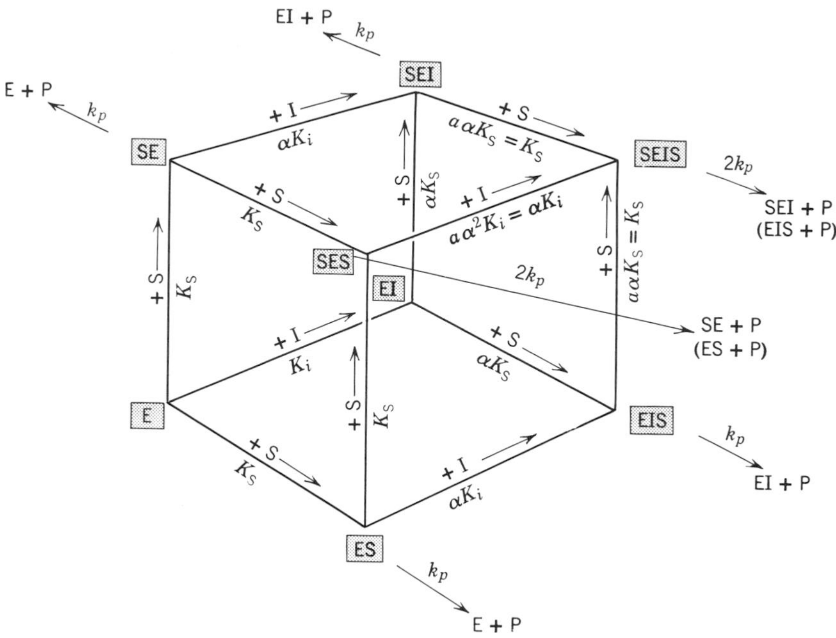

Fig. VII-20. System B3b; equilibria among enzyme species for an enzyme with two identical noncooperative substrate sites and one inhibitor site. I increases the dissociation constant for S to αK_S. The first molecule of S that binds overcomes the effect of I.

dissociation constant of the vacant site to aK_S) regardless of the presence of the inhibitor, the equilibria will be those shown in the rear cube of Figure VII-21. In effect, the substrate interaction factor is changed by the inhibitor from a to ba, where b (in this case) equals $1/\alpha$. The velocity equation is:

$$\frac{v}{V_{\max}} = \frac{\dfrac{[S]}{K_S} + \dfrac{[S]^2}{aK_S} + \dfrac{[S][I]}{\alpha K_S K_i} + \dfrac{[S]^2[I]}{a\alpha K_S^2 K_i}}{1 + \dfrac{2[S]}{K_S} + \dfrac{[S]^2}{aK_S^2} + \dfrac{2[S][I]}{\alpha K_S K_i} + \dfrac{[S]^2[I]}{a\alpha K_S^2 K_i} + \dfrac{[I]}{K_i}} \quad \text{(VII-34)}$$

where $V_{\max} = 2k_p[E]_t$. At an infinitely high inhibitor concentration, all the enzyme will be driven to those complexes shown on the back face of the rear

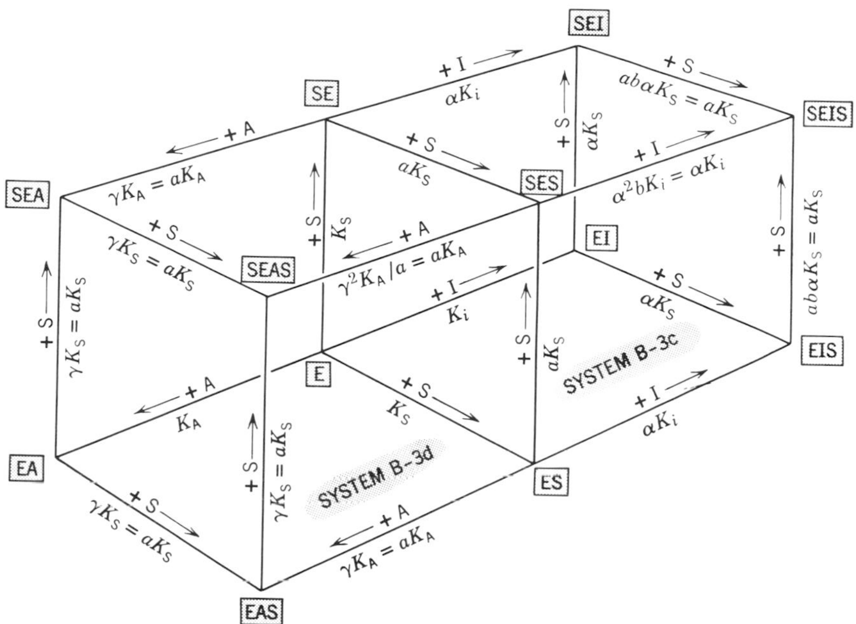

Fig. VII-21. System B3c (rear cube); S binds cooperatively; I increases the dissociation constant for S to αK_S. The first molecule of S changes the dissociation constant of the vacant S site to αK_S whether or not I is bound. System B3d (forward cube); A is an activator that excludes I and induces cooperative S binding; $\alpha > 1, b < 1, a < 1, \gamma = a, b = 1/\alpha$.

cube and the equation reduces to:

$$\frac{v}{V_{\max}} = \frac{\dfrac{[S]}{\alpha K_S} + \dfrac{[S]^2}{a\alpha K_S^2}}{1 + \dfrac{2[S]}{\alpha K_S} + \dfrac{[S]^2}{a\alpha K_S^2}} \qquad \text{(VII-35)}$$

The velocity curve is shown in Figure VII-22. In the presence of inhibitor, $[S]_{0.5}$ and the cooperativity increase. For $\alpha = 4$ and $a = 0.2$, the $[S]_{0.9}/[S]_{0.1}$ ratio at $[I] \gg K_i$ is about 16. The slope of the Hill plot is about 1.58.

System B3d. Activator Mimics Substrate and Excludes Inhibitor

The forward cube of Figure VII-21 describes a system in which an activator binds to a cooperative enzyme and increases the affinities of both vacant

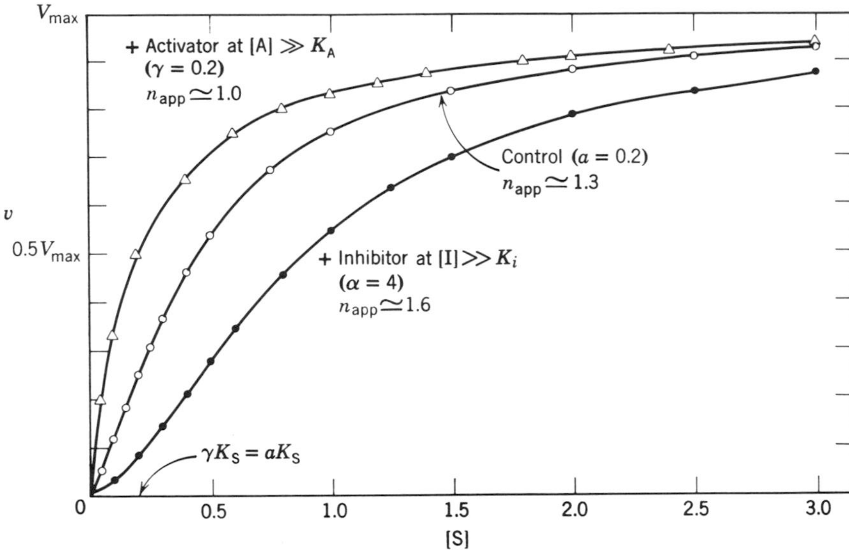

Fig. VII-22. v versus [S] plot for System B3c and System B3d.

substrate sites. The substrate dissociation constant changes from K_S to γK_S, where $\gamma < 1$. If the conformational change induced by the activator is identical to that induced by the substrate, then $\gamma = a$ and the first molecule of substrate bound has no further effect on K_S. It is assumed that the activator excludes the inhibitor either by binding to the inhibitor site, or by steric hinderance, or by having a binding site that overlaps or utilizes part of the inhibitor site. The velocity equation in the absence of inhibitor is:

$$\frac{v}{V_{max}} = \frac{\dfrac{[S]}{K_S} + \dfrac{[S]^2}{aK_S^2} + \dfrac{[S][A]}{aK_S K_A} + \dfrac{[S]^2[A]}{a^2K_S^2K_A}}{1 + \dfrac{2[S]}{K_S} + \dfrac{[S]^2}{aK_S^2} + \dfrac{2[S][A]}{aK_S K_A} + \dfrac{[S]^2[A]}{a^2K_S^2K_A} + \dfrac{[A]}{K_A}} \qquad \text{(VII-36)}$$

where $V_{max} = 2k_p[E]_t$. In the presence of an infinitely high activator concentration all the enzyme is driven to the complexes shown on the front face of the forward cube and the velocity equation reduces to:

$$\frac{v}{V_{\max}} = \frac{\dfrac{[S]}{aK_S} + \dfrac{[S]^2}{a^2K_S^2}}{1 + \dfrac{2[S]}{aK_S} + \dfrac{[S]^2}{a^2K_S^2}} = \frac{\dfrac{[S]}{aK_S}}{1 + \dfrac{[S]}{aK_S}} \qquad \text{(VII-37)}$$

Thus the velocity curve becomes hyperbolic as shown in Figure VII-22. In the presence of activator and inhibitor the velocity equation contains all the terms of equations VII-34 and VII-36.

$$\frac{v}{V_{\max}} = \frac{\left(\dfrac{[S]}{K_S} + \dfrac{[S]^2}{aK_S^2}\right) + \dfrac{[I]}{K_i}\left(\dfrac{[S]}{\alpha K_S} + \dfrac{[S]^2}{\alpha a K_S^2}\right) + \dfrac{[A]}{K_A}\left(\dfrac{[S]}{aK_S} + \dfrac{[S]^2}{a^2K_S^2}\right)}{\left(1 + \dfrac{2[S]}{K_S} + \dfrac{[S]^2}{aK_S^2}\right) + \dfrac{[I]}{K_i}\left(1 + \dfrac{2[S]}{\alpha K_S} + \dfrac{[S]^2}{\alpha a K_S^2}\right) + \dfrac{[A]}{K_A}\left(1 + \dfrac{2[S]}{aK_S} + \dfrac{[S]^2}{a^2K_S^2}\right)} \qquad \text{(VII-38)}$$

In the form above the expressions for the various types of complexes are set off by the parentheses. Each parenthetical term and the accompanying $[I]/K_i$ or $[A]/K_A$ factor relate to one of the vertical faces of the equilibrium cube viewed from the front. The $[S]_{0.5}$ and apparent n value will depend on the $[A]/[I]$ ratio.

System B3e. Inhibitor Eliminates Substrate Cooperativity

An inhibitor may "freeze" an enzyme in its original conformation and thereby prevent the substrate-induced or activator-induced increase in the affinity of vacant substrate sites. In this case, the inhibitor has no effect on the affinity of the enzyme for the first substrate molecule. The equilibria for this type of system is shown in Figure VII-23. The velocity equation is:

$$\frac{v}{V_{\max}} = \frac{\dfrac{[S]}{K_S} + \dfrac{[S]^2}{aK_S^2} + \dfrac{[S][I]}{K_S K_i} + \dfrac{[S]^2[I]}{K_S^2 K_i}}{1 + \dfrac{2[S]}{K_S} + \dfrac{[S]^2}{aK_S^2} + \dfrac{2[S][I]}{K_S K_i} + \dfrac{[S]^2[I]}{K_S^2 K_i} + \dfrac{[I]}{K_i}} \qquad \text{(VII-39)}$$

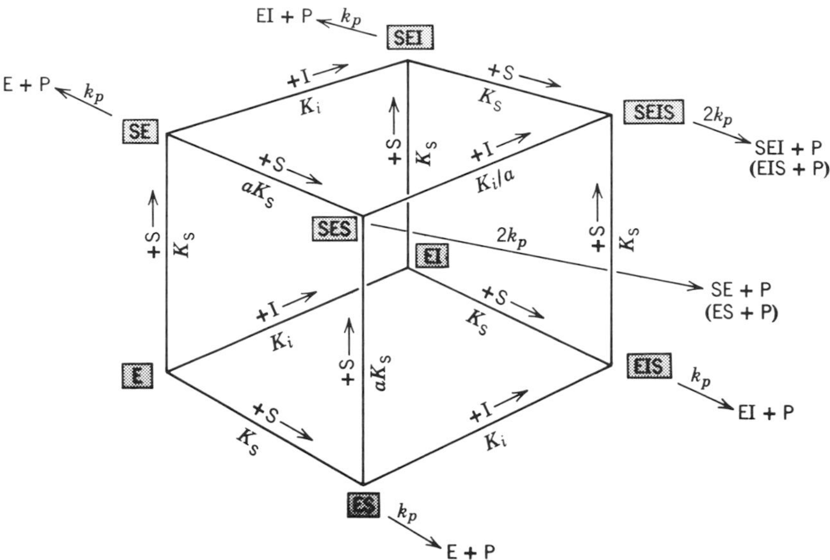

Fig. VII-23. Equilibria among enzyme species where I prevents the interaction between the two substrate sites. In the absence of I S binds cooperatively with an interaction factor a.

where $V_{max} = 2k_p[\mathrm{E}]_t$. Now only the $[\mathrm{S}]^2$ term contains the interaction factor. At an infinitely high inhibitor concentration the equation reduces to:

$$\frac{v}{V_{max}} = \frac{\dfrac{[\mathrm{S}]}{K_S} + \dfrac{[\mathrm{S}]^2}{K_S^2}}{1 + \dfrac{2[\mathrm{S}]}{K_S} + \dfrac{[\mathrm{S}]^2}{K_S^2}} = \frac{\dfrac{[\mathrm{S}]}{K_S}}{1 + \dfrac{[\mathrm{S}]}{K_S}}$$

Thus in the presence of inhibitor $[\mathrm{S}]_{0.5}$ increases and the velocity curve becomes less cooperative, with a limiting n_{app} value of 1.0.

System B4. Substrate Activation

So far, we have considered multisite enzymes in which all substrate binding sites are catalytically active. It is possible for an enzyme to possess substrate binding sites that play a regulatory role. Two examples are described below.

System B4a. Ordered Binding

Consider an enzyme with two substrate sites. One site is an effector site and is always accessible to the substrate, but has no catalytic activity. The catalytic site becomes accessible only after the substrate binds to the effector site and induces a conformational change in the enzyme. System B4a is an extreme example of cooperative binding as the catalytic site has no affinity at all for the substrate until the effector site is occupied. The equilibria describing this system are shown below.

$$\begin{array}{l} E + S \underset{}{\overset{K_{S_1}=K_S}{\rightleftharpoons}} SE \\ \qquad\qquad\quad + \\ \qquad\qquad\quad S \\ \qquad\qquad\quad \Big\downarrow\!\!\Big\uparrow K_{S_2}=\alpha K_S \\ \qquad\qquad SES \xrightarrow{k_p} SE + P \end{array}$$

The K_{S_1} (or K_S) is an intrinsic dissociation constant because only one site on E is available for the first molecule of S. Similarly, K_{S_2} (or αK_S) is the intrinsic dissociation constant of the substrate-catalytic site complex. The equilibria and the velocity equation are identical to those for a rapid equilibrium essential activation system in which the substrate can bind only to the enzyme-activator complex. Now the activator is another substrate molecule. The enzyme could, in fact, be a monomer. The velocity equation is:

$$\boxed{\frac{v}{V_{max}} = \frac{\dfrac{[S]^2}{\alpha K_S^2}}{1 + \dfrac{[S]}{K_S} + \dfrac{[S]^2}{\alpha K_S^2}}} \qquad \text{(VII-40)}$$

The equation differs from that for an enzyme with two identical catalytic sites in that there is no $[S]/K_S$ term in the numerator (because SE does not produce product and ES does not form), and there is no factor of 2 in the $[S]/K_S$ term in the denominator (only one singly occupied enzyme-substrate complex forms). The velocity curve is sigmoidal even if $\alpha = 1$. In this case, the $[S]_{0.9}/[S]_{0.1}$ ratio is 25, which corresponds to an apparent n value of 1.36 (if the data are analyzed by the Hill equation). The reciprocal plot is curved and intercepts the $1/v$-axis at $1/V_{max}$. The Hill plot is also curved with limiting slopes of 1.0 (at very high substrate concentrations) and 2.0 (at very low substrate concentrations). The slope between the points corresponding to $0.9\,V_{max}$ and $0.1\,V_{max}$ is 1.36. If S binding is measured (rather than v), then in place of v/V_{max} we can plot Y_S (fraction of total sites occupied) versus [S]. In this case, the numerator of equation VII-40 would contain an additional $[S]/2K_S$ term. The denominator would be unchanged. The binding curve is sigmoidal with an n_{app} of about 1.25. As α decreases, the curves become more sigmoidal and n_{app} approaches 2.

An inhibitor that competes only for the effector site but does not promote substrate binding increases $[S]_{0.5}$ and the apparent n value. An inhibitor that combines at both sites (but does not mimic S at the effector site) increases $[S]_{0.5}$ but decreases the apparent n value. An inhibitor that combines only at the catalytic site behaves similarly. If I mimics S at the effector site by promoting the binding of S to the catalytic site, then I is really an activator; $[S]_{0.5}$ and n_{app} decrease. If I binds to both sites (and mimics S at the effector site), then at very low substrate concentrations the activation by I may more than compensate for the inhibition resulting from competition at the catalytic site. In this case, the v versus [I] plot will rise to a maximum, then fall. The v versus [S] plot in the presence of I will rise more rapidly then the control plot, but after a certain point, the usual inhibition is observed.

System B4b. Random Binding

If S binds randomly and independently (noncooperatively) to the effector and catalytic sites, the equilibria are:

$$
\begin{array}{ccccc}
E + S & \xrightleftharpoons{K_{S_1}=K_S} & ES & & \\
+ & & + & & \\
S & & S & & \\
K_{S_2}=\alpha K_S \updownarrow & & \updownarrow K_{S_2}=\alpha K_S & & \\
SE + S & \xrightleftharpoons{K_{S_1}=K_S} & SES & \xrightarrow{k_p} & SE + P
\end{array}
$$

The velocity equation is given by:

$$
\frac{v}{V_{max}} = \frac{\dfrac{[S]^2}{\alpha K_S^2}}{1 + \dfrac{[S]}{K_S} + \dfrac{[S]}{\alpha K_S} + \dfrac{[S]^2}{\alpha K_S^2}} \qquad \text{(VII-41)}
$$

It may be difficult to distinguish System B4b from System B4a. System B4b also yields sigmoidal velocity curves even when $\alpha = 1$. In this case, the $[S]_{0.9}/[S]_{0.1}$ ratio is 40. As α decreases, the $[S]_{0.9}/[S]_{0.1}$ ratio decreases, the curve becomes more sigmoidal, and n_{app} approaches 2.

We might expect the sigmoidicity of the velocity curve to increase as the number of noncatalytic effector sites increase. However, if S binding to the multiple sites is noncooperative, the sigmoidicity of the velocity curve remains low (i.e., the inflection point remains at a low specific velocity—only $[S]_{0.5}$ increases). This phenomenon is described in more detail in System B5.

System B5. Multiple Essential Activator Sites

If multiple essential activator sites must be filled before any catalytic activity is possible, and the binding of one ligand has no effect on the binding of another, the velocity equations are:

Ordered: n molecules of A *bind randomly before* S

$$\frac{v}{V_{max}} = \frac{\dfrac{[S][A]^n}{K_S K_A^n}}{\left(1 + \dfrac{[A]}{K_A}\right)^n + \dfrac{[S][A]^n}{K_S K_A^n}} \qquad \text{(VII-42)}$$

Ordered: S *binds first, then n molecules of* A *bind randomly*

$$\frac{v}{V_{max}} = \frac{\dfrac{[S][A]^n}{K_S K_A^n}}{1 + \dfrac{[S]}{K_S}\left(1 + \dfrac{[A]}{K_A}\right)^n} \qquad \text{(VII-43)}$$

Random binding of S *and n molecules of* A

$$\frac{v}{V_{max}} = \frac{\dfrac{[S][A]^n}{K_S K_A^n}}{\left(1 + \dfrac{[S]}{K_S}\right)\left(1 + \dfrac{[A]}{K_A}\right)^n} \qquad \text{(VII-44)}$$

In all cases, the single numerator term represents the concentration of the single catalytically active species, ESA_n, relative to free E. The denominator, when multiplied out, would contain a term for each enzyme species that forms. The A could just as well be a cosubstrate that binds to $n-1$ activator sites and one catalytic site. The v versus [A] curves are sigmoidal for $n>1$. In general, as n increases, $[A]_{0.5}$ increases. For the random or ordered model (S before nA) at saturating [S]:

$$[A]_{0.5} = \frac{\sqrt[n]{0.5}}{1 - \sqrt[n]{0.5}} K_A$$

Although the velocity curve is sigmoidal, the inflection occurs at rather low specific velocities and throughout most of the v' range the curve decreases in slope (Fig. VII-24 for $n=10$). Specifically the inflection point occurs at:

$$[A] = \frac{n-1}{2} K_A, \qquad v = \left(\frac{n-1}{n+1}\right)^n V_{max}$$

where n is the actual number of activator sites. The Hill plot for $n=10$ is curved and has limiting slopes of 1.0 (at very high [A]) and 10 (at very low [A]). In the region of $v=0.3$ to $0.7\,V_{max}$, the plot is reasonably linear with a slope of about 1.3. Multiple essential activation would not provide much in

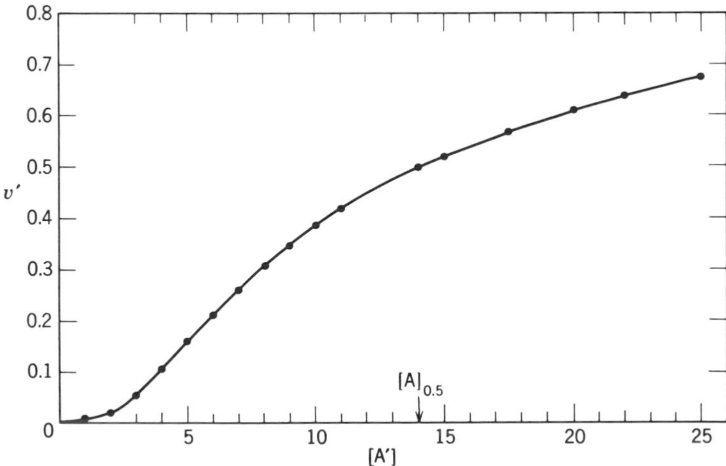

Fig. VII-24. The v versus [A] plot at $[S] \gg K_S$ for a Rapid Equilibrium Random system involving one molecule of S and 10 molecules of A. Only ESA_{10} is catalytically active.

the way of an "off-on" switch unless the activator displayed cooperative binding. In this case, as few as two or four activator sites can result in rather steep sigmoidal velocity curves with the v' coordinate of the inflection point approaching $0.5\,V_{\max}$ (see below).

System B6. Cooperative Essential Activation

Multiple essential cooperative activation can be considered in exactly the same manner as cooperative substrate binding. For example, suppose one molecule of S and n molecules of A bind to E randomly and only ESA_n is catalytically active. Let us further assume that A binds cooperatively ($a < 1$, $b < 1$, etc.), but the binding of A has no effect on the binding of S (i.e., S binds equally well to E, EA, EA_2, etc.). The velocity equation would be:

$$\frac{v}{V_{\max}} = \frac{\dfrac{[S][A]^n}{K_S K'}}{\left(1 + \dfrac{[S]}{K_S}\right)\left(1 + \cdots \dfrac{[A]^n}{K'}\right)} \qquad \text{(VII-45)}$$

where K' represents the product of K_A^n and $(n-1)$ interaction factors (raised to appropriate powers). If A binding is strongly cooperative, the velocity equation can be simplified (as shown) by ignoring all the complexes containing less than n molecules of A. When S is the varied ligand:

$$\boxed{\frac{v}{V_{\max}} = \frac{[S]}{K_S\left(1 + \dfrac{K'}{[A]^n}\right) + [S]\left(1 + \dfrac{K'}{[A]^n}\right)}} \qquad \text{(VII-46)}$$

The plots are hyperbolic and the reciprocal plots are linear although both $slope_{1/S}$ and $1/V_{\max_{app}}$ versus $1/[A]$ replots are markedly concave up. When [A] is varied, the velocity can be expressed by a modified Hill equation:

$$\boxed{\frac{v}{V_{\max}} = \frac{\dfrac{[A]^n}{K'}}{\left(1 + \dfrac{K_S}{[S]}\right)\left(1 + \dfrac{[A]^n}{K'}\right)}} \qquad \text{(VII-47)}$$

or

$$\boxed{\frac{v}{V_{\max_{\rm app}}} = \frac{\dfrac{[\mathrm{A}]^n}{K'}}{1+\dfrac{[\mathrm{A}]^n}{K'}} = \frac{[\mathrm{A}]^n}{K'+[\mathrm{A}]^n}} \qquad \text{(VII-48)}$$

where

$$V_{\max_{\rm app}} = \frac{V_{\max}}{\left(1+\dfrac{K_S}{[\mathrm{S}]}\right)} = \text{an apparent maximal velocity at the fixed [S]}$$

The v versus [A] plots are sigmoidal. The inflection point occurs at:

$$\boxed{[\mathrm{A}] = \frac{n-1}{n+1}K_{\mathrm{A}}, \qquad v = \frac{n-1}{2n}V_{\max_{\rm app}}} \qquad \text{(VII-49)}$$

Unlike the multiple essential noncooperative activation system described in System B5, variations in [A] will provide a very sensitive control over v' at any fixed [S]; that is, within a certain [A] range, a small increment in [A] will cause a large change in v'. Also in contrast to the noncooperative system, $[\mathrm{A}]_{0.5}$ at any fixed [S] decreases as n increases.

C. THE GENERAL SEQUENTIAL INTERACTION MODEL OF KOSHLAND, NEMETHY, AND FILMER (RESTRICTED INTERACTIONS BETWEEN SITES)

In our previous discussion of site-site interactions, we assumed that the affinities of all vacant sites are affected equally by an occupied site. Thus when the first site of an enzyme with four identical sites is occupied, the dissociation constant of all three vacant sites was shown changed from K_S to aK_S. However, if the electronic and chemical environments of the vacant sites differ, the dissociation constants can differ. For example, a vacant site situated between another vacant site and an occupied site might be subject to forces different from those exerted on a vacant site situated between two other vacant sites, or between two occupied sites. Koshland, Nemethy, and Filmer (1966) have described several models and the corresponding saturation equations for proteins with different permissable site-site interactions.

The substrate binding sites are assumed to reside on different subunits. Each subunit is assumed to exist in either of two distinct conformations. These are designated the *A conformation* when the subunit is vacant, and *B conformation* when the subunit is occupied. The substrate binding constant, K_S, which represents the intrinsic affinity of an individual subunit is given as:

$$\boxed{K_S = \frac{[BS]}{[B][S]}} \tag{VII-50}$$

A transformation constant, K_t, represents the equilibrium constant for the change from conformation A to conformation B.

$$\boxed{K_t = \frac{[B]}{[A]}} \tag{VII-51}$$

The overall binding sequence is described by the product of the two equilibrium constants.

$$S + A \rightleftharpoons BS$$

$$\boxed{K_S K_t = \frac{[BS]}{[S][A]}} \tag{VII-52}$$

The sequence can be visualized in three ways (Fig. VII-25): (*a*) The substrate binds to the subunit in the A conformation with a binding constant

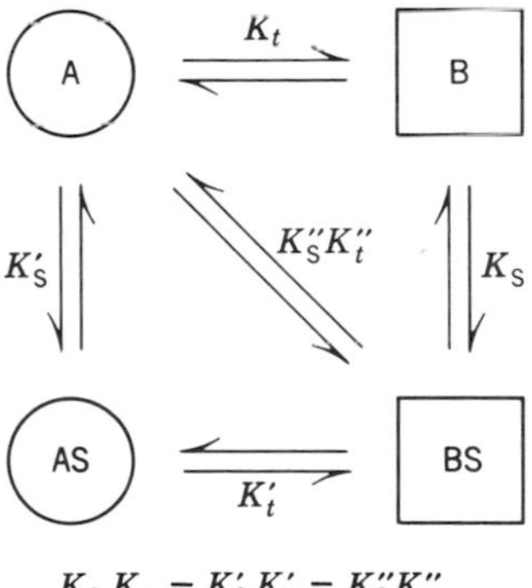

$$K_S K_t = K'_S K'_t = K''_S K''_t$$

Fig. VII-25. Overall conversion of an unoccupied A subunit to an occupied B subunit as induced by substrate binding. The overall, combined binding transformation constant is $K_S K_t$ regardless of the sequence.

K'_S, and thereby induces a conformational change to the B form:

$$\begin{array}{llcll} & A+S & \rightleftharpoons & AS & K_{eq}=K'_S \\ & AS & \rightleftharpoons & BS & K_{eq}=K'_t \\ \hline \text{Sum:} & A+S & \rightleftharpoons & BS & K_{eq}=K'_S K'_t \end{array}$$

(*b*) An equilibrium preexists between conformations A and B. The substrate binds to conformation B and pulls the coupled reactions to the right (i.e., the B conformation is stabilized by the substrate):

$$\begin{array}{llcll} & A & \rightleftharpoons & B & K_{eq}=K_t \\ & B+S & \rightleftharpoons & BS & K_{eq}=K_S \\ \hline \text{Sum:} & A+S & \rightleftharpoons & BS & K_{eq}=K_S K_t \end{array}$$

(*c*) The binding and transformation occur simultaneously:

$$S+A \rightleftharpoons BS \qquad K_{eq}=K''_S K''_t$$

In all cases $K_S K_t$ is the overall intrinsic *binding* constant. Thus $1/K_S K_t$ is equivalent to the intrinsic *dissociation* constant, K_S, used in our earlier discussion.

The interactions between subunits that affect the binding or dissociation constants of the vacant sites are indicated as K_{AA}, K_{AB}, and K_{BB}. The K_{AA} represents the interaction constant between two vacant sites and is taken as unity. (A choice of any other value simply changes the standard state but does not change the relative magnitudes of other interactions.) The interaction between an occupied and a vacant site is indicated by the constant K_{AB}. The K_{AB} is, in a sense, an equilibrium constant expressing the strength of an AB interaction relative to an AA interaction:

$$\boxed{K_{AB}=\frac{[AB][A]}{[AA][B]}} \qquad \text{(VII-53)}$$

where [AB] and [AA] are interacting subunits, and [A] and [B] are noninteracting subunits. The latter are included so that the numerator and denominator of the K_{AB} expression contain the same total number of subunits. A K_{AB} value >1 indicates that an interaction between an A and a B subunit is more favorable than an interaction between two A subunits. In other words, if $K_{AB}>1$, complexes with a large number of AB interactions are more stable than complexes with fewer AB interactions, and complexes with some AB interactions (i.e., partially filled complexes) are more stable than the complex with no AB interactions (free A_4). The interaction between two occupied subunits is indicated by the constant K_{BB}, where:

$$\boxed{K_{BB}=\frac{[BB][A][A]}{[AA][B][B]}} \qquad \text{(VII-54)}$$

A K_{BB} value >1 indicates that complexes with BB interactions are more stable than complexes with only AA interactions. The constants K_{AB} and K_{BB} have the same effect on the binding constants of vacant sites and on the relative distribution of complexes as the interaction factors a, b, and c used earlier. For example, if $K_{AB} \geqslant 1$ and $K_{BB} > K_{AB}$, the binding constants of vacant sites increase as sites are occupied resulting in a sigmoidal velocity curve. If $K_{AB} > 1$ but $K_{BB} \leqslant 1$, then the intermediate, partially filled complexes (with the greatest AB interactions) are more stable than either the free enzyme (A_4) or the completely filled enzyme (B_4S_4). The partially filled complexes will persist as [S] increases and we will observe negative cooperativity.

Dimer Model

To illustrate the model, let us consider first a simple dimer. The binding sequence can be visualized as follows:

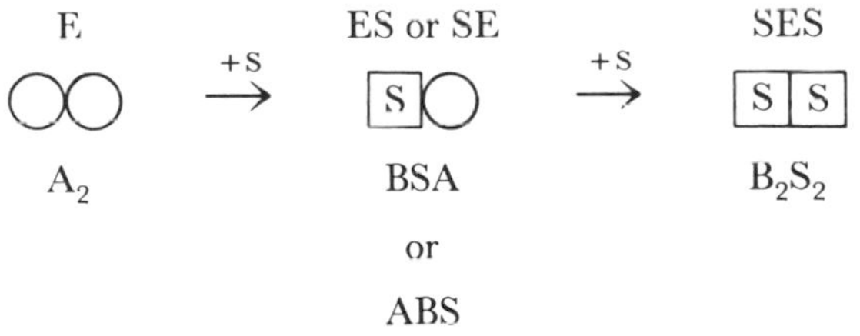

E		ES or SE		SES
OO	$\xrightarrow{+S}$	[S]O	$\xrightarrow{+S}$	[S|S]
A_2		BSA		B_2S_2
		or		
		ABS		

Note that there are really two different singly occupied complexes, BSA and ABS. If we write out the total reaction pattern showing each site separately, the statistical factor of 2 will be automatically taken care of and we can also directly write the overall binding constant for a given reaction. For example, the binding constant for the reaction $BSA+S \rightleftharpoons B_2S_2$ is composed of the factors (a) K_t (transformation), (b) K_S (binding), and (c) K_{BB}/K_{AB} (the introduction of a new BB interaction in place of the previous AB interaction). This latter factor may be considered as the equilibrium constant for the change: AB interaction $\rightleftharpoons$BB interaction, that is a measure of the relative strengths of the two kinds of interactions.

$$
\begin{array}{ccccccc}
A_2 + S & \underset{}{\overset{K_iK_SK_{AB}/K_{AA}}{\rightleftharpoons}} & BSA & \xrightarrow{k_p} & A_2 + P \\
(E) & & (ES) & & \\
+ & & + & & \\
S & & S & & \\
\frac{K_iK_SK_{AB}}{K_{AA}} \downarrow\uparrow & & \uparrow\downarrow \frac{K_iK_SK_{BB}}{K_{AB}} & & \\
A_2 + P \xleftarrow{k_p} ABS + S & \underset{K_iK_SK_{BB}/K_{AB}}{\rightleftharpoons} & B_2S_2 & \xrightarrow{k_p} & ABS + P \\
(SE) & & (SES) & & \\
 & & \downarrow k_p & & \\
 & & BSA + P & &
\end{array}
$$

The concentration of any given species relative to free enzyme (A_2) is simply the product of all binding constants and ligand concentrations between that species and the free enzyme. For example:

$$[B_2S_2] = \left(\frac{K_iK_SK_{AB}[S]}{K_{AA}}\right)\left(\frac{K_iK_SK_{BB}[S]}{K_{AB}}\right)[A_2]$$

$$= K_i^2K_S^2K_{BB}[S]^2[A_2] \qquad (\text{where } K_{AA}=1)$$

Similarly:

$$[BSA] = K_iK_SK_{AB}[S][A_2]$$

$$[ABS] = K_iK_SK_{AB}[S][A_2]$$

while $[E]_t = [A_2] + [BSA] + [ABS] + [B_2S_2]$. The proportion of the total enzyme represented by each species is:

$$\frac{[A_2]}{[E]_t} = \frac{1}{1 + 2K_iK_SK_{AB}[S] + K_i^2K_S^2K_{BB}[S]^2}, \qquad \frac{[BSA]}{[E]_t} = \frac{K_iK_SK_{AB}[S]}{\text{same denominator}}$$

$$\frac{[ABS]}{[E]_t} = \frac{K_iK_SK_{AB}[S]}{\text{same denominator}}, \qquad \frac{[B_2S_2]}{[E]_t} = \frac{K_i^2K_S^2K_{BB}[S]^2}{\text{same denominator}}$$

The velocity equation is obtained in the usual manner:

$$v = k_p[\text{BSA}] + k_p[\text{ABS}] + 2k_p[\text{B}_2\text{S}_2]$$

$$\frac{v}{[\text{E}]_t} = \frac{k_p 2K_t K_\text{S} K_\text{AB}[\text{S}] + 2k_p K_t^2 K_\text{S}^2 K_\text{BB}[\text{S}]^2}{1 + 2K_t K_\text{S} K_\text{AB}[\text{S}] + K_t^2 K_\text{S}^2 K_\text{BB}[\text{S}]^2}$$

or

$$\boxed{\frac{v}{V_\text{max}} = \frac{K_t K_\text{S} K_\text{AB}[\text{S}] + K_t^2 K_\text{S}^2 K_\text{BB}[\text{S}]^2}{1 + 2K_t K_\text{S} K_\text{AB}[\text{S}] + K_t^2 K_\text{S}^2 K_\text{BB}[\text{S}]^2}} \qquad \text{(VII-55)}$$

where $2k_p[\text{E}]_t = V_\text{max}$.

The relative species concentrations can be written directly without showing the individual reactions at each site (which can be complicated for multisite enzymes). The procedure is as follows: (*a*) Indicate the various species present. (*b*) Determine the number of ways of arranging each species. This introduces the necessary statistical factors. For the dimer example, there are two ways of arranging the singly occupied species: [S]○ (BSA or ES) and ○[S] (ABS or SE) and only one way of arranging the doubly occupied species. (*c*) Indicate the product of the transformation and binding constants for the overall conversion of free enzyme to the given species. This product is then multiplied by the concentration of each ligand raised to a power equal to the number of molecules of ligand in the species. (*d*) Determine the kinds of subunit interactions possible in each species. Indicate these interactions by the appropriate constant (K_AA, K_AB, K_BB, etc.). Each constant is then raised to a power equal to the number of specific interactions of that kind. The various factors obtained for a given species by this procedure are then multiplied to obtain a term for that species. This term turns out to be the concentration of the species relative to the concentration of free enzyme. The procedure is illustrated below for the dimer.

	A_2	ABS / BSA	B_2S_2
Species	○○	○[S] / [S]○	[S][S]
Factor for number of arrangements	1	2	1
Factor for product of transformation and binding constants and the number of molecules of ligand	—	$K_t K_S[S]$	$K_t^2 K_S^2[S]^2$
Number of AA interactions	1	0	0
Factor	$K_{AA} = 1$	—	—
Number of AB interactions	0	1	0
Factor	—	K_{AB}	—
Number of BB interactions	0	0	1
Factor	—	—	K_{BB}
Term for species	1	$2K_t K_S K_{AB}[S]$	$K_t^2 K_S^2 K_{BB}[S]^2$
Effective k_p value	—	k_p	$2k_p$

Once we have the terms for each species, the velocity equation can be written directly taking into account the effective k_p values. Comparing velocity equation VII-54 with equation VII-6 derived earlier for the simple sequential interaction model, we find that in place of the simple K_S (dissociation constant) we now have the more complex $K_S K_t$ (combined binding-transformation constant). In place of the simple interaction factor a, we now have two kinds of interaction factors, K_{AB} and K_{BB}. If there are no AB interactions ($K_{AB} = K_{AA} = 1$), the two equations become equivalent when:

$$K_{S_{Dis}} = \frac{1}{K_t K_{S_{Bind}}} \qquad \text{and} \qquad a = \frac{1}{K_{BB}}$$

The general sequential interaction model of Koshland, Nemethy, and Filmer takes into account a variety of possible subunit interactions that may affect affinities of vacant sites. Consequently, the general model is far more versatile than the simple sequential interaction model.

Tetramer Models

If we consider a tetramer, then there are three possible subunit interaction models: "tetrahedral" (each subunit can interact with the other three, Fig. VII-26a), "linear" (two of the subunits can interact with two others, while two of the subunits can interact with only one other, Fig. VII-26b), and "square" (each subunit can interact with only two of the other three). The terms tetrahedral, linear, and square do not necessarily mean that the subunits are arranged spatially in these relationships. The square geometry model is illustrated in Figure VII-26c. If we indicate the subunits in a square arrangement, then horizontal and vertical interactions are permissable, but no interactions across the diagonal occur. Note that there are two possible modes of arranging the four subunits as an ES_2 ($A_2B_2S_2$) complex. In one mode the two B subunits are adjacent ("cis" arrangement), while in the second mode the two B subunits lie across the diagonal ("trans" arrangement). There are four possible "cis" arrangements and two possible "trans" arrangements. The "cis" arrangement permits two possible AB interactions and one BB interaction. The restriction on "trans" interactions permits the "trans" arrangement four possible AB interactions (1–2, 1–4, 3–2, and 3–4), but no BB interactions. The interaction factors for the "cis" and "trans" ES_2

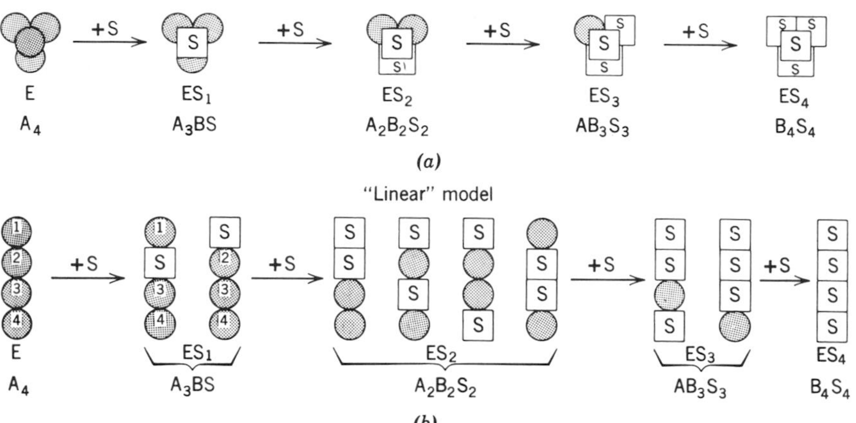

Fig. VII-26. (*a*) "Tetrahedral" model of a four-site enzyme (*b*) "Linear" model.

complexes must be indicated separately. The relative distribution of enzyme species is:

$$[\mathrm{ES}]=[\mathrm{A_3BS}]=4K_{\mathrm{AB}}^2(K_\mathrm{S}K_t[\mathrm{S}])[\mathrm{A_4}]$$

$$[\mathrm{ES_2}]=[\mathrm{A_2B_2S_2}]=(4K_{\mathrm{AB}}^2K_{\mathrm{BB}}+2K_{\mathrm{AB}}^4)(K_\mathrm{S}K_t[\mathrm{S}])^2[\mathrm{A_4}]$$

$$[\mathrm{ES_3}]=[\mathrm{AB_3S_3}]=4K_{\mathrm{AB}}^2K_{\mathrm{BB}}^2(K_\mathrm{S}K_t[\mathrm{S}])^3[\mathrm{A_4}]$$

$$[\mathrm{ES_4}]=[\mathrm{B_4S_4}]=K_{\mathrm{BB}}^4(K_\mathrm{S}K_t[\mathrm{S}])^4[\mathrm{A_4}]$$

The velocity equation is obtained from:

$$v=k_p[\mathrm{ES}]+2k_p[\mathrm{ES_2}]+3k_p[\mathrm{ES_3}]+4k_p[\mathrm{ES_4}] \qquad \text{and} \qquad V_{\max}=4k_p[\mathrm{E}]_t$$

$$\therefore \quad \boxed{\frac{v}{V_{\max}}=\frac{K_{\mathrm{AB}}^2(K_\mathrm{S}K_t[\mathrm{S}])+(2K_{\mathrm{AB}}^2K_{\mathrm{BB}}+K_{\mathrm{AB}}^4)(K_\mathrm{S}K_t[\mathrm{S}])^2+3K_{\mathrm{AB}}^2K_{\mathrm{BB}}^2(K_\mathrm{S}K_t[\mathrm{S}])^3+K_{\mathrm{BB}}^4(K_\mathrm{S}K_t[\mathrm{S}])^4}{1+4K_{\mathrm{AB}}^2(K_\mathrm{S}K_t[\mathrm{S}])+(4K_{\mathrm{AB}}^2K_{\mathrm{BB}}+2K_{\mathrm{AB}}^4)(K_\mathrm{S}K_t[\mathrm{S}])^2+4K_{\mathrm{AB}}^2K_{\mathrm{BB}}^2(K_\mathrm{S}K_t[\mathrm{S}])^3+K_{\mathrm{BB}}^4(K_\mathrm{S}K_t[\mathrm{S}])^4}} \tag{VII-56}$$

"Square" model

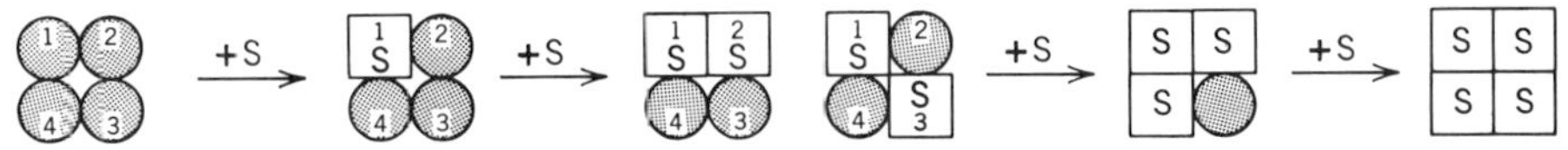

	E	ES₁	ES₂		ES₃	ES₄
Complex	A_4	A_3BS	$A_2B_2S_2$		AB_3S_3	B_4S_4
Number of modes of binding S		1	2		1	1
Number of ways of arranging S		4	4*	2**	4	1
Number of possible AB interactions		2	2	4	2	0
Number of possible BB interactions		0	1	0	2	4
Term for species (concentration relative to A_4)	1	$4K_{AB}^2\,(K_S K_t[S])$	$(4K_{AB}^2 K_{BB} + 2K_{AB}^4)\,(K_S K_t[S])^2$		$4K_{AB}^2 K_{BB}^2\,(K_S K_t[S])^3$	$K_{BB}^4\,(K_S K_t[S])^4$
Effective catalytic rate constant of complex (k_p)		$1k_p$	$2k_p$		$3k_p$	$4k_p$

*Four "cis" arrangements of ES_2

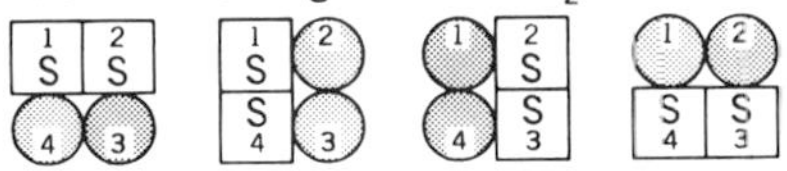

**Two "trans" arrangements of ES_2

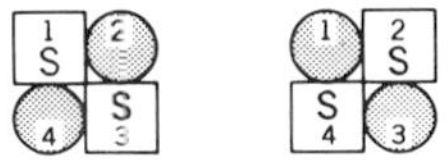

(c)

Fig. VII-26. (*Cont.*) (*c*) "Square" model.

As noted earlier, increasing the value of K_{AB} increases the stability of intermediate complexes and, hence, flattens the velocity curve, while increasing K_{BB} tends to increase the sigmodicity of the velocity curve. Thus with the proper combination of interaction factors, an enzyme with strong subunit interactions could yield a hyperbolic velocity curve. For example if $K_{AB} = \sqrt{K_{BB}}$ in the "square" model, the velocity equation reduces to:

$$\boxed{\frac{v}{V_{max}} = \frac{K_{BB}K_SK_t[S]}{1 + K_{BB}K_SK_t[S]}} \quad \text{(VII-57)}$$

where $V_{max} = 4k_p[E]_t$.

Steep velocity curves (sigmoidal on a linear scale) can result even if $K_{BB} = 1$ provided $K_{AB} < 1$. In this case there are no BB interactions, but the instability of partially filled complexes promotes the formation of complexes with B subunits. Figure VII-27 shows the effect of varying K_{AB} at a constant K_SK_t and $K_{BB} = 1$ for the square model. As K_{AB} increases, the partially filled

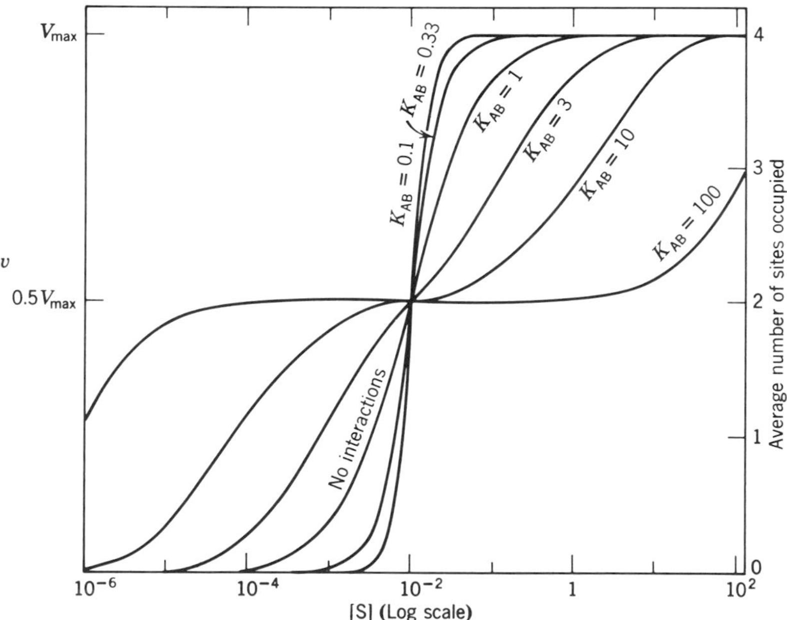

Fig. VII-27. Effect of different K_{AB} values on the v versus log [S] plots for the "square" model; $K_{BB} = 1$, $K_SK_t = 100$. [Redrawn with permission from Koshland, Jr., D. E., Nemethy, G. and Filmer, D., *Biochemistry* **5**, 365 (1966). Copyright by the American Chemical Society.]

complexes persist over a wider range of [S] and the velocity curve tends to flatten at velocities below V_{max}. As [S] increases, the vacant sites are eventually filled and the velocity approaches V_{max}. Similar negatively co-operative velocity curves will be obtained if $K_{AB}=1$ and $K_{BB}<1$ (Fig. VII-28). In this case there are no AB interactions but the unfavorable BB interactions tend to stabilize the partially filled complexes.

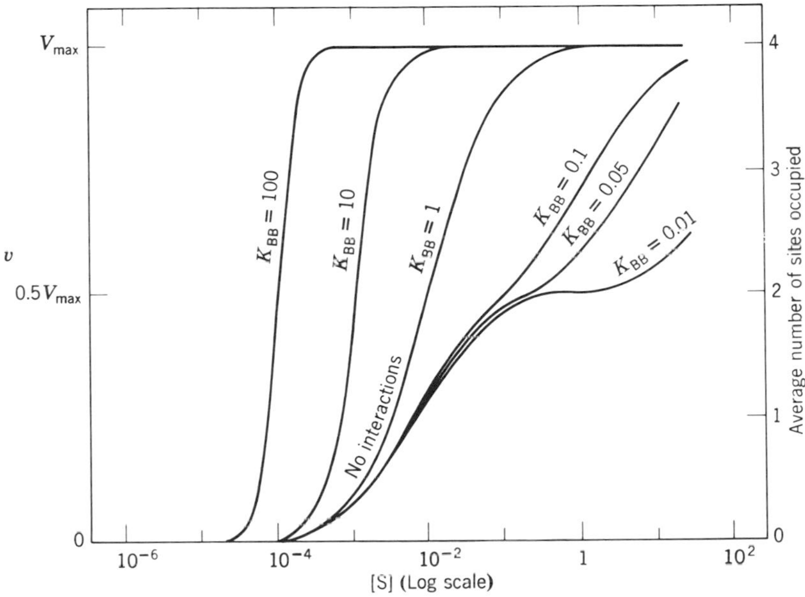

Fig. VII-28. Effect of different K_{BB} values on the v versus log [S] plots for the square model; $K_{AB}=1$, $K_SK_t=100$. [Redrawn with permission from Koshland Jr., D. E., Nemethy, G. and Filmer, D., *Biochemistry* **5**, 365 (1966). Copyright by the American Chemical Society.]

Nonidentical Subunits

If the enzyme is made of nonidentical subunits, the number and kind of possible interactions increase. For example, consider a tetramer composed of two α subunits and two β subunits. The first molecule of substrate could bind either to an α subunit or a β subunit converting the subunit to the B form. The intrinsic subunit binding constants, $K_{S\alpha}$ and $K_{S\beta}$, could be the same or different. Similarly, the transition constants, $K_{t\alpha}$ and $K_{t\beta}$, may be identical or different. Two kinds of AB interactions are possible in ES, one between the α subunit in the A form and a β subunit in the B form, the other between a β subunit in the A form and an α subunit in the B form.

Thus there are at least two K_{AB} constants ($K_{A\alpha B\beta}$ and $K_{A\beta B\alpha}$) which may be identical or different. Similarly, there are at least three kinds of BB interactions in the higher complexes ($K_{B\alpha}K_{B\alpha}, K_{B\alpha}K_{B\beta}, K_{B\beta}K_{B\beta}$).

Inhibition and Activation—A Dimer Model

Figure VII-29 illustrates the application of the general sequential interaction model to a system involving an activator and an inhibitor. For simplicity, the enzyme is assumed to be a dimer, and each subunit is assumed to possess a single site for each ligand. The first column shows the complexes that can form in the presence of the substrate (S). As before, it is assumed that the binding of S induces a conformational change (from the A form to the B form) in the subunit carrying S. The mathematical term shown beneath each species represents the concentration of that species relative to A_2. The second column shows the additional complexes that can form in the presence of substrate and activator (J). It is assumed that the activator induces the same conformational change as the substrate, but binds to a site distinct from the substrate binding site. The mathematical terms for species containing J now contain an additional K_J factor (the binding constant for J). For example, the species B_2S_2J is described by $2K_{t_{AB}}^2 K_{BB} K_S^2 K_J [S]^2 [J]$. The 2 indicates the two ways of forming the species ($\boxed{SJ|S}$ and $\boxed{S|SJ}$); $K_{t_{AB}}^2$ indicates that two subunits have undergone an A-B transition; K_{BB} indicates that there is one B-B subunit-subunit interaction; K_S^2 and $[S]^2$ indicate that two molecules of S are bound; K_J and [J] indicate that only one molecule of J is bound. The third column shows the additional species that can form in the presence of substrate, activator, and inhibitor (I). It is assumed that the inhibitor induces a change to a new conformation (C) that cannot bind either S or J. The transformation constant for this change is $K_{t_{AC}}$. The inhibitor binding constant is K_i. The terms for the species containing I now include K_{AC}, or K_{BC}, or K_{CC} representing the new interactions between A and C, B and C, or two C subunits, respectively.

The distribution of complexes can be expressed in the usual manner. For example, the concentration of B_2S_2J in terms of $[A_2]$ is:

$$[B_2S_2J] = 2K_{t_{AB}}^2 K_{BB} K_S^2 K_J [S]^2 [J][A_2]$$

The fraction of the total enzyme concentration that is present as B_2S_2J is:

$$\frac{[B_2S_2J]}{[E]_t} = \frac{2K_{t_{AB}}^2 K_{BB} K_S^2 K_J [S]^2 [J]}{1 + \text{sum of the terms for all complexes}}$$

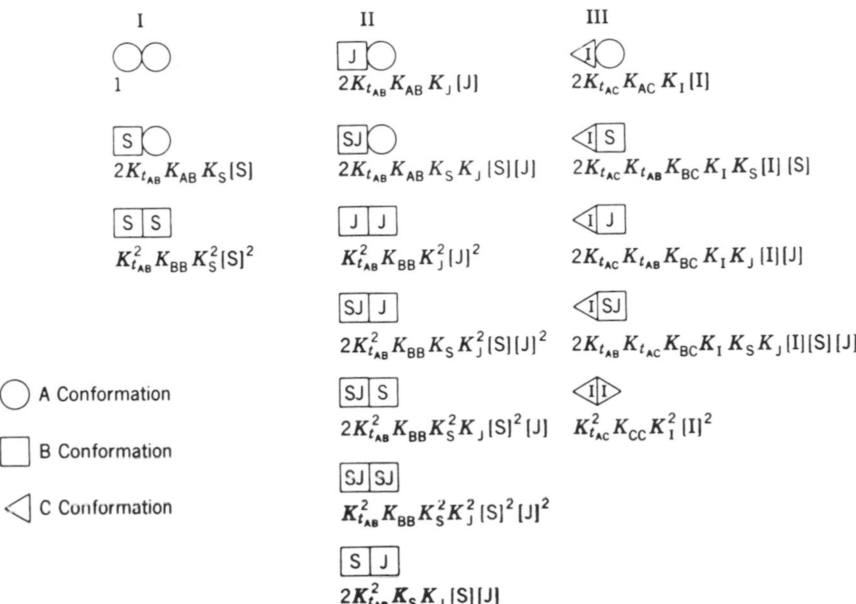

Fig. VII-29. Dimer model; J is an activator that induces the same conformational change as S but binds to a distinct site; I is an inhibitor that induces a conformational change different from that induced by S or J; I and S or I and J are mutually exclusive. Column I: enzymespecies in the presence of S. Collumn II: additional enzyme species in the presence of S and J. Column III: additional enzyme species in the presence of S, J, and I. The transition from an A conformation to a B conformation is described by $K_{t_{AB}}$; the transition from A conformation to a C conformation is described by $K_{t_{AC}}$.

The terms for the different complexes (which represent their concentrations relative to free A_2) are those shown in Figure VII-29. The velocity of the reaction, also obtained in the usual way, is the sum of the concentrations of all complexes containing S, each multiplied by the appropriate k_p value (k_p for species containing one S, $2k_p$ for species containing two molecules of S). The $2k_p[E]_t$ term can be factored out as V_{max}. The final velocity equation will have the usual form:

$$\frac{v}{V_{max}} = \frac{K_{t_{AB}}K_{AB}K_S[S] + K_{t_{AB}}^2K_{BB}K_S^2[S]^2 + K_{t_{AB}}K_{AB}K_SK_J[S][J] + K_{t_{AB}}^2K_{BB}K_SK_J^2[S][J]^2 \ldots \text{etc.}}{1 + 2K_{t_{AB}}K_{AB}K_S[S] + K_{t_{AB}}^2K_{BB}K_S^2[S]^2 + 2K_{t_{AB}}K_{AB}K_J[J] \ldots \text{etc.}}$$

The velocity equation in the presence of just S and I would be:

$$\frac{v}{V_{max}} = \frac{K_{t_{AB}}K_{AB}K_S[S] + K_{t_{AB}}^2 K_{BB}K_S^2[S]^2 + K_{t_{AC}}K_{t_{AB}}K_{BC}K_SK_i[S][I]}{1 + 2K_{t_{AB}}K_{AB}K_S[S] + K_{t_{AB}}^2 K_{BB}K_S^2[S]^2 + 2K_{t_{AC}}K_{t_{AB}}K_{BC}K_SK_i[S][I] + 2K_{t_{AC}}K_{AC}K_i[I] + K_{t_{AC}}^2 K_{CC}K_i^2[I]^2}$$

(VII-58)

Equation VII-58 represents the most general form of the equations derived for the pure competitive Systems B2a through System B2e. System B2 is shown below in terms of the general interaction model. The overall binding constant for any reaction is determined in the manner described earlier. For example, the reaction from ES to IES includes (*a*) a transformation of an A subunit to a C subunit ($K_{t_{AC}}$), (*b*) the binding of one molecule of inhibitor to the C subunit (K_i), and (*c*) the introduction of a new BC interaction in place of an AB interaction (K_{BC}/K_{AB}).

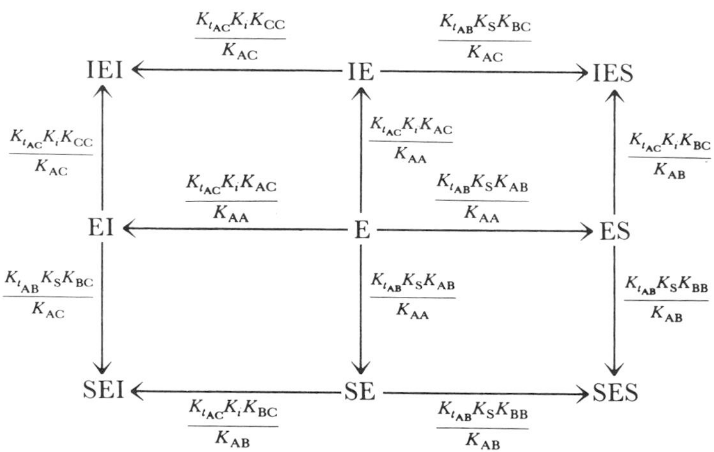

If there are no interactions between an A subunit and any other subunit ($K_{AA} = K_{AB} = K_{AC} = 1$) then we obtain System B2a where $a = 1/K_{CC} =$

$1/K_{BC}=1/K_{BB}$ (i.e., $K_{CC}=K_{BC}=K_{BB}>K_{AA}$). If the only interactions that occur are between two B subunits, then we have System B2b where $a=1/K_{BB}$.

Independent Binding Model

If the substrate and a second ligand (L) are not mutually exclusive, complexes containing both ligands on the same subunit can form. Together S and L might induce a conformational change in the subunit that is different from the conformation induced by either ligand alone. Thus we have the possibility of four different subunit conformations (Fig. VII-30a) and the attendant interactions (K_{AA}, K_{AB}, K_{AC}, K_{AD}, K_{BB}, K_{BC}, K_{BD}, K_{CC}, K_{CD}, K_{DD}). The overall equilibrium constant for the conversion of an A subunit to a D subunit must be the same regardless of the path taken. Thus $K_{t_{AB}}K_{S_B}K_{t_{BD}}K_{L_D}=K_{t_{AC}}K_{L_C}K_{t_{CD}}K_{S_D}$. One or two of the conformations may be equivalent with respect to their interactions with other subunits. For example, the subunit conformation D carrying both S and L may behave identically to subunit conformation B (i.e., $K_{AD}=K_{AB}$, $K_{CD}=K_{BC}$). The independent binding of the two ligands, S and L, can then be represented by Figure VII-30b. If, in addition, subunit conformation C behaves identically to subunit conformation A, then the independent binding of the two ligands can be represented by Figure VII-30c. In effect, the binding of S induces the A-B conformational change, but the binding of L does not. Another way of stating this is that S binds only to a B subunit, but L binds equally well to an A or B subunit. A fourth possibility is illustrated in Figure VII-30d. Here it is assumed that either S or L can induce the A-B conformational change, i.e., both ligands bind only to the B subunit. The relative concentrations of species is determined in the usual manner. This is illustrated in Figure VII-31 for a dimer of the model shown in Figure VII-30b; L=I, a noncompetitive inhibitor. In any of the systems described above, ligand L could be either a cosubstrate, an activator, or an inhibitor. If L is a cosubstrate or essential activator, then the numerator of the velocity equation will contain only those terms corresponding to species with both S and L present on the same subunit. If L is a noncompetitive inhibitor (i.e., subunits containing S and L are catalytically inactive), then the effective k_p values will vary accordingly. For example, the effective catalytic rate constant of the

| SI | S | complex will be k_p, rather than $2k_p$, since only one of the two subunits will be active.

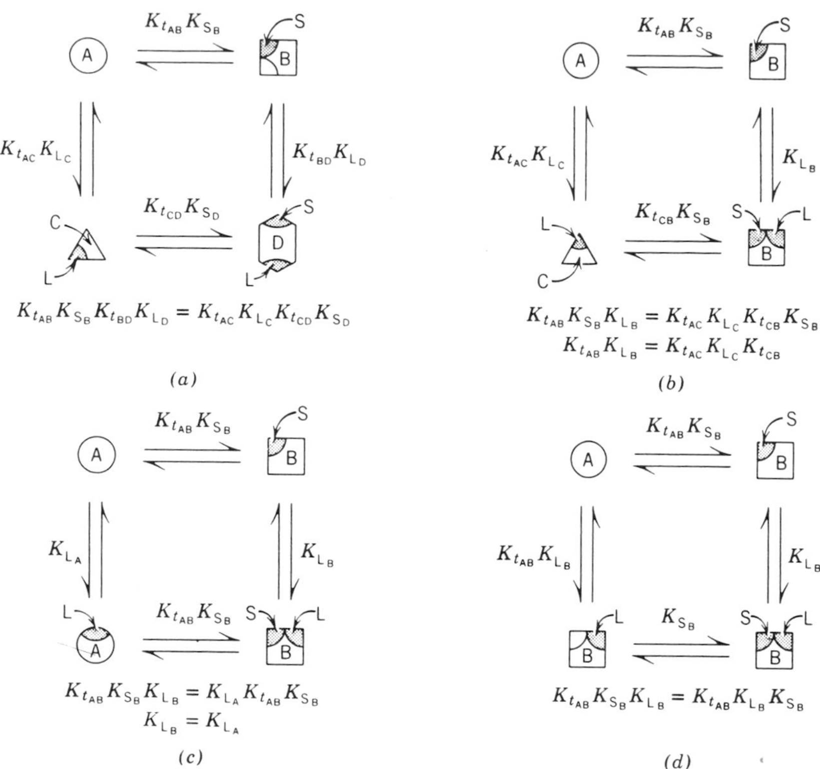

Fig. VII-30. Subunit conformations where S and L are not mutually exclusive. (*a*) S and L induce different conformations and when S and L are both bound, a fourth conformation results. (*b*)–(*d*) Two or more of the conformations shown in *a* are equivalent with respect to their effective binding constants and interactions with other subunits. The constants, K_{S_B}, K_{L_D}, K_{L_B}, etc. are binding constants for the respective ligands to the indicated conformation. The constants $K_{t_{AB}}$, $K_{t_{CB}}$, $K_{t_{AC}}$, etc., are transformation constants for the indicated conformational change.

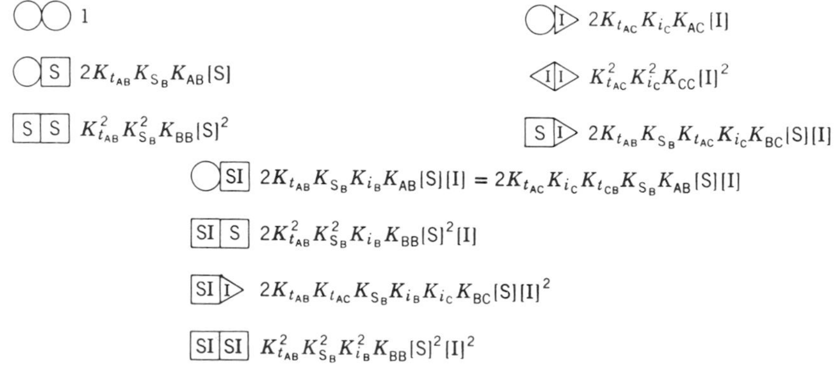

Fig. VII-31. Enzyme species of the dimer model of VII-30*b*; I induces conformation C; S and I are not mutually exclusive; S induces conformation B whether or not I is bound.

Summary

The Koshland-Nemethy-Filmer model of allosteric enzymes represents an extension of the "induced fit" or "flexible enzyme" hypothesis of Koshland (1958) and the simple sequential interaction models of Adair (1925) and Pauling (1935). The model is based on the assumptions that (*a*) hybrid conformational states of a protein can exist; that is, that each subunit of an enzyme (identical or nonidentical) can exist in two or more different conformations. (*b*) The conformation of a subunit depends on the number and nature of ligands bound. (*c*) The change from one conformation to another occurs as a result of ligand binding. In effect, the ligand binds preferentially or exclusively to only one (or some, but not all) of the conformations. (*d*) A change in the conformation of one subunit can affect the relative stability of conformations of neighboring subunits. (*e*) The affinity of a subunit for a ligand depends on whether the subunit conformation will be more or less stable after ligand binding. The model differs from the simple sequential interaction described earlier in that the binding of a ligand to one site (subunit) does not necessarily result in an identical change in the affinities of all the remaining vacant sites. The model differs from the concerted-symmetry model of Monod, Wyman, and Changeux in that the symmetry of the protein need not be retained after ligand binding. The concerted-symmetry model requires that all subunits in a protein exist in the same conformation at a given time and that they all change conformation simultaneously as a result of ligand binding to any one subunit. Usually, it is easier to analyze the behavior of an allosteric enzyme in terms of the concerted-symmetry model where the sigmoidicity of ligand binding curves or velocity curves can be directly related to just a few constants which can be obtained by graphical means. In contrast, the general sequential interaction model can be applied only by fitting the experimental data to theoretical curves. Nevertheless, the sequential interaction models are more general than the concerted-symmetry model and frequently provide a better explanation for experimental data.

D. THE SYMMETRY MODEL OF ALLOSTERIC ENZYMES

(The Concerted Transition Model of Monod, Wyman, and Changeux)

In 1965, Monod, Wyman, and Changeux proposed a unique model for allosteric proteins. The features of this model are as follows: (*a*) allosteric proteins are polymeric ("oligomers") containing identical minimal units ("protomers") arranged in a symmetrical fashion. (*b*) Each identical protomer possesses one, and only one, binding site for any given ligand (substrate, inhibitor, activator). (*c*) The oligomer can exist in at least two different conformations which are in equilibrium. The different conformations can arise from a rearrangement of the quaternary structure or from a change in the tertiary structure of the protomers (or both). The transition between one conformation and another is an all-or-nothing event; that is, the symmetry of the oligomer is conserved in the transition. Thus there are no hybrid states where some protomers have rearranged in space or changed in conformation while others have not. (*d*) The affinity of a binding site for a given ligand depends on the conformation of the protomer (hence on the conformation of the oligomer). Some ligands bind preferentially to one oligomer conformation, while other ligands bind preferentially to the other oligomer conformation. The binding of a ligand to one particular conformation will cause the equilibrium between oligomer conformations to shift in favor of the conformation with the bound ligand. Because each oligomer possesses more than one ligand binding site (one per protomer) and the transition from the lower affinity to the higher affinity conformation occurs simultaneously for all protomers, the number of higher affinity binding sites made available by the transition exceeds the one used up. As a result, the ligand binding curve or velocity curve is sigmoidal.

Derivation of the General Velocity Equation

Figure VII-32 illustrates the concerted-symmetry model for a tetramer. The T ("taut" or "tight") state represents the conformation with the lower affinity for the ligand, S. The R ("relaxed") state represents the conformation with the higher affinity for S. The equilibrium constant (or allosteric constant) for the transition $R_0 \rightleftharpoons T_0$ is designated L. The L represents the ratio of T to R states in the absence of any ligand.

$$\boxed{\mathrm{L} = \frac{[\mathrm{T}_0]}{[\mathrm{R}_0]}} \qquad \text{(VII-59)}$$

The intrinsic dissociation constant for the S binding site on a protomer in the T state is designated K_T. The intrinsic dissociation constant for the S binding site on a protomer in the R state is designated K_R. (Later, when we deal with multiple ligands, the dissociation constants for S will be designated K_{S_R} and K_{S_T}, or just K_S if only one state binds S). The equilibria are shown

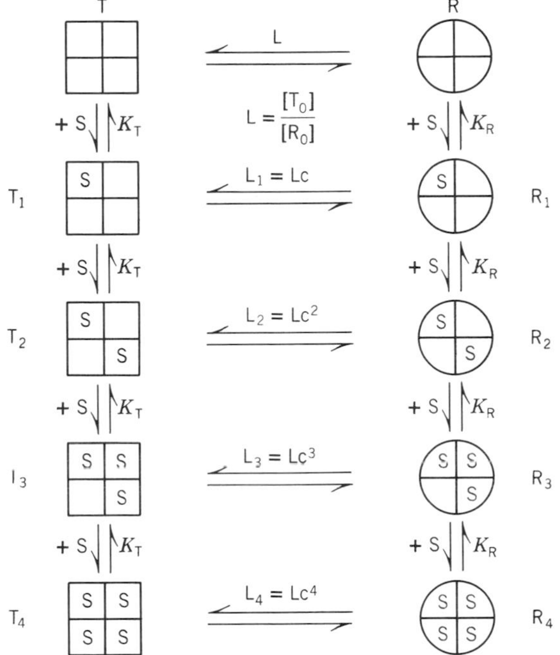

Fig. VII-32. The concerted-symmetry model of Monod, Wyman, and Changeux; T represents a low affinity form of an oligomeric enzyme which is in equilibrium with R, a high affinity form of the enzyme.

below.

$$T_0 \overset{L}{\rightleftharpoons} R_0$$

$$T_0 + S \overset{K_T}{\rightleftharpoons} T_1 = TS \qquad R_0 + S \overset{K_R}{\rightleftharpoons} R_1 = RS$$

$$T_1 + S \overset{K_T}{\rightleftharpoons} T_2 = TS_2 \qquad R_1 + S \overset{K_R}{\rightleftharpoons} R_2 = RS_2$$

$$T_2 + S \overset{K_T}{\rightleftharpoons} T_3 = TS_3 \qquad R_2 + S \overset{K_R}{\rightleftharpoons} R_3 = RS_3$$

$$T_3 + S \overset{K_T}{\rightleftharpoons} T_4 = TS_4 \qquad R_3 + S \overset{K_R}{\rightleftharpoons} R_4 = RS_4$$

The distribution of complexes can be written in the usual manner, taking into account the statistical factor (i.e., the number of different ways of arranging S in any given complex). The distribution of R complexes is shown in Table VII-1. The first column gives the concentration of each complex in general terms for an enzyme with n sites. The concentration of each complex is given in terms of the preceding complex. The second column gives the general distribution in terms of $[R_0]$. The third column gives the distribution in terms of $[R_0]$ for an enzyme with four sites. The fourth column repeats the third using the notation $\alpha = [S]/K_R$. Thus α represents the specific (reduced, normalized, relative) concentration ($[S']$).

Table VII-1 Distribution of R Complexes

In Terms of Each other	In Terms of $[R_0]$	In Terms of $[R_0]$ Where $n=4$	
$[R_1] = \frac{n[S]}{K_R}[R_0]$	$= \frac{n[S]}{K_R}[R_0]$	$= \frac{4[S]}{K_R}[R_0]$	$= 4\alpha[R_0]$
$[R_2] = \frac{(n-1)[S]}{2K_R}[R_1]$	$= \frac{n(n-1)[S]^2}{2K_R^2}[R_0]$	$= \frac{6[S]^2}{K_R^2}[R_0]$	$= 6\alpha^2[R_0]$
$[R_3] = \frac{(n-2)[S]}{3K_R}[R_2]$	$= \frac{n(n-1)(n-2)[S]^3}{6K_R^3}[R_0]$	$= \frac{4[S]^3}{K_R^3}[R_0]$	$= 4\alpha^3[R_0]$
$[R_4] = \frac{(n-3)[S]}{4K_R}[R_3]$	$= \frac{n(n-1)(n-2)(n-3)[S]^4}{24K_R^4}[R_0]$	$= \frac{[S]^4}{K_R^4}[R_0]$	$= \alpha^4[R_0]$
$\vdots$			
$[R_n] = \frac{[S]}{nK_R}[R_{n-1}]$	$= \frac{[S]^n}{K_R^n}[R_0]$		

The distribution of T complexes is shown in Table VII-2 in terms of T_0 and in terms of R_0. The conversion from T_0 to R_0 notation makes use of the relationships:

$$[T_0] = L[R_0]$$

and

$$c = \frac{K_R}{K_T} \qquad \text{or} \qquad K_T = \frac{K_R}{c}$$

where c = the nonexclusive binding coefficient, that is, the ratio of dissociation constants for the R and T states (the ratio of affinities of S for the T and R states). The fraction of total protein in the R state (called the "function of

Table VII-2 Distribution of T Complexes

In Terms of Each Other	In Terms of $[T_0]$	In Terms of $[R_0]$	In Terms of $[R_0]$ Where $n=4$
$[T_1]=\frac{n[S]}{K_T}[T_0]$	$=\frac{n[S]}{K_T}[T_0]$	$=\frac{cn[S]}{K_R}L[R_0]$	$=\frac{4c[S]}{K_R}L[R_0] \quad =4c\alpha L[R_0]$
$[T_2]=\frac{(n-1)[S]}{2K_T}[T_1]$	$=\frac{n(n-1)[S]^2}{2K_T^2}[T_0]$	$=\frac{c^2n(n-1)[S]^2}{2K_R^2}L[R_0]$	$=\frac{6c^2[S]^2}{K_R^2}L[R_0] \quad =6c^2\alpha^2L[R_0]$
$[T_3]=\frac{(n-2)[S]}{3K_T}[T_2]$	$=\frac{n(n-1)(n-2)[S]^3}{6K_T^3}[T_0]$	$=\frac{c^3n(n-1)(n-2)[S]^3}{6K_R^3}L[R_0]$	$=\frac{4c^3[S]^3}{K_R^3}L[R_0] \quad =4c^3\alpha^3L[R_0]$
$[T_4]=\frac{(n-3)[S]}{4K_T}[T_3]$	$=\frac{n(n-1)(n-2)(n-3)[S]^4}{24K_T^4}[T_0]$	$=\frac{c^4n(n-1)(n-2)(n-3)[S]^4}{24K_R^4}L[R_0]$	$=\frac{c^4[S]^4}{K_R^4}L[R_0] \quad =c^4\alpha^4L[R_0]$
$\vdots$	$\vdots$	$\vdots$	
$[T_n]=\frac{[S]}{nK_T}[T_{n-1}]$	$=\frac{[S]^n}{K_T^n}[T_0]$	$=\frac{c^n[S]^n}{K_R^n}L[R_0]$	

state" $\overline{\mathrm{R}}$) is:

$$\overline{\mathrm{R}}=\frac{[\mathrm{R}_0]+[\mathrm{R}_1]+[\mathrm{R}_2]+[\mathrm{R}_3]+[\mathrm{R}_4]\cdots+[\mathrm{R}_n]}{([\mathrm{T}_0]+[\mathrm{T}_1]+[\mathrm{T}_2]+[\mathrm{T}_3]+[\mathrm{T}_4]\cdots+[\mathrm{T}_n])+([\mathrm{R}_0]+[\mathrm{R}_1]+[\mathrm{R}_2]+[\mathrm{R}_3]+[\mathrm{R}_4]\cdots+[\mathrm{R}_n])} \tag{VII-60}$$

The fraction of total sites occupied by S (the "saturation function") designated $\overline{Y}_{\mathrm{S}}$ is:

$$\overline{Y}_{\mathrm{S}}=\frac{([\mathrm{T}_1]+2[\mathrm{T}_2]+3[\mathrm{T}_3]+4[\mathrm{T}_4]\cdots+n[\mathrm{T}_n])+([\mathrm{R}_1]+2[\mathrm{R}_2]+3[\mathrm{R}_3]+4[\mathrm{R}_4]\cdots+n[\mathrm{R}_n])}{n([\mathrm{T}_0]+[\mathrm{T}_1]+[\mathrm{T}_2]+[\mathrm{T}_3]+[\mathrm{T}_4]\cdots+[\mathrm{T}_n])+n([\mathrm{R}_0]+[\mathrm{R}_1]+[\mathrm{R}_2]+[\mathrm{R}_3]+[\mathrm{R}_4]\cdots+[\mathrm{R}_n])} \tag{VII-61}$$

The numerator expresses the fact that there is one occupied site on T_1 and R_1, two occupied sites on T_2 and R_2, and so on. The denominator represents the sum of all sites on all R and T complexes. The expression for $\overline{\mathrm{R}}$ can be simplified and expressed in terms of α, c, and L. For the four-site enzyme:

$$\overline{\mathrm{R}}=\frac{(1+\alpha)^4}{\mathrm{L}(1+\mathrm{c}\alpha)^4+(1+\alpha)^4} \tag{VII-62}$$

In general:

$$\overline{\mathrm{R}}=\frac{(1+\alpha)^n}{\mathrm{L}(1+\mathrm{c}\alpha)^n+(1+\alpha)^n} \tag{VII-63}$$

Similarly, the "function of state" $\overline{\mathrm{T}}$ is:

$$\boxed{\overline{\mathrm{T}} = \frac{\mathrm{L}(1+\mathrm{c}\alpha)^n}{\mathrm{L}(1+\mathrm{c}\alpha)^n + (1+\alpha)^n}} \qquad \text{(VII-64)}$$

The ratio of the fraction of enzyme in the R state to the fraction of enzyme in the T state (the "quotient function") designated $\overline{\mathrm{Q}}$ is:

$$\boxed{\overline{\mathrm{Q}} = \frac{\overline{\mathrm{R}}}{\overline{\mathrm{T}}} = \frac{\overline{\mathrm{R}}}{1-\overline{\mathrm{R}}} = \frac{(1+\alpha)^n}{\mathrm{L}(1+\mathrm{c}\alpha)^n}} \qquad \text{(VII-65)}$$

For an enzyme under rapid equilibrium conditions, the fraction of total sites occupied is equivalent to $v/V_{\max}$; therefore:

$$\boxed{\overline{Y}_{\mathrm{S}} = \frac{v}{V_{\max}} = \frac{\mathrm{L}\mathrm{c}\alpha(1+\mathrm{c}\alpha)^{n-1} + \alpha(1+\alpha)^{n-1}}{\mathrm{L}(1+\mathrm{c}\alpha)^n + (1+\alpha)^n}} \qquad \text{(VII-66)}$$

and

$$\boxed{\overline{Y}_{\mathrm{S}} = \frac{\mathrm{c}\alpha}{1+\mathrm{c}\alpha}(1-\overline{\mathrm{R}}) + \frac{\alpha}{1+\alpha}\overline{\mathrm{R}}} \qquad \text{(VII-67)}$$

When $\mathrm{c}=1$ (i.e., the affinity of both states for S is the same), or when L is very small (i.e., the enzyme exists predominantly in the R state), the velocity equation reduces to the usual Henri-Michaelis-Menten equation.

Exclusive Ligand Binding

The simplest concerted-symmetry model assumes that c is very small; that is, the T state has essentially no affinity for S. Under this condition:

$$\boxed{\overline{\mathrm{R}} = \frac{(1+\alpha)^n}{\mathrm{L} + (1+\alpha)^n}} \qquad \text{(VII-68)}$$

and the velocity equation reduces to:

$$\bar{Y}_S = \frac{v}{V_{max}} = \frac{\alpha}{(1+\alpha)}\bar{R} = \frac{\alpha(1+\alpha)^{n-1}}{L+(1+\alpha)^n} \qquad \text{(VII-69)}$$

The expanded form of equation VII-69 for a four-site enzyme is:

$$\frac{v}{V_{max}} = \frac{\frac{[S]}{K_S} + \frac{3[S]^2}{K_S^2} + \frac{3[S]^3}{K_S^3} + \frac{[S]^4}{K_S^4}}{L+1+\frac{4[S]}{K_S} + \frac{6[S]^2}{K_S^2} + \frac{4[S]^3}{K_S^3} + \frac{[S]^4}{K_S^4}} \qquad \text{(VII-70)}$$

where $K_S = K_R$. Dividing the numerator and denominator of the resulting equation by $(L+1) = M$ yields:

$$\frac{v}{V_{max}} = \frac{\frac{[S]}{MK_S} + \frac{3[S]^2}{MK_S^2} + \frac{3[S]^3}{MK_S^3} + \frac{[S]^4}{MK_S^4}}{1+\frac{4[S]}{MK_S} + \frac{6[S]^2}{MK_S^2} + \frac{4[S]^3}{MK_S^3} + \frac{[S]^4}{MK_S^4}} \qquad \text{(VII-71)}$$

According to the simple sequential interaction model, equation VII-71 gives the velocity for a four-site enzyme where the first substrate molecule binds with a dissociation constant MK_S, and subsequent substrate molecules bind with dissociation constants K_S. In other words, the interaction factor a is equivalent to $1/M$. As shown later, the two models can be made formally equivalent for positive cooperativity even when $c \neq 0$ (nonexclusive substrate binding).

Effect of L and c on Cooperativity

The cooperativity of substrate binding depends on L and c. The velocity curves become more sigmoidal as L increases (i.e., as the $R_0 \rightleftharpoons T_0$ equilibrium favors T_0) and as c decreases (i.e., as the affinity of the T state decreases relative to the affinity of the R state for S). The effects of varying L and c are shown in Figure VII-33 *a* and *b*. Allosteric inhibitors are assumed to bind preferentially to the T state thereby displacing the $T_0 \rightleftharpoons R_0$ equilibrium in favor of T_0. In effect, the allosteric constant, L, increases and the velocity curves become more sigmoidal with n_{app} approaching the actual number of sites. An activator is assumed to bind preferentially to the R state and thus mimics the substrate by shifting the $T_0 \rightleftharpoons R_0$ equilibrium to the right. As a result, the velocity curves become less sigmoidal. At an infinitely high activator concentration, all the enzyme will be driven to the R state and the velocity curve will be hyperbolic (Fig. VII-33*c*). (See System B3c and System B3d for a simple sequential interaction model that yields the same results.) Systems in which the T and R states have different affinities for S and for effectors, but the same intrinsic catalytic activities (i.e., $k_{p_R} = k_{p_T}$), are designated *K* systems because effectors alter the apparent "K_m" (really $[S]_{0.5}$), but V_{max} is unaffected.

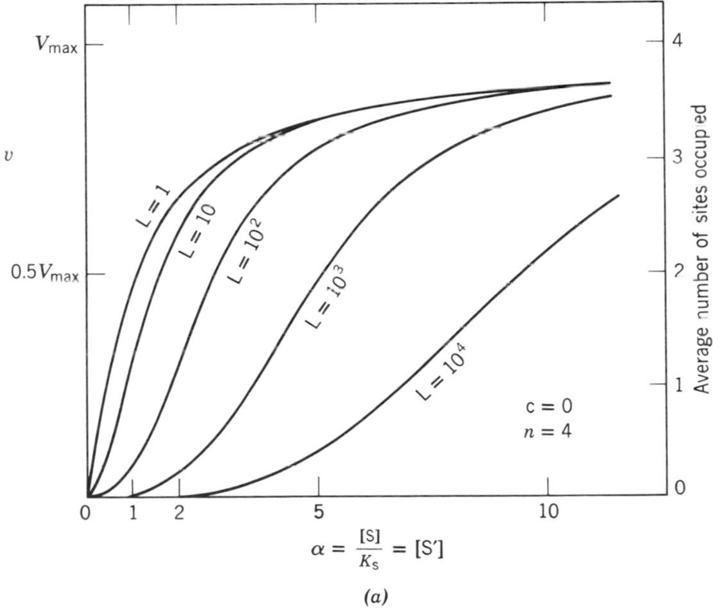

(*a*)

Fig. VII-33. (*a*) Effect of different values of the allosteric constant, L, on the v versus [S] plot of a four-site enzyme. The nonexclusive binding coefficient, c, is assumed to be zero; that is, S binds only to the R state. As L increases, the plots become more sigmoidal.

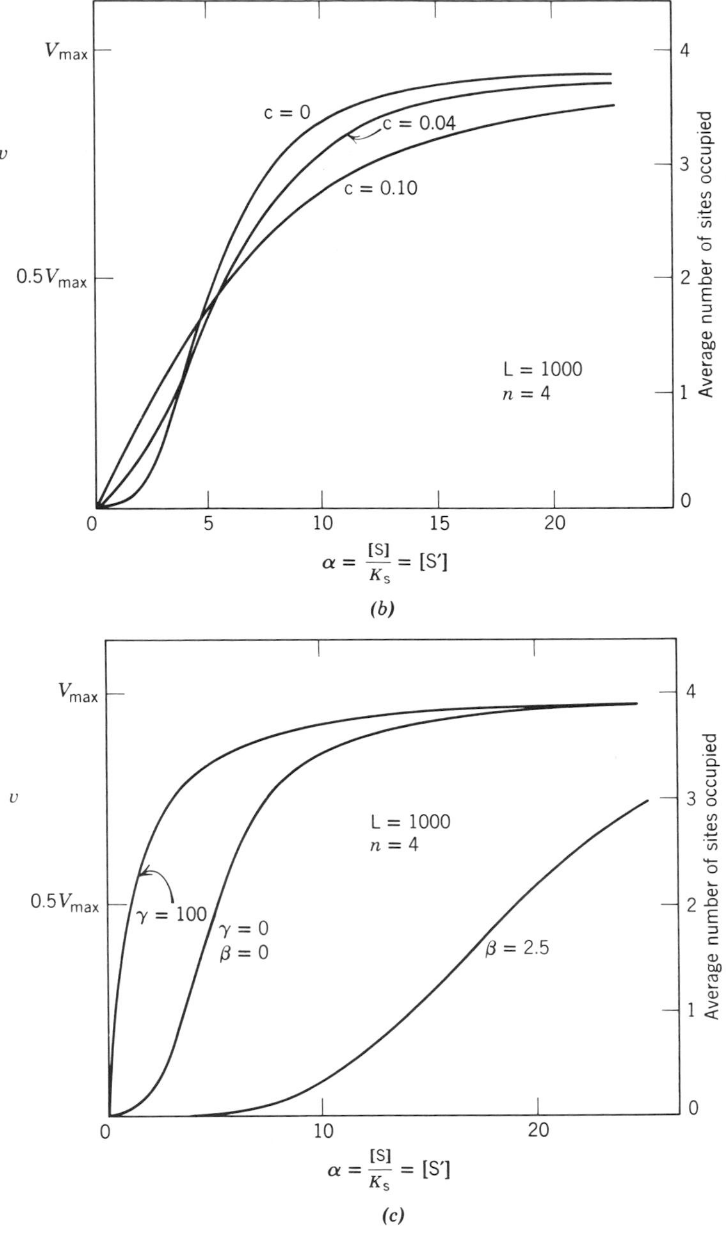

Fig. VII-33. (*Cont.*) (*b*) Effect of different values of c on the v versus [S] plot of a four-site enzyme; L = 1000. As c decreases, the plots become more sigmoidal. (*c*) Effect of an inhibitor and an activator on the v versus [S] plot of a four-site enzyme. The inhibitor binds exclusively to the T state, while the activator and the substrate bind exclusively to the R state; $\beta = [I]/K_{i_T}$, $\gamma = [A]/K_{A_R}$; L = 1000. [Redrawn from Monod, J., Wyman, J. and Changeux, J. P., *J. Mol. Biol.* **12**, 88 (1965).]

V Systems

Systems in which the T and R states have the same affinity for S, but different intrinsic catalytic activities ($k_{p_R} > k_{p_T}$), are designated V systems (the V_{max} of the two states differ). The velocity equation for a V system can be derived in the usual manner.

$$v = \frac{L\alpha(1+\alpha)^{n-1} gnk_p[E]_t + \alpha(1+\alpha)^{n-1} nk_p[E]_t}{L(1+\alpha)^n + (1+\alpha)^n} \qquad \text{(VII-72)}$$

Simplifying:

$$v = \frac{\alpha(1+\alpha)^{n-1}(1+gL)nk_p[E]_t}{(1+\alpha)^n(1+L)} = \frac{\alpha(1+gL)nk_p[E]_t}{(1+\alpha)(1+L)}$$

or

$$\frac{v}{V_{max}} = \frac{\alpha(1+gL)}{(1+\alpha)(1+L)} \qquad \text{(VII-73)}$$

where $V_{max} = nk_p[E]_t$. In writing equation VII-73, it is assumed that the R state has the higher catalytic activity (i.e., $g < 1$). If $g > 1$, the true maximal velocity will be $gnk_p[E]_t = gV_{max}$. The velocity curve for a V system is hyperbolic, just as would be observed for a mixture of two enzymes with the same K_S but different V_{max} values. An effector that preferentially binds to the R state shifts the $T_0 \rightleftharpoons R_0$ equilibrium toward the higher activity R state and thereby behaves as an activator. The V_{max} will increase, but the velocity curves remain hyperbolic with a constant $[S]_{0.5}$. An effector that binds preferentially to the lower activity T state shifts the equilibrium toward the T state and thereby acts as an inhibitor. If the T state has absolutely no catalytic activity ($g = 0$), the activator will be essential and the inhibitor will act in a pure noncompetitive manner. If both T and R states are catalytically active, A will be a nonessential activator, while I will behave as a partial noncompetitive inhibitor. Although the v versus [S] curves are hyperbolic in the presence of effector, the v versus [I] curve at a fixed [A] and [S] can be sigmoidal, and, similarly, the v versus [A] curve at a fixed [I] and [S] can be sigmoidal. The sigmoidicity of the v versus [varied effector] curve increases as the concentration of fixed effector increases.

Mixed K and V Systems

If the R and T states have different affinities for S ($c \neq 1$) as well as different catalytic activities ($k_{p_R} \neq k_{p_T}$ or $g \neq 1$), the velocity equation becomes:

$$v = \frac{Lc\alpha(1+c\alpha)^{n-1} gnk_p[E]_t + (1+\alpha)^{n-1} nk_p[E]_t}{L(1+c\alpha)^n + (1+\alpha)^n} \qquad \text{(VII-74)}$$

If $g < 1$ and $c < 1$, the R state has both the higher affinity for S as well as the higher catalytic activity. Alternately it is possible for the higher affinity state to have the lower catalytic activity ($c < 1, g > 1$) leading to substrate inhibition. If $c \neq 1$ and $g \neq 1$, effectors that bind preferentially to one state will now affect both $[S]_{0.5}$ and V_{max}.

The kinetics of activation and inhibition in terms of the concerted-symmetry model, and the consequences of nonexclusive ligand binding are described in more detail later. In fact, it is only through the use of inhibitors and activators that display preferential binding that all the parameters K_R, K_T (hence c), n, and L can be estimated.

Comparison and Formal Equivalence of the Sequential and Concerted Models

For any finite values of K_R, L, and c of the concerted symmetry model, we can calculate the corresponding K_S, a, b, c, and so on, values of the simple sequential interaction model. In other words, a sigmoidal substrate saturation curve that fits the velocity equation of the former model can be shown to fit the velocity equation of the latter model. For a four-site enzyme, the equations and equivalent values are shown below:

Simple sequential interaction model

$$\frac{v}{V_{max}} = \frac{\dfrac{[S]}{K_S} + \dfrac{3[S]^2}{aK_S^2} + \dfrac{3[S]^3}{a^2bK_S^3} + \dfrac{[S]^4}{a^3b^2cK_S^4}}{1 + \dfrac{4[S]}{K_S} + \dfrac{6[S]^2}{aK_S^2} + \dfrac{4[S]^3}{a^2bK_S^3} + \dfrac{[S]^4}{a^3b^2cK_S^4}} \qquad \text{(VII-75)}$$

or

$$\frac{v}{V_{\max}} = \frac{\dfrac{[\mathrm{S}]}{K_{\mathrm{S}}} + 3i\left(\dfrac{[\mathrm{S}]}{K_{\mathrm{S}}}\right)^2 + 3i^2j\left(\dfrac{[\mathrm{S}]}{K_{\mathrm{S}}}\right)^3 + i^3j^2k\left(\dfrac{[\mathrm{S}]}{K_{\mathrm{S}}}\right)^4}{1 + 4\left(\dfrac{[\mathrm{S}]}{K_{\mathrm{S}}}\right) + 6i\left(\dfrac{[\mathrm{S}]}{K_{\mathrm{S}}}\right)^2 + 4i^2j\left(\dfrac{[\mathrm{S}]}{K_{\mathrm{S}}}\right)^3 + i^3j^2k\left(\dfrac{[\mathrm{S}]}{K_{\mathrm{S}}}\right)^4} \qquad \text{(VII-76)}$$

Concerted-symmetry model

$$\frac{v}{V_{\max}} = \frac{\mathrm{Lc}\alpha(1+\alpha\mathrm{c})^3 + \alpha(1+\alpha)^3}{\mathrm{L}(1+\alpha\mathrm{c})^4 + (1+\alpha)^4} \qquad \text{(VII-77)}$$

or

$$\frac{v}{V_{\max}} = \frac{\mathrm{Lc}\dfrac{[\mathrm{S}]}{K_{\mathrm{R}}}\left(1 + \mathrm{c}\dfrac{[\mathrm{S}]}{K_{\mathrm{R}}}\right)^3 + \dfrac{[\mathrm{S}]}{K_{\mathrm{R}}}\left(1 + \dfrac{[\mathrm{S}]}{K_{\mathrm{R}}}\right)^3}{\mathrm{L}\left(1 + \mathrm{c}\dfrac{[\mathrm{S}]}{K_{\mathrm{R}}}\right)^4 + \left(1 + \dfrac{[\mathrm{S}]}{K_{\mathrm{R}}}\right)^4} \qquad \text{(VII-78)}$$

Equations for both models yield the same $v/V_{\max}$ values when:

$$K_{\mathrm{S}} = K_{\mathrm{R}}\left(\frac{1+\mathrm{L}}{1+\mathrm{Lc}}\right)$$

$$\frac{1}{a} = i = 1 + \frac{\dfrac{(1-\mathrm{c})^2}{2\mathrm{c}}}{1 + \dfrac{1}{2\mathrm{Lc}} + \dfrac{\mathrm{Lc}}{2}}$$

$$\frac{1}{b} = j = 1 + \frac{\dfrac{(1-\mathrm{c})^2}{2\mathrm{c}}}{1 + \dfrac{1}{2\mathrm{Lc}^2} + \dfrac{\mathrm{Lc}^2}{2}}$$

$$\frac{1}{c} = k = 1 + \frac{\dfrac{(1-\mathrm{c})^2}{2\mathrm{c}}}{1 + \dfrac{1}{2\mathrm{Lc}^3} + \dfrac{\mathrm{Lc}^3}{2}}$$

The interaction factors a, b, and c are factors by which the dissociation constants of vacant sites are progressively changed as sites are filled. The factors i, j, and k are the corresponding reciprocals and can be considered as factors by which the binding constants are changed.

The simple sequential interaction model is more general than the concerted-symmetry model in that there are many combinations of values for a, b, and c for which there are no equivalent values of L and c. Even values of $a = b = c \neq 1$ are excluded from the concerted-symmetry model. Furthermore, the concerted-symmetry model explicitly excludes negative cooperativity, a property displayed by many allosteric enzymes. Nevertheless, the concerted-symmetry model does provide explanations for many properties of allosteric enzymes including (a) partial allosteric inhibition, (b) increased, decreased, or constant substrate cooperativity in the presence of effectors, and (c) activation by low concentrations of competitive inhibitors (see below). Most importantly, the concerted-symmetry model provides a single, easily visualized physical explanation for all cooperative effects (*viz.*, the displacement of the $T \rightleftharpoons R$ equilibrium by preferential ligand binding).

Inhibition in Exclusive Binding *K* Systems

In the concerted-symmetry model, an allosteric inhibitor is assumed to have a much higher affinity for the T state than for the R state. As a result, an allosteric inhibitor will displace the $T_0 \rightleftharpoons R_0$ equilibrium in favor of T_0. If we assume that S binds exclusively to the R state (i.e., c = 0) and I binds exclusively to the T state (i.e., $K_{i_R}/K_{i_T} = f \gg 1$) the kinds of complexes that can form are those shown in Figure VII-34. The velocity equation can be derived as described earlier except now all the ES complexes exist solely in the R state, while the T state is composed of T_0 and the various complexes containing I. The equations become:

$$\boxed{\bar{R} = \frac{(1+\alpha)^n}{L(1+\beta)^n + (1+\alpha)^n}} \qquad \text{(VII-79)}$$

$$\boxed{\frac{v}{V_{max}} = \frac{\alpha}{(1+\alpha)}\bar{R} = \frac{\alpha(1+\alpha)^{n-1}}{L(1+\beta)^n + (1+\alpha)^n}} \qquad \text{(VII-80)}$$

where

$$\alpha = \frac{[S]}{K_{S_R}} = \frac{[S]}{K_S} \qquad \text{and} \qquad \beta = \frac{[I]}{K_{i_T}} = \frac{[I]}{K_i}$$

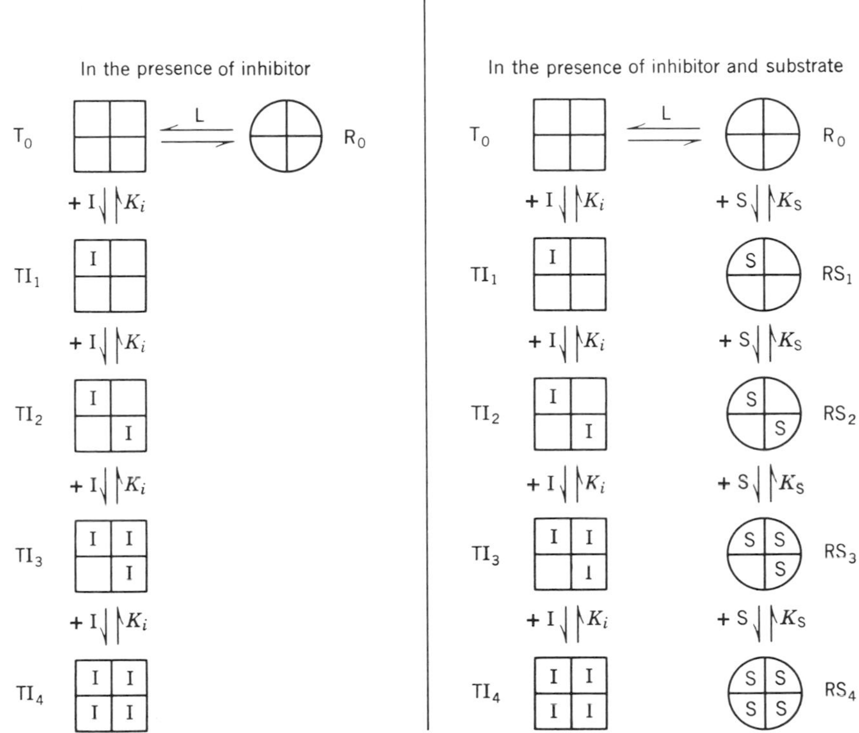

Fig. VII-34. Enzyme species in the presence of S and I; S binds exclusively to the R state; I binds exclusively to the T state.

or

$$\frac{v}{V_{\max}} = \frac{\alpha(1+\alpha)^{n-1}}{\mathrm{L}' + (1+\alpha)^n} \qquad \text{(VII-81)}$$

where

$$\mathrm{L}' = \text{an apparent allosteric constant} = \mathrm{L}(1+\beta)^n = \mathrm{L}\left(1 + \frac{[\mathrm{I}]}{K_i}\right)^n$$

As [I] increases, L′ increases and the velocity curves become more sigmoidal (Fig. VII-33*a* and *c*). The n_{app} (determined from the slope of the Hill plot) approaches the actual number of substrate binding sites.

In the expanded form, the equation for a four-site enzyme would be:

$$\frac{v}{V_{\max}} = \frac{\dfrac{[S]}{K_S} + \dfrac{3[S]^2}{K_S^2} + \dfrac{3[S]^3}{K_S^3} + \dfrac{[S]^4}{K_S^4}}{L\left(1 + \dfrac{4[I]}{K_i} + \dfrac{6[I]^2}{K_i^2} + \dfrac{4[I]^3}{K_i^3} + \dfrac{[I]^4}{K_i^4}\right) + \left(1 + \dfrac{4[S]}{K_S} + \dfrac{6[S]^2}{K_S^2} + \dfrac{4[S]^3}{K_S^3} + \dfrac{[S]^4}{K_S^4}\right)} \quad \text{(VII-82)}$$

Note that there are no terms containing both [I] and [S] because of the assumption of exclusive binding of each ligand to only one of the two states. As [S] increases, the v versus [I] curves become more sigmoidal (Fig. VII-35) and n_{app} approaches the actual number of inhibitor sites (which equals the number of substrate sites in the simplest concerted-symmetry model).

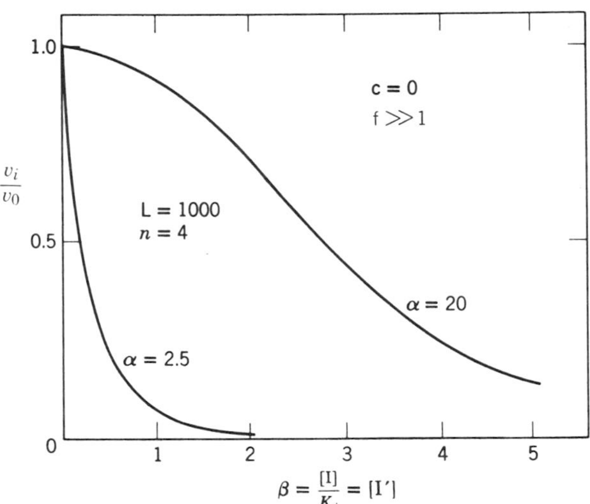

Fig. VII-35. The v versus [I] plot for a four-site enzyme at two different fixed substrate concentrations; L = 1000; S binds exclusively to the R state (i.e., c = 0), while I binds exclusively to the T state. [Redrawn from Monod, J., Wyman, J. and Changeux, J. P., *J. Mol. Biol.* **12**, 88 (1965).]

Determination of K_i and L

As shown earlier:

$$\bar{Y}_S = \frac{v}{V_{max}} = \frac{c\alpha}{1+c\alpha}(1-\bar{R}) + \frac{\alpha}{1+\alpha}\bar{R} \qquad \text{(VII-83)}$$

When c is very small, equation VII-83 reduces to:

$$\bar{Y}_S = \frac{v}{V_{max}} = \frac{\alpha}{1+\alpha}\bar{R} \qquad \text{(VII-84)}$$

Equation VII-84 expresses the fact that in an exclusive binding K system (i.e., c = 0) the variation in the specific velocity at constant specific substrate concentration is a function of the fraction of the enzyme in the R state. The $\bar{R}$ in turn will vary with the concentrations of effectors which bind exclusively to either the R state or the T state. Rearranging equation VII-84:

$$\bar{R} = \frac{v}{V_{max}}\frac{(1+\alpha)}{\alpha} = \frac{v}{V'} \qquad \text{(VII-85)}$$

where $V' = V_{max}\left(\frac{\alpha}{1+\alpha}\right)$

= the velocity at the fixed substrate concentration that would be observed if all the enzyme were in the R state

= the velocity at the fixed substrate concentration in the presence of a saturating concentration of an activator

In the presence of an allosteric inhibitor, the ratio between the R and T states, $\bar{Q}$, is (from equation VII-65):

$$\bar{Q} = \frac{\bar{R}}{\bar{T}} = \frac{\bar{R}}{1-\bar{R}} = \frac{\dfrac{(1+\alpha)^n}{L(1+\beta)^n + (1+\alpha)^n}}{1 - \dfrac{(1+\alpha)^n}{L(1+\beta)^n + (1+\alpha)^n}} = \frac{(1+\alpha)^n}{L(1+\beta)^n}$$

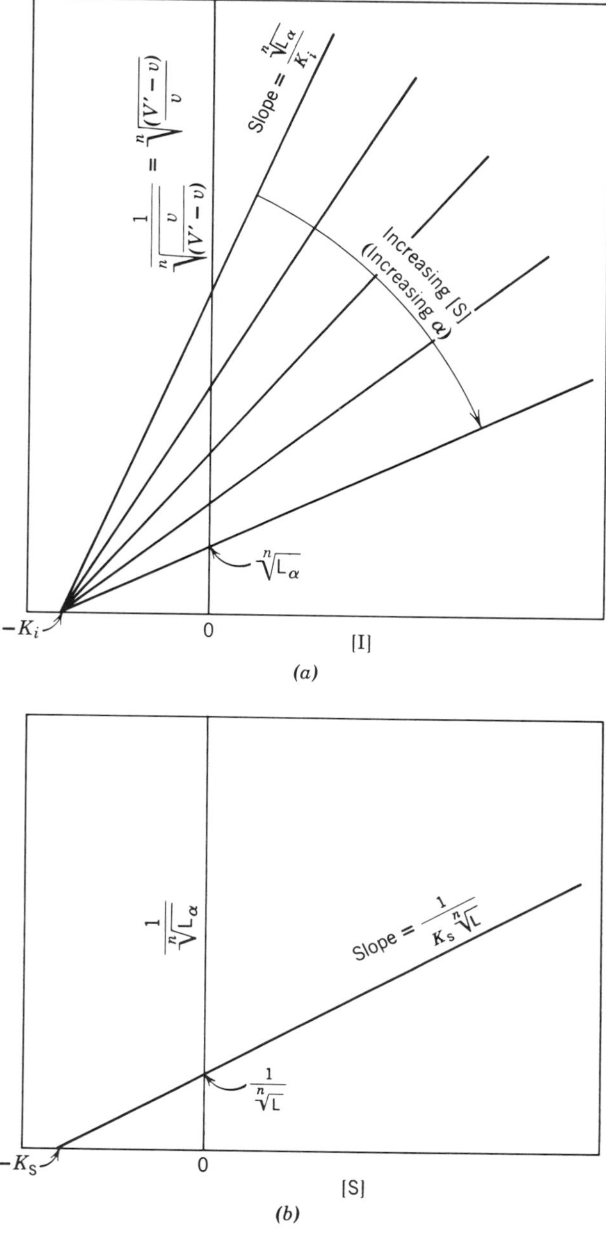

Fig. VII-36. (*a*) Plot of $\sqrt[n]{(V'-v)/v}$ versus [I] at different fixed substrate concentrations. (*b*) Replot of the reciprocal vertical-axis intercepts versus [S]. The plots determine K_i (i.e., K_{i_T}) and L if n (the actual number of sites) is known.

If α is held constant so that $\overline{Q}$ depends only on β:

$$\boxed{\overline{Q} = \frac{1}{L_\alpha(1+\beta)^n}} \tag{VII-86}$$

where

$$L_\alpha = \frac{L}{(1+\alpha)^n} = \text{an apparent allosteric constant at a constant } \alpha$$

From equations VII-65 and VII-85, we obtain:

$$\overline{Q} = \frac{\overline{R}}{1-\overline{R}} = \frac{v}{V'-v} \tag{VII-87}$$

Substituting for $\overline{Q}$ in equation VII-86, we obtain:

$$\boxed{\frac{v}{V'-v} = \frac{1}{L_\alpha(1+\beta)^n}} \tag{VII-88}$$

Equation VII-88 may be converted to a linear form. Taking the nth root of both sides:

$$\sqrt[n]{\frac{v}{(V'-v)}} = \frac{1}{\sqrt[n]{L_\alpha}\left(1+\frac{[I]}{K_i}\right)} = \frac{K_i}{\sqrt[n]{L_\alpha}\,K_i + \sqrt[n]{L_\alpha}\,[I]}$$

Inverting and separating terms:

$$\boxed{\sqrt[n]{\frac{(V'-v)}{v}} = \frac{\sqrt[n]{L_\alpha}}{K_i}[I] + \sqrt[n]{L_\alpha}} \tag{VII-89}$$

Thus a plot of $\sqrt[n]{(V'-v)/v}$ versus [I] at a fixed [S] is a straight line with a slope of $\sqrt[n]{L_\alpha}/K_i$ and intercepts of $\sqrt[n]{L_\alpha}$ and K_i (Fig. VII-36*a*). The L may be calculated from the vertical-axis intercept (if $\alpha = [S]/K_S$ is known) or, alternately, the reciprocal of the intercepts for different fixed [S] may be replotted as shown in Figure VII-36*b*.

$$\boxed{\frac{1}{\sqrt[n]{L_\alpha}} = \frac{1}{K_S \sqrt[n]{L}}[S] + \frac{1}{\sqrt[n]{L}}} \qquad \text{(VII-90)}$$

Thus the replot is linear (Fig. VII-36*b*) with a slope of $1/K_S \sqrt[n]{L}$ and an intercept on the vertical axis of $1/\sqrt[n]{L}$. When $1/\sqrt[n]{L_\alpha} = 0$, the intercept on the [S]-axis gives $-K_S$ (i.e., $-K_{S_R}$).

To determine K_i and L as described above we must know V' and n. Even if $c = 0$, the n_{app} determined from the slope of the Hill plot will be less than n unless L is very large. However, as noted earlier, an inhibitor that binds exclusively to the T form increases the effective allosteric constant; hence n_{app}. Thus a minimum value for n may be estimated from Hill plots of v versus [S] data obtained in the presence of very high inhibitor concentrations. For example, if n_{app} increases from 2.4 (in the absence of inhibitor) to a maximum of 3.3 (in the presence of saturating inhibitor), then n is at least 4. Alternately, n may be estimated from ligand binding studies by means of equilibrium dialysis (n = moles ligand bound per mole of oligomer at saturating ligand concentration). If the $\sqrt[n]{(V' - v)/v}$ versus [I] plots are nonlinear and do not intersect on the [I]-axis then the assumed value of n is either too high or too low. If n cannot be estimated by either of the methods above then the data can be plotted according to equation VII-89 with assumed values of n. The acceptable value would be that which yields linear plots. To obtain V' we need an activator that drives all the enzyme to the R state.

Activation in Exclusive Binding *K* Systems

Activators are assumed to bind preferentially to the R state at sites distinct from the substrate binding sites (i.e., $K_{A_R}/K_{A_T} = e \ll 1$). An activator mimics the substrate by displacing the $T_0 \rightleftharpoons R_0$ equilibrium in the direction of R_0. Because both S and A can occupy an R subunit independently, there are a large number of possible R complexes. Figure VII-37 shows the various R complexes that can exist for an enzyme with two subunits (i.e., two S sites and two A sites). The velocity equation can be derived in the usual manner. For the two-site enzyme where S and A bind only to the R state the equation is:

$$\frac{v}{V_{\max}} = \frac{\dfrac{[S]}{K_S} + \dfrac{[S]^2}{K_S^2} + \dfrac{2[S][A]}{K_S K_A} + \dfrac{2[S]^2[A]}{K_S^2 K_A} + \dfrac{[S][A]^2}{K_S K_A^2} + \dfrac{[S]^2[A]^2}{K_S^2 K_A^2}}{L + 1 + \dfrac{2[S]}{K_S} + \dfrac{[S]^2}{K_S^2} + \dfrac{2[A]}{K_A} + \dfrac{[A]^2}{K_A^2} + \dfrac{4[S][A]}{K_S K_A} + \dfrac{2[S]^2[A]}{K_S^2 K_A} + \dfrac{2[S][A]^2}{K_S K_A^2} + \dfrac{[S]^2[A]^2}{K_S^2 K_A^2}} \tag{VII-91}$$

where $V_{\max} = 2k_p[E]_t$. The equation can be simplified to:

$$\frac{v}{V_{\max}} = \frac{\dfrac{[S]}{K_S}\left(1 + \dfrac{[S]}{K_S}\right)\left(1 + \dfrac{[A]}{K_A}\right)^2}{L + \left(1 + \dfrac{[S]}{K_S}\right)^2\left(1 + \dfrac{[A]}{K_A}\right)^2} = \frac{\alpha(1+\alpha)(1+\gamma)^2}{L + (1+\alpha)^2(1+\gamma)^2} \tag{VII-92}$$

where

$$\alpha = \frac{[S]}{K_S} = \frac{[S]}{K_{S_R}} \qquad \text{and} \qquad \gamma = \frac{[A]}{K_A} = \frac{[A]}{K_{A_R}}$$

Dividing numerator and denominator by $(1+\gamma)^2$:

$$\frac{v}{V_{\max}} = \frac{\alpha(1+\alpha)}{\dfrac{L}{(1+\gamma)^2} + (1+\alpha)^2} \tag{VII-93}$$

Or, in general:

$$\frac{v}{V_{\max}} = \frac{\alpha(1+\alpha)^{n-1}}{L' + (1+\alpha)^n} \tag{VII-94}$$

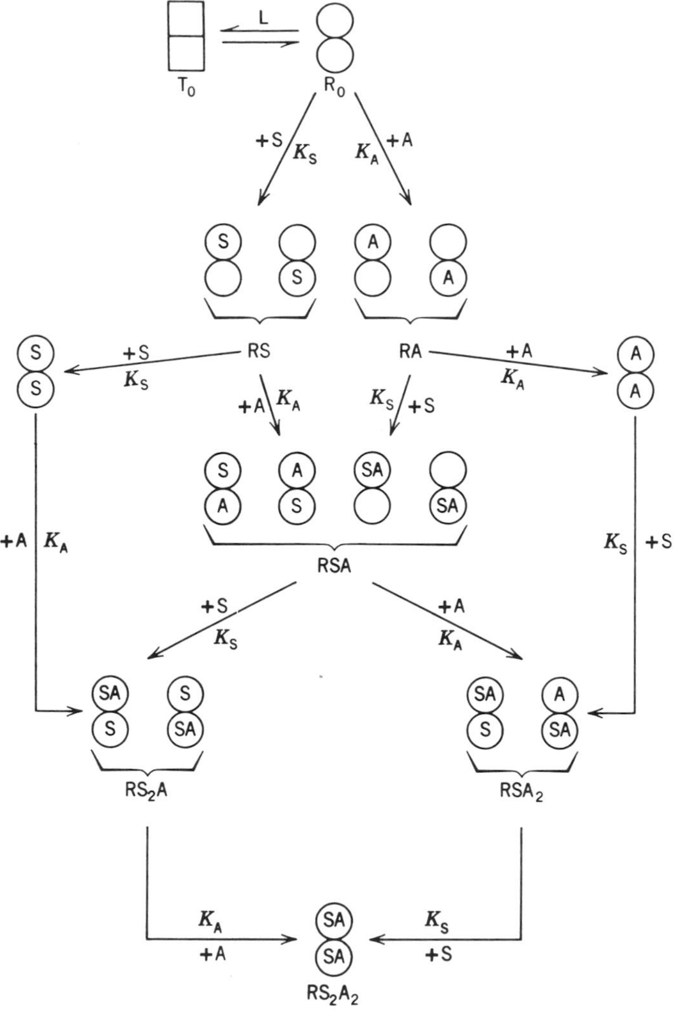

Fig. VII-37. Distribution of enzyme species in the presence of the substrate and an activator, both of which bind to the R state.

where L′ = an apparent allosteric constant = $L/(1+\gamma)^n$. As [A] increases, L′ decreases. When [A] is infinitely high, L′ is zero and the velocity curve reduces to:

$$\frac{v}{V_{\max}} = \frac{\alpha(1+\alpha)^{n-1}}{(1+\alpha)^n} = \frac{\alpha}{1+\alpha} = \frac{\frac{[S]}{K_S}}{1+\frac{[S]}{K_S}} = \frac{[S]}{K_S+[S]}$$

Thus in the presence of an infinitely high activator concentration all the enzyme is driven to the R state and the velocity curve becomes hyperbolic with $[S]_{0.5} = K_S$.

In the presence of an inhibitor and an activator, the velocity equation contains terms representing all the RA, RS, and RSA complexes shown in Figure VII-37, as well as terms representing all the T complexes containing I shown in Figure VII-34. The velocity equation simplifies to:

$$\boxed{\frac{v}{V_{max}} = \frac{\alpha(1+\alpha)^{n-1}}{L\left(\dfrac{(1+\beta)^n}{(1+\gamma)^n}\right) + (1+\alpha)^n}} \tag{VII-95}$$

or

$$\boxed{\frac{v}{V_{max}} = \frac{\alpha(1+\alpha)^{n-1}}{L' + (1+\alpha)^n}} \tag{VII-96}$$

where L′, the apparent allosteric constant $= L(1+\beta)^n/(1+\gamma)^n$.

Because A and I have opposite effects on the $T_0 \rightleftharpoons R_0$ equilibrium, plots of v versus [I] at fixed [S] and [A] (Fig. VII-38*a*) and v versus [A] at fixed [S] and [I] (Fig. VII-38*b*) are sigmoidal with n_{app} values approaching the actual number of sites as the fixed effector concentration increases. When the inhibitor and activator are maintained at equal specific concentrations (i.e., $\beta = \gamma$) $L' = L$ and the specific velocity at any given substrate concentration is constant, regardless of the actual concentrations of inhibitor or activator (equation VII-95).

Determination of K_A and L

When the activator and substrate both bind exclusively to the R state, then:

$$\bar{R} = \frac{(1+\alpha)^n(1+\gamma)^n}{L + (1+\alpha)^n(1+\gamma)^n}$$

and

$$\bar{Q} = \frac{\bar{R}}{\bar{T}} = \frac{\bar{R}}{1-\bar{R}} = \frac{(1+\alpha)^n(1+\gamma)^n}{L}$$

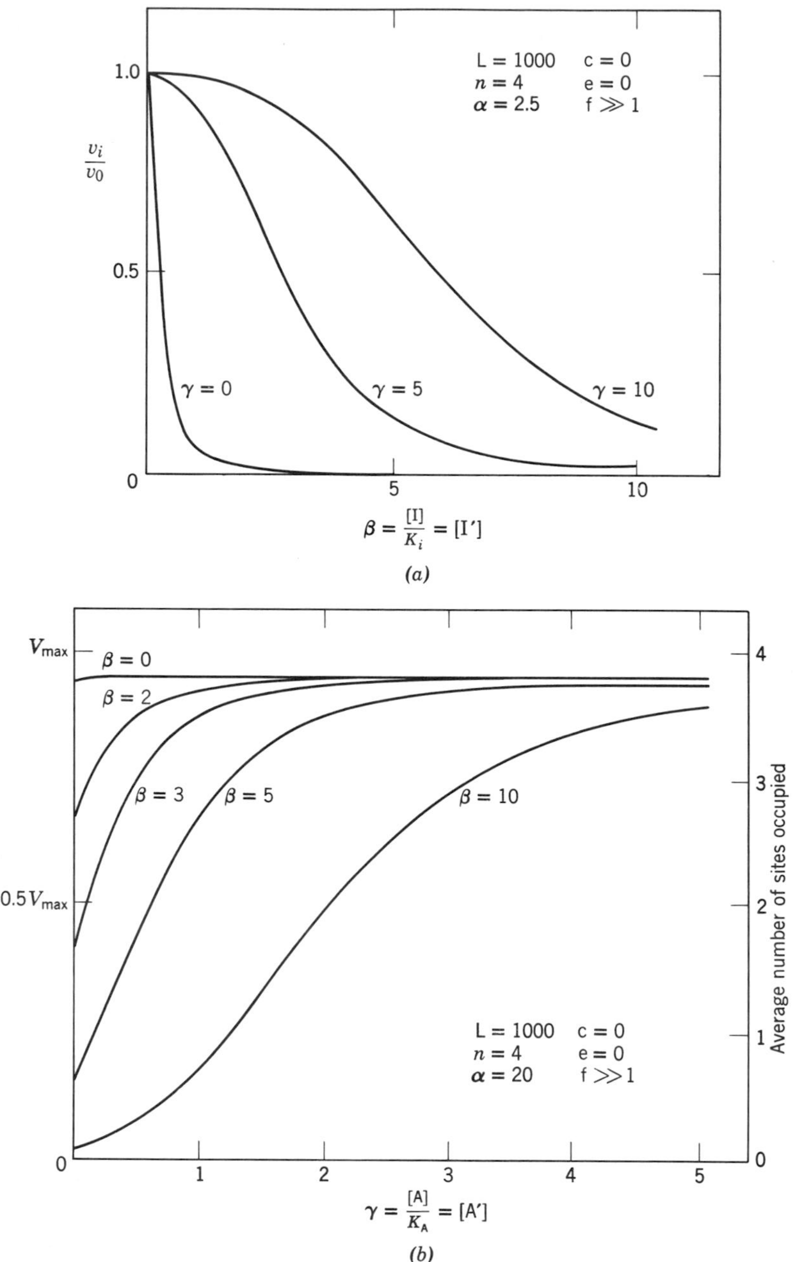

Fig. VII-38. *(a)* v_i/v versus [I] plot in the presence of different fixed concentrations of activator and a constant substrate concentration; $n=4$; L=1000; $\alpha=[S]/K_{S_R}=2.5$. *(b)* v versus [A] plot in the presence of different fixed concentrations of inhibitor and a constant substrate concentration; $n=4$; L=1000; $\alpha=20$. [Redrawn from Monod, J., Wyman, J. and Changeux, J. P., *J. Mol. Biol.* **12**, 88 (1965).]

At a constant α:

$$\bar{Q} = \frac{(1+\gamma)^n}{L_\alpha}$$

where

$$L_\alpha = \frac{L}{(1+\alpha)^n}$$

As shown earlier (equation VII-87):

$$\bar{Q} = \frac{v}{V' - v}$$

$$\therefore \quad \boxed{\frac{v}{V'-v} = \frac{(1+\gamma)^n}{L_\alpha}} \qquad \text{(VII-97)}$$

where V' = the velocity at the fixed substrate concentration that would be observed if all the enzyme were in the R state; that is, the velocity at the fixed substrate concentration and a saturating concentration of activator.

Equation VII-97 can be converted to a linear form. Taking the nth root of both sides and separating terms:

$$\boxed{\sqrt[n]{\frac{v}{(V'-v)}} = \frac{1}{K_A \sqrt[n]{L_\alpha}}[A] + \frac{1}{\sqrt[n]{L_\alpha}}} \qquad \text{(VII-98)}$$

The plots and the replot of the vertical-axis intercepts are shown in Figure VII-39. If n and V' are known, K_A and L may be determined.

Horn-Börnig Plot to Determine n and L (when c = 0)

Horn and Börnig (1969) have described a linear plot that allows n and L (or L′) to be determined in exclusive binding systems (c = 0). The linear transformation of the exclusive binding equation is described below.

$$\bar{Y}_S = \frac{v}{V_{max}} = \frac{\alpha(1+\alpha)^{n-1}}{L' + (1+\alpha)^n}$$

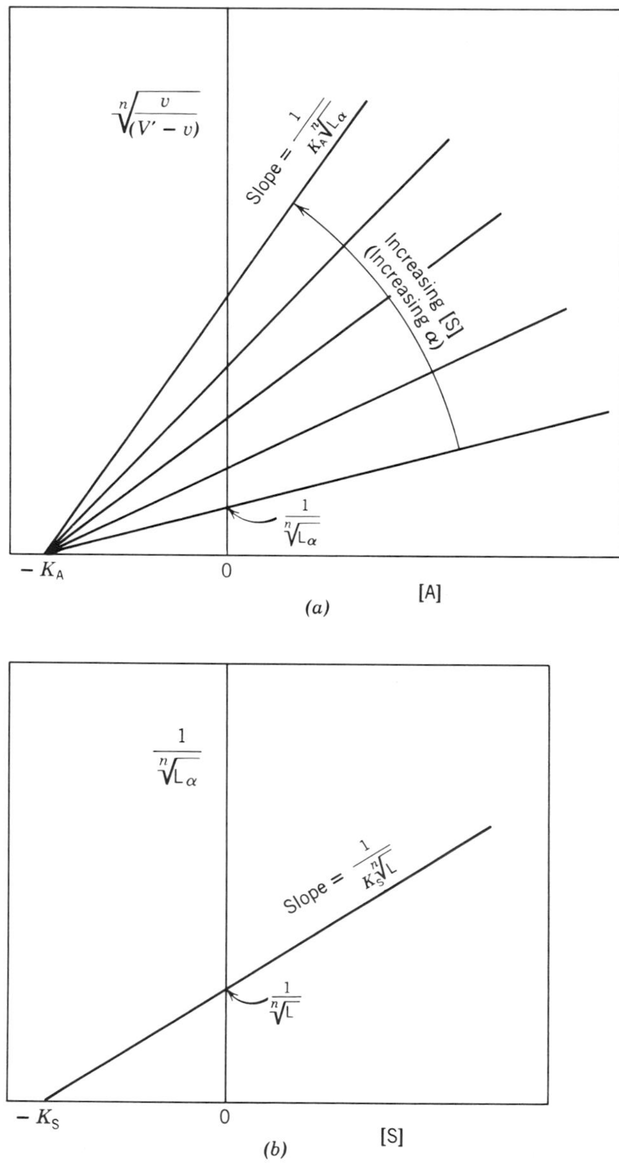

Fig. VII-39. (*a*) Plot of $\sqrt[n]{v/(V'-v)}$ versus [A] at different fixed substrate concentrations. (*b*) Replot of the vertical-axis intercepts versus [S]. The plots determine K_A (i.e., K_{A_R}) and L if *n* (the actual number of sites) is known.

Inverting:

$$\frac{V_{max}}{v} = \frac{L' + (1+\alpha)^n}{\alpha(1+\alpha)^{n-1}}$$

Dividing numerator and denominator of the right side of the equation by $(1+\alpha)^{n-1}$:

$$\frac{V_{max}}{v} = \frac{\dfrac{L'}{(1+\alpha)^{n-1}} + 1 + \alpha}{\alpha} = \frac{1}{\alpha}\left(\frac{L'}{(1+\alpha)^{n-1}} + 1 + \alpha\right)$$

Rearranging:

$$\frac{\alpha V_{max}}{v} - \alpha - 1 = \frac{L'}{(1+\alpha)^{n-1}}$$

Taking logarithms of both sides:

$$\log\left(\frac{\alpha V_{max}}{v} - \alpha - 1\right) = \log L' - (n-1)\log(1+\alpha)$$

or

$$\boxed{\log\left(\frac{\alpha V_{max}}{v} - \alpha - 1\right) = (1-n)\log(1+\alpha) + \log L'} \qquad \text{(VII-99)}$$

Thus a plot of $\log(\alpha V_{max}/v - \alpha - 1)$ versus $\log(1+\alpha)$ is a straight line with a slope of $(1-n)$ and an intercept on the vertical axis of $\log L'$. The intercept on the horizontal axis gives $\log L'/(n-1)$. The plot is illustrated in Figure VII-40. The L is determined from data obtained in the absence of effectors; L′ is determined from data obtained in the presence of effectors. Inhibitors increase the effective allosteric constant from L to L′ and thus displace the line above and parallel to the control line. In the presence of an inhibitor, the allosteric constant is given by:

$$L' = L(1+\beta)^n \quad \text{or} \quad \log L' = \log L + n\log(1+\beta)$$

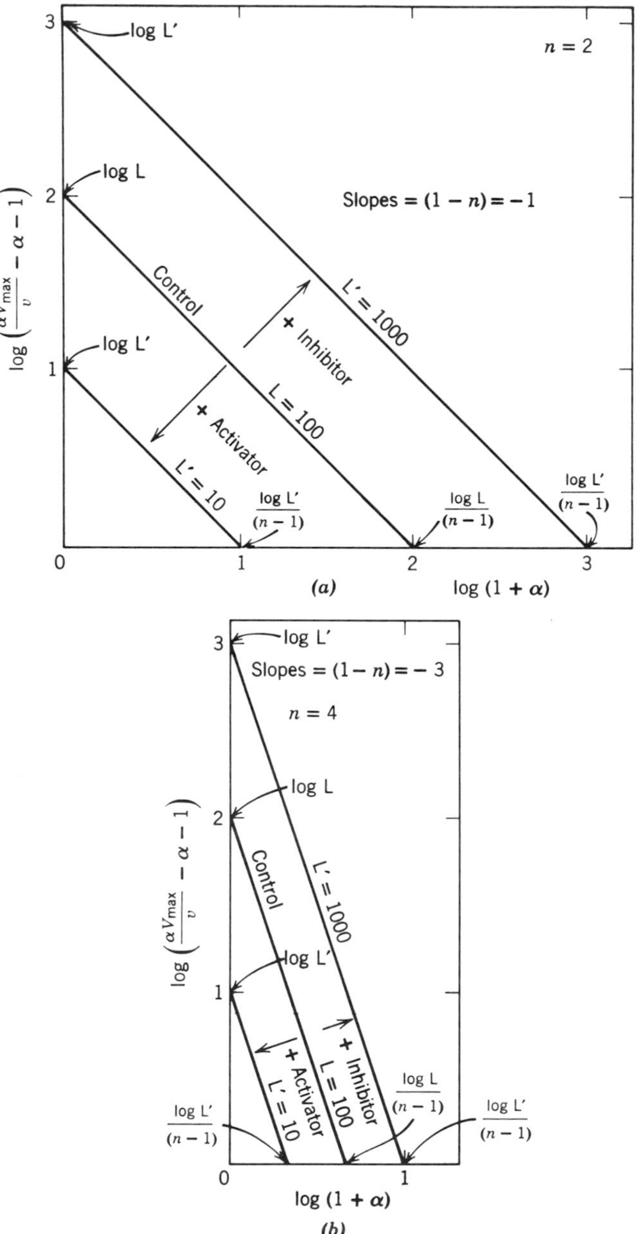

Fig. VII-40. Horn-Börnig plot of $\log[(\alpha V_{max}/v) - \alpha - 1]$ versus $\log(1+\alpha)$ for exclusive binding systems. The plot determines n and L or L′ (hence K_{A_R} and K_{i_T}) if α is known. (*a*) Plot for $n=2$. (*b*) Plot for $n=4$.

Thus the displacement is equivalent to $n \log(1+\beta)$. Since L and the concentration of inhibitor will be known, β and K_{i_T} can be calculated from L′. Similarly, activators decrease the effective allosteric constant from L to L′ and thus displace the line below and parallel to the control line. In the presence of an activator, the allosteric constant is given by:

$$L' = \frac{L}{(1+\gamma)^n} \quad \text{or} \quad \log L' = \log L - n \log(1+\gamma)$$

Thus the displacement is equivalent to $n \log(1+\gamma)$. Since L and the concentration of activator will be known, γ and K_{A_R} can be calculated from L′. Construction of the Horn-Börnig plot requires a knowledge of α, and, therefore, of K_{S_R}. Consequently, it is necessary to have an activator that will drive all the enzyme to the R state so that K_{S_R} can be determined from the hyperbolic kinetics. The Horn-Börnig plot works only for exclusive binding systems (where $c=0$). Allosteric enzymes with nonexclusive ligand binding ($c>0$) give sigmoidal curves.

Combinations of Alternate Effectors

An enzyme may respond to either of two (or more) alternate effectors. If the effectors bind to the same site, they will compete with each other. For example, if the effectors are activators that bind exclusively to the R state, the velocity equation becomes:

$$\boxed{\frac{v}{V_{max}} = \frac{\alpha(1+\alpha)^{n-1}}{\dfrac{L}{(1+\gamma+\delta)^n} + (1+\alpha)^n}} \qquad \text{(VII-100)}$$

where $\gamma = \dfrac{[A]}{K_A}$ (the normalized or specific concentration of one activator)

$\delta = \dfrac{[A']}{K_{A'}}$ (the normalized or specific concentration of the other activator)

If the two effectors were inhibitors, the L′ term in the denominator would be $L(1+\gamma+\delta)^n$. The additive notation indicates that the two effectors bind in an "either-or" (competitive) manner. For example, if $n=2$, the expansion of $(1+\gamma+\delta)^2$ equals $1+2\gamma+2\delta+\gamma^2+\delta^2+2\delta\gamma$. Thus among the R forms, we have free R (indicated by the 1), two ways of arranging R singly occupied with one activator (the 2γ), or the other activator (the 2δ), one way of arranging R doubly occupied with one activator (the γ^2), or the other activator (the δ^2), and two ways of arranging R with both activators (the $2\delta\gamma$). There are no $\gamma^2\delta$, $\delta^2\gamma$, or $\gamma^2\delta^2$ terms because there are no complexes containing more than two molecules of effector per protomer.

If the two activators bind to different sites, the velocity equation becomes:

$$\boxed{\frac{v}{V_{\max}} = \frac{\alpha(1+\alpha)^{n-1}}{\dfrac{L}{(1+\gamma)^n(1+\delta)^n} + (1+\alpha)^n}} \qquad \text{(VII-101)}$$

If the two effectors are inhibitors, the L′ term becomes $L(1+\gamma)^n(1+\delta)^n$. Now if $n=2$, the expansion contains additional terms $(2\gamma^2\delta, 2\gamma\delta^2, \delta^2\gamma^2)$ corresponding to the RA_2A', RA'_2A, and $RA_2A'_2$ complexes. The expansion also contains a $4\delta\gamma$ term (in place of the $2\delta\gamma$) because the independent binding of the two effectors provides four ways of arranging A and A′ on two protomers.

Competitive Inhibition

A compound that competes with the substrate for the R-state site can have a dual effect. At low substrate concentrations, low concentrations of the inhibitor may stimulate the velocity by promoting the T$\rightleftharpoons$R transition. At higher concentrations, the inhibitor will act in a pure competitive manner. The velocity equation describing this situation is:

$$\boxed{\frac{v}{V_{\max}} = \frac{\alpha(1+\alpha+\theta)^{n-1}}{L+(1+\alpha+\theta)^n}} \qquad \text{(VII-102)}$$

where $\theta = \dfrac{[I]}{K_{i_R}}$ = the normalized or specific inhibitor concentration

Note that the true competitive inhibitor acts in a manner completely different from an allosteric inhibitor. If the competitive inhibitor binds equally well to both the T and R states (but the true substrate binds exclusively to the R state) the velocity equation becomes:

$$\frac{v}{V_{\max}} = \frac{\alpha(1+\alpha+\theta)^{n-1}}{L(1+\theta)^n + (1+\alpha+\theta)^n} \qquad \text{(VII-103)}$$

In this case, the inhibitor will not activate at any substrate or inhibitor concentration, nor will the inhibitor interfere with substrate cooperativity or display homotropic cooperativity (in v versus [I] curves at constant [S]). The system is analogous to System B3d.

Nonexclusive Substrate and Effector Binding

If the substrate and effector bind to both the T and R states and both states are equally active, the complete velocity equation is:

$$\frac{v}{V_{\max}} = \frac{Lc\alpha(1+c\alpha)^{n-1}(1+e\gamma)^n + \alpha(1+\alpha)^{n-1}(1+\gamma)^n}{L(1+c\alpha)^n(1+e\gamma)^n + (1+\alpha)^n(1+\gamma)^n} \qquad \text{(VII-104)}$$

The numerator now represents the contributions of all possible T and R complexes containing S. The equation can be written:

$$\frac{v}{V_{\max}} = \frac{L'c\alpha(1+c\alpha)^{n-1} + \alpha(1+\alpha)^{n-1}}{L'(1+c\alpha)^n + (1+\alpha)^n} \qquad \text{(VII-105)}$$

where L' = an apparent allosteric constant $= L\left(\frac{(1+e\gamma)^n}{(1+\gamma)^n}\right)$

γ = the specific effector concentration, $\frac{[A]}{K_A}$ or $\frac{[I]}{K_i}$

e = the ratio of effector dissociation constants of the two states, $\frac{K_{i_R}}{K_{i_T}}$ or $\frac{K_{A_R}}{K_{A_T}}$

If $c < 1$ and $e < 1$, the effector is an activator; that is, the effector and the substrate bind preferentially to the R state. If $c < 1$ and $e > 1$, the effector is an inhibitor; that is, the effector binds preferentially to the T state while the substrate binds preferentially to the R state. If both an inhibitor and activator are present:

$$\boxed{\mathrm{L}'=\mathrm{L}\frac{(1+\mathrm{e}\gamma)^n(1+\mathrm{f}\beta)^n}{(1+\gamma)^n(1+\beta)^n}=\mathrm{L}\left(\frac{(1+\mathrm{e}\gamma)(1+\mathrm{f}\beta)}{(1+\gamma)(1+\beta)}\right)^n} \qquad \text{(VII-106)}$$

where

$$\mathrm{e}=K_{\mathrm{A_R}}/K_{\mathrm{A_T}} \qquad \text{and} \qquad \mathrm{f}=K_{i_\mathrm{R}}/K_{i_\mathrm{T}}.$$

Determination of $K_{\mathrm{S_T}}$, $K_{\mathrm{S_R}}$, and c in Nonexclusive K Systems

Consider a system in which the substrate has an appreciable affinity for the T state although c is still significantly less than unity. The velocity curve will still be sigmoidal because the preferential binding of S to the R state will displace the T$\rightleftharpoons$R equilibrium in favor of R and, thereby, create more S binding sites than the one used up. However, since the T state is now contributing significantly to the observed velocity, we can expect that the velocity curve will be less sigmoidal than that for a similar enzyme (same n and same L) where c = 0. Suppose now we add an inhibitor that binds exclusively to the T state. At a saturating inhibitor concentration, all the enzyme will be driven to the T state. Because the T state binds S and is catalytically active, we can obtain a velocity curve that is characteristic of the lower affinity state. As long as the inhibitor concentration is high enough to "freeze" the enzyme in the T state throughout the substrate concentration range used, the resulting velocity curve will be hyperbolic ($n_{\mathrm{app}}=1, [\mathrm{S}]_{0.9}/[\mathrm{S}]_{0.1}=81$) with an $[\mathrm{S}]_{0.5}$ equivalent to $K_{\mathrm{S_T}}$.

In the presence of a saturating concentration of an activator that binds exclusively to the R state, the enzyme will be driven entirely to the R state. Again the resulting v versus [S] curve will be hyperbolic. This time $[\mathrm{S}]_{0.5}$ is equivalent to $K_{\mathrm{S_R}}$. From these two experiments, c may be calculated from the definition:

$$\boxed{\mathrm{c}=\frac{K_{\mathrm{S_R}}}{K_{\mathrm{S_T}}}} \qquad \text{(VII-107)}$$

Figure VII-41a shows the velocity curves for phosphofructokinase (v versus [fructose-6-phosphate] at fixed saturating ATP) in the presence of

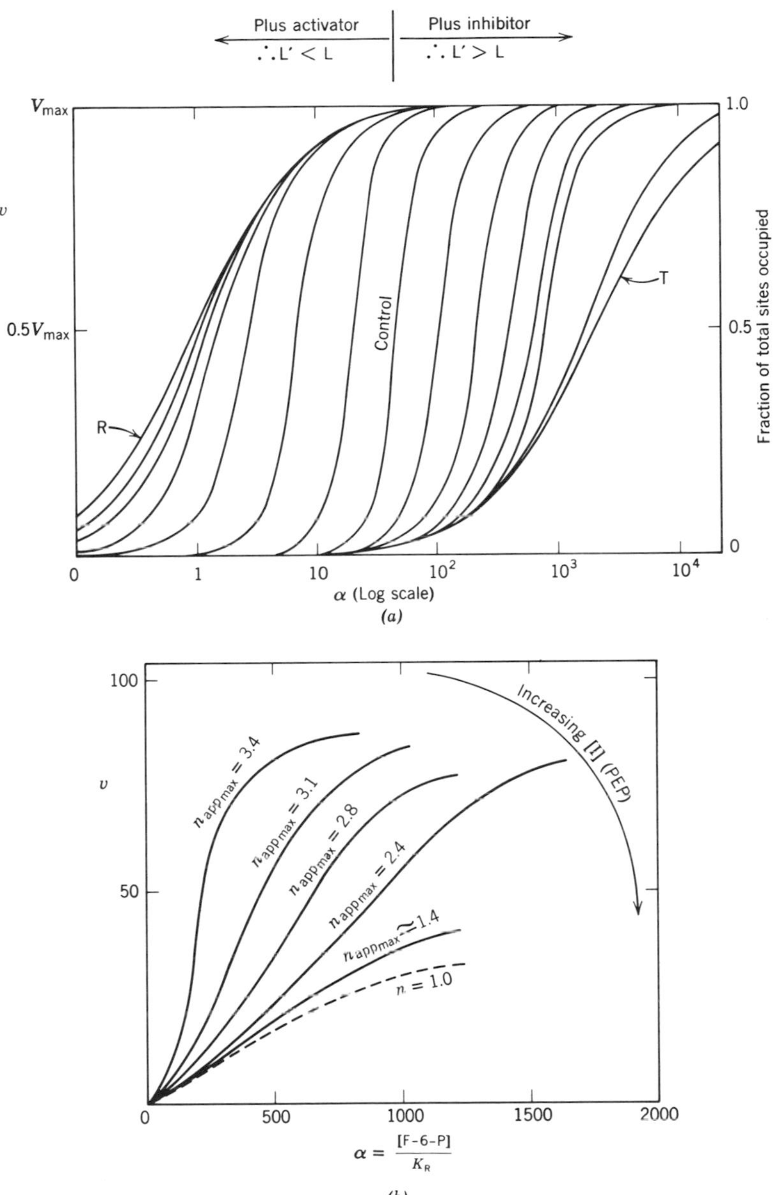

Fig. VII-41. Nonexclusive ligand binding by *E. coli* phosphofructokinase. (*a*) v versus log[S] plot in the presence of different fixed concentrations of an inhibitor or different fixed concentrations of an activator; I binds preferentially to the T state; A binds preferentially to the R state; S binds preferentially to the R state. S=F−6−P; I=PEP; A=ADP. [ATP] is constant and saturating. (*b*) v versus [F−6−P] for *E. coli* phosphofructokinase at different fixed concentrations of PEP and saturating ATP; PEP binds preferentially to the T state. At saturating PEP, the plot becomes near-hyperbolic. [Redrawn from Blangy, D., Buc, H. and Monod, J., *J. Mol. Biol.* **31**, 13 (1968).]

different fixed concentrations of activator (ADP) or inhibitor (PEP). One curve represents the control. The curves to the left of the control curve (i.e., with decreasing $[S]_{0.5}$ values) were obtained with different ADP concentrations; curves to the right of the control curve (i.e., with increasing $[S]_{0.5}$ values) were obtained in the presence of different PEP concentrations. Because the effective allosteric constant, L′, depends on the concentrations of effectors, we can consider the family of curves shown in Figure VII-41*a* as representing the velocity curves at different fixed values of L′. Although the curves are plotted on a semi-log scale, the decrease in cooperativity at either extreme is obvious. The decrease in cooperativity with increasing inhibitor concentrations is better illustrated in the linear plot shown in Figure VII-41*b*. (Note that these results are quite different from those obtained for an enzyme where $c=0$.) The fact that the limiting plots shown in Figure VII-41*a* and *b* are not quite hyperbolic (i.e., n_{app} is slightly greater than 1.0) suggests that the effectors do not exhibit completely exclusive binding to only one state.

Determination of *n*

A Hill plot can be constructed for each of the curves shown in Figure VII-41*a*. Each plot will yield an n_{app} which depends on the values of L′ and c (just as in the sequential interaction models n_{app} depends on the values of the interaction factors). The Hill equation can be expressed in terms of the concerted-symmetry model by solving the velocity equation (equation VII-66) for the ratio $\bar{Y}_S/(1-\bar{Y}_S)$ or $v/(V_{max}-v)$. After substituting for $\bar{Y}_S$ or v/V_{max}, rearranging terms, expanding the power functions, and condensing them again, we obtain:

$$\boxed{\frac{\bar{Y}_S}{1-\bar{Y}_S}=\frac{v}{V_{max}-v}=\frac{L'c\alpha(1+c\alpha)^{n-1}+\alpha(1+\alpha)^{n-1}}{L'(1+c\alpha)^{n-1}+(1+c\alpha)^{n-1}}} \qquad \text{(VII-108)}$$

The slope of the log-log plot gives n_{app}:

$$\boxed{n_{app}=\frac{d\log\left(\dfrac{v}{V_{max}-v}\right)}{d\log\alpha}} \qquad \text{(VII-109)}$$

As shown earlier, the Hill plots may deviate from linearity and the maximum slope may not occur at $v=0.5\,V_{max}$ [i.e., when $v/(V_{max}-v)=1.0$].

Qualitatively, we can see from Figure VII-41*a* that $n_{app_{max}}$ will be 1.0 for the limiting plots and will pass through a maximum at some intermediary value of L′ (not necessarily at $L'=L$). Thus we can correlate $n_{app_{max}}$ with L′ or (experimentally) with $\alpha_{1/2}$ (i.e., $[S]_{0.5}$) since L′ and $\alpha_{1/2}$ vary in the same direction. Figure VII-42*a* shows the theoretical variation of $n_{app_{max}}$ with L′ as computed by Rubin and Changeux (1966) for a four-site enzyme. The curves are symmetrical and bell-shaped for values of $c>0$. The maximum value of $n_{app_{max}}$ ($n^{max}_{app_{max}}$) occurs at $L'=c^{-n/2}$. When $c=0$, $n_{app_{max}}$ approaches n (the actual number of sites) as L′ increases. Note that $n^{max}_{app_{max}}$ can deviate significantly from n if $c>0$. For example, even if the T state has only 10% of the affinity of the R state for S ($c=0.10$), $n^{max}_{app_{max}}$ is still less than 2 when $n=4$.

In general, L′ will not be known. So under experimental conditions we would determine $n_{app_{max}}$ for each fixed effector concentration (from the Hill plots) and then replot $n_{app_{max}}$ versus the corresponding $\alpha_{1/2}$ or $[S]_{0.5}$. The experimental curve for phosphofructokinase is shown in Figure VII-42*b*. In this case, c is relatively low, so the value of $n^{max}_{app_{max}}$ of 3.8 strongly suggests that $n=4$. However, even if c is not very small it is still possible to calculate n from the relationship:

$$\boxed{n^{max}_{app_{max}}-1=(n-1)\left(\frac{1-\sqrt{c}}{1+\sqrt{c}}\right)^2} \qquad \text{(VII-110)}$$

Determination of L′ and L

Once c and n are known, L′ or L can be calculated by substituting any known value of $\bar{Y}_S$ or v/V_{max} and the corresponding value of α into equation VII-66. For example, by setting $\bar{Y}_S$ or v/V_{max} equal to 0.5 and substituting the value of $\alpha_{1/2}$ for α, equation VII-66 can be rearranged and solved for L′:

$$\boxed{L'=\frac{(\alpha_{1/2}-1)(\alpha_{1/2}+1)^{n-1}}{(1-c\alpha_{1/2})(1+c\alpha_{1/2})^{n-1}}} \qquad \text{(VII-111)}$$

The L would be calculated using the value of $\alpha_{1/2}$ of the control plot.

Consequences of Nonexclusive Ligand Binding

Many of the properties of regulatory enzymes can be explained by the concerted-symmetry model, especially with the added consequences of non-

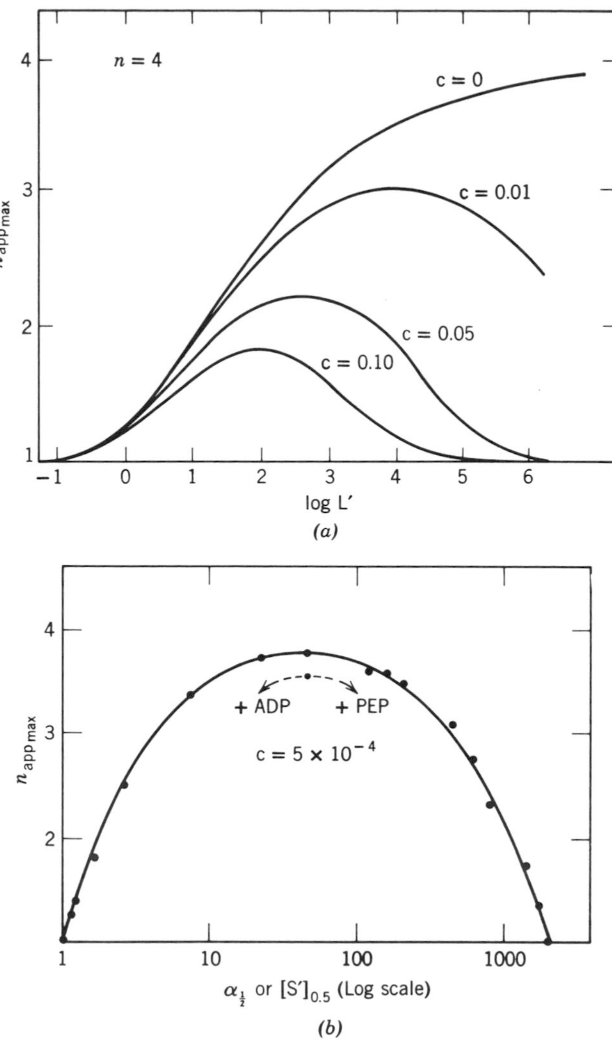

Fig. VII-42. (*a*) $n_{app_{max}}$ values as a function of L′ at different values of c. When c=0, $n_{app_{max}}$ obtained from Hill plots approaches the actual number of sites. [Redrawn from Rubin, M. M. and Changeux, J. P., *J. Mol. Biol.* **21**, 265 (1966).] (*b*) $n_{app_{max}}$ versus log$[S]_{0.5}$. [Redrawn from Blangy, D., Buc, H. and Monod, J., *J. Mol. Biol.* **31**, 13 (1968).] In general, as c increases $n^{max}_{app_{max}}$ decreases and log$\alpha_{1/2}$ at the peak decreases.

exclusive ligand binding. For example, we have seen that an allosteric inhibitor may either increase or decrease substrate cooperativity, depending on the value of c and the intrinsic value of L. The concerted-symmetry model can also explain how an allosteric effector could change $[S]_{0.5}$ markedly without affecting n_{app} or $n_{app_{max}}$. In this case it is assumed that the values of c and L for these systems are such that substrate cooperativity is already close to maximal; that is, $n_{app_{max}}$ (Fig. VII-42*b*) is close to $n_{app_{max}}^{max}$. Partial inhibition can also be explained on the basis of nonexclusive binding of either the substrate or the inhibitor. Thus if the T state binds substrate and is catalytically active, an infinitely high concentration of inhibitor (that binds preferentially or exclusively to the T state) will decrease the velocity to that fixed by K_{S_T} and V_{max}, but not to zero (Fig. VII-43*a*). Alternately, c may be equal to zero but if the inhibitor displays nonexclusive binding, an infinitely high inhibitor concentration will not be able to drive all the enzyme to the T state; hence v cannot be driven to zero (Fig. VII-43*b*). Nonexclusive effector binding can also explain why high concentrations of activators are unable to overcome the effects of inhibitors.

General and Hybrid Models

The concerted-symmetry and sequential interaction models may be considered extreme cases of the more general model shown in Figure VII-44*a*. Even this general model represents a simplification, since states with more than one bound ligand can take up different geometrical configurations. For example, $T_3R(S)_2$ can exist as the nonequivalent complexes shown in Figure VII-44*b*. All together, there are 44 nonequivalent complexes for the four-site enzyme. The concerted symmetry model can be extended further if we include ligand-induced dissociation-association of polymeric enzymes as a possible allosteric interaction. For example, consider that (*a*) association-dissociation is concerted (i.e., no intermediate oligomers containing less than n monomers exist), and (*b*) the monomer and oligomer have different affinities for a given ligand or different intrinsic catalytic activities. The monomer might represent the R state and the oligomer might represent the T state. This system has the added versatility in that its kinetic properties would vary with enzyme concentration. At low enzyme concentrations, dissociation would be favored, while at high enzyme concentrations the equilibrium would be shifted in favor of the polymer.

The velocity equation for a "dimer" that undergoes rapid reversible dissociation has been derived by Frieden (1970):

$$\frac{v}{V_{max}} = \frac{\alpha(1+\alpha)^{n-1}+2K_{eq}[E]c\alpha(1+c\alpha)^{2n-1}}{(1+\alpha)^{n}+2K_{eq}[E](1+c\alpha)^{2n}} \qquad \text{(VII-112)}$$

The dissociation under consideration could actually be hexamer⇌2 trimer, tetramer⇌2 dimer, and so forth. Thus n, which represents the number of sites on the "monomer" is $\geqslant 1$, while the dimer contains $2n$ sites. $V_{max} = n k_p [E]_t$.

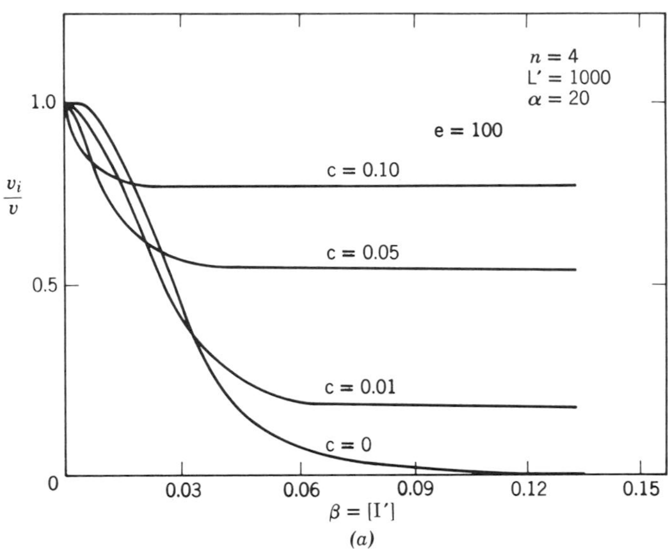

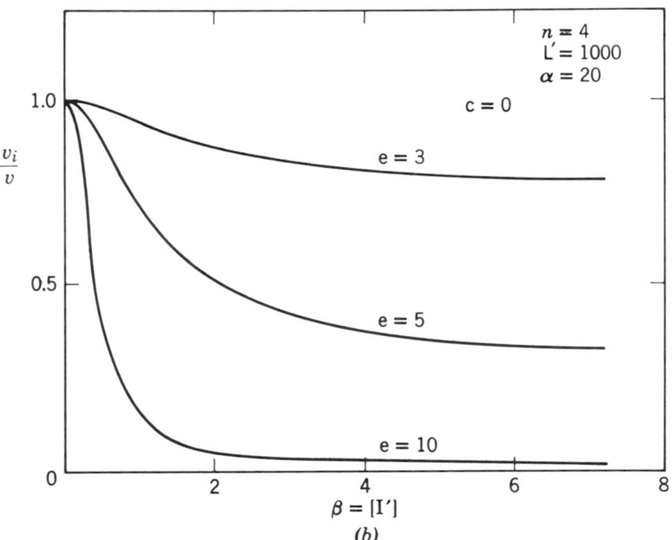

Fig. VII-43. Consequences of nonexclusive ligand binding. (*a*) v_i/v versus [I] plot for different values of c; e=100 (i.e., I binds almost exclusively to the T state). Saturating [I] will not drive v to zero if c>0 (i.e., if the T state has some affinity for S). (*b*) v_i/v plot for different values of e; c=0. Saturating [I] will not drive v to zero unless I binds exclusively to the T state (i.e., unless e≫1). L′ is an apparent allosteric constant at the fixed concentrations of substrate and activators. [Redrawn from Rubin, M. M. and Changeux, J. P., *J. Mol. Biol.* **21**, 265 (1966).]

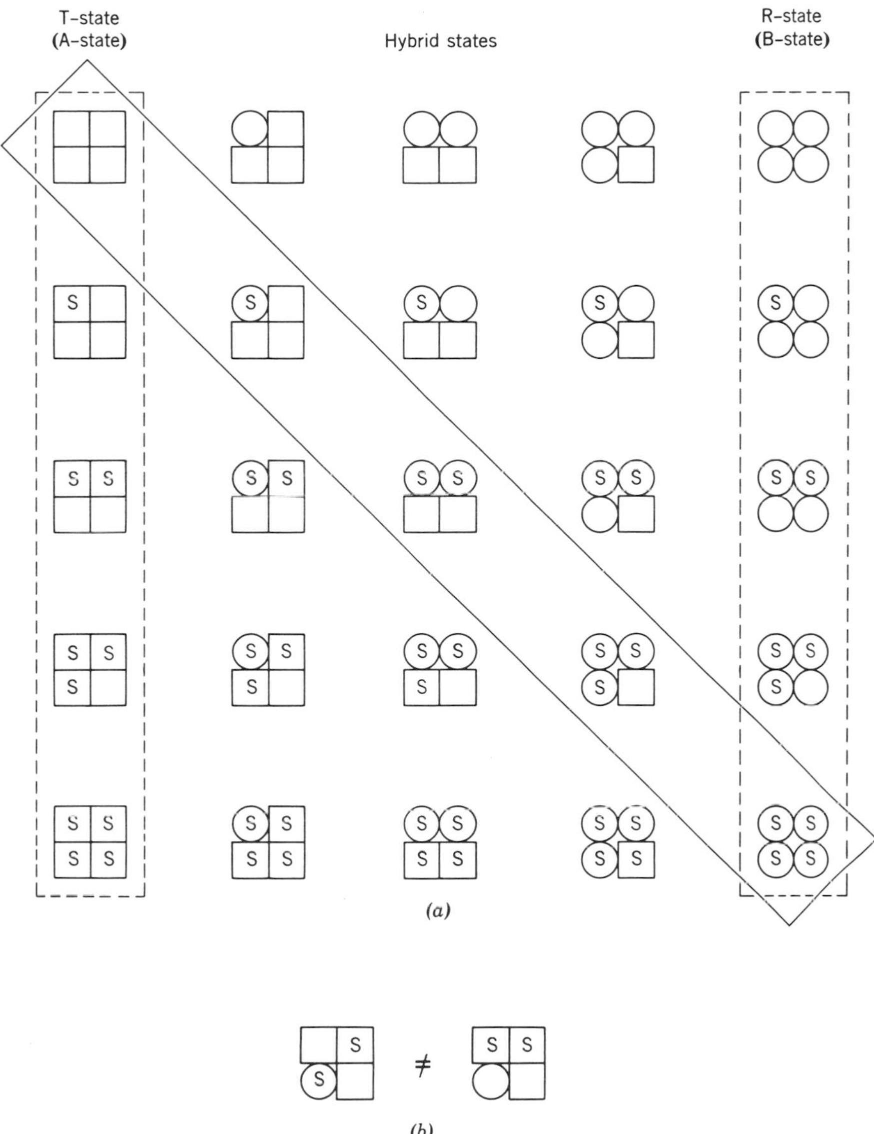

Fig. VII-44. (*a*) General model for a four-site allosteric enzyme involving hybrid oligomers. The first and fourth columns represent the concerted symmetry model. The diagonal represents the sequential interaction model. As shown, there are 25 different types of enzyme forms. (*b*) If the potential nonequivalent complexes are included (e.g., two different T_3RS_2 forms), then there are 44 possible enzyme forms. [Redrawn from Hammes, G. G. and Wu, C-W., *Science* **172**, 1205 (1971). Copyright by the American Association for the Advancement of Science.]

$\alpha = [S]/K_S$ for the "monomer," c is the ratio of "monomer" to "dimer" substrate dissociation constants, [E] is the concentration of free "monomer," and K_{eq} is the equilibrium constant for the dimerization, $[E_2]/[E]^2$. The concentration of free "monomer" is given by:

$$[E] = \frac{-(1+\alpha)^n + \sqrt{(1+\alpha)^{2n} + 8[E]_t K_{eq}(1+c\alpha)^{2n}}}{4K_{eq}(1+c\alpha)^{2n}} \tag{VII-113}$$

where $[E]_t$ is the total enzyme concentration in terms of the "monomer." It is assumed that the intrinsic k_p value of the catalytic site is the same in the "monomer" and "dimer." For higher degrees of polymerization, the equations become far more complex. For polymerization beyond a "tetramer," there is no way of solving for [E]. The situation becomes even more complex if the number of binding sites does not increase linearly with the number of monomers; that is, some sites become inaccessible as a result of subunit polymerization. Further complexities are introduced if a number of oligomers coexist in equilibrium (i.e., E, E_2, E_3, etc.).

E. ALTERNATIVE KINETIC EXPLANATIONS FOR SIGMOIDAL RESPONSES

Sigmoidal responses can arise in two-substrate (or substrate plus effector) reactions in which there is a preferred (but not exclusive) kinetic pathway to a ternary complex. In such systems, there is no need to assume subunit-subunit interactions or ligand-induced conformational changes. For example, consider the random bireactant pathway shown below.

$$\begin{array}{ccccc} & & EA & & \\ & {\scriptstyle k_1}\nearrow\swarrow{\scriptstyle k_{-1}} & & {\scriptstyle k_3}\searrow\nwarrow{\scriptstyle k_{-3}} & \\ E & & & & EAB \xrightarrow{k_5} \text{products} \\ & {\scriptstyle k_2}\searrow\nwarrow{\scriptstyle k_{-2}} & & {\scriptstyle k_4}\nearrow\swarrow{\scriptstyle k_{-4}} & \\ & & EB & & \end{array}$$

where the rate constants (in arbitrary units) are: $k_1 = 10^5$, $k_{-1} = 10^{-1}$, $k_2 = 10^4$, $k_{-2} = 10^{-2}$, $k_3 = 10^5$, $k_{-3} = 10^{-1}$, $k_4 = 10^4$, $k_{-4} = 10^{-2}$, $k_5 = 10^6$. The rate constants were chosen to fulfill the requirement that $k_1 k_3 k_{-4} k_{-2}$

$= k_2 k_4 k_{-3} k_{-1}$ (i.e., $K_1 K_3 = K_2 K_4$). Note that k_5 is greater than any of the other forward rate constants so that we cannot make the usual rapid equilibrium assumptions. The product $k_1 k_3$ is greater than $k_2 k_4$. Consequently, the overall reaction has a kinetically preferred pathway ($E \xrightarrow{+A} EA \xrightarrow{+B} EAB \longrightarrow$ products), although the alternate route via EB is not forbidden. The dissociation constants K_1, K_2, K_3, and K_4 are all equal to 10^{-6}, so there is no thermodynamic preference for any step. The steady-state velocity equation for the system above is rather complex (equation IX-182). However, we can visualize the substrate response curves in a qualitative way. Suppose we measure the initial velocity as a function of [A] in the presence of a fixed unsaturating concentration of B. At zero [A], the enzyme will be present as E and EB. Now let us increase [A] starting at very low A concentrations. At first, EAB will form faster via the $EB \xrightarrow{+A} EAB$ reaction than via the $E \xrightarrow{+A} EA \xrightarrow{+B} EAB$ route. As [A] is increased further, the faster sequence gradually takes over and the slope of the velocity curve increases sharply. At some A concentration, the $E \xrightarrow{+A} A \xrightarrow{+B} EAB$ route becomes the predominant pathway of EAB formation. As [A] is increased still further, we observe the usual sloping off of the velocity curve as it approaches V_{max}. The overall velocity curve is sigmoidal. (See Figure IX-37*a*.)

Now consider the converse experiment: [A] is held constant at some fixed concentration and the velocity is measured at various increasing concentrations of B. At first, the velocity curve will rise in the usual way as EAB is produced via the kinetically favored sequence. However, when the concentration of B becomes high enough to overcome the kinetic factors favoring the $E \xrightarrow{+A} EA \xrightarrow{+B} EAB$ pathway, a greater proportion of the reaction flux will proceed via the $E \xrightarrow{+B} EB \xrightarrow{+A} EAB$ route. As a result, the initial velocity will pass through a maximum and then decrease to a limiting plateau velocity dictated by the rate constants of the kinetically less favorable route. (See Figure IX-37*b*.) Jensen and Trentini (1970) have interpreted the kinetics of the 3-deoxy-D-arabino-heptulosonate-7-phosphate synthetase of *Rhodomicrobium vannielii* in terms of this kinetic model. The two substrates, phosphoenolpyruvate and erythrose-4-phosphate, are equivalent to A and B, respectively.

The kinetic argument will not be valid for rapid equilibrium systems (i.e., systems in which the breakdown of EAB is rate-limiting), nor for systems in which both routes to the ternary complex are equally favorable, nor for systems in which A and B add in an obligate order. Furthermore, there is no need to invoke a kinetic model if equilibrium ligand binding curves are sigmoidal. The kinetics of a random bireactant system and other steady-state systems capable of yielding sigmoidal velocity curves are described in more detail in Chapter Nine.

REFERENCES

General

Atkinson, D. E., "Regulation of Enzyme Activity," in *Ann. Rev. Biochem.* **35**, 85 (1966).

Bohr, C., *Zentr. Physiol.* **17**, 682 (1903).

Hammes, G. G. and Wu, C-W., *Science* **172**, 1205 (1971).

Hammes, G. G. and Wu, C-W., "Kinetics of Allosteric Enzymes," in *Ann. Rev. Biophys. Bioeng.* **3**, 1 (1974).

Hill, A. V., *Biochem. J.* **7**, 471 (1913).

Koshland, D. E., Jr., *Proc. Nat. Acad. Sci. U.S.* **44**, 98 (1958).

Koshland, D. E., Jr., "The Active Site and Enzyme Action," in *Advances in Enzymology*, Vol. 22, F. F. Nord, ed., Interscience, 1960, p. 45.

Koshland, D. E. and Neet, K. E., "The Catalytic and Regulatory Properties of Enzymes," in *Ann. Rev. Biochem.* **37**, 359 (1968).

Monod, J., Changeux, J-P. and Jacob, F., *J. Mol. Biol.* **6**, 306 (1963).

Reiner, J. M., *Behavior of Enzyme Systems*, 2nd ed., Van Nostrand-Reinhold, 1969, p. 93.

Stadtman, E. R., "Allosteric Regulation of Enzyme Activity," in *Advances in Enzymology* Vol. 28, F. F. Nord, ed., Wiley-Interscience, 1966, p. 41.

Walter, C., *J. Biol. Chem.* **249**, 699 (1974).

Webb, J. L., *Enzyme and Metabolic Inhibitors*, Vol. 1, Academic Press, 1963, p. 36.

Wong, J. T-F. and Endrenyi, L., *Can. J. Biochem.* **49**, 568 (1970).

Wyman, J., "On Allosteric Models," in *Current Topics in Cellular Regulation*, Vol. 6, B. L. Horecker, and E. R. Stadtman, eds., Academic Press, 1972, p. 209.

Simple Sequential Interaction Model

Adair, G. S., *J. Biol. Chem.* **63**, 529 (1925).

Atkinson, D. E., Hathaway, J. A. and Smith, E. C., *J. Biol. Chem.* **240**, 2682 (1965) (Isocitrate dehydrogenase).

Frieden, C., *J. Biol. Chem.* **242**, 4045 (1967).

Pauling, L., *Proc. Nat. Acad. Sci. U.S.* **21**, 186 (1935).

General Sequential Interaction Model

Haber, J. E. and Koshland, D. E., Jr., *Proc. Nat. Acad. Sci. U.S.* **58**, 2087 (1967).

Kirtley, M. E. and Koshland, D. E., Jr., *J. Biol. Chem.* **242**, 4192 (1967).

Koshland, D. E., Jr. and Neet, K. E., "The Catalytic and Regulatory Properties of Enzymes," in *Ann. Rev. Biochem.* **37**, 359 (1968).

Koshland, D. E., Jr., "The Molecular Basis for Enzyme Regulation," in *The Enzymes*, Vol. 1, 3rd ed., P. Boyer, ed., Academic Press, 1970, p. 461.

Koshland, D. E., Jr., Nemethy, G. and Filmer, D., *Biochemistry* **5**, 365 (1966).

Negative Cooperativity

Cook, R. A. and Koshland, D. E., Jr., *Biochemistry* **9**, 3337 (1970).

Levitski, A. and Koshland, D. E., Jr., *Proc. Nat. Acad. Sci. U.S.* **62**, 1121 (1969).

Levitski, A. and Koshland, D. E., Jr., *FEBS Symposium* **19**, 263 (1969).

Substrate Inhibition

Webb, J. L., *Enzyme and Metabolic Inhibitors*, Vol. 1, Academic Press, 1963, Ch. 4.

Concerted-Symmetry Model

Blangy, D., *Biochimie* **53**, 135 (1971).

Blangy, D., Buc, H. and Monod, J., *J. Mol. Biol.* **31**, 13 (1968).

Endrenyi, L., Chan, M-S. and Wong, J. T-F., *Can. J. Biochem.* **49**, 581 (1971).

Gerhart, J. C., "Regulatory Properties of Aspartic Transcarbamylase," in *Current Topics in Cellular Regulation*, Vol. 2, B. L. Horecker and E. R. Stadtman, eds., Academic Press, 1970, p. 276.

Horn, A. and Börnig, H., *FEBS Letters* **3**, 325 (1969).

Johannes, K.-J. and Hess, B., *J. Mol. Biol.* **76**, 181 (1973).

Markus, G., McClintock, D. K. and Bussel, J. B., *J. Biol. Chem.* **246**, 762 (1971).

Monod, J., Wyman, J., Changeux, J-P., *J. Mol. Biol.* **12**, 88 (1965).

Rubin, M. M. and Changeux, J-P., *J. Mol. Biol.* **21**, 265 (1966).

Allosteric Ping Pong Enzymes

Sumi, T. and Ui, M., *Biochim. Biophys. Acta* **276**, 12 (1972).

Ligand-Exclusion Model

Fisher, H. F., Gates, R. E. and Cross, D. G., *Nature* **228**, 247 (1970).

Monomeric Allosteric Enzyme

Panagou, D., Orr, M. D., Dunstone, J. R. and Blakley, R. L., *Biochemistry* **11**, 2378 (1972).

Rubsamen, H., Khandker, R. and Witzel, H., *Hoppe Seyler Zeit. Physiol. Chem.*, **355**, 687 (1974).

Flip-Flop Mechanisms and Half-Site Enzymes

Lazdunski, M., "Flip-Flop Mechanisms and Half-Site Enzymes," in *Current Topics in Cellular Regulation*, Vol. 6, B. L. Horecker and E. R. Stadtman, eds., Academic Press, 1972, p. 267.

Hysteretic Enzymes

Frieden, C., *J. Biol. Chem.* **245**, 5788 (1970).

Frieden, C., "Protein-Protein Interaction and Enzymatic Activity," in *Ann. Rev. Biochem.* **40**, 653 (1971).

Smith, T. E. and Perry, M., *Arch. Biochem. Biophys.* **156**, 448 (1973).

Effector Changes V_{max}

Fritzson, P., *Eur. J. Biochem.* **38**, 408 (1973) (Nucleotidase).

Ribereau-Gayon, G., Sabraw, A., Lammel, C. and Preiss, J., *Arch. Biochem. Biophys.* **142**, 675 (1971) (ADPG pyrophosphorylase).

Schramm, V. L. and Morrison, J. F., *Biochemistry* **10**, 2272 (1971) (Nucleoside diphosphatase).

Alternative Kinetic Models for Allosteric Enzymes

Ferdinand, W., *Biochem. J.* **98**, 278 (1966).

Fisher, J. R., *Arch. Biochem. Biophys.* **152**, 638 (1972).

Griffen, C. C. and Brand, L., *Arch. Biochem. Biophys.* **126**, 856 (1968).

Jensen, R. A. and Trentini, W. C., *J. Biol. Chem.* **245**, 2018 (1970).

Plowman, K. M., *Enzyme Kinetics*, McGraw-Hill, 1972, Ch. 5.

Rabin, B. R., *Biochem. J.* **102**, 22C (1967).

Sanwal, B. G. and Cook, R. A., *Biochemistry* **5**, 886 (1966).

Sweeny, J. R. and Fisher, J. R., *Biochemistry* **7**, 561 (1968).

Weber, G., "The Binding of Small Molecules to Proteins," in *Molecular Biophysics*, B. Pullman and M. Weissbluth, eds., Academic Press, 1965.

Association-Dissociation

Frieden, C., *J. Biol. Chem.* **242**, 4045 (1967).

Frieden, C., "Protein-Protein Interaction and Enzymatic Activity," in *Ann. Rev. Biochem.* **40**, 653 (1971).

Gerlt, J. A., Rabinowitz, K. W., Dunne, C. P. and Wood, W. A., *J. Biol. Chem.* **248**, 8200 (1973).

Klapper, M. H. and Klotz, I. M., *Biochemistry* **7**, 223 (1968).

Nichol, L. W., Jackson, W. J. H. and Winzor, D. J., *Biochemistry* **6**, 2449 (1967).

Sigmoidal Velocity Curves with Inflection Points at Low v/V_{max} Values

Bisswanger, H., *Eur. J. Biochem.* **48**, 377 (1974). (Pyruvate dehydrogenase in terms of the concerted-symmetry model).

CHAPTER EIGHT

MULTIPLE INHIBITION ANALYSIS

Many kinetics studies result in a list of compounds that inhibit a particular enzyme, but an extremely important question goes unanswered: How many distinct inhibitor sites are present on the enzyme? Phrased differently the question is: Are the inhibitors all mutually exclusive, or can two or more occupy the enzyme simultaneously? A multiple inhibition analysis can tell us something about the different binding sites on an enzyme, and may also provide information on the interactions of physiologically important feedback inhibitors. In this chapter we examine the kinetics of enzymes with (*a*) multiple sites for a given inhibitor, (*b*) a single site capable of binding different inhibitors, and (*c*) multiple sites each specific for a different inhibitor. For simplicity, we assume that only a single substrate is involved. If the system is in fact multireactant, then the calculated constants are apparent constants for the given concentration of cosubstrate which is held constant. However, the general properties of the system are unchanged. (See equations IX-348 to IX-352.)

A. MULTIPLE SITES FOR A GIVEN INHIBITOR

Consider a unireactant enzyme capable of binding two molecules of a competitive inhibitor, I, in a random manner. The binding of I to either site is sufficient to exclude S. The equilibria are:

$$\begin{array}{ccccccccc}
IE & \overset{K_i}{\rightleftharpoons} & I + E & + & S & \overset{K_S}{\rightleftharpoons} & ES & \overset{k_p}{\longrightarrow} & E + P \\
+ & & + & & & & & & \\
I & & I & & & & & & \\
K_i \upharpoonleft\downharpoonright & & K_i \upharpoonleft\downharpoonright & & & & & & \\
IEI & \overset{K_i}{\rightleftharpoons} & I + EI & & & & & &
\end{array}$$

The velocity equation can be derived from rapid equilibrium assumptions in the usual manner.

$$\frac{v}{V_{max}} = \frac{\dfrac{[S]}{K_S}}{1 + \dfrac{[S]}{K_S} + \dfrac{2[I]}{K_i} + \dfrac{[I]^2}{K_i^2}} \qquad \text{(VIII-1)}$$

or

$$\frac{v}{V_{max}} = \frac{[S]}{K_S\left(1 + \dfrac{[I]}{K_i}\right)^2 + [S]} \qquad \text{(VIII-2)}$$

If the binding of the first inhibitor molecule changes the intrinsic dissociation constant of the vacant site by a factor a, then the velocity equation becomes:

$$\frac{v}{V_{max}} = \frac{\dfrac{[S]}{K_S}}{1 + \dfrac{[S]}{K_S} + \dfrac{2[I]}{K_i} + \dfrac{[I]^2}{aK_i^2}} \qquad \text{(VIII-3)}$$

Two-site, pure competitive inhibition will not be distinguishable from one-site, pure competitive inhibition, or one-site, partial competitive inhibition by the usual v versus [S] curves (in presence and absence of inhibitor) nor by the corresponding reciprocal plots, which are linear. This is understandable as all three systems can be described by the same equations.

$$\frac{v}{V_{max}} = \frac{[S]}{K_{S_{app}} + [S]} \qquad \text{and} \qquad \frac{1}{v} = \frac{K_{S_{app}}}{V_{max}} \frac{1}{[S]} + \frac{1}{V_{max}}$$

However, the slope and $K_{S_{app}}$ are a different function of [I] in each system. Theoretically, the three types of competitive systems can be distinguished by plotting v versus [I] at a fixed [S]. In partial competitive systems, the curve plateaus at a limiting velocity. In both the one-site and two-site pure competitive systems, the velocity can be driven to zero by increasing the inhibitor concentration sufficiently (but this may be hard to demonstrate). Inspection of Figure VIII-1*a* shows that the $[I]_{0.9}/[I]_{0.1}$ ratio for the non-cooperative two-site system is 18 compared to 81 for the one-site system. The two-site system is not noticeably sigmoidal, although the $[I]_{0.9}/[I]_{0.1}$ ratio corresponds to an n value of 1.5. The best indication of the nature of the inhibition is obtained from the replots of the slopes of the reciprocal plot versus the corresponding [I]. The reciprocal form of the velocity equation is:

$$\boxed{\frac{1}{v} = \frac{K_S}{V_{max}}\left(1 + \frac{[I]}{K_i}\right)^2 \frac{1}{[S]} + \frac{1}{V_{max}}} \qquad \text{(VIII-4)}$$

The replot of $slope_{1/S}$ versus [I] is a parabola, as shown in Figure VIII-1*b*. One-site pure competitive and one-site partial competitive systems yield linear and hyperbolic replots, respectively. Because of the shape of the replot, the two-site, pure competitive system can be called slope-parabolic competitive inhibition. Replots of $K_{S_{app}}$ versus [I] and $1/v$ versus [I] are also parabolic. The K_i cannot be determined directly from the Dixon plot, the reciprocal plot, or from the usual replots. The "K_i" calculated from the slope of the reciprocal plot is actually $K_{i_{slope}}$, which is not the K_i for the dissociation of an EI complex, but rather a more complex function of K_i which varies with [I]:

$$\boxed{slope_{1/S} = \frac{K_S}{V_{max}}\left(1 + \frac{[I]}{K_i}\right)^2 = \frac{K_S}{V_{max}}\left(1 + \frac{[I]}{K_{i_{slope}}}\right)} \qquad \text{(VIII-5)}$$

where

$$1 + \frac{[I]}{K_{i_{slope}}} = 1 + \frac{2[I]}{K_i} + \frac{[I]^2}{K_i^2}$$

$$\frac{[I]}{K_{i_{slope}}} = \frac{2[I]}{K_i} + \frac{[I]^2}{K_i^2} = \frac{2K_i[I] + [I]^2}{K_i^2}$$

$$\boxed{K_{i_{slope}} = \frac{K_i^2}{2K_i + [I]}} \qquad \text{(VIII-6)}$$

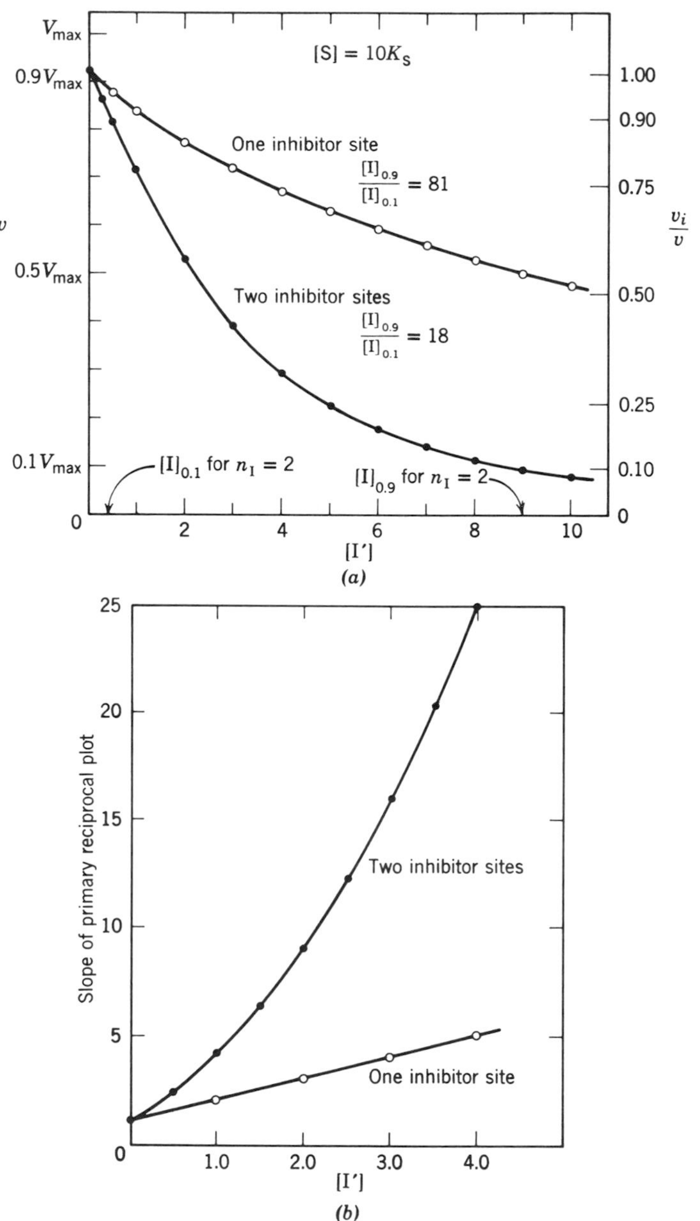

Fig. VIII-1. (*a*) v versus [I] at $[S]=10K_S$. (*b*) $slope_{1/S}$ versus [I] replot for one-site and two-site competitive inhibition. The two molecules of I are assumed to bind randomly and independently (no interaction between sites).

The expression for $K_{i_{slope}}$ can be rearranged to a linear form:

$$\boxed{\frac{1}{K_{i_{slope}}} = \frac{1}{K_i^2}[\mathrm{I}] + \frac{2}{K_i}} \qquad \text{(VIII-7)}$$

Thus if the reciprocal of the $K_{i_{slope}}$ determined from each reciprocal plot is replotted versus the corresponding [I], the replot will be a straight line with a slope of $1/K_i^2$ and a $1/K_{i_{slope}}$-axis intercept of $2/K_i$. When $1/K_{i_{slope}} = 0$, the intercept on the [I]-axis gives $-2K_i$. If there is cooperative binding, the slope of the replot is $1/aK_i^2$, the $1/K_{i_{slope}}$-axis intercept is still $2/K_i$, and the [I]-axis intercept is $-2aK_i$. Nonlinear replots indicate that the $K_{i_{slope}}$ term contains higher powers of [I], or, in other words, that there are more than two inhibitor binding sites.

The v versus [I] curve for a four-site pure competitive inhibitor at $[\mathrm{S}] = 10K_{\mathrm{S}}$ is shown in Figure VIII-2. The curve was generated from the velocity equation:

$$\boxed{\frac{v}{V_{max}} = \frac{[\mathrm{S}]}{K_{\mathrm{S}}\left(1 + \frac{[\mathrm{I}]}{K_i}\right)^4 + [\mathrm{S}]}} \qquad \text{(VIII-8)}$$

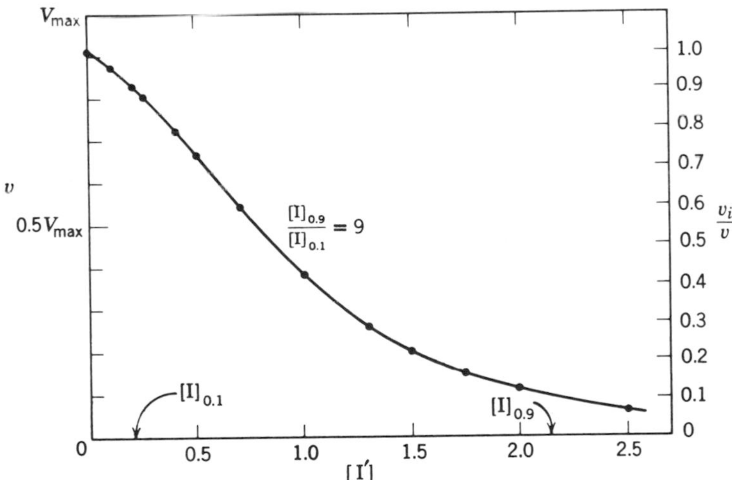

Fig. VIII-2. The v versus [I] plot at $[\mathrm{S}] = 10K_{\mathrm{S}}$ for an enzyme with four identical inhibitor sites; I is competitive with S. The four molecules of I bind randomly and independently.

where K_i is the intrinsic inhibitor dissociation constant. Now the curve is noticeably sigmoidal at low inhibitor concentrations, yet no true cooperative binding was assumed; that is, the binding of one inhibitor molecule had no effect on the intrinsic dissociation constants of the vacant inhibitor sites. For $[S]=10K_S$, the $[I]_{0.9}/[I]_{0.1}$ ratio is 9. The ratio is not constant, but varies with [S]. For example, at $[S]=2K_S$, the ratio increases to about 16. At $[S]=K_S$, the ratio is about 22. As [S] decreases, the v versus [I] curves become less sigmoidal and the $[I]_{0.9}/[I]_{0.1}$ ratio approaches 81 (i.e., the v versus [I] curve approaches a hyperbola as the fixed [S] decreases). A replot of the slopes of the reciprocal plot versus the corresponding [I] is concave up. However, the replot of $1/K_{i_{\text{slope}}}$ versus [I] is not linear because of the $[I]^3$ and $[I]^4$ terms.

Hill Equation and Hill Plots for Multisite Inhibition

Initial velocity versus [I] data for multisite inhibition can be expressed in terms of a Hill equation, although the constant, K', will have different meanings for different types of inhibition systems and different values at different substrate concentrations. For example, the equation for multisite competitive inhibition with n inhibitor sites and m substrate sites would be:

$$\frac{v_i}{V_{\max}}=\frac{\dfrac{[S]}{K_{S_1}}+\dfrac{[S]^2}{K_{S_1}K_{S_2}}+\cdots\dfrac{[S]^m}{K_{S_1}K_{S_2}\cdots K_{S_m}}}{1+\dfrac{[S]}{K_{S_1}}+\dfrac{[S]^2}{K_{S_1}K_{S_2}}+\cdots\dfrac{[S]^m}{K_{S_1}K_{S_2}\cdots K_{S_m}}+\dfrac{[I]}{K_{i_1}}+\dfrac{[I]^2}{K_{i_1}K_{i_2}}+\cdots\dfrac{[I]^n}{K_{i_1}K_{i_2}\cdots K_{i_n}}}$$

where K_{S_1}, K_{S_2}, K_{i_1}, K_{i_2}, and so on, are effective dissociation constants. At any fixed substrate concentration, all the terms containing [S] and K_S can be combined into a single constant, K. If the inhibitor exhibits cooperative binding, all terms containing less than n molecules of I can be omitted. (This, of course, is a simplification, and immediately we can predict nonintegral values of n for the inhibitor if the binding is not strongly cooperative.) The equation can then be rewritten:

$$\frac{v_i}{V_{\max}}=\frac{K}{(1+K)+\dfrac{[I]^n}{a^{n-1}b^{n-2}\cdots K_i^n}}$$

where K_i = the intrinsic dissociation constant for the EI complex
a, b, etc. = the interaction factors

In the absence of inhibitor, the velocity at any substrate concentration is:

$$v_0 = \frac{K V_{max}}{(1+K)}$$

Therefore, the relative velocity at any fixed [S] is:

$$\frac{v_i}{v_0} = \frac{\dfrac{K V_{max}}{(1+K) + \dfrac{[I]^n}{a^{n-1} b^{n-2} \ldots K_i^n}}}{\dfrac{K V_{max}}{(1+K)}}$$

$$= \frac{(1+K)}{(1+K) + \dfrac{[I]^n}{a^{n-1} b^{n-2} \ldots K_i^n}}$$

If the group of constants $a^{n-1} b^{n-2} \ldots K_i^n$ are also combined into a single constant, Z, and numerator and denominator multiplied by this constant, the equation becomes:

$$\frac{v_i}{v_0} = \frac{Z(1+K)}{Z(1+K) + [I]^n} = \frac{K'}{K' + [I]^n} \qquad \text{(VIII-9)}$$

The equation can be converted to the linear form shown below.

$$\boxed{\log \frac{v_i}{v_0 - v_i} = -n \log [I] + \log K'} \qquad \text{(VIII-10)}$$

Thus a plot of $\log v_i/(v_0 - v_i)$ versus $\log [I]$ will be a straight line with a slope of $-n$. When $v_i = 0.5 v_0$, $\log v_i/(v_0 - v_i) = 0$, that is, $v_i/(v_0 - v_i) = 1$ on a log scale. The corresponding position on the $\log [I]$ axis gives $[I]_{0.5}$; $[I]_{0.5}$ is related to K' as:

$$\boxed{n \log [I]_{0.5} = \log K' \qquad \text{or} \qquad [I]_{0.5}^n = K'} \qquad \text{(VIII-11)}$$

Note that the plot for inhibitors involves $v_i/(v_0 - v_i)$, not $v_i/(V_{max} - v_i)$.

The Hill plot for two-site competitive inhibition with no cooperativity described in Figure VIII-1 is shown in Figure VIII-3. The plot is curved with limiting slopes of -1.0 (at very low inhibitor concentrations) and -2.0 (at very high inhibitor concentrations). If a straight line is fitted to the points obtained at intermediate inhibitor concentrations [e.g., for values of $v_i/(v_0-v_i)$ between 0.1 and 10, such as might be obtained under normal experimental conditions], a nonintegral value of n between 1 and 2 will be obtained. The n_{app} value of, for example, 1.5 indicates more than one inhibitor binding site, but it does not indicate cooperative binding of the inhibitor in the usual sense. The Hill plot for the four-site competitive inhibition system will be curved with limiting slopes of -1 and -4. The

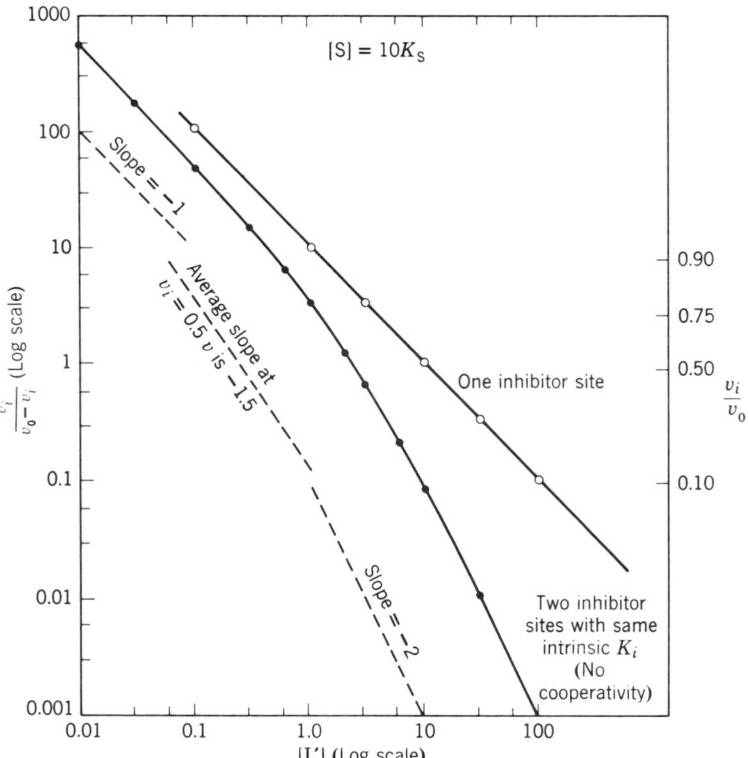

Fig. VIII-3. Hill plots for competitive inhibitors: $\log v_i/(v_0-v_i)$ versus $\log$ [I]. If only one I site is available, the slope of the Hill plot is -1. The plot for an enzyme with two identical, noncooperative inhibitor sites is curved with limiting slopes of -1 at very low [I] and -2 at very high [I].

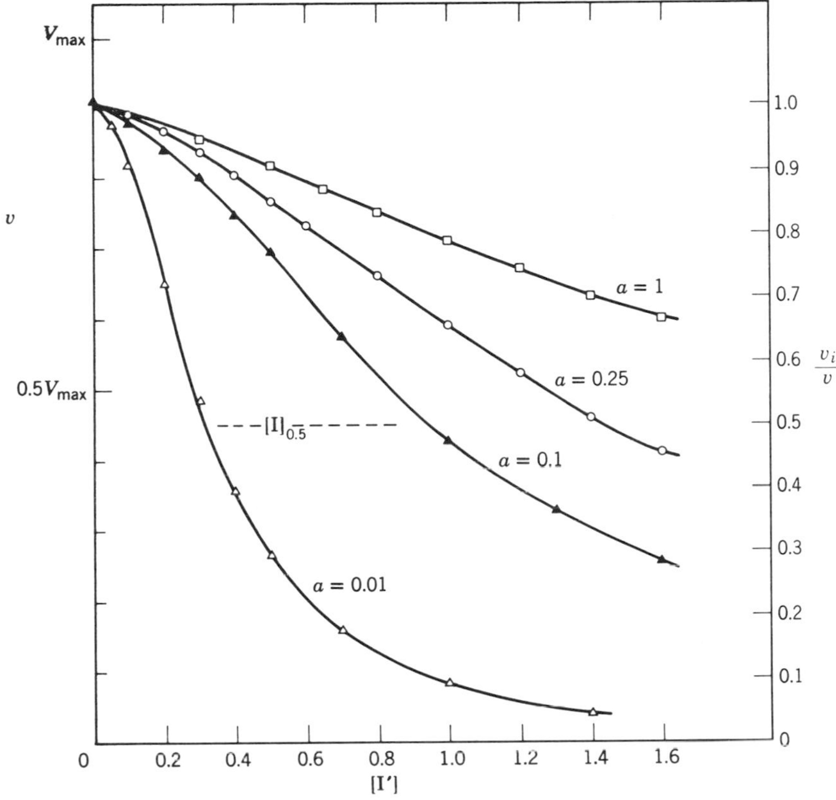

Fig. VIII-4. Effect of the interaction factor, a, on the shape of the v versus [I] plot for an enzyme with two identical inhibitor sites. $[S]=10K_S$. As the interaction factor, a, decreases, the curves become more sigmoidal and $[I]_{0.5}$ decreases.

average slope in the region between 10% and 90% inhibition is -2 at $[S]=10K_S$. Thus we see that inhibitors, unlike substrates, can yield n values greater than 1.0 in the absence of true cooperative binding. If the inhibitor does display cooperative binding, the velocity versus [I] curves become more sigmoidal (Fig. VIII-4) and the apparent n value determined from the $[I]_{0.9}/[I]_{0.1}$ ratio or the Hill plot approaches the actual number of inhibitor sites. For example, if in the two-site system the binding of the first inhibitor molecule increases the affinity of the vacant site by a factor of 10 ($a=0.1$), the $[I]_{0.9}/[I]_{0.1}$ ratio decreases from 18 ($n_{app}=1.5$) to about 11.6 ($n_{app}=1.8$). The average slope of the Hill plot in the region of $v_i=0.5v$ will be 1.8. The Hill plot also becomes more linear at intermediate $v_i/(v-v_i)$ regions.

B. INHIBITION BY MIXTURES OF DIFFERENT INHIBITORS

System B1. Pure Competitive Inhibition by Two Different Exclusive Inhibitors

The equilibria for an enzyme subject to competitive inhibition by two different compounds, I and X, are shown below. In writing the equilibria, it is assumed that the enzyme can combine with either I or X, but not both; I and X are competitive with respect to each other.

$$\begin{array}{ccccccccc}
EX & & & & & & & & \\
K_X \Big\downarrow\!\Big\uparrow & & & & & & & & \\
X & & & & & & & & \\
+ & & & & & & & & \\
E & + & S & \overset{K_S}{\rightleftharpoons} & ES & \overset{k_p}{\longrightarrow} & E & + & P \\
+ & & & & & & & & \\
I & & & & & & & & \\
K_i \Big\uparrow\!\Big\downarrow & & & & & & & & \\
EI & & & & & & & &
\end{array}$$

The velocity equation is:

$$\boxed{\frac{v}{V_{max}} = \frac{\dfrac{[S]}{K_S}}{1 + \dfrac{[S]}{K_S} + \dfrac{[I]}{K_i} + \dfrac{[X]}{K_X}}} \qquad \text{(VIII-12)}$$

or

$$\boxed{\frac{v}{V_{max}} = \frac{[S]}{K_S\left(1 + \dfrac{[I]}{K_i} + \dfrac{[X]}{K_X}\right) + [S]}} \qquad \text{(VIII-13)}$$

The system behaves identically to single-inhibitor pure competitive inhibition. Either inhibitor alone can reduce the velocity to zero. The slope replot for either inhibitor alone is linear (there are no squared terms in the velocity equation). Mixtures of I and X are no more inhibitory than is an equal total specific concentration of either inhibitor alone. For example, the degree of inhibition obtained at $[\mathrm{I}]=2K_i$ or $[\mathrm{X}]=2K_{\mathrm{X}}$ is identical to the degree of inhibition obtained with $[\mathrm{I}]=K_i$ plus $[\mathrm{X}]=K_{\mathrm{X}}$. This observation in itself tells us that I and X are mutually exclusive. This result is not obtained if both inhibitors can combine with the enzyme to form an IEX complex (see System B3). Another diagnostic test for mutual exclusivity is provided by Dixon plots of $1/v$ versus the concentration of one inhibitor at a constant concentration of S and different fixed concentrations of the other inhibitor. The equation for the Dixon plot can be rearranged to show either I or X as the varied inhibitor. When [I] is varied, the equation is:

$$\frac{1}{v}=\frac{K_{\mathrm{S}}}{[\mathrm{S}]\,V_{\max}K_i}[\mathrm{I}]+\frac{1}{V_{\max}}\left(1+\frac{K_{\mathrm{S}}}{[\mathrm{S}]}+\frac{K_{\mathrm{S}}[\mathrm{X}]}{[\mathrm{S}]K_{\mathrm{X}}}\right) \tag{VIII-14}$$

When [X] is varied:

$$\frac{1}{v}=\frac{K_{\mathrm{S}}}{[\mathrm{S}]\,V_{\max}K_{\mathrm{X}}}[\mathrm{X}]+\frac{1}{V_{\max}}\left(1+\frac{K_{\mathrm{S}}}{[\mathrm{S}]}+\frac{K_{\mathrm{S}}[\mathrm{I}]}{[\mathrm{S}]K_i}\right) \tag{VIII-15}$$

Thus the plot of $1/v$ versus [I] at a fixed [S] and [X] is linear with a slope of $K_{\mathrm{S}}/[\mathrm{S}]\,V_{\max}K_i$. The slope is independent of [X]. Consequently, the family of curves obtained at different fixed concentrations of X are parallel (Fig. VIII-5*a*). The curves obtained when $1/v$ versus [X] is plotted at different fixed concentrations of I are also parallel (Fig. VIII-5*b*). The curves will not be parallel if both I and X can combine simultaneously with the enzyme. The intercept on the $1/v$-axis represents the effect of varying concentrations of the second (fixed) inhibitor in the absence of the varied inhibitor. Thus the replot of the $1/v$-axis intercepts versus the corresponding concentration of fixed inhibitor gives the usual Dixon plot for the fixed inhibitor.

Another diagnostic plot for the mutual exclusivity has been suggested by Yagi and Ozawa (1960) and is obtained from the equation for the relative

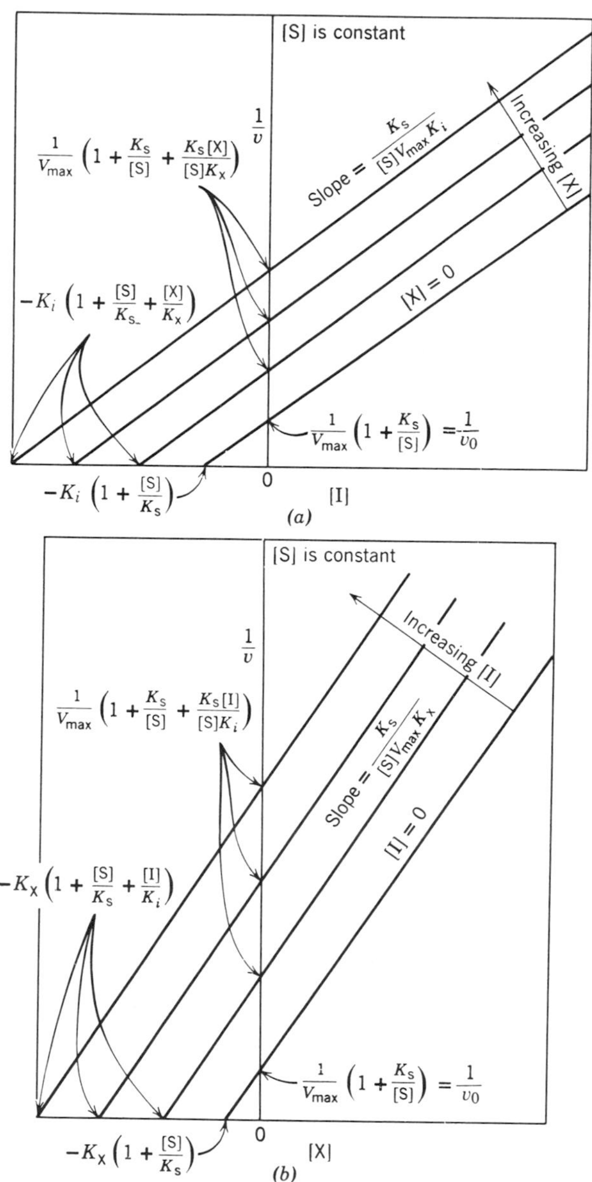

Fig. VIII-5. Dixon plots of (*a*) $1/v$ versus [I] at different fixed concentrations of linear competitive inhibitor X and a constant [S], (*b*) $1/v$ versus [X] at different fixed concentrations of linear competitive inhibitor I and a constant [S]. The parallel plots show that the two inhibitors, I and X, are mutually exclusive. A replot of the $1/v$-axis intercepts would give the usual Dixon plot for the changing fixed inhibitor.

velocities, v_0/v_i:

$$\frac{v_0}{v_i}=\frac{\dfrac{[\mathrm{S}]V_{\max}}{K_\mathrm{S}+[\mathrm{S}]}}{\dfrac{[\mathrm{S}]V_{\max}}{K_\mathrm{S}\left(1+\dfrac{[\mathrm{I}]}{K_i}+\dfrac{[\mathrm{X}]}{K_\mathrm{X}}\right)+[\mathrm{S}]}}=\frac{K_\mathrm{S}\left(1+\dfrac{[\mathrm{I}]}{K_i}+\dfrac{[\mathrm{X}]}{K_\mathrm{X}}\right)+[\mathrm{S}]}{K_\mathrm{S}+[\mathrm{S}]}$$

$$\boxed{\frac{v_0}{v_i}=\frac{1}{\left(1+\dfrac{[\mathrm{S}]}{K_\mathrm{S}}\right)}\left(\frac{[\mathrm{I}]}{K_i}+\frac{[\mathrm{X}]}{K_\mathrm{X}}\right)+1} \qquad \text{(VIII-16)}$$

or

$$\boxed{\frac{v_0}{v_i}=\left(1-\frac{v_0}{V_{\max}}\right)\left(\frac{[\mathrm{I}]}{K_i}+\frac{[\mathrm{X}]}{K_\mathrm{X}}\right)+1} \qquad \text{(VIII-17)}$$

where v_0 = the uninhibited velocity at the fixed [S]
v_i = the velocity at the fixed [S] in the presence of the inhibitors

The plot is constructed by varying [I] and [X] together, maintaining the two inhibitors at a constant ratio, $[\mathrm{X}]=n[\mathrm{I}]$. The plot is a straight line if the two inhibitors are mutually exclusive. In effect, this is the same as saying that a mixture of the two inhibitors gives the same inhibition as the same total specific concentration of either inhibitor alone. If the two inhibitors are not mutually exclusive, the plot will be concave upward (parabolic), as in System B3 described in a following section.

The Dixon plot and the Yagi-Ozawa plot have been combined into a single plot by Yonetani and Theorell (1964). In this plot v_0/v_i is plotted against [I] at different fixed concentrations of [X] (or vice versa). The equation for the plot is obtained by multiplying out and rearranging equation VIII-16:

$$\boxed{\frac{v_0}{v_i}=\frac{1}{K_i\left(1+\dfrac{[\mathrm{S}]}{K_\mathrm{S}}\right)}[\mathrm{I}]+\left[1+\frac{[\mathrm{X}]}{K_\mathrm{X}\left(1+\dfrac{[\mathrm{S}]}{K_\mathrm{S}}\right)}\right]} \qquad \text{(VIII-18)}$$

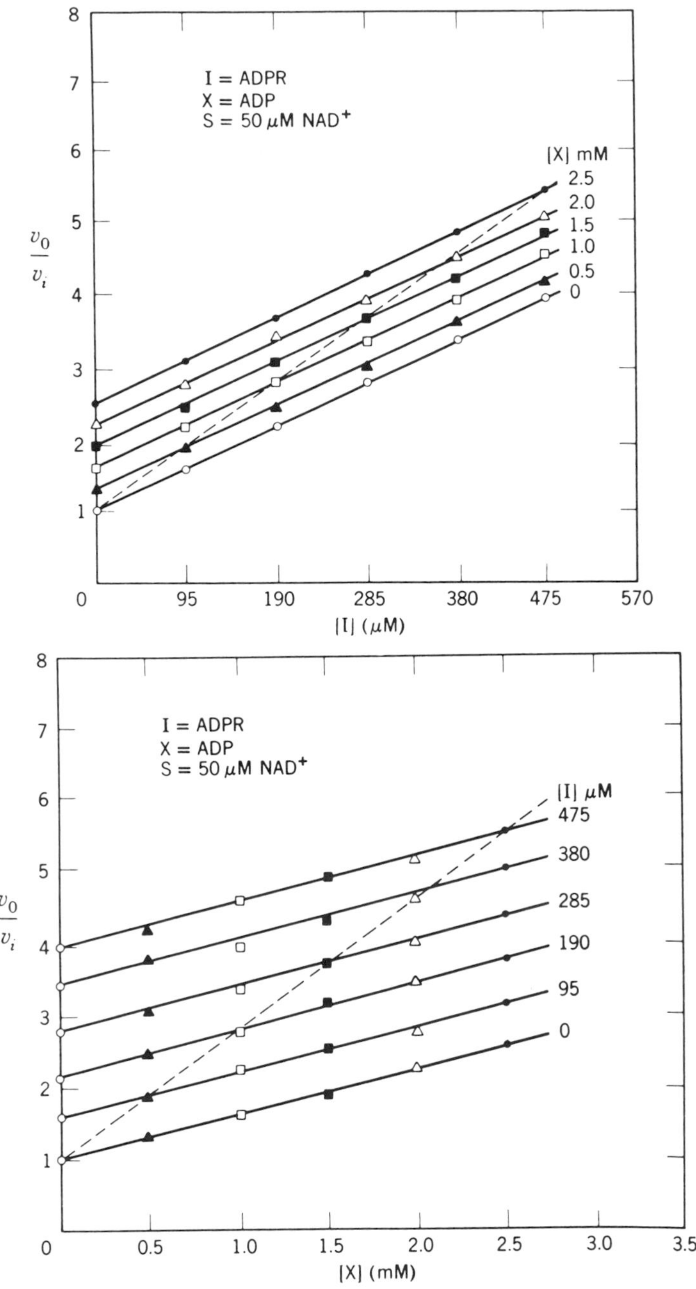

Fig. VIII-6. Yonetani-Theorell plot of v_0/v_i versus [I] and v_0/v_i versus [X] for liver alcohol dehydrogenase. The varied inhibitor and the changing fixed inhibitor are increased by the same factor permitting the superimposition of the linear Yagi-Ozawa plot. [Redrawn from Yonetani, T. and Theorell, H., *Arch. Biochem. Biophys.* **106**, 243 (1964).]

The family of curves are parallel and the plot for $[\mathrm{X}]=0$ intercepts the v_0/v_i-axis at 1. If the fixed concentrations of X are varied by the same factor as the increments of I, then the Yagi-Ozawa plot can be obtained by drawing a line between the points corresponding to the varying $[\mathrm{I}]+[\mathrm{X}]$ (where [I] and [X] are each increased by the same factor). The Dixon plot and the Yonetani-Theorell plot can be used for multisubstrate enzymes where I and X are competitive with respect to one of the substrates. The equations will be more complex, but the characteristics of the plots remain the same (i.e., parallel lines if I and X are mutually exclusive). The Yonetani-Theorell plot is illustrated in Figure VIII-6 for the mutually exclusive competitive inhibition of liver alcohol dehydrogenase by ADP and ADPR (adenosine diphosphate ribose). The inhibitors compete with NAD^+ for the enzyme.

System B2. Noncompetitive and Mixed-Type Inhibition by Two Different Mutually Exclusive Inhibitors

The equilibria shown below describe a system in which two different mutually exclusive inhibitors, I and X, form nonproductive ESX and ESI complexes.

$$
\begin{array}{ccccc}
\mathrm{EX} + \mathrm{S} & \underset{}{\overset{\beta K_S}{\rightleftharpoons}} & \mathrm{ESX} & & \\
K_X \updownarrow & & \updownarrow \beta K_X & & \\
\mathrm{X} & & \mathrm{X} & & \\
+ & & + & & \\
\mathrm{E} + \mathrm{S} & \overset{K_S}{\rightleftharpoons} & \mathrm{ES} & \overset{k_p}{\longrightarrow} & \mathrm{E} + \mathrm{P} \\
+ & & + & & \\
\mathrm{I} & & \mathrm{I} & & \\
K_i \updownarrow & & \updownarrow \alpha K_i & & \\
\mathrm{EI} + \mathrm{S} & \overset{\alpha K_S}{\rightleftharpoons} & \mathrm{ESI} & &
\end{array}
$$

The system may be purely noncompetitive if $\alpha=\beta=1$, "completely" mixed if $\alpha\neq 1$ and $\beta\neq 1$, or "partially" mixed if $\alpha=1$ and $\beta\neq 1$, or $\alpha\neq 1$ and

$\beta = 1$. The velocity equation in the presence of I and X is:

$$\frac{v}{V_{max}} = \frac{\dfrac{[S]}{K_S}}{1 + \dfrac{[S]}{K_S} + \dfrac{[I]}{K_i} + \dfrac{[X]}{K_X} + \dfrac{[S][I]}{\alpha K_S K_i} + \dfrac{[S][X]}{\beta K_S K_X}} \qquad \text{(VIII-19)}$$

or

$$\frac{v}{V_{max}} = \frac{[S]}{K_S\left(1 + \dfrac{[I]}{K_i} + \dfrac{[X]}{K_X}\right) + [S]\left(1 + \dfrac{[I]}{\alpha K_i} + \dfrac{[X]}{\beta K_X}\right)} \qquad \text{(VIII-20)}$$

If $\alpha = \beta$ (whether or not $\alpha = \beta = 1$), the degree of inhibition by mixtures of I and X will be identical to that observed in the presence of the same total specific concentration of I or X alone. If $\alpha \neq \beta$, then the degree of inhibition by mixtures of I and X will be more or less than the same total specific concentration of either inhibitor, depending on the values of α and β. Either inhibitor alone can drive the velocity to zero. The values of K_i, K_X, α, and β can be obtained by treating each inhibitor separately as described earlier for linear noncompetitive and mixed-type inhibitions. The mutual exclusivity of I and X will be evident from the Dixon plots of $1/v$ versus the concentration of one inhibitor at a fixed [S] and various fixed concentrations of the other inhibitor. The equations for the Dixon plot are:

$$\frac{1}{v} = \frac{\left(1 + \dfrac{\alpha K_S}{[S]}\right)}{\alpha K_i V_{max}}[I] + \frac{1}{V_{max}}\left(1 + \frac{K_S}{[S]} + \frac{[X]}{\beta K_X} + \frac{K_S[X]}{[S]K_X}\right) \qquad \text{(VIII-21)}$$

and

$$\frac{1}{v} = \frac{\left(1 + \dfrac{\beta K_S}{[S]}\right)}{\beta K_X V_{max}}[X] + \frac{1}{V_{max}}\left(1 + \frac{K_S}{[S]} + \frac{[I]}{\alpha K_i} + \frac{K_S[I]}{[S]K_i}\right) \qquad \text{(VIII-22)}$$

Thus the slopes of the plots are independent of the concentration of the second (fixed) inhibitor. Consequently, the family of curves are parallel as shown for System B1. The intercepts on the $1/v$-axis represent the effect of varying concentrations of the fixed inhibitor in the absence of the varied inhibitor. Thus a replot of $1/v$-axis intercept versus the corresponding concentration of fixed inhibitor gives the usual Dixon plot for that inhibitor. Note that the Dixon plots are linear only because the ESX and ESI complexes are not catalytically active (as in mixed-type inhibition System C1, Chapter Four).

System B3. Cooperative (Synergistic) Pure Competitive Inhibition by Two Different Nonexclusive Inhibitors

The equilibria shown below describe a system in which two different inhibitors, I and X, bind to the enzyme at different sites and the binding of either inhibitor prevents the substrate from binding. The two inhibitors may compete for different portions of the substrate binding site, or they may combine with the enzyme at specific sites in such a way as to distort the substrate binding site. The binding of one inhibitor may ($\alpha \neq 1$) or may not ($\alpha = 1$) affect the binding of the other.

EX + I $\overset{\alpha K_i}{\rightleftharpoons}$ EIX

E + X $\overset{K_X}{\rightleftharpoons}$ EX

E + I $\overset{K_i}{\rightleftharpoons}$ EI

EI + X $\overset{\alpha K_X}{\rightleftharpoons}$ EIX

$$\mathrm{E} + \mathrm{S} \overset{K_S}{\rightleftharpoons} \mathrm{ES} \overset{k_p}{\longrightarrow} \mathrm{E} + \mathrm{P}$$

The velocity equation is:

$$\frac{v}{V_{\max}} = \frac{\dfrac{[\mathrm{S}]}{K_S}}{1 + \dfrac{[\mathrm{S}]}{K_S} + \dfrac{[\mathrm{I}]}{K_i} + \dfrac{[\mathrm{X}]}{K_X} + \dfrac{[\mathrm{I}][\mathrm{X}]}{\alpha K_i K_X}} \qquad \text{(VIII-23)}$$

or

$$\boxed{\frac{v}{V_{max}} = \frac{[S]}{K_S\left(1 + \frac{[I]}{K_i} + \frac{[X]}{K_X} + \frac{[I][X]}{\alpha K_i K_X}\right) + [S]}} \qquad \text{(VIII-24)}$$

The v versus [I] or v versus [X] curve is identical to that for single-site, pure competitive inhibition. When both I and X are present, the degree of inhibition is always greater than that observed at the same total specific concentration of I or X alone. For example, when $\alpha = 1$, $[S] = 10K_S$, $[X] = 6K_X$, and $[I] = 0$, v is $0.588\,V_{max}$ (i.e., 35.3% inhibition). When $[X] = 3K_X$ and $[I] = 3K_i$, v is $0.384\,V_{max}$ (i.e., 57.7% inhibition). The inhibition is clearly synergistic in spite of the fact that no cooperative binding in the usual sense was assumed. The apparent cooperativity results from the formation of the additional EIX complex. (In the presence of both inhibitors, the denominator of the velocity equation contains an additional term thereby decreasing v below that which would be observed with I or X alone.) If the binding of one inhibitor increases the binding constant for the other inhibitor (i.e., $\alpha < 1$), the synergistic effect will be much more pronounced. The enzyme glutamine : PRPP amido transferase, which catalyzes the first reaction of the purine ribonucleotide biosynthesis pathway, shows synergistic or cooperative feedback inhibition by adenine nucleotides plus guanine nucleotides (Caskey *et al.*, 1964; Nierlich and Magasanik, 1965). In this case, the inhibition is competitive with respect to PRPP and mixed-type with respect to glutamine.

The nonexclusivity of I and X binding can be shown by Dixon plots of $1/v$ versus the concentration of one inhibitor at a fixed [S] and different fixed concentrations of the second inhibitor. The equations for the Dixon plots are:

$$\boxed{\frac{1}{v} = \frac{K_S}{[S]\,V_{max}K_i}\left(1 + \frac{[X]}{\alpha K_X}\right)[I] + \frac{1}{V_{max}}\left(1 + \frac{K_S}{[S]} + \frac{K_S[X]}{[S]K_X}\right)} \qquad \text{(VIII-25)}$$

and

$$\boxed{\frac{1}{v} = \frac{K_S}{[S]\,V_{max}K_X}\left(1 + \frac{[I]}{\alpha K_i}\right)[X] + \frac{1}{V_{max}}\left(1 + \frac{K_S}{[S]} + \frac{K_S[I]}{[S]K_i}\right)} \qquad \text{(VIII-26)}$$

The slope of the plot now depends on the concentration of the second inhibitor. Consequently, the family of curves obtained at different fixed concentrations of one inhibitor will not be parallel (Fig. VIII-7). The intercepts on the $1/v$-axis represent the effect of varying concentrations of the fixed inhibitor in the absence of the varied inhibitor. As before, the replot of the $1/v$-axis intercepts versus the corresponding concentration of fixed inhibitor represents the usual linear Dixon plot for that inhibitor alone.

If we set $1/v$ for $[X]_1$ equal to $1/v$ for $[X]_2$, we find that the family of curves intersect to the left of the $1/v$-axis at $[I] = -\alpha K_i$ (Fig. VIII-7*a*). The corresponding $1/v$ coordinate is:

$$\boxed{\frac{1}{V_{max}}\left(1 + \frac{K_S(1-\alpha)}{[S]}\right) \quad \text{at} \quad [I] = -\alpha K_i} \qquad \text{(VIII-27)}$$

Thus when $\alpha = 1$, the curves intersect at a height of $1/V_{max}$ above the [I]-axis. If $\alpha > 1$ (mutually hindered inhibitor binding), the curves intersect below $1/V_{max}$ and also below the horizontal axis if $\alpha > (1 + [S]/K_S)$. If $\alpha < 1$ (mutually cooperative inhibitor binding), the curves intersect above $1/V_{max}$. The $1/v$ versus [X] plots give analogous results (Fig. VIII-7*b*). Thus both αK_i and αK_X can be obtained from the two plots. The values can also be obtained from a single plot. For example, the $1/v$ versus [X] plot gives $-\alpha K_X$ as the intersection point. The αK_i can be obtained from the ratio of slope with I to slope without I:

$$\frac{\textit{slope} \text{ at } [I] > 0}{\textit{slope} \text{ at } [I] = 0} = \frac{\dfrac{K_S}{[S] V_{max} K_X}\left(1 + \dfrac{[I]}{\alpha K_i}\right)}{\dfrac{K_S}{[S] V_{max} K_X}} = \left(1 + \frac{[I]}{\alpha K_i}\right)$$

Alternately, the slopes of the $1/v$ versus [X] plots can be replotted against the corresponding [I]:

$$\boxed{\textit{slope} = \frac{K_S}{\alpha K_i K_X [S] V_{max}}[I] + \frac{K_S}{K_X [S] V_{max}}} \qquad \text{(VIII-28)}$$

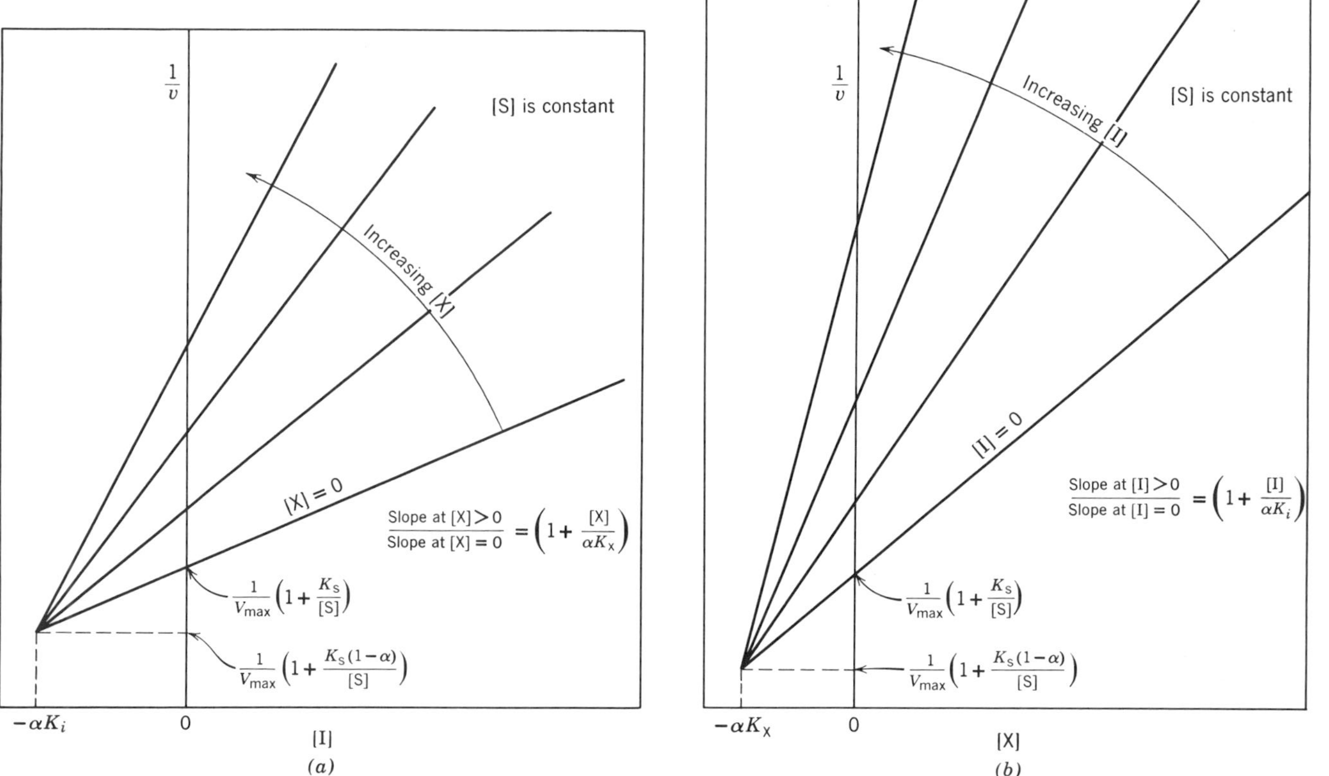

Fig. VIII-7. Dixon plots of (*a*) $1/v$ versus [I] at different fixed concentrations of X and a constant [S] and (*b*) $1/v$ versus [X] at different fixed concentrations of I and a constant [S]. I and X are both linear competitive inhibitors with respect to S. The intersecting lines show that I and X are not mutually exclusive. The families of plots may intersect above, on, or below the horizontal axis, depending on the values of α and $[S]/K_S$. A replot of the $1/v$-axis would give the usual Dixon plot for the changing fixed inhibitor. If K_X and K_i are relatively large but $\alpha \ll 1$, I and X will appear to be concerted competitive inhibitors. The plots will then intersect almost on the $1/v$ axis

The intercept on the [I]-axis gives $-\alpha K_i$. The K_i and K_X can be obtained independently from suitable plots treating each inhibitor separately (or replots of the $1/v$-axis intercepts, noted above). With K_i and K_X known, α can be calculated. The relationship $\alpha K_i K_X = \alpha K_X K_i$ can also be used to calculate any one dissociation constant if the other three are known.

The equation for the Yagi-Ozawa plot is obtained from the equation for the relative velocities:

$$\frac{v_0}{v_i} = \left(1 - \frac{v_0}{V_{max}}\right)\left(\frac{[I]}{K_i} + \frac{[X]}{K_X} + \frac{[I][X]}{\alpha K_i K_X}\right) + 1 \qquad \text{(VIII-29)}$$

If I and X are maintained at a constant ratio, $[X] = n[I]$, so that both inhibitors are varied by the same factor, the equation can be written:

$$\frac{v_0}{v_i} = \left(1 - \frac{v_0}{V_{max}}\right)\left(\frac{[I]}{K_i} + \frac{n[I]}{K_X} + \frac{n[I]^2}{\alpha K_i K_X}\right) + 1 \qquad \text{(VIII-30)}$$

The plot of v/v_i versus [I] or [X] will be parabolic (i.e., concave upward).

The equation for the Yonetani-Theorell plot is:

$$\frac{v_0}{v_i} = \frac{\left(1 + \frac{[X]}{\alpha K_X}\right)}{K_i\left(1 + \frac{[S]}{K_S}\right)}[I] + \left[1 + \frac{[S]}{K_X\left(1 + \frac{[S]}{K_S}\right)}\right] \qquad \text{(VIII-31)}$$

The Yonetani-Theorell plot for the competitive inhibition of liver alcohol dehydrogenase by ADP and o-phenanthroline is shown in Figure VIII-8. The curves intersect to the left of the v_0/v_i-axis indicating that the two inhibitors can occupy the enzyme simultaneously. This conclusion is confirmed by the superimposed Yagi-Ozawa plot, which is concave upward. If I and X have very little affinity for E, no inhibition will be observed in the presence of either compound alone. However, if $\alpha \ll 1$, mixtures of I and X will be strongly inhibitory because of their mutually cooperative binding. This limiting case of synergistic binding is called concerted inhibition. The plots of $1/v$ versus [I] or [X] will intersect very close to the $1/v$-axis.

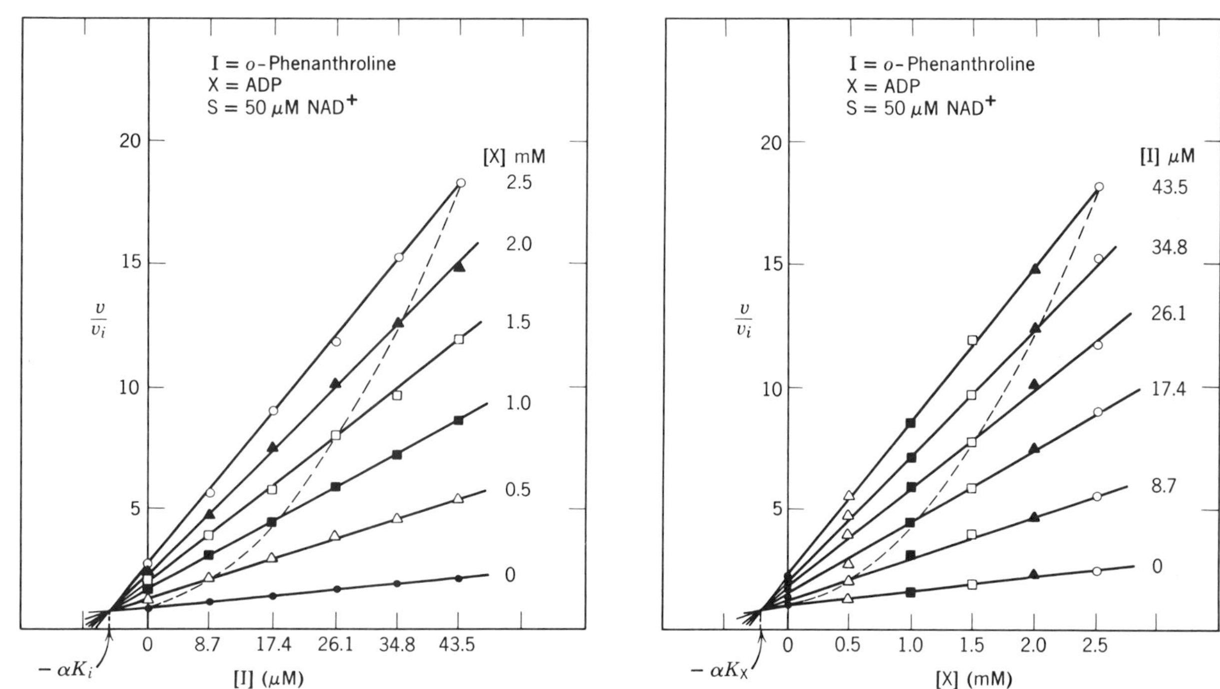

Fig. VIII-8. Yonetani-Theorell plots with superimposed parabolic Yagi-Ozawa plots for the inhibition of liver alcohol dehydrogenase by *o*-phenanthroline. [Redrawn from Yonetani, T. and Theorell, H., *Arch. Biochem. Biophys.* **106**, 243 (1964).]

The β-aspartylkinases of *Bacillus polymyxa* and *Rhodopseudomonas capsulata* are examples of enzymes subject to concerted feedback inhibition. β-Aspartylkinase catalyzes the first "committed" reaction leading to the biosynthesis of lysine, threonine, methionine, and isoleucine. The enzyme is inhibited by the concerted action of lysine and threonine. (In *Escherichia coli*, three different β-aspartylkinases are present. One is specifically inhibited and repressed by lysine, one is specifically inhibited by homoserine, a methionine precursor, while the third is specifically inhibited by threonine.)

True cooperative or concerted inhibition by two feedback inhibitors is never additive. That is, if I at $6K_i$ yields 35.3% inhibition, and X at $6K_X$ yields 35.3% inhibition, then I at $6K_i$ *plus* X at $6K_X$ will not yield 70.6% inhibition. Additive inhibition implies the presence of two distinct enzymes or sites, each sensitive to only one of the inhibitors. Each inhibitor alone appears to act in a partial manner.

Ordered Inhibitor Binding

If X does not bind to free E but does bind to EI, the results would be similar to those described above. X alone will have no effect on the velocity, but X plus I will be more inhibitory than I alone. The equilibria and equations are:

$$\begin{array}{ccccc} E + S & \overset{K_S}{\rightleftharpoons} & ES & \overset{k_p}{\longrightarrow} & E + P \\ + & & & & \\ I & & & & \\ K_i \updownarrow & & & & \\ EI + X & \overset{K_X}{\rightleftharpoons} & EIX & & \end{array}$$

$$\frac{v}{V_{max}} = \frac{\dfrac{[S]}{K_S}}{1 + \dfrac{[S]}{K_S} + \dfrac{[I]}{K_i} + \dfrac{[I][X]}{K_i K_X}} = \frac{[S]}{K_S\left(1 + \dfrac{[I]}{K_i} + \dfrac{[I][X]}{K_i K_X}\right) + [S]} \qquad \text{(VIII-32)}$$

$$\boxed{\frac{1}{v} = \frac{K_S}{V_{max}K_i[S]}\left(1 + \frac{[X]}{K_X}\right)[I] + \frac{1}{V_{max}}\left(1 + \frac{K_S}{[S]}\right)} \tag{VIII-33}$$

$$\boxed{\frac{1}{v} = \frac{K_S[I]}{V_{max}K_iK_X[S]}[X] + \frac{1}{V_{max}}\left(1 + \frac{K_S}{[S]} + \frac{K_S[I]}{[S]K_i}\right)} \tag{VIII-34}$$

Both families of Dixon plots are intersecting. The $1/v$ versus [I] plots at different fixed concentrations of X and constant [S] intersect on the $1/v$-axis. The K_X can be obtained as the horizontal intercept of the slope versus [X] replot. The plots of $1/v$ versus [X] intersect to the left of the $1/v$-axis. The slope versus [I] replot goes through the origin.

System B4. Cooperative (Synergistic) Noncompetitive Inhibition by Two Different Nonexclusive Inhibitors

Figure VIII-9 shows a general scheme for an enzyme with one substrate site and two nonexclusive effector sites. I and X could be two different inhibitors, two different activators, or one could be an inhibitor and the other an activator. The effectors could act in a pure, partial, or mixed fashion, depending on the values of the interaction factors. Suppose I and X are two pure noncompetitive inhibitors; that is, the inhibitors have no effect on substrate binding ($\alpha = \beta = 1$) and the ESI, ESX, and ESXI complexes are not catalytically active ($x = y = z = 0$).The velocity equation is:

$$\boxed{\frac{v}{V_{max}} = \frac{[S]}{K_S\left(1 + \frac{[I]}{K_i} + \frac{[X]}{K_X} + \frac{[I][X]}{\gamma K_iK_X}\right) + [S]\left(1 + \frac{[I]}{K_i} + \frac{[X]}{K_X} + \frac{[I][X]}{\gamma K_iK_X}\right)}} \tag{VIII-35}$$

where $V_{max} = k_p[E]_t$. As expected, the equation is similar to that for System B3, except now both K_S and [S] are multiplied by the parenthetical term. The inhibition caused by mixtures of I and X is always greater than that obtained by the same total specific concentration of I and X alone, regardless of the value of γ. The synergism is evident from equation VIII-35 where

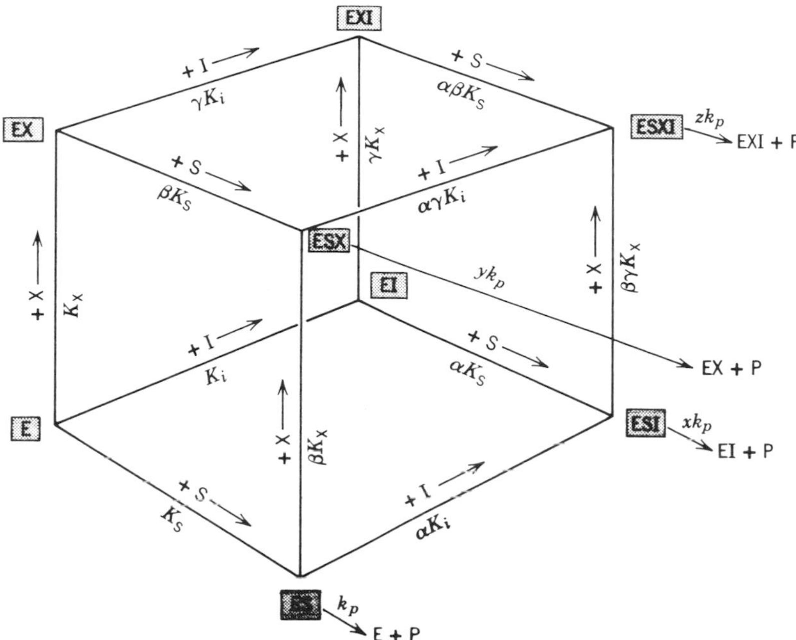

Fig. VIII-9. Equilibria among enzyme species where I and X are two nonexclusive inhibitors.

we see that the simultaneous presence of I and X introduces additional terms to the denominator. If K_i and K_X are very large but $\gamma \ll 1$, then I and X will act as concerted noncompetitive inhibitors. The values of K_i and K_X can be determined in the usual manner by treating each inhibitor separately if each alone is significantly inhibitory. The inhibitor interaction factor, γ, can be obtained from Dixon plots of $1/v$ versus the concentration of one inhibitor at a fixed substrate concentration and various fixed concentrations of the other inhibitor. The equations for the Dixon plots are:

$$\boxed{\frac{1}{v} = \frac{\left(1 + \frac{K_S}{[S]}\right)}{K_i V_{max}}\left(1 + \frac{[X]}{\gamma K_X}\right)[I] + \frac{1}{V_{max}}\left(1 + \frac{K_S}{[S]}\right)\left(1 + \frac{[X]}{K_X}\right)} \qquad \text{(VIII-36)}$$

and

$$\frac{1}{v}=\frac{\left(1+\frac{K_S}{[S]}\right)}{K_X V_{max}}\left(1+\frac{[I]}{\gamma K_i}\right)[X]+\frac{1}{V_{max}}\left(1+\frac{K_S}{[S]}\right)\left(1+\frac{[I]}{K_i}\right) \qquad \text{(VIII-37)}$$

The family of curves for equation VIII-36 or equation VIII-37 intersect at $[I]=-\gamma K_i$ or $[X]=-\gamma K_X$. The $1/v$ coordinate of the intersection point is given by:

$$\frac{1}{v_i}=\frac{1}{V_{max}}(1-\gamma)\left(1+\frac{K_S}{[S]}\right) \quad \text{at} \quad \begin{cases}[I]=-\gamma K_i\\ [X]=-\gamma K_X\end{cases} \qquad \text{(VIII-38)}$$

or

$$\frac{1}{v_i}=\frac{1}{v_0}(1-\gamma) \qquad \text{(VIII-39)}$$

where v = the uninhibited velocity at the fixed [S].

The plots for I as the varied inhibitor are shown in Figure VIII-10. An analogous series of plots are obtained when X is the varied inhibitor. When $\gamma=1$ (i.e., the binding of one inhibitor has no effect on the binding of the other), the curves intersect on the horizontal axis. When $\gamma<1$ (i.e., cooperative inhibitor binding), the curves intersect above the horizontal axis. If γ is very small and K_X relatively large (i.e., concerted inhibition), the family of Dixon plots will intersect very close to the $1/v$-axis. When $\gamma>1$ (i.e., one inhibitor hinders the binding of the other), the curves intersect below the horizontal axis. The values of γ and K_i are obtained from the intersection coordinates. The γK_X can be obtained from the ratio of slope with X to slope without X:

$$\frac{\textit{slope}\text{ at }[X]>0}{\textit{slope}\text{ at }[X]=0}=\frac{\dfrac{\left(1+\frac{K_S}{[S]}\right)\left(1+\frac{[X]}{\gamma K_X}\right)}{K_i V_{max}}}{\dfrac{\left(1+\frac{K_S}{[S]}\right)}{K_i V_{max}}}=\left(1+\frac{[X]}{\gamma K_X}\right)$$

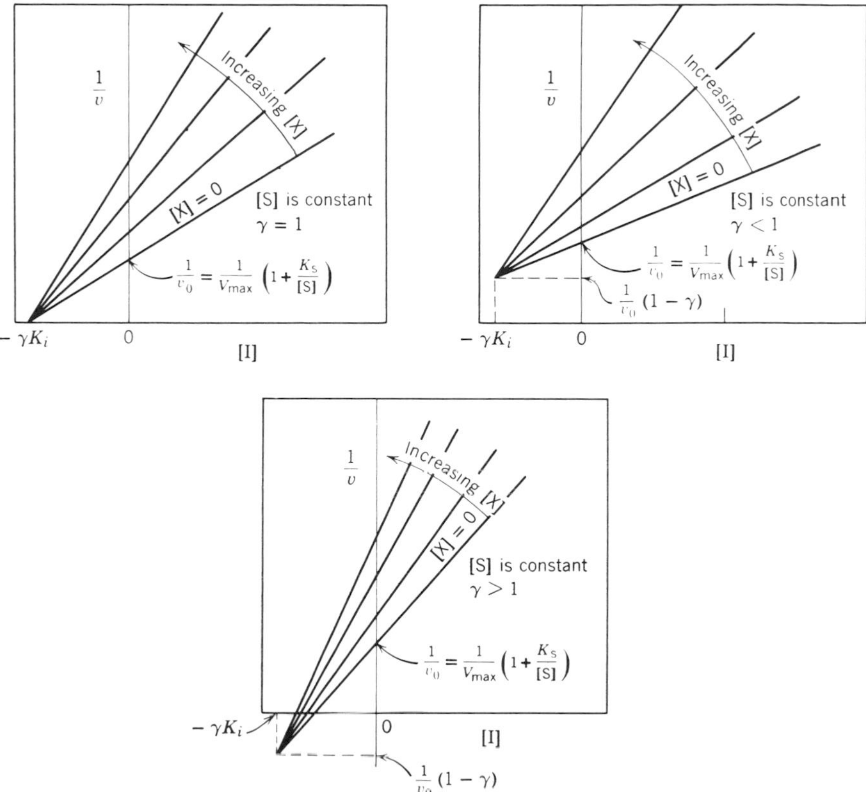

Fig. VIII-10. Dixon plots of $1/v$ versus [I] at different fixed concentrations of X and a constant [S]; I and X are both linear noncompetitive inhibitors with respect to S. The intersecting plots show that I and X can both occupy the enzyme simultaneously. The $1/v$ coordinate of the intersection point depends only on the value of the interaction factor, γ. The plots of $1/v$ versus [X] are symmetrical to those shown for varied [I]. A replot of the $1/v$-axis intercepts would give the usual Dixon plot for the changing fixed inhibitor. If K_X and K_i are relatively high but $\gamma \ll 1$, I and X will be concerted noncompetitive inhibitors. The plots will then intersect very close to the $1/v$-axis.

Alternately, the slopes of the $1/v$ versus [I] plots can be replotted against the corresponding [X]:

$$slope = \frac{\left(1 + \dfrac{K_S}{[S]}\right)}{\gamma K_X K_i V_{max}}[X] + \frac{\left(1 + \dfrac{K_S}{[S]}\right)}{K_i V_{max}} \qquad \text{(VIII-40)}$$

The intercept on the [X]-axis gives $-\gamma K_X$. The replot of $1/v$-axis intercept versus [X] represents the ordinary Dixon plot for X in the absence of I and permits K_X to be determined.

If X and I are linear mixed-type inhibitors, the equation for the Dixon plot would be more complex, but the general characteristics of the family of plots would be the same: they intersect at $-\gamma K_i$ or $-\gamma K_X$. The β-aspartylkinase of *Pseudomonas fluorescens* is subject to noncompetitive and mixed-type cooperative feedback inhibition (approaching concerted inhibition) by threonine plus lysine and by threonine plus methionine (Dungan and Datta, 1973). Physical evidence suggests that threonine and lysine induce the formation of inactive oligomers of the enzyme.

System B5. Cooperative (Synergistic) Uncompetitive Inhibitors

If I and X, two uncompetitive inhibitors, can both occupy the enzyme simultaneously, the equilibria and velocity equations are:

$$\begin{array}{ccccccccc} & & & & & & \text{ESX} & & \\ & & & & & \overset{K_X}{\nearrow\!\!\swarrow} & & + & \\ & & & & \text{X} & & & & \text{I} \\ & & & + & & & & & \searrow\!\!\nwarrow \alpha K_i \\ \text{E} + \text{S} & \underset{}{\overset{K_S}{\rightleftharpoons}} & \text{ES} & \overset{k_p}{\longrightarrow} & \text{E} & + & \text{P} & & \text{ESIX} \\ & & & + & & & & & \nearrow\!\!\swarrow \alpha K_X \\ & & & & \text{I} & & & & \text{X} \\ & & & & & \underset{K_i}{\searrow\!\!\nwarrow} & & + & \\ & & & & & & \text{ESI} & & \end{array}$$

$$\boxed{\frac{v}{V_{max}} = \frac{[S]}{K_S + [S]\left(1 + \frac{[I]}{K_i} + \frac{[X]}{K_X} + \frac{[I][X]}{\alpha K_i K_X}\right)}} \qquad \text{(VIII-41)}$$

Mixtures of I and X will be more inhibitory than the same total specific concentration of I or X alone because of the formation of the additional

ESIX complex. The Dixon plots are described by:

$$\boxed{\frac{1}{v}=\frac{1}{K_iV_{\max}}\left(1+\frac{[\mathrm{X}]}{\alpha K_{\mathrm{X}}}\right)[\mathrm{I}]+\frac{1}{V_{\max}}\left(1+\frac{K_{\mathrm{S}}}{[\mathrm{S}]}+\frac{[\mathrm{X}]}{K_{\mathrm{X}}}\right)} \qquad \text{(VIII-42)}$$

and

$$\boxed{\frac{1}{v}=\frac{1}{K_{\mathrm{X}}V_{\max}}\left(1+\frac{[\mathrm{I}]}{\alpha K_i}\right)[\mathrm{X}]+\frac{1}{V_{\max}}\left(1+\frac{K_{\mathrm{S}}}{[\mathrm{S}]}+\frac{[\mathrm{I}]}{K_i}\right)} \qquad \text{(VIII-43)}$$

The family of plots intersect at a common point above, on, or below the horizontal axis depending on whether α is less than, equal to, or greater than $(1+K_{\mathrm{S}}/[\mathrm{S}])$. The intersection coordinates are:

$$[\mathrm{X}]=-\alpha K_{\mathrm{X}} \qquad \text{(plotting } 1/v \text{ versus } [\mathrm{X}])$$

$$[\mathrm{I}]=-\alpha K_i \qquad \text{(plotting } 1/v \text{ versus } [\mathrm{I}])$$

$$\frac{1}{v}=\frac{1}{V_{\max}}\left(1+\frac{K_{\mathrm{S}}}{[\mathrm{S}]}-\alpha\right)=\frac{1}{v_0}\left[1-\frac{\alpha}{1+\dfrac{K_{\mathrm{S}}}{[\mathrm{S}]}}\right] \qquad \text{(VIII-44)}$$

The constants, α, K_{X}, and K_i can be obtained from the intersection coordinates or, alternately, from replots of the slopes and $1/v$-axis intercepts. As before, a single family of plots is sufficient to obtain all three inhibitor constants. For example, the plot of $1/v$ versus [X] at different fixed concentrations of I will allow α and K_{X} to be determined. The value of αK_i can be calculated from the ratio of *slope* at $[\mathrm{I}]>0$/*slope* at $[\mathrm{I}]=0$. As before, if K_i and K_{X} are relatively large but $\alpha \ll 1$, I and X will be concerted inhibitors.

System B6. I Is Competitive and X Is Noncompetitive With Respect to S

If I is a linear competitive inhibitor and X is a noncompetitive or linear mixed-type inhibitor, then mixtures of I and X will not yield the same degree of inhibition as the same total specific concentration of either alone although the Dixon plots will be parallel. The equilibria assuming I and X are mutually exclusive are shown below.

$$
\begin{array}{ccccccc}
\mathrm{EI} & \underset{}{\overset{K_i}{\rightleftharpoons}} & \mathrm{I} + \mathrm{E} + \mathrm{S} & \overset{K_S}{\leftrightharpoons} & \mathrm{ES} & \overset{k_p}{\longrightarrow} & \mathrm{E} + \mathrm{P} \\
 & & + & & + & & \\
 & & \mathrm{X} & & \mathrm{X} & & \\
 & & K_X \updownarrows & & \downuparrows \alpha K_X & & \\
 & & \mathrm{EX} + \mathrm{S} & \overset{\alpha K_S}{\leftrightharpoons} & \mathrm{ESX} & &
\end{array}
$$

The velocity equation and Dixon plots are given by:

$$\frac{v}{V_{max}} = \frac{[S]}{K_S\left(1 + \frac{[I]}{K_i} + \frac{[X]}{K_X}\right) + [S]\left(1 + \frac{[X]}{\alpha K_X}\right)} \tag{VIII-45}$$

$$\frac{1}{v} = \frac{K_S}{V_{max} K_i [S]}[I] + \frac{1}{V_{max}}\left[\left(1 + \frac{[X]}{\alpha K_X}\right) + \frac{K_S}{[S]}\left(1 + \frac{[X]}{K_X}\right)\right] \tag{VIII-46}$$

and

$$\frac{1}{v} = \frac{\left(1 + \frac{\alpha K_S}{[S]}\right)}{\alpha K_X V_{max}}[X] + \frac{1}{V_{max}}\left[1 + \frac{K_S}{[S]}\left(1 + \frac{[I]}{K_i}\right)\right] \tag{VIII-47}$$

If I and X are not mutually exclusive, then it is likely that the only additional complex that can form will be EIX (as long as I and S remain mutually exclusive). This introduces an $[I][X]/\beta K_i K_X$ denominator term which becomes part of the slope factor of the velocity equation. The equations for the Dixon plots are:

$$\frac{1}{v} = \frac{K_S}{V_{max} K_i [S]}\left(1 + \frac{[X]}{\beta K_X}\right)[I] + \frac{1}{V_{max}}\left[\left(1 + \frac{[X]}{\alpha K_X}\right) + \frac{K_S}{[S]}\left(1 + \frac{[X]}{K_X}\right)\right]$$

(VIII-48)

and

$$\frac{1}{v} = \frac{1}{\alpha K_X V_{max}}\left[1 + \frac{\alpha K_S}{[S]}\left(1 + \frac{[I]}{\beta K_i}\right)\right][X] + \frac{1}{V_{max}}\left[1 + \frac{K_S}{[S]}\left(1 + \frac{[I]}{K_i}\right)\right]$$

(VIII-49)

The families of plots are now intersecting. The inhibitor interaction factor, β, can be obtained easily from the slope replot of equation VIII-48 ($-\beta K_X$ is the [X]-axis intercept). The slope replot of equation VIII-49 gives $-\beta K_i(1 + [S]/\alpha K_S)$ as the [I]-axis intercept.

System B7. Two Partial Inhibitors

System B7a. I and X Are Partial Competitive Inhibitors

Suppose I and X shown in Figure VIII-9 are two different partial competitive inhibitors. The binding of I changes K_S by a factor α; the binding of X changes K_S by a factor β. The inhibitors have no effect on the binding of each other (i.e., $\gamma = 1$), nor on the catalytic rate constant, k_p (i.e., $x = y = z = 1$). At any given substrate concentration, increasing [I] (in the absence of X) will decrease the initial velocity hyperbolically to a limit given by:

$$\frac{v}{V_{max}} = \frac{[S]}{\alpha K_S + [S]}$$

Increasing [X] (in the absence of I) will decrease the initial velocity hyperbolically to a limit given by:

$$\frac{v}{V_{max}} = \frac{[S]}{\beta K_S + [S]}$$

If I and X are both increased (together, or first one then the other), the enzyme will be driven to the EXI and ESXI forms, and the velocity will decrease hyperbolically to a limit given by:

$$\frac{v}{V_{max}} = \frac{[S]}{\alpha\beta K_S + [S]}$$

Figure VIII-11 shows the effect of increasing I followed by increasing X at $[S] = 10K_S$, where $\alpha = 10$, and $\beta = 10$. The curve was calculated from the velocity equation:

$$\frac{v}{V_{max}} = \frac{\frac{[S]}{K_S} + \frac{[S][I]}{\alpha K_S K_i} + \frac{[S][X]}{\beta K_S K_X} + \frac{[S][I][X]}{\alpha\beta K_S K_i K_X}}{1 + \frac{[S]}{K_S} + \frac{[I]}{K_i} + \frac{[X]}{K_X} + \frac{[S][I]}{\alpha K_S K_i} + \frac{[S][X]}{\beta K_S K_X} + \frac{[I][X]}{K_i K_X} + \frac{[S][I][X]}{\alpha\beta K_S K_i K_X}} \tag{VIII-50}$$

We see that either inhibitor alone can never reduce the velocity below $0.5\,V_{max}$ (the plateau velocity will vary with different values of α, β, and [S]).

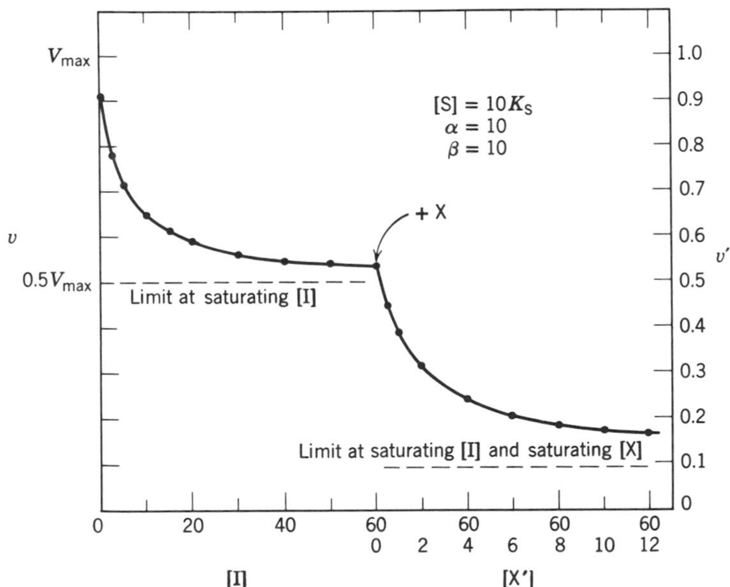

Fig. VIII-11. Plot of v versus [I] followed by [X]; I and X are two partial competitive inhibitors ($\alpha = 10, \beta = 10$); $[S] = 10K_S$.

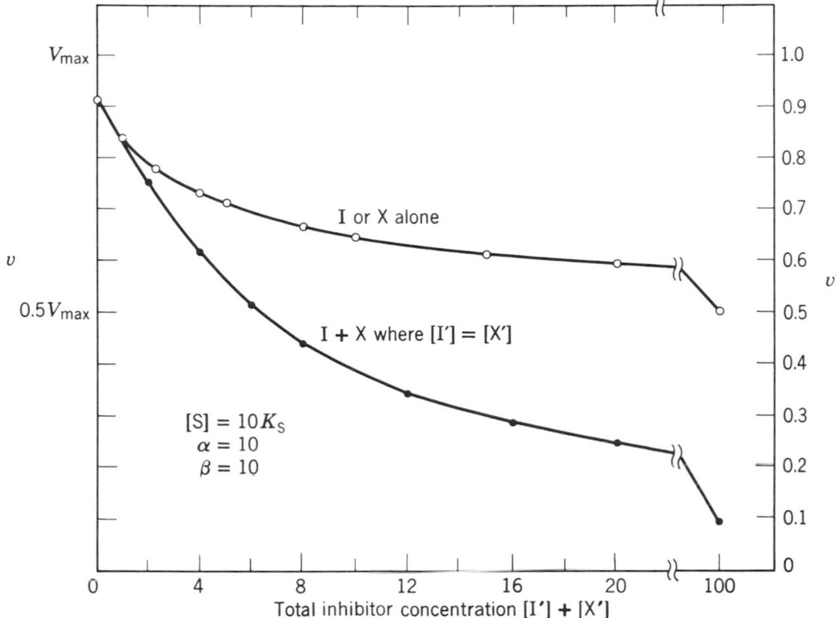

Fig. VIII-12. The v versus [I]+[X] plot where [I′]=[X′]; I and X are two partial competitive inhibitors ($\alpha=10, \beta=10$); [S]=$10K_S$. Mixtures of I and X are more inhibitory than the same total specific concentration of either inhibitor alone.

Yet, in combination the velocity can be supressed to $0.091\,V_{max}$. Figure VIII-12 illustrates this synergistic action of the two inhibitors. For example, when either I or X is present alone at $8K_i$, $v=0.667\,V_{max}$ (i.e., 26.7% inhibition). When I and X are present together at the same total specific concentration ($[I]=4K_i$ plus $[X]=4K_X$), $v=0.44\,V_{max}$ (i.e., 51.5% inhibition).

System B7b. I and X Are Partial Noncompetitive Inhibitors (Cumulative Inhibition)

If I and X have no effect on S binding but $x<1, y<1$, and $z<1$, the velocity equation is:

$$\frac{v}{V_{max}} = \frac{\dfrac{[S]}{K_S} + \dfrac{x[S][I]}{K_S K_i} + \dfrac{y[S][X]}{K_S K_X} + \dfrac{z[S][I][X]}{K_S K_i K_X}}{1 + \dfrac{[S]}{K_S} + \dfrac{[I]}{K_i} + \dfrac{[X]}{K_X} + \dfrac{[S][I]}{K_S K_i} + \dfrac{[S][X]}{K_S K_X} + \dfrac{[I][X]}{K_i K_X} + \dfrac{[S][I][X]}{K_S K_i K_X}} \qquad \text{(VIII-51)}$$

If the individual fractional effects of I and X on k_p are unchanged when both inhibitors bind, then $z = xy$ and I + X will act in a cumulative fashion at all substrate concentrations (see below). If $z \neq xy$, cumulative inhibition will not be observed.The cumulative action of two different inhibitors is illustrated below. Inhibitor X reduces the initial velocity to 60% of the original value when present alone at some concentration. Inhibitor I reduces the initial velocity to 50% of the original value when present alone at some concentration. When X and I are both present, X inhibits 40% and I inhibits the residual 60% by 50% (or vice versa) yielding a total of 70% inhibition. (The net activity is 0.6×0.5, or 0.3 of the original activity).

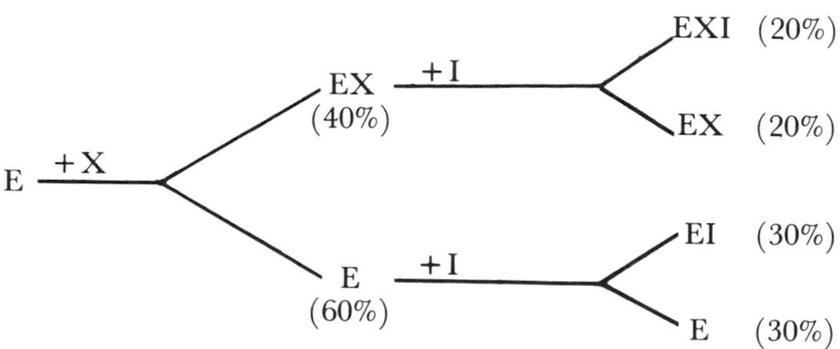

Glutamine synthetase of *E. coli* is subject to feedback inhibition by eight different compounds that derive their nitrogen from glutamine. L-Alanine, AMP, and carbamyl phosphate are partially noncompetitive with respect to any of the substrates (NH_4^+, glutamate, and $MgATP^{2-}$). Glycine, CTP, and L-tryptophan are partially competitive with respect to glutamate, while L-histidine and glucosamine-6-phosphate are partially competitive with respect to NH_4^+. Stadtman and co-workers (1968) have shown that the inhibitors act independently and (under certain assay conditions) in a cumulative fashion.

System B8. Concerted (Multivalent) Inhibition by Two Different Inhibitors

The equilibria shown below describe an enzyme that is subject to inhibition only by the concerted action of two different inhibitors. Neither X nor I alone has any effect (i.e., EX and EI have the same affinity for S as free E and the EXS and EIS complexes are as catalytically active as ES). The EIX complex, however, either cannot bind substrate (competitive action, as shown) or forms an inactive EIXS complex (a noncompetitive variation of this scheme, not shown). The binding of one inhibitor may ($\alpha \neq 1$) or may not $(\alpha=1)$ affect the binding of the other. In the competitive model, the binding of any two ligands induces a conformational change in the enzyme that excludes the third ligand.

$$
\begin{array}{ccccccc}
 & & \mathrm{EX} + \mathrm{S} & \overset{K_S}{\rightleftharpoons} & \mathrm{EXS} & \overset{k_p}{\longrightarrow} & \mathrm{EX} + \mathrm{P} \\
 & \mathrm{I} \; + & \uparrow\downarrow K_X & & \uparrow\downarrow K_X & & \\
 \alpha K_i \nearrow\!\!\swarrow & & \mathrm{X} & & \mathrm{X} & & \\
 & & + & & + & & \\
 \mathrm{EIX} & & \mathrm{E} + \mathrm{S} & \overset{K_S}{\rightleftharpoons} & \mathrm{ES} & \overset{k_p}{\longrightarrow} & \mathrm{E} + \mathrm{P} \\
 & & + & & + & & \\
 \alpha K_X \nwarrow\!\!\searrow & & \mathrm{I} & & \mathrm{I} & & \\
 & \mathrm{X} \; + & \downarrow\uparrow K_i & & \downarrow\uparrow K_i & & \\
 & & \mathrm{EI} + \mathrm{S} & \overset{K_S}{\rightleftharpoons} & \mathrm{ESI} & \overset{k_p}{\longrightarrow} & \mathrm{EI} + \mathrm{P}
\end{array}
$$

The velocity equation for the competitive situation shown is:

$$
\frac{v}{V_{\max}} = \frac{\dfrac{[\mathrm{S}]}{K_S} + \dfrac{[\mathrm{X}][\mathrm{S}]}{K_X K_S} + \dfrac{[\mathrm{I}][\mathrm{S}]}{K_i K_S}}{1 + \dfrac{[\mathrm{S}]}{K_S} + \dfrac{[\mathrm{X}]}{K_X} + \dfrac{[\mathrm{I}]}{K_i} + \dfrac{[\mathrm{X}][\mathrm{S}]}{K_X K_S} + \dfrac{[\mathrm{I}][\mathrm{S}]}{K_i K_S} + \dfrac{[\mathrm{X}][\mathrm{I}]}{\alpha K_X K_i}} \qquad \text{(VIII-52)}
$$

For the noncompetitive situation, the denominator would contain an additional $[\mathrm{X}][\mathrm{I}][\mathrm{S}]/\alpha K_X K_i K_S$ term corresponding to the EIXS complex. Other forms of the velocity equation for the competitive system are:

$$
\frac{v}{V_{\max}} = \frac{[\mathrm{S}]}{K_S \dfrac{\left(1 + \dfrac{[\mathrm{X}]}{K_X} + \dfrac{[\mathrm{I}]}{K_i} + \dfrac{[\mathrm{X}][\mathrm{I}]}{\alpha K_X K_i}\right)}{\left(1 + \dfrac{[\mathrm{X}]}{K_X} + \dfrac{[\mathrm{I}]}{K_i}\right)} + [\mathrm{S}]} \qquad \text{(VIII-53)}
$$

or

$$\boxed{\frac{v}{V_{\max}} = \frac{[S]}{K_S\left(1 + \dfrac{[X]}{\alpha K_X\left(1 + \dfrac{K_i}{[I]}\right) + \dfrac{\alpha K_i[X]}{[I]}}\right) + [S]}} \qquad \text{(VIII-54)}$$

or

$$\boxed{\frac{v}{V_{\max}} = \frac{[S]}{K_S\left(1 + \dfrac{[I]}{\alpha K_i\left(1 + \dfrac{K_X}{[X]}\right) + \dfrac{\alpha K_X[I]}{[X]}}\right) + [S]}} \qquad \text{(VIII-55)}$$

In the forms above, the equation for the noncompetitive situation would have both K_S and [S] multiplied by the parenthetical term in the denominator. If either I or X is absent, the parenthetical factor reduces to unity (i.e., no inhibition). In the presence of a saturating concentration of I, X behaves as a linear competitive inhibitor with a dissociation constant of αK_X. (When [I] is infinitely high, the parenthetical factor reduces to $(1+[X]/\alpha K_X)$. In this case, saturating [X] will drive the velocity to zero.) Similarly, in the presence of saturating [X], I behaves as a linear competitive inhibitor with a dissociation constant of αK_i. In the presence of an unsaturating concentration of one inhibitor, the other will act as a partial competitive inhibitor (unlike the concerted case of System B3 where v can be driven to zero). This is illustrated in Figure VIII-13 for the competitive and noncompetitive systems where $\alpha = 1$. At the fixed unsaturating concentration of X, an infinitely high concentration of I drives the enzyme to a mixture of the active EI and inactive EIX forms. The v versus [I] curves are hyperbolic.

The values of K_X, K_i, and α can be obtained in a manner similar to that described earlier for partial inhibition systems. Several families of reciprocal plots are constructed, each family at a different constant unsaturating [I]. The lines within each family represent the $1/v$ versus $1/[S]$ plots at different

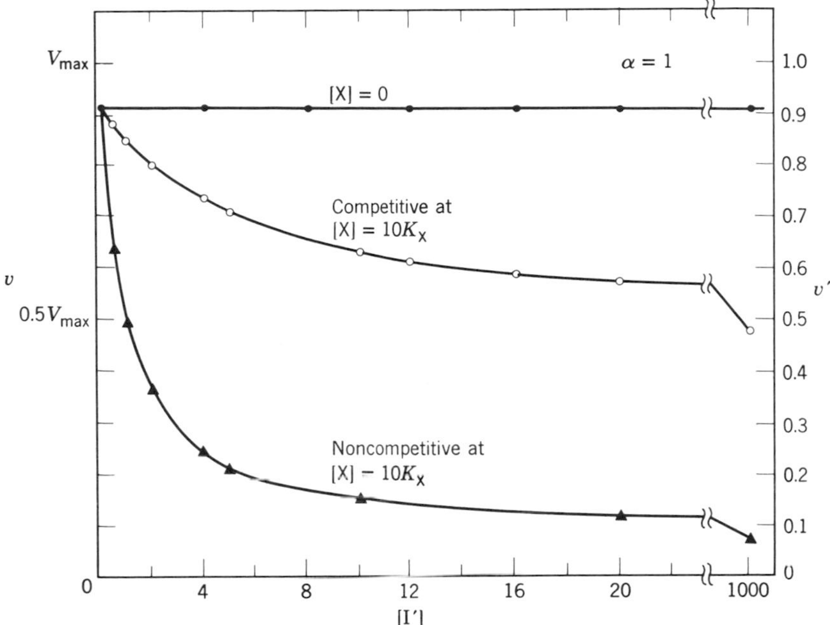

Fig. VIII-13. System B8; concerted or multivalent inhibition: v versus [I] at $[S]=10K_S$ in the absence and in the presence of X; $\alpha=1$.

fixed concentrations of X (including zero). The slope of any plot is given by:

$$slope_{1/S} = \frac{K_S}{V_{max}} \frac{\left(1 + \frac{[X]}{K_X} + \frac{[I]}{K_i} + \frac{[X][I]}{\alpha K_X K_i}\right)}{\left(1 + \frac{[X]}{K_X} + \frac{[I]}{K_i}\right)} = \frac{K_{S_{app}}}{V_{max}}$$

The difference between the slope of the control plot and the slope of any other within a family (Δ *slope*) is given by:

$$\Delta\, slope_{1/S} = \frac{K_{S_{app}}}{V_{max}} - \frac{K_S}{V_{max}}$$

The reciprocal is given by:

$$\boxed{\frac{1}{\Delta\, slope} = \frac{\alpha K_X V_{max}}{K_S}\left(1 + \frac{K_i}{[I]}\right)\frac{1}{[X]} + \frac{\alpha K_i V_{max}}{K_S[I]}} \qquad \text{(VIII-56)}$$

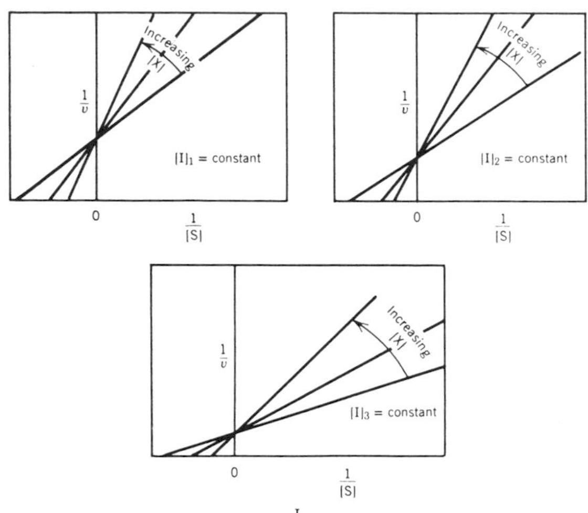

Fig. VIII-14. Procedure for obtaining K_i, K_X, and α where I+X excludes S but neither compound alone has any effect on S binding or catalytic activity. I. Plot $1/v$ versus $1/[S]$. II. Replot $1/\Delta$ *slope* versus $1/[X]$. III. Replot slopes of plot II versus $1/[I]$.

Thus for each family of reciprocal plots, a replot of $1/\Delta$ *slope* versus the corresponding $1/[X]$ is linear. The αK_i can be calculated from the vertical-axis intercept. The slopes of the replots are a function of the constant [I] for each family. A secondary replot of $slope_{family}$ against the corresponding $1/[I]$ permits K_i and αK_X to be determined:

$$slope_{family} = \frac{\alpha K_X K_i V_{max}}{K_S} \frac{1}{[I]} + \frac{\alpha K_X V_{max}}{K_S}$$

The horizontal-axis intercept gives $-1/K_i$. The αK_X can be calculated from the vertical-axis intercept. With K_i and αK_i known, α and K_X can be calculated. The procedure is illustrated in Figure VIII-14.

REFERENCES

Cumulative Feedback Inhibition (Glutamine Synthetase)

Shapiro, B. M. and Stadtman, E. R., "The Regulation of Glutamine Synthetase in Microorganisms," in *Ann. Rev. Microbiol.* **24**, 501 (1970).

Stadtman, E. R., Shapiro, B. M., Kingdon, H. S., Woolfolk, C. A. and Hubbard, J. S., "Cellular Regulation of Glutamine Synthetase Activity in *Escherichia coli*," in *Advances in Enzyme Regulation*, Vol. 6, G. Weber, ed., Pergamon Press, 1968, p. 257.

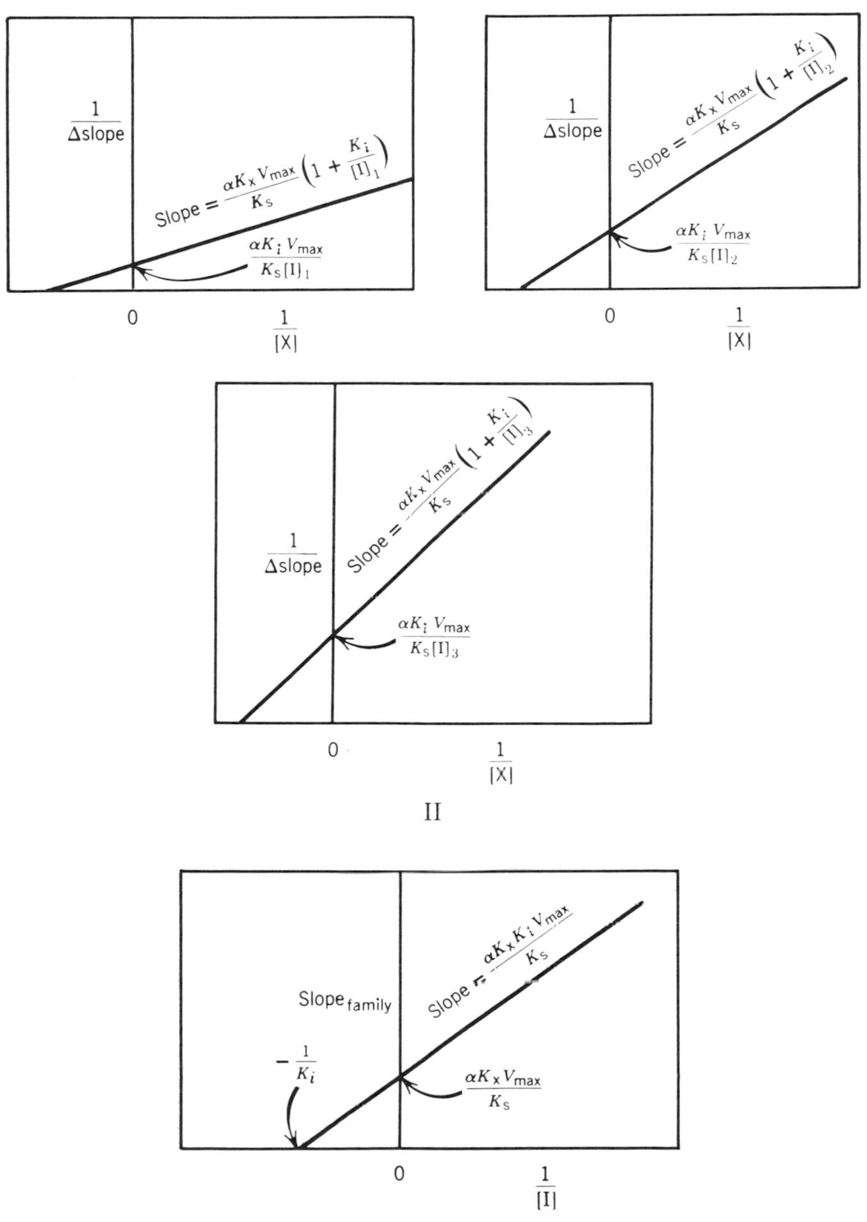

Fig. VIII-14. (*Cont.*)

Concerted (Multivalent) and Cooperative (Synergistic) Feedback Inhibition

Brush, A. and Paulus, H., *Biochem. Biophys. Res. Commun.* **45**, 735 (1971) (*O*-acetylhomoserine synthetase from *Bacillus subtilis*).

Caskey, C. T., Ashton, D. M. and Wyngaarden, J. B., *J. Biol. Chem.* **239**, 2570 (1964) (5′-Phosphoribosylamine synthetase from pigeon liver).

Datta, P. and Gest, H., *Proc. Natl. Acad. Sci. U.S.* **52**, 1004 (1964) (Aspartokinase from *Rhodopseudomonas capsulatus*).

Dungan, S. M. and Datta P., *J. Biol. Chem.* **248**, 8541 (1973) (Aspartokinase from *Pseudomonas fluorescens*).

Haas, D. and Leisinger, T., *Biochem. Biophys. Res. Commun.* **60**, 42 (1974) (N-acetylglutamate synthetase from *Pseudomonas aeruginosa*).

Jensen, R. A. and Trentini, W. C., *J. Biol. Chem.* **245**, 2018 (1970) (DHAP synthetase of *Rhodomicrobium vannielii*).

Klungsøyr, L. and Atkinson, D. E., *Biochemistry* **9**, 2021 (1970) (PRibATP synthetase from *E. coli*).

Lee, L, Ravel, J. N. and Shive, W., *J. Biol. Chem.* **241**, 5479 (1966) (*O*-succinyl homoserine synthetase from *E. coli*).

Marr, J. J. and Weber, M. M., *Biochem. Biophys. Res. Commun.* **45**, 1019 (1971) (Isocitrate dehydrogenase from *Neurospora crassa* and *Saccharomyces cerevisiae*).

Miflin, B. J., *Arch. Biochem. Biophys.* **146**, 542 (1971) (Acetohydroxy acid synthetase of barley).

Nierlich, D. P. and Magasanik, B., *J. Biol. Chem.* **240**, 358 (1965) (5′-Phosphoribosylamine synthetase of *Aerobacter aerogenes*).

Patte, J. C., Loviny, T. and Cohen, G. N., *Biochim. Biophys. Acta* **99**, 523 (1965) (Aspartokinase of *E. coli*).

Paulus, H. and Gray, E., *J. Biol. Chem.* **239** PC4008 (1964) (Aspartokinase of *Bacillus polymyxa*).

Shiio, I. and Ozaki, M., *J. Biochem.* (*Tokyo*) **64**, 45 (1968) (Isocitrate dehydrogenase from *Brevibacterium flavum*).

Shiio, I., Sugimoto, S. and Miyajima, R., *J. Biochem. (Tokyo)* **75**, 987 (1974) (DHAP synthetase from *Brevibacterium flavum*).

Multiple Inhibition Analysis

Colombo, V. E. and Semenza, G., *Biochim. Biophys. Acta* **288**, 145 (1972).

Loftfield, R. B. and Eigner, E. A., *Science.* **164**, 305 (1969).

Lucas, J. J., Burchiel, S. W. and Segel, I. H., *Arch. Biochem. Biophys.* **153**, 664 (1972).

Northrup, D. B. and Cleland, W. W., *J. Biol. Chem.* **249**, 2928 (1974).

Semenza, G. and von Balthazar, A-K. *Eur. J. Biochem.* **41**, 149 (1974).

Yagi, K. and Ozawa, T., *Biochem. Biophys. Acta* **39**, 304 (1960).

Yagi, K. and Ozawa, T., *Biochem. Biophys. Acta* **42**, 381 (1960).

Yonetani, T. and Theorell, H., *Arch. Biochem. Biophys.* **106**, 243 (1964).

CHAPTER NINE

STEADY-STATE KINETICS OF MULTIREACTANT ENZYMES

A great many multisubstrate systems cannot be described adequately by velocity equations derived from rapid equilibrium assumptions. In many systems, the isomerization of the central complex and subsequent product release steps are so rapid that E, EA, EAB, and so on, never attain equilibrium. The distribution of enzyme species will depend on the rate constants of all of the steps, including those involving product release. The transitory enzyme-product complexes may now represent a significant fraction of the total enzyme even in the absence of added products. Velocity equations for nonrapid equilibrium multisubstrate systems can be derived algebraically from steady-state assumptions. However, as the number of enzyme species increase, the algebraic manipulations become increasingly more complicated. The derivation can be shortened considerably by using the schematic method of King and Altman (1956). This method provides a shortcut to expressions for the relative distribution of enzyme species (e.g., $[EA]/[E]_t$, $[EAB]/[E]_t$, etc.) in terms of the substrate concentrations and the various rate constants. A velocity equation is obtained by inserting the expressions into an appropriate equation for the net steady-state velocity (e.g., $v = k_3[EAB] - k_{-3}[P][EQ]$ for an Ordered Bi Bi reaction). The resulting equation is still relatively useless until the rate constants are grouped into kinetic constants that can be experimentally determined (e.g., V_{max}, K_m, and K_i constants). While several methods of grouping the individual rate constants have been suggested, the method most commonly used is that of W. W. Cleland of the University of Wisconsin. Cleland's elegant approach and nomenclature, published in 1963, has brought the subject of steady-state enzyme kinetics within the grasp of the nonspecialist. The King-Altman and Cleland methods are described in the following sections for the most common steady-state systems. The nomenclature of Cleland will be used in most places. However, to remain consistent with earlier chapters, rate constants for forward reactions will be numbered consecutively with positive subscripts; rate constants for the reverse reactions will be numbered with corresponding negative subscripts. Michaelis constants will be indicated by the symbol K_m (e.g., K_{m_A}, K_{m_P}). In the Cleland nomenclature Michaelis constants are indicated as K_a, K_p, and so on.

A. THE KING-ALTMAN METHOD OF DERIVING STEADY-STATE VELOCITY EQUATIONS

Uni Uni Reaction

To illustrate the King-Altman method, let us consider the simple Uni Uni reaction shown below:

$$\mathrm{E}+\mathrm{A} \underset{k_{-1}}{\overset{k_1}{\rightleftharpoons}} \mathrm{EA} \underset{k_{-2}}{\overset{k_2}{\rightleftharpoons}} \mathrm{EP} \underset{k_{-3}}{\overset{k_3}{\rightleftharpoons}} \mathrm{E}+\mathrm{P}$$

The velocity equation is obtained as follows.

1. Arrange the different enzyme species into a geometric figure, with each species as one corner of the figure. For the Uni Uni reaction shown above, there are three different enzyme species, E, EA, and EP. Therefore the basic figure is a triangle.

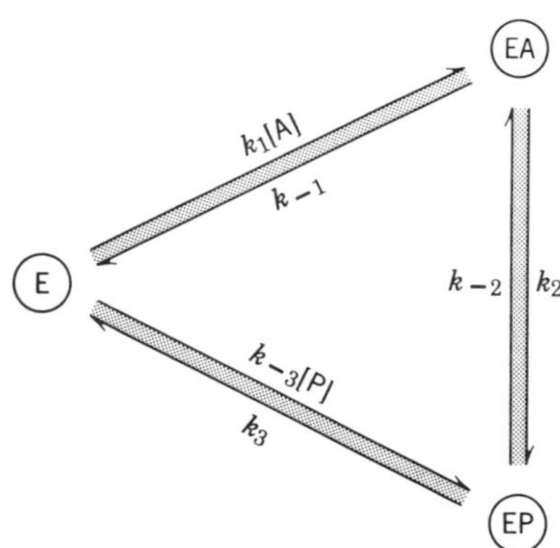

The line connecting any two enzyme species represents the unimolecular or biomolecular step by which the two species are interconverted. The lines are labeled with the appropriate rate constant or rate constant × ligand concentration for each direction. The lines leading to any one species represent all the ways that species can be formed. For example, EP can be made from

E by the reaction of E with A to yield EA, followed by the unimolecular isomerization reaction from EA to EP. EP can also be formed from E by the direct reaction of E with P.

2. The second step is to write all possible patterns containing one less line than the number of enzyme species. The basic figure contains three corners. The possible patterns containing two lines are as follows.

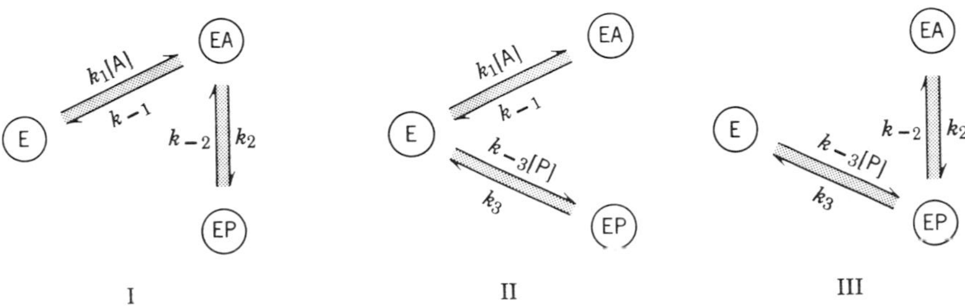

The patterns containing one less line than the basic figure represent all the ways any one enzyme species can be formed from the others by paths not involving closed loops. The total number of interconversion patterns is given by:

$$\text{number of patterns with } (n-1) \text{ lines} = \frac{m!}{(n-1)!(m-n+1)!} \qquad \text{(IX-1)}$$

where n = the number of enzyme species
= the number of corners in the basic figure
m = the number of lines in the basic figure

For the simple Uni Uni reaction, none of the $(n-1)$-lined interconversion patterns calculated by equation IX-1 contain closed loops. However, equation IX-1 gives the maximum number of $(n-1)$-lined patterns, which may include closed loops for more complex systems. For example, suppose we have a system involving five enzyme species which can be interconverted as shown by the basic figure below. (The pattern could represent linear mixed-type inhibition.)

$m = 6$
$n = 5$

Equation IX-1 predicts 15 patterns with $(n-1)$ lines:

$$\text{number of patterns with four lines} = \frac{6!}{(5-1)!(6-5+1)!} = 15$$

However, we can see that there are four patterns containing four lines that include closed loops:

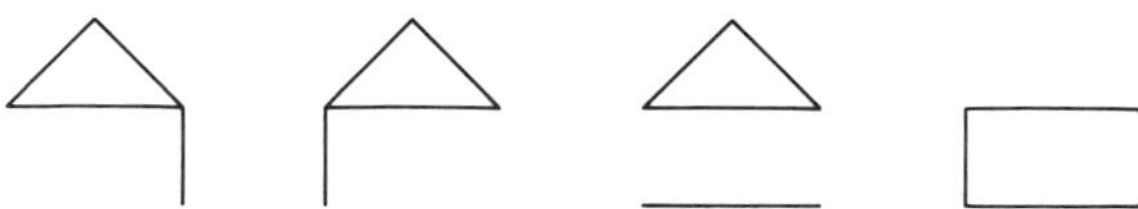

These must be subtracted from the total of 15 leaving 11 possible four-lined interconversion patterns (Fig. IX-1). The number of patterns unacceptable because of closed loops can be calculated ahead of time from equation IX-2 given below.

$$\begin{array}{l}\text{number of patterns with } (n-1) \text{ lines} \\ \text{containing a given closed loop of } r \text{ sides}\end{array} = \frac{(m-r)!}{(n-1-r)!(m-n+1)!} \tag{IX-2}$$

Basic figure $n = 5$ $m = 6$

Interconversion patterns containing $(n-1)$ lines and no closed loops

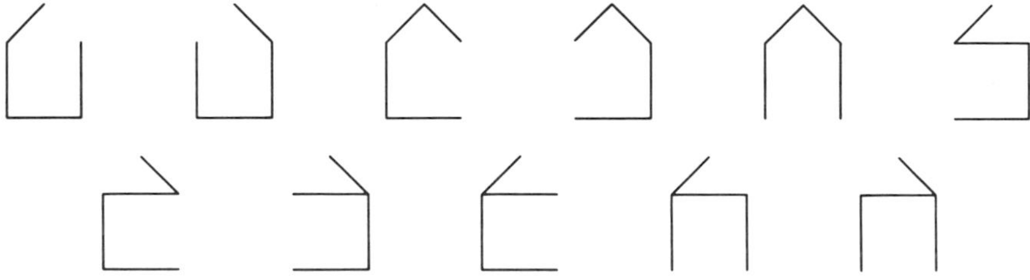

Fig. IX-1. Eleven valid four-lined King-Altman interconversion patterns for the basic five-cornered figure shown.

Thus for $m=6$ and $n=5$:

$$\text{number of four-lined patterns containing a closed loop of three sides} = \frac{(6-3)!}{(5-1-3)!(6-5+1)!} = 3$$

and

$$\text{number of four-lined patterns containing a closed loop of four sides} = \frac{(6-4)!}{(5-1-4)!(6-5+1)!} = 1$$

(Note that $0! = 1$.) The number of unacceptable patterns must be calculated for all possible values of r equal to and less than $n-1$. (In the example above, there are no possible closed loops of two sides.) When calculating the number of unacceptable patterns, care should be taken not to subtract any pattern more than once. For example, a three-sided loop subtracted once may be included in some of the unacceptable patterns containing four- or five-sided loops. Also note that equation IX-2 predicts the number of times a *given* closed loop appears. For example, consider the basic pattern shown below. There are five corners and six lines:

$$\left.\begin{matrix} n=5 \\ m=6 \end{matrix}\right\} \xrightarrow{\therefore} \frac{(6\times5\times4\times3\times2\times1)}{(4\times3\times2\times1)(2\times1)} = 15$$

Thus there are 15 possible four-sided patterns. However, there are three *different* four-sided patterns containing closed loops:

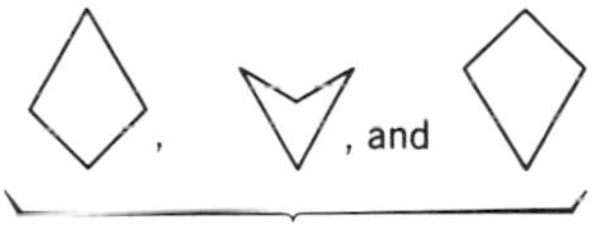

, , and

$$\downarrow \therefore$$

$$\frac{(6-4)!}{(5-1-4)!(6-5+1)!} = \frac{(2\times1)}{(0)!(2)!} = \frac{2\times1}{2\times1} = 1$$

Equation IX-2 predicts that *each one* of the closed loops appears only once. Therefore, a total of three is subtracted from the 15 possible four-sided interconversion patterns yielding 12 possible interconversion patterns with four lines and no closed loops.

Once we have all the interconversion patterns, the relative distribution of enzyme species is determined as described below.

3. The relative proportion of the total enzyme, $[E]_t$, represented by any given species, [EX], is given by:

$$\frac{[EX]}{[E]_t} = \frac{\text{the sum of } z \text{ terms}}{\text{a denominator}}$$

where z is the number of acceptable $(n-1)$ lined patterns. That is, the numerator will contain the terms for the formation of EX by each King-Altman interconversion pattern. The denominator will be the same for all enzyme species and represents the sum of all numerator terms for all enzyme species. The same interconversion patterns are used for each enzyme species, although the direction in which they are read will vary. The numerator term for any one interconversion pattern is the product of all rate constants and associated ligand concentrations read along the lines *leading to* the species in question. This is illustrated below for the Uni Uni reaction.

$$\frac{[E]}{[E]_t} = \frac{k_{-2}k_{-1} \quad + \quad k_{-1}k_3 \quad + \quad k_2k_3}{\text{denominator}} \tag{IX-3}$$

$$\frac{[EA]}{[E]_t} = \frac{k_{-2}k_1[A] \quad + \quad k_3k_1[A] \quad + \quad k_{-3}[P]k_{-2}}{\text{denominator}} \tag{IX-4}$$

$$\frac{[EP]}{[E]_t} = \frac{k_1[A]k_2 \quad + \quad k_{-1}k_{-3}[P] \quad + \quad k_2k_{-3}[P]}{\text{denominator}} \tag{IX-5}$$

The denominator is equal to the sum of all nine numerator terms. Grouping similar terms:

$$\text{denominator} = [E]_t = (k_{-2}k_{-1} + k_{-1}k_3 + k_2k_3) + k_1(k_{-2} + k_2 + k_3)[A]$$
$$+ k_{-3}(k_{-1} + k_{-2} + k_2)[P]$$

4. Now that we have expressions for the concentration of all enzyme species in terms of rate constants and ligand concentrations, we can proceed to write the velocity equation. The steady-state net velocity is given by the difference between the forward and reverse velocities of any step. For example:

$$v = k_2[EA] - k_{-2}[EP] \tag{IX-6}$$

$$v = \frac{k_2(k_{-2}k_1[A] + k_3k_1[A] + k_{-3}k_{-2}[P])[E]_t}{\text{denominator}} - \frac{k_{-2}(k_1k_2[A] + k_{-1}k_{-3}[P] + k_2k_{-3}[P])[E]_t}{\text{denominator}} \tag{IX-7}$$

or

$$\frac{v}{[E]_t} = \frac{k_1k_2k_3[A] \quad k_{-1}k_{-2}k_{-3}[P]}{(k_{-1}k_{-2} + k_{-1}k_3 + k_2k_3) + k_1(k_{-2} + k_2 + k_3)[A] + k_{-3}(k_{-1} + k_{-2} + k_2)[P]} \tag{IX-8}$$

When there are no alternate reaction paths, the numerator of the velocity equation will always contain the difference between two terms. The first term will be the product of all rate constants and substrate concentrations for the forward reaction, while the second term will be the product of all rate constants and product concentrations for the reverse reaction.

Equation IX-8 can be written in the coefficient form of Cleland:

$$v = \frac{\text{num}_1[A] - \text{num}_2[P]}{\text{const} + \text{Coef}_A[A] + \text{Coef}_P[P]} \tag{IX-9}$$

where num_1 = the coefficient of the positive numerator term
= the product of $[E]_t$ and the rate constants that multiply [A] in the numerator

num_2 = the coefficient of the negative numerator term
= the product of $[E]_t$ and the rate constants that multiply [P] in the numerator

const = the constant term in the denominator
= the group of rate constants in the denominator which is not associated with any ligand concentration term

Coef_A = the coefficient of [A] in the denominator

Coef_P = the coefficient of [P] in the denominator.

5. The final velocity equation is obtained by redefining the coefficients of equation IX-9 as kinetic constants that can be determined experimentally. The following definitions will allow us to simplify the velocity equation to the usual Henri-Michaelis-Menten form.

$$V_{\max_f} = \frac{\text{num}_1}{\text{Coef}_\text{A}}, \qquad K_{m_\text{A}} = \frac{\text{const}}{\text{Coef}_\text{A}}, \qquad V_{\max_r} = \frac{\text{num}_2}{\text{Coef}_\text{P}}, \qquad K_{m_\text{P}} = \frac{\text{const}}{\text{Coef}_\text{P}}$$

We can also see that the ratio of $\text{num}_1/\text{num}_2$ is equivalent to K_{eq}:

$$\frac{\text{num}_1}{\text{num}_2} = \frac{k_1 k_2 k_3}{k_{-1} k_{-2} k_{-3}} = K_{eq}$$

Now, multiplying numerator and denominator of equation IX-9 by $\text{num}_2/(\text{Coef}_\text{A})(\text{Coef}_\text{P})$, we obtain:

$$v = \frac{\dfrac{(\text{num}_1)(\text{num}_2)[\text{A}]}{(\text{Coef}_\text{A})(\text{Coef}_\text{P})} - \dfrac{(\text{num}_2)(\text{num}_2)(\text{num}_1)[\text{P}]}{(\text{Coef}_\text{A})(\text{Coef}_\text{P})(\text{num}_1)}}{\dfrac{(\text{const})(\text{num}_2)}{(\text{Coef}_\text{A})(\text{Coef}_\text{P})} + \dfrac{(\text{num}_2)(\text{Coef}_\text{A})[\text{A}]}{(\text{Coef}_\text{A})(\text{Coef}_\text{P})} + \dfrac{(\text{num}_2)(\text{Coef}_\text{P})(\text{num}_1)[\text{P}]}{(\text{Coef}_\text{A})(\text{Coef}_\text{P})(\text{num}_1)}}$$

Note that the [P] terms in the numerator and denominator have also been multiplied by $\text{num}_1/\text{num}_1$. Substituting the definitions given above, the velocity equation becomes:

$$v = \frac{V_{\max_f} V_{\max_r} [\text{A}] - \dfrac{V_{\max_r} V_{\max_f} [\text{P}]}{K_{eq}}}{K_{m_\text{A}} V_{\max_r} + V_{\max_r} [\text{A}] + \dfrac{V_{\max_f} [\text{P}]}{K_{eq}}} \qquad \text{(IX-10)}$$

or

$$v = \frac{V_{\max_f} V_{\max_r} \left([\text{A}] - \dfrac{[\text{P}]}{K_{eq}} \right)}{K_{m_\text{A}} V_{\max_r} + V_{\max_r} [\text{A}] + \dfrac{V_{\max_f} [\text{P}]}{K_{eq}}} \qquad \text{(IX-11)}$$

When [P]=0, the equation reduces to the usual Henri-Michaelis-Menten equation for the forward velocity. When [A]=0, we expect equation IX-11 to reduce to the Henri-Michaelis-Menten equation for the velocity in the reverse direction with P as the substrate. That is:

$$v=\frac{-V_{\max_r}[\mathrm{P}]}{K_{m_{\mathrm{P}}}+[\mathrm{P}]} \tag{IX-12}$$

(The minus sign means that the observed net velocity is from P to A.) However, when [A]=0, equation IX-11 becomes:

$$v=\frac{-V_{\max_f}V_{\max_r}\dfrac{[\mathrm{P}]}{K_{\mathrm{eq}}}}{K_{m_{\mathrm{A}}}V_{\max_r}+\dfrac{V_{\max_f}[\mathrm{P}]}{K_{\mathrm{eq}}}}=\frac{-V_{\max_r}[\mathrm{P}]}{\dfrac{K_{m_{\mathrm{A}}}V_{\max_r}K_{\mathrm{eq}}}{V_{\max_f}}+[\mathrm{P}]} \tag{IX-13}$$

Equations IX-12 and IX-13 must be identical. Therefore:

$$\frac{K_{m_{\mathrm{A}}}V_{\max_r}K_{\mathrm{eq}}}{V_{\max_f}}=K_{m_{\mathrm{P}}} \tag{IX-14}$$

This is also seen when we rewrite the constant term in the denominator of equation IX-13 in coefficient form:

$$\frac{K_{m_{\mathrm{A}}}V_{\max_r}K_{\mathrm{eq}}}{V_{\max_f}}=\frac{(\mathrm{const})(\mathrm{num}_2)(\mathrm{num}_1)(\mathrm{Coef}_{\mathrm{A}})}{(\mathrm{Coef}_{\mathrm{A}})(\mathrm{Coef}_{\mathrm{P}})(\mathrm{num}_2)(\mathrm{num}_1)}=\frac{\mathrm{const}}{\mathrm{Coef}_{\mathrm{P}}}=K_{m_{\mathrm{P}}}$$

Equation IX-14 can be rearranged to:

$$\boxed{\frac{K_{m_{\mathrm{P}}}V_{\max_f}}{K_{m_{\mathrm{A}}}V_{\max_r}}=K_{\mathrm{eq}}} \tag{IX-15}$$

which we see is the Haldane equation, which was derived by a slightly different method in Chapter Two. Every kinetic system has at least one Haldane equation relating the kinetic constants to the thermodynamic equilibrium constant. The "Haldanes" permit us to eliminate K_{eq} from the complete velocity equation. For example, substituting for K_{eq} in the denominator of equation IX-11, we obtain after cancelling V_{max_r} and factoring:

$$\boxed{\frac{v}{V_{max_f}} = \frac{[A] - \dfrac{[P]}{K_{eq}}}{K_{m_A}\left(1 + \dfrac{[P]}{K_{m_P}}\right) + [A]}} \qquad \text{(IX-16)}$$

As pointed out in Chapter Two, the numerator of equation IX-16 reflects the thermodynamic driving force (i.e., the displacement of the system from equilibrium), while the denominator reflects the competitive inhibition by the product (i.e., the fact that some of the enzyme is tied up with P). Equation IX-16 gives the *net* forward velocity in the presence of A and P. If v is measured as the initial rate at which radioactivity from A appears in P, then P can be treated as an ordinary competitive inhibitor. In this case the velocity equation reduces to:

$$\boxed{\frac{v}{V_{max_f}} = \frac{[A]}{K_{m_A}\left(1 + \dfrac{[P]}{K_{m_P}}\right) + [A]}} \qquad \text{(IX-17)}$$

A $slope_{1/A}$ versus [P] replot can be used to obtain K_{m_P}. If A is not radioactive, then plots of $1/v$ (actually, $1/v_{net}$) versus $1/[A]$ in the presence of P may be nonlinear and concave up unless K_{eq} is $\gg 1$. Linear plots can be obtained by plotting $1/v_{net}$ versus $1/\Delta A$ (see equation II-21). The reciprocal of the intercept on the $1/\Delta A$-axis is given by:

$$-\Delta A = K_{m_A}\left(1 + \frac{[P]}{K_{m_P}}\right) + \frac{[P]}{K_{eq}}$$

Isomerization of Central Complexes

In the derivation of the velocity equation for the simple Uni Uni reaction, we indicated the two central complexes, EA and EP, as separate entities with rate constants for their interconversion. However, the final form of the velocity equation is the same as that derived earlier when the reaction sequence was considered to be simply $E + A \rightleftharpoons EA \rightleftharpoons E + P$. As a general rule, isomerizations of central complexes have no effect on the final form of the velocity equation. Consequently, in applying the King-Altman method to more complicated systems, corresponding pairs of central complexes will usually be considered together as a single corner of the interconversion pattern.

B. SIMPLIFICATION OF COMPLEX KING-ALTMAN PATTERNS

The number of King-Altman interconversion patterns increases as the number of enzyme species increases. For some complicated mechanisms, the King-Altman method can be simplified. Two examples are described below.

Addition of Multiple Lines Connecting Two Corners

Consider the single substrate-single modifier system shown below where M could be either a partial inhibitor or a nonessential activator of the reaction $S \rightarrow P$.

$$\begin{array}{ccccc}
E + S & \underset{k_{-1}}{\overset{k_1}{\rightleftharpoons}} & ES & \underset{k_{-2}}{\overset{k_2}{\rightleftharpoons}} & E + P \\
+ & & + & & \\
M & & M & & \\
k_3 \downarrow\uparrow k_{-3} & & k_{-5} \uparrow\downarrow k_5 & & \\
EM + S & \underset{k_4}{\overset{k_{-4}}{\leftrightharpoons}} & EMS & \underset{k_{-6}}{\overset{k_6}{\rightleftharpoons}} & EM + P
\end{array}$$

The basic King-Altman figure is shown below.

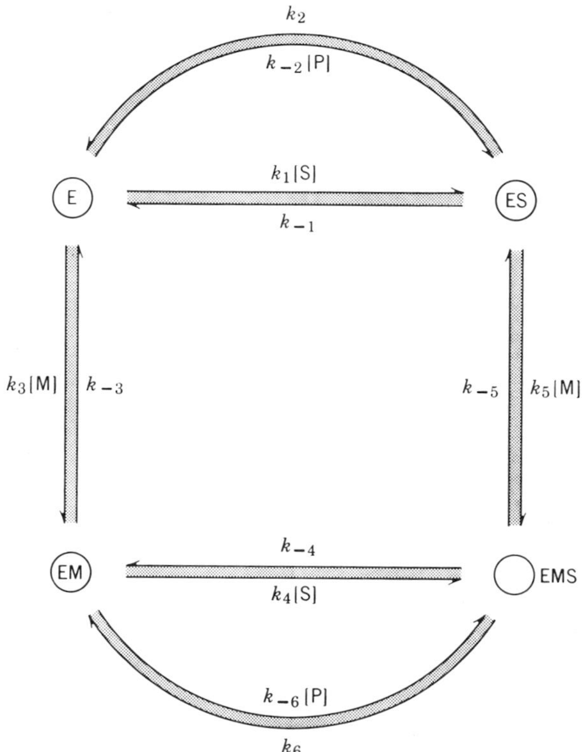

The basic figure contains four corners ($n = 4$) and six lines ($m = 6$). Equation IX-1 predicts 20 patterns with three lines. However, the 20 patterns include all the three-lined patterns containing the two different two-line loops shown below.

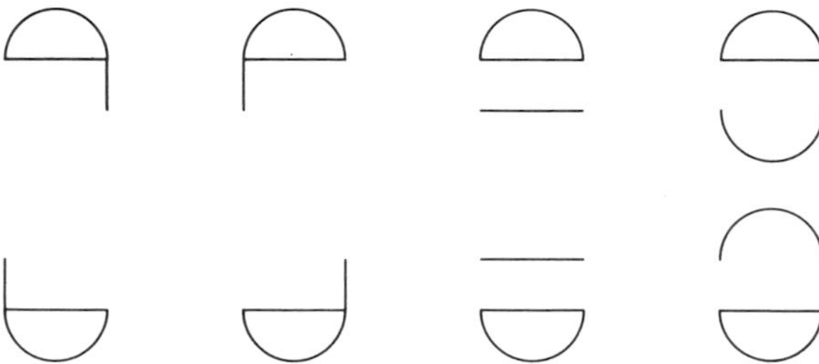

Thus by inspection or by using equation IX-2, we would eliminate four three-lined patterns for each kind of loop (i.e., we reject eight patterns). The 12 valid interconversion patterns are shown below.

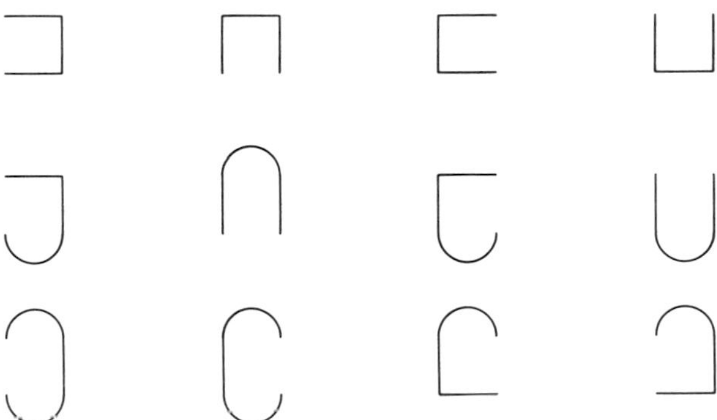

Instead of working with 12 patterns, the basic figure can be simplified by adding the lines connecting E with ES and EM with EMS. In general, when multiple lines connect two enzyme species (corners) in one or both directions, the lines can be added. Thus the basic figure can be simplified to:

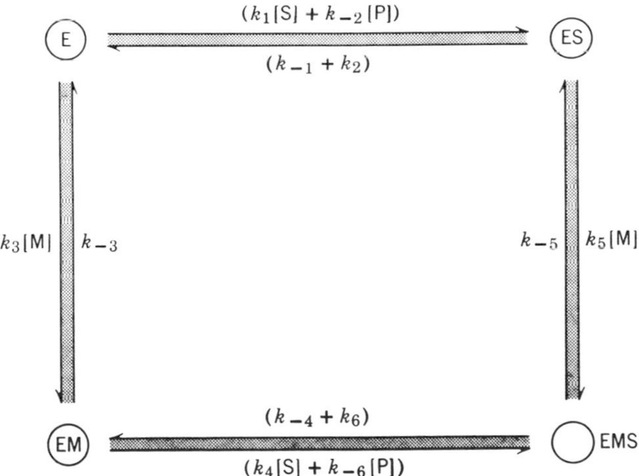

The velocity equation can now be derived using four different three-lined patterns:

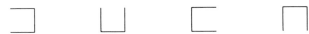

Reduction of Matrices to a Point

In the reaction scheme shown below, I is an alternate substrate of enzyme E which can substitute for A in the Ping Pong Bi Bi mechanism. The first product released, P, is common to both A and I, but the second product released is different. A and I could represent two esters of the same alcohol which undergo hydrolysis (B is H_2O, P the common alcohol, Q and S the different free acids).

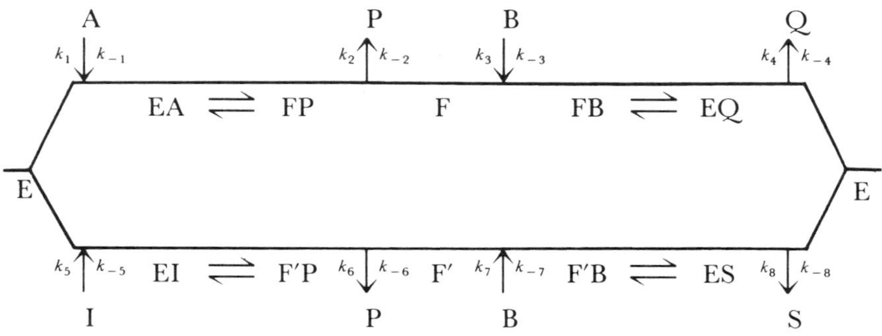

There are seven different enzyme species present in the presence of both I and A. These can be arranged into the basic figure shown below.

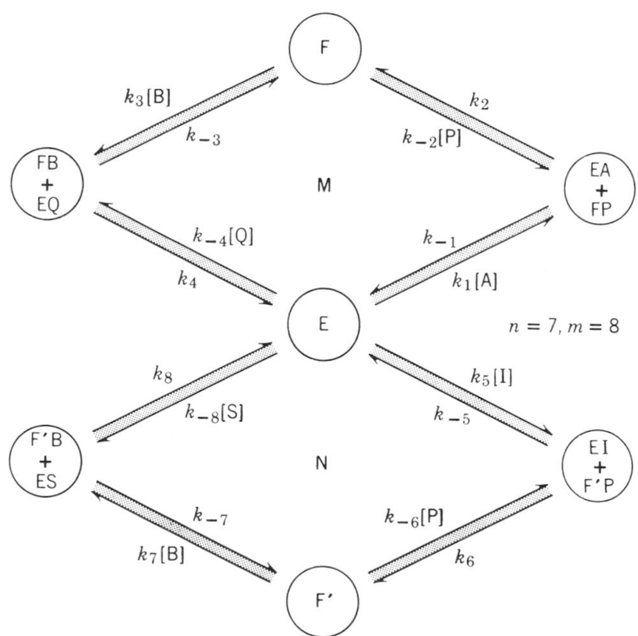

The velocity equation can be worked out from the 16 possible six-lined valid interconversion patterns, but a shortcut becomes obvious if we recognize that the 16 patterns are composed of the four possible three-lined patterns obtained for the top four-sided figure (matrix M), each combined with the four possible three-lined patterns of the bottom four-sided figure (matrix N).

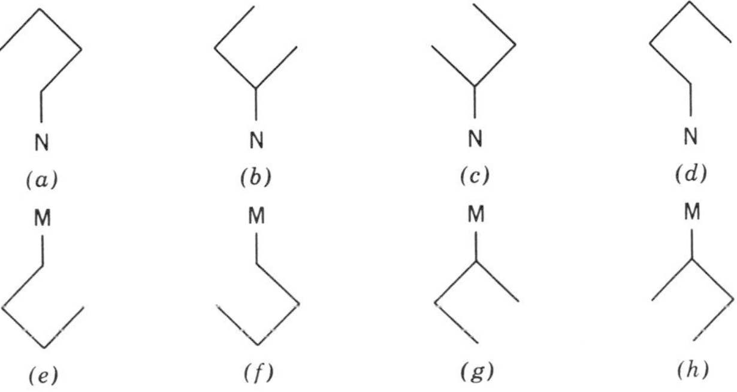

(a) (b) (c) (d)

(e) (f) (g) (h)

The basic figure can be considered as two separate figures connected at a common point. The fractional concentration of an enzyme form in matrix M is given by N times the usual M-matrix solution for that form. The fractional concentration of an enzyme form in matrix N is given by M times the usual N-matrix solution for that form. (The matrix solution is the sum of all rate constants and concentration factors read along the lines leading to the enzyme form.) The fractional concentration of the point common to both matrices, [E], is given by M × N, where

$$\mathrm{M} = \text{the matrix solution for } \frac{[\mathrm{E}]}{[\mathrm{E}]_t} \text{ determined in matrix M}$$

$$\mathrm{N} = \text{the matrix solution for } \frac{[\mathrm{E}]}{[\mathrm{E}]_t} \text{ determined in matrix N}$$

Thus:

$$\frac{[\mathrm{E}]}{[\mathrm{E}]_t} = \frac{\begin{array}{c}(k_{-3}k_{-2}[\mathrm{P}]k_{-1} + k_3[\mathrm{B}]k_4k_{-1} + k_4k_{-2}[\mathrm{P}]k_{-1} + k_2k_3[\mathrm{B}]k_4) \\ \times\,(k_6k_7[\mathrm{B}]k_8 + k_{-7}k_{-6}[\mathrm{P}]k_{-5} + k_{-5}k_7[\mathrm{B}]k_8 + k_8k_{-6}[\mathrm{P}]k_{-5})\end{array}}{\text{denominator}}$$

(IX-18)

where M is the first parenthetical numerator factor and N is the second parenthetical numerator factor.

$$\frac{[EA]}{[E]_t} = \frac{N(k_{-3}k_{-2}[P]k_1[A] + k_3[B]k_4k_1[A] + k_{-2}[P]k_4k_1[A] + k_{-4}[Q]k_{-3}k_{-2}[P])}{\text{denominator}} \qquad \text{(IX-19)}$$

$$\frac{[F]}{[E]_t} = \frac{N(k_{-3}k_1[A]k_2 + k_{-1}k_{-4}[Q]k_{-3} + k_4k_1[A]k_2 + k_2k_{-4}[Q]k_{-3})}{\text{denominator}} \qquad \text{(IX-20)}$$

$$\frac{[FB+EQ]}{[E]_t} = \frac{N(k_1[A]k_2k_3[B] + k_3[B]k_{-1}k_{-4}[Q] + k_{-2}[P]k_{-1}k_{-4}[Q] + k_{-4}[Q]k_2k_3[B])}{\text{denominator}} \qquad \text{(IX-21)}$$

$$\frac{[EI]}{[E]_t} = \frac{M(k_{-8}[S]k_{-7}k_{-6}[P] + k_5[I]k_{-7}k_{-6}[P] + k_7[B]k_8k_5[I] + k_{-6}[P]k_8k_5[I])}{\text{denominator}} \qquad \text{(IX-22)}$$

$$\frac{[F']}{[E]_t} = \frac{M(k_6k_{-8}[S]k_{-7} + k_{-7}k_5[I]k_6 + k_{-5}k_{-8}[S]k_{-7} + k_8k_5[I]k_6)}{\text{denominator}} \qquad \text{(IX-23)}$$

$$\frac{[F'B+ES]}{[E]_t} = \frac{M(k_{-8}[S]k_6k_7[B] + k_5[I]k_6k_7[B] + k_7[B]k_{-5}k_{-8}[S] + k_{-6}[P]k_{-5}k_{-8}[S])}{\text{denominator}} \qquad \text{(IX-24)}$$

Steady-State Systems with Rapid Equilibrium Segments

When one or more portions of a multistep reaction sequence is significantly faster than the overall reaction, the reactions within such segments must reach near equilibrium as the overall reaction attains a steady-state. Cha (1968) has shown that the derivation of the complete velocity equation by the King-Altman method can be greatly simplified by considering all the enzyme forms within the rapid equilibrium segment as a single entity (i.e., as a single corner of the basic King-Altman figure). The distribution of species within the rapid equilibrium segment is obtained by the usual rules for rapid equilibrium systems. For example, consider the Ordered Bi Bi system in which the binding and dissociation of the first substrate and last product are both significantly faster than the other reactions. The basic interconversion figure is shown below. The shaded area represents the rapid equilibrium segment, which we can call X.

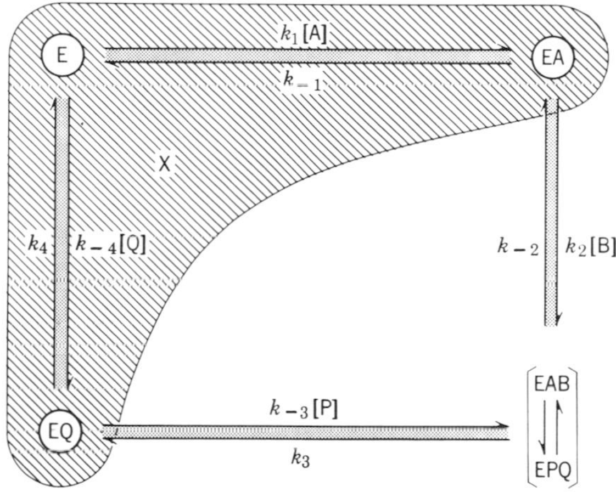

The basic figure can now be condensed to only two lines connecting X and the central complexes EAB + EPQ.

$$[X]=[E]+[EA]+[EQ]$$

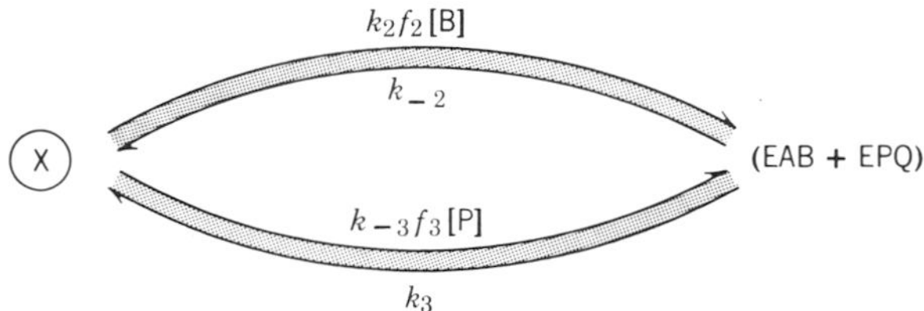

The interconversion patterns used for the derivation are:

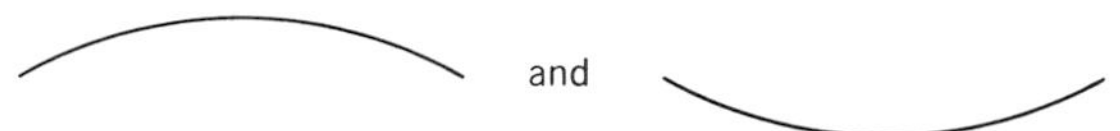

The symbols f_2 and f_3 represent *fractional concentrations* and stand for the relative proportion of the rapid equilibrium segment, X, that actually is involved in the given reaction. That is, of the species comprising X, it is EA that reacts with B (with a rate constant k_2) to yield EAB. The symbol f_2 represents the proportion of [X] that is [EA]:

$$f_2 = \frac{[\mathrm{EA}]}{[\mathrm{X}]} = \frac{\dfrac{[\mathrm{A}]}{K_\mathrm{A}}}{1 + \dfrac{[\mathrm{A}]}{K_\mathrm{A}} + \dfrac{[\mathrm{Q}]}{K_\mathrm{Q}}} = \frac{\dfrac{k_1[\mathrm{A}]}{k_{-1}}}{1 + \dfrac{k_1[\mathrm{A}]}{k_{-1}} + \dfrac{k_{-4}[\mathrm{Q}]}{k_4}}$$

$$f_2 = \frac{k_1 k_4[\mathrm{A}]}{k_{-1}k_4 + k_1 k_4[\mathrm{A}] + k_{-1}k_{-4}[\mathrm{Q}]} \qquad \text{(IX-25)}$$

Similarly:

$$f_3 = \frac{[\mathrm{EQ}]}{[\mathrm{X}]} = \frac{\dfrac{[\mathrm{Q}]}{K_\mathrm{Q}}}{1 + \dfrac{[\mathrm{A}]}{K_\mathrm{A}} + \dfrac{[\mathrm{Q}]}{K_\mathrm{Q}}} = \frac{\dfrac{k_{-4}[\mathrm{Q}]}{k_4}}{1 + \dfrac{k_1[\mathrm{A}]}{k_{-1}} + \dfrac{k_{-4}[\mathrm{Q}]}{k_4}}$$

$$f_3 = \frac{k_{-1}k_{-4}[\mathrm{Q}]}{k_{-1}k_4 + k_1 k_4[\mathrm{A}] + k_{-1}k_{-4}[\mathrm{Q}]} \qquad \text{(IX-26)}$$

Thus $k_2 f_2$ and $k_{-3} f_3$ represent effective or apparent rate constants. The King-Altman treatment yields:

$$\frac{[\mathrm{X}]}{[\mathrm{E}]_t} = \frac{k_{-2} + k_3}{\text{denominator}} \quad \text{and} \quad \frac{[\mathrm{EAB} + \mathrm{EPQ}]}{[\mathrm{E}]_t} = \frac{k_2 f_2[\mathrm{B}] + k_{-3} f_3[\mathrm{P}]}{\text{denominator}}$$

where the denominator is the sum of both numerators. The velocity equation is obtained by substituting the relationships above into the steady-state equation:

$$v = k_2[\text{EA}][\text{B}] - k_{-2}[\text{EAB} + \text{EPQ}]$$

where $[\text{EA}] = f_2[\text{X}]$. The final velocity equation and its relationship to the total rapid equilibrium and total steady-state equations are discussed at the end of Section G describing the Ordered Bi Bi mechanism.

C. GENERAL RULES FOR DEFINING KINETIC CONSTANTS AND DERIVING VELOCITY EQUATIONS

Cleland Nomenclature

The conversion of the velocity equation from the coefficient form to a form composed entirely of kinetic constants was relatively straightforward for the Uni Uni reaction. The general rules of Cleland (1963) outlined below can be applied to the more complicated nonrandom steady-state systems described in the following sections.

Maximum Velocities and K_{eq}

The V_{max} in either direction will equal the numerator coefficient for that direction (num_1 or num_2) divided by the coefficient of the denominator term which contains the product of the concentrations of all substrates for the reaction in that direction. Thus if we are considering a Ter Ter reaction ($\text{A} + \text{B} + \text{C} \rightleftharpoons \text{P} + \text{Q} + \text{R}$):

$$V_{\max_f} = \frac{\text{num}_1}{\text{Coef}_{\text{ABC}}}, \qquad V_{\max_r} = \frac{\text{num}_2}{\text{Coef}_{\text{PQR}}}$$

The K_{eq} will always equal $\text{num}_1/\text{num}_2$.

Michaelis Constants

The Michaelis constant for any substrate or product will equal the ratio of two denominator coefficients. The denominator of the ratio is the same as that used in defining V_{max} for the given direction. The numerator of the ratio is chosen so that, after canceling subscript letters, the only letter remaining as a denominator subscript is that corresponding to the substrate or product in question. Thus the numerator of the ratio must contain subscripts corresponding to all substrates in the given direction *except* the one whose K_m is desired. For a Ter Ter reaction:

$$K_{m_A} = \frac{\text{Coef}_{BC}}{\text{Coef}_{ABC}}, \qquad K_{m_B} = \frac{\text{Coef}_{AC}}{\text{Coef}_{ABC}}, \qquad K_{m_C} = \frac{\text{Coef}_{AB}}{\text{Coef}_{ABC}}$$

$$K_{m_P} = \frac{\text{Coef}_{QR}}{\text{Coef}_{PQR}}, \qquad K_{m_Q} = \frac{\text{Coef}_{PR}}{\text{Coef}_{PQR}}, \qquad K_{m_R} = \frac{\text{Coef}_{PQ}}{\text{Coef}_{PQR}}$$

The Michaelis constants defined above are limiting constants in that they equal the concentration of the particular substrate that yields half-maximal velocity when all the other substrates in the given direction are saturating. The $K_{m_{app}}$ for a given substrate at unsaturating cosubstrate concentrations can be greater than, less than, or equal to the limiting value, depending on the kinetic mechanism.

Inhibition Constants

Other constants associated with a given ligand will equal the ratio of two denominator coefficients. The ratio must be chosen so that, after canceling subscript letters, the letter corresponding to the ligand associated with the constant remains as a subscript in the denominator of the ratio. A constant so defined is written as K_{ia}, K_{ip}, and so on, and is called an inhibition constant because it is related to the "K_i" for the given ligand as a product inhibitor. In many cases, the K_i will be identical to the actual dissociation constant of the substrate or product (which in steady-state systems is distinct from the Michaelis constant and often appears together with the corresponding Michaelis constant in velocity equations). For some systems, it will be possible to find quite a few ratios of coefficients that satisfy the cancellation requirements above. Some of the ratios may turn out to be identical in terms of their component rate constants. For example, a definition of K_{ic} that permits an Ordered Ter Ter reaction to be simplified is:

$$K_{ic} = \frac{\text{Coef}_{ABP}}{\text{Coef}_{ABCP}} = \frac{\text{Coef}_{ABPQ}}{\text{Coef}_{ABCPQ}} = \frac{\text{Coef}_{BPQR}}{\text{Coef}_{BCPQR}} = \frac{k_{-3}}{k_3}$$

The three definitions above reduce to the same value in terms of rate constants. Other definitions that might have been chosen on the basis of the preceding rules are:

$$\frac{\text{const}}{\text{Coef}_C} = \frac{\text{Coef}_R}{\text{Coef}_{CR}} = \frac{\text{Coef}_A}{\text{Coef}_{AC}} = \frac{\text{Coef}_{QR}}{\text{Coef}_{CQR}} = \frac{k_{-2}(k_{-3}+k_4)}{k_3 k_4}$$

However, these definitions would not permit the coefficient form of the velocity equation to simplify to a form containing only kinetic constants and K_{eq}. For bireactant mechanisms, there will be two suitable definitions for some K_i values, but for mechanisms that are terreactant or greater in either direction, only one definition (or set of equivalent definitions) works.

Isoinhibition Constants

If a stable enzyme form isomerizes in a step that is partially rate-limiting, additional terms occur in the denominator of the velocity equation, and it becomes necessary to define additional constants called isoinhibition constants and written as K_{iip}, K_{iiq}, and so on. For example, if after the release of Q in an Ordered Bi Bi sequence, the free enzyme exists as form E′, which must isomerize to E before reacting with substrate A, additional APQ, ABQ, and ABPQ terms are present in the velocity equation. The K_{ii} for a given ligand that combines with an isomer of a stable form is defined as the denominator coefficient used to define the maximum velocity for the formation of that ligand, divided by the coefficient of the denominator term containing the given ligand as well as the ligands in the numerator of the ratio. Thus after canceling ligands, the letter corresponding to the ligand that combines with the stable enzyme form remains in the denominator. For the Iso Ordered Bi Bi system:

$$K_{iia} = \frac{\text{Coef}_{\text{PQ}}}{\text{Coef}_{\text{APQ}}}, \qquad K_{iiq} = \frac{\text{Coef}_{\text{AB}}}{\text{Coef}_{\text{ABQ}}}$$

Velocity Equation for the Forward Direction

The complete velocity equation is obtained by multiplying the equation in coefficient form by a factor composed of (num_2) divided by the product of the two denominator coefficients which contain as subscripts all the substrates for the reaction in each direction (i.e., the two denominator coefficients used to define $V_{\max_f}$ and $V_{\max_r}$).

$$\text{factor} = \frac{\text{num}_2}{(\text{Coef}_{\text{ABC}\cdots})(\text{Coef}_{\text{PQR}\cdots})}$$

After multiplication by this factor, the above rules are used to convert coefficient ratios to kinetic constants. Some ratios will not be immediately definable. In this case we multiply the ratio by $\text{num}_1/\text{num}_1$. In some cases, we will have to substitute other coefficients to obtain an expression that reduces to kinetic constants. For example, a ratio $\text{Coef}_{\text{BQ}}/\text{Coef}_{\text{AB}}$ cannot be defined, but it may be possible to substitute $\text{Coef}_{\text{B}}/K_{m_\text{A}}$ for Coef_{AB} (from the definition of K_{m_A}). We can then define the resulting $\text{Coef}_{BQ}/\text{Coef}_B$ as K_{iq}. In searching for substitutions, care must be taken not to define any given constant more than one way, unless the alternate ways are identical in terms of rate constants. For example, if K_{ip} is defined as $\text{Coef}_{AB}/\text{Coef}_{ABP}$ in one part of the equation, we cannot define K_{ip} as $\text{Coef}_A/\text{Coef}_{AP}$ in another part unless the two definitions are equivalent. After the conversion to kinetic constants, the numerator of the velocity equation has the form:

$$\boxed{V_{\max_f} V_{\max_r}\left([\text{A}][\text{B}][\text{C}]\ldots - \frac{[\text{P}][\text{Q}][\text{R}]\ldots}{K_{eq}}\right)} \tag{IX-27}$$

while the denominator terms contain $V_{\max_r}$ as a factor of some terms and $V_{\max_f}/K_{eq}$ as a factor of others. If numerator and denominator are divided by $V_{\max_r}$, the equation is put into a form giving the net steady-state velocity in the forward direction. After the division, some denominator terms will contain the factor $V_{\max_f}/V_{\max_r}K_{eq}$, which can be converted to K_m and K_i constants by substitution of the proper Haldane equation. The numerator of the velocity equation after dividing by $V_{\max_r}$ has the form

$$V_{\max_f}\left([\text{A}][\text{B}][\text{C}]\ldots - \frac{[\text{P}][\text{Q}][\text{R}]\ldots}{K_{eq}}\right) \tag{IX-28}$$

Velocity Equation for the Reverse Direction

If the equation in the form of IX-27 is divided by $V_{\max_f}/K_{eq}$ (instead of $V_{\max_r}$), we will obtain a form giving the net steady-state velocity in the reverse direction. Both sides of the resulting equation are multiplied by -1 to put the numerator in the form:

$$\boxed{V_{\max_r}\left([\text{P}][\text{Q}][\text{R}]\ldots - \frac{[\text{A}][\text{B}][\text{C}]\ldots}{K'_{eq}}\right)} \tag{IX-29}$$

The K'_{eq} is the equilibrium constant for the *reverse* reaction; $K'_{eq} = 1/K_{eq}$. The $-v$ that is obtained on the left-hand side of the equation represents the initial steady-state velocity in the reverse direction and is usually just written as v. The denominator of the equation at this point contains a number of terms that have K_{eq} as a factor. One of these terms may be a constant which can be used to derive a Haldane equation for the system.

Haldane Equations

Every kinetic mechanism has at least one Haldane equation relating K_{eq} to the kinetic constants. The Haldane equations have the form:

$$\boxed{K_{eq} = \left(\frac{V_{\max_f}}{V_{\max_r}}\right)^n \frac{K_{(p)}K_{(q)}K_{(r)}\cdots}{K_{(a)}K_{(b)}K_{(c)}\cdots}} \qquad \text{(IX-30)}$$

The K's will be either inhibition or Michaelis constants. The choice is not arbitrary, but, rather, is set for any given mechanism. There will always be at least one Haldane equation where $n=1$ which can be used to eliminate K_{eq} from the velocity equation. If the velocity equation contains a constant term in the denominator, we can conveniently obtain a Haldane equation with $n=1$ by rearranging the equation to express the net velocity of the reverse reaction, as described above. The constant denominator term containing K_{eq} is then written out in coefficient form. After canceling identical coefficients and substituting equivalent terms for others, the entire constant term will be expressed in kinetic constants. Setting this new expression equal to the original constant denominator term allows us to solve for K_{eq}. For example, in the Ordered Bi Bi reaction, the constant denominator term for the reverse reaction can be written as:

$$\frac{K_{ia}K_{m_B}V_{\max_r}K_{eq}}{V_{\max_f}} = \frac{(\text{const})(\text{Coef}_A)(\text{num}_2)(\text{num}_1)(\text{Coef}_{AB})}{(\text{Coef}_A)(\text{Coef}_{AB})(\text{Coef}_{PQ})(\text{num}_2)(\text{num}_1)}$$

$$= \frac{(\text{const})}{(\text{Coef}_{PQ})} = \frac{(\text{const})K_{m_P}}{(\text{Coef}_Q)} = K_{iq}K_{m_P}$$

$$\therefore \quad K_{eq} = \frac{V_{\max_f}K_{iq}K_{m_P}}{V_{\max_r}K_{ia}K_{m_B}}$$

Other Haldane equations for a given system might be found by writing out in coefficient form the Haldane equation obtained as described above and then recombining coefficients into new definitions. If there is no constant term in the denominator, then the same procedure can be followed using one of the other denominator terms of the original velocity equation containing K_{eq}. We can also examine the definitions for the various constants to see which combinations, when expressed as rate constants, reduce to $k_1k_2k_3\ldots/k_{-1}k_{-2}k_{-3}$. The numerator of the Haldane equation must contain a constant for each product. The denominator must contain a constant for each substrate.

A Shortcut for Obtaining Velocity Equations

The conversion of the velocity equation from coefficient form to a form composed entirely of kinetic constants can often be shortened by dividing numerator and denominator by $\text{Coef}_{\text{ABC}\ldots}$(i.e., the coefficient of the denominator term containing all the substrates for the reaction in the desired direction). This procedure is useful when we wish to obtain the equation for the forward velocity in the absence of any products, or in the presence of only one product at a time.

Rate Constants

The values of individual rate constants can be calculated for the ordered portions of many systems of Bi or greater reactancy by combining the equations for the definitions of kinetic constants. For example, in the Ordered Bi Bi system:

$$V_{\max_f} = \frac{\text{num}_1}{\text{Coef}_{\text{AB}}} = \frac{k_1k_2k_3k_4[\text{E}]_t}{k_1k_2(k_4+k_3)} \quad \text{and} \quad K_{m_{\text{A}}} = \frac{\text{Coef}_{\text{B}}}{\text{Coef}_{\text{AB}}} = \frac{k_2k_3k_4}{k_1k_2(k_4+k_3)}$$

$$\therefore \quad \frac{V_{\max_f}}{K_{m_{\text{A}}}[\text{E}]_t} = k_1$$

Some calculations assume that transitory complexes do not isomerize. If the isomerization is taken into account, the form of the velocity equation is unchanged, but some of the calculated rate constants will not be valid. (The calculated rate constant will be a composite of several rate constants.)

Distribution Equations

The King-Altman method yields equations for the distribution of enzyme species in terms of rate constants and ligand concentrations. The equations can be expressed in terms of kinetic constants, K_{eq}, and ligand concentrations by multiplying numerator and denominator by the same factor used to convert the coefficient form of the velocity equation to a form composed of kinetic constants, that is, $(\text{num}_2)/(\text{Coef}_{\text{ABC}\ldots})(\text{Coef}_{\text{PQR}\ldots})$. The denominator of the distribution equations is the same as the denominator of the velocity equation and consequently can be expressed in terms of kinetic constants and K_{eq} as described earlier. The numerators of the distribution equations sometimes contain entire denominator terms which present no problem, but in many cases only part of a denominator term is present. For example, in the Ordered Bi Bi system:

$$\frac{[EAB+EPQ]}{[E]_t} = \frac{k_1k_2k_4[A][B] + k_1k_2k_{-3}[A][B][P] + k_2k_{-3}k_{-4}[B][P][Q] + k_{-1}k_{-3}k_{-4}[P][Q]}{\text{denominator}}$$

The group of rate constants multiplying [A][B][P] is identical to Coef_{ABP}, and the group of rate constants multiplying [B][P][Q] is identical to Coef_{BPQ}. However, the rate constants multiplying [A][B] and [P][Q] are only parts of Coef_{AB} and Coef_{PQ}, respectively. (The rest of Coef_{AB} comes from the $[EQ]/[E]_t$ equation; the rest of Coef_{PQ} comes from the $[EA]/[E]_t$ equation.) The [A][P][B] and [B][P][Q] terms can be converted to kinetic constants in the usual way. The [A][B] term, after multiplying by the usual factor, becomes:

$$\frac{k_1k_2k_4(\text{num}_2)[A][B]}{(\text{Coef}_{AB})(\text{Coef}_{PQ})} = \frac{k_1k_2k_4V_{\max_r}[A][B]}{k_1k_2(k_3+k_4)} = \frac{V_{\max_r}[A][B]}{1+\dfrac{k_3}{k_4}}$$

The [A][B] term can be expressed entirely in terms of kinetic constants because it is possible to evaluate k_3 and k_4 in terms of kinetic constants:

$$\frac{k_3}{k_4} = \frac{\dfrac{1}{k_4}}{\dfrac{1}{k_3}} = \frac{\dfrac{K_{m_Q}[E]_t}{K_{iq}V_r}}{\dfrac{[E]_t}{V_f} - \dfrac{K_{m_Q}[E]_t}{K_{iq}V_r}} = \frac{1}{\dfrac{K_{iq}V_r}{K_{m_Q}V_f} - 1}$$

Substituting, the [A][B] term becomes:

$$\left(V_r - \frac{K_{m_Q}V_f}{K_{iq}}\right)[A][B]$$

In a similar manner, the [P][Q] term can be expressed entirely in terms of kinetic constants.

Alternative Nomenclature

Although the Cleland nomenclature is now in general use, other methods of defining groups of rate constants have been used. These are illustrated below for the Ordered Bi Bi reaction. As we shall see later, this mechanism yields a steady-state velocity equation in the absence of products which can be written as:

$$v = \frac{f_1[\mathrm{E}]_t[\mathrm{A}][\mathrm{B}]}{f_2 + f_3[\mathrm{A}] + f_4[\mathrm{B}] + f_5[\mathrm{A}][\mathrm{B}]} \quad \text{(IX-31)}$$

where f_1, f_2, and so on, are different groups of rate constants.

Alberty (1953)

Numerator and denominator of equation IX-31 are divided by f_5 and the resulting constant terms defined as shown below.

$$v = \frac{V_f[\mathrm{A}][\mathrm{B}]}{K_{\mathrm{AB}} + K_{\mathrm{B}}[\mathrm{A}] + K_{\mathrm{A}}[\mathrm{B}] + [\mathrm{A}][\mathrm{B}]} \quad \text{(IX-32)}$$

Or, in reciprocal form:

$$\frac{V_f}{v} = 1 + \frac{K_{\mathrm{A}}}{[\mathrm{A}]} + \frac{K_{\mathrm{B}}}{[\mathrm{B}]} + \frac{K_{\mathrm{AB}}}{[\mathrm{A}][\mathrm{B}]} \quad \text{(IX-33)}$$

where

$$V_f = \frac{f_1[\mathrm{E}]_t}{f_5}, \qquad K_{\mathrm{AB}} = \frac{f_2}{f_5}, \qquad K_{\mathrm{B}} = \frac{f_3}{f_5}, \qquad \text{and} \qquad K_{\mathrm{A}} = \frac{f_4}{f_5}$$

The K_{A} and K_{B} are limiting Michaelis constants, while K_{AB} is a complex constant with dimensions of M^2.

Dalziel (1957)

Numerator and denominator of equation IX-31 are divided by f_1 giving:

$$v = \frac{[E]_t[S_1][S_2]}{\phi_{12} + \phi_2[S_1] + \phi_1[S_2] + \phi_0[S_1][S_2]} \tag{IX-34}$$

Or, in reciprocal form:

$$\frac{[E]_t}{v} = \phi_0 + \frac{\phi_1}{[S_1]} + \frac{\phi_2}{[S_2]} + \frac{\phi_{12}}{[S_1][S_2]} \tag{IX-35}$$

where the two substrates, A and B, are indicated as S_1 and S_2, and $\phi_{12} = f_2/f_1$, $\phi_2 = f_3/f_1$, $\phi_1 = f_4/f_1$, and $\phi_0 = f_5/f_1$. As in the Alberty system, the dimensions of the different constants are not all the same; ϕ has the dimensions of, for example, min; ϕ_1 and ϕ_2 are in $M \times \text{min}$, while ϕ_{12} is in $M^2 \times \text{min}$.

Bloomfield, Peller, and Alberty (1962)

Dividing numerator and denominator of equation IX-31 by f_2, we obtain:

$$v = \frac{\dfrac{V_{AB}}{K_{AB}}[A][B]}{1 + \dfrac{[A]}{K_A} + \dfrac{[B]}{K_B} + \dfrac{[A][B]}{K_{AB}}} \tag{IX-36}$$

or

$$v = \frac{\dfrac{V_{AB}}{K_{AB}}[A][B]}{K_{AB} + \dfrac{K_{AB}[A]}{K_A} + \dfrac{K_{AB}[B]}{K_B} + [A][B]} \tag{IX-37}$$

Or, in reciprocal form:

$$\boxed{\frac{V_{AB}}{v} = 1 + \frac{K_{AB}}{K_B[A]} + \frac{K_{AB}}{K_A[B]} + \frac{K_{AB}}{[A][B]}} \qquad \text{(IX-38)}$$

where

$$V_{AB} = \frac{f_1[E]_t}{f_5}, \qquad K_A = \frac{f_2}{f_3}, \qquad K_B = \frac{f_2}{f_4}, \qquad \text{and} \qquad K_{AB} = \frac{f_2}{f_5}$$

The Bloomfield-Peller-Alberty system leads to relatively uncomplicated looking equations. However, a different kinetic constant is defined for each term in the denominator of equation IX-36, none of which correspond to actual Michaelis constants.

Cleland (1963)

Dividing equation IX-22 by f_5, we obtain:

$$\boxed{v = \frac{V_1[A][B]}{K_{ia}K_b + K_b[A] + K_a[B] + [A][B]}} \qquad \text{(IX-39)}$$

Or, in reciprocal form:

$$\boxed{\frac{V_1}{v} = 1 + \frac{K_a}{[A]} + \frac{K_b}{[B]} + \frac{K_{ia}K_b}{[A][B]}} \qquad \text{(IX-40)}$$

where

$$V_1 = \frac{f_1[E]_t}{f_5}, \qquad K_a = \frac{f_4}{f_5}, \qquad K_b = \frac{f_3}{f_5}, \qquad \text{and} \qquad K_{ia} = \frac{f_2}{f_3}$$

A comparison of the symbols used in the different nomenclature systems is shown in Table IX-1 for an Ordered Bi Bi reation.

Table IX-1 Different Symbols Used for Kinetic Constants of an Ordered Bi Bi Reaction

Definition	Alberty	Dalziel	Bloomfield *et al.*	Cleland	Mahler and Cordes	Enzyme Commission	This Book
Limiting Michaelis constant for A	K_A	$\frac{\phi_1}{\phi_0}$	$\frac{K_{AB}}{K_B}$	K_a	K_a	K_m^A	K_{m_A}
Limiting Michaelis constant for B	K_B	$\frac{\phi_2}{\phi_0}$	$\frac{K_{AB}}{K_A}$	K_b	K_b	K_m^B	K_{m_B}
Dissociation constant for A	$\frac{K_{AB}}{K_B}$	$\frac{\phi_{12}}{\phi_2}$	K_A	K_{ia}	$\bar{K}_a$	K_S^A	$K_{ia}(K_A)$[a]
Limiting maximum velocity	V_f	$\frac{e}{\phi_0}$	V_{AB}	V_1	V_1	V	V_{max}
Turnover number	$\frac{V_f}{E_0}$	$\frac{1}{\phi_0}$	$\frac{V_{AB}}{E_0}$	$\frac{V_1}{E_t}$	$\frac{V_1}{e_0}$	—	$\frac{V_{max}}{[E]_t}$

[a]K_{ia} is used for steady-state systems; K_A is used for rapid equilibrium systems.

K_{slope} and K_{int} Nomenclature

As we have seen in Chapter Six, the K_{slope} of a multiligand system, represents an apparent constant expressing the effect of a changing fixed ligand on the slope of the reciprocal plot of $1/v$ versus $1/[\text{varied ligand}]$. In some cases, K_{slope} is a limiting constant but more often, it is a function of the limiting constant and the concentrations of the other ligands that are held constant. Similarly, K_{int} represents an apparent constant expressing the effect of a changing fixed ligand on the $1/v$-axis intercept of the reciprocal plot. These apparent constants are identified according to the ligands which are varied. For example, $K_{i_{slope}}^{P/B}$ is the apparent inhibitor constant expressing the effect of P on the slope of the $1/v$ versus $1/[B]$ plots. The latter is indicated as $slope_{1/B}$. The apparent constant is obtained by replotting $slope_{1/B}$ versus [P]. Similarly, $K_{int}^{C/A}$ is an apparent activation constant expressing the effect of substrate C on the $1/v$-axis intercept of the $1/v$ versus $1/[A]$ plots. The intercepts are indicated as $1/V_{max_{app}}$ or $int_{1/A}$. In this case, $K_{int}^{C/A}$ is obtained by replotting the intercepts versus $1/[C]$.

D. ISO UNI UNI SYSTEM (MOBILE CARRIER MODEL OF MEMBRANE TRANSPORT)

We have seen that the general form of the velocity equation is unchanged by the introduction of additional central complexes. Thus the sequences $E + A \rightleftharpoons EA \rightleftharpoons E + P$ and $E + A \rightleftharpoons EA \rightleftharpoons EP \rightleftharpoons E + P$ both yield the same final velocity equation. In general, the isomerization of any transitory complex has no effect on the form of the velocity equation. However, if a stable enzyme form undergoes isomerization, additional terms appear in the denominator of the velocity equation. For example, consider the reaction sequence shown below.

$$E + A \underset{k_{-1}}{\overset{k_1}{\rightleftharpoons}} EA \underset{k_{-2}}{\overset{k_2}{\rightleftharpoons}} E'P \underset{k_{-3}}{\overset{k_3}{\rightleftharpoons}} P + E' \qquad E' \underset{k_{-4}}{\overset{k_4}{\rightleftharpoons}} E$$

The E′ represents a stable enzyme form that is produced when the central complex breaks down liberating P. E′ must isomerize to E before another reaction sequence can occur. Iso mechanisms may be uncommon in ordinary soluble enzyme systems, but membrane transport systems might be approximated by an Iso Uni Uni sequence. For example, let us rewrite the sequence as a basic King-Altman figure:

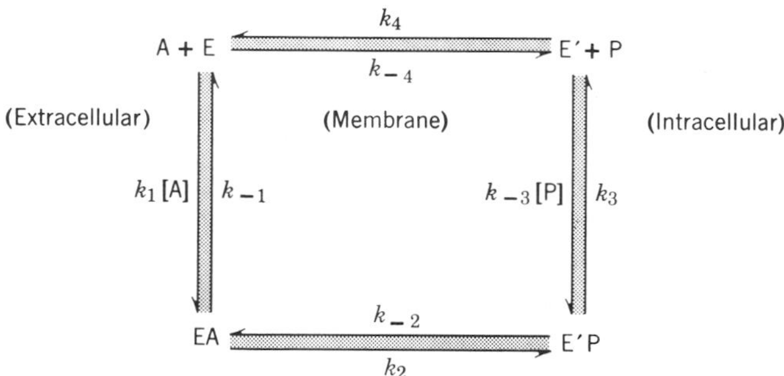

The E could represent a binding protein or carrier molecule that combines with the external substrate, A, at the outer surface of the membrane. The EA complex is then translocated (or undergoes a conformational change) bringing the complex to the cytoplasm-membrane boundary, where the substrate is released. Although the "product" of the permease system is the intracellular, unchanged substrate, we will use the symbol P to distinguish intracellular substrate from the extracellular substrate. After the release of P, the free carrier molecule is not in a conformation or location that is immediately accessible to A, and thus is indicated by the symbol E′. E′ is converted back to E by the kinetically significant reaction indicated by rate constants k_4/k_{-4}. We have four enzyme species. Hence the King-Altman interconversion patterns contain three lines from which we obtain:

$$\frac{[\mathrm{E}]}{[\mathrm{E}]_t} = \frac{k_{-1}k_3k_4 + k_2k_3k_4 + k_{-3}[\mathrm{P}]k_{-2}k_{-1} + k_4k_{-2}k_{-1}}{\text{denominator}} \quad \text{(IX-41)}$$

$$\frac{[\mathrm{EA}]}{[\mathrm{E}]_t} = \frac{k_3k_4k_1[\mathrm{A}] + k_{-4}k_{-3}[\mathrm{P}]k_{-2} + k_1[\mathrm{A}]k_{-3}[\mathrm{P}]k_{-2} + k_4k_1[\mathrm{A}]k_{-2}}{\text{denominator}} \quad \text{(IX-42)}$$

$$\frac{[\mathrm{E'P}]}{[\mathrm{E}]_t} = \frac{k_{-1}k_{-4}k_{-3}[\mathrm{P}] + k_{-4}k_{-3}[\mathrm{P}]k_2 + k_1[\mathrm{A}]k_2k_{-3}[\mathrm{P}] + k_4k_1[\mathrm{A}]k_2}{\text{denominator}} \quad \text{(IX-43)}$$

$$\frac{[\mathrm{E'}]}{[\mathrm{E}]_t} = \frac{k_{-1}k_{-4}k_3 + k_{-4}k_2k_3 + k_1[\mathrm{A}]k_2k_3 + k_{-2}k_{-1}k_{-4}}{\text{denominator}} \quad \text{(IX-44)}$$

The denominator is the sum of all 16 numerator terms. Substituting into an equation for the steady-state net forward velocity:

$$v = k_2[\mathrm{EA}] - k_{-2}[\mathrm{E'P}]$$

we obtain after grouping denominator terms:

$$\frac{v}{[\mathrm{E}]_t} = \frac{k_1k_2k_3k_4[\mathrm{A}] - k_{-1}k_{-2}k_{-3}k_{-4}[\mathrm{P}]}{(k_4+k_{-4})(k_{-1}k_3+k_{-1}k_{-2}+k_2k_3) + k_1(k_2k_3+k_2k_4+k_{-2}k_4+k_3k_4)[\mathrm{A}] + k_{-3}(k_{-1}k_{-2}+k_{-1}k_{-4}+k_2k_{-4}+k_{-2}k_{-4})[\mathrm{P}] + (k_1k_{-3})(k_2+k_{-2})[\mathrm{A}][\mathrm{P}]} \quad \text{(IX-45)}$$

or

$$v = \frac{\text{num}[A] - \text{num}_2[P]}{\text{const} + \text{Coef}_A[A] + \text{Coef}_P[P] + \text{Coef}_{AP}[A][P]}$$

Since the number of central complexes does not influence the final form of the velocity equation, we could have drawn the overall reaction sequence as a three-lined basic figure:

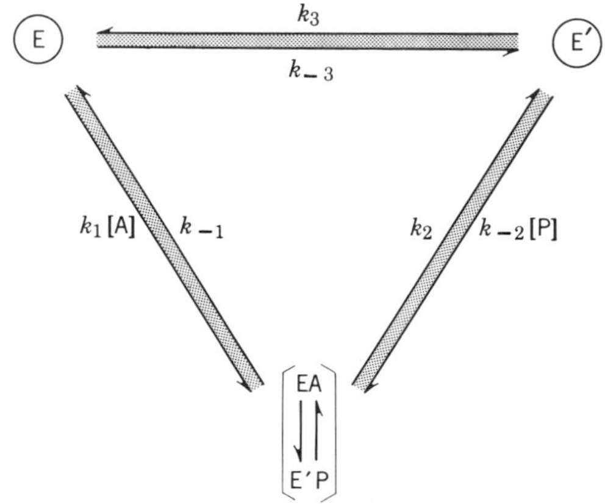

The King-Altman interconversion patterns would contain two lines:

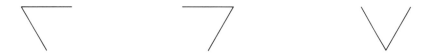

The final equation would be the same in coefficient form as IX-45. Only the rate constants comprising the coefficients would be different.

Compared to equation IX-9 for the ordinary Uni Uni reaction, we see that the isomerization of the stable enzyme form introduces an [A][P] term to the denominator. To obtain a velocity equation in terms of kinetic constants, we divide numerator and denominator of equation IX-45 by Coef_A:

$$v = \frac{\dfrac{\text{num}_1}{\text{Coef}_A}[A] - \dfrac{\text{num}_2}{\text{Coef}_A}[P]}{\dfrac{\text{Const}}{\text{Coef}_A} + \dfrac{\text{Coef}_A[A]}{\text{Coef}_A} + \dfrac{\text{Coef}_P[P]}{\text{Coef}_A} + \dfrac{\text{Coef}_{AP}[A][P]}{\text{Coef}_A}}$$

Substituting:

$$\frac{\text{num}_1}{\text{Coef}_\text{A}} = V_{\text{max}_f}, \qquad \frac{\text{const}}{\text{Coef}_\text{A}} = K_{m_\text{A}}$$

$$\frac{\text{Coef}_\text{P}}{\text{Coef}_\text{A}} = \frac{\text{Coef}_\text{P} K_{m_\text{A}}}{\text{const}} = \frac{K_{m_\text{A}}}{K_{m_\text{P}}},$$

$$\frac{\text{num}_2}{\text{Coef}_\text{A}} = \frac{\text{num}_1}{K_\text{eq}\text{Coef}_\text{A}} = \frac{V_{\text{max}_f}}{K_\text{eq}}$$

we obtain:

$$\frac{v}{V_{\text{max}_f}} = \frac{[\text{A}] - \dfrac{[\text{P}]}{K_\text{eq}}}{K_{m_\text{A}} + [\text{A}] + \dfrac{K_{m_\text{A}}[\text{P}]}{K_{m_\text{P}}} + \dfrac{\text{Coef}_\text{AP}[\text{A}][\text{P}]}{\text{Coef}_\text{A}}} \tag{IX-46}$$

The denominator still contains a number of rate constants grouped as $\text{Coef}_\text{AP}/\text{Coef}_\text{A}$. This ratio must be defined as a kinetic constant before the velocity equation is useable:

$$\frac{\text{Coef}_\text{A}}{\text{Coef}_\text{AP}} = K_{iip}$$

When $1/K_{iip}$ is substituted for $\text{Coef}_\text{AP}/\text{Coef}_\text{A}$, the velocity equation simplifies to:

$$\boxed{\frac{v}{V_{\text{max}_f}} = \frac{[\text{A}] - \dfrac{[\text{P}]}{K_\text{eq}}}{K_{m_\text{A}}\left(1 + \dfrac{[\text{P}]}{K_{m_\text{P}}}\right) + [\text{A}]\left(1 + \dfrac{[\text{P}]}{K_{iip}}\right)}} \tag{IX-47}$$

where we see that K_{iip} is an inhibition constant, specifically an isoinhibition constant. If K_eq is very large, or if the system is far from the equilibrium point, the velocity equation reduces to:

$$\boxed{\frac{v}{V_{\max}} = \frac{[\mathrm{A}]}{K_{m_\mathrm{A}}\left(1 + \frac{[\mathrm{P}]}{K_{m_\mathrm{P}}}\right) + [\mathrm{A}]\left(1 + \frac{[\mathrm{P}]}{K_{iip}}\right)}} \qquad \text{(IX-48)}$$

Thus we see that intracellular A (i.e., P) acts as a mixed-type inhibitor with respect to the further transport of extracellular A. Both the apparent K_{m_A} and $V_{\max}$ are affected:

$$K_{m_{\mathrm{A(app)}}} = K_{m_\mathrm{A}} \frac{\left(1 + \frac{[\mathrm{P}]}{K_{m_\mathrm{P}}}\right)}{\left(1 + \frac{[\mathrm{P}]}{K_{iip}}\right)}, \qquad V_{\max_{\mathrm{app}}} = \frac{V_{\max}}{\left(1 + \frac{[\mathrm{P}]}{K_{iip}}\right)}$$

The reciprocal form of the velocity equation is:

$$\boxed{\frac{1}{v} = \frac{K_{m_\mathrm{A}}}{V_{\max}}\left(1 + \frac{[\mathrm{P}]}{K_{m_\mathrm{P}}}\right)\frac{1}{[\mathrm{A}]} + \frac{1}{V_{\max}}\left(1 + \frac{[\mathrm{P}]}{K_{iip}}\right)} \qquad \text{(IX-49)}$$

The plot is shown in Figure IX-2. The family of curves intersect above or below the $1/[\mathrm{A}]$-axis, depending on the relative values of K_{m_P} and K_{iip}. The $1/[\mathrm{A}]$ coordinate is given by:

$$\frac{1}{[\mathrm{A}]} = -\frac{1}{K_{m_\mathrm{A}}}\frac{(K_{i_{\mathrm{slope}}})}{(K_{i_{\mathrm{int}}})} = -\frac{K_{m_\mathrm{P}}}{K_{m_\mathrm{A}} K_{iip}}$$

The $1/v$ coordinate is given by:

$$\frac{1}{v} = \frac{1}{V_{\max}}\left(1 - \frac{K_{i_{\mathrm{slope}}}}{K_{i_{\mathrm{int}}}}\right) = \frac{1}{V_{\max}}\left(1 - \frac{K_{m_\mathrm{P}}}{K_{iip}}\right)$$

If, coincidentally, $K_{m_\mathrm{P}} = K_{iip}$, P will act as a pure noncompetitive inhibitor.

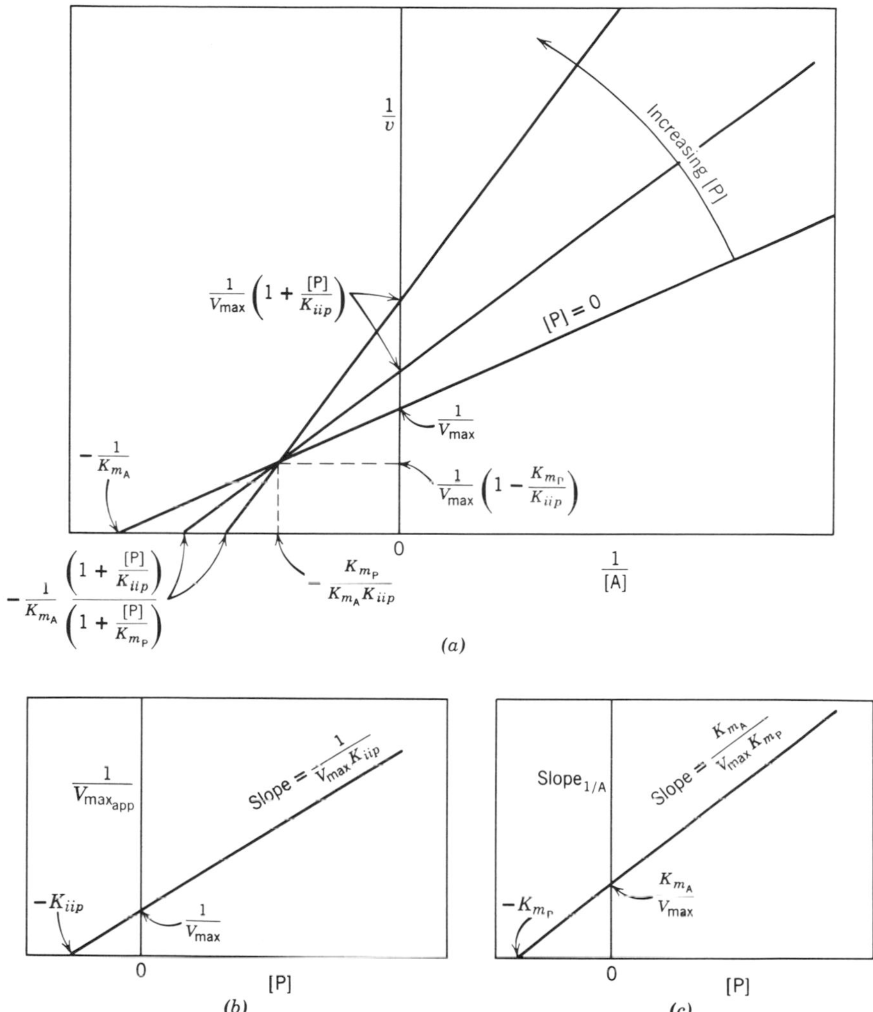

Fig. IX-2. Iso Uni Uni system. It is assumed that K_{eq} is very large so that v represents the actual forward velocity, uncomplicated by the reverse reaction. (V_{max_r} might be very low.) (*a*) $1/v$ versus $1/[A]$ in the presence of different fixed concentrations of P. (*b*) $1/v$-axis intercept replot. (*c*) $Slope_{1/A}$ replot.

The K_{iip} can be determined from replots of $1/v$-axis intercept versus [P]:

$$\frac{1}{V_{\max_{app}}} = \frac{1}{V_{\max} K_{iip}}[\mathrm{P}] + \frac{1}{V_{\max}}$$

The K_{m_P} can be determined from replots of $slope_{1/A}$ versus [P]:

$$slope_{1/\mathrm{A}} = \frac{K_{m_A}}{V_{\max} K_{m_P}}[\mathrm{P}] + \frac{K_{m_A}}{V_{\max}}$$

If K_{eq} is not $\gg 1$, plots of $1/v$ (actually $1/v_{net}$) versus 1/[A] in the presence of P are nonlinear and concave up. Linear plots can be obtained by plotting $1/v_{net}$ versus $1/\Delta\mathrm{A}$ where $\Delta\mathrm{A} = [\mathrm{A}] - [\mathrm{P}]/K_{eq}$. The velocity and reciprocal equations are:

$$\frac{v_{net}}{V_{\max_f}} = \frac{\Delta \mathrm{A}}{K_{m_A}\left(1 + \frac{[\mathrm{P}]}{K_{m_P}}\right) + \frac{[\mathrm{P}]}{K_{eq}}\left(1 + \frac{[\mathrm{P}]}{K_{iip}}\right) + \Delta\mathrm{A}\left(1 + \frac{[\mathrm{P}]}{K_{iip}}\right)} \quad \text{(IX-50)}$$

$$\frac{1}{v_{net}} = \left[\frac{K_{m_A}}{V_{\max_f}}\left(1 + \frac{[\mathrm{P}]}{K_{m_P}}\right) + \frac{[\mathrm{P}]}{V_{\max_f} K_{eq}}\left(1 + \frac{[\mathrm{P}]}{K_{iip}}\right)\right]\frac{1}{\Delta\mathrm{A}} + \frac{1}{V_{\max_f}}\left(1 + \frac{[\mathrm{P}]}{K_{iip}}\right)$$

(IX-51)

The family of plots do not intersect at a common point. The reciprocal of the horizontal-axis intercept gives:

$$-\Delta\mathrm{A} = K_{m_A}\frac{\left(1 + \frac{[\mathrm{P}]}{K_{m_P}}\right)}{\left(1 + \frac{[\mathrm{P}]}{K_{iip}}\right)} + \frac{[\mathrm{P}]}{K_{eq}}$$

The $1/v_{net}$-axis intercept replot is linear (and gives $1/V_{\max_f}$ and K_{iip}) but the $slope_{1/A}$ versus [P] replot is parabolic.

"Transinhibition," that is, the inhibition of further transport of an extracellular substrate after preloading cells with the same substrate (or another substrate of the same permease system), has been observed in a number of microbial transport systems.

We can see that the overall reaction sequence, $A+E \rightleftharpoons EA \rightleftharpoons E'P \rightleftharpoons E'+P$, is symmetrical. Therefore, we can predict that A will be a mixed-type inhibitor with respect to P. If we write the reaction backwards and go through the same procedure as outlined above (dividing by Coef_P instead of Coef_A), we will obtain the equation for the net steady-state velocity in the direction P→A:

$$\boxed{\frac{v}{V_{\max_r}} = \frac{[P] - \dfrac{[A]}{K'_{eq}}}{K_{m_P}\left(1 + \dfrac{[A]}{K_{m_A}}\right) + [P]\left(1 + \dfrac{[A]}{K_{iia}}\right)}} \qquad \text{(IX-52)}$$

where

$$K_{iia} = \frac{\text{Coef}_P}{\text{Coef}_{AP}}, \qquad K'_{eq} = \frac{1}{K_{eq}}$$

In deriving equations IX-47 and IX-52, we have taken a shortcut by dividing the coefficient form of the velocity equation (equation IX-45) by Coef_A (for the forward reaction) or Coef_P (for the reverse reaction). This procedure keeps K_{eq} from appearing in the denominator and eliminates the need for a Haldane equation. Let us backtrack a bit and rederive the velocity equations, this time following the general procedure outlined earlier. Equation IX-45 is multiplied by the factor $(\text{num}_2)/(\text{Coef}_A\text{Coef}_P)$. The [P] terms are also multiplied by $\text{num}_1/\text{num}_1$. We obtain:

$$\boxed{v = \frac{V_{\max_f}\left([A] - \dfrac{[P]}{K_{eq}}\right)}{K_{m_A} + [A] + \dfrac{V_{\max_f}[P]}{V_{\max_r}K_{eq}} + \dfrac{[A][P]}{K_{iip}}}} \qquad \text{(IX-53)}$$

If we set $[P]=0$, the equation reduces to the usual Henri-Michaelis-Menten equation for the forward velocity. To obtain the equation for the net reverse steady-state velocity, we divide numerator and denominator of equation

IX-53 by $V_{\max_f}/K_{eq}$. We obtain (neglecting the negative sign):

$$v = \frac{V_{\max_r}\left([\mathrm{P}] - \dfrac{[\mathrm{A}]}{K'_{eq}}\right)}{\dfrac{K_{m_A} V_{\max_r} K_{eq}}{V_{\max_f}} + \dfrac{V_{\max_r}[\mathrm{A}]K_{eq}}{V_{\max_f}} + [\mathrm{P}] + \dfrac{V_{\max_r}[\mathrm{A}][\mathrm{P}]K_{eq}}{K_{iip} V_{\max_f}}} \qquad \text{(IX-54)}$$

where $[\mathrm{A}]/K'_{eq} = [\mathrm{A}]K_{eq}$. If we set $[\mathrm{A}]=0$, equation IX-54 reduces to:

$$v = \frac{V_{\max_r}[\mathrm{P}]}{\dfrac{K_{m_A} V_{\max_r} K_{eq}}{V_{\max_f}} + [\mathrm{P}]}$$

which must equal

$$v = \frac{V_{\max_r}[\mathrm{P}]}{K_{m_P} + [\mathrm{P}]}$$

$$\therefore \quad \frac{K_{m_A} V_{\max_r} K_{eq}}{V_{\max_f}} = K_{m_P} \qquad \text{or} \qquad K_{eq} = \frac{K_{m_P} V_{\max_f}}{K_{m_A} V_{\max_r}}$$

Thus the Haldane equation is the same as for the simple Uni Uni reaction. Now that we can express K_{eq} in terms of kinetic constants, we can substitute into equation IX-54 to obtain:

$$\frac{v}{V_{\max_r}} = \frac{\left([\mathrm{P}] - \dfrac{[\mathrm{A}]}{K'_{eq}}\right)}{K_{m_P}\left(1 + \dfrac{[\mathrm{A}]}{K_{m_A}}\right) + [\mathrm{P}]\left(1 + \dfrac{[\mathrm{A}]K_{m_P}}{K_{iip}K_{m_A}}\right)} \qquad \text{(IX-55)}$$

Equation IX-55 must be identical to equation IX-52. Therefore:

$$\frac{K_{m_P}}{K_{iip}K_{m_A}} = \frac{1}{K_{iia}} \qquad \text{or} \qquad \frac{K_{m_P}}{K_{m_A}} = \frac{K_{iip}}{K_{iia}}$$

The identity is also seen when we write the kinetic constants in coefficient form:

$$\frac{K_{m_P}}{K_{iip}K_{m_A}}=\frac{(\text{const})(\text{Coef}_{AP})(\text{Coef}_A)}{(\text{Coef}_P)(\text{Coef}_A)(\text{const})}=\frac{\text{Coef}_{AP}}{\text{Coef}_P}=\frac{1}{K_{iia}}$$

If A is radioactive and P is not, the equation for the rate of labeled A uptake can be derived from Cleland's rules for isotope exchange described in Chapter Ten. The procedure yields:

$$v^*=\frac{k_1k_2k_3[\text{A}^*][\text{E}]}{k_{-1}k_3+k_{-1}k_{-2}+k_2k_3}$$

Substituting for [E] from IX-41 and defining terms in the usual way, we obtain:

$$\frac{v^*}{V_{\max_f}}=\frac{[\text{A}^*]\left(1+\frac{[\text{P}]}{K_\text{I}}\right)}{K_{m_A}\left(1+\frac{[\text{P}]}{K_{m_P}}\right)+[\text{A}]\left(1+\frac{[\text{P}]}{K_{iip}}\right)}=\frac{[\text{A}^*]}{K_{m_A}\frac{\left(1+\frac{[\text{P}]}{K_{m_P}}\right)}{\left(1+\frac{[\text{P}]}{K_\text{I}}\right)}+[\text{A}]\frac{\left(1+\frac{[\text{P}]}{K_{iip}}\right)}{\left(1+\frac{[\text{P}]}{K_\text{I}}\right)}} \tag{IX-56}$$

where

$$K_\text{I}=\frac{k_4(k_{-1}k_3+k_{-1}k_{-2}+k_2k_3)}{k_{-1}k_{-2}k_{-3}}$$

If $K_\text{I}>K_{m_P}$ and $>K_{iip}$, P will act as an inhibitor. If $K_\text{I}<K_{m_P}$ and $<K_{iip}$, P will activate. In effect, the efflux of unlabeled P provides an alternate route for the regeneration of free E. If this route ($\text{P}+\text{E}'\rightarrow\text{E}'\text{P}\rightarrow\text{EA}\rightarrow\text{A}+\text{E}$) is faster than the rate of E regeneration via the influx route ($\text{E}'\rightarrow\text{E}$), P will activate and transstimulation will be observed. If the efflux route is slower than the influx route (which is likely if $K_{eq}>1$), transinhibition will be observed. Equation IX-56 predicts hyperbolic slope and intercept replots. That is, as [P] increases, v^* approaches a limit. If K_I is much larger than K_{m_P} and K_{iip}, equation IX-56 reduces to equation IX-48 and the replots will appear to be linear in a [P] range in the neighborhood of K_{m_P} and K_{iip}. (See Chapter Ten for a further discussion of isotope exchange in Iso Uni Uni systems.)

E. ORDERED UNI BI AND ORDERED BI UNI SYSTEMS

The Ordered Uni Bi reaction can be written:

$$E + A \underset{k_{-1}}{\overset{k_1}{\rightleftharpoons}} EA \underset{k_{-p}}{\overset{k_p}{\rightleftharpoons}} EPQ \underset{k_{-2}}{\overset{k_2}{\rightleftharpoons}} P + EQ \underset{k_{-3}}{\overset{k_3}{\rightleftharpoons}} E + Q$$

There are four enzyme species: E, EA, EPQ, and EQ. However, since the isomerization of the central complex has no effect on the final form of the velocity equation, we can combine EA and EPQ and show the reaction as:

$$E + A \underset{k_{-1}}{\overset{k_1}{\rightleftharpoons}} (EA \rightleftharpoons EPQ) \underset{k_{-2}}{\overset{k_2}{\rightleftharpoons}} P + EQ \underset{k_{-3}}{\overset{k_3}{\rightleftharpoons}} E + Q$$

Or, in the notation of Cleland:

$$\begin{array}{ccccccc} & A & & P & & Q & \\ & k_1 \downarrow k_{-1} & & k_2 \uparrow k_{-2} & & k_3 \uparrow k_{-3} & \\ \hline E & & (EA \rightleftharpoons EPQ) & & EQ & & E \end{array}$$

Now we effectively have only three enzyme species which yield the basic King-Altman figure shown below.

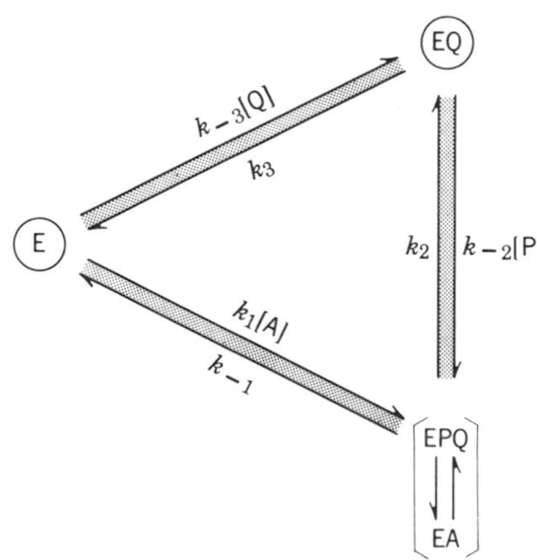

The distribution of enzyme species is obtained from the two-lined interconversion patterns shown below.

$$\frac{[E]}{[E]_t} = \frac{k_2k_3 \quad + \quad k_{-2}[P]k_{-1} \quad + \quad k_3k_{-1}}{\text{denominator}} \tag{IX-57}$$

$$\frac{[EA+EPQ]}{[E]_t} = \frac{k_{-3}[Q]k_{-2}[P] \quad + \quad k_1[A]k_{-2}[P] \quad + \quad k_3k_1[A]}{\text{denominator}} \tag{IX-58}$$

$$\frac{[EQ]}{[E]_t} = \frac{k_{-3}[Q]k_2 \quad + \quad k_1[A]k_2 \quad + \quad k_{-1}k_{-3}[Q]}{\text{denominator}} \tag{IX-59}$$

$$v = k_1[E][A] - k_{-1}[EA+EPQ]$$

$$v = \frac{k_1(k_2k_3 + k_{-2}[P]k_{-1} + k_3k_{-1})[A][E]_t - k_{-1}(k_{-3}[Q]k_{-2}[P] + k_1[A]k_{-2}[P] + k_3k_1[A])[E]_t}{\text{denominator}}$$

Grouping denominator terms and simplifying the numerator:

$$\frac{v}{[E]_t} = \frac{k_1k_2k_3[A] - k_{-1}k_{-2}k_{-3}[P][Q]}{k_3(k_2+k_{-1}) + k_1(k_2+k_3)[A] + k_{-1}k_{-2}[P] + k_{-3}(k_2+k_{-1})[Q] + k_1k_{-2}[A][P] + k_{-2}k_{-3}[P][Q]} \tag{IX-60}$$

When $[P]=0$ and $[Q]=0$, we obtain after dividing numerator and denominator by Coef_A:

$$v = \frac{\dfrac{\text{num}_1}{\text{Coef}_A}[A]}{\dfrac{\text{const}}{\text{Coef}_A} + [A]} = \frac{V_{max}[A]}{K_{m_A} + [A]}$$

When [A] = 0, we can obtain the velocity equation for the reverse Bi Uni reaction by dividing numerator and denominator by Coef_{PQ}:

$$v = \frac{\frac{\text{num}_2}{\text{Coef}_{PQ}}[P][Q]}{\frac{\text{const}}{\text{Coef}_{PQ}} + \frac{\text{Coef}_P}{\text{Coef}_{PQ}}[P] + \frac{\text{Coef}_Q}{\text{Coef}_{PQ}}[Q] + \frac{\text{Coef}_{PQ}}{\text{Coef}_{PQ}}[P][Q]} \qquad \text{(IX-61)}$$

Defining the ratios in the usual way:

$$v = \frac{V_{\max}[P][Q]}{\frac{\text{const}}{\text{Coef}_{PQ}} + K_{m_Q}[P] + K_{m_P}[Q] + [P][Q]} \qquad \text{(IX-62)}$$

The term $\text{const}/\text{Coef}_{PQ}$ can be converted to kinetic constants in two ways, depending on what we substitute for Coef_{PQ}. The two possibilities based on allowable definitions for inhibition constants are shown below:

(a) $$K_{m_Q} = \frac{\text{Coef}_P}{\text{Coef}_{PQ}} \qquad \therefore \quad \text{Coef}_{PQ} = \frac{\text{Coef}_P}{K_{m_Q}}$$

$$\frac{\text{const}}{\text{Coef}_{PQ}} = \frac{\text{const}\, K_{m_Q}}{\text{Coef}_P} = K'_{ip} K_{m_Q}$$

where

$$K'_{ip} = \frac{\text{const}}{\text{Coef}_P} = \frac{k_3(k_2 + k_{-1})}{k_{-1}k_{-2}}$$

(b) $$K_{m_P} = \frac{\text{Coef}_Q}{\text{Coef}_{PQ}} \qquad \therefore \quad \text{Coef}_{PQ} = \frac{\text{Coef}_Q}{K_{m_P}}$$

$$\frac{\text{const}}{\text{Coef}_{PQ}} = \frac{\text{const}\, K_{m_P}}{\text{Coef}_Q} = K_{iq} K_{m_P}$$

where

$$K_{iq} = \frac{\text{const}}{\text{Coef}_Q} = \frac{k_3(k_2 + k_{-1})}{k_{-3}(k_2 + k_{-1})} = \frac{k_3}{k_{-3}}$$

The velocity equation can then be written as:

$$\text{(a)} \quad \boxed{v = \frac{V_{\max}[\mathrm{P}][\mathrm{Q}]}{K'_{ip}K_{m_\mathrm{Q}} + K_{m_\mathrm{Q}}[\mathrm{P}] + K_{m_\mathrm{P}}[\mathrm{Q}] + [\mathrm{P}][\mathrm{Q}]}} \tag{IX-63}$$

or

$$\text{(b)} \quad \boxed{v = \frac{V_{\max}[\mathrm{P}][\mathrm{Q}]}{K_{iq}K_{m_\mathrm{P}} + K_{m_\mathrm{Q}}[\mathrm{P}] + K_{m_\mathrm{P}}[\mathrm{Q}] + [\mathrm{P}][\mathrm{Q}]}} \tag{IX-64}$$

Thus $K'_{ip}K_{m_\mathrm{Q}} = K_{iq}K_{m_\mathrm{P}}$ for the definitions of K'_{ip} and K_{iq} that we have used. Note that $\mathrm{Coef}_\mathrm{A}/\mathrm{Coef}_\mathrm{AP}$ is another valid definition of a K_{ip}, although numerically it is not equivalent to K'_{ip} defined as $\mathrm{Coef}_\mathrm{P}/\mathrm{Coef}_\mathrm{PQ}$. The alternate definition is not needed here, but is the definition of choice when we consider the complete velocity equation (i.e., all substrates and products present). In that case an [A][P] denominator term is simplified using the alternate definition.

When equation IX-64 is rearranged to show either P or Q as the varied substrate, we obtain:

$$\boxed{\frac{v}{V_{\max}} = \frac{[\mathrm{P}]}{K_{m_\mathrm{P}}\left(1 + \dfrac{K_{iq}}{[\mathrm{Q}]}\right) + [\mathrm{P}]\left(1 + \dfrac{K_{m_\mathrm{Q}}}{[\mathrm{Q}]}\right)}} \tag{IX-65}$$

and

$$\boxed{\frac{v}{V_{\max}} = \frac{[\mathrm{Q}]}{K_{m_\mathrm{Q}}\left(1 + \dfrac{K_{iq}K_{m_\mathrm{P}}}{K_{m_\mathrm{Q}}[\mathrm{P}]}\right) + [\mathrm{Q}]\left(1 + \dfrac{K_{m_\mathrm{P}}}{[\mathrm{P}]}\right)}} \tag{IX-66}$$

When [P] is varied:

$$K_{m_{\mathrm{P}\,\mathrm{app}}} = K_{m_\mathrm{P}}\frac{\left(1 + \dfrac{K_{iq}}{[\mathrm{Q}]}\right)}{\left(1 + \dfrac{K_{m_\mathrm{Q}}}{[\mathrm{Q}]}\right)}, \qquad V_{\max_{\mathrm{app}}} = \frac{V_{\max}}{\left(1 + \dfrac{K_{m_\mathrm{Q}}}{[\mathrm{Q}]}\right)}$$

When [Q] is varied:

$$K_{m_{\mathrm{Q}app}} = K_{m_{\mathrm{Q}}} \frac{\left(1 + \dfrac{K_{iq}K_{m_{\mathrm{P}}}}{K_{m_{\mathrm{Q}}}[\mathrm{P}]}\right)}{\left(1 + \dfrac{K_{m_{\mathrm{P}}}}{[\mathrm{P}]}\right)}, \qquad V_{\max_{app}} = \frac{V_{\max}}{\left(1 + \dfrac{K_{m_{\mathrm{P}}}}{[\mathrm{P}]}\right)}$$

The reciprocal forms of equations IX-65 and IX-66 are:

$$\boxed{\frac{1}{v} = \frac{K_{m_{\mathrm{P}}}}{V_{\max}}\left(1 + \frac{K_{iq}}{[\mathrm{Q}]}\right)\frac{1}{[\mathrm{P}]} + \frac{1}{V_{\max}}\left(1 + \frac{K_{m_{\mathrm{Q}}}}{[\mathrm{Q}]}\right)} \qquad \text{(IX-67)}$$

$$\boxed{\frac{1}{v} = \frac{K_{m_{\mathrm{Q}}}}{V_{\max}}\left(1 + \frac{K_{iq}K_{m_{\mathrm{P}}}}{K_{m_{\mathrm{Q}}}[\mathrm{P}]}\right)\frac{1}{[\mathrm{Q}]} + \frac{1}{V_{\max}}\left(1 + \frac{K_{m_{\mathrm{P}}}}{[\mathrm{P}]}\right)} \qquad \text{(IX-68)}$$

The plots are shown in Figure IX-3*a* and Figure IX-4*a*. The coordinates of the intersection points are given by:

$$\frac{1}{[\mathrm{X}]} = -\frac{1}{K_{m_{\mathrm{X}}}}\left(\frac{K_{\mathrm{int}}}{K_{\mathrm{slope}}}\right) \qquad \text{and} \qquad \frac{1}{v} = \frac{1}{V_{\max}}\left(1 - \frac{K_{\mathrm{int}}}{K_{\mathrm{slope}}}\right)$$

where X is a substrate and K_{int} and K_{slope} are apparent intercept and slope constants. Thus for P as the varied substrate, the coordinates are:

$$\frac{1}{[\mathrm{P}]} = -\frac{1}{K_{m_{\mathrm{P}}}}\left(\frac{K_{m_{\mathrm{Q}}}}{K_{iq}}\right) \qquad \text{and} \qquad \frac{1}{v} = \frac{1}{V_{\max}}\left(1 - \frac{K_{m_{\mathrm{Q}}}}{K_{iq}}\right)$$

When [Q] is varied, $K_{\mathrm{int}} = K_{m_{\mathrm{P}}}$ and $K_{\mathrm{slope}} = K_{iq}K_{m_{\mathrm{P}}}/K_{m_{\mathrm{Q}}}$. The intersection coordinates are:

$$\frac{1}{[\mathrm{Q}]} = -\frac{1}{K_{m_{\mathrm{Q}}}}\left(\frac{K_{m_{\mathrm{P}}}K_{m_{\mathrm{Q}}}}{K_{iq}K_{m_{\mathrm{P}}}}\right) = -\frac{1}{K_{iq}}$$

and

$$\frac{1}{v} = \frac{1}{V_{\max}}\left(1 - \frac{K_{m_{\mathrm{P}}}K_{m_{\mathrm{Q}}}}{K_{iq}K_{m_{\mathrm{P}}}}\right) = \frac{1}{V_{\max}}\left(1 - \frac{K_{m_{\mathrm{Q}}}}{K_{iq}}\right)$$

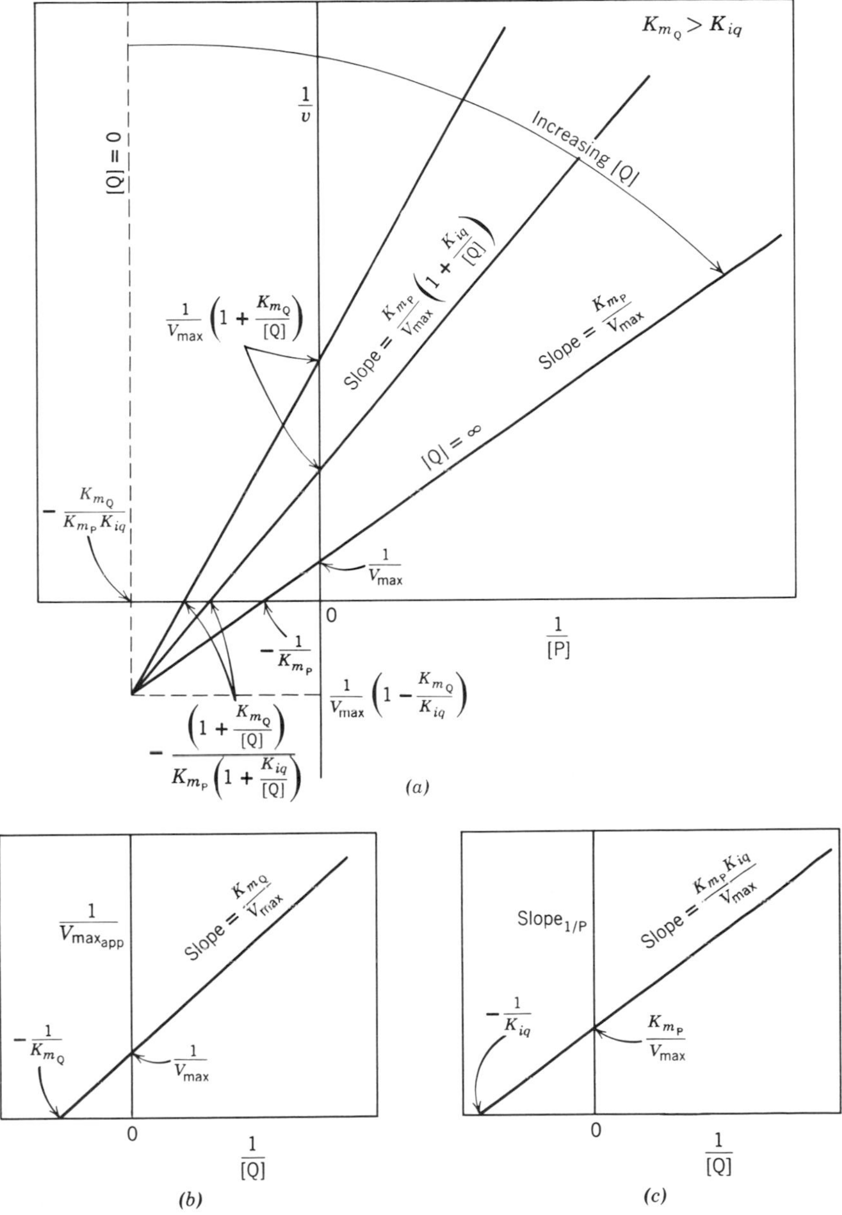

Fig. IX-3. Bi Uni system: P + Q ⟶ A. (*a*) $1/v$ versus $1/[P]$ in the presence of different fixed concentrations of cosubstrate Q. (*b*) $1/v$-axis intercept replot. (*c*) $Slope_{1/P}$ replot. The plots are shown for $K_{m_Q} > K_{iq}$.

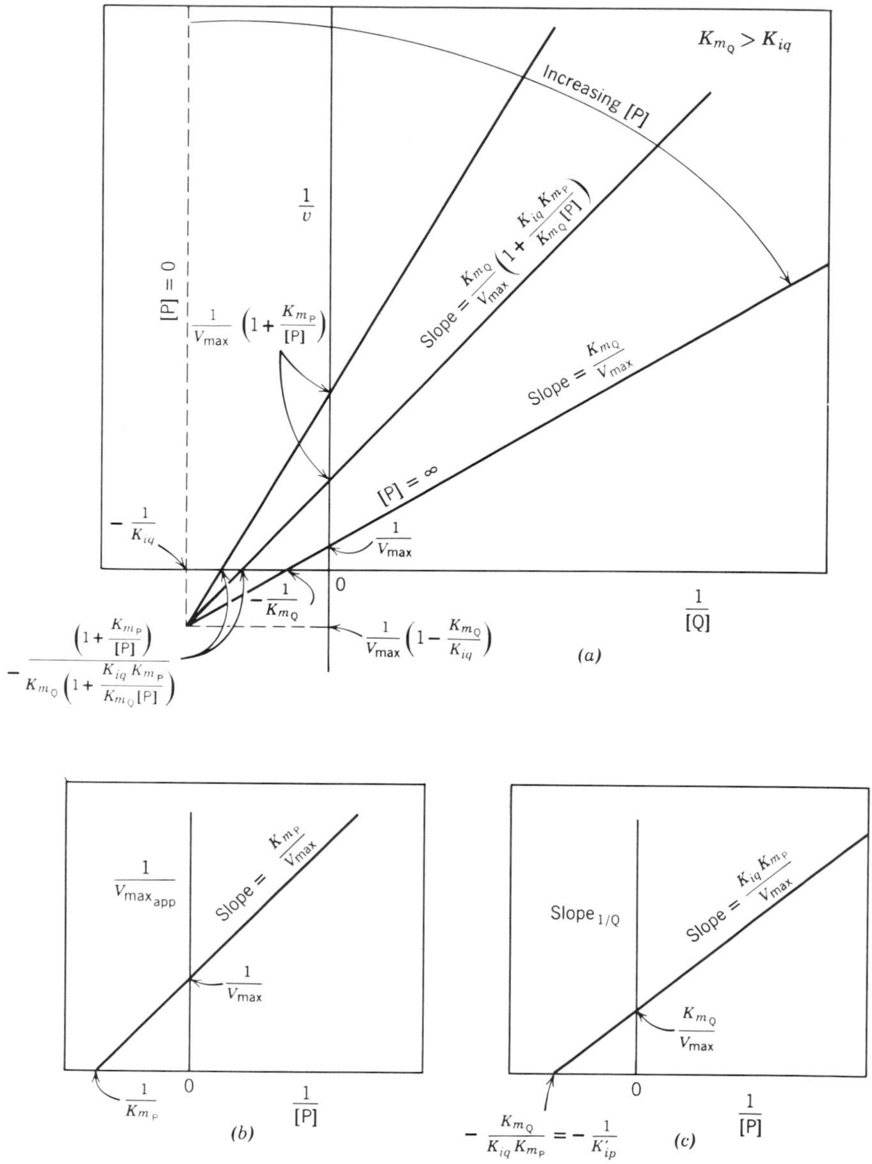

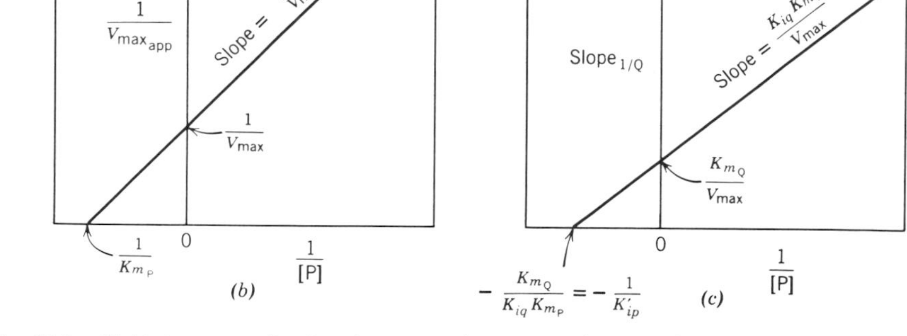

Fig. IX-4. Bi Uni system: P + Q ⟶ A. (*a*) $1/v$ versus $1/[Q]$ in the presence of different fixed concentrations of cosubstrate P. (*b*) $1/v$-axis intercept replot. (*c*) $Slope_{1/Q}$ replot.

The intersection points will be above or below the horizontal axis, depending on the relative values of K_{m_Q} and K_{iq}. If coincidentally, $K_{m_Q} = K_{iq}$, the family of curves will intersect on the horizontal axis. Replots of $1/v$-axis intercepts permit K_{m_Q} and K_{m_P} to be determined (Fig. IX-3*b* and Fig. IX-4*b*). Slope replots permit K_{iq} and $K_{iq}K_{m_P}/K_{m_Q}$ (i.e., K'_{ip}) to be determined (Fig. IX-3*c* and Fig. IX-4*c*). Equations IX-67 and IX-68 are identical in form. Consequently it is impossible to identify which substrate is P and which is Q from the initial velocity studies alone.

Complete Velocity Equation—Product Inhibition

P and Q can be identified from product inhibition studies. To derive the required equations, we apply the general rules for obtaining the complete velocity equation in terms of kinetic constants. Multiplying numerator and denominator of equation IX-60 by $(\text{num}_2)/(\text{Coef}_A)(\text{Coef}_{PQ})$, and the [P], [Q], and [P][Q] terms by $(\text{num}_1)/(\text{num}_1)$, we obtain:

$$v = \frac{V_{\max_f} V_{\max_r}[\mathrm{A}] - \dfrac{V_{\max_f} V_{\max_r}[\mathrm{P}][\mathrm{Q}]}{K_{eq}}}{V_{\max_r} K_{m_A} + V_{\max_r}[\mathrm{A}] + \dfrac{V_{\max_f} K_{m_Q}[\mathrm{P}]}{K_{eq}} + \dfrac{V_{\max_f} K_{m_P}[\mathrm{Q}]}{K_{eq}} + \dfrac{V_{\max_r}[\mathrm{A}][\mathrm{P}]}{K_{ip}} + \dfrac{V_{\max_f}[\mathrm{P}][\mathrm{Q}]}{K_{eq}}} \tag{IX-69}$$

where

$$K_{m_A} = \frac{\text{const}}{\text{Coef}_A}, \qquad K_{m_Q} = \frac{\text{Coef}_P}{\text{Coef}_{PQ}}, \qquad K_{m_P} = \frac{\text{Coef}_Q}{\text{Coef}_{PQ}}$$

$$K_{ip} = \frac{\text{Coef}_{AP}}{\text{Coef}_A}, \qquad V_{\max_f} = \frac{\text{num}_1}{\text{Coef}_A}, \qquad V_{\max_r} = \frac{\text{num}_2}{\text{Coef}_{PQ}}$$

$$K_{eq} = \frac{\text{num}_1}{\text{num}_2}$$

Note that K_{ip} is not the same as K'_{ip} defined earlier. Other constants, not needed for the simplification above, but useful elsewhere, are:

$$K_{iq} = \frac{\text{const}}{\text{Coef}_Q} = \frac{k_3}{k_{-3}}, \qquad K_{ia} = \frac{\text{Coef}_P}{\text{Coef}_{AP}} = \frac{k_{-1}}{k_1}$$

To put the equation in a form suitable for consideration of the forward velocity we divide numerator and denominator by $V_{\max_r}$. This gives:

$$v = \frac{V_{\max_f}\left([\mathrm{A}] - \dfrac{[\mathrm{P}][\mathrm{Q}]}{K_{\mathrm{eq}}}\right)}{K_{m_\mathrm{A}} + [\mathrm{A}] + \dfrac{V_{\max_f}K_{m_\mathrm{Q}}[\mathrm{P}]}{V_{\max_r}K_{\mathrm{eq}}} + \dfrac{V_{\max_f}K_{m_\mathrm{P}}[\mathrm{Q}]}{V_{\max_r}K_{\mathrm{eq}}} + \dfrac{[\mathrm{A}][\mathrm{P}]}{K_{ip}} + \dfrac{V_{\max_f}[\mathrm{P}][\mathrm{Q}]}{V_{\max_r}K_{\mathrm{eq}}}}$$

When both $[\mathrm{P}]=0$ and $[\mathrm{Q}]=0$, we obtain the usual Henri-Michaelis-Menten equation for the forward velocity. To obtain an equation suitable for consideration of the reverse reaction, the numerator and denominator of equation IX-69 are multiplied by $K_{\mathrm{eq}}/V_{\max_f}$. We obtain:

$$\boxed{v = \frac{V_{\max_r}\left([\mathrm{P}][\mathrm{Q}] - \dfrac{[\mathrm{A}]}{K'_{\mathrm{eq}}}\right)}{\dfrac{V_{\max_r}K_{m_\mathrm{A}}K_{\mathrm{eq}}}{V_{\max_f}} + \dfrac{V_{\max_r}[\mathrm{A}]K_{\mathrm{eq}}}{V_{\max_f}} + K_{m_\mathrm{Q}}[\mathrm{P}] + K_{m_\mathrm{P}}[\mathrm{Q}] + \dfrac{V_{\max_r}[\mathrm{A}][\mathrm{P}]K_{\mathrm{eq}}}{V_{\max_f}K_{ip}} + [\mathrm{P}][\mathrm{Q}]}} \tag{IX-70}$$

where $K'_{\mathrm{eq}} = 1/K_{\mathrm{eq}}$. The constant term in the denominator is used to derive the Haldane equation which we will then use to eliminate K_{eq}. Writing the constant term in coefficient form:

$$\frac{V_{\max_r}K_{m_\mathrm{A}}K_{\mathrm{eq}}}{V_{\max_f}} = \frac{(\mathrm{num}_2)(\mathrm{const})(\mathrm{num}_1)(\mathrm{Coef_A})}{(\mathrm{Coef_{PQ}})(\mathrm{Coef_A})(\mathrm{num}_2)(\mathrm{num}_1)} = \frac{\mathrm{const}}{\mathrm{Coef_{PQ}}}$$

Substituting $\mathrm{Coef_Q}/K_{m_\mathrm{P}}$ for $\mathrm{Coef_{PQ}}$:

$$\frac{V_{\max_r}K_{m_\mathrm{A}}K_{\mathrm{eq}}}{V_{\max_f}} = \frac{(\mathrm{const})K_{m_\mathrm{P}}}{(\mathrm{Coef_Q})} = K_{iq}K_{m_\mathrm{P}}$$

$$\therefore \quad \boxed{K_{\mathrm{eq}} = \frac{V_{\max_f}K_{iq}K_{m_\mathrm{P}}}{V_{\max_r}K_{m_\mathrm{A}}}} \tag{IX-71}$$

Another Haldane equation can be obtained by rewriting equation IX-71 in coefficient form:

$$K_{eq}=\frac{V_{max_f}K_{iq}K_{m_P}}{V_{max_r}K_{m_A}}=\frac{V_{max_f}(\text{const})(\text{Coef}_Q)(\text{Coef}_A)}{V_{max_r}(\text{Coef}_Q)(\text{Coef}_{PQ})(\text{const})}=\frac{V_{max_f}(\text{Coef}_A)}{V_{max_r}(\text{Coef}_{PQ})}$$

Substituting $K_{ip}(\text{Coef}_{AP})$ for Coef_A, and Coef_P/K_{m_Q} for Coef_{PQ}:

$$\boxed{K_{eq}=\frac{V_{max_f}K_{ip}K_{m_Q}}{V_{max_r}K_{ia}}} \qquad \text{(IX-72)}$$

From the two Haldane equations, we see that:

$$K_{iq}K_{m_P}K_{ia}=K_{ip}K_{m_Q}K_{m_A}$$

To determine the effect of any one product as an inhibitor, we set the concentration of the other product equal to zero and eliminate K_{eq} with a Haldane equation. Thus the velocity equation for P as an inhibitor of the forward velocity in the absence of Q is:

$$\boxed{\frac{v}{V_{max}}=\frac{[\text{A}]}{K_{m_A}\left(1+\dfrac{K_{m_Q}[\text{P}]}{K_{iq}K_{m_P}}\right)+[\text{A}]\left(1+\dfrac{[\text{P}]}{K_{ip}}\right)}} \qquad \text{(IX-73)}$$

Thus P acts as a mixed-type inhibitor with respect to A. The reciprocal plot is shown in Figure IX-5*a*. The intersection point has coordinates of:

$$\frac{1}{[\text{A}]}=-\frac{1}{K_{m_A}}\left(\frac{K_{i_{slope}}}{K_{i_{int}}}\right)=-\frac{K_{iq}K_{m_P}}{K_{m_A}K_{m_Q}K_{ip}}=-\frac{1}{K_{ia}}$$

$$\frac{1}{v}=\frac{1}{V_{max}}\left(1-\frac{K_{i_{slope}}}{K_{i_{int}}}\right)=\frac{1}{V_{max}}\left(1-\frac{K_{iq}K_{m_P}}{K_{m_Q}K_{ip}}\right)=\frac{1}{V_{max}}\left(1-\frac{K_{m_A}}{K_{ia}}\right)$$

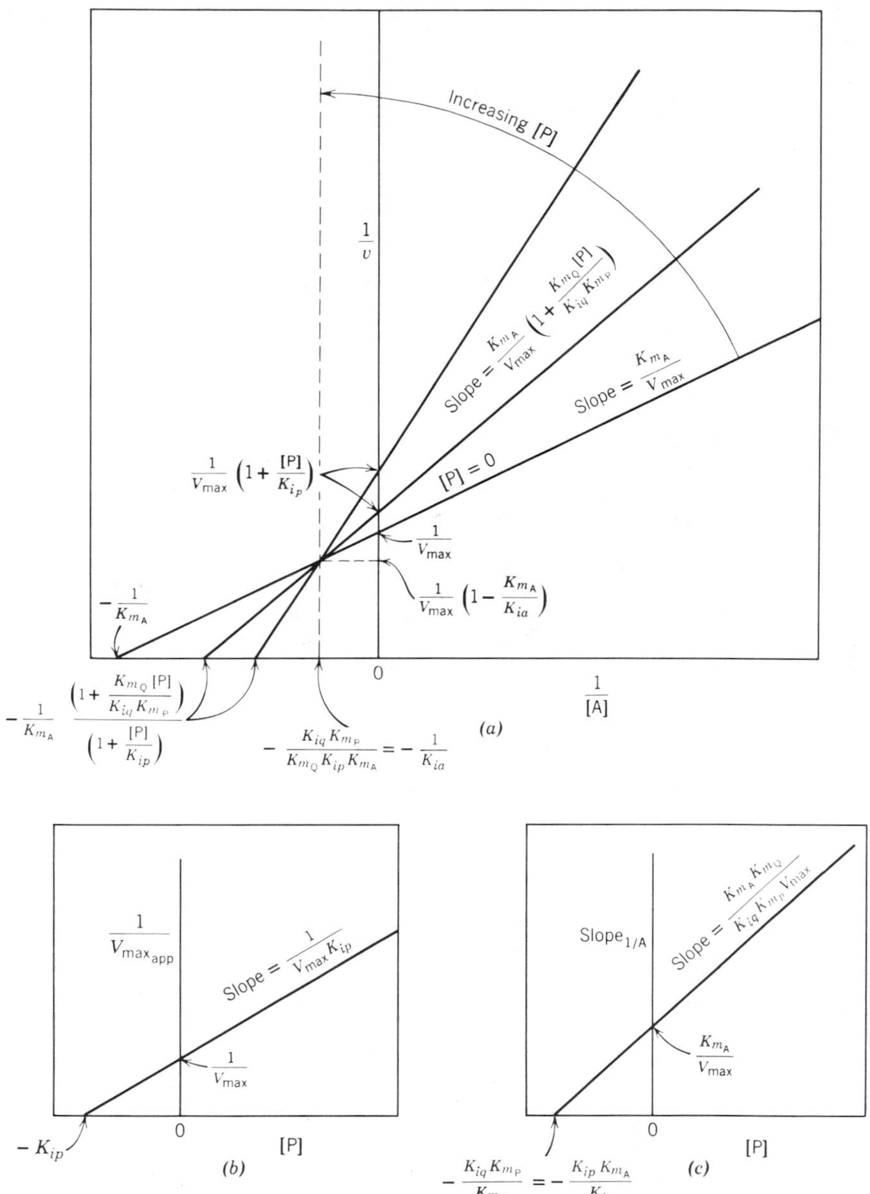

Fig. IX-5. Product inhibition by P in a Uni Bi system: A ⟶ P + Q. (*a*) $1/v$ versus $1/[A]$ in the presence of different fixed concentrations of P; Q is absent. (*b*) $1/v$-axis intercept replot. (*c*) *Slope*$_{1/A}$ replot.

When Q is the product inhibitor, the velocity equation is:

$$\frac{v}{V_{\max}} = \frac{[A]}{K_{m_A}\left(1+\frac{[Q]}{K_{iq}}\right)+[A]} \qquad \text{(IX-74)}$$

Thus, Q acts as a competitive inhibitor with respect to A (Fig. IX-6*a*). The different inhibition patterns permit P and Q to be identified. The product inhibition studies do not allow K_{m_P} and K_{m_Q} to be determined, but their ratio can be calculated from the relationship:

$$\frac{K_{m_Q}}{K_{m_P}} = \frac{K_{iq}K_{ia}}{K_{ip}K_{m_A}} \qquad \text{(IX-75)}$$

Product inhibition studies of the P+Q→A Bi Uni reverse reaction can be complicated by the reaction in the direction A→P+Q. (Normally, we don't have to worry about the reverse reaction because one or more of the products is omitted from the reaction mixture.) If v is measured as the rate of appearance of radioactivity from Q in A, then the measurements present no problem. This is, in effect, an isotope exchange study (Chapter Ten) but the initial forward velocity and the isotope exchange velocity are identical. If the assay is based on a P*-A exchange, then nonlinear reciprocal plots may be obtained when Q is varied at a constant [P] and different fixed concentrations of A. When [P] is varied at a constant [Q], plots may be linear, but A will act as a partial inhibitor. (The equation for this situation can be derived using the procedures outlined in Chapter Ten.) For the moment, let us assume that the assay is based on the incorporation of label from Q into A or that the reaction A→P+Q is negligible at the A concentrations used (i.e., $[A]/K'_{eq}$ is extremely small). The velocity equations for varied [P] and [Q] are:

$$\frac{v}{V_{\max}} = \frac{[P]}{K_{m_P}\left(1+\frac{K_{iq}}{[Q]}+\frac{K_{iq}[A]}{K_{m_A}[Q]}\right)+[P]\left(1+\frac{K_{m_Q}}{[Q]}+\frac{K_{m_Q}[A]}{K_{ia}[Q]}\right)} \qquad \text{(IX-76)}$$

$$\frac{v}{V_{\max}} = \frac{[Q]}{K_{m_Q}\left(1+\frac{K_{m_A}K_{ip}}{K_{ia}[P]}+\frac{K_{ip}[A]}{K_{ia}[P]}+\frac{[A]}{K_{ia}}\right)+[Q]\left(1+\frac{K_{m_P}}{[P]}\right)} \qquad \text{(IX-77)}$$

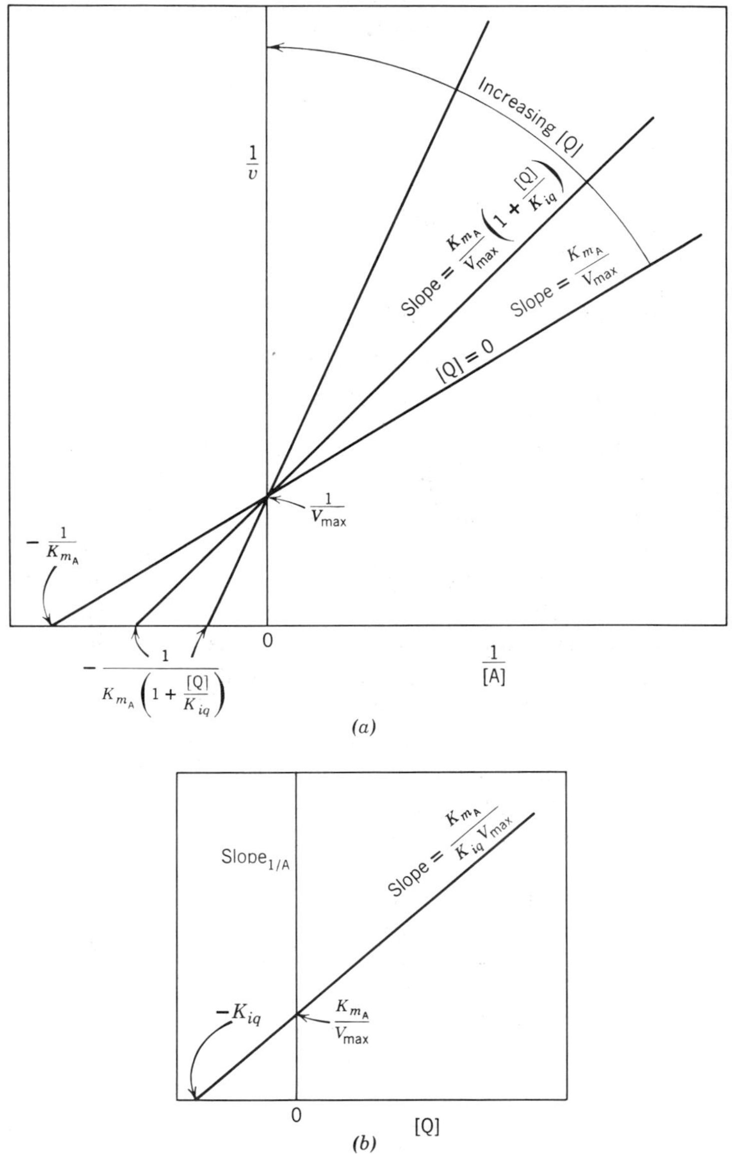

Fig. IX-6. Product inhibition by Q in a Uni Bi system: $A \longrightarrow P + Q$. (*a*) $1/v$ versus $1/[A]$ in the presence of different fixed concentrations of Q; P is absent. (*b*) $Slope_{1/A}$ replot.

We see that A acts as a mixed-type inhibitor with respect to P at unsaturating Q concentrations. The inhibition is overcome by saturating [Q] (i.e., the slope and intercept factors reduce to unity when the terms containing [Q] go to zero at infinitely high [Q]). On the other hand, A is a competitive inhibitor with respect to Q at all P concentrations. Saturating [P] changes $V_{\max_{app}}$ to the true $V_{\max}$, but the apparent K_{m_Q} still depends on [A] (i.e., the slope factor still contains an $[A]/K_{ia}$ term at saturating [P]). The various constants can be obtained from suitable replots. For example, the $1/v$-axis intercept of the $1/v$ versus $1/[P]$ plot at a constant unsaturating [Q] and different fixed concentrations of A is given by:

$$\frac{1}{V_{\max_{app}}} = \frac{K_{m_Q}}{V_{\max} K_{ia} [Q]} [A] + \frac{1}{V_{\max}} \left(1 + \frac{K_{m_Q}}{[Q]}\right) \qquad \text{(IX-78)}$$

The slope is given by:

$$slope_{1/P} = \frac{K_{m_P} K_{iq}}{V_{\max} K_{m_A} [Q]} [A] + \frac{K_{m_P}}{V_{\max}} \left(1 + \frac{K_{iq}}{[Q]}\right) \qquad \text{(IX-79)}$$

The reciprocal plot and replots are shown in Figure IX-7. The constants K_{m_Q}, K_{m_P}, K_{iq}, and K_{ip} can all be determined from control plots in the absence of A. The replots allow K_{m_A} and K_{ia} to be determined. The reciprocal plot and slope replot for Q as the varied substrate is shown in Figure IX-8. If [P] is saturating, the horizontal intercept of the $slope_{1/Q}$ replot gives $-K_{ia}$. The constants K_{ia} and K_{iq} represent true dissociation constants:

$$K_{ia} = \frac{\text{Coef}_P}{\text{Coef}_{AP}} = \frac{k_{-1}}{k_1}, \qquad K_{iq} = \frac{\text{const}}{\text{Coef}_Q} = \frac{k_3}{k_{-3}}$$

Thus equilibrium binding studies should yield a dissociation constant for Q equivalent to K_{iq} determined kinetically. Binding studies with A may be impossible because the reaction to P+Q will take place.

Calculation of Rate Constants

The individual rate constants can be calculated from the kinetic constants as shown below:

$$k_1 = \frac{V_{\max_r}}{K_{ia}[E]_t}, \qquad k_{-1} = \frac{V_{\max_r}}{[E]_t}, \qquad \frac{1}{k_2} = \frac{[E]_t}{V_{\max_f}} - \frac{1}{k_3}$$

$$k_{-2} = \frac{k_{-1} + k_2}{K_{m_P}}, \qquad k_3 = \frac{K_{iq} V_{\max_r}}{K_{m_Q}[E]_t}, \qquad k_{-3} = \frac{V_{\max_r}}{K_{m_Q}[E]_t}$$

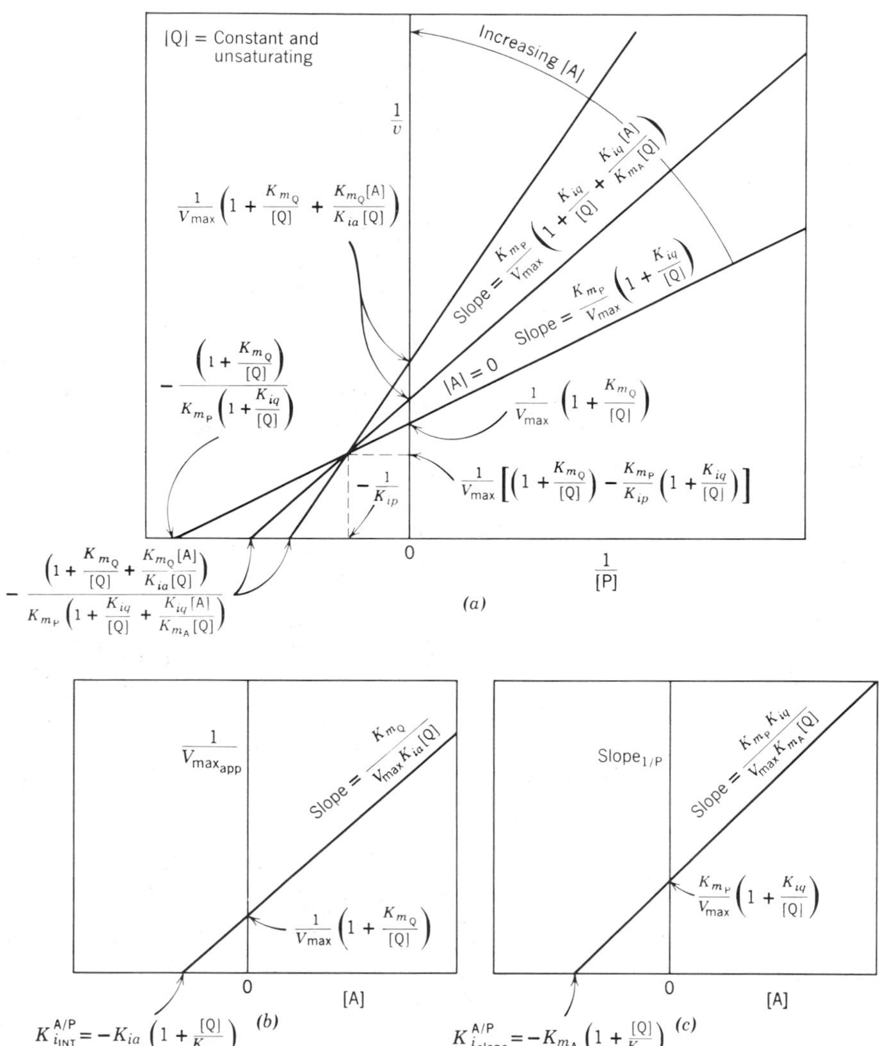

Fig. IX-7. Product inhibition in a Bi Uni system: P + Q ⟶ A. It is assumed that the reverse reaction does not occur to any significant extent (see text). (*a*) $1/v$ versus 1/[P] at different fixed concentrations of A and a constant, unsaturating Q. (*b*) $1/v$-axis intercept replot. (*c*) $Slope_{1/P}$ replot.

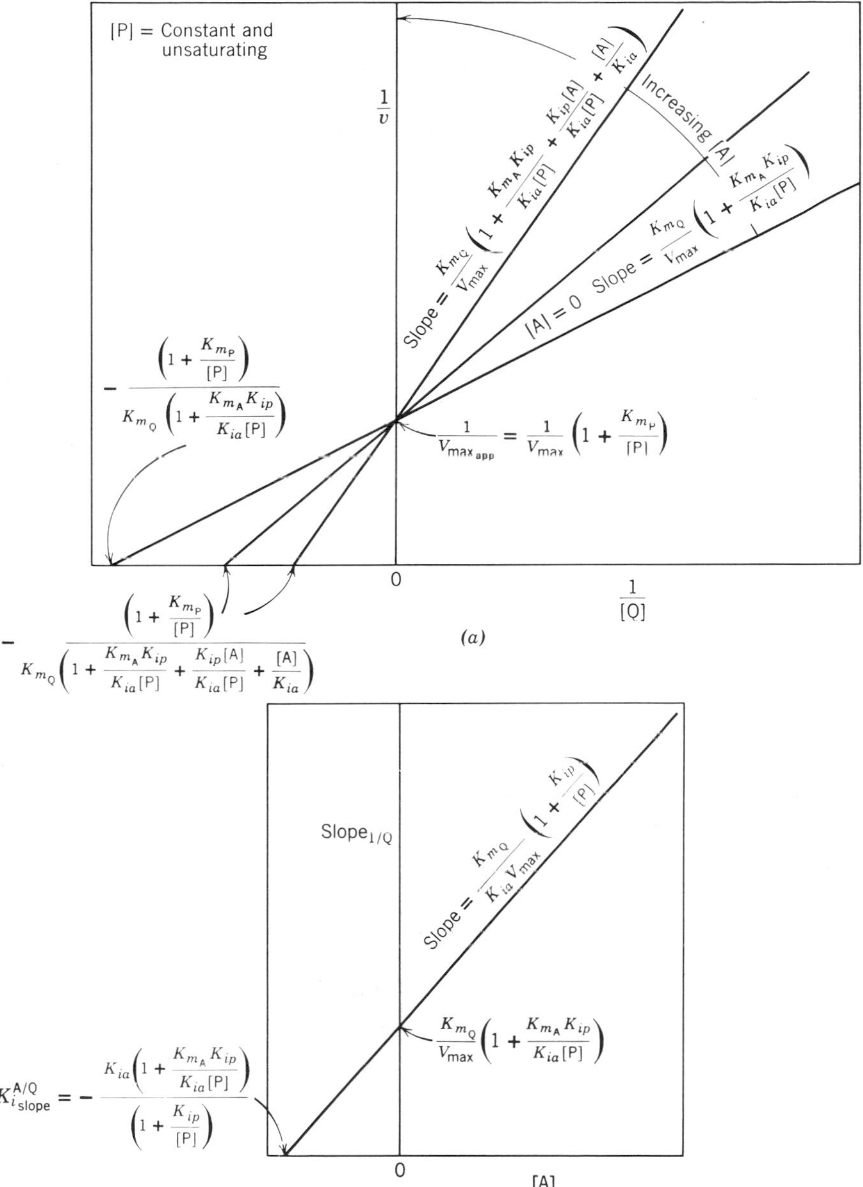

Fig. IX-8. Product inhibition in a Bi Uni system: P + Q ⟶ A. (*a*) $1/v$ versus $1/[Q]$ at different fixed concentrations of A and a constant, unsaturating concentration of P. (*b*) $Slope_{1/Q}$ replot.

Distribution Equations

$$\frac{[E]}{[E]_t} = \frac{K_{m_A} V_r + \dfrac{K_{m_Q} V_f [P]}{K_{eq}}}{\text{denominator of velocity equation}} \qquad \text{(IX-80)}$$

$$\frac{[EA + EPQ]}{[E]_t} = \frac{\left(V_r - \dfrac{K_{m_Q} V_f}{K_{iq}}\right)[A] + \dfrac{V_r [A][P]}{K_{ip}} + \dfrac{V_f [P][Q]}{K_{eq}}}{\text{denominator of velocity equation}} \qquad \text{(IX-81)}$$

$$\frac{[EQ]}{[E]_t} = \frac{\dfrac{K_{m_P} V_f [Q]}{K_{eq}} + \dfrac{K_{m_Q} V_f [Q]}{K_{iq}}}{\text{denominator of velocity equation}} \qquad \text{(IX-82)}$$

Effect of Isomerizations

The validity of the distribution equations and the calculations of rate constants depends on the assumption that transitory complexes do not isomerize. For example, if in the Ordered Uni Bi system, we show the isomerization of the central complexes as $EA \rightleftharpoons EPQ$ and incorporate the isomerization step into the derivation of the velocity equation, then the calculations of k_1, k_{-1}, k_2, and k_{-2} (as originally defined) are invalid. If EQ isomerizes to EQ′, then all the rate constant calculations are invalid, and among the distribution equations only $[E]/[E]_t$ may be calculated. As noted earlier, isomerization of a transitory complex has no effect on the form of the velocity equation.

F. ORDERED BI BI SYSTEM

An ordered reaction that is bireactant in both directions is:

$$E + A \underset{k_{-1}}{\overset{k_1}{\rightleftharpoons}} EA \qquad EQ \underset{k_{-4}}{\overset{k_4}{\rightleftharpoons}} E + Q$$

$$EA + B \underset{k_{-2}}{\overset{k_2}{\rightleftharpoons}} (EAB \underset{k_{-p}}{\overset{k_p}{\rightleftharpoons}} EPQ) \underset{k_{3}}{\overset{k_{-3}}{\rightleftharpoons}} EQ + P$$

or

$$\begin{array}{ccccccccc} & A & & B & & & P & & Q & \\ & k_1 \downarrow k_{-1} & & k_2 \downarrow k_{-2} & & & k_3 \uparrow k_{-3} & & k_4 \uparrow k_{-4} & \\ \hline E & & EA & & (EAB \rightleftharpoons EPQ) & & & EQ & & E \end{array}$$

The basic King-Altman figure has four corners:

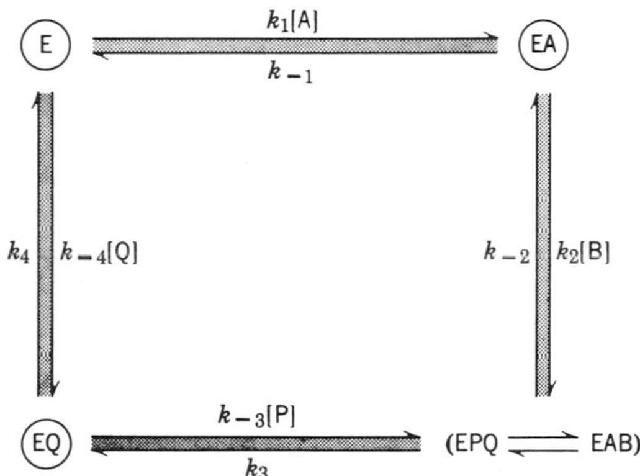

The distribution of enzyme species is obtained from four three-lined interconversion patterns.

$$\frac{[E]}{[E]_t} = \frac{k_4k_{-1}k_{-2} + k_{-3}[P]k_{-2}k_{-1} + k_2[B]k_3k_4 + k_{-1}k_3k_4}{\text{denominator}} \tag{IX-83}$$

$$\frac{[EA]}{[E]_t} = \frac{k_4k_1[A]k_{-2} + k_1[A]k_{-3}[P]k_{-2} + k_{-4}[Q]k_{-3}[P]k_{-2} + k_3k_4k_1[A]}{\text{denominator}} \tag{IX-84}$$

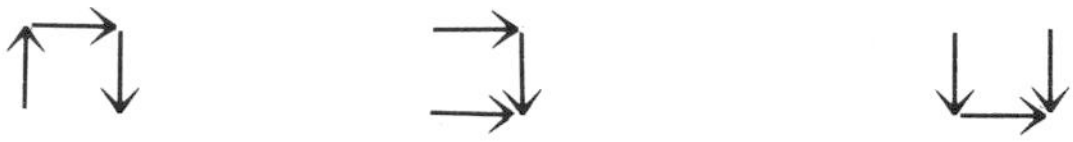

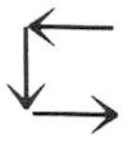

$$\frac{[\mathrm{EAB}+\mathrm{EPQ}]}{[\mathrm{E}]_t}=\frac{k_4k_1[\mathrm{A}]k_2[\mathrm{B}]+k_1[\mathrm{A}]k_2[\mathrm{B}]k_{-3}[\mathrm{P}]+k_{-4}[\mathrm{Q}]k_{-3}[\mathrm{P}]k_2[\mathrm{B}]+k_{-1}k_{-4}[\mathrm{Q}]k_{-3}[\mathrm{P}]}{\text{denominator}}\tag{IX-85}$$

$$\frac{[\mathrm{EQ}]}{[\mathrm{E}]_t}=\frac{k_{-2}k_{-1}k_{-4}[\mathrm{Q}]+k_1[\mathrm{A}]k_2[\mathrm{B}]k_3+k_{-4}[\mathrm{Q}]k_2[\mathrm{B}]k_3+k_{-1}k_{-4}[\mathrm{Q}]k_3}{\text{denominator}}\tag{IX-86}$$

Letting $v=k_1[\mathrm{E}][\mathrm{A}]-k_{-1}[\mathrm{EA}]$ (or any other difference between the forward and reverse velocities of a given step), we obtain after grouping denominator terms:

$$\begin{aligned}\frac{v}{[\mathrm{E}]_t}=\;&\frac{k_1k_2k_3k_4[\mathrm{A}][\mathrm{B}]-k_{-1}k_{-2}k_{-3}k_{-4}[\mathrm{P}][\mathrm{Q}]}{k_{-1}k_4(k_{-2}+k_3)+k_1k_4(k_{-2}+k_3)[\mathrm{A}]+k_2k_3k_4[\mathrm{B}]+k_{-1}k_{-2}k_{-3}[\mathrm{P}]}\\&+k_{-1}k_{-4}(k_{-2}+k_3)[\mathrm{Q}]+k_1k_2(k_3+k_4)[\mathrm{A}][\mathrm{B}]+k_1k_{-2}k_{-3}[\mathrm{A}][\mathrm{P}]\\&+k_{-3}k_{-4}(k_{-1}+k_{-2})[\mathrm{P}][\mathrm{Q}]+k_2k_3k_{-4}[\mathrm{B}][\mathrm{Q}]+k_1k_2k_{-3}[\mathrm{A}][\mathrm{B}][\mathrm{P}]\\&+k_2k_{-3}k_{-4}[\mathrm{B}][\mathrm{P}][\mathrm{Q}]\end{aligned}\tag{IX-87}$$

The numerator and denominator of the equation above are multiplied by $(\mathrm{num}_2)/(\mathrm{Coef}_{\mathrm{AB}})(\mathrm{Coef}_{\mathrm{PQ}})$. After this multiplication, the AB numerator term and the A, B, AB, and ABP denominator terms can be defined directly in terms of kinetic constants. The PQ numerator term and the P, Q, PQ, and BPQ denominator terms can be reduced to kinetic constants (and K_{eq}) after additional multiplication by $\mathrm{num}_1/\mathrm{num}_1$. The AP term simplifies after multiplication by $\mathrm{num}_1/\mathrm{num}_1$ and substitution of $\mathrm{Coef}_{\mathrm{P}}/K_{ia}$ for $\mathrm{Coef}_{\mathrm{AP}}$. The BQ term simplifies after substitution of $\mathrm{Coef}_{\mathrm{B}}/K_{m_{\mathrm{A}}}$ for $\mathrm{Coef}_{\mathrm{AB}}$. The constant term in the denominator is simplified by substituting $\mathrm{Coef}_{\mathrm{A}}/K_{m_{\mathrm{B}}}$ for $\mathrm{Coef}_{\mathrm{AB}}$. We did not use $\mathrm{Coef}_{\mathrm{B}}/K_{m_{\mathrm{A}}}$ here because this would have converted the constant term to $V_{\max_r}(\mathrm{const})K_{m_{\mathrm{A}}}/\mathrm{Coef}_{\mathrm{B}}$. While $(\mathrm{const})/\mathrm{Coef}_{\mathrm{B}}$ is a good definition for K_{ib}, it is not identical to K_{ib} as $\text{Coef}_{PQ}/\text{Coef}_{BPQ}$ which is used in simplifying the BPQ term. The equation can now be written in terms of kinetic constants and K_{eq}. For simplicity, we will indicate the forward and reverse maximal velocities as V_f and V_r, respectively.

$$v=\frac{V_f V_r\left([A][B]-\dfrac{[P][Q]}{K_{eq}}\right)}{\begin{array}{l} V_r K_{ia} K_{m_B} + V_r K_{m_B}[A] + V_r K_{m_A}[B] + \dfrac{V_f K_{m_Q}[P]}{K_{eq}} + \dfrac{V_f K_{m_P}[Q]}{K_{eq}} + V_r[A][B] \\ \quad + \dfrac{V_f K_{m_Q}[A][P]}{K_{eq} K_{ia}} + \dfrac{V_f [P][Q]}{K_{eq}} + \dfrac{V_r K_{m_A}[B][Q]}{K_{iq}} + \dfrac{V_r [A][B][P]}{K_{ip}} \\ \quad + \dfrac{V_f [B][P][Q]}{K_{ib} K_{eq}} \end{array}} \qquad \text{(IX-88)}$$

where

$$K_{m_A}=\frac{\text{Coef}_B}{\text{Coef}_{AB}}, \qquad K_{m_B}=\frac{\text{Coef}_A}{\text{Coef}_{AB}}, \qquad K_{ia}=\frac{\text{Coef}_P}{\text{Coef}_{AP}}=\frac{\text{const}}{\text{Coef}_A}=\frac{k_{-1}}{k_1}$$

$$K_{ib}=\frac{\text{Coef}_{PQ}}{\text{Coef}_{BPQ}}=\frac{k_{-1}+k_{-2}}{k_2}, \qquad K_{m_P}=\frac{\text{Coef}_Q}{\text{Coef}_{PQ}}, \qquad K_{m_Q}=\frac{\text{Coef}_P}{\text{Coef}_{PQ}}$$

$$K_{ip}=\frac{\text{Coef}_{AB}}{\text{Coef}_{ABP}}=\frac{k_3+k_4}{k_{-3}}, \qquad K_{iq}=\frac{\text{Coef}_B}{\text{Coef}_{BQ}}=\frac{\text{const}}{\text{Coef}_Q}=\frac{k_4}{k_{-4}}$$

$$V_f=\frac{\text{num}_1}{\text{Coef}_{AB}}, \qquad V_r=\frac{\text{num}_2}{\text{Coef}_{PQ}}, \qquad K_{eq}=\frac{\text{num}_1}{\text{num}_2}$$

Note that K_{ib} and K_{ip} are not the dissociation constants for B and P, respectively, from the central complexes. The dissociation constants can be expressed in terms of the other kinetic constants (Ainslie and Cleland, 1972):

$$K_B=\frac{k_{-2}}{k_2}=\frac{K_{ib}K_{m_A}V_{\max_r}}{K_{ia}V_{\max_f}}, \qquad K_P=\frac{k_3}{k_{-3}}=\frac{K_{ip}K_{m_Q}V_{\max_f}}{K_{iq}V_{\max_r}} \qquad \text{(IX-88a)}$$

By way of comparison, Table IX-2 gives the definitions of the kinetic constants in terms of rate constants where the interconversion of EAB and EPQ is taken into account ($\text{EAB} \underset{k_{-p}}{\overset{k_p}{\rightleftharpoons}} \text{EPQ}$) (Plapp, 1973).

Table IX-2 Definition of Kinetic Constants for an Ordered Bi Bi System

$K_{ia} = \dfrac{k_{-1}}{k_1}$	$K_{iq} = \dfrac{k_4}{k_{-4}}$
$K_{m_A} = \dfrac{k_3k_4k_p}{k_1(k_3k_4 + k_3k_p + k_4k_p + k_4k_{-p})}$	$K_{m_Q} = \dfrac{k_{-1}k_{-2}k_{-p}}{k_{-4}(k_{-1}k_{-2} + k_{-1}k_p + k_{-1}k_{-p} + k_{-2}k_{-p}}$
$K_{ib} = \dfrac{k_{-1}k_{-2} + k_{-1}k_p + k_{-1}k_{-p} + k_{-2}k_{-p}}{k_2(k_p + k_{-p})}$	$K_{ip} = \dfrac{k_3k_4 + k_3k_p + k_4k_p + k_4k_{-p}}{k_{-3}(k_p + k_{-p})}$
$K_{m_B} = \dfrac{k_4(k_{-2}k_3 + k_{-2}k_{-p} + k_3k_p)}{k_2(k_3k_4 + k_3k_p + k_4k_p + k_4k_{-p})}$	$K_{m_P} = \dfrac{k_{-1}(k_{-2}k_3 + k_{-2}k_{-p} + k_3k_p)}{k_{-3}(k_{-1}k_{-2} + k_{-1}k_p + k_{-1}k_{-p} + k_{-2}k_{-p}}$
$\dfrac{V_{\max_f}}{[\mathrm{E}]_t} = \dfrac{k_3k_4k_p}{k_3k_4 + k_3k_p + k_4k_p + k_4k_{-p}}$	$\dfrac{V_{\max_r}}{[\mathrm{E}]_t} = \dfrac{k_{-1}k_{-2}k_{-p}}{k_{-1}k_{-2} + k_{-1}k_p + k_{-1}k_{-p} + k_{-2}k_{-p}}$

Initial Forward Velocity in the Absence of Products

In the absence of products, the velocity equation is given (after dividing out V_r) by:

$$\frac{v}{V_{\max}} = \frac{[\mathrm{A}][\mathrm{B}]}{K_{ia}K_{m_B} + K_{m_B}[\mathrm{A}] + K_{m_A}[\mathrm{B}] + [\mathrm{A}][\mathrm{B}]} \quad \text{(IX-89)}$$

The equation is identical to that for the Rapid Equilibrium Random Bi Bi system. If k_3 is very small compared to the other rate constants, K_{m_A}, which is given by:

$$K_{m_A} = \frac{\mathrm{Coef_B}}{\mathrm{Coef_{AB}}} = \frac{k_2k_3k_4}{k_1k_2(k_4 + k_3)}$$

reduces to zero. In this case, the $K_{m_A}[\mathrm{B}]$ term is absent and the equation reduces to that shown earlier for the Rapid Equilibrium Ordered Bi Bi system (Chapter Six) with $K_{m_B} \cong K_\mathrm{B} = k_{-2}/k_2$.

When [A] is varied:

$$\frac{v}{V_{\max}} = \frac{[\mathrm{A}]}{K_{m_A}\left(1 + \dfrac{K_{ia}K_{m_B}}{K_{m_A}[\mathrm{B}]}\right) + [\mathrm{A}]\left(1 + \dfrac{K_{m_B}}{[\mathrm{B}]}\right)} \quad \text{(IX-90)}$$

$$\frac{1}{v} = \frac{K_{m_A}}{V_{\max}}\left(1 + \frac{K_{ia}K_{m_B}}{K_{m_A}[\mathrm{B}]}\right)\frac{1}{[\mathrm{A}]} + \frac{1}{V_{\max}}\left(1 + \frac{K_{m_B}}{[\mathrm{B}]}\right) \quad \text{(IX-91)}$$

When [B] is varied:

$$\frac{v}{V_{max}} = \frac{[B]}{K_{m_B}\left(1+\frac{K_{ia}}{[A]}\right)+[B]\left(1+\frac{K_{m_A}}{[A]}\right)} \quad \text{(IX-92)}$$

$$\frac{1}{v} = \frac{K_{m_B}}{V_{max}}\left(1+\frac{K_{ia}}{[A]}\right)\frac{1}{[B]}+\frac{1}{V_{max}}\left(1+\frac{K_{m_A}}{[A]}\right) \quad \text{(IX-93)}$$

The reciprocal plots and replots (Fig. IX-9 and Fig. IX-10) are identical to those shown earlier for the Bi Uni system. The reciprocal plots may intersect above, on, or below the horizontal axis, depending on the ratio of K_{m_A}/K_{ia}. If K_{ia} is very small compared to K_{m_A}, the slopes of the plots become insensitive to changes in the concentration of the cosubstrate and the family of plots will be essentially parallel.

Other Methods of Plotting Data

Although the Lineweaver-Burk double reciprocal plot is the most commonly used linear plot for analyzing initial velocity data, any of the other linear plots described in Chapter Four will serve as well. Figure IX-11 to IX-15 show the three other types of linear plots for a steady-state ordered bireactant system. Note that the [S]/v versus [S] plots (Figs. IX-11 and IX-12) are quite similar to the 1/v versus 1/[S] plots (the slopes and intercepts are interchanged). The [S]/v coordinate of the intersection point (above, on, or below the [S]-axis) depends on whether K_{ia} is smaller than, equal to, or greater than K_{m_A}. The velocity equations are:

$$\frac{[A]}{v} = \frac{\left(1+\frac{K_{m_B}}{[B]}\right)}{V_{max}}[A]+\frac{K_{m_A}}{V_{max}}\left(1+\frac{K_{ia}K_{m_B}}{K_{m_A}[B]}\right) \quad \text{(IX-94)}$$

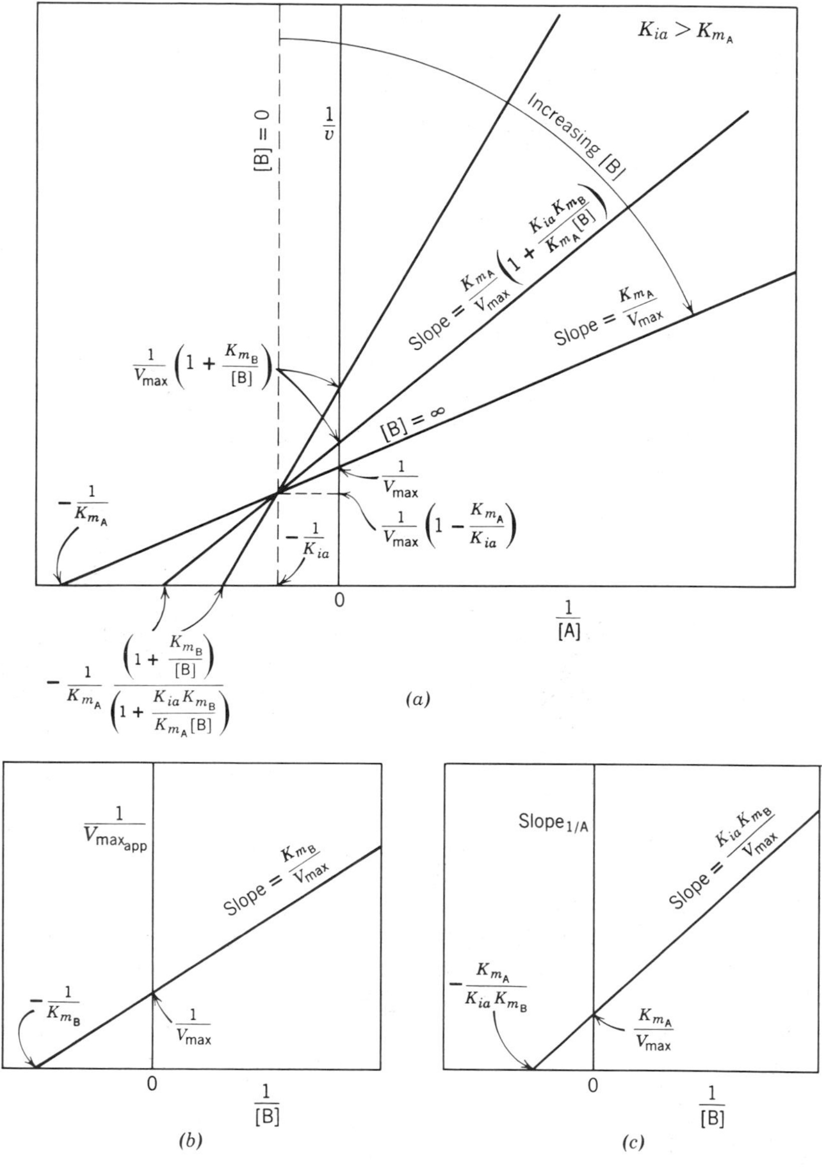

Fig. IX-9. Ordered Bi Bi system. (*a*) $1/v$ versus $1/[A]$ at different fixed B concentrations. (*b*) $1/v$-axis intercept replot. (*c*) $Slope_{1/A}$ replot. The plots are shown for $K_{ia} > K_{m_A}$.

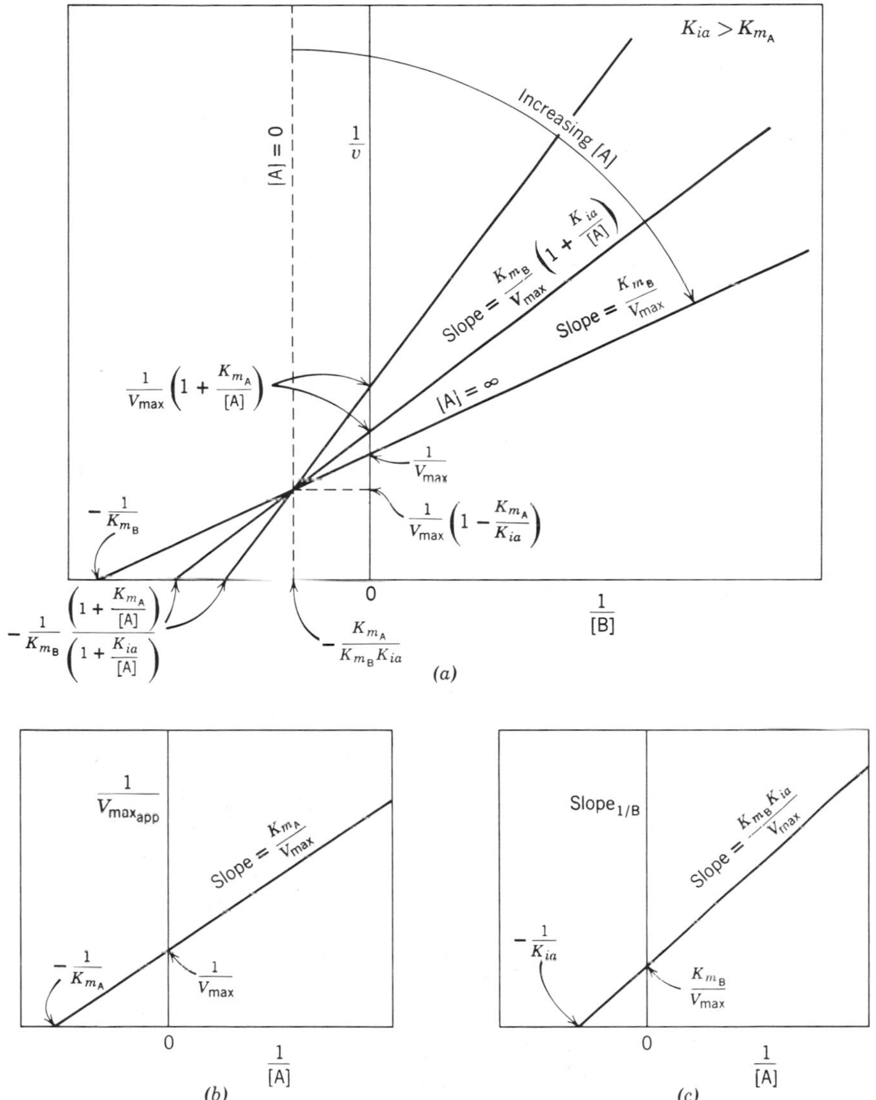

Fig. IX-10. Ordered Bi Bi system. (*a*) $1/v$ versus $1/[B]$ at different fixed A concentrations. (*b*) $1/v$-axis intercept replot. (*c*) $Slope_{1/B}$ replot.

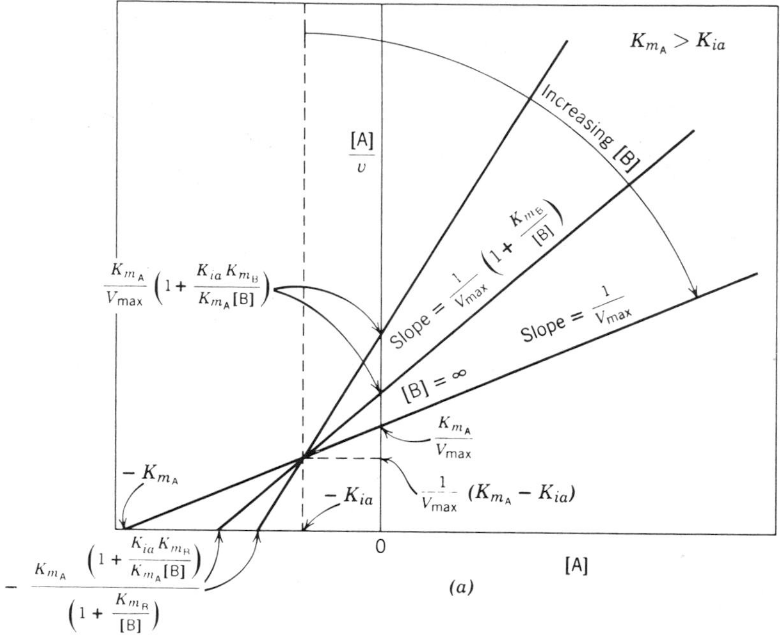

(a)

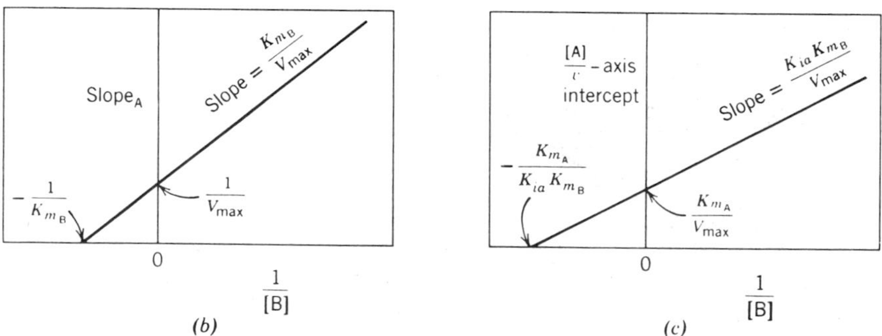

(b) *(c)*

Fig. IX-11. Ordered Bi Bi system. Plot of $[A]/v$ versus [A] and the *slope*$_A$ and $[A]/v$-axis intercept replots. The plots are shown for $K_{m_A} > K_{ia}$.

and

$$\boxed{\frac{[B]}{v} = \frac{\left(1 + \frac{K_{m_A}}{[A]}\right)}{V_{max}}[B] + \frac{K_{m_B}}{V_{max}}\left(1 + \frac{K_{ia}}{[A]}\right)} \qquad \text{(IX-95)}$$

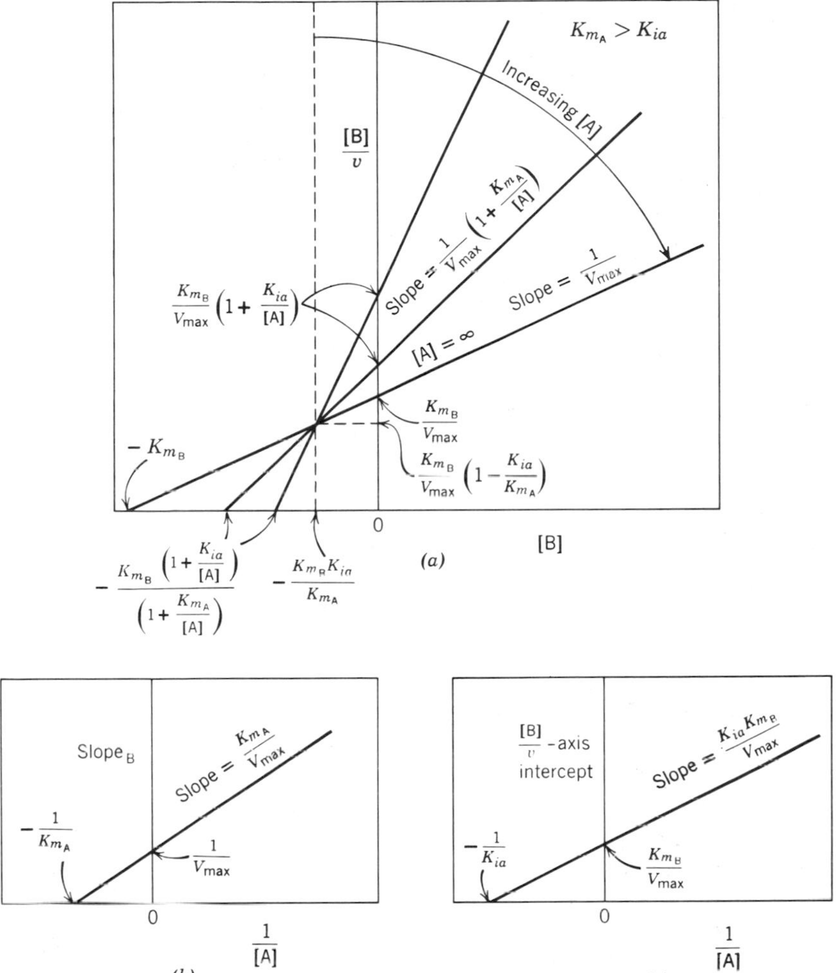

Fig. IX-12. Ordered Bi Bi system. Plot of $[B]/v$ versus [B] and the *slope*$_B$ and $[B]/v$-axis intercept replots.

The equations for the $v/[S]$ versus v plots are:

$$\frac{v}{[A]} = -\frac{\left(1+\frac{K_{m_B}}{[B]}\right)}{K_{m_A}\left(1+\frac{K_{ia}K_{m_B}}{K_{m_A}[B]}\right)}v + \frac{V_{max}}{K_{m_A}\left(1+\frac{K_{ia}K_{m_B}}{K_{m_A}[B]}\right)} \qquad \text{(IX-96)}$$

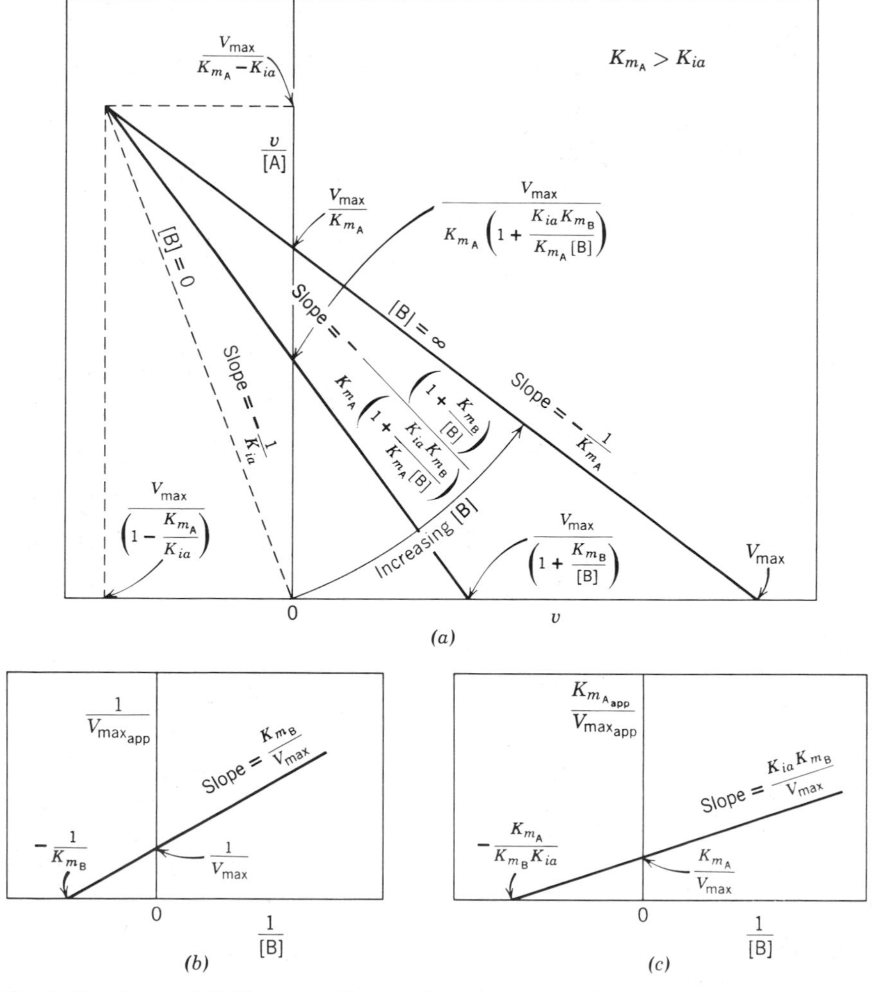

Fig. IX-13. Ordered Bi Bi system. Plot of $v/[A]$ versus v and the replots of the intercepts. The plots are shown for $K_{m_A} > K_{ia}$.

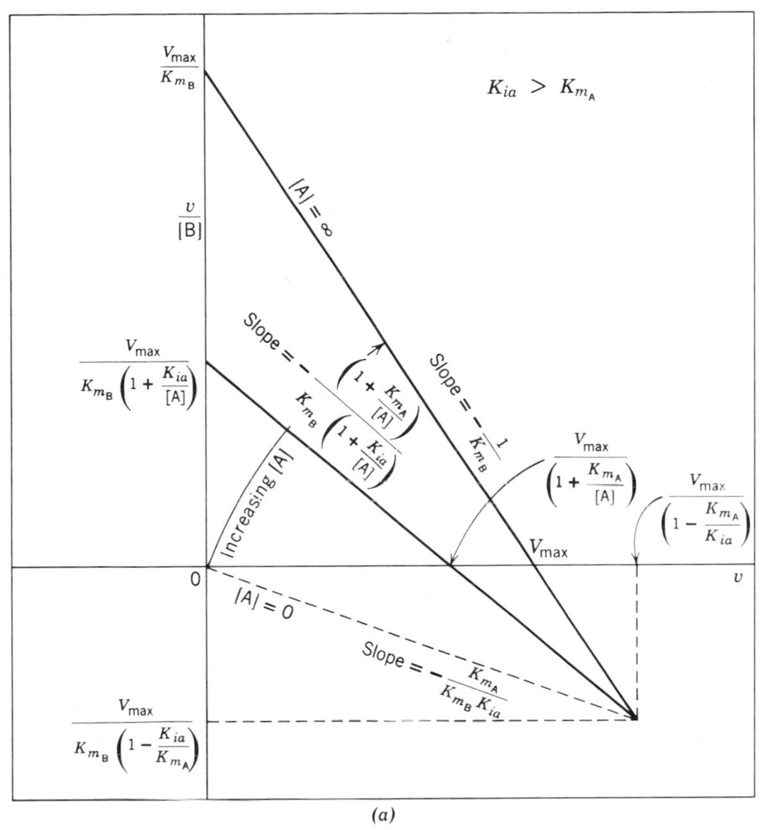

(a)

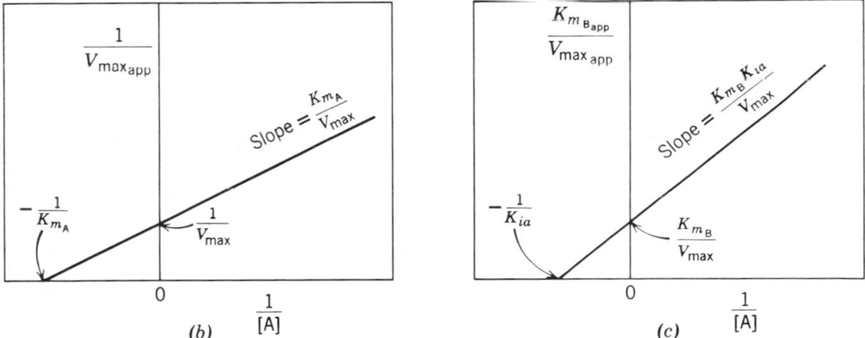

Fig. IX-14. Ordered Bi Bi system. Plot of $v/[B]$ versus v and the replots of the intercepts. The plots are shown for $K_{ia} > K_{m_A}$.

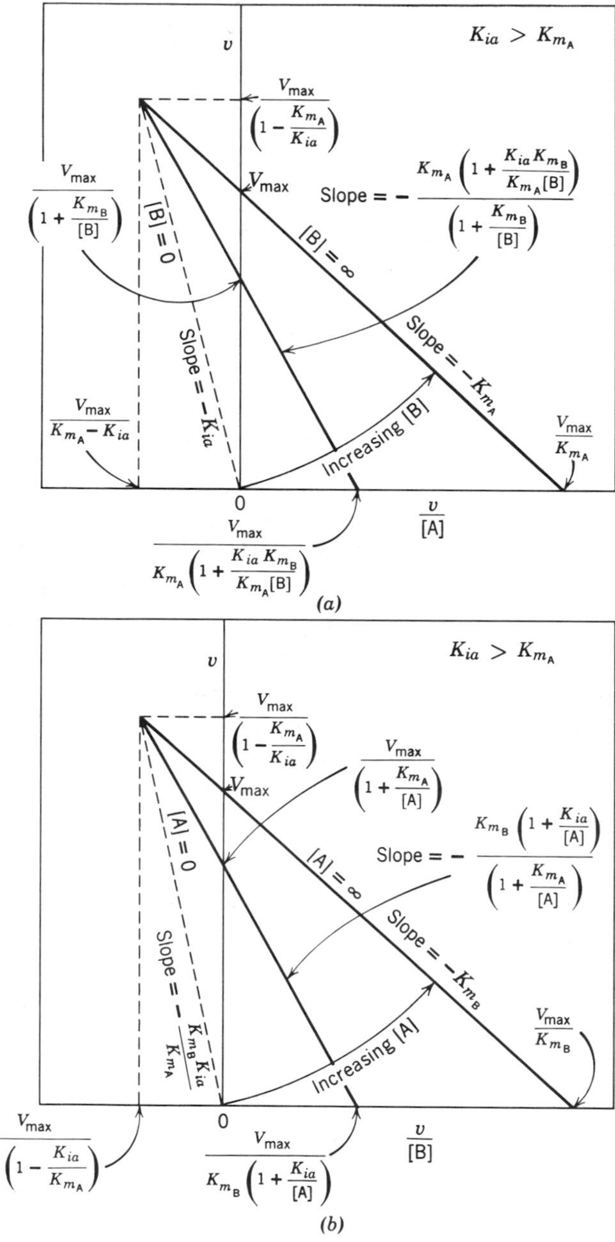

Fig. IX-15. Ordered Bi Bi system. Plots of v versus $v/[A]$ and v versus $v/[B]$ where $K_{ia} > K_{m_A}$. The replots are identical to those shown for the $v/[A]$ versus v and $v/[B]$ versus v plots.

and

$$\frac{v}{[\mathrm{B}]} = -\frac{\left(1+\frac{K_{m_\mathrm{A}}}{[\mathrm{A}]}\right)}{K_{m_\mathrm{B}}\left(1+\frac{K_{ia}}{[\mathrm{A}]}\right)}v + \frac{V_{\max}}{K_{m_\mathrm{B}}\left(1+\frac{K_{ia}}{[\mathrm{A}]}\right)} \qquad \text{(IX-97)}$$

If $K_{ia} = K_{m_\mathrm{A}}$, the slopes of the plots are constant at all concentrations of the nonvaried substrate. (The plots are parallel.) If $K_{m_\mathrm{A}} > K_{ia}$, the plots will intersect to the left of the $v/[\mathrm{S}]$-axis, above the v-axis. This is shown in Figure IX-13*a* for the $v/[\mathrm{A}]$ versus v plot. If $K_{m_\mathrm{A}} < K_{ia}$, the plots intersect below the v-axis and to the right of the $v/[\mathrm{S}]$-axis. This is shown in Figure IX-14*a* for the $v/[\mathrm{B}]$ versus v plot. The replots (Fig. IX-13*b,c* and Fig. IX-14*b,c*) are constructed from the reciprocals of the two axes intercepts.

The equations for the v versus $v/[\mathrm{S}]$ plots are:

$$v = \frac{-K_{m_\mathrm{A}}\left(1+\frac{K_{ia}K_{m_\mathrm{B}}}{K_{m_\mathrm{A}}[\mathrm{B}]}\right)}{\left(1+\frac{K_{m_\mathrm{B}}}{[\mathrm{B}]}\right)}\frac{v}{[\mathrm{A}]} + \frac{V_{\max}}{\left(1+\frac{K_{m_\mathrm{B}}}{[\mathrm{B}]}\right)} \qquad \text{(IX-98)}$$

and

$$v = \frac{-K_{m_\mathrm{B}}\left(1+\frac{K_{ia}}{[\mathrm{A}]}\right)}{\left(1+\frac{K_{m_\mathrm{A}}}{[\mathrm{A}]}\right)}\frac{v}{[\mathrm{B}]} + \frac{V_{\max}}{\left(1+\frac{K_{m_\mathrm{A}}}{[\mathrm{A}]}\right)} \qquad \text{(IX-99)}$$

The plots are shown in Figure IX-15a,b for $K_{ia} > K_{m_A}$. If $K_{ia} = K_{m_A}$, the plots would be parallel. If $K_{ia} < K_{m_A}$, the plots would intersect below the $v/[S]$-axes, to the right of the v-axis. The replots are identical to those shown for the $v/[S]$ versus v plots.

Complete Velocity Equation—Product Inhibition

To obtain the velocity equation for the reverse reaction, we multiply numerator and denominator of equation IX-88 by $K_{eq}/V_{\max_f}$. This puts the equation in the form:

$$v = \frac{V_{\max_r}([P][Q] - [A][B]K_{eq})}{\dfrac{V_{\max_r}K_{ia}K_{m_B}K_{eq}}{V_{\max_f}} + \cdots}$$

For the moment, all we need is the constant term in the denominator from which the Haldane equation is obtained. Writing the constant term in coefficient form, using the same definitions given earlier:

$$\frac{V_{\max_r}K_{ia}K_{m_B}K_{eq}}{V_{\max_f}} = \frac{(\text{num}_2)(\text{const})(\text{Coef}_A)(\text{num}_1)(\text{Coef}_{AB})}{(\text{Coef}_{PQ})(\text{Coef}_A)(\text{Coef}_{AB})(\text{num}_2)(\text{num}_1)}$$

$$= \frac{(\text{const})}{(\text{Coef}_{PQ})} = \frac{(\text{const})K_{m_P}}{(\text{Coef}_Q)} = K_{iq}K_{m_P}$$

Solving for K_{eq}:

$$\boxed{K_{eq} = \frac{V_{\max_f}K_{iq}K_{m_P}}{V_{\max_r}K_{ia}K_{m_B}}} \qquad \text{(IX-100)}$$

Another Haldane equation is:

$$\boxed{K_{eq} = \left(\frac{V_{\max_f}}{V_{\max_r}}\right)^2 \frac{K_{ip}K_{m_Q}}{K_{ib}K_{m_A}}} \qquad \text{(IX-101)}$$

Equation IX-100 can be used to eliminate K_{eq} in velocity equations rearranged to express P or Q as product inhibitors with respect to either A or B. When P is present and [Q] = 0, the velocity equation becomes:

$$\frac{v}{V_{\max_f}} = \frac{[A][B]}{K_{ia}K_{m_B} + K_{m_B}[A] + K_{m_A}[B] + \dfrac{V_f K_{m_Q}[P]}{V_r K_{eq}} + [A][B] + \dfrac{V_f K_{m_Q}[A][P]}{V_r K_{ia} K_{eq}} + \dfrac{[A][B][P]}{K_{ip}}}$$

The denominator P term becomes:

$$\frac{V_f K_{m_Q}[P]}{V_r K_{eq}} = \frac{V_f K_{m_Q}[P]\, V_r K_{ia} K_{m_B}}{V_r V_f K_{iq} K_{m_P}} = \frac{K_{m_Q} K_{m_B} K_{ia}[P]}{K_{iq} K_{m_P}}$$

The denominator AP term becomes:

$$\frac{V_f K_{m_Q}[A][P]}{V_r K_{ia} K_{eq}} = \frac{V_f K_{m_Q}[A][P]\, V_r K_{ia} K_{m_B}}{V_r K_{ia} V_f K_{iq} K_{m_P}} = \frac{K_{m_Q} K_{m_B}[A][P]}{K_{iq} K_{m_P}}$$

Inserting the expressions above for the P and AP terms, dividing numerator and denominator by [B] to obtain an equation for A as the varied substrate, and factoring, we obtain:

$$\boxed{\frac{v}{V_{\max}} = \frac{[A]}{K_{m_A}\left[1 + \dfrac{K_{ia}K_{m_B}}{K_{m_A}[B]}\left(1 + \dfrac{K_{m_Q}[P]}{K_{iq}K_{m_P}}\right)\right] + [A]\left[1 + \dfrac{K_{m_B}}{[B]}\left(1 + \dfrac{K_{m_Q}[P]}{K_{iq}K_{m_P}}\right) + \dfrac{[P]}{K_{ip}}\right]}}$$

(IX-102)

We see that (*a*) P will act as a mixed-type inhibitor with respect to A at unsaturating [B] (Fig. IX-16*a*), and (*b*) the inhibition cannot be overcome by saturation with B. Saturating [B] reduces the slope factor unity, but the $1/v$-axis intercept factor retains the $[P]/K_{ip}$ term. Thus P acts as an uncompetitive inhibitor with respect to A at saturating [B] (Fig. IX-17*a*). When [B] is varied, the equation becomes:

$$\boxed{\frac{v}{V_{\max}} = \frac{[B]}{K_{m_B}\left(1 + \dfrac{K_{ia}}{[A]}\right)\left(1 + \dfrac{K_{m_Q}[P]}{K_{iq}K_{m_P}}\right) + [B]\left(1 + \dfrac{K_{m_A}}{[A]} + \dfrac{[P]}{K_{ip}}\right)}} \quad \text{(IX-103)}$$

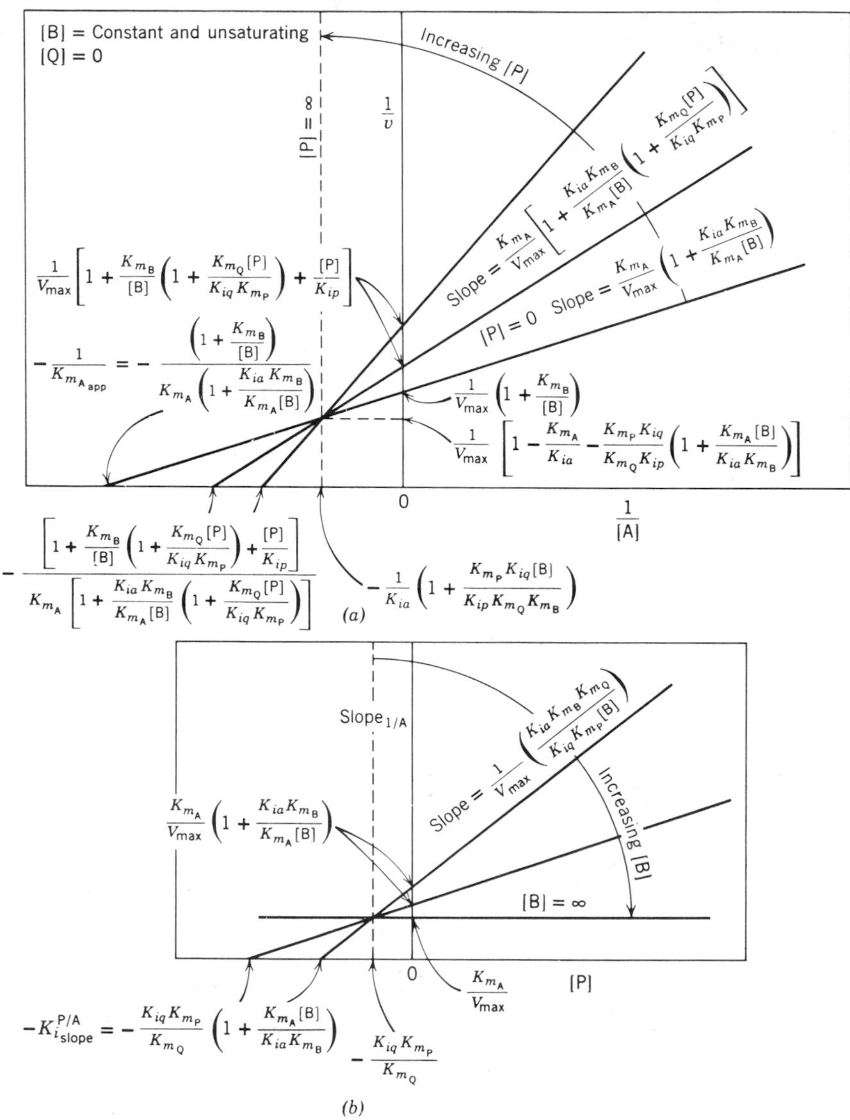

Fig. IX-16. Product inhibition by P in an Ordered Bi Bi system. (*a*) $1/v$ versus $1/[A]$ in the presence of different fixed P concentrations and a constant, unsaturating B concentration; Q is absent. (*b*) $Slope_{1/A}$ replot. Each replot corresponds to a different constant B concentration. The intercept on the [P]-axis gives $K_{i_{slope}}^{P/A}$. (*c*) Secondary replot of $K_{i_{slope}}$ versus [B]. (*d*) $1/v$-axis intercept replot. Each replot corresponds to a different constant B concentration. The intercept on the [P]-axis gives $K_{i_{int}}^{P/A}$. (*e*) Secondary replot of $K_{i_{int}}^{P/A}$ versus [B].

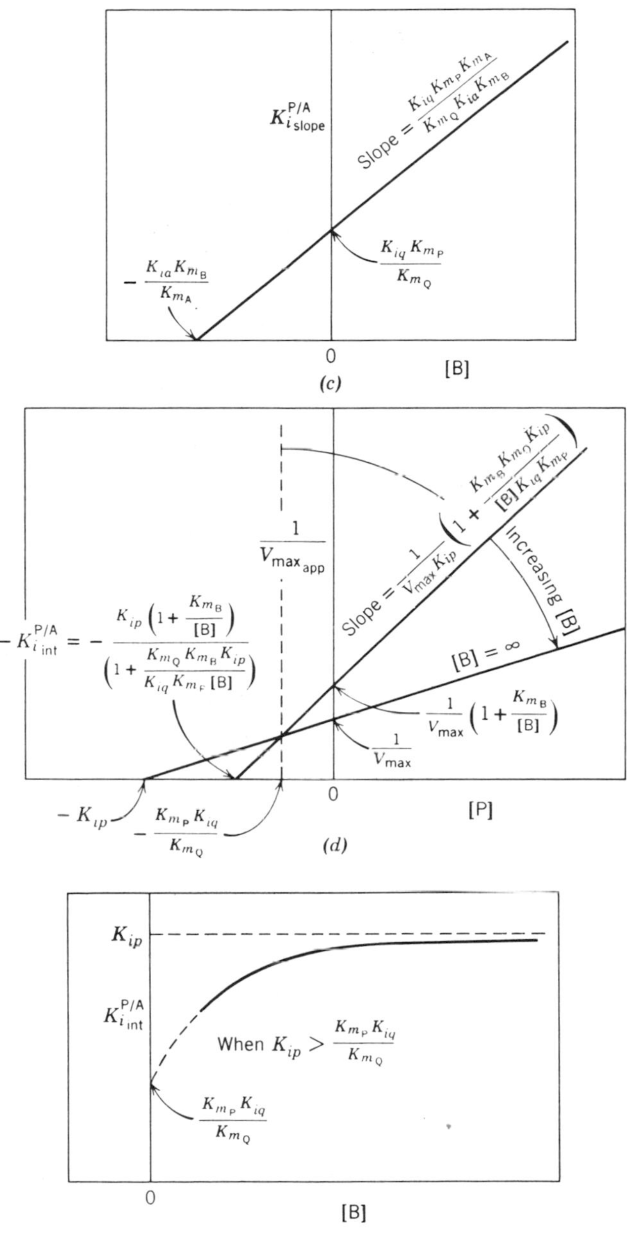

Fig. IX-16. (*Cont.*)

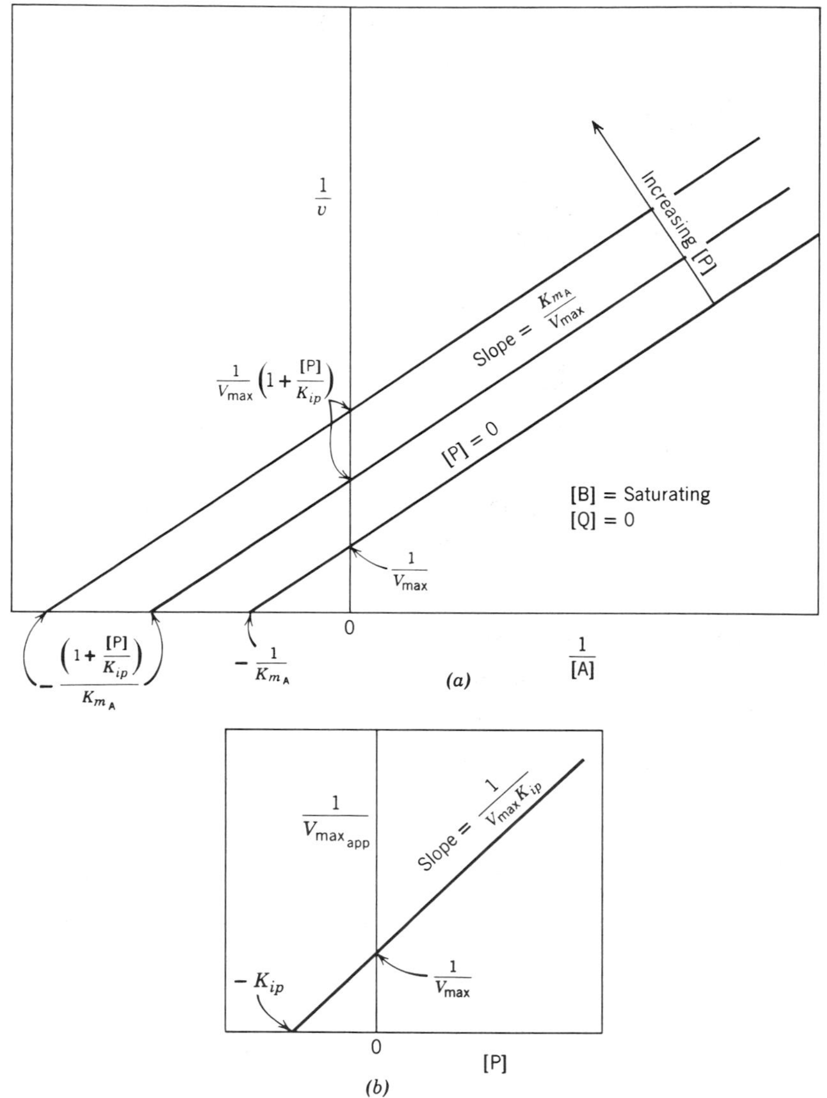

Fig. IX-17. Product inhibition by P in an Ordered Bi Bi system. (*a*) $1/v$ versus $1/[A]$ plot at different fixed P concentrations and a constant, saturating B concentration; Q is absent. (*b*) $1/v$-axis intercept replot.

P behaves as a mixed-type inhibitor with respect to B at all concentrations of A (Fig. IX-18). (Saturating [A] leaves [P] terms in the slope and intercept factors.) The intersection point may be above, on, or below the horizontal axis, depending on the ratio of $K_{m_P}K_{iq}/K_{m_Q}K_{ip}$. The horizontal coordinate of the intersection point is given by:

$$-\frac{K_{m_P}K_{iq}}{K_{m_B}K_{m_Q}K_{ip}\left(1+\frac{K_{ia}}{[A]}\right)}=-\frac{k_2(k_p+k_{-p})}{k_{-2}k_{-p}\left(1+\frac{K_{ia}}{[A]}\right)}$$

For the special case where $K_{ip} \gg K_{m_P}K_{iq}/K_{m_Q}$ or $K_B \gg K_{m_B}$ (because $k_{-2} \gg k_4$), the intersection point will fall very close to the vertical axis suggesting that P is competitive with B. Porter, Modebe, and Stark (1969) have suggested that this is the situation with the catalytic subunit of aspartate transcarbamylase where A = carbamylphosphate, B = asparate, P = carbamylaspartate, and Q = inorganic phosphate. Carbamylaspartate appears to be competitive with aspartate and a mixed-type inhibitor with respect to carbamylphosphate. Inorganic phosphate is competitive with carbamylphosphate and a mixed-type inhibitor with respect to aspartate. However, more recent studies by Heyde, Nagabhusanam, and Morrison (1973) suggest a random mechanism with three dead-end complexes (EBQ, EBP, and EAP).

When Q is present and $[P]=0$, equation IX-88 becomes:

$$\frac{v}{V_{max}}=\frac{[A][B]}{K_{ia}K_{m_B}+K_{m_B}[A]+K_{m_A}[B]+\frac{V_{max_f}K_{m_P}[Q]}{V_{max_r}K_{eq}}+[A][B]+\frac{K_{m_A}[B][Q]}{K_{iq}}}$$

The K_{eq} is eliminated from the Q term:

$$\frac{V_{max_f}K_{m_P}[Q]}{V_{max_r}K_{eq}}=\frac{V_{max_f}K_{m_P}[Q]\,V_{max_r}K_{ia}K_{m_B}}{V_{max_r}V_{max_f}K_{iq}K_{m_P}}=\frac{K_{ia}K_{m_B}[Q]}{K_{iq}}$$

When [A] is varied,

$$\boxed{\frac{v}{V_{max}}=\frac{[A]}{K_{m_A}\left(1+\frac{[Q]}{K_{iq}}\right)\left(1+\frac{K_{ia}K_{m_B}}{K_{m_A}[B]}\right)+[A]\left(1+\frac{K_{m_B}}{[B]}\right)}} \qquad \text{(IX-104)}$$

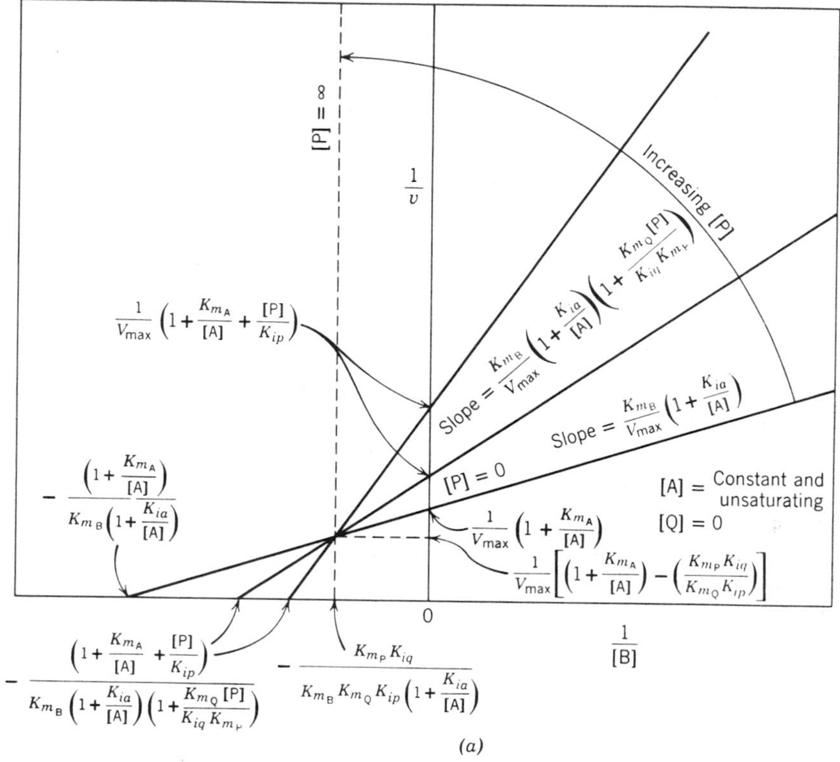

Fig. IX-18. Product inhibition by P in an Ordered Bi Bi system. (*a*) $1/v$ versus $1/[B]$ at different fixed P concentrations and a constant, unsaturating A concentration; Q is absent. (*b*) $Slope_{1/B}$ replot. Each replot corresponds to a different constant A concentration. The intercept on the [P]-axis gives $-K_{i_{slope}}^{P/B}$, which is constant for all A concentrations. (*c*) $1/v$-axis intercept replot. Each replot corresponds to a different constant A concentration. The intercept on the [P]-axis gives $K_{i_{int}}^{P/B}$. (*d*) Secondary replot of $K_{i_{int}}^{P/B}$.

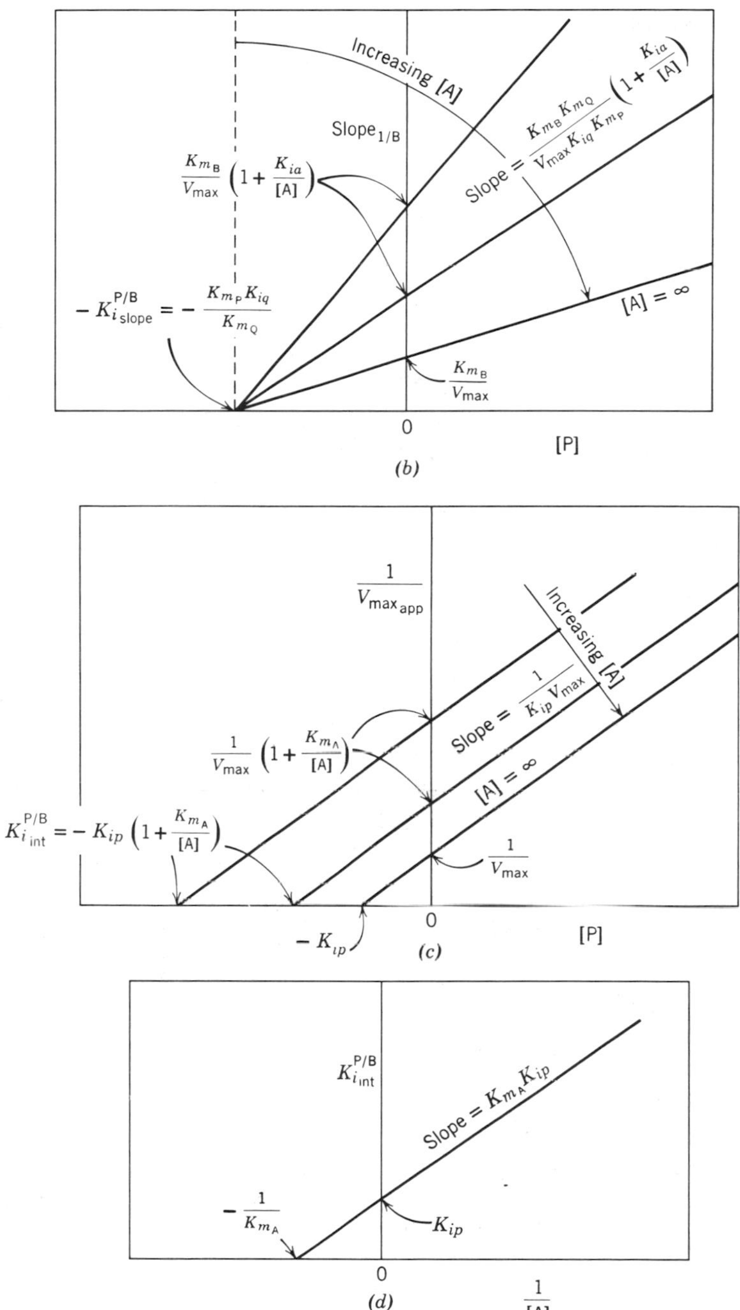

Fig. IX-18. (*Cont.*)

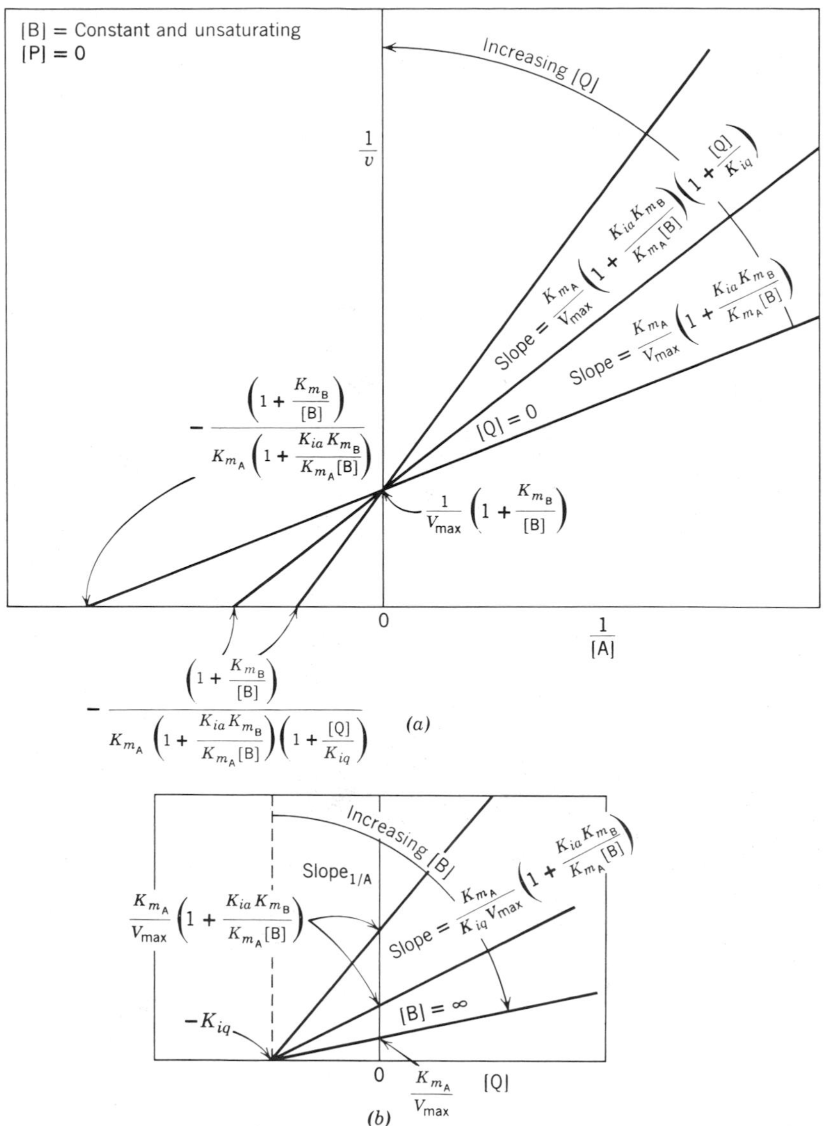

Fig. IX-19. Product inhibition by Q in an Ordered Bi Bi system. (*a*) $1/v$ versus $1/[A]$ at different fixed Q concentrations and a constant, unsaturating B; P is absent. (*b*) $Slope_{1/A}$ replot. Each replot represents a different constant B concentration. The intercept on the [Q]-axis gives $K_{i_{slope}}^{Q/A}$, which equals K_{iq} for all B concentrations.

We see that Q acts as a competitive inhibitor with respect to A at all B concentrations (Fig. IX-19*a*). When [B] is varied:

$$\frac{v}{V_{max}} = \frac{[B]}{K_{m_B}\left[1+\frac{K_{ia}}{[A]}\left(1+\frac{[Q]}{K_{iq}}\right)\right]+[B]\left[1+\frac{K_{m_A}}{[A]}\left(1+\frac{[Q]}{K_{iq}}\right)\right]} \qquad \text{(IX-105)}$$

The Q acts as a mixed-type inhibitor with respect to B at unsaturating A concentrations (Fig. IX-20*a*). Saturation with A overcomes the inhibition. Keep in mind, however, that saturation does not simply mean ten-times the Michaelis constant. In the present example, [A] must be very large compared to $K_{ia}[Q]/K_{iq}$, and $K_{m_A}[Q]/K_{iq}$ for these [Q] terms to disappear from the slope and intercept factors. It is noteworthy that the intersection coordinates of Figure IX-20*a* are the same as those for the $1/v$ versus $1/[B]$ plots at different fixed A concentrations. This is not surprising as Q is competitive with A. Thus increasing [Q] has the same effect as decreasing [A]. The product inhibition patterns permit us to establish the order of substrate addition and product release; that is, we can tell A from B and P from Q. The patterns also distinguish the Ordered Bi Bi system from the Rapid Equilibrium Random Bi Bi system.

K_{ip} can be determined from $1/v$-axis intercept replots of the $1/v$ versus $1/[B]$ data obtained at a constant [A] and different fixed [P]. The intercept can be written as:

$$\frac{1}{V_{max_{app}}} = \frac{1}{V_{max}K_{ip}}[P] + \frac{1}{V_{max}}\left(1+\frac{K_{m_A}}{[A]}\right) \qquad \text{(IX-106)}$$

The horizontal-axis intercept of the replot gives $K_{i_{int}}^{P/B} = K_{ip}(1+K_{m_A}/[A])$ (Fig. IX-18*c*). If [A] is saturating, the intercept gives K_{ip}. If several families of $1/v$ versus $1/[B]$ plots are constructed, each at a different constant unsaturating [A], then $K_{i_{int}}^{P/B}$ can be replotted against the corresponding $1/[A]$ to yield K_{ip} (Fig. IX-18*d*). K_{iq} can be obtained from the $slope_{1/A}$ replot (Fig. IX-19*b*) or the intercept replot of $1/v$ versus $1/[B]$ data with Q as a product inhibitor. For the latter replot:

$$\frac{1}{V_{max_{app}}} = \frac{K_{m_A}}{K_{iq}V_{max}[A]}[Q] + \frac{1}{V_{max}}\left(1+\frac{K_{m_A}}{[A]}\right) \qquad \text{(IX-107)}$$

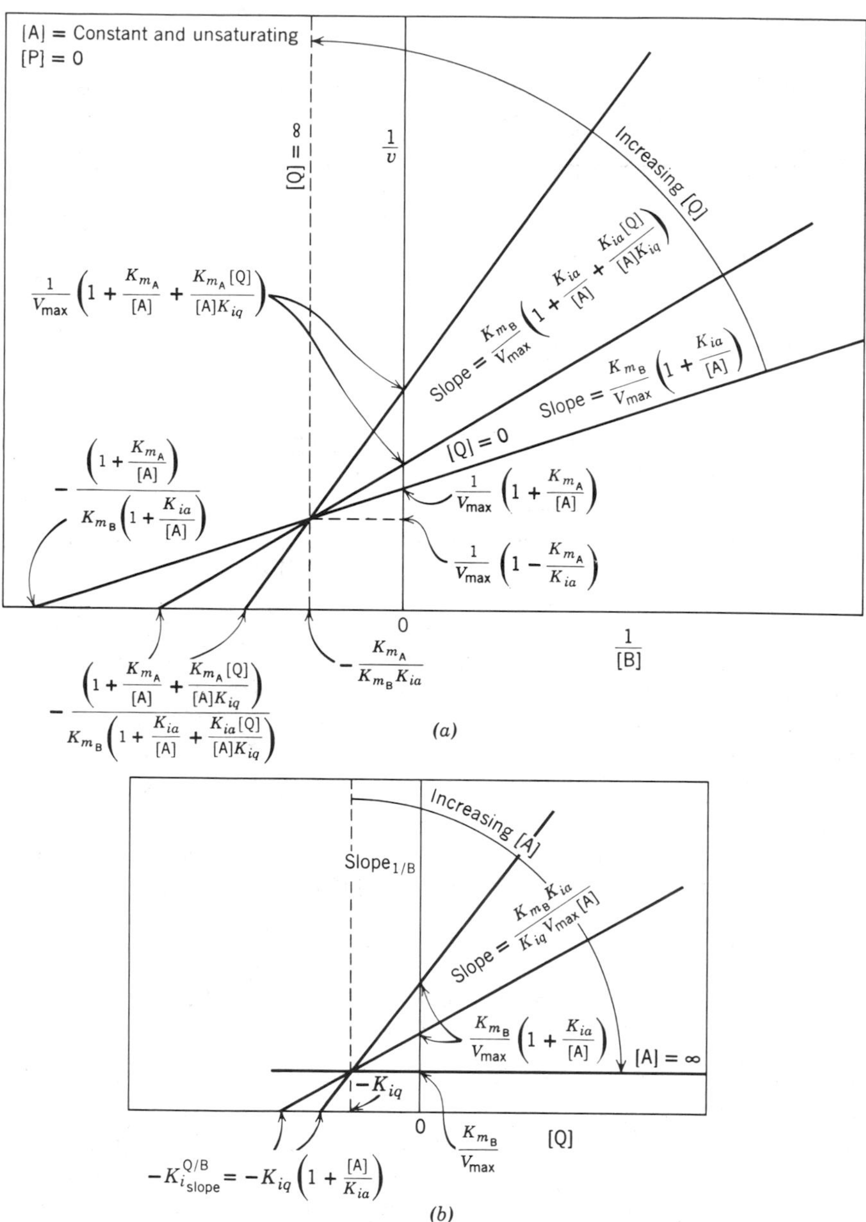

Fig. IX-20. Product inhibition by Q in an Ordered Bi Bi system. (*a*) $1/v$ versus $1/[B]$ at different fixed Q concentrations and a constant, unsaturating A; P is absent. (*b*) $Slope_{1/B}$ replot. Each replot represents a diffeerent constant A concentration. The intercept on the [Q]-axis gives $K_{i_{slope}}^{Q/B}$. (*c*) Secondary replot of $K_{i_{slope}}^{Q/B}$ versus [A]. (*d*) $1/v$-axis intercept replot. Each replot represents a different constant A concentration. The intercept on the [Q]-axis gives $K_{i_{int}}^{Q/B}$. (*e*) Secondary replot of $K_{i_{int}}^{Q/B}$ versus [A].

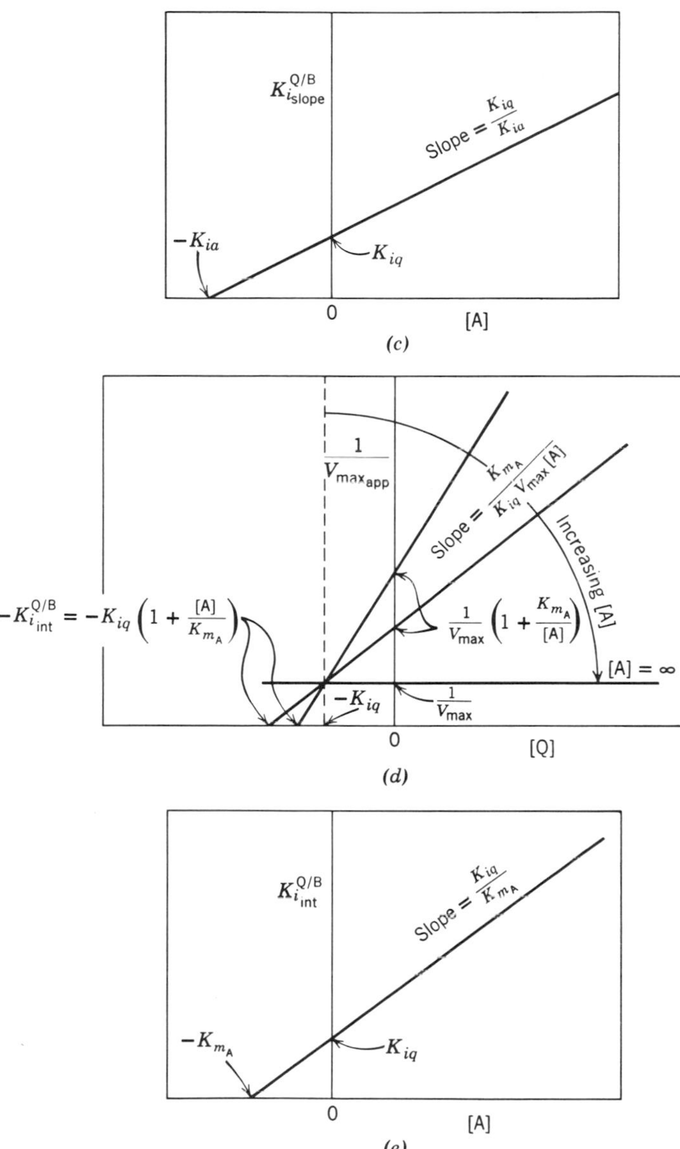

Fig. IX-20. (*Cont.*)

The horizontal intercept of the replot gives $K_{i_{\text{int}}}^{\text{Q/B}} = K_{iq}(1 + [\text{A}]/K_{m_\text{A}})$ (Fig. IX-20*b*). If several families of plots at different constant A concentrations are obtained, the slopes or horizontal intercepts of the replots can be replotted against an appropriate function of A to obtain a better evaluation of K_{iq} (Fig. IX-20*d*,*e*). If the reaction can be assayed in the reverse direction, K_{m_Q} and K_{m_P} can be obtained in the usual way (and also, K_{ib} from studies with B as a product inhibitor). If the reaction cannot be conveniently assayed in the reverse direction, the ratio of $K_{m_\text{P}}/K_{m_\text{Q}}$ can still be determined from the slopes (or slope replots) when P is the product inhibitor (Fig. IX-16*b*,*d* and Fig. IX-18*b*).

$K_{i_{\text{slope}}}$ and $K_{i_{\text{int}}}$

As shown in Figures IX-16 to IX-20, the K_i's calculated from the slopes and intercepts of the reciprocal plots are not necessarily equivalent to K_{ip} and K_{iq}. We can determine the composition of these K_i's directly from the original velocity equations. For example, the reciprocal plot for P as the product inhibitor with A as the varied substrate can be rearranged to:

$$\frac{1}{v} = \frac{K_{m_\text{A}}}{V_{\max}}\left(1 + \frac{K_{ia}K_{m_\text{B}}}{K_{m_\text{A}}[\text{B}]}\right)\left[1 + \frac{[\text{P}]}{\dfrac{K_{m_\text{P}}K_{iq}}{K_{m_\text{Q}}}\left(1 + \dfrac{K_{m_\text{A}}[\text{B}]}{K_{ia}K_{m_\text{B}}}\right)}\right]\frac{1}{[\text{A}]}$$

$$+ \frac{1}{V_{\max}}\left(1 + \frac{K_{m_\text{B}}}{[\text{B}]}\right)\left[1 + \frac{[\text{P}]}{\dfrac{K_{ip}\left(1 + \dfrac{K_{m_\text{B}}}{[\text{B}]}\right)}{\left(1 + \dfrac{K_{m_\text{Q}}K_{m_\text{B}}K_{ip}}{K_{m_\text{P}}K_{iq}[\text{B}]}\right)}}\right] \tag{IX-108}$$

Thus the K_i values for P calculated from the slopes and intercepts of the $1/[A]$ plots (indicated as $K_{i_{\text{slope}}}^{P/A}$ and $K_{i_{\text{int}}}^{P/A}$), are given by:

$$K_{i_{\text{slope}}}^{P/A} = \frac{K_{m_P} K_{iq}}{K_{m_Q}} \left(1 + \frac{K_{m_A}[B]}{K_{ia} K_{m_B}}\right) \qquad \text{and} \qquad K_{i_{\text{int}}}^{P/A} = \frac{K_{ip}\left(1 + \dfrac{K_{m_B}}{[B]}\right)}{\left(1 + \dfrac{K_{m_Q} K_{m_B} K_{ip}}{K_{m_P} K_{iq}[B]}\right)}$$

$K_{i_{\text{slope}}}^{P/A}$ increases linearly to infinity as [B] increases (Fig. IX-16*c*); $K_{i_{\text{int}}}^{P/A}$ approaches a limit of K_{ip} as [B] increases (Fig. IX-16*e*). However, we may observe an increase, decrease, or no change in $K_{i_{\text{int}}}^{P/A}$, depending on the relative values of K_{ip}, K_{m_P}, and K_{iq}. For example, if the ratio $K_{m_Q} K_{ip}/K_{m_P} K_{iq}$ is unity, then $K_{i_{\text{int}}}^{P/A} = K_{ip}$ at all B concentrations.

When [B] is varied, the reciprocal equation can be written:

$$\frac{1}{v} = \frac{K_{m_B}}{V_{\max}}\left(1 + \frac{K_{ia}}{[A]}\right)\left[1 + \frac{[P]}{\dfrac{K_{m_P} K_{iq}}{K_{m_Q}}}\right]\frac{1}{[B]} + \frac{1}{V_{\max}}\left(1 + \frac{K_{m_A}}{[A]}\right)\left[1 + \frac{[P]}{K_{ip}\left(1 + \dfrac{K_{m_A}}{[A]}\right)}\right] \qquad \text{(IX-109)}$$

Thus:

$$K_{i_{\text{slope}}}^{P/B} = \frac{K_{m_P} K_{iq}}{K_{m_Q}} \qquad \text{and} \qquad K_{i_{\text{int}}}^{P/B} = K_{ip}\left(1 + \frac{K_{m_A}}{[A]}\right)$$

$K_{i_{\text{slope}}}^{P/B}$ is constant but not equal to K_{ip}. The $K_{i_{\text{int}}}^{P/B}$ decreases with increasing [A] approaching K_{ip} as a limit (Fig. IX-18*d*). When Q is the product inhibitor, the reciprocal equations can be written:

$$\frac{1}{v} = \frac{K_{m_A}}{V_{\max}}\left(1 + \frac{K_{ia} K_{m_B}}{K_{m_A}[B]}\right)\left(1 + \frac{[Q]}{K_{iq}}\right)\frac{1}{[A]} + \frac{1}{V_{\max}}\left(1 + \frac{K_{m_B}}{[B]}\right) \qquad \text{(IX-110)}$$

$$\therefore \quad K_{i_{\text{slope}}}^{Q/A} = K_{iq} \qquad \text{at all B concentrations}$$

$$\frac{1}{v}=\frac{K_{m_{\mathrm{B}}}}{V_{\max}}\left(1+\frac{K_{ia}}{[\mathrm{A}]}\right)\left[1+\frac{[\mathrm{Q}]}{K_{iq}\left(1+\frac{[\mathrm{A}]}{K_{ia}}\right)}\right]\frac{1}{[\mathrm{B}]}+\frac{1}{V_{\max}}\left(1+\frac{K_{m_{\mathrm{A}}}}{[\mathrm{A}]}\right)\left[1+\frac{[\mathrm{Q}]}{K_{iq}\left(1+\frac{[\mathrm{A}]}{K_{m_{\mathrm{A}}}}\right)}\right] \tag{IX-111}$$

$$\therefore \quad K_{i_{\text{slope}}}^{\mathrm{Q}/\mathrm{B}}=K_{iq}\left(1+\frac{[\mathrm{A}]}{K_{ia}}\right) \quad \text{and} \quad K_{i_{\text{int}}}^{\mathrm{Q}/\mathrm{B}}=K_{iq}\left(1+\frac{[\mathrm{A}]}{K_{m_{\mathrm{A}}}}\right)$$

Thus both $K_{i_{\text{slope}}}^{\mathrm{Q}/\mathrm{B}}$ and $K_{i_{\text{int}}}^{\mathrm{Q}/\mathrm{B}}$ are functions of [A] (Fig. IX-20*d*, *e*).

Calculation of Rate Constants

The individual rate constants can be calculated from the kinetic constants as shown below.

$$k_1=\frac{V_{\max_f}}{K_{m_{\mathrm{A}}}[\mathrm{E}]_t}, \qquad k_{-1}=\frac{V_{\max_f}K_{ia}}{K_{m_{\mathrm{A}}}[\mathrm{E}]_t}, \qquad k_2=\frac{V_{\max_f}}{K_{m_{\mathrm{B}}}[\mathrm{E}]_t}\left(1+\frac{k_{-2}}{k_3}\right)$$

$$\frac{1}{k_{-2}}=\frac{[\mathrm{E}]_t}{V_{\max_r}}-\frac{1}{k_{-1}}, \qquad \frac{1}{k_3}=\frac{[\mathrm{E}]_t}{V_{\max_f}}-\frac{1}{k_4}, \qquad k_{-3}=\frac{V_{\max_r}}{K_{m_{\mathrm{P}}}[\mathrm{E}]_t}\left(1+\frac{k_3}{k_{-2}}\right)$$

$$k_4=\frac{V_{\max_r}K_{iq}}{K_{m_{\mathrm{Q}}}[\mathrm{E}]_t}, \qquad k_{-4}=\frac{V_{\max_r}}{K_{m_{\mathrm{Q}}}[\mathrm{E}]_t}$$

If the isomerization of the central complexes is included in the reaction sequence (Table IX-2), then the kinetic constants cannot be used to calculate k_2, k_{-2}, k_3, and k_{-3}. However, the calculations of k_1, k_{-1}, k_4, and k_{-4} are still valid because the additional rate constants (k_p and k_{-p}) cancel out (e.g., they appear in the same way in $V_{\max_f}/[\mathrm{E}]_t$ and $K_{m_{\mathrm{A}}}$, and in $V_{\max_r}/[\mathrm{E}]_t$ and $K_{m_{\mathrm{Q}}}$).

Distribution Equations

$$\frac{[\mathrm{E}]}{[\mathrm{E}]_t} = \frac{K_{ia}K_{m_\mathrm{B}}V_r + \dfrac{K_{m_\mathrm{Q}}V_f[\mathrm{P}]}{K_{\mathrm{eq}}} + K_{m_\mathrm{A}}V_r[\mathrm{B}]}{\text{denominator of velocity equation}} \tag{IX-112}$$

$$\frac{[\mathrm{EA}]}{[\mathrm{E}]_t} = \frac{K_{m_\mathrm{B}}V_r[\mathrm{A}] + \dfrac{K_{m_\mathrm{Q}}V_f[\mathrm{A}][\mathrm{P}]}{K_{ia}K_{\mathrm{eq}}} + \dfrac{K_{m_\mathrm{A}}V_r[\mathrm{P}][\mathrm{Q}]}{K_{ia}K_{\mathrm{eq}}}}{\text{denominator of velocity equation}} \tag{IX-113}$$

$$\frac{[\mathrm{EAB+EPQ}]}{[\mathrm{E}]_t} = \frac{\left(V_r - \dfrac{V_fK_{m_\mathrm{Q}}}{K_{iq}}\right)[\mathrm{A}][\mathrm{B}] + \left(\dfrac{V_f}{K_{\mathrm{eq}}} - \dfrac{V_rK_{m_\mathrm{A}}}{K_{\mathrm{eq}}K_{ia}}\right)[\mathrm{P}][\mathrm{Q}] + \dfrac{V_r[\mathrm{A}][\mathrm{B}][\mathrm{P}]}{K_{ip}} + \dfrac{V_f[\mathrm{B}][\mathrm{P}][\mathrm{Q}]}{K_{ib}K_{\mathrm{eq}}}}{\text{denominator of velocity equation}} \tag{IX-114}$$

$$\frac{[\mathrm{EQ}]}{[\mathrm{E}]_t} = \frac{\dfrac{K_{m_\mathrm{P}}V_f[\mathrm{Q}]}{K_{\mathrm{eq}}} + \dfrac{K_{m_\mathrm{Q}}V_f[\mathrm{A}][\mathrm{B}]}{K_{iq}} + \dfrac{K_{m_\mathrm{A}}V_r[\mathrm{B}][\mathrm{Q}]}{K_{iq}}}{\text{denominator of velocity equation}} \tag{IX-115}$$

Effect of Isomerizations

EAB + EPQ: The distribution equations shown above are still valid and k_1, k_{-1}, k_4, and k_{-4} can be calculated.

EA: Only $[\mathrm{E}]/[\mathrm{E}]_t$, $[\mathrm{EQ}]/[\mathrm{E}]_t$, k_4, and k_{-4} can be calculated.

EA and EQ: Only $[\mathrm{E}]/[\mathrm{E}]_t$ can be calculated.

Agreement of Kinetic Data With the Haldane Equations

The values of the experimentally determined kinetic constants can be checked for consistency against the known or experimentally determined equilibrium constant by means of the Haldane equations. In attempting such a check, care should be taken to determine $V_{\max_f}$ and $V_{\max_r}$ with the same preparation, as close together in time as possible. The values of the Michaelis constants and inhibition constants are independent of the enzyme concentration, but $V_{\max}$ values are a direct measurement of the amount of enzyme present. Generally, we use a fixed volume of enzyme solution in each assay. However, the actual concentration of active enzyme may change from day to day as a result of slow denaturation. (Other "aging" phenomena such as aggregation or dissociation may affect the K values as well as V_{max} values.) The V_{max} values can be conveniently obtained by increasing all substrates together, maintaining them at a constant ratio. The resulting reciprocal plot will be concave upward (parabolic for the Ordered Bi Bi reaction). Extrapolation to the vertical axis gives V_{max} for the given direction (Fig. IX-21). The equation when $[B]=x[A]$ is:

$$\frac{v}{V_{max}} = \frac{x[A]^2}{K_{ia}K_{m_B} + K_{m_B}[A] + K_{m_A}[xA] + x[A]^2}$$

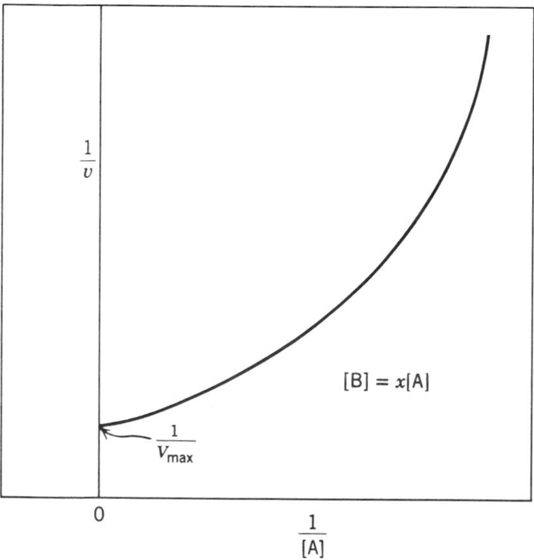

Fig. IX-21. Ordered Bi Bi system; $1/v$ versus $1/[A]$ where [B] is varied along with [A] at $[B]=x[A]$.

or

$$\boxed{\frac{1}{v} = \frac{K_{m_A}}{V_{max}}\left(1 + \frac{K_{ia}K_{m_B}}{xK_{m_A}[A]} + \frac{K_{m_B}}{xK_{m_A}}\right)\frac{1}{[A]} + \frac{1}{V_{max}}} \qquad \text{(IX-116)}$$

G. PARTIAL RAPID EQUILIBRIUM ORDERED BI BI SYSTEM

The Steady-State Ordered Bi Bi system reduces to a rapid equilibrium system when k_3 is much smaller than the other rate constants (Chapter Six). In this case, K_{m_A} reduces to zero and the other kinetic constants reduce to dissociation constants. A somewhat similar situation arises if k_{-1} is much larger than certain other rate constants. Let us write out the definitions of some kinetic constants:

$$K_{m_A} = \frac{\text{Coef}_B}{\text{Coef}_{AB}} = \frac{k_3 k_4}{k_1(k_3 + k_4)}, \qquad K_{ia} = \frac{k_{-1}}{k_1}$$

$$\frac{V_{\max_f}}{[E]_t} = \frac{\text{num}_1}{\text{Coef}_{AB}} = \frac{k_3 k_4}{(k_3 + k_4)}$$

We see that the K_{m_A} term will drop out of the velocity equation without losing other terms when:

$$K_{ia} \gg K_{m_A} \quad \text{or} \quad \frac{k_{-1}}{k_1} \gg \frac{k_3 k_4}{k_1(k_3 + k_4)} \quad \text{or} \quad k_{-1} \gg \frac{k_3 k_4}{(k_3 + k_4)}$$

In other words, when the rate constant for the dissociation of A is greater than the maximal velocity in the forward direction ($k_{-1} \gg V_{\max_f}/[E]_t$) the K_{m_A} term is eliminated from the denominator of the velocity equation, but the K_{ia} and other terms remain. If, in addition, k_4 is much greater than $V_{\max_r}/[E]_t$, the K_{m_Q} term will drop out and the system will have the same characteristics as a Rapid Equilibrium Ordered Bi Bi system in both directions. We could approach the situation in a different manner by deriving the complete velocity equation for an Ordered Bi Bi system in which the first ligand to add in either direction is in equilibrium with free E and the respective complexes, but EAB and EPQ are not at equilibrium. If we use Cha's method for deriving the velocity equation (described in Section B), we obtain

$$v = k_2[EA][B] - k_{-2}[EAB + EPQ] = k_2 f_2[X][B] - k_{-2}[EAB + EPQ]$$

$$\frac{v}{[E]_t} = \frac{k_2 k_3 f_2[B] - k_{-2} k_{-3} f_3[P]}{k_{-2} + k_3 + k_2 f_2[B] + k_{-3} f_3[P]}$$

where f_2 and f_3 are the fractional concentrations of EA and EQ, respectively.

Substituting for f_2 and f_3 from equations IX-25 and IX-26 and simplifying, we obtain:

$$v = \frac{V_f V_r \left([A][B] - \dfrac{[P][Q]}{K_{eq}} \right)}{V_r K_{ia} K_{m_B} + V_r K_{m_B}[A] + V_r [A][B] + \dfrac{V_f K_{m_P}[Q]}{K_{eq}} + \dfrac{V_f [P][Q]}{K_{eq}}} \tag{IX-117}$$

The velocity equation has the same form as that for the total Rapid Equilibrium Ordered Bi Bi system except the constants associated with B and P are K_m's rather than dissociation constants. P will not be a product inhibitor unless a dead-end EAP complex forms; Q will be competitive with A and B provided only the normal EQ complex forms (i.e., no dead-end EQB complex).

Another situation that yields $slope_{1/A}$ replots that go through the origin is shown below. Here A is an activator (e.g., a metal ion) which must add to E before the substrate, B, adds. However, once A adds, it need not dissociate from the enzyme during each catalytic cycle.

A: $k_1 \downarrow k_{-1}$ B: $k_2 \downarrow k_{-2}$ P: $k_3 \uparrow k_{-3}$ B: $k_2 \downarrow k_{-2}$

E EA (EAB ⇌ EAP) EA (EAB ⇌

E, A, and EA will be at equilibrium in the steady-state. The King-Altman pattern is:

$$\begin{array}{c} E \\ k_1[A] \downarrow \uparrow k_{-1} \\ EA \end{array} \quad \underset{k_{-2}+k_3}{\overset{k_2[B]+k_{-3}[P]}{\rightleftharpoons}} \quad \begin{pmatrix} EAB \\ + \\ EAP \end{pmatrix}$$

The resulting velocity equation is:

$$\frac{v}{V_{\max}} = \frac{[A][B]}{K_{ia}K_{m_B} + K_{m_B}[A] + [A][B] + \dfrac{K_{m_B}[A][P]}{K_{ip}}} \qquad \text{(IX-118)}$$

In the absence of P, the equation is identical to that derived for a Rapid Equilibrium Ordered Bireactant system with $K_{m_B} = (k_3 + k_{-2})/k_2$ in place of K_B. P will be competitive with B and uncompetitive with A.

H. THEORELL-CHANCE BI BI SYSTEM

Theorell and Chance (1951) have proposed an Ordered Bi Bi mechanism without a ternary central complex for alcohol dehydrogenase. The hit-and-run reaction sequence can be written:

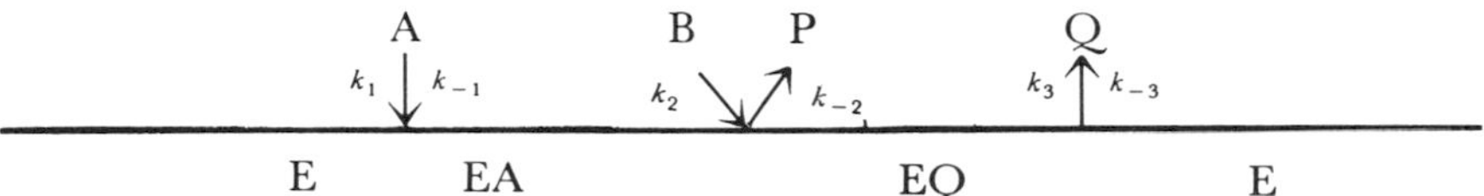

The basic King-Altman figure is:

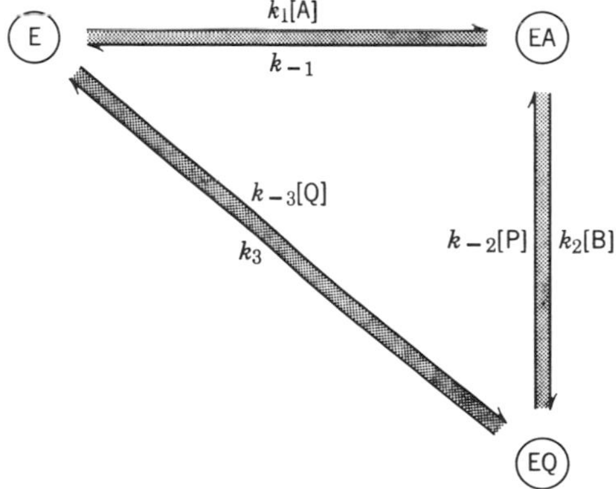

The distribution of enzyme forms is obtained from three two-lined interconversion patterns.

$$\frac{[E]}{[E]_t} = \frac{k_{-2}[P]k_{-1} + k_{-1}k_3 + k_2[B]k_3}{\text{denominator}} \qquad \text{(IX-119)}$$

$$\frac{[EA]}{[E]_t} = \frac{k_1[A]k_{-2}[P] + k_3k_1[A] + k_{-3}[Q]k_{-2}[P]}{\text{denominator}} \qquad \text{(IX-120)}$$

$$\frac{[EQ]}{[E]_t} = \frac{k_1[A]k_2[B] + k_{-1}k_{-3}[Q] + k_{-3}[Q]k_2[B]}{\text{denominator}} \qquad \text{(IX-121)}$$

Letting $v = k_1[E][A] - k_{-1}[EA]$ and grouping denominator terms, we obtain:

$$\frac{v}{[E]_t} = \frac{k_1k_2k_3[A][B] - k_{-1}k_{-2}k_{-3}[P][Q]}{k_{-1}k_3 + k_1k_3[A] + k_2k_3[B] + k_{-1}k_{-2}[P] + k_{-1}k_{-3}[Q] + k_1k_2[A][B] + k_1k_{-2}[A][P] + k_2k_{-3}[B][Q] + k_{-2}k_{-3}[P][Q]} \qquad \text{(IX-122)}$$

Compared to the usual Ordered Bi Bi, the equation for the Theorell-Chance mechanism lacks ABP and BPQ terms in the denominator. After substituting and defining coefficient ratios as described for the Ordered Bi Bi system, we obtain:

$$v = \frac{V_fV_r\left([A][B] - \dfrac{[P][Q]}{K_{eq}}\right)}{V_rK_{ia}K_{m_B} + V_rK_{m_B}[A] + V_rK_{m_A}[B] + \dfrac{V_fK_{m_Q}[P]}{K_{eq}} + \dfrac{V_fK_{m_P}[Q]}{K_{eq}} + V_r[A][B] + \dfrac{V_fK_{m_Q}[A][P]}{K_{eq}K_{ia}} + \dfrac{V_rK_{m_A}[B][Q]}{K_{iq}} + \dfrac{V_f[P][Q]}{K_{eq}}} \qquad \text{(IX-123)}$$

In the absence of products, the velocity equation for the net steady-state forward velocity is the same as that for the usual Ordered Bi Bi reaction.

$$\boxed{\frac{v}{V_{max}} = \frac{[A][B]}{K_{ia}K_{m_B} + K_{m_B}[A] + K_{m_A}[B] + [A][B]}} \qquad \text{(IX-124)}$$

Thus initial velocity studies alone will not distinguish between the two systems. However, the missing ABP and BPQ denominator terms lead to different product inhibition patterns.

Product Inhibition

There are sixteen Haldane equations for the Theorell-Chance Bi Bi system, including the four shown below with V_{max_f}/V_{max_r} to the first power.

$$K_{eq}=\frac{V_{max_f}K_{m_P}K_{iq}}{V_{max_r}K_{ia}K_{m_B}}=\frac{V_{max_f}K_{m_P}K_{iq}}{V_{max_r}K_{m_A}K_{ib}}=\frac{V_{max_f}K_{ip}K_{m_Q}}{V_{max_r}K_{m_A}K_{ib}}=\frac{V_{max_f}K_{ip}K_{mQ}}{V_{max_r}K_{ia}K_{m_B}} \qquad \text{(IX-125)}$$

where the new definitions are:

$$K_{ib}=\frac{\text{Coef}_Q}{\text{Coef}_{BQ}}=\frac{\text{const}}{\text{Coef}_B}=\frac{k_{-1}}{k_2}$$

$$K_{ip}=\frac{\text{Coef}_A}{\text{Coef}_{AP}}=\frac{\text{const}}{\text{Coef}_P}=\frac{k_3}{k_{-2}}$$

Also from the Haldane equations we see that:

$$K_{ia}K_{m_B}=K_{m_A}K_{ib}, \qquad K_{m_P}K_{iq}=K_{ip}K_{m_Q} \qquad \text{(IX-126)}$$

When P is the product inhibitor:

$$\frac{v}{V_{max}}=\frac{[A]}{K_{m_A}\left(1+\frac{K_{ib}}{[B]}+\frac{K_{ib}[P]}{K_{ip}[B]}\right)+[A]\left(1+\frac{K_{m_B}}{[B]}+\frac{K_{m_B}[P]}{K_{ip}[B]}\right)} \qquad \text{(IX-127)}$$

When [B] is varied:

$$\frac{v}{V_{max}} = \frac{[B]}{K_{m_B}\left(1+\frac{K_{ia}}{[A]}\right)\left(1+\frac{[P]}{K_{ip}}\right)+[B]\left(1+\frac{K_{m_A}}{[A]}\right)} \qquad \text{(IX-128)}$$

Thus P is a mixed-type inhibitor with respect to A at unsaturating [B]. The inhibition is overcome by saturating [B]; P is a competitive inhibitor with respect to B at all concentrations of A.

When Q is the product inhibitor:

$$\frac{v}{V_{max}} = \frac{[A]}{K_{m_A}\left(1+\frac{K_{ib}}{[B]}\right)\left(1+\frac{[Q]}{K_{iq}}\right)+[A]\left(1+\frac{K_{m_B}}{[B]}\right)} \qquad \text{(IX-129)}$$

$$\frac{v}{V_{max}} = \frac{[B]}{K_{m_B}\left(1+\frac{K_{ia}}{[A]}+\frac{K_{ia}[Q]}{K_{iq}[A]}\right)+[B]\left(1+\frac{K_{m_A}}{[A]}+\frac{K_{m_A}[Q]}{K_{iq}[A]}\right)} \qquad \text{(IX-130)}$$

Q is competitive with respect to A at all B concentrations and is a mixed-type inhibitor with respect to B at unsaturating [A]. Saturating [A] overcomes the inhibition. Thus the product inhibition patterns are different from those of the usual Ordered Bi Bi system. The reciprocal plots and replots are shown in Figures IX-22 to IX-25. Table IX-3 gives the expressions for $K_{i_{slope}}$ and $K_{i_{int}}$. Note that the product inhibition patterns are symmetrical. Thus product inhibition studies only identify A–Q and B–P pairs and do not disclose the order of substrate addition and product release. Binding studies will permit identification of A and Q (only these ligands bind to free E). Once A, B, P, and Q are identified, all the constants except K_{m_P} and K_{m_Q} can be determined. The ratio of K_{m_P}/K_{m_Q} can be calculated from $K_{i_{slope}}^{P/A}$ or $K_{i_{int}}^{P/A}$.

Table IX-3 K_i Expressions for Product Inhibition in Theorell-Chance Bi Bi Systems

Product Inhibitor	Varied Substrate A: $K_{i_{slope}}$ Unsat B	$K_{i_{slope}}$ Sat B	$K_{i_{int}}$ Unsat B	$K_{i_{int}}$ Sat B
P	$K_{ip}\left(1+\frac{[B]}{K_{ib}}\right)$	∞	$K_{ip}\left(1+\frac{[B]}{K_{m_B}}\right)$	∞
Q	K_{iq}	K_{iq}	—	—

Product Inhibitor	Varied Substrate B: $K_{i_{slope}}$ Unsat A	$K_{i_{slope}}$ Sat A	$K_{i_{int}}$ Unsat A	$K_{i_{int}}$ Sat A
P	$K_{ip}=\frac{K_{m_P}K_{iq}}{K_{m_Q}}$	$K_{ip}=\frac{K_{m_P}K_{iq}}{K_{m_Q}}$	—	—
Q	$K_{iq}\left(1+\frac{[A]}{K_{ia}}\right)$	∞	$K_{iq}\left(1+\frac{[A]}{K_{m_A}}\right)$	∞

Rate Constants

The individual rate constants for the Theorell-Chance mechanism are related to the kinetic constants in the following way: k_1, k_{-1}, k_3, and k_{-3} are the same as k_1, k_{-1}, k_4, and k_{-4} shown earlier for the usual Ordered Bi Bi system. In addition:

$$k_2=\frac{V_{\max_f}}{K_{m_B}[E]_t}, \qquad k_{-2}=\frac{V_{\max_r}}{K_{m_P}[E]_t}$$

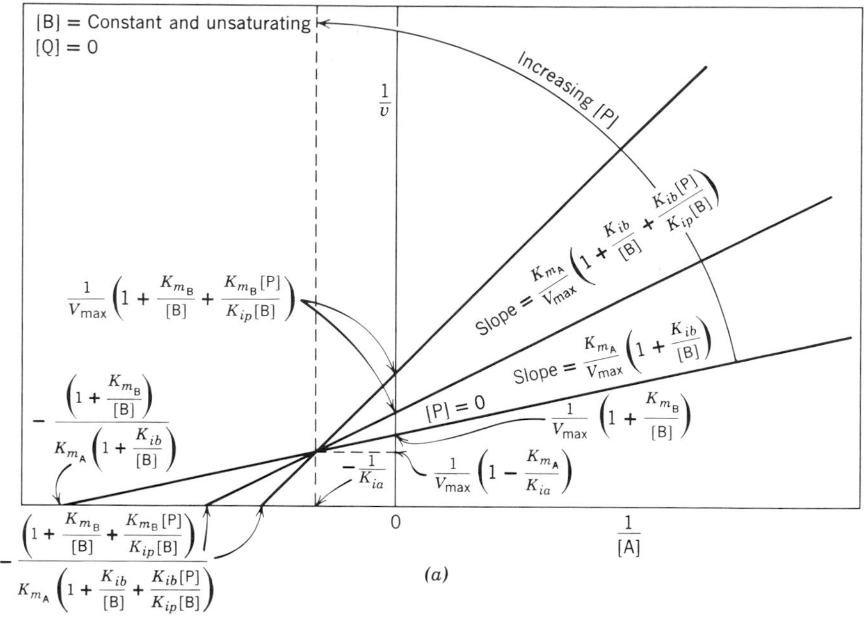

(a)

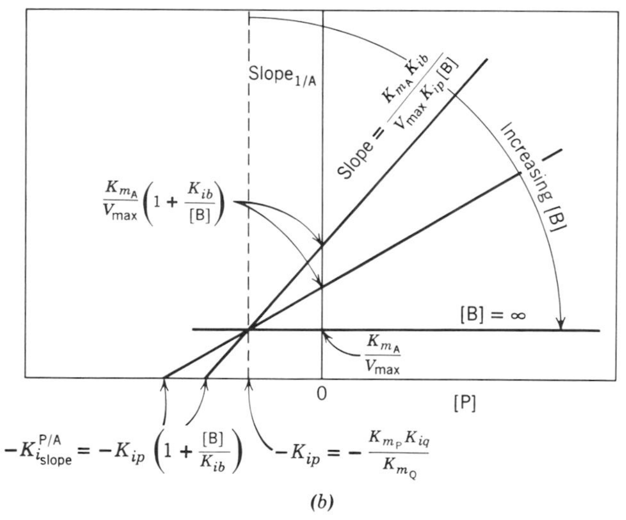

(b)

Fig. IX-22. Product inhibition by P in a Theorell-Chance Bi Bi system. (*a*) $1/v$ versus $1/[A]$ at different fixed P concentrations and a constant, unsaturating B; Q is absent. (*b*)–(*e*) *Slope*$_{1/A}$, and $1/v$-axis intercept replots and the secondary replots of $K_{i_{slope}}^{P/A}$ and $K_{i_{int}}^{P/A}$.

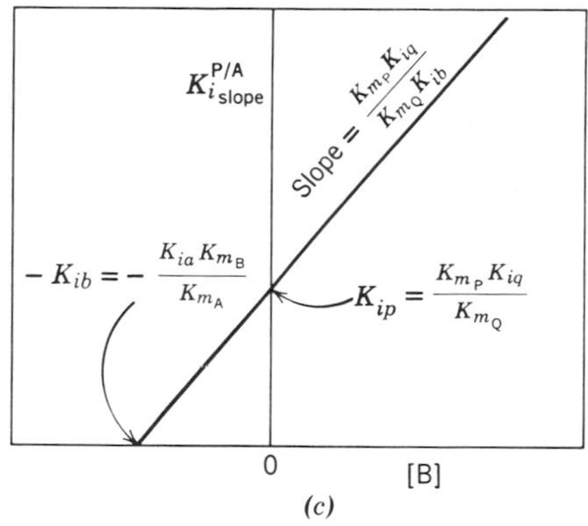

(c)

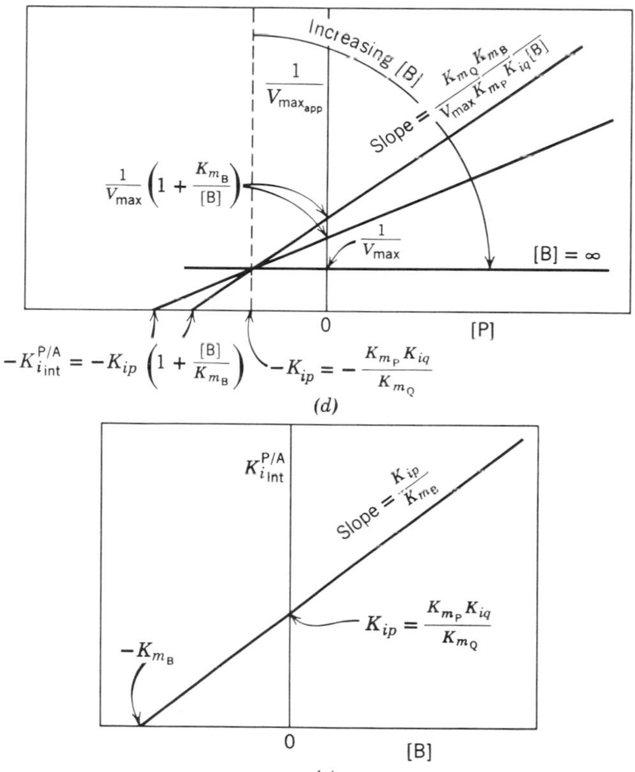

(d)

(e)

Fig. IX-22. (*Cont.*)

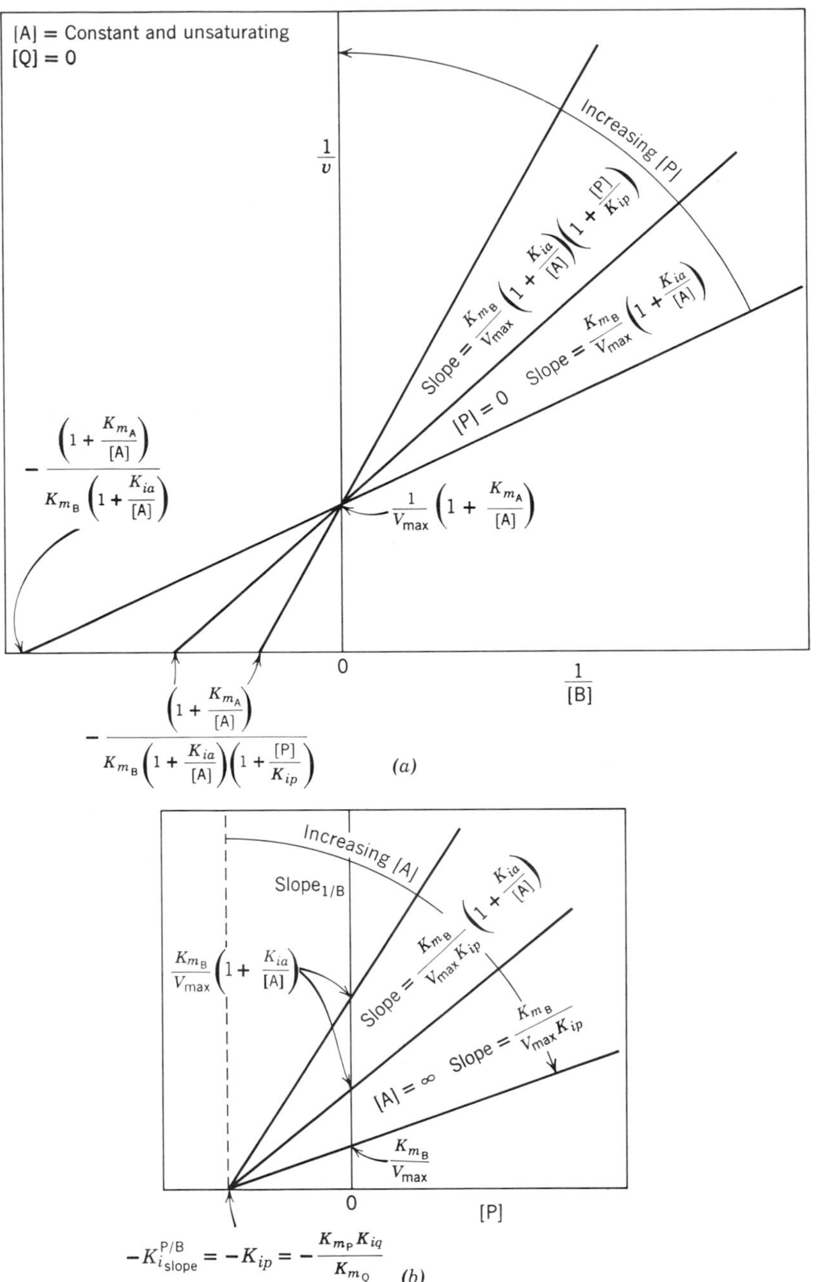

Fig. IX-23. Product inhibition by P in a Theorell-Chance Bi Bi system. (*a*) $1/v$ versus $1/[B]$ at different fixed P concentrations and a constant unsaturating A; Q is absent. (*b*) $Slope_{1/B}$ replots at different constant A concentrations.

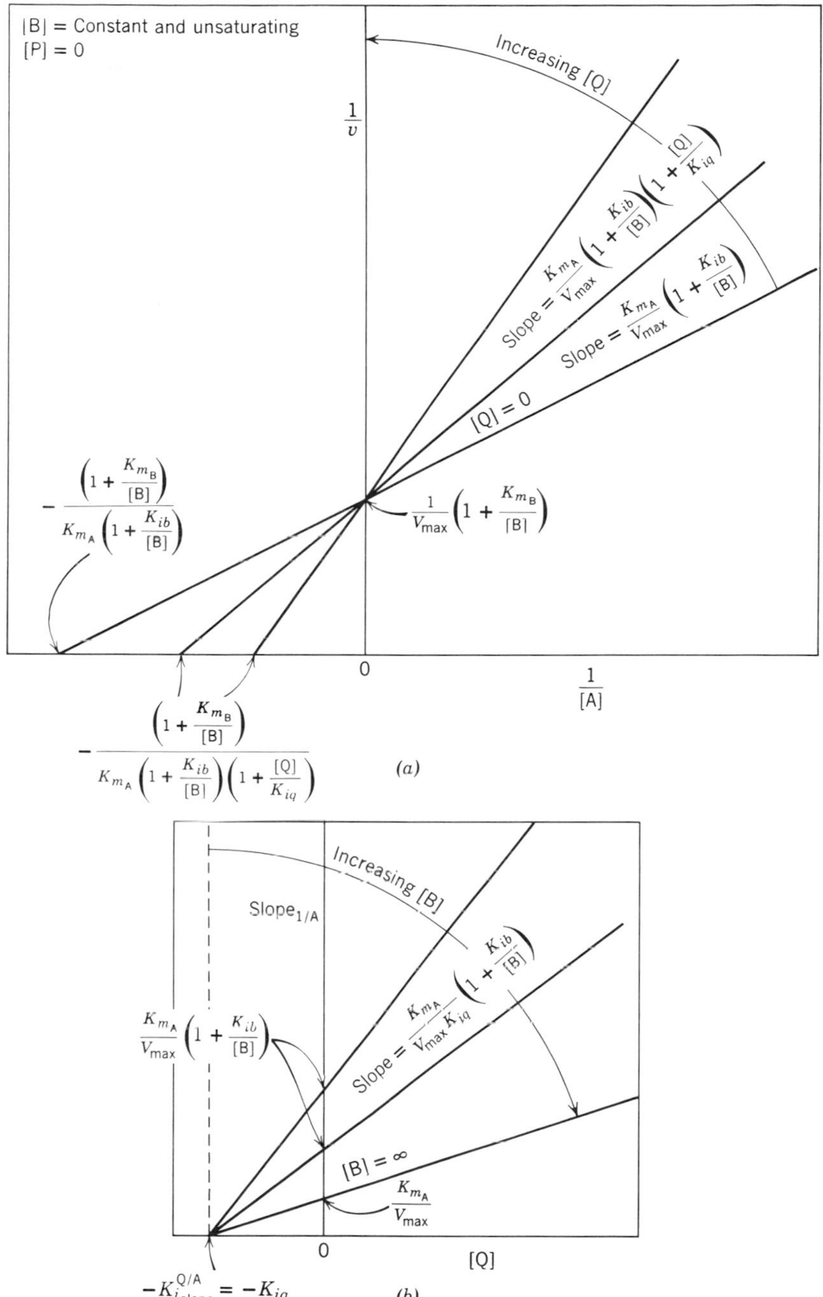

Fig. IX-24. Product inhibition by Q in a Theorell-Chance Bi Bi system. (*a*) $1/v$ versus $1/[A]$ at different fixed Q concentrations and a constant, unsaturating B concentration; P is absent. (*b*) $Slope_{1/A}$ replots at different constant B concentrations.

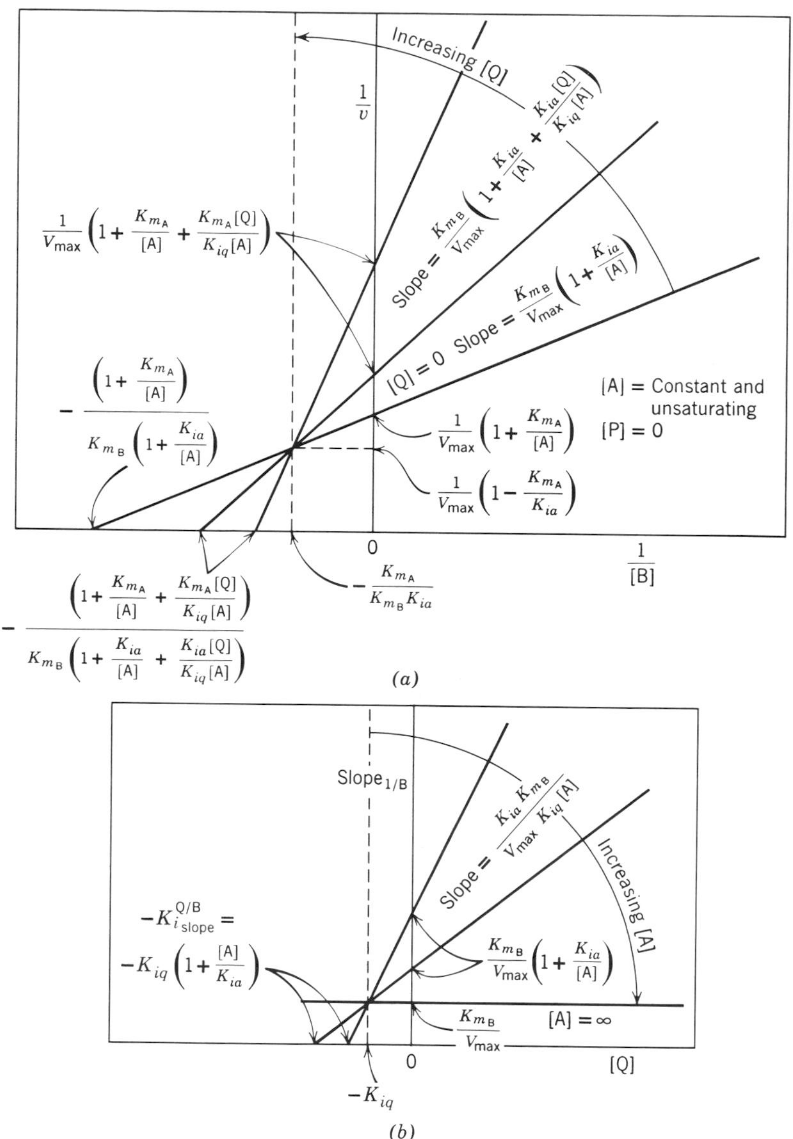

Fig. IX-25. Product inhibition by Q in a Theorell-Chance Bi Bi system. (*a*) $1/v$ versus $1/[B]$ at different fixed Q concentrations and a constant, unsaturating A concentration; P is absent. (*b*)–(*e*) Replots and secondary replots of the primary data. Each replot represents a different constant A concentration.

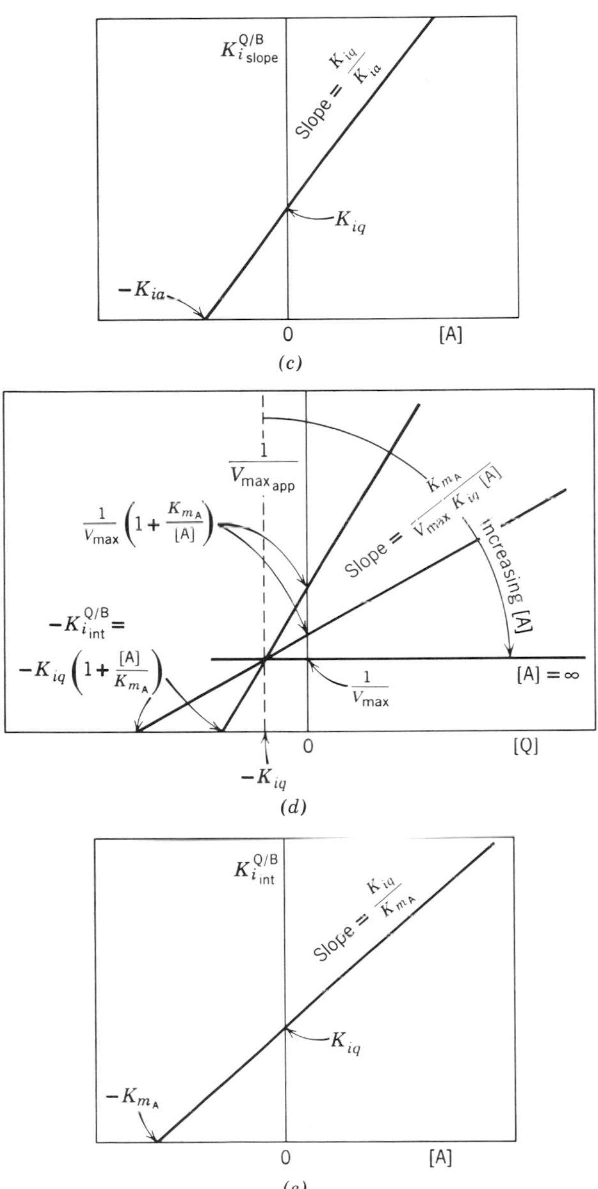

Fig. IX-25. (*Cont.*)

Distribution Equations

$$\frac{[\mathrm{E}]}{[\mathrm{E}]_t} = \frac{K_{ia}K_{m_\mathrm{B}}V_r + \dfrac{K_{m_\mathrm{Q}}V_f[\mathrm{P}]}{K_{\mathrm{eq}}} + K_{m_\mathrm{A}}V_r[\mathrm{B}]}{\text{denominator of velocity equation}} \tag{IX-131}$$

$$\frac{[\mathrm{EA}]}{[\mathrm{E}]_t} = \frac{K_{m_\mathrm{B}}V_r[\mathrm{A}] + \dfrac{K_{m_\mathrm{Q}}V_f[\mathrm{A}][\mathrm{P}]}{K_{ia}K_{\mathrm{eq}}} + \dfrac{V_f[\mathrm{P}][\mathrm{Q}]}{K_{\mathrm{eq}}}}{\text{denominator of velocity equation}} \tag{IX-132}$$

$$\frac{[\mathrm{EAB}+\mathrm{EPQ}]}{[\mathrm{E}]_t} \cong 0 \tag{IX-133}$$

$$\frac{[\mathrm{EQ}]}{[\mathrm{E}]_t} = \frac{\dfrac{K_{m_\mathrm{P}}V_f[\mathrm{Q}]}{K_{\mathrm{eq}}} + V_r[\mathrm{A}][\mathrm{B}] + \dfrac{K_{m_\mathrm{A}}V_r[\mathrm{B}][\mathrm{Q}]}{K_{iq}}}{\text{denominator of velocity equation}} \tag{IX-134}$$

Effect of Isomerizations

EA: k_{-2}, k_3, and k_{-3} may still be calculated. In addition, the rate constant for the conversion of EA to EA′ in the forward direction (k') is given by:

$$\frac{1}{k'} = \frac{[\mathrm{E}]_t}{V_{\max_f}} - \frac{1}{k_3}$$

The $[\mathrm{E}]/[\mathrm{E}]_t$ and $[\mathrm{EQ}]/[\mathrm{E}]_t$ are given by the same equations as shown earlier for Ordered Bi Bi. Also:

$$\frac{[\mathrm{EA}+\mathrm{EA'}]}{[\mathrm{E}]_t} = \frac{K_{m_\mathrm{B}}V_r[\mathrm{A}] + \dfrac{V_f[\mathrm{P}][\mathrm{Q}]}{K_{\mathrm{eq}}} + \left(V_r - \dfrac{V_fK_{m_\mathrm{Q}}}{K_{iq}}\right)[\mathrm{A}][\mathrm{B}] + \dfrac{K_{m_\mathrm{Q}}V_f[\mathrm{A}][\mathrm{P}]}{K_{ia}K_{\mathrm{eq}}}}{\text{denominator of velocity equation}} \tag{IX-135}$$

EA and EQ: No rate constants can be calculated. Only $[\mathrm{E}]/[\mathrm{E}]_t$ can be determined.

Reduction of Ordered Bi Bi to Theorell-Chance

The Theorell-Chance mechanism is a special case of the Ordered Bi Bi mechanism in which the concentrations of the central EAB and EPQ complexes are essentially zero. The velocity equations are identical except that the $[P]/K_{ip}$ term (as defined for Ordered Bi Bi) is missing from the $1/v$-axis intercept factor when P is the product inhibitor. (In the reverse direction, the $[B]/K_{ib}$ term, as defined for Ordered Bi Bi, is missing.) Thus the Ordered Bi Bi mechanism reduces to the Theorell-Chance mechanism when $K_{ip}=\infty$ and $K_{ib}=\infty$. Specifically, the Ordered Bi Bi mechanism reduces to the Theorell-Chance mechanism when certain rate constants are very large compared to others. The conditions are: (*a*) $k_3 \gg k_4$, (*b*) $k_{-2} \gg k_{-1}$, (*c*) $k_3 \gg k_{-3}$, and (*d*) $k_{-2} \gg k_2$. Condition *a* eliminates the [A][B] term in the distribution equation for $[EAB+EPQ]/[E]_t$, but does not eliminate the [A][B] term in the distribution equation for $[EQ]/[E]_t$. Condition *b* eliminates the [P][Q] term from the distribution equation for $[EAB+EPQ]/[E]_t$, but does not eliminate the [P][Q] term from the distribution equation for $[EA]/[E]_t$. Condition *c* eliminates the [B][P][Q] term from the distribution equation for $[EAB+EPQ]/[E]_t$, but does not eliminate the [B][Q] term from the distribution equation for $[EQ]/[E]_t$. Condition *d* eliminates the [A][B][P] term from the distribution equation for $[EAB+EPQ]/[E]_t$, but does not eliminate the [A][P] term from $[EA]/[E]_t$. Thus, when all four conditions are met, $[EAB+EPQ]/[E]_t$ is essentially zero, but the concentrations of all the other species are finite. In effect, conditions *c* and *d* state that the central complexes breakdown as rapidly as they form. It is unlikely that a true Theorell-Chance system exists.

Evaluating the Kinetic Significance of the Central Complexes

Janson and Cleland (1974) have presented a simple way of evaluating the kinetic importance of the central complexes in an Ordered Bi Bi system where the reaction can be measured in both directions. The ratio R is calculated, where R equals the sum of the vertical intersection coordinates of the original reciprocal plots in both directions, divided by the sum of the reciprocal V_{max} values in both directions:

$$R=\frac{\sum \text{vertical intersection coordinates}}{\sum \frac{1}{V_{max}}}=\frac{\frac{1}{V_{max_f}}\left(1-\frac{K_{m_A}}{K_{ia}}\right)+\frac{1}{V_{max_r}}\left(1-\frac{K_{m_Q}}{K_{iq}}\right)}{\frac{1}{V_{max_f}}+\frac{1}{V_{max_r}}} \quad \text{(IX-135a)}$$

The value of R varies from zero for a Theorell-Chance mechanism (where the second product release step is solely rate limiting) to 1.0 for a Rapid Equilibrium Ordered system (where catalysis and/or the first product release step is solely rate limiting). If V_{max_f} and V_{max_r} are unequal, R indicates the rate limiting step in the slower direction only. In this case R gives the fraction of the total enzyme present as the central complexes when both substrates for the slower direction are saturating. The ratio $R/(1-R)$ then gives the ratio of the central complexes to all other enzyme species present.

I. PING PONG BI BI SYSTEM

A mechanism in which a product is released between the addition of two substrates is called Ping Pong. Ping Pong mechanisms are common in group transfer or "substituted enzyme" reactions. For example, consider a transaminase reaction:

$$E + H_2N—\overset{\overset{\displaystyle R}{|}}{C}H—COOH \rightleftharpoons E—NH_2 + O=\overset{\overset{\displaystyle R}{|}}{C}—COOH$$

$$E—NH_2 + O=\overset{\overset{\displaystyle R'}{|}}{C}—COOH \rightleftharpoons E + H_2N—\overset{\overset{\displaystyle R'}{|}}{C}—COOH$$

In this case, the keto acid product is released before the keto acid substrate adds. If we show the individual steps using the Cleland notation, the reaction can be written as:

$$\begin{array}{ccccccccc} & A & & P & & B & & Q & \\ & k_1 \downarrow k_{-1} & & k_2 \uparrow k_{-2} & & k_3 \downarrow k_{-3} & & k_4 \uparrow k_{-4} & \\ \hline E & & (EA \rightleftharpoons FP) & & F & & (FB \rightleftharpoons EQ) & & E \end{array}$$

where E = free enzyme
EA = the enzyme-substrate (amino $acid_1$) complex
FP = the amino enzyme-product (keto $acid_1$) complex
F = the amino enzyme
FB = the amino enzyme-substrate (keto $acid_2$) complex
EQ = the enzyme-product (amino $acid_2$) complex

Thus the mechanism involves two different stable enzyme forms, and two sets of central complexes. Since each substrate addition is followed by a product release, the mechanism could be called Uni Uni Uni Uni Ping Pong, or Tetra Uni Ping Pong. The basic King-Altman figure is shown below.

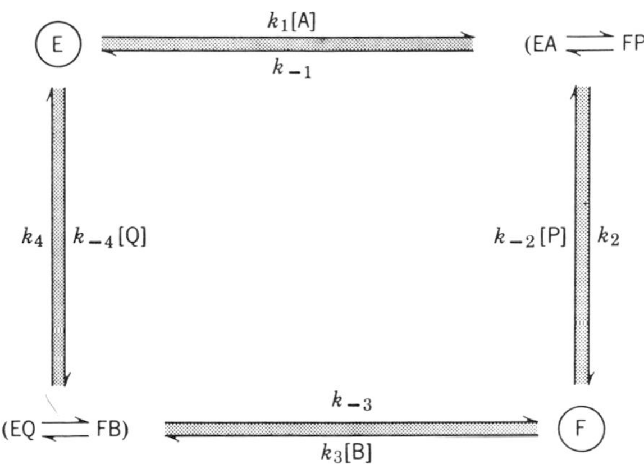

The three-lined interconversion patterns yield the distribution equations shown below.

$$\frac{[\mathrm{E}]}{[\mathrm{E}]_t} = \frac{k_4k_{-1}k_{-2}[\mathrm{P}] + k_2k_3[\mathrm{B}]k_4 + k_3[\mathrm{B}]k_4k_{-1} + k_{-3}k_{-2}[\mathrm{P}]k_{-1}}{\text{denominator}} \qquad \text{(IX-136)}$$

$$\frac{[\mathrm{EA}+\mathrm{FP}]}{[\mathrm{E}]_t} = \frac{k_4k_1[\mathrm{A}]k_{-2}[\mathrm{P}] + k_{-4}[\mathrm{Q}]k_{-3}k_{-2}[\mathrm{P}] + k_3[\mathrm{B}]k_4k_1[\mathrm{A}] + k_{-3}k_{-2}[\mathrm{P}]k_1[\mathrm{A}]}{\text{denominator}} \qquad \text{(IX-137)}$$

$$\frac{[\mathrm{F}]}{[\mathrm{E}]_t} = \frac{k_4k_1[\mathrm{A}]k_2 + k_{-4}[\mathrm{Q}]k_{-3}k_2 + k_{-1}k_{-4}[\mathrm{Q}]k_{-3} + k_{-3}k_1[\mathrm{A}]k_2}{\text{denominator}} \qquad \text{(IX-138)}$$

$$\frac{[\mathrm{FB}+\mathrm{EQ}]}{[\mathrm{E}]_t} = \frac{k_{-2}[\mathrm{P}]k_{-1}k_{-4}[\mathrm{Q}] + k_{-4}[\mathrm{Q}]k_2k_3[\mathrm{B}] + k_{-1}k_{-4}[\mathrm{Q}]k_3[\mathrm{B}] + k_1[\mathrm{A}]k_2k_3[\mathrm{B}]}{\text{denominator}} \qquad \text{(IX-139)}$$

Letting $v = k_1[\mathrm{E}][\mathrm{A}] - k_{-1}[\mathrm{EA+FP}]$ and grouping denominator terms, we obtain:

$$v = \frac{k_1k_2k_3k_4[\mathrm{E}]_t[\mathrm{A}][\mathrm{B}] - k_{-1}k_{-2}k_{-3}k_{-4}[\mathrm{E}]_t[\mathrm{P}][\mathrm{Q}]}{\begin{array}{l} k_1k_2(k_{-3}+k_4)[\mathrm{A}] + k_3k_4(k_2+k_{-1})[\mathrm{B}] + k_{-1}k_{-2}(k_{-3}+k_4)[\mathrm{P}] \\ + k_{-3}k_{-4}(k_{-1}+k_2)[\mathrm{Q}] + k_1k_3(k_2+k_4)[\mathrm{A}][\mathrm{B}] + k_1k_{-2}(k_{-3}+k_4)[\mathrm{A}][\mathrm{P}] \\ + k_{-2}k_{-4}(k_{-1}+k_{-3})[\mathrm{P}][\mathrm{Q}] + k_3k_{-4}(k_{-1}+k_2)[\mathrm{B}][\mathrm{Q}] \end{array}} \quad \text{(IX-140)}$$

The equation has no ABP, BPQ, nor constant term in the denominator. Multiplying numerator and denominator by the usual factor, $\mathrm{num}_2/(\mathrm{Coef}_{\mathrm{AB}})(\mathrm{Coef}_{\mathrm{PQ}})$, and the P, Q, AP, and PQ terms by $\mathrm{num}_1/\mathrm{num}_1$ as well, we obtain:

$$v = \frac{V_f V_r\left([\mathrm{A}][\mathrm{B}] - \dfrac{[\mathrm{P}][\mathrm{Q}]}{K_{\mathrm{eq}}}\right)}{\begin{array}{l} V_r K_{m_\mathrm{B}}[\mathrm{A}] + V_r K_{m_\mathrm{A}}[\mathrm{B}] + \dfrac{V_f K_{m_\mathrm{Q}}[\mathrm{P}]}{K_{\mathrm{eq}}} + \dfrac{V_f K_{m_\mathrm{P}}[\mathrm{Q}]}{K_{\mathrm{eq}}} + V_r[\mathrm{A}][\mathrm{B}] \\ + \dfrac{V_f K_{m_\mathrm{Q}}[\mathrm{A}][\mathrm{P}]}{K_{\mathrm{eq}}K_{ia}} + \dfrac{V_f[\mathrm{P}][\mathrm{Q}]}{K_{\mathrm{eq}}} + \dfrac{V_r K_{m_\mathrm{A}}[\mathrm{B}][\mathrm{Q}]}{K_{iq}} \end{array}} \quad \text{(IX-141)}$$

where the definitions of the K_m's, K_{ia}, and K_{iq} are the same as those given for the Ordered Bi Bi system. The definitions of K_{ib} and K_{ip} used later in deriving the Haldane equations are:

$$K_{ib} = \frac{\mathrm{Coef}_\mathrm{Q}}{\mathrm{Coef}_{\mathrm{BQ}}} = \frac{k_{-3}}{k_3}, \qquad K_{ip} = \frac{\mathrm{Coef}_\mathrm{A}}{\mathrm{Coef}_{\mathrm{AP}}} = \frac{k_2}{k_{-2}}$$

Initial Forward Velocity in the Absence of Products

In the absence of P and Q, the forward velocity is given by:

$$\boxed{\frac{v}{V_{\max}} = \frac{[\mathrm{A}][\mathrm{B}]}{K_{m_\mathrm{B}}[\mathrm{A}] + K_{m_\mathrm{A}}[\mathrm{B}] + [\mathrm{A}][\mathrm{B}]}} \quad \text{(IX-142)}$$

Rearranged to show each substrate as the varied substrate:

$$\frac{v}{V_{max}} = \frac{[A]}{K_{m_A} + [A]\left(1 + \frac{K_{m_B}}{[B]}\right)} \quad \text{(IX-143)}$$

$$\frac{v}{V_{max}} = \frac{[B]}{K_{m_B} + [B]\left(1 + \frac{K_{m_A}}{[A]}\right)} \quad \text{(IX-144)}$$

In reciprocal form, the equations become:

$$\frac{1}{v} = \frac{K_{m_A}}{V_{max}} \frac{1}{[A]} + \frac{1}{V_{max}}\left(1 + \frac{K_{m_B}}{[B]}\right) \quad \text{(IX-145)}$$

and

$$\frac{1}{v} = \frac{K_{m_B}}{V_{max}} \frac{1}{[B]} + \frac{1}{V_{max}}\left(1 + \frac{K_{m_A}}{[A]}\right) \quad \text{(IX-146)}$$

The reciprocal plots and replots are shown in Figure IX-26 for A as the varied substrate. The plots for B as the varied substrate are symmetrical to those for A. While the system is symmetrical with respect to A and B (and also with respect to the A–Q, B–P pairs, as shown later), there is generally no difficulty in identifying which substrate is A and which is B. A must be the substrate with the group to be transferred and B the acceptor. Similarly, P must be the product of A while Q is the product of B plus the group transferred from A. If [A] and [B] are varied together while maintaining their concentrations at a constant ratio, $[B] = x[A]$, the reciprocal equation becomes:

$$\frac{1}{v} = \frac{K_{m_A}}{V_{max}}\left(1 + \frac{K_{m_B}}{xK_{m_A}}\right)\frac{1}{[A]} + \frac{1}{V_{max}} \quad \text{(IX-147)}$$

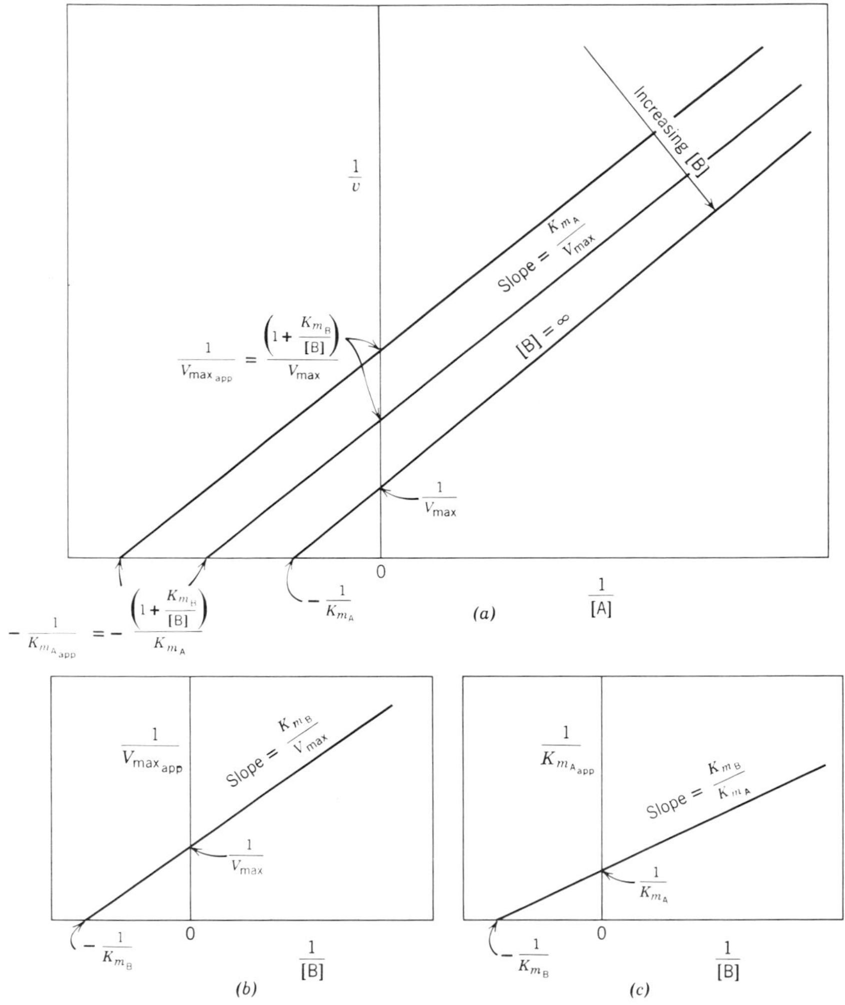

Fig. IX-26. Ping Pong Bi Bi system. (*a*) $1/v$ versus $1/[A]$ at different fixed B concentrations. (*b*) $1/v$-axis intercept replot. (*c*) $1/K_{m_{A\,app}}$ replot. The plots and replots for B as the varied substrate are symmetrical to those shown for varied [A].

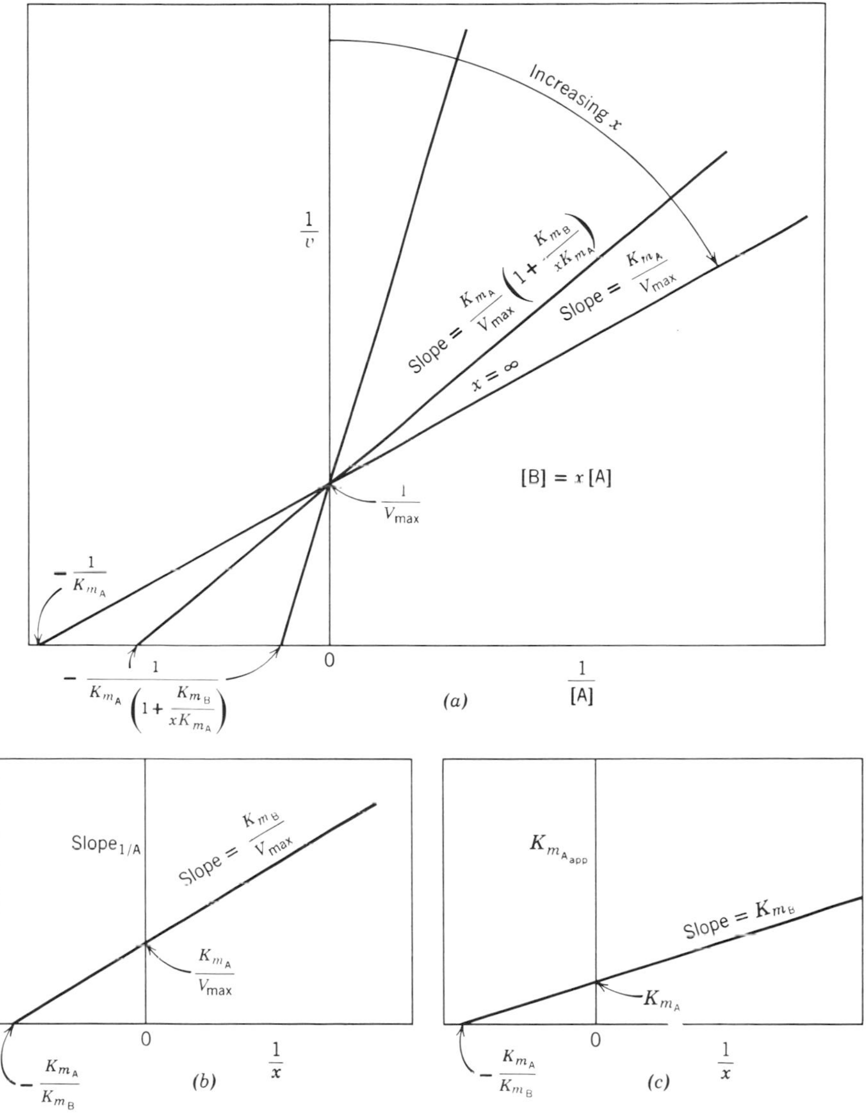

Fig. IX-27. Ping Pong Bi Bi system. (*a*) $1/v$ versus $1/[A]$ where the [B]/[A] ratio is maintained constant. (*b*) Replot of the *slope*$_{1/A}$ versus $1/x$, where $x = [B]/[A]$. (*c*) Replot of $K_{m_{A_{app}}}$ versus $1/x$.

The reciprocal plot is linear (Fig. IX-27*a*) unlike the plot for the Ordered Bi Bi systems. The $1/v$-axis intercept gives the reciprocal of the true V_{max}. The apparent K_{m_A} is a function of the ratio constant, x. If a series of plots are constructed at different values of x, K_{m_A} and K_{m_B} can be determined from slope or $K_{m_{A(app)}}$ replots (Fig. IX-27*b* and *c*).

$$\boxed{slope_{1/A} = \frac{K_{m_B}}{V_{max}}\frac{1}{x} + \frac{K_{m_A}}{V_{max}}} \qquad \text{(IX-148)}$$

$$\boxed{K_{m_{A(app)}} = K_{m_B}\frac{1}{x} + K_{m_A}} \qquad \text{(IX-149)}$$

The family of parallel lines (Fig. IX-26*a*) and the linear reciprocal plot when [A] and [B] are varied together (Fig. IX-27*a*) are generally sufficient to identify a Ping Pong Bi Bi system. However, the Ordered Bi Bi system will yield the same results if K_{ia} is very small compared to K_{m_A}. Product inhibition studies are then necessary to distinguish between the two systems.

Figures IX-28, IX-29, and IX-30 show the alternate linear plots for a Ping Pong bireactant system where [A] is varied. The plots for varied [B] are symmetrical to those shown for [A].

Haldane Equations

There is no constant term in the denominator, but we can obtain a Haldane equation from the AP term:

$$\frac{V_f K_{m_Q}[A][P]}{K_{eq}K_{ia}} = \frac{(\text{num}_1)(\text{Coef}_P)(\text{num}_2)(\text{Coef}_{AP})[A][P]}{(\text{Coef}_{AB})(\text{Coef}_{PQ})(\text{num}_1)(\text{Coef}_P)} = \frac{(\text{num}_2)(\text{Coef}_{AP})[A][P]}{(\text{Coef}_{PQ})(\text{Coef}_{AB})}$$

$$= \frac{V_r(\text{Coef}_A)K_{m_B}[A][P]}{K_{ip}(\text{Coef}_A)} = \frac{V_r K_{m_B}[A][P]}{K_{ip}}$$

$$\therefore \quad \boxed{K_{eq} = \frac{V_f K_{m_Q} K_{ip}}{V_r K_{ia} K_{m_B}}} \qquad \text{(IX-150)}$$

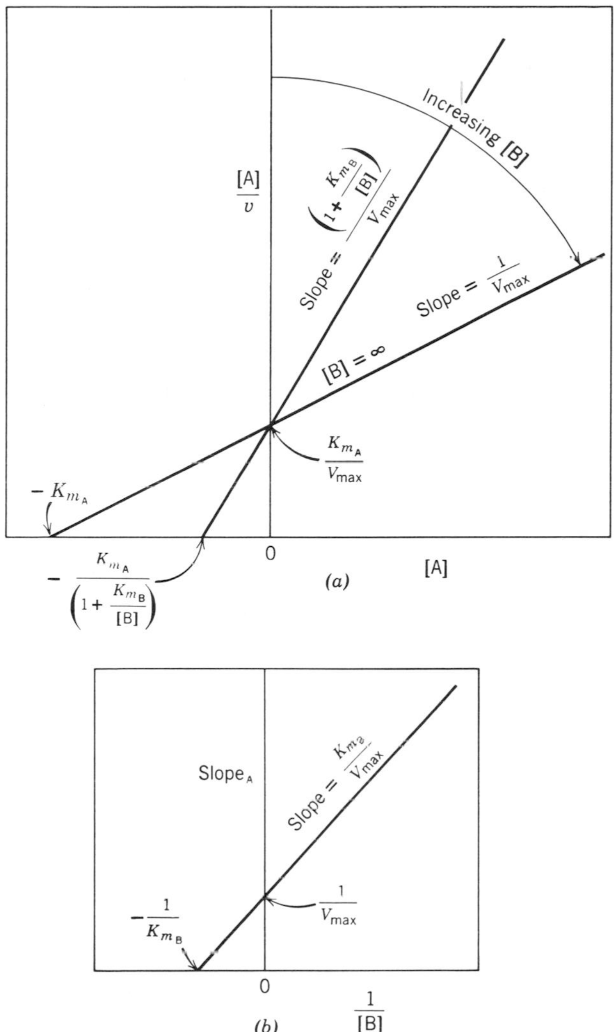

Fig. IX-28. Ping Pong Bi Bi system. (*a*) $[A]/v$ versus [A] at different fixed B concentrations. (*b*) $Slope_A$ replot. The plots for B as the varied substrate are symmetrical to those shown for varied [A].

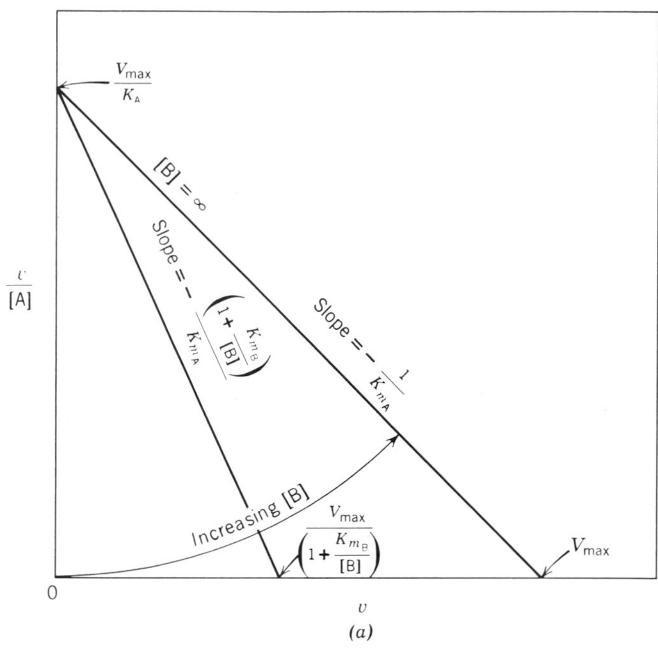

(a)

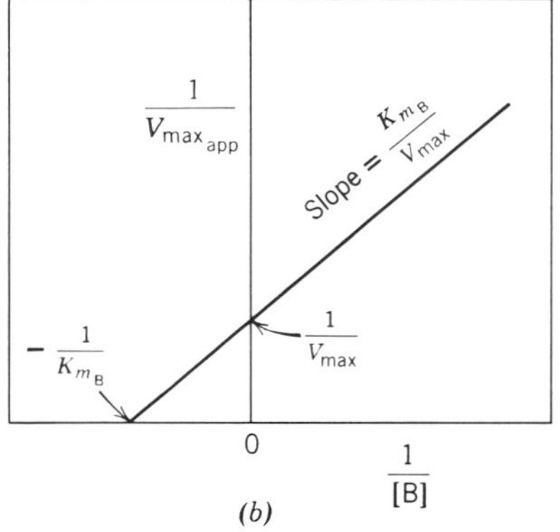

(b)

Fig. IX-29. Ping Pong Bi Bi system. (*a*) $v/[A]$ versus v at different fixed B concentrations. (*b*) v-axis intercept replot. The plots for varied [B] are symmetrical to those shown for varied [A].

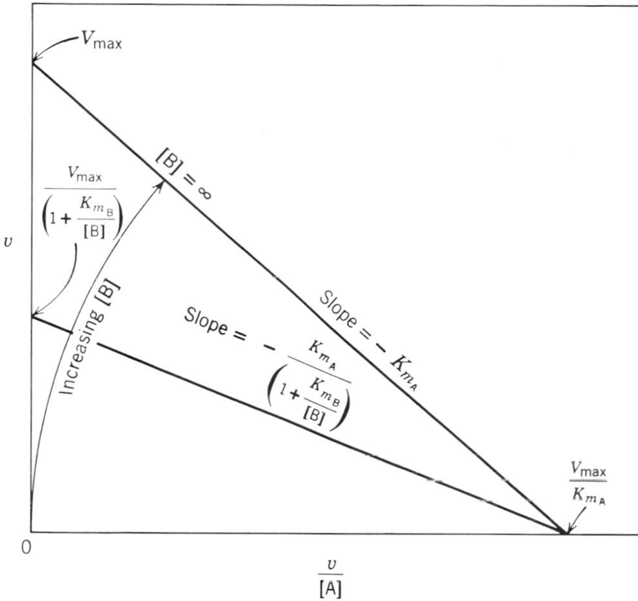

Fig. IX-30. Ping Pong Bi Bi system; v versus $v/[A]$. The replots are the same as those shown for the $v/[A]$ versus v plot. The plots for varied [B] are symmetrical to those shown for varied [A].

Similarly, from the Q term we obtain:

$$\frac{V_f K_{m_P}[Q]}{K_{eq}} = \frac{(num_1)(num_2)(Coef_Q)[Q]}{(Coef_{AB})(num_1)(Coef_{PQ})} = \frac{V_r Coef_Q[Q]}{Coef_{AB}} = \frac{V_r K_{ib} Coef_{BQ}[Q]}{Coef_{AB}}$$

$$= \frac{V_r K_{ib} Coef_B[Q]}{K_{iq} Coef_{AB}} = \frac{V_r K_{ib} K_{m_A}[Q]}{K_{iq}}$$

$$\therefore \quad \boxed{K_{eq} = \frac{V_f K_{m_P} K_{iq}}{V_r K_{ib} K_{m_A}}} \qquad \text{(IX-151)}$$

From equations IX-150 and IX-151, we see that:

$$\frac{K_{m_Q} K_{ip}}{K_{ia} K_{m_B}} = \frac{K_{m_P} K_{iq}}{K_{ib} K_{m_A}} \qquad \text{(IX-152)}$$

Other Haldane equations for the Ping Pong Bi Bi system are:

$$\boxed{K_{eq} = \frac{K_{ip}K_{iq}}{K_{ia}K_{ib}}} \quad \text{and} \quad \boxed{K_{eq} = \left(\frac{V_f}{V_r}\right)^2 \frac{K_{m_P}K_{m_Q}}{K_{m_A}K_{m_B}}} \qquad \text{(IX-153)}$$

Also:

$$\frac{V_{\max_f}K_{ia}}{K_{m_A}} = \frac{V_{\max_r}K_{ip}}{K_{m_P}} = \frac{k_{-1}k_2}{(k_{-1}+k_2)} \quad \text{and} \quad \frac{V_{\max_f}K_{ib}}{K_{m_B}} = \frac{V_{\max_r}K_{iq}}{K_{m_Q}} = \frac{k_{-3}k_4}{(k_{-3}+k_4)}$$

Product Inhibition

Eliminating K_{eq} from the velocity equation by means by equations IX-150 and IX-152 we obtain for varied [A] in the presence of P:

$$\boxed{\frac{v}{V_{\max}} = \frac{[A]}{K_{m_A}\left(1 + \frac{K_{ia}K_{m_B}[P]}{K_{m_A}K_{ip}[B]}\right) + [A]\left(1 + \frac{K_{m_B}[P]}{K_{ip}[B]} + \frac{K_{m_B}}{[B]}\right)}} \qquad \text{(IX-154)}$$

In the presence of Q:

$$\boxed{\frac{v}{V_{\max}} = \frac{[A]}{K_{m_A}\left(1 + \frac{K_{ib}[Q]}{K_{iq}[B]} + \frac{[Q]}{K_{iq}}\right) + [A]\left(1 + \frac{K_{m_B}}{[B]}\right)}} \qquad \text{(IX-155)}$$

When [B] is varied, the equations are:

$$\boxed{\frac{v}{V_{\max}} = \frac{[B]}{K_{m_B}\left(1 + \frac{K_{ia}[P]}{K_{ip}[A]} + \frac{[P]}{K_{ip}}\right) + [B]\left(1 + \frac{K_{m_A}}{[A]}\right)}} \qquad \text{(IX-156)}$$

$$\boxed{\frac{v}{V_{\max}} = \frac{[B]}{K_{m_B}\left(1 + \frac{K_{ib}K_{m_A}[Q]}{K_{m_B}K_{iq}[A]}\right) + [B]\left(1 + \frac{K_{m_A}}{[A]} + \frac{K_{m_A}[Q]}{K_{iq}[A]}\right)}} \qquad \text{(IX-157)}$$

Thus P is a mixed-type inhibitor with respect to A at unsaturating [B] and competitive with B at all A concentrations. Q is competitive with A at all B concentrations and a mixed-type inhibitor with respect to B at unsaturating [A]. The reciprocal plots and replots are shown in Figures IX-31 and IX-32 for A as the varied substrate. The plots for varied [B] are symmetrical to those for A. All the constants except K_{m_P} and K_{m_Q} can be determined from the reciprocal plots and replots. The ratio K_{m_P}/K_{m_Q} can be obtained from the $slope_{1/B}$ replot with Q as the inhibitor, or from $K_{i_{slope}}^{Q/B}$. The apparent K_i values are given in Table IX-4. Individual rate constants cannot be calculated from the kinetic constants.

Table IX-4 K_i Expressions for Product Inhibition in Ping Pong Bi Bi Systems

Product Inhibitor	Varied Substrate A: $K_{i_{slope}}$ Unsat B	A: $K_{i_{slope}}$ Sat B	A: $K_{i_{int}}$ Unsat B	A: $K_{i_{int}}$ Sat B	B: $K_{i_{slope}}$ Unsat A	B: $K_{i_{slope}}$ Sat A	B: $K_{i_{int}}$ Unsat A	B: $K_{i_{int}}$ Sat A
P	$\frac{K_{ip}K_{m_A}[B]}{K_{ia}K_{m_B}}$ or $\frac{K_{m_P}K_{iq}[B]}{K_{m_Q}K_{ib}}$	∞	$K_{ip}\left(1+\frac{[B]}{K_{m_B}}\right)$	∞	$\frac{K_{ip}}{\left(1+\frac{K_{ia}}{[A]}\right)}$	K_{ip}	—	—
Q	$\frac{K_{iq}}{\left(1+\frac{K_{ib}}{[B]}\right)}$	K_{iq}	—	—	$\frac{K_{m_Q}K_{ip}[A]}{K_{m_P}K_{ia}}$ or $\frac{K_{iq}K_{m_B}[A]}{K_{ib}K_{m_A}}$	∞	$K_{iq}\left(1+\frac{[A]}{K_{m_A}}\right)$	∞

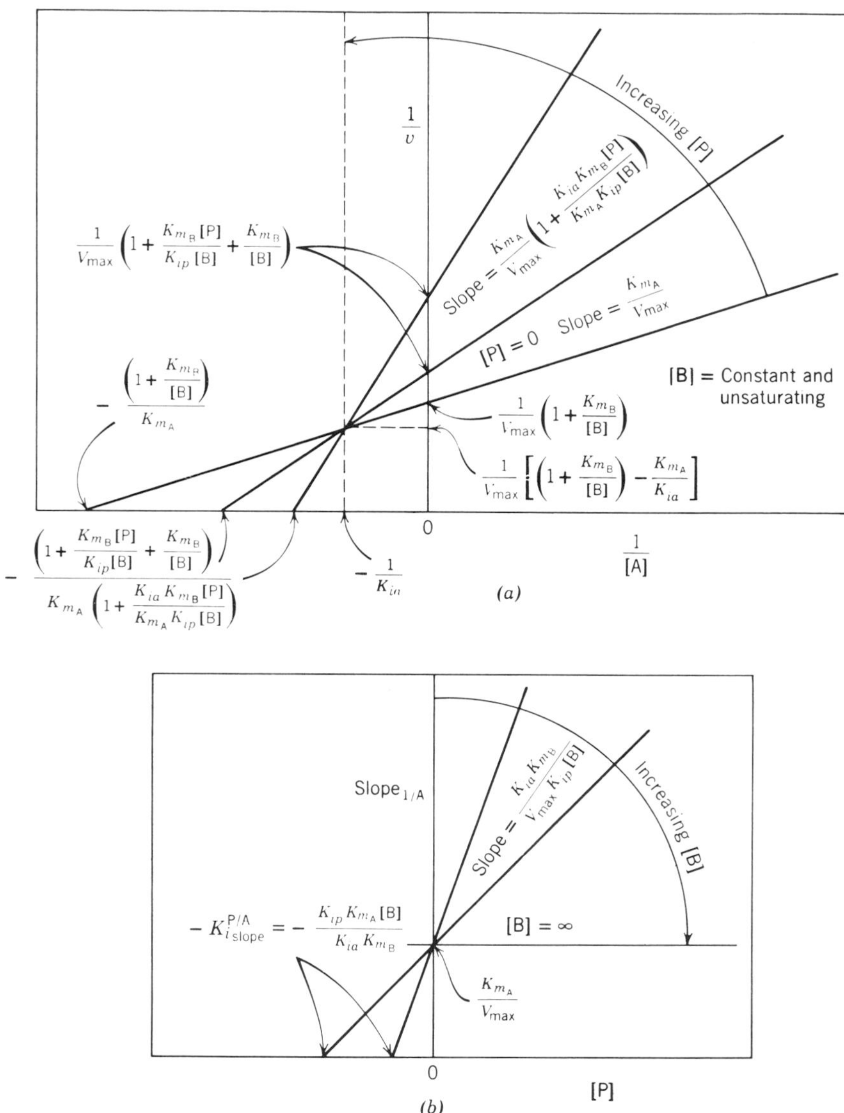

Fig. IX-31. Product inhibition by P in a Ping Pong Bi Bi system. (*a*) $1/v$ versus $1/[A]$ at different fixed P concentrations and a constant, unsaturating B concentration; Q is absent. (*b*)–(*e*) $Slope_{1/A}$ and $1/v$-axis intercept replots and secondary replots for different constant B concentrations.

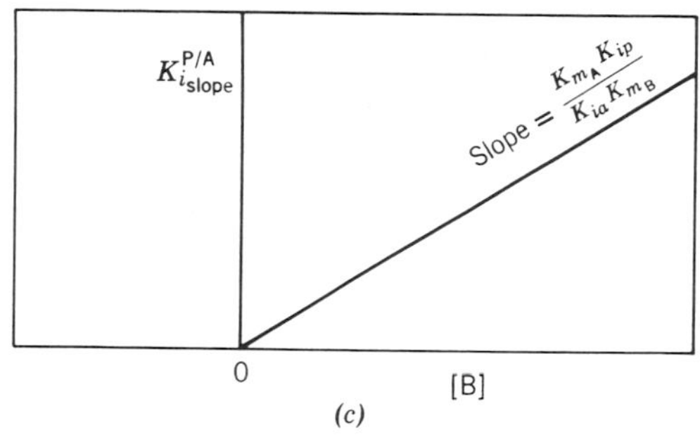

(c)

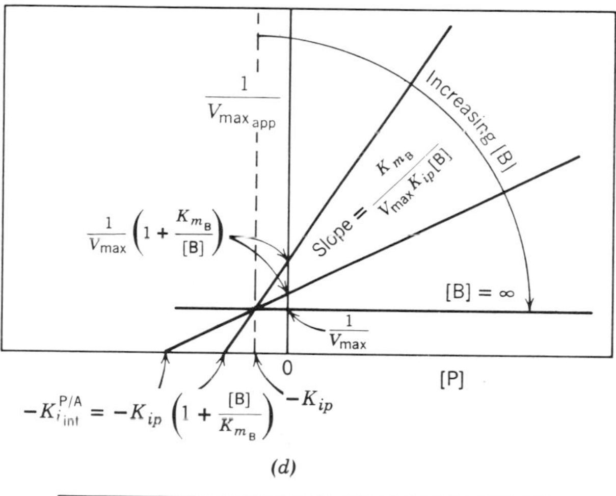

(d)

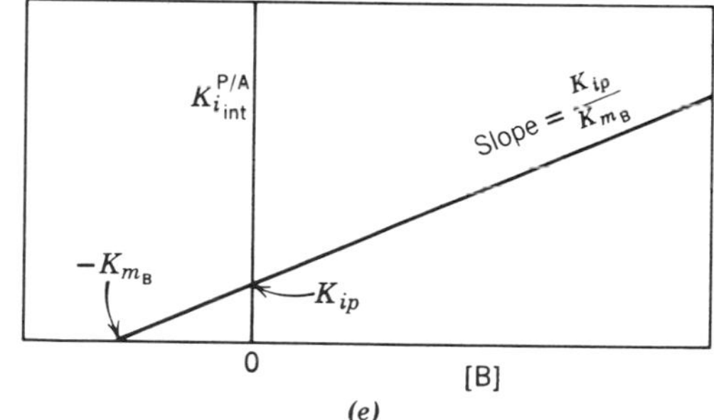

(e)

Fig. IX-31. *(Cont.)*

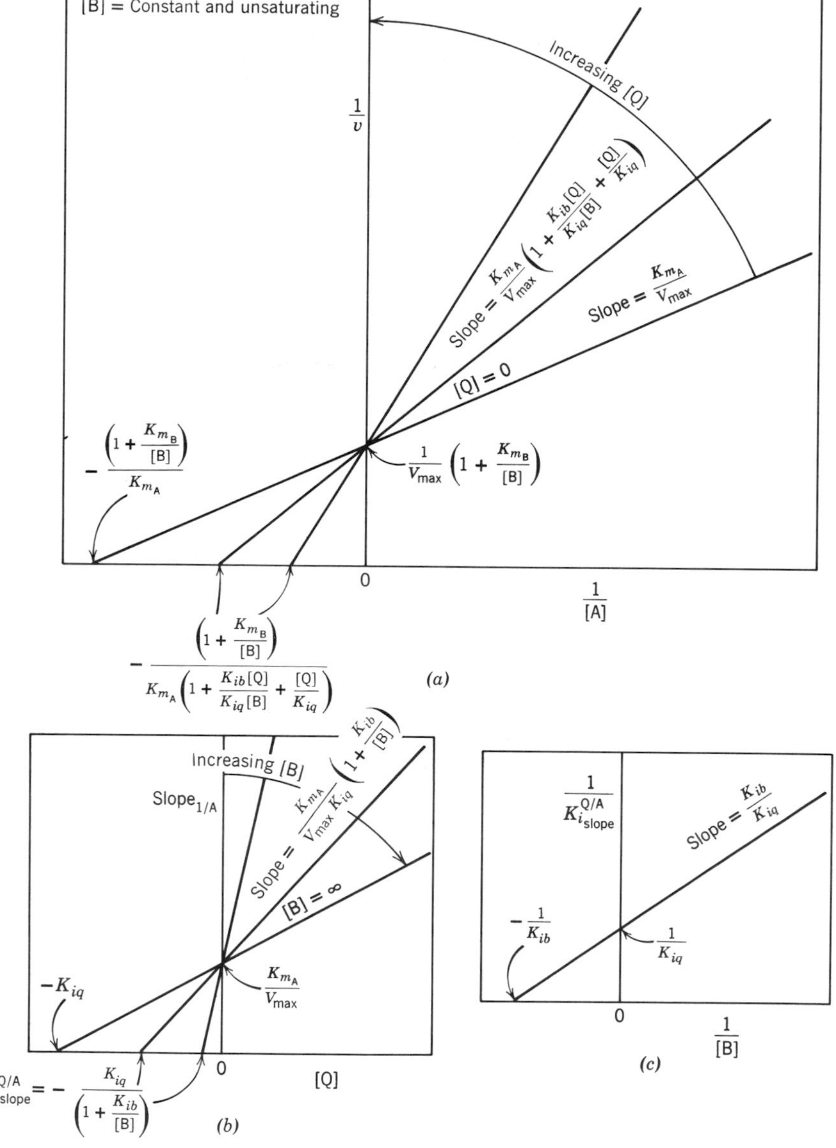

Fig. IX-32. Product inhibition by Q in a Ping Pong Bi Bi system. (*a*) $1/v$ versus $1/[A]$ at different fixed Q concentrations and a constant, unsaturating B concentration; P is absent. (*b*) $Slope_{1/A}$ replots for different constant B concentrations. (*c*) Secondary replot of $1/K_{i_{slope}}^{Q/A}$ versus $1/[B]$. The plots for varied [B] are symmetrical to those shown in Figs. IX-31 and IX-32 for varied [A].

Distribution Equations

$$\frac{[E]}{[E]_t} = \frac{K_{m_A} V_r [B] + \dfrac{K_{m_Q} V_f [P]}{K_{eq}}}{\text{denominator of velocity equation}} \qquad \text{(IX-158)}$$

$$\frac{[F]}{[E]_t} = \frac{K_{m_B} V_r [A] + \dfrac{K_{m_P} V_f [Q]}{K_{eq}}}{\text{denominator of velocity equation}} \qquad \text{(IX-159)}$$

The fraction of the total enzyme present as EA + FP and FB + EQ cannot be determined.

Effect of Isomerizations

Isomerizations of transitory complexes (including central complexes) have no effect on the two distributions that can be calculated.

Effect of Impure Substrates

Noninhibitory impurities in the varied substrate will have no effect on the nature of the reciprocal plot family. Intersecting plots remain intersecting, and parallel plots remain parallel. The intercepts on the $1/v$-axis will be unchanged, but the slopes of the plots will increase if the impurities represent a significant fraction of the varied substrate. For example, if the substrate is only 75% pure, the horizontal coordinates of the velocity points will be misplotted, and the slopes and apparent K_m values will all be 33.3% greater than the actual values (see equation II-35). If the impurity in the varied substrate is competitive with the varied substrate, then both the slopes and the $1/v$-axis intercepts will be in error (equation III-42), but the plots will still be linear and the nature of the family unchanged. Inhibitors present in one substrate that compete with another substrate yield reciprocal plot families identical to those obtained for the corresponding substrate inhibition systems (described in Section HH).

Presteady-State "Burst" Phenomenon with Ping Pong Enzymes

Hydrolytic enzymes often have a Ping Pong Bi Bi mechanism. At the constant water concentration, the reaction can be regarded as Ordered Uni Bi. For example, consider the hydrolysis of *p*-nitrophenylphosphate by a phosphatase. The reaction sequence is shown below.

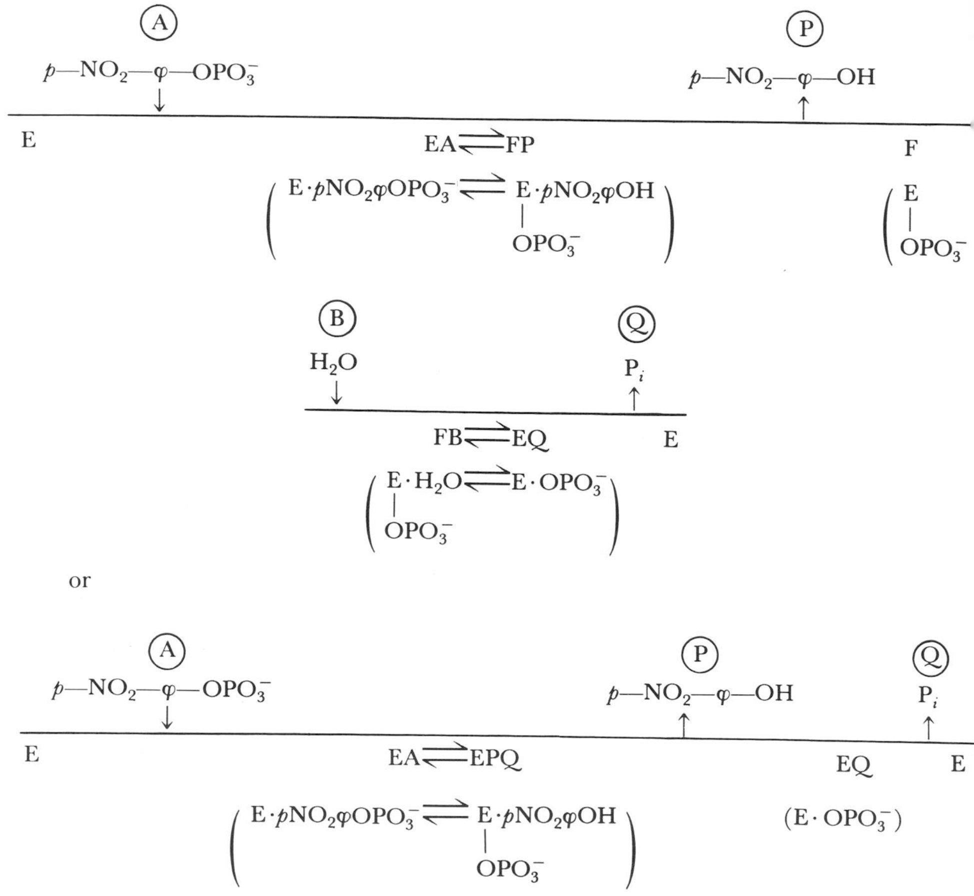

When the reaction is monitored by following the rate of P formation, we may see an initial "burst" of P followed by a slower linear appearance (Fig. IX-33). The "burst" results from a rapid, presteady-state reaction of A with free E in which P is released and an acyl-enzyme is formed. (At the start of the reaction, all the enzyme is present as free E, available for combination with A.) The slower, linear appearance of P occurs after a steady-state is attained. In the steady-state, the rate of P or Q appearance is determined by the slow, rate-limiting hydrolysis of the acyl-enzyme. (The slow step controls the subsequent level of E available for reaction with more A.) The presteady-state phase may last for only a few seconds or less. To follow the

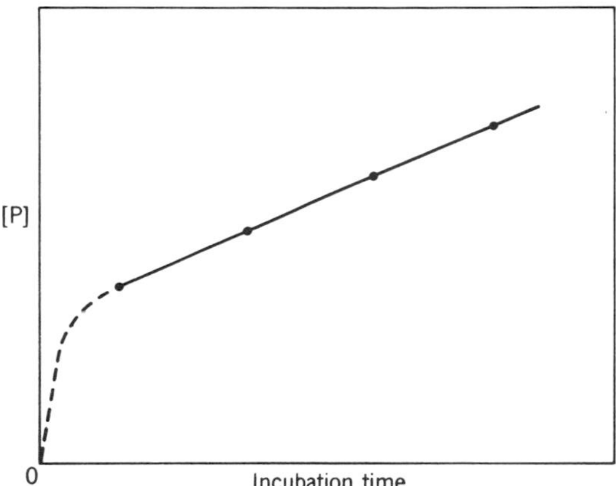

Fig. IX-33. "Burst" phenomenon in a Ping Pong hydrolysis reaction.

appearance of P during this phase, rapid reaction techniques (e.g., stopped flow methods) must be used. Under normal assay conditions, where very low enzyme concentrations are used, the "burst" of P may be undetectable. At relatively high enzyme concentrations, the extrapolated linear portion of the progress curve may intersect the vertical axis at a point significantly above zero. When working at high enzyme concentrations, it is necessary to ensure that $[A]_0 \gg [E]_t$. If $[A]_0$ is not very much greater than $[E]_t$, the appearance of P will not be linear with time in the steady-state. ([A] will decrease very rapidly and the reaction will not be pseudo-zero-order over the assay time.) If the release of Q is not rate-limiting, then even at high enzyme and substrate concentrations, the [P] versus time curve will not be biphasic.

Ordered Bi Bi Systems That Appear to be Ping Pong

It would seem that there should be no difficulty in deciding between a Ping Pong mechanism and an ordered mechanism (or another that yields intersecting reciprocal plots). However, there are some conditions where lines that seem parallel really are not. Figure IX-34 shows a family of reciprocal plots for an Ordered Bi Bi system in which K_{ia} is very much smaller than K_{m_A}; that is, the enzyme has a very high affinity for A but because of the relative values of the other rate constants, K_{m_A} is much larger than K_{ia}. The family of plots intersect far to the left of the $1/v$-axis and far below the $1/[A]$-axis. The same thing is observed when [B] is varied. Allowing for the usual small errors in determining v, it is quite possible that the "best fitting" lines drawn through the points will appear to be parallel. However, the product inhibition patterns will not conform to those expected for a Ping

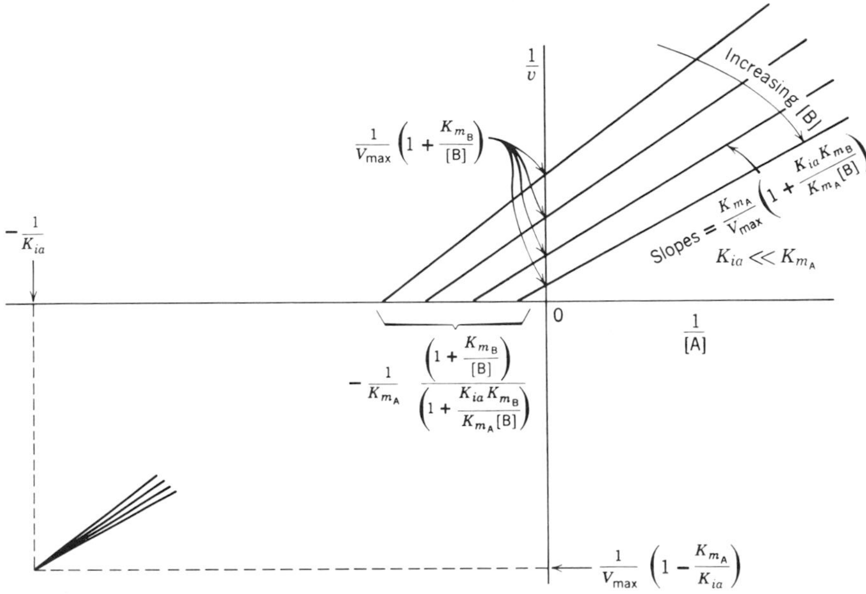

Fig. IX-34. Ordered Bi Bi where $K_{ia} \ll K_{m_A}$. The reciprocal plots may appear to be parallel. If, in addition, $K_{ip} \gg K_{mp}K_{iq}/K_{m_Q}$, then P will appear to be competitive with B.

Pong mechanism. For example, consider a kinase reaction. Suppose that reciprocal plots of $1/v$ versus 1/[ATP] and $1/v$ versus 1/[B] at different fixed concentrations of the nonvaried substrate appear parallel. We would then assume that the sequence is that shown below.

$$\begin{array}{l} \text{ATP} = \text{A} \downarrow \quad \text{ADP} = \text{P} \uparrow \quad \text{B} \downarrow \quad \text{BPO}_4 = \text{Q} \uparrow \\ \text{E} \quad \left(\text{E—ATP} \rightleftharpoons \underset{\text{PO}_4}{\text{E—ADP}}\right) \quad \underset{\text{PO}_4}{\text{E}} \quad \left(\underset{\text{PO}_4}{\text{EB}} \rightleftharpoons \text{E—BPO}_4\right) \quad \text{E} \end{array}$$

However, we find that ADP is competitive with ATP (not a mixed-type inhibitor as expected), and uncompetitive with respect to B (not competitive as expected). Similarly, the inhibition patterns for BPO_4 are different from those expected if the system were truly Ping Pong. We find that BPO_4 appears to be uncompetitive with respect to ATP (instead of competitive).

The patterns conform to an Ordered Bi Bi mechanism where $K_{ia} \ll K_{m_A}$.

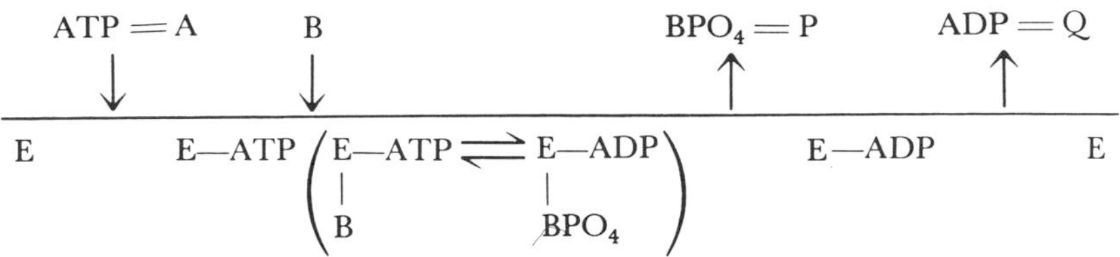

If $K_{ip} \gg K_{mP}K_{iq}/K_{mQ}$ (because $K_{-2} \gg K_4$), BPO_4 will appear to be competitive with respect to B (the intersection point will fall very close to the $1/v$-axis). In this case, substrate-product pairs can be identified, but the order of substrate addition and product release cannot be established (the product-substrate inhibition patterns are symmetrical). If K_{ip} is not very large, BPO_4 will be a mixed-type inhibitor with respect to B. The phosphofructokinase of *Dictostelium discoideum* shows parallel reciprocal plots when either ATP or F-6-P is varied (Baumann and Wright, 1968). However, ADP is competitive with ATP and uncompetitive with F-6-P, while FDP is competitive with F-6-P and uncompetitive with ATP.

Seemingly parallel initial velocity patterns can also occur in rapid equilibrium random systems where the binding of one substrate strongly inhibits the binding of the other (i.e., $\alpha \gg 1$). A product will be competitive with whichever substrate occupies the same binding site and a noncompetitive or mixed-type inhibitor with whichever substrate it forms a dead-end complex. The noncompetitive or mixed-type patterns will appear uncompetitive if the binding of the product does not strongly inhibit the binding of the co-substrate and vice versa; that is, if the interaction factor β in System A2a or γ in System A4a (Chapter Six) is $\ll \alpha$.

A nonrapid equilibrium random system will also yield seemingly parallel plots if the rate constants for the release of A and B are lower than V_{max_f} (Bar-Tana and Cleland, 1974). In this case, the plots are really concave up, but will appear parallel in the usual range of [A] and [B] plotted (i.e., in the region of K_{m_A} and K_{m_B}). If lower substrate concentrations are plotted, the lines will appear to intersect. If the substrate release rate constants are significantly lower than V_{max_f}, the plots in the region of K_{m_A} and K_{m_B} will appear to intersect to the right of the $1/v$-axis.

J. PARTIAL RAPID EQUILIBRIUM PING PONG BI BI SYSTEM

Suppose that the substrate addition and product release steps are much faster than the interconversions of the central complexes so that E, EA, and

EQ are at equilibrium and F, FB, and FP are at equilibrium. The basic King-Altman figure can be represented as:

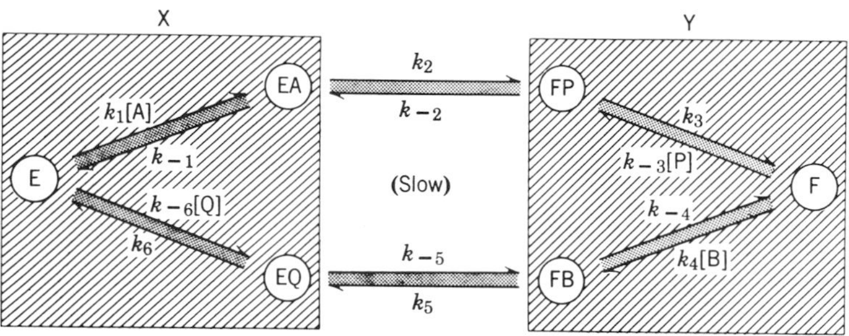

The shaded areas represent the two rapid equilibrium segments, X and Y. Note that the steps EA$\rightleftharpoons$FP and FB$\rightleftharpoons$EQ are now included in the basic figure. Consequently, we can now apply Cha's simplification and indicate the figure as:

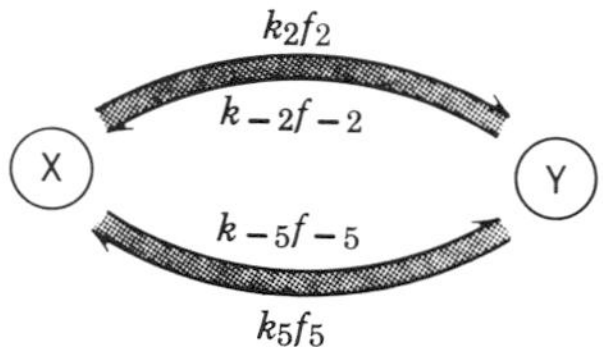

where f_2, f_{-2}, f_5, and f_{-5} represent the fractional concentrations of EA, FP, FB, and EQ, respectively. Although the rapid equilibrium segments affect the composition of the kinetic constants (in terms of rate constants), the final form of the velocity equation is unchanged. Thus the reciprocal plots and product inhibition patterns are unchanged.

K. HYBRID PING PONG—RAPID EQUILIBRIUM RANDOM (TWO-SITE) BI BI SYSTEMS

Transcarboxylase (methylmalonyl-*S*-CoA : pyruvate carboxytransferase) catalyzes the reversible reaction:

$$\underset{\text{methylmalonyl-}S\text{-CoA}}{CH_3{-}\overset{COOH}{\underset{H}{C}}{-}\underset{O}{\overset{\|}{C}}{-}S{-}CoA} + \underset{\text{pyruvate}}{CH_3{-}\underset{O}{\overset{\|}{C}}{-}COOH} \rightleftharpoons \underset{\text{propionyl-}S\text{-CoA}}{CH_3{-}CH_2{-}\underset{O}{\overset{\|}{C}}{-}S{-}CoA} + \underset{\text{oxalacetate}}{\overset{COOH}{CH_2}{-}\underset{O}{\overset{\|}{C}}{-}COOH}$$

The overall reaction is the sum of two partial reactions in which protein-bound biotin acts as a carboxyl-carrying intermediate:

$$\text{methylmalonyl-}S\text{-CoA} + \text{E–biotin} \rightleftharpoons \text{propionyl-}S\text{-CoA} + \text{E–biotin}{\sim}CO_2$$

$$\text{E–biotin}{\sim}CO_2 + \text{pyruvate} \rightleftharpoons \text{E–biotin} + \text{oxalacetate}$$

Initial velocity studies yield a series of parallel lines when either substrate is varied at different fixed levels of the other. The patterns are consistent with a classical Ping Pong Bi Bi sequence:

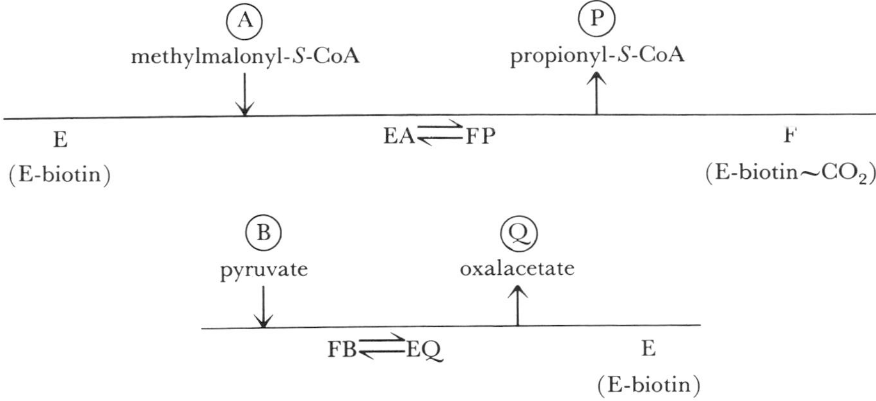

However, the product inhibition patterns are just the opposite to those expected: propionyl-*S*-CoA and methylamlonyl-*S*-CoA are competitive; oxalacetate and pyruvate are competitive; propionyl-*S*-CoA is a mixed-type inhibitor with respect to pyruvate (and vice versa); oxalacetate is a mixed-type inhibitor with respect to methylmalonyl-*S*-CoA (and vice versa). α-Ketobutyrate, a dead-end inhibitor competitive with the α-keto acids, was uncompetitive with methylmalonyl-*S*-CoA. Free CoA-SH was competitive with the thioesters but was uncompetitive with the α-keto acids. A classical Ping Pong sequence predicts mixed-type patterns instead of the observed uncompetitive patterns. To account for these observations, Northrup (1969) has proposed that the enzyme possesses two separate and independent sites (as opposed to the classical single-site model shown in Figure IX-56). One site binds the thioesters, and the other site binds the α-keto acids. Decarboxylation-carboxylation occurs at both sites, the carboxyl group being carried between the two sites by biotin attached in ϵ-peptide linkage to lysine. It is assumed that all substrates and products bind and dissociate independently in a Rapid Equilibrium Random fashion; that is, a thioester can combine at site 1 before or after an α-keto acid binds at site 2 and the binding of one ligand has no effect on the binding of another. Similarly, the rates of carboxyl transfer and biotin migration are unaffected by the binding of any ligand. Since the mechanism involves a substituted enzyme (E-biotin $\sim CO_2$), the mechanism is hybrid Ping Pong—Rapid Equilibrium Random. The sequence can be diagrammed as shown below.

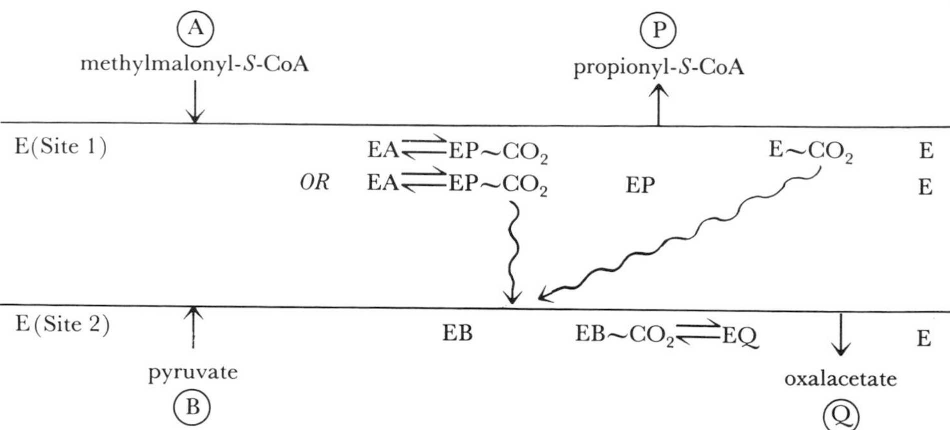

Thus one possible sequence is true Ping Pong in which methylmalonyl-*S*-CoA adds, CO_2 is transferred to biotin, and propionyl-*S*-CoA leaves. Then pyruvate adds, picks up the CO_2, and leaves as oxalacetate. But another possible sequence involves the random addition of both substrates to their respective sites, the transfer of CO_2 from methylmalonyl-*S*-CoA to biotin and from biotin to pyruvate followed by the random release of both products. Also possible is a hybrid sequence in which both substrates add randomly, CO_2 is transferred from methylmalonyl-*S*-CoA to biotin and then propionyl-*S*-CoA leaves before the CO_2 is transferred to pyruvate. A full steady-state treatment of the proposed mechanism would be impossibly complex. However, a combined rapid equilibrium-steady-state treatment employing the method of Cha is relatively simple. The basic King-Altman figure is shown on the following page. The four corners are four Rapid Equilibrium Random segments. The ${}^{0}E$ represents all enzyme forms containing uncarboxylated biotin at the thioester site; ${}^{CO_2}E$ represents all enzyme forms containing carboxylated biotin at the thioester site; E^{CO_2} represents all enzyme forms containing carboxylated biotin at the α-keto acid site; E^0 represents all enzyme forms containing uncarboxylated biotin at the α-keto acid site. The reversible steady-state steps are (*a*) the transfer of CO_2 from methylmalonyl-*S*-CoA to biotin (the $k_1 f_1/k_{-1} f_{-1}$ step), (*b*) the migration of carboxylated biotin from the thioester site to the α-keto acid site (the k_2/k_{-2}

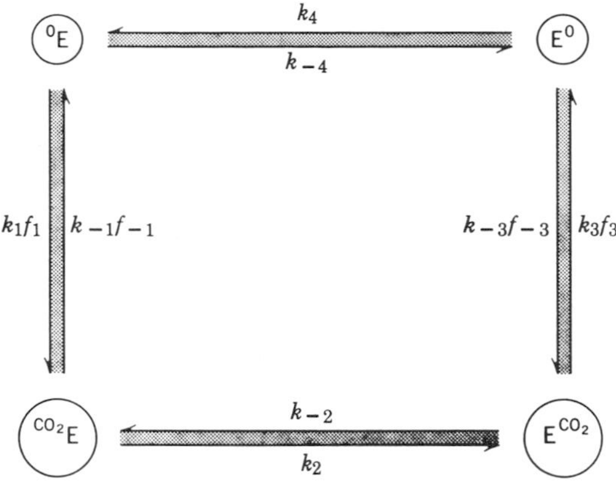

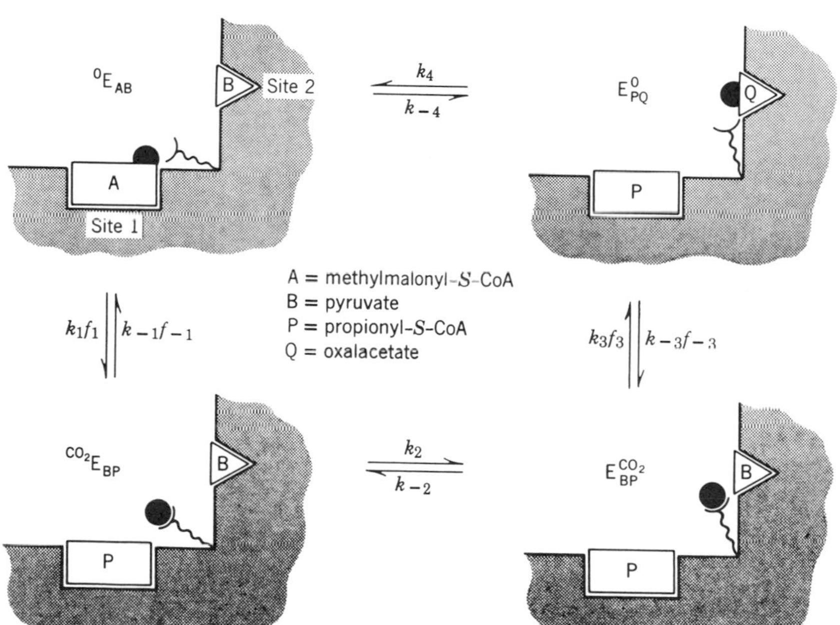

Fig. IX-35. Hybrid Ping Pong-Rapid Equilibrium Random model for methylmalonyl-S-CoA: pyruvate transcarboxylase. Substrates A and B bind randomly and independently at distinct sites. The CO_2 is carried from one site to another by a mobile, biotin-containing peptide.

step), (c) the transfer of CO_2 from biotin to pyruvate (the $k_3f_3/k_{-3}f_{-3}$ step), and the migration of the uncarboxylated biotin back to the thioester site (the k_4/k_{-4} step). (The final equation is unchanged if the steps involving the migration of the biotin are omitted. However, it is difficult to imagine two independent sites without some migration or conformational change that allows the biotin to exchange the carboxyl group.) The mechanism is shown in Figure IX-35. Only one complex of each rapid equilibrium segment is shown (${}^{0}E_{AB}$ for ${}^{0}E$, ${}^{CO_2}E_{BP}$ for ${}^{CO_2}E$, etc.), but a total of nine different enzyme forms are possible within each segment if all four ligands are present. The different forms and their interconversions are shown below. The shaded area represents the enzyme forms in the presence of P as a product inhibitor. The area bounded by the broken line represents enzyme forms in the presence of Q as a product inhibitor. Note that A and P are mutually exclusive. Similarly, B and Q are mutually exclusive.

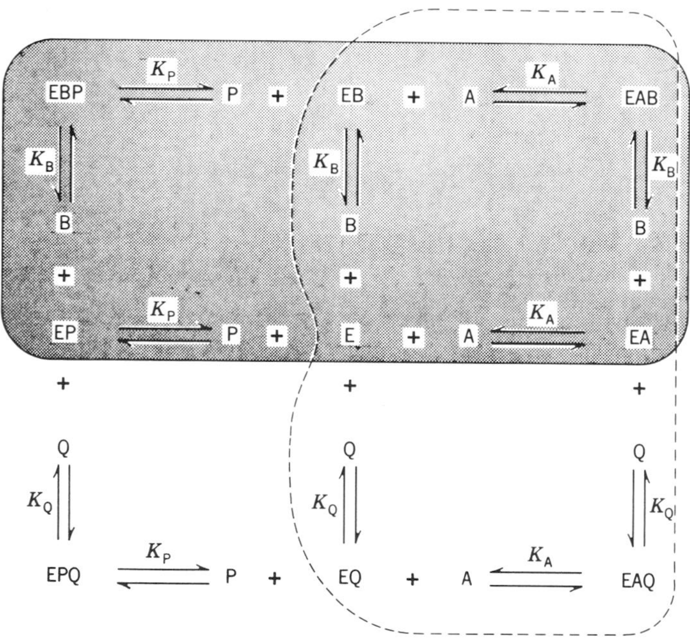

It is assumed that the binding of one ligand has no effect on the binding of another (i.e., $\alpha = 1$, $\beta = 1$). It is also assumed that a carboxylated substrate and carboxylated biotin at the same site are not mutually exclusive. The fractional concentration factors, f_1, f_{-1}, f_3, and f_{-3}, represent the fraction of the respective rapid equilibrium segment capable of undergoing the particular step shown. Thus f_1 represents that fraction of ^0E that can transfer CO_2 to biotin (i.e., the complexes containing A). Similarly, f_{-1} is the fraction of ^{CO_2}E that can accept CO_2 from biotin (i.e., the complexes containing P), f_3 is the fraction of E^{CO_2} capable of accepting CO_2 from biotin (i.e., the complexes containing B), and f_{-3} is the fraction of E^0 capable of donating CO_2 to biotin (i.e., those complexes containing Q). Thus:

$$f_1 = \frac{[EA + EAQ + EAB]}{[^0E]} = \frac{[\text{“E”A}]}{[^0E]}$$

$$= \frac{\dfrac{[A]}{K_A} + \dfrac{[A][B]}{K_A K_B} + \dfrac{[A][Q]}{K_A K_Q}}{1 + \dfrac{[A]}{K_A} + \dfrac{[B]}{K_B} + \dfrac{[A][B]}{K_A K_B} + \dfrac{[P]}{K_P} + \dfrac{[Q]}{K_Q} + \dfrac{[A][Q]}{K_A K_Q} + \dfrac{[B][P]}{K_B K_P} + \dfrac{[P][Q]}{K_P K_Q}}$$

$$= \frac{\dfrac{[A]}{K_A}\left(1 + \dfrac{[B]}{K_B} + \dfrac{[Q]}{K_Q}\right)}{\left(1 + \dfrac{[B]}{K_B} + \dfrac{[Q]}{K_Q}\right) + \dfrac{[A]}{K_A}\left(1 + \dfrac{[B]}{K_B} + \dfrac{[Q]}{K_Q}\right) + \dfrac{[P]}{K_P}\left(1 + \dfrac{[B]}{K_B} + \dfrac{[Q]}{K_Q}\right)}$$

$$= \frac{\dfrac{[A]}{K_A}}{1 + \dfrac{[A]}{K_A} + \dfrac{[P]}{K_P}} = \frac{\dfrac{[A]}{K_A}}{d_1} \qquad \text{(IX-160)}$$

Thus, f_1, the fractional A occupancy of sites, is the same as if we were dealing with a simple case of competition between A and P for a common site. (The expression for f_1 would not be so simple if the binding of one ligand affected the binding of another.) In a similar manner, we can show that:

$$f_{-1} = \frac{[\text{“E”P}]}{[^{CO_2}E]} = \frac{\dfrac{[P]}{K_P}}{1 + \dfrac{[A]}{K_A} + \dfrac{[P]}{K_P}} = \frac{\dfrac{[P]}{K_P}}{d_1} \qquad \text{(IX-161)}$$

$$f_3 = \frac{[\text{"E"B}]}{[\text{E}^{\text{CO}_2}]} = \frac{\dfrac{[\text{B}]}{K_\text{B}}}{1 + \dfrac{[\text{B}]}{K_\text{B}} + \dfrac{[\text{Q}]}{K_\text{Q}}} = \frac{\dfrac{[\text{B}]}{K_\text{B}}}{d_2} \qquad \text{(IX-162)}$$

$$f_{-3} = \frac{[\text{"E"Q}]}{[\text{E}^0]} = \frac{\dfrac{[\text{Q}]}{K_\text{Q}}}{1 + \dfrac{[\text{B}]}{K_\text{B}} + \dfrac{[\text{Q}]}{K_\text{Q}}} = \frac{\dfrac{[\text{Q}]}{K_\text{Q}}}{d_2} \qquad \text{(IX-163)}$$

The velocity can be taken as the net rate through any steady-state step. For example:

$$v = k_2[{}^{\text{CO}_2}\text{E}] - k_{-2}[\text{E}^{\text{CO}_2}]$$

Dividing by $[\text{E}]_t$:

$$\frac{v}{[\text{E}]_t} = \frac{k_2[{}^{\text{CO}_2}\text{E}] - k_{-2}[\text{E}^{\text{CO}_2}]}{[{}^0\text{E}] + [{}^{\text{CO}_2}\text{E}] + [\text{E}^{\text{CO}_2}] + [\text{E}^0]}$$

Expressions for $[{}^0\text{E}]/[\text{E}]_t$, $[{}^{\text{CO}_2}\text{E}]/[\text{E}]_t$, and so on, are obtained from four three-lined King-Altman interconversion patterns. After grouping similar terms, we obtain:

$$\frac{v}{[\text{E}]_t} = \frac{k_1k_2k_3k_4f_1f_3 - k_{-1}k_{-2}k_{-3}k_{-4}f_{-1}f_{-3}}{\begin{array}{l} k_1k_4(k_2+k_{-2})f_1 + k_2k_3(k_4+k_{-4})f_3 + k_1k_3(k_2+k_4)f_1f_3 + k_{-1}k_{-2}(k_4+k_{-4})f_{-1} \\ + k_{-3}k_{-4}(k_2+k_{-2})f_{-3} + k_{-1}k_3(k_4+k_{-4})f_{-1}f_3 + k_1k_{-3}(k_2+k_{-2})f_1f_{-3} \\ + k_{-1}k_{-3}(k_{-2}+k_{-4})f_{-1}f_{-3} \end{array}} \qquad \text{(IX-164)}$$

After substituting for the fractional concentration factors and multiplying numerator and denominator by d_1d_2, the equation in the presence of one product at a time becomes:

$$\boxed{\frac{v}{V_{\max}} = \frac{[\text{A}][\text{B}]}{K_{m_\text{B}}[\text{A}] + K_{m_\text{A}}[\text{B}] + [\text{A}][\text{B}] + \dfrac{K_{m_\text{A}}K_{ib}[\text{P}]}{K_{ip}} + \dfrac{K_{ia}K_{m_\text{B}}[\text{Q}]}{K_{iq}} + \dfrac{K_{m_\text{A}}[\text{B}][\text{P}]}{K_{ip}} + \dfrac{K_{m_\text{B}}[\text{A}][\text{Q}]}{K_{iq}}}}$$

(IX-165)

where

$$K_{ia}=\frac{\text{Coef}_{\text{Q}}}{\text{Coef}_{\text{AQ}}},\qquad K_{ib}=\frac{\text{Coef}_{\text{P}}}{\text{Coef}_{\text{BP}}},\qquad K_{ip}=\frac{\text{Coef}_{\text{B}}}{\text{Coef}_{\text{BP}}},\qquad \text{and}\qquad K_{iq}=\frac{\text{Coef}_{\text{A}}}{\text{Coef}_{\text{AQ}}}$$

The Haldane equations are:

$$K_{\text{eq}}=\frac{V_{\max_f}K_{m_{\text{P}}}K_{iq}}{V_{\max_r}K_{ia}K_{m_{\text{B}}}}=\frac{V_{\max_f}K_{ip}K_{m_{\text{Q}}}}{V_{\max_r}K_{m_{\text{A}}}K_{ib}}=\frac{K_{ip}K_{iq}}{K_{ia}K_{ib}}=\left(\frac{V_{\max_f}}{V_{\max_r}}\right)^2\frac{K_{m_{\text{P}}}K_{m_{\text{Q}}}}{K_{m_{\text{A}}}K_{m_{\text{B}}}} \qquad \text{(IX-166)}$$

The last two Haldane expressions are identical to those of the classical Ping Pong Bi Bi system. In the absence of products, the velocity equation is identical to that for the classical Ping Pong Bi Bi system. The opposite product inhibition patterns stem from the [B][P] and [A][Q] terms in place of the [B][Q] and [A][P] terms of the classical system. In the presence of P, the velocity equations are:

$$\frac{v}{V_{\max}}=\frac{[\text{A}]}{K_{m_{\text{A}}}\left(1+\frac{K_{ib}[\text{P}]}{K_{ip}[\text{B}]}+\frac{[\text{P}]}{K_{ip}}\right)+[\text{A}]\left(1+\frac{K_{m_{\text{B}}}}{[\text{B}]}\right)} \qquad \text{(IX-167)}$$

and

$$\frac{v}{V_{\max}}=\frac{[\text{B}]}{K_{m_{\text{B}}}\left(1+\frac{K_{m_{\text{A}}}K_{ib}[\text{P}]}{K_{m_{\text{B}}}K_{ip}[\text{A}]}\right)+[\text{B}]\left(1+\frac{K_{m_{\text{A}}}}{[\text{A}]}+\frac{K_{m_{\text{A}}}[\text{P}]}{K_{ip}[\text{A}]}\right)} \qquad \text{(IX-168)}$$

In the presence of Q, the equations are:

$$\frac{v}{V_{\max}}=\frac{[\text{A}]}{K_{m_{\text{A}}}\left(1+\frac{K_{ia}K_{m_{\text{B}}}[\text{Q}]}{K_{m_{\text{A}}}K_{iq}[\text{B}]}\right)+[\text{A}]\left(1+\frac{K_{m_{\text{B}}}}{[\text{B}]}+\frac{K_{m_{\text{B}}}[\text{Q}]}{K_{iq}[\text{B}]}\right)} \qquad \text{(IX-169)}$$

and

$$\frac{v}{V_{max}} = \frac{[B]}{K_{m_B}\left(1+\frac{K_{ia}[Q]}{K_{iq}[A]}+\frac{[Q]}{K_{iq}}\right)+[B]\left(1+\frac{K_{m_A}}{[A]}\right)} \qquad \text{(IX-170)}$$

A dead-end inhibitor (substrate analog) that binds to the A–P site will introduce an $[I]/K_{i_1}$ term to the denominator of f_1 and f_{-1}. Upon simplification this results in the [B] and [Q] terms in the denominator of equation IX-165 being multiplied by the factor $(1+[I]/K_{i_1})$. When [A] is varied, the equation is:

$$\frac{v}{V_{max}} = \frac{[A]}{K_{m_A}\left(1+\frac{[I]}{K_{i_1}}\right)+[A]\left(1+\frac{K_{m_B}}{[B]}\right)} \qquad \text{(IX-171)}$$

As expected, I is competitive with A. When [B] is varied:

$$\frac{v}{V_{max}} = \frac{[B]}{K_{m_B}+[B]\left(1+\frac{K_{m_A}}{[A]}+\frac{K_{m_A}[I]}{[A]K_{i_1}}\right)} \qquad \text{(IX-172)}$$

I is uncompetitive with respect to B. Similarly, a dead-end inhibitor combining at the B–Q site adds an $[I]/K_{i_2}$ term to the denominator of f_3 and f_{-3}. Upon simplification, this introduces a $(1+[I]/K_{i_2})$ factor to the denominator [A] and [P] terms. I will be competitive with B and uncompetitive with respect to A.

L. ISO BI BI SYSTEMS

The isomerization of a transitory complex has no effect on the complete velocity equation, but if a stable enzyme form isomerizes, new terms appear in the denominator of the velocity equation (e.g., the [A][P] term in the Iso Uni Uni system described earlier). Isomerization of a stable enzyme form does not affect the velocity equation in the absence of products, but the product inhibition patterns are changed. The new denominator terms for several iso systems are summarized below.

Table IX-5 New Denominator Terms in the Complete Velocity Equation for Iso Systems

System	New Denominator Terms
Iso Uni Uni	[A][P]
Iso Ordered Uni Bi	[A][Q] and [A][P][Q]
Iso Random Uni Bi	$[A][P]^2$, $[A][Q]^2$, $[A][P]^2[Q]$, and $[A][P][Q]^2$
Iso Ordered Bi Bi	[A][P][Q], [A][B][Q], and [A][B][P][Q]
Iso Theorell Chance Bi Bi	[A][P][Q] and [A][B][Q]
Iso Tetra-Uni Ping Pong (E isomerizes)	[A][Q], [A][B][Q], and [A][P][Q]
Uni Uni Iso Uni Uni Ping Pong (F isomerizes)	[B][P], [A][B][P], and [B][P][Q]
Di-Iso Ping Pong Bi Bi (E and F isomerize)	[A][Q], [A][B][Q], [A][P][Q], [B][P], [A][B][P], and [B][P][Q]

Since the products are used one at a time in product inhibition studies, the extra denominator terms containing either [P] or [Q] (but not both) are the ones that influence the inhibition patterns. (In the reverse direction, the denominator terms containing either [A] or [B] are the important ones.) To express the velocity equations for the nonrandom iso systems in terms of kinetic constants, two new inhibition constants (isoinhibition constants) must be defined for each stable enzyme form that isomerizes. These isoinhibition constants have subscripts corresponding to the ligands combining with the isomers. Thus for an Iso Ordered Bi Bi system in which E isomerizes and A combines with E while Q combines with E′, we have K_{iia} and K_{iiq}. The Iso Ordered Bi Bi system is described below.

Iso Ordered Bi Bi

The reaction sequence is:

$$
\begin{array}{ccccccc}
 & A & B & & P & Q & \\
 & k_1 \downarrow k_{-1} & k_2 \downarrow k_{-2} & & k_3 \uparrow k_{-3} & k_4 \uparrow k_{-4} & \\
\hline
E & & EA & (EAB \rightleftharpoons E'PQ) & & E'Q & E' \underset{k_{-5}}{\overset{k_5}{\rightleftharpoons}} E
\end{array}
$$

The basic King-Altman figure is:

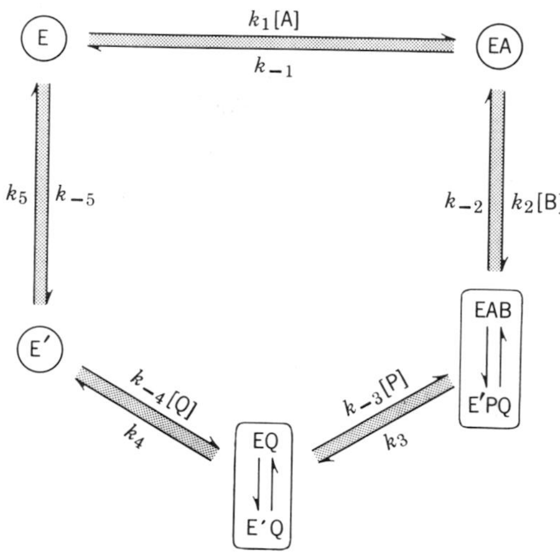

The distribution of enzyme species is:

$$\frac{[E]}{[E]_t} = \frac{k_{-4}[Q]k_{-3}[P]k_{-2}k_{-1} + k_2[B]k_3k_4k_5 + k_{-1}k_3k_4k_5 + k_{-2}k_{-1}k_4k_5 + k_5k_{-3}[P]k_{-2}k_{-1}}{\text{denominator}} \quad \text{(IX-173)}$$

$$\frac{[EA]}{[E]_t} = \frac{\begin{array}{l} k_1[A]k_{-4}[Q]k_{-3}[P]k_{-2} + k_{-5}k_{-4}[Q]k_{-3}[P]k_{-2} + k_3k_4k_5k_1[A] \\ \quad + k_{-2}k_4k_5k_1[A] + k_5k_1[A]k_{-3}[P]k_{-2} \end{array}}{\text{denominator}} \quad \text{(IX-174)}$$

$$\frac{[EAB + E'PQ]}{[E]_t} = \frac{\begin{array}{l} k_1[A]k_2[B]k_{-4}[Q]k_{-3}[P] + k_2[B]k_{-5}k_{-4}[Q]k_{-3}[P] + k_{-1}k_{-5}k_{-4}[Q]k_{-3}[P] \\ \quad + k_4k_5k_1[A]k_2[B] + k_{-3}[P]k_5k_1[A]k_2[B] \end{array}}{\text{denominator}} \quad \text{(IX-175)}$$

$$\frac{[E'Q]}{[E]_t} = \frac{\begin{array}{l} k_{-4}[Q]k_1[A]k_2[B]k_3 + k_{-5}k_{-4}[Q]k_2[B]k_3 + k_3k_{-1}k_{-5}k_{-4}[Q] \\ \quad + k_{-2}k_{-1}k_{-5}k_{-4}[Q] + k_5k_1[A]k_2[B]k_3 \end{array}}{\text{denominator}} \quad \text{(IX-176)}$$

$$\frac{[E']}{[E]_t} = \frac{k_1[A]k_2[B]k_3k_4 + k_{-5}k_2[B]k_3k_4 + k_{-1}k_{-5}k_3k_4 + k_4k_{-2}k_{-1}k_{-5} + k_{-3}[P]k_{-2}k_{-1}k_{-5}}{\text{denominator}} \quad \text{(IX-177)}$$

The velocity equation is:

$$v=\frac{\text{num}_1[\text{A}][\text{B}]-\text{num}_2[\text{P}][\text{Q}]}{\text{same as Ordered Bi Bi }+\text{Coef}_{\text{ABQ}}[\text{A}][\text{B}][\text{Q}]+\text{Coef}_{\text{APQ}}[\text{A}][\text{P}][\text{Q}]+\text{Coef}_{\text{ABPQ}}[\text{A}][\text{B}][\text{P}][\text{Q}]} \quad \text{(IX-178)}$$

Except for the three new terms, the coefficient form of the denominator is the same as that of the equation for Ordered Bi Bi. (However, the coefficients are composed of different rate constants.) Proceeding as usual, the [A][B][Q] term is converted to kinetic constants by multiplying by $(\text{num}_2)/(\text{Coef}_{\text{AB}})(\text{Coef}_{\text{PQ}})$. The [A][P][Q] term is multiplied by $\text{num}_1/\text{num}_1$ as well.

$$\frac{(\text{Coef}_{\text{ABQ}})(\text{num}_2)[\text{A}][\text{B}][\text{Q}]}{(\text{Coef}_{\text{AB}})(\text{Coef}_{\text{PQ}})}=\frac{V_{\text{max}_r}[\text{A}][\text{B}][\text{Q}]}{K_{iiq}}$$

and

$$\frac{(\text{Coef}_{\text{APQ}})(\text{num}_2)(\text{num}_1)[\text{A}][\text{P}][\text{Q}]}{(\text{Coef}_{\text{AB}})(\text{Coef}_{\text{PQ}})(\text{num}_1)}=\frac{V_{\text{max}_f}[\text{A}][\text{P}][\text{Q}]}{K_{iia}K_{\text{eq}}}$$

where the new isoinhibition constants for A and Q are defined as shown below:

$$K_{iiq}=\frac{\text{Coef}_{\text{AB}}}{\text{Coef}_{\text{ABQ}}}, \qquad K_{iia}=\frac{\text{Coef}_{\text{PQ}}}{\text{Coef}_{\text{APQ}}}$$

The [A][B][P][Q] term, after multiplication by the usual factor, becomes:

$$\frac{(\text{Coef}_{\text{ABPQ}})(\text{num}_2)[\text{A}][\text{B}][\text{P}][\text{Q}]}{(\text{Coef}_{\text{AB}})(\text{Coef}_{\text{PQ}})}$$

Substituting for Coef_{AB} from the definition for K_{ip}, the term becomes:

$$\frac{(\text{Coef}_{\text{ABPQ}})V_{\text{max}_r}}{(\text{Coef}_{\text{ABP}})K_{ip}}$$

The ratio $\text{Coef}_{\text{ABPQ}}/\text{Coef}_{\text{ABP}}$ does not correspond to any of the defined kinetic constants for Q, nor can it be converted to the defined constants by substitution since none of the definitions include a $\text{Coef}_{\text{ABPQ}}$. We could define a new inhibition constant for Q, but this is not necessary. If we substitute the equivalent rate constants, $\text{Coef}_{ABPQ}/\text{Coef}_{ABP}$ reduces to k_{-4}/k_5, which is equivalent to $\text{Coef}_{APQ}/\text{Coef}_{AP}$. The latter by substitution can be converted to defined kinetic constants. The term then becomes:

$$\frac{K_{ia}V_{\max_r}[A][B][P][Q]}{K_{iia}K_{m_Q}K_{ip}}$$

Of the three new terms, only the [A][B][Q] term has an effect on the product inhibition patterns (when each product is used singly). Since there are no new [P] terms that do not also contain [Q], the product inhibition pattern by P is the same as in Ordered Bi Bi. The velocity equations for Q as a product inhibitor are the same as those for Ordered Bi Bi with an additional $[Q]/K_{iiq}$ term in the $1/v$-axis intercept factor. Thus Q is a mixed-type inhibitor with respect to A at all B concentrations; Q is a mixed-type inhibitor with respect to B at unsaturating [A], but is an uncompetitive inhibitor at saturating [A]. (The $K_{ia}[Q]/[A]K_{iq}$ and $K_{m_A}[Q]/[A]K_{iq}$ terms in the slope and intercept factors, respectively, go to zero at saturating [A], but the $[Q]/K_{iiq}$ term in the intercept factor remains.) The product inhibition patterns of P and Q are symmetrical, and consequently only substrate-product pairs, but not the order of addition and release of ligands, can be established without binding studies.

Iso Theorell-Chance

The velocity equation for the Iso Theorell-Chance system contains new [A][P][Q] and [A][B][Q] terms in the denominator, of which only the latter is important in product inhibition studies where only one product at a time is used. The [A][B][Q] term becomes $V_{\max_r}[A][B][Q]/K_{iiq}$ after multiplication by the usual factor. The term is incorporated into the $1/v$-axis intercept factor when either A or B is the varied substrate, and remains in the factor at saturating cosubstrate. Thus when B is the varied substrate, saturating [A] reduces the slope factor to unity, but the $[Q]/K_{iiq}$ term remains in the intercept factor and Q becomes an uncompetitive inhibitor. (In the normal Theorell-Chance system, both slope and intercept factors reduce to unity at saturating [A] and the inhibition disappears.) The product inhibition patterns of the Iso Theorell-Chance are symmetrical to those of the the Ordered Bi Bi system, but since we do not know which substrate is A and which is B, nor which product is P and which is Q, the inhibition patterns will not distinguish between the two systems. Ligand binding studies are necessary. The ligands that bind to free E and E′ are A and Q. If Q yields competitive product inhibition with respect to A, the system is Ordered Bi Bi. In the Iso Theorell-Chance system, Q yields mixed-type inhibition patterns, or, to put it another way, the product yielding competitive patterns will not bind to free E and E′.

Iso Ping Pong

The three Iso Ping Pong Bi Bi systems are:

1. Iso Tetra Uni Ping Pong (E isomerizes)

$$\begin{array}{ccccccccc} & \mathrm{A} & & \mathrm{P} & & \mathrm{B} & & \mathrm{Q} & \\ & k_1\downarrow k_{-1} & & k_2\uparrow k_{-2} & & k_3\downarrow k_{-3} & & k_4\uparrow k_{-4} & \\ \hline \mathrm{E} & & (\mathrm{EA}\rightleftharpoons \mathrm{FP}) & & \mathrm{F} & & (\mathrm{FB}\rightleftharpoons \mathrm{E'Q}) & & \mathrm{E'}\underset{k_{-5}}{\overset{k_5}{\rightleftharpoons}}\mathrm{E} \end{array}$$

2. Uni Uni Iso Uni Uni Ping Pong (F isomerizes)

$$\begin{array}{ccccccccc} & \mathrm{A} & & \mathrm{P} & & \mathrm{B} & & \mathrm{Q} & \\ & k_1\downarrow k_{-1} & & k_2\uparrow k_{-2} & & k_4\downarrow k_{-4} & & k_5\uparrow k_{-5} & \\ \hline \mathrm{E} & & (\mathrm{EA}\rightleftharpoons \mathrm{F'P}) & & \mathrm{F'}\underset{k_{-3}}{\overset{k_3}{\rightleftharpoons}}\mathrm{F} & & (\mathrm{FB}\rightleftharpoons \mathrm{EQ}) & & \mathrm{E} \end{array}$$

3. Di-Iso Ping Pong (E and F isomerize)

$$\begin{array}{ccccccccc} & \mathrm{A} & & \mathrm{P} & & \mathrm{B} & & \mathrm{Q} & \\ & k_1\downarrow k_{-1} & & k_2\uparrow k_{-2} & & k_4\downarrow k_{-4} & & k_5\uparrow k_{-5} & \\ \hline \mathrm{E} & & (\mathrm{EA}\rightleftharpoons \mathrm{F'P}) & & \mathrm{F'}\underset{k_{-3}}{\overset{k_3}{\rightleftharpoons}}\mathrm{F} & & (\mathrm{FB}\rightleftharpoons \mathrm{E'Q}) & & \mathrm{E'}\underset{k_{-6}}{\overset{k_6}{\rightleftharpoons}}\mathrm{E} \end{array}$$

Four isoinhibition constants must be defined for the Di-Iso Ping Pong mechanism because there are two stable enzyme forms that isomerize, and two ligands that bind to each stable form.

$$K_{iia}=\frac{\mathrm{Coef}_{\mathrm{PQ}}}{\mathrm{Coef}_{\mathrm{APQ}}},\qquad K_{iiq}=\frac{\mathrm{Coef}_{\mathrm{AB}}}{\mathrm{Coef}_{\mathrm{ABQ}}},\qquad K_{iib}=\frac{\mathrm{Coef}_{\mathrm{PQ}}}{\mathrm{Coef}_{\mathrm{BPQ}}},\qquad K_{iip}=\frac{\mathrm{Coef}_{\mathrm{AB}}}{\mathrm{Coef}_{\mathrm{ABP}}}$$

The product inhibition patterns are summarized in Table IX-6.

M. HYBRID THEORELL-CHANCE PING PONG (AND ISO PING PONG) SYSTEMS

Three hybrid bireactant systems combining the substituted enzyme feature of the Ping Pong sequence with the hit-and-run feature of the Theorell-

Chance mechanism can be devised. These are really limiting cases of the classical Ping Pong Bi Bi system in which one or two central complexes have extremely short lives. The reaction sequences and product inhibition patterns are shown below

1.

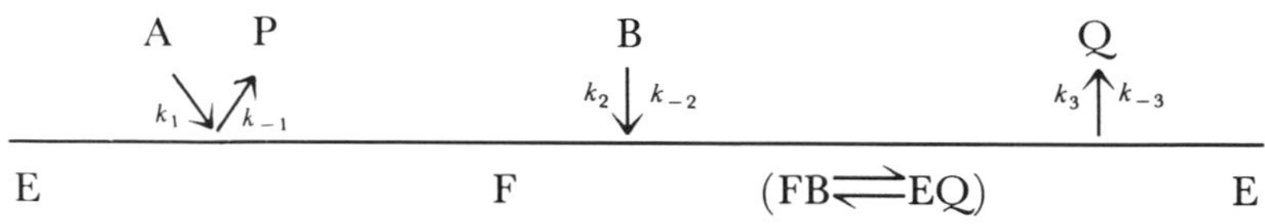

Product Inhibitor	A Varied Unsat B	A Varied Sat B	B Varied Unsat A	B Varied Sat A
P	C	—	C	—
Q	C	C	MT	—

2.

A (k_1, k_{-1})   P (k_2, k_{-2})   B Q (k_3, k_{-3})

E   (EA⇌FP)   F   E

Product Inhibitor	A Varied Unsat B	A Varied Sat B	B Varied Unsat A	B Varied Sat A
P	MT	—	C	C
Q	C	—	C	—

3.

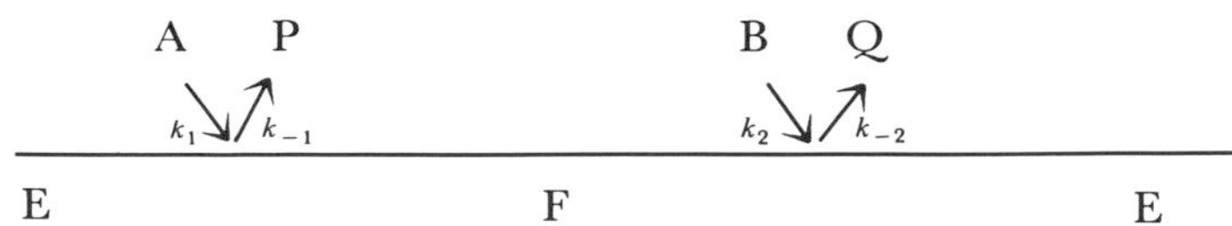

Product Inhibitor	A Varied: Unsat B	A Varied: Sat B	B Varied: Unsat A	B Varied: Sat A
P	C	—	C	—
Q	C	—	C	—

In the absence of products, all initial velocity patterns are parallel.

If we include isomerization of the stable enzyme forms, then for each of the systems above there are three iso systems (isomerization of E, isomerization of F, and isomerization of both E and F). The initial velocity and product inhibition patterns for these nine other systems can be predicted from the general rules outlined in Section CC. The most interesting of these hybrid systems is the hybrid Di-Iso Ping Pong-Di-Theorell-Chance system called Di Uni Iso Ping Pong by Hurst (1969).

(P∼X) (B∼X)

A P (k_1, k_{-1}) B Q (k_3, k_{-3})

$$E \qquad F' \underset{k_{-2}}{\overset{k_2}{\rightleftharpoons}} F \qquad E' \underset{k_{-4}}{\overset{k_4}{\rightleftharpoons}} E$$

(E–) (E–X) (X–E) (–E) (E–)

The isomerization of F′ to F could represent a conformational change that brings the transferred group, X, into a position that permits the hit-and-run attack by B. The isomerization of E′ to E returns the enzyme to a conformation that permits the acceptor group to receive X from A in another

Theorell-Chance step. The basic King-Altman figure for this system is shown below.

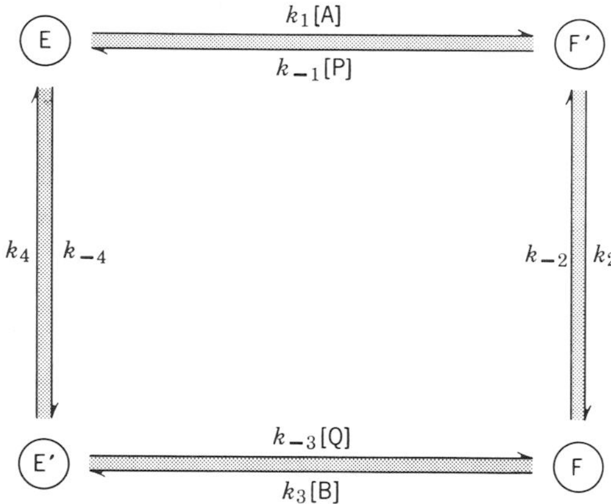

The velocity equation for the forward reaction in the presence of one product at a time is:

$$\frac{v}{V_{\max}} = \frac{[A][B]}{K_{m_B}[A] + K_{m_A}[B] + [A][B] + \dfrac{K_{ib}K_{m_A}[P]}{K_{ip}} + \dfrac{K_{ia}K_{m_B}[Q]}{K_{iq}} + \dfrac{K_{m_B}[A][Q]}{K_{iq}} + \dfrac{K_{m_A}[B][P]}{K_{ip}}} \tag{IX-179}$$

where

$$K_{ia} = \frac{\text{Coef}_Q}{\text{Coef}_{AQ}}, \qquad K_{ib} = \frac{\text{Coef}_P}{\text{Coef}_{BP}}, \qquad K_{ip} = \frac{\text{Coef}_B}{\text{Coef}_{BP}}, \qquad \text{and} \qquad K_{iq} = \frac{\text{Coef}_A}{\text{Coef}_{AQ}}$$

The equation is identical in form to that of the classical Ping Pong Bi Bi system except that the product inhibition terms are reversed. Initial velocity reciprocal plot patterns are parallel. P will be competitive with A at all B concentrations and a mixed-type inhibitor with respect to B at unsaturating [A]. Saturating [A] overcomes the inhibition. Q will be competitive with B at all A concentrations and a mixed-type inhibitor with respect to A at unsaturating [B]. The product inhibition patterns are symmetrical, but it is obvious that A is the group-donating substrate, B the acceptor, P the residue of A, and Q the product of B with the group from A attached. This brings to six the total number of bireactant kinetic mechanisms which yield parallel lines or seemingly parallel lines on reciprocal plots: (*a*) Classical Ping Pong Bi Bi, (*b*) Hybrid Ping Pong—Rapid Equilibrium Random Bi Bi (transcarboxylase model), (*c*) Hybrid Di Iso Ping Pong—Di Theorell Chance (Di Uni Iso Ping Pong), (*d*) Ordered Bi Bi where $K_{ia} \ll K_{m_A}$, (*e*) Rapid Equilibrium Random Bi Bi where $\alpha \gg 1$, and (*f*) a random (nonrapid equilibrium) bireactant system where the rate constants for the release of the substrates are smaller than $V_{\max_f}$ (Bar-Tana and Cleland, 1974). Systems *a* through *c* yield true parallel lines. Systems *b* and *c* yield product inhibition patterns opposite to those of system *a*. No single type of experiment will distinguish between all six systems. A combination of product inhibition studies, dead-end inhibition studies, and perhaps isotope exchange and equilibrium ligand binding studies may be necessary.

N. RAPID EQUILIBRIUM RANDOM BI BI SYSTEMS

The overall reaction sequence for a Rapid Equilibrium Random Bi Bi system is shown below.

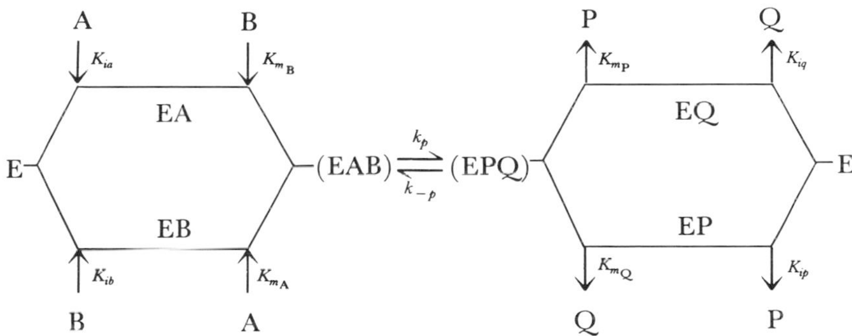

where $K_{ia} = K_A$, $K_{ib} = K_B$, $K_{m_B} = \alpha K_B$, $K_{m_A} = \alpha K_A$, $K_{ip} = K_P$, $K_{iq} = K_Q$, $K_{m_P} = \beta K_P$, and $K_{m_Q} = \beta K_Q$ in the notation used in Chapter Six. The characteristics and product inhibition patterns have been described in System A4 of Chapter Six. As noted in these sections, a variety of product inhibition patterns can be obtained, depending on the type of dead-end complexes that can form (and there usually will be at least one dead-end complex). Generally, the Rapid Equilibrium Random system can be distinguished from the different Steady-State Ordered systems if a complete product inhibition analysis is carried out. However, there are situations where one of the products may show negligible inhibition at reasonable concentrations. In this case it may be difficult to decide the kinetic mechanism. For example, suppose the reciprocal plots for two substrates yield intersecting patterns and a product, Z, is competitive with one substrate, S, at all concentrations of the second substrate, S′, and is a mixed-type inhibitor with respect to S′ at unsaturating concentrations of S (no inhibition at saturating [S]). Z could be Q with S=A and S′=B in an Odered Bi Bi or Theorell-Chance mechanism (Ping Pong Bi Bi is excluded by the intersecting pattern); Z and S could be P and B in a Theorell-Chance or Iso Theorell-Chance mechanism, and so on. The same product inhibition pattern is observed in Rapid Equilibrium Random mechanisms where Z forms a dead-end complex. While we cannot distinguish between the various Steady-State Ordered mechanisms, we may be able to tell whether the mechanism is ordered or random from the slope replots of the competitive product inhibition pattern. Several families of $1/v$ versus $1/[S]$ plots at various fixed concentrations of Z are constructed, each family at a different constant [S′]. The slopes within each family are replotted against the corresponding [Z]. In an ordered system, $K_{i_{slope}}$ (the intercept on the [Z]-axis) is constant for all S′ concentrations, while in a random system, $K_{i_{slope}}$ may vary with [S′]. For example, in a random system with a dead-end EBQ complex (System A4a, Chapter Six, with EBQ=ES′Z and A=S), the slope of the $1/v$ versus $1/[A]$ plots is given by:

$$slope_{1/A} = \frac{\alpha K_A}{V_{max}}\left(1 + \frac{K_B}{[B]} + \frac{K_B[Q]}{[B]K_Q} + \frac{[Q]}{\gamma K_Q}\right), \quad \text{and} \quad K_{i_{slope}}^{Q/A} = \gamma K_Q \frac{\left(1 + \dfrac{K_B}{[B]}\right)}{\left(1 + \dfrac{\gamma K_B}{[B]}\right)}$$

The $K_{i_{slope}}^{Q/A}$ will approach γK_Q as [B] increases. This involves an increase in $K_{i_{slope}}$ as [B] increases if $\gamma > 1$, or a decrease in $K_{i_{slope}}$ if $\gamma < 1$. If $\gamma = 1$, then $K_{i_{slope}}$ will not vary as [B] varies and no conclusions as to whether the system is ordered or random can be made. The slope and $K_{i_{slope}}$ replots for a Steady-State Ordered Bi Bi and for a Rapid Equilibrium Random Bi Bi system with a dead-end EBQ complex are shown in Figure IX-36. The plots for a Rapid Equilibrium Random systems with a product inhibitor that excludes both substrates are shown in Figure VI-12 (System A4a, Chapter Six). Here too, $K_{i_{slope}}$ varies with the concentration of the fixed substrate, but linearly rather than hyperbolically.

Haldane Equations

The Haldane equations for any mechanism must fit the general equation IX-21. Keeping in mind that $K_{ia}K_{m_B} = K_{ib}K_{m_A}$ and $K_{iq}K_{m_P} = K_{ip}K_{m_Q}$ for a

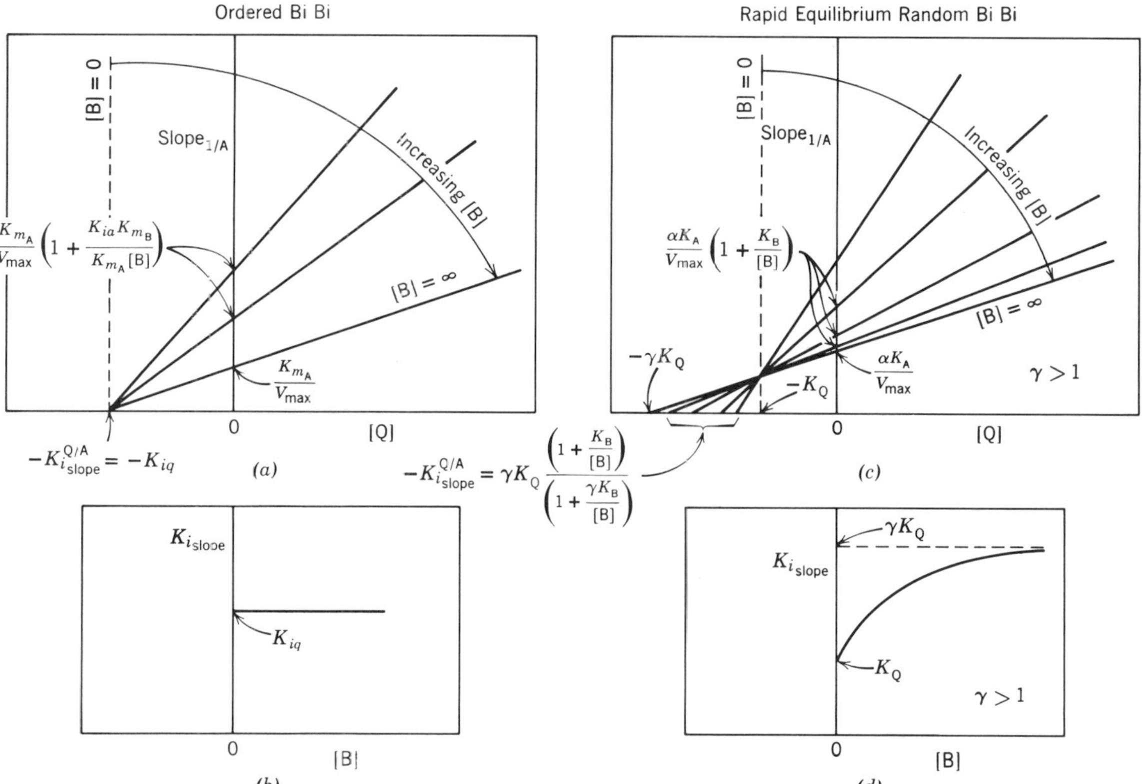

Fig. IX-36. Discriminating between an Ordered Bi Bi and a Rapid Equilibrium Random Bi Bi by means of product inhibition; Q is competitive with A. Several families of $1/v$ versus $1/[A]$ plots at different fixed Q concentrations and a constant, unsaturating B concentration are constructed. Each family represents a different constant B concentration. *Slope*$_{1/A}$ versus [Q] replots at different constant B concentrations are then constructed. (*a*) and (*b*) In the ordered system, $K_{i_{slope}}^{Q/A}$ is independent of [B]. (*c*) and (*d*) In the random system, $K_{i_{slope}}^{Q/A}$ may vary with [B] if $\gamma \neq 1$.

Rapid Equilibrium Random system, we can write the following Haldane equations:

$$K_{\mathrm{eq}} = \frac{V_f K_{m_{\mathrm{P}}} K_{iq}}{V_r K_{ia} K_{m_{\mathrm{B}}}} = \frac{V_f K_{ip} K_{m_{\mathrm{Q}}}}{V_r K_{ia} K_{m_{\mathrm{B}}}} = \frac{V_f K_{ip} K_{m_{\mathrm{Q}}}}{V_r K_{m_{\mathrm{A}}} K_{ib}} = \frac{V_f K_{m_{\mathrm{P}}} K_{iq}}{V_r K_{m_{\mathrm{A}}} K_{ib}} \qquad \text{(IX-180)}$$

The Haldane equations are identical to those of the Theorell-Chance mechanism.

O. STEADY-STATE RANDOM MECHANISMS

Bi Uni System—Velocity Equation

If the breakdown of the central complex is not the sole rate-limiting step, then the velocity equation for a random bireactant system becomes quite complex. For example, consider the Random Bi Uni system shown below.

$$\begin{array}{ccccccc}
\mathrm{E} + \mathrm{A} & \underset{k_{-1}}{\overset{k_1}{\rightleftharpoons}} & \mathrm{EA} & & & & \\
+ & & + & & & & \\
\mathrm{B} & & \mathrm{B} & & & & \\
k_2 \downarrow\uparrow k_{-2} & & k_{-3} \uparrow\downarrow k_3 & & & & \\
\mathrm{EB} + \mathrm{A} & \underset{k_4}{\overset{k_{-4}}{\leftrightharpoons}} & (\mathrm{EAB} \rightleftharpoons \mathrm{EP}) & \underset{k_{-5}}{\overset{k_5}{\rightleftharpoons}} & \mathrm{E} + \mathrm{P} & &
\end{array}$$

The basic King-Altman figure and the eight interconversion patterns are shown below.

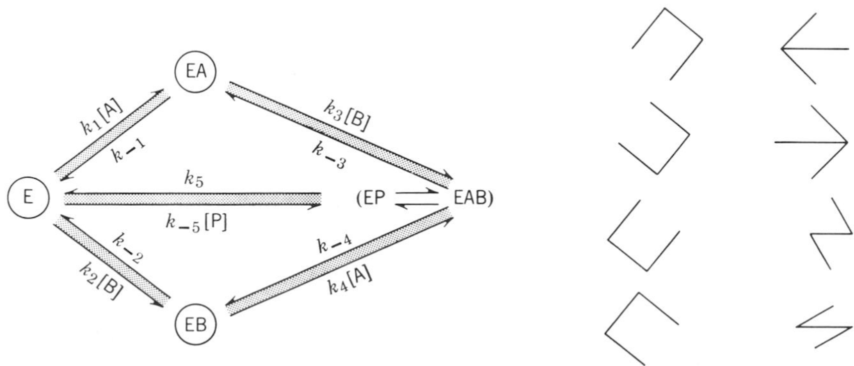

The complete velocity equation, after grouping similar terms is:

$$\frac{v}{[\mathrm{E}]_t}=\frac{K_1[\mathrm{A}][\mathrm{B}]+K_2[\mathrm{A}]^2[\mathrm{B}]+K_3[\mathrm{A}][\mathrm{B}]^2-K_4[\mathrm{P}]-K_5[\mathrm{A}][\mathrm{P}]-K_6[\mathrm{B}][\mathrm{P}]}{K_7+K_8[\mathrm{A}]+K_9[\mathrm{B}]+K_{10}[\mathrm{A}][\mathrm{B}]+K_{11}[\mathrm{A}]^2+K_{12}[\mathrm{B}]^2+K_{13}[\mathrm{A}]^2[\mathrm{B}]}$$

$$+K_{14}[\mathrm{A}][\mathrm{B}]^2+K_{15}[\mathrm{P}]+K_{16}[\mathrm{A}][\mathrm{P}]+K_{17}[\mathrm{B}][\mathrm{P}]+K_{18}[\mathrm{A}][\mathrm{B}][\mathrm{P}] \qquad \text{(IX-181)}$$

where K_1 through K_{18} represent combinations of rate constants:

$K_1=k_1k_{-2}k_3k_5+k_{-1}k_2k_4k_5,\quad K_2=k_1k_3k_4k_5,\quad K_3=k_2k_3k_4k_5,$

$K_4=k_{-1}k_{-2}k_{-3}k_{-5}+k_{-1}k_{-2}k_{-4}k_{-5},\quad K_5=k_{-1}k_{-3}k_4k_{-5},$

$K_6=k\ {}_2k_3k_{-4}k_{-5},\quad K_7=k_{-1}k_{-2}k_{-3}+k_{-1}k_{-2}k_{-4}+k_{-1}k_{-2}k_5,$

$K_8=k_1k_{-2}k_{-3}+k_1k_{-2}k_{-4}+k_1k_{-2}k_5+k_{-1}k_{-3}k_4+k_{-1}k_4k_5,$

$K_9=k_{-1}k_2k_{-3}+k_{-1}k_2k_{-4}+k_{-1}k_2k_5+k_{-2}k_3k_{-4}+k_{-2}k_3k_5,$

$K_{10}=k_1k_{-2}k_3+k_{-1}k_2k_4+k_1k_3k_{-4}+k_2k_{-3}k_4+k_3k_4k_5,$

$K_{11}=k_1k_{-3}k_4+k_1k_4k_5,\quad K_{12}=k_2k_3k_{-4}+k_2k_3k_5,\quad K_{13}=k_1k_3k_4,$

$K_{14}=k_2k_3k_4,\quad K_{15}=k_{-1}k_{-2}k_{-5}+k_{-1}k_{-4}k_{-5}+k_{-2}k_{-3}k_{-5},$

$K_{16}=k_{-1}k_4k_{-5}+k_{-3}k_4k_{-5},\quad K_{17}=k_{-2}k_3k_{-5}+k_3k_{-4}k_{-5},\quad K_{18}=k_3k_4k_{-5}$

The velocity equation for the forward reaction in the absence of P contains two new numerator terms and four new denominator terms in addition to those in the equation for the Ordered Bi Uni system. The equation does not describe a hyperbola and, theoretically, the reciprocal plots are not linear, unless one substrate is saturating. (In this case, all the reaction flux proceeds via one path to EAB and the randomness disappears.) The groups of rate constants cannot be combined into convenient kinetic constants. If one substrate (e.g., B) is held constant, the equation for the forward velocity can be written as:

$$\boxed{\frac{v}{[E]_t} = \frac{i[A]^2 + j[A]}{k + l[A]^2 + m[A]}} \qquad \text{(IX-182)}$$

where

$$i = K_2[B], \qquad j = K_1[B] + K_3[B]^2, \qquad k = K_7 + K_9[B] + K_{12}[B]^2,$$

$$l = K_{11} + K_{13}[B], \qquad \text{and} \qquad m = K_8 + K_{10}[B] + K_{14}[B]^2$$

When [A] is held constant and [B] is varied, the equation becomes:

$$\boxed{\frac{v}{[E]_t} = \frac{i'[B]^2 + j'[B]}{k' + l'[B]^2 + m'[B]}} \qquad \text{(IX-183)}$$

where

$$i' = K_3[A], \qquad j' = K_1[A] + K_2[A]^2, \qquad k' = K_7 + K_8[A] + K_{11}[A]^2,$$

$$l' = K_{12} + K_{14}[A], \qquad \text{and} \qquad m' = K_9 + K_{10}[A] + K_{13}[A]^2$$

Although the reciprocal plots are nonlinear, the departure from linearity may be impossible to detect if both routes to EAB are about equally favorable. The curvature of the reciprocal plot will occur close to the $1/v$-axis. The system will appear to be Rapid Equilibrium Random although the K_m values may not equal the K_S values determined from binding studies. Equations IX-182 and IX-183 are known as "2/1" functions because in reciprocal form (after dividing numerator and denominator by $[S]^2$), $1/[S]^2$ appears in the numerator, while $1/[S]$ appears in the denominator.

Reverse Reaction—Uni Bi

In the reverse direction, the velocity equation in the absence of A and B is:

$$\frac{v_r}{[E]_t} = \frac{K_4[P]}{K_7 + K_{15}[P]} \qquad \text{or} \qquad \frac{v}{V_{max}} = \frac{[P]}{K_{m_P} + [P]}$$

where

$$V_{\max} = \frac{K_4[\mathrm{E}]_t}{K_{15}} \qquad \text{and} \qquad K_{m_\mathrm{P}} = \frac{K_7}{K_{15}} = \frac{\text{const}}{\text{Coef}_\mathrm{P}}$$

Thus in the Uni Bi direction, the velocity is described by the usual Henri-Michaelis-Menten equation. A and B theoretically yield mixed-type inhibition kinetics even in the absence of a dead-end complex. The slopes and intercepts are not linear functions of product inhibitor concentration.

Random Bi Bi Systems

Random Bi Bi systems yield even more complicated velocity equations with 37 new terms in the denominator, in addition to those of the Ordered Bi Bi equation. As above, the velocity curve is not a hyperbola and reciprocal plots are nonlinear unless one substrate is saturating. P and Q yield mixed-type inhibition kinetics. At a fixed unsaturating cosubstrate concentration, $1/v$-axis intercept replots are hyperbolic, while the slope replots are more complex functions of the product inhibitor concentrations. At saturating cosubstrate, both slope and intercept replots are hyperbolic. If there are no unusual rate constants, the Random Bi Bi systems will appear to be a Rapid Equilibrium Random system.

P. PARTIAL RAPID EQUILIBRIUM RANDOM BI BI SYSTEM

If the dissociation of the first ligand in either direction is much faster than the subsequent bimolecular reactions, then A, B, P, and Q will be in equilibrium with E, EA, EB, EP, and EQ. The overall velocity is determined by the slower steady-state rates of second ligand addition and the interconversion of the central complex. The repeating sequence is shown below. The shaded area represents the rapid equilibrium segment.

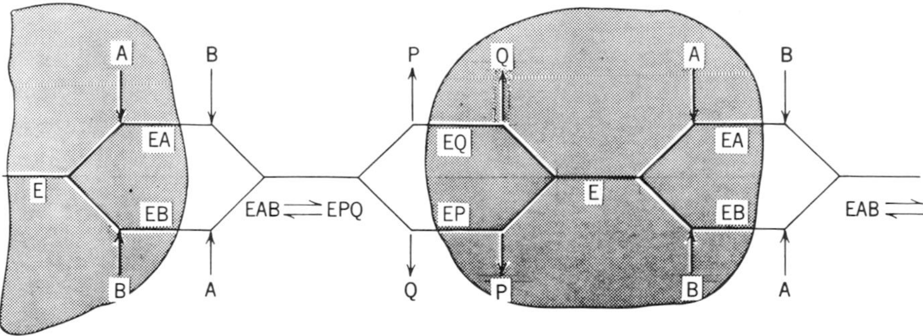

The rapid equilibrium segment can be considered as a single corner, X, in the basic King-Altman figure and the method of Cha used to derive the velocity equation. The basic figure and the simplified figure are shown below.

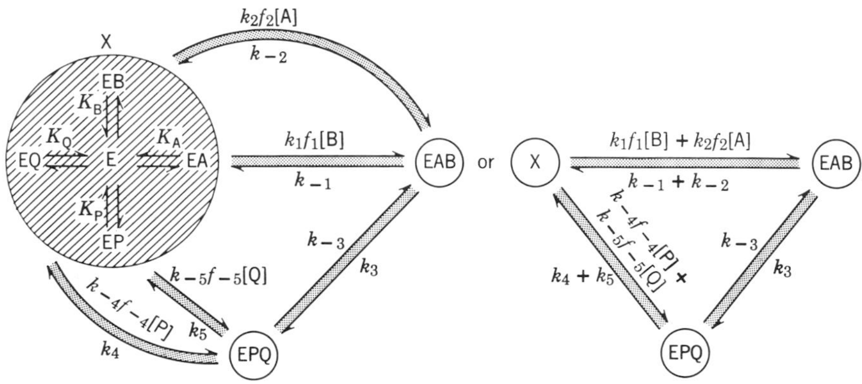

The factors f_1, f_2, f_{-4}, and f_{-5} stand for the fractions of segment X represented by EA, EB, EQ, and EP, respectively. The velocity can be taken as the net rate through a steady-state step. Solving for $[X]/[E]_t$, $[EAB]/[E]_t$, and $[EPQ]/[E]_t$ and substituting into $v = k_3[EAB] - k_{-3}[EPQ]$ we obtain:

$$v = \frac{\text{num}_1[A][B] - \text{num}_2[P][Q]}{\text{const} + \text{Coef}_A[A] + \text{Coef}_B[B] + \text{Coef}_{AB}[A][B] + \text{Coef}_P[P] + \text{Coef}_Q[Q] + \text{Coef}_{PQ}[P][Q]} \tag{IX-184}$$

Even though all the enzyme forms are not at equilibrium, the complete velocity equation has the same form as that of the total Rapid Equilibrium Random system. However, the rate constant composition of the coefficients are different, and, consequently, some of the definitions of kinetic constants are K_m's. The products P and Q are both competitive inhibitors with respect to A and B. (The derivation assumes that no dead-end complexes form. Otherwise, each product will be competitive with whichever substrate occupies the same binding site and a mixed-type inhibitor with respect to the substrate with which it forms a dead-end complex.)

Another situation in which linear reciprocal plots are expected when one of the substrates is varied even though all enzyme forms are not at equilibrium is shown below. Here it is assumed that only one of the substrates adds in a rapid equilibrium fashion while the addition of the other substrate, the interconversion of the central complex, and the release of products are in steady-state.

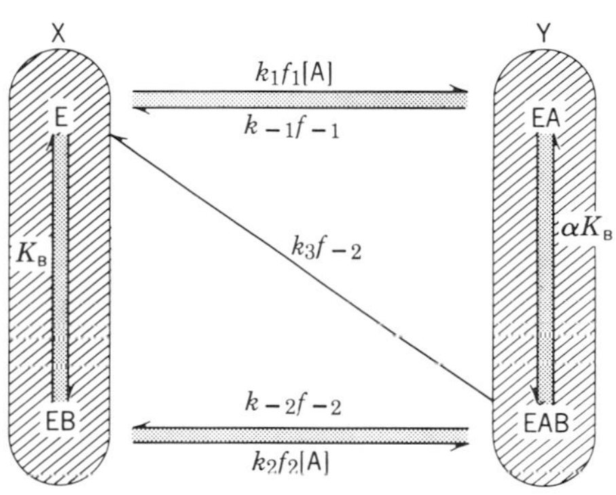

$$v = k_3[\mathrm{EAB}] = k_3 f_{-2}[\mathrm{Y}]$$

$$f_1 = \frac{[\mathrm{E}]}{[\mathrm{X}]} = \frac{1}{1 + \dfrac{[\mathrm{B}]}{K_\mathrm{B}}} = \frac{K_\mathrm{B}}{K_\mathrm{B} + [\mathrm{B}]}$$

$$f_{-1} = \frac{[\mathrm{EA}]}{[\mathrm{Y}]} = \frac{1}{1 + \dfrac{[\mathrm{B}]}{\alpha K_\mathrm{B}}} = \frac{\alpha K_\mathrm{B}}{\alpha K_\mathrm{B} + [\mathrm{B}]}$$

$$f_2 = \frac{[\mathrm{EB}]}{[\mathrm{X}]} = \frac{\dfrac{[\mathrm{B}]}{K_\mathrm{B}}}{1 + \dfrac{[\mathrm{B}]}{K_\mathrm{B}}} = \frac{[\mathrm{B}]}{K_\mathrm{B} + [\mathrm{B}]}$$

$$f_{-2} = \frac{[\mathrm{EAB}]}{[\mathrm{Y}]} = \frac{\dfrac{[\mathrm{B}]}{\alpha K_\mathrm{B}}}{1 + \dfrac{[\mathrm{B}]}{\alpha K_\mathrm{B}}} = \frac{[\mathrm{B}]}{\alpha K_\mathrm{B} + [\mathrm{B}]}$$

The X and Y represent two rapid equilibrium segments; f_1 and f_2 are the fractional concentrations of the two components of segment X; f_{-1} and f_{-2} are the fractional concentrations of the two components of segment Y (relative to EA in this case). For simplicity, it is assumed that no products are present. Since all the lines of the King-Altman figure connect the same two points, the ones going in the same direction may be added and the figure simplified to:

$$\text{(X)} \underset{k_{-1}f_{-1} + k_{-2}f_{-2} + k_3 f_{-2}}{\overset{k_1 f_1[\text{A}] + k_2 f_2[\text{A}]}{\rightleftharpoons}} \text{(Y)}$$

Thus:

$$\frac{[\text{X}]}{[\text{E}]_t} = \frac{k_{-1}f_{-1} + (k_{-2} + k_3)f_{-2}}{\text{denominator}} \quad \text{and} \quad \frac{[\text{Y}]}{[\text{E}]_t} = \frac{(k_1 f_1 + k_2 f_2)[\text{A}]}{\text{denominator}}$$

where the denominator is the sum of the two numerators. The forward velocity is given by:

$$\frac{v}{[\text{E}]_t} = \frac{\left(k_1 k_3 K_{\text{B}}[\text{B}] + k_2 k_3 [\text{B}]^2\right)[\text{A}]}{\alpha k_{-1} K_{\text{B}}^2 + (\alpha k_{-1} + k_{-2} + k_3) K_{\text{B}}[\text{B}] + (k_{-2} + k_3)[\text{B}]^2 + (\alpha k_2 + k_1) K_{\text{B}}[\text{A}][\text{B}] + k_2[\text{A}][\text{B}]^2 + \alpha k_1 K_{\text{B}}^2[\text{A}]} \qquad \text{(IX-185)}$$

or

$$\frac{v}{[\text{E}]_t} = \frac{\left(a[\text{B}] + b[\text{B}]^2\right)[\text{A}]}{\left(c + d[\text{B}] + e[\text{B}]^2\right) + \left(f + g[\text{B}] + h[\text{B}]^2\right)[\text{A}]} \qquad \text{(IX-186)}$$

where a–h are conbinations of rate constants and K_{B}. At any constant [B], the equation has the form:

$$v = \frac{\text{num}\,[\text{A}]}{\text{const} + \text{Coef}_{\text{A}}\,[\text{A}]}$$

Thus the $1/v$ versus $1/[\text{A}]$ plots are linear. However, the $slope_{1/\text{A}}$ and $1/v$-axis intercept are complex functions of [B]. Plots of $1/v$ versus $1/[\text{B}]$ are theoretically nonlinear.

Table IX-6 summarizes the product inhibition patterns for a variety of bireactant systems.

Mechanism	Product Inhibitor	Varied Substrate A: Unsaturated with B	Varied Substrate A: Saturated with B[b]	Varied Substrate B: Unsaturated with A	Varied Substrate B: Saturated with A[b]
Ordered Uni Bi	P	MT			
	Q	C			
Ordered Bi Uni	P	C	C	MT	—
Iso Ordered Uni Bi	P	MT			
	Q	MT			
Iso Ordered Bi Uni	P	MT	MT	MT	MT
Ordered Bi Bi[c]	P	MT	UC	MT	MT
	Q	C	C	MT	—
Theorell-Chance[d,g]	P	MT	—	C	C
	Q	C	C	MT	—
Ping Pong (Tetra-Uni Ping Pong)	P	MT	—	C	C
	Q	C	C	MT	—
Iso Ordered Bi Bi[d]	P	MT	UC	MT	MT
	Q	MT	MT	MT	UC
Iso Theorell-Chance[c]	P	MT	—	C	C
	Q	MT	MT	MT	UC
Iso Tetra Uni Ping Pong (E isomerizes)	P	MT	—	C	C
	Q	MT	MT	MT	MT

Table IX-6 (*Continued*)

Mechanism	Product Inhibitor	Varied Substrate A: Unsaturated with B	A: Saturated with B[b]	B: Unsaturated with A	B: Saturated with A[b]
Uni Uni Iso Uni Uni Ping Pong (F isomerizes)	P	MT	MT	MT	MT
	Q	C	C	MT	—
Di-Iso Ping Pong (E and F isomerize)	P	MT	MT	MT	MT
	Q	MT	MT	MT	MT
Ordered Bi Ter	P	MT	UC	MT	MT
	Q	UC	UC	UC	UC
	R	C	C	MT	—
Uni Uni Uni Bi Ping Pong Bi Ter	P	MT	—	C	C
	Q	UC	UC	MT	MT
	R	C	C	UC	—
Uni Bi Uni Uni Ping Pong Bi Ter (Equivalent to Uni Uni Uni Bi Ping Pong Bi Ter)	P	MT	MT	UC	UC
	Q	UC	—	C	C
	R	C	C	MT	—
Rapid Equilibrium Ordered Bi Bi (No Dead-End Complexes)	P	—	—	—	—
	Q	C	—	C	—
Rapid Equilibrium Ordered Bi Bi with Dead-End EBQ	P	—	—	—	—
	Q	C	C	MT	—

Rapid Equilibrium Ordered	P	[illegible]				
Bi Bi with Dead-End EAP	Q	C		—	C	—
Rapid Equilibrium Ordered Bi Uni	P	C		—	C	—
Rapid Equilibrium Ordered Uni Bi	P	—				
	Q	C				
Random Uni Bi	P		MT^f			
	Q		MT^f			
Random Bi Uni	P	MT^e		MT^f	MT^e	MT^f
Random Bi Bi	P	MT^e		MT^f	MT^e	MT^f
	Q	MT^e		MT^f	MT^e	MT^f
Rapid Equilibrium Random Bi Bi	P	C		—	C	—
	Q	C		—	C	—
Rapid Equilibrium Random Bi Bi with Dead-End EAP[h]	P	MT		—	C	C
	Q	C		—	C	—
Rapid Equilibrium Random Bi Bi with Dead-End EBQ[h]	P	C		—	C	—
	Q	C		C	MT	—
Rapid Equilibrium Random Bi Bi with Dead-End EBP[h]	P	C		C	MT	—
	Q	C		—	C	—
Rapid Equilibrium Random Bi Bi with Dead-End EAP and EBQ[g]	P	MT		—	C	C
	Q	C		C	MT	—
Rapid Equilibrium Random Bi Uni	P	C		—	C	—
Rapid Equilibrium Random Uni Bi	P		C			
	Q		C			

Table IX-6 (*Continued*)

[a] C = competitive (only slope changes); UC = uncompetitive (only $1/v$-axis changes); MT = mixed-type (both slope and $1/v$-axis change). In the Cleland nomenclature, mixed-type inhibition patterns are called noncompetitive. The patterns shown for steady-state mechanisms assume that no dead-end complexes form. A horizontal line means no inhibition.
[b] "Saturation" with the fixed cosubstrate may require more than 100 times its K_m value. This may be difficult to achieve because of solubility, or ionic strength problems, or expense. In practice, the demonstration that the K_i's increase linearly with increasing concentration of fixed substrate is sufficient indication that saturation will overcome the inhibition.
[c] The product inhibition patterns are symmetrical. The mechanisms can be distinguished by identifying A and Q via ligand binding studies.
[d] The product inhibition patterns identify substrate-product pairs, but do not disclose the order of addition or release. Binding studies are necessary to identify A and Q.
[e] Reciprocal plots are theoretically nonlinear although the curvature may be difficult to see. The $1/v$-axis intercepts are hyperbolic functions of product inhibitor concentration while the slopes are more complex functions of inhibitor concentration.
[f] Reciprocal plots are linear. Slope and intercept replots are hyperbolic functions of product inhibitor concentration.
[g] The product inhibition patterns are identical. The mechanisms can be distinguished only by binding or isotope exchange studies. In a Rapid Equilibrium Random system, A, B, P, and Q all bind to free E. In a Theorell-Chance system, only A and Q bind to free E.
[h] In a random system, the designations A, B and P, Q are arbitrary (see Chapter Six).

Q. VARIETIES OF NONHYPERBOLIC VELOCITY CURVES

Random Bi Bi Systems

If E, EA, EB, A, and B of a Random Bi Bi System are not at equilibrium, and one route to EAB is significantly more favorable than the other, then a variety of interesting nonhyperbolic velocity curves can be obtained. One possibility is a sigmoidal curve which was discussed qualitatively in Chapter Seven, Section E. The mathematical basis for nonhyperbolic responses, as described by Ferdinand (1966), is given below.

The first derivative of equation IX-182 is given by:

$$\frac{dv}{d[\mathrm{A}]}=\frac{[\mathrm{A}]^2(im-jl)+2ki[\mathrm{A}]+kj}{\left(k+l[\mathrm{A}]^2+m[\mathrm{A}]\right)^2} \qquad \text{(IX-187)}$$

The velocity curve will have maxima or minima when $dv/d[\mathrm{A}]=0$. This occurs at $[\mathrm{A}]=\pm\infty$, which is of no particular interest, and when the numerator of the first derivative is zero; that is, at:

$$[\mathrm{A}]=-\frac{ki\pm\sqrt{k^2i^2-kj(im-jl)}}{(im-jl)} \qquad \text{(IX-188)}$$

The value of [A] at the maxima or minima will be negative (and again, of no practical interest) except when $im<jl$. In this case, one of the roots will be positive. The second derivative of equation IX-182 is given by:

$$\frac{d^2v}{d[\mathrm{A}]^2}=\frac{2\left[l(im-jl)[\mathrm{A}]^3-3kil[\mathrm{A}]^2-3kjl[\mathrm{A}]+k(ki-mj)\right]}{\left(k+l[\mathrm{A}]^2+m[\mathrm{A}]\right)^3} \qquad \text{(IX-189)}$$

The velocity curve will have an inflection point when $d^2v/d[\mathrm{A}]^2=0$. This occurs at $[\mathrm{A}]=\pm\infty$, or when the numerator of equation IX-189 equals zero; that is, when

$$[\mathrm{A}]^3+\frac{3ki}{(im-jl)}[\mathrm{A}]^2+\frac{3kj}{(im-jl)}[\mathrm{A}]+\frac{k(mj-ki)}{l(im-jl)}=0$$

Positive roots occur under either of two conditions:

(*a*) $im>jl$ and $ki>mj$ (Fig. IX-37*a*), (*b*) $im<jl$

Under condition *b*, the velocity curve will also have a maximum at a positive

[A]. If, at the same time, $ki < mj$, there will be only one inflection point (Fig. IX-37*b*), but if it is possible for $ki > mj$, there will be two inflection points (Fig. IX-37*c*). (Two inflection points may be impossible with real, positive rate constants.) If $ki < mj$ and $im > jl$, the velocity curve will rise to a limit of V_{max}. There is no inflection point. However, the reciprocal plot may curve downward as it approaches the $1/v$-axis (Figure IX-37*d*). When the data of Figure IX-37*a* are replotted as $\log(V_{max} - v)/v$ versus $\log[A]$ (Hill plot) a slope (i.e., n value) of >1 is obtained. Thus the possibility that a sigmoidal velocity curve arises as a consequence of a random path to the central complex should be considered seriously by investigators dealing with "allosteric" bireactant or terreactant enzymes. Note that equation IX-182 has the same form as the general velocity equation for an allosteric enzyme with two sites (equation VII-24). The equivalent terms are:

$$i = \frac{2bk_p}{aK_S^2}, \qquad j = \frac{2k_p}{K_S}, \qquad k = 1$$

$$l = \frac{1}{aK_S^2}, \qquad \text{and} \qquad m = \frac{2}{K_S}$$

Unireactant Systems

An equation identical to equation IX-182 can be obtained for a single substrate-single site enzyme capable of existing in two forms if steady-state rather than rapid equilibrium conditions prevail. (See references by Weber, by Rabin, and by Plowman at the end of Chapter Seven.) The reaction sequence and King-Altman figure are shown below.

$$\begin{array}{ccccccc}
 & \mathrm{T} & \underset{k_{-1}}{\overset{k_1}{\rightleftharpoons}} & \mathrm{R} & \\
 & + & & + & \\
 & \mathrm{S} & & \mathrm{S} & \\
 & k_4 \downarrow\uparrow k_{-4} & & k_{-2} \uparrow\downarrow k_2 & \\
\mathrm{T+P} \underset{k_{-5}}{\overset{k_5}{\rightleftharpoons}} & \mathrm{TS} & \underset{k_6}{\overset{k_{-6}}{\rightleftharpoons}} & \mathrm{RS} & \underset{k_{-3}}{\overset{k_3}{\rightleftharpoons}} \mathrm{P+R}
\end{array}$$

$$v = k_3[\mathrm{RS} + \mathrm{RP}] + k_5[\mathrm{TS} + \mathrm{TP}]$$

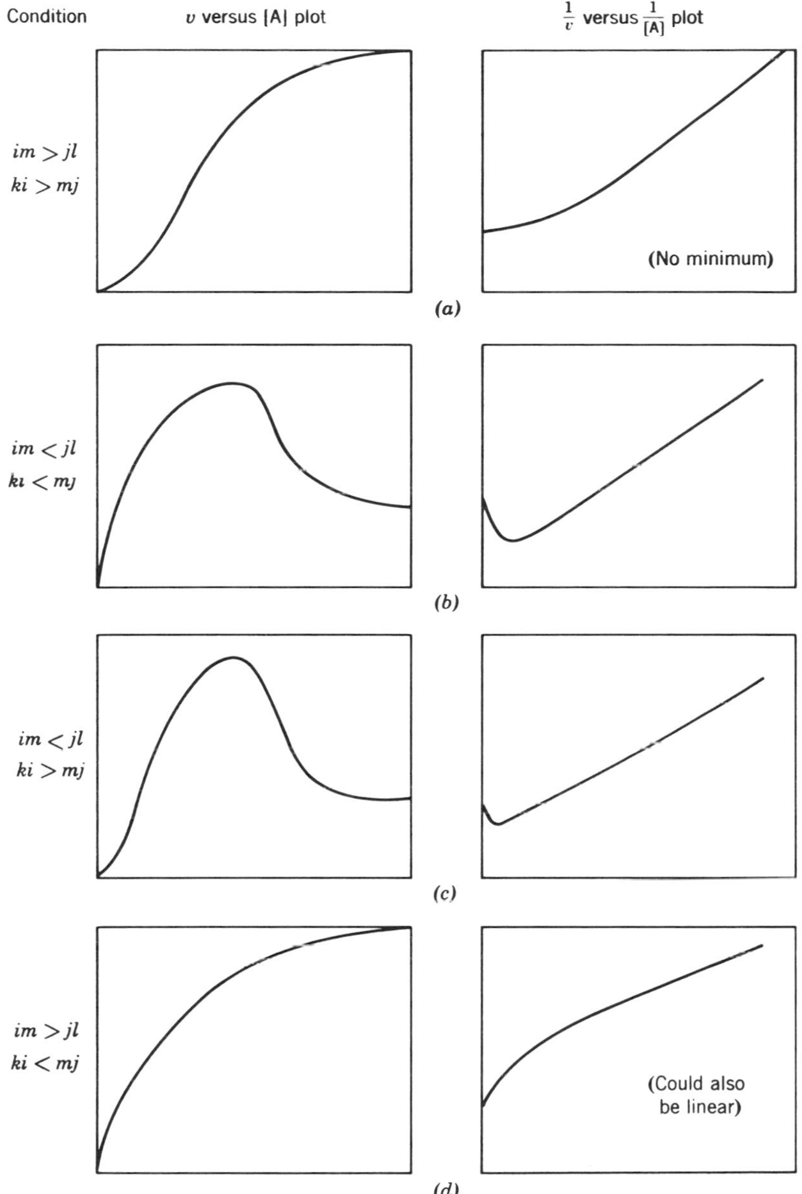

Fig. IX-37. Varieties of velocity curves and reciprocal plots possible in a Nonrapid Equilibrium Random Bireactant system.

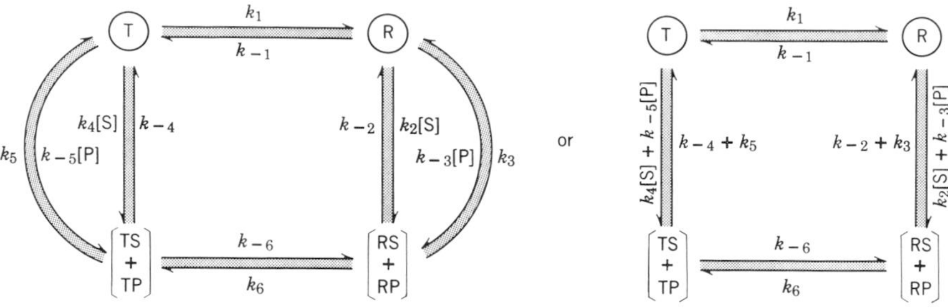

The resulting equation in the absence of P is:

$$\frac{v}{[\mathrm{E}]_t} = \frac{i[\mathrm{S}]^2 + j[\mathrm{S}]}{k + l[\mathrm{S}]^2 + m[\mathrm{S}]} \tag{IX-190}$$

where i, j, and so on, are combinations of rate constants. The equation has the same form if only one of the central complexes (RS or TS) is catalytically active.

Hybrid Ping Pong-Ordered and Ping Pong-Random Bi Bi Systems

There are a number of possible hybrids of the classical Ping Pong Bi Bi mechanism and a Steady-State Ordered or Random mechanism. These systems are of interest because they provide alternate explanations for nonhyperbolic velocity curves (including sigmoidal curves) that are not tied to rapid equilibrium assumptions. We have already seen how a Steady-State Random Bi Bi system can yield sigmoidal velocity curves if certain relationships between groups of rate constants exist. In general, nonhyperbolic velocity curves (and nonlinear reciprocal plots) can arise whenever the velocity equation contains ligand concentration terms to a power greater than unity. Some examples of hybrid Ping Pong systems are described below.

Hybrid Ping Pong-Ordered

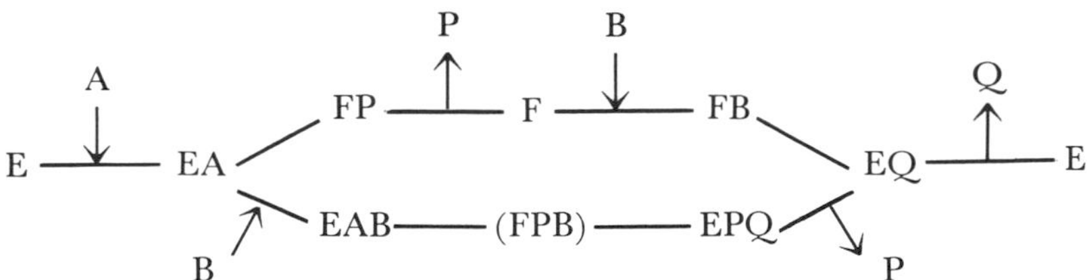

In the reaction mechanism shown above, A, the first substrate to add in ordered sequence is capable of participating in a Ping Pong sequence by donating a group to E, which is subsequently pased on to B. However, it is also possible for B to bind to EA and accept the group directly. Alternately, the group may be passed from A to E, and then from E to B making the true sequence of the ordered segment, $EAB \rightleftharpoons FPB \rightleftharpoons EPQ$. The important point is that there are two different sequences by which A and B are converted to P and Q. Substrate A adds once in the total sequence, but B adds twice. The resulting velocity equation contains terms in $[B]^2$ and thus can yield nonhyperbolic v versus [B] plots. The plots for A as the varied substrate are normal although replots of $slope_{1/A}$ and $1/V_{max_{app}}$ versus $1/[B]$ can be nonlinear. Mannervik (1973) has suggested a hybrid Ping Pong-Ordered mechanism for glutathione reductase.

Hybrid Ping Pong-Semirandom

In the mechanism shown below, either A or B can add to free E. If A adds first, EA participates in a Ping Pong reaction in which P is released before B adds. If B adds first, then the reaction proceeds in an ordered fashion.

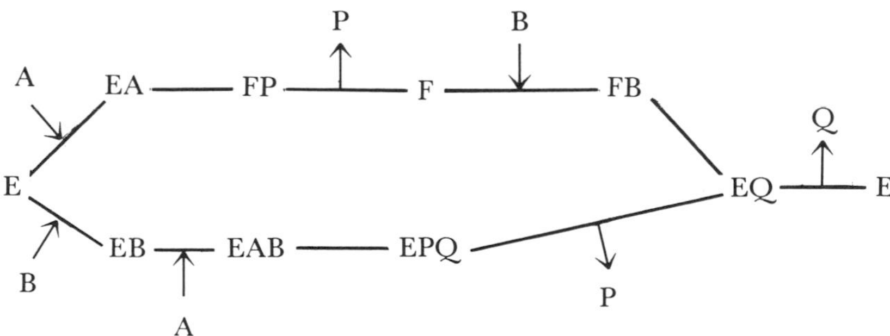

The velocity equation contains $[A]^2$ and $[B]^2$ terms. An extension of the system above allows B to add to FP yielding FPB which then can isomerize to EPQ.

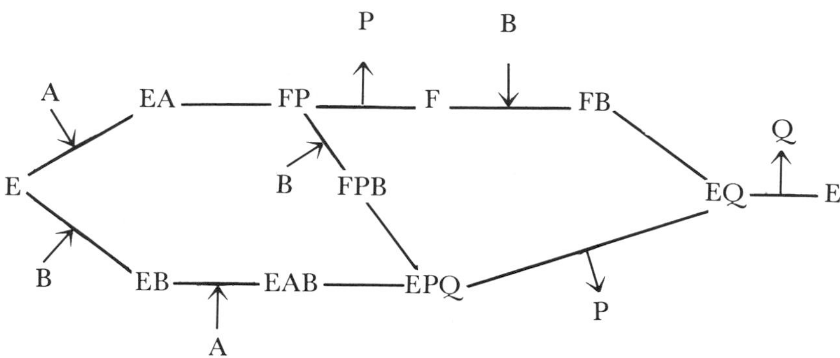

Now there are three B-adding steps and the velocity equation will contain a $[B]^3$ term. Griffen and Brand (1968) have shown that the system above can yield highly sigmoidal v versus [B] plots with Hill plots of slope approaching 3.0.

Hybrid Ping Pong-Random

In the more general scheme shown below, A and B add randomly to form a ternary EAB complex, but EA can still undergo a partial reaction.

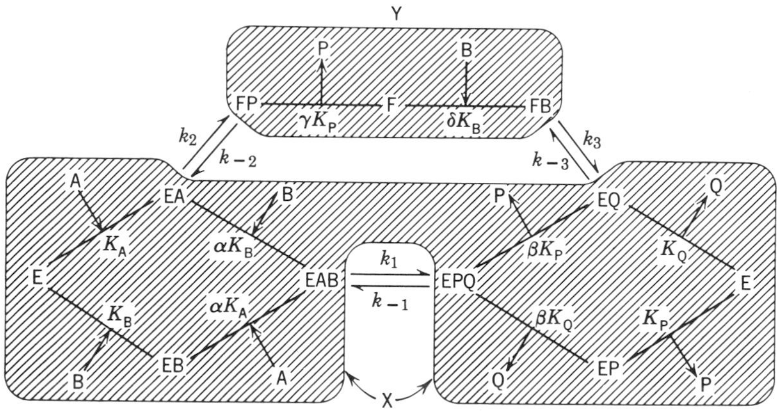

The velocity equation will contain $[A]^2$ and $[B]^3$ terms. If the ligands add in a rapid equilibrium fashion with rate limiting steps at the catalytic reactions

EA$\rightleftharpoons$FP, FB$\rightleftharpoons$EQ, and EAB$\rightleftharpoons$EPQ, then the equation will be considerably simpler. The velocity equation for the rapid equilibrium situation is obtained from the basic King-Altman figure shown below.

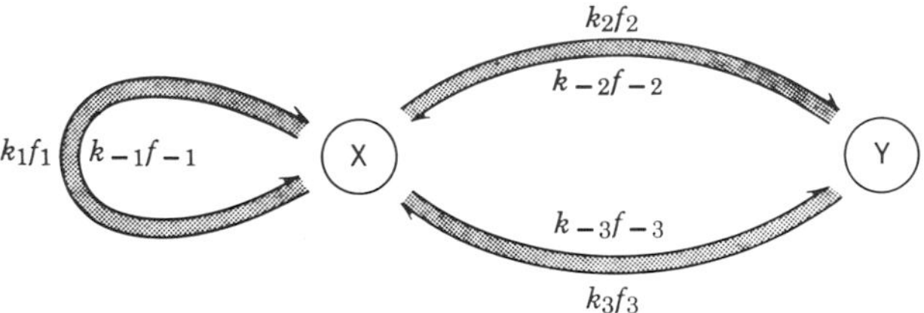

where X represents the Random Bi Bi segment and Y represents the Ordered Uni Uni segment of the Ping Pong sequence. The internal $k_1 f_1/k_{-1} f_{-1}$ loop in segment X is not included in the King-Altman solutions. This reaction represents the rate limiting steps in which the EAB and EPQ portions of X are interconverted. The total net velocity is given by:

$$v=(k_1[\mathrm{EAB}]-k_{-1}[\mathrm{EPQ}])+(k_2[\mathrm{EA}]-k_{-2}[\mathrm{FP}])$$

where the terms in the second parentheses represent that portion of the net velocity through the Ping Pong segment. The equation can be written as:

$$v=(k_1 f_1[\mathrm{X}]-k_{-1}f_{-1}[\mathrm{X}])+(k_2 f_2[\mathrm{X}]-k_{-2}f_{-2}[\mathrm{Y}]) \qquad \text{(IX-191)}$$

where:

$$f_1=\frac{[\mathrm{EAB}]}{[\mathrm{X}]}=\frac{\dfrac{[\mathrm{A}][\mathrm{B}]}{\alpha K_{\mathrm{A}}K_{\mathrm{B}}}}{1+\dfrac{[\mathrm{A}]}{K_{\mathrm{A}}}+\dfrac{[\mathrm{B}]}{K_{\mathrm{B}}}+\dfrac{[\mathrm{A}][\mathrm{B}]}{\alpha K_{\mathrm{A}}K_{\mathrm{B}}}+\dfrac{[\mathrm{P}]}{K_{\mathrm{P}}}+\dfrac{[\mathrm{Q}]}{K_{\mathrm{Q}}}+\dfrac{[\mathrm{P}][\mathrm{Q}]}{\beta K_{\mathrm{P}}K_{\mathrm{Q}}}}=\frac{\dfrac{[\mathrm{A}][\mathrm{B}]}{\alpha K_{\mathrm{A}}K_{\mathrm{B}}}}{d_1}$$

$$f_{-1}=\frac{[\mathrm{EPQ}]}{[\mathrm{X}]}=\frac{\dfrac{[\mathrm{P}][\mathrm{Q}]}{\beta K_{\mathrm{P}}K_{\mathrm{Q}}}}{d_1}, \qquad f_2=\frac{[\mathrm{EA}]}{[\mathrm{X}]}=\frac{\dfrac{[\mathrm{A}]}{K_{\mathrm{A}}}}{d_1}$$

$$f_{-2}=\frac{[\mathrm{FP}]}{[\mathrm{Y}]}=\frac{\dfrac{[\mathrm{P}]}{\gamma K_{\mathrm{P}}}}{1+\dfrac{[\mathrm{B}]}{\delta K_{\mathrm{B}}}+\dfrac{[\mathrm{P}]}{\gamma K_{\mathrm{P}}}}=\frac{\dfrac{[\mathrm{P}]}{\gamma K_{\mathrm{P}}}}{d_2}, \qquad f_3=\frac{[\mathrm{FB}]}{[\mathrm{Y}]}=\frac{\dfrac{[\mathrm{B}]}{\delta K_{\mathrm{B}}}}{d_2}$$

$$f_{-3}=\frac{[\mathrm{EQ}]}{[\mathrm{X}]}=\frac{\dfrac{[\mathrm{Q}]}{K_{\mathrm{Q}}}}{d_1}$$

The King-Altman interconversion patterns and solutions are:

$$\frac{[X]}{[E]_t} = \frac{k_{-2}f_{-2} + k_3 f_3}{k_{-2}f_{-2} + k_3 f_3 + k_2 f_2 + k_{-3}f_{-3}} \qquad \text{(IX-192)}$$

$$\frac{[Y]}{[E]_t} = \frac{k_2 f_2 + k_{-3}f_{-3}}{k_{-2}f_{-2} + k_3 f_3 + k_2 f_2 + k_{-3}f_{-3}} \qquad \text{(IX-193)}$$

where $[E]_t = [X] + [Y]$. From equation IX-191 and the King-Altman solutions for [X] and [Y], the velocity equation for the forward reaction in the absence of products can be written as:

$$v = \frac{\dfrac{[B]}{\delta K_B}\left(\dfrac{V_{max_1}[A][B]}{\alpha K_A K_B} + \dfrac{V_{max_2}[A]}{K_A}\right)}{\dfrac{[B]}{\delta K_B}\left(1 + \dfrac{[A]}{K_A} + \dfrac{[B]}{K_B} + \dfrac{[A][B]}{\alpha K_A K_B}\right) + \dfrac{k_2[A]}{k_3 K_A}\left(1 + \dfrac{[B]}{\delta K_B}\right)} \qquad \text{(IX-194)}$$

where

$$V_{max_1} = k_1[E]_t = \text{the maximal velocity through EAB}$$

$$V_{max_2} = k_3[E]_t = \text{the maximal velocity through FB}$$

Thus even under rapid equilibrium conditions, the equation contains $[B]^2$ terms although when [A] is varied normal hyperbolic plots and linear reciprocal plots are observed. When the flux through F is negligible, the equation reduces to that for a Rapid Equilibrium Random system. When the flux through EAB is negligible, the equation reduces to that for a Rapid Equilibrium Ping Pong Bi Bi system.

General Hybrid System

Figure IX-38 shows the most general hybrid Ping Pong—Random mechanism (which includes the four systems described above). There are two steps in which A adds and four steps in which B adds. The velocity equation will contain $[A]^2$ and $[B]^4$ terms providing for a variety of nonhyperbolic plots, including sigmoidal curves.

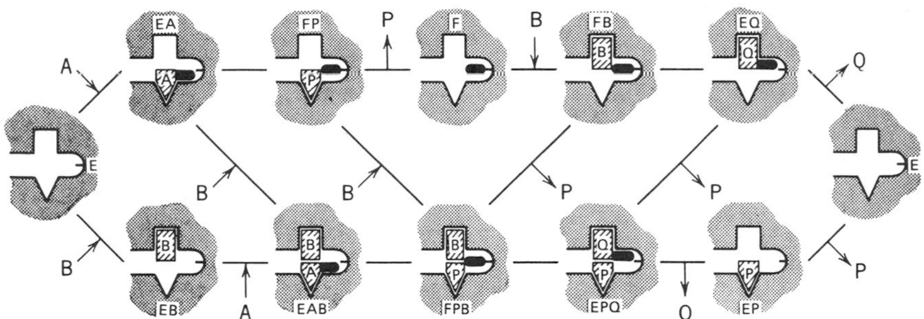

Fig. IX-38. General hybrid Ping Pong-Random mechanism. The overall reaction involves a group transfer from A to B producing P (the residue of A) and Q (B + group transferred from A). In all possible routes, the transfer involves the intermediary formation of an enzyme-group complex.

R. ORDERED TER BI SYSTEM

Velocity Equation, Kinetic Constants, and Haldane Equations

The Ordered Ter Bi system can be represented as:

$$\begin{array}{cccccc} & \text{A} & \text{B} & \text{C} & \text{P} & \text{Q} \\ & k_1 \downarrow k_{-1} & k_2 \downarrow k_{-2} & k_3 \downarrow k_{-3} & k_4 \uparrow k_{-4} & k_5 \uparrow k_{-5} \\ \hline \text{E} & \text{EA} & \text{EAB} & (\text{EABC} \rightleftharpoons \text{EPQ}) & \text{EQ} & \text{E} \end{array}$$

The velocity equation obtained from five four-lined King-Altman interconversion patterns is:

$$\frac{v}{[\mathrm{E}]_t} = \frac{k_1k_2k_3k_4k_5[\mathrm{A}][\mathrm{B}][\mathrm{C}] - k_{-1}k_{-2}k_{-3}k_{-4}k_{-5}[\mathrm{P}][\mathrm{Q}]}{k_{-1}k_{-2}k_5(k_{-3}+k_4) + k_1k_{-2}k_5(k_{-3}+k_4)[\mathrm{A}] + k_{-1}k_3k_4k_5[\mathrm{C}]}$$

$$+ k_1k_2k_5(k_{-3}+k_4)[\mathrm{A}][\mathrm{B}] + k_1k_3k_4k_5[\mathrm{A}][\mathrm{C}] + k_2k_3k_4k_5[\mathrm{B}][\mathrm{C}]$$

$$+ k_1k_2k_3(k_4+k_5)[\mathrm{A}][\mathrm{B}][\mathrm{C}] + k_{-1}k_{-2}k_{-3}k_{-4}[\mathrm{P}] + k_{-1}k_{-2}k_{-5}(k_{-3}+k_4)[\mathrm{Q}]$$

$$+ k_{-4}k_{-5}(k_{-2}k_{-3} + k_{-1}k_{-3} + k_{-1}k_{-2})[\mathrm{P}][\mathrm{Q}] + k_1k_{-2}k_{-3}k_{-4}[\mathrm{A}][\mathrm{P}]$$

$$+ k_1k_2k_{-3}k_{-4}[\mathrm{A}][\mathrm{B}][\mathrm{P}] + k_1k_2k_3k_{-4}[\mathrm{A}][\mathrm{B}][\mathrm{C}][\mathrm{P}]$$

$$+ k_{-1}k_3k_4k_{-5}[\mathrm{C}][\mathrm{Q}] + k_2k_3k_4k_{-5}[\mathrm{B}][\mathrm{C}][\mathrm{Q}] + k_{-1}k_3k_{-4}k_{-5}[\mathrm{C}][\mathrm{P}][\mathrm{Q}]$$

$$+ k_2k_{-3}k_{-4}k_{-5}[\mathrm{B}][\mathrm{P}][\mathrm{Q}] + k_2k_3k_{-4}k_{-5}[\mathrm{B}][\mathrm{C}][\mathrm{P}][\mathrm{Q}] \qquad \text{(IX-195)}$$

Multiplying by the usual factors, the equation becomes:

$$v = \frac{V_f V_r\left([\mathrm{A}][\mathrm{B}][\mathrm{C}] - \dfrac{[\mathrm{P}][\mathrm{Q}]}{K_{\mathrm{eq}}}\right)}{\begin{aligned}&V_r K_{ia} K_{ib} K_{m_\mathrm{C}} + V_r K_{ib} K_{m_\mathrm{C}}[\mathrm{A}] + V_r K_{ia} K_{m_\mathrm{B}}[\mathrm{C}] + V_r K_{m_\mathrm{C}}[\mathrm{A}][\mathrm{B}] \\ &+ K_{m_\mathrm{B}}[\mathrm{A}][\mathrm{C}] + V_r K_{m_\mathrm{A}}[\mathrm{B}][\mathrm{C}] + V_r[\mathrm{A}][\mathrm{B}][\mathrm{C}] + \frac{V_f K_{m_\mathrm{Q}}[\mathrm{P}]}{K_{\mathrm{eq}}} \\ &+ \frac{V_f K_{m_\mathrm{P}}[\mathrm{Q}]}{K_{\mathrm{eq}}} + \frac{V_f[\mathrm{P}][\mathrm{Q}]}{K_{\mathrm{eq}}} + \frac{V_f K_{m_\mathrm{Q}}[\mathrm{A}][\mathrm{P}]}{K_{ia}K_{\mathrm{eq}}} + \frac{V_f K_{m_\mathrm{Q}}[\mathrm{A}][\mathrm{B}][\mathrm{P}]}{K_{ia}K_{ib}K_{\mathrm{eq}}} \\ &+ \frac{V_f K_{m_\mathrm{Q}}[\mathrm{A}][\mathrm{B}][\mathrm{C}][\mathrm{P}]}{K_{ia}K_{ib}K_{ic}K_{\mathrm{eq}}} + \frac{V_r K_{ia} K_{m_\mathrm{B}}[\mathrm{C}][\mathrm{Q}]}{K_{iq}} + \frac{V_r K_{m_\mathrm{A}}[\mathrm{B}][\mathrm{C}][\mathrm{Q}]}{K_{iq}} \\ &+ \frac{V_r K_{ia} K_{m_\mathrm{B}}[\mathrm{C}][\mathrm{P}][\mathrm{Q}]}{K_{ip}K_{iq}} + \frac{V_r K_{m_\mathrm{A}} K_{ic}[\mathrm{B}][\mathrm{P}][\mathrm{Q}]}{K_{ip}K_{iq}} + \frac{V_r K_{m_\mathrm{A}}[\mathrm{B}][\mathrm{C}][\mathrm{P}][\mathrm{Q}]}{K_{ip}K_{iq}}\end{aligned}} \qquad \text{(IX-196)}$$

where

$$K_{m_\mathrm{A}} = \frac{\mathrm{Coef}_{\mathrm{BC}}}{\mathrm{Coef}_{\mathrm{ABC}}}, \qquad K_{m_\mathrm{B}} = \frac{\mathrm{Coef}_{\mathrm{AC}}}{\mathrm{Coef}_{\mathrm{ABC}}}, \qquad K_{m_\mathrm{C}} = \frac{\mathrm{Coef}_{\mathrm{AB}}}{\mathrm{Coef}_{\mathrm{ABC}}}$$

$$K_{m_\mathrm{P}} = \frac{\mathrm{Coef}_{\mathrm{Q}}}{\mathrm{Coef}_{\mathrm{PQ}}} \qquad K_{m_\mathrm{Q}} = \frac{\mathrm{Coef}_{\mathrm{P}}}{\mathrm{Coef}_{\mathrm{PQ}}}$$

$$K_{ia} = \frac{\mathrm{const}}{\mathrm{Coef}_{\mathrm{A}}} = \frac{\mathrm{Coef}_{\mathrm{P}}}{\mathrm{Coef}_{\mathrm{AP}}} = \frac{\mathrm{Coef}_{\mathrm{C}}}{\mathrm{Coef}_{\mathrm{AC}}} = \frac{k_{-1}}{k_1}$$

$$K_{ib} = \frac{\mathrm{Coef}_{\mathrm{A}}}{\mathrm{Coef}_{\mathrm{AB}}} = \frac{\mathrm{Coef}_{\mathrm{AP}}}{\mathrm{Coef}_{\mathrm{ABP}}} = \frac{k_{-2}}{k_2}$$

$$K_{ic} = \frac{\mathrm{Coef}_{\mathrm{BPQ}}}{\mathrm{Coef}_{\mathrm{BCPQ}}} = \frac{\mathrm{Coef}_{\mathrm{ABP}}}{\mathrm{Coef}_{\mathrm{ABCP}}} = \frac{k_{-3}}{k_3}$$

$$K_{ip} = \frac{\mathrm{Coef}_{\mathrm{CQ}}}{\mathrm{Coef}_{\mathrm{CPQ}}} = \frac{\mathrm{Coef}_{\mathrm{BCQ}}}{\mathrm{Coef}_{\mathrm{BCPQ}}} = \frac{k_4}{k_{-4}}$$

$$K_{iq} = \frac{\mathrm{const}}{\mathrm{Coef}_{\mathrm{Q}}} = \frac{\mathrm{Coef}_{\mathrm{C}}}{\mathrm{Coef}_{\mathrm{CQ}}} = \frac{\mathrm{Coef}_{\mathrm{BC}}}{\mathrm{Coef}_{\mathrm{BCQ}}} = \frac{k_5}{k_{-5}}$$

Note that all the inhibition constants represent true dissociation constants for the given ligand. Equation IX-196 can be expressed entirely in terms of kinetic constants by substituting for K_{eq} using the Haldane equations.

$$\boxed{K_{eq} = \frac{V_{\max_f} K_{m_P} K_{iq}}{V_{\max_r} K_{ia} K_{ib} K_{m_C}} = \frac{K_{ip} K_{iq}}{K_{ia} K_{ib} K_{ic}}} \qquad \text{(IX-197)}$$

Rate Constants

$$k_1 = \frac{V_{\max_f}}{K_{m_A}[\mathrm{E}]_t}, \qquad k_{-1} = \frac{V_{\max_f} K_{ia}}{K_{m_A}[\mathrm{E}]_t}, \qquad k_2 = \frac{V_{\max_f}}{K_{m_B}[\mathrm{E}]_t}$$

$$k_{-2} = \frac{V_{\max_f} K_{ib}}{K_{m_B}[\mathrm{E}]_t}, \qquad k_3 = \left(1 + \frac{k_{-3}}{k_4}\right) \frac{V_{\max_f}}{K_{m_C}[\mathrm{E}]_t}$$

$$\frac{1}{k_{-3}} = \frac{[\mathrm{E}]_t}{V_{\max_r}} - \frac{1}{k_{-1}} - \frac{1}{k_{-2}}, \qquad \frac{1}{k_4} = \frac{[\mathrm{E}]_t}{V_{\max_f}} - \frac{1}{k_5}$$

$$k_{-4} = \left(1 + \frac{k_4}{k_{-3}}\right) \frac{V_{\max_r}}{K_{m_P}[\mathrm{E}]_t}, \qquad k_5 = \frac{V_{\max_r} K_{iq}}{K_{m_Q}[\mathrm{E}]_t}$$

$$k_{-5} = \frac{V_{\max_r}}{K_{m_Q}[\mathrm{E}]_t}$$

Distribution Equations

$$\frac{[\mathrm{E}]}{[\mathrm{E}]_t} = \frac{\dfrac{K_{m_Q} V_f [\mathrm{P}]}{K_{eq}} + K_{m_A} V_r [\mathrm{B}][\mathrm{C}] + K_{ia} K_{m_B} V_r [\mathrm{C}] + K_{ia} K_{ib} K_{m_C} V_r}{\text{denominator of velocity equation}} \qquad \text{(IX-198)}$$

$$\frac{[\mathrm{EA}]}{[\mathrm{E}]_t} = \frac{\dfrac{K_{m_Q} V_f [\mathrm{A}][\mathrm{P}]}{K_{ia} K_{eq}} + \dfrac{K_{m_A} V_r [\mathrm{P}][\mathrm{Q}]}{K_{ia} K_{eq}} + K_{m_B} V_r [\mathrm{A}][\mathrm{C}] + K_{ib} K_{m_C} V_r [\mathrm{A}]}{\text{denominator of velocity equation}}$$

$$\text{(IX-199)}$$

$$\frac{[\mathrm{EAB}]}{[\mathrm{E}]_t} = \frac{\dfrac{K_{m_\mathrm{Q}} V_f [\mathrm{A}][\mathrm{B}][\mathrm{P}]}{K_{ia} K_{ib} K_{\mathrm{eq}}} + \dfrac{K_{m_\mathrm{A}} K_{ic} V_r [\mathrm{B}][\mathrm{P}][\mathrm{Q}]}{K_{ip} K_{iq}} + \dfrac{K_{m_\mathrm{B}} V_r [\mathrm{P}][\mathrm{Q}]}{K_{ib} K_{\mathrm{eq}}} + K_{m_\mathrm{C}} V_r [\mathrm{A}][\mathrm{B}]}{\text{denominator of velocity equation}} \quad \text{(IX-200)}$$

$$\frac{[\mathrm{EABC+EPQ}]}{[\mathrm{E}]_t} = \frac{\dfrac{K_{m_\mathrm{Q}} V_f [\mathrm{A}][\mathrm{B}][\mathrm{C}][\mathrm{P}]}{K_{ia} K_{ib} K_{ic} K_{\mathrm{eq}}} + \dfrac{K_{m_\mathrm{A}} V_r [\mathrm{B}][\mathrm{C}][\mathrm{P}][\mathrm{Q}]}{K_{ip} K_{iq}} + \dfrac{K_{ia} K_{m_\mathrm{B}} V_r [\mathrm{C}][\mathrm{P}][\mathrm{Q}]}{K_{ip} K_{iq}} + \left[\dfrac{V_f}{K_{\mathrm{eq}}} - \left(\dfrac{K_{m_\mathrm{A}}}{K_{ia}} + \dfrac{K_{m_\mathrm{B}}}{K_{ib}} \right) \dfrac{V_r}{K_{\mathrm{eq}}} \right] [\mathrm{P}][\mathrm{Q}] + \left(\dfrac{V_r - K_{m_\mathrm{Q}} V_f}{K_{iq}} \right) [\mathrm{A}][\mathrm{B}][\mathrm{C}]}{\text{denominator of velocity equation}} \quad \text{(IX-201)}$$

$$\frac{[\mathrm{EQ}]}{[\mathrm{E}]_t} = \frac{\dfrac{K_{m_\mathrm{Q}} V_f [\mathrm{A}][\mathrm{B}][\mathrm{C}]}{K_{iq}} + \dfrac{K_{m_\mathrm{A}} V_r [\mathrm{B}][\mathrm{C}][\mathrm{Q}]}{K_{iq}} + \dfrac{K_{ia} K_{m_\mathrm{B}} V_r [\mathrm{C}][\mathrm{Q}]}{K_{iq}} + \dfrac{K_{m_\mathrm{P}} V_f [\mathrm{Q}]}{K_{\mathrm{eq}}}}{\text{denominator of velocity equation}} \quad \text{(IX-202)}$$

Effect of Isomerizations

EABC + EPQ: The calculations of k_3, k_{-3}, k_4, and k_{-4} are invalid.

EA: [E]/[E]$_t$, [EAB]/[E]$_t$, [EQ]/[E]$_t$, k_{-2}, k_5, and k_{-5} may still be calculated.

EAB: [E]/[E]$_t$, [EA]/[E]$_t$, [EQ]/[E]$_t$, k_1, k_{-1}, k_5, and k_{-5} may still be calculated.

EQ: [E]/[E]$_t$, [EA]/[E]$_t$, [EAB]/[E]$_t$, k_1, k_{-1}, k_2, and k_{-2} may still be calculated.

Initial Velocity Studies in the Forward Direction

The velocity equation for the forward reaction in the absence of products is:

$$\frac{v}{V_{\max}} = \frac{[A][B][C]}{K_{ia}K_{ib}K_{m_C} + K_{ib}K_{m_C}[A] + K_{ia}K_{m_B}[C] + K_{m_C}[A][B] + K_{m_B}[A][C] + K_{m_A}[B][C] + [A][B][C]}$$

(IX-203)

or

$$\frac{v}{V_{\max}} = \frac{[A]}{K_{m_A}\left(1 + \frac{K_{ia}K_{ib}K_{m_C}}{K_{m_A}[B][C]} + \frac{K_{ia}K_{m_B}}{K_{m_A}[B]}\right) + [A]\left(1 + \frac{K_{ib}K_{m_C}}{[B][C]} + \frac{K_{m_C}}{[C]} + \frac{K_{m_B}}{[B]}\right)}$$

(IX-204)

$$\frac{v}{V_{\max}} = \frac{[B]}{K_{m_B}\left(1 + \frac{K_{ia}}{[A]}\right)\left(1 + \frac{K_{ib}K_{m_C}}{K_{m_B}[C]}\right) + [B]\left(1 + \frac{K_{m_A}}{[A]} + \frac{K_{m_C}}{[C]}\right)}$$ (IX-205)

$$\frac{v}{V_{\max}} = \frac{[C]}{K_{m_C}\left(1 + \frac{K_{ia}K_{ib}}{[A][B]} + \frac{K_{ib}}{[B]}\right) + [C]\left(1 + \frac{K_{ia}K_{m_B}}{[A][B]} + \frac{K_{m_B}}{[B]} + \frac{K_{m_A}}{[A]}\right)}$$

(IX-206)

The usual procedure would be to treat each substrate as the varied substrate at different fixed concentrations of another substrate, maintaining a constant concentration of the third substrate (which can be changed for a different family of plots). All such plots intersect to the left of the $1/v$-axis, except when [B] is saturating. In this case, the $slope_{1/A}$ and $slope_{1/C}$ factors reduce to unity and the families of $1/v$ versus $1/[A]$ plots at different fixed [C] and $1/v$ versus $1/[C]$ plots at different fixed [A] are parallel. When either [A] or [C] is saturating, the reciprocal plot families still intersect. Thus B, the second substrate to add, can be identified. However, the initial velocity studies do not allow A and C to be distinguished from each other. (Binding studies, however, will identify A as it is the only substrate which binds to free E.) Often it is not practical to maintain the substrates at saturation (i.e., ca.

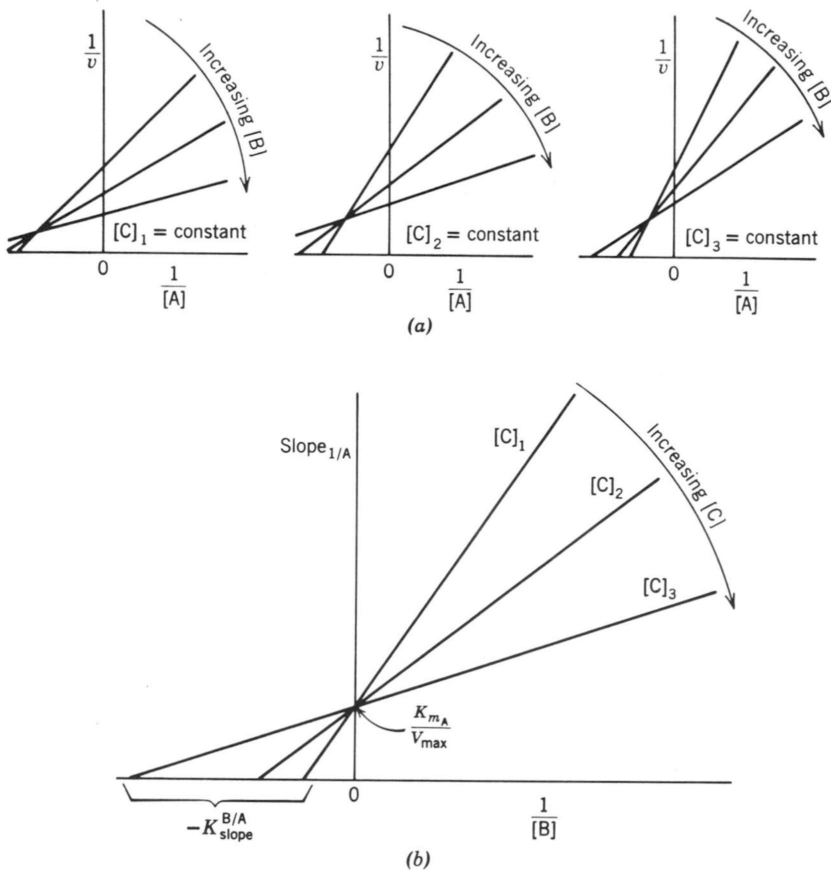

Fig. IX-39. Ordered terreactant system; B can be identified as that changing fixed substrate which yields a family of *slope* replots that intersect on the *slope*-axis at all concentrations of the constant substrate: (*a*) $1/v$ versus $1/[A]$ plots at different fixed B concentrations and a constant, unsaturating C concentration (changed for each family of plots). (*b*) $Slope_{1/A}$ replots. (*c*) [A] is varied as in *a*, except now C is the changing fixed substrate and B is the constant substrate. (*d*) $Slope_{1/A}$ replots.

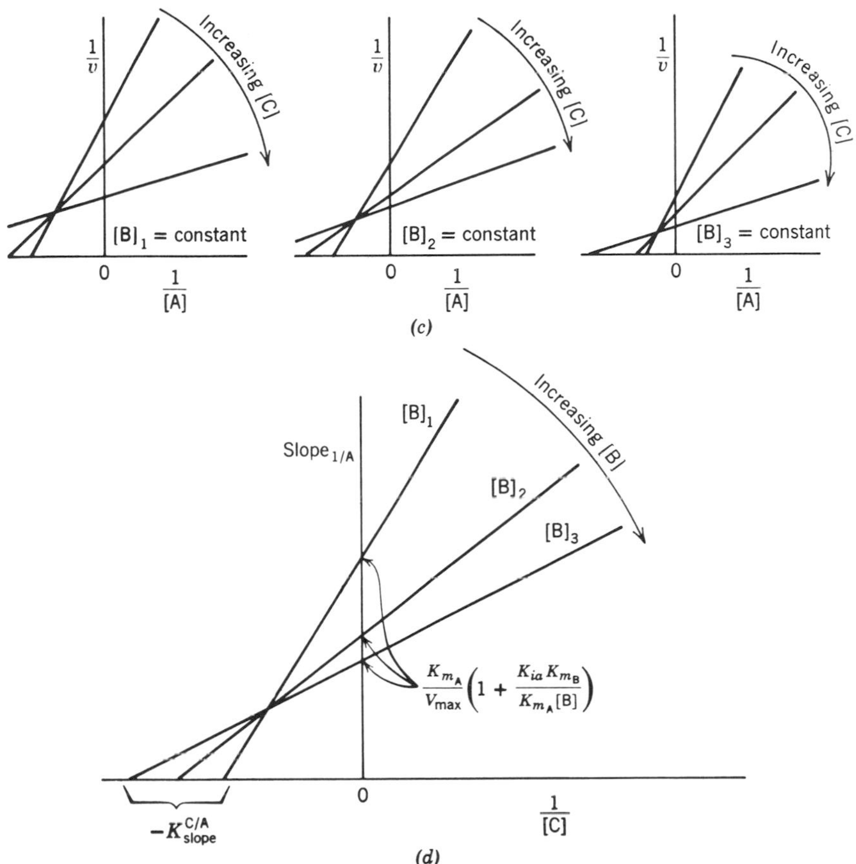

Fig. IX-39. (*Cont.*)

100 times the K_m value). In this case, B can still be identified from slope replots. For example, when B is the changing fixed substrate and C the constant substrate (Fig. IX-39*a*), the $slope_{1/A}$ replot is given by:

$$slope_{1/A} = \frac{K_{ia}K_{m_B}}{V_{max}}\left(1 + \frac{K_{ib}K_{m_C}}{K_{m_B}[C]}\right)\frac{1}{[B]} + \frac{K_{m_A}}{V_{max}} \tag{IX-207}$$

Thus the intercept on the slope-axis (i.e., the slope at an infinitely high [B]) is the same for all constant C concentrations (Fig. IX-39*b*). On the other hand, when C is the changing fixed substrate and B is the constant substrate

(Fig. IX-39*c*) the $slope_{1/\mathrm{A}}$ is:

$$slope_{1/\mathrm{A}} = \frac{K_{ia}K_{ib}K_{m_\mathrm{C}}}{V_{\max}[\mathrm{B}]}\frac{1}{[\mathrm{C}]} + \frac{K_{m_\mathrm{A}}}{V_{\max}}\left(1 + \frac{K_{ia}K_{m_\mathrm{B}}}{K_{m_\mathrm{A}}[\mathrm{B}]}\right) \qquad \text{(IX-208)}$$

In this case, the $slope_{1/\mathrm{A}}$ replots for different constant concentrations of B do not intersect on the vertical axis (Fig. IX-39*d*). The $slope_{1/\mathrm{C}}$ versus $1/[\mathrm{B}]$ replots are analogous to those for $slope_{1/\mathrm{A}}$: the family of replots when A is the constant substrate intersect on the vertical axis at $K_{m_\mathrm{C}}/V_{\max}$. When B is the constant substrate, both families of $slope_{1/\mathrm{C}}$ replots intersect to the left of the vertical axis. When [B] is varied and A or C is the changing fixed substrate (Fig. IX-40*a,c*), the $slope_{1/\mathrm{B}}$ replots intersect to the left of the $slope_{1/\mathrm{B}}$-axis but on the $1/[\mathrm{A}]$ or $1/[\mathrm{C}]$-axis (Fig. IX-40*b,d*).

$$K_{\text{slope}}^{\mathrm{A/B}} = K_{ia} \text{ at all constant } [\mathrm{C}], \qquad K_{\text{slope}}^{\mathrm{C/B}} = \frac{K_{m_\mathrm{B}}}{K_{ib}K_{m_\mathrm{C}}} \text{ at all constant } [\mathrm{A}]$$

Plots With Two Changing Fixed Substrates

A rapid way to identify B is to vary each substrate separately at different fixed concentrations of the other two, maintaining the fixed substrates at a constant ratio (Rudolph, Purich, and Fromm, 1968). Any two lines within each family of reciprocal plots intersect to the left of the $1/v$-axis. The intercept replots of $1/V_{\max_{\text{app}}}$ versus $1/[\mathrm{B}]$ or $1/[\mathrm{C}]$ (when [A] is varied) and $1/V_{\max_{\text{app}}}$ versus $1/[\mathrm{A}]$ or $1/[\mathrm{B}]$ (when [C] is varied) are both parabolic. On the other hand the replot of $1/V_{\max_{\text{app}}}$ versus $1/[\mathrm{A}]$ or $1/[\mathrm{C}]$ (when [B] is varied) is linear. All three intercept replots extrapolate to the true $1/V_{\max}$. All three slope replots will be parabolic when one substrate is varied at different fixed concentrations of the other two, but the intercept on the vertical axis yields the ratio of K_m (for the varied substrate) to $V_{\max}$. For example, when [C] is varied at different fixed concentrations of A and B, where $[\mathrm{A}] = x[\mathrm{B}]$, the slope is given by:

$$slope_{1/\mathrm{A}} = \frac{K_{m_\mathrm{C}}K_{ib}}{V_{\max}}\left(1 + \frac{K_{ia}K_{m_\mathrm{B}}}{xK_{ib}[\mathrm{B}]}\right)\frac{1}{[\mathrm{B}]} + \frac{K_{m_\mathrm{C}}}{V_{\max}} \qquad \text{(IX-209)}$$

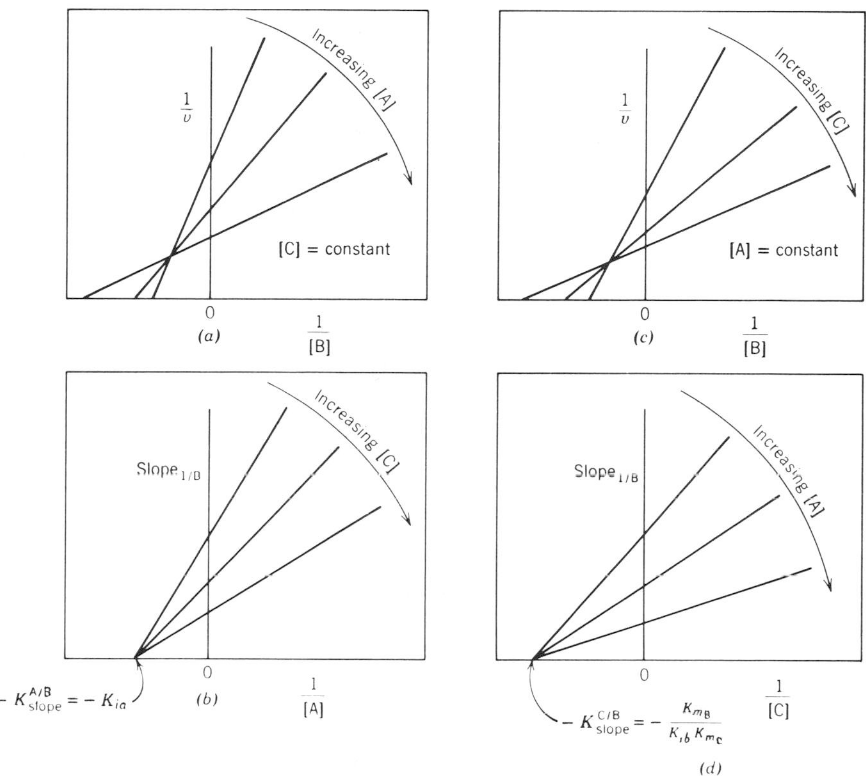

Fig. IX-40. Ordered terreactant system. B can be identified as that varied substrate which yields a family of *slope* replots that intersect on the horizontal-axis at all concentrations of the constant substrate: (*a*) [B] is varied at different fixed concentrations of A and a constant, unsaturating C concentration (changed for each family of plots). (*b*) The $slope_{1/B}$ replot gives K_{ia} at all constant C concentrations. (*c*) [B] is varied at different fixed C concentrations and a constant, unsaturating A concentration (changed for each family of plots). (*d*) The $slope_{1/B}$ replot gives $-K_{slope}^{C/B} = -K_{m_B}/K_{ib}K_{m_C}$ at all constant A concentrations.

Thus once V_{max} is known, all three K_m's can be determined. The value of K_{ib} can be calculated from $K_{slope}^{C/B}$; K_{ic} does not appear in the velocity equation for the forward reaction in the absence of products.

To predict the shape of the slope or intercept replot when one substrate is varied at different fixed concentrations of the other two, we need only examine the slope or intercept factors in the original velocity or reciprocal equation. If the factor contains a product of two concentration terms (for example, [A][B] in the intercept factor when [C] is varied, or [B][C] in the slope factor when [A] is varied), then the replot will be parabolic. (The product of concentration terms becomes a squared term.)

Varying Two Substrates Together

An alternate method of determining V_{max} and the K_m values of the three substrates is to vary two substrates together at a constant ratio at different fixed concentrations of the third substrate. For example, when [A] and [B] are varied together at $[B]=x[A]$, the reciprocal plot is given by:

$$\frac{1}{v}=\frac{K_{m_A}}{V_{max}}\left(1+\frac{K_{ia}K_{ib}K_{m_C}}{xK_{m_A}[A][C]}+\frac{K_{ib}K_{m_C}}{xK_{m_A}[C]}+\frac{K_{ia}K_{m_B}}{xK_{m_A}[A]}+\frac{K_{m_B}}{xK_{m_A}}\right)\frac{1}{[A]}+\frac{1}{V_{max}}\left(1+\frac{K_{m_C}}{[C]}\right)$$

(IX-210)

When A and C are varied together maintaining [C] at $x[A]$:

$$\frac{1}{v}=\frac{K_{m_A}}{V_{max}}\left(1+\frac{K_{ia}K_{ib}K_{m_C}}{xK_{m_A}[A][B]}+\frac{K_{ib}K_{m_C}}{xK_{m_A}[B]}+\frac{K_{ia}K_{m_B}}{K_{m_A}[B]}+\frac{K_{m_C}}{xK_{m_A}}\right)\frac{1}{[A]}+\frac{1}{V_{max}}\left(1+\frac{K_{m_B}}{[B]}\right)$$

(IX-210a)

The families of reciprocal plots are all parabolic at unsaturating concentrations of the fixed substrate. If, however, A and C are varied together at saturating [B], the plot will be linear. (Saturating [B] reduces the $K_{ia}K_{ib}K_{m_C}/xK_{m_A}[A][B]$ term to zero; hence, there is no $[A]^2$ term in the reciprocal equation IX-210a.) In all cases, the $1/v$-axis intercept replots give V_{max} and the K_m value of the changing fixed substrate.For example, when [A] and [B] are varied together at different fixed concentrations of C, the $1/v$-axis intercept is given by:

$$\frac{1}{V_{max_{app}}}=\frac{K_{m_C}}{V_{max}}\frac{1}{[C]}+\frac{1}{V_{max}} \qquad \text{(IX-211)}$$

The vertical-axis intercept of the replot gives $1/V_{max}$ while the horizontal-axis intercept gives $-1/K_{m_C}$.

Product Inhibition

In the presence of P, the velocity equation contains additional [P], [A][P], [A][B][P], and [A][B][C][P] terms in the denominator which may be expressed entirely in terms of kinetic constants by substitution for K_{eq} from the first Haldane equation. When [A] is varied, the four new terms appear in the denominator as shown below.

$$\frac{v}{V_{max}} = \frac{[A]}{K_{m_A}\left(1 + \text{original terms} + \dfrac{K_{ia}K_{ib}K_{m_C}K_{m_Q}[P]}{K_{m_A}K_{m_P}K_{iq}[B][C]}\right) + [A]\left(1 + \text{original terms} + \dfrac{K_{ib}K_{m_C}K_{m_Q}[P]}{K_{m_P}K_{iq}[B][C]} + \dfrac{K_{m_C}K_{m_Q}[P]}{K_{m_P}K_{iq}[C]} + \dfrac{K_{m_C}K_{m_Q}[P]}{K_{m_P}K_{iq}K_{ic}}\right)}$$

(IX-212)

Thus P changes both the slope and intercept of the $1/v$ versus $1/[A]$ plots (at fixed unsaturating concentrations of B and C). Saturation with either B or C eliminates the slope effect, but the intercept effect remains and P becomes an uncompetitive inhibitor.

When [B] is varied:

$$\frac{v}{V_{max}} = \frac{[B]}{K_{m_B}\left(1 + \text{original terms} + \dfrac{K_{ia}K_{ib}K_{m_C}K_{m_Q}[P]}{K_{m_B}K_{m_P}K_{iq}[A][C]} + \dfrac{K_{ib}K_{m_C}K_{m_Q}[P]}{K_{m_B}K_{m_P}K_{iq}[C]}\right) + [B]\left(1 + \text{original terms} + \dfrac{K_{m_C}K_{m_Q}[P]}{K_{m_P}K_{iq}[C]} + \dfrac{K_{m_C}K_{m_Q}[P]}{K_{m_P}K_{iq}K_{ic}}\right)}$$

(IX-213)

P affects both the slope and intercept at fixed unsaturating [A] and [C]. Saturation with A does not change the pattern; saturation with C eliminates the slope effect, but not the intercept effect.

When [C] is varied:

$$\frac{v}{V_{\max}}=\frac{[\mathrm{C}]}{K_{m_{\mathrm{C}}}\left(1+\text{original terms}+\dfrac{K_{ia}K_{ib}K_{m_{\mathrm{Q}}}[\mathrm{P}]}{K_{m_{\mathrm{P}}}K_{iq}[\mathrm{A}][\mathrm{B}]}+\dfrac{K_{ib}K_{m_{\mathrm{Q}}}[\mathrm{P}]}{K_{m_{\mathrm{P}}}K_{iq}[\mathrm{B}]}+\dfrac{K_{m_{\mathrm{Q}}}[\mathrm{P}]}{K_{m_{\mathrm{P}}}K_{iq}}\right)+[\mathrm{C}]\left(1+\text{original terms}+\dfrac{K_{m_{\mathrm{Q}}}K_{m_{\mathrm{C}}}[\mathrm{P}]}{K_{m_{\mathrm{P}}}K_{ic}K_{iq}}\right)}$$

(IX-214)

P affects both the slope and intercept at all A and B concentrations. The effect of P at saturating concentrations of two cosubstrates is shown in Figure IX-41. The value of K_{ic} can be obtained from the intersection point of the $1/v$ versus $1/[\mathrm{C}]$ plots at saturating [A] and [B]. If [A] is not saturating but [B] is, the intersection point is still $-1/K_{ic}$. If only [A] is saturating, the intersection occurs at $1/[\mathrm{C}]=-1/K_{ic}(1+K_{ib}/[\mathrm{B}])$.

If it is impractical to maintain the nonvaried substrates at saturation, it is still possible to determine whether saturation with a given substrate will eliminate the slope effect. For example, $slope_{1/\mathrm{A}}$ and $slope_{1/\mathrm{B}}$ in the presence of P are given by:

$$slope_{1/\mathrm{A}}=\frac{K_{ia}K_{ib}K_{m_{\mathrm{C}}}K_{m_{\mathrm{Q}}}}{V_{\max}K_{m_{\mathrm{P}}}K_{iq}[\mathrm{B}][\mathrm{C}]}[\mathrm{P}]+\frac{K_{m_{\mathrm{A}}}}{V_{\max}}\left[1+\frac{K_{ia}K_{m_{\mathrm{B}}}}{K_{m_{\mathrm{A}}}[\mathrm{B}]}\left(1+\frac{K_{m_{\mathrm{C}}}}{[\mathrm{C}]}\right)\right]$$

(IX-215)

$$slope_{1/\mathrm{B}}=\frac{K_{ib}K_{m_{\mathrm{C}}}K_{m_{\mathrm{Q}}}}{V_{\max}K_{m_{\mathrm{P}}}K_{iq}[\mathrm{C}]}\left(1+\frac{K_{ia}}{[\mathrm{A}]}\right)[\mathrm{P}]+\frac{K_{m_{\mathrm{B}}}}{V_{\max}}\left(1+\frac{K_{ia}}{[\mathrm{A}]}\right)\left(1+\frac{K_{ib}K_{m_{\mathrm{C}}}}{K_{m_{\mathrm{B}}}[\mathrm{C}]}\right)$$

(IX-216)

Thus several families of $1/v$ versus $1/[\mathrm{A}]$ plots at different fixed concentrations of P and a constant [B] and [C] are constructed. The constant [B] (or the constant [C]) is changed for each family. The slopes of each family are replotted against [P] (equation IX-215). Finally, the slopes of the $slope_{1/\mathrm{A}}$ replots are plotted against $1/[\mathrm{B}]$ or $1/[\mathrm{C}]$ (whichever substrate was changed for each family). The secondary replot will go through the origin indicating

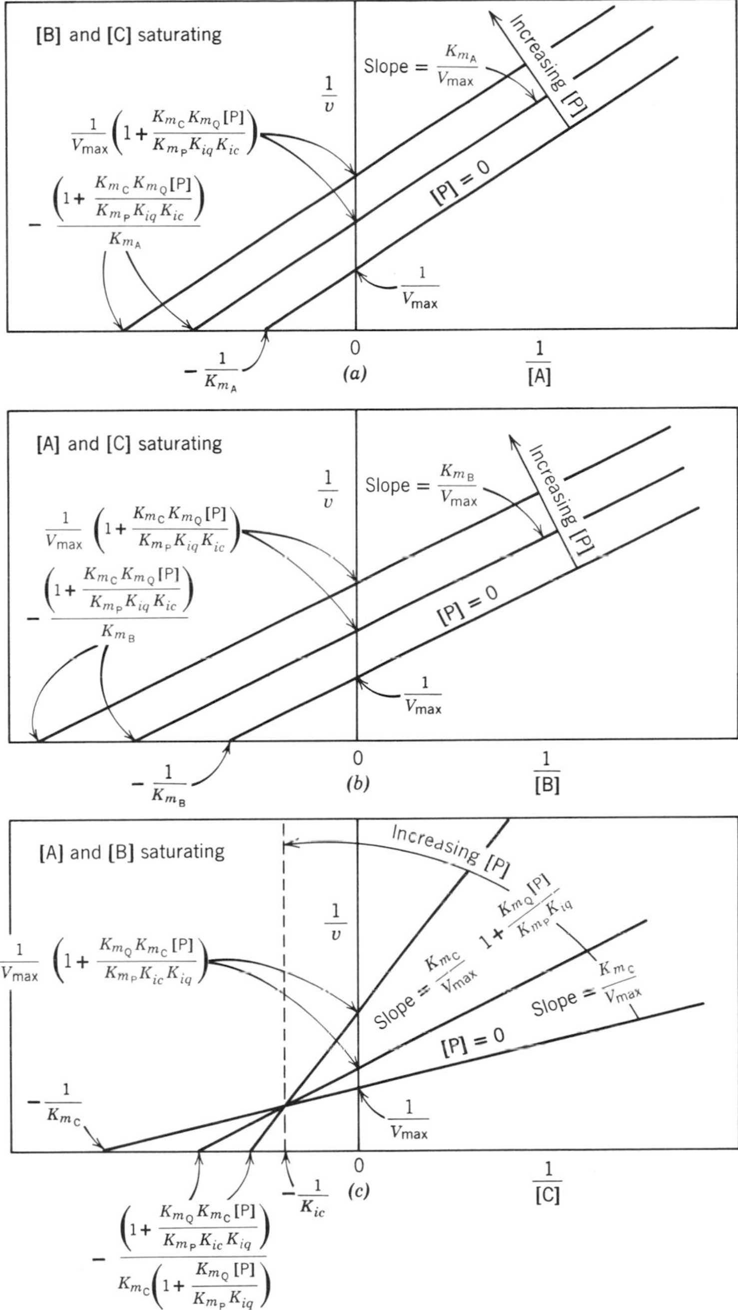

Fig. IX-41. Product inhibition by P in an Ordered Ter Bi system. (*a*) [A] is varied; [B] and [C] are maintained constant and saturating. (*b*) [B] is varied; [A] and [C] are maintained constant and saturating. (*c*) [C] is varied; [A] and [B] are maintained constant and saturating.

that at saturating [B] or saturating [C], P has no effect on $slope_{1/A}$. The same procedure can be used to determine the effect of saturating [A] and saturating [C] on $slope_{1/B}$ (equation IX-216). In this case, a secondary replot against 1/[C] goes through the origin, but if A is changed for each family, the secondary replot against 1/[A] has finite intercepts; that is, saturating [A] will not eliminate the effect of P on $slope_{1/B}$. Both secondary replots of $slope_{1/C}$ versus 1/[A] and versus 1/[B] have finite intercepts.

When Q is the product inhibitor, the velocity equation contains additional [Q], [C][Q], and [B][C][Q] terms which appear in the denominator of the velocity equation as shown below. When [A] is varied:

$$\frac{v}{V_{\max}} = \frac{[\mathrm{A}]}{K_{m_\mathrm{A}}\left(1 + \frac{K_{ia}K_{ib}K_{m_\mathrm{C}}}{K_{m_\mathrm{A}}[\mathrm{B}][\mathrm{C}]} + \frac{K_{ia}K_{m_\mathrm{B}}}{K_{m_\mathrm{A}}[\mathrm{B}]}\right)\left(1 + \frac{[\mathrm{Q}]}{K_{iq}}\right) + [\mathrm{A}]\left(1 + \frac{K_{ib}K_{m_\mathrm{C}}}{[\mathrm{B}][\mathrm{C}]} + \frac{K_{m_\mathrm{C}}}{[\mathrm{C}]} + \frac{K_{m_\mathrm{B}}}{[\mathrm{B}]}\right)} \tag{IX-217}$$

Thus $K_{i_{\text{slope}}}^{\mathrm{Q}/\mathrm{A}} = K_{iq}$ at all fixed B and C concentrations. When [B] is varied:

$$\frac{v}{V_{\max}} = \frac{[\mathrm{B}]}{K_{m_\mathrm{B}}\left(1 + \text{original terms} + \frac{K_{ia}[\mathrm{Q}]}{K_{iq}[\mathrm{A}]} + \frac{K_{ia}K_{ib}K_{m_\mathrm{C}}[\mathrm{Q}]}{K_{iq}K_{m_\mathrm{B}}[\mathrm{A}][\mathrm{C}]}\right) + [\mathrm{B}]\left(1 + \text{original terms} + \frac{K_{m_\mathrm{A}}[\mathrm{Q}]}{K_{iq}[\mathrm{A}]}\right)} \tag{IX-218}$$

When [C] is varied:

$$\frac{v}{V_{\max}} = \frac{[\mathrm{C}]}{K_{m_\mathrm{C}}\left(1 + \text{original terms} + \frac{K_{ia}K_{ib}[\mathrm{Q}]}{[\mathrm{A}][\mathrm{B}]K_{iq}}\right) + [\mathrm{C}]\left(1 + \text{original terms} + \frac{K_{ia}K_{m_\mathrm{B}}[\mathrm{Q}]}{K_{iq}[\mathrm{A}][\mathrm{B}]} + \frac{K_{m_\mathrm{A}}[\mathrm{Q}]}{K_{iq}[\mathrm{A}]}\right)} \tag{IX-219}$$

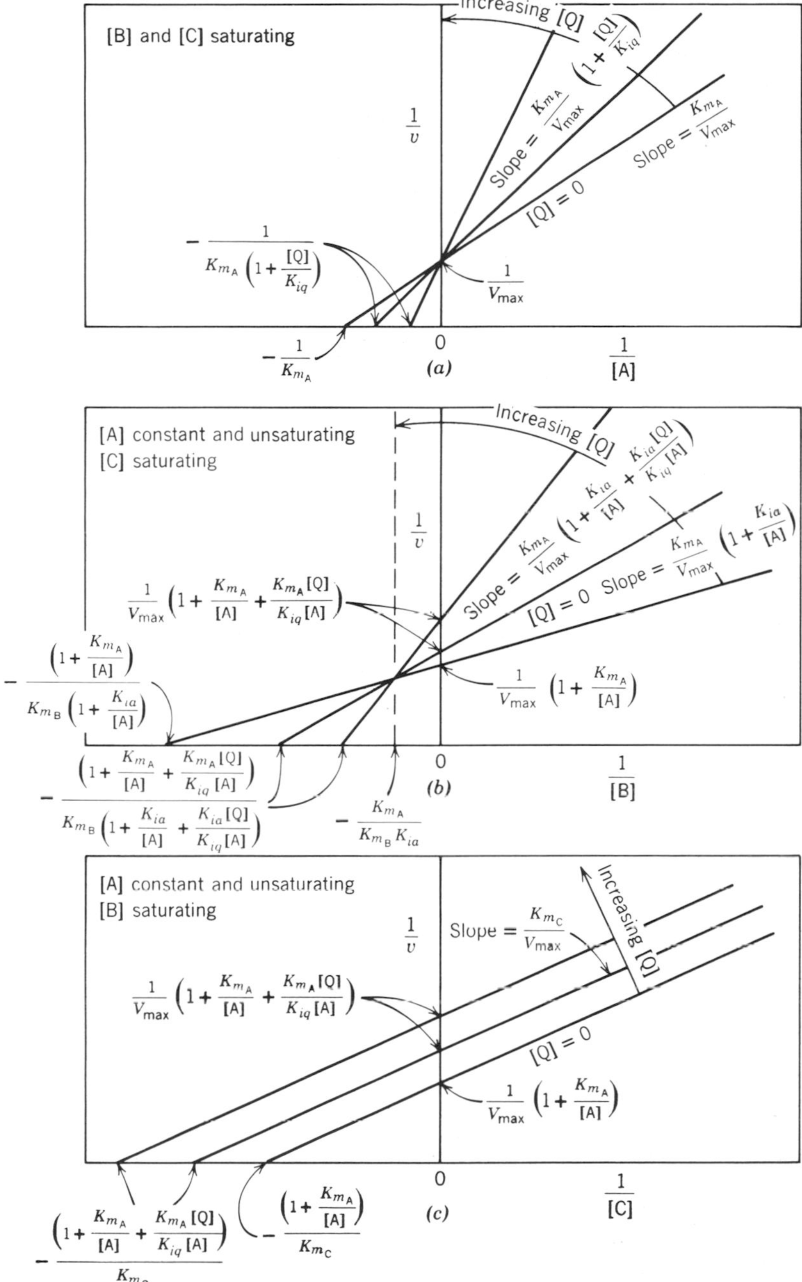

Fig. IX-42. Product inhibition by Q in an Ordered Ter Bi system. Reciprocal plots in the presence of different fixed Q concentrations and saturating concentrations of the nonvaried substrate (where possible without eliminating the inhibition).

Q is competitive with A at all B and C concentrations and is a mixed-type inhibitor with respect to B at all C concentrations and unsaturating [A]. Saturation with A eliminates the inhibition. Q is a mixed-type inhibitor with respect to C at unsaturating A and B concentrations. Saturation with A eliminates the inhibition. Saturation with B eliminates the slope effect, but the intercept effect remains and Q becomes an uncompetitive inhibitor. The product inhibition studies easily identify the A–Q pair (the only competitive pattern). With B already known, C and P can be identified. Figure IX-42 shows the effect of Q at fixed concentrations of two cosubstrates (saturating, where possible without eliminating the inhibition). If [C] is constant but unsaturating, the $1/v$ versus $1/[B]$ plots intersect at:

$$-\frac{K_{m_A}}{K_{ia}K_{m_B}\left(1+\dfrac{K_{ib}K_{m_C}}{K_{m_B}[C]}\right)}$$

If it is possible to construct all the plots described, then most of the kinetic constants can be determined from appropriate replots. The K_{m_Q} and K_{m_P} cannot be determined from forward velocity data, but their ratio can be calculated; K_{ip} does not appear in the velocity equation for the forward reaction, but can be calculated from the Haldane equation (IX-197) if K_{eq} is known.

Reverse Direction—Ordered Bi Ter

In the reverse direction the velocity equation is, of course, the same as that shown earlier for the Ordered Bi Uni or Ordered Bi Bi systems. But in the presence of products (A, B, or C), new terms appear in the slope or intercept factors, as shown below.

When [P] is varied:

$$\frac{v}{V_{\max}}=\frac{[P]}{K_{m_P}\left(1+\dfrac{K_{iq}}{[Q]}+\dfrac{K_{iq}[A]}{K_{ia}[Q]}\right)+[P]\left(1+\dfrac{K_{m_Q}}{[Q]}+\dfrac{K_{m_Q}[A]}{K_{ia}[Q]}\right)} \qquad \text{(IX-220)}$$

$$\frac{v}{V_{\max}} = \frac{[\mathrm{P}]}{K_{m_\mathrm{P}}\left(1 + \frac{K_{iq}}{[\mathrm{Q}]}\right) + [\mathrm{P}]\left(1 + \frac{K_{m_\mathrm{Q}}}{[\mathrm{Q}]} + \frac{K_{m_\mathrm{A}} K_{ic} K_{m_\mathrm{P}}[\mathrm{B}]}{K_{ia} K_{ib} K_{m_\mathrm{C}} K_{ip}}\right)} \qquad \text{(IX-221)}$$

$$\frac{v}{V_{\max}} = \frac{[\mathrm{P}]}{K_{m_\mathrm{P}}\left(1 + \frac{K_{iq}}{[\mathrm{Q}]}\right)\left(1 + \frac{K_{m_\mathrm{B}}[\mathrm{C}]}{K_{ib} K_{m_\mathrm{C}}}\right) + [\mathrm{P}]\left(1 + \frac{K_{m_\mathrm{Q}}}{[\mathrm{Q}]} + \frac{K_{m_\mathrm{B}} K_{m_\mathrm{P}}[\mathrm{C}]}{K_{ib} K_{ip} K_{m_\mathrm{C}}}\right)} \qquad \text{(IX-222)}$$

Thus $K_{i_{\text{slope}}}^{\mathrm{C/P}} = K_{ib}K_{m_\mathrm{C}}/K_{m_\mathrm{B}}$ at all Q concentrations; A and C are mixed-type inhibitors with respect to P. Saturation with [Q] eliminates the inhibition by A, but not by C; B is an uncompetitive inhibitor with respect to P at all Q concentrations.

When [Q] is varied:

$$\frac{v}{V_{\max}} = \frac{[\mathrm{Q}]}{K_{m_\mathrm{Q}}\left(1 + \frac{K_{m_\mathrm{P}} K_{iq}}{K_{m_\mathrm{Q}}[\mathrm{P}]}\right)\left(1 + \frac{[\mathrm{A}]}{K_{ia}}\right) + [\mathrm{Q}]\left(1 + \frac{K_{m_\mathrm{P}}}{[\mathrm{P}]}\right)} \qquad \text{(IX-223)}$$

$$\frac{v}{V_{\max}} = \frac{[\mathrm{Q}]}{K_{m_\mathrm{Q}}\left(1 + \frac{K_{m_\mathrm{P}} K_{iq}}{K_{m_\mathrm{Q}}[\mathrm{P}]}\right) + [\mathrm{Q}]\left(1 + \frac{K_{m_\mathrm{P}}}{[\mathrm{P}]} + \frac{K_{m_\mathrm{A}} K_{ic} K_{m_\mathrm{P}}[\mathrm{B}]}{K_{ia} K_{ib} K_{m_\mathrm{C}} K_{ip}}\right)} \qquad \text{(IX-224)}$$

$$\frac{v}{V_{\max}} = \frac{[\mathrm{Q}]}{K_{m_\mathrm{Q}}\left(1 + \frac{K_{m_\mathrm{P}} K_{iq}}{K_{m_\mathrm{Q}}[\mathrm{P}]} + \frac{K_{m_\mathrm{P}} K_{iq} K_{m_\mathrm{B}}[\mathrm{C}]}{K_{m_\mathrm{Q}} K_{ib} K_{m_\mathrm{C}}[\mathrm{P}]}\right) + [\mathrm{Q}]\left(1 + \frac{K_{m_\mathrm{P}}}{[\mathrm{P}]} + \frac{K_{m_\mathrm{B}} K_{m_\mathrm{P}}[\mathrm{C}]}{K_{ib} K_{m_\mathrm{C}}[\mathrm{P}]} + \frac{K_{m_\mathrm{B}} K_{m_\mathrm{P}}[\mathrm{C}]}{K_{ib} K_{ip} K_{m_\mathrm{C}}}\right)} \qquad \text{(IX-225)}$$

Thus A is competitive with Q, and $K_{i_{\text{slope}}}^{A/Q} = K_{ia}$ at all P concentrations. B is uncompetitive with respect to Q at all P concentrations; C is a mixed-type inhibitor with respect to Q at unsaturating [P], and an uncompetitive inhibitor at saturating [P]. The K_{m_P}, K_{m_Q}, and K_{iq} can be determined as described earlier for ordered bireactant systems. If K_{m_B} and K_{m_C} are known, K_{ib} can be calculated from $K_{i_{\text{slope}}}^{C/P}$. With K_{ib} known, K_{ip} can be calculated from $K_{i_{\text{int}}}^{C/P}$, where:

$$K_{i_{\text{int}}}^{C/P} = \frac{K_{ib}K_{ip}K_{m_C}}{K_{m_B}K_{m_P}}\left(1 + \frac{K_{m_Q}}{[Q]}\right)$$

The K_{ic} can then be calculated from $K_{i_{\text{int}}}^{B/P}$ or $K_{i_{\text{int}}}^{B/Q}$, where:

$$K_{i_{\text{int}}}^{B/P} = \frac{K_{ia}K_{ib}K_{m_C}K_{ip}}{K_{m_A}K_{ic}K_{m_P}}\left(1 + \frac{K_{m_Q}}{[Q]}\right) \quad \text{and} \quad K_{i_{\text{int}}}^{B/Q} = \frac{K_{ia}K_{ib}K_{m_C}K_{ip}}{K_{m_A}K_{ic}K_{m_P}}\left(1 + \frac{K_{m_P}}{[P]}\right)$$

Other apparent K_i values that may be of use are:

$$K_{i_{\text{slope}}}^{A/P} = K_{ia}\left(1 + \frac{[Q]}{K_{iq}}\right) \quad \text{and} \quad K_{i_{\text{int}}}^{A/P} = K_{ia}\left(1 + \frac{[Q]}{K_{m_Q}}\right)$$

$$K_{i_{\text{slope}}}^{C/Q} = \frac{K_{ib}K_{m_C}}{K_{m_B}}\left(1 + \frac{K_{m_Q}[P]}{K_{iq}K_{m_P}}\right) \quad \text{and} \quad K_{i_{\text{int}}}^{C/Q} = \frac{\left(\dfrac{K_{ib}K_{m_C}}{K_{m_B}}\right)\left(1 + \dfrac{[P]}{K_{m_P}}\right)}{\left(1 + \dfrac{[P]}{K_{ip}}\right)}$$

Each of the $K_{i_{\text{slope}}}$ and $K_{i_{\text{int}}}$ values shown above can be calculated from one or two velocity points, or preferably, from appropriate slope and intercept replots. (The $K_{i_{\text{slope}}}$ or $K_{i_{\text{int}}}$ appear as the horizontal-axis intercept.) The relationship shown below (obtained from the Haldane equations) may also be useful in calculating an unknown kinetic constant, or if $V_{\max_f}/V_{\max_r}$ is known with confidence, to check the value of a K_i or K_m.

$$\frac{V_{\max_f}}{V_{\max_r}} = \frac{K_{ip}K_{m_C}}{K_{m_P}K_{ic}} \tag{IX-226}$$

Reduction to Rapid Equilibrium Ordered Ter Bi

The kinetic constants for the forward reaction are:

$$K_{m_A} = \frac{\text{Coef}_{BC}}{\text{Coef}_{ABC}} = \frac{k_4 k_5}{k_1(k_4 + k_5)}, \qquad K_{ia} = \frac{\text{const}}{\text{Coef}_A} = \frac{k_{-1}}{k_1}$$

$$K_{m_B} = \frac{\text{Coef}_{AC}}{\text{Coef}_{ABC}} = \frac{k_4 k_5}{k_2(k_4 + k_5)}, \qquad K_{ib} = \frac{\text{Coef}_A}{\text{Coef}_{AB}} = \frac{k_{-2}}{k_2}$$

$$\frac{V_{max_f}}{[E]_t} = \frac{\text{num}_1}{\text{Coef}_{ABC}} = \frac{k_4 k_5}{(k_4 + k_5)}$$

Thus when the rate constants for the dissociations of A and B (k_{-1} and k_{-2}) are very large compared to the maximal forward velocity, the K_{m_A} and K_{m_B} terms drop out, but the K_{ia} and K_{ib} terms remain in the velocity equation. Under this condition, the velocity equation has the same form, and the reciprocal plots have the same characteristics as those for a completely Rapid Equilibrium Ordered Ter Bi system. However, the constant associated with C is still K_{m_C}.

$$K_{m_C} = \frac{\text{Coef}_{AB}}{\text{Coef}_{ABC}} = \frac{k_5(k_{-3} + k_4)}{k_3(k_4 + k_5)}$$

If k_4 is very small compared to k_{-3} and k_5, then the constant reduces to $K_{ic} = k_{-3}/k_3$.

If only A is in equilibrium with E and EA, the velocity equation for the forward reaction becomes:

$$\frac{v}{V_{max}} = \frac{[A][B][C]}{K_{ia}K_{m_B}K_{m_C} + K_{ib}K_{m_C}[A] + K_{ia}K_{m_B}[C] + K_{m_C}[A][B] + K_{m_B}[A][C] + [A][B][C]}$$

(IX-227)

Now, only the K_{m_A} denominator term is missing. Morrison and Ebner (1971) have proposed this partial Rapid Equilibrium Ter Bi mechanism for galactotransferase where A = Mn^{2+}, B = UDP-Gal, and C = *N*-acetylglucosamine. The data suggest that Mn^{2+} adds first but need not dissociate after each catalytic cycle. (See equation IX-118 for an Ordered Bi Uni example. Also see equation IX-271 for the complete velocity equation for a partial Rapid Equilibrium Ter Ter system.)

S. BI UNI UNI UNI PING PONG TER BI SYSTEM

Reaction Sequence

At first glance, there appears to be two Ter Bi Ping Pong systems: Bi Uni Uni Uni Ping Pong:

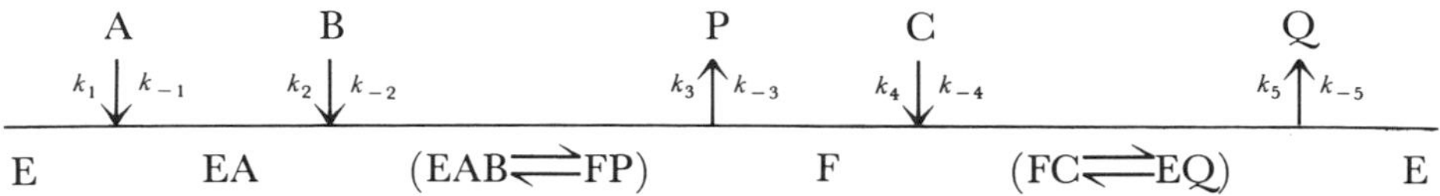

and Uni Uni Bi Uni Ping Pong:

A (k_1, k_{-1}) P (k_2, k_{-2}) B (k_3, k_{-3}) C (k_4, k_{-4}) Q (k_5, k_{-5})

E (EA⇌FP) F FB (FBC⇌EQ) E

The two reaction sequences, however, are identical and if the Bi Uni Uni Uni sequence is started at F, it becomes Uni Uni Bi Uni. This is more easily seen if we show the repeating sequence.

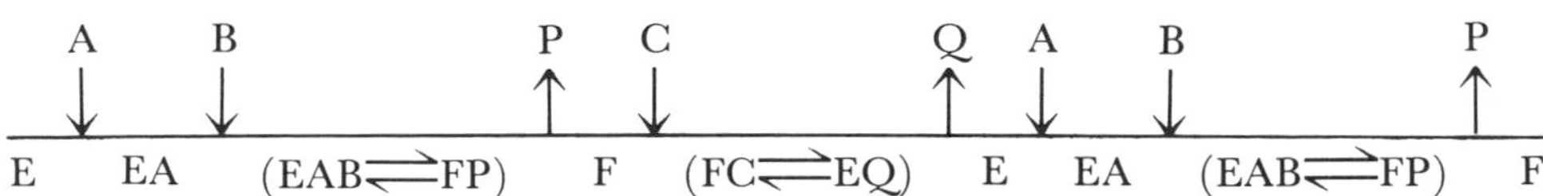

The velocity equations for the sequence written both ways have the same form. The reciprocal plots and product inhibition patterns will be identical. Only the letters designating the five ligands differ. The "correct," or "better," sequence depends on the chemistry of the reaction.

Velocity Equation, Kinetic Constants, and Haldane Equations

The velocity equation for the Bi Uni Uni Uni sequence obtained from five four-sided King-Altman interconversion patterns is:

$$\frac{v}{[\mathrm{E}]_t} = \frac{k_1k_2k_3k_4k_5[\mathrm{A}][\mathrm{B}][\mathrm{C}] - k_{-1}k_{-2}k_{-3}k_{-4}k_{-5}[\mathrm{P}][\mathrm{Q}]}{\begin{aligned}&k_{-1}k_4k_5(k_{-2}+k_3)[\mathrm{C}] + k_1k_2k_3(k_{-4}+k_5)[\mathrm{A}][\mathrm{B}] + k_1k_4k_5(k_{-2}+k_3)[\mathrm{A}][\mathrm{C}]\\ &+ k_2k_3k_4k_5[\mathrm{B}][\mathrm{C}] + k_1k_2k_4(k_3+k_5)[\mathrm{A}][\mathrm{B}][\mathrm{C}] + k_1k_{-2}k_{-3}(k_{-4}+k_5)[\mathrm{A}][\mathrm{P}]\\ &+ k_1k_2k_{-3}(k_{-4}+k_5)[\mathrm{A}][\mathrm{B}][\mathrm{P}] + k_{-1}k_{-2}k_{-3}(k_{-4}+k_5)[\mathrm{P}]\\ &+ k_{-1}k_{-4}k_{-5}(k_{-2}+k_3)[\mathrm{Q}] + k_{-3}k_{-5}(k_{-2}k_{-4}+k_{-1}k_{-4}+k_{-1}k_{-2})[\mathrm{P}][\mathrm{Q}]\\ &+ k_2k_3k_{-4}k_{-5}[\mathrm{B}][\mathrm{Q}] + k_{-1}k_4k_{-5}(k_{-2}+k_3)[\mathrm{C}][\mathrm{Q}] + k_2k_{-3}k_{-4}k_{-5}[\mathrm{B}][\mathrm{P}][\mathrm{Q}]\\ &+ k_2k_3k_4k_{-5}[\mathrm{B}][\mathrm{C}][\mathrm{Q}]\end{aligned}} \tag{IX-228}$$

In terms of kinetic constants, the equation is:

$$v = \frac{V_{\max_f}V_{\max_r}\left([\mathrm{A}][\mathrm{B}][\mathrm{C}] - \dfrac{[\mathrm{P}][\mathrm{Q}]}{K_{\mathrm{eq}}}\right)}{\begin{aligned}&V_rK_{ia}K_{m_\mathrm{B}}[\mathrm{C}] + V_rK_{m_\mathrm{C}}[\mathrm{A}][\mathrm{B}] + V_rK_{m_\mathrm{B}}[\mathrm{A}][\mathrm{C}] + V_rK_{m_\mathrm{A}}[\mathrm{B}][\mathrm{C}] + V_r[\mathrm{A}][\mathrm{B}][\mathrm{C}]\\ &+ \frac{V_fK_{m_\mathrm{Q}}[\mathrm{A}][\mathrm{P}]}{K_{ia}K_{\mathrm{eq}}} + \frac{V_fK_{m_\mathrm{Q}}[\mathrm{A}][\mathrm{B}][\mathrm{P}]}{K_{ia}K_{ib}K_{\mathrm{eq}}} + \frac{V_fK_{m_\mathrm{Q}}[\mathrm{P}]}{K_{\mathrm{eq}}} + \frac{V_fK_{m_\mathrm{P}}[\mathrm{Q}]}{K_{\mathrm{eq}}}\\ &+ \frac{V_f[\mathrm{P}][\mathrm{Q}]}{K_{\mathrm{eq}}} + \frac{V_rK_{m_\mathrm{A}}K_{ic}[\mathrm{B}][\mathrm{Q}]}{K_{iq}} + \frac{V_rK_{ia}K_{m_\mathrm{B}}[\mathrm{C}][\mathrm{Q}]}{K_{iq}}\\ &+ \frac{V_rK_{m_\mathrm{A}}K_{ic}[\mathrm{B}][\mathrm{P}][\mathrm{Q}]}{K_{ip}K_{iq}} + \frac{V_rK_{m_\mathrm{A}}[\mathrm{B}][\mathrm{C}][\mathrm{Q}]}{K_{iq}}\end{aligned}} \tag{IX-229}$$

where

$$K_{m_\mathrm{A}} = \frac{\mathrm{Coef_{BC}}}{\mathrm{Coef_{ABC}}}, \quad K_{m_\mathrm{B}} = \frac{\mathrm{Coef_{AC}}}{\mathrm{Coef_{ABC}}}, \quad K_{m_\mathrm{C}} = \frac{\mathrm{Coef_{AB}}}{\mathrm{Coef_{ABC}}},$$

$$K_{m_\mathrm{P}} = \frac{\mathrm{Coef_{Q}}}{\mathrm{Coef_{PQ}}}, \quad K_{m_\mathrm{Q}} = \frac{\mathrm{Coef_{P}}}{\mathrm{Coef_{PQ}}}, \quad K_{ia} = \frac{\mathrm{Coef_{P}}}{\mathrm{Coef_{AP}}} = \frac{\mathrm{Coef_{C}}}{\mathrm{Coef_{AC}}} = \frac{k_{-1}}{k_1},$$

$$K_{ib} = \frac{\mathrm{Coef_{AP}}}{\mathrm{Coef_{ABP}}} = \frac{k_{-2}}{5_2}, \quad K_{ic} = \frac{\mathrm{Coef_{Q}}}{\mathrm{Coef_{CQ}}} = \frac{\mathrm{Coef_{BQ}}}{\mathrm{Coef_{BCQ}}} = \frac{k_{-4}}{k_4},$$

$$K_{ip} = \frac{\mathrm{Coef_{BQ}}}{\mathrm{Coef_{BPQ}}} = \frac{\mathrm{Coef_{AB}}}{\mathrm{Coef_{ABP}}} = \frac{k_3}{k_{-3}}, \quad K_{iq} = \frac{\mathrm{Coef_{C}}}{\mathrm{Coef_{CQ}}} = \frac{\mathrm{Coef_{BC}}}{\mathrm{Coef_{BCQ}}} = \frac{k_5}{k_{-5}}$$

The K_{eq} may be eliminated by substituting from either of the first two Haldane equations.

$$\boxed{K_{eq} = \frac{V_{max_f} K_{ip} K_{m_Q}}{V_{max_r} K_{ia} K_{ib} K_{m_C}} = \frac{V_{max_f} K_{m_P} K_{iq}}{V_{max_r} K_{ia} K_{m_B} K_{ic}} = \frac{K_{ip} K_{iq}}{K_{ia} K_{ib} K_{ic}} = \left(\frac{V_f}{V_r}\right)^2 \frac{K_{m_P} K_{m_Q}}{K_{ia} K_{m_B} K_{m_C}}}$$

(IX-230)

From the first two Haldane equations, we see that:

$$\boxed{\frac{K_{ip} K_{m_Q}}{K_{ib} K_{m_C}} = \frac{K_{m_P} K_{iq}}{K_{m_B} K_{ic}}} \qquad \text{(IX-231)}$$

Rate Constants

$$k_1 = \frac{V_{max_f}}{K_{m_A}[E]_t}, \qquad k_{-1} = \frac{V_{max_f} K_{ia}}{K_{m_A}[E]_t}$$

The other rate constants cannot be determined.

Distribution Equations

$\frac{[E]}{[E]_t}$ = the same as that shown for Ordered Ter Bi without the constant term in the numerator

$\frac{[EA]}{[E]_t}$ = the same as that shown for Ordered Ter Bi without the [A] term in the numerator

$$\frac{[F]}{[E]_t} = \frac{V_r K_{m_C}[A][B] + \dfrac{V_f K_{m_P}[Q]}{K_{eq}} + \dfrac{V_r K_{m_A} K_{ic}[B][Q]}{K_{iq}}}{\text{denominator of velocity equation}} \qquad \text{(IX-232)}$$

The proportion of the enzyme as EAB + FP and FC + EQ cannot be determined.

Effect of Isomerizations

EAB + FP and/or FC + EQ: The rate constants and distributions shown above can still be calculated.

EA: Only $[E]/[E]_t$ and $[F]/[E]_t$ can be calculated.

Initial Velocity Studies in the Forward Direction

In the absence of products, the velocity equation is:

$$\frac{v}{V_{\max}} = \frac{[A][B][C]}{K_{ia}K_{m_B}[C] + K_{m_C}[A][B] + K_{m_B}[A][C] + K_{m_A}[B][C] + [A][B][C]} \quad \text{(IX-233)}$$

or

$$\frac{v}{V_{\max}} = \frac{[A]}{K_{m_A}\left(1 + \frac{K_{ia}K_{m_B}}{K_{m_A}[B]}\right) + [A]\left(1 + \frac{K_{m_B}}{[B]} + \frac{K_{m_C}}{[C]}\right)} \quad \text{(IX-234)}$$

$$\frac{v}{V_{\max}} = \frac{[B]}{K_{m_B}\left(1 + \frac{K_{ia}}{[A]}\right) + [B]\left(1 + \frac{K_{m_A}}{[A]} + \frac{K_{m_C}}{[C]}\right)} \quad \text{(IX-235)}$$

$$\frac{v}{V_{\max}} = \frac{[C]}{K_{m_C} + [C]\left(1 + \frac{K_{m_A}}{[A]} + \frac{K_{m_B}}{[B]} + \frac{K_{ia}K_{m_B}}{[A][B]}\right)} \quad \text{(IX-236)}$$

The substrate that adds between two product release steps, C, can be identified as the only varied substrate that yields parallel reciprocal plots when either of the other two substrates is the changing fixed substrate (Fig. IX-45). When [A] is varied, the family of reciprocal plots intersect when B is

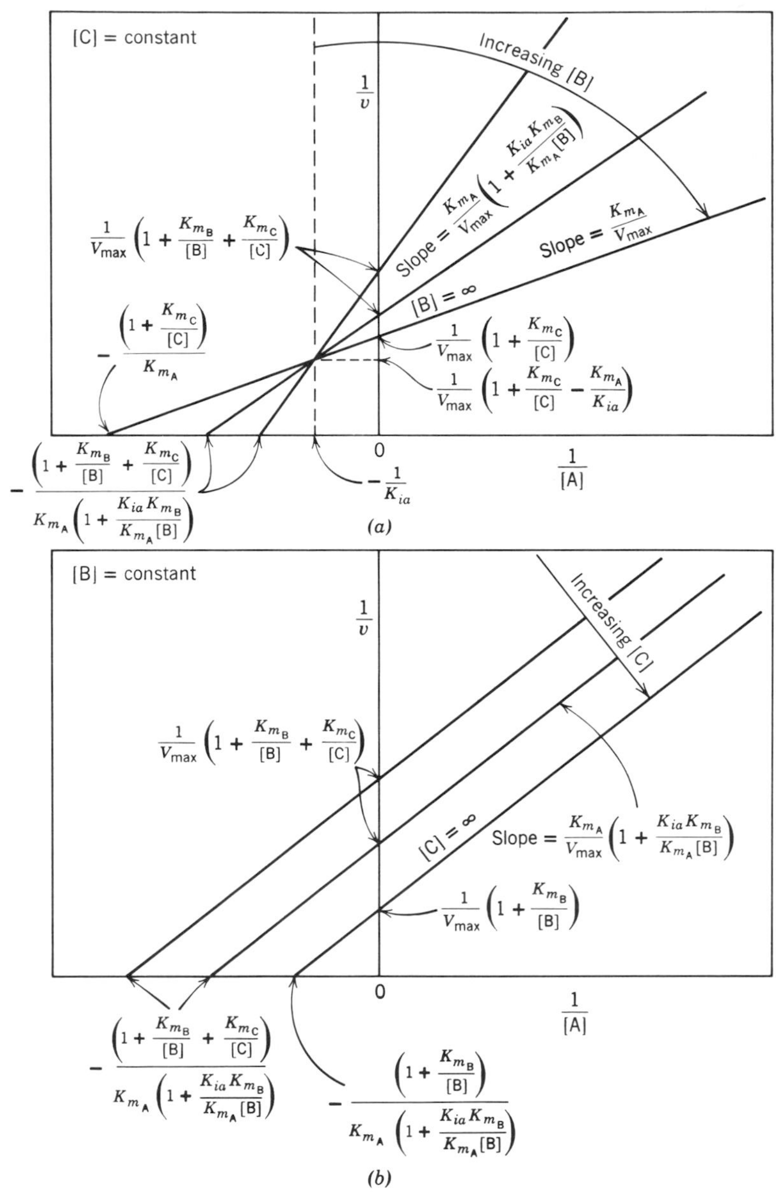

Fig. IX-43. Bi Uni Uni Uni Ping Pong Ter Bi system. (*a*) $1/v$ versus 1/[A] at different fixed B concentrations and a constant C concentration. (*b*) $1/v$ versus 1/[A] at different fixed C concentrations and a constant B concentration.

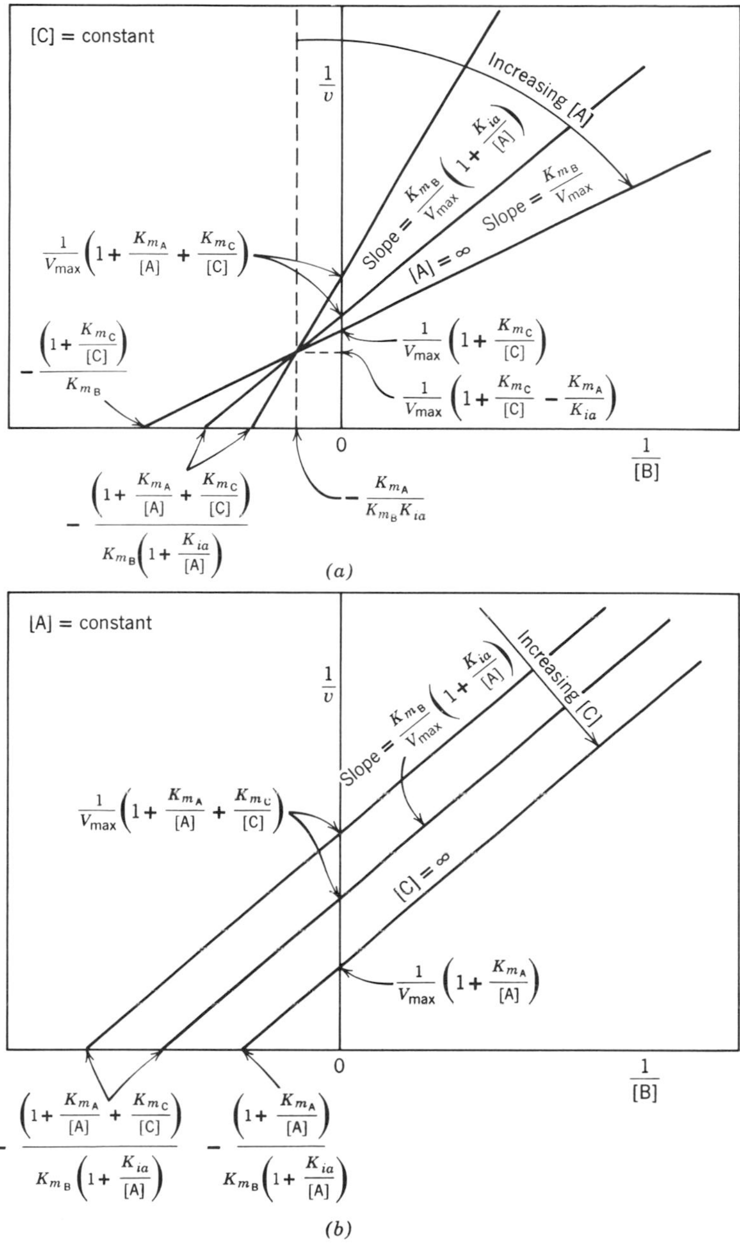

Fig. IX-44. Bi Uni Uni Uni Ping Pong Ter Bi system; $1/v$ versus $1/[B]$. (*a*) A is the changing fixed substrate and C is the constant substrate. (*b*) C is the changing fixed substrate and A is the constant substrate. The plots are symmetrical to the $1/v$ versus $1/[A]$ plots. Consequently, initial velocity studies will not distinguish between A and B.

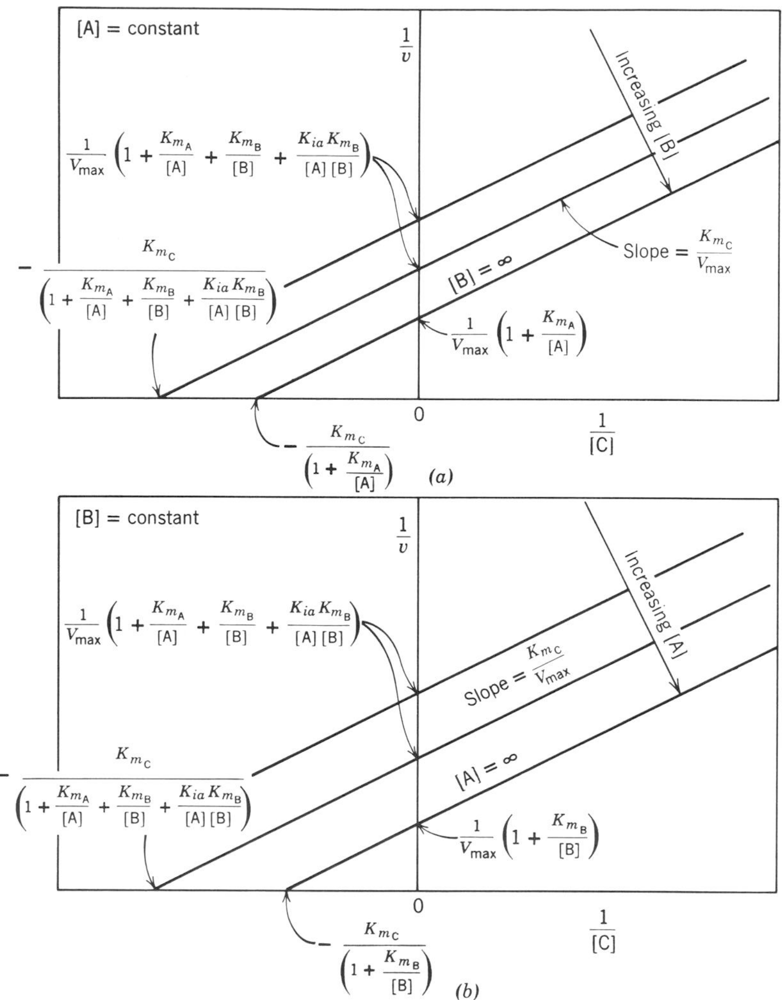

Fig. IX-45. Bi Uni Uni Uni Ping Pong Ter Bi system; $1/v$ versus $1/[C]$. (*a*) B is the changing fixed substrate and A is the constant substrate. (*b*) A is the changing fixed substrate and B is the constant substrate.

the changing fixed substrate and C is the constant substrate. The plots are parallel when C is the changing fixed substrate and [B] is held constant (Fig. IX-43). A similar pair of plots are obtained when [B] is varied (Fig. IX-44).

Plots With Two Changing Fixed Substrates

Instead of six families of plots, three families can be constructed. In each family, one substrate is varied at different fixed concentrations of the other two which are maintained at a constant ratio. The family of $1/v$ versus $1/[C]$ plots are parallel, while the $1/v$ versus $1/[A]$ and $1/v$ versus $1/[B]$ plots intersect. The $1/V_{max_{app}}$ replot of the $1/v$ versus $1/[C]$ plot is parabolic (because of the $[A][B]=[B]^2$ term), while the corresponding replots of $1/v$ versus $1/[A]$ and $1/v$ versus $1/[B]$ data are linear. All three replots extrapolate to the true $1/V_{max}$. For example, when [A] is varied and $[B]=x[C]$, the reciprocal plot is given by:

$$\frac{1}{v}=\frac{K_{m_A}}{V_{max}}\left(1+\frac{K_{ia}K_{m_B}}{K_{m_A}x[C]}\right)\frac{1}{[A]}+\frac{1}{V_{max}}\left(1+\frac{K_{m_B}}{x[C]}+\frac{K_{m_C}}{[C]}\right) \quad \text{(IX-237)}$$

The intercept replot is given by:

$$\frac{1}{V_{max_{app}}}=\frac{K_{m_C}}{V_{max}}\left(1+\frac{K_{m_B}}{xK_{m_C}}\right)\frac{1}{[C]}+\frac{1}{V_{max}} \quad \text{(IX-238)}$$

Varying Two Substrates Together

When [A] and [C] are varied at different fixed concentrations of B (and $[C]=x[A]$), the plot of $1/v$ versus $1/[A]$ is given by:

$$\frac{1}{v}=\frac{K_{m_A}}{V_{max}}\left(1+\frac{K_{ia}K_{m_B}}{K_{m_A}[B]}+\frac{K_{m_C}}{xK_{m_A}}\right)\frac{1}{[A]}+\frac{1}{V_{max}}\left(1+\frac{K_{m_B}}{[B]}\right) \quad \text{(IX-239)}$$

The plots are linear and intersect to the left of the $1/v$-axis at $-1/K_{ia}$. Similarly, when [B] and [C] are varied together at $[C]=x[B]$, the reciprocal plots at different fixed concentrations of A are linear and intersect at $-K_{m_A}/K_{m_B}K_{ia}$. However, when [A] and [B] are varied together at $[B]=x[A]$, the reciprocal plot is given by:

$$\frac{1}{v}=\frac{K_{m_A}}{V_{max}}\left(1+\frac{K_{ia}K_{m_B}}{xK_{m_A}[A]}+\frac{K_{m_B}}{x}\right)\frac{1}{[A]}+\frac{1}{V_{max}}\left(1+\frac{K_{m_C}}{[C]}\right) \quad \text{(IX-240)}$$

The reciprocal plots are now parabolic. Thus there are several ways of identifying C, but initial velocity studies do not distinguish between A and B. Binding or product inhibition studies are necessary. Once A and B are identified, the various kinetic constants for the forward reaction can be obtained easily. The K_{m_C} can be obtained from the slope of the $1/v$ versus $1/[C]$ plot (Fig. IX-45) if V_{max} is known; K_{ia} can be obtained from the $1/[A]$ coordinate of the $1/v$ versus $1/[A]$ plot (Fig. IX-44) or as $K_{slope}^{A/B}$ at all constant C concentrations. The K_{m_B} can be obtained from $1/v$-axis intercept replots of $1/v$ versus $1/[A]$ plots. Similarly, K_{m_A} can be obtained from the $1/v$ versus $1/[B]$ intercept replots. The value of $K_{slope}^{B/A}$, which equals $K_{ia}K_{m_B}/K_{m_A}$ at all constant C concentrations can be used to calculate one of the kinetic constants if the other two are known. Replots of $1/v$-axis intercepts when two substrates are varied together allow V_{max}, K_{m_A}, K_{m_B}, and K_{m_C} to be determined. The K_{ib} and K_{ic} do not appear in the velocity equations for the forward reaction in the absence of products.

Product Inhibition Studies

In the presence of P, three new terms are present in the denominator of the velocity equation. After conversion to kinetic constants by substitution from the first Haldane equation, these become:

$$\frac{V_{max_r}K_{ib}K_{m_C}[A][P]}{K_{ip}}, \qquad \frac{V_{max_r}K_{m_C}[A][B][P]}{K_{ip}}, \qquad \text{and} \qquad \frac{V_{max_r}K_{ia}K_{ib}K_{m_C}[P]}{K_{ip}}$$

Dividing by V_{max_r} and the two nonvaried substrate concentration terms and then factoring, the velocity equations become as follows.

When [A] is varied:

$$\frac{v}{V_{max}} = \frac{[A]}{K_{m_A}\left(1 + \frac{K_{ia}K_{m_B}}{K_{m_A}[B]} + \frac{K_{ia}K_{ib}K_{m_C}[P]}{K_{m_A}K_{ip}[B][C]}\right) + [A]\left(1 + \frac{K_{m_B}}{[B]} + \frac{K_{m_C}}{[C]} + \frac{K_{ib}K_{m_C}[P]}{K_{ip}[B][C]} + \frac{K_{m_C}[P]}{K_{ip}[C]}\right)} \qquad \text{(IX-241)}$$

The product P is a mixed-type inhibitor with respect to A at unsaturating B and C concentrations. Saturating [B] eliminates the [P] term in the slope factor, but the $K_{m_C}[P]/K_{ip}[C]$ term remains in the $1/v$-axis intercept factor. Thus P becomes an uncompetitive inhibitor. Saturating [C] overcomes the inhibition.

When [B] is varied:

$$\frac{v}{V_{\max}} = \frac{[B]}{K_{m_B}\left(1 + \frac{K_{ia}}{[A]}\right)\left(1 + \frac{K_{ib}K_{m_C}[P]}{K_{m_B}K_{ip}[C]}\right) + [B]\left(1 + \frac{K_{m_A}}{[A]} + \frac{K_{m_C}}{[C]} + \frac{K_{m_C}[P]}{K_{ip}[C]}\right)}$$

(IX-242)

P is a mixed-type inhibitor with respect to B at all A concentrations. Saturating [C] eliminates the inhibition. The $K_{i_{slope}}^{P/B}$ is independent of [A] but depends on [C]:

$$K_{i_{slope}}^{P/B} = \frac{K_{m_B}K_{ip}[C]}{K_{ib}K_{m_C}}$$

The ratio K_{ip}/K_{ib} can be calculated from the value of $K_{i_{slope}}^{P/B}$ at fixed [C] if K_{m_B} and K_{m_C} are known. The $K_{i_{slope}}^{P/B}$ can be obtained as the [P]-axis intercept of a $slope_{1/B}$ replot:

$$slope_{1/B} = \frac{K_{ib}K_{m_C}}{V_{\max}K_{ip}[C]}\left(1 + \frac{K_{ia}}{[A]}\right)[P] + \frac{K_{m_B}}{V_{\max}}\left(1 + \frac{K_{ia}}{[A]}\right) \qquad \text{(IX-243)}$$

The $1/v$-axis intercept replot is given by:

$$\frac{1}{V_{\max_{app}}} = \frac{K_{m_C}}{V_{\max}K_{ip}[C]}[P] + \frac{1}{V_{\max}}\left(1 + \frac{K_{m_A}}{[A]} + \frac{K_{m_C}}{[C]}\right) \qquad \text{(IX-244)}$$

The intercept on the [P]-axis gives:

$$K_{i_{int}}^{P/B} = -K_{ip}\left[1 + \frac{[C]}{K_{m_C}}\left(1 + \frac{K_{m_A}}{[A]}\right)\right]$$

If K_{m_C} is known, K_{ip} may be calculated from the slope of the replot. If K_{m_A} is also known, K_{ip} can be calculated from the $K_{i_{\text{int}}}^{\text{P/B}}$. With K_{ip} known, K_{ib} can be calculated from $K_{i_{\text{slope}}}^{\text{P/B}}$.

When [C] is varied:

$$\frac{v}{V_{\max}} = \frac{[\text{C}]}{K_{m_{\text{C}}}\left(1 + \dfrac{K_{ib}[\text{P}]}{K_{ip}[\text{B}]} + \dfrac{[\text{P}]}{K_{ip}} + \dfrac{K_{ia}K_{ib}[\text{P}]}{K_{ip}[\text{A}][\text{B}]}\right) + [\text{C}]\left(1 + \dfrac{K_{m_{\text{A}}}}{[\text{A}]} + \dfrac{K_{m_{\text{B}}}}{[\text{B}]} + \dfrac{K_{ia}K_{m_{\text{B}}}}{[\text{A}][\text{B}]}\right)}$$

(IX-245)

P is competitive with respect to C at all A and B concentrations.

In the presence of Q, four additional denominator terms appear in the velocity equation. One of them, the $V_f K_{m_{\text{P}}}[\text{Q}]/K_{\text{eq}}$ term transforms to $K_{m_{\text{P}}}K_{ia}K_{ib}K_{m_{\text{C}}}[\text{Q}]/K_{ip}K_{m_{\text{Q}}}$ if we use the first Haldane equation to eliminate K_{eq}. If the second Haldane equation is used, we obtain the simpler $K_{ia}K_{m_{\text{B}}}K_{ic}[\text{Q}]/K_{iq}$.

When A is the varied substrate, all the [Q] terms appear in the slope factor. The velocity equation can be written:

$$\frac{v}{V_{\max}} = \frac{[\text{A}]}{K_{m_{\text{A}}}\left(1 + \dfrac{K_{ia}K_{m_{\text{B}}}}{K_{m_{\text{A}}}[\text{B}]}\right)\left[1 + \dfrac{[\text{Q}]}{K_{iq}\left(1 + \dfrac{K_{ic}}{[\text{C}]}\right)}\right] + [\text{A}]\left(1 + \dfrac{K_{m_{\text{B}}}}{[\text{B}]} + \dfrac{K_{m_{\text{C}}}}{[\text{C}]}\right)}$$

(IX-246)

Thus $K_{i_{\text{slope}}}^{\text{Q/A}}$ depends on [C] but is constant at all B concentrations:

$$K_{i_{\text{slope}}}^{\text{Q/A}} = \frac{K_{iq}}{\left(1 + \dfrac{K_{ic}}{[\text{C}]}\right)}$$

The $K_{i_{\text{slope}}}^{\text{Q/A}}$ is obtained as the horizontal-axis intercept of the *slope*$_{1/\text{A}}$ replot at constant [B] and [C]. If $K_{i_{\text{slope}}}^{\text{Q/A}}$ is determined at several different constant C concentrations, a secondary replot can be constructed:

$$\frac{1}{K_{i_{\text{slope}}}^{\text{Q/A}}} = \frac{K_{ic}}{K_{iq}}\frac{1}{[\text{C}]} + \frac{1}{K_{iq}} \qquad \text{(IX-247)}$$

The vertical-axis intercept ($1/K_{i_{slope}}^{Q/A}$ at saturating [C]) gives $1/K_{iq}$; K_{ic} can be calculated from the slope of the secondary replot.

When [B] is varied:

$$\frac{v}{V_{max}} = \frac{[B]}{K_{m_B}\left(1+\frac{K_{ia}}{[A]}+\frac{K_{ia}K_{ic}[Q]}{K_{iq}[A][C]}+\frac{K_{ia}[Q]}{K_{iq}[A]}\right)+[B]\left(1+\frac{K_{m_A}}{[A]}+\frac{K_{m_C}}{[C]}+\frac{K_{m_A}K_{ic}[Q]}{K_{iq}[A][C]}+\frac{K_{m_A}[Q]}{K_{iq}[A]}\right)} \quad \text{(IX-248)}$$

Q is a mixed-type inhibitor with respect to B at all C concentrations and unsaturating A concentrations. Saturating [A] overcomes the inhibition. From the slope factor, we see that:

$$K_{i_{slope}}^{Q/B} = K_{iq}\frac{\left(1+\frac{[A]}{K_{ia}}\right)}{\left(1+\frac{K_{ic}}{[C]}\right)}$$

With K_{ia} and K_{iq} known, the relationship above provides another way of determining K_{ic}. The relationship can be written as:

$$\frac{1}{K_{i_{slope}}^{Q/B}} = \frac{K_{ic}}{K_{iq}\left(1+\frac{[A]}{K_{ia}}\right)}\frac{1}{[C]}+\frac{1}{K_{iq}\left(1+\frac{[A]}{K_{ia}}\right)} \quad \text{(IX-249)}$$

Thus replots of $1/K_{i_{slope}}^{Q/B}$ versus $1/[C]$ (at a constant [A]) gives $-1/K_{ic}$ as the horizontal-axis intercept.

When [C] is varied:

$$\frac{v}{V_{max}} = \frac{[C]}{K_{m_C}\left(1+\frac{K_{m_B}K_{ic}K_{ia}[Q]}{K_{m_C}K_{iq}[A][B]}+\frac{K_{m_A}K_{ic}[Q]}{K_{m_C}K_{iq}[A]}\right)+[C]\left(1+\frac{K_{m_A}}{[A]}+\frac{K_{m_B}}{[B]}+\frac{K_{ia}K_{m_B}[Q]}{K_{iq}[A][B]}+\frac{K_{m_A}[Q]}{K_{iq}[A]}\right)} \quad \text{(IX-250)}$$

Q is a mixed-type inhibitor with respect to C at all B concentrations and unsaturating [A]. Saturation with A eliminates the inhibition.

The competitive patterns identify the C–P and A–Q pairs. Since initial velocity studies in the absence of products identifies C, the sequence of substrate addition and product release can be established. Most of the kinetic constants can be calculated. The ratio K_{m_P}/K_{m_Q} can be obtained from equation IX-231.

Reverse Direction—Uni Uni Uni Bi Ping Pong Bi Ter

In the absence of A, B, and C, the reverse reaction is the usual Ping Pong Bi Bi mechanism. In the presence of A, the denominator of the velocity equation contains an [A][P] term. After dividing equation IX-229 by $V_{\max_f}/K_{eq}$ to put it in a form suitable for consideration of the reverse reaction, and then dividing by either [P] or [Q], we obtain:

$$\frac{v}{V_{\max}} = \frac{[P]}{K_{m_P} + [P]\left(1 + \frac{K_{m_Q}}{[Q]} + \frac{K_{m_Q}[A]}{K_{ia}[Q]}\right)} \qquad \text{(IX-251)}$$

and

$$\frac{v}{V_{\max}} = \frac{[Q]}{K_{m_Q}\left(1 + \frac{[A]}{K_{ia}}\right) + [Q]\left(1 + \frac{K_{m_P}}{[P]}\right)} \qquad \text{(IX-252)}$$

Thus A is uncompetitive with P at unsaturating [Q], and competitive with Q at all P concentrations.

$$K_{i_{\text{int}}}^{A/P} = K_{ia}\left(1 + \frac{[Q]}{K_{m_Q}}\right), \qquad K_{i_{\text{slope}}}^{A/Q} = K_{ia}$$

In the presence of B, the denominator of equation IX-229 contains [B][Q] and [B][P][Q] terms, which after substitution from the second Haldane equation yields:

$$\frac{v}{V_{\max}} = \frac{[P]}{K_{m_P}\left(1 + \frac{K_{m_A}[B]}{K_{ia}K_{m_B}}\right) + [P]\left(1 + \frac{K_{m_Q}}{[Q]} + \frac{K_{m_A}K_{m_P}[B]}{K_{ia}K_{ip}K_{m_B}}\right)} \qquad \text{(IX-253)}$$

and

$$\boxed{\frac{v}{V_{\max}} = \frac{[\mathrm{Q}]}{K_{m_{\mathrm{Q}}} + [\mathrm{Q}]\left(1 + \dfrac{K_{m_{\mathrm{P}}}}{[\mathrm{P}]} + \dfrac{K_{m_{\mathrm{A}}}K_{m_{\mathrm{P}}}[\mathrm{B}]}{K_{ia}K_{m_{\mathrm{B}}}[\mathrm{P}]} + \dfrac{K_{m_{\mathrm{A}}}K_{m_{\mathrm{P}}}[\mathrm{B}]}{K_{ia}K_{ip}K_{m_{\mathrm{B}}}}\right)}} \qquad \text{(IX-254)}$$

B is a mixed-type inhibitor with respect to P at all Q concentrations and an uncompetitive inhibitor with respect to Q at all P concentrations.

$$K_{i_{\text{slope}}}^{\mathrm{B/P}} = \frac{K_{ia}K_{m_{\mathrm{B}}}}{K_{m_{\mathrm{A}}}}, \qquad K_{i_{\text{int}}}^{\mathrm{B/P}} = \frac{K_{ia}K_{ip}K_{m_{\mathrm{B}}}\left(1 + \dfrac{K_{m_{\mathrm{Q}}}}{[\mathrm{Q}]}\right)}{K_{m_{\mathrm{A}}}K_{m_{\mathrm{P}}}}$$

In the presence of C, equation IX-229 contains a [C] and a [C][Q] term. The velocity equation transforms to:

$$\boxed{\frac{v}{V_{\max}} = \frac{[\mathrm{P}]}{K_{m_{\mathrm{P}}}\left(1 + \dfrac{[\mathrm{C}]}{K_{ic}} + \dfrac{K_{iq}[\mathrm{C}]}{K_{ic}[\mathrm{Q}]}\right) + [\mathrm{P}]\left(1 + \dfrac{K_{m_{\mathrm{Q}}}}{[\mathrm{Q}]}\right)}} \qquad \text{(IX-255)}$$

and

$$\boxed{\frac{v}{V_{\max}} = \frac{[\mathrm{Q}]}{K_{m_{\mathrm{Q}}}\left(1 + \dfrac{K_{m_{\mathrm{P}}}K_{iq}[\mathrm{C}]}{K_{m_{\mathrm{Q}}}K_{ic}[\mathrm{P}]}\right) + [\mathrm{Q}]\left(1 + \dfrac{K_{m_{\mathrm{P}}}}{[\mathrm{P}]} + \dfrac{K_{m_{\mathrm{P}}}[\mathrm{C}]}{K_{ic}[\mathrm{P}]}\right)}} \qquad \text{(IX-256)}$$

C is competitive with respect to P at all Q concentrations and a mixed-type inhibitor with respect to Q at unsaturating [P]. Saturation with P overcomes the inhibition.

$$K_{i_{\text{slope}}}^{\mathrm{C/P}} = \frac{K_{ic}}{\left(1 + \dfrac{K_{iq}}{[\mathrm{Q}]}\right)}, \qquad K_{i_{\text{slope}}}^{\mathrm{C/Q}} = \frac{K_{m_{\mathrm{Q}}}K_{ic}[\mathrm{P}]}{K_{m_{\mathrm{P}}}K_{iq}}, \qquad K_{i_{\text{int}}}^{\mathrm{C/Q}} = K_{ic}\left(1 + \frac{[\mathrm{P}]}{K_{m_{\mathrm{P}}}}\right)$$

The product inhibition studies allow the order of substrate addition and product release to be determined. The K_{m_P} and K_{m_Q} can be determined in the usual way from initial velocity studies in the absence of products; K_{ia} is obtained as $K_{i_{slope}}^{A/Q}$. With K_{ia} known, the ratio K_{m_B}/K_{m_A} can be calculated from $K_{i_{slope}}^{B/P}$, and then K_{ip} from $K_{i_{int}}^{B/P}$. The K_{ic} can be obtained from $K_{i_{int}}^{C/Q}$. With K_{ic} known, K_{iq} can be calculated from $K_{i_{slope}}^{C/Q}$ or $K_{i_{slope}}^{C/P}$. Each of the $K_{i_{slope}}$ and $K_{i_{int}}$ values above can be calculated from a few velocity points, or, preferably, from the appropriate slope or intercept replots. The K_{ib} and K_{m_C} do not appear in the velocity equation for the reverse reaction in the presence of A or B or C although the ratio $K_{m_B}/K_{ib}K_{m_C}$ can be calculated from equation IX-231.

Multiple Inhibition Studies

In a Bi Ter reaction, two of the three products can be present simultaneously without introducing the complications of the back reaction. For example, if both A and B are present, the [A][B] and [A][B][P] terms are incorporated into the velocity equation yielding:

$$\frac{v}{V_{max}} = \frac{[P]}{K_{m_P}\left(1 + \frac{K_{m_A}[B]}{K_{ia}K_{m_B}} + \frac{K_{ip}K_{m_Q}[A][B]}{K_{m_P}K_{ia}K_{ib}[Q]}\right) + [P]\left(1 + \frac{K_{m_Q}}{[Q]} + \frac{K_{m_Q}[A]}{K_{ia}[Q]} + \frac{K_{m_A}K_{m_P}[B]}{K_{ia}K_{ip}K_{m_B}} + \frac{K_{m_Q}[A][B]}{K_{ia}K_{ib}[Q]}\right)} \tag{IX-257}$$

$$\frac{v}{V_{max}} = \frac{[Q]}{K_{m_Q}\left(1 + \frac{[A]}{K_{ia}} + \frac{K_{ip}[A][B]}{K_{ia}K_{ib}[P]} + \frac{[A][B]}{K_{ia}K_{ib}}\right) + [Q]\left(1 + \frac{K_{m_P}}{[P]} + \frac{K_{m_A}K_{m_P}[B]}{K_{ia}K_{m_B}[P]} + \frac{K_{m_A}K_{m_P}[B]}{K_{ia}K_{ip}K_{m_B}}\right)} \tag{IX-258}$$

If [B] and [Q] are held constant, and $1/v$ versus $1/[P]$ plotted at different fixed concentrations of A, the $slope_{1/P}$ replot is given by:

$$\boxed{slope_{1/P} = \frac{K_{ip}K_{m_Q}[B]}{V_{max}K_{ia}K_{ib}[Q]}[A] + \frac{K_{m_P}}{V_{max}}\left(1 + \frac{K_{m_A}[B]}{K_{ia}K_{m_B}}\right)} \qquad (IX\text{-}259)$$

Thus K_{ib} can be calculated from the slope of the replot. If [A] and [P] are held constant and $1/v$ versus $1/[Q]$ plotted at different fixed concentrations of B, the $slope_{1/Q}$ replot is given by:

$$\boxed{slope_{1/Q} = \frac{K_{m_Q}[A]}{V_{max}K_{ia}K_{ib}}\left(1 + \frac{K_{ip}}{[P]}\right)[B] + \frac{K_{m_Q}}{V_{max}}\left(1 + \frac{[A]}{K_{ia}}\right)} \qquad (IX\text{-}260)$$

The K_{ib} can be calculated from the slope of this replot also. With K_{ib} known, the ratio of K_{m_B}/K_{m_C} can be calculated from equation IX-231. The ratio of K_{m_A}/K_{m_C} can also be calculated if K_{m_B}/K_{m_A} is known.

Alternate Designation—Uni Bi Uni Uni Ping Pong Bi Ter

If the reverse reaction is started with stable form F, the reaction becomes Uni Bi Uni Uni. As in the forward direction, the velocity equations, reciprocal plots and product inhibition patterns for the two designations have the same forms. Only the letters representing the five ligands differ.

T. ORDERED TER TER SYSTEM

Velocity Equation, Kinetic Constants, and Haldane Equations

The Ordered Ter Ter sequence is:

$$\begin{array}{ccccccccccccc}
 & A & & B & & C & & & P & & Q & & R & \\
 & k_1 \downarrow k_{-1} & & k_2 \downarrow k_{-2} & & k_3 \downarrow k_{-3} & & & k_4 \uparrow k_{-4} & & k_5 \uparrow k_{-5} & & k_6 \uparrow k_{-6} & \\
E & & EA & & EAB & & (EABC \rightleftharpoons EPQR) & & & EQR & & ER & & E
\end{array}$$

The velocity equation obtained from six five-sided King-Altman interconversion patterns contains 27 terms as shown below.

$$\frac{v}{[\mathrm{E}]_t}=\frac{k_1k_2k_3k_4k_5k_6[\mathrm{A}][\mathrm{B}][\mathrm{C}]-k_{-1}k_{-2}k_{-3}k_{-4}k_{-5}k_{-6}[\mathrm{P}][\mathrm{Q}][\mathrm{R}]}{\begin{array}{l}k_{-1}k_{-2}k_5k_6(k_{-3}+k_4)+k_1k_{-2}k_5k_6(k_{-3}+k_4)[\mathrm{A}]+k_{-1}k_3k_4k_5k_6[\mathrm{C}]\\
+k_1k_2k_5k_6(k_{-3}+k_4)[\mathrm{A}][\mathrm{B}]+k_1k_3k_4k_5k_6[\mathrm{A}][\mathrm{C}]\\
+k_2k_3k_4k_5k_6[\mathrm{B}][\mathrm{C}]+k_1k_2k_3(k_4k_5+k_4k_6+k_5k_6)[\mathrm{A}][\mathrm{B}][\mathrm{C}]\\
+k_{-1}k_{-2}k_{-3}k_{-4}k_6[\mathrm{P}]+k_{-1}k_{-2}k_5k_{-6}(k_{-3}+k_4)[\mathrm{R}]\\
+k_{-1}k_{-2}k_{-3}k_{-4}k_{-5}[\mathrm{P}][\mathrm{Q}]+k_{-1}k_{-2}k_{-3}k_{-4}k_{-6}[\mathrm{P}][\mathrm{R}]\\
+k_{-1}k_{-2}k_{-5}k_{-6}(k_{-3}+k_4)[\mathrm{Q}][\mathrm{R}]\\
+k_{-4}k_{-5}k_{-6}(k_{-1}k_{-2}+k_{-1}k_{-3}+k_{-2}k_{-3})[\mathrm{P}][\mathrm{Q}][\mathrm{R}]\\
+k_1k_{-2}k_{-3}k_{-4}k_6[\mathrm{A}][\mathrm{P}]+k_{-1}k_3k_4k_5k_{-6}[\mathrm{C}][\mathrm{R}]\\
+k_1k_2k_{-3}k_{-4}k_6[\mathrm{A}][\mathrm{B}][\mathrm{P}]+k_2k_3k_4k_5k_{-6}[\mathrm{B}][\mathrm{C}][\mathrm{R}]\\
+k_1k_{-2}k_{-3}k_{-4}k_{-5}[\mathrm{A}][\mathrm{P}][\mathrm{Q}]+k_{-1}k_3k_4k_{-5}k_{-6}[\mathrm{C}][\mathrm{Q}][\mathrm{R}]\\
+k_1k_2k_3k_{-4}k_6[\mathrm{A}][\mathrm{B}][\mathrm{C}][\mathrm{P}]+k_1k_2k_3k_4k_{-5}[\mathrm{A}][\mathrm{B}][\mathrm{C}][\mathrm{Q}]\\
+k_1k_2k_{-3}k_{-4}k_{-5}[\mathrm{A}][\mathrm{B}][\mathrm{P}][\mathrm{Q}]+k_2k_3k_4k_{-5}k_{-6}[\mathrm{B}][\mathrm{C}][\mathrm{Q}][\mathrm{R}]\\
+k_2k_{-3}k_{-4}k_{-5}k_{-6}[\mathrm{B}][\mathrm{P}][\mathrm{Q}][\mathrm{R}]+k_{-1}k_3k_{-4}k_{-5}k_{-6}[\mathrm{C}][\mathrm{P}][\mathrm{Q}][\mathrm{R}]\\
+k_1k_2k_3k_{-4}k_{-5}[\mathrm{A}][\mathrm{B}][\mathrm{C}][\mathrm{P}][\mathrm{Q}]+k_2k_3k_{-4}k_{-5}k_{-6}[\mathrm{B}][\mathrm{C}][\mathrm{P}][\mathrm{Q}][\mathrm{R}]\end{array}} \tag{IX-261}$$

In terms of kinetic constants, the equation becomes:

$$v=\frac{V_fV_r\left([\mathrm{A}][\mathrm{B}][\mathrm{C}]-\dfrac{[\mathrm{P}][\mathrm{Q}][\mathrm{R}]}{K_{\mathrm{eq}}}\right)}{\begin{array}{l}V_rK_{ia}K_{ib}K_{m_\mathrm{C}}+V_rK_{ib}K_{m_\mathrm{C}}[\mathrm{A}]+V_rK_{ia}K_{m_\mathrm{B}}[\mathrm{C}]+V_rK_{m_\mathrm{C}}[\mathrm{A}][\mathrm{B}]+V_rK_{m_\mathrm{B}}[\mathrm{A}][\mathrm{C}]\\
+V_rK_{m_\mathrm{A}}[\mathrm{B}][\mathrm{C}]+V_r[\mathrm{A}][\mathrm{B}][\mathrm{C}]+\dfrac{V_fK_{ir}K_{m_\mathrm{Q}}[\mathrm{P}]}{K_{\mathrm{eq}}}+\dfrac{V_fK_{iq}K_{m_\mathrm{P}}[\mathrm{R}]}{K_{\mathrm{eq}}}\\
+\dfrac{V_fK_{m_\mathrm{R}}[\mathrm{P}][\mathrm{Q}]}{K_{\mathrm{eq}}}+\dfrac{V_fK_{m_\mathrm{Q}}[\mathrm{P}][\mathrm{R}]}{K_{\mathrm{eq}}}+\dfrac{V_fK_{m_\mathrm{P}}[\mathrm{Q}][\mathrm{R}]}{K_{\mathrm{eq}}}+\dfrac{V_f[\mathrm{P}][\mathrm{Q}][\mathrm{R}]}{K_{\mathrm{eq}}}\end{array}}$$

$$+\frac{V_f K_{m_Q} K_{ir}[A][P]}{K_{ia}K_{eq}}+\frac{V_r K_{ia} K_{m_B}[C][R]}{K_{ir}}+\frac{V_f K_{m_Q} K_{ir}[A][B][P]}{K_{ia}K_{ib}K_{eq}}$$

$$+\frac{V_r K_{m_A}[B][C][R]}{K_{ir}}+\frac{V_f K_{m_R}[A][P][Q]}{K_{ia}K_{eq}}+\frac{V_r K_{ia} K_{m_B}[C][Q][R]}{K_{iq}K_{ir}}$$

$$+\frac{V_f K_{ir} K_{m_Q}[A][B][C][P]}{K_{ia}K_{ib}K_{ic}K_{eq}}+\frac{V_f K_{ip} K_{m_R}[A][B][C][Q]}{K_{ia}K_{ib}K_{ic}K_{eq}}+\frac{V_f K_{m_R}[A][B][P][Q]}{K_{ia}K_{ib}K_{eq}}$$

$$+\frac{V_r K_{m_A}[B][C][Q][R]}{K_{iq}K_{ir}}+\frac{V_r K_{m_A} K_{ic}[B][P][Q][R]}{K_{ip}K_{iq}K_{ir}}+\frac{V_r K_{ia} K_{m_B}[C][P][Q][R]}{K_{ip}K_{iq}K_{ir}}$$

$$+\frac{V_f K_{m_R}[A][B][C][P][Q]}{K_{ia}K_{ib}K_{ic}K_{eq}}+\frac{V_r K_{m_A}[B][C][P][Q][R]}{K_{ip}K_{iq}K_{ir}} \qquad \text{(IX-262)}$$

where

$$K_{m_A}=\frac{\text{Coef}_{\text{BC}}}{\text{Coef}_{\text{ABC}}}, \qquad K_{m_B}=\frac{\text{Coef}_{\text{AC}}}{\text{Coef}_{\text{ABC}}}, \qquad K_{m_C}=\frac{\text{Coef}_{\text{AB}}}{\text{Coef}_{\text{ABC}}}$$

$$K_{m_P}=\frac{\text{Coef}_{\text{QR}}}{\text{Coef}_{\text{PQR}}}, \qquad K_{m_Q}=\frac{\text{Coef}_{\text{PR}}}{\text{Coef}_{\text{PQR}}}, \qquad K_{m_R}=\frac{\text{Coef}_{\text{PQ}}}{\text{Coef}_{\text{PQR}}}$$

$$K_{ia}=\frac{\text{const}}{\text{Coef}_{\text{A}}}=\frac{\text{Coef}_{\text{C}}}{\text{Coef}_{\text{AC}}}=\frac{\text{Coef}_{\text{P}}}{\text{Coef}_{\text{AP}}}=\frac{\text{Coef}_{\text{PQ}}}{\text{Coef}_{\text{APQ}}}=\frac{k_{-1}}{k_1}$$

$$K_{ib}=\frac{\text{Coef}_{\text{A}}}{\text{Coef}_{\text{AB}}}=\frac{\text{Coef}_{\text{AP}}}{\text{Coef}_{\text{ABP}}}=\frac{\text{Coef}_{\text{APQ}}}{\text{Coef}_{\text{ABPQ}}}=\frac{k_{-2}}{k_2}$$

$$K_{ic}=\frac{\text{Coef}_{\text{ABP}}}{\text{Coef}_{\text{ABCP}}}=\frac{\text{Coef}_{\text{ABPQ}}}{\text{Coef}_{\text{ABCPQ}}}=\frac{\text{Coef}_{\text{BPQR}}}{\text{Coef}_{\text{BCPQR}}}=\frac{k_{-3}}{k_3}$$

$$K_{ip}=\frac{\text{Coef}_{\text{CQR}}}{\text{Coef}_{\text{CPQR}}}=\frac{\text{Coef}_{\text{ABCQ}}}{\text{Coef}_{\text{ABCPQ}}}=\frac{\text{Coef}_{\text{BCQR}}}{\text{Coef}_{\text{BCPQR}}}=\frac{k_4}{k_{-4}}$$

$$K_{iq}=\frac{\text{Coef}_{\text{R}}}{\text{Coef}_{\text{QR}}}=\frac{\text{Coef}_{\text{CR}}}{\text{Coef}_{\text{CQR}}}=\frac{\text{Coef}_{\text{BCR}}}{\text{Coef}_{\text{BCQR}}}=\frac{k_5}{k_{-5}}$$

$$K_{ir}=\frac{\text{const}}{\text{Coef}_{\text{R}}}=\frac{\text{Coef}_{\text{P}}}{\text{Coef}_{\text{PR}}}=\frac{\text{Coef}_{\text{C}}}{\text{Coef}_{\text{CR}}}=\frac{\text{Coef}_{\text{BC}}}{\text{Coef}_{\text{BCR}}}=\frac{k_6}{k_{-6}}$$

The Haldane equations are:

$$K_{eq} = \frac{V_{max_f} K_{m_P} K_{iq} K_{ir}}{V_{max_r} K_{ia} K_{ib} K_{m_C}} = \frac{K_{ip} K_{iq} K_{ir}}{K_{ia} K_{ib} K_{ic}} \tag{IX-263}$$

Rate Constants

The equations for k_1, k_{-1}, k_2, k_{-2}, k_3, k_{-3}, k_{-4}, k_5, and k_{-5} are the same as those shown for the Ordered Ter Bi system. In addition:

$$\frac{1}{k_4} = \frac{[E]_t}{V_{max_f}} - \frac{1}{k_5} - \frac{1}{k_6}, \qquad k_6 = \frac{V_{max_r} K_{ir}}{[E]_t K_{m_R}}, \qquad k_{-6} = \frac{V_{max_r}}{[E]_t K_{m_R}}$$

Distribution Equations

$$\frac{[E]}{[E]_t} = \frac{K_{m_A} V_r [B][C] + K_{ia} K_{m_B} V_r [C] + K_{ia} K_{ib} K_{m_C} V_r + \dfrac{K_{ir} K_{m_Q} V_f [P]}{K_{eq}} + \dfrac{K_{m_R} V_f [P][Q]}{K_{eq}}}{\text{denominator of velocity equation}} \tag{IX-264}$$

$$\frac{[EA]}{[E]_t} = \frac{\dfrac{K_{m_A} V_r [P][Q][R]}{K_{ia} K_{eq}} + K_{m_B} V_r [A][C] + K_{ib} K_{m_C} V_r [A] + \dfrac{K_{m_Q} K_{ir} V_f [A][P]}{K_{ia} K_{eq}} + \dfrac{K_{m_R} V_f [A][P][Q]}{K_{ia} K_{eq}}}{\text{denominator of velocity equation}} \tag{IX-265}$$

$$\frac{[EAB]}{[E]_t} = \frac{\dfrac{K_{m_A} K_{ic} V_r [B][P][Q][R]}{K_{ip} K_{iq} K_{ir}} + \dfrac{K_{m_B} V_r [P][Q][R]}{K_{ib} K_{eq}} + K_{m_C} V_r [A][B] + \dfrac{K_{m_Q} K_{ir} [A][B][P]}{K_{ia} K_{ib} K_{eq}} + \dfrac{K_{m_R} V_f [A][B][P][Q]}{K_{ia} K_{ib} K_{eq}}}{\text{denominator of velocity equation}} \tag{IX-266}$$

$$\frac{[\mathrm{EABC}+\mathrm{EPQR}]}{[\mathrm{E}]_t}=\frac{\begin{array}{c}\dfrac{K_{m_\mathrm{A}}V_r[\mathrm{B}][\mathrm{C}][\mathrm{P}][\mathrm{Q}][\mathrm{R}]}{K_{ip}K_{iq}K_{ir}}+\dfrac{K_{ia}K_{m_\mathrm{B}}V_r[\mathrm{C}][\mathrm{P}][\mathrm{Q}][\mathrm{R}]}{K_{ip}K_{iq}K_{ir}}\\+\dfrac{K_{m_\mathrm{Q}}K_{ir}V_f[\mathrm{A}][\mathrm{B}][\mathrm{C}][\mathrm{P}]}{K_{ia}K_{ib}K_{ic}K_{\mathrm{eq}}}+\dfrac{K_{m_\mathrm{R}}V_f[\mathrm{A}][\mathrm{B}][\mathrm{C}][\mathrm{P}][\mathrm{Q}]}{K_{ia}K_{ib}K_{ic}K_{\mathrm{eq}}}\\+\left[\dfrac{V_f}{K_{\mathrm{eq}}}-\dfrac{V_r}{K_{\mathrm{eq}}}\left(\dfrac{K_{m_\mathrm{A}}}{K_{ia}}+\dfrac{K_{m_\mathrm{B}}}{K_{ib}}\right)\right][\mathrm{P}][\mathrm{Q}][\mathrm{R}]\\+\left[V_r-V_f\left(\dfrac{K_{m_\mathrm{Q}}}{K_{iq}}+\dfrac{K_{m_\mathrm{B}}}{K_{ir}}\right)\right][\mathrm{A}][\mathrm{B}][\mathrm{C}]\end{array}}{\text{denominator of velocity equation}}$$

(IX-267)

$$\frac{[\mathrm{EQR}]}{[\mathrm{E}]_t}=\frac{\begin{array}{c}\dfrac{K_{m_\mathrm{A}}V_r[\mathrm{B}][\mathrm{C}][\mathrm{Q}][\mathrm{R}]}{K_{iq}K_{ir}}+\dfrac{K_{ia}K_{m_\mathrm{B}}V_r[\mathrm{C}][\mathrm{Q}][\mathrm{R}]}{K_{iq}K_{ir}}+\dfrac{K_{m_\mathrm{P}}V_f[\mathrm{Q}][\mathrm{R}]}{K_{\mathrm{eq}}}\\+\dfrac{K_{m_\mathrm{Q}}V_f[\mathrm{A}][\mathrm{B}][\mathrm{C}]}{K_{iq}}+\dfrac{K_{ip}K_{m_\mathrm{R}}V_f[\mathrm{A}][\mathrm{B}][\mathrm{C}][\mathrm{Q}]}{K_{ia}K_{ib}K_{ic}K_{\mathrm{eq}}}\end{array}}{\text{denominator of velocity equation}}$$

(IX-268)

$$\frac{[\mathrm{ER}]}{[\mathrm{E}]_t}=\frac{\begin{array}{c}\dfrac{K_{m_\mathrm{A}}V_r[\mathrm{B}][\mathrm{C}][\mathrm{R}]}{K_{ir}}+\dfrac{K_{ia}K_{m_\mathrm{B}}V_r[\mathrm{C}][\mathrm{R}]}{K_{ir}}+\dfrac{K_{m_\mathrm{P}}K_{iq}V_f[\mathrm{R}]}{K_{\mathrm{eq}}}\\+\dfrac{K_{m_\mathrm{Q}}V_f[\mathrm{P}][\mathrm{R}]}{K_{\mathrm{eq}}}+\dfrac{K_{m_\mathrm{R}}V_f[\mathrm{A}][\mathrm{B}][\mathrm{C}]}{K_{ir}}\end{array}}{\text{denominator of velocity equation}}$$

(IX-269)

Effect of Isomerizations

EABC + EPQR: All the distributions shown above are valid but the calculations of k_3, k_{-3}, k_4, and k_{-4} are invalid.

EA: The calculations of k_1, k_{-1}, k_2, k_3, k_{-3}, k_4, k_{-4}, [EA]/[E]$_t$, and [EABC + EPQR]/[E]$_t$ are all invalid.

EAB: The calculations of k_2, k_{-2}, k_3, k_{-3}, k_4, k_{-4}, [EAB]/[E]$_t$, and [EABC + EPQR]/[E]$_t$ are all invalid.

Initial Velocity in the Forward and Reverse Directions

The velocity equation for the reaction in the forward direction in the absence of products is that given earlier for the Ordered Ter Bi System. The same equation gives the velocity in the reverse direction (with R replacing A, Q replacing B, and P replacing C).

Product Inhibition Studies

In the presence of P, the velocity equations and inhibition patterns are identical to those given earlier for the Ordered Ter Bi system. (The extra K_{ir} that appears in the [P], [A][P], and [A][P][B] terms drops out when K_{eq} is eliminated by means of the first Haldane equation.) In the presence of R, the velocity equations and inhibition patterns are identical to those given for the Ordered Ter Bi system in the presence of Q (with [R] and K_{ir} replacing [Q] and K_{iq}). The K_{ir} can be obtained as $K_{i_{\text{slope}}}^{\text{R/A}}$. In the presence of Q, a single new term is added to the denominator of the velocity equation. This is:

$$\frac{V_f K_{ip} K_{m_R}[A][B][C][Q]}{K_{ia} K_{ib} K_{ic} K_{eq}} = \frac{K_{ip} K_{m_R} K_{m_C}[A][B][C][Q]}{K_{ic} K_{m_P} K_{iq} K_{ir}}$$

The term becomes part of the $1/v$-axis intercept factor when either [A], [B], or [C] is varied. For example, when [B] is varied, the velocity equation is:

$$\frac{v}{V_{\max}} = \frac{[B]}{K_{m_B}\left(1+\frac{K_{ia}}{[A]}\right)\left(1+\frac{K_{ib}K_{m_C}}{K_{m_B}[C]}\right)+[B]\left(1+\frac{K_{m_A}}{[A]}+\frac{K_{m_C}}{[C]}+\frac{K_{ip}K_{m_R}K_{m_C}[Q]}{K_{ic}K_{m_P}K_{iq}K_{ir}}\right)} \tag{IX-270}$$

The ratio of $K_{ip}K_{m_R}/K_{m_P}K_{iq}$ can be obtained from $1/v$-axis intercept replots. The remaining kinetic constants can be obtained from initial velocity and product inhibition studies in the reverse direction (which is symmetrical to the forward direction).

U. PARTIAL RAPID EQUILIBRIUM ORDERED TERREACTANT SYSTEMS

There are a number of limiting cases in which one or two of the three substrates (and/or one or two of the three products) are in equilibrium with their respective complexes. In some cases the system behaves as a full rapid equilibrium system even though only some of the enzyme forms are at equilibrium (e.g., the Ordered Ter Bi system described earlier in which A and B are in equilibrium with E, EA, and EAB). In other cases, the system acts intermediate between the full steady-state and the full rapid equilibrium systems. For example, consider the Ordered Ter Ter sequence in which the rate constant for the dissociation of the first ligand to add in either direction is very large compared to the $V_{\max}/[\mathrm{E}]_t$ for that direction. (Another way of stating the condition is that $k_{-1} \gg k_2[\mathrm{B}]$ and $k_6 \gg k_{-5}[\mathrm{Q}]$.) In this case, K_{m_A} and K_{m_R} are very small compared to the other kinetic constants and can be neglected. Alternately, the velocity equation can be derived directly by the method of Cha, in which E, EA, and ER are taken as a single corner of the basic King-Altman figure. The resulting equation for the forward direction is the same as that for the full Steady-State Ordered Ter Ter system with all denominator terms containing K_{m_A} or K_{m_R} omitted. If we consider only the forward velocity in the presence of only one of the products at a time, the equation becomes:

$$\frac{v}{V_{\max}} = \frac{[\mathrm{A}][\mathrm{B}][\mathrm{C}]}{\begin{aligned} &K_{ia}K_{ib}K_{m_\mathrm{C}} + K_{ia}K_{m_\mathrm{C}}[\mathrm{A}] + K_{ia}K_{m_\mathrm{B}}[\mathrm{C}] + K_{m_\mathrm{C}}[\mathrm{A}][\mathrm{B}] + K_{m_\mathrm{B}}[\mathrm{A}][\mathrm{C}] + [\mathrm{A}][\mathrm{B}][\mathrm{C}] \\ &+ \frac{K_{ia}K_{ib}K_{m_\mathrm{C}}K_{m_\mathrm{Q}}[\mathrm{P}]}{K_{m_\mathrm{P}}K_{iq}} + \frac{K_{ib}K_{m_\mathrm{C}}K_{m_\mathrm{Q}}[\mathrm{A}][\mathrm{P}]}{K_{m_\mathrm{P}}K_{iq}} + \frac{K_{m_\mathrm{C}}K_{m_\mathrm{Q}}[\mathrm{A}][\mathrm{B}][\mathrm{P}]}{K_{m_\mathrm{P}}K_{iq}} \\ &+ \frac{K_{m_\mathrm{C}}K_{m_\mathrm{Q}}[\mathrm{A}][\mathrm{B}][\mathrm{C}][\mathrm{P}]}{K_{ic}K_{m_\mathrm{P}}K_{iq}} + \frac{K_{ia}K_{ib}K_{m_\mathrm{C}}[\mathrm{R}]}{K_{ir}} + \frac{K_{ia}K_{m_\mathrm{B}}[\mathrm{C}][\mathrm{R}]}{K_{ir}} \end{aligned}} \tag{IX-271}$$

The missing denominator [B][C] term affects the characteristics of the initial velocity plots in the absence of products. When [A] is varied at different fixed concentrations of B and C (maintained at a constant ratio), a mixed-type pattern results as usual, but the parabolic $slope_{1/\mathrm{A}}$ versus $1/[\mathrm{B}]$ or $1/[\mathrm{C}]$ replots go through the origin (i.e., at saturating [B] or [C], the slope of the $1/v$ versus $1/[\mathrm{A}]$ plot is zero). The equation contains no [Q] terms. Hence Q is not an inhibitor. (The [A][B][C][Q] term present in the equation for the full steady-state system is missing.) Also missing is the [B][C][R] denominator term. As a result, R is competitive with both A and B (instead of just with A); R is still a mixed-type inhibitor with respect to C.

As shown earlier, if both A and B are at equilibrium with E, EA, and EAB, the system behaves as a full rapid equilibrium system even though the reaction EAB + C$\rightleftharpoons$EABC is in steady-state. *Slope*$_{1/A}$ and *slope*$_{1/B}$ replots have zero intercepts, *intercept*$_{1/B}$ replots are linear, and $1/v$ versus $1/[C]$ plots intersect on the $1/v$-axis. If in the reverse direction, E, ER, and EQR are at equilibrium, then R will be the only product inhibitor (competitive with respect to A, B, and C).

V. ORDERED TERREACTANT SYSTEMS WITH RAPID EQUILIBRIUM RANDOM SEQUENCES

Two ligands in an otherwise ordered terreactant sequence may add randomly. Several possibilities were described earlier for completely rapid equilibrium systems (Tables VI-2 to VI-4). Random sequences in steady-state systems introduce squared concentration terms into the velocity equation, and, theoretically, the reciprocal plots are nonlinear, although the curvature may not be obvious. There are several conditions in which linear reciprocal plots are expected even though all the enzyme forms may not be at equilibrium. Some examples are described below.

Random A–B, Ordered C

Consider the Random A–B system shown below in which the first ligands to add in either direction are at equilibrium with their respective complexes and free E; that is, the dissociations of EA and EB are very rapid compared to the rates of the forward steps.

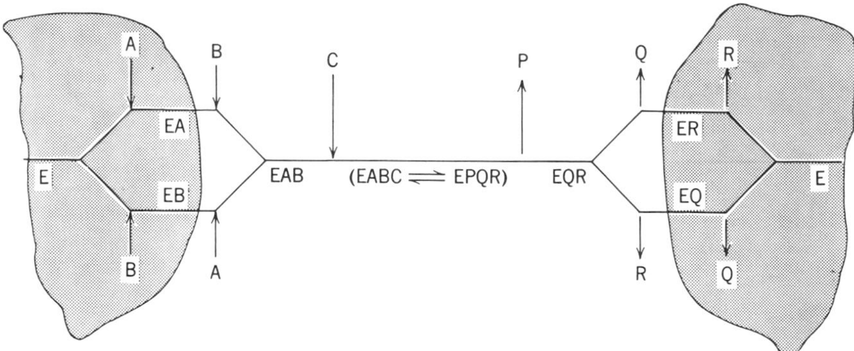

For simplicity, let us consider only the forward reaction. The basic King-Altman figure is shown below with E, EA, and EB taken as a single corner, X. Note that EPQR is shown going to E in a single step. The step labeled k_4 is really several product release steps leading from EPQR back to E. The line in the reverse direction labeled k_{-4} ["P"] represents several product addition steps leading from E back to EPQR. Since the product release and addition steps are omitted, the basic figure could describe any ordered terreactant system (Ter Ter, Ter Bi, etc.).

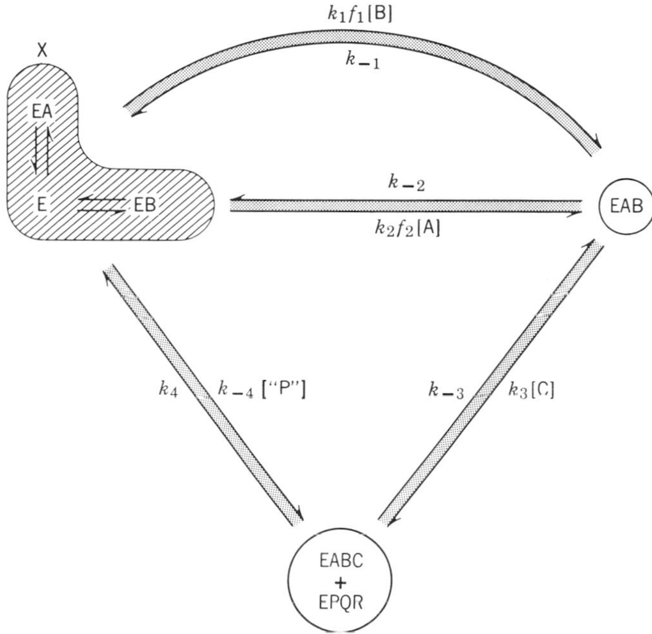

Combining the two lines connecting X and EAB:

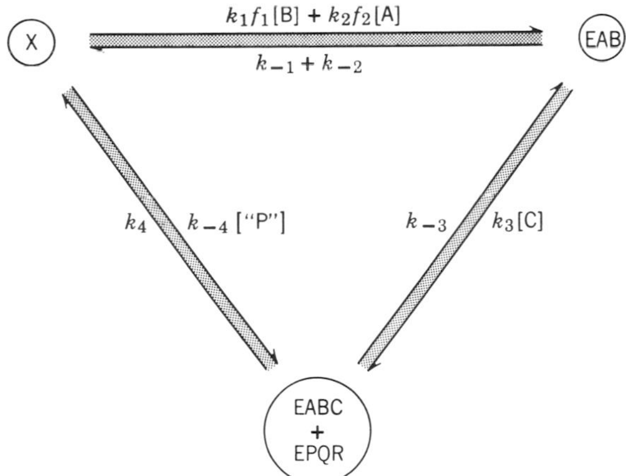

$$f_1 = \frac{[EA]}{[X]} = \frac{\dfrac{[A]}{K_A}}{1 + \dfrac{[A]}{K_A} + \dfrac{[B]}{K_B}}$$

$$f_2 = \frac{[EB]}{[X]} = \frac{\dfrac{[B]}{K_B}}{1 + \dfrac{[A]}{K_A} + \dfrac{[B]}{K_B}}$$

$$v = k_3[C][EAB] - k_{-3}[EABC + EPQR]$$

Substitution and simplification yields:

$$\frac{v}{V_{max}} = \frac{[A][B][C]}{K_{ia}K_{m_B}K_{ic} + K_{m_B}K_{ic}[A] + K_{m_A}K_{ic}[B] + K_{ia}K_{m_B}[C] + K_{m_B}[A][C] + K_{m_A}[B][C] + K_{m_C}[A][B] + [A][B][C]} \qquad \text{(IX-272)}$$

where

$$K_{ia} = \frac{\text{const}}{\text{Coef}_A} = \frac{\text{Coef}_C}{\text{Coef}_{AC}} = K_A, \qquad K_{ib} = \frac{\text{const}}{\text{Coef}_B} = \frac{\text{Coef}_C}{\text{Coef}_{BC}} = K_B$$

$$K_{ic} = \frac{\text{const}}{\text{Coef}_C} = \frac{\text{Coef}_B}{\text{Coef}_{BC}} = \frac{\text{Coef}_A}{\text{Coef}_{AC}}, \qquad \text{and} \qquad K_{ia}K_{m_B} = K_{ib}K_{m_A}$$

If we choose the same definition for K_{ia} but define:

$$K_{ib} = \frac{\text{Coef}_A}{\text{Coef}_{AB}} \neq K_B$$

the equation becomes:

$$\frac{v}{V_{\max}} = \frac{[A][B][C]}{K_{ia}K_{ib}K_{m_C} + K_{ib}K_{m_C}[A] + K_{ia}K_{m_C}[B] + K_{ia}K_{m_B}[C] + K_{m_B}[A][C] + K_{m_A}[B][C] + K_{m_C}[A][B] + [A][B][C]}$$

(IX-273)

The equations contain new [C], [A][C], and [B][C] terms compared to the equation for the total Rapid Equilibrium Random A–B system and a new [B] term compared to the equation for the steady-state fully ordered terreactant system. All reciprocal plots yield mixed-type patterns when one substrate is varied at different fixed concentrations of another and a constant concentration of the third. Mixed-type patterns also result if two substrates are treated together as the changing fixed substrates. In this case, all slope and $1/v$-axis intercept replots are parabolic with finite intercepts on the vertical axis. The patterns are identical to those expected for a rapid equilibrium completely random terreactant system. Product inhibition studies may differentiate between the two possibilities. In the Random A–B, Q–R system, Q and R are competitive with A and B and are mixed-type inhibitors with respect to C; P is a mixed-type inhibitor with respect to all three substrates. In the completely random system, all three products are competitive with all three substrates, since they all combine with free E. However, one or more of the products may form dead-end complexes resulting in mixed-type inhibition patterns. If the likely dead-end complexes can be predicted, then it may be possible to distinguish between the two possible mechanisms. If it is assumed that all reactions between E and EAB are at equilibrium, while the reaction EAB + C$\rightleftharpoons$EABC and the interconversion of the central complex are in steady-state, then the final velocity equation has the same form as that for the full Rapid Equilibrium Random A–B system (Chapter Six). If it is assumed that E is not part of the rapid equilibrium segment, but EA, EB, and EAB are, or if it is assumed that only one of the first complexes (e.g., EA or EB, but not both) is in equilibrium with E and EAB, then the final velocity equation will contain squared concentration terms.

Ordered A, Random B–C

If EAB, EAC, and EA are at equilibrium with B and C while the reactions E + A⇌EA, EAB + C⇌EABC, EAC + B⇌EABC, and the interconversion of the central complex are in steady-state, then the velocity equation for the forward reaction becomes:

$$\frac{v}{V_{\max}} = \frac{[A][B][C]}{K_{ia}K_{m_B}K_{ic} + K_{m_B}K_{ic}[A] + K_{m_C}[A][B] + K_{m_B}[A][C] + K_{m_A}[B][C] + [A][B][C]}$$

(IX-274)

where the definitions of the kinetic constants are the same (where possible) as those given for the Random A–B system. The same equation is obtained if, in addition, it is assumed that EABC is in equilibrium with EA, EAB, EAC, B, and C. The equation contains a new denominator [B][C] term compared to the full Rapid Equilibrium Random B–C system. Compared to the full Steady-State Ordered Terreactant system, equation IX-274 is missing the denominator [C] term. Equation IX-274 is rearranged below to show A, B, and C as the varied substrate.

$$\frac{v}{V_{\max}} = \frac{[A]}{K_{m_A}\left(1 + \frac{K_{ia}K_{m_B}K_{ic}}{K_{m_A}[B][C]}\right) + [A]\left(1 + \frac{K_{m_C}}{[C]} + \frac{K_{m_B}}{[B]} + \frac{K_{m_B}K_{ic}}{[B][C]}\right)}$$

(IX-275)

$$\frac{v}{V_{\max}} = \frac{[B]}{K_{m_B}\left(1 + \frac{K_{ic}}{[C]} + \frac{K_{ia}K_{ic}}{[A][C]}\right) + [B]\left(1 + \frac{K_{m_A}}{[A]} + \frac{K_{m_C}}{[C]}\right)}$$

(IX-276)

$$\frac{v}{V_{\max}} = \frac{[C]}{K_{m_C}\left(1 + \frac{K_{ia}K_{m_B}K_{ic}}{K_{m_C}[A][B]} + \frac{K_{m_B}K_{ic}}{K_{m_C}[B]}\right) + [C]\left(1 + \frac{K_{m_A}}{[A]} + \frac{K_{m_B}}{[B]}\right)}$$

(IX-277)

When [A] is varied at different fixed concentrations of B and C (maintained at a constant ratio), a mixed-type pattern results with parabolic slope and intercept replots which have finite intercpets. When [B] is varied at different fixed concentrations of A and C, the pattern is again mixed-type. Slope replots are parabolic with finite intercepts, but the intercept replot (against 1/[C]) is linear with a finite intercept. If [C] is held constant and A is the changing fixed substrate, the family of plots intersect on the $1/v$-axis. When [C] is varied at different fixed concentrations of A and B, the pattern is mixed-type with a parabolic slope replot and a linear $1/v$-axis intercept replot, both with finite intercepts.

If EAB, EAC, and EABC are assumed to be in equilibrium with B and C, while all other steps are in steady-state, the velocity equation contains squared concentration terms and reciprocal plots are (theoretically) non-linear. Table IX-7 lists some of the properties of the various terreactant systems involving a quaternary EABC complex. (See also Table VI-3 for some completely rapid equilibrium systems.)

W. BI UNI UNI BI PING PONG TER TER SYSTEM

Reaction Sequence

There are three basic Ter Ter Ping Pong mechanisms. One is the Bi Uni Uni Bi Ping Pong system shown below.

A ($k_1 \downarrow k_{-1}$) B ($k_2 \downarrow k_{-2}$) P ($k_3 \uparrow k_{-3}$) C ($k_4 \downarrow k_{-4}$) Q ($k_5 \uparrow k_{-5}$) R ($k_6 \uparrow k_{-6}$)

E EA (EAB $\rightleftharpoons$ FP) F (FC $\rightleftharpoons$ EQR) ER E

The Bi Uni Uni Bi system is equivalent to the Uni Bi Bi Uni system shown below if the sequence starts with stable form F.

A ($k_1 \downarrow k_{-1}$) P ($k_2 \uparrow k_{-2}$) Q ($k_3 \uparrow k_{-3}$) B ($k_4 \downarrow k_{-4}$) C ($k_5 \downarrow k_{-5}$) R ($k_6 \uparrow k_{-6}$)

E (EA $\rightleftharpoons$ FPQ) FQ F FB (FBC $\rightleftharpoons$ ER) E

The velocity equations have the same forms, with different letters representing the six ligands. The Uni Bi Bi Uni Ping Pong system is an unlikely mechanism. It requires that a single substrate be converted to three fragments—two that are released and one that stays on the enzyme in covalent linkage to yield form F. Then two additional substrates must add, and both substrates plus the covalent fragment leaves as a single product, R.

Table IX-7 Slope and Intercept Replots of some Terreactant Systems Involving a Quaternary EABC Complex[a]

Mechansim	Rapid Equilibrium Segment	Varied Substrate	Reciprocal Plot Pattern[b]	Slope Replot[c]	$1/v$-Axis Intercept[c] Replot
Completely Random	All enzyme forms	A	MT	P[d]	P
		B	MT	P	P
		C	MT	P	P
Completely Ordered	None	A	MT	P	P
		B	MT	P	L[e]
		C	MT	P	P
Completely Ordered	E-EA	A	MT	P(INT = 0)[f]	P
		B	MT	P	L
		C	MT	P	P
Completely Ordered	EA-EAB	A	MT	P	P
		B	MT	P(INT = 0)	L
		C	MT	P	L
Completely Ordered	All enzyme forms	A	MT	P(INT = 0)	P
	or	B	MT	P(INT = 0)	L
	E-EA-EAB	C	C[g]	P	—
Random A–B	All enzyme forms	A	MT	P(INT = 0)	P
	or	B	MT	P(INT = 0)	P
	E, EA, EB, and EAB	C	C	P	—

Random A–B	E-EA — EB	A	MT	P	P
		B	MT	P	P
		C	MT	P	P
Random B–C	All enzyme forms	A	MT	P(INT=0)	P
		B	MT	P	L
		C	MT	P	L
Random B–C	EA-EAC — EAB (±EABC)	A	MT	P	P
		B	MT	P	L
		B	MT	P	L
Random A–C	All enzyme forms	A	MT	P	P
		B	MT	P(INT=0)	L
		C	MT	P	P

[a]Each substrate is varied at different fixed concentrations of the other two (maintained at a constant ratio).
[b]MT = mixed-type: any two lines of the family of plots intersect to the left of the $1/v$-axis, above, below, or coincidentally on the horizontal axis. All the lines do not necessarily intersect at a common point.
[c]The slope and $1/v$-axis intercepts are replotted against the reciprocal concentration of one of the changing fixed substrates.
[d]P = parabolic.
[e]L = linear.
[f]INT = 0 means that the replot goes through the origin.
[g]C = competitive: the family of reciprocal plots intersect on the $1/v$-axis.

Velocity Equation, Kinetic Constants, and Haldane Equations

The velocity equation for the Bi Uni Uni Bi sequence is:

$$\frac{v}{[\mathrm{E}]_t} = \frac{k_1k_2k_3k_4k_5k_6[\mathrm{A}][\mathrm{B}][\mathrm{C}] - k_{-1}k_{-2}k_{-3}k_{-4}k_{-5}k_{-6}[\mathrm{P}][\mathrm{Q}][\mathrm{R}]}{k_{-1}k_4k_5k_6(k_{-2}+k_3)[\mathrm{C}] + k_1k_2k_3k_6(k_{-4}+k_5)[\mathrm{A}][\mathrm{B}] + k_1k_4k_5k_6(k_{-2}+k_3)[\mathrm{A}][\mathrm{C}]}$$

$$+ k_2k_3k_4k_5k_6[\mathrm{B}][\mathrm{C}] + k_1k_2k_4(k_3k_5 + k_3k_6 + k_5k_6)[\mathrm{A}][\mathrm{B}][\mathrm{C}]$$

$$+ k_{-1}k_{-2}k_{-3}k_6(k_{-4}+k_5)[\mathrm{P}] + k_{-1}k_{-2}k_{-3}k_{-4}k_{-5}[\mathrm{P}][\mathrm{Q}]$$

$$+ k_{-1}k_{-2}k_{-3}k_{-6}(k_{-4}+k_5)[\mathrm{P}][\mathrm{R}] + k_{-1}k_{-4}k_{-5}k_{-6}(k_{-2}+k_3)[\mathrm{Q}][\mathrm{R}]$$

$$+ k_{-3}k_{-5}k_{-6}(k_{-1}k_{-2} + k_{-1}k_{-4} + k_{-2}k_{-4})[\mathrm{P}][\mathrm{Q}][\mathrm{R}]$$

$$+ k_1k_{-2}k_{-3}k_6(k_{-4}+k_5)[\mathrm{A}][\mathrm{P}]$$

$$+ k_{-1}k_4k_5k_{-6}(k_{-2}+k_3)[\mathrm{C}][\mathrm{R}] + k_1k_2k_{-3}k_6(k_{-4}+k_5)[\mathrm{A}][\mathrm{B}][\mathrm{P}]$$

$$+ k_1k_2k_3k_{-4}k_{-5}[\mathrm{A}][\mathrm{B}][\mathrm{Q}] + k_2k_3k_4k_5k_{-6}[\mathrm{B}][\mathrm{C}][\mathrm{R}]$$

$$+ k_1k_{-2}k_{-3}k_{-4}k_{-5}[\mathrm{A}][\mathrm{P}][\mathrm{Q}]$$

$$+ k_2k_3k_{-4}k_{-5}k_{-6}[\mathrm{B}][\mathrm{Q}][\mathrm{R}] + k_{-1}k_4k_{-5}k_{-6}(k_{-2}+k_3)[\mathrm{C}][\mathrm{Q}][\mathrm{R}]$$

$$+ k_1k_2k_3k_4k_{-5}[\mathrm{A}][\mathrm{B}][\mathrm{C}][\mathrm{Q}] + k_1k_2k_{-3}k_{-4}k_{-5}[\mathrm{A}][\mathrm{B}][\mathrm{P}][\mathrm{Q}]$$

$$+ k_2k_3k_4k_{-5}k_{-6}[\mathrm{B}][\mathrm{C}][\mathrm{Q}][\mathrm{R}] + k_2k_{-3}k_{-4}k_{-5}k_{-6}[\mathrm{B}][\mathrm{P}][\mathrm{Q}][\mathrm{R}]$$

(IX-278)

In terms of kinetic constants, the velocity equation is the same as that shown for the Ordered Ter Ter system except there are no constant, [A], [R], [A][B][C][P], [C][P][Q][R], [A][B][C][P][Q], or [B][C][P][Q][R] terms.

The denominator contains two new terms in [A][B][Q] and [B][Q][R].

$$v = \frac{V_f V_r\left([A][B][C] - \dfrac{[P][Q][R]}{K_{eq}}\right)}{V_r K_{ia} K_{m_B}[C] + V_r K_{m_C}[A][B] + V_r K_{m_B}[A][C] + V_r K_{m_A}[B][C]}$$

$$+ V_r[A][B][C] + \frac{V_f K_{ir} K_{m_Q}[P]}{K_{eq}} + \frac{V_f K_{m_A}[P][Q]}{K_{eq}} + \frac{V_f K_{m_Q}[P][R]}{K_{eq}}$$

$$+ \frac{V_f K_{m_P}[Q][R]}{K_{eq}} + \frac{V_f [P][Q][R]}{K_{eq}} + \frac{V_f K_{m_Q} K_{ir}[A][P]}{K_{ia} K_{eq}}$$

$$+ \frac{V_r K_{ia} K_{m_B}[C][R]}{K_{ir}} + \frac{V_f K_{m_Q} K_{ir}[A][B][P]}{K_{ia} K_{ib} K_{eq}} + \frac{V_f K_{ip} K_{m_R}[A][B][Q]}{K_{ia} K_{ib} K_{eq}}$$

$$+ \frac{V_r K_{m_A}[B][C][R]}{K_{ir}} + \frac{V_f K_{m_R}[A][P][Q]}{K_{ia} K_{eq}} + \frac{V_r K_{m_A} K_{ia}[B][Q][R]}{K_{iq} K_{ir}} + \frac{V_r K_{ia} K_{m_B}[C][Q][R]}{K_{iq} K_{ir}}$$

$$+ \frac{V_f K_{ip} K_{m_R}[A][B][C][Q]}{K_{ia} K_{ib} K_{ic} K_{eq}} + \frac{V_f K_{m_R}[A][B][P][Q]}{K_{ia} K_{ib} K_{eq}} + \frac{V_r K_{m_A}[B][C][Q][R]}{K_{iq} K_{ir}}$$

$$+ \frac{V_r K_{m_A} K_{ic}[B][P][Q][R]}{K_{ip} K_{iq} K_{ir}} \qquad \text{(IX-279)}$$

where K_{m_A}, K_{m_B}, K_{m_C}, K_{m_P}, K_{m_Q}, K_{m_R}, $V_{\max_f}$, and $V_{\max_r}$ are defined in the usual way. Also:

$$K_{ia} = \frac{\text{Coef}_{\text{C}}}{\text{Coef}_{\text{AC}}} = \frac{\text{Coef}_{\text{P}}}{\text{Coef}_{\text{AP}}} = \frac{\text{Coefp}_{\text{PQ}}}{\text{Coef}_{\text{APQ}}} = \frac{k_{-1}}{k_1}$$

$$K_{ib} = \frac{\text{Coef}_{\text{AP}}}{\text{Coef}_{\text{ABP}}} = \frac{\text{Coef}_{\text{APQ}}}{\text{Coef}_{\text{ABPQ}}} = \frac{k_{-2}}{k_2}$$

$$K_{ic} = \frac{\text{Coef}_{\text{QR}}}{\text{Coef}_{\text{CQR}}} = \frac{\text{Coef}_{\text{ABQ}}}{\text{Coef}_{\text{ABCQ}}} = \frac{\text{Coef}_{\text{BQR}}}{\text{Coef}_{\text{BCQR}}} = \frac{k_{-4}}{k_4}$$

$$K_{ip} = \frac{\text{Coef}_{\text{AB}}}{\text{Coef}_{\text{ABP}}} = \frac{\text{Coef}_{\text{ABQ}}}{\text{Coef}_{\text{ABPQ}}} = \frac{\text{Coef}_{\text{BQR}}}{\text{Coef}_{\text{BPQR}}} = \frac{k_3}{k_{-3}}$$

$$K_{iq} = \frac{\text{Coef}_{\text{CR}}}{\text{Coef}_{\text{CQR}}} = \frac{\text{Coef}_{\text{BCR}}}{\text{Coef}_{\text{BCQR}}} = \frac{k_5}{k_{-5}}$$

$$K_{ir} = \frac{\text{Coef}_{\text{P}}}{\text{Coef}_{\text{PR}}} = \frac{\text{Coef}_{\text{C}}}{\text{Coef}_{\text{CR}}} = \frac{\text{Coef}_{\text{BC}}}{\text{Coef}_{\text{BCR}}} = \frac{k_6}{k_{-6}}$$

The Haldane equations are:

$$K_{eq} = \frac{V_f K_{ip} K_{m_Q} K_{ir}}{V_r K_{ia} K_{ib} K_{m_C}} = \frac{V_f K_{m_P} K_{iq} K_{ir}}{V_r K_{ia} K_{m_B} K_{ic}} = \frac{K_{ip} K_{iq} K_{ir}}{K_{ia} K_{ib} K_{ic}} = \left(\frac{V_f}{V_r}\right)^2 \frac{K_{m_P} K_{m_Q} K_{ir}}{K_{ia} K_{m_B} K_{m_C}}$$

(IX-280)

Rate Constants

The rate constants k_1, k_{-1}, k_6, and k_{-6} are given by the same equations shown for Ordered Ter Ter. The other rate constants cannot be determined.

Distribution Equations

$$\frac{[E]}{[E]_t} = \text{same as shown for Ordered Ter Ter without the numerator constant term}$$

$$\frac{[EA]}{[E]_t} = \text{same as shown for Ordered Ter Ter without the numerator [A] term}$$

$$\frac{[ER]}{[E]_t} = \text{same as shown for Ordered Ter Ter without the numerator R term}$$

$$\frac{[F]}{[E]_t} = \frac{V_r K_{m_C}[A][B] + \dfrac{V_f K_{ip} K_{m_R}[A][B][Q]}{K_{ia} K_{ib} K_{eq}} + \dfrac{V_f K_{m_P}[Q][R]}{K_{eq}} + \dfrac{V_r K_{m_A} K_{ic}[B][Q][R]}{K_{iq} K_{ir}}}{\text{denominator of velocity equation}}$$

(IX-281)

The fraction of the total enzyme present as central complexes cannot be determined.

Effect of Isomerizations

EAB + FP and/or FC + EQR: All the distributions and rate constants shown above may be calculated.

EA: The calculations of k_1, k_{-1}, and $[EA]/[E]_t$ are invalid.

ER: The calculations of k_6, k_{-6}, and $[ER]/[E]_t$ are invalid.

Initial Velocity Studies in the Forward Direction

In the absence of products, the velocity equation for the forward reaction is the same as that shown earlier for the Bi Uni Uni Uni Ping Pong Ter Bi system. Thus C may be identified and V_{max}, K_{m_A}, K_{ia}, K_{m_B}, and K_{m_C} determined as described earlier.

Product Inhibition Studies

In the presence of P, the velocity equation is identical to that shown for the Bi Uni Uni Uni Ping Pong Ter Bi system (when the first Haldane equation is used to eliminate K_{eq}). In the presence of Q, two new terms appear in the denominator of the velocity equation. These appear as shown below.

When [A] is varied:

$$\frac{v}{V_{max}} = \frac{[A]}{K_{m_A}\left(1 + \frac{K_{ia}K_{m_B}}{K_{m_A}[B]}\right) + [A]\left(1 + \frac{K_{m_B}}{[B]} + \frac{K_{m_C}}{[C]} + \frac{K_{m_C}K_{m_R}[Q]}{K_{m_Q}K_{ir}[C]} + \frac{K_{m_R}K_{m_C}[Q]}{K_{ic}K_{m_Q}K_{ir}}\right)}$$

(IX-282)

Thus Q is an uncompetitive inhibitor with respect to A at all concentrations of B and C.

When [B] is varied:

$$\frac{v}{V_{max}} = \frac{[B]}{K_{m_B}\left(1 + \frac{K_{ia}}{[A]}\right) + [B]\left(1 + \frac{K_{m_A}}{[A]} + \frac{K_{m_C}}{[C]} + \frac{K_{m_C}K_{m_R}[Q]}{K_{m_Q}K_{ir}[C]} + \frac{K_{m_C}K_{m_R}[Q]}{K_{ic}K_{m_Q}K_{ir}}\right)}$$

(IX-283)

Q is an uncompetitive inhibitor with respect to B at all A and C concentrations.

When [C] is varied:

$$\frac{v}{V_{max}} = \frac{[C]}{K_{m_C}\left(1 + \frac{K_{m_R}[Q]}{K_{m_Q}K_{ir}}\right) + [C]\left(1 + \frac{K_{m_A}}{[A]} + \frac{K_{m_B}}{[B]} + \frac{K_{ia}K_{m_B}}{[A][B]} + \frac{K_{m_C}K_{m_R}[Q]}{K_{ic}K_{m_Q}K_{ir}}\right)}$$

(IX-284)

Q is a mixed-type inhibitor with respect to C at all A and B concentrations; $K_{i_{\text{slope}}}^{\text{Q}/\text{C}}$ is constant:

$$K_{i_{\text{slope}}}^{\text{Q}/\text{C}} = \frac{K_{m_\text{Q}} K_{ir}}{K_{m_\text{R}}}$$

In the presence of R, two new terms appear in the denominator of the velocity equations.

When [A] is varied:

$$\frac{v}{V_{\max}} = \frac{[\text{A}]}{K_{m_\text{A}}\left(1 + \frac{K_{ia}K_{m_\text{B}}}{K_{m_\text{A}}[\text{B}]} + \frac{K_{ia}K_{m_\text{B}}[\text{R}]}{K_{m_\text{A}}K_{ir}[\text{B}]} + \frac{[\text{R}]}{K_{ir}}\right) + [\text{A}]\left(1 + \frac{K_{m_\text{B}}}{[\text{B}]} + \frac{K_{m_\text{C}}}{[\text{C}]}\right)}$$

(IX-285)

Thus R is competitive with A at unsaturating [B] and all C concentrations.

When [B] is varied:

$$\frac{v}{V_{\max}} = \frac{[\text{B}]}{K_{m_\text{B}}\left(1 + \frac{K_{ia}}{[\text{A}]} + \frac{K_{ia}[\text{R}]}{K_{ir}[\text{A}]}\right) + [\text{B}]\left(1 + \frac{K_{m_\text{A}}}{[\text{A}]} + \frac{K_{m_\text{C}}}{[\text{C}]} + \frac{K_{m_\text{A}}[\text{R}]}{K_{ir}[\text{A}]}\right)}$$

(IX-286)

R is a mixed-type inhibitor with respect to B at unsaturating [A] and all C concentrations.

When [C] is varied:

$$\frac{v}{V_{\max}} = \frac{[\text{C}]}{K_{m_\text{C}} + [\text{C}]\left(1 + \frac{K_{m_\text{A}}}{[\text{A}]} + \frac{K_{m_\text{B}}}{[\text{B}]} + \frac{K_{ia}K_{m_\text{B}}}{[\text{A}][\text{B}]} + \frac{K_{ia}K_{m_\text{B}}[\text{R}]}{K_{ir}[\text{A}][\text{B}]} + \frac{K_{m_\text{A}}[\text{R}]}{K_{ir}[\text{A}]}\right)}$$

(IX-287)

R is uncompetitive with C at unsaturating [A] and all B concentrations. The product inhibition studies permit each substrate and product to be identified. The K_{ip} and K_{ib} can be obtained from studies with P as a product inhibitor, as described earlier for the Bi Uni Uni Uni Ping Pong Ter Bi system; K_{ir} can be obtained from $K_{i_{\text{slope}}}^{\text{R/A}}$:

$$K_{i_{\text{slope}}}^{\text{R/A}} = K_{ir}\left(1 + \frac{K_{m_{\text{A}}}[\text{B}]}{K_{ia}K_{m_{\text{B}}}}\right)$$

If $K_{i_{\text{slope}}}^{\text{R/A}}$ is determined at several B concentrations, the data can be replotted against [B]:

$$K_{i_{\text{slope}}}^{\text{R/A}} = \frac{K_{ir}K_{m_{\text{A}}}}{K_{ia}K_{m_{\text{B}}}}[\text{B}] + K_{ir} \qquad \text{(IX-288)}$$

The intercept on the vertical axis gives K_{ir}. The ratio $K_{m_{\text{Q}}}/K_{m_{\text{R}}}$ can be calculated from $K_{i_{\text{spe}}}^{\text{Q/C}}$. The individual kinetic constants, K_{ic}, $K_{m_{\text{P}}}$, $K_{m_{\text{Q}}}$, $K_{m_{\text{R}}}$, and K_{iq}, can be obtained from initial velocity and product inhibition studies of the reverse reaction, which is symmetrical to the forward reaction.

X. BI BI UNI UNI PING PONG TER TER SYSTEM

Reaction Sequence

A second Ter Ter Ping Pong mechanism is the Bi Bi Uni Uni mechanism shown below.

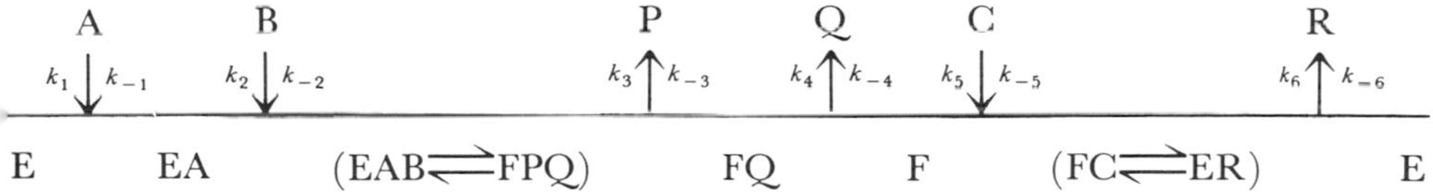

A (k_1 ↓ k_{-1}) B (k_2 ↓ k_{-2}) P (k_3 ↑ k_{-3}) Q (k_4 ↑ k_{-4}) C (k_5 ↓ k_{-5}) R (k_6 ↑ k_{-6})

E EA (EAB ⇌ FPQ) FQ F (FC ⇌ ER) E

The sequence of the reverse reaction and the sequence of the forward reaction starting with stable form F is Uni Uni Bi Bi:

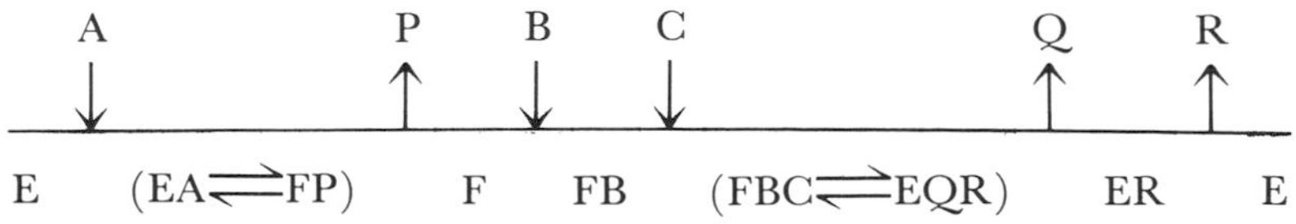

A ↓ P ↑ B ↓ C ↓ Q ↑ R ↑

E (EA ⇌ FP) F FB (FBC ⇌ EQR) ER E

Thus the two sequences are mechanistically and formally equivalent.

Velocity Equation, Kinetic Constants, and Haldane Equations

The velocity equation of the Bi Bi Uni Uni sequence is:

$$\frac{v}{[\mathrm{E}]_t} = \frac{k_1k_2k_3k_4k_5k_6[\mathrm{A}][\mathrm{B}][\mathrm{C}] - k_{-1}k_{-2}k_{-3}k_{-4}k_{-5}k_{-6}[\mathrm{P}][\mathrm{Q}][\mathrm{R}]}{k_{-1}k_4k_5k_6(k_{-2}+k_3)[\mathrm{C}] + k_1k_2k_3k_4(k_{-5}+k_{-6})[\mathrm{A}][\mathrm{B}]}$$

$$+ k_1k_4k_5k_6(k_{-2}+k_3)[\mathrm{A}][\mathrm{C}] + k_2k_3k_4k_5k_6[\mathrm{B}][\mathrm{C}]$$

$$+ k_1k_2k_5(k_3k_6+k_3k_4+k_4k_6)[\mathrm{A}][\mathrm{B}][\mathrm{C}] + k_{-1}k_4k_{-5}k_{-6}(k_{-2}+k_3)[\mathrm{R}]$$

$$+ k_{-1}k_{-2}k_{-3}k_{-4}(k_{-5}+k_6)[\mathrm{P}][\mathrm{Q}] + k_{-1}k_2k_{-3}k_{-5}k_{-6}[\mathrm{P}][\mathrm{R}]$$

$$+ k_{-1}k_{-4}k_{-5}k_{-6}(k_{-2}+k_3)[\mathrm{Q}][\mathrm{R}]$$

$$+ k_{-3}k_{-4}k_{-6}(k_{-1}k_{-2}+k_{-1}k_{-5}+k_{-2}k_{-5})[\mathrm{P}][\mathrm{Q}][\mathrm{R}]$$

$$+ k_2k_3k_4k_{-5}k_{-6}[\mathrm{B}][\mathrm{R}] + k_{-1}k_4k_5k_{-6}(k_{-2}+k_3)[\mathrm{C}][\mathrm{R}] + k_{-1}k_{-2}k_{-3}k_5k_6[\mathrm{C}][\mathrm{P}]$$

$$+ k_1k_2k_3k_{-4}(k_{-5}+k_6)[\mathrm{A}][\mathrm{B}][\mathrm{Q}] + k_1k_{-2}k_{-3}k_5k_6[\mathrm{A}][\mathrm{C}][\mathrm{P}]$$

$$+ k_2k_3k_4k_5k_{-6}[\mathrm{B}][\mathrm{C}][\mathrm{R}] + k_1k_{-2}k_{-3}k_{-4}(k_{-5}+k_6)[\mathrm{A}][\mathrm{P}][\mathrm{Q}]$$

$$+ k_{-1}k_{-2}k_{-3}k_5k_{-6}[\mathrm{C}][\mathrm{P}][\mathrm{R}] + k_2k_3k_{-4}k_{-5}k_{-6}[\mathrm{B}][\mathrm{Q}][\mathrm{R}]$$

$$+ k_1k_2k_{-3}k_5k_6[\mathrm{A}][\mathrm{B}][\mathrm{C}][\mathrm{P}] + k_1k_2k_{-3}k_{-4}(k_{-5}+k_6)[\mathrm{A}][\mathrm{B}][\mathrm{P}][\mathrm{Q}]$$

$$+ k_2k_{-3}k_{-4}k_{-5}k_{-6}[\mathrm{B}][\mathrm{P}][\mathrm{Q}][\mathrm{R}]$$

(IX-289)

In terms of kinetic constants, the equation becomes:

$$
= \frac{V_f V_r\left([A][B][C] - \frac{[P][Q][R]}{K_{eq}}\right)}{\begin{aligned}& V_r K_{ia} K_{m_B}[C] + V_r K_{m_C}[A][B] + V_r K_{m_B}[A][C] + V_r K_{m_A}[B][C] \\
& + V_r[A][B][C] + \frac{V_f K_{iq} K_{m_P}[R]}{K_{eq}} + \frac{V_f K_{m_R}[P][Q]}{K_{eq}} + \frac{V_f K_{m_Q}[P][R]}{K_{eq}} + \frac{V_f K_{m_P}[Q][R]}{K_{eq}} \\
& + \frac{V_f[P][Q][R]}{K_{eq}} + \frac{V_r K_{m_A} K_{ic}[B][R]}{K_{ir}} + \frac{V_r K_{ia} K_{m_B}[C][R]}{K_{ir}} + \frac{V_f K_{m_Q} K_{ir}[C][P]}{K_{ic} K_{eq}} \\
& + \frac{V_f K_{ip} K_{m_R}[A][B][Q]}{K_{ia} K_{ib} K_{eq}} + \frac{V_f K_{m_Q} K_{ir}[A][C][P]}{K_{ia} K_{ic} K_{eq}} + \frac{V_r K_{m_A}[B][C][R]}{K_{ir}} \\
& + \frac{V_f K_{m_R}[A][P][Q]}{K_{ia} K_{eq}} + \frac{V_f K_{m_Q}[C][P][R]}{K_{ic} K_{eq}} + \frac{V_r K_{m_A} K_{ic}[B][Q][R]}{K_{iq} K_{ir}} \\
& + \frac{V_f K_{ir} K_{m_Q}[A][B][C][P]}{K_{ia} K_{ib} K_{ic} K_{eq}} + \frac{V_f K_{m_R}[A][B][P][Q]}{K_{ia} K_{ib} K_{eq}} + \frac{V_r K_{m_A} K_{ic}[B][P][Q][R]}{K_{ip} K_{iq} K_{ir}}\end{aligned}}
\tag{IX-290}
$$

where K_{m_A}, K_{m_B}, K_{m_C}, K_{m_P}, K_{m_Q}, K_{m_R}, $V_{\max_f}$, and $V_{\max_r}$ are defined in the usual way and

$$K_{ia} = \frac{\text{Coef}_{C}}{\text{Coef}_{AC}} = \frac{\text{Coef}_{PQ}}{\text{Coef}_{APQ}} = \frac{\text{Coef}_{CP}}{\text{Coef}_{ACP}} = \frac{k_{-1}}{k_1}$$

$$K_{ib} = \frac{\text{Coef}_{APQ}}{\text{Coef}_{ABPQ}} = \frac{\text{Coef}_{ACP}}{\text{Coef}_{ABCP}} = \frac{k_{-2}}{k_2}$$

$$K_{ic} = \frac{\text{Coef}_{BR}}{\text{Coef}_{BCR}} = \frac{\text{Coef}_{R}}{\text{Coef}_{CR}} = \frac{\text{Coef}_{PR}}{\text{Coef}_{CPR}} = \frac{k_{-5}}{k_5}$$

$$K_{ip} = \frac{\text{Coef}_{ABQ}}{\text{Coef}_{ABPQ}} = \frac{\text{Coef}_{BQR}}{\text{Coef}_{BPQR}} = \frac{k_3}{k_{-3}}$$

$$K_{iq} = \frac{\text{Coef}_{R}}{\text{Coef}_{QR}} = \frac{\text{Coef}_{BR}}{\text{Coef}_{BQR}} = \frac{\text{Coef}_{AB}}{\text{Coef}_{ABQ}} = \frac{k_4}{k_{-4}}$$

$$K_{ir} = \frac{\text{Coef}_{C}}{\text{Coef}_{CR}} = \frac{\text{Coef}_{BC}}{\text{Coef}_{BCR}} = \frac{\text{Coef}_{CP}}{\text{Coef}_{CPR}} = \frac{k_6}{k_{-6}}$$

As in the previous terreactant systems considered, all the K_i constants are true dissociation constants.

The Haldane equations are:

$$K_{eq} = \frac{V_f K_{m_P} K_{iq} K_{ir}}{V_r K_{ia} K_{m_B} K_{ic}} = \frac{V_f K_{ip} K_{iq} K_{m_R}}{V_r K_{ia} K_{ib} K_{m_C}} = \frac{K_{ip} K_{iq} K_{ir}}{K_{ia} K_{ib} K_{ic}} = \left(\frac{V_f}{V_r}\right)^2 \frac{K_{m_P} K_{iq} K_{m_R}}{K_{ia} K_{m_B} K_{m_C}}$$

(IX-291)

From the first two Haldane equations we see that:

$$\frac{K_{ip} K_{m_R}}{K_{ib} K_{m_C}} = \frac{K_{m_P} K_{ir}}{K_{m_B} K_{ic}} \qquad \text{(IX-292)}$$

Rate Constants

$$k_1 = \frac{V_{\max_f}}{[E]_t K_{m_A}}, \qquad k_{-1} = \frac{V_{\max_f} K_{ia}}{[E]_t K_{m_A}}, \qquad k_4 = \frac{V_{\max_r} K_{iq}}{[E]_t K_{m_Q}}, \qquad k_{-4} = \frac{V_{\max_r}}{[E]_t K_{m_Q}}$$

The remaining rate constants cannot be calculated from the kinetic constants.

Distribution Equations

$$\frac{[E]}{[E]_t} = \frac{V_r K_{ia} K_{m_B}[C] + V_r K_{m_A}[B][C] + \dfrac{V_f K_{m_R}[P][Q]}{K_{eq}} + \dfrac{V_f K_{m_Q} K_{ir}[C][P]}{K_{ic} K_{eq}}}{\text{denominator of velocity equation}}$$

(IX-293)

$$\frac{[EA]}{[E]_t} = \frac{V_r K_{m_B}[A][C] + \dfrac{V_f K_{m_A}[P][Q][R]}{K_{ia} K_{eq}} + \dfrac{V_f K_{m_Q} K_{ir}[A][C][P]}{K_{ia} K_{ic} K_{eq}} + \dfrac{V_f K_{m_R}[A][P][Q]}{K_{ia} K_{eq}}}{\text{denominator of velocity equation}}$$

(IX-294)

$$\frac{[FQ]}{[E]_t} = \frac{\dfrac{V_f K_{m_Q}[A][B][C]}{K_{iq}} + \dfrac{V_f K_{m_P}[Q][R]}{K_{eq}} + \dfrac{V_f K_{ip} K_{m_R}[A][B][Q]}{K_{ia} K_{ib} K_{eq}} + \dfrac{V_r K_{m_A} K_{ic}[B][Q][R]}{K_{iq} K_{ir}}}{\text{denominator of velocity equation}}$$

(IX-295)

$$\frac{[F]}{[E]_t} = \frac{V_r K_{m_C}[A][B] + \dfrac{V_f K_{iq} K_{m_P}[R]}{K_{eq}} + \dfrac{V_f K_{m_Q}[P][R]}{K_{eq}} + \dfrac{V_r K_{m_A} K_{ic}[B][R]}{K_{ir}}}{\text{denominator of velocity equation}}$$

(IX-296)

The proportion of the total enzyme present as central complexes cannot be determined.

Effect of Isomerizations

EAB + FPQ and/or FC + ER: All the rate constants and distributions shown can be calculated.

EA: The calculations of k_1, k_{-1}, and $[EA]/[E]_t$ are invalid.

FQ: The calculations of k_4, k_{-4}, and $[FQ]/[E]_t$ are invalid.

Initial Velocity Studies in the Forward Direction

In the absence of products, the velocity equation for the forward reaction is the same as that for the Bi Uni Uni Bi Ping Pong Ter Ter system. The two Ping Pong systems can be distinguished from each other by product inhibition studies.

Product Inhibition Studies

In the presence of P, the denominator of the velocity equation contains three new terms, which, after eliminating K_{eq} by means of the first Haldane equation, appear as shown below.

When [A] is varied:

$$\frac{v}{V_{max}} = \frac{[A]}{K_{m_A}\left(1 + \frac{K_{ia}K_{m_B}}{K_{m_A}[B]} + \frac{K_{ia}K_{m_B}K_{m_Q}[P]}{K_{m_A}K_{m_P}K_{iq}[B]}\right) + [A]\left(1 + \frac{K_{m_B}}{[B]} + \frac{K_{m_C}}{[C]} + \frac{K_{m_B}K_{m_Q}[P]}{K_{m_P}K_{iq}[B]} + \frac{K_{m_B}K_{m_Q}[P]}{K_{ib}K_{m_P}K_{iq}}\right)}$$

(IX-297)

When [B] is varied:

$$\frac{v}{V_{max}} = \frac{[B]}{K_{m_B}\left(1 + \frac{K_{ia}}{[A]}\right)\left(1 + \frac{K_{m_B}K_{m_Q}[P]}{K_{m_A}K_{m_P}K_{iq}}\right) + [B]\left(1 + \frac{K_{m_A}}{[A]} + \frac{K_{m_C}}{[C]} + \frac{K_{m_B}K_{m_Q}[P]}{K_{ib}K_{m_P}K_{iq}}\right)}$$

(IX-298)

When [C] is varied:

$$\frac{v}{V_{\max}} = \frac{[\mathrm{C}]}{K_{m_\mathrm{C}} + [\mathrm{C}]\left(1 + \frac{K_{m_\mathrm{A}}}{[\mathrm{A}]} + \frac{K_{m_\mathrm{B}}}{[\mathrm{B}]} + \frac{K_{ia}K_{m_\mathrm{B}}}{[\mathrm{A}][\mathrm{B}]} + \frac{K_{ia}K_{m_\mathrm{B}}K_{m_\mathrm{Q}}[\mathrm{P}]}{K_{m_\mathrm{P}}K_{iq}[\mathrm{A}][\mathrm{B}]} + \frac{K_{m_\mathrm{B}}K_{m_\mathrm{Q}}[\mathrm{P}]}{K_{m_\mathrm{P}}K_{iq}[\mathrm{B}]} + \frac{K_{m_\mathrm{B}}K_{m_\mathrm{Q}}[\mathrm{P}]}{K_{ib}K_{m_\mathrm{P}}K_{iq}}\right)} \tag{IX-299}$$

Thus P is a mixed-type inhibitor with respect to A at unsaturating [B] and all C concentrations, and an uncompetitive inhibitor at saturating [B]; P is a mixed-type inhibitor with respect to B at all A and C concentrations and an uncompetitive inhibitor with respect to C at all A and B concentrations. The last inhibition pattern distinguishes the Bi Bi Uni Uni Ping Pong system from the Bi Uni Uni Bi system. In the latter system, no product yields uncompetitive patterns with respect to C at all concentrations of A and B, while in the Bi Bi Uni Uni system no product yields a mixed-type pattern with respect to C at all A and B concentrations. (Keep in mind that C can be identified from initial velocity studies in the absence of products.)

In the presence of Q, a single new term appears in the velocity equation. When A is the varied substrate and the second Haldane equation is used to eliminate K_{eq}, the term is:

$\dfrac{K_{m_\mathrm{C}}[\mathrm{A}][\mathrm{Q}]}{K_{iq}[\mathrm{C}]}$, which appears as part of the intercept factor when [A] is factored out

When B is the varied substrate, the term is

$\dfrac{K_{m_\mathrm{C}}[\mathrm{B}][\mathrm{Q}]}{K_{iq}[\mathrm{C}]}$, which appears as part of the intercept factor when [B] is factored out

When C is the varied substrate, the term is

$\dfrac{K_{m_\mathrm{C}}[\mathrm{Q}]}{K_{iq}}$, which introduces a $[\mathrm{Q}]/K_{iq}$ term into the slope factor

Thus Q is an uncompetitive inhibitor with respect to A at all B concentrations. Saturation with C overcomes the inhibition; Q is also uncompetitive with respect to B at all A concentrations. Again, saturation with C overcomes the inhibition; Q is competitive with resepct to C at all A and B concentrations. The $K_{i_{slope}}^{Q/C}$ always equals K_{iq}. With K_{iq} known (and also K_{m_A}, K_{m_B}, and K_{ia}), the ratio K_{m_P}/K_{m_Q} can be calculated from $K_{i_{slope}}^{P/A}$ or $K_{i_{slope}}^{P/B}$.

$$K_{i_{slope}}^{P/A} = \frac{K_{m_P} K_{iq}}{K_{m_Q}} \left(1 + \frac{K_{m_A}[B]}{K_{ia} K_{m_B}} \right) \quad \text{and} \quad K_{i_{slope}}^{P/B} = \frac{K_{m_A} K_{m_P} K_{iq}}{K_{m_B} K_{m_Q}}$$

With the ratio of K_{m_P}/K_{m_Q} known, K_{ib} can be calculated from the slope of the intercept replot of $1/v$ versus $1/[B]$ data obtained in the presence of P. The intercept is given by:

$$\frac{1}{V_{max_{app}}} = \frac{K_{m_B} K_{m_Q}}{K_{ib} K_{m_P} K_{iq} V_{max}} [P] + \frac{1}{V_{max}} \left(1 + \frac{K_{m_A}}{[A]} + \frac{K_{m_C}}{[C]} \right) \qquad \text{(IX-300)}$$

The intercept on the horizontal axis gives $K_{i_{int}}^{P/B}$:

$$K_{i_{int}}^{P/B} = \frac{K_{ib} K_{m_P} K_{iq} \left(1 + \frac{K_{m_A}}{[A]} + \frac{K_{m_C}}{[C]} \right)}{K_{m_B} K_{m_Q}}$$

In the presence of R, four new terms appear in the denominator of the velocity equation. The first Haldane equation is used to eliminate K_{eq}.

When [A] is varied, all four [R] terms appear in the slope factor. The equation can be factored to:

$$\frac{v}{V_{max}} = \frac{[A]}{K_{m_A} \left(1 + \frac{K_{ia} K_{m_B}}{K_{m_A}[B]} \right) \left(1 + \frac{K_{ic}[R]}{K_{ir}[C]} + \frac{[R]}{K_{ir}} \right) + [A] \left(1 + \frac{K_{m_B}}{[B]} + \frac{K_{m_C}}{[C]} \right)} \qquad \text{(IX-301)}$$

Thus R is competitive with respect to A at all B and C concentrations and

$$K_{i_{slope}}^{R/A} = \frac{K_{ir}}{\left(1 + \frac{K_{ic}}{[C]} \right)}$$

A replot of $1/K_{i_{slope}}^{R/A}$ versus $1/[C]$ is given by:

$$\frac{1}{K_{i_{slope}}^{R/A}} = \frac{K_{ic}}{K_{ir}} \frac{1}{[C]} + \frac{1}{K_{ir}} \qquad \text{(IX-302)}$$

Thus K_{ir} and K_{ic} can be determined.

When [B] is varied:

$$\frac{v}{V_{\max}} = \frac{[B]}{K_{m_B}\left(1 + \frac{K_{ia}}{[A]} + \frac{K_{ia}K_{m_B}K_{ic}[R]}{K_{m_A}K_{ir}[A][C]} + \frac{K_{ia}K_{m_B}[R]}{K_{m_A}K_{ir}[A]}\right) + [B]\left(1 + \frac{K_{m_A}}{[A]} + \frac{K_{m_C}}{[C]} + \frac{K_{m_A}K_{ic}[R]}{K_{ir}[A][C]} + \frac{K_{m_A}[R]}{K_{ir}[A]}\right)} \tag{IX-303}$$

R is a mixed-type inhibitor with respect to B at all C concentrations and unsaturating [A]. Saturation with A eliminates the inhibition.

When [C] is varied:

$$\frac{v}{V_{\max}} = \frac{[C]}{K_{m_C}\left(1 + \frac{K_{ia}K_{m_B}K_{ic}[R]}{K_{m_C}K_{ir}[A][B]} + \frac{K_{m_A}K_{ic}[R]}{K_{m_C}K_{ir}[A]}\right) + [C]\left(1 + \frac{K_{m_A}}{[A]} + \frac{K_{m_B}}{[B]} + \frac{K_{ia}K_{m_B}}{[A][B]} + \frac{K_{ia}K_{m_B}[R]}{K_{ir}[A][B]} + \frac{K_{m_A}[R]}{K_{ir}[A]}\right)} \tag{IX-304}$$

R is a mixed-type inhibitor with respect to C to all B concentrations and unstaurating [A]. Saturation with A eliminates the inhibition. The remaining kinetic constants can be obtained from studies of the reverse reaction. The ratio $K_{ip}K_{m_R}/K_{m_P}$ can be calculated from equation IX-292.

Reverse Direction—Uni Uni Bi Bi Ping Pong Ter Ter

The velocity equations for the forward and reverse reactions are shown below.

Forward (Bi Bi Uni Uni)

$$\frac{v}{V_{\max}} = \frac{[A][B][C]}{K_{ia}K_{m_B}[C] + K_{m_C}[A][B] + K_{m_B}[A][C] + K_{m_A}[B][C] + [A][B][C]} \tag{IX-305}$$

Reverse (Uni Uni Bi Bi)

$$\frac{v}{V_{max}} = \frac{[P][Q][R]}{K_{iq}K_{m_P}[R] + K_{m_R}[P][Q] + K_{m_Q}[P][R] + K_{m_P}[Q][R] + [P][Q][R]}$$

(IX-306)

If the Uni Uni Bi Bi sequence is written as "substrates" (i.e., A, B, C) going to "products" (i.e., P, Q, R), the equation becomes:

$$\frac{v}{V_{max}} = \frac{[A][B][C]}{K_{ib}K_{m_C}[A] + K_{m_A}[B][C] + K_{m_B}[A][C] + K_{m_C}[A][B] + [A][B][C]}$$

(IX-307)

All three equations have exactly the same form, even though the forward and reverse reactions are not symmetrical. The forms are identical because, as pointed out earlier, the Uni Uni Bi Bi sequence is equivalent to the Bi Bi Uni Uni sequence if the repeating pattern is started with stable form F. The K_m and K_i values can be determined by the same procedures as described for the "forward" reaction. For example, when P, Q, or R is the substrate, A is the product that introduces the single new term to the denominator of the velocity equation and $K_{i_{slope}}^{A/R} = K_{ia}$. Thus product A in the reverse direction is equivalent to product Q in the forward direction. (Both are products released immediately before the Uni Uni steps.) Substrate R in the reverse direction is equivalent to substrate C of the forward direction. (Both are substrates that add immediately after the Bi Bi steps.) This is more easily seen if we write the repeating sequence.

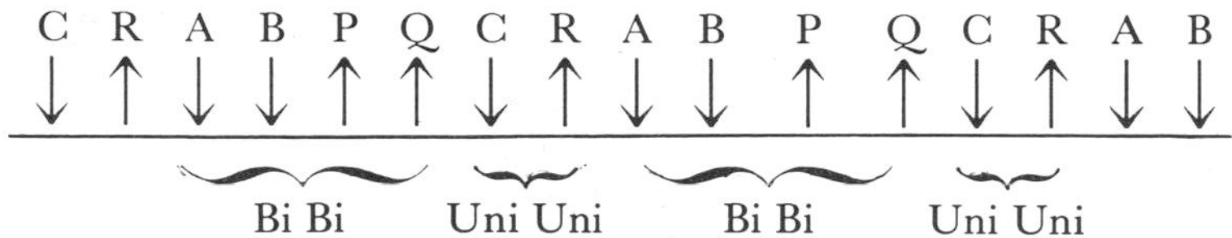

Y. HEXA UNI PING PONG SYSTEM

Velocity Equation, Kinetic Constants, and Haldane Equations

The third basic Ter Ter Ping Pong mechanism is Hexa Uni Ping Pong. The reaction sequence is the same regardless of which stable form is chosen as the start. A product is released after each substrate addition.

$$\mathrm{E}\xrightarrow[k_{-1}]{k_1\,\mathrm{A}\downarrow}(\mathrm{EA}\rightleftharpoons\mathrm{FP})\xrightarrow[k_{-2}]{k_2\,\mathrm{P}\uparrow}\mathrm{F}\xrightarrow[k_{-3}]{k_3\,\mathrm{B}\downarrow}(\mathrm{FB}\rightleftharpoons\mathrm{GQ})\xrightarrow[k_{-4}]{k_4\,\mathrm{Q}\uparrow}\mathrm{G}\xrightarrow[k_{-5}]{k_5\,\mathrm{C}\downarrow}(\mathrm{GC}\rightleftharpoons\mathrm{ER})\xrightarrow[k_{-6}]{k_6\,\mathrm{R}\uparrow}\mathrm{E}$$

The complete velocity equation contains 17 terms:

$$\frac{v}{[\mathrm{E}]_t}=\frac{k_1k_2k_3k_4k_5k_6[\mathrm{A}][\mathrm{B}][\mathrm{C}]-k_{-1}k_{-2}k_{-3}k_{-4}k_{-5}k_{-6}[\mathrm{P}][\mathrm{Q}][\mathrm{R}]}{\begin{array}{l}k_1k_2k_3k_4(k_{-5}+k_6)[\mathrm{A}][\mathrm{B}]+k_1k_2k_5k_6(k_{-3}+k_4)[\mathrm{A}][\mathrm{C}]+k_3k_4k_5k_6(k_{-1}+k_2)[\mathrm{B}][\mathrm{C}]\\ +k_1k_3k_5(k_2k_4+k_2k_6+k_4k_6)[\mathrm{A}][\mathrm{B}][\mathrm{C}]+k_{-1}k_{-2}k_{-3}k_{-4}(k_{-5}+k_6)[\mathrm{P}][\mathrm{Q}]\\ +k_{-1}k_{-2}k_{-5}k_{-6}(k_{-3}+k_4)[\mathrm{P}][\mathrm{R}]+k_{-3}k_{-4}k_{-5}k_{-6}(k_{-1}+k_2)[\mathrm{Q}][\mathrm{R}]\\ +k_{-2}k_{-4}k_{-6}(k_{-1}k_{-3}+k_{-1}k_{-5}+k_{-3}k_{-5})[\mathrm{P}][\mathrm{Q}][\mathrm{R}]+k_1k_2k_{-3}k_{-4}(k_{-5}+k_6)[\mathrm{A}][\mathrm{Q}]\\ +k_3k_4k_{-5}k_{-6}(k_{-1}+k_2)[\mathrm{B}][\mathrm{R}]+k_{-1}k_{-2}k_5k_6(k_{-3}+k_4)[\mathrm{C}][\mathrm{P}]\\ +k_1k_2k_3k_{-4}(k_{-5}+k_6)[\mathrm{A}][\mathrm{B}][\mathrm{Q}]+k_1k_{-2}k_5k_6(k_{-3}+k_4)[\mathrm{A}][\mathrm{C}][\mathrm{P}]\\ +k_3k_4k_5k_{-6}(k_{-1}+k_2)[\mathrm{B}][\mathrm{C}][\mathrm{R}]+k_1k_{-2}k_{-3}k_{-4}(k_{-5}+k_6)[\mathrm{A}][\mathrm{P}][\mathrm{Q}]\\ +k_3k_{-4}k_{-5}k_{-6}(k_{-1}+k_2)[\mathrm{B}][\mathrm{Q}][\mathrm{R}]+k_{-1}k_{-2}k_5k_{-6}(k_{-3}+k_4)[\mathrm{C}][\mathrm{P}][\mathrm{R}]\end{array}} \tag{IX-308}$$

In terms of kinetic constants, the velocity equation is:

$$v=\frac{V_fV_r\left([\mathrm{A}][\mathrm{B}][\mathrm{C}]-\dfrac{[\mathrm{P}][\mathrm{Q}][\mathrm{R}]}{K_{\mathrm{eq}}}\right)}{\begin{array}{l}V_rK_{m_\mathrm{C}}[\mathrm{A}][\mathrm{B}]+V_rK_{m_\mathrm{B}}[\mathrm{A}][\mathrm{C}]+V_rK_{m_\mathrm{A}}[\mathrm{B}][\mathrm{C}]+V_r[\mathrm{A}][\mathrm{B}][\mathrm{C}]\\ +\dfrac{V_fK_{m_\mathrm{R}}[\mathrm{P}][\mathrm{Q}]}{K_{\mathrm{eq}}}+\dfrac{V_fK_{m_\mathrm{Q}}[\mathrm{P}][\mathrm{R}]}{K_{\mathrm{eq}}}+\dfrac{V_fK_{m_\mathrm{P}}[\mathrm{Q}][\mathrm{R}]}{K_{\mathrm{eq}}}+\dfrac{V_f[\mathrm{P}][\mathrm{Q}][\mathrm{R}]}{K_{\mathrm{eq}}}\\ +\dfrac{V_fK_{ip}K_{m_\mathrm{R}}[\mathrm{A}][\mathrm{Q}]}{K_{ia}K_{\mathrm{eq}}}+\dfrac{V_rK_{m_\mathrm{A}}K_{ic}[\mathrm{B}][\mathrm{R}]}{K_{ir}}+\dfrac{V_fK_{m_\mathrm{Q}}K_{ir}[\mathrm{C}][\mathrm{P}]}{K_{ic}K_{\mathrm{eq}}}\\ +\dfrac{V_fK_{ip}K_{m_\mathrm{R}}[\mathrm{A}][\mathrm{B}][\mathrm{Q}]}{K_{ia}K_{ib}K_{\mathrm{eq}}}+\dfrac{V_fK_{m_\mathrm{Q}}K_{ir}[\mathrm{A}][\mathrm{C}][\mathrm{P}]}{K_{ia}K_{ic}K_{\mathrm{eq}}}+\dfrac{V_rK_{m_\mathrm{A}}[\mathrm{B}][\mathrm{C}][\mathrm{R}]}{K_{ir}}\\ +\dfrac{V_fK_{m_\mathrm{R}}[\mathrm{A}][\mathrm{P}][\mathrm{Q}]}{K_{ia}K_{\mathrm{eq}}}+\dfrac{V_rK_{m_\mathrm{A}}K_{ic}[\mathrm{B}][\mathrm{Q}][\mathrm{R}]}{K_{iq}K_{ir}}+\dfrac{V_fK_{m_\mathrm{Q}}[\mathrm{C}][\mathrm{P}][\mathrm{R}]}{K_{ic}K_{\mathrm{eq}}}\end{array}}$$

where the K_m values are defined in the usual way and

$$K_{ia} = \frac{\text{Coef}_{\text{CP}}}{\text{Coef}_{\text{ACP}}} = \frac{\text{Coef}_{\text{PQ}}}{\text{Coef}_{\text{APQ}}} = \frac{k_{-1}}{k_1}, \qquad K_{ib} = \frac{\text{Coef}_{\text{AQ}}}{\text{Coef}_{\text{ABQ}}} = \frac{\text{Coef}_{\text{QR}}}{\text{Coef}_{\text{BQR}}} = \frac{k_{-3}}{k_3}$$

$$K_{ic} = \frac{\text{Coef}_{\text{BR}}}{\text{Coef}_{\text{BCR}}} = \frac{\text{Coef}_{\text{PR}}}{\text{Coef}_{\text{CPR}}} = \frac{k_{-5}}{k_5}, \qquad K_{ip} = \frac{\text{Coef}_{\text{AC}}}{\text{Coef}_{\text{ACP}}} = \frac{\text{Coef}_{\text{AQ}}}{\text{Coef}_{\text{APQ}}} = \frac{k_2}{k_{-2}}$$

$$K_{iq} = \frac{\text{Coef}_{\text{AB}}}{\text{Coef}_{\text{ABQ}}} = \frac{\text{Coef}_{\text{BR}}}{\text{Coef}_{\text{BQR}}} = \frac{k_4}{k_{-4}}$$

As in the other terreactant systems, all the inhibition constants represent true dissociation constants.

Some of the Haldane equations are:

$$K_{\text{eq}} = \frac{V_f K_{ip} K_{m_\text{Q}} K_{ir}}{V_r K_{ia} K_{m_\text{B}} K_{ic}} = \frac{V_f K_{ip} K_{iq} K_{m_\text{R}}}{V_r K_{ia} K_{ib} K_{m_\text{C}}} = \frac{V_f K_{m_\text{P}} K_{iq} K_{ir}}{V_r K_{m_\text{A}} K_{ib} K_{ic}} = \frac{K_{ip} K_{iq} K_{ir}}{K_{ia} K_{ib} K_{ic}} \qquad \text{(IX-310)}$$

We see from the first three Haldane equations that:

$$\frac{K_{ip} K_{m_\text{R}}}{K_{ia} K_{m_\text{C}}} = \frac{K_{m_\text{P}} K_{ir}}{K_{m_\text{A}} K_{ic}}, \qquad \frac{K_{m_\text{P}} K_{iq}}{K_{m_\text{A}} K_{ib}} = \frac{K_{ip} K_{m_\text{Q}}}{K_{ia} K_{m_\text{B}}}, \qquad \text{and} \qquad \frac{K_{iq} K_{m_\text{R}}}{K_{ib} K_{m_\text{C}}} = \frac{K_{m_\text{Q}} K_{ir}}{K_{m_\text{B}} K_{ic}}$$

(IX-311)

Rate Constants

None of the rate constants can be calculated.

Distribution Equations

$$\frac{[\text{E}]}{[\text{E}]_t} = \frac{V_r K_{m_\text{A}}[\text{B}][\text{C}] + \dfrac{V_f K_{m_\text{Q}} K_{ir}[\text{C}][\text{P}]}{K_{ic} K_{\text{eq}}} + \dfrac{V_f K_{m_\text{R}}[\text{P}][\text{Q}]}{K_{\text{eq}}}}{\text{denominator of velocity equation}} \qquad \text{(IX-312)}$$

$$\frac{[\text{F}]}{[\text{E}]_t} = \frac{V_r K_{m_\text{B}}[\text{A}][\text{C}] + \dfrac{V_f K_{ip} K_{m_\text{R}}[\text{A}][\text{Q}]}{K_{ia} K_{\text{eq}}} + \dfrac{V_f K_{m_\text{P}}[\text{Q}][\text{R}]}{K_{\text{eq}}}}{\text{denominator of velocity equation}} \qquad \text{(IX-313)}$$

$$\frac{[\text{G}]}{[\text{E}]_t} = \frac{V_r K_{m_\text{C}}[\text{A}][\text{B}] + \dfrac{V_r K_{m_\text{A}} K_{ic}[\text{B}][\text{R}]}{K_{ir}} + \dfrac{V_f K_{m_\text{Q}}[\text{P}][\text{R}]}{K_{\text{eq}}}}{\text{denominator of velocity equation}} \qquad \text{(IX-314)}$$

The proportion of the enzyme existing as central complexes cannot be calculated.

Effect of Isomerizations

Isomerization of any of the central complexes has no effect on the distribution equations.

Initial Velocity Studies in the Forward Direction

In the absence of products, the velocity equation is:

$$\frac{v}{V_{\max}} = \frac{[\mathrm{A}][\mathrm{B}][\mathrm{C}]}{K_{m_\mathrm{C}}[\mathrm{A}][\mathrm{B}] + K_{m_\mathrm{B}}[\mathrm{A}][\mathrm{C}] + K_{m_\mathrm{A}}[\mathrm{B}][\mathrm{C}] + [\mathrm{A}][\mathrm{B}][\mathrm{C}]} \qquad \text{(IX-315)}$$

When [A] is varied:

$$\frac{v}{V_{\max}} = \frac{[\mathrm{A}]}{K_{m_\mathrm{A}} + [\mathrm{A}]\left(1 + \dfrac{K_{m_\mathrm{B}}}{[\mathrm{B}]} + \dfrac{K_{m_\mathrm{C}}}{[\mathrm{C}]}\right)} \qquad \text{(IX-316)}$$

When [B] is varied:

$$\frac{v}{V_{\max}} = \frac{[\mathrm{B}]}{K_{m_\mathrm{B}} + [\mathrm{B}]\left(1 + \dfrac{K_{m_\mathrm{A}}}{[\mathrm{A}]} + \dfrac{K_{m_\mathrm{C}}}{[\mathrm{C}]}\right)} \qquad \text{(IX-317)}$$

When [C] is varied:

$$\frac{v}{V_{\max}} = \frac{[\mathrm{C}]}{K_{m_\mathrm{C}} + [\mathrm{C}]\left(1 + \dfrac{K_{m_\mathrm{A}}}{[\mathrm{A}]} + \dfrac{K_{m_\mathrm{B}}}{[\mathrm{B}]}\right)} \qquad \text{(IX-318)}$$

Thus all three substrates yield parallel reciprocal plots at different fixed concentrations of another substrate and a constant concentration of the third substrate. If one substrate is varied at different fixed concentrations of the other two (maintained at a constant ratio), all the plots are parallel with linear intercept replots. For example when [A] is varied and [C] $= x$[B] (Fig. IX-46*a*), the reciprocal plot is given by:

$$\boxed{\frac{1}{v} = \frac{K_{m_A}}{V_{max}}\frac{1}{[A]} + \frac{1}{V_{max}}\left(1 + \frac{xK_{m_B} + K_{m_C}}{x[B]}\right)} \qquad \text{(IX-319)}$$

The intercept replot extrapolates to $1/V_{max}$ as the vertical-axis intercept (Fig. IX-46*b*). With V_{max} known, the K_m values of each substrate can be calculated from the slopes of the reciprocal plots. As an alternate procedure, two substrates can be varied together (at constant ratio) at different fixed concentrations of the third substrate. Again, all three families of reciprocal plots are parallel with linear intercept replots. For example, when A and B are varied together maintaining [B] $= x$[A] (Fig. IX-47*a,b*), the reciprocal equation is:

$$\boxed{\frac{1}{v} = \frac{K_{m_A}}{V_{max}}\left(1 + \frac{K_{m_B}}{xK_{m_A}}\right)\frac{1}{[A]} + \frac{1}{V_{max}}\left(1 + \frac{K_{m_C}}{[C]}\right)} \qquad \text{(IX-320)}$$

The intercept replot gives K_{m_C} and V_{max}. The other K_m values can be obtained in the same way. If all three substrates are varied together maintaining [B] at x[A] and [C] at y[A], the resulting $1/v$ versus $1/[A]$ plot is linear and extrapolates to $1/V_{max}$. The reciprocal equation is:

$$\boxed{\frac{1}{v} = \frac{K_{m_A}}{V_{max}}\left(1 + \frac{K_{m_B}}{xK_{m_A}} + \frac{K_{m_C}}{yK_{m_A}}\right)\frac{1}{[A]} + \frac{1}{V_{max}}} \qquad \text{(IX-321)}$$

Product Inhibition Studies

In the presence of P, the denominator [C][P] and [A][C][P] terms are incorporated into the velocity equation as shown below when the first

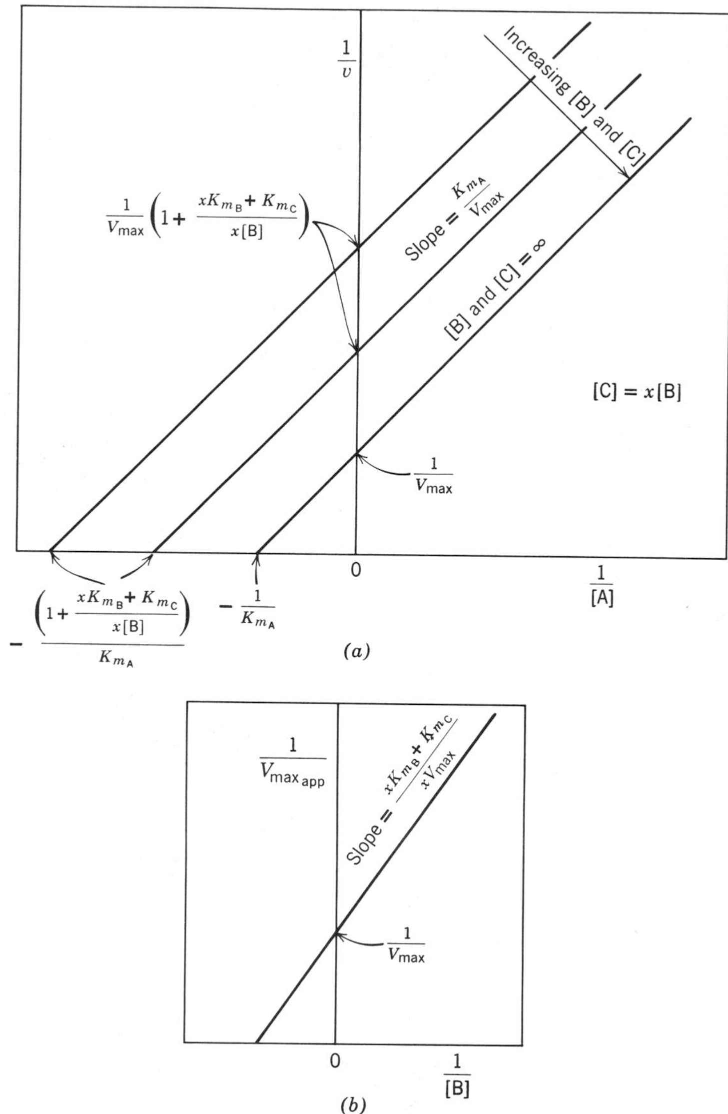

Fig. IX-46. Hexa Uni Ping Pong terreactant system. (*a*) $1/v$ versus 1/[A] at different fixed concentrations of B and C where the ratio of [C]/[B] is maintained constant. (*b*) Intercept replot.

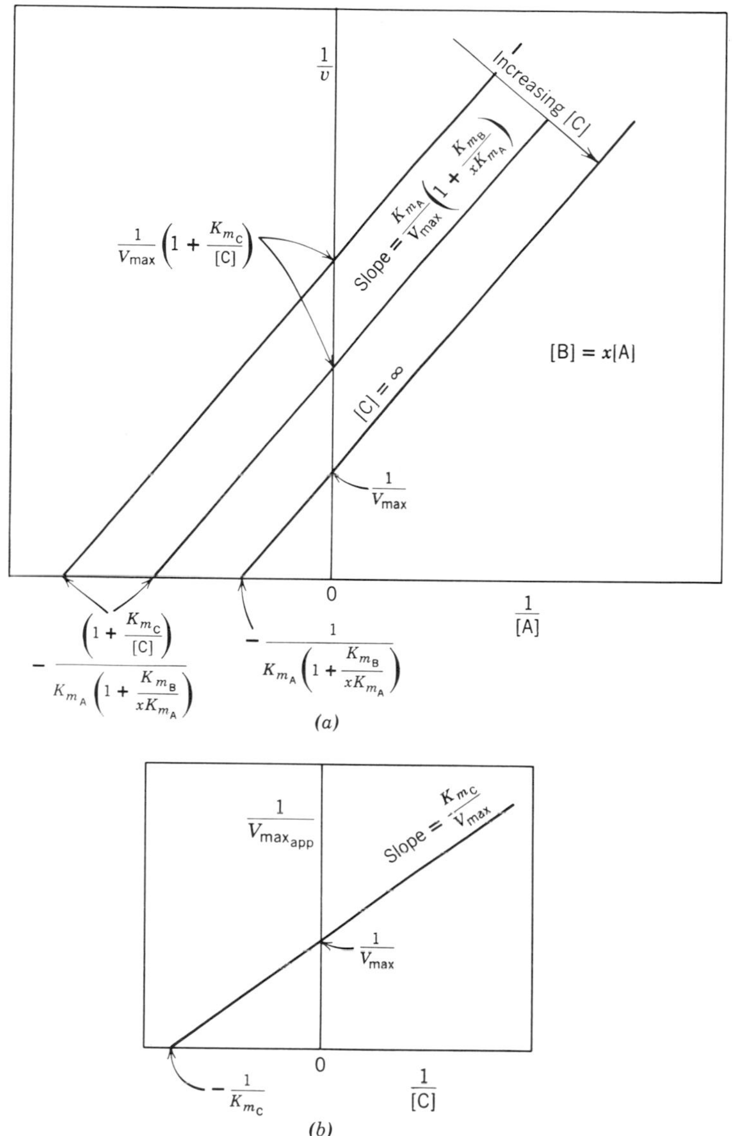

Fig. IX-47. Hexa Uni Ping Pong terreactant system. (*a*) $1/v$ versus 1/[A] at different fixed C concentrations. [B] is varied along with [A] maintaining the [B]/[A] ratio constant. (*b*) Intercept replot.

Haldane equation is used to eliminate K_{eq}.

$$\frac{v}{V_{max}} = \frac{[A]}{K_{m_A}\left(1 + \frac{K_{ia}K_{m_B}[P]}{K_{m_A}K_{ip}[B]}\right) + [A]\left(1 + \frac{K_{m_B}}{[B]} + \frac{K_{m_C}}{[C]} + \frac{K_{m_B}[P]}{K_{ip}[B]}\right)} \quad \text{(IX-322)}$$

$$\frac{v}{V_{max}} = \frac{[B]}{K_{m_B}\left(1 + \frac{K_{ia}[P]}{K_{ip}[A]} + \frac{[P]}{K_{ip}}\right) + [B]\left(1 + \frac{K_{m_A}}{[A]} + \frac{K_{m_C}}{[C]}\right)} \quad \text{(IX-323)}$$

$$\frac{v}{V_{max}} = \frac{[C]}{K_{m_C} + [C]\left(1 + \frac{K_{m_A}}{[A]} + \frac{K_{m_B}}{[B]} + \frac{K_{ia}K_{m_B}[P]}{K_{ip}[A][B]} + \frac{K_{m_B}[P]}{K_{ip}[B]}\right)} \quad \text{(IX-324)}$$

P is a mixed-type inhibitor with respect to A at all C concentrations and unsaturating [B]. Saturation with B eliminates the inhibition. P is competitive with respect to B at all A and B concentrations; P is an uncompetitive inhibitor with respect to C at all A concentrations and unsaturating [B]. The inhibition is overcome when [B] is saturating.

When Q is the product inhibitor and the second Haldane equation is used to eliminate K_{eq}, the equations for A, B, and C as varied substrates are:

$$\frac{v}{V_{max}} = \frac{[A]}{K_{m_A} + [A]\left(1 + \frac{K_{m_B}}{[B]} + \frac{K_{m_C}}{[C]} + \frac{K_{ib}K_{m_C}[Q]}{K_{iq}[B][C]} + \frac{K_{m_C}[Q]}{K_{iq}[C]}\right)} \quad \text{(IX-325)}$$

$$\frac{v}{V_{max}} = \frac{[B]}{K_{m_B}\left(1 + \frac{K_{ib}K_{m_C}[Q]}{K_{m_B}K_{iq}[C]}\right) + [B]\left(1 + \frac{K_{m_A}}{[A]} + \frac{K_{m_C}}{[C]} + \frac{K_{m_C}[Q]}{K_{iq}[C]}\right)} \quad \text{(IX-326)}$$

$$\frac{v}{V_{\max}} = \frac{[\mathrm{C}]}{K_{m_{\mathrm{C}}}\left(1+\frac{K_{ib}[\mathrm{Q}]}{K_{iq}[\mathrm{B}]}+\frac{[\mathrm{Q}]}{K_{iq}}\right)+[\mathrm{C}]\left(1+\frac{K_{m_{\mathrm{A}}}}{[\mathrm{A}]}+\frac{K_{m_{\mathrm{B}}}}{[\mathrm{B}]}\right)} \qquad \text{(IX-327)}$$

Q is an uncompetitive inhibitor with respect to A at all B concentrations and unsaturating [C]. Saturation with C eliminates the inhibition. Q is a mixed-type inhibitor with respect to B at all A concentrations and unsaturating [C]. Again, saturation with C eliminates the inhibition; Q is competitive with respect to C at all A and B concentrations.

In the presence of R, the equations are:

$$\frac{v}{V_{\max}} = \frac{[\mathrm{A}]}{K_{m_{\mathrm{A}}}\left(1+\frac{K_{ic}[\mathrm{R}]}{K_{ir}[\mathrm{C}]}+\frac{[\mathrm{R}]}{K_{ir}}\right)+[\mathrm{A}]\left(1+\frac{K_{m_{\mathrm{B}}}}{[\mathrm{B}]}+\frac{K_{m_{\mathrm{C}}}}{[\mathrm{C}]}\right)} \qquad \text{(IX-328)}$$

$$\frac{v}{V_{\max}} = \frac{[\mathrm{B}]}{K_{m_{\mathrm{B}}}+[\mathrm{B}]\left(1+\frac{K_{m_{\mathrm{A}}}}{[\mathrm{A}]}+\frac{K_{m_{\mathrm{C}}}}{[\mathrm{C}]}+\frac{K_{m_{\mathrm{A}}}K_{ic}[\mathrm{R}]}{K_{ir}[\mathrm{A}][\mathrm{C}]}+\frac{K_{m_{\mathrm{A}}}[\mathrm{R}]}{K_{ir}[\mathrm{A}]}\right)} \qquad \text{(IX-329)}$$

$$\frac{v}{V_{\max}} = \frac{[\mathrm{C}]}{K_{m_{\mathrm{C}}}\left(1+\frac{K_{m_{\mathrm{A}}}K_{ic}[\mathrm{R}]}{K_{m_{\mathrm{C}}}K_{ir}[\mathrm{A}]}\right)+[\mathrm{C}]\left(1+\frac{K_{m_{\mathrm{A}}}}{[\mathrm{A}]}+\frac{K_{m_{\mathrm{B}}}}{[\mathrm{B}]}+\frac{K_{m_{\mathrm{A}}}[\mathrm{R}]}{K_{ir}[\mathrm{A}]}\right)} \qquad \text{(IX-330)}$$

R is competitive with respect to A at all B and C concentrations; R is an uncompetitive inhibitor with respect to B at all C concentrations and unsaturating [A]. Saturation with A eliminates the inhibition; R is a mixed-type inhibitor with respect to C at all B concentrations and unsaturating [A]. Saturation with A overcomes the inhibition.

The product inhibition patterns clearly identify substrate-product pairs, but do not indicate the order of binding and release. (For each substrate, one product yields a competitive pattern, another product yields an uncompetitive pattern eliminated by saturation with one of the nonvaried substrates, and a third product yields a mixed-type pattern eliminated by saturation with a nonvaried substrate.) However, since the reaction sequence is the same regardless of which stable form is taken as the "start," the order can be established. The preferred way of indicating the sequence and designating the substrates and products should also be obvious from the chemistry of the reaction.

The K_{ip} can be obtained from $K_{i_{int}}^{P/A}$. The intercept replot is given by:

$$\boxed{\frac{1}{V_{max_{app}}} = \frac{K_{m_B}}{V_{max} K_{ip}[B]}[P] + \frac{1}{V_{max}}\left(1 + \frac{K_{m_B}}{[B]} + \frac{K_{m_C}}{[C]}\right)} \qquad \text{(IX-331)}$$

$$K_{i_{int}}^{P/A} = K_{ip}\left(1 + \frac{[B]}{K_{m_B}} + \frac{[B]K_{m_C}}{K_{m_B}[C]}\right)$$

With K_{ip} known, K_{ia} can be calculated from $K_{i_{slope}}^{P/A}$:

$$K_{i_{slope}}^{P/A} = \frac{K_{ip}K_{m_A}[B]}{K_{ia}K_{m_B}}$$

In a similar manner, K_{iq} can be calculated from $K_{i_{int}}^{Q/B}$ and then K_{ib} can be calculated from $K_{i_{slope}}^{Q/B}$; K_{ir} can be calculated from $K_{i_{int}}^{R/C}$ and then K_{ic} from $K_{i_{slope}}^{R/C}$. The values can be checked using other K_i values. For example:

$$K_{i_{slope}}^{P/B} = \frac{K_{ip}}{\left(1 + \frac{K_{ia}}{[A]}\right)}, \qquad K_{i_{slope}}^{Q/C} = \frac{K_{iq}}{\left(1 + \frac{K_{ib}}{[B]}\right)}, \qquad K_{i_{slope}}^{R/A} = \frac{K_{ir}}{\left(1 + \frac{K_{ic}}{[C]}\right)}$$

The ratios K_{m_P}/K_{m_R}, K_{m_Q}/K_{m_P}, and K_{m_Q}/K_{m_R} can be calculated from equation IX-311; K_{m_P}, K_{m_Q}, K_{m_R}, as well as all the K_i values can be obtained from the kinetics of the reverse reaction, which is symmetrical to the forward reaction.

Z. SUMMARY OF NONRANDOM TERREACTANT SYSTEMS

There are two basic nonrandom Ter Bi systems and four basic nonrandom Ter Ter systems. A preliminary distinction between three of the four nonrandom Ter Ter systems can be made from initial velocity studies in which two substrates are varied together (as the varied substrates or as the changing fixed substrates). The patterns are summarized in Table IX-8. The product inhibition patterns are summarized in Table IX-9.

Table IX-8 Summary of Nonrandom Terreactant Systems: Reciprocal Plot Patterns Observed When Two Substrates Are Varied

Procedure	Ordered Ter Ter (and Ordered Ter Bi)	Bi Bi Uni Uni Ping Pong Ter Ter, Bi Uni Uni Bi Ping Pong Ter Ter (and Bi Uni Uni Uni Ping Pong Ter Bi)	Hexa Uni Ping Pong Ter Ter
Vary each substrate separately at different fixed concentrations of the other two (maintained at a constant ratio).	All families are linear and intersecting.[a] Two of the $1/v$-axis intercept replots are parabolic. One replot is linear.[b]	One family of plots consists of parallel lines which yields a parabolic replot of $1/v$-axis intercept. Two families are linear and intersecting with linear intercept replots.[b]	All three families of plots consist of parallel lines with linear $1/v$-axis intercept replots.[b]
Vary two substrates at a time (maintained at a constant ratio) at different fixed concentrations of the third.[c]	All families are parabolic with linear replots of $1/v$-axis intercepts.[d]	Two families are linear and intersecting. One family is parabolic. All yield linear $1/v$-axis intercept replots.[d]	All three families of plots consist of parallel lines with linear $1/v$-axis intercept replots.[d]

[a] Any two lines of the family intersect but all the lines may not intersect at a common point (as in Fig. IX-56).
[b] The intercepts are replotted against the reciprocal concentration of one of the changing fixed substrates.
[c] The $1/v$ is plotted against the reciprocal concentration of one of the varied substrates.
[d] The intercepts are replotted against the reciprocal concentration of the changing fixed substrate.

Table IX-9 Product Inhibition Patterns for Some Terreactant Mechanisms[a]

Mechanism	Product Inhibitor	Varied Substrate A: Unsaturated	A: Saturated with B	A: Saturated with C	B: Unsaturated	B: Saturated with A	B: Saturated with C	C: Unsaturated	C: Saturated with A	C: Saturated with B
Ordered Ter Bi	P	MT	UC	UC	MT	MT	UC	MT	MT	MT
	Q	C	C	C	MT	—	MT	MT	—	UC
Ordered Bi Ter	P	MT	UC		MT	MT				
	Q	UC	UC		UC	UC				
	R	C	C		MT	—				
Uni Uni Uni Bi Ping Pong Bi Ter	P	MT	—		C	C				
	Q	UC	UC		MT	MT				
	R	C	C		UC	—				
Uni Bi Uni Uni Ping Pong Bi Ter (equivalent to Uni Uni Uni Bi)	P	MT	MT		UC	UC				
	Q	UC	—		C	C				
	R	C	C		MT	—				
Bi Uni Uni Uni Ping Pong Ter Bi (reverse of Uni Uni Uni Bi)	P	MT	UC	—	MT	MT	—	C	C	C
	Q	C	C	C	MT	—	MT	MT	—	MT
Uni Uni Bi Uni Ping Pong Ter Bi (reverse of Uni Bi Uni Uni and equivalent to Bi Uni Uni Uni)	P	MT	—	MT	C	C	C	MT	MT	—
	Q	C	C	C	MT	—	UC	MT	—	MT

Ordered Ter Ter	P	MT	UC	UC	MT	MT	UC	MT	MT	MT
	Q	UC	UC	UC	UC	UC	UC	UC	UC	UC
	R	C	C	C	MT	—	MT	MT	—	UC
Bi Uni Uni Bi Ping Pong Ter Ter	P	MT	UC	—	MT	MT	—	C	C	C
	Q	UC	UC	UC	UC	UC	UC	MT	MT	MT
	R	C	C	C	MT	—	MT	UC	—	UC
Uni Bi Bi Uni Ping Pong Ter Ter (equivalent to Bi Uni Uni Bi)	P	MT	MT	MT	UC	UC	UC	UC	UC	UC
	Q	UC	—	UC	C	C	C	MT	MT	—
	R	C	C	C	MT	—	UC	MT	—	MT
Bi Bi Uni Uni Ping Pong Ter Ter	P	MT	UC	MT	MT	MT	MT	UC	UC	UC
	Q	UC	UC	—	UC	UC	—	C	C	C
	R	C	C	C	MT	—	MT	MT	—	MT
Uni Uni Bi Bi Ping Pong Ter Ter (reverse of and equivalent to Bi Bi Uni Uni)	P	MT	—	MT	C	C	C	MT	MT	—
	Q	UC	UC	UC	MT	MT	UC	MT	MT	MT
	R	C	C	C	UC	—	UC	UC	—	UC
Hexa Uni Ping Pong Ter Ter	P	MT	—	MT	C	C	C	UC	UC	—
	Q	UC	UC	—	MT	MT	—	C	C	C
	R	C	C	C	UC	—	UC	MT	—	MT

[a] The product inhibition patterns assume that no dead-end complexes form and that there are no rapid equilibrium steps. Dead-end complexes and rapid equilibrium segments will change the patterns (see text).

AA. OTHER POSSIBLE TERREACTANT SYSTEMS

Theorell-Chance Systems

The terreactant systems listed in Table IX-9 and described earlier do not exhaust all the possibilities. For most of the steady-state systems listed, we could devise limiting Theorell-Chance type cases in which the steady-state levels of one or more central complexes are essentially zero.

Ordered Ter Ter, Theorell-Chance C–P

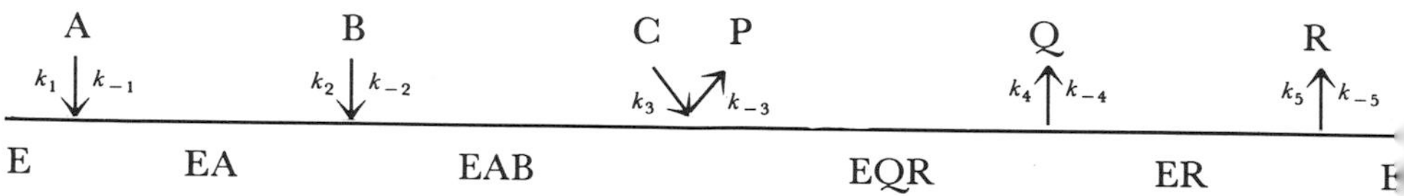

Bi Uni Uni Bi Ping Pong Ter Ter, Theorell-Chance B–P

A B P C Q R

k_1 k_{-1} k_2 k_{-2} k_3 k_{-3} k_4 k_{-4} k_5 k_{-5}

E EA F (FC ⇌ EQR) ER E

Uni Uni Bi Bi Ping Pong Ter Ter, Theorell-Chance C–Q

A P B C Q R

k_1 k_{-1} k_2 k_{-2} k_3 k_{-3} k_4 k_{-4} k_5 k_{-5}

E (EA ⇌ FP) F FB ER E

The absence of a central complex changes the inhibition patterns for the product released in the Theorell-Chance step. For example, in the Bi Uni Uni Bi system above, saturating [B] overcomes the inhibition by P when either A or C is the varied substrate. Also, P acts as a competitive inhibitor with respect to B at unsaturating [A] and [C] and at unsaturating [C] and saturating [A]. In the usual Bi Uni Uni Bi system, P is a mixed-type inhibitor under these conditions.

Terreactant Ping Pong Systems with Rapid Equilibrium Segments

When some steps of a reaction sequence are at equilibrium, the velocity equation will be modified. The numerator is unaffected, but certain terms may be missing from the denominator. The missing terms may alter the characteristics of the initial velocity plots in the absence of products, the product inhibition patterns, or both. A number of partial rapid equilibrium systems involving a quaternary EABC complex have been described earlier. Rapid equilibrium segments can also occur within Ping Pong sequences. For example, consider the Bi Uni Uni Bi Ping Pong Ter Ter system in which the first ligand that binds in either direction and their respective complexes are at equilibrium. The overall reaction rate is set by the slower steady-state steps between EA and ER.

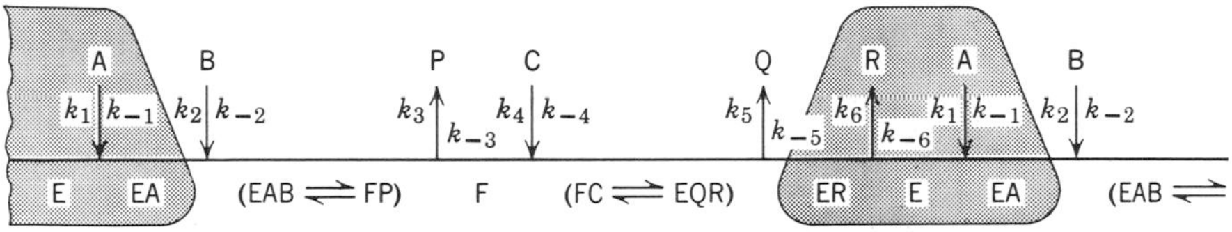

A B P C Q R A B
k_1 k_{-1} k_2 k_{-2} k_3 k_{-3} k_4 k_{-4} k_5 k_{-5} k_6 k_{-6} k_1 k_{-1} k_2 k_{-2}
E EA (EAB ⇌ FP) F (FC ⇌ EQR) ER E EA (EAB ⇌

If we consider only the forward reaction in the absence of products, the equation becomes:

$$\frac{v}{V_{\max}} = \frac{[A][B][C]}{K_{ia}K_{m_B}[C] + K_{m_C}[A][B] + K_{m_B}[A][C] + [A][B][C]} \qquad \text{(IX-332)}$$

Compared to the equivalent velocity equation for the full steady-state system, equation IX-332 is missing the $K_{m_A}[B][C]$ term.

When [A] is varied:

$$\frac{v}{V_{\max}} = \frac{[A]}{\frac{K_{ia}K_{m_B}}{[B]} + [A]\left(1 + \frac{K_{m_B}}{[B]} + \frac{K_{m_C}}{[C]}\right)} \qquad \text{(IX-333)}$$

Because of the missing term $slope_{1/A}$ versus $1/[B]$ replots go through the origin distinguishing this system from the full steady-state system. When [B] is varied, the equation is:

$$\frac{v}{V_{max}} = \frac{[B]}{K_{m_B}\left(1 + \frac{K_{ia}}{[A]}\right) + [B]\left(1 + \frac{K_{m_C}}{[C]}\right)} \qquad \text{(IX-334)}$$

The $K_{m_A}/[A]$ term is missing from the intercept factor. Thus the $1/v$-axis intercept of $1/v$ versus $1/[B]$ plots at constant [C] is independent of [A] providing another distinguishing feature of this system. When [C] is varied:

$$\frac{v}{V_{max}} = \frac{[C]}{K_{m_C} + [C]\left(1 + \frac{K_{m_B}}{[B]} + \frac{K_{ia}K_{m_B}}{[A][B]}\right)} \qquad \text{(IX-335)}$$

The $1/v$-axis intercept still depends on both [A] and [B].

The product inhibition patterns are also changed because the [B][Q], [A][B][Q], and [B][C][R] terms are missing. (A number of other terms are also missing, but these contain two product concentration terms and are not considered for normal product inhibition patterns when only one product at a time is used.) Of the existing denominator terms containing product concentrations, only [P], [A][P], [A][B][P], and [C][R] are of interest. Thus we see immediately that Q is not an inhibitor. Rearranging the equation for each varied substrate, we find that R is competitive with both A and B, and uncompetitive with C, while P is competitive with C and a mixed-type inhibitor with respect to A and B. The unusual product inhibition patterns are a result of the rapid equilibrium segment, and are predictable by the general rules for rapid equilibrium systems described in Chapter Six.

Terreactant Ping Pong Systems with Rapid Equilibrium Random Segments

Other possible terreactant Ping Pong systems include those with random sequences. Random sequences introduce squared terms into the velocity equation and the reciprocal plots are theoretically nonlinear. However, the curvature occurs close to the $1/v$-axis and usually cannot be detected. These systems can be treated as if rapid equilibrium conditions prevail. Three examples are shown below.

Random A–B, Random Q–R Bi Uni Uni Bi Ping Pong

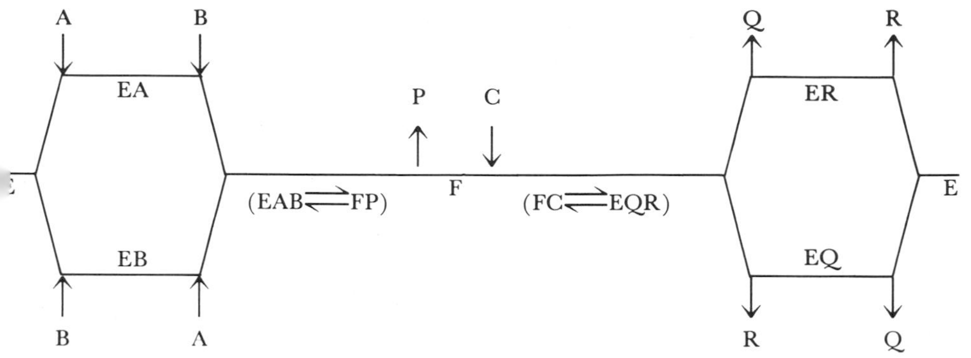

Random B–C Uni Uni Bi Bi Ping Pong

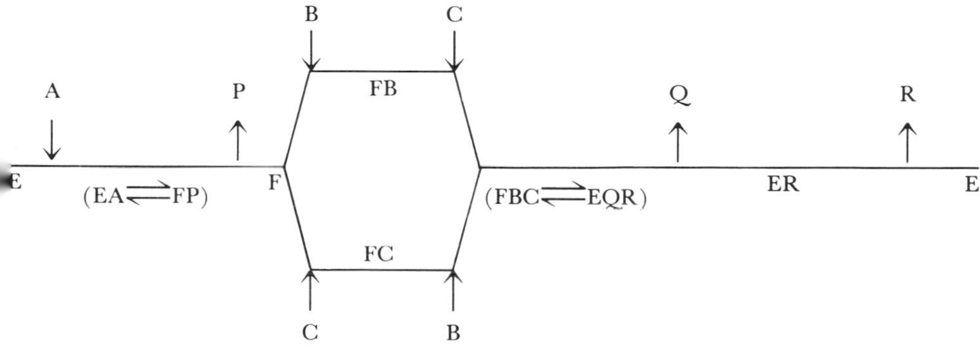

Random A–B, Random P–Q Bi Bi Uni Uni Ping Pong

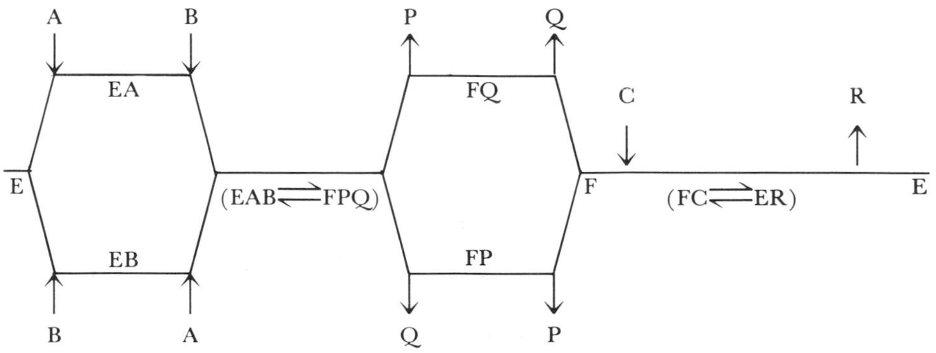

If the complexes comprising the random segment are assumed to be in equilibrium, the method of Cha can be used to derive the velocity equation. In the absence of products, the velocity equation for each of the partially random Ping Pong systems has the same form as that of the analogous nonrandom system. Consequently, initial velocity studies in the absence of products would give no clue as to the presence of random sequences. However, the product inhibition patterns are different. The exact patterns can be predicted from the rules outlined in a following section.

Hybrid (Two-Site) Rapid Equilibrium Random-Ping Pong Bi Bi Uni Uni System

Pyruvic carboxylase catalyzes the overall reaction:

$$\text{pyruvate} + \text{MgATP} + \text{HCO}_3^- \rightleftharpoons \text{oxalacetate} + \text{MgADP} + \text{P}_i$$

which occurs via two partial reactions:

$$(a)\quad \text{E-biotin} + \text{MgATP} + \text{HCO}_3^- \rightleftharpoons \text{E-biotin}{\sim}\text{CO}_2 + \text{MgADP} + \text{P}_i$$

$$(b)\quad \text{E-biotin}{\sim}\text{CO}_2 + \text{pyruvate} \rightleftharpoons \text{E-biotin} + \text{oxalacetate}$$

Barden, Fung, Utter, and Scrutton (1972) observed the initial velocity patterns shown in Table IX-10 for the chicken liver enzyme.

Table IX-10 Initial Velocity Patterns for Chicken Liver Pyruvic Carboxylase

Varied Substrate	Changing Fixed Substrate	Observed Pattern
MgATP	HCO_3^-	Intersecting
	Pyruvate	Parallel
HCO_3^-	MgATP	Intersecting
	Pyruvate	Parallel
Pyruvate	HCO_3^-	Parallel
	MgATP	Parallel
MgADP	Phosphate	Intersecting
	Oxalacetate	Parallel
Phosphate	MgADP	Intersecting
	Oxalacetate	Parallel
Oxalacetate	Phosphate	Parallel
	MgADP	Parallel

The patterns are consistent with a Bi Bi Uni Uni Ping Pong sequence:

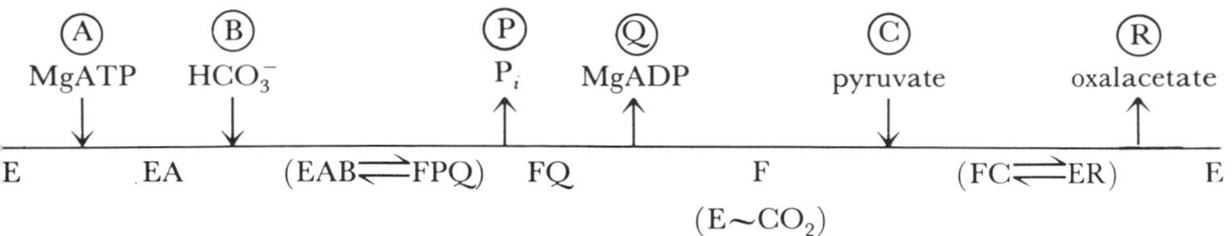

Table IX-11 Product Inhibition Patterns for Various Bi Bi Uni Uni Ping Pong Systems

System	Product Inhibitor	Varied Substrate at Unsaturating Cosubstrates		
		A	B	C
Observed for pyruvic carboxylase	P	C	MT	U
	Q	C	MT	U
	R	MT	MT	C
Classical Bi Bi Uni Uni	P	MT	MT	U
	Q	U	U	C
	R	C	MT	MT
Rapid Equilibrium Random A–B and P–Q (no dead-end complexes)	P	U	U	C
	Q	U	U	C
	R	C	C	MT
Rapid Equilibrium Random A–B and P–Q with dead-end EBP and EBQ	P	MT	MT	MT
	Q	MT	MT	MT
	R	C	C	MT
Hybrid (Two-Site) Rapid Equilibrium Random A–B and P–Q (no dead-end complexes)	P	C	C	U
	Q	C	C	U
	R	MT	MT	C
Hybrid (Two-Site) Rapid Equilibrium Random A–B and P–Q with dead-end EBP and EBQ	P	C	MT	U
	Q	C	MT	U
	R	MT	MT	C

However, the product inhibition patterns did not conform to the classical Ping Pong sequence. For example, the classical sequence predicts competitive interactions between MgATP and oxalacetate, and also between MgADP and pyruvate, because in both cases the varied substrate and product inhibitor combine at the same site with the same enzyme form. Instead, the patterns shown in the first row of Table IX-11 were obtained. For comparison, the patterns expected for several other Bi Bi Uni Uni Ping Pong systems are also shown. The observed patterns are consistent with a hybrid, two-site model similar to that described earlier for transcarboxylase. The assumptions made in deriving the velocity equation for this model are: (*a*) the two partial reactions occur at two different and independent sites linked by a mobile biotin residue; (*b*) A and B (ATP and HCO_3^-) and the products P and Q (ADP and P_i) bind at site 1 in a Rapid Equilibrium Random fashion; the binding of one ligand at site 1 may affect the binding of another ligand at this site. The binding of any ligand to site 1 has no effect on the binding of pyruvate and oxalacetate at site 2, and vice versa. Dead-end EBP and EBQ complexes can form at site 1. (*c*) The rate of biotin carboxylation reaction at site 1 is unaffected by occupancy at site 2, and vice versa. Similarly, the rate of biotin migration between the two sites is unaffected by substrates or products binding to either site. (*d*) The steady-state steps are the carboxylation-decarboxylation of biotin and the migration of the biotin from one site to another. The basic King-Altman figure is identical to that shown earlier for transcarboxylase. The fractional concentration factors are (in the presence of all ligands):

$$f_1 = \frac{[\text{“E”AB}]}{[{}^0\text{E}]} = \frac{\dfrac{[\text{A}][\text{B}]}{\alpha K_\text{A} K_\text{B}}}{1 + \dfrac{[\text{A}]}{K_\text{A}} + \dfrac{[\text{B}]}{K_\text{B}} + \dfrac{[\text{A}][\text{B}]}{\alpha K_\text{A} K_\text{B}} + \dfrac{[\text{P}]}{K_\text{P}} + \dfrac{[\text{Q}]}{K_\text{Q}} + \dfrac{[\text{P}][\text{Q}]}{\beta K_\text{P} K_\text{Q}} + \dfrac{[\text{B}][\text{P}]}{\gamma K_\text{B} K_\text{P}} + \dfrac{[\text{B}][\text{Q}]}{\delta K_\text{B} K_\text{Q}}}$$

$$f_{-1} = \frac{[\text{“E”PQ}]}{[{}^{\text{CO}_2}\text{E}]} = \frac{\dfrac{[\text{P}][\text{Q}]}{\beta K_\text{P} K_\text{Q}}}{d_1 = \text{same denominator as in } f_1}$$

$$f_3 = \frac{[\text{“E”C}]}{[\text{E}^{\text{CO}_2}]} = \frac{\dfrac{[\text{C}]}{K_\text{C}}}{1 + \dfrac{[\text{C}]}{K_\text{C}} + \dfrac{[\text{R}]}{K_\text{R}}}$$

$$f_{-3} = \frac{[\text{“E”R}]}{[\text{E}^0]} = \frac{\dfrac{[\text{R}]}{K_\text{R}}}{d_2 = \text{same denominator as in } f_3}$$

Note that f_1 represents all the complexes in ^{0}E containing bound A and B, including those complexes with C or R at site 2. Similarly, f_3 represents all C-containing complexes of E^{CO_2}, including all the complexes with various ligands at site 1. However, as described earlier for transcarboxylase, the expressions simplify to the usual distribution equation when the terms corresponding to the complexes of the "other" site are factored and cancelled. The King-Altman solutions yield equation IX-164 shown earlier for transcarboxylase. Substituting for the fractional concentration factors, the equation becomes:

$$\frac{v}{[E]_t} = \frac{K_1\frac{[A][B][C]}{d_1d_2} - K_2\frac{[P][Q][R]}{d_1d_2}}{\frac{K_3[C]}{d_2} + \frac{K_4[A][B]}{d_1} + \frac{K_5[A][B][C]}{d_1d_2} + \frac{K_6[A][B][R]}{d_1d_2} + \frac{K_7[R]}{d_2} + \frac{K_8[P][Q]}{d_2} + \frac{K_9[C][P][Q]}{d_1d_2} + \frac{[P][Q][R]}{d_1d_2}} \quad \text{(IX-336)}$$

where K_1, K_2, and so on, are groups of rate constants. Multiplying numerator and denominator by d_1d_2 and defining terms in the usual manner, the complete velocity equation is:

$$v = \frac{V_fV_r\left([A][B][C] - \frac{[P][Q][R]}{K_{eq}}\right)}{\begin{gathered}V_rK_{ia}K_{m_B}[C] + V_rK_{m_C}[A][B] + V_rK_{m_A}[B][C] + V_rK_{m_B}[A][C] + V_r[A][B][C] \\ + \frac{V_fK_{ip}K_{m_Q}[R]}{K_{eq}} + \frac{V_rK_{ia}K_{m_B}[C][P]}{K_{ip}} + \frac{V_rK_{ia}K_{m_B}[C][Q]}{K_{iq}} \\ + \frac{V_fK_{iq}K_{m_P}[B][R]}{K_{ib}K_{eq}} + \frac{V_fK_{iq}K_{m_P}[A][R]}{K_{ia}K_{eq}} + \frac{V_rK_{m_C}[A][B][R]}{K_{ir}} \\ + \frac{V_rK_{m_A}[B][C][Q]}{K_{iq}} + \frac{V_rK_{m_A}[B][C][P]}{K_{ip}} + \frac{V_fK_{m_Q}[P][R]}{K_{eq}} \\ + \frac{V_fK_{m_P}[Q][R]}{K_{eq}} + \frac{V_fK_{m_R}[P][Q]}{K_{eq}} + \frac{V_f[P][Q][R]}{K_{eq}} + \frac{V_fK_{m_R}[C][P][Q]}{K_{ic}K_{eq}} \\ + \frac{V_fK_{m_P}[B][Q][R]}{K_{ib}K_{eq}} + \frac{V_fK_{m_Q}[B][P][R]}{K_{ib}K_{eq}}\end{gathered}}$$

(IX-337)

where the K_m's are defined as usual and:

$$K_{ia} = \frac{\text{Coef}_\text{C}}{\text{Coef}_\text{AC}} \qquad K_{ip} = \frac{\text{Coef}_\text{C}}{\text{Coef}_\text{CP}}$$

$$K_{ib} = \frac{\text{Coef}_\text{QR}}{\text{Coef}_\text{BQR}} = \frac{\text{Coef}_\text{PR}}{\text{Coef}_\text{BPR}} \qquad K_{iq} = \frac{\text{Coef}_\text{C}}{\text{Coef}_\text{CQ}}$$

$$K_{ic} = \frac{\text{Coef}_\text{PQ}}{\text{Coef}_\text{CPQ}} \qquad K_{ir} = \frac{\text{Coef}_\text{AB}}{\text{Coef}_\text{ABR}}$$

Additional inhibition constants must be defined for the ligands of the Bi Bi partial reaction to eliminate K_eq.

$$K'_{ia} = \frac{\text{Coef}_\text{BR}}{\text{Coef}_\text{ABR}}, \qquad K'_{ib} = \frac{\text{Coef}_\text{AR}}{\text{Coef}_\text{ABR}}, \qquad K'_{ip} = \frac{\text{Coef}_\text{CQ}}{\text{Coef}_\text{CPQ}}, \qquad K'_{iq} = \frac{\text{Coef}_\text{CP}}{\text{Coef}_\text{CPQ}}$$

The Haldane equations are:

$$\boxed{K_\text{eq} = \frac{V_f K'_{ip} K_{iq} K_{m_\text{R}}}{V_r K_{ia} K_{m_\text{B}} K_{ic}} = \frac{V_f K_{ip} K'_{iq} K_{m_\text{R}}}{V_r K_{ia} K_{m_\text{B}} K_{ic}} = \frac{V_f K_{m_\text{P}} K_{iq} K_{ir}}{V_r K'_{ia} K_{ib} K_{m_\text{C}}} = \frac{V_f K_{m_\text{P}} K_{iq} K_{ir}}{V_r K_{ia} K'_{ib} K_{m_\text{C}}}} \tag{IX-338}$$

On the basis of the equations derived for the two-site models for transcarboxylase and pyruvic carboxylase, the product inhibition patterns conform to the following rules:

> *Competitive patterns are observed only when the product inhibitor and the varied substrate are members of the same partial reaction. Mixed-type and uncompetitive patterns are observed when the product inhibitor and the varied substrate are members of different partial reactions. The pattern is uncompetitive if the product inhibitor is involved in a bireactant (or higher reactancy) random reaction; the pattern is mixed-type if the product inhibitor is a member of a unireactant partial reaction. Substrate analogs that act as true dead-end inhibitors are competitive with respect to the substrate they resemble and uncompetitive with respect to the reactants of the other partial reaction.*

In contrast to the chicken liver enzyme, Warren and Tipton (1974) find intersecting patterns for all initial velocity studies with the pig liver enzyme. These authors propose a Rapid Equilibrium Ordered A, Ordered B–C, Theorell-Chance D–P mechanism where A = Mg^{2+}, B = MgATP, C = HCO_3^-, D = pyruvate, and P = oxalacetate. Products Q and R (MgADP and P_i) are assumed to leave in a Rapid Equilibrium Random fashion.

BB. QUADREACTANT SYSTEMS

Quadreactant systems are not as common as bireactant and terreactant systems, but some do exist. There are four possible Quad Bi systems, of which three are really different. There are ten possible Quad Ter systems, of which five are different. Of the 20 possible Quad Quad systems, ten are different. The velocity equations for these systems can be derived in the usual manner and the various ligands identified as described earlier for the bireactant and terreactant systems. The analyses, of course, will be more complicated, but the procedures are the same. For initial velocity studies, two of the substrates will have to be held constant while one is varied at different fixed concentrations of another. The two constant substrates are maintained at a fixed ratio, which can be changed for a different series of reciprocal plot families for the same varied and fixed substrates. If we are only interested in determining the mechanism (and not the various kinetic constants), then some shortcuts can be taken. For example, two substrates can be varied together as the "varied substrate" or as the "changing fixed substrates." The general rules described below can be used to diagnose the mechanism without knowing the velocity equation.

CC. GENERAL RULES FOR PREDICTING INITIAL VELOCITY PATTERNS

Cleland has formulated a series of general rules for predicting the effect of a compound (e.g., nonvaried substrate, a normal or alternate product, or a dead-end inhibitor) on the slope and intercept of reciprocal plots for a given varied substrate. Intercept and slope effects are predicted separately and then combined to obtain the final reciprocal plot pattern. If only intercept effects are predicted, the family of reciprocal plots will consist of a series of parallel lines. If only slope effects are predicted, the family of plots will intersect on the $1/v$-axis. If both slope and intercept effects are predicted, the plots will intersect to the left of the $1/v$-axis, above, on, or below the horizontal axis.

Intercept Effects

> *A compound affects the $1/v$-axis intercept of a reciprocal plot if it and the varied substrate combine with different enzyme forms.*

The intercept rule is based on the fact that the $1/v$-axis intercept represents the reciprocal velocity at an infinitely high concentration of varied substrate. The concentration of the enzyme form (e.g., E′) that combines with the varied substrate (e.g., S), will be driven to zero by the reaction $E' + S \rightleftharpoons E'S$. If the compound in question still affects the velocity at saturating [S], it must do so by combining with some enzyme form other than E′. This rule applies to all steady-state systems and to Rapid Equilibrium Random systems. If in the Rapid Equilibrium Random system, A is the varied substrate and B the changing fixed substrate, B will affect the $1/v$-axis intercept of the $1/v$ versus $1/[A]$ plot even though B and A can both combine with free E (the same enzyme form) because B also can combine with EA (and A cannot combine with EA).

Exceptions to the Intercept Rule

Exceptions to this rule include Rapid Equilibrium Ordered systems. For example, consider the Rapid Equilibrium Ordered Bi Bi system (Chapter Six). Here, A and B combine with different enzyme forms but saturation with B eliminates the dependence of v on [A] by forcing the equilibrium $E + A \rightleftharpoons EA$ completely to the right as EA reacts with B to form EAB. As a result, the $1/v$ versus $1/[B]$ plots intersect on the $1/v$-axis at all A concentrations (where $[A] > [E]_t$). Other exceptions are encountered in steady-state mechanisms without a central complex, such as the Theorell-Chance Bi Bi system:

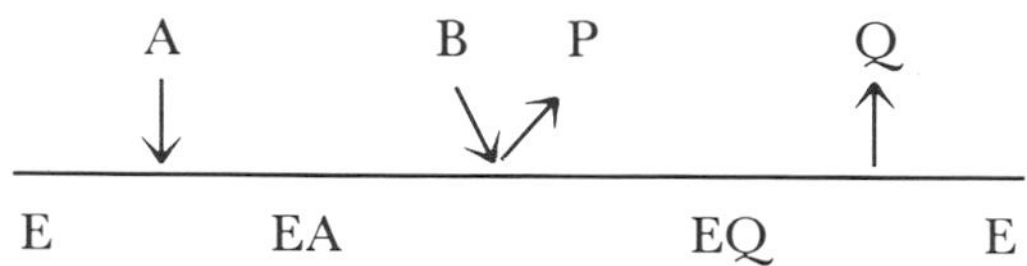

Here, B and P combine with two different enzyme forms (B with EA, and P with EQ). However, P is competitive with B (i.e., no intercept effect). At unsaturating [B], P inhibits by backing up the reaction sequence to reform EA: $EQ + P \rightleftharpoons EA + B$. However, when [B] is infinitely high, P cannot push the reaction back and, consequently, no inhibition is observed. In the Ordered Bi Bi system, P can back up the reaction sequence by reacting with

EQ to form EPQ. In this case, saturating [B] cannot prevent the back up into EPQ, and, consequently, an intercept effect is observed. (Saturating [B] will only increase the steady-state level of EAB.) Thus the rule can be restated in a more complete manner for product inhibition.

> *A compound affects the $1/v$-axis intercept of a reciprocal plot if it and the varied substrate combine with different enzyme forms and saturation with the varied substrate cannot overcome the inhibition.*

Product inhibition is discussed in more detail in a following section.

Slope Effects

> *A cosubstrate or product affects the slope of a reciprocal plot if it and the varied substrate combine reversibly with the same enzyme form or with different forms that are separated in the reaction sequence by freely reversible steps.*

The slope rule can be appreciated if we recall the fact that the slope of a reciprocal plot, K_m/V_{max}, is the reciprocal of the apparent first-order rate constant, k, for the reaction of the varied substrate (when its concentration is very low).

$$v = \frac{V_{max}[S]}{K_m + [S]} \xrightarrow{[S] \ll K_m} \frac{V_{max}}{K_m}[S] = k[S]$$

$$\therefore \quad slope = \frac{K_m}{V_{max}} = \frac{1}{k}$$

Suppose S combines with enzyme form E′ to yield E′S. The k at any fixed $[E]_t$ depends on the relative levels of E′ and E′S. Any factor that lowers the steady-state level of E′ (or increases the steady-state level of E′S) will decrease the apparent first-order rate constant, or, in other words, increase the slope of the $1/v$ versus $1/[S]$ plot. Similarly, any factor that increases E′/E′S decreases the slope. Thus a product that combines with E′ (thereby lowering the E′ level) will increase the slope, while a cosubstrate that reacts with E′S (thereby lowering the E′S level) will decrease the slope. If, for example, E′S is actually EA in an Ordered Bi Bi system, increasing [B] would decrease [EA] relative to [E] causing the slope of the $1/v$ versus $1/[A]$ plot to decrease. When B is the varied substrate, increasing [A] increases [EA] relative to [EAB + EPQ]. Consequently, the slope of the $1/v$ versus $1/[B]$ plot decreases as [A] increases.

Effect of Irreversible Sequences

The slope rule states that a compound affects the slope if it and the varied substrate combine with the same enzyme form or with different enzyme forms *that are reversibly connected.*

> *A reaction sequence becomes irreversible if either (a) the two points under consideration are separated by the addition of a substrate that is present at a saturating concentration, or (b) the two points under consideration are separated by a product release step and that product is not present at a significant concentration.*

For example, consider the Ordered Ter Ter reaction system.

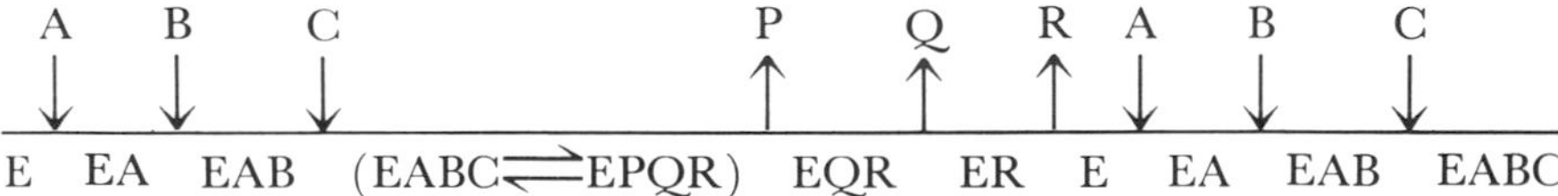

The enzyme forms adding A and C are reversibly connected through the addition or dissociation of B. Consequently, C will affect the $slope_{1/A}$. That is, the family of $1/v$ versus $1/[A]$ plots at a constant unsaturating [B] will have a different slope for each fixed concentration of C (A and C combine with different enzyme forms so the $1/v$-axis intercepts will also vary and the family of plots intersect to the left of the $1/v$-axis). Specifically, as [C] increases, the steady-state distribution of enzyme forms is forced to the right so that the [EAB]/[EA] and [EA]/[E] ratios at any constant [B] decrease resulting in a decrease in $slope_{1/A}$. When [B] is saturating, the reversible connection between the forms combining with A and C is broken (assuming P, Q, and R are not present at a concentration high enough to make the reversible connection downstream). Now, C will have no effect on $slope_{1/A}$

and the family of $1/v$ versus $1/[A]$ plots at different fixed C concentrations will consist of a series of parallel lines. (Saturating [B] reduces the steady-state level of EA to zero so that the [EAB]/[EA] and [EA]/[E] ratios are essentially unaffected by changing [C].) The completely random terreactant system (Chapter Six) will show intersecting patterns regardless of which substrate is saturating. If one substrate must add first, but the next two add randomly (random B–C) then parallel reciprocal plots will be observed when either B or C is saturating.

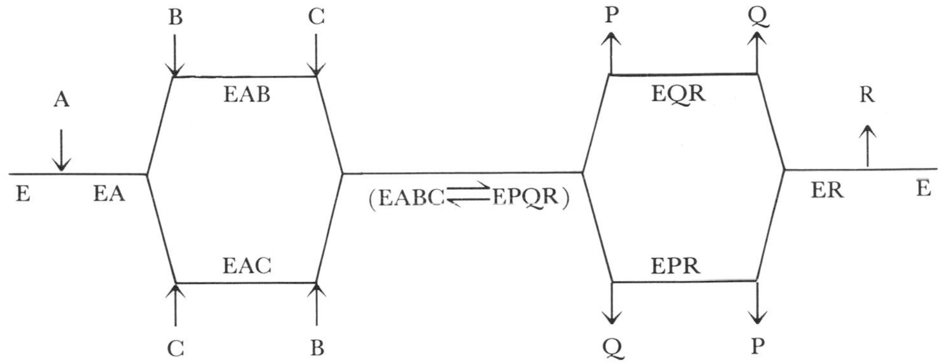

In this case saturation with B converts all the EA to EAB and the reaction is diverted to the ordered E→EA→EAB→EABC pathway. Similarly, saturation with C converts the random B–C sequence to an ordered E→EA→EAC→EABC sequence.

Consider the Bi Bi Uni Uni Ping Pong Ter Ter system:

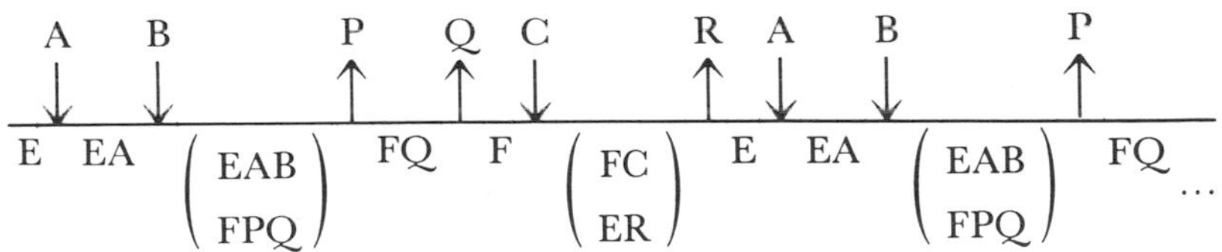

A and B combine with different enzyme forms which are reversibly connected. Therefore, plots of $1/v$ versus $1/[A]$ at different fixed levels of B and a constant [C] and plots of $1/v$ versus $1/[B]$ at different fixed levels of A and a constant [C] are intersecting (i.e., we observe both intercept and slope effects). However, the A–C plots at constant [B] and B–C plots at constant [A] will be a series of parallel lines because the points of addition of A and C or B and C are separated by product release steps (i.e., no slope effect). If P and Q are present at significant concentrations, a reversible connection will be established upstream between A and C or B and C. The families of A–C and B–C plots will now intersect to the left of the $1/v$-axis. Similarly, if R is present, a reversible connection is established downstream and the A–C and B–C plots intersect. The intersecting patterns can be converted back to parallel patterns by interrupting the reversible sequence. For example, if P and Q are present, but [B] is saturating, the points of addition of A and C will no longer be reversibly connected and a parallel pattern is observed again. If R is present and [A] is saturating, the downstream reversible connection between the points of addition of C and B would be eliminated and the B–C pattern would be parallel again. In general then:

> *When any two substrates are varied (one the varied substrate and one the changing fixed substrate), the series of reciprocal plots obtained at constant levels of the other substrates will be parallel if the points of addition of the two varied substrates are separated by one or more irreversible steps (e.g., a product release step or saturation with a substrate that adds between the two varied substrates). The series of reciprocal plots will intersect if the points of addition of the two varied substrates are connected by reversible steps (e.g., presence of unsaturating substrates that add between the two varied substrates or presence of products released between the points of addition of the two varied substrates).*

The general rules can be applied directly to iso systems. For example, consider the Iso Ordered Bi Bi system shown below.

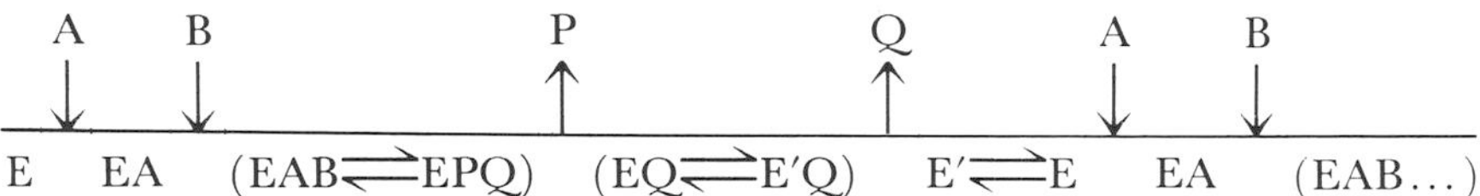

Compared to the regular Ordered Bi Bi system, the inhibition patterns by P are unchanged. However, now Q is a mixed-type inhibitor with respect to A at all B concentrations (A and Q combine with different enzyme forms that are reversibly connected). Q is a mixed-type inhibitor with respect to B at unsaturating [A]. But now saturating [A] converts Q to an uncompetitive inhibitor. (Saturating [A] breaks the downstream reversible connection between E′ and EA.)

As a corollary to the general rule, we can add:

> *When two substrates are varied together as the changing fixed substrates (maintaining them at a constant ratio), the resulting series of reciprocal plots will be (a) parallel if neither of the changing fixed substrates is reversibly connected to the varied substrate, or (b) intersecting (any two lines) if either of the changing fixed substrates is linked reversibly to the varied substrate. (All plots may not intersect at a common point as in Fig. IX-56.)*

Thus in the Bi Bi Uni Uni Ping Pong Ter Ter System, plots of $1/v$ versus $1/[C]$ at different fixed concentrations of A and B (where $[A] = x[B]$) are parallel. But, plots of $1/v$ versus $1/[A]$ at different fixed concentrations of B and C and plots of $1/v$ versus $1/[B]$ at different fixed concentrations of A and C are intersecting.

We can also predict the shape of intercept and slope replots when two substrates are varied together as the changing fixed substrates:

> *The $1/v$-axis intercept replots will be parabolic if both changing fixed substrates individually affect the $1/v$-axis intercept and are reversibly connected at saturating concentrations of the varied substrate. Intercept replots will be linear if the two changing fixed substrates are not reversibly connected, even though each affects the intercept individually. Slope replots are parabolic if both changing fixed substrates affect the slope individually and are reversibly connected, and linear if the two changing fixed substrates are not reversibly connected.*

For example, in the Bi Uni Uni Bi Ping Pong system, the plots of $1/v$ versus $1/[C]$ at different fixed concentrations of A and B are parallel. The replot of $1/v$-axis intercept versus $1/[A]$ or $1/[B]$ will be parabolic because both A and B affect the intercept and the two fixed substrates are reversibly connected (even at saturating [C]). When [A] is varied at different fixed levels of B and C, or when [B] is varied at different fixed levels of A and C, the intercept replots are linear because neither B and C nor A and C are reversibly connected. Slope replots are also linear. Essentially, a parabolic replot means that the appropriate slope or intercept factor contains squared concentrations terms (i.e., the product of the concentrations of the two changing fixed substrates; for example, see equation IX-236). As another example, consider the Ordered Ter Ter system. Here all three substrates are reversibly connected so we might expect parabolic slope and intercept replots regardless of which pair of substrates are varied together as the changing fixed substrates. But keep in mind that the $1/v$-axis intercept represents the reciprocal velocity at a saturating concentration of the varied substrate. Thus when [B] is saturating, the reversible connection between A and C is broken. Consequently, intercept replots of $1/v$ versus 1/[B] data obtained at different fixed concentrations of A and C are linear. (Examination of equation IX-205 shows no [A][C] term in the intercept factor.) When either B and C or A and B are the changing fixed substrates, intercept replots are parabolic. All slope replots for the Ordered Ter Ter system are parabolic.

A similar rule predicts the results when two or more substrates are treated as the varied substrates (maintaining their concentrations at a constant ratio):

> *When two substrates are varied together (at a constant ratio), the reciprocal plots will be (a) linear if there is no reversible connection between the two varied substrates or (b) parabolic if the two substrates are linked reversibly.*

Thus for the Bi Bi Uni Uni Ping Pong Ter Ter system, plots of $1/v$ versus 1/[A] at different fixed concentrations of B where $[C]=x[A]$ and plots of $1/v$ versus 1/[B] at different fixed concentrations of A where $[C]=x[B]$ are linear. However, plots of $1/v$ versus 1/[A] at different fixed concentrations of C where $[B]=x[A]$ are parabolic. For an Ordered Ter Ter system, all reciprocal plots are parabolic when two substrates are varied together at a constant ratio. However, if [B] is saturating, the reversible connection between the points of addition of A and C is broken and plots of $1/v$ versus 1/[A] or 1/[C] are linear.

Substrates That Add Twice

The rules above can be used to predict reciprocal plot patterns for systems in which a substrate adds twice. For example, consider the following sequences:

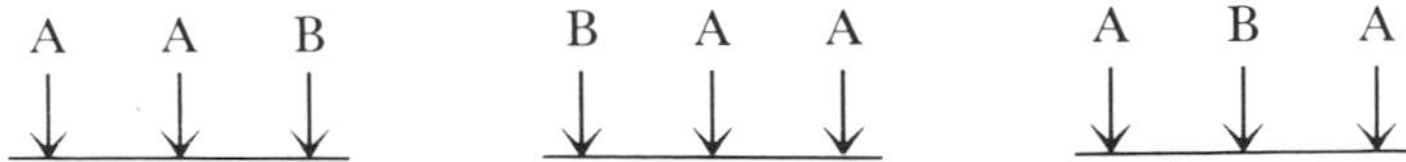

When A is the varied substrate, the reciprocal plots at different fixed levels of B are parabolic in all three sequences. However, the plot for the A–B–A sequence would become linear as [B] approaches saturation (the reversible connection between the points of addition of A will be broken). When B is the varied substrate and A the changing fixed substrate, the plots are linear. Slope replots are parabolic in all three sequences. Intercept replots for the A–A–B and B–A–A sequences are parabolic, but in the A–B–A sequence, the intercept replot is linear.

Product Inhibition

The effect of different product inhibitors on the reciprocal plots can be similarly predicted. For example, consider the Ordered Ter Ter system. P and A are reversibly connected in the presence of unsaturating [B] and [C]. Consequently P increases $slope_{1/\text{A}}$. Since P and A combine with different enzyme forms, P also affects the $1/v$-axis intercept. Thus at unsaturating [B] and [C], P is a mixed-type inhibitor with respect to A. When either [B] or [C] is saturating, the reversible connection between P and A is broken. P can no longer increase the [EA]/[E] ratio and consequently has no effect on $slope_{1/\text{A}}$. The $1/v$-axis intercept is still affected though; thus P acts as an uncompetitive inhibitor with respect to A at saturating [B] or [C]. Similarly, we would predict that P will affect the intercept and $slope_{1/\text{B}}$ at unsaturating [A] and [C]. Saturating [A] has no effect (P remains a mixed-type inhibitor with respect to B), but saturating [C] breaks the connection (and P becomes an uncompetitive inhibitor). P is a mixed-type inhibitor with respect to C at all A and B concentrations because (*a*) P and C combine with different enzyme forms, (*b*) the enzyme forms combining with the two different ligands are reversibly connected, and (*c*) saturation with [A] or [B] does not break the reversible connection. Q is not reversibly connected to A or B or C (in the absence of P and R). On the downstream side, the connection is broken by the release of R while the release of P on the upstream side prevents Q from backing up the reaction. (Although some P and R are produced in the reaction, their concentrations during the course of the assay remain essentially zero.) Thus Q is an uncompetitive inhibitor with respect to all three substrates at all concentrations of the nonvaried substrates. R and A combine with the same enzyme form. Therefore, R affects the $slope_{1/A}$ and is competitive with A at all B and C concentrations. If we write the Ordered Ter Ter reaction sequence only once, the points of combination of R and B or R and C appear not to be reversibly connected (in the absence of P and Q). But, if the repeating sequence is written out, we see that R and B or R and C are reversibly connected further downstream. Consequently, R will affect $slope_{1/B}$ and $slope_{1/C}$. Ligands R, B, and C all combine with different enzyme forms, so R will act as a mixed-type inhibitor with respect to the two substrates. Specifically, R reacts with E thereby decreasing [E] relative to [EA], [EA] relative to [EAB], and [EAB] relative to [EABC + EPQR]. The increase in the [EAB]/[EA] and [EABC]/[EAB] ratios result in increases in $slope_{1/B}$ and $slope_{1/C}$. When [A] is saturating, the reversible connection between R and B or R and C is broken. Saturating [A] reduces the steady-state level of E to zero, so R has nothing to combine with. In this case, R is not an inhibitor. Saturating [B] will eliminate the reversible connection between the points of combination of R and C. Consequently, R has no effect on $slope_{1/C}$ at saturating [B] (i.e., R becomes an uncompetitive inhibitor). Saturating [C] will not change the effect of R on $slope_{1/B}$ (R and B remain reversibly connected through A) and consequently R is a mixed-type inhibitor with respect to B at all C concentrations. The effect of saturating [A] suggests another rule to keep in mind.

> *A compound (product, inhibitor) has no effect on the reciprocal plot if the steady-state level of the enzyme form it combines with is zero (because of saturation with another ligand that combines with the same enzyme form). Stated differently: A compound cannot combine with an enzyme form whose steady-state level is already zero.*

Establishing a Reversible Connection by Adding Another Product

Earlier we saw how the addition of a significant concentration of one or more products can establish a reversible connection between the points of addition of two substrates and thereby change parallel initial velocity reciprocal plots to intersecting ones. The same procedure can be used in analyzing product inhibition patterns. For example, consider the Ordered Bi

Quad system shown below.

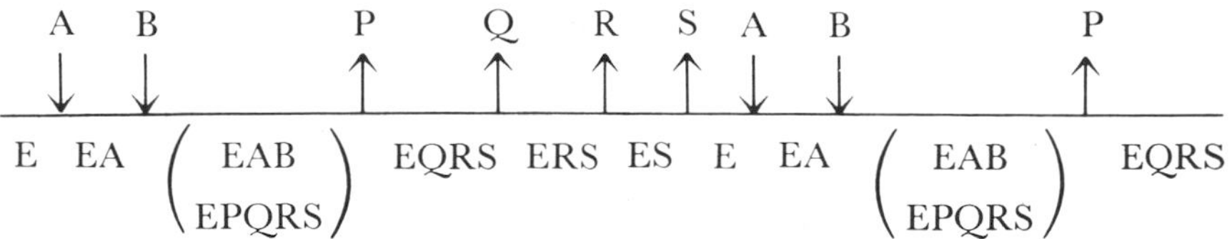

The general rules will permit A, B, P, and S to be identified: S is competitive with A at all B concentrations and a mixed-type inhibitor with respect to B at unsaturating [A]. Saturation with A overcomes the inhibition; P is a mixed-type inhibitor with respect to A at unsaturating [B], an uncompetitive inhibitor with respect to A at saturating [B], and a mixed-type inhibitor with respect to B at all A concentrations. The products Q and R cannot be distinguished from each other by the usual product inhibition studies. Both products are uncompetitive inhibitors with respect to both substrates at all concentrations of the nonvaried substrate. However, if a significant concentration of S is added, a reversible connection is established between R and A or B. Now R will act as a mixed-type inhibitor with respect to both substrates; Q remains an uncompetitive inhibitor. The addition of both R and S or of P will convert Q from an uncompetitive inhibitor to a mixed-type inhibitor.

Table IX-12 illustrates the use of Cleland's rules in predicting reciprocal plot patterns for the Bi Bi Uni Uni Ping Pong Ter Ter system in the presence of different product inhibitors. In practice, of course, we work in the opposite direction (the patterns are used to predict the mechanism).

Modification of Slope and Intercept Rules for Steady-State Systems with Rapid Equilibrium Segments

Care must be exercised in applying the general slope and intercept rules when rapid equilibrium segments are present. For example, consider the Bi Uni Uni Bi Ping Pong Ter Ter system described earlier in which E, EA, and ER are at equilibrium with A and R.

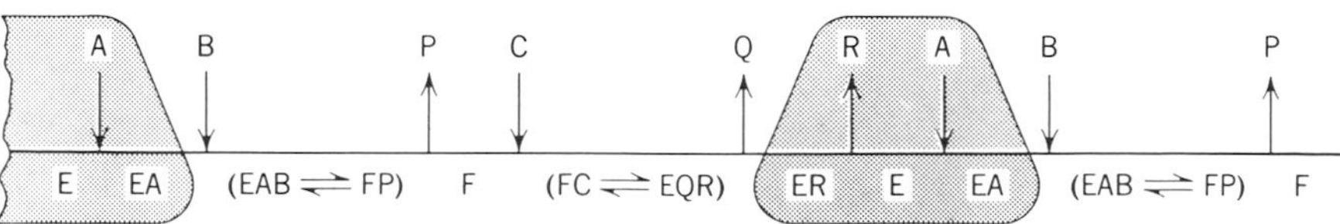

Table IX-12 Predicting Reciprocal Plot Patterns for Product Inhibition in a Bi Bi Uni Uni Ping Pong Ter Ter System[a]

$$\begin{array}{ccccccccccccc} & A\downarrow & B\downarrow & P\uparrow & Q\uparrow & C\downarrow & R\uparrow & A\downarrow & B\downarrow & P\uparrow & Q\uparrow & C\downarrow & R\uparrow \\ E & EA & \begin{pmatrix} EAB \\ FPQ \end{pmatrix} & FQ & F & \begin{pmatrix} FC \\ ER \end{pmatrix} & E & EA & \begin{pmatrix} EAB \\ FPQ \end{pmatrix} & FQ & F & \begin{pmatrix} FC \\ ER \end{pmatrix} & E \end{array}$$

Product Inhibitor	Varied Substrate	Saturating Substrate	Additional Product at Constant Level[b]	Reasoning and Prediction
P	A	—	—	P and A combine with different enzyme forms. Therefore, P affects the $1/v$-axis intercept. The points of attachment (FQ and E) are reversibly connected via the sequence $E \rightleftharpoons EA \rightleftharpoons EAB \rightleftharpoons EPQ \rightleftharpoons FQ$. Therefore, P affects the slope of the $1/v$ versus $1/[A]$ plot; P then is a mixed-type inhibitor with respect to A.
P	A	B	—	The intercept effect is unaffected. The slope effect is eliminated because saturating [B] reduces the steady-state level of EA to zero breaking the reversible connection between FQ and E; P now acts as an uncompetitive inhibitor.
P	A	C	—	Saturating [C] reduces the steady-state level of F to zero, but this has no effect on the sequence between FQ and E; P is still a mixed-type inhibitor.

P	B	—	—	P and B combine with different enzyme forms. Therefore, P affects the $1/v$-axis intercept. The sequence between the points of attachment (FQ and EA) is reversible. Therefore, the $slope_{1/B}$ is affected; P is a mixed-type inhibitor.
P	B	A	—	Saturating [A] reduces the steady-state level of E to zero but this has no effect on the reversible sequence between FQ and EA; P is still a mixed-type inhibitor.
P	B	C	—	Saturating [C] reduces the steady-state level of F to zero, but this has no effect on the reversible sequence between FQ and EA; P is still a mixed-type inhibitor.
P	C	—	—	P and C combine with different enzyme forms. Therefore, P will affect the $1/v$-axis intercept. The points of attachment of P and C (FQ and F) are not reversibly connected. Downstream (to the right), the connection is broken by the release of Q (but Q is not present at a significant concentration). Upstream, the connection is broken by the release of R (also not present at a significant concentration). Thus P is an uncompetitive inhibitor.
P	C	—	Q	If a significant level of Q is present, a reversible connection between FQ and F will be established. The plots of $1/v$ versus $1/[C]$ at different fixed concentrations of P (and constant [A], [B], and [Q]) will now intersect to the left of the $1/v$-axis; that is, P now affects both $slope_{1/C}$ and $1/v$-axis intercept. (In the presence of Q, P can back up the reaction decreasing the level of F relative to FC + ER.)

Table IX-12 (*Continued*)

Product Inhibitor	Varied Substrate	Saturating Substrate	Additional Product at Constant Level[b]	Reasoning and Prediction
P	C	—	R	The presence of a significant constant level of R establishes a reversible connection between F and FQ in the downstream direction. As above, the plots at different fixed concentrations of P intersect. (In the presence of R, P can back up the reaction increasing the level of FC+ER relative to F.)
P	C	A	—	Saturating [A] has no effect on pattern; P is still an uncompetitive inhibitor.
P	C	A	Q	The finite constant level of Q establishes the reversible connection upstream between F and FQ. Pattern is still intersecting as in P+Q at unsaturating [A].
P	C	A	R	A and R both combine with E. Saturating [A] reduces the steady-state level of E to zero. Thus there is nothing for R to combine with and the finite unsaturating concentration of R cannot establish a reversible connection downstream between F and FQ; P is still an uncompetitive inhibitor.
P	C	B	—	Saturating [B] has no effect on the situation; P is still an uncompetitive inhibitor with respect to C.

Q	A	—	—	Q and A combine with different enzyme forms (F and E, respectively). Therefore, there will be an intercept effect. The two forms are not reversibly linked in the absence of P or R. Therefore, Q is uncompetitive with respect to A.
Q	A	—	P or R	A significant constant level of P or R establish a reversible connection between F and E; Q will behave as a mixed-type inhibitor. Plots of $1/v$ versus $1/[A]$ at different fixed concentrations of Q (and a constant [B], [C], and [P] or [R]) intersect to the left of the $1/v$-axis.
Q	A	B	—	Saturating [B] reduces the steady-state level of EA to zero, but A combines with E and Q combines with F, whose levels remain finite. Therefore, there is no change; Q is still an uncompetitive inhibitor with respect to A.
Q	A	B	P	Although P establishes an upstream reversible connection between the points of Q and A addition, saturating [B] interrupts the connection. Thus the plots are still parallel.
Q	A	C	—	Saturating [C] reduces the steady-state level of F to zero. Q has nothing to combine with; hence the inhibition is overcome.
Q	B	—	—	Q and B combine with different enzyme forms that are not reversibly linked in the absence of P. Therefore, there is an intercept effect but no slope effect; Q is uncompetitive with respect to B.

Table IX-12 (*Continued*)

Product Inhibitor	Varied Substrate	Saturating Substrate	Additional Product at Constant Level[b]	Reasoning and Prediction
Q	B	—	P or R	A significant level of P or R establishes a reversible connection between F and EA and thereby changes the uncompetitive parallel pattern to a mixed-type intersecting pattern.
Q	B	A	R	The reversible downstream connection between F and EA established by R is broken when [A] is saturating. Thus the pattern becomes parallel again.
Q	B	A	—	Saturating [A] does not change the original situation; Q is still uncompetitive with respect to B.
Q	B	C	—	Saturating [C] overcomes the inhibition by Q by reducing the steady-state level of F to zero.
Q	C	—	—	Q and C combine with the same enzyme form (F). Thus we will see a slope effect, but no intercept effect. Q is a competitive inhibitor with respect to C.
Q	C	A or B	—	Saturating [A] or [B] has no effect; Q is still competitive.
R	A	—	—	R and A both combine with form E. Therefore, R is a competitive inhibitor with respect to A.

R	A	B or C	—	Saturating [B] or [C] has no effect; R is still competitive.
R	B	—	—	R and B combine with different enzyme forms (E and EA, respectively). The two forms are reversibly linked in the presence of A. Therefore, we observe intercept and slope effects; R is a mixed-type inhibitor with respect to B.
R	B	A	—	Saturating [A] reduces the steady-state level of E to zero. R has nothing to combine with. Therefore, the inhibition is overcome.
R	B	C	—	Saturating [C] has no effect; R remains a mixed-type inhibitor with respect to B.
R	C	—	—	R and C combine with different enzyme forms which are reversibly linked. Therefore, R is a mixed-type inhibitor with respect to C.
R	C	A	—	Saturating [A] overcomes the inhibition by driving the steady-state level of E to zero; R has nothing to combine with.
R	C	B	—	Saturating [B] has no effect; R is still a mixed-type inhibitor with respect to C.

[a]The analyses assume that only normal complexes form in the presence of products (i.e., no dead-end complexes form).

[b]The conversion of a parallel (uncompetitive) pattern to an intersecting (mixed-type) pattern by the addition of a second product can also be seen from the complete velocity equation. For example, if the complete velocity equation is rearranged to show C as the varied substrate without dropping any terms, we would find no [P] term in the slope factor. Hence P alone has no effect on the $slope_{1/C}$. The slope factor will, however, contain terms in [P][Q] and [P][R]. Thus in the presence of P and Q or P and R, the $slope_{1/C}$ is increased.

P is competitive with C and a mixed-type inhibitor with respect to A and B; R is obviously competitive with A and uncompetitive with C. Normally, we would expect R to be a mixed-type inhibitor with respect to B, and Q to be uncompetitive with A and B, and a mixed-type inhibitor with respect to C. Instead, we find (from the velocity equation derived by the method of Cha) that R is competitive with B while Q is not an inhibitor at all. The fact that R is competitive with B stems from the fact that ER, E, and EA are at equilibrium. Saturation with B will drive all the EA to EAB. Since E is at equilibrium with EA, all the E is also driven to EAB, leaving nothing for R to combine with. Q does not inhibit because in the absence of a finite concentration of R, the steady-state level of ER is essentially zero so Q has nothing to combine with. If a finite concentration of R is present, Q will be competitive with A and B; that is, saturating [A] or [B] will drive all the ER to EAB and, once again, Q has nothing to combine with. Also, with Q and R present together, both products now behave as mixed-type inhibitors with respect to C. In general, a product will not act as an inhibitor if it must bind to an enzyme form that is part of the rapid equilibrium segment *unless* it competes directly with a substrate or it forms a dead-end complex together with one of the substrates. If we keep the rules regarding rapid equilibrium systems in mind, then product inhibition patterns can be predicted quite easily. For example, consider the Random A–B, Random Q–R Bi Uni Uni Bi Ping Pong Ter Ter system shown below in which A and B add and Q and R leave in a Rapid Equilibrium Random fashion.

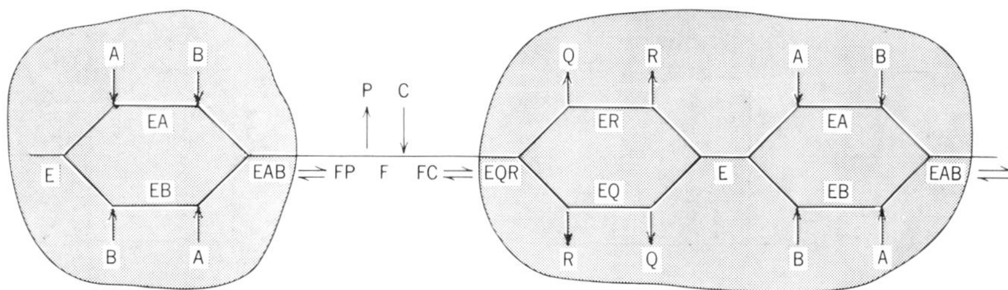

P is competitive with C and a mixed-type inhibitor with respect to A and B (i.e., the usual steady-state prediction); Q and R are both competitive with A and B. The Q that adds to ER and the R that adds to EQ are ignored; we consider only the Q and R that are inhibitory (the ones adding to E). Thus Q and R are uncompetitive with respect to C. (The point of addition, E, is not reversibly connected to F.) There are product release steps in both directions. These predictions assume that the products add only to the enzyme forms that they normally react with as substrates in the reverse direction. Asparagine synthetase has the kinetic mechanism above where A and B are aspartate and MgATP, C is NH_3, Q and R are AMP and asparagine, and P is $MgPP_i$ (Ceder and Schwartz, 1969). AMP is competitive with ATP but a mixed-type inhibitor with respect to aspartate. Asparagine is competitive with aspartate, but a mixed-type inhibitor with respect to MgATP; $MgPP_i$ is a mixed-type inhibitor with respect to MgATP and aspartate. The results suggest that AMP and asparagine form dead-end complexes with AMP occupying the MgATP site and asparagine occupying the aspartate site. Neither product interferes with the binding of the substrate to the vacant site. $MgPP_i$ also forms an EP and dead-end EPB complex by binding to the MgATP site.

DD. DEAD-END INHIBITION

A dead-end inhibitor is a compound that reacts with one or more enzyme forms to yield a complex that cannot participate in the reaction. The velocity equation in the presence of a dead-end inhibitor can be derived in the usual manner by the King-Altman method. The basic figure will contain additional corners corresponding to each I-containing complex that forms. However, if we know the velocity equation for the uninhibited reaction, then we can easily write the new velocity equation as modified by the inhibitor without going through an entire derivation. The effect of a dead-end inhibitor is to multiply certain terms in the denominator of the uninhibited velocity equation (written in the form $v/V_{max} = [A][B][C]\cdots/\text{denominator}$) by $(1+[I]/K_i)$. The terms multiplied by the factor are those representing the enzyme form combining with the inhibitor. The K_i is the constant for the dissociation of I from the specific enzyme form-inhibitor complex. In rapid equilibrium systems, the relative concentration of any particular enzyme form is given by a single denominator term (e.g., $[A]/K_A$ represents the EA complex and $[A][B]/K_AK_B$ represents the EAB complex). However, in steady-state systems, the relative concentration of a particular enzyme form may be expressed by several denominator terms, and, often, a particular denominator term represents more than one enzyme form. The denominator terms multiplied by the $(1+[I]/K_i)$ factor are those which appear in the numerator of the distribution equation for the enzyme form combining with the inhibitor. If a distribution equation in terms of kinetic constants cannot be written for the enzyme form combining with the inhibitor, then K_i cannot be evaluated kinetically because only some of the rate constants comprising certain denominator terms will be multiplied by the factor. Even if a distribution equation in terms of kinetic constants can be written, K_i will still be undeterminable by kinetic means if the same numerator terms (of the distribution equation) are used in the numerator of the distribution equation for another enzyme form. To see how the velocity equations are affected consider a Ping Pong Bi Bi system in which an inhibitor, I, combines in a dead-end fashion with E.

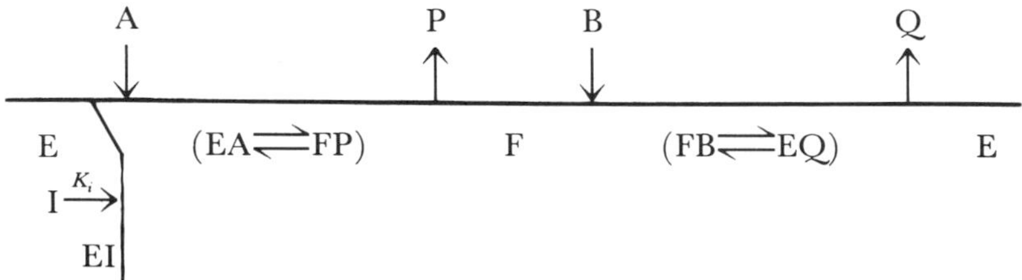

The velocity equation for the forward reaction in the absence of any products or inhibitor is:

$$\frac{v}{V_{max}} = \frac{[A][B]}{K_{m_B}[A] + K_{m_A}[B] + [A][B]} \tag{IX-339}$$

The I combines with E. The distribution equation for $[E]/[E]_t$ in the absence of P and Q is (from equation IX-136 or IX-158):

$$\frac{[E]}{[E]_t} = \frac{K_{m_A}[B]}{K_{m_B}[A] + K_{m_A}[B] + [A][B]} \tag{IX-340}$$

Thus in the presence of I, the $K_{m_A}[B]$ term in the denominator of the velocity equation is multiplied by $(1+[I]/K_i)$:

$$\boxed{\frac{v}{V_{max}} = \frac{[A][B]}{K_{m_B}[A] + K_{m_A}[B]\left(1 + \frac{[I]}{K_i}\right) + [A][B]}} \tag{IX-341}$$

When [A] is varied, the equation is:

$$\boxed{\frac{v}{V_{max}} = \frac{[A]}{K_{m_A}\left(1 + \frac{[I]}{K_i}\right) + [A]\left(1 + \frac{K_{m_B}}{[B]}\right)}} \tag{IX-342}$$

When [B] is varied:

$$\boxed{\frac{v}{V_{max}} = \frac{[B]}{K_{m_B} + [B]\left(1 + \frac{K_{m_A}}{[A]} + \frac{K_{m_A}[I]}{K_i[A]}\right)}} \qquad \text{(IX-343)}$$

Thus I is competitive with respect to A, and $K_{i_{slope}} = K_i$ (Fig. IX-48); I is uncompetitive with respect to B at unsaturating [A] (Fig. IX-49). The K_i can be evaluated by any convenient plot and replot (e.g., $slope_{1/A}$ versus [I]).

Suppose that I reacts with stable enzyme form F. The reaction sequence is:

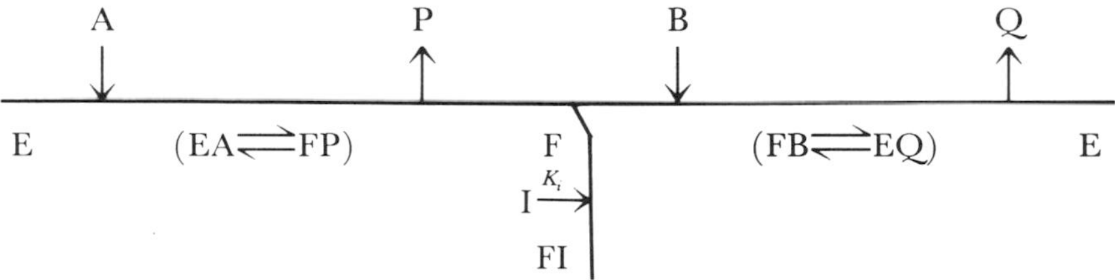

The distribution equation for form F in the absence of P and Q is given by equation IX-159, which can be written as:

$$\frac{[F]}{[E]_t} = \frac{K_{m_B}[A]}{K_{m_B}[A] + K_{m_A}[B] + [A][B]} \qquad \text{(IX-344)}$$

Consequently, the $K_{m_B}[A]$ term in the denominator will be multiplied by $(1 + [I]/K_i)$. The resulting equations are symmetrical to equations IX-341, IX-342, and IX-343. I will be competitive with B and uncompetitive with A. If I combines with both E and F with dissociation constants K_{i_1} and K_{i_2}, respectively, the velocity equation becomes:

$$\frac{v}{V_{max}} = \frac{[A][B]}{K_{m_B}[A]\left(1 + \frac{[I]}{K_{i_2}}\right) + K_{m_A}[B]\left(1 + \frac{[I]}{K_{i_1}}\right) + [A][B]} \qquad \text{(IX-345)}$$

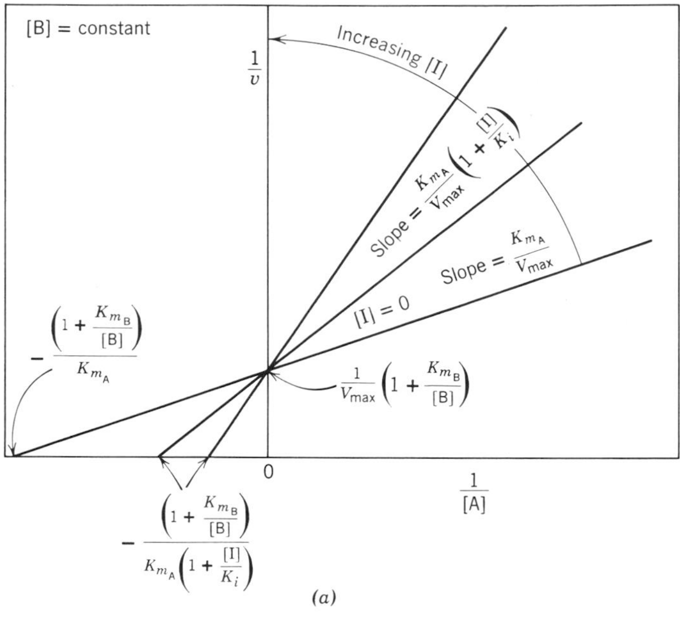

(*a*)

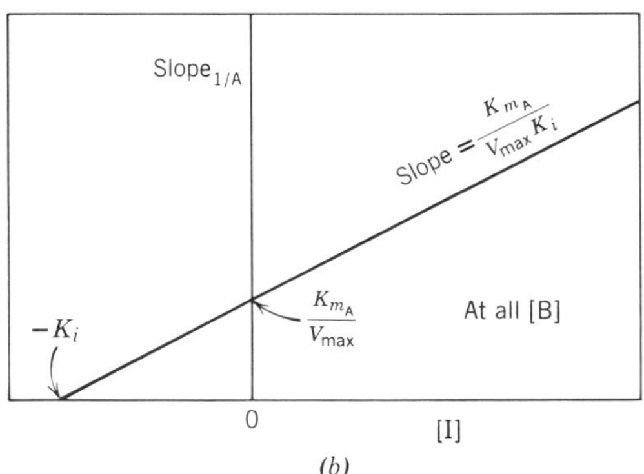

(*b*)

Fig. IX-48. Dead-end inhibition in a bireactant Ping Pong system; I combines with E. (*a*) $1/v$ versus $1/[A]$ plot at different fixed I concentrations and a constant B concentration. (*b*) $Slope_{1/A}$ replot.

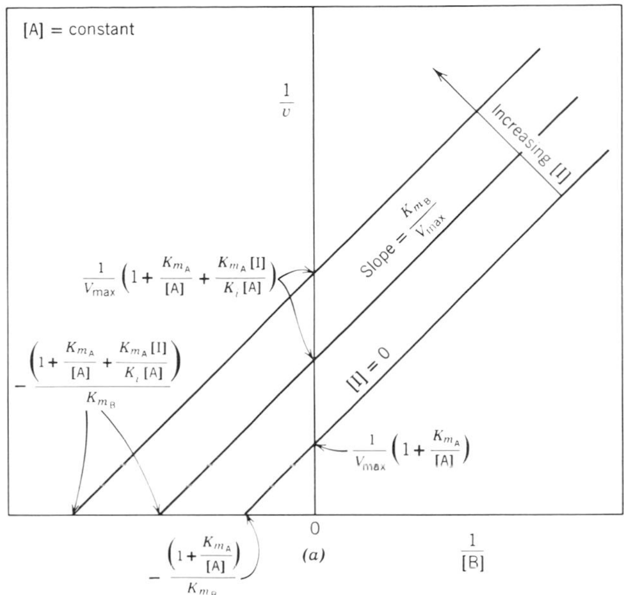

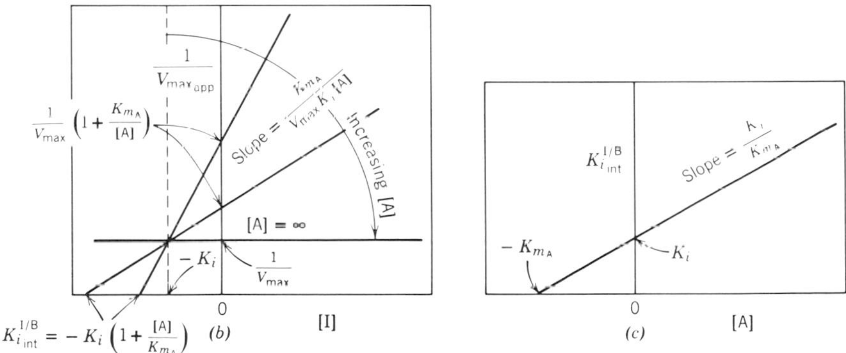

Fig. IX-49. Dead-end inhibition in a bireactant Ping Pong system; I combines with E. (*a*) $1/v$ versus $1/[B]$ plot at different fixed I concentrations and a constant A concentration. (*b*) $1/v$-axis intercept replot. Each replot represents a different constant A concentration. (*c*) Secondary replot of $K_{i_{int}}^{I/B}$ versus [A].

When [A] is varied:

$$\frac{v}{V_{\max}} = \frac{[\mathrm{A}]}{K_{m_{\mathrm{A}}}\left(1+\frac{[\mathrm{I}]}{K_{i_1}}\right)+[\mathrm{A}]\left(1+\frac{K_{m_{\mathrm{B}}}}{[\mathrm{B}]}\right)\left[1+\frac{[\mathrm{I}]}{K_{i_2}\left(1+\frac{[\mathrm{B}]}{K_{m_{\mathrm{B}}}}\right)}\right]} \tag{IX-346}$$

When [B] is varied:

$$\frac{v}{V_{\max}} = \frac{[\mathrm{B}]}{K_{m_{\mathrm{B}}}\left(1+\frac{[\mathrm{I}]}{K_{i_2}}\right)+[\mathrm{B}]\left(1+\frac{K_{m_{\mathrm{A}}}}{[\mathrm{A}]}\right)\left[1+\frac{[\mathrm{I}]}{K_{i_1}\left(1+\frac{[\mathrm{A}]}{K_{m_{\mathrm{A}}}}\right)}\right]} \tag{IX-347}$$

Thus I is a mixed-type inhibitor with respect to the varied substrate at unsaturating concentrations of the nonvaried substrate. When the nonvaried substrate is saturating, I acts as a competitive inhibitor (because the [I] term in the intercept factor goes to zero). K_{i_1} can be obtained from $slope_{1/\mathrm{A}}$ versus [I] replots or from $1/V_{\max_{app}}$ versus [I] replots when B is the varied substrate; K_{i_2} can be obtained from $slope_{1/\mathrm{B}}$ versus [I] replots or from $1/V_{\max_{app}}$ versus [I] replots when A is the varied substrate. The reciprocal plots for A as the varied substrate are shown in Figure IX-50. The plots for B as the varied substrate are symmetrical to those shown for A.

It is unlikely that an inhibitor would combine with either of the two central complexes, since all ligand binding sites are occupied. If such a combination did occur though, K_i could not be determined kinetically. For example, suppose I combined with EA + FP. If we examine the King-Altman solution for $[\mathrm{EA}+\mathrm{FP}]/[\mathrm{E}]_t$, we find an [A][B] term. However, the [A][B] term in the denominator of the velocity equation does not arise solely from the King-Altman solution for $[\mathrm{EA}+\mathrm{FP}]/[\mathrm{E}]_t$, but also from the solution for $[\mathrm{FB}+\mathrm{EQ}]/[\mathrm{E}]_t$. Therefore, we cannot multiply the denominator [A][B] term by $(1+[\mathrm{I}]/K_i)$. (In contrast, only the King-Altman solution for $[\mathrm{E}]/[\mathrm{E}]_t$ contains a numerator [B] term, and the [A] term comes solely from the King-Altman solution for $[\mathrm{F}]/[\mathrm{E}]_t$.)

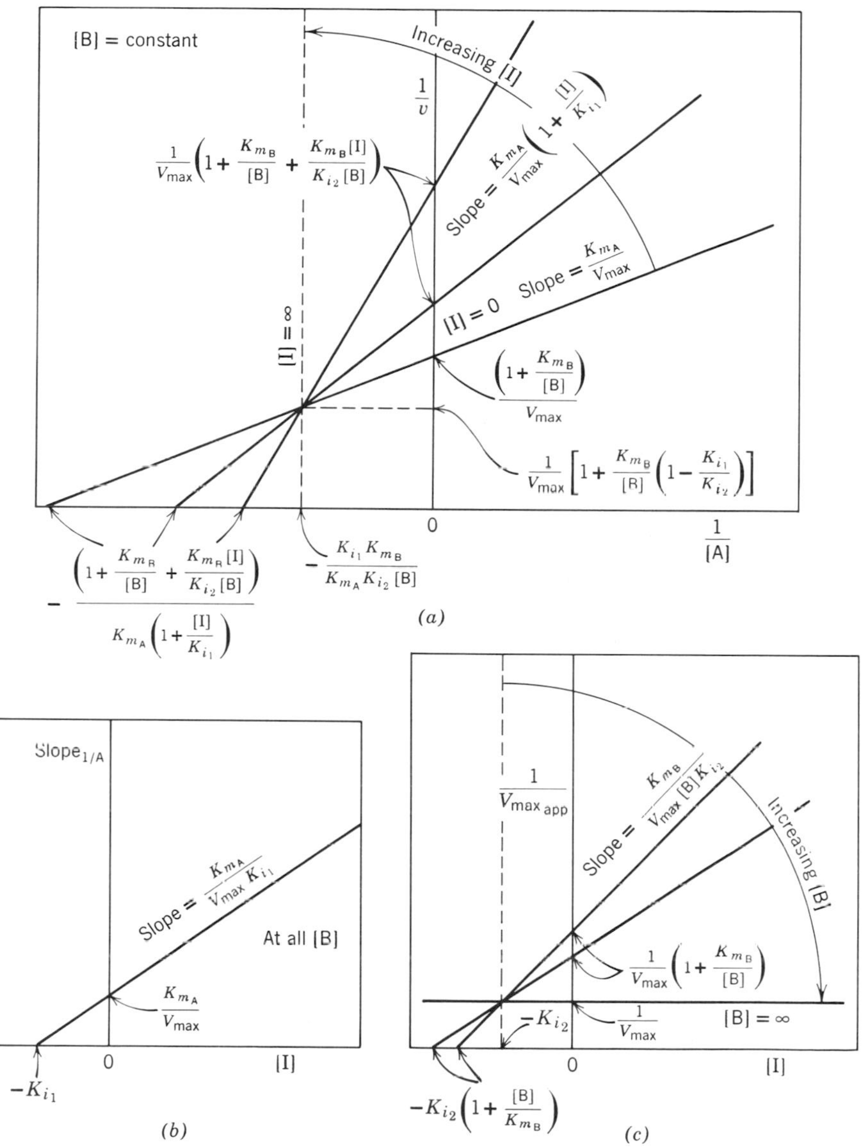

Fig. IX-50. Dead-end inhibition in a bireactant Ping Pong system; I combines with both E and F. (*a*) $1/v$ versus $1/[A]$ at different fixed I concentrations and a constant B concentration. (*b*) $Slope_{1/A}$ replot. (*c*) $1/v$-axis intercept replot.

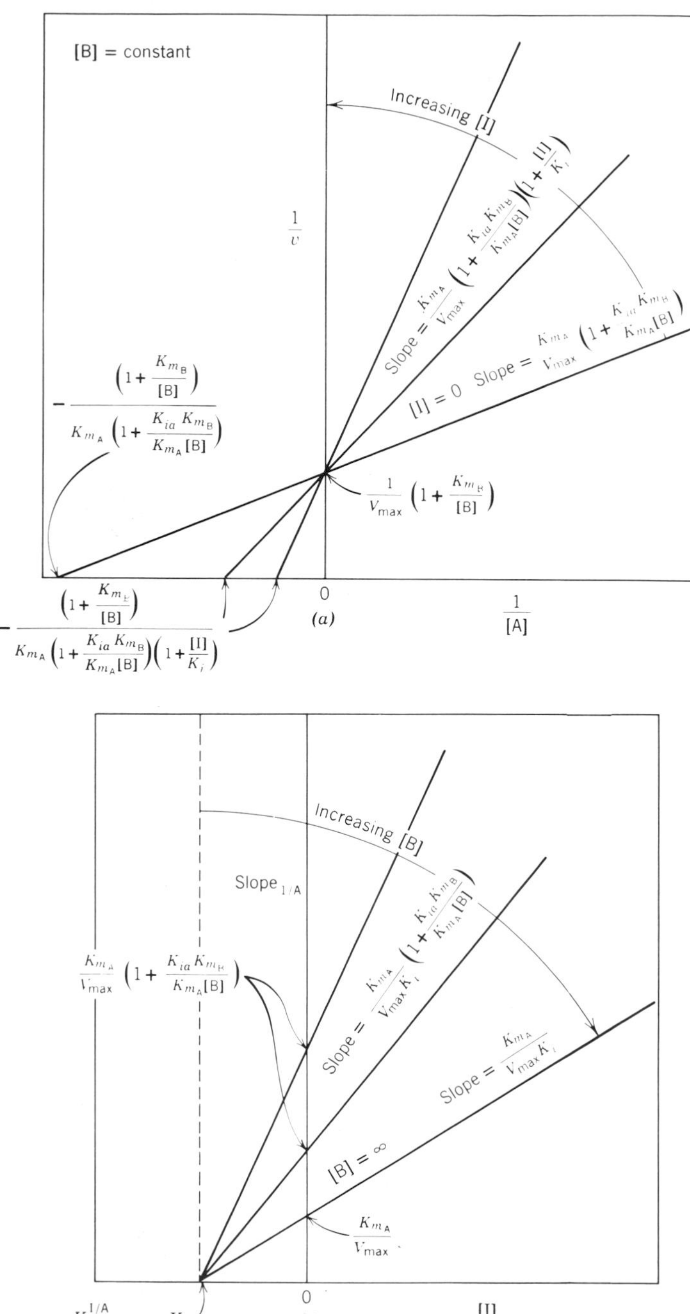

Fig. IX-51. Dead-end inhibition in an ordered bireactant system; I combines with E. (*a*) $1/v$ versus $1/[A]$. (*b*) $Slope_{1/A}$ replot.

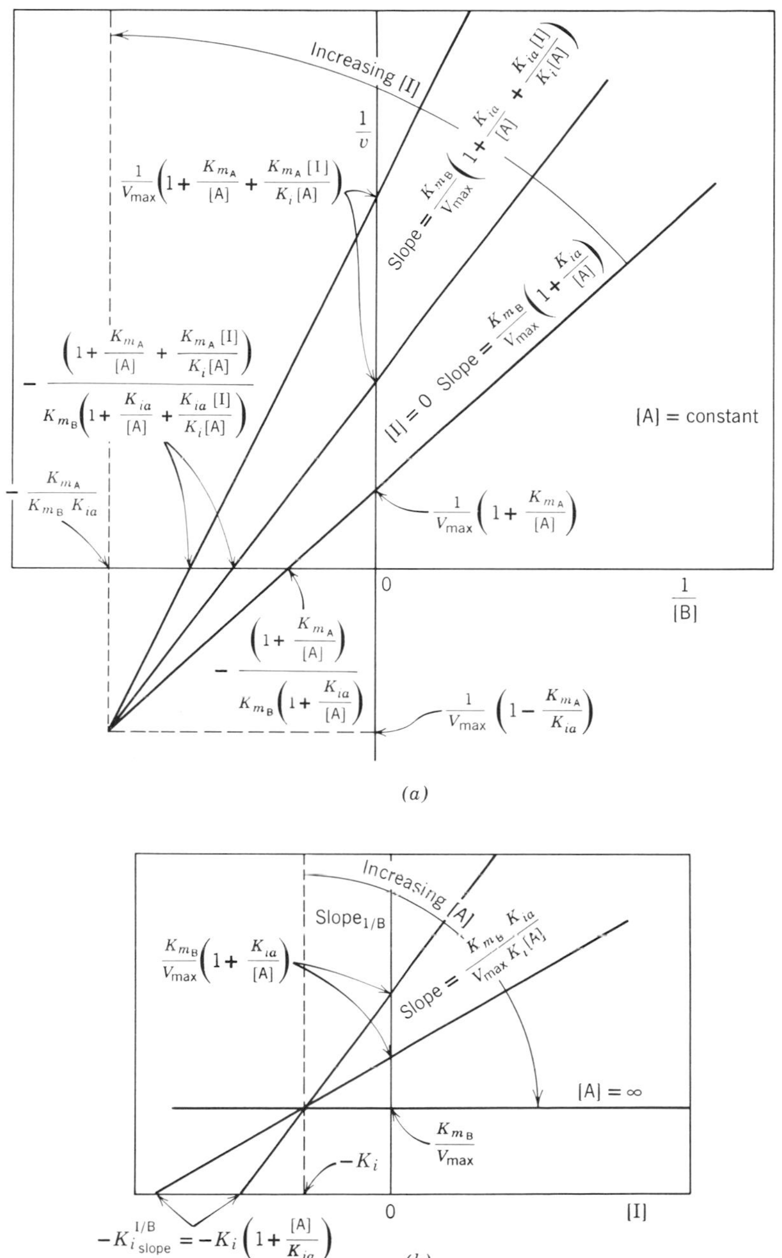

Fig. IX-52. Dead-end inhibition in an ordered bireactant system; I combines with E; $1/v$ versus $1/[B]$ and appropriate replots.

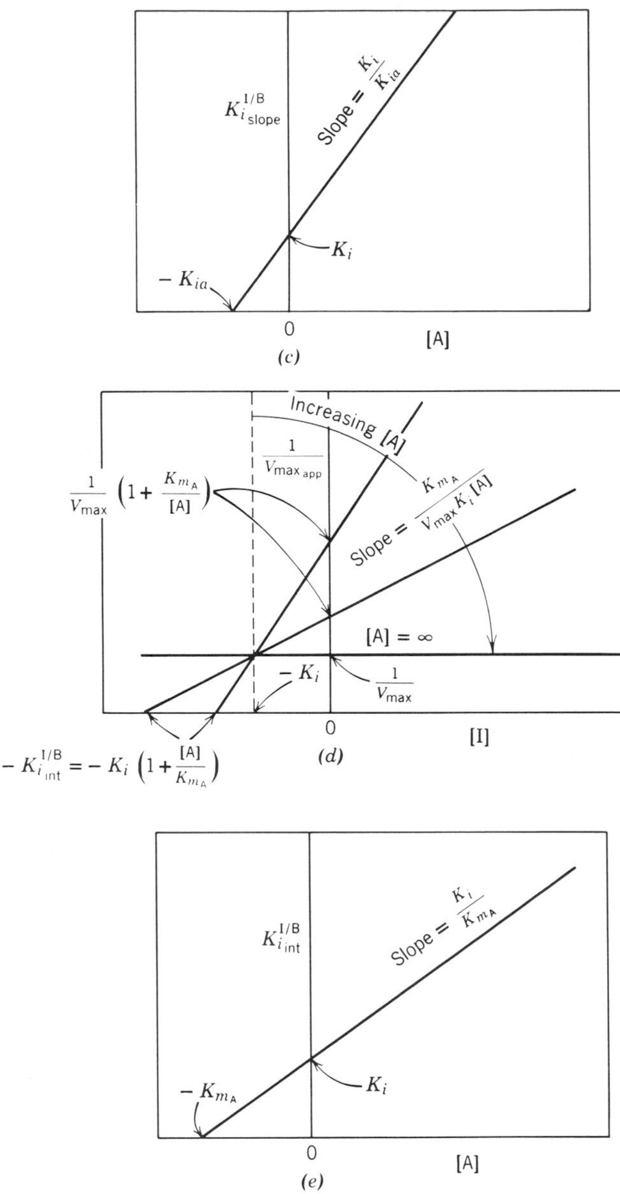

Fig. IX-52. (*Cont.*)

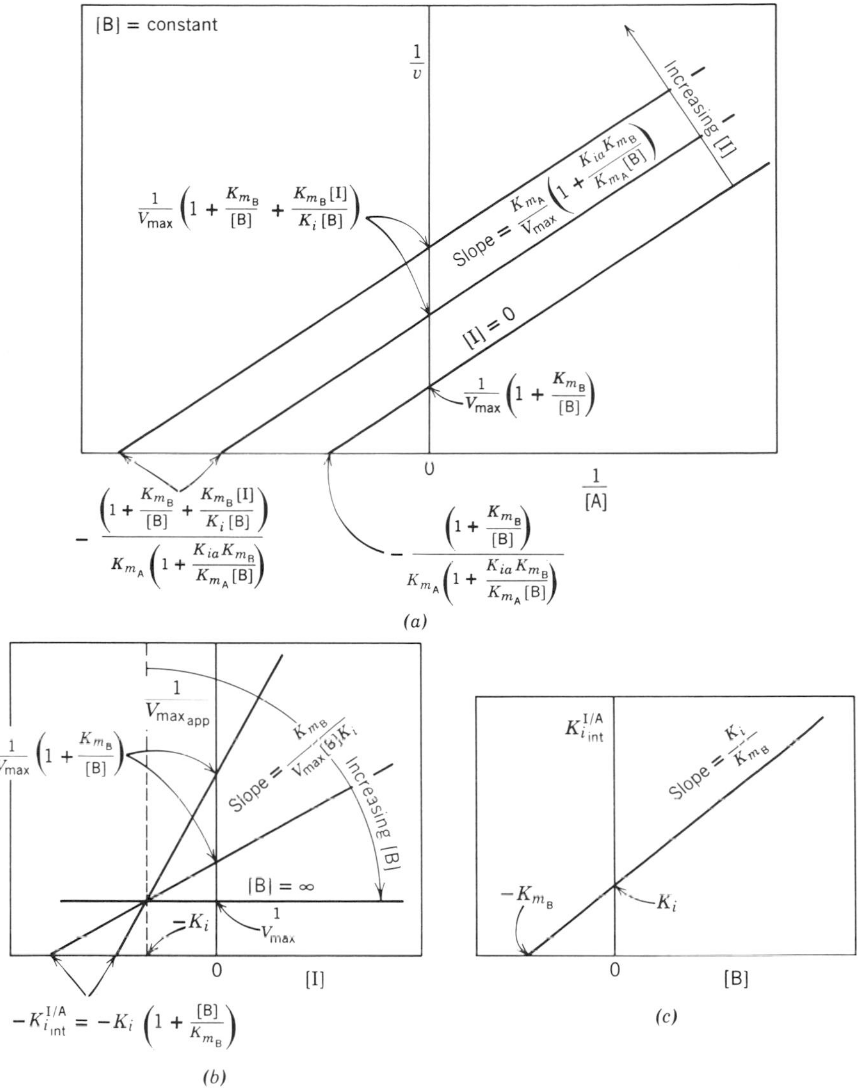

Fig. IX-53. Dead-end inhibition in an ordered bireactant system; I combines with EA; $1/v$ versus $1/[A]$ plot and appropriate replots.

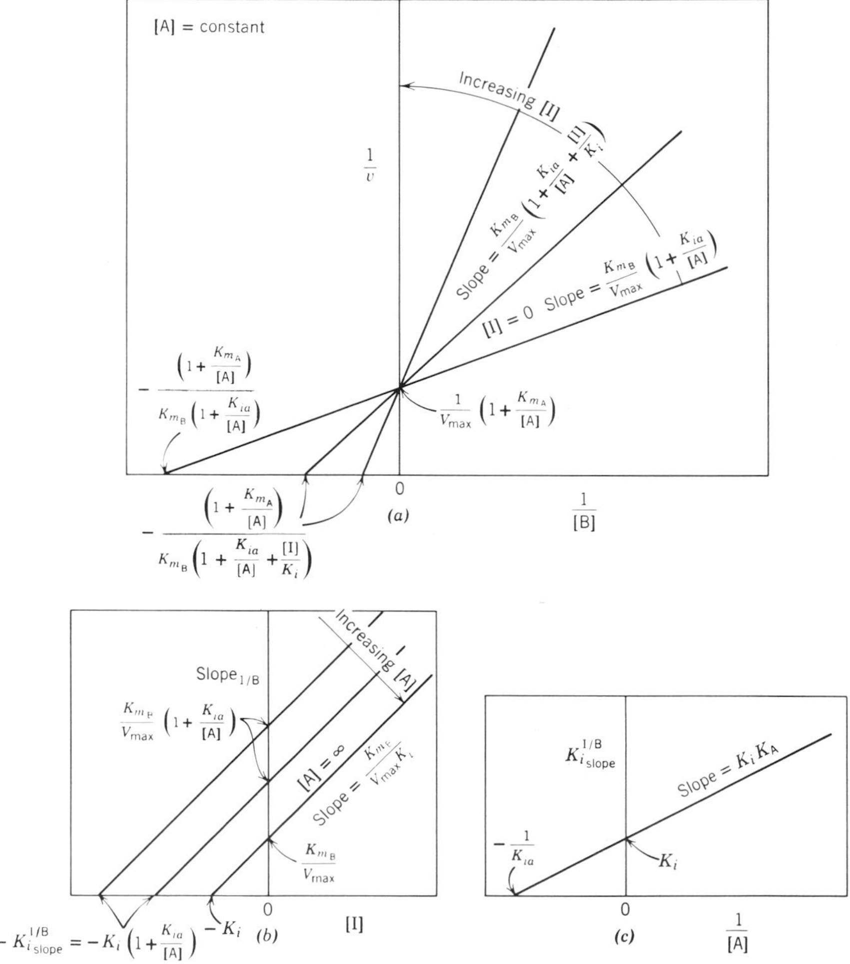

Fig. IX-54. Dead-end inhibition in an ordered bireactant system; I combines with EA; $1/v$ versus $1/[B]$ plot and appropriate replots.

As another example, consider the Ordered Bi Bi system. Inspection of the distribution equations or of the King-Altman solutions for the various enzyme forms shows that, in the absence of products, the constant and [B] terms in the denominator of the velocity equation represent free E, the [A] term represents the EA complex, while the [A][B] term represents the sum of the central complexes, [EAB + EPQ], and the EQ complex. Thus a dead-end inhibitor combining with free E introduces a $(1+[I]/K_i)$ factor to both the $K_{ia}K_{m_B}$ term and the K_{m_A}[B] term. A dead-end inhibitor combining with EA multiplies the K_{m_B}[A] term by $(1+[I]/K_i)$. The reciprocal plots and replots are shown in Figures IX-51 to IX-54. A dead-end inhibitor combining with EAB + EPQ (an unlikely event) or with EQ affects only part of the [A][B] term (i.e., some of the rate constants of $Coef_{AB}$ come from the King-Altman solution for [EAB + EPQ] and some come from the solution for [EQ]). Thus we cannot multiply the [A][B] term by $(1+[I]/K_i)$. A K_i determined kinetically will be an apparent value. The inhibition will be uncompetitive. Dead-end inhibitors have exactly the same effects on the Theorell-Chance system, since E and EA are represented by the same terms as in the Ordered Bi Bi system.

General Rules

The effect of a dead-end inhibitor on the $1/v$-axis intercept is predicted by the same rule described earlier (which is applicable to both product inhibitors and dead-end inhibitors):

> *The $1/v$-axis intercept of a reciprocal plot is affected when the inhibitor and varied substrate combine with different enzyme forms and saturation with the varied substrate cannot overcome the inhibition.*

The rule predicting slope effects by dead-end inhibitors is slightly different from the rule used for product inhibitors, because dead-end inhibitors, unlike products, are not normal participants in a reaction and, therefore, cannot back up the reaction sequence. Consequently, the slope is unaffected if the dead-end inhibitor adds after the varied substrate.

> *A dead-end inhibitor affects the slope of the reciprocal plot if (a) it and the varied substrate combine with the same enzyme form, or (b) it combines* ***before*** *(upstream from) the varied substrate with an enzyme form that is reversibly connected to the form combining with the varied substrate (i.e., the varied substrate must add after the inhibitor with no irreversible steps between the points of addition of the two ligands).*

Thus in the Ordered Bi Bi system shown below,

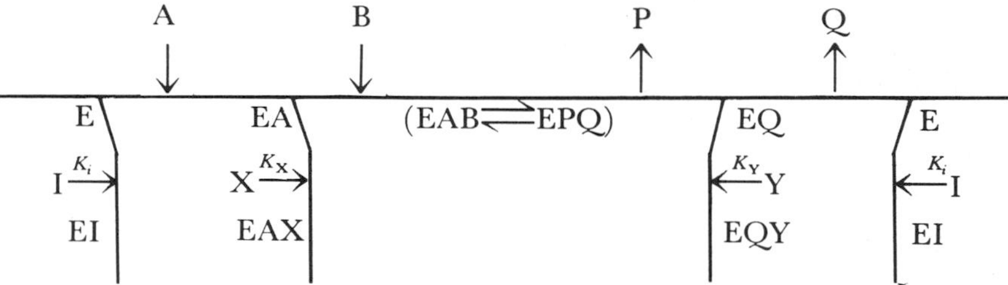

dead-end inhibitor I is competitive with A, and a mixed-type inhibitor with respect to B. Inhibitor X is competitive with B and uncompetitive with A (X adds after A; thus there is no slope effect). Inhibitor Y is uncompetitive with respect to both A and B. In the reverse direction, Y is competitive with P and uncompetitive with Q. I is competitive with Q and a mixed-type inhibitor with respect to P. In general, all reciprocal plots in the presence of dead-end inhibitors are linear. Similarly, all slope and $1/v$-axis intercept replots are linear except when more than one molecule of dead-end inhibitor combines with the same enzyme form. In this case, replots will be parabolic (if two molecules of I add) or higher ordered (if more than two molecules of I add). (See Chapter Eight, Section A.)

It is often difficult to distinguish between a random bireactant system and an ordered bireactant system. Dead-end inhibitors may help. Suppose we find that product Q is competitive with A and a mixed-type inhibitor with respect to B. If the system is ordered, A adds first and Q leaves last. A dead-end inhibitor, X, competitive with B will be uncompetitive with A. If the system is really random, then X will be a mixed-type inhibitor with respect to A. It is important to establish that X is really a dead-end inhibitor and not an alternate substrate (described in Section FF). An alternate substrate for B that promotes the reaction sequence (thereby regenerating EA) will act as a mixed-type inhibitor with respect to A in steady-state systems. (In Rapid Equilibrium Ordered systems an alternate substrate for B will be uncompetitive with A.)

Multiple Inhibition Analysis

If two different dead-end inhibitors, I and X, combine with the same enzyme form but are mutually exclusive, the denominator terms representing that enzyme form are multiplied by the factor $(1+[\mathrm{I}]/K_i+[\mathrm{X}]/K_\mathrm{X})$. For the situation where I and X combine with E in an Ordered Bi Bi system, the equation becomes:

$$\frac{v}{V_{max}} = \frac{[A][B]}{K_{ia}K_{m_B}\left(1+\frac{[I]}{K_i}+\frac{[X]}{K_X}\right)+K_{m_B}[A]+K_{m_A}[B]\left(1+\frac{[I]}{K_i}+\frac{[X]}{K_X}\right)+[A][B]}$$

(IX-348)

A mixture of I and X has the same effect as the same total specific concentration of either inhibitor alone. Regrouping terms for a Dixon plot of $1/v$ versus [I]:

$$\boxed{\frac{1}{v} = \frac{1}{V_{max}K_i}\left(\frac{K_{ia}K_{m_B}}{[A][B]}+\frac{K_{m_A}}{[A]}\right)[I]+\frac{1}{V_{max}}\left[\left(1+\frac{K_{m_B}}{[B]}\right)+\left(\frac{K_{ia}K_{m_B}}{[A][B]}+\frac{K_{m_A}}{[A]}\right)\left(1+\frac{[X]}{K_X}\right)\right]}$$

(IX-349)

Thus at a constant [A] and [B], the slope of the $1/v$ versus [I] plot is independent of [X]. The family of plots obtained at different fixed concentrations of X are parallel.

If I and X are not mutually exclusive, the factor becomes:

$$\left(1+\frac{[I]}{K_i}+\frac{[X]}{K_X}+\frac{[I][X]}{\alpha K_i K_X}\right)$$

where α is an interaction factor, that is, a factor by which the binding of one inhibitor changes the dissociation constant for the other. The reciprocal equation rearranged for a Dixon plot of $1/v$ versus [I] is:

$$\boxed{\begin{aligned}\frac{1}{v} &= \frac{1}{V_{max}K_i}\left(\frac{K_{ia}K_{m_B}}{[A][B]}+\frac{K_{m_A}}{[A]}\right)\left(1+\frac{[X]}{\alpha K_X}\right)[I]\\ &+\frac{1}{V_{max}}\left[\left(1+\frac{K_{m_B}}{[B]}\right)+\left(\frac{K_{ia}K_{m_B}}{[A][B]}+\frac{K_{m_A}}{[A]}\right)\left(1+\frac{[X]}{K_X}\right)\right]\end{aligned}}$$

(IX-350)

Now the slope of the plot varies with the fixed concentration of X. The same type of analysis can be carried out for multiple dead-end inhibitors that combine with any enzyme form in any of the systems described. The resulting plots are identical to those described in Chapter Eight for a single substrate system, except that the equations are more complex. In general, mutually exclusive inhibitors yield a series of parallel lines on a Dixon plot. Inhibitors that are not mutually exclusive yield a family of plots that intersect to the left of the $1/v$-axis at $-\alpha K_i$. The $1/v$ coordinate of the intersection point is given by:

$$\boxed{\frac{1}{v} = \frac{1}{V^*_{\max}}\left(1 + \frac{K^*_m(1-\alpha)}{[\mathrm{S}]}\right)}$$

where S is the competitive substrate, $V^*_{\max}$ is the apparent $V_{\max}$ at saturating competitive substrate and the constant concentrations of the not-competitive substrates, and K^*_m is the apparent K_m of the competitive substrate at the constant concentrations of cosubstrates in the absence of inhibitors. For the Ordered Bi Bi system described above where I and X combine with E (and compete with A):

$$V^*_{\max} = \frac{V_{\max}}{\left(1 + \frac{K_{m_\mathrm{B}}}{[\mathrm{B}]}\right)}, \qquad K^*_m = \frac{K_{m_\mathrm{A}}\left(1 + \frac{K_{ia}K_{m_\mathrm{B}}}{K_{m_\mathrm{A}}[\mathrm{B}]}\right)}{\left(1 + \frac{K_{m_\mathrm{B}}}{[\mathrm{B}]}\right)}$$

The $1/v$ intersection coordinate is:

$$\boxed{\frac{1}{v} = \frac{1}{V_{\max}}\left[\left(1 + \frac{K_{m_\mathrm{B}}}{[\mathrm{B}]}\right) + \frac{K_{m_\mathrm{A}}\left(1 + \frac{K_{ia}K_{m_\mathrm{B}}}{K_{m_\mathrm{A}}[\mathrm{B}]}\right)(1-\alpha)}{[\mathrm{A}]}\right]}$$

The equation of the Dixon plot for two nonmutually exclusive inhibitors that combine with enzyme form F in a Ping Pong Bi Bi system is:

$$\frac{1}{v} = \frac{K_{m_B}}{V_{max}K_i[B]}\left(1+\frac{[X]}{\alpha K_X}\right)[I] + \frac{1}{V_{max}}\left[\left(1+\frac{K_{m_A}}{[A]}\right)+\frac{K_{m_B}}{[B]}\left(1+\frac{[X]}{\alpha K_X}\right)\right]$$

(IX-351)

The inhibitors are competitive with B. Therefore, the $1/v$ coordinate of the intersection point is given by:

$$\frac{1}{v} = \frac{1}{V_{max}}\left(1+\frac{K_{m_A}}{[A]}+\frac{K_{m_B}(1-\alpha)}{[B]}\right) \qquad \text{(IX-352)}$$

Dead-End Inhibitors Versus Alternate Substrates

In order for an inhibitor to bind to a particular enzyme form, it must bear some resemblance to the substrate with which it competes. If the inhibitor resembles the substrate sufficiently, it may promote the binding of other substrates and, perhaps, participate in the overall reaction to yield an alternate product. If the inhibitor promotes cosubstrate binding, it will not act as a true dead-end inhibitor (which by definition only undergoes dissociation back to the original enzyme form). The velocity equations for inhibitors that act completely or partially as alternate substrates are different from the equations for true dead-end inhibitors. Several examples are discussed in Section FF.

EE. MIXED DEAD-END AND PRODUCT INHIBITION

Products usually bear some structural resemblance to the substrates. Consequently, it is not surprising to find that products frequently combine with the "wrong" enzyme form (at the substrate binding site) to yield dead-end complexes. A number of examples have been discussed earlier in Chapter Six for rapid equilibrium systems. A common example of mixed dead-end and product inhibition observed in bireactant steady state systems is the combination of P with EA in an Ordered Bi Bi system. In this case, A combines with the enzyme and induces the formation of the binding site for B. However, P, which bears a structural similarity to B, binds instead of B.

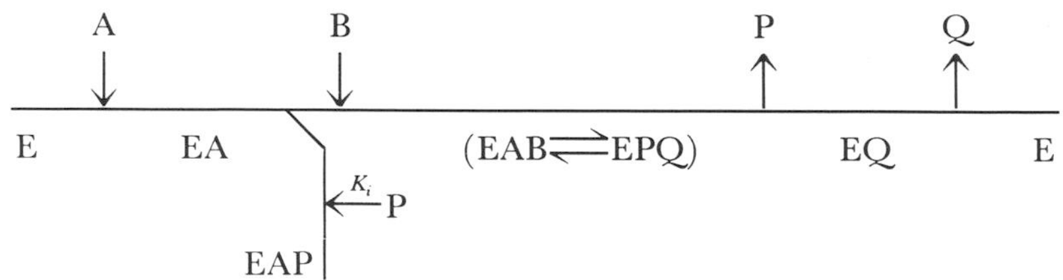

When [A] is varied at a fixed unsaturating [B], the combination of P with EQ (to form the normal EPQ complex) results in both slope and intercept effects. The combination of P with EA to yield the dead-end EAP complex adds only an intercept effect (because P combines after A in the reaction sequence). Thus P behaves as a mixed-type inhibitor. The inhibition can be distinguished from normal product inhibition by the fact that the families of reciprocal plots obtained at different fixed concentrations of P do not intersect at a common point. The mixed dead-end plus normal product inhibition is further distinguished by the $1/v$-axis intercept replots. The intercept replot is parabolic because P combines at two points that are reversibly connected in the proper sequence (see below) and combination at both points results in intercept effects. Specifically, the combination of P with EQ increases the steady-state level of the central complexes, which in turn increases the level of EA available for combination with P. In other words, P promotes the formation of EAP by two means (backup of the normal reaction sequence and direct combination with EA). In the example above, only one point of P addition predicts slope effects. Consequently, the slope replot is linear. Saturation with B will overcome the dead-end inhibition by reducing the level of EA to zero. In this case, only the normal uncompetitive inhibition by P is seen and $1/v$-axis intercept replots are linear. The system can be described as *S*-linear, *I*-parabolic (nonintersecting) mixed-type inhibition when A is the varied substrate and [B] is unsaturating, changing to *I*-linear uncompetitive inhibition when [B] is saturating. When [B] is varied, the combination of P with EQ predicts both slope and intercept effects, while the combination with EA predicts only a slope effect (P and B combine with the same enzyme form). Thus the inhibition is mixed-type and the $1/v$-axis intercept replot is linear, but now the slope replot is parabolic. Saturation with A has no effect on the patterns. The system can be described as *S*-parabolic, *I*-linear (nonintersecting) mixed-type inhibition at all A concentrations. As general rules:

When a product combines at two points in a reaction sequence, the effects for each point are predicted separately by the usual rules. A parabolic replot is expected if the specific effect is predicted for both combination points, and product combination at one point increase the steady-state level of the enzyme form combining with the second molecule of product. That is, replots are parabolic only if the usual point of product combination is reversibly connected to the dead-end point of combination upstream.

The definitions of reversibility are the same as given earlier: no release of another product or saturation with a substrate between the two points of inhibitory product addition.

Another common example of mixed dead-end and product inhibition occurs when a product combines with the "wrong" stable enzyme form in a Ping Pong system (Fig. IX-55). For example, consider what happens when P combines with E to form a dead-end EP complex in a Ping Pong Bi Bi system.

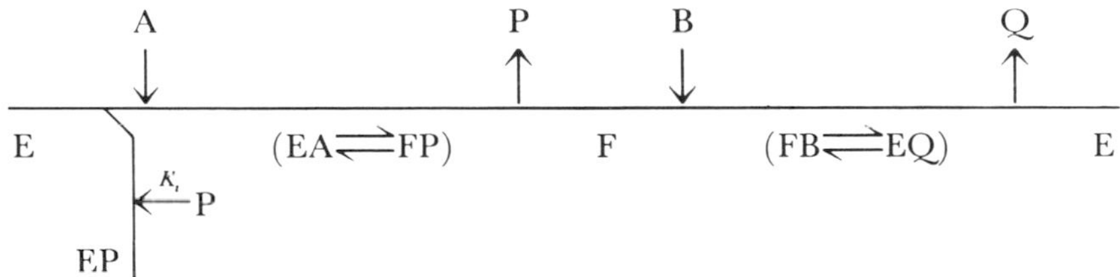

When [A] is varied at unsaturating [B], the combination of P with F introduces slope and intercept effects. The combination of P with E introduces a slope effect. The two points of P addition are reversibly connected. Therefore, P acts as a mixed-type inhibitor with parabolic slope replots and linear $1/v$-axis intercept replots (S-parabolic, I-linear, nonintersecting, mixed-type inhibition). Saturating [B] reduces the steady-state level of F to zero. Consequently, P can combine only with E. Under this condition P yields S-linear competitive inhibition. When [B] is varied at unsaturating [A], the combination of P with F introduces a slope effect, while the combination of P with E introduces both a slope and intercept effect. Thus P yields S-parabolic, I-linear (nonintersecting) mixed-type inhibition. Satura-

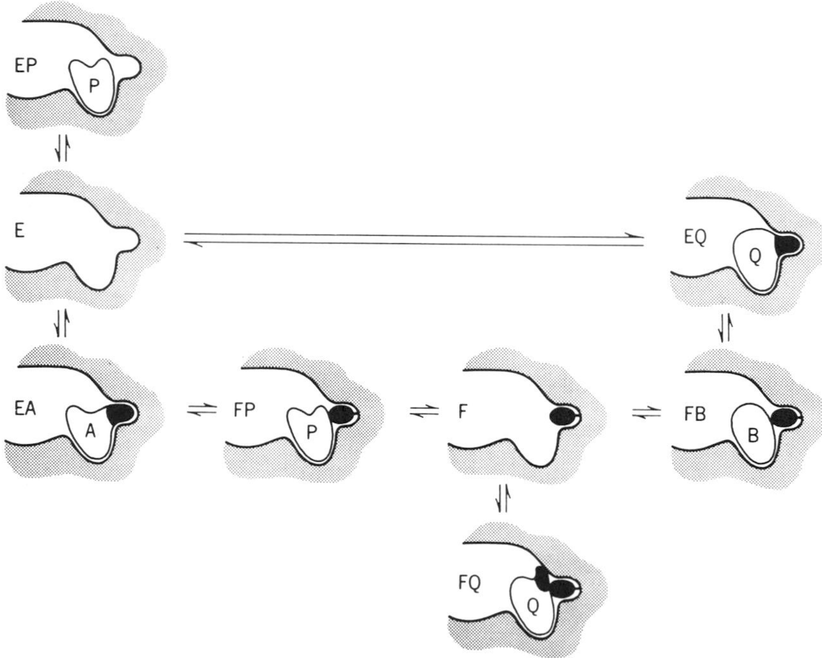

Fig. IX-55. Mixed product and dead-end inhibition in a Bi Bi Ping Pong system; Q combines with form F as well as with form E; P combines with form E as well as with form F.

tion with A prevents P from combining with E and the inhibition with respect to B changes to *S*-linear competitive. The velocity equation is obtained by multiplying the [B] and [P] terms in the denominator of the complete velocity equation by $(1+[P]/K_i)$. (See equations IX-356 and 357).

Care must be taken in predicting parabolic intercept replots because the $1/v$-axis intercept represents saturation with the varied substrate. In this regard, the following general rule should be kept in mind:

> *The addition of the varied substrate between the two points of product addition acts as an irreversible step with regard to predicting parabolic intercept effects.*

For example, consider the Bi Uni Uni Uni Ping Pong Ter Bi system shown below in which P combines with E to form a dead-end EP complex:

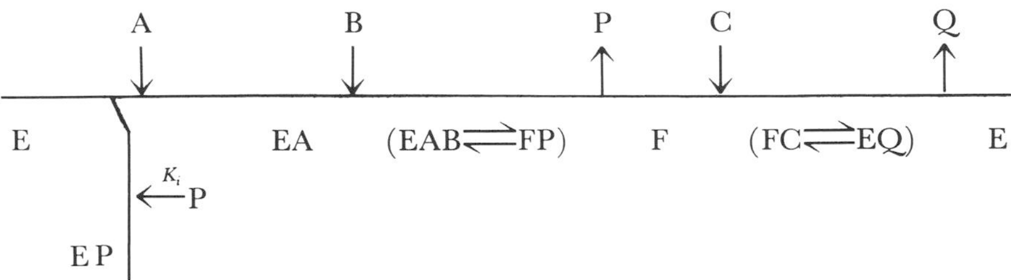

The combination of P with F predicts slope and intercept effects when A is the varied substrate and [B] and [C] are unsaturating. The combination of P with E predicts only a slope effect. Thus P yields *S*-parabolic, *I*-linear (nonintersecting) mixed-type inhibition. When [B] is saturating, the two points of P addition are no longer reversibly connected. Combination of P with F now yields only an intercept effect. Combination of P with E still yields only a slope effect. Thus the inhibition changes to *S*-linear, *I*-linear (intersecting) mixed-type. Saturation with C reduces the level of F to zero; P still can combine with E and so we observe *S*-linear competitive inhibition. When B is the varied substrate, and [A] and [C] are unsaturating, combination of P with F introduces slope and intercept effects. Combination with E also introduces slope and intercept effects. However, the $1/v$-axis intercept represents saturation with B. Thus only the slope replot is parabolic and the inhibition is *S*-parabolic, I-linear (nonintersecting) mixed-type. When [A] is saturating, P cannot combine with E and we observe the usual *S*-linear, *I*-linear (intersecting) mixed-type inhibition by P. Saturation with C eliminates the combination of P with F and we observe only the dead-end combination with E which yields *S*-linear, *I*-linear (intersecting) mixed-type inhibition. When [C] is varied at unsaturating [A] and [B], combination of P with F yields a slope effect while combination of P with E yields a slope and intercept effect. The inhibition is *S*-parabolic, *I*-linear (nonintersecting) mixed-type. Saturation with B interrupts the reversible connection between E and F and the inhibition becomes *S*-linear, *I*-linear (intersecting) mixed-type. Saturation with [A] eliminates the dead-end combination of P with E yielding the usual *S*-linear competitive inhibition.

As another example, consider the Ordered Ter Ter system in which one of the products reacts with EA to yield a dead-end complex. The reaction sequence is written to emphasize the repeating nature of the process.

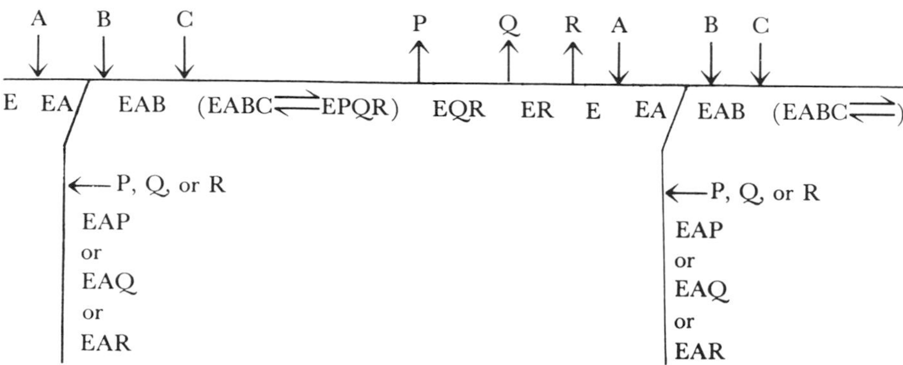

Suppose that [C] is varied at unsaturating [A] and [B]. P adding to EQR introduces slope and intercept effects; P adding to EA also introduces slope and intercept effects. The two points of P addition are reversibly connected but since C adds between the two points, the connection becomes irreversible with regard to multiplying intercept effects. Thus P yields *S*-parabolic, *I*-linear (nonintersecting) mixed-type inhibition. Q adding to ER introduces only an intercept effect; Q adding to EA introduces slope and intercept effects. There is no reversible connection in the absence of P. Therefore, the inhibition by Q is *S*-linear, *I*-linear (intersecting) mixed-type. R adding to E introduces slope and intercept effects. (The points of R and C addition are reversibly connected in the downstream direction.) R adding to EA also introduces slope and intercept effects. The two points of R addition are reversibly connected in the downstream direction but the effects are not multiplied because the addition of R to E does not increase the relative level of EA (the complex combining in dead-end fashion with the second molecule of R). Thus we observe *S*-linear, *I*-linear mixed-type inhibition. These predictions can be verified by examining the distribution and velocity equations as described below. For the Ordered Ter Ter system with dead-end complexes made from EA, only the formation of EAP introduces a squared term into the velocity equation.

Velocity Equations

When a product combines with an enzyme form in addition to the one it normally combines with as a substrate in the reverse direction, certain terms in the denominator of the normal velocity equation are multiplied by the factor $(1+[\mathrm{P}]/K_i)$, or $(1+[\mathrm{Q}]/K_i)$, and so on, where K_i is the dissociation constant of the product from the dead-end complex. If the denominator terms that are multiplied already contain a [P] or [Q] term, the resulting inhibition will be parabolic. For example, consider the Ordered Bi Bi system in which P reacts with EA to form a dead-end EAP complex. The denominator terms that are multiplied by $(1+[\mathrm{P}]/K_i)$ are those corresponding to the relative concentration of the EA complex. These terms are found in the numerator of the distribution equation for $[\mathrm{EA}]/[\mathrm{E}]_t$ (equation IX-84 or equation IX-113). In the presence of A, B, and P (and absence of Q), the numerator of the distribution equation for $[\mathrm{EA}]/[\mathrm{E}]_t$ contains an [A] term and an [A][P] term. Consequently, the velocity equation in the presence of P is (from equation IX-89):

$$\frac{v}{V_{\max}}=\frac{[\mathrm{A}][\mathrm{B}]}{K_{ia}K_{m_\mathrm{B}}+K_{m_\mathrm{B}}\left(1+\frac{[\mathrm{P}]}{K_i}\right)[\mathrm{A}]+K_{m_\mathrm{A}}[\mathrm{B}]+\frac{K_{m_\mathrm{Q}}K_{m_\mathrm{B}}K_{ia}[\mathrm{P}]}{K_{iq}K_{m_\mathrm{P}}}+[\mathrm{A}][\mathrm{B}]+\frac{K_{m_\mathrm{Q}}K_{m_\mathrm{B}}}{K_{iq}K_{m_\mathrm{P}}}\left(1+\frac{[\mathrm{P}]}{K_i}\right)[\mathrm{A}][\mathrm{P}]+\frac{[\mathrm{A}][\mathrm{B}][\mathrm{P}]}{K_{ip}}}$$

(IX-353)

When [A] is varied, the resulting $[\mathrm{P}]^2$ terms appear in the $1/v$-axis intercept factor. When B is the varied substrate, the resulting $[\mathrm{P}]^2$ terms appear in the slope factor. The velocity equations can be written as shown below.

$$\frac{v}{V_{\max}}=\frac{[\mathrm{A}]}{K_{m_\mathrm{A}}\left[1+\frac{K_{ia}K_{m_\mathrm{B}}}{K_{m_\mathrm{A}}[\mathrm{B}]}\left(1+\frac{K_{m_\mathrm{Q}}[\mathrm{P}]}{K_{iq}K_{m_\mathrm{P}}}\right)\right]+[\mathrm{A}]\left[1+\frac{[\mathrm{P}]}{K_{ip}}+\frac{K_{m_\mathrm{B}}}{[\mathrm{B}]}\left(1+\frac{[\mathrm{P}]}{K_i}\right)\left(1+\frac{K_{m_\mathrm{Q}}[\mathrm{P}]}{K_{iq}K_{m_\mathrm{P}}}\right)\right]}$$

(IX-354)

$$\frac{v}{V_{\max}}=\frac{[\mathrm{B}]}{K_{m_\mathrm{B}}\left(1+\frac{[\mathrm{P}]}{K_i}\right)\left[1+\frac{K_{ia}}{[\mathrm{A}]\left(1+\frac{[\mathrm{P}]}{K_i}\right)}\right]\left(1+\frac{K_{m_\mathrm{Q}}[\mathrm{P}]}{K_{iq}K_{m_\mathrm{P}}}\right)+[\mathrm{B}]\left(1+\frac{K_{m_\mathrm{A}}}{[\mathrm{A}]}+\frac{[\mathrm{P}]}{K_{ip}}\right)}$$

(IX-355)

The linear or parabolic nature of the replots is seen better if the denominator of equations IX-354 and IX-355 are multiplied out and the terms regrouped. The slopes and $1/v$-axis intercepts then have the forms:

$$slope_{1/\mathrm{A}} = \frac{K_{m_\mathrm{A}}}{V_{\max}}(a + b[\mathrm{P}]) \qquad \text{i.e., a straight line}$$

$$int_{1/\mathrm{A}} = \frac{1}{V_{\max}}\left(a + b[\mathrm{P}] + c[\mathrm{P}]^2\right) \qquad \text{i.e., a parabola}$$

$$slope_{1/\mathrm{B}} = \frac{K_{m_\mathrm{B}}}{V_{\max}}\left(a + b[\mathrm{P}] + c[\mathrm{P}]^2\right) \qquad \text{i.e., a parabola}$$

$$int_{1/\mathrm{B}} = \frac{1}{V_{\max}}(a + b[\mathrm{P}]) \qquad \text{i.e., a striaght line}$$

For the system above, the constant for the dissociation of I from the dead-end complex is not easily obtained. In some cases, however, K_i appears by itself in a slope or intercept factor. For example, when P combines in a dead-end fashion with E in a Ping Pong Bi Bi system, the velocity equations become:

$$\frac{v}{V_{\max}} = \frac{[\mathrm{A}]}{K_{m_\mathrm{A}}\left(1 + \frac{[\mathrm{P}]}{K_i} + \frac{K_{ia}K_{m_\mathrm{B}}[\mathrm{P}]}{K_{m_\mathrm{A}}K_{ip}[\mathrm{B}]} + \frac{K_{ia}K_{m_\mathrm{B}}[\mathrm{P}]^2}{K_{m_\mathrm{A}}K_{ip}K_i[\mathrm{B}]}\right) + [\mathrm{A}]\left(1 + \frac{K_{m_\mathrm{B}}}{[\mathrm{B}]} + \frac{K_{m_\mathrm{B}}[\mathrm{P}]}{K_{ip}[\mathrm{B}]}\right)} \tag{IX-356}$$

$$\frac{v}{V_{\max}} = \frac{[\mathrm{B}]}{K_{m_\mathrm{B}}\left(1 + \frac{K_{ia}[\mathrm{P}]}{K_{ip}[\mathrm{A}]} + \frac{[\mathrm{P}]}{K_{ip}} + \frac{K_{ia}[\mathrm{P}]^2}{K_{ip}K_i[\mathrm{A}]}\right) + [\mathrm{B}]\left(1 + \frac{K_{m_\mathrm{A}}}{[\mathrm{A}]} + \frac{K_{m_\mathrm{A}}[\mathrm{P}]}{K_i[\mathrm{A}]}\right)} \tag{IX-357}$$

The patterns are *S*-parabolic, *I*-linear (nonintersecting) mixed-type for both substrates (Fig. IX-56). Replots of $int_{1/\mathrm{A}}$ will allow K_{ip} to be determined, while replots of $int_{1/\mathrm{B}}$ allow K_i to be determined.

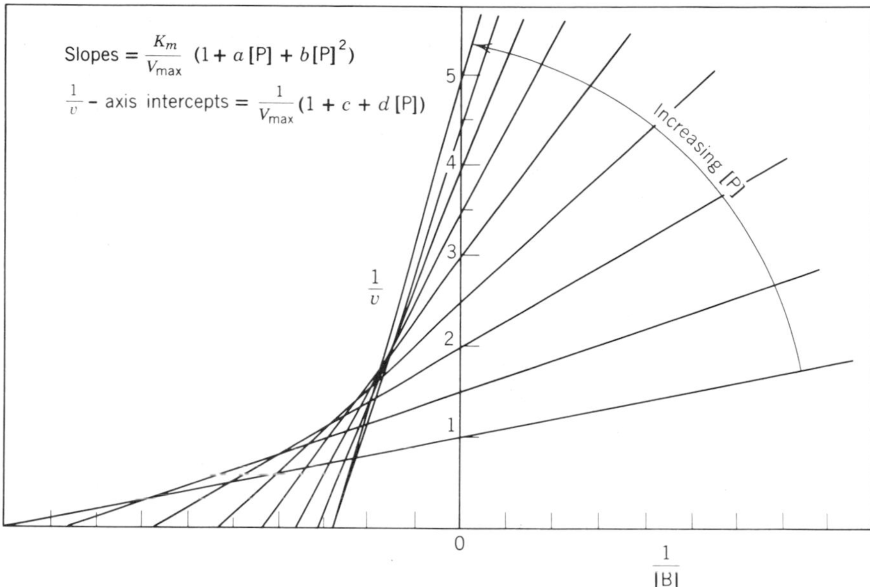

Fig. IX-56. Mixed product and dead-end inhibition in an Ordered Bi Bi system; P reacts with EA as well as with EQ. The plot of $1/v$ versus $1/[B]$ is slope-parabolic, intercept-linear.

Dead-End Complexes with the Normal Enzyme Form

A product may react with one enzyme form to yield two different kinds of complexes, one of which is a normal complex of the reaction sequence, while the other is a dead-end complex. For example, suppose P reacts in two ways with form F in a Ping Pong Bi Bi system:

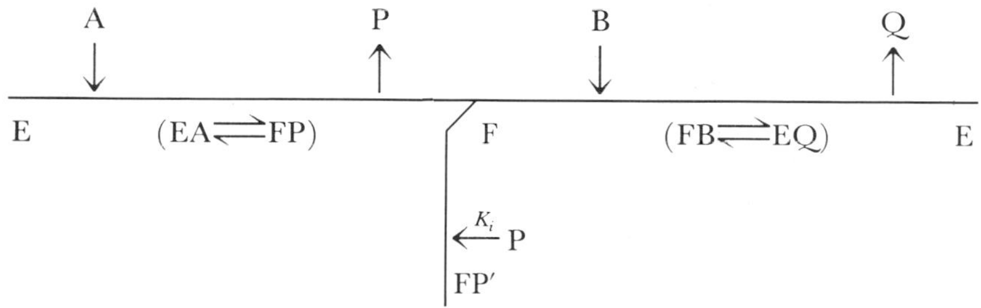

If we follow the usual rules we might predict that when [A] is varied, combination of P with F to yield the normal FP introduces slope and intercept effects while combination with F to yield the dead end FP′ introduces only an intercept effect. Since the points of P addition are reversibly connected (they are in fact, the same point), we might expect a parabolic intercept effect. However, examination of the distribution equation for $[F]/[E]_t$ shows only an [A] term in the numerator when A, B, and P are present and Q is absent. Thus the $K_{m_B}[A]$ term in the denominator of the velocity equation is multiplied by $(1+[P]/K_i)$. There are no [P]-containing terms that are multiplied; hence no $[P]^2$ terms are generated. The inhibition remains *S*-linear, *I*-linear mixed-type. Similarly, when [B] is varied, there is no parabolic slope effect; the inhibition remains *S*-linear competitive. As a general rule then:

> *If a product combines with one enzyme form to yield two different kinds of complexes (a normal complex and a dead-end complex), the inhibition is still linear.*

For the example above, the velocity equations become:

$$\frac{v}{V_{max}} = \frac{[A]}{K_{m_A}\left(1+\dfrac{K_{ia}K_{m_B}[P]}{K_{m_A}K_{ip}[B]}\right)+[A]\left(1+\dfrac{K_{m_B}[P]}{K_{ip}[B]}+\dfrac{K_{m_B}}{[B]}+\dfrac{K_{m_B}[P]}{K_i[B]}\right)} \tag{IX-358}$$

$$\frac{v}{V_{max}} = \frac{[B]}{K_{m_B}\left(1+\dfrac{K_{ia}[P]}{K_{ip}[A]}+\dfrac{[P]}{K_{ip}}+\dfrac{[P]}{K_i}\right)+[B]\left(1+\dfrac{K_{m_A}}{[A]}\right)} \tag{IX-359}$$

The extra $[P]/K_i$ term appears in the intercept factor when [A] is varied, and in the slope factor when [B] is varied. The presence of the dead-end complex can be missed easily, because the product inhibition patterns are not changed. However, the K_{ip} calculated from $slope_{1/A}$ replots will not give the same value of K_{ip} calculated from $int_{1/A}$ replots. The former is the true K_{ip}, but the latter is an apparent value composed of K_{ip} and K_i.

$$K_{i_{int}}^{P/A} = \left(1+\frac{[B]}{K_{m_B}}\right)\frac{K_{ip}K_i}{K_{ip}+K_i}, \qquad K_{ip}^{app} = \frac{K_{ip}K_i}{(K_{ip}+K_i)} = \frac{K_{ip}}{\left(1+\dfrac{K_{ip}}{K_i}\right)}$$

The same apparent K_{ip} is calculated from $slope_{1/B}$ replots when [A] is saturating. With normal experimental error, the difference between K_{ip} calculated from $K_{i_{slope}}^{P/A}$ and $K_{i_{int}}^{P/A}$ (or $K_{i_{slope}}^{P/B}$ at saturating [A]) may not be obvious.

FF. INHIBITION BY ALTERNATIVE SUBSTRATES

Substrate analogs are often used as inhibitors in diagnosing reaction mechanisms or determining enzyme specificity. The resulting inhibition patterns depend on the extent to which the analog participates in the reaction sequence. An analog may act simply as a dead-end inhibitor, or it may act as an alternative substrate yielding an alternative product. In between these extremes, we have an analog that mimics the substrate it replaces by promoting the binding of one or more cosubstrates, but the resulting complex is unable to undergo further reaction. We cannot always directly determine whether an analog is an alternative substrate. For example, the assay may be based on the appearance of radioactive P from radioactive A. The analog, I, may be unavailable in labeled form. Nevertheless, by determining the inhibition patterns caused by I we may be able to deduce whether it is a substrate, partial substrate, or dead-end inhibitor. If I is an alternative substrate, then it is frequently possible to determine K_{m_I} and K_{ii} from appropriate plots and replots where the analog is used as an inhibitor.

Ordered Bi Bi with Alternate A

In the reaction sequence shown below, I is an alternative substrate for A. A and I yield a common first product, P, but different second products, Q (from A) or S (from I).

A (k_1, k_{-1}) B (k_2, k_{-2}) P (k_3, k_{-3}) Q (k_4, k_{-4})

E — EA — (EAB ⇌ EPQ) — EQ — E

E — EI — (EIB ⇌ EPS) — ES — E

I (k_5, k_{-5}) B (k_6, k_{-6}) P (k_7, k_{-7}) S (k_8, k_{-8})

The basic King-Altman figure is:

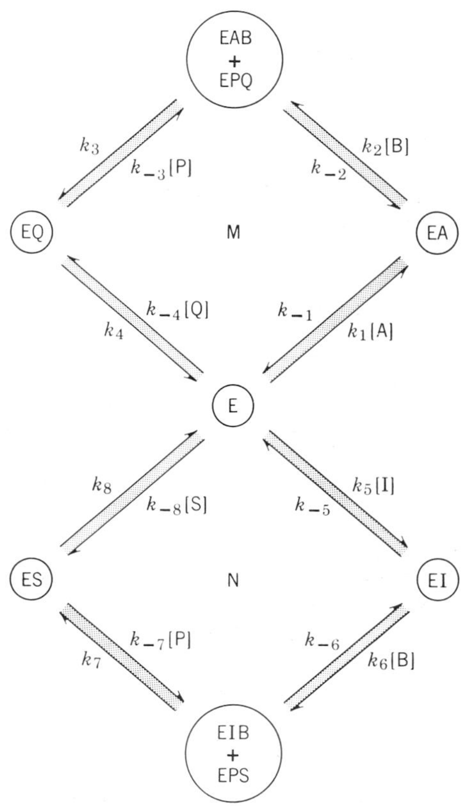

The distribution of enzyme forms can be obtained by the simplified procedure described in Section C (in which matrices M and N are handled separately and only eight interconversion patterns are necessary). If the velocity of the reaction is taken as the rate of Q production (i.e., the rate of appearance of the unique product of A), then in the absence of products:

$$v = \frac{d[\mathrm{Q}]}{dt} = k_4[\mathrm{EQ}]$$

$$\frac{v}{[\mathrm{E}]_t} = \frac{k_4[\mathrm{EQ}]}{[\mathrm{E}] + [\mathrm{EA}] + [\mathrm{EAB} + \mathrm{EPQ}] + [\mathrm{EQ}] + [\mathrm{EI}] + [\mathrm{EIB} + \mathrm{EPS}] + [\mathrm{ES}]}$$

Substituting from the King-Altman solutions:

$$\frac{v}{[\mathrm{E}]_t} = \frac{k_4K_1[\mathrm{A}][\mathrm{B}] + k_4K_2[\mathrm{A}][\mathrm{B}]^2}{K_3 + K_4[\mathrm{B}] + K_6[\mathrm{A}] + K_7[\mathrm{A}][\mathrm{B}] + K_5[\mathrm{B}]^2 + K_8[\mathrm{A}][\mathrm{B}]^2 + K_9[\mathrm{I}] + K_{10}[\mathrm{I}][\mathrm{B}] + K_{11}[\mathrm{I}][\mathrm{B}]^2} \quad \text{(IX-360)}$$

where K_1 through K_{11} are groups of rate constants. (In the absence of I, only the first numerator term and the first four denominator terms would be present.) At any fixed [A] and [I], the equation can be written:

$$\boxed{\frac{v}{[\mathrm{E}]_t} = \frac{K_{12}[\mathrm{B}] + K_{13}[\mathrm{B}]^2}{K_{14} + K_{15}[\mathrm{B}] + K_{16}[\mathrm{B}]^2}} \quad \text{(IX-361)}$$

Equation IX-361 does not have the same form as the Henri-Michaelis-Menten equation and, consequently, we might expect nonlinear reciprocal plots when [B] is varied. This would distinguish the Ordered Bi Bi system from a Rapid Equilibrium Random Bi Bi system which yields linear mixed-type inhibition plots when [B] is varied in the presence of an alternative substrate for A. However, equation IX-361 has the same form as that for a Steady-State Random mechanism (equation IX-183). Computer analyses of random systems using reasonable values for rate constants have shown that the reciprocal plots will usually appear linear. The nonlinear portion of the plot occurs close to the $1/v$-axis and would not be observed under the usual experimental conditions. If one path of the random sequence has rate constants that are significantly different from those of the other path, then the primary velocity curve can have an inflection point, or pass through a maximum and then decrease (substrate inhibition), or both. The reciprocal plot will then be nonlinear. An analogous situation arises when an alternative substrate for A is used in an Ordered Bi Bi system. The alternative substrate, I, provides an alternative point for the addition of B. If the rate constants of the alternative path are identical or close to those of the path with A as the first substrate (which is what we might expect), then the reciprocal plots will not show a nonlinear portion. Consequently, care should be taken in interpreting kinetic data obtained with an alternative substrate. Linear plots cannot be used to exclude a mechanism. A mechanism may be excluded only if a reciprocal plot exhibits obvious and significant deviations from linearity and the varied substrate does not display substrate inhibition in the absence of I. Substrate inhibition by itself causes nonlinear reciprocal plots in steady-state and many rapid equilibrium systems. An additional deviation from linearity caused by I may be masked.

At a fixed concentration of B, equation IX-360 can be written as:

$$\frac{v}{[\mathrm{E}]_t} = \frac{K_{17}[\mathrm{A}]}{K_{18} + K_{19}[\mathrm{A}] + K_{20}[\mathrm{I}]} \tag{IX-362}$$

where K_{17} through K_{20} represent combinations of the original rate constants and the fixed concentration of B. Dividing numerator and denominator by K_{19} (i.e., $\mathrm{Coef_A}$) and defining the ratios in the usual manner:

$$\frac{v}{V_{\max_{\mathrm{app}}}} = \frac{[\mathrm{A}]}{K_{m_{\mathrm{A(app)}}}\left(1 + \frac{[\mathrm{I}]}{K_{i_{\mathrm{app}}}}\right) + [\mathrm{A}]} \tag{IX-363}$$

Thus at any given [B], the reciprocal plots for A as the varied substrate are linear and I acts as a competitive inhibitor with respect to A. The $slope_{1/\mathrm{A}}$ is a linear function of [I] but a complex function of [B].

Suppose that I yielded a different first product but the same second product as A. The reaction sequence would then be:

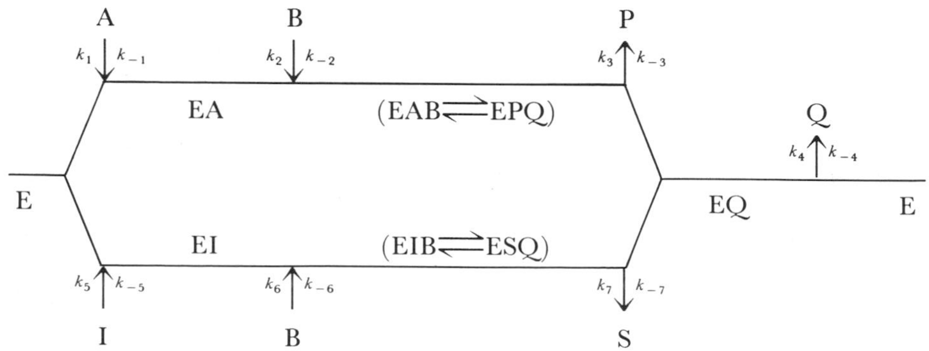

The basic King-Altman figure and the 15 valid five-lined interconversion patterns are shown below.

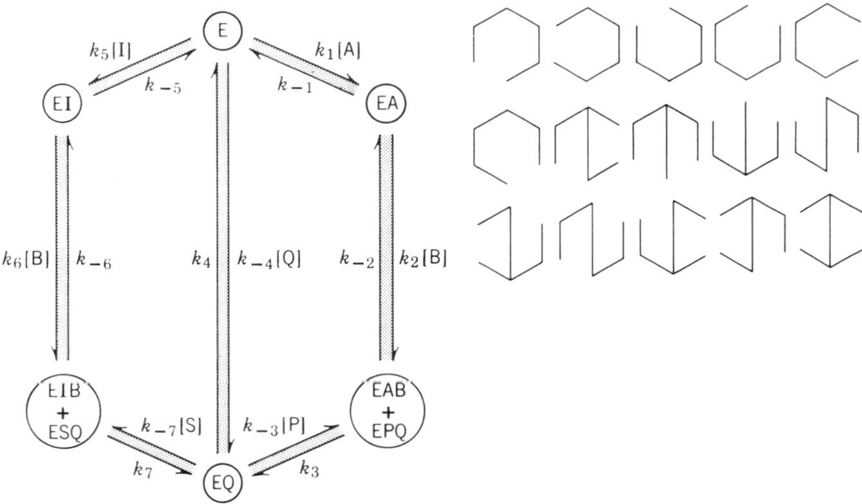

If v is taken as the rate of P formation (i.e., the rate of unique product formation from substrate A):

$$v = \frac{d[\mathrm{P}]}{dt} = k_3[\mathrm{EAB} + \mathrm{EPQ}]$$

and

$$\frac{v}{[\mathrm{E}]_t} = \frac{k_3[\mathrm{EAB} + \mathrm{EPQ}]}{[\mathrm{E}] + [\mathrm{EA}] + [\mathrm{EAB} + \mathrm{EPQ}] + [\mathrm{EQ}] + [\mathrm{EI}] + [\mathrm{EIB} + \mathrm{ESQ}]}$$

The velocity equation obtained from the five-lined interconversion patterns is identical to equation IX-360 (except the K constants are composed of different rate constants). Thus I is competitive with A and may induce nonlinear reciprocal plots when B is the varied substrate (or I may simply appear to be a linear mixed-type inhibitor). In general:

> *The velocity equation in the presence of an alternative substrate is the same regardless of which product is the unique product of the alternative substrate, as long as the velocity is taken as the rate of appearance of the product of the regular substrate.*

Ordered Bi Bi with Alternative B

If I substitutes for B to yield a different first product, S, but the same second product, the reaction sequence is:

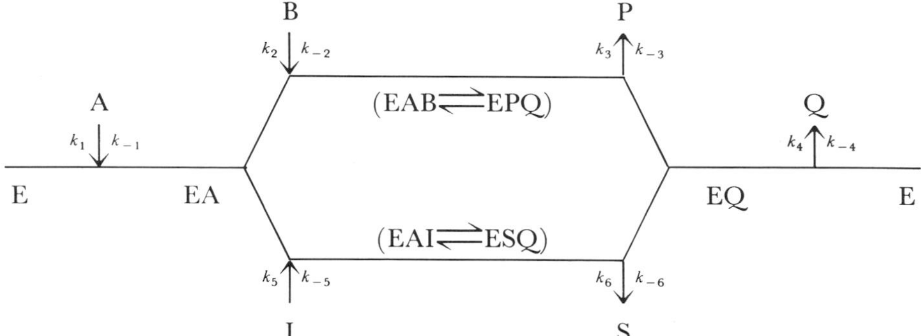

The King-Altman figure and interconversion patterns are shown below.

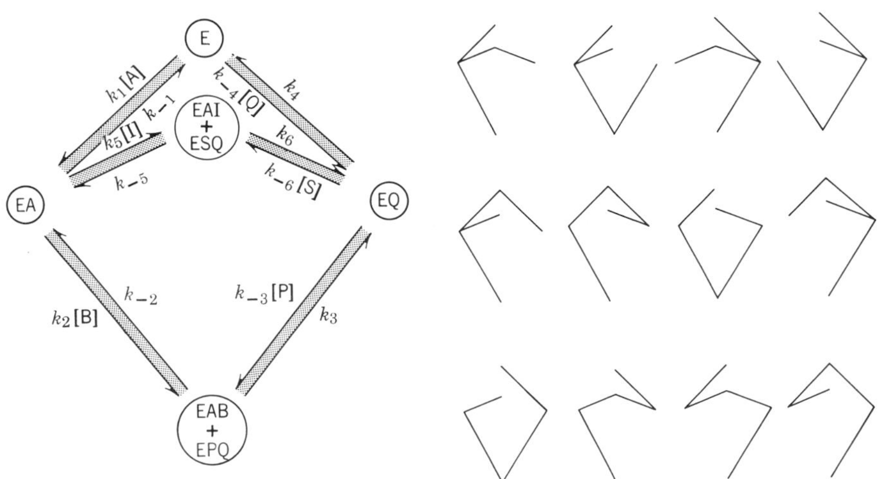

In the absence of products, P, Q, and S, the velocity of P formation is given by:

$$\frac{v}{V_{\max}} = \frac{[\mathrm{A}]}{K_{m_\mathrm{A}}\left(1 + \frac{K_{ia}K_{m_\mathrm{B}}}{K_{m_\mathrm{A}}[\mathrm{B}]} + \frac{K_{ia}K_{m_\mathrm{B}}[\mathrm{I}]}{K_{m_\mathrm{A}}K_{ii}[\mathrm{B}]}\right) + [\mathrm{A}]\left(1 + \frac{K_{m_\mathrm{B}}}{[\mathrm{B}]} + \frac{K_{m_\mathrm{B}}[\mathrm{I}]}{K_{m_\mathrm{I}}[\mathrm{B}]}\right)}$$

(IX-364)

$$\boxed{\frac{v}{V_{\max}} = \frac{[B]}{K_{m_B}\left(1 + \frac{K_{ia}}{[A]} + \frac{[I]}{K_{m_I}} + \frac{K_{ia}[I]}{K_{ii}[A]}\right) + [B]\left(1 + \frac{K_{m_A}}{[A]}\right)}} \qquad \text{(IX-365)}$$

where $K_{m_I} = \dfrac{\text{Coef}_A}{\text{Coef}_{AI}}$ = the K_m for I as a substrate in place of B

$K_{ii} = \dfrac{\text{const}}{\text{Coef}_I}$ = an inhibition constant for I (but not defined analogously to K_{ib}).

As expected, I affects only the slope when [B] is varied.The competitive $K_{i_{\text{slope}}}$ is equivalent to the apparent K_m for I as a substrate (in place of B) at the fixed [A]:

$$K_{i_{\text{slope}}}^{I/B} = \frac{K_{m_I}\left(1 + \frac{K_{ia}}{[A]}\right)}{\left(1 + \frac{K_{ia}K_{m_I}}{K_{ii}[A]}\right)}$$

where $\dfrac{K_{ia}K_{m_I}}{K_{ii}} = K'_{m_A}$ = the K_m for A when I is the cosubstrate

When [A] is varied, I acts as a mixed-type inhibitor. Note that I affects the slope of the $1/v$ versus $1/[A]$ plot even though it adds after A in the reaction sequence. Thus I as an alternative substrate clearly does not act in the same way as a dead-end inhibitor. This is understandable if we keep in mind that I promotes a reaction sequence leading to the regeneration of E whereas a dead-end inhibitor binding to EA does not. The K_{m_I} and K_{ii} can be obtained from appropriate plots and replots (Figs. IX-57 and IX-58). Of course, if S can be measured, both constants can be obtained from initial velocity and product inhibition studies with A and I as substrates (in the absence of B) and with S and Q as product inhibitors.

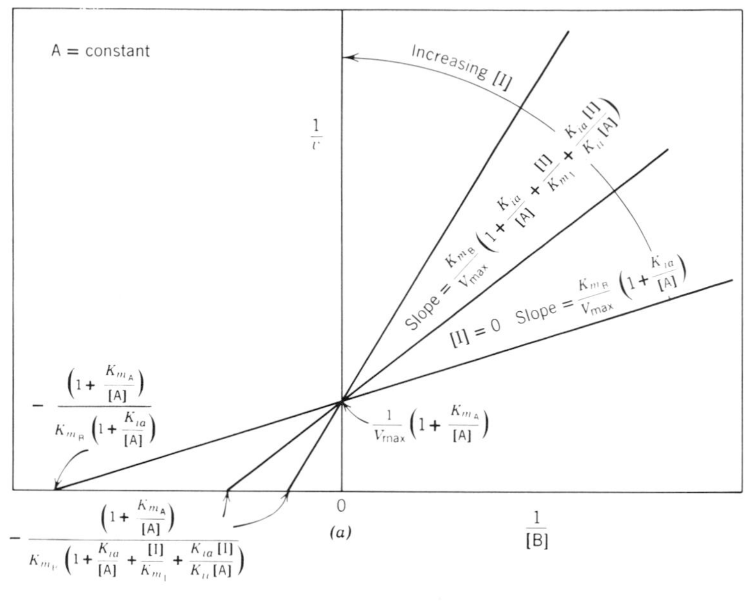

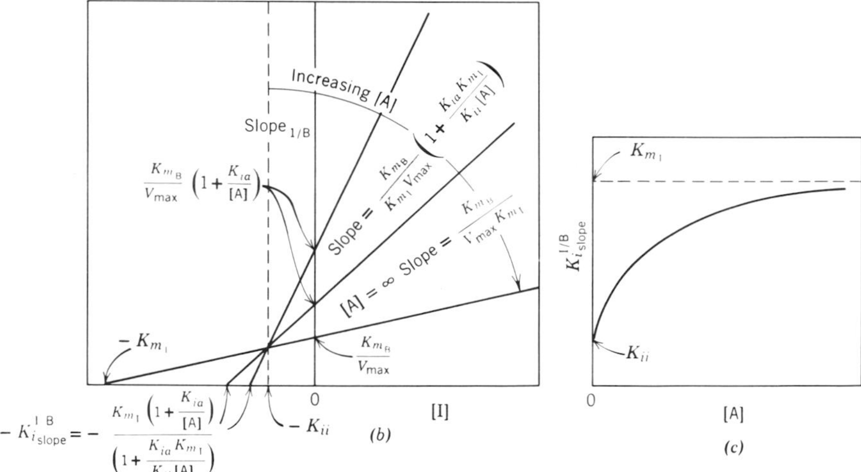

Fig. IX-57. Inhibition by an alternative substrate for B in an ordered bireactant system; $1/v$ versus $1/[B]$ plot and appropriate replots.

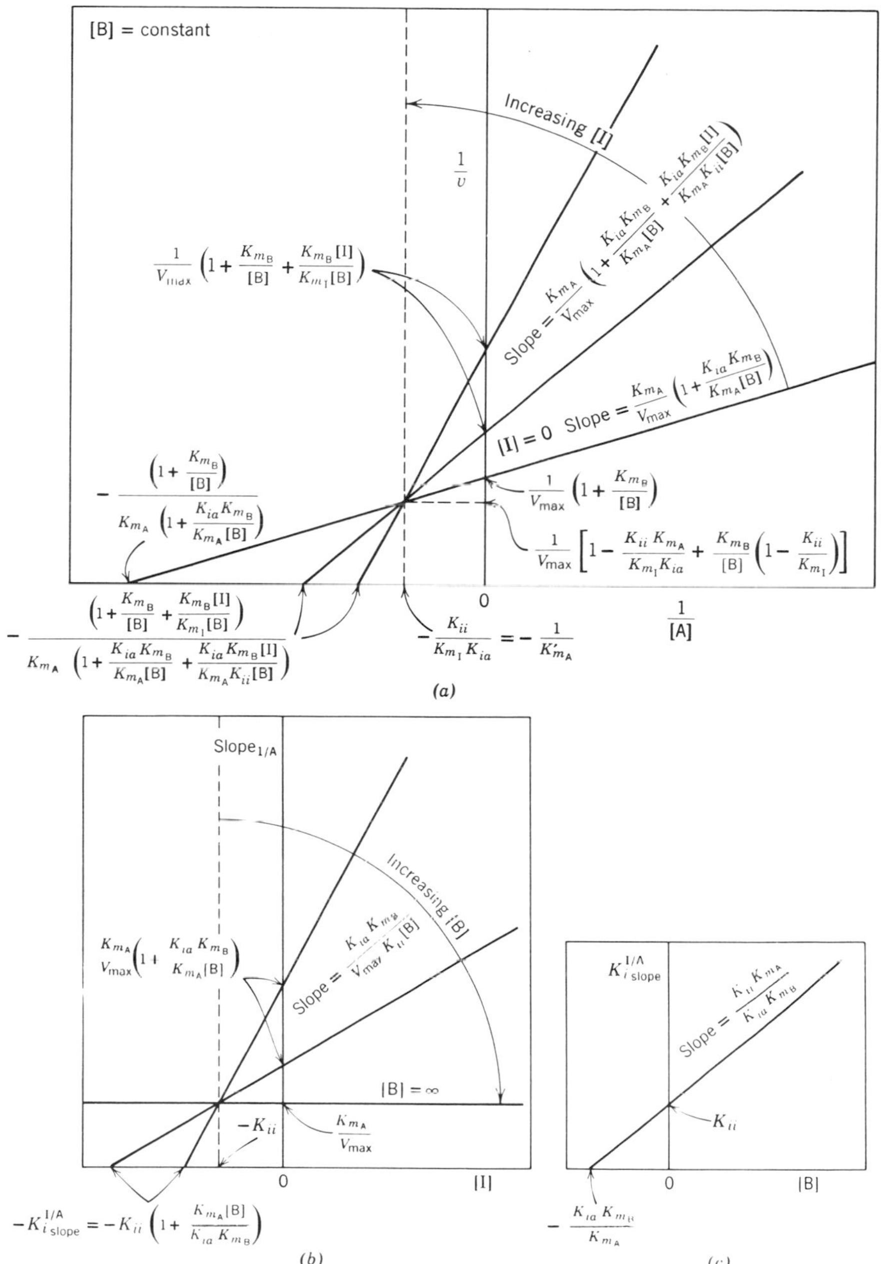

Fig. IX-58. Inhibition by an alternative substrate for B in an ordered bireactant system; $1/v$ versus $1/[A]$ plot and appropriate replots.

Alternative Substrates That Promote a Partial Reaction in an Ordered Sequence

Compound I, an analog of B that binds to EA to yield a catalytically inactive EAI complex, is an ordinary dead-end inhibitor; I will be competitive with B and uncompetitive with A. Similarly, I, an analog of A that yields an EI complex that cannot add B, is a dead-end inhibitor; I will be competitive with A and a mixed-type inhibitor with respect to B. Suppose, however, that the binding of I to E induces the proper conformational change in the enzyme that permits B to bind, but the resulting EIB complex is catalytically inactive. In this case, I is neither a true alternative substrate nor a true dead-end inhibitor. The reaction sequence is shown below.

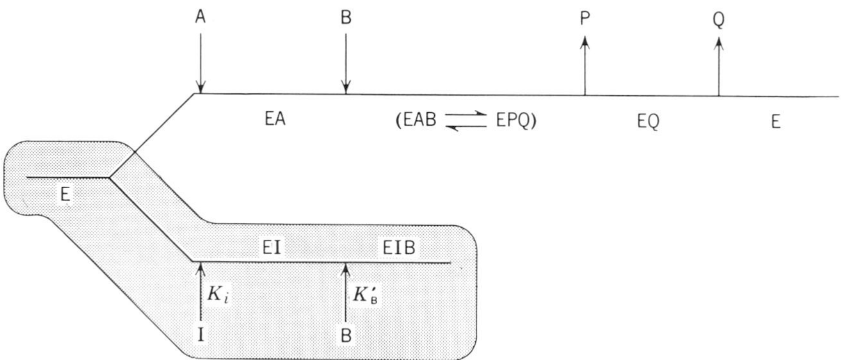

The shaded area represents a rapid equilibrium segment, since EI and EIB can regenerate E only by dissociation. The velocity equation can be derived in the usual manner taking E, EI, and EIB as a single corner, X, of the basic King-Altman figure. Alternatively, we can start with the velocity equation for the uninhibited reaction and multiply the denominator terms representing free E by the factor $(1+[\mathrm{I}]/K_i+[\mathrm{I}][\mathrm{B}]/K_iK'_{\mathrm{B}})$, which represents the distribution of enzyme forms within the rapid equilibrium segment. Free E is represented by the $K_{ia}K_{m_{\mathrm{B}}}$ and $K_{m_{\mathrm{A}}}[\mathrm{B}]$ terms. The resulting equations are:

$$\frac{v}{V_{\max}}=\frac{[\mathrm{A}]}{K_{m_{\mathrm{A}}}\left(1+\frac{K_{ia}K_{m_{\mathrm{B}}}}{K_{m_{\mathrm{A}}}[\mathrm{B}]}\right)\left(1+\frac{[\mathrm{I}]}{K_i}+\frac{[\mathrm{I}][\mathrm{B}]}{K_iK'_{\mathrm{B}}}\right)+[\mathrm{A}]\left(1+\frac{K_{m_{\mathrm{B}}}}{[\mathrm{B}]}\right)} \tag{IX-366}$$

and

$$\frac{v}{V_{\max}} = \frac{[\mathrm{B}]}{K_{m_\mathrm{B}}\left[1 + \frac{K_{ia}}{[\mathrm{A}]}\left(1 + \frac{[\mathrm{I}]}{K_i} + \frac{[\mathrm{I}][\mathrm{B}]}{K_i K'_\mathrm{B}}\right)\right] + [\mathrm{B}]\left[1 + \frac{K_{m_\mathrm{A}}}{[\mathrm{A}]}\left(1 + \frac{[\mathrm{I}]}{K_i} + \frac{[\mathrm{I}][\mathrm{B}]}{K_i K'_\mathrm{B}}\right)\right]}$$

(IX-367)

I is competitive with A at any constant [B]. When [A] is varied at a constant [I] and different fixed B concentrations, B acts as a competitive substrate inhibitor as its concentration becomes significant compared to K'_B. The $1/v$ versus $1/[\mathrm{B}]$ plots at any fixed [A] and [I] are nonlinear at high [B] (i.e., low $1/[\mathrm{B}]$). At low B concentrations the plots may appear linear. In this region, I acts as a mixed-type inhibitor at any constant unsaturating [A] (i.e., the extrapolated asymptotes intersect to the left of the $1/v$-axis).

Ping Pong Bi Bi With Alternative Substrates

Two Ping Pong Bi Bi systems with an alternative substrate for A are shown below.

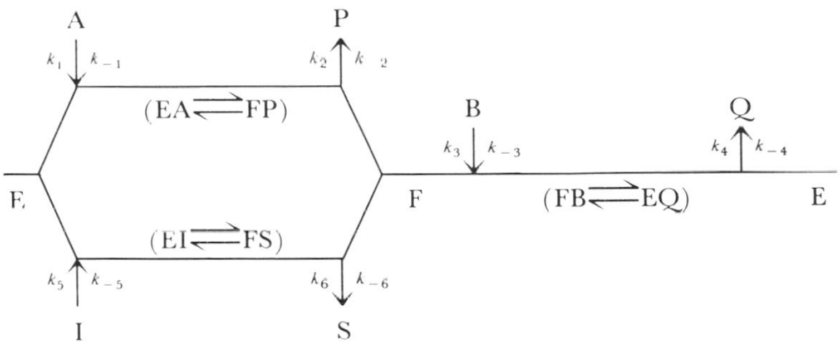

A P B Q
k_1 k_{-1} k_2 k_{-2} k_3 k_{-3} k_4 k_{-4}
(EA⇌FP) F (FB⇌EQ)
E E
(EI⇌F′P) F′ (F′B⇌ES)
k_5 k_{-5} k_6 k_{-6} k_7 k_{-7} k_8 k_{-8}
I P B S

In one case, I yields an alternative first product and v can be taken as k_2[EA + FP]. This sequence could represent a transaminase working on two different amino acids (A and I) releasing two different α-keto acids (P and S). B could be a common acceptor such as α-ketoglutarate while Q is the common amino acid product, glutamate. In the other case, I yields an alternative second product and v can be taken as k_4[FB + EQ]. The velocity equation for the first system is obtained from the basic King-Altman figure shown below.

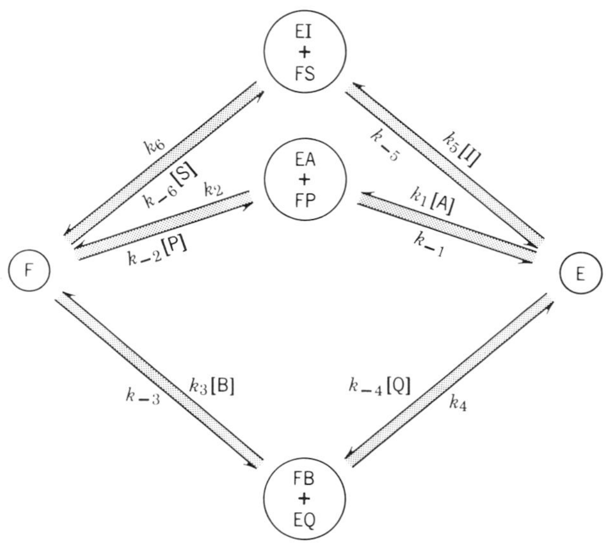

The denominator of the complete velocity equation, representing $[E]_t$, contains 15 terms: [A], [B], [A][B], [A][P], [A][S], [P], [Q], [P][Q], [B][Q], [I], [I][B], [I][P], [S], [I][S], and [S][Q]. In the absence of products, the equation becomes:

$$v = \frac{\text{num}[A][B]}{\text{Coef}_A[A] + \text{Coef}_B[B] + \text{Coef}_{AB}[A][B] + \text{Coef}_I[I] + \text{Coef}_{IB}[I][B]} \tag{IX-368}$$

The King-Altman solution for the second system yields an equation containing an $[A][B]^2$ numerator term and [A][B], $[B]^2$, $[A][B]^2$, [I][B], and $[I][B]^2$ denominator terms. Canceling [B] in the numerator and denominator yields, as expected, the same equation IX-368. Dividing numerator and denominator by Coef_{AB} and defining terms in the usual manner, we obtain:

$$\boxed{\frac{v}{V_{max}} = \frac{[A]}{K_{m_A}\left(1+\frac{K'_{m_B}[I]}{K_{m_I}[B]}+\frac{[I]}{K_{m_I}}\right)+[A]\left(1+\frac{K_{m_B}}{[B]}\right)}} \quad \text{(IX-369)}$$

$$\boxed{\frac{v}{V_{max}} = \frac{[B]}{K_{m_B}\left(1+\frac{K'_{m_B}K_{m_A}[I]}{K_{m_B}K_{m_I}[A]}\right)+[B]\left(1+\frac{K_{m_A}}{[A]}+\frac{K_{m_A}[I]}{K_{m_I}[A]}\right)}} \quad \text{(IX-370)}$$

where $K_{m_I} = \dfrac{\text{Coef}_B}{\text{Coef}_{IB}}$ = the K_m for I as a substrate

$K'_{m_B} = \dfrac{\text{Coef}_I}{\text{Coef}_{IB}}$ = the K_m for B when I is the substrate

Thus I is competitive with A at all B concentrations and a mixed-type inhibitor with respect to B at unsaturating [A]. Slope and $1/v$-axis intercept replots are linear. The K_{m_I} and K'_{m_B} can be obtained from appropriate plots and replots (Figs. IX-59 and IX-60). The fact that I has a slope effect on $1/v$ versus $1/[B]$ plots may be unexpected; B adds after a product release step and consequently we might have predicted that I would be uncompetitive with respect to B. To understand why I affects the $slope_{1/B}$, we must keep in mind that the slope of the reciprocal plot is the reciprocal of the apparent first-order rate constant at low [B] for the reaction of B to yield the measured product. When I is present, some additions of B occur in catalytic sequences that result in the production of S rather than P. Thus increasing [I] causes a decrease in the apparent first-order rate constant and an increase in the $slope_{1/B}$. The slope effect shows that I is not a dead-end inhibitor. As usual, the competitive $K_{i_{slope}}$ is equal to the apparent K_m for I as a substrate (in place of A) at the fixed [B]:

$$K_{i_{slope}}^{I/A} = \frac{K_{m_I}}{\left(1+\frac{K'_{m_B}}{[B]}\right)}$$

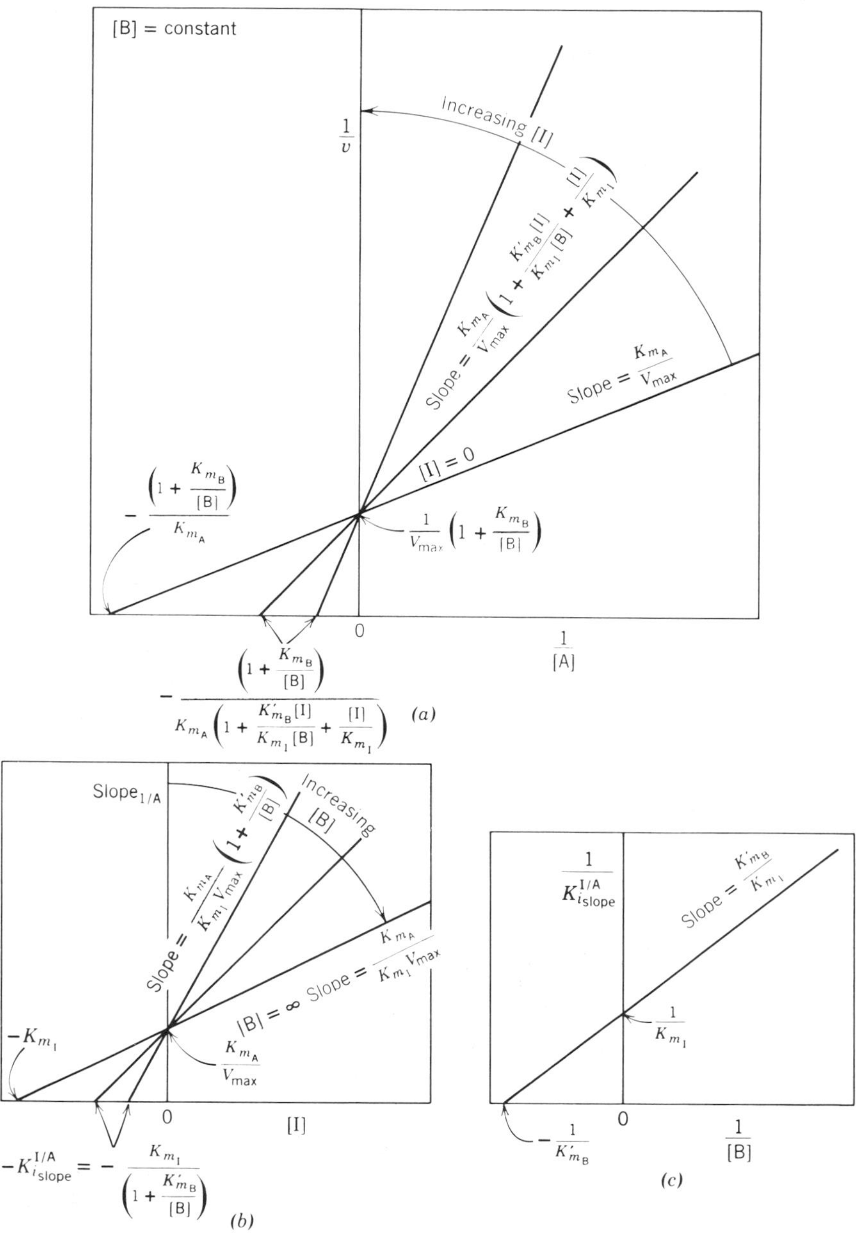

Fig. IX-59. Inhibition by an alternative substrate for A in a Ping Pong bireactant system; $1/v$ versus $1/[A]$ plot and appropriate replots.

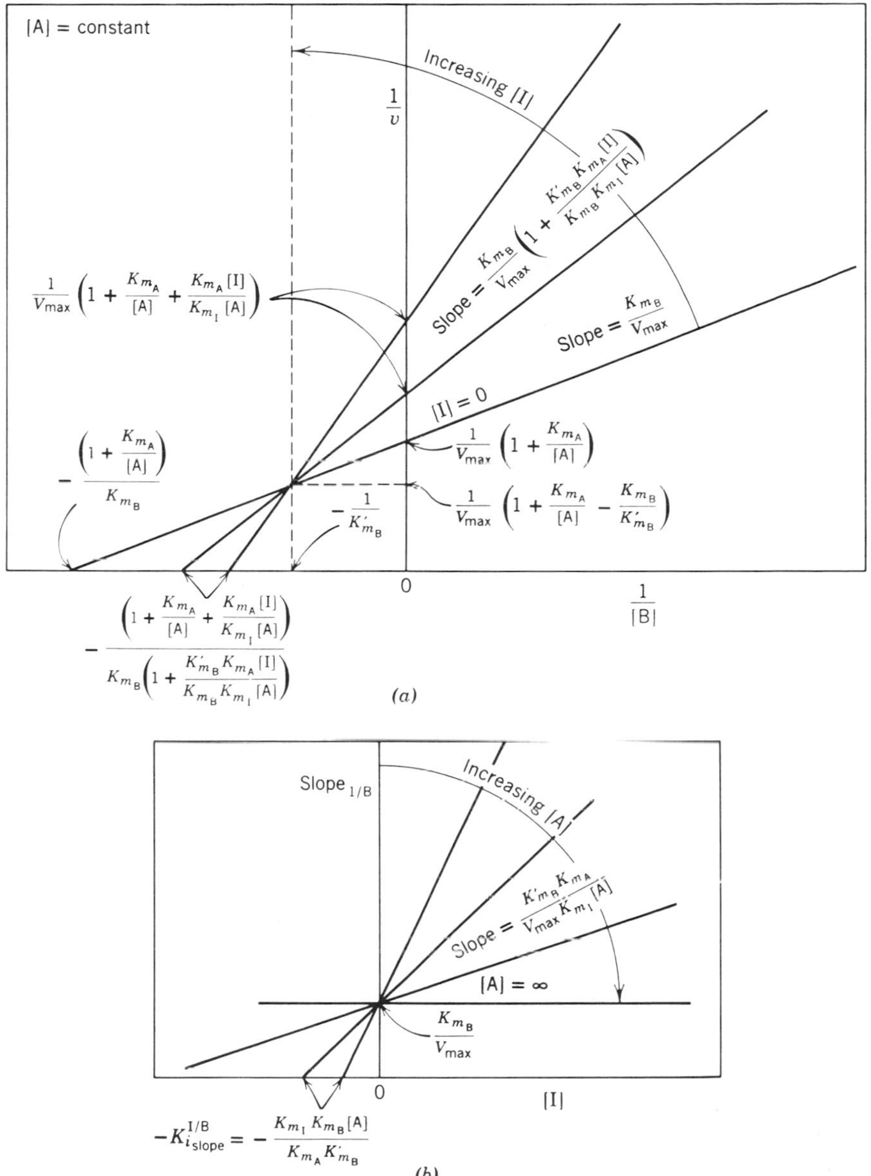

Fig. IX-60. Inhibition by an alternative substrate for A in a Ping Pong bireactant system; $1/v$ versus $1/[B]$ plot and appropriate replots.

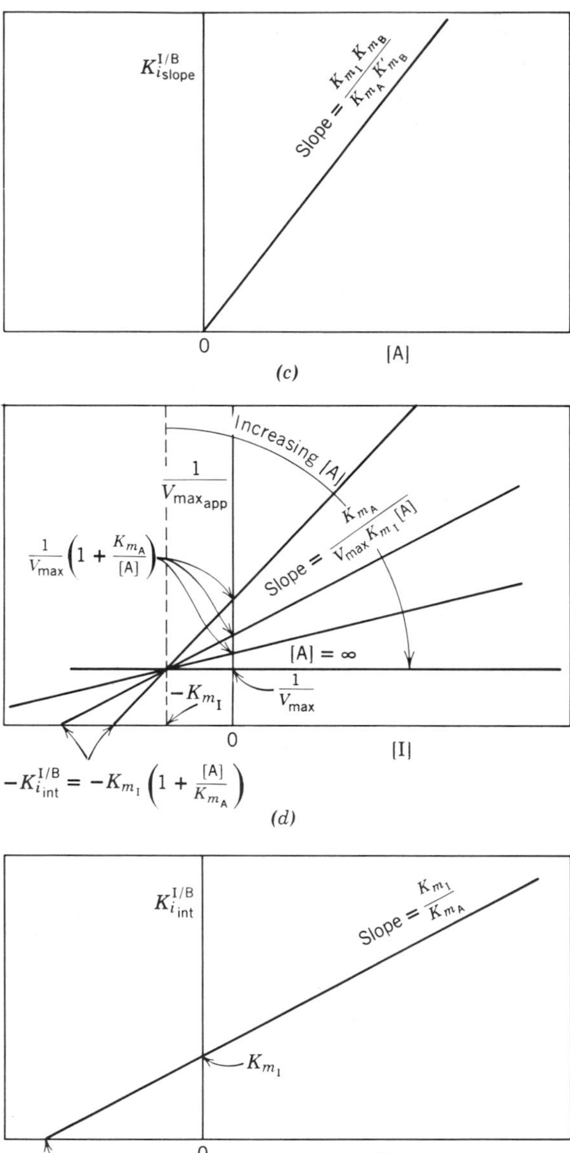

Fig. IX-60. (*Cont.*)

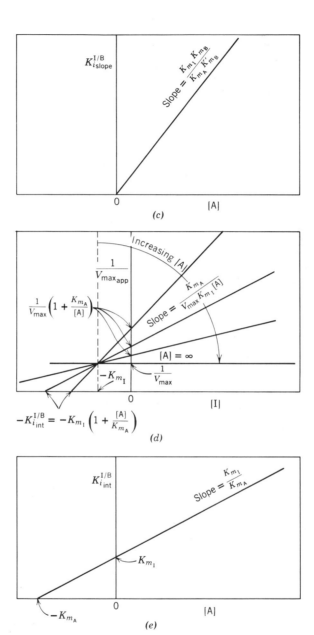

When alternative substrate I substitutes for B to yield alternative product S, the reaction sequence is:

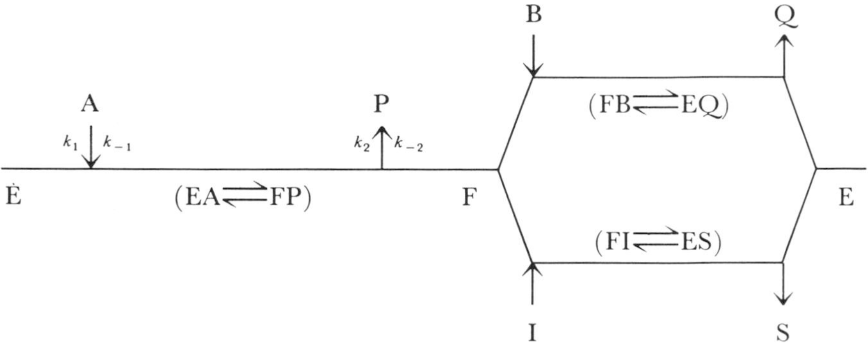

The velocity equations are symmetrical to equations IX-369 and IX-370.

$$\frac{v}{V_{\max}} = \frac{[A]}{K_{m_A}\left(1 + \frac{K'_{m_A}K_{m_B}[I]}{K_{m_A}K_{m_I}[B]}\right) + [A]\left(1 + \frac{K_{m_B}}{[B]} + \frac{K_{m_B}[I]}{K_{m_I}[B]}\right)} \qquad \text{(IX-371)}$$

$$\frac{v}{V_{\max}} = \frac{[B]}{K_{m_B}\left(1 + \frac{K'_{m_A}[I]}{K_{m_I}[A]} + \frac{[I]}{K_{m_I}}\right) + [B]\left(1 + \frac{K_{m_A}}{[A]}\right)} \qquad \text{(IX-372)}$$

where $K'_{m_A} = \frac{\text{Coef}_I}{\text{Coef}_{AI}}$ = the K_m for A when I is the second substrate

Thus I is competitive with B and a mixed-type inhibitor with respect to A. If I is a dead-end inhibitor instead of an alternative substrate, we would observe an uncompetitive pattern in place of the mixed-type pattern.

In general, for full steady-state systems:

> *An alternative substrate is always competitive with the substrate it replaces and a mixed-type inhibitor with respect to the other substrates. Reciprocal plots for a cosubstrate* ***may*** *be nonlinear if that substrate adds more than once in the overall parallel reaction sequence at points of addition that are reversibly connected. Reciprocal plots are always linear if the points of cosubstrate addition are separated by a product release step. Alternative substrates do not yield uncompetitive patterns in full steady-state systems. The competitive* $K_{i_{\text{slope}}}$ *is equivalent to the apparent* K_m *of the alternative substrate at the fixed concentrations of cosubstrates.*

Alternative Substrates That Promote a Partial Ping Pong Reaction Sequence

In the reaction sequences shown below, I is an alternative first substrate that promotes a partial Ping Pong reaction sequence. In scheme 1, the alternative stable enzyme form, F′, is unable to bind B. In scheme 2, F′ binds B but the F′B complex is unable to isomerize to ES.

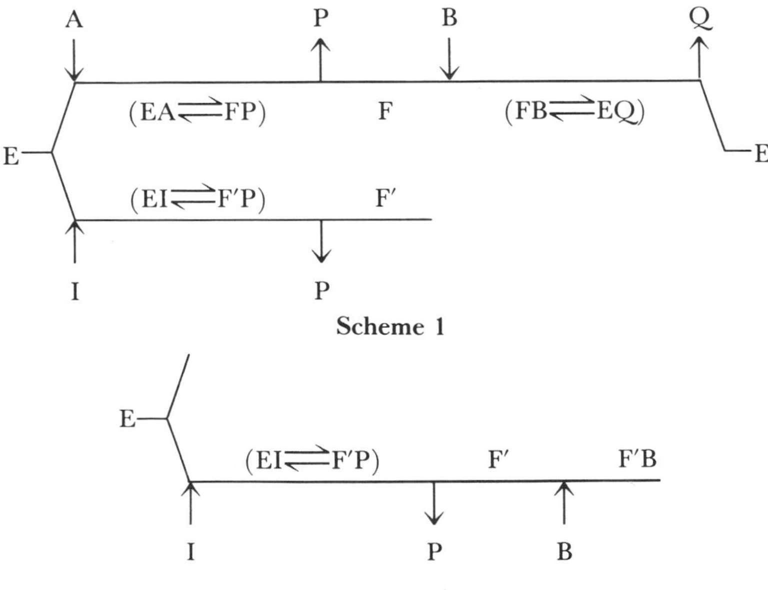

Scheme 1

Scheme 2

In both schemes, the enzyme would be stoichiometrically titrated to an inactive F′ (or F′ + F′B). In the absence of P the reaction is irreversible, and I behaves as an irreversible noncompetitive inhibitor (i.e., an active site label).

Measurement of Common Product

In most studies with alternative substrates, the rate of unique product appearance is measured. If the rate of formation of the common product is measured, then the velocity equations are more complicated (because the velocity is now the sum of the velocities of two parallel reaction sequences, one of which may be faster or slower than the other). For example, if we measure Q in the Ping Pong Bi Bi system where I is an alternative first substrate yielding S, an alternative first product, then v is given by $v = k_4[\mathrm{FB} + \mathrm{EQ}]$ or:

$$v = \frac{V_{\max_A}[\mathrm{A}][\mathrm{B}] + \dfrac{\mathrm{num_{IB}}}{\mathrm{Coef_{AB}}}[\mathrm{I}][\mathrm{B}]}{K_{m_B}[\mathrm{A}] + K_{m_A}[\mathrm{B}] + [\mathrm{A}][\mathrm{B}] + \dfrac{K_{m_A}K'_{m_B}[\mathrm{I}]}{K_{m_I}} + \dfrac{K_{m_A}[\mathrm{I}][\mathrm{B}]}{K_{m_I}}} \qquad \text{(IX-373)}$$

The ratio $\mathrm{num_{IB}}/\mathrm{Coef_{AB}}$ can be converted to kinetic constants as shown below.

$$\frac{\mathrm{num_{IB}}}{\mathrm{Coef_{AB}}} = \frac{V_{\max_I}\mathrm{Coef_{IB}}}{\mathrm{Coef_{AB}}} = \frac{V_{\max_I}K_{m_A}\mathrm{Coef_{IB}}}{\mathrm{Coef_B}} = \frac{V_{\max_I}K_{m_A}}{K_{m_I}}$$

where

$V_{\max_A}$ = the maximal velocity with A as the first substrate

$V_{\max_I}$ = the maximal velocity with I as the first substrate

The velocity equation can now be written as:

$$\frac{v}{V_{\max_A}} = \frac{[\mathrm{A}][\mathrm{B}]\left(1 + \dfrac{V_{\max_I}K_{m_A}[\mathrm{I}]}{V_{\max_A}K_{m_I}[\mathrm{A}]}\right)}{\text{denominator}}$$

When [A] is varied:

$$\frac{v}{V_{\max_A}} = \frac{[A]}{K_{m_A}\dfrac{\left(1+\dfrac{K'_{m_B}[I]}{K_{m_I}[B]}+\dfrac{[I]}{K_{m_I}}\right)}{\left(1+\dfrac{V_{\max_I}K_{m_A}[I]}{V_{\max_A}K_{m_I}[A]}\right)}+[A]\dfrac{\left(1+\dfrac{K_{m_B}}{[B]}\right)}{\left(1+\dfrac{V_{\max_I}K_{m_A}[I]}{V_{\max_A}K_{m_I}[A]}\right)}}$$

(IX-374)

The reciprocal plots are nonlinear. This is understandable, since Q is formed from I whether A is present or absent. As [A] decreases (i.e., as $1/[A]$ increases), the reciprocal plot approaches a horizontal line corresponding to the $1/v$ for the fixed [B] and [I]. As [A] increases, the plot curves as it approaches the $1/v$-axis. The $1/v$-axis intercept is above the horizontal asymptote (i.e., the plot curves up) if $V_{\max_A}$ at the fixed [B] is less than $v_{Q/I}$, the velocity of Q formation from I alone at the fixed [B]. The plot curves down as it approaches the $1/v$-axis if $V_{\max_A}$ is greater than $v_{Q/I}$ at the fixed [B]. When [B] is varied:

$$\frac{v}{V_{\max_A}} = \frac{[B]}{K_{m_B}\dfrac{\left(1+\dfrac{K'_{m_B}K_{m_A}[I]}{K_{m_B}K_{m_I}[A]}\right)}{\left(1+\dfrac{V_{\max_I}K_{m_A}[I]}{V_{\max_A}K_{m_I}[A]}\right)}+[B]\dfrac{\left(1+\dfrac{K_{m_A}}{[A]}+\dfrac{K_{m_A}[I]}{K_{m_I}[A]}\right)}{\left(1+\dfrac{V_{\max_I}K_{m_A}[I]}{V_{\max_A}K_{m_I}[A]}\right)}}$$

(IX-375)

The reciprocal plots are linear ([B] does not appear in either the slope or intercept factor); I acts as a partial inhibitor or activator at unsaturating [A]. Slope and intercept replots are hyperbolic. As [I] increases, the terms not containing [I] become insignificant and the slope and $1/v$-axis intercept approach limits given by:

$$slope_{1/B}\ (\text{at } [\text{I}]=\infty)=\frac{K_{m_B}}{V_{\max_A}}\left(\frac{K'_{m_B}V_{\max_A}}{K_{m_B}V_{\max_I}}\right)=\frac{K'_{m_B}}{V_{\max_I}}=\frac{\alpha}{\beta}\frac{K_{m_B}}{V_{\max_A}}$$

$$int_{1/B}\ (\text{at } [\text{I}]=\infty)=\frac{1}{V_{\max_A}}\left(\frac{V_{\max_A}}{V_{\max_I}}\right)=\frac{1}{V_{\max_I}}=\frac{1}{\beta V_{\max_A}}$$

where $\alpha=K'_{m_B}/K_{m_B}$ and $\beta=V_{\max_I}/V_{\max_A}$. These are the same relationships derived earlier for the rapid equilibrium, hyperbolic mixed-type inhibition systems (Chapter Four).

GG. INHIBITION BY ALTERNATIVE PRODUCTS

Cleland has suggested the use of alternative products as a way of diagnosing or confirming a kinetic mechanism. An alternative product is a substance that would be produced if an alternative substrate had been used. For example, consider the Ping Pong Bi Bi reaction where S is an alternative first product of I, an alternative first substrate.

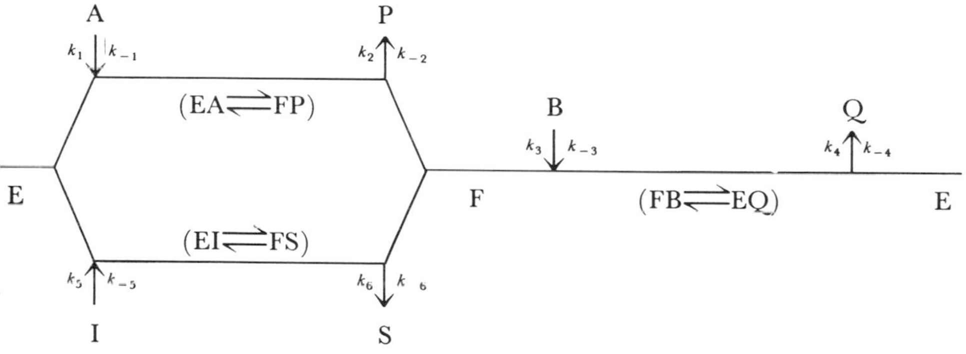

The basic King-Altman figure shown earlier yields a complete velocity equation with 15 denominator terms. The numerator depends on which product is being assayed. If we measure the rate of P formation from A in the presence of S, then:

$$v^{S}_{P/A}=\frac{d[\text{P}]}{dt}=k_2[\text{EA}+\text{FP}]$$

Substituting from the King-Altman solution, we obtain:

$$v_{\mathrm{P/A}}^{\mathrm{S}} = \frac{\mathrm{num_{AB}[A][B] + num_{AS}[A][S]}}{\mathrm{Coef_A[A] + Coef_B[B] + Coef_{AB}[A][B] + Coef_{AS}[A][S] + Coef_S[S]}} \tag{IX-376}$$

In the absence of I, P, and Q, 10 of the denominator terms drop out (but their coefficients are still used to define kinetic constants, as shown below). There are two numerator terms because there are two overall sequences by which P is produced from A. One is the usual sequence in which B reacts with F to form Q and regenerate E. The other is the sequence in which S reacts with F to produce I and regenerate E. Factoring and dividing numerator and denominator by $\mathrm{Coef_{AB}}$:

$$v_{\mathrm{P/A}}^{\mathrm{S}} = \frac{\dfrac{\mathrm{num_{AB}}}{\mathrm{Coef_{AB}}}\mathrm{[A][B]}\left(1 + \dfrac{\mathrm{num_{AS}}}{\mathrm{num_{AB}}}\dfrac{\mathrm{[S]}}{\mathrm{[B]}}\right)}{\dfrac{\mathrm{Coef_A}}{\mathrm{Coef_{AB}}}\mathrm{[A]} + \dfrac{\mathrm{Coef_B}}{\mathrm{Coef_{AB}}}\mathrm{[B]} + \mathrm{[A][B]} + \dfrac{\mathrm{Coef_{AS}}}{\mathrm{Coef_{AB}}}\mathrm{[A][S]} + \dfrac{\mathrm{Coef_S}}{\mathrm{Coef_{AB}}}\mathrm{[S]}} \tag{IX-377}$$

The ratios of coefficients can be defined as kinetic constants where:

$$K_{m_{\mathrm{A}}} = \frac{\mathrm{Coef_B}}{\mathrm{Coef_{AB}}}, \qquad K_{m_{\mathrm{B}}} = \frac{\mathrm{Coef_A}}{\mathrm{Coef_{AB}}}, \qquad K_{m_{\mathrm{I}}} = \frac{\mathrm{Coef_B}}{\mathrm{Coef_{IB}}}$$

$K'_{m_{\mathrm{B}}} = \dfrac{\mathrm{Coef_I}}{\mathrm{Coef_{IB}}}$ = the K_m for B when I is the first substrate

$K_{is} = \dfrac{\mathrm{Coef_I}}{\mathrm{Coef_{IS}}}$ = an inhibition constant for product S, analogous to K_{ip} of the usual Ping Pong Bi Bi system

$K'_{is} = \dfrac{\mathrm{Coef_A}}{\mathrm{Coef_{AS}}}$ = a different inhibition constant for product S when A is the substrate

$K_{ii} = \dfrac{\mathrm{Coef_S}}{\mathrm{Coef_{IS}}}$ = an inhibition constant for substrate I, analogous to K_{ia} of the usual Ping Pong Bi Bi system

$K'_{ia} = \dfrac{\mathrm{Coef_S}}{\mathrm{Coef_{AS}}}$ = an inhibition constant for substrate A when S is the product inhibitor

Using the definitions above:

$$\frac{\text{Coef}_{\text{S}}}{\text{Coef}_{\text{AB}}} = \frac{\text{Coef}_{\text{S}} K_{m_{\text{B}}}}{\text{Coef}_{\text{A}}} = \frac{\text{Coef}_{\text{S}} K_{m_{\text{B}}}}{\text{Coef}_{\text{AS}} K'_{is}} = \frac{K'_{ia} K_{m_{\text{B}}}}{K'_{is}}$$

or

$$\frac{\text{Coef}_{\text{S}}}{\text{Coef}_{\text{AB}}} = \frac{\text{Coef}_{\text{S}} K_{m_{\text{A}}}}{\text{Coef}_{\text{B}}} = \frac{\text{Coef}_{\text{S}} K_{m_{\text{A}}}}{\text{Coef}_{\text{IB}} K_{m_{\text{I}}}} = \frac{\text{Coef}_{\text{S}} K_{m_{\text{A}}} K'_{m_{\text{B}}}}{\text{Coef}_{\text{I}} K_{m_{\text{I}}}} = \frac{\text{Coef}_{\text{S}} K_{m_{\text{A}}} K'_{m_{\text{B}}}}{K_{is} K_{m_{\text{I}}} \text{Coef}_{\text{IS}}} = \frac{K_{ii} K_{m_{\text{A}}} K'_{m_{\text{B}}}}{K_{is} K_{m_{\text{I}}}}$$

and

$$\frac{\text{Coef}_{\text{AS}}}{\text{Coef}_{\text{AB}}} = \frac{\text{Coef}_{\text{AS}} K_{m_{\text{B}}}}{\text{Coef}_{\text{A}}} = \frac{K_{m_{\text{B}}}}{K'_{is}}$$

From the two ways of expressing $\text{Coef}_{\text{S}}/\text{Coef}_{\text{AB}}$, we see the equality:

$$\frac{K'_{ia} K_{m_{\text{B}}}}{K'_{is} K_{m_{\text{A}}}} = \frac{K_{ii} K'_{m_{\text{B}}}}{K_{is} K_{m_{\text{I}}}}$$

The velocity equation can now be written:

$$\frac{v^{\text{S}}_{\text{P/A}}}{V_{\max_{\text{P/A}}}} = \frac{[\text{A}][\text{B}]\left(1 + \dfrac{\text{num}_{\text{AS}}}{\text{num}_{\text{AB}}} \dfrac{[\text{S}]}{[\text{B}]}\right)}{K_{m_{\text{B}}}[\text{A}] + K_{m_{\text{A}}}[\text{B}] + [\text{A}][\text{B}] + \dfrac{K_{m_{\text{B}}}[\text{A}][\text{S}]}{K'_{is}} + \dfrac{K_{ii} K_{m_{\text{A}}} K'_{m_{\text{B}}}[\text{S}]}{K_{is} K_{m_{\text{I}}}}}$$

(IX-378)

where $V_{\max_{\text{P/A}}}$ is the maximal velocity of P formation from A defined in the usual way. The undefined ratio $\text{num}_{\text{AS}}/\text{num}_{\text{AB}}$ reduces to:

$$\frac{\text{num}_{\text{AS}}}{\text{num}_{\text{AB}}} = \frac{(k_{-5}k_{-6})(k_{-3} + k_4)}{(k_3 k_4)(k_{-5} + k_6)}$$

Searching for an equivalent result, we find that:

$$\frac{(k_{-5}k_{-6})(k_{-3} + k_4)}{(k_3 k_4)(k_{-5} + k_6)} = \frac{\text{Coef}_{\text{S}}}{\text{Coef}_{\text{B}}} = \frac{K_{ii} K'_{m_{\text{B}}}}{K_{is} K_{m_{\text{I}}}}$$

When [A] is varied:

$$\frac{v^{S}_{P/A}}{V_{\max_{P/A}}} = \frac{[A]}{K_{m_A} + [A]\left[\dfrac{1 + \dfrac{K_{m_B}}{[B]}\left(1 + \dfrac{[S]}{K'_{is}}\right)}{1 + \dfrac{K_{ii}K'_{m_B}[S]}{K_{is}K_{m_I}[B]}}\right]} \qquad \text{(IX-379)}$$

Thus when P is measured, S acts as a partial uncompetitive inhibitor or activator at unsaturating [B]. The replot of $1/v$-axis intercept versus [S] is hyperbolic. The situation is not the same as when alternative substrate I is present. In the presence of S (and absence of a finite concentration of I), all the reaction flux goes through the E→EA→FP→F sequence. Only the steps from F back to E are affected. Thus S has no effect on $slope_{1/A}$ because each catalytic cycle produces P regardless of whether Q or I is the other product formed. At saturating [S], all the P formed is produced by the sequence E→EA→FP→F→FS→EI→E. The effect of S on the $1/v$-axis intercept depends on whether the path by which I is formed is faster or slower than the path by which Q is formed. If the rate of I formation is slower, S will be a hyperbolic uncompetitive inhibitor. If the rate of I formation is faster, S will be a hyperbolic uncompetitive activator. If both paths have the same overall rate, then S will have no effect.

If the rate of Q formation from A is taken as the velocity, then:

$$v^{S}_{Q/A} = \frac{d[Q]}{dt} = k_4[EQ]$$

Substitution from the King-Altman solutions yields:

$$\frac{v^{S}_{Q/A}}{V_{\max_{Q/A}}} = \frac{[A]}{K_{m_A}\left(1 + \dfrac{K_{ii}K'_{m_B}[S]}{K_{is}K_{m_I}[B]}\right) + [A]\left(1 + \dfrac{K_{m_B}}{[B]} + \dfrac{K_{m_B}[S]}{K'_{is}[B]}\right)} \qquad \text{(IX-380)}$$

Thus when Q is measured, S is a linear mixed-type inhibitor.

When I is used as the substrate and S as a product inhibitor, the velocity is described by the usual equation for a Ping Pong Bi Bi system with the first product as inhibitor:

$$v = \frac{d[\text{Q}]}{dt} = \frac{d[\text{S}]}{dt} = k_6[\text{FS}] = k_4[\text{EQ}]$$

$$\boxed{\frac{v}{V_{\max_{\text{I}}}} = \frac{[\text{I}]}{K_{m_{\text{I}}}\left(1 + \dfrac{K_{ii}K'_{m_{\text{B}}}[\text{S}]}{K_{is}K_{m_{\text{I}}}[\text{B}]}\right) + [\text{I}]\left(1 + \dfrac{K'_{m_{\text{B}}}}{[\text{B}]} + \dfrac{K'_{m_{\text{B}}}[\text{S}]}{K_{is}[\text{B}]}\right)}} \qquad \text{(IX-381)}$$

Suppose P and S combine with free E to form a dead-end EP or ES complex (a not uncommon event in Ping Pong systems). When P is used as a product inhibitor with A as the substrate, and also when S is used as a product inhibitor with I as the substrate, $slope_{1/\text{A}}$ and $slope_{1/\text{I}}$ replots will be parabolic (equation IX-356). The dissociation constant for the dead-end complex, K_i, appears as part of the $slope_{1/\text{A}}$ and $slope_{1/\text{I}}$ factors and as part of the $int_{1/\text{B}}$ factor. The K_i can be obtained from $int_{1/\text{B}}$ replots. However, in some cases, [B] cannot be varied. For example, B might be water in a Ping Pong hydrolysis reaction. In this case, K_i would have to be extracted from $slope_{1/\text{A}}$ or $slope_{1/\text{I}}$ replots by various curve fitting procedures. An easier way is to use S as an alternative product inhibitor with A as the substrate (and similarly, P as an alternative product inhibitor with I as the substrate). In the presence of S, which forms a dead-end complex with E, the denominator $K_{m_{\text{A}}}[\text{B}]$ term is multiplied by $(1 + [\text{S}]/K_i)$. (The denominator $K_{m_{\text{A}}}[\text{B}]$ is the only denominator term representing [E] in the absence of a finite concentration of P.) The rate of P formation when [A] is varied is given by:

$$\boxed{\frac{v^{\text{S}}_{\text{P/A}}}{V_{\max_{\text{P/A}}}} = \frac{[\text{A}]}{K_{m_{\text{A}}}\left(1 + \dfrac{[\text{S}]}{K_i}\right) + [\text{A}]\left[\dfrac{1 + \dfrac{K_{m_{\text{B}}}}{[\text{B}]}\left(1 + \dfrac{[\text{S}]}{K'_{is}}\right)}{1 + \dfrac{K_{ii}K'_{m_{\text{B}}}[\text{S}]}{K_{is}K_{m_{\text{I}}}[\text{B}]}}\right]}} \qquad \text{(IX-382)}$$

Now S causes *S*-linear inhibition, *I*-hyperbolic inhibition, or activation, and the K_i for the dissociation of S from ES can be easily obtained from $slope_{1/\text{A}}$ replots. Similarly, the K_i for the dissociation of P from EP will appear in the $slope_{1/\text{I}}$ factor when P is used as an alternative product inhibitor with I as the substrate and v is taken as the rate of S appearance. If B were water, then the system behaves as if it were Ordered Uni Bi. The velocity equations are the same as those shown above with all the $K_{m_B}/[B]$ and $K'_{m_B}/[B]$ terms missing. For example, $v^S_{P/A}$ is given by:

$$\frac{v^S_{P/A}}{V_{\max_{P/A}}} = \frac{[A]}{K_{m_A}\left(1 + \frac{[S]}{K_i}\right) + [A]\left[\dfrac{1 + \dfrac{[S]}{K'_{is}}}{1 + \dfrac{K_{ii}[S]}{K_{is}K_{m_I}}}\right]} \qquad \text{(IX-383)}$$

In the Ordered Uni Bi system, [E] is represented by the constant in the denominator (which is K_{m_A}). Since the formation of a dead-end enzyme-product complex results in slope effects, it is obvious that if we observe only hyperbolic intercept effects, we can be certain that S or P does not form a dead-end complex with E.

The approach above is not limited to Ordered Uni Bi hydrolyses or Ping Pong Bi Bi systems but can also be used with Ordered Bi Bi enzymes where an alternative second substrate yields an alternative first product. When the second substrate is varied and the rate of the unique first product appearance is taken as v, an alternative first product should yield hyperbolic uncompetitive inhibition or activation. Any slope effect results from dead-end combination of the alternative first product with EA or E (the latter can be avoided by maintaining a reasonably high A concentration).

HH. SUBSTRATE INHIBITION

Most substrate inhibitions result from the combination of a substrate with the wrong enzyme form and, in general, are apparent only at high substrate concentrations and/or when the reaction is studied in the nonphysiological direction. The best way to study substrate inhibition is to vary a noninhibitory substrate at several fixed levels of the inhibitory substrate and then to determine whether the slope, the intercept, or both are affected. Sometimes we may wish to vary the inhibitory substrate. In this case, the reciprocal plot will pass through a minimum and bend up as it approaches the $1/v$-axis. The plot approaches a straight line asymptote only at very high values of $1/[S]$. The position of the asymptote may not be easy to establish, but if the position of the minimum is known, the slope and $1/v$-axis intercept of the asymptote can be established by adjusting the line until its slope is half that of the line from the intercept to the minimum point, as described in Figure VI-17. Some rapid equilibrium examples of substrate inhibition have been described in Chapter Six. Some steady-state examples are described below.

Substrate Inhibition in an Ordered Bireactant System

Dead-End EB Complex (Competitive Substrate Inhibition)

When the two substrates are structurally similar, B may react with free E to yield a dead-end EB complex:

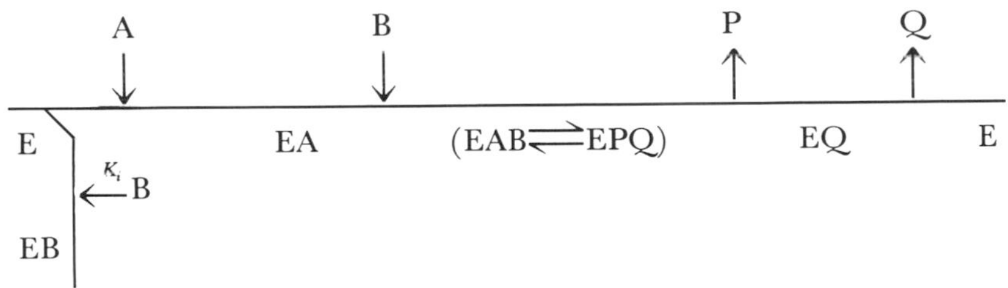

The velocity equation is obtained in the usual manner: those terms in the denominator representing free E are multiplied by $(1+[B]/K_i)$, where K_i is the dissociation constant of EB. The equation becomes:

$$\frac{v}{V_{max}} = \frac{[A][B]}{K_{ia}K_{m_B}\left(1+\frac{[B]}{K_i}\right) + K_{m_A}[B]\left(1+\frac{[B]}{K_i}\right) + K_{m_B}[A] + [A][B]} \tag{IX-384}$$

When [A] is varied, the equation can be written:

$$\frac{v}{V_{max}} = \frac{[A]}{K_{m_A}\left(1+\frac{K_{ia}K_{m_B}}{K_{m_A}[B]}\right)\left(1+\frac{[B]}{K_i}\right) + [A]\left(1+\frac{K_{m_B}}{[B]}\right)} \tag{IX-385}$$

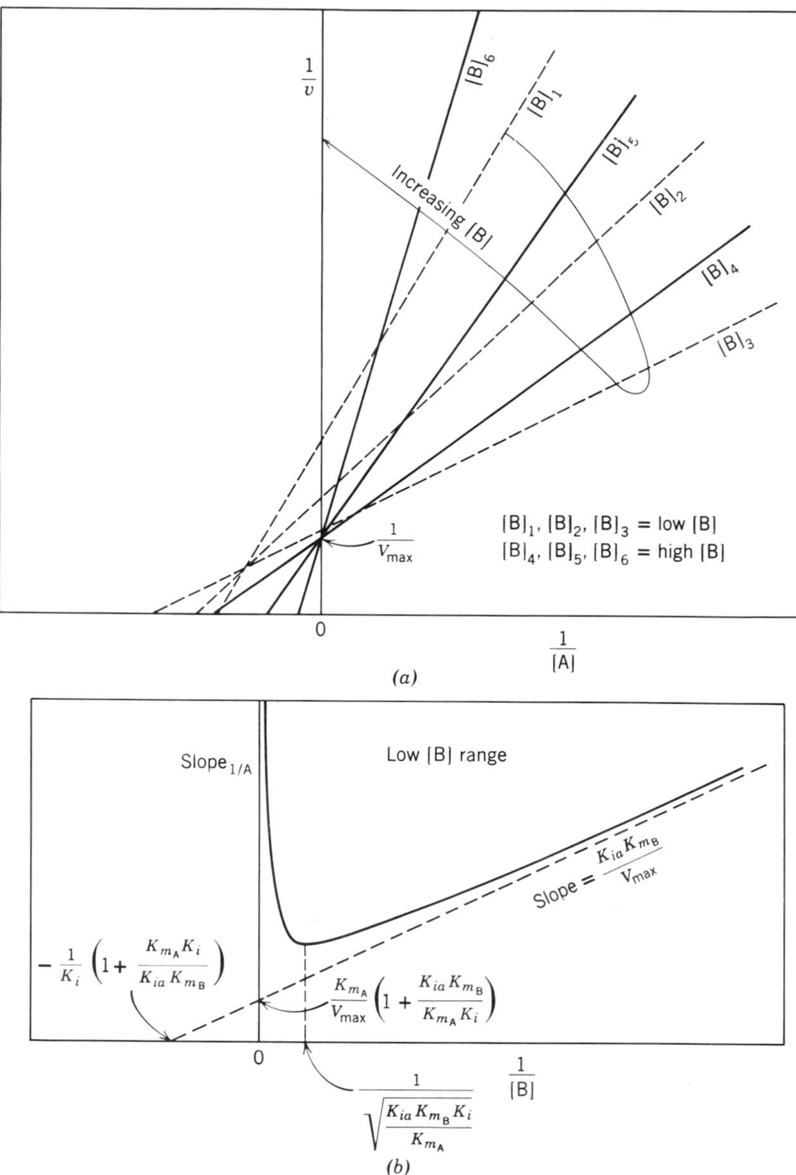

Fig. IX-61. Substrate inhibition in an ordered bireactant system. B combines with E as well as with EA. (*a*) Reciprocal plots at low and high fixed B concentrations. (*b*) $Slope_{1/A}$ replot in a low range of B concentrations. (B acts mainly as a substrate.) (*c*) $Slope_{1/A}$ replot in a high range of B concentrations. (B acts mainly as an inhibitor.) (*d*) $1/v$-axis intercept replot.

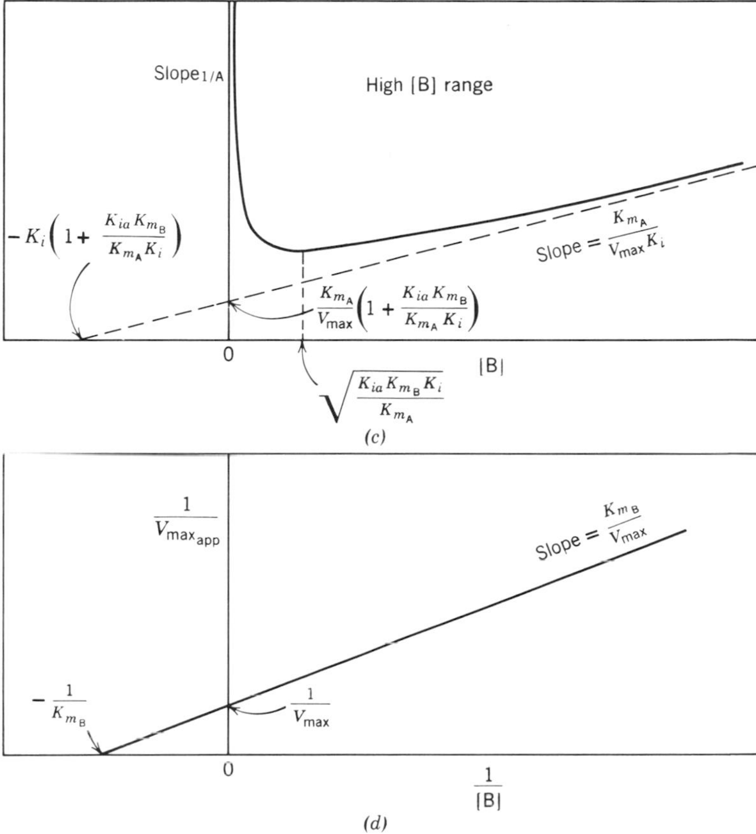

Fig. IX-61. (*Cont.*)

The $1/v$ versus $1/[A]$ plots are linear at all B concentrations; B as an inhibitor affects only the $slope_{1/A}$. At very low fixed B concentrations, the family of reciprocal plots may appear normal, intersecting to the left of the $1/v$-axis, above, on, or below the $1/[A]$-axis. As the fixed [B] is increased, the substrate inhibition becomes more obvious: the $1/v$-axis intercept decreases as usual (to a limit of $1/V_{max}$) but the slope passes through a minimum and then increases again (Fig. IX-61*a*). At very high fixed B concentrations, the family of reciprocal plots intersect on the $1/v$-axis at $1/V_{max}$. The $slope_{1/A}$ is given by:

$$slope_{1/A} = \frac{K_{m_A}}{V_{max}K_i}[B] + \frac{K_{ia}K_{m_B}}{V_{max}}[B]^{-1} + \frac{K_{m_A}}{V_{max}}\left(1 + \frac{K_{ia}K_{m_B}}{K_{m_A}K_i}\right) \quad \text{(IX-386)}$$

Differentiating and setting $d\ slope/d[\mathrm{B}]$ equal to zero, we find that the minimum slope occurs at:

$$[\mathrm{B}]_{\min}=\sqrt{\frac{K_{ia}K_{m_\mathrm{B}}K_i}{K_{m_\mathrm{A}}}}$$

If data are available over a sufficiently large range of fixed B concentrations, two $slope_{1/\mathrm{A}}$ replots can be constructed (Fig. IX-61*b,c*). The intercept replot (Fig. IX-61*d*) is normal and allows K_{m_B} and $V_{\max}$ to be determined. With these values, K_{ia} and the ratio K_{m_A}/K_i can be calculated from the slopes of the asymptotes. The ratio can then be used to calculate K_i or K_{m_A} from the horizontal-axis intercept of one of the $slope_{1/\mathrm{A}}$ replots. With one of these constants known, the other may be calculated.

When [B] is varied, the plots will be linear only at high values of 1/[B]. As the plots approach the $1/v$-axis, they pass through a minimum and curve up. The velocity equation is:

$$\frac{v}{V_{\max}}=\frac{[\mathrm{B}]}{K_{m_\mathrm{B}}\left(1+\dfrac{K_{ia}}{[\mathrm{A}]}\right)+[\mathrm{B}]\left(1+\dfrac{K_{ia}K_{m_\mathrm{B}}}{K_i[\mathrm{A}]}+\dfrac{K_{m_\mathrm{A}}}{[\mathrm{A}]}+\dfrac{K_{m_\mathrm{A}}[\mathrm{B}]}{K_i[\mathrm{A}]}\right)} \quad \text{(IX-387)}$$

The minimum occurs at:

$$[\mathrm{B}]_{\min}=\sqrt{\frac{K_{m_\mathrm{B}}K_i([\mathrm{A}]+K_{ia})}{K_{m_\mathrm{A}}}}$$

As the fixed [A] increases, the minimum moves to higher [B] (lower 1/[B]) until at saturating [A], there is no minimum and the plot is linear (Fig. IX-62).

Dead-End EA_2 Complex (Competitive Substrate Inhibition)

Substrate A may compete with B and react with EA to form a dead-end EA_2 complex.

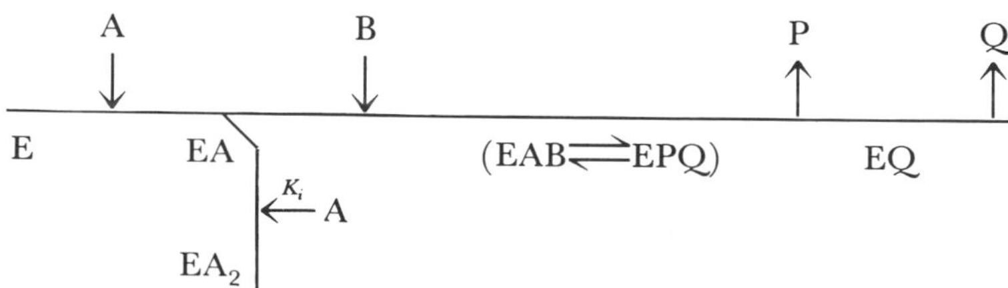

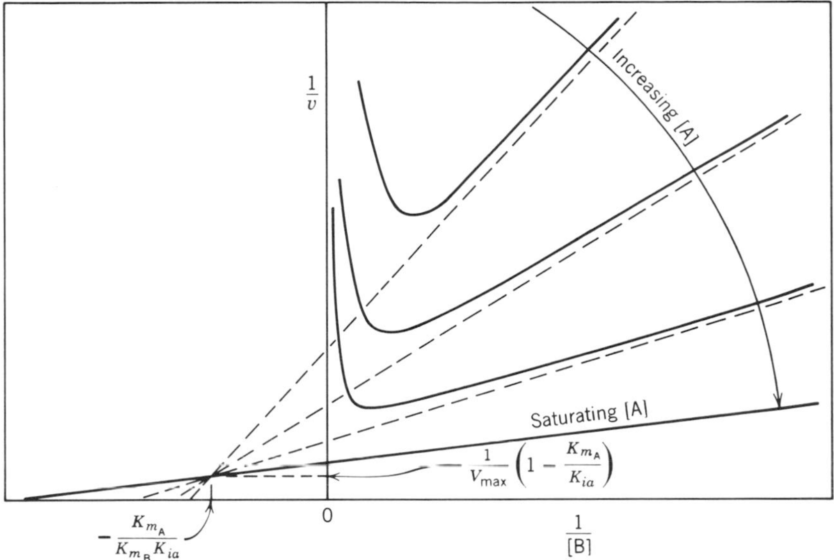

Fig. IX-62. Substrate inhibition in an ordered bireactant system; B reacts with E as well as with EA; $1/v$ versus $1/[B]$ plot.

The velocity equation is:

$$\frac{v}{V_{max}} = \frac{[A][B]}{K_{ia}K_{m_B} + K_{m_A}[B] + K_{m_B}[A]\left(1 + \frac{[A]}{K_i}\right) + [A][B]} \quad \text{(IX-388)}$$

The $K_{m_B}[A]$ term in the denominator (representing the EA complex) is multiplied by $(1 + [A]/K_i)$. The inhibition by A will be obvious from the reciprocal plots of $1/v$ versus $1/[A]$, which pass through a minimum and then curve up as they approach the $1/v$-axis. The velocity equation when [A] is varied is:

$$\frac{v}{V_{max}} = \frac{[A]}{K_{m_A}\left(1 + \frac{K_{ia}K_{m_B}}{K_{m_A}[B]}\right) + [A]\left[1 + \frac{K_{m_B}}{[B]}\left(1 + \frac{[A]}{K_i}\right)\right]} \quad \text{(IX-389)}$$

The minimum occurs at a point corresponding to:

$$[\mathrm{A}]_{\min} = \sqrt{K_{ia}K_i\left(1 + \frac{K_{m_\mathrm{A}}[\mathrm{B}]}{K_{ia}K_{m_\mathrm{B}}}\right)}$$

Thus as the fixed [B] increases, the minimum point moves closer to the 1/[A]-axis. When [B] is saturating, the inhibition by A is overcome and the plots are linear. The plots are similar to those shown in Figure IX-62.

When [B] is varied the velocity equation is:

$$\boxed{\frac{v}{V_{\max}} = \frac{[\mathrm{B}]}{K_{m_\mathrm{B}}\left(1 + \frac{[\mathrm{A}]}{K_i} + \frac{K_{ia}}{[\mathrm{A}]}\right) + [\mathrm{B}]\left(1 + \frac{K_{m_\mathrm{A}}}{[\mathrm{A}]}\right)}} \qquad \text{(IX-390)}$$

The plots are linear at all fixed A concentrations. The $1/v$-axis intercept decreases normally as the fixed [A] increases. Replots of $1/V_{\max_{\mathrm{app}}}$ versus 1/[A] are linear and allow $V_{\max}$ and K_{m_A} to be determined (Fig. IX-63*a*). The $slope_{1/\mathrm{B}}$ replot passes through a minimum and then bends up as it approaches the vertical (Fig. IX-63*b*,*c*). If data over a large range of fixed A concentrations are available, both types of $slope_{1/\mathrm{B}}$ replots can be constructed and K_i as well as K_{ia} can be determined.

Dead-End EBQ Complex (Uncompetitive Substrate Inhibition)

Uncompetitive substrate inhibition is characteristic of ordered systems and occurs when B reacts in dead-end fashion with EQ.

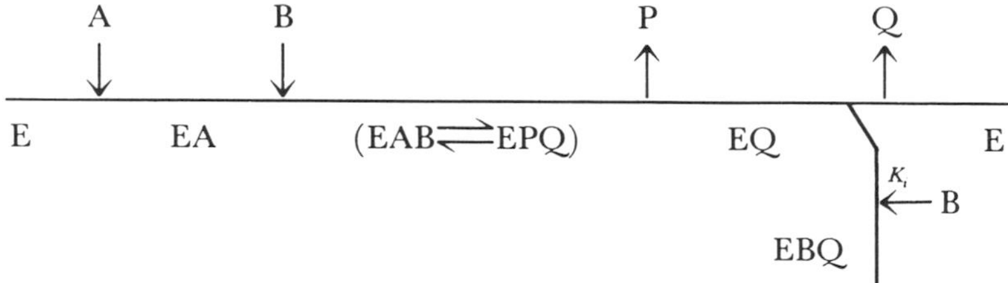

There is no rapid equilibrium counterpart, since in the absence of added Q the level of EQ is negligible. In Steady-State Ordered Bi Bi systems, there are no complete denominator terms that represent solely the EQ complex.

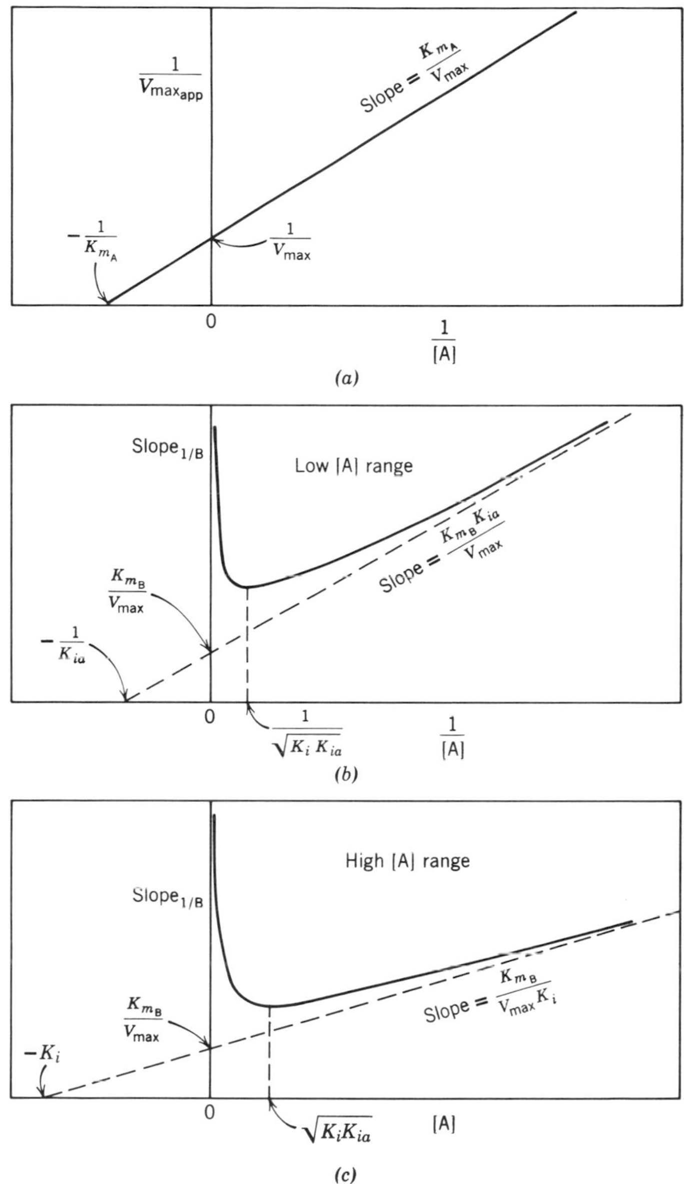

Fig. IX-63. Substrate inhibition in an ordered bireactant system; A reacts with EA as well as with E. Replots of the $1/v$ versus $1/[B]$ plot.

The [A][B] term represents both EQ and the central complexes. Consequently, a velocity equation cannot be written and the dissociation constant for the EBQ complex cannot be determined. Nevertheless, uncompetitive substrate inhibition by B can be easily recognized from plots of $1/v$ versus $1/[A]$ over a wide range of fixed B concentrations (Fig. IX-64). As [B] is increased, the $slope_{1/A}$ decreases normally to a limit of K_{m_A}/V_{max}. The $1/v$-axis intercept decreases to a minimum and then increases again so that at high [B] the plots become parallel. The intercept replot passes through a minimum and then bends up as it approaches the $1/V_{max_{app}}$-axis. The constants obtained by extrapolation of the asymptote are only apparent constants.

Substrate Inhibition in Ping Pong Systems

Dead-End EB Complex (Competitive Substrate Inhibition)

Competitive substrate inhibition resulting from the combination of B with E as well as with F is characteristic of Ping Pong systems.

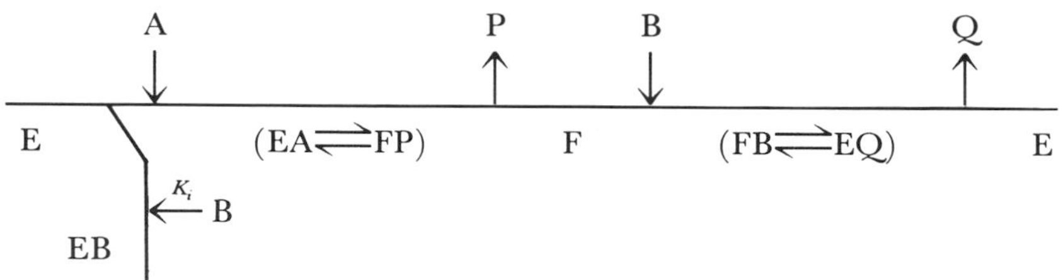

The $K_{m_A}[B]$ term in the denominator of the velocity equation represents [E]. Consequently, this term is multiplied by $(1+[B]/K_i)$:

$$\frac{v}{V_{max}} = \frac{[A][B]}{K_{m_A}[B]\left(1+\frac{[B]}{K_i}\right)+K_{m_B}[A]+[A][B]} \tag{IX-391}$$

When [A] is varied:

$$\frac{v}{V_{max}} = \frac{[A]}{K_{m_A}\left(1+\frac{[B]}{K_i}\right)+[A]\left(1+\frac{K_{m_B}}{[B]}\right)} \tag{IX-392}$$

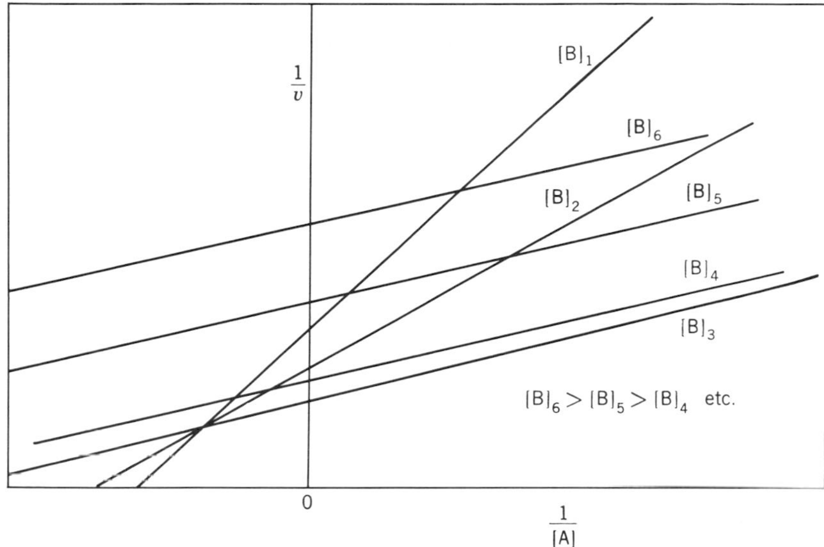

Fig. IX-64. Uncompetitive substrate inhibition by B in an Ordered Bi Bi system; B reacts with EQ to form a dead-end EBQ complex.

At low fixed B concentrations, the plots may appear parallel, as long as $K_i > K_{m_B}$. As the fixed [B] is increased, the $1/v$-axis intercept decreases as usual to a limit of $1/V_{max}$. The $slope_{1/A}$ may appear constant for several low B concentrations, but eventually, as the fixed [B] increases, the slope increases. When the fixed B concentrations are very high compared to K_i, the plots will intersect on the $1/v$-axis at $1/V_{max}$ (Fig. IX-65*a*). Since the variation in $slope_{1/A}$ results only from the competitive inhibition by B, a replot of $slope_{1/A}$ versus [B] is linear and yields K_i directly (Fig. IX-65*b*). The same type of plots are obtained if substrate B is contaminated with an inhibitor that is competitive with A.

When [B] is varied, the velocity equation is given by:

$$\frac{v}{V_{max}} = \frac{[B]}{K_{m_B} + [B]\left[1 + \frac{K_{m_A}}{[A]}\left(1 + \frac{[B]}{K_i}\right)\right]} \qquad \text{(IX-393)}$$

The plots appear parallel at high $1/[B]$ but pass through a minimum and

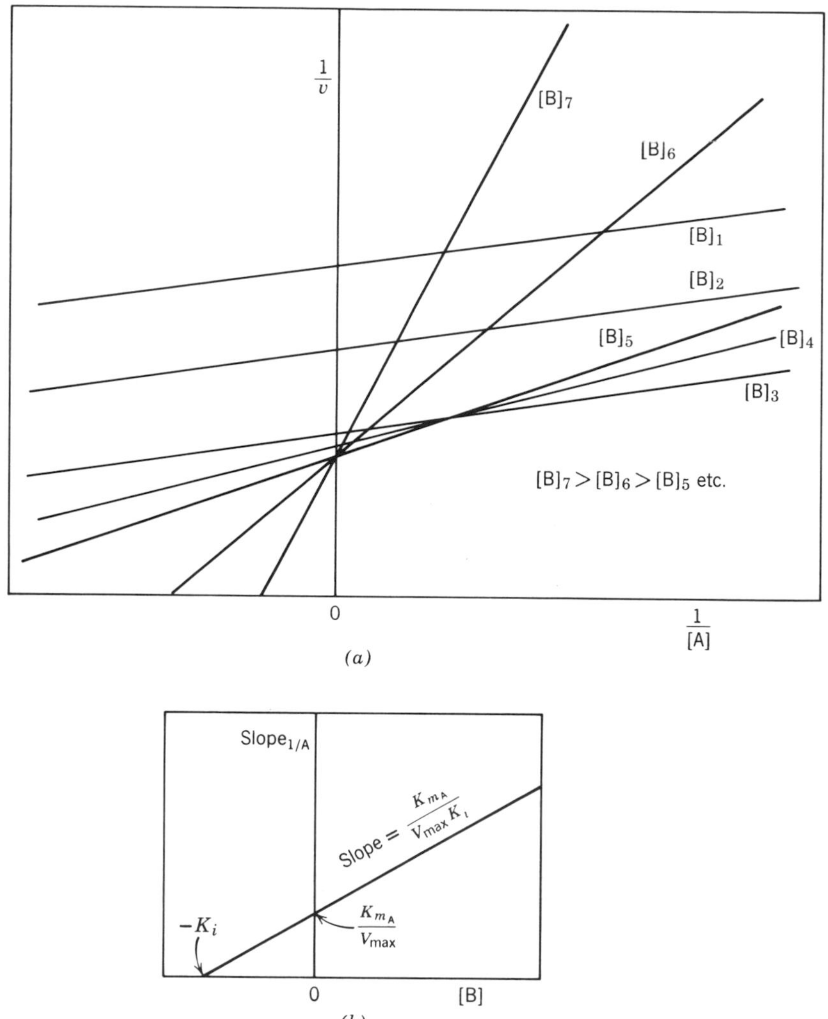

Fig. IX-65. Competitive substrate inhibition by B in a Ping Pong bireactant system; B reacts with E to form a dead-end EB complex. (*a*) $1/v$ versus 1/[A] and (*b*) *slope*$_{1/\mathrm{A}}$ replot.

bend up as they approach the $1/v$-axis (Fig. IX-66). The minimum occurs at:

$$[\mathrm{B}]_{\min}=\sqrt{\frac{K_{m_\mathrm{B}}K_i[\mathrm{A}]}{K_{m_\mathrm{A}}}}$$

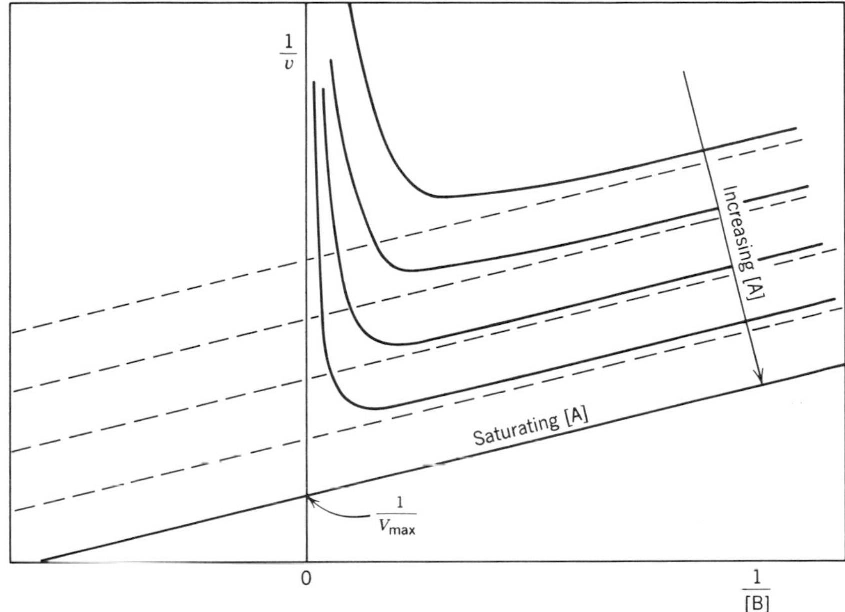

Fig. IX-66. Substrate inhibition by B in a Ping Pong bireactant system; B reacts with E; $1/v$ versus $1/[B]$ plot.

As the fixed [A] increases, the position of the minimum moves closer to the $1/v$-axis and disappears when [A] is saturating (i.e., saturating [A] overcomes the inhibition by B).

Dead-End FA Complex (Competitive Substrate Inhibition)

If A combines with F to form a dead-end FA complex, then the denominator $K_{m_B}[A]$ is multiplied by $(1+[A]/K_i)$. The plots are symmetrical to those described above.

Dead-End EB and FA Complexes (Double Competitive Substrate Inhibition)

Double competitive substrate inhibition where both substrates form dead-end complexes with the wrong enzyme forms is not uncommon in Ping Pong systems.

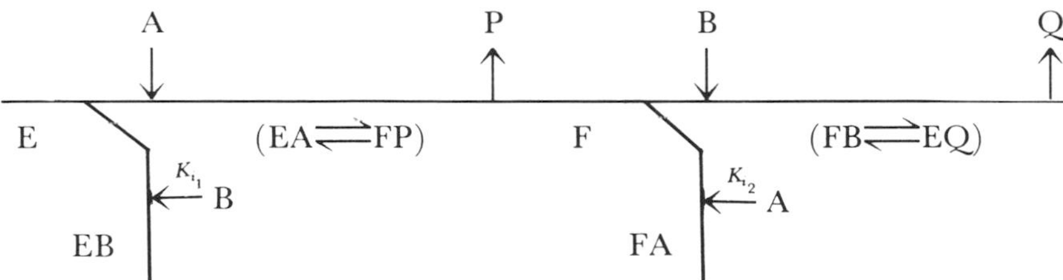

The velocity equation is:

$$\frac{v}{V_{\max}} = \frac{[\mathrm{A}][\mathrm{B}]}{K_{m_\mathrm{A}}[\mathrm{B}]\left(1+\frac{[\mathrm{B}]}{K_{i_1}}\right)+K_{m_\mathrm{B}}[\mathrm{A}]\left(1+\frac{[\mathrm{A}]}{K_{i_2}}\right)+[\mathrm{A}][\mathrm{B}]} \tag{IX-394}$$

The reciprocal plot patterns for the two substrates are symmetrical. When either substrate is varied at different fixed levels of the other, the plots curve up as they approach the $1/v$-axis. As the concentration of the fixed substrate increases, the minimum point moves closer to the $1/v$-axis (as the fixed substrate overcomes the inhibition by the varied substrate) and the slope of the reciprocal plot increases (as the fixed substrate introduces its own inhibitory effect). Figure IX-67 shows a typical double competitive substrate inhibition pattern. It may be difficult to determine all the constants without curve-fitting procedures. If K_{i_1} and K_{i_2} are relatively large compared to K_{m_A} and K_{m_B} (which is not unexpected), then K_{m_A} and K_{m_B} might be obtainable from replots of the $1/v$-axis intercepts of the parallel asymptotes obtained at several low concentrations of the fixed substrate (e.g., $[\mathrm{B}]_1$ and $[\mathrm{B}]_2$ on Fig. IX-67). The dissociation constants for the dead-end complexes might be obtainable from slope replots of data obtained at high concentrations of the fixed substrate (e.g., $[\mathrm{B}]_7$ and $[\mathrm{B}]_6$ on Fig. IX-67).

II. A REVIEW OF INHIBITION SYSTEMS

In the preceding sections we have encountered a number of different types of inhibitors, including dead-end inhibitors, alternate substrates, normal products, and alternate products. The inhibitors have been classified as (*a*) *competitive* if they affect only the slope of the reciprocal plot, (*b*) *uncompetitive* if

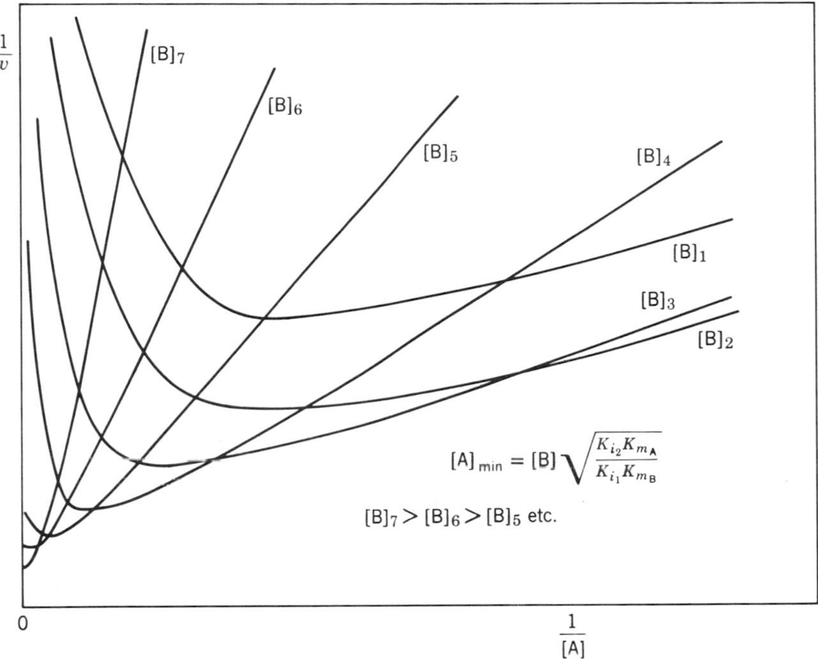

Fig. IX-67. Double substrate inhibition in a Ping Pong bireactant system; B reacts with E forming a dead-end EB complex; A reacts with F forming a dead-end FA complex.

they affect only the $1/v$-axis intercept of the reciprocal plot, and (c) *mixed-type* (or *noncompetitive*) if both the slope and $1/v$-axis intercept are affected. The type of inhibition has been further classified as (a) *linear* if the replot of slope and/or intercept versus [I] is a straight line, (b) *parabolic* if the slope and/or intercept replot is a parabola, and (c) *hyperbolic* if the replot is a hyperbola.

Linear Inhibition

In linear mixed-type inhibitions, the velocity equation has the form:

$$\frac{v}{V_{\max}} = \frac{[\mathrm{S}]}{K_{m_\mathrm{A}}(a+b[\mathrm{I}])+[\mathrm{S}](c+d[\mathrm{I}])} \tag{IX-395}$$

where S is one of the substrates and a and c are all the non-I-containing terms composed of kinetic constants and concentration terms of the other constant ligands. Sometimes it is convenient to factor out a and c and write the equation as:

$$\frac{v}{V^*_{\max}} = \frac{[S]}{K^*_{m_S}(1+e[I])+[S](1+f[I])} \tag{IX-396}$$

where $V^*_{\max} = \dfrac{V_{\max}}{c}$ = the limiting maximal velocity at the fixed concentrations of all the nonvaried ligands in the absence of I

$K^*_{m_S} = \dfrac{aK_{m_S}}{c}$ = the limiting K_m for S at the fixed concentrations of all the nonvaried ligands in the absence of I

$$e = \frac{b}{a} \quad \text{and} \quad f = \frac{d}{c}$$

Equation IX-396 can be written:

$$\frac{v}{V^*_{\max}} = \frac{[S]}{K^*_{m_S}\left(1+\dfrac{[I]}{K_{i_{slope}}}\right)+[S]\left(1+\dfrac{[I]}{K_{i_{int}}}\right)} \tag{IX-397}$$

where $K_{i_{slope}} = 1/e$ and $K_{i_{int}} = 1/f$.

The family of reciprocal plots obtained at different fixed concentrations of I intersect to the left of the $1/v$-axis, above the $1/[S]$-axis if $K_{i_{slope}} < K_{i_{int}}$, on the $1/[S]$-axis if $K_{i_{slope}} = K_{i_{int}}$, and below the $1/[S]$-axis if $K_{i_{slope}} > K_{i_{int}}$. The coordinates of the intersection point are:

$$-\frac{1}{K^*_{m_S}}\left(\frac{K_{i_{slope}}}{K_{i_{int}}}\right) \quad \text{and} \quad \frac{1}{V^*_{\max}}\left(1-\frac{K_{i_{slope}}}{K_{i_{int}}}\right) \tag{IX-398}$$

The equations for the slope and intercept replots have the form of a straight line.

$$slope = \frac{K_{m_S}}{V_{\max}}(a+b[I]) = \frac{K^*_{m_S}}{V^*_{\max}}(1+e[I]) \tag{IX-399}$$

$$int = \frac{1}{V_{\max}}(c+d[I]) = \frac{1}{V^*_{\max}}(1+f[I]) \tag{IX-400}$$

The replots yield $K_{i_{slope}}$ and $K_{i_{int}}$ for the constant concentrations of the other ligands. The true K_i can be obtained from secondary replots of $K_{i_{slope}}$ and $K_{i_{int}}$ against an appropriate concentration function of the constant ligands.

Parabolic Inhibition

In parabolic mixed-type or noncompetitive inhibition systems, the velocity equation contains $[I]^2$ terms resulting from the combination of I with two enzyme forms. The velocity equation has the form:

$$\frac{v}{V_{max}} = \frac{[S]}{K_{m_S}\left(a+b[I]+c[I]^2\right)+[S]\left(d+e[I]+f[I]^2\right)} \tag{IX-401}$$

or

$$\frac{v}{V_{max}} = \frac{[S]}{K^*_{m_S}\left(1+g[I]+h[I]^2\right)+[S]\left(1+j[I]+k[I]^2\right)} \tag{IX-402}$$

The replots are given by:

$$slope = \frac{K^*_{m_S}}{V^*_{max}}\left(1+g[I]+h[I]^2\right) \tag{IX-403}$$

$$int = \frac{1}{V^*_{max}}\left(1+j[I]+k[I]^2\right) \tag{IX-404}$$

The family of reciprocal plots at different fixed concentrations of I will not intersect at a common point unless the ratio of g/h in the slope factor is the same as the ratio of j/k in the intercept factor. In this case, the plots intersect at:

$$-\frac{1}{K^*_{m_S}}\left(\frac{j}{g}\right) \quad \text{and} \quad \frac{1}{V^*_{max}}\left(1-\frac{j}{g}\right) \tag{IX-405}$$

The intersection point is above, on, or below the $1/[S]$-axis depending on whether j is less than, equal to, or greater than g, respectively. In systems where one factor (slope or intercept) is a parabolic function of [I] and the other factor is a linear function of [I], the family of plots do not intersect at a common point.

Hyperbolic Inhibition

Hyperbolic systems arise from alternate paths to product formation; I can be a partial inhibitor or a partial (nonessential) activator. The velocity equation has the form:

$$\frac{v}{V_{max}} = \frac{[S]}{K_{m_S}\left(\frac{a+b[I]}{c+d[I]}\right)+[S]\left(\frac{e+f[I]}{g+h[I]}\right)} \tag{IX-406}$$

or

$$\frac{v}{V^*_{\max}} = \frac{[S]}{K^*_{m_S}\left(\dfrac{1+i[I]}{1+j[I]}\right)+[S]\left(\dfrac{1+k[I]}{1+l[I]}\right)} \tag{IX-407}$$

where i and j represent $1/K_{i_{slope}}^{num}$ and $1/K_{i_{slope}}^{denom}$ respectively, and k and l represent $1/K_{i_{int}}^{num}$ and $1/K_{i_{int}}^{denom}$, respectively. If both K_i^{denom}'s are the same (i.e., $j = l$), the family of reciprocal plots will intersect at a common point. The coordinates are:

$$-\frac{1}{K^*_{m_S}}\left(\frac{k-j}{i-j}\right) \quad \text{and} \quad \frac{1}{V^*_{\max}}\left(1-\frac{k-j}{i-j}\right) \tag{IX-408}$$

If $K_{i_{slope}}^{num}$ is less than $K_{i_{int}}^{num}$ (i.e., $i > k$), the lines cross above the $1/[S]$-axis. If $i < k$, the lines cross below the $1/[S]$-axis. If $i = k$, the lines cross on the $1/[S]$-axis. If the numerator and denominator K_i's are combined into artificial, [I]-dependent $K_{i_{slope}}$ and $K_{i_{int}}$ (as shown in Chapter Four for Rapid Equilibrium Random partial and mixed-type inhibition), then the intersection coordinates are given by equation IX-398. The lines do not intersect at a common point in mixed linear, hyperbolic inhibition systems.

The slope and intercept in an S-hyperbolic, I-hyperbolic system are given by:

$$slope = \frac{K_{m_S}}{V_{\max}}\left(\frac{a+b[I]}{c+d[I]}\right) = \frac{K^*_{m_S}}{V^*_{\max}}\left(\frac{1+i[I]}{1+j[I]}\right) \tag{IX-409}$$

$$int = \frac{1}{V_{\max}}\left(\frac{e+f[I]}{g+h[I]}\right) = \frac{1}{V^*_{\max}}\left(\frac{1+k[I]}{1+l[I]}\right) \tag{IX-410}$$

The replots are hyperbolic with limits of:

$$\frac{K^*_{m_S}}{V^*_{\max}}\left(\frac{i}{j}\right) = \frac{K^*_{m_S}}{V^*_{\max}}\,\frac{K_{i_{slope}}^{denom}}{K_{i_{slope}}^{num}}$$

and

$$\frac{1}{V^*_{\max}}\left(\frac{k}{l}\right) = \frac{K_{i_{int}}^{denom}}{V^*_{\max}K_{i_{int}}^{num}}$$

Thus I is an inhibitor if $K_i^{denom} > K_i^{num}$ and an activator if $K_i^{num} > K_i^{denom}$ Also, as [I] increases, the slope might increase while the intercept decreases, or vice versa, depending on the relative values of the four K_i's. The $K_{i_{\text{slope}}}^{\text{num}}$ and so on, can be determined from replots of $1/\Delta$ *slope* versus $1/[\text{I}]$ or $1/\Delta$ *int* versus $1/[\text{I}]$:

$$\Delta\,slope = \frac{K_{m_S}^*}{V_{\max}^*}\frac{\left(1+\dfrac{[\text{I}]}{K_{i_{\text{slope}}}^{\text{num}}}\right)}{\left(1+\dfrac{[\text{I}]}{K_{i_{\text{slope}}}^{\text{denom}}}\right)} - \frac{K_{m_S}^*}{V_{\max}^*}$$

$$\boxed{\frac{1}{\Delta\,slope} = \frac{K_{i_{\text{slope}}}^{\text{num}}K_{i_{\text{slope}}}^{\text{denom}}V_{\max}^*}{K_S^*\left(K_{i_{\text{slope}}}^{\text{denom}} - K_{i_{\text{slope}}}^{\text{num}}\right)}\frac{1}{[\text{I}]} + \frac{V_{\max}^*K_{i_{\text{slope}}}^{\text{num}}}{K_S^*\left(K_{i_{\text{slope}}}^{\text{denom}} - K_{i_{\text{slope}}}^{\text{num}}\right)}} \qquad \text{(IX-411)}$$

The intercept on the $1/[\text{I}]$-axis gives $-1/K_{i_{\text{slope}}}^{\text{denom}}$ The $K_{i_{\text{slope}}}^{\text{num}}$ can be calculated from the intercept on the $1/\Delta$ *slope*-axis. Similarly:

$$\boxed{\frac{1}{\Delta\,int} = \frac{K_{i_{\text{int}}}^{\text{num}}K_{i_{\text{int}}}^{\text{denom}}V_{\max}^*}{\left(K_{i_{\text{int}}}^{\text{denom}} - K_{i_{\text{int}}}^{\text{num}}\right)}\frac{1}{[\text{I}]} + \frac{V_{\max}^*K_{i_{\text{int}}}^{\text{num}}}{\left(K_{i_{\text{int}}}^{\text{denom}} - K_{i_{\text{int}}}^{\text{num}}\right)}} \qquad \text{(IX-412)}$$

For the Rapid Equilibrium Random partial mixed-type system described in Chapter Four (see equation IV-24):

$$K_{i_{\text{int}}}^{\text{num}} = \alpha K_i, \qquad K_{i_{\text{int}}}^{\text{denom}} = \frac{\alpha}{\beta}K_i, \qquad K_{i_{\text{slope}}}^{\text{num}} = K_i, \qquad \text{and} \qquad K_{i_{\text{slope}}}^{\text{denom}} = \frac{\alpha}{\beta}K_i$$

For more complex systems, $K_{i_{\text{slope}}}$ and $K_{i_{\text{int}}}$ are apparent constants for the fixed concentrations of the other ligands (cosubstrates, etc.). For example, in the Ping Pong Bi Bi system, alternate product S is a hyperbolic uncompetitive effector when [A] is varied. The intercept replot is given by:

$$int_{1/\text{A}} = \frac{1}{V_{\max}}\frac{\left(1+\dfrac{K_{m_B}}{[\text{B}]}+\dfrac{K_{m_B}[\text{S}]}{K_{is}'[\text{B}]}\right)}{\left(1+\dfrac{K_{ii}K_{m_B}'[\text{S}]}{K_{is}K_{m_1}[\text{B}]}\right)} = \frac{1}{V_{\max}}\left(1+\frac{K_{m_B}}{[\text{B}]}\right)\frac{\left[1+\dfrac{[\text{S}]}{K_{is}'\left(1+\dfrac{[\text{B}]}{K_{m_B}}\right)}\right]}{\left[1+\dfrac{[\text{S}]}{\dfrac{K_{is}K_{m_1}[\text{B}]}{K_{ii}K_{m_B}}}\right]}$$

(IX-413)

Thus

$$K_{i_{\text{int}}}^{\text{num}} = K_{is}'\left(1 + \frac{[\text{B}]}{K_{m_\text{B}}}\right), \qquad K_{i_{\text{int}}}^{\text{denom}} = K_{is}\left(\frac{K_{m_\text{I}}[\text{B}]}{K_{ii}K_{m_\text{B}}'}\right)$$

and

$$\frac{1}{V_{\max}^*} = \frac{1}{V_{\max}}\left(1 + \frac{K_{m_\text{B}}}{[\text{B}]}\right) = \text{the intercept at the fixed [B] when [I]} = 0$$

The replots would determine $K_{i_{\text{int}}}^{\text{num}}$ and $K_{i_{\text{int}}}^{\text{denom}}$; K_{is}' and K_{is} would have to be calculated from the known values of K_{m_I}, K_{ii}, and K_{m_B}'.

More Complex Types of Nonlinear Inhibition Systems

Inhibition systems where the slopes or intercepts are more complicated functions of [I] than those described above are quite possible, especially in systems containing alternate reaction sequences. For example, consider the hit-and-run substrate-inhibitor displacement system shown below (Reiner, 1969). There are two routes to ES, one of which involves the binding and displacement of I and which may be more or less favorable than the direct route.

$$\text{E} + \text{S} \underset{k_{-1}}{\overset{k_1}{\rightleftharpoons}} (\text{ES} \rightleftharpoons \text{EP}) \underset{k_{-2}}{\overset{k_2}{\rightleftharpoons}} \text{E} + \text{P}$$

$$\text{E} + \text{I} \underset{k_{-3}}{\overset{k_3}{\rightleftharpoons}} \text{EI}$$

$$\text{EI} + \text{S} \underset{k_{-4}}{\overset{k_4}{\rightleftharpoons}} (\text{ES} \rightleftharpoons \text{EP}) + \text{I}$$

The velocity equation for this system will contain $[\text{S}]^2$ and $[\text{I}]^2$ terms and, theoretically, reciprocal plots of $1/v$ versus $1/[\text{S}]$ in the presence of I are nonlinear. If, however, E and EI are at equilibrium, then the reciprocal plots will be linear. In this case the King-Altman figure for the forward reaction is:

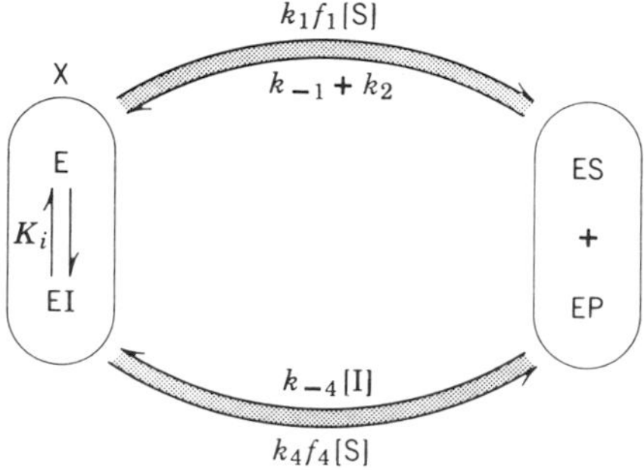

$$f_1 = \frac{[\mathrm{E}]}{[\mathrm{X}]} = \frac{1}{1+\frac{[\mathrm{I}]}{K_i}} = \frac{K_i}{K_i+[\mathrm{I}]}$$

$$f_4 = \frac{[\mathrm{EI}]}{[\mathrm{X}]} = \frac{\frac{[\mathrm{I}]}{K_i}}{1+\frac{[\mathrm{I}]}{K_i}} = \frac{[\mathrm{I}]}{K_i+[\mathrm{I}]}$$

The forward velocity is given by:

$$\frac{v}{[\mathrm{E}]_t} = \frac{k_2(k_1K_i + k_4[\mathrm{I}])[\mathrm{S}]}{(k_{-1}+k_2)K_i + (k_{-1}+k_2+k_4)[\mathrm{I}] + k_{-4}[\mathrm{I}]^2 + (k_1K_i + k_4[\mathrm{I}])[\mathrm{S}]} \qquad \text{(IX-414)}$$

or, dividing numerator and denominator by $\mathrm{Coef_S}$:

$$\boxed{\frac{v}{V_{\max}} = \frac{[\mathrm{S}]}{\frac{a+b[\mathrm{I}]+c[\mathrm{I}]^2}{d+e[\mathrm{I}]}+[\mathrm{S}]}} \qquad \text{(IX-415)}$$

Factoring out the non-[I]-containing terms:

$$\boxed{\frac{v}{V_{\max}} = \frac{[\mathrm{S}]}{\frac{a\left(1+f[\mathrm{I}]+g[\mathrm{I}]^2\right)}{d(1+h[\mathrm{I}])}+[\mathrm{S}]}} \qquad \text{(IX-416)}$$

where

$$\frac{a}{d} = \frac{(k_{-1}+k_2)}{k_1} = K_{m_\mathrm{S}}, \qquad f=\frac{b}{a}, \qquad g=\frac{c}{a}, \qquad \text{and} \qquad h=\frac{e}{d}$$

I behaves as a pure competitive inhibitor. The slope of the reciprocal plot is given by:

$$slope = \frac{K_{m_\mathrm{S}}}{V_{\max}} \frac{\left(1+f[\mathrm{I}]+g[\mathrm{I}]^2\right)}{(1+h[\mathrm{I}])} \qquad \text{(IX-417)}$$

The slope is a "2/1" function of [I]. The replot is a hyperbola with nonhorizontal asymptotes. As [I] increases, the replot approaches a straight line with a slope of $gK_{m_S}/hV_{\max}$ (which equals $k_{-4}/k_4V_{\max}$ in the present case). Thus v can be driven to zero by an infinitely high [I]. As [I] decreases, the replot attains another limiting slope. The slope of the replot near the slope-axis (when $[\mathrm{I}]=0$) is $K_{m_S}(f-h)/V_{\max}$, which may be greater than, equal to, or less than the limiting slope at high [I], depending on the ratio of $(g/h)/(f-h)$.

Another system involving alternate routes to product formation is the general modifier system shown below where I is either a partial inhibitor or a nonessential activator. It is assumed that I binds in a rapid equilibrium fashion.

$$\begin{array}{ccccccc}
\mathrm{E} & + \;\mathrm{S} & \underset{k_{-1}}{\overset{k_1}{\rightleftharpoons}} & \mathrm{ES} & \overset{k_3}{\longrightarrow} & \mathrm{E} & + \;\mathrm{P} \\
+ & & & + & & & \\
\mathrm{I} & & & \mathrm{I} & & & \\
K_i \upharpoonleft\!\downharpoonright & & & \downharpoonleft\!\upharpoonright \alpha K_i & & & \\
\mathrm{EI} & + \;\mathrm{S} & \underset{k_{-2}}{\overset{k_2}{\rightleftharpoons}} & \mathrm{ESI} & \overset{k_4}{\longrightarrow} & \mathrm{EI} & + \;\mathrm{P}
\end{array}$$

The basic King-Altman figure for the forward reaction is:

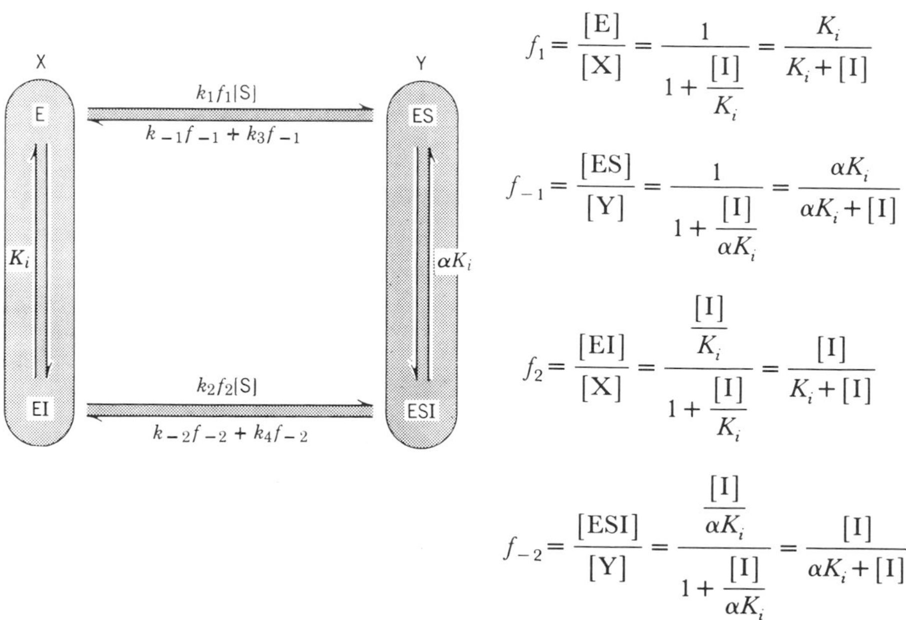

$$f_1 = \frac{[\mathrm{E}]}{[\mathrm{X}]} = \frac{1}{1+\dfrac{[\mathrm{I}]}{K_i}} = \frac{K_i}{K_i+[\mathrm{I}]}$$

$$f_{-1} = \frac{[\mathrm{ES}]}{[\mathrm{Y}]} = \frac{1}{1+\dfrac{[\mathrm{I}]}{\alpha K_i}} = \frac{\alpha K_i}{\alpha K_i+[\mathrm{I}]}$$

$$f_2 = \frac{[\mathrm{EI}]}{[\mathrm{X}]} = \frac{\dfrac{[\mathrm{I}]}{K_i}}{1+\dfrac{[\mathrm{I}]}{K_i}} = \frac{[\mathrm{I}]}{K_i+[\mathrm{I}]}$$

$$f_{-2} = \frac{[\mathrm{ESI}]}{[\mathrm{Y}]} = \frac{\dfrac{[\mathrm{I}]}{\alpha K_i}}{1+\dfrac{[\mathrm{I}]}{\alpha K_i}} = \frac{[\mathrm{I}]}{\alpha K_i+[\mathrm{I}]}$$

or

$$\text{(X)} \underset{(k_{-1}+k_3)f_{-1}+(k_{-2}+k_4)f_{-2}}{\overset{(k_1 f_1+k_2 f_2)[\mathrm{S}]}{\rightleftharpoons}} \text{(Y)}$$

The forward velocity is given by:

$$\frac{v}{[\mathrm{E}]_t}=\frac{\alpha k_1 k_3 K_i^2[\mathrm{S}]+(\alpha k_2 k_3 K_i+k_1 k_4 K_i)[\mathrm{I}][\mathrm{S}]+k_2 k_4[\mathrm{I}]^2[\mathrm{S}]}{\alpha K_i^2(k_{-1}+k_3)+[\alpha K_i(k_{-1}+k_3)+(k_{-2}+k_4)K_i][\mathrm{I}]+(k_{-2}+k_4)[\mathrm{I}]^2 + \alpha k_1 K_i^2[\mathrm{S}]+[k_1 K_i+\alpha k_2 K_i][\mathrm{I}][\mathrm{S}]+k_2[\mathrm{I}]^2[\mathrm{S}]} \qquad \text{(IX-418)}$$

Factoring [S]:

$$\frac{v}{[\mathrm{E}]_t}=\frac{[\mathrm{S}]\left(a+b[\mathrm{I}]+c[\mathrm{I}]^2\right)}{\left[d+e[\mathrm{I}]+f[\mathrm{I}]^2\right]+[\mathrm{S}]\left[g+h[\mathrm{I}]+i[\mathrm{I}]^2\right]} \qquad \text{(IX-419)}$$

Or, factoring the non-[I]-containing terms:

$$\boxed{\frac{v}{V_{\max}}=\frac{[\mathrm{S}]\left(1+j[\mathrm{I}]+k[\mathrm{I}]^2\right)}{K_{m_S}\left(1+l[\mathrm{I}]+m[\mathrm{I}]^2\right)+[\mathrm{S}]\left(1+n[\mathrm{I}]+o[\mathrm{I}]^2\right)}} \qquad \text{(IX-420)}$$

where $V_{\max}=\frac{a}{g}[\mathrm{E}]_t=k_3[\mathrm{E}]_t=$ the maximal velocity in the absence of I

$K_{m_S}=\frac{d}{g}=\frac{k_{-1}+k_3}{k_1}$

$j=\frac{b}{a}$, $k=\frac{c}{a}$, $l=\frac{e}{d}$, $m=\frac{f}{d}$, $n=\frac{h}{g}$, and $o=\frac{i}{g}$.

The equation can be written as:

$$\boxed{\frac{v}{V_{\max}}=\frac{[\mathrm{S}]}{K_{m_S}\left(\frac{1+l[\mathrm{I}]+m[\mathrm{I}]^2}{1+j[\mathrm{I}]+k[\mathrm{I}]^2}\right)+[\mathrm{S}]\left(\frac{1+n[\mathrm{I}]+o[\mathrm{I}]^2}{1+j[\mathrm{I}]+k[\mathrm{I}]^2}\right)}} \qquad \text{(IX-421)}$$

As [I] increases, the kinetic constants approach limiting values. Replots are hyperbolic. At saturating [I], the apparent $K_{m_S}=mK_{m_S}/o=(k_{-2}+k_4)/k_2$.

The apparent $V_{max} = kV_{max}/o = k_4[E]_t$. These limiting values may be higher or lower than the values in the absence of I.

If only ES forms product, the velocity would be given by:

$$v = k_3 f_{-1}[Y] \qquad \text{and} \qquad \frac{v}{[E]_t} = \frac{k_3 f_{-1}[Y]}{[X]+[Y]}$$

The numerator of the velocity equation would contain fewer terms (no $[I]^2$ term), but the denominator, which represents all the enzyme forms, remains the same (except now $k_4 = 0$). The velocity equation can be written:

$$\boxed{\frac{v}{V_{max}} = \frac{[S]}{K_{m_S}\left(\dfrac{1+l[I]+m[I]^2}{1+j[I]}\right)+[S]\left(\dfrac{1+n[I]+o[I]^2}{1+j[I]}\right)}} \qquad \text{(IX-422)}$$

The coefficients c and k are missing since $k_4 = 0$. The slope and intercept are hyperbolic functions of [I] with nonhorizontal asymptotes, as described for the substrate-inhibitor displacement system. Unlike the preceding system in which both ES and ESI form product, saturating [I] will drive the velocity to zero as all the enzyme is driven to the nonproductive EI and ESI forms. The slope and $1/v$-axis intercept approach infinity as [I] increases because the asymptote of the hyperbolic replot is not a horizontal line, but rather a straight line of positive slope.

If only ESI forms product, then I becomes a cosubstrate or essential activator. In this case:

$$v = k_4 f_{-2}[Y] \qquad \text{and} \qquad \frac{v}{[E]_t} = \frac{k_4 f_{-2}[Y]}{[X]+[Y]}$$

The velocity equation can be written:

$$\boxed{\frac{v}{[E]_t} = \frac{[S]}{\left(\dfrac{d+e[I]+f[I]^2}{b[I]+c[I]^2}\right)+[S]\left(\dfrac{g+h[I]+i[I]^2}{b[I]+c[I]^2}\right)}} \qquad \text{(IX-423)}$$

Notice that a is missing since $k_3 = 0$. The apparent K_{m_S} and V_{max} are complex functions of [I]. At saturating [I], the limiting kinetic constants are:

$$K_{m_S} = \frac{f}{i} = \frac{k_{-2} + k_4}{k_2} \qquad \text{and} \qquad V_{\max} = \frac{c[\mathrm{E}]_t}{i} = k_4[\mathrm{E}]_t$$

Replots of $slope_{1/S}$ or $int_{1/S}$ versus $1/[\mathrm{I}]$ are theoretically nonlinear, although they might appear so. In this case, the system is most conveniently analyzed as a Rapid Equilibrium Random Bireactant system.

REFERENCES

Cleland Nomenclature and General Rules

Cleland, W. W., *Biochim. Biophys. Acta* **67**, 104, 173, and 188 (1963).

Cleland, W. W., "Steady State Kinetics," in *The Enzymes*, Vol. 2, 3rd ed., P. D. Boyer, ed., Academic Press, 1970, p. 1.

Cleland, W. W., "Enzyme Kinetics," *Ann. Rev. Biochem.* **36**, 77 (1967).

Plapp, B. V., *Arch. Biochem. Biophys.* **156**, 112 (1973).

Plowman, K. M., *Enzyme Kinetics*, McGraw-Hill, 1972.

Rudolph, F. B., Purich, D. L., and Fromm, H. J., *J. Biol. Chem.* **243**, 5539 (1968).

Other Nomenclature

Alberty, R. A., *J. Am. Chem. Soc.* **75**, 1928 (1953).

Bloomfield, V., Peller, L. and Alberty, R. A., *J. Am. Chem. Soc.* **84**, 4367 (1962).

Dalziel, K., *Acta Chem. Scand.* **11**, 1706 (1957).

Laidler, K. J. and Bunting, P. S., *The Chemical Kinetics of Enzyme Action*, 2nd ed., Oxford, 1973.

Reiner, J. M., *Behavior of Enzyme Systems*, 2nd ed., Van Nostrand-Reinhold, 1969.

Wong, J. T-F. and Hanes, C. S., *Can. J. Biochem. Physiol.* **40**, 763 (1962).

King-Altman Method and Modifications

Cha, S., *J. Biol. Chem.* **243**, 820 (1968).

Fromm, H. J., *Biochem. Biophys. Res. Commun.* **40**, 692 (1970).

King, E. L. and Altman, C., *J. Phys. Chem.* **60**, 1375 (1956).

Volkenstein, M. V. and Goldstein, B. N., *Biochim. Biophys. Acta* **155**, 471 (1966).

Iso Uni Uni and Membrane Transport Systems

Britton, H. G., *Biochem. J.* **133**, 255 (1973).

Crabeel, M. and Grenson, M., *Eur. J. Biochem.* **14**, 197 (1970).

Cuppoletti, J. and Segel, I. H., *J. Membrane Biol.* **17**, 239 (1974).

Hunter, D. R. and Segel, I. H., *Arch. Biochem. Biophys.* **154**, 387 (1973).

Nearne, K. D. and Richards, T. G., *Elementary Kinetics of Membrane Carrier Transport*, John Wiley & Sons, 1972.

Pall, M. L. and Kelly, K. A., *Biochem. Biophys. Res. Commun.* **42**, 940 (1971).

Pall, M. L., *Biochim. Biophys. Acta* **233**, 201 (1971).

Ring, K., Gross, K. and Heinz, E., *Arch. Biochem. Biophys.* **137**, 243 (1970).

Schachter, H., "The Use of the Steady State Assumption to Derive Kinetic Formulations for the Transport of a Solute Across a Membrane," in *Metabolic Pathways*, Vol. 6, L. E. Hokin ed., Academic Press, 1972, p. 1.

Theorell-Chance Mechanism

Theorell, H. and Chance, B., *Acta Chem. Scand.* **5**, 1127 (1951).

Evaluating the Kinetic Significance of the Central Complexes

Janson, C. A. and Cleland, W. W., *J. Biol. Chem.* **249**, 2562 (1974) (Glycerokinase).

Allosteric Ping Pong Systems

Sumi, T. and Ui, M., *Biochim. Biophys. Acta* **276**, 12 (1972).

Uyeda, K. and Kurooka, S., *J. Biol. Chem.* **245**, 3315 (1970).

Two-Site and Three-Site Ping Pong Mechanisms

Barden, R. E., Fung, C-H., Utter, M. and Scrutton, M. C., *J. Biol. Chem.* **247**, 1323 (1972).

Cleland, W. W., *J. Biol. Chem.* **248**, 8353 (1973).

Northrop, D. B., *J. Biol. Chem.* **244**, 5808 (1969).

Hybrid Ping Pong-Theorell Chance Systems ("Di-Uni Iso Ping Pong")

Hurst, R. O., *Can. J. Biochem.* **47**, 111 (1969).

Hybrid Ping Pong-Random, Ping Pong-Ordered Systems Yielding Sigmoidal Velocity Curves

Griffen, C. C. and Brand, L., *Arch. Biochem. Biophys.* **126**, 856 (1968).

Sweeny, J. R. and Fisher, J. R., *Biochem.* **7**, 561 (1968).

Steady-State Random Bireactant Systems

Dixon, M. and Webb, E. C., *Enzymes*, 2nd ed., Academic Press, 1964, p. 95.

Ferdinand, W., *Biochem. J.* **98**, 278 (1966).

Fisher, J. R., *Arch. Biochem. Biophys.* **152**, 638 (1972).

Alternate Substrates for Bireactant and Terreactant Systems

Fromm, H. J., *Biochim. Biophys. Acta* **81**, 413 (1964).

Hamagishi, Y. and Yoshida, H., *J. Biochem. (Tokyo)* **76**, 81 (1974).

Hinberg, I. and Laidler, K., *Can. J. Biochem.* **50**, 1334 and 1360 (1972).

Ricardi, J., Noat, G., Got, C. and Borel, M., *Eur. J. Biochem.* **31**, 14 (1972).

Rudolph, F. B. and Fromm, H. J., *Biochemistry* **9**, 4660 (1970).

Rudolph, F. B., and Fromm, H. J., *Arch. Biochem. Biophys.* **147**, 515 (1971).

Zewe, V., Fromm, H. J. and Fabiano, R., *J. Biol. Chem.* **239**, 1625 (1964).

Measurement of Common Product

Cha, S., *Mol. Pharmacol.* **4**, 621 (1968).

Pocklington, T. and Jeffery, J., *Biochem. J.* **112**, 331 (1969).

Dead-End Inhibitors

Fromm, H. J., *Biochim. Biophys. Acta* **139**, 221 (1967).

Partial Rapid Equilibrium Ordered System

Morrison, J. F. and Ebner, K. E., *J. Biol. Chem.* **246**, 3977 (1971) (Galactotransferase—Mn^{2+} adds first and does not dissociate after each catalytic cycle).

Partially Random Systems

Dalziel, K., *Biochem. J.* **114**, 547 (1969) (Criticism of partially random systems).

Rudolph, F. B. and Fromm, H. J., *J. Theoret. Biol.* **39**, 363 (1973) (Support of partially random systems).

Unexpected Product Inhibition Patterns and Parallel Plots That Do Not Indicate Ping Pong Kinetics

Bar-Tana, J. and Cleland, W. W., *J. Biol. Chem.*, **249**, 1271 (1974) (Non Rapid Equilibrium Random System where the rate constants for the release of substrates is less than V_{max}).

Baumann, P. and Wright, B. E., *Biochemistry* **7**, 3653 (1968) (Phosphofructokinase).

DeLeo, A. B., Dayan, J. and Sprinson, D. B., *J. Biol. Chem.* **248**, 2344 (1973) (DAHP synthetase).

Gold, M. and Segel, I. H., *J. Biol. Chem.* **249**, 2417 (1974) (Protein kinase).

Hanson, T. L. and Fromm, H. J., *J. Biol. Chem.* **240**, 4133 (1965) (Hexokinase).

Henderson, J. F., Brox, L. W., Kelley, W. N., Rosenbloom, F. M. and Seegmiller, J. E., *J. Biol. Chem.* **243**, 2514 (1968) (Hypoxanthine-guanine phosphoribosyltransferase).

Lucas, J. J., Burchiel, S. W. and Segel, I. H., *Arch. Biochem. Biophys.* **153**, 664 (1972) (Choline sulfatase).

Mannervik, B., *Biochem. Biophys. Res. Commun.* **53**, 1151 (1973) (Glutathione reductase).

Porter, R. W., Modebe, M. O. and Stark, G. P., *J. Biol. Chem.* **244**, 1846 (1969) (Aspartate transcarbamylase catalytic subunit).

Substrate Inhibition and Varying Inhibitor and Substrate at Constant Ratio

Cleland, W. W., Gross, M. and Folk, J. E., *J. Biol. Chem.* **248**, 6541 (1973).

Some Other Specific Examples

Ainslie, G. R., Jr. and Cleland, W. W., *J. Biol. Chem.* **247**, 946 (1972) (Liver alcohol dehydrogenase).

Bar-Tana, J. and Rose, G., *Biochem. J.* **109**, 275 (1968) (Liver fatty acyl CoA synthetase: Bi Uni Uni Bi Ping Pong).

Brostrom, M. and Browning, E. T., *J. Biol. Chem.* **248**, 2364 (1973) (Yeast choline kinase: Random Bi Bi).

Buzdygon, B. E., Braginski, J. E. and Chung, A. E., *Arch. Biochem. Biophys.* **159**, 400 (1973) (Isocitrate dehydrogenase from *Rhodopseudomonas spheroides*: Ordered Bi Bi with one dead-end complex).

Ceder, H. and Schwartz, J. H., *J. Biol. Chem.* **244**, 4122 (1969) [Asparagine synthetase, Bi (random) Uni Uni Bi (random)].

Coward, J. K., Slisz, E. P. and Wu, F. Y-H., *Biochemistry* **12**, 2291 (1973) (Catechol *O*-methyltransferase: Random Bi Bi with two modes of catechol binding).

Dela Fuenta, G. and Sols, A., *Eur. J. Biochem.* **16**, 234 (1970) (Yeast hexokinase: Ordered Bi Bi).

Din, G. A. and Suzuki, I., *Can. J. Biochem.* **45**, 1547 (1967) (Fe^{++}-cytochrome *c* reductase of *Ferrobacillus ferrooxidans*: Ping Pong Bi Bi).

Duggleby, R. G. and Dennis, D. T., *J. Biol. Chem.* **249**, 167 (1974) (Glyceraldehyde-3-phosphate dehydrogenase: Bi Uni Uni Uni Ping Pong Ter Bi).

Fromm, H. J., *Eur. J. Biochem.* **7**, 385 (1969) (Yeast hexokinase: Random Bi Bi).

Garces, E. and Cleland, W. W., *Biochemistry* **8**, 633 (1969) (Nucleoside diphosphate kinase: Ping Pong Bi Bi with competitive substrate inhibition).

Gillham, B., *Biochem. J.* **135**, 797 (1973) (Glutathione-*S*-transferase of rat liver: Ordered Bi Bi).

Greene, R. C., *Biochemistry* **8**, 2255 (1969) (Yeast *S*-adenosylmethionine synthetase: Ordered Bi Ter).

Gulbinsky, J. S. and Cleland, W. W., *Biochemistry* **7**, 566 (1966) (*E. coli* galactokinase: Random Bi Bi).

Heyde, E. and Ainsworth, S., *J. Biol. Chem.* **243**, 2413 (1968) (Malate dehydrogenase: Ordered Bi Bi).

Heyde, E., Nagabhushanam, A. and Morrison, J. F., *Biochemistry* **12**, 4718 (1973) (Aspartate transcarbamylase catalytic subunit: Rapid Equilibrium Random Bi Bi with three possible dead-end complexes).

Houslay, M. D. and Tipton, K. F., *Biochem. J.* **135**, 735 (1973) (Monoamine oxidase of rat liver mitochondria: Ping Pong Bi Bi).

Hsu, R., Lardy, H. A. and Cleland, W. W., *J. Biol. Chem.* **242**, 5315 (1967) (Pigeon liver malic enzyme: Ordered Bi Ter).

Kuhn, E. and Brand, K., *Biochemistry* **12**, 5217 (1973) (Yeast transaldolase: Ping Pong; bovine transaldolase: Random).

Mannervik, B., Bartafi, T. and Gorna-Hall, B., *J. Biol. Chem.* **249**, 901 (1974) (Glyoxalase I from yeast).

Mannervik, B., Gorna-Hall, B. and Bartafi, T., *Eur. J. Biochem.* **37**, 270 (1973) (Glyoxalase I from porcine erythrocytes: Random Bi Uni).

McGregor, W. G., Phillips, J. and Suelter, C. H., *J. Biol. Chem.* **249**, 3152 (1974) (Glycerol dehydrogenase: Ordered Bi Bi).

McVerrey, P. and Kim, K-H., *Biochemistry* **13**, 3505 (1974) (Glycogen synthetase: Random Bi Bi with nonlinear reciprocal plots for varied glycogen).

Middleton, B., *Biochem. J.* **139**, 109 (1974) (Acetoacetyl CoA Thiolase Ping Pong Bi Bi where A = B).

Nazario, M. and Evans, J. A., *J. Biol. Chem.* **249** 4934 (1974) (Arginyl *t*-RNA ligase: Ordered Ter with random release of P and Q).

Orsi, B. A. and Cleland, W. W., *Biochemistry* **11**, 102 (1972) (Glyceraldehyde-3-phosphate dehydrogenase: Ordered Bi Bi).

Paule, M. R. and Preiss, J., *J. Biol. Chem.* **246**, 4602 (1971) (ADPG pyrophosphorylase).

Preuveneers, M. J., Peacock, D., Crook, E. M., Clark, J. B. and Brocklehurst, K., *Biochem. J.* **133**, 133 (1973) (*D*-3-hydroxybutyrate dehydrogenase from *Rhodopseudomonas spheroides*: Ordered Bi Bi).

Thorner, J. W. and Paulus, H., *J. Biol. Chem.* **248**, 3922 (1973) (Glycerol kinase from *E. coli*: Ordered Bi Bi with unusual MgATP velocity curves).

Tsai, C. S., Burgett, M. W. and Reed, L. J., *J. Biol. Chem.* **248**, 8348 (1973) (α-Keto acid dehydrogenase complexes: Three-site Ping Pong).

Uhr, M. L., Thompson, V. W. and Cleland, W. W., *J. Biol. Chem.* **249**, 2920 (1974) (Pig heart isocitrate dehydrogenase: Random Bi Ter).

Warren, G. B. and Tipton, K. F., *Biochem. J.* **139**, 311 and 321 (1974) (Pig liver pyruvate carboxylase: Rapid Equilibrium (Ordered A, Random B–C) Ordered (Theorell-Chance) D for OAA decarboxylation).

Yoshida, H., *J. Biochem. (Tokyo)* **75**, 905 (1974) (Phosphodiesterase-phosphomonoesterase: Apparent Uni Ter).

Errors in Spectrophotometer Assays

Cavalieri, R. L. and Sable, H. Z., *Anal Biochem.* **59**, 122 (1974) (Spurious inhibition patterns).

Eyzaguirre, J., *Biochem. Biophys. Res. Commun.* **60**, 35 (1974) (Stray light effects).

CHAPTER TEN

ISOTOPE EXCHANGE

All of the studies described earlier were concerned with the observed *net* velocity of a chemical reaction. This net velocity is the difference between the absolute forward and reverse velocities of any step. The unidirectional velocity of a step or series of steps may be considerably faster than the observed net velocity. For example, in a Uni Uni reaction at chemical (thermodynamic) equilibrium (Chapter Two), the net velocity is zero, yet the rate at which S is converted to P certainly is not zero. The net velocity is zero because the $S \rightarrow P$ rate equals the $P \rightarrow S$ rate. The actual $S \rightarrow P = P \rightarrow S$ rate ("exchange rate") can be measured by adding a small amount of radioactively labeled S (designated S*) to the equilibrium mixture and measuring the initial velocity at which label appears in P. If the mass of labeled S* is extremely small compared to the $[S]_{eq}$ present, the system will remain at equilibrium. In multireactant systems, the unidirectional rates of any two segments need not be equal. For example, in an Ordered Bi Bi sequence with both substrates and both products present, the net rate of P formation must equal the net rate of Q formation, but the initial velocity at which label from A* appears in P can be quite different from the rate at which label from B* appears in Q. The A*–P exchange rate might be 1000 in arbitrary units. If the system is not at chemical equilibrium, the P*–A exchange rate (i.e., the rate at which label from P* appears in A) might be 900. The observed net rate of P formation is 100. The net rate of Q appearance must also be 100 but the B*–Q exchange rate might be 500. The Q*–B exchange rate has to be 400. Thus isotope exchange studies provide a way of measuring the unidirectional rates of certain steps within a reaction sequence. Isotope exchange studies are usually conducted at chemical equilibrium, but they need not be. In all cases one or more products must be present. (If no products are present, the initial velocity of label appearance in P or Q simply equals v, the initial reaction velocity.) Of course, it is necessary to have a convenient method of separating the members of an exchange pair to measure the rate of label transfer from one to another. In most cases the velocity equations for isotope exchange are far too complicated to permit the determination of the usual kinetic constants. Nevertheless, the technique can be used in a qualitative way. By observing the effects of varying reactant concentrations on the exchange rates, the kinetic mechanism can be deduced or confirmed. The key questions asked in diagnosing multireactant kinetic mechanisms are: (*a*) will the enzyme catalyze an exchange between a given substrate-product pair in the absence of all other reactants, and (*b*) is substrate inhibition observed when a substrate-product pair is varied, and if so, is it total or partial? The design of isotope exchange experiments and the interpretation of the results are described below for some common multireactant systems.

A. ORDERED BI BI

Exchange can be measured between any substrate-product pair that share a common part. For example, consider the Ordered Bi Bi PRPP synthetase reaction (Switzer, 1970) shown below.

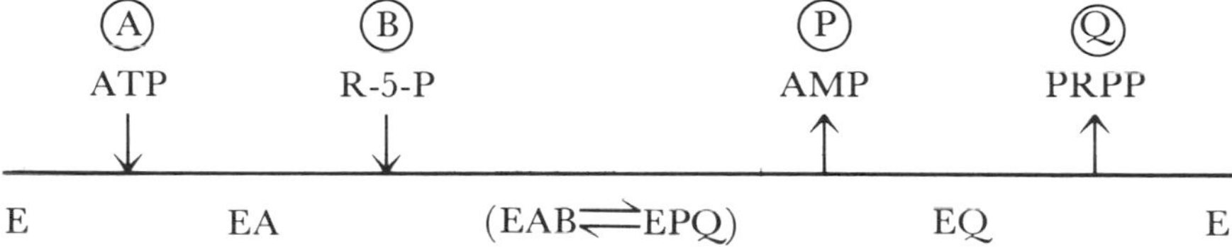

The A*–P (or P*–A) exchange can be measured if the nucleotide is labeled with C^{14} in the adenine or ribose part; A*–Q (or Q*–A) exchange can be measured if the β or γ phosphate of ATP (or pyrophosphate of PRPP) is labeled with P^{32}; B*–Q (or Q*–B) exchange can be measured if the ribose-5-phosphate or PRPP is labeled with C^{14} in the ribose part or P^{32} in the 5′-phosphate. No B*–P (or P*–B) exchange occurs. In the hypothetical dehydrogenase reaction shown below, we could measure A*–Q, B*–P, and B*–Q exchanges, but not A*–P.

If the exchange rates are measured at chemical equilibrium, then the rate in one direction (e.g., A*–P) must equal the rate in the opposite direction

(P*–A). The choice of which exchange is measured depends on which reactant can be obtained in labeled form with a reasonably high specific activity. Exchange rates are measured at several different concentrations of some substrate-product pair. For example, the equilibrium A*–P exchange rate can be measured at several different concentrations of A* and P (maintaining [B] and [Q] constant), or at several different concentrations of B and Q (maintaining [A*] and [P] constant), or at several different concentrations of B and P (maintaining [A*] and [Q] constant). The varied pair is maintained at a constant ratio so that the system remains at equilibrium. Thus if the equilibrium ratio of [Q]/[B] is 50, the ratio must remain 50 as [Q] and [B] are varied. If the system is not at equilibrium, the varied reactants must still be maintained at their fixed ratio so that the system will remain at the same distance from equilibrium at all concentrations of varied reactants. The general procedure is to mix A, B, P, and Q at the desired concentrations (chosen to yield $[P][Q]/[A][B] = K_{eq}$ or a fixed multiple of K_{eq}). A small amount (i.e., an essentially massless amount) of labeled reactant is added. The reaction is then started by adding enzyme. Alternately, if exchange is to be measured at equilibrium, the reactants and enzyme can be preincubated for a short time (to establish equilibrium in case the original reactant ratios were not quite at equilibrium) and then the exchange reaction can be started by adding a small amount of high specific activity labeled reactant (containing so little mass that the equilibrium ratios are not affected). If the reaction is to be started by adding enzyme, then there is no reason why the label cannot be mixed directly with the unlabeled reactant beforehand to make a stock solution of known concentration and specific activity. This procedure is preferred when one of the varied reactants is the labeled reactant, or if the commercially available labeled reactant contains a significant amount of mass.

The reactants can be identified from the results of two series of exchange studies: the A*–Q (or Q*–A) exchange at different fixed concentrations of B and P, and the B*–P (or P*–B) exchange at different fixed concentrations of A and Q. The A*–Q exchange shows total substrate inhibition as [B] and [P] are increased (Fig. X-1*a*, *b*), while the B*–P exchange rate rises in a normal hyperbolic fashion (Fig. X-1*c*, *d*). However, if we do not already know which reactant is which, then we might have chosen A and P or B and Q as the substrate-product pairs, so let us examine in a qualitative way the results of several possible exchange studies.

A*–Q Exchange

As the concentrations of B and P are raised, the exchange velocity increases *at first* simply because B is needed for the reaction in which part of A

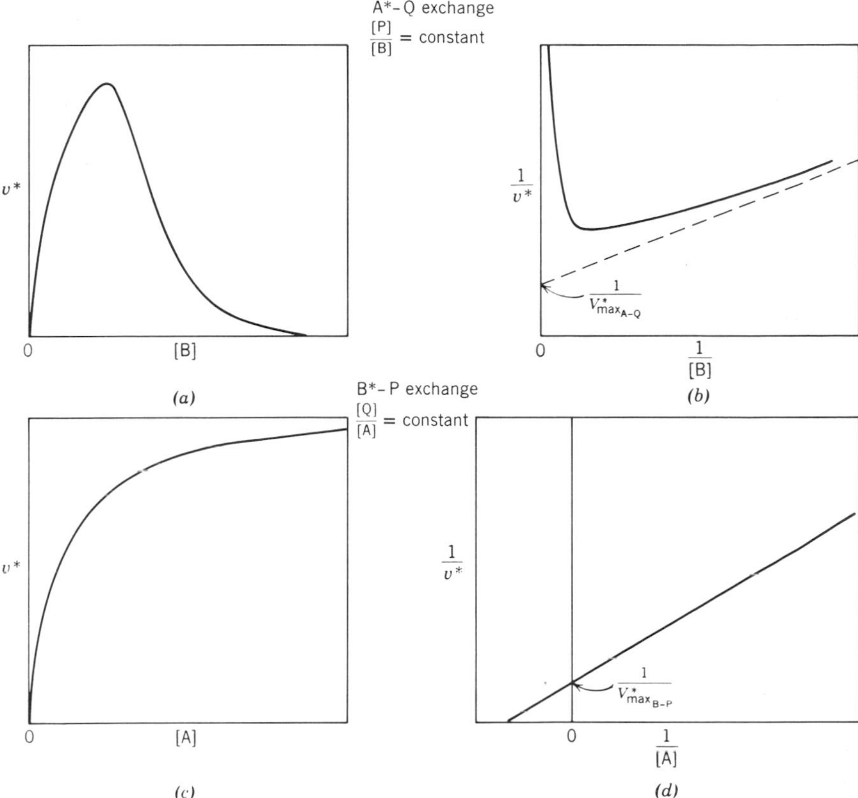

Fig. X-1. Isotope exchange in an Ordered Bi Bi reaction at chemical equilibrium. (*a*) Velocity of the A*–Q exchange as [B] is increased; [P] is increased along with [B]. maintaining the [P]/[B] ratio constant at the equilibrium ratio. (*b*) Reciprocal plot of $1/v^*$ versus 1/[B]. (*c*) Velocity of the B*–P exchange as [A] is increased; [Q] is increased along with [A], maintaining the [Q]/[A] ratio constant at the equilibrium ratio; B and P can be identified as the inner substrate-product pair by the total inhibition of A*–Q exchange at high [B] and [P]. (*d*) Reciprocal plot of $1/v^*$ versus 1/[A].

becomes part of Q. The EA* that is formed must react with B in order to pass through the central complexes and form EQ*, which liberates the labeled product. However, after a certain point, a further increase in [B] and [P] causes the exchange velocity to decrease. The reason for the substrate inhibition is as follows: The rate at which labeled Q appears depends on the level of EQ*: $v^* = k_4[\text{EQ}^*]$. The system is at equilibrium. Consequently, as [P] increases, the level of EQ* decreases (E and EQ* are driven to the central complexes). At saturating [P], the level of EQ* is zero and total substrate inhibition is observed. Similarly, saturating [B] drives all the E and EA to the central complexes. If no E is present, there is nothing for A* to react with and we observe total substrate inhibition. The Q*–A exchange behaves identically. At first, the increase in [P] promotes the reaction, but eventually saturating [B] and [P] leaves no EA* to dissociate or E for combination with Q*.

B*–P Exchange

In order for label from B to appear in P, B* must react with EA. Thus increasing [A] promotes the exchange by increasing the level of EA. Increasing [Q] along with [A] simply increases the level of EQ, which has no effect on the rate of P* release from EP*Q. Thus the initial velocity of the B*–P exchange increases normally as [A] and [Q] are increased. All the enzyme is driven to EA, EAB + EPQ, and EQ, which maximizes the exchange rate. No inhibition is observed. Similarly, the P*–B exchange rate increases without any substrate inhibition as [A] and [Q] are increased.

A*–P Exchange

Increasing [B] and [Q] will promote the reaction *at first* because B is necessary for the reaction (B reacts with EA* to yield EP*Q which dissociates to form P*). However, for label from A* to appear in P, A* must react with E. But as [B] and [Q] increase, the level of E decreases. (Increasing [B] forces E and EA into the central complexes; increasing [Q] forces E to EQ.) Consequently, after a certain point, substrate inhibition is observed and, at saturating [B] and [Q], the velocity is driven to zero. (Remember that [A*], which competes with Q, is maintained constant.) Similarly, the P*–A exchange increases at first because increasing [Q] forms more EQ for P* to react with, but increasing [B] eventually forces all the EA to the central complexes. If there is no EA, then A* cannot form ($v^* = k_{-1}[\text{EA}^*]$).

B*–Q Exchange

B* must react with EA for label to appear in Q. Thus increasing [A] and [P] promotes the reaction *at first*. But as [P] increases, EQ is forced into the central complexes; thus after a point, total substrate inhibition is observed (no EQ* to liberate Q*). Similarly, the Q*–B exchange rate increases at first (P is needed for the reaction) and then drops to zero as all the E (needed for combination with Q*) is driven to EA.

Cleland (1970) has formulated a simple rule by which we can predict whether substrate inhibition will be observed in *reactions with no alternate sequences*.

> *Total substrate inhibition is observed if either of the varied reactants (of a pair) adds between the points of combination of the exchange reactants (and both varied reactants do no add to the same stable enzyme form).*

In addition, keep in mind that anything that inhibits exchange at equilibrium in one direction will inhibit equally in the opposite direction. Thus B will show total substrate inhibition of exchange in either direction between A and P and between A and Q. P will show total substrate inhibition of exchange in either direction between A and Q and between B and Q. Consequently, A and Q, the two outer reactants can be identified as that pair which causes no inhibition of exchange between the other substrate-product pair.

In the exchange studies described above, the concentrations of the exchange pair were held constant while the other substrate-product pair was varied. Exchange rates can also be measured at different fixed concentrations of one or both of the exchange reactants. For example, the A*–Q exchange rate can be measured at different concentrations of (*a*) A* and Q, (*b*) A* and P, or (*c*) B and Q. The same rule applies. Exchange *a* shows no substrate inhibition. The Q reacts with E; however, so does A*; thus the exchange velocity simply rises to a maximum value and the reciprocal plot is linear. Exchange *b* and *c* ultimately show substrate inhibition, since in both cases, one of the varied reactants adds between A and Q. When A*–P exchange is measured at different concentrations of (*a*) A* and P, (*b*) A* and Q, and (*c*) B and P, we see no inhibition in case *a* or case *b* because neither A, P, nor Q adds *between* A* and P. (In the repeating sequence, Q adds between P and A if we look downstream, but Q and A combine with the same stable enzyme form. Increasing [A] along with [Q] prevents Q from forcing all the E to EQ.) Increasing [B] and [P] causes inhibition. The B*–P exchange is not inhibited as (*a*) [B*] and [P], or (*b*) [B*] and [Q], or (*c*) [A] and [P] are raised.

Although varying [A] and [Q] together and [B] and [P] together gives all the information required to determine the reaction sequence, varying [A] and [P] together and [B] and [Q] together can be used to confirm the presence or absence of possible dead-end complexes. For example, in the malate dehydrogenase reaction, increasing oxalacetate and NAD^+ together inhibits the NADH–NAD^+ exchange but not the oxalacetate-malate exchange, which are the expected results if NADH adds before oxalacetate and malate leaves before NAD^+. When malate and NADH are increased together, both exchanges are inhibited. We would expect the NADH*–NAD^+ exchange to be inhibited, since NADH = A, oxalacetate = B, malate = P, and NAD^+ = Q, but the B–P exchange should not be. The results suggest that a dead-end E-NADH-malate complex forms.

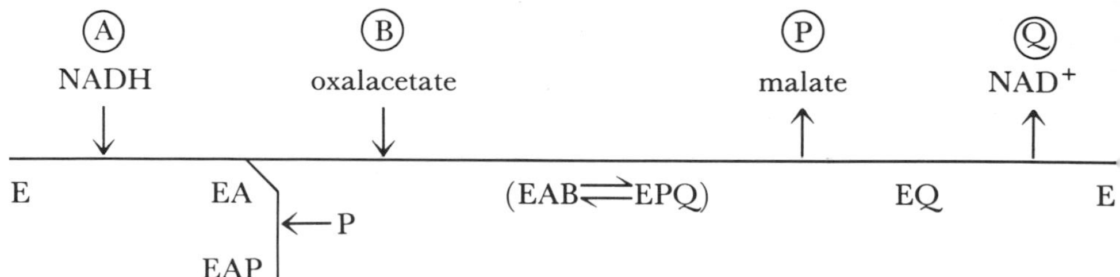

There is no E-NAD^+-oxalacetate (EQB) dead-end complex, since the B*–P exchange is not inhibited when [B] and [Q] are increased. The formation of EQB would decrease the level of EQ available for combination with P and thereby inhibit the P–B exchange. (Note that even if the reaction being measured is B*→P, anything that would inhibit the P*→B reaction would inhibit the B*→P reaction identically. At equilibrium, the rates in both directions are equal and at a fixed distance from equilibrium; the initial exchange rate in one direction is a fixed multiple of the rate in the opposite direction.)

When [A] and [Q] are varied together and the A*–Q exchange rate is measured, the intercept on the $1/v^*$-axis gives $1/V^*_{max_{A-Q}}$. Similarly, when [B] and [P] are varied together and B*–P exchange is measured, the intercept gives $1/V^*_{max_{B-P}}$. The V^*_{max} values may be greater than, equal to, or smaller than the V_{max_f} or V_{max_r} of the reaction determined from the usual initial velicity studies. The relative values of $V^*_{max_{A-Q}}$ and $V^*_{max_{B-P}}$ give an indication of the rate-limiting step. If the interconversion of the central complexes or the release of P is rate-limiting, both V^*_{max} values will be about the same. If, however, the release of Q is rate-limiting, then $V^*_{max_{B-P}}$ will be faster than $V^*_{max_{A-Q}}$. In a Theorell-Chance mechanism, where presumably there is no central complex, the reciprocal plot for B*–P exchange as [B] and [P] are varied goes through the origin (i.e., $V^*_{max_{B-P}}$ appears to be infinite). It is doubtful that a true Theorell-Chance system exists.

Even when substrate inhibition is not observed, reciprocal plots for some exchange patterns may not be linear. The plots are concave up or concave down near the $1/v^*$-axis and intersect the $1/v^*$-axis at a finite value. However, there is no minimum. The plots are known as 2/1 functions because 1/[A] or 1/[B] appears as a squared term in the numerator and to the power 1 in the denominator of the reciprocal equation. The plots may resemble those shown in Figures IX-37*a* and IX-37*d*.

B. RANDOM BI BI

One of the major uses of isotope exchange studies is in discriminating between Ordered Bi Bi and Random Bi Bi systems. In a Random Bi Bi sequence *with no preferred pathway*, the A*–Q or Q*–A exchange increases in a normal hyperbolic fashion as [B] and [P] are increased. Similarly, the B*–P or P*–B exchange shows normal hyperbolic kinetics.

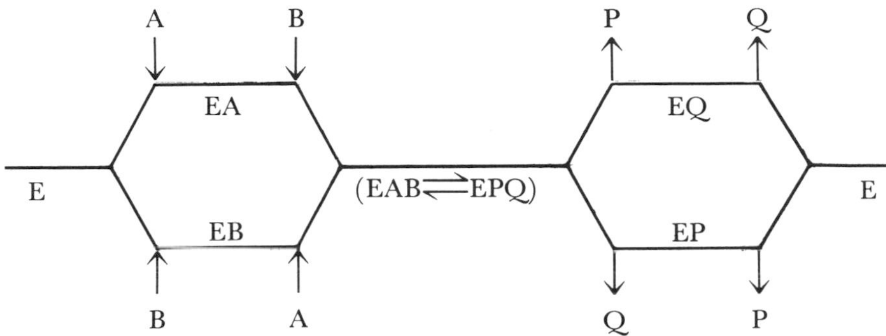

Saturating [B] and [P] simply forces all the A–Q flux through the lower pathway. Saturating [A] and [Q] forces all the flux through the upper pathway. Thus the exchange pair become the inner pair of an Ordered Bi Bi system when the concentrations of the varied pair are infinitely high. If one of the pathways is kinetically less favorable, the reciprocal plots for one exchange will be nonlinear near the $1/v^*$-axis, but the plot intersects the axis at a finite value. In other words, no total substrate inhibition is observed in random pathways (*unless dead-end complexes form*), since no reactant can prevent another from dissociating. In general, if like reactants are varied (e.g., MgATP and MgADP as A and Q and galactose and galactose-1-phosphate as B and P), no total substrate inhibition is observed for any exchange. If unlike reactants *which can form dead-end complexes* are varied, then total substrate inhibition is observed for all exchanges. In the above example, varying [galactose] and [MgADP] would show total substrate inhibition of (*a*) galactose–galactose-1-phosphate, (*b*) MgATP–MgADP, and (*c*) MgATP–galactose-1-phosphate exchanges because of the formation of a dead-end E-galactose-MgADP complex. Similarly, varying [MgATP] and [galactose-1-phosphate] would show total substrate inhibition of all three exchanges if a dead-end E-MgATP–galactose-1-phosphate complex formed. In a truly random system we would never observe total substrate inhibition of one exchange and not another.

Isotope exchange is the only way to determine whether a random system is a rapid equilibrium one. If the interconversion of the central complexes is the sole rate limiting step, then all exchanges *in a given direction* must proceed at the same rate at any given set of reactant concentrations. Thus $v^*_{A-Q} = v^*_{A-P} = v^*_{B-P}$ and $v^*_{Q-A} = v^*_{P-A} = v^*_{P-B}$. If exchange is measured at chemical equilibrium, then all exchange rates are equal at any given set of equilibrium reactant concentrations (because $v^*_{A-Q} = v^*_{Q-A}$, $v^*_{A-P} = v^*_{P-A}$, and $v^*_{B-P} = v^*_{P-B}$). The equality of exchange rates holds even if dead-end complexes form.

Isotope exchange can also be useful in detecting partially random segments in what otherwise appears to be an ordered sequence. For example, if increasing [B] and [P] of an Ordered Bi Bi system yields significant substrate inhibition, but the exchange velocity cannot be driven to zero (i.e., the plot intersects the $1/v^*$-axis at a finite value), then we might suspect that some degree of randomness is present.

C. ISOTOPE EXCHANGE DURING A NET REACTION

Most isotope exchange studies are conducted at chemical equilibrium, but sometimes it is profitable to measure exchange between a product and a substrate while the reaction is proceeding. For example, consider an Ordered Bi Bi system.

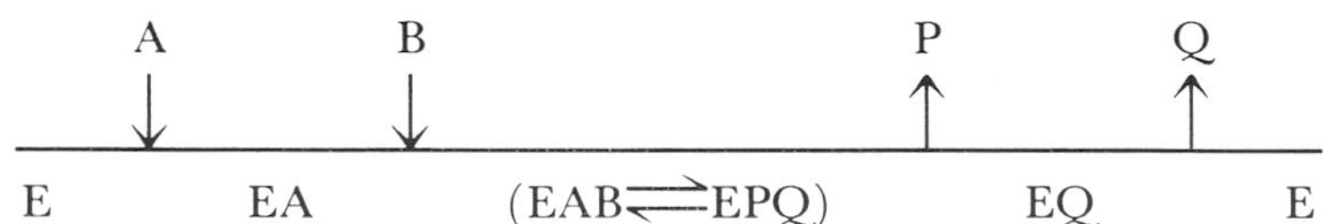

Label from P can appear in the corresponding substrate in the absence of Q. All that is required is a reasonable level of EQ in the steady-state and a significant concentration of P* (not just the P produced as the reaction occurs). The exchange of label from Q into a substrate will not occur unless a significant concentration of P is present. Both products should be tested individually at levels at or above their inhibition constants. Thus the order of product release can be established: the first product released is that which exchanges with a substrate in the absence of the other product. If both products show exchange, then their release must be random. If the product release steps are not at all rate-limiting (i.e., the system is rapid equilibrium), then there will be essentially no EQ for P* to combine with and consequently, no exchange of label from P* will be seen in the absence of Q. The same approach can be used to identify the first product released in a hydrolytic Ping Pong Bi Bi system (which appears as Ordered Uni Bi at the constant water concentration).

D. PING PONG SYSTEMS

Isotope exchange is a powerful tool for probing Ping Pong systems, because isolated portions of the overall reaction sequence can be studied. For example, in a Ping Pong Bi Bi system, the A*–P and P*–A exchange can be measured in the absence of B and Q.

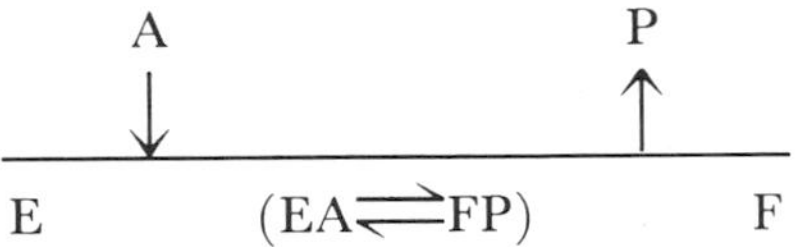

Almost immediately after mixing E, A, and P, the system will come to equilibrium at essentially the established [P]/[A] ratio. (The levels of E and F adjust accordingly.) Similarly, when E, B, and Q are mixed, the other partial reaction is automatically established at chemical equilibrium.

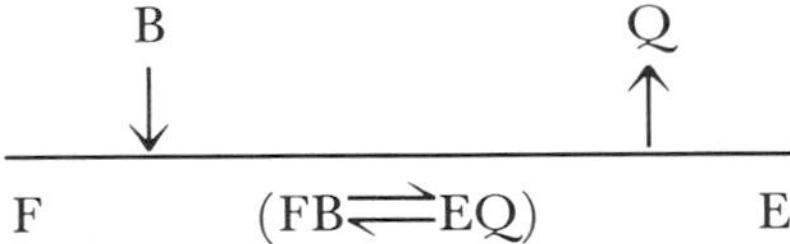

If no exchange takes place between the suspected A–P and B–Q pairs unless all reactants are present, then the system is probably ordered or random. The exchange velocity equation for an isolated portion of a Ping Pong system is relatively simple compared to those for complete random and ordered systems (as described later). For example, v^* for an A*–P exchange is given by:

$$\boxed{\frac{v^*}{V^*_{\max_{A\text{-}P}}} = \frac{[A^*][P]}{K_{ia}[P] + K_{ip}[A] + [A][P]}} \tag{X-1}$$

where

$$V^*_{\max_{A\text{-}P}} = \frac{V_{\max_f} K_{ia}}{K_{m_A}} = \frac{V_{\max_r} K_{ip}}{K_{m_P}}$$

Note that the maximal A*–P or P*–A exchange rate may be greater than, equal to, or less than $V_{\max_f}$ or $V_{\max_r}$, depending on the ratio of K_{ia}/K_{m_A} and K_{ip}/K_{m_P}, respectively. Both [A*] and [A] represent the concentration of A. The asterisk simply means that [A*] can be expressed as cpm/ml and v^* as cpm$\times$ml$^{-1}\times$min^{-1}. Equation X-1 has the same form as the equation for the initial velocity with A and B. Thus if [A*] is varied at different fixed levels of [P], a parallel pattern of reciprocal plots is obtained (Fig. X-2*a*). The intercept replot gives K_{ip} and $V^*_{\max_{A-P}}$ (Fig. X-2*b*). If P*–A exchange is measured, the plots give K_{ia} and $V^*_{\max_{P-A}}$. The equation for the B*–P exchange is:

$$\boxed{\frac{v^*}{V^*_{\max_{B-P}}} = \frac{[B^*][P]}{K_{ib}[Q] + K_{iq}[B] + [B][Q]}} \tag{X-2}$$

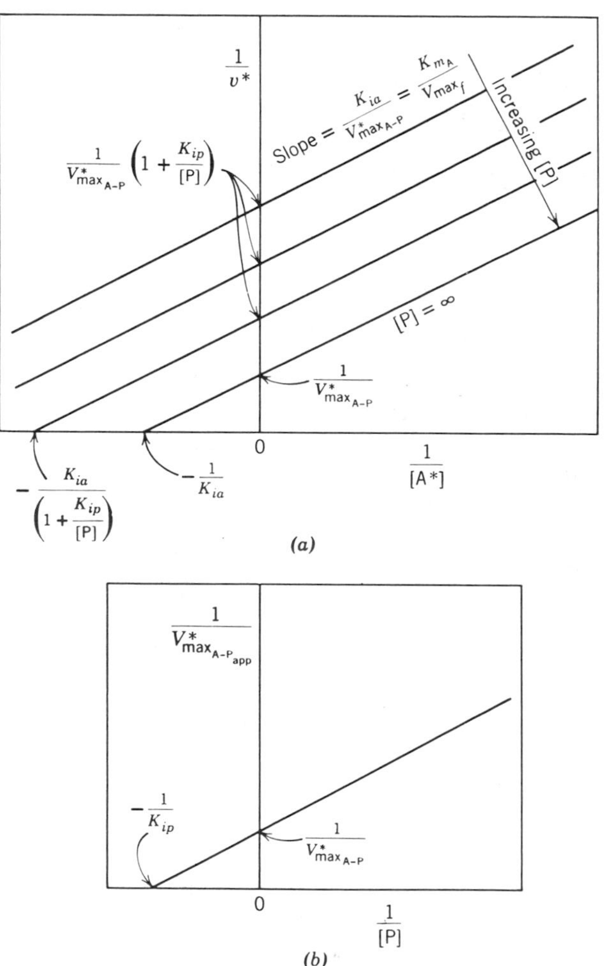

Fig. X-2. The A*–P exchange in an isolated Ping Pong segment (only A* and P present). (*a*) $1/v$ versus $1/[A^*]$ at different fixed concentrations of P. (*b*) Replot of $1/v^*$-axis intercept versus $1/[P]$.

If $V_{\max_f}$ and $V_{\max_r}$ are known, K_{m_A} and K_{m_B} can be calculated from the $V^*_{\max}$ values.

The reciprocal plot patterns for isotope exchange in Ping Pong systems can be predicted from general rules similar to those used in predicting initial velocity patterns. For the A*–P exchange, we recognize that the $1/v^*$-axis intercepts represent the reciprocals of the apparent maximal exchange velocities at an infinitely high concentration of A*. At saturating [A*], increasing [P] increases the exchange rate of P with the central complexes. As a result the $1/v^*$-axis intercepts are decreased. The slopes represent the reciprocals of the apparent first-order rate constant for the exchange reaction at very low [A*]. Since the isolated reaction segment is automatically at equilibrium, essentially all of the enzyme will be present as free E at very low [A*] (i.e., essentially no central complexes are present). Thus at very low [A*], changing [P] will have no effect on v^*. Consequently, the slopes are unaffected. In general, the following intercept and slope rules are combined to predict the overall effects.

The $1/v^*$-axis intercept decreases as the concentration of the changing fixed reactant increases provided the fixed reactant adds (a) on the opposite side of the central complexes or (b) on the same side of the central complexes as the varied reactant, but after the varied reactant in sequence. There is no intercept effect if the changing fixed reactant adds before the varied reactant (i.e., we see the Rapid Equilibrium Ordered pattern).

The slope decreases as the concentration of the changing fixed reactant increases if the changing fixed reactant adds on the same side of the central complexes as the varied reactant.

If the changing fixed reactant adds within the exchange path, total substrate inhibition is superimposed on the above patterns at high concentrations: the slope and/or intercept increases as the concentration of the changing fixed reactant increases. The inhibition is mixed type if the changing fixed reactant and the varied reactant are on opposite sides of the central complex. The inhibition is uncompetitive if the changing fixed reactant and the varied reactant are on the same side of the central complex. If the varied reactant adds within the exchange path, the reciprocal plots bend up as they approach the $1/v^*$-axis.

Let us see how these rules can be applied to a slightly more complex Ping Pong sequence. For example, consider the Bi Uni portion of a Bi Uni Uni Bi system:

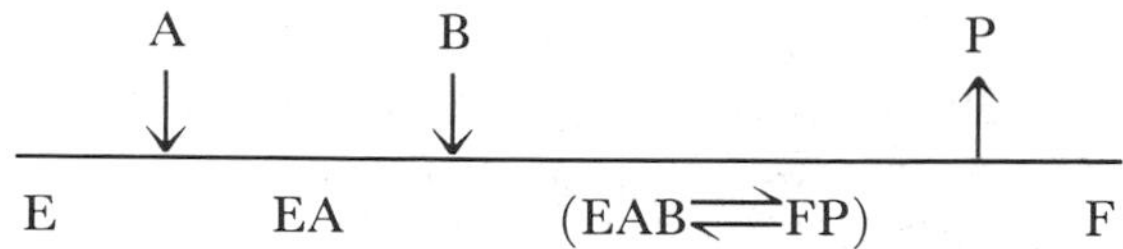

In most real situations, P will pick up label from A or B, but not both. Consider the A*–P exchange. We can (*a*) vary [A*] at different fixed concentrations of P and constant [B], (*b*) vary [B] at different fixed concentrations of P and constant [A*], or (*c*) vary [A*] at different fixed concentrations of B and constant [P]. The same information would be obtained if the varied and changing fixed reactants were reversed. Also, the same families of curves could be plotted for the P*–A exchange (which must equal the A*–P exchange at chemical equilibrium). Family *a* will consist of a series of parallel lines because A* and P are on opposite sides of the central complexes. Competitive substrate inhibition will be observed only if P combines in dead-end fashion with E or A* with F. Family *b* will also be parallel because B and P are on opposite sides of the central complexes. Superimposed on the parallel pattern will be total substrate inhibition at high [B] because B adds within the exchange path. Thus the $1/v^*$ versus 1/[B] plots will curve up as they approach the vertical axis. The $1/v$ versus 1/[P] plots at different fixed concentrations of B will start out parallel and then display total mixed-type substrate inhibition as [B] becomes very high and drives all the EA and E to the central complexes. Note that B as an inhibitor causes both a slope and intercept effect, because it is on the opposite side of the central complexes from P. Family *c* is intersecting when [A*] is varied because B adds between A* and the central complexes (intercept effect) and on the same side of the central complexes as A* (slope effect). At high fixed concentrations of B, total uncompetitive substrate inhibition is observed. When [B] is varied at different fixed concentrations of A*, the plots are intersecting but curve up as they approach the $1/v^*$-axis. (The asymptotes intersect on the $1/v^*$-axis.) If A and B add randomly, the A–B patterns will be intersecting, but no total substrate inhibition will be observed.

If B exchanges with P, the B–P and A–P patterns will be parallel with no substrate inhibition unless dead-end complexes form. Plots of $1/v^*$ versus 1/[A] at different fixed concentrations of B* will be intersecting. At saturating [B*], the plot is a horizontal line at $1/V^*_{\max_{B\text{-}P}}$. The replot of $slope_{1/A}$ versus 1/[B] will go through the origin just as in the initial velocity study of a Rapid Equilibrium Ordered Bireactant system. At saturating [B*], all the E is driven to EAB regardless of the concentration of A (as long as there is at least one molecule of A for every molecule of E present). Plots of $1/v$ versus

$1/[B^*]$ at different fixed concentrations of A will intersect on the $1/v^*$-axis. (In other words v^* is independent of [A] at saturating [B], as noted above). Thus by measuring the exchange of P with its corresponding substrate, we can determine just from the A–B pattern whether the pattern is random or ordered, and, if ordered, whether the exchange substrate is A or B.

Another common partial Ping Pong sequence is the Bi Bi portion of a Bi Bi Uni Uni system.

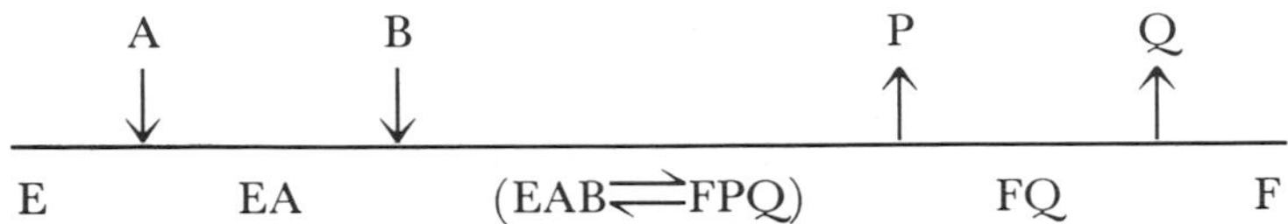

When the changing fixed reactant and varied reactant are on opposite sides of the central complexes, the reciprocal plot patterns will be parallel. Total substrate inhibition is observed whenever one of the reactants adds within the exchange path. The A–B and P–Q patterns will be intersecting as for the Bi Uni system.

Isotope exchange can be used to establish the presence of alternate Ping Pong sequences in what is predominately a sequential system. For example, suppose that most of the reaction flux occurs via a Random Bi Bi sequence, but A can be slowly converted to P (and P released) without B being present. The enzyme will catalyze an A*–P exchange in the absence of B and Q, but the rate will be quite low compared to the exchange rate in the presence of all reactants. Smith and Morrison (1969) have observed such a situation with arginine kinase. Of course, an alternate explanation for an A*–P exchange in the absence of all reactants is that the enzyme preparation is impure and contains small amounts of another enzyme that catalyzes a Ping Pong sequence between A and P. For this reason, enzyme preparations used for isotope exchange studies might have to be purified to a greater extent than those used for initial velocity studies. This is especially true for enzymes utilizing ATP. A great many enzymes catalyze ATP–ADP, ATP–AMP, and ATP–PP_i exchanges.

Isotope exchange at chemical equilibrium can be measured in Ping Pong systems with all reactants present. For example, consider the Ping Pong Bi Bi system:

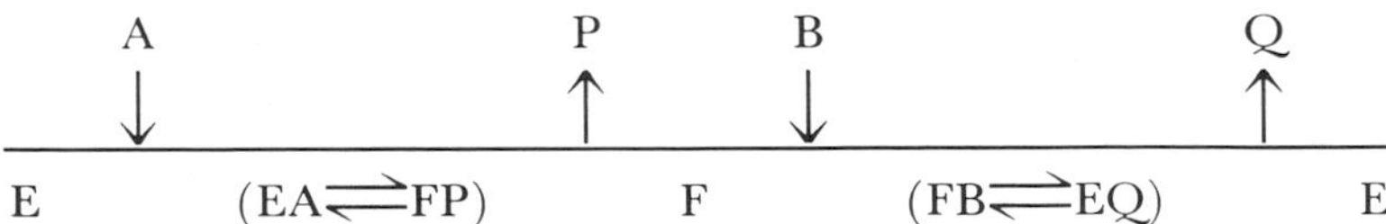

The A*–P or P*–A exchange rate increases normally as [A*] and [P] (or [A] or [P*]) are increased at constant [B] and [Q]. Increasing [B] and [Q] at constant [A*] and [P] will show only total substrate inhibition of the A*–P or P*–A exchange; B and Q are not needed for the A*–P reaction. Increasing their concentrations drives all the E and F to a central complex, leaving nothing for A* or P* to combine with. Increasing [A*] and [Q] at constant [B] and [P] does not lead to substrate inhibition because both A* and Q combine with the same enzyme form (i.e., increasing [A*] along with [Q] prevents all the E from being tied up as EQ). Similarly increasing [B] and [P] will not lead to substrate inhibition. B and P combine with the same enzyme form. Increasing [B] would drive all the F to FB+EQ, but this is prevented by simultaneously increasing [P]. The B*–Q or Q*–B exchange shows substrate inhibition only when [A] and [P] are increased. In some systems, it is possible to measure exchange between A and Q. In such systems, increasing [A] and [Q] or [B] and [P] shows no inhibition, while increasing [A] and [P] or [B] and [Q] ultimately leads to total substrate inhibition.

E. DETERMINING EXCHANGE VELOCITIES

The initial exchange velocity at chemical equilibrium, v^*, can be determined in two ways. The first way is simply to measure the actual initial rate at which label appears in the coreactant. The sample for analysis must be taken before the exchange has proceeded to any great extent, just as in the usual initial velocity studies. The second method is to take at least one sample at a known time during the approach to isotopic equilibrium. The transfer of label from one reactant to another follows an exponential curve. If we start with all the radioactivity in, for example, A*, then at isotopic equilibrium the distribution of label between A* and, for example, P* will be proportional to their equilibrium concentrations (which presumably are known). Thus we can calculate the fraction of exchange equilibrium that has occurred from a sample taken at any time. The initial exchange velocity is given by:

$$v^* = -\frac{[A][P]}{[A]+[P]}\frac{2.3}{t}\log(1-F) \qquad \text{(X-3)}$$

where F = the fraction of isotopic equilibrium attained at time t. For example, if we have [A] = 8 mM and [P] = 2 mM as the equilibrium concentrations of A and P, then at isotopic equilibrium P* would contain 20% of the label originally added as A*. If at $t = 3$ min, we find 5% of the original label in P*, then F is 0.25. The fraction F can be written as:

$$\boxed{F = \frac{A^* - A_0^*}{A_\infty^* - A_0^*} = \frac{P^* - P_0^*}{P_\infty^* - P_0^*}} \qquad \text{(X-4)}$$

where A* and P* = the label (radioactivity) in A and P, respectively, at any time

A_0^* and P_0^* = the label in A and P, respectively, at zero-time

A_∞^* and P_∞^* = the label in A and P, respectively, at isotopic equilibrium

If P has no label at $t = 0$:

$$\boxed{F = \frac{P^*}{P_\infty^*}} \qquad \text{(X-5)}$$

P* and P_∞^* can be expressed in cpm or in specific activities (cpm/μmole). Thus in the example above, if $A_0^* = 10{,}000$ cpm and $P_0^* = 0$, then $A_\infty^* = 8000$ cpm, and $P_\infty^* = 2000$ cpm. When P* = 500 cpm, A* must equal 9500 cpm and

$$F = \frac{9500 - 10{,}000}{8000 - 10{,}000} = \frac{500 - 0}{2000 - 0} = 0.25$$

Twenty-five percent of isotopic equilibrium has been attained. If $t = 3$ min:

$$v^* = \frac{-(8)(2)}{(8) + (2)} \frac{2.3}{3} \log(1 - 0.25)$$

$$= \frac{-(16)(2.3)}{(10)(3)} \times -0.125 = 0.153\,\mu\text{moles} \times \text{ml}^{-1} \times \text{min}^{-1}$$

In general, several time samples should be taken as a check on the experimental procedure. If the data are plotted on a semi-log scale (Fig. X-3), a straight line passing through unity at $t = 0$ should be obtained. The half-

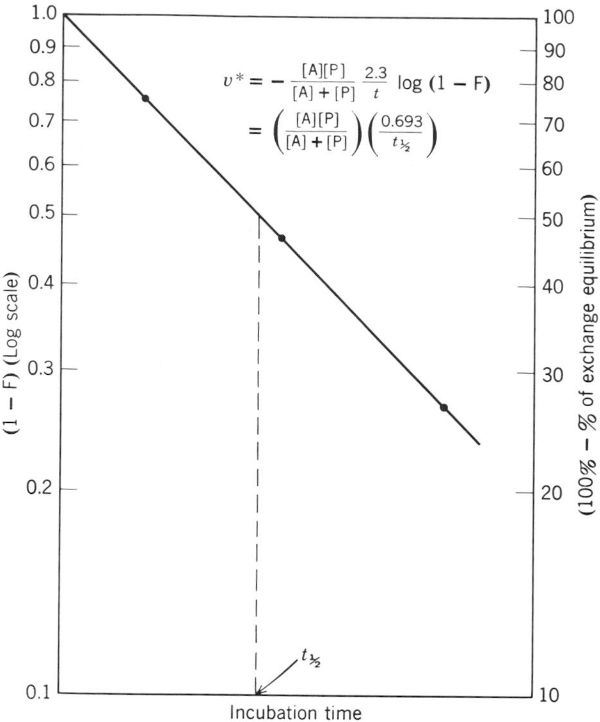

Fig. X-3. Plot of log $(1-F)$ versus incubation time where F = the fraction of isotopic equilibrium achieved at any time. The plot can be used to calculate the exchange velocity, v^*, from several samples taken during the approach to isotopic equilibrium.

time, $t_{1/2}$, can be read off the plot and v^* calculated from equation X-6.

$$v^* = \left(\frac{[A][P]}{[A]+[P]}\right)\left(\frac{0.693}{t_{1/2}}\right) \tag{X-6}$$

Equation X-3 is derived in the following way: At chemical equilibrium the rate at which A is converted to P (in terms of μmoles per liter per minute) is constant throughout the incubation period. However, as unlabeled P is converted to A, the specific activity of A decreases with time. Consequently, the rate at which label appears in P (in terms of $\text{cpm} \times 1^{-1} \times \text{min}^{-1}$) decreases with time. At any time:

$$\frac{dP^*}{dt} = v^*(S.A._A - S.A._P) \tag{X-7}$$

where $S.A._A$ = the specific activity of A (cpm/ μmole)
$S.A._P$ = the specific activity of P (cpm/ μmole)
P^* = the label in P (cpm)
v^* = the constant exchange velocity (μmoles $\times l^{-1} \times min^{-1}$)

Thus if the initial velocity at which P becomes labeled is measured immediately after adding A*:

$$v^* = \frac{dP^*/dt}{S.A._A} \tag{X-8}$$

Equation X-7 can be written as:

$$\frac{dP^*}{dt} = v^*\left[\frac{A^*}{[A]} - \frac{P^*}{[P]}\right] = \frac{v^*}{[A][P]}(A^*[P] - P^*[A]) \tag{X-9}$$

where [A] and [P] are the equilibrium concentrations of A and P (μmoles/ l). The total amount of label present at any time is constant.

$$A^* + P^* = A^*_\infty + P^*_\infty \qquad \therefore \quad A^* = (A^*_\infty + P^*_\infty - P^*) \tag{X-10}$$

At isotopic equilibrium, the distribution of label is porportional to the equilibrium concentrations:

$$\frac{P^*_\infty}{A^*_\infty} = \frac{[P]}{[A]} \qquad \therefore \quad A^*_\infty = \frac{P^*_\infty[A]}{[P]} \tag{X-11}$$

Substituting for A* in equation X-9 from equations X-10 and X-11:

$$\frac{dP^*}{dt} = \frac{v^*}{[A][P]}(P^*_\infty[A] + P^*_\infty[P] - P^*[P] - P^*[A])$$

$$= \frac{v^*}{[A][P]}([A] + [P])(P^*_\infty - P^*)$$

Rearranging and integrating between $t = 0$ (when $P^*_\infty - P^* = P^*_\infty$ because $P^* = 0$) and any other time:

$$\int_{P^*_\infty}^{P^*_\infty - P^*} \frac{dP^*}{(P^*_\infty - P^*)} = \frac{v^*}{[A][P]}([A] + [P])\int_0^t dt$$

$$-\ln\frac{P^*_\infty - P^*}{P^*} = \frac{v^* t}{[A][P]}([A] + [P])$$

or

$$v^* = -\frac{[A][P]}{[A]+[P]}\frac{2.3}{t}\log\left(1-\frac{P^*}{P^*_\infty}\right) \qquad \text{(X-12)}$$

where $P^*/P^*_\infty = F$.

F. DETERMINING K_{eq}

To conduct isotope exchange experiments at chemical equilibrium, we must know K_{eq}. Even if K_{eq} is given in the literature, it is advisable to run a few experiments under the specific assay conditions we are using. (The literature value of K_{eq} might be for a different pH or temperature.) One way to determine K_{eq} is to allow the reaction to proceed to equilibrium from both directions (starting with known initial concentrations of A, B, etc., and P, Q, etc.) and then measure or calculate the equilibrium concentrations of all reactants. This method may be impractical if a long incubation time is required to attain equilibrium. (The enzyme might not be stable under prolonged incubation times at the assay temperature.) A quicker method is to set up reaction mixtures containing known concentrations of all reactants at some ratio close to the suspected equilibrium ratio. We then determine $\Delta[A]$ or $\Delta[P]$, and so on, over a short time period. If $\Delta[P]$ is positive (i.e., [P] increases), then the initial ratios are less than K_{eq} (the reaction mixture is not yet at equilibrium). If $\Delta[P]$ is negative, then the $[P][Q]\ldots/[A][B]\ldots$ ratio is greater than K_{eq} (the net reaction proceeds from right to left). When no change takes place, the reactant ratios equal K_{eq}. Alternately, $\Delta[P]$ can be plotted against the $[P][Q]\ldots/[A][P]\ldots$ ratio (Fig. X-4). The K_{eq} is obtained as the value where the $\Delta[P]$ line crosses the horizontal axis.

G. DERIVATION OF ISOTOPE EXCHANGE VELOCITY EQUATIONS

A*–Q Exchange in a Bi Bi Uni Uni System

The rate at which label of a substrate (e.g., A*) appears in a product (e.g., Q*) is given by:

$$v^*_{A-Q} = \frac{dQ^*}{dt} = k[EX^*] + k'[EY^*] + \ldots \text{etc.}$$

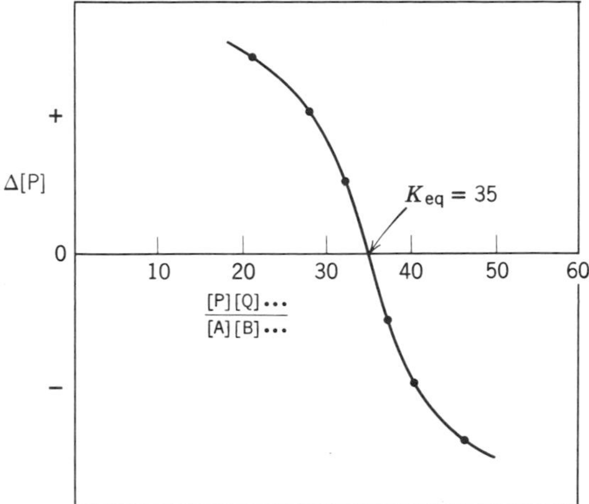

Fig. X-4. Procedure for establishing the exact product/substrate ratio at equilibrium. Several initial ratios are set up and the change in one of the reactants is measured. The ΔP is plotted against the initial ratio. When ΔP = 0, the ratio equals K_{eq}.

where EX*, EY*, and so on represent the different labeled enzyme forms that yield Q* by dissociation, and k and k' are the respective rate constants. In unbranched pathways, there is usually only one labeled enzyme form that yields the labeled product. In branched pathways, the labeled product may be produced by the dissociation of several complexes. For example, in a Random Bi Bi system:

$$v^*_{A-Q} = k_6[EQ^*] + k_7[EA^*B + EPQ^*] \qquad (X\text{-}13)$$

EX*, EY*, and so on can be evaluated algebraically by solving the appropriate set of simultaneous differential steady-state equations. (See the reference by Switzer, 1970, for an example.) Alternately, the King-Altman method can be used treating the labeled and unlabeled species as separate entities and then omitting negligible concentration terms. (See the reference by Schacter, 1972, for the use of this procedure for an Iso Uni Uni system. Fifteen five-lined patterns must be used for each species.) Cleland (1967) has formulated a much simpler schematic method for evaluating [EX*] based on the King-Altman method. The expression for any complex (e.g., EQ*) is given by

$$[EQ^*] = \frac{N_{EQ^*}}{D} \qquad (X\text{-}14)$$

where D is the same for all labeled enzyme species and N_{EQ} is a numerator specific for EQ*. The D is obtained in the following way: (*a*) draw the basic King-Altman figure. Locate the steps in which the exchange pair add or dissociate. Indicate with an asterisk all enzyme forms which become labeled during the exchange. Also indicate with some special marking all steps which involve interconversions of *unlabeled* enzyme forms. This is illustrated below for the A*–Q exchange in a Bi Bi Uni Uni Ping Pong Ter Ter system (where $v^*_{\text{A–Q}} = d\text{Q}^*/dt = k_4[\text{FQ}^*]$).

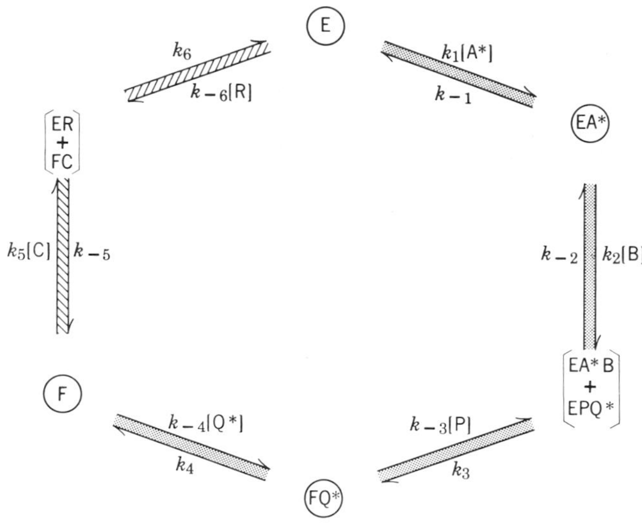

(*b*) Next draw the interconversion patterns which contain one less line than the basic figure. Only those patterns in which the omitted line is one that touches a labeled enzyme form are used in evaluating N_{FQ^*} and D. In other words, the acceptable patterns must contain all the specially marked steps between unlabeled enzyme forms. Thus for the present system there are only four acceptable patterns:

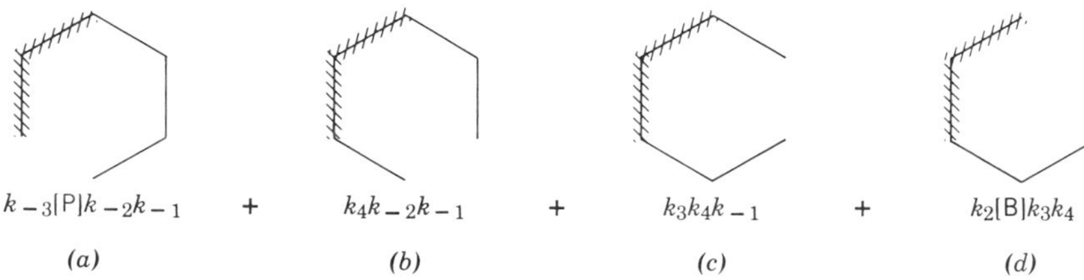

$k_{-3}[\text{P}]k_{-2}k_{-1}$ + $k_4k_{-2}k_{-1}$ + $k_3k_4k_{-1}$ + $k_2[\text{B}]k_3k_4$
(a)
(b)
(c)
(d)

In general, if the basic figure has m lines and n corners, the interconversion patterns have $n-1$ lines. If p is the number of labeled enzyme forms, then the maximum number of patterns is:

$$\frac{(m-n+1+p)!}{p!(m-n+1)!}$$

As usual, patterns containing closed loops must be omitted. (c) D is evaluated in the following way: in each acceptable pattern, follow along the lines which do not connect unlabeled enzyme forms and write the product of the rate constants and reactant concentration terms for the paths by which labeled enzyme forms are converted to unlabeled forms. The resulting terms (shown below the patterns) are added. Combining terms:

$$\mathrm{D}=k_{-1}k_4(k_{-2}+k_3)+k_2k_3k_4[\mathrm{B}]+k_{-1}k_{-2}k_{-3}[\mathrm{P}]$$

The expression for D contains only rate constants and chemical concentrations of unlabeled reactants. (d) Each N is evaluated in the following way: select from among the acceptable interconversion patterns those which include paths by which the labeled substrate adds to an unlabeled enzyme form and is converted to the labeled enzyme form which releases the labeled product. For each pattern chosen, read the product of all rate constants and reactant concentration terms along the lines by which the labeled substrate and all labeled enzyme forms are converted to the enzyme complex which releases the labeled product. Do not read any terms from the specially marked lines which connect unlabeled enzyme forms. If the labeled substrate (e.g., A*) can add to several unlabeled enzyme forms (e.g., E and EB), then some of the patterns may involve dissociation of A* (e.g., from EA*), passage through unlabeled enzyme forms, and then recombination of A* with EB. In evaluating N, the enzyme form which adds the labeled substrate is considered as a reactant and included in the term. In the Bi Bi Uni Uni example, only interconversion pattern a is used. Thus:

$$\mathrm{N}_{\mathrm{FQ}^*}=k_1[\mathrm{A}^*][\mathrm{E}]k_2[\mathrm{B}]k_3$$

and consequently:

$$v^*_{\mathrm{A\text{-}Q}}=k_4[\mathrm{FQ}^*]=\frac{k_4\mathrm{N}_{\mathrm{FQ}^*}}{\mathrm{D}}=\frac{k_1k_2k_3k_4[\mathrm{A}^*][\mathrm{B}][\mathrm{E}]}{k_{-1}k_4(k_{-2}+k_3)+k_2k_3k_4[\mathrm{B}]+k_{-1}k_{-2}k_{-3}[\mathrm{P}]} \qquad \text{(X-15)}$$

If we compare the groups of rate constants to the definitions of the usual kinetic constants, we can write (after multiplying numerator and denominator of equation X-15 by k_5k_6 and then dividing by Coef_{ABC}):

$$v^*_{A\text{-}Q} = \frac{V_{\max_f}[A^*][B]\dfrac{[E]}{[E]_t}}{K_{ia}K_{m_B} + K_{m_A}[B] + \dfrac{K_{ia}K_{m_B}K_{m_Q}[P]}{K_{m_P}K_{iq}}} \qquad \text{(X-16)}$$

The completed velocity equation is obtained when $[E]/[E]_t$ is expressed in terms of reactant concentrations and kinetic constants. In the absence of C and R, the Ping Pong segment is at chemical equilibrium and we can write:

$$\frac{[E]}{[E]_t} = \frac{1}{1 + \dfrac{[A]}{K_{ia}} + \dfrac{[A][B]}{K_{ia}K_{ib}} + \dfrac{[A][B]K_{ip}}{K_{ia}K_{ib}[P]} + \dfrac{[A][B]K_{ip}K_{iq}}{K_{ia}K_{ib}[P][Q]}} \qquad \text{(X-17)}$$

The last two denominator terms represent, respectively, the concentrations of FQ and F in terms of E. Substituting into equation X-16:

$$v^*_{A\text{-}Q} = \frac{V_{\max}[A^*][B]}{\left(K_{ia}K_{m_B} + K_{m_A}[B] + \dfrac{K_{ia}K_{m_B}K_{m_Q}[P]}{K_{m_P}K_{iq}}\right) \times \left(1 + \dfrac{[A]}{K_{ia}} + \dfrac{[A][B]}{K_{ia}K_{ib}} + \dfrac{[A][B]K_{ip}}{K_{ia}K_{ib}[P]} + \dfrac{[A][B]K_{ip}K_{iq}}{K_{ia}K_{ib}[P][Q]}\right)} \qquad \text{(X-18)}$$

The equation confirms the general rules of Cleland. First let us see which reactants must be present in order for label from A* to appear in Q. Obviously, if $[B]=0$, then $v^*_{A\text{-}Q}=0$. If the denominator is multiplied out, it would contain $K_{m_B}K_{ip}[A][B]/K_{ib}[P]$ and $K_{m_B}K_{ip}K_{iq}[A][B]/K_{ib}[P][Q]$ terms, among others. If either $[P]=0$ or $[Q]=0$, at least one of these terms would become infinitely high and, again, $v^*_{A\text{-}Q}=0$. Thus to observe exchange, all four reactants must be present. The denominator also contains

$$\frac{K_{ia}K_{m_B}K_{m_Q}[P]}{K_{m_P}K_{iq}}, \quad \frac{K_{m_B}K_{m_Q}[A][P]}{K_{m_P}K_{iq}}, \quad \text{and} \quad \frac{K_{m_B}K_{m_Q}[A][B][P]}{K_{ib}K_{m_P}K_{iq}}$$

terms. If [P] becomes infinitely high, the whole denominator becomes infinitely high and again $v^*_{\text{A-Q}}=0$. Thus if [P] is increased as a changing fixed reactant or along with some substrate, we eventually observe total substrate inhibition. The denominator also contains $[B]^2$ terms. Thus at some point as [B] is increased, the denominator increases faster than the numerator, and we see substrate inhibition. Note that if [Q] becomes infinitely high, no inhibition is observed. (There are no denominator terms with [Q] in the numerator.) Also, there are no $[A]^2$ denominator terms, so that no substrate inhibition is observed when [A*] is varied. If all reactants of the system are present, equation X-16 would be unchanged, but the expression for $[E]/[E]_t$ would be more complex. If the system is at chemical equilibrium, the expression for $[E]/[E]_t$ would contain an additional $[R]/K_{ir}$ term representing the (ER + FC) central complex. This central complex could also be expressed as $[A][B][C]K_{ip}K_{iq}/K_{ia}K_{ib}K_{ic}[P][Q]$ because at chemical equilibrium:

$$\frac{[P][Q][R]}{[A][B][C]} = K_{\text{eq}} = \frac{K_{ip}K_{iq}K_{ir}}{K_{ia}K_{ib}K_{ic}}$$

Similarly, stable form F could be given as $[R]K_{ic}/K_{ir}[C]$. If all reactants are present but the system is not at chemical equilibrium, $[E]/[E]_t$ can be expressed by the distribution equation IX-293 which is derived for steady-state conditions with no assumptions of chemical equilibrium.

A*–Q Exchange in an Ordered Bi Bi System

Equation X-16 also describes the A*–Q exchange for an Ordered Bi Bi system. The derivation would be identical to that shown for the Bi Bi Uni Uni system except there would be no specially marked lines (all enzyme complexes become labeled and all four three-lined interconversion patterns would be used in obtaining D). The final velocity equation would differ from equation X-18 because the expression for $[E]/[E]_t$ is different. At chemical equilibrium:

$$\frac{[E]}{[E]_t} = \frac{1}{1 + \frac{[A]}{K_{ia}} + \frac{[A][B]}{K_{ia}K_B} + \frac{[Q]}{K_{iq}}} \tag{X-19}$$

Note that K_B represents the dissociation constant for B from the central complexes, which is not the same as K_{ib} defined for this system (see equation IX-88a). The expression for the (EAB + EPQ) central complex can also be written as $[P][Q]/K_{iq}K_P$, where K_P is the dissociation constant for P from (EAB + EPQ) and is not the same as the K_{ip} defined for this system. We could also indicate the EQ complex as $[A][B]K_P/K_{ia}K_B[P]$. (At chemical equilibrium $[P][Q]/[A][B] = K_{iq}K_P/K_{ia}K_B = K_{eq}$.) The exchange velocity equation is:

$$v^*_{A-Q} = \frac{V_{max_f}[A^*][B]}{\left(K_{ia}K_{m_B} + K_{m_A}[B] + \dfrac{K_{ia}K_{m_B}K_{m_Q}[P]}{K_{m_P}K_{iq}}\right)\left(1 + \dfrac{[A]}{K_{ia}} + \dfrac{[A][B]}{K_{ia}K_B} + \dfrac{[Q]}{K_{iq}}\right)} \quad \text{(X-20)}$$

Keep in mind that we are at chemical equilibrium with all reactants present. If multiplied out, the denominator would contain $[B]^2$ terms, thus confirming that when [B] is varied along with [P] or [Q], substrate inhibition is observed. The denominator also contains [A][P] and [A][B][P] terms. Thus when [A] and [P] are varied at $[P] = x[A]$, we have $[A]^2$ terms in the denominator and just an [A] term in the numerator. Thus substrate inhibition is observed. When [A] and [Q] are varied at $[Q] = x[A]$, no $[A]^2$ terms appear in the denominator. Thus there is no substrate inhibition.

If the system is not at chemical equilibrium, $[E]/[E]_t$ is given by distribution equation IX-112. The resulting exchange velocity equation is far more complex than equation X-20 but the predictions are the same.

A*–P Exchange in an Ordered Bi Bi System

As another example, consider the A*–P exchange in an Ordered Bi Bi system; v^* is given by:

$$v^*_{A-P} = \frac{dP^*}{dt} = k_3[EA^*B + EPQ] = \frac{k_3 N_{EA^*B+EP^*Q}}{D} \quad \text{(X-21)}$$

The basic King-Altman figure is shown below.

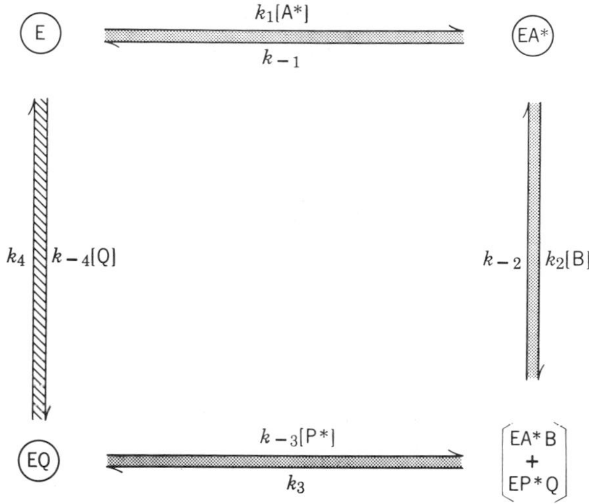

The valid interconversion patterns and the resulting expressions for D are:

$k_{-1}k_3$ + k_2[B]k_3 + $k_{-2}k_{-1}$

(a) *(b)* *(c)*

The N_{EQ} is evaluated from pattern *c*:

$$N_{EA^*B+EP^*Q} = k_1[A^*][E]k_2[B]$$

Therefore:

$$v^*_{A\text{–}P} = \frac{k_1k_2k_3[A^*][B][E]}{k_{-1}(k_{-2}+k_3)+k_2k_3[B]} = \frac{V_{\max_f}[A^*][B]\dfrac{[E]}{[E]_t}}{K_{ia}K_{m_B}+K_{m_A}[B]} \qquad \text{(X-22)}$$

The second form is obtained by multiplying the numerator and denominator of the first by k_4 and then defining the groups of rate constants in the usual manner. At chemical equilibrium, $[E]/[E]_t$ is given by equation X-19.

Substituting, we obtain:

$$v^*_{\text{A-P}} = \frac{V_{\max_f}[\text{A}^*][\text{B}]}{K_{ia}K_{m_\text{B}} + K_{m_\text{B}}[\text{A}] + \frac{K_{m_\text{B}}[\text{A}][\text{B}]}{K_\text{B}} + \frac{K_{ia}K_{m_\text{B}}[\text{Q}]}{K_{iq}} + K_{m_\text{A}}[\text{B}] + \frac{K_{m_\text{A}}[\text{A}][\text{B}]}{K_{ia}} + \frac{K_{m_\text{A}}[\text{A}][\text{B}]^2}{K_{ia}K_\text{B}} + \frac{K_{m_\text{A}}[\text{B}][\text{Q}]}{K_{iq}}}$$

(X-23)

We can see directly that varying [B] along with either [P] or [Q] leads to total substrate inhibition because of the $[\text{B}]^2$ terms that appear in the denominator. Varying [A] and [Q] where $[\text{Q}] = x[\text{A}]$ does not introduce any $[\text{A}]^2$ terms into the denominator and, consequently, no substrate inhibition is observed. No [P] terms appear in the equation, so we migh predict that varying [A] and [P] would not lead to substrate inhibition. This is confirmed by substituting $[\text{P}][\text{Q}]/K_{iq}K_\text{P}$ for $[\text{A}][\text{B}]/K_{ia}K_\text{B}$, and $[\text{A}][\text{B}]K_\text{P}/K_{ia}K_\text{B}[\text{P}]$ for $[\text{Q}]/K_{iq}$ in the expression for $[\text{E}]/[\text{E}]_t$. This leads to a final exchange velocity equation containing denominator [P] terms. However, there will be no [A][P] terms to become $[\text{A}]^2$ when $[\text{P}] = x[\text{A}]$.

Equation X-23 rearranges to:

$$\frac{v^*_{\text{A-P}}}{V_{\max_f}} = \frac{[\text{A}^*]}{K_{m_\text{A}}\left(1 + \frac{[\text{Q}]}{K_{iq}}\right)\left(1 + \frac{K_{ia}K_{m_\text{B}}}{K_{m_\text{A}}[\text{B}]}\right) + [\text{A}]\left(\frac{K_{m_\text{B}}}{K_\text{B}} + \frac{K_{m_\text{B}}}{[\text{B}]} + \frac{K_{m_\text{A}}}{K_{ia}} + \frac{K_{m_\text{A}}[\text{B}]}{K_{ia}K_\text{B}}\right)}$$

(X-24)

Thus when [A] and [Q] are varied at different fixed concentrations of B and P, we eventually see uncompetitive inhibition by the changing fixed reactants. When [A] and [P] are varied at different fixed concentrations of B and Q, we eventually see mixed-type inhibition by the changing fixed reactants.

B*–P Exchange in an Ordered Bi Bi System

The same procedure for the B*–P exchange yields:

$$v^*_{\text{B-P}} = k_3[\text{EAB}^* + \text{EP}^*\text{Q}] = \frac{k_3\text{N}_{\text{EAB+EPQ}}}{\text{D}} \quad \text{(X-24a)}$$

$$\mathrm{N} = k_2[\mathrm{B}^*][\mathrm{EA}], \qquad \mathrm{D} = k_{-2} + k_3$$

$$v^*_{\mathrm{B-P}} = \frac{k_2 k_3[\mathrm{B}^*][\mathrm{EA}]}{(k_{-2} + k_3)} = \frac{k_1 k_2 k_3 k_4[\mathrm{B}^*][\mathrm{EA}]}{k_1 k_4(k_{-2} + k_3)}$$

Substituting for [EA] and defining the groups of rate constants in the usual way ($\mathrm{Coef_A}/\mathrm{Coef_{AB}} = K_{m_\mathrm{B}}$), we obtain:

$$\boxed{\frac{v^*_{\mathrm{B-P}}}{V_{\max_f}} = \frac{[\mathrm{A}][\mathrm{B}^*]}{K_{ia}K_{m_\mathrm{B}}\left(1 + \dfrac{[\mathrm{A}]}{K_{ia}} + \dfrac{[\mathrm{A}][\mathrm{B}]}{K_{ia}K_\mathrm{B}} + \dfrac{[\mathrm{Q}]}{K_{iq}}\right)}} \qquad \text{(X-24b)}$$

The lack of inhibition by any reactant is obvious.

A*–P Exchange in a Ping Pong Bi Bi System

As a further example, consider the A*–P exchange in a Ping Pong Bi Bi system. The exchange velocity is given by:

$$v^*_{\mathrm{A-P}} = \frac{d\mathrm{P}^*}{dt} = k_2[\mathrm{EA}^* + \mathrm{FP}^*] = \frac{k_2 \mathrm{N}_{\mathrm{EA+FP}}}{\mathrm{D}} \qquad \text{(X-25)}$$

The basic King-Altman figure is:

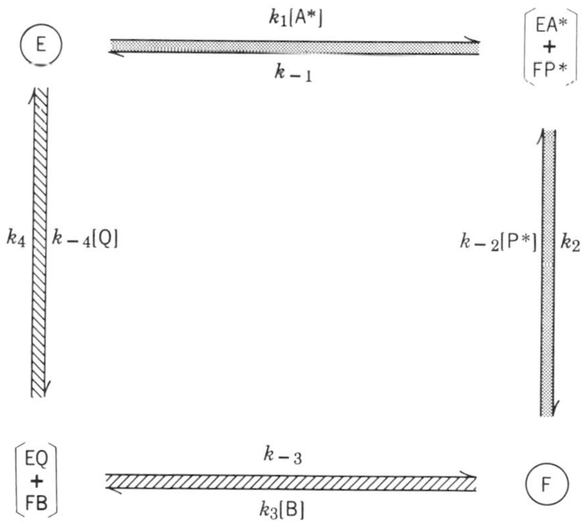

E ⇄ (EA* + FP*): $k_1[\mathrm{A}^*]$, k_{-1}

(EA* + FP*) ⇄ F: k_2, $k_{-2}[\mathrm{P}^*]$

F ⇄ (EQ + FB): $k_3[\mathrm{B}]$, k_{-3}

(EQ + FB) ⇄ E: k_4, $k_{-4}[\mathrm{Q}]$

The interconversion patterns and expressions for D and N_{EA+FP} are:

k_{-1} + k_2

$$D = k_{-1} + k_2, \qquad N_{EA^*+FP^*} = k_1[A^*][E]$$

$$\therefore \quad v^*_{A-P} = \frac{k_1k_2[A^*][E]}{k_{-1}+k_2} = \frac{k_1k_2k_3k_4[A^*][E]}{k_3k_4(k_{-1}+k_2)} = \frac{V_{max_f}[A^*]\dfrac{[E]}{[E]_t}}{K_{m_A}} \tag{X-26}$$

At chemical equilibrium:

$$\frac{[E]}{[E]_t} = \frac{1}{1 + \dfrac{[A]}{K_{ia}} + \dfrac{[A]K_{ip}}{K_{ia}[P]} + \dfrac{[A]K_{ip}[B]}{K_{ia}[P]K_{ib}}} \tag{X-27}$$

Substituting:

$$v^*_{A-P} = \frac{V_{max_f}[A^*]}{K_{m_A} + \dfrac{K_{m_A}[A]}{K_{ia}} + \dfrac{K_{m_A}K_{ip}[A]}{K_{ia}[P]} + \dfrac{K_{m_A}K_{ip}[A][B]}{K_{ia}K_{ib}[P]}} \tag{X-28}$$

B is obviously inhibitory when varied along with Q but not when it is varied along with P at constant ratio. There is no [B] term in the numerator confirming that B is not required for the A*–P exchange. If [B]=0, the equation rearranges to:

$$v^*_{A-P} = \frac{V_{max_f}\dfrac{K_{ia}}{K_{m_A}}[A^*][P]}{K_{ia}[P] + K_{ip}[A] + [A][P]} \quad \text{or} \quad \frac{v^*_{A-P}}{V^*_{max_{A-P}}} = \frac{[A^*][P]}{K_{ia}[P] + K_{ip}[A] + [A][P]} \tag{X-29}$$

where $V^*_{max_{A-P}} = V_{max_f}K_{ia}/K_{m_A}$.

Exchange Equations for a Random Bi Bi System

As a more complicated example, consider the A*–P exchange in a Random Bi Bi system at chemical equilibrium.

$$v^*_{A-P} = \frac{dP^*}{dt} = k_8[EP^*] + k_5[EA^*B + EPQ^*] = k_8\frac{N_{EP}}{D} + k_5\frac{N_{EAB+EPQ}}{D} \qquad \text{(X-30)}$$

The basic eight-lined King-Altman figure is shown on page 876. There are two specially marked lines which connect only unlabeled enzyme forms. Of the 32 possible five-lined interconversion patterns, only the 12 shown are used to determine D. After grouping terms D is given by:

$$\begin{aligned} D = {} & (k_{-1}k_8)(k_{-2} + k_{-4} + k_5 + k_7) + k_2k_8(k_{-4} + k_5 + k_7)[B] \\ & + k_{-1}k_{-7}(k_{-2} + k_4 + k_5)[Q] + k_2k_{-7}(k_{-4} + k_5)[B][Q] \end{aligned} \qquad \text{(X-31)}$$

The N_{EP*} is obtained from patterns *b*, *d*, and *f* because only these contain paths by which A* adds to an unlabeled enzyme form and is converted to EP*.

$$N_{EP*} = k_{-1}k_4k_7[A^*][EB] + k_1k_2k_7[A^*][B][E] + k_2k_4k_7[A^*][B][EB]$$

The $N_{EA*B+EP*Q}$ is obtained from patterns *b*, *d*, *f*, *g*, *h*, and *l*.

$$\begin{aligned} N_{EA*B+EP*Q} = {} & k_{-1}k_4k_{-7}[A^*][Q][EB] + k_1k_2k_{-7}[A^*][B][Q][E] \\ & + k_2k_4k_{-7}[A^*][B][Q][EB] + k_2k_4k_8[A^*][B][EB] \\ & + k_1k_2k_8[A^*][B][E] + k_{-1}k_4k_8[A^*][EB] \end{aligned} \qquad \text{(X-32)}$$

The final exchange velocity equation is obtained by substituting for N_{EP*} and $N_{EA*B+EP*Q}$ into equation X-30, factoring, substituting $[B][E]/K_B$ for [EB], and then substituting the usual distribution equation for [E]. Needless to say, the final equation is rather complicated and can be written in several

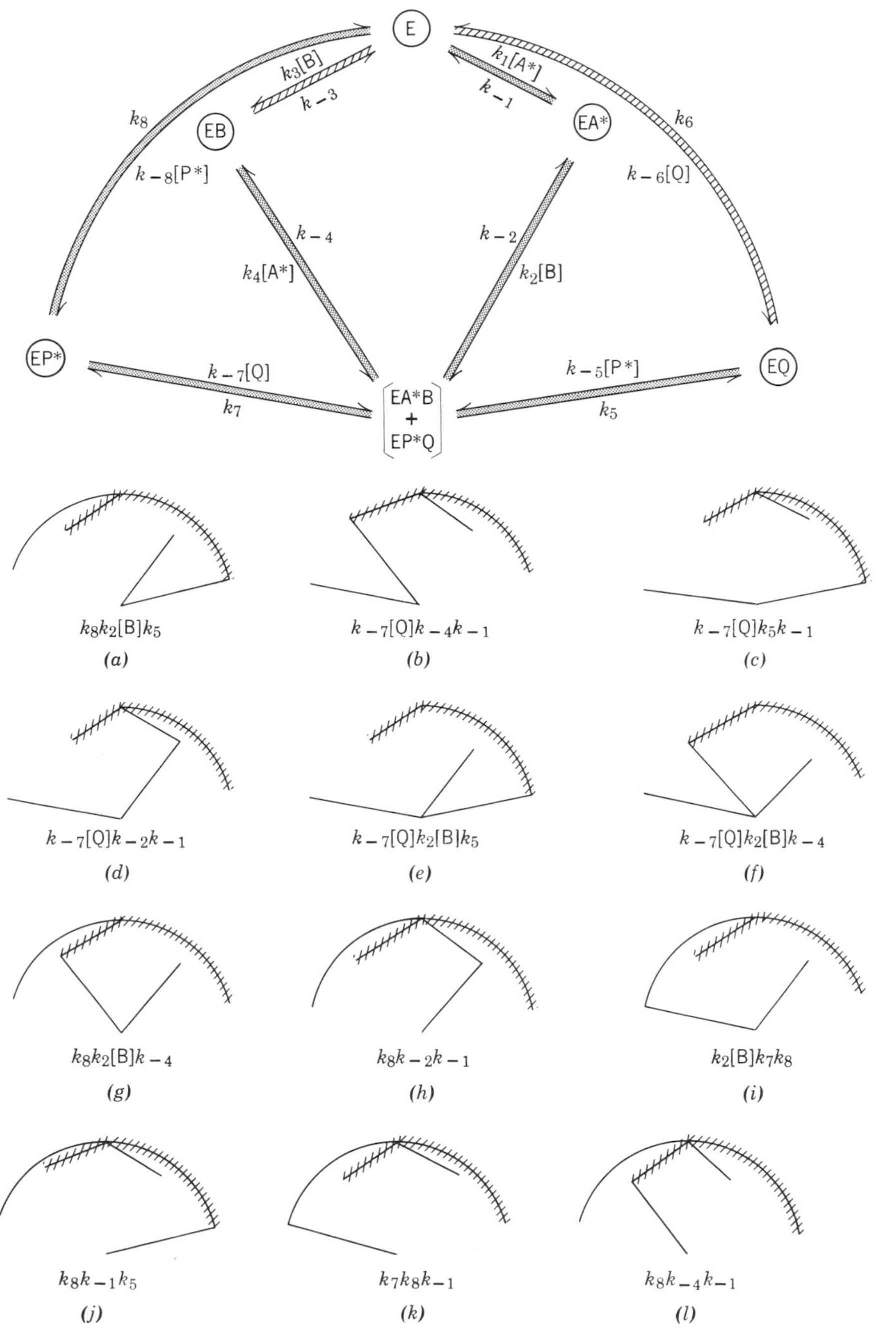
E
k_3[B]
k_{-3}
k_1[A*]
k_{-1}
k_8
EB
EA*
k_6
k_{-8}[P*]
k_{-6}[Q]
k_{-4}
k_{-2}
k_4[A*]
k_2[B]
EP*
EQ
k_{-7}[Q]
k_{-5}[P*]
EA*B
+
EP*Q
k_7
k_5
k_8k_2[B]k_5
(a)
k_{-7}[Q]$k_{-4}k_{-1}$
(b)
k_{-7}[Q]k_5k_{-1}
(c)
k_{-7}[Q]$k_{-2}k_{-1}$
(d)
k_{-7}[Q]k_2[B]k_5
(e)
k_{-7}[Q]k_2[B]k_{-4}
(f)
k_8k_2[B]k_{-4}
(g)
$k_8k_{-2}k_{-1}$
(h)
k_2[B]k_7k_8
(i)
$k_8k_{-1}k_5$
(j)
$k_7k_8k_{-1}$
(k)
$k_8k_{-4}k_{-1}$
(l)

forms. One form given by Plowman (1972) is shown below.

$$v^*_{\text{A-P}} = \frac{(k_7k_8 + k_5k_8 + k_5k_{-7}[\text{Q}])\left(k_1k_2 + \dfrac{k_3k_4}{k_{-3}}(k_{-1} + k_2[\text{B}])\right)[\text{A}^*][\text{B}][\text{E}]_t}{\text{D}\left(1 + \dfrac{[\text{A}]}{K_\text{A}} + \dfrac{[\text{B}]}{K_\text{B}} + \dfrac{[\text{A}][\text{B}]}{K_\text{A}K_\text{B}} + \dfrac{[\text{P}]}{K_\text{P}} + \dfrac{[\text{Q}]}{K_\text{Q}}\right)} \tag{X-33}$$

where D represents the denominator terms (equation X-31) and $k_{-3}/k_3 = K_\text{B}$.

Boyer and Silverstein (1963) have presented the exchange velocity equation X-34 for a random system in which the step $\text{EAP} \underset{k_{-p}}{\overset{k_p}{\rightleftharpoons}} \text{EPQ}$ is included in the reaction sequence.

$$v^*_{\text{A-P}} = \frac{[\text{E}]_t}{\left(1 + \dfrac{k_{-p}(k_8 + k_{-7}[\text{Q}])}{k_7k_8 + k_5(k_8 + k_{-7}[\text{Q}])} + \dfrac{k_p(k_{-1} + k_2[\text{B}])}{k_{-1}k_{-2} + k_{-4}(k_{-1} + k_2[\text{B}])}\right)}$$
$$\times\left[\frac{1}{k_{-p}}\left(1 + \frac{K_\text{Q}}{[\text{Q}]} + \frac{K_\text{P}}{[\text{P}]}\right) + \frac{1}{k_p}\left(1 + \frac{K_\text{A}}{[\text{A}]} + \frac{K_\text{B}}{[\text{B}]} + \frac{K_\text{A}K_\text{B}}{[\text{A}][\text{B}]}\right)\right] \tag{X-34}$$

When [B] and [Q] are saturating, equation X-34 becomes:

$$v^*_{\text{A-P}} = \frac{[\text{E}]_t}{\left(1 + \dfrac{k_{-p}}{k_5} + \dfrac{k_p}{k_{-4}}\right)\left[\dfrac{1}{k_{-p}}\left(1 + \dfrac{K_\text{P}}{[\text{P}]}\right) + \dfrac{1}{k_p}\left(1 + \dfrac{K_\text{A}}{[\text{A}]}\right)\right]} \tag{X-35}$$

When all reactants are saturating:

$$v^*_{\text{A-P}} = \frac{[\text{E}]_t}{\left(1 + \dfrac{k_{-p}}{k_5} + \dfrac{k_p}{k_{-4}}\right)\left(\dfrac{1}{k_{-p}} + \dfrac{1}{k_p}\right)} \tag{X-36}$$

If k_p and k_{-p} are very small compared to the rate constants for reactant dissociations, the interconversion of EAB and EPQ becomes rate-limiting (i.e., the system is a Rapid Equilibrium Random one) and equation X-36 becomes:

$$v^*_{\text{A-P}} = \frac{[\text{E}]_t}{\dfrac{1}{k_{-p}} + \dfrac{1}{k_p}} = \frac{k_p k_{-p}[\text{E}]_t}{k_p + k_{-p}}$$

or

$$\boxed{v^*_{\text{A-P}} = \frac{V_{\max_f} V_{\max_r}}{V_{\max_f} + V_{\max_r}}} \qquad \text{(X-37)}$$

The exchange velocity for a true rapid equilibrium system at any set of reactant concentrations is identical to the initial unidirectional velocity. Thus for a Rapid Equilibrium Random system v^* would be given by equation VI-46 with the [P][Q] numerator term omitted. [See also equation 12 of Gulbinsky and Cleland (1968).]

A*–P Exchange in Iso Uni Uni (Membrane Transport) Systems

Finally, let us return to the simple Iso Uni Uni system for membrane transport (equation IX-47). Unlike the regular Uni Uni mechanism, the exchange velocity equation is not identical to the initial velocity equation. It is noteworthy that most studies on the effect of internal substrate on the rate of further external substrate transport (transinhibition) are in fact isotope exchange studies where the internal substrate, P, is unlabeled and the external substrate, A, is radioactive. The basic figure and interconversion patterns are shown below.

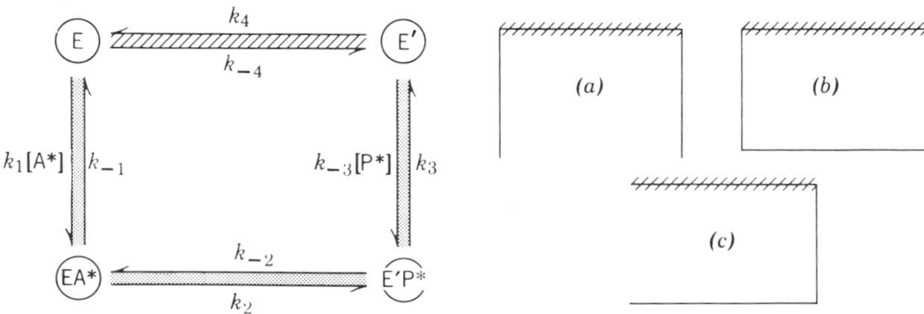

$$v^*_{\text{A-P}} = \frac{d\text{P}^*}{dt} = k_3[\text{E'P}^*] = \frac{k_3 \text{N}_{\text{E'P}^*}}{\text{D}}$$

Pattern b yields: $\text{N}_{\text{E'P}^*} = k_1[\text{A}^*][\text{E}]k_2$
Patterns a, b, and c yield: $\text{D} = k_{-1}k_3 + k_{-1}k_{-2} + k_2k_3$

$$\therefore \quad v^*_{\text{A-P}} = \frac{k_1k_2k_3[\text{A}^*][\text{E}]}{k_{-1}k_3 + k_{-1}k_{-2} + k_2k_3} \tag{X-38}$$

Substituting for [E] from equation IX-41 and then dividing numerator and denominator by $(k_{-1}k_3 + k_{-1}k_{-2} + k_2k_3)$ and factoring, we obtain:

$$\frac{v^*_{\text{A-P}}}{[\text{E}]_t} = \frac{\text{num}_1[\text{A}^*]\left(1 + \dfrac{k_{-1}k_{-2}k_{-3}[\text{P}]}{k_4(k_{-1}k_3 + k_{-1}k_{-2} + k_2k_3)}\right)}{\text{const} + \text{Coef}_{\text{A}}[\text{A}] + \text{Coef}_{\text{P}}[\text{P}] + \text{Coef}_{\text{AP}}[\text{A}][\text{P}]}$$

The equation can be expressed entirely in terms of kinetic constants by employing the usual definitions, and defining a new constant:

$$K_{\text{I}} = \frac{k_4(k_{-1}k_3 + k_{-1}k_{-2} + k_2k_3)}{k_{-1}k_{-2}k_{-3}}$$

$$\boxed{\frac{v^*_{\text{A-P}}}{V_{\max}} = \frac{[\text{A}^*]\left(1 + \dfrac{[\text{P}]}{K_{\text{I}}}\right)}{K_{m_{\text{A}}}\left(1 + \dfrac{[\text{P}]}{K_{m_{\text{P}}}}\right) + [\text{A}]\left(1 + \dfrac{[\text{P}]}{K_{iip}}\right)}} \tag{X-39}$$

Equation X-39 will predict the initial velocity of A* transport in the presence of any fixed P concentration. Reciprocal plots of $1/v^*$ versus $1/[\text{A}^*]$ are linear but slope and intercept replots are hyperbolic (Fig. X-5). At an infinitely high [P], the limiting $K_{m_{\text{A(app)}}}$ is $\alpha K_{m_{\text{A}}}/\beta$, where $\alpha = K_{\text{I}}/K_{m_{\text{P}}}$, and $\beta = K_{\text{I}}/K_{iip}$. The limiting $V_{\max_{f(\text{app})}}$ is $V_{\max_f}/\beta$ (Fig. X-5). However, in practice it may be impossible to preload cells to a sufficiently high internal level to observe these limiting constants.

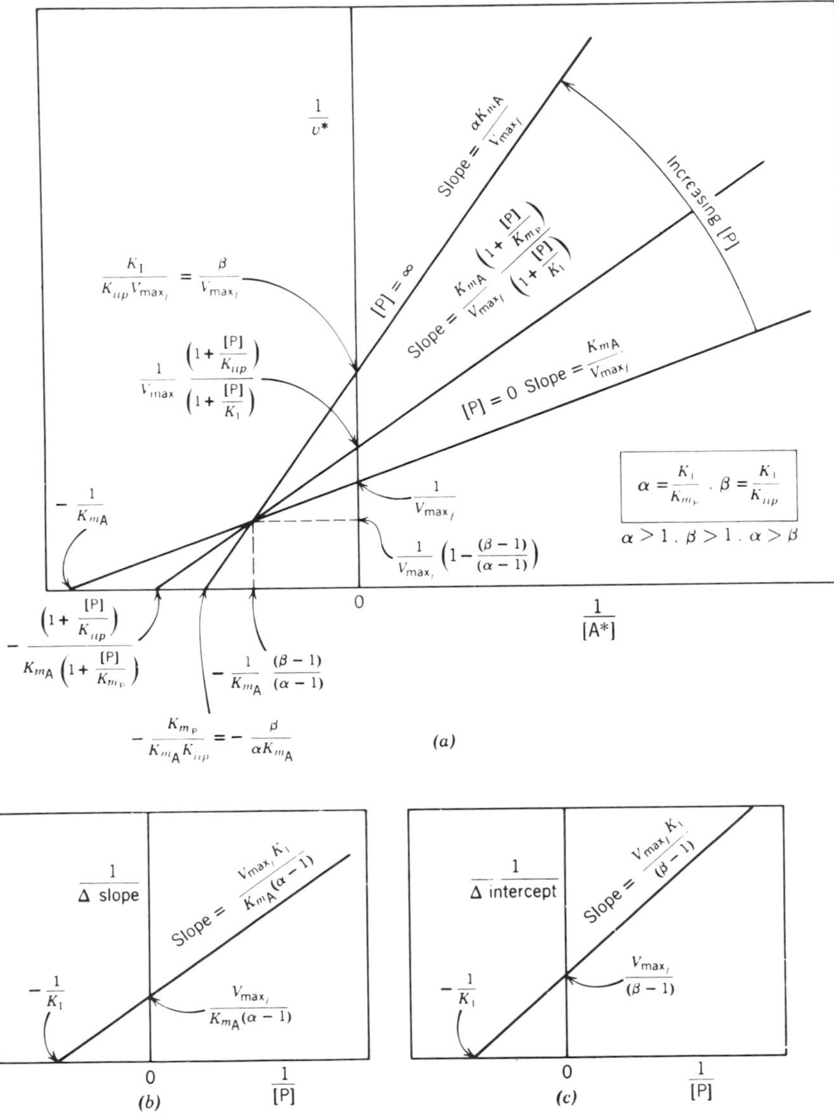

Fig. X-5. Reciprocal plots and replots for isotope exchange in an Iso Uni Uni (membrane transport) system. [A*] is varied at different fixed concentrations of P.

The K_I can be determined from replots of $1/\Delta slope$ and $1/\Delta intercept$ (Chapter Four). The equations for the replots are:

$$\frac{1}{\Delta slope} = \frac{V_{\max_f} K_I}{K_{m_A}(\alpha - 1)} \frac{1}{[P]} + \frac{V_{\max_f}}{K_{m_A}(\alpha - 1)} \tag{X-40}$$

$$\frac{1}{\Delta intercept} = \frac{V_{\max_f} K_I}{(\beta - 1)} \frac{1}{[P]} + \frac{V_{\max_f}}{(\beta - 1)} \tag{X-41}$$

The intercept on the $1/[P]$-axis gives $-1/K_I$ on both replots. With K_{m_A} and $V_{\max_f}$ known from control plots, α and β (hence K_{m_P} and K_{iip}) can be calculated from the vertical-axis intercepts of the $1/\Delta$ replots. Because equations X-40 and X-41 assume that internal P acts as a transinhibitor, $\Delta slope$ and $\Delta intercept$ are taken as (*slope* or *intercept* in the presence of P) − (*slope* or *intercept* of the control). If $K_I \gg K_{m_P}$ and $\gg K_{iip}$, the $1/\Delta$ replots will go through the origin. The slopes of the replots reduce to $V_{\max_f} K_{m_P}/K_{m_A}$ (plotting $1/\Delta slope$) and $V_{\max_f} K_{iip}$ (plotting $1/\Delta intercept$). In this case, the system can be treated as linear inhibition and analyzed by the usual slope versus [P] and intercept versus [P] replots to obtain K_{m_P} and K_{iip}, respectively.

If $K_I < K_{m_P}$ and $< K_{iip}$, P can be treated as a mixed-type, nonessential activator (Chapter Five). The same replots are used, only now $\Delta slope$ or $\Delta intercept$ is taken as (*slope* or *intercept* of the control) − (*slope* or *intercept* in the presence of P). This changes the $(K_I/K_{m_P} - 1)$ and $(K_I/K_{iip} - 1)$ terms to $(1 - K_I/K_{m_P})$ and $(1 - K_I/K_{iip})$, respectively.

In general, if internal, unlabeled P acts as a transinhibitor of external, labeled A uptake, then external, unlabeled A will act as a transstimulator of internal, labeled P efflux. Analysis of A-stimulated efflux of labeled P will permit K_{iia}, K_{m_A}, and the analogous K_I' to be determined ($K_I/K_I' = K_{eq}$).

In a soluble system, varying [A*] and [P] together at $[P] = [A^*]K_{eq}$ converts equation X-39 to:

$$\frac{v^*}{V_{\max_f}} = \frac{\dfrac{K_{eq}}{K_I}[A^*]^2 + [A^*]}{K_{m_A} + \dfrac{K_{eq}}{K_{iip}}[A^*]^2 + \left(1 + \dfrac{K_{m_A} K_{eq}}{K_{m_P}}\right)[A^*]} \tag{X-42}$$

Equation X-42 has the same form as equations IX-182 and VII-24. Thus the shape of the v^* versus [A*] plot will depend on the relative values of the kinetic constants and can be hyperbolic or sigmoidal, and may show a maximum followed by partial substrate inhibition (Fig. IX-37).

In contrast to the Iso Uni Uni system, the simple Uni Uni system always yields a normal hyperbolic plot as [A*] and [P] are varied together. Following the rules outlined earlier, we obtain:

$$v^*_{\mathrm{A-P}} = \frac{V_{\max_f}[\mathrm{A}^*]\dfrac{[\mathrm{E}]}{[\mathrm{E}]_t}}{K_{m_\mathrm{A}}} \tag{X-43}$$

where

$$\frac{[\mathrm{E}]}{[\mathrm{E}]_t} = \frac{1}{\left(1+\dfrac{[\mathrm{A}]}{K_\mathrm{A}}+\dfrac{[\mathrm{P}]}{K_\mathrm{P}}\right)} = \frac{1}{\left(1+\dfrac{[\mathrm{A}]}{K_{m_\mathrm{A}}}+\dfrac{[\mathrm{P}]}{K_{m_\mathrm{P}}}\right)} \tag{X-44}$$

The two parenthetical terms are equal even though $K_\mathrm{A} \neq K_{m_\mathrm{A}}$ and $K_\mathrm{P} \neq K_{m_\mathrm{P}}$. Substituting for $[\mathrm{E}]/[\mathrm{E}]_t$ in equation X-43 we obtain (as expected):

$$\frac{v^*}{V_{\max_f}} = \frac{[\mathrm{A}^*]}{K_{m_\mathrm{A}}\left(1+\dfrac{[\mathrm{P}]}{K_{m_\mathrm{P}}}\right)+[\mathrm{A}]} \tag{X-45}$$

If $[\mathrm{P}] = [\mathrm{A}^*]K_{\mathrm{eq}} = [\mathrm{A}^*]\,V_{\max_f}K_{m_\mathrm{P}}/V_{\max_r}K_{m_\mathrm{A}}$, equation X-45 can be rearranged to:

$$\boxed{\frac{v^*_{\mathrm{A-P}}}{V_{\max_f}} = \frac{[\mathrm{A}^*]}{K_{m_\mathrm{A}}+[\mathrm{A}^*]\left(1+\dfrac{V_{\max_f}}{V_{\max_r}}\right)}} \tag{X-46}$$

Thus, the reciprocal plot is linear with intercepts of:

$$\frac{1}{V^*_{\max_{\mathrm{A-P}}}} = \left(\frac{1}{V_{\max_f}}+\frac{1}{V_{\max_r}}\right) \quad \text{and} \quad -\frac{1}{K_{m_{\mathrm{A(app)}}}} = -\frac{\left(1+\dfrac{V_{\max_f}}{V_{\max_r}}\right)}{K_{m_\mathrm{A}}}$$

REFERENCES

General

Boyer, P. D., *Arch. Biochem. Biophys.* **82**, 387 (1959).

Boyer, P. D. and Silverstein, E., *Acta Chem. Scand.* **17**, Suppl. 1, 195 (1963).

Cleland, W. W., "Enzyme Kinetics," *Ann. Rev. Biochem.* **36**, 77 (1967).

Cleland, W. W., "Steady State Kinetics," in *The Enzymes*, Vol. 2, 3rd ed., P. D. Boyer, ed., Academic Press, 1970, p. 1.

Darvey, I. G., *J. Theoret. Biol.* **42**, 55 (1973).

Darvey, I. G., *Biochem J.* **135**, 861 (1973).

Hass, L. F. and Byrne, W. L., *J. Am. Chem. Soc.* **82**, 947 (1960).

Morrison, J. F., *Aust. J. Sci.* **27**, 317 (1965).

Myers, O. E. and Prestwood, R. J., in *Radioactivity Applied to Chemistry*, A. C. Wahl and N. A. Bonner, eds., John Wiley and Sons, 1951, Ch. 1.

Plowman, K., *Enzyme Kinetics*, McGraw-Hill, 1972, Ch. 4.

Wedler, F. C. and Boyer, P. D., *J. Theoret. Biol.* **38**, 539 (1973).

Specific Examples

Ainslie, G. R., Jr. and Cleland, W. W., *J. Biol. Chem.* **247**, 946 (1972) (Liver alcohol dehydrogenase).

Balinsky, D., Dennis, A. W. and Cleland, W. W., *Biochemistry* **10**, 1947 (1971) (Shikimate dehydrogenase).

Cuppoletti, J. and Segel, I. H., *J. Membrane Biol.* **17**, 239 (1974) (Sulfate transport system).

Davis, J. S., Balinsky, J. B., Harington, J. S. and Shepherd, J. B., *Biochem. J.* **133**, 667 (1973) (γ-Glutamylcysteine synthetase).

Engers, H. D., Bridger, W. A. and Madsen, N. B., *J. Biol. Chem.* **244**, 5936 (1969) (Glycogen phosphorylase).

Farrar, Y. J. K. and Plowman, K. M., *J. Biol. Chem.* **246**, 3783 (1971) (Citrate cleavage enzyme).

Fromm, H. J., Silverstein, E. and Boyer, P. D., *J. Biol. Chem.* **239**, 3645 (1964) (Yeast hexokinase).

Gold, A. M., Johnson, R. M. and Tseng, J. K., *J. Biol. Chem.* **245**, 2564 (1970) (Glycogen phosphorylase).

Gulbinsky, J. S. and Cleland, W. W., *Biochemistry* **7**, 566 (1968) (Galactokinase).

Hansen, J. N., Dinovo, E. L. and Boyer, P. D., *J. Biol. Chem.* **244**, 6270 (1969) (Fumarase).

Moffet, F. J. and Bridger, W. A., *Can. J. Biochem.* **51**, 44 (1973) (Succinyl CoA synthetase).

Papas, T. S. and Peterkofsky, A., *Biochemistry* **11**, 4602 (1972) (Arginyl *t*-RNA synthetase).

Reed, J. K., *J. Biol. Chem.* **248**, 4834 (1973) (Lipoamide dehydrogenase).

Schachter, H., "The Use of the Steady State Assumption to Derive Kinetic Formulations for the Transport of a Solute Across a Membrane," in *Metabolic Pathways*, Vol. 6, L. E. Hokin, ed., Academic Press, 1972, (Iso Uni Uni—membrane transport).

Silverstein, E. and Sulebele, G., *Biochemistry* **8**, 2543 (1969) (Malate dehydrogenase).

Silverstein, E. and Sulebele, G., *Biochemistry* **12**, 2164 (1973) (Glutamate dehydrogenase).

Smith, E. and Morrison, J. F., *J. Biol. Chem.* **244**, 4224 (1969) (Arginine kinase).

Switzer, R. L., *J. Biol. Chem.* **245**, 483 (1970) (PRPP synthetase).

Uhr, M. L., Thompson, V. W., and Cleland, W. W., *J. Biol. Chem.*, **249** 2920 (1974) (Pig heart isocitrate dehydrogenase).

Wedler, F. C. and Boyer, P. D., *J. Biol. Chem.* **247**, 984 (1972) (Glutamine synthetase).

Young, O. A. and Anderson, J. W., *Biochem. J.* **138**, 435 (1974) (Fatty acyl CoA synthetase. Ordered first partial reaction).

CHAPTER ELEVEN

EFFECTS OF pH AND TEMPERATURE

A. EFFECT OF pH

It is not surprising that pH influences the velocity of an enzyme-catalyzed reaction. The active sites on enzymes are frequently composed of ionizable groups which must be in the proper ionic form to maintain the conformation of the active site, bind the substrates, or catalyze the reaction. Furthermore, one or more of the substrates themselves may contain ionizable groups, and only one ionic form of that substrate may bind to the enzyme or undergo catalysis. The pK values of the prototropic groups of the active site can often be determined by measuring the pH-dependence of $slope_{1/S}$ and $V_{\max_{app}}$. Once the pK values are known, we can make an educated guess as to the identities of the groups involved. Table XI-1 lists some common prototropic groups that may be involved at the active site. A range of pK values is shown for each group because the exact pK is strongly influenced by its environment. For example, a γ-carboxyl group of p$K_a = 4$ in the free amino acid may become a much weaker acid in a highly nonpolar environment.

Effect of pH on Enzyme Stability

The pH also affects the stability of an enzyme and this must be taken into account in any study of the effect of pH on substrate binding and catalysis. Figure XI-1 shows an experimental v versus pH curve for an enzyme (curve A).

Table XI-1 Prototropic Groups which may be Present at the Active Site of an Enzyme

Group	Ionization Reaction	pK_a	ΔH_{ion} (kcal/mole)
α-Carboxyl (at end of polypeptide chain)		3.0–3.2	
	$—COOH \rightleftharpoons —COO^- + H^+$		±1.5
β-or γ-Carboxyl (of aspartic or glutamic acid)		3.0–5.0	
Imidazolium (of histidine)	imidazolium ring (HN … $\overset{+}{N}H$) $\rightleftharpoons$ imidazole ring (HN … N) $+ H^+$	5.5–7.0	6.9–7.5
α-Amino (at end of polypeptide chain)		7.5–8.5	
	$—\overset{+}{N}H_3 \rightleftharpoons —NH_2 + H^+$		10–13
ε-Amino (of lysine)		9.5–10.6	
Sulfhydryl (of cysteine)	$SH \rightleftharpoons —S^- + H^+$	8.0–8.5	6.5–7.0
Phenolic OH (of tyrosine)	$—C_6H_4—OH \rightleftharpoons —C_6H_4—O^- + H^+$	9.8–10.5	6
Guanidinium (of arginine)	$—N(H)—C(=\overset{+}{N}H_2)—NH_2 \rightleftharpoons —N(H)—C(=NH)—NH_2 + H^+$	11.6–12.6	12–13

The pH "optimum" is at 6.8. Curve *A* gives no indication why the velocity declines above and below pH 6.8. The decline could result from the formation of an improper ionic form of the substrate or enzyme (or both), or from inactivation of the enzyme, or from a combination of these effects. Curve *B* shows the effect of pH on enzyme stability. We see that preincubation of the enzyme at pH 5 or pH 8 has no effect on the activity measured at

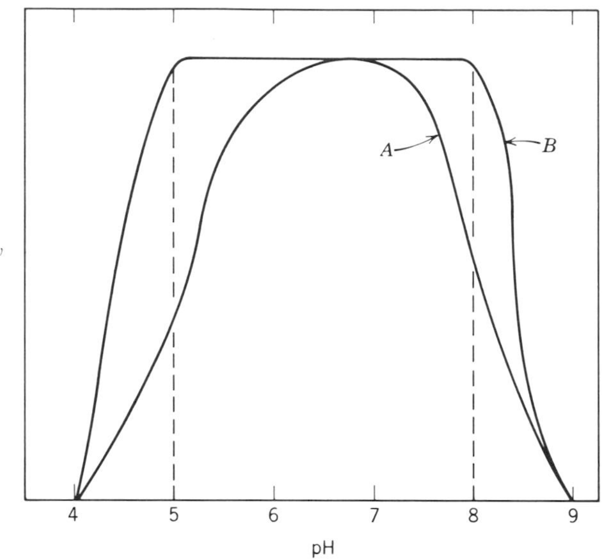

Fig. XI-1. Effect of pH on activity and stability of an enzyme. Curve *A*: Observed activity versus pH plot. Curve *B*: Activity at pH 6.8 after preincubating the enzyme at the indicated pH values. The decline in activity between pH 6.8 and 5.0 and between pH 6.8 and 8.0 can be ascribed to the effect of pH on ionizable groups of the active site or substrate. The decline in activity above pH 8.0 and below pH 5.0 can be ascribed in part to irreversible denaturation of the enzyme.

pH 6.8. Thus the decline in activity between pH 6.8 and 8 and between pH 6.8 and 5 must result from the formation of an improper ionic form of the enzyme and/or substrate. When the enzyme is preincubated at pH > 8 or pH < 5, full activity is not regained at pH 6.8. Thus part of the decline in activity above pH 8 and below pH 5 results from irreversible enzyme inactivation. A pH stability study, such as that shown in curve *B*, is an essential part of any enzyme characterization. Unfortunately, it is frequently omitted and only the curve *A* data presented. The stability curve *B* can be obtained by preincubating the enzyme at the indicated pH for a time at least as long as the usual assay time. Enzyme activity is then measured at the optimum pH. A more complete study is shown in Figure XI-2*a*. The slope of the inactivation curve represents the first-order rate constant for enzyme inactivation (Figure XI-2*b*). The pH stability of an enzyme depends on many factors including (*a*) temperature, (*b*) ionic strength, (*c*) chemical nature of the buffer, (*d*) concentration of various preservatives (e.g., glycerol, sulfhydryl compounds), (*e*) concentration of contaminating metal ions, (*f*) concentration of substrates or cofactors of the enzyme, and (*g*) enzyme

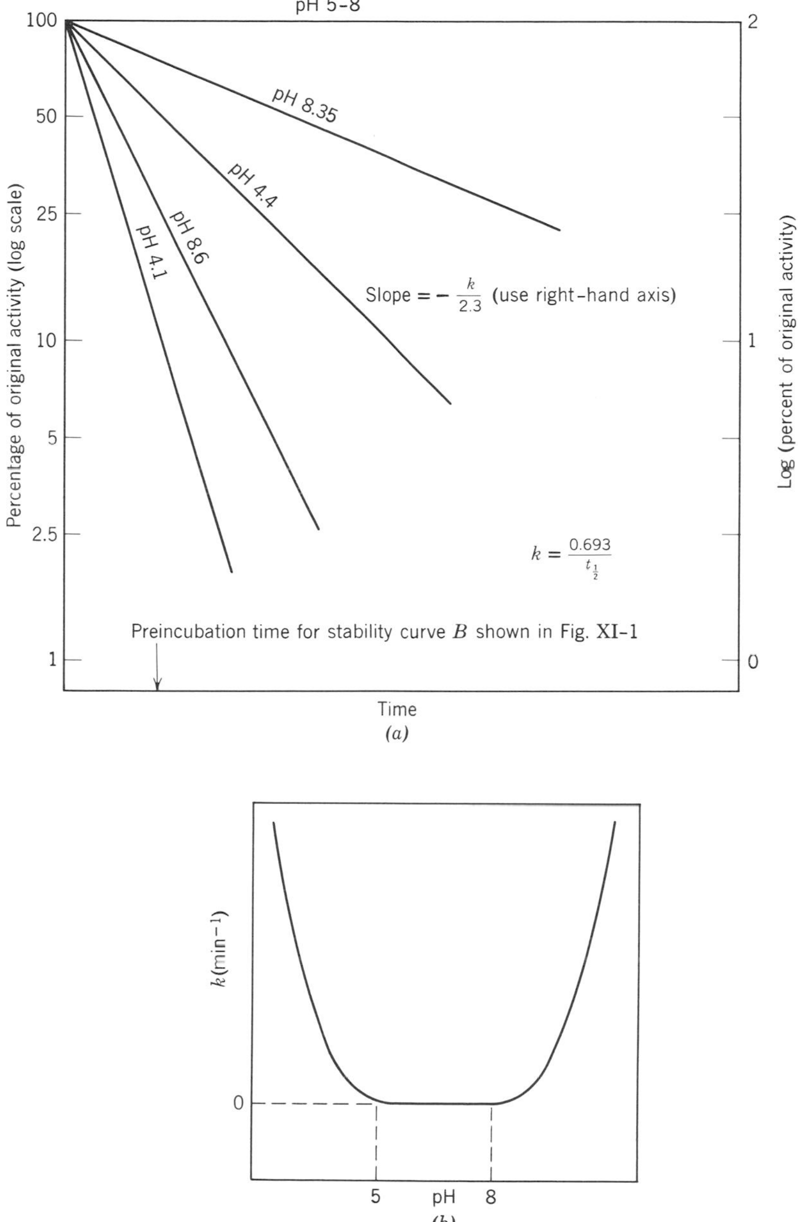

Fig. XI-2. Enzyme inactivation. (*a*) Log percentage of original activity versus preincubation time at different pH values. Activity is measured at pH 6.8. (*b*) First-order rate constant for enzyme inactivation versus pH.

concentration. In many cases a substrate may induce a conformational change in the enzyme to a form that is more resistant or less resistant to pH or temperature denaturation. The concentration of the enzyme itself may be a factor. At low concentrations, the enzyme may dissociate into smaller oligomers or monomers, which may be less stable than the original oligomer. It should also be kept in mind that an enzyme may be more stable over a long period of time at a pH significantly different from the "optimum" used in the assay.

In the following discussion of the effects of pH on enzyme activity it is assumed that preliminary studies have established that the enzyme is stable over the pH range studied.

System A1. All Forms of "E" Bind S; Only E^nS Yields Product

The active site of an enzyme may contain several ionizable groups. Some may be involved in substrate binding and some in catalysis. A simple rapid equilibrium diprotic model is shown below. The various enzyme species are indicated in a general way. Thus E^{n+1}, E^n, and E^{n-1} might represent EH_2, EH^-, and E^{2-}, respectively, or alternately, EH_2^+, EH^0, E^-, or EH_2^{2+}, EH^+, and E^0, and so on.

$$
\begin{array}{ccccccc}
E^{n+1} & + & S & \overset{\alpha K_S}{\rightleftharpoons} & E^{n+1}S & & E^{n+1} \\
K_{e_1}\updownarrow & & & & \updownarrow K_{es_1}=\alpha K_{e_1} & & \updownarrow K_{e_1} \\
H^+ & & & & H^+ & & H^+ \\
+ & & & & + & & + \\
E^n & + & S & \overset{K_S}{\rightleftharpoons} & E^nS & \overset{k_p}{\longrightarrow} & E^n + P \\
K_{e_2}\updownarrow & & & & \updownarrow K_{es_2}=\frac{K_{e_2}}{\beta} & & \updownarrow K_{e_2} \\
H^+ & & & & H^+ & & H^+ \\
+ & & & & + & & + \\
E^{n-1} & + & S & \overset{\beta K_S}{\rightleftharpoons} & E^{n-1}S & & E^{n-1}
\end{array}
$$

It is assumed that all three forms of "E" are able to bind substrate but only E^nS is catalytically active. Only one ionic form of the substrate is shown. Thus it is implied that either (*a*) S is an uncharged substance, or (*b*) S has essentially the same charge at all pH values or at least over the pH range studied, or (*c*) all ionic forms of S are equivalent with respect to their binding to a given form of "E" or (*d*) the total concentration of S is adjusted at each pH so that the concentration of the active ionic form remains constant. For most enzymes the vertical reactions involving the addition or dissociation of H^+ will be very fast compared to the catalytic step. Consequently, we can assume that the H^+ steps are at equilibrium. The horizontal reactions involving the addition and dissociation of S may or may not be rapid compared to the catalytic step. But to obtain a reasonable velocity equation without $[H^+]^2$ terms, we have to assume that the catalytic step is relatively slow and rate-limiting. The velocity is obtained from:

$$v = k_p[E^nS]$$

$$\frac{v}{[E]_t} = \frac{k_p[E^nS]}{[E^n]+[E^{n+1}]+[E^{n-1}]+[E^nS]+[E^{n+1}S]+[E^{n-1}S]} \qquad \text{(XI-1)}$$

The final equation can be written easily using the usual rules for rapid equilibrium systems. The concentration of each enzyme species is given by the product of the concentrations of all ligands adding between the reference state and the given species, divided by the product of all dissociation constants between the reference state and the given species. If a step is a dissociation instead of an addition, the fraction for that step is inverted, giving dissociation constant over ligand concentration. Taking E^n as the reference state:

$$\boxed{\frac{v}{V_{max}} = \frac{\dfrac{[S]}{K_S}}{1+\dfrac{[H^+]}{K_{e_1}}+\dfrac{K_{e_2}}{[H^+]}+\dfrac{[S]}{K_S}+\dfrac{[H^+][S]}{K_{e_1}\alpha K_S}+\dfrac{K_{e_2}[S]}{[H^+]\beta K_S}}} \qquad \text{(XI-2)}$$

or

$$\boxed{\frac{v}{V_{max}} = \frac{\dfrac{[S]}{K_S}}{1+\dfrac{[H^+]}{K_{e_1}}+\dfrac{K_{e_2}}{[H^+]}+\dfrac{[S]}{K_S}+\dfrac{[S][H^+]}{K_S K_{es_1}}+\dfrac{[S]K_{es_2}}{K_S[H^+]}}} \qquad \text{(XI-3)}$$

Or, multiplying numerator and denominator by K_S and factoring:

$$\frac{v}{V_{max}} = \frac{[S]}{K_S\left(1+\frac{[H^+]}{K_{e_1}}+\frac{K_{e_2}}{[H^+]}\right)+[S]\left(1+\frac{[H^+]}{K_{es_1}}+\frac{K_{es_2}}{[H^+]}\right)} \tag{XI-4}$$

where $V_{max} = k_p[E]_t$ = a theoretical maximal velocity that would be observed if all the enzyme could exist as E^nS. The parenthetical factor of K_S represents f_E, the Michaelis pH function for "E." The parenthetical factor of [S] represents f_{ES}, the Michaelis pH function for "E"S. The function f_E is just the denominator of the equation expressing the effect of $[H^+]$ on the distribution of $["E"]_t$ between E^n, E^{n+1}, and E^{n-1}. For example:

$$\frac{[E^{n+1}]}{["E"]_t} = \frac{[E^{n+1}]}{[E^n]+[E^{n+1}]+[E^{n-1}]} = \frac{\frac{[H^+]}{K_{e_1}}[E^n]}{[E^n]+\frac{[H^+]}{K_{e_1}}[E^n]+\frac{K_{e_2}}{[H^+]}[E^n]}$$

$$= \frac{\frac{[H^+]}{K_{e_1}}}{1+\frac{[H^+]}{K_{e_1}}+\frac{K_{e_2}}{[H^+]}} \tag{XI-5}$$

where E^n is taken as the reference state. Similarly f_{ES} is the denominator of the distribution equation for $["E"S]_t$ where E^nS is taken as the reference state. For example:

$$\frac{[E^{n-1}S]}{["E"S]_t} = \frac{[E^{n-1}S]}{[E^nS]+[E^{n+1}S]+[E^{n-1}S]} = \frac{\frac{K_{es_2}}{[H^+]}[E^nS]}{[E^nS]+\frac{[H^+]}{K_{es_1}}[E^nS]+\frac{K_{es_2}}{[H^+]}[E^nS]}$$

$$= \frac{\frac{K_{es_2}}{[H^+]}}{1+\frac{[H^+]}{K_{es_1}}+\frac{K_{es_2}}{[H^+]}} \tag{XI-6}$$

If E^{n+1} is chosen as the reference state, the distribution equation for E^{n+1} would be:

$$\frac{[E^{n+1}]}{[\text{"E"}]_t} = \frac{1}{1 + \frac{K_{e_1}}{[H^+]} + \frac{K_{e_1}K_{e_2}}{[H^+]^2}} \qquad \text{(XI-7)}$$

If E^{n-1} is chosen as the reference state, the distribution equations would be:

$$\frac{[E^{n+1}]}{[\text{"E"}]_t} = \frac{\frac{[H^+]^2}{K_{e_1}K_{e_2}}}{1 + \frac{[H^+]}{K_{e_2}} + \frac{[H^+]^2}{K_{e_1}K_{e_2}}}, \qquad \frac{[E^n]}{[\text{"E"}]_t} = \frac{\frac{[H^+]}{K_{e_2}}}{1 + \frac{[H^+]}{K_{e_2}} + \frac{[H^+]^2}{K_{e_1}K_{e_2}}}$$

$$\frac{[E^{n-1}]}{[\text{"E"}]_t} = \frac{1}{1 + \frac{[H^+]}{K_{e_2}} + \frac{[H^+]^2}{K_{e_1}K_{e_2}}}$$

Using E^{n-1} as the reference state, the velocity equation obtained after substitution into equation XI-1 is:

$$\frac{v}{V_{\max}} = \frac{\frac{[H^+][S]}{K_{e_2}K_S}}{\underset{\uparrow \atop E^{n-1}}{1} + \underset{\uparrow \atop E^{n}}{\frac{[H^+]}{K_{e_2}}} + \underset{\uparrow \atop E^{n+1}}{\frac{[H^+]^2}{K_{e_1}K_{e_2}}} + \underset{\uparrow \atop E^{n-1}S}{\frac{[S]}{\beta K_S}} + \underset{\uparrow \atop E^{n}S}{\frac{[H^+][S]}{K_{e_2}K_S}} + \underset{\uparrow \atop E^{n+1}S}{\frac{[H^+]^2[S]}{K_{e_2}K_S K_{es_1}}}} \qquad \text{(XI-8)}$$

The $[S]/\beta K_S$ term is equal to $[H^+][S]K_{es_2}/K_{e_2}K_S[H^+]$ because the overall K_{eq} between E^{n-1} and $E^{n-1}S$ must be the same regardless of the path. Making this substitution and then multiplying numerator and denominator by $K_{e_2}/[H^+]$ we obtain equation XI-3. Thus the same final equation is obtained regardless of the choice of reference state.

At any fixed $[H^+]$, the observed kinetic constants are:

$$V_{max_{app}} = \frac{V_{max}}{\left(1 + \frac{[H^+]}{K_{es_1}} + \frac{K_{es_2}}{[H^+]}\right)}, \qquad K_{S_{app}} = \frac{K_S\left(1 + \frac{[H^+]}{K_{e_1}} + \frac{K_{e_2}}{[H^+]}\right)}{\left(1 + \frac{[H^+]}{K_{es_1}} + \frac{K_{es_2}}{[H^+]}\right)}$$

and

$$slope_{1/S} = \frac{K_S}{V_{max}}\left(1 + \frac{[H^+]}{K_{e_1}} + \frac{K_{e_2}}{[H^+]}\right)$$

If the binding of S has no effect on the dissociation constants of the active site groups, K_{e_1} will equal K_{es_1}, and K_{e_2} will equal K_{es_2}. In this case varying pH will have no effect on K_S. If the corresponding dissociation constants of "E" and "E"S are not identical, $K_{S_{app}}$ will vary from:

$$\frac{K_S K_{es_1}}{K_{e_1}} = \alpha K_S \quad \text{at low pH} \qquad \text{to} \qquad \frac{K_S K_{e_2}}{K_{es_2}} = \beta K_S \quad \text{at high pH}$$

The H^+ ion acts as a substrate that displays substrate inhibition. The reciprocal plots of $1/v$ versus $1/[S]$ at different fixed H^+ concentrations are linear. At very low $[H^+]$, increasing $[H^+]$ increases the reaction velocity by promoting the formation of E^n and E^nS. In this region the $[H^+]/K_{e_1}$ and $[H^+]/K_{es_1}$ terms can be ignored. The family of reciprocal plots at different fixed $[H^+]$ intersect to the left of the $1/v$-axis above, on, or below the $1/[S]$-axis. The intersection coordinates are:

$$\frac{1}{[S]} = -\frac{K_{es_2}}{K_S K_{e_2}} = -\frac{1}{\beta K_S} \qquad \text{and} \qquad \frac{1}{v} = \frac{1}{V_{max}}\left(1 - \frac{K_{es_2}}{K_{e_2}}\right) = \frac{1}{V_{max}}\left(1 - \frac{1}{\beta}\right)$$

As $[H^+]$ increases, the velocity will pass through a maximum (at the optimum pH) and then decrease as H^+ becomes inhibitory by promoting the formation of the inactive (or less active) E^{n-1} and inactive $E^{n-1}S$ forms. At high $[H^+]$ the $K_{e_2}/[H^+]$ and $K_{es_2}/[H^+]$ terms can be ignored. In this region the family of reciprocal plots intersect at:

$$\frac{1}{[S]} = -\frac{K_{e_1}}{K_S K_{es_1}} = -\frac{1}{\alpha K_S} \qquad \text{and} \qquad \frac{1}{v} = \frac{1}{V_{max}}\left(1 - \frac{K_{e_1}}{K_{es_1}}\right) = \frac{1}{V_{max}}\left(1 - \frac{1}{\alpha}\right)$$

If $K_{e_1}/K_{es_1} = K_{es_2}/K_{e_2}$ the plots will intersect at a common point at all pH values (on the 1/[S]-axis if $K_{e_1} = K_{es_1}$ and $K_{e_2} = K_{es_2}$).

If the H^+ dissociation constants of the enzyme are separated by a factor of 100 or more the values of K_{e_1}, K_{e_2}, and K_S/V_{max} can be obtained from slope replots. Similarly, K_{es_1}, K_{es_2}, and $1/V_{max}$ can be obtained from $1/v$-axis intercept (i.e., $1/V_{max_{app}}$) replots. If the $[H^+]$ range plotted is chosen properly the replots will be essentially linear. The equations for the replots are:

$$\boxed{slope_{1/S} = \frac{K_S}{V_{max}K_{e_1}}[H^+] + \frac{K_S K_{e_2}}{V_{max}}\frac{1}{[H^+]} + \frac{K_S}{V_{max}}} \tag{XI-9}$$

$$\boxed{\frac{1}{V_{max_{app}}} = \frac{1}{V_{max}K_{es_1}}[H^+] + \frac{K_{es_2}}{V_{max}}\frac{1}{[H^+]} + \frac{1}{V_{max}}} \tag{XI-10}$$

Figure XI-3*a* and *b* show the replots at relatively high $[H^+]$ (i.e., in the region of 0.1 to 10 times K_{e_1} and K_{es_1}). In this region, the $1/[H^+]$ terms of the equations for the slope and $1/V_{max_{app}}$ replots will be very small and can be ignored. Figure XI-3*c* and *d* show the replots at relatively low $[H^+]$ (i.e., in the region of 0.1 to 10 times K_{e_2} and K_{es_2}). In this region, the $[H^+]$ terms of equations XI-9 and XI-10 will be very small and can be ignored. If the H^+ dissociation constants are relatively close, or if the $[H^+]$ range plotted is in the region of the optimum $[H^+]$, then we can not omit a term from the slope and intercept equations. The replots will not be linear and will have an obvious minimum. Figure XI-3*e* shows such a slope replot. The minimum occurs at $\sqrt{K_{e_1}K_{e_2}}$ (or the reciprocal if $1/[H^+]$ is plotted). The minimum of the $1/V_{max_{app}}$ versus $[H^+]$ replot occurs at $\sqrt{K_{es_1}K_{es_2}}$ (or the reciprocal if $1/[H^+]$ is plotted).

Plots of $V_{max_{app}}$ Versus pH and 1/*slope* versus pH

When the pK values are separated by at least 3.5 pH units (i.e., $K_1 \geqslant 3162K_2$) plots of $V_{max_{app}}$ versus pH and $V_{max_{app}}/K_{m_{app}}$ (i.e., $1/slope_{1/S}$) versus pH will be bell-shaped with a maximum very close to V_{max} or V_{max}/K_m. The pK values can be read off directly as the pH values at $\frac{1}{2}V_{max_{app}}$ or $\frac{1}{2}V_{max_{app}}/K_{m_{app}}$. Figure XI-4*a* shows a $V_{max_{app}}$ plot for p$K_{es_1} = 4.0$

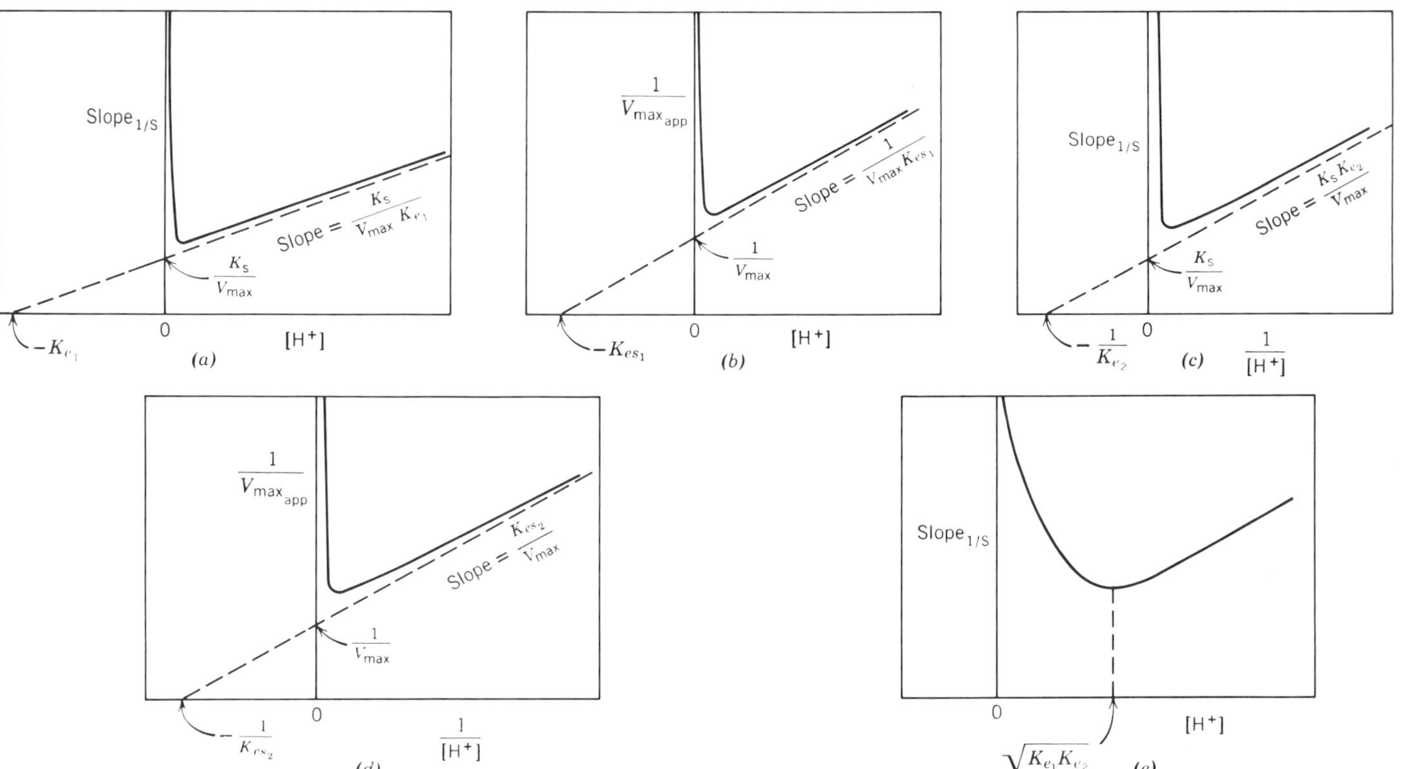

Fig. XI-3. Diprotic model; E^n, E^{n-1}, and E^{n+1} all bind S but only E^nS is catalytically active. (*a*) and (*b*) *Slope*$_{1/S}$ and $1/v$-axis intercept replots in an [H$^+$] range near K_{e_1} and K_{es_1}. (*c*) and (*d*) *Slope*$_{1/S}$ and $1/v$-axis intercept replots in an [H$^+$] range near K_{e_2} and K_{es_2}. (*e*) *Slope*$_{1/S}$ replot in an [H$^+$] range intermediate between K_{e_1} and K_{e_2}. It is assumed that the pK values of "E" and "ES" are well separated and that S is uncharged, or has the same charge throughout the pH range tested, or that all ionic forms of S that exist in the pH range tested are equal with respect to binding and catalytic activity.

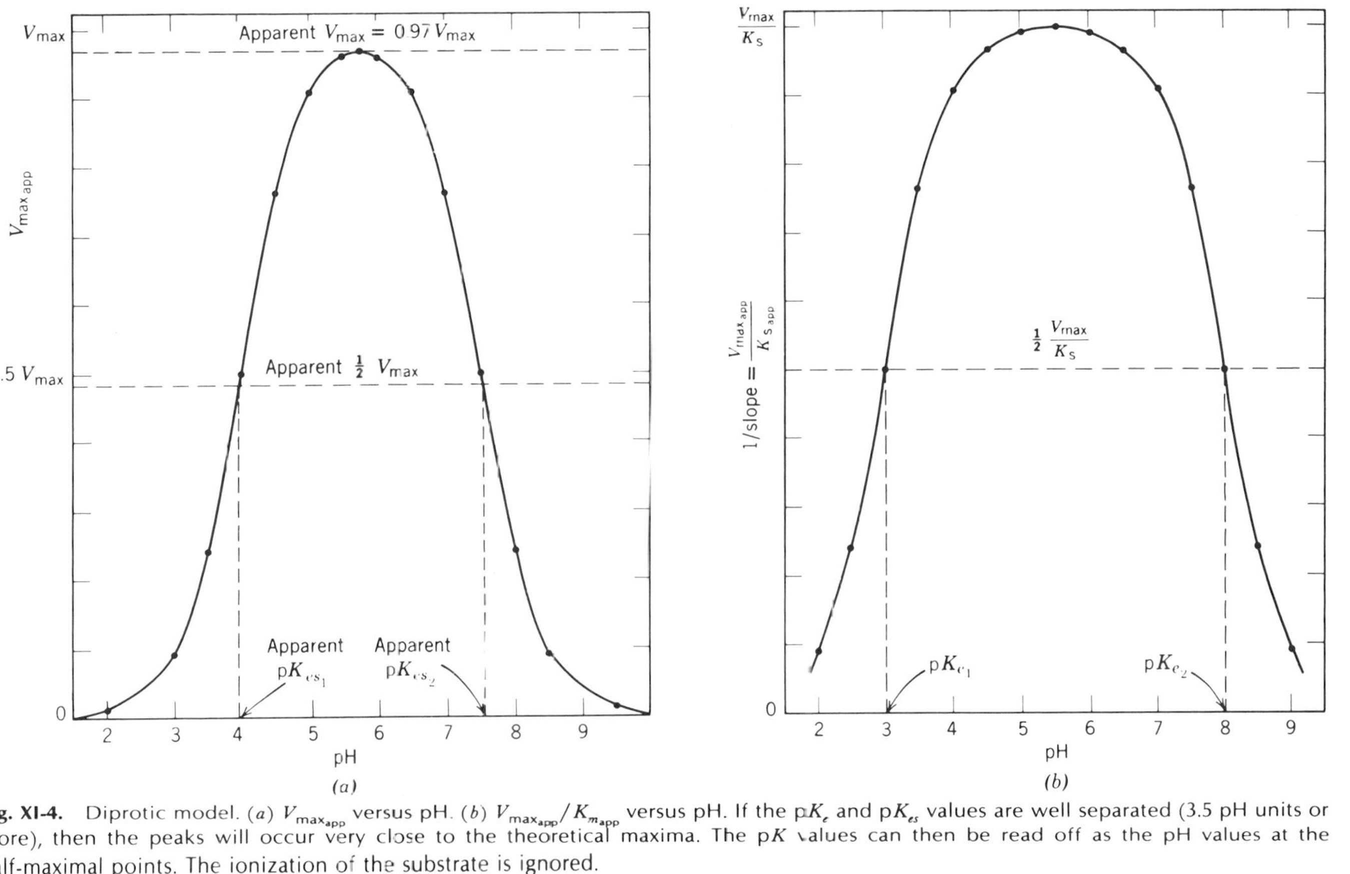

Fig. XI-4. Diprotic model. (*a*) $V_{max_{app}}$ versus pH. (*b*) $V_{max_{app}}/K_{m_{app}}$ versus pH. If the pK_e and pK_{es} values are well separated (3.5 pH units or more), then the peaks will occur very close to the theoretical maxima. The pK values can then be read off as the pH values at the half-maximal points. The ionization of the substrate is ignored.

and $pK_{es_2}=7.5$. Figure XI-4*b* shows a $V_{max_{app}}/K_{S_{app}}$ plot for $pK_{e_1}=3.0$ and $pK_{e_2}=8.0$. If the two p*K* values are separated by less than 3.5 pH units, the maximum of the plot will occur at a value significantly lower than the theoretical maximum and consequently the pH values at the half-maximum points will not correspond to the p*K* values. The pK_{es} values chosen for Figure XI-4*a* are exactly 3.5 pH units apart. The $V_{max_{app}}$ versus pH plot peaks at 97% of the theoretical V_{max}. The pK_{es} values read off at the half-maximum points are about 3.95 and 7.55 which are well within experimental error. The pK_e values chosen for Figure XI-4*b* are 5 pH units apart. Here the plot passes through a maximum at better than 99.9% of the theoretical V_{max}. The pH values at half-maximum $V_{max_{app}}$ give pK_{e_1} and pK_{e_2} exactly.

Dixon-Webb Log Plots

Another plot suitable for systems in which the p*K* values are separated by 3.5 pH units or more is the Dixon-Webb (1964) plot of log $V_{max_{app}}$ versus pH and log $V_{max_{app}}/K_{S_{app}}$ versus pH. Log $V_{max_{app}}$ is given by:

$$\log V_{max_{app}} = \log V_{max} - \log\left(1 + \frac{[H^+]}{K_{es_1}} + \frac{K_{es_2}}{[H^+]}\right) \qquad \text{(XI-11)}$$

At low pH values, the $K_{es_2}/[H^+]$ term of f_{ES} may be neglected. Also, the ratio $[H^+]/K_{es_1}$ will be $\gg 1$, so the 1 can be ignored. The resulting equation is:

$$\log V_{max_{app}} = \log V_{max} - \log\frac{[H^+]}{K_{es_1}} = \log V_{max} - \left(\log[H^+] - \log K_{es_1}\right)$$

$$\log V_{max_{app}} = \log V_{max} + pH - pK_{es_1} \qquad \text{(XI-12)}$$

Equation XI-12 can be written:

$$\boxed{\log\frac{V_{max_{app}}}{V_{max}} = pH - pK_{es_1}} \qquad \text{(XI-13)}$$

Thus at low pH values, the plot starts off as a straight line with a slope of +1. At a point on this line where $V_{max_{app}} = V_{max}$, $\log V_{max_{app}}/V_{max}=0$, and $pH = pK_{es_1}$. To obtain pK_{es_1}, we need only draw a horizontal line at V_{max} and a line of *slope* = 1 that is tangent to the plot at low pH. The horizontal

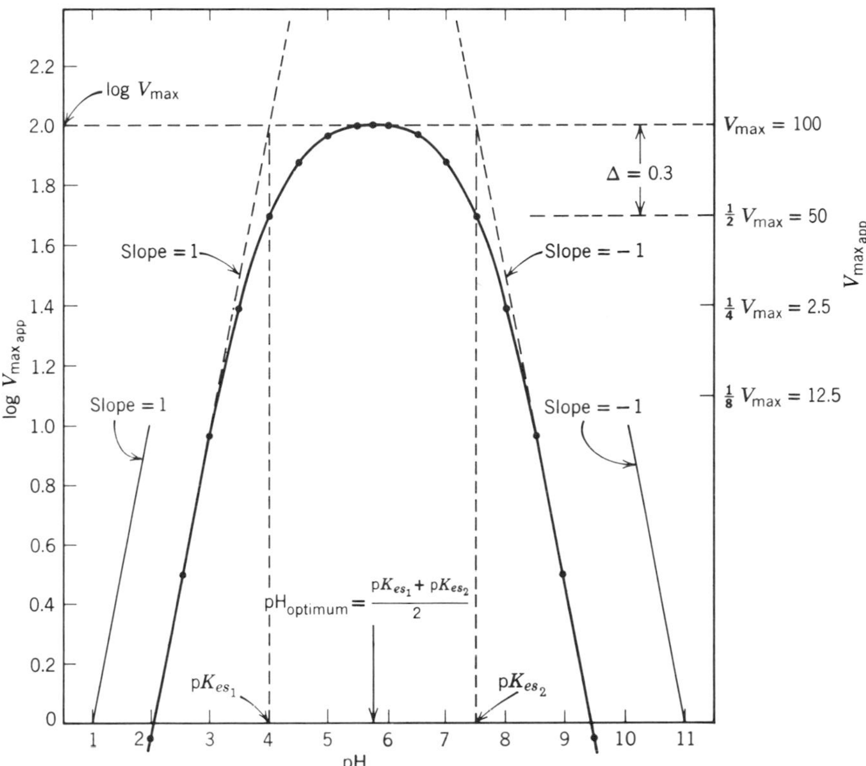

Fig. XI-5. Dixon-Webb plot: Log $V_{max_{app}}$ versus pH. Lines of slope=1 and slope=−1 drawn tangent to the curve at very low log $V_{max_{app}}$ values intersect the horizontal log V_{max} line at pK_{es_1} and pK_{es_2}. A similar plot of log $V_{max_{app}}/K_{m_{app}}$ versus pH will yield pK_{e_1} and pK_{e_2}.

coordinate of the intersection point gives pK_{es_1}. The plot is shown in Figure XI-5 for $V_{max} = 100$, $pK_{es_1} = 4.0$, and $pK_{es_2} = 7.5$. At high pH, the logarithmic form of equation XI-13 becomes:

$$\log V_{max_{app}} = \log V_{max} - \log \frac{K_{es_2}}{[H^+]} = \log V_{max} - \left(\log K_{es_2} - \log [H^+]\right)$$

$$= \log V_{max} + pK_{es_2} - pH \qquad \text{(XI-14)}$$

A line having a slope of −1 drawn tangent to the plot at high pH intersects the horizontal line at V_{max} at a pH equal to pK_{es_2}. A similar plot of $\log V_{max_{app}}/K_{S_{app}}$ versus pH can be used to obtain pK_{e_1} and pK_{e_2}. If the

$V_{\max_{app}}$ or $V_{\max_{app}}/K_{m_{app}}$ values are plotted directly on semilog paper (against pH on the horizontal linear scale), then the lines drawn tangent to the plot are such that they rise one decade for each pH unit.

Correction For Ionization of the Substrate

If the substrate has acid dissociation groups with pK_a values in the pH range studied, then the velocity equation becomes:

$$\frac{v}{V_{\max}} = \frac{\dfrac{[\mathrm{S}]_t}{\left(1+\dfrac{[\mathrm{H}^+]}{K_{a_1}}+\dfrac{K_{a_2}}{[\mathrm{H}^+]}\right)}}{K_{\mathrm{S}}\left(1+\dfrac{[\mathrm{H}^+]}{K_{e_1}}+\dfrac{K_{e_2}}{[\mathrm{H}^+]}\right)+[\mathrm{S}]_t\dfrac{\left(1+\dfrac{[\mathrm{H}^+]}{K_{es_1}}+\dfrac{K_{es_2}}{[\mathrm{H}^+]}\right)}{\left(1+\dfrac{[\mathrm{H}^+]}{K_{a_1}}+\dfrac{K_{a_2}}{[\mathrm{H}^+]}\right)}}$$

$$= \frac{\dfrac{[\mathrm{S}]_t}{f_{\mathrm{S}}}}{K_{\mathrm{S}} f_{\mathrm{E}} + \dfrac{[\mathrm{S}]_t f_{\mathrm{ES}}}{f_{\mathrm{S}}}} = \frac{[\mathrm{S}]_t}{K_{\mathrm{S}} f_{\mathrm{E}} f_{\mathrm{S}} + [\mathrm{S}]_t f_{\mathrm{ES}}} \qquad \textbf{(XI-15)}$$

where f_{S} is the Michaelis pH function for the substrate. The substrate is assumed to be diprotic with the S^n form as the active species and reference state:

$$\frac{[\mathrm{S}^n]}{[\mathrm{S}]_t} = \frac{[\mathrm{S}^n]}{[\mathrm{S}^n]+[\mathrm{S}^{n+1}]+[\mathrm{S}^{n-1}]} = \frac{[\mathrm{S}^n]}{[\mathrm{S}^n]+\dfrac{[\mathrm{H}^+]}{K_{a_1}}[\mathrm{S}^n]+\dfrac{K_{a_2}}{[\mathrm{H}^+]}[\mathrm{S}^n]}$$

$$= \frac{1}{1+\dfrac{[\mathrm{H}^+]}{K_{a_1}}+\dfrac{K_{a_2}}{[\mathrm{H}^+]}} \qquad \therefore \quad [\mathrm{S}^n] = \frac{[\mathrm{S}]_t}{1+\dfrac{[\mathrm{H}^+]}{K_{a_1}}+\dfrac{K_{a_2}}{[\mathrm{H}^+]}} = \frac{[\mathrm{S}]_t}{f_{\mathrm{S}}}$$

The apparent kinetic constants are:

$$V_{\max_{app}} = \frac{V_{\max}}{f_{ES}}, \qquad K_{S_{app}} = K_m \frac{f_E f_S}{f_{ES}}$$

$$slope = \frac{K_m f_E f_S}{V_{\max}}, \qquad \frac{V_{\max_{app}}}{K_{S_{app}}} = \frac{1}{slope} = \frac{V_{\max}}{K_m f_E f_S}$$

To restrict our studies to the effect of pH on the enzyme, we must either

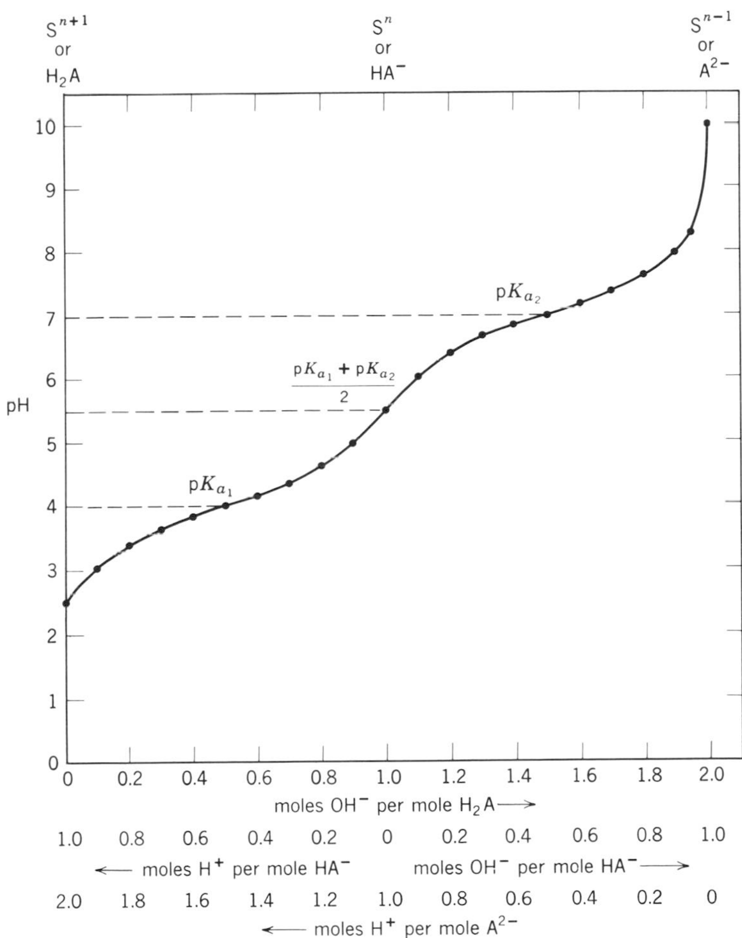

Fig. XI-6. Titration curve of a diprotic substrate, H_2A.

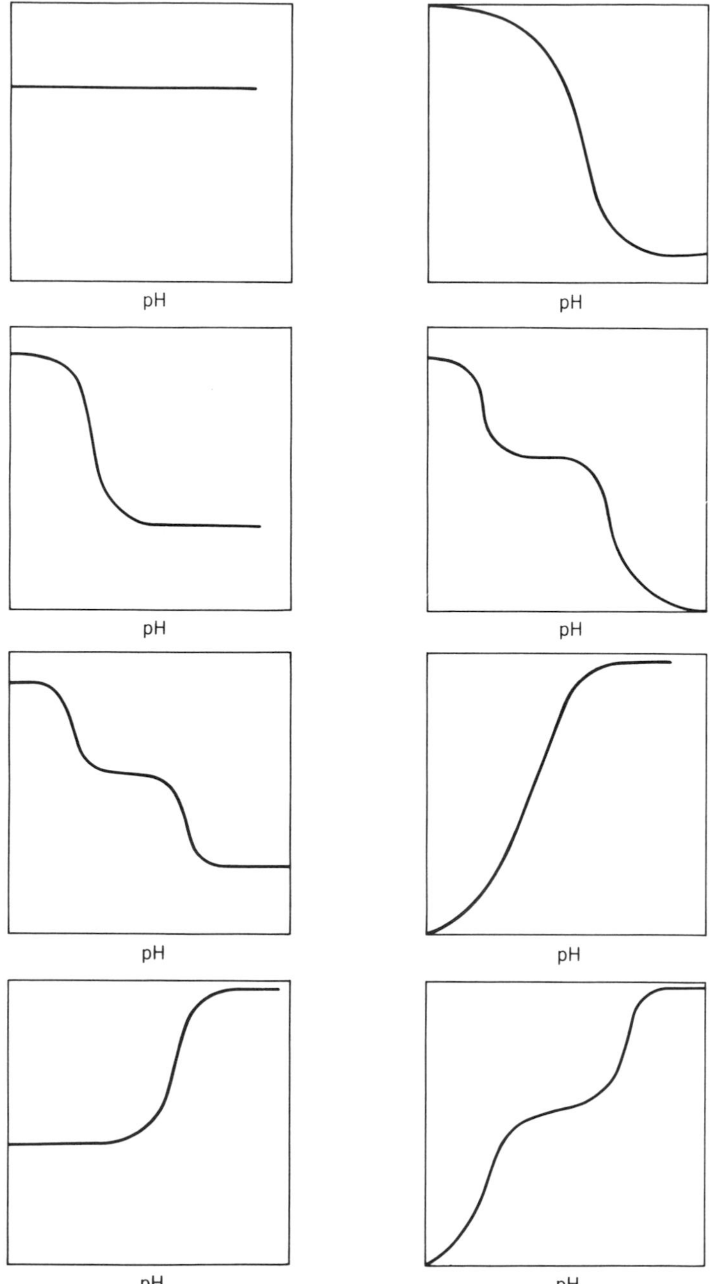

Fig. XI-7. Varieties of $V_{\max_{app}}$ versus pH or $V_{\max_{app}}/K_{m_{app}}$ versus pH plots that may be observed for a diprotic system. A limiting nonzero plateau indicates that more than one ionic form of the enzyme is active.

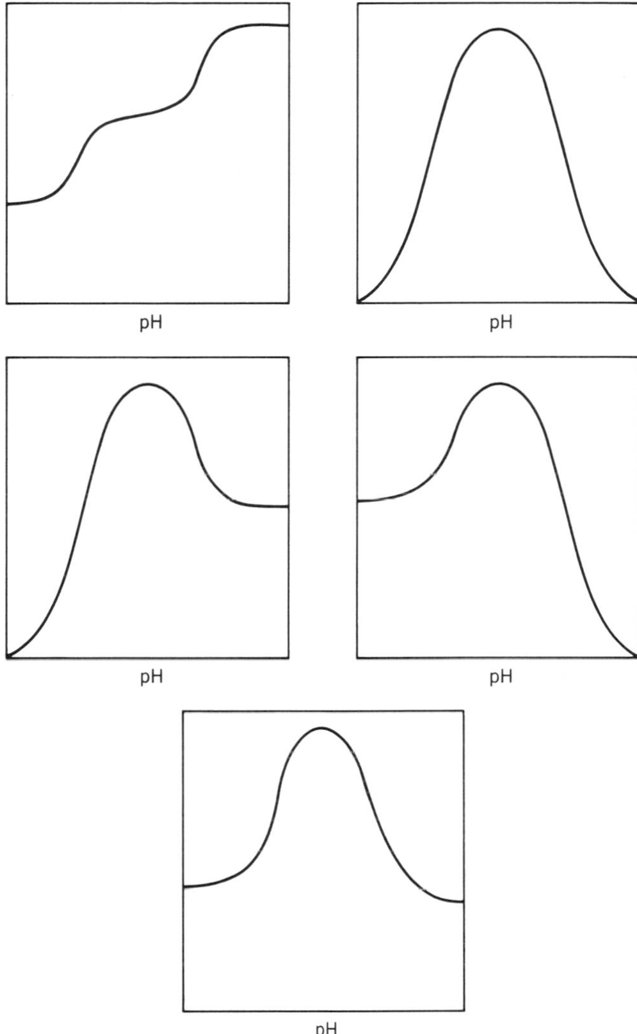

Fig. XI-7. (*Cont.*)

adjust $[S]_t$ at each pH so that $[S^n]$ is constant, or else correct the observed kinetic constants for f_S. To do either, we must know the pK_a values of the substrate. If the values are not known, a sample of the substrate can be titrated and the pK_a's determined as the pH values at the midpoints of the buffering regions. A titration curve for a diprotic substrate with pK_a values of 4.0 and 7.0 is shown in Figure XI-6. If the two pK_a values are only two pH units apart (or less), the titration curve will appear almost linear over a broad pH range from about 0.5 pH unit below pK_{a_1} to about 0.5 pH unit above pK_{a_2}.

Assuming that we can look up or determine the pK_a values of the substrate, we must next decide which ionic form of the substrate is active. Here, an educated guess may help. An enzyme that displays good activity over a given pH range probably accepts the predominant ionic form (or forms) of the substrate in that pH range. Otherwise, a good deal of the catalytic potential of the enzyme would be wasted.

Varieties of pH Responses

For many enzymes, a good deal of important information can be obtained by determining the effect of pH on $V_{\max_{app}}$, slope, and $K_{S_{app}}$. Unfortunately, most research papers simply report a pH optimum determined at a single substrate concentration. Plots of $V_{\max_{app}}$, $1/slope_{1/S}$, or $1/K_{S_{app}}$ versus pH (and log plots of these constants) can have a variety of shapes. Some are shown in Figure XI-7. The combination of responses tells us which ionic form of "E" binds S and which ionic form of "E"S yields product. The pK_e and pK_{es} values give an indication of the amino acid residues that might be involved. In the simple diprotic model described above, $V_{\max_{app}}$ and $1/slope_{1/S}$ versus pH plots display maxima, while the plot of $1/K_{S_{app}}$ versus pH would rise, fall, or remain constant, depending on the relative values of K_{e_1}/K_{es_1} (i.e., α) and K_{es_2}/K_{e_2} (i.e., β). Some other simple systems yielding different pH responses are described below.

System A2. Only E^n Binds S; Only E^nS Yields Product

System A2 illustrated in Figure XI-8 is a limiting case of System A1 in which $\alpha \gg 1$ and $\beta \gg 1$. Thus E^{n+1} and E^{n-1} have essentially no affinity for S. The Y^- group alone cannot bind S and the XH^+ group cannot dissociate when S is bound. (This also follows from the rules of equilibrium: if $\alpha \gg 1$ and $\beta \gg 1$ then K_{es_1}, which equals αK_{e_1}, must be $\gg K_{e_1}$, and K_{es_2}, which equals K_{e_2}/β, must be $\ll K_{e_2}$). To put it another way: if E^nS picks up or gives off an H^+, then S immediately dissociates so that there are no $E^{n+1}S$ and $E^{n-1}S$

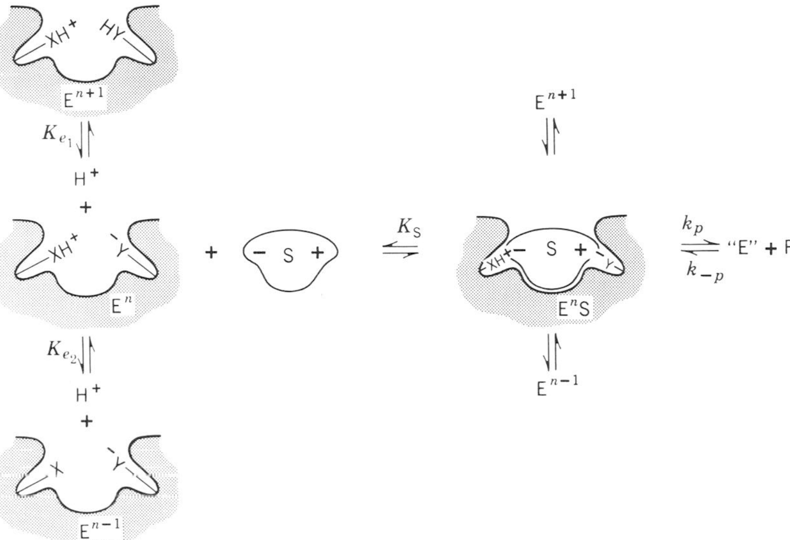

Fig. XI-8. Diprotic model. "E" can exist in three ionic forms but only E^n binds S.

forms. The velocity equation is:

$$\frac{v}{V_{\max}} = \frac{\dfrac{[S]}{K_S}}{1 + \dfrac{[S]}{K_S} + \dfrac{[H^+]}{K_{e_1}} + \dfrac{K_{e_2}}{[H^+]}} \tag{XI-16}$$

or

$$\frac{v}{V_{\max}} = \frac{[S]}{K_S\left(1 + \dfrac{[H^+]}{K_{e_1}} + \dfrac{K_{e_2}}{[H^+]}\right) + [S]} \tag{XI-17}$$

where $V_{\max} = k_p[E]_t$, and E^n is the reference state. Thus $V_{\max_{app}} = V_{\max}$ at all pH values (there is no $1/v$-axis intercept factor). Plots of $V_{\max_{app}}/K_{S_{app}}$ versus pH and $1/K_{S_{app}}$ versus pH (or Dixon-Webb log plots) rise, pass through a maximum at $pH_{optimum} = (pK_{e_1} + pK_{e_2})/2$, and then fall. (Plot $1/K_{S_{app}}$ or $-\log K_{S_{app}}$ to obtain a rising curve. If $K_{S_{app}}$ or $\log K_{S_{app}}$ versus pH is plotted, the curve will fall, pass through a minimum at $pH_{optimum}$, and then rise again.) The pK_{e_1} and pK_{e_2} can be obtained from $slope_{1/S}$ versus $[H^+]$ and $slope_{1/S}$ versus $1/[H^+]$ replots, or linear or log plots of $1/slope_{1/S}$ versus pH as described earlier.

Treating H^+ as a Substrate

Usually, H^+ is treated as the changing fixed ligand in a biligand or triligand system. Under some circumstances it may be of interest to consider H^+ as a varied substrate. For example, if E^{n-1} of System A2 is taken as the reference state then one of the H^+ ions and S can be considered as two substrates that add in an ordered sequence. The H^+ of the K_{e_1} ionization can be considered as an inhibitor competitive with S. The velocity equation is:

$$\frac{v}{V_{max}} = \frac{\dfrac{[H^+][S]}{K_{e_2}K_S}}{1 + \dfrac{[H^+]}{K_{e_2}} + \dfrac{[H^+][S]}{K_{e_2}K_S} + \dfrac{[H^+]^2}{K_{e_2}K_{e_1}}} \qquad \text{(XI-18)}$$

where $V_{max} = k_p[EH^+S]$. When $[H^+]$ is varied at different fixed concentrations of S, the equation becomes:

$$\frac{v}{V_{max}} = \frac{[H^+]}{K_{e_2}\left(\dfrac{K_S}{[S]}\right) + [H^+]\left(1 + \dfrac{K_S}{[S]} + \dfrac{K_S[H^+]}{[S]K_{e_1}}\right)} \qquad \text{(XI-19)}$$

The $K_S[H^+]/[S]K_{e_1}$ term can be neglected at high pH values, that is, in a range of $[H^+]$ very low compared to K_{e_1}. This is the region of $[H^+]$ on the right-hand limb of the "pH optimum" curve where H^+ behaves as a cosubstrate. The system behaves as a simple Rapid Equilibrium Ordered Bireactant system. Plots of $1/v$ versus $1/[H^+]$ at different fixed concentrations of S intersect to the left of the $1/v$-axis at $-1/K_{e_2}$. The limiting slope at saturating [S] is zero. $Slope_{1/H^+}$ versus $1/[S]$ replots go through the origin. Plots of $1/v$ versus $1/[S]$ at different fixed H^+ concentrations intersect on the $1/v$-axis at $1/V_{max}$.

System A3. E^n and E^{n+1} Bind S; Only E^nS Yields Product

In System A3, S binds to two forms of "E" but only one form of "E"S is catalytically active. The binding of S masks the K_{e_2} group, or if E^nS ionizes,

S immediately dissociates.

$$
\begin{array}{ccccccc}
E^{n+1} & + & S & \underset{}{\overset{\alpha K_S}{\rightleftharpoons}} & E^{n+1}S & & E^{n+1} \\
K_{e_1}\downarrow\uparrow & & & & \uparrow\downarrow K_{es_1} = \alpha K_{e_1} & & \uparrow\downarrow K_{e_1} \\
H^+ & & & & H^+ & & H \\
+ & & & & + & & + \\
E^n & + & S & \overset{K_S}{\rightleftharpoons} & E^nS & \overset{k_p}{\longrightarrow} & E^n \;+\; P \\
K_{e_2}\downarrow\uparrow & & & & & & \uparrow\downarrow K_{e_2} \\
H^+ & & & & & & H^+ \\
+ & & & & & & + \\
E^{n-1} & & & & & & E^{n-1}
\end{array}
$$

The velocity equation is:

$$
\boxed{\frac{v}{V_{\max}} = \frac{\dfrac{[S]}{K_S}}{1 + \dfrac{[H^+]}{K_{e_1}} + \dfrac{K_{e_2}}{[H^+]} + \dfrac{[S]}{K_S} + \dfrac{[S][H^+]}{K_S K_{es_1}}}} \qquad \text{(XI-20)}
$$

or

$$
\boxed{\frac{v}{V_{\max}} = \frac{[S]}{K_S\left(1 + \dfrac{[H^+]}{K_{e_1}} + \dfrac{K_{e_2}}{[H^+]}\right) + [S]\left(1 + \dfrac{[H^+]}{K_{es_1}}\right)}} \qquad \text{(XI-21)}
$$

The $1/v$-axis intercept factor (i.e., f_{ES}) has the usual form of an inhibition factor. Thus an intercept replot is perfectly linear and yields $V_{\max}$ and K_{es_1} directly (Fig. XI-9*a*). Linear or log plots of $V_{\max_{app}}$ versus pH rise and approach $V_{\max}$ as a alimit (Fig. XI-9*b*,*c*). Linear or log plots of $1/slope_{1/S}$ $= V_{\max_{app}}/K_{S_{app}}$ versus pH pass through a maximum (Fig. XI-9*d*,*e*). The

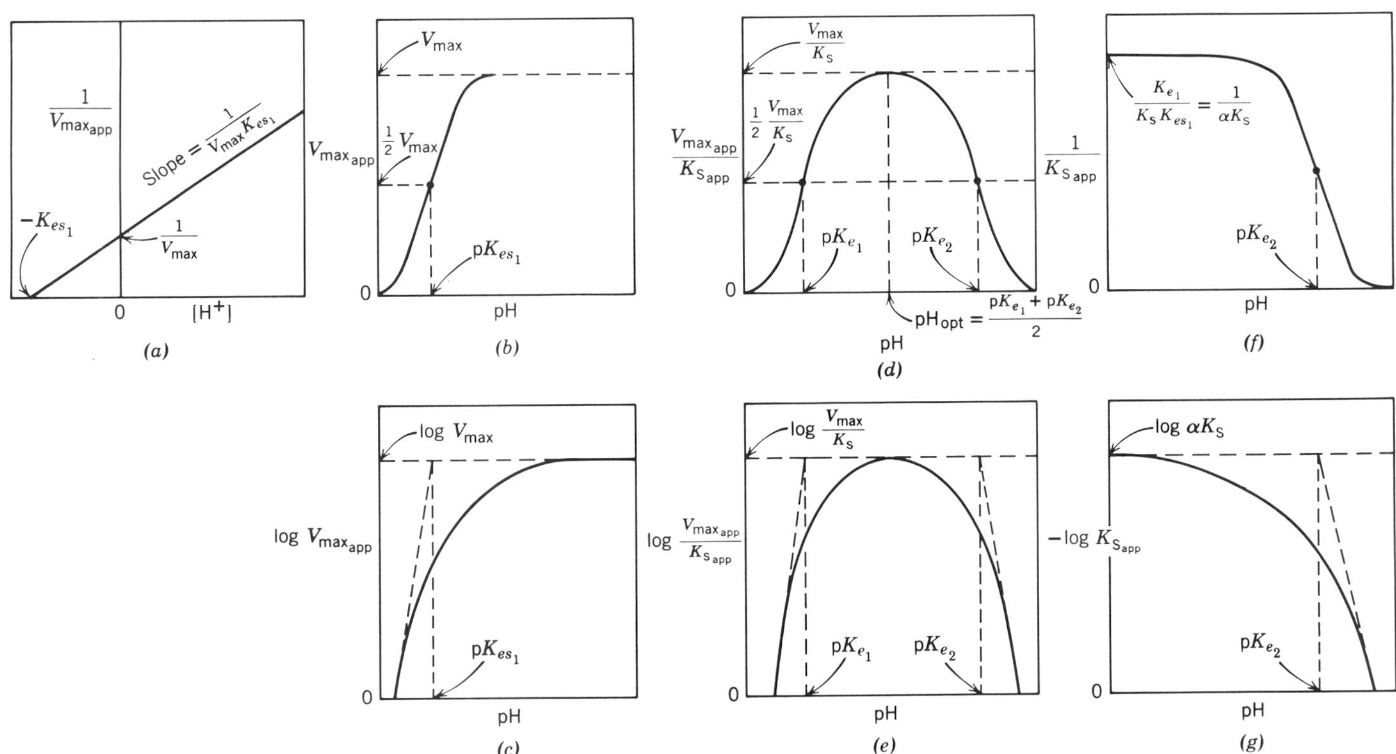

Fig. XI-9. Diagnostic plots for a diprotic model in which E^n and E^{n+1} bind S to form E^nS and $E^{n+1}S$ but only E^nS is catalytically active.

$K_{S_{app}}$ is given by:

$$K_{S_{app}} = K_S \frac{\left(1 + \dfrac{[H^+]}{K_{e_1}} + \dfrac{K_{e_2}}{[H^+]}\right)}{\left(1 + \dfrac{[H^+]}{K_{es_1}}\right)}$$

As $[H^+]$ increases, the $K_{e_2}/[H^+]$ term becomes insignificant and $K_{S_{app}}$ approaches a limit of $K_S K_{es_1}/K_{e_1} = \alpha K_S$. As $[H^+]$ decreases, the $[H^+]/K_{e_1}$ and $[H^+]/K_{es_1}$ terms approach zero and $K_{S_{app}}$ increases to infinity (i.e., all the "E" is driven to E^{n-1} which has no affinity for S). The plots of $1/K_{S_{app}}$ versus pH and $\log 1/K_{S_{app}} = -\log K_{S_{app}}$ versus pH are shown in Figure XI-9*f*,*g*. The plots are redundant since all three enzyme dissociation constants can be obtained from the $V_{max_{app}}$ and $1/slope_{1/S}$ plots.

System A4. All Forms of "E" Bind S; Only E^nS and $E^{n-1}S$ Yield Product

System A4 is similar to System A1 with the added complication that both E^nS and $E^{n-1}S$ are catalytically active.

$$\begin{array}{ccccccccc}
E^{n+1} & + & S & \overset{\alpha K_S}{\rightleftharpoons} & E^{n+1}S & & E^{n+1} & & \\
K_{e_1}\updownarrow & & & & \updownarrow K_{es_1} = \alpha K_{e_1} & & \updownarrow K_{e_1} & & \\
H^+ & & & & H^+ & & H^+ & & \\
+ & & & & + & & + & & \\
E^n & + & S & \overset{K_S}{\rightleftharpoons} & E^nS & \overset{k_p}{\longrightarrow} & E^n & + & P \\
K_{e_2}\updownarrow & & & & \updownarrow K_{es_2} = \dfrac{K_{e_2}}{\beta} & & \updownarrow K_{e_2} & & \\
H^+ & & & & H^+ & & H^+ & & \\
+ & & & & + & & + & & \\
E^{n-1} & + & S & \overset{\beta K_S}{\rightleftharpoons} & E^{n-1}S & \overset{\gamma k_p}{\longrightarrow} & E^{n-1} & + & P
\end{array}$$

The velocity equation is obtained from:

$$\frac{v}{[\mathrm{E}]_t} = \frac{k_p\frac{[\mathrm{S}]}{K_\mathrm{S}} + \gamma k_p\frac{[\mathrm{S}]K_{es_2}}{K_\mathrm{S}[\mathrm{H}^+]}}{1 + \frac{[\mathrm{S}]}{K_\mathrm{S}} + \frac{[\mathrm{H}^+]}{K_{e_1}} + \frac{K_{e_2}}{[\mathrm{H}^+]} + \frac{[\mathrm{S}][\mathrm{H}^+]}{K_\mathrm{S}K_{es_1}} + \frac{[\mathrm{S}]K_{es_2}}{K_\mathrm{S}[\mathrm{H}^+]}} \quad \text{(XI-22)}$$

where E^n is taken as the reference state. Factoring and taking $k_p[\mathrm{E}]_t$ as $V_{\max}$, the equation rearranges to:

$$\frac{v}{V_{\max}} = \frac{[\mathrm{S}]}{K_\mathrm{S}\dfrac{\left(1 + \dfrac{[\mathrm{H}^+]}{K_{e_1}} + \dfrac{K_{e_2}}{[\mathrm{H}^+]}\right)}{\left(1 + \dfrac{\gamma K_{es_2}}{[\mathrm{H}^+]}\right)} + [\mathrm{S}]\dfrac{\left(1 + \dfrac{[\mathrm{H}^+]}{K_{es_1}} + \dfrac{K_{es_2}}{[\mathrm{H}^+]}\right)}{\left(1 + \dfrac{\gamma K_{es_2}}{[\mathrm{H}^+]}\right)}} \quad \text{(XI-23)}$$

If the successive pK values are well separated (3.5 pH units or more), a plot of $V_{\max_{\mathrm{app}}}$ versus pH will rise, approach $V_{\max}$ asymptotically, and then fall (or, if $\gamma > 1$, rise) to a limit of $\gamma V_{\max}$. The plot of $V_{\max_{\mathrm{app}}}/K_{\mathrm{S}_{\mathrm{app}}}$ will be similarly biphasic (if $\beta \neq \gamma$) with an intermediate limit of $V_{\max}/K_\mathrm{S}$ and a final limit of $\gamma V_{\max}/\beta K_\mathrm{S}$. The initial phase of the curves can be used to determine K_{e_1} and K_{es_1}, as shown earlier. The second phase of the $V_{\max_{\mathrm{app}}}$ plot can be used to obtain K_{es_2} (Fig. XI-10). The data of the second phase can also be treated in the manner described in Chapters Four and Five for hyperbolic mixed-type or partial inhibition or nonessential activation systems. For example, suppose $\gamma < 1$ and $\beta > 1$. The reciprocal plot is shown in Figure XI-11*a*. $1/V_{\max_{\mathrm{app}}}$ is given by:

$$\frac{1}{V_{\max_{\mathrm{app}}}} = \frac{1}{V_{\max}}\frac{\left(1 + \dfrac{[\mathrm{H}^+]}{K_{es_1}} + \dfrac{K_{es_2}}{[\mathrm{H}^+]}\right)}{\left(1 + \dfrac{\gamma K_{es_2}}{[\mathrm{H}^+]}\right)}$$

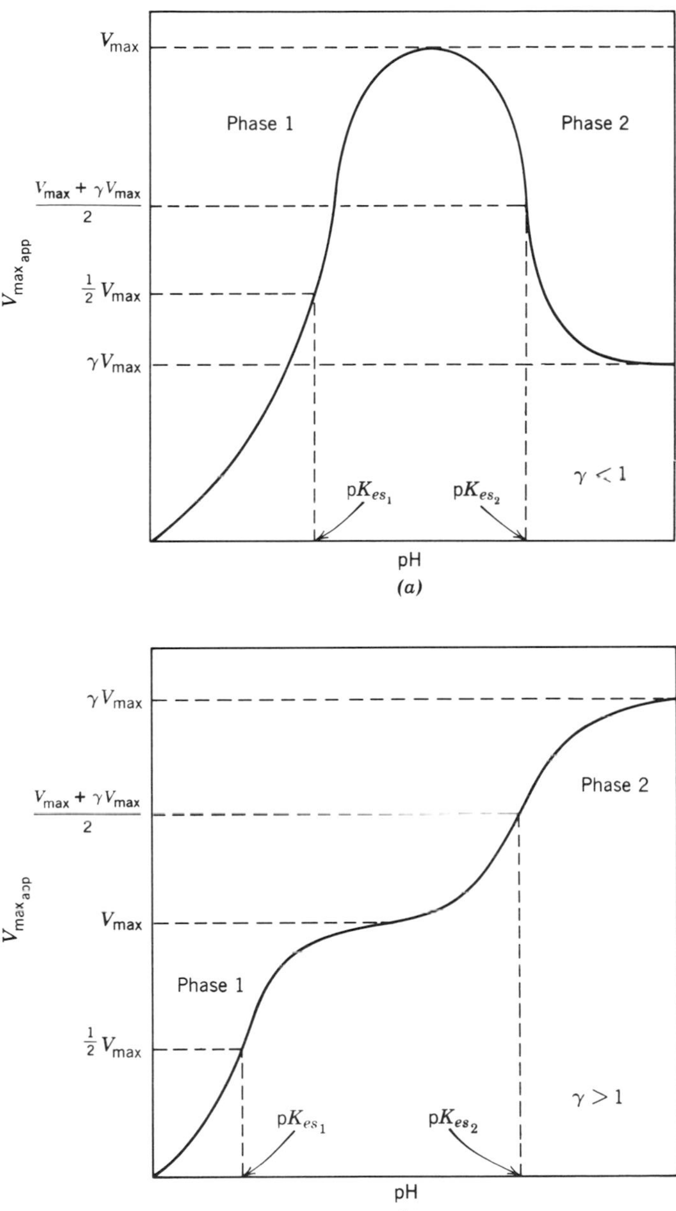

Fig. XI-10. The $V_{max_{app}}$ versus pH plot for a diprotic model in which all three forms of "E" bind S but only E^nS and $E^{n-1}S$ are catalytically active. (*a*) $\gamma < 1$; that is, the effective k_p for $E^{n-1}S$ is less than that of E^nS. (*b*) $\gamma > 1$; that is, $E^{n-1}S$ is catalytically more active than E^nS.

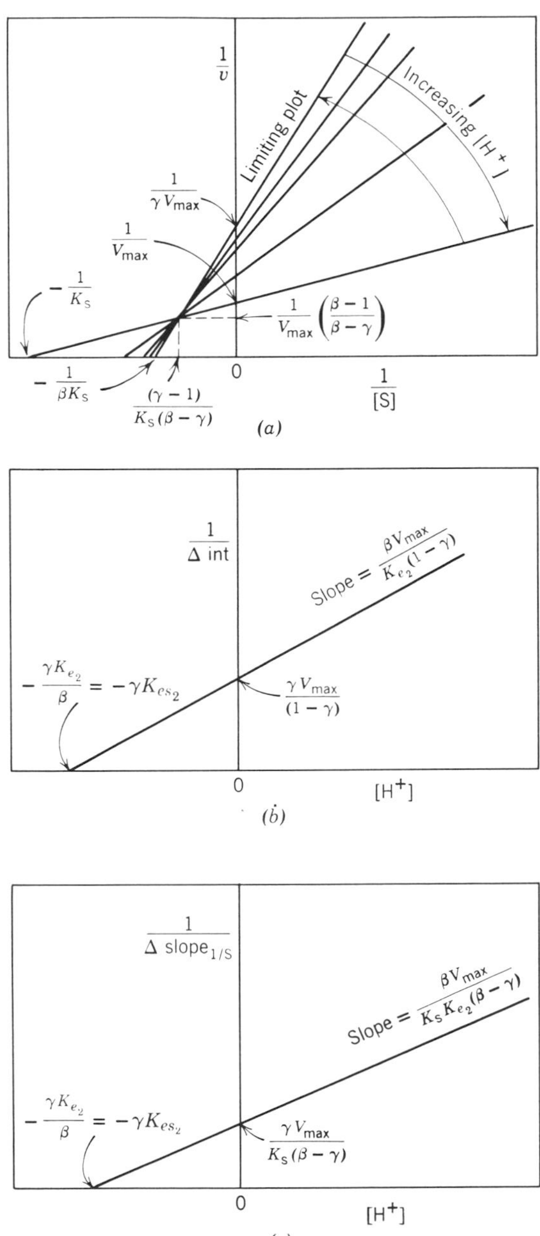

Fig. XI-11. Analysis of System A4 in the region of K_{e_2} and K_{es_2} in which increasing pH is treated as hyperbolic (mixed-type) inhibition (or H^+ is treated as a nonessential activator) ($\gamma < 1$, $\beta > 1$).

In the region between $V_{\max}$ and $\gamma V_{\max}$, the pH will be high compared to pK_{es_1} ($[H^+]$ is very low compared to K_{es_1}). Neglecting the $[H^+]/K_{es_1}$ term, $1/V_{\max_{app}}$ can be written:

$$\frac{1}{V_{\max_{app}}} = \frac{1}{V_{\max}} \frac{\left(1+\dfrac{K_{es_2}}{[H^+]}\right)}{\left(1+\dfrac{\gamma K_{es_2}}{[H^+]}\right)}$$

$$= \frac{1}{V_{\max}}\left(\frac{[H^+]+K_{es_2}}{[H^+]+\gamma K_{es_2}}\right) = \frac{1}{V_{\max}}\left(\frac{\beta[H^+]+K_{e_2}}{\beta[H^+]+\gamma K_{e_2}}\right)$$

$$\Delta\,\text{int} = \frac{1}{V_{\max_{app}}} - \frac{1}{V_{\max}} = \frac{\beta[H^+]+K_{e_2}-\beta[H^+]-\gamma K_{e_2}}{V_{\max}(\beta[H^+]+\gamma K_{e_2})}$$

$$= \frac{K_{e_2}(1-\gamma)}{V_{\max}(\beta[H^+]+\gamma K_{e_2})}$$

Inverting:

$$\frac{1}{\Delta\,\text{int}} = \frac{V_{\max}\beta[H^+]+V_{\max}\gamma K_{e_2}}{K_{e_2}(1-\gamma)}$$

or

$$\boxed{\frac{1}{\Delta\,\text{int}} = \frac{\beta V_{\max}}{K_{e_2}(1-\gamma)}[H^+] + \frac{\gamma V_{\max}}{(1-\gamma)}} \qquad \text{(XI-24)}$$

Thus a plot of $1/\Delta\,\text{int}$ versus $[H^+]$ is linear (Fig. XI-11*b*). The intercept on the horizontal axis gives $-\gamma K_{e_2}/\beta = -\gamma K_{es_2}$. γ can be calculated from the vertical-axis intercept or from the limiting values of $V_{\max}$ and $\gamma V_{\max}$ on the

V_{max} versus pH plot. Thus K_{es_2} can be calculated. The slope of the reciprocal plots in the second phase is given by (ignoring the $[\mathrm{H}^+]/K_{e_1}$ term):

$$slope = \frac{K_{\mathrm{S_{app}}}}{V_{max_{app}}} = \frac{K_{\mathrm{S}}}{V_{max}} \frac{\left(1+\dfrac{K_{e_2}}{[\mathrm{H}^+]}\right)}{\left(1+\dfrac{\gamma K_{es_2}}{[\mathrm{H}^+]}\right)} = \frac{K_{\mathrm{S}}}{V_{max}}\left(\frac{[\mathrm{H}^+]+K_{e_2}}{[\mathrm{H}^+]+\gamma K_{es_2}}\right)$$

$$= \frac{K_{\mathrm{S}}}{V_{max}}\left(\frac{\beta[\mathrm{H}^+]+\beta K_{e_2}}{\beta[\mathrm{H}^+]+\gamma K_{e_2}}\right)$$

$$\Delta\, slope = \frac{K_{\mathrm{S_{app}}}}{V_{max_{app}}} - \frac{K_{\mathrm{S}}}{V_{max}} = \frac{\beta K_{\mathrm{S}}[\mathrm{H}^+]+\beta K_{\mathrm{S}}K_{e_2} - \beta K_{\mathrm{S}}[\mathrm{H}^+] - \gamma K_{\mathrm{S}}K_{e_2}}{V_{max}(\beta[\mathrm{H}^+]+\gamma K_{e_2})}$$

$$= \frac{K_{\mathrm{S}}K_{e_2}(\beta-\gamma)}{V_{max}(\beta[\mathrm{H}^+]+\gamma K_{e_2})}$$

Inverting and separating terms:

$$\boxed{\frac{1}{\Delta\, slope} = \frac{\beta V_{max}}{K_{\mathrm{S}}K_{e_2}(\beta-\gamma)}[\mathrm{H}^+] + \frac{\gamma V_{max}}{K_{\mathrm{S}}(\beta-\gamma)}} \qquad \text{(XI-25)}$$

A plot of $1/\Delta\, slope$ versus $[\mathrm{H}^+]$ is linear (Fig. XI-11*c*). The horizontal intercept gives $-\gamma K_{es_2}$. With γ, K_{S}, and V_{max} known, β can be calculated from the vertical-axis intercept (or from βK_{S}) and then K_{e_2} can be calculated from the slope of the replot or from the value of $\gamma K_{es_2} = \gamma K_{e_2}/\beta$.

If $\gamma = 1$ but $\beta \neq 1$, the $V_{max_{app}}$ versus pH plot will rise and approach V_{max} as a limit. The two parenthetical terms of the $1/v$-axis intercept factor of equation XI-23 will cancel at pH values $\gg \mathrm{p}K_{es_1}$. The $V_{max_{app}}/K_{\mathrm{S_{app}}}$ versus pH plot will still be biphasic. The $1/\Delta\, slope$ plot can be used to determine β and K_{e_2}. If $\beta = 1$, but $\gamma \neq 1$, the system will resemble partial noncompetitive inhibition or nonessential activation. (When $\beta = 1$, $K_{e_2} = K_{es_2}$. The slope and $1/v$-axis intercept factors of equation XI-23 will be identical.) Then $\mathrm{p}K_{e_2} = \mathrm{p}K_{es_2}$ can be determined as the pH half way between V_{max} and γV_{max} on the $V_{max_{app}}$ versus pH plot, or from a similar biphasic $V_{max_{app}}/K_{\mathrm{S_{app}}}$ plot, or from $1/\Delta\, slope$ or $1/\Delta\mathrm{int}$ replots. If $\beta = 1$ and $\gamma = 1$, the numerator and denominator portions of both the slope and $1/v$-axis intercept factors are identical at high pH and cancel. The linear or log plots of $V_{max_{app}}$ and $V_{max_{app}}/K_{S_{app}}$ show a single phase (rising to the limits of V_{max} and V_{max}/K_S), from which $K_{e_1} = K_{es_1}$ can be determined; $K_{e_2} = K_{es_2}$ will not show up in the data. This is the same result that would be obtained if (a) E^{n-1} and $E^{n-1}S$ did not form, or (b) if E^{n-1} cannot bind S, but once S binds to E^n, the resulting E^nS can dissociate to $E^{n-1}S$ without causing S to dissociate.

System A5. General System: All Forms of "E" Bind S; All Forms of "E"S Yield Product

The most general rapid equilibrium system is shown below. Here it is assumed that all three forms of "E" bind S and all three form of "E"S yield product.

$$
\begin{array}{ccccccccc}
E^{n+1} & + & S & \underset{}{\overset{\alpha K_S}{\rightleftharpoons}} & E^{n+1}S & \overset{\gamma k_p}{\longrightarrow} & E^{n+1} & + & P \\
K_{e_1}\downarrow\uparrow & & & & \uparrow\downarrow K_{es_1}=\alpha K_{e_1} & & \uparrow\downarrow K_{e_1} & & \\
H^+ & & & & H^+ & & H^+ & & \\
+ & & & & + & & + & & \\
E^n & + & S & \overset{K_S}{\rightleftharpoons} & E^nS & \overset{k_p}{\longrightarrow} & E^n+P & & \\
K_{e_2}\downarrow\uparrow & & & & \uparrow\downarrow K_{es_2}=\dfrac{K_{e_2}}{\beta} & & \uparrow\downarrow K_{e_2} & & \\
H^+ & & & & H^+ & & H^+ & & \\
+ & & & & + & & + & & \\
E^{n-1} & + & S & \overset{\beta K_S}{\rightleftharpoons} & E^{n-1}S & \overset{\delta k_p}{\longrightarrow} & E^{n-1} & + & P
\end{array}
$$

The velocity is given by:

$$
\frac{v}{[E]_t} = \frac{k_p\dfrac{[S]}{K_S} + \gamma k_p\dfrac{[S][H^+]}{K_S K_{es_1}} + \delta k_p\dfrac{[S]K_{es_2}}{K_S[H^+]}}{1 + \dfrac{[H^+]}{K_{e_1}} + \dfrac{K_{e_2}}{[H^+]} + \dfrac{[S]}{K_S} + \dfrac{[S][H^+]}{K_S K_{es_1}} + \dfrac{[S]K_{es_2}}{K_S[H^+]}} \qquad \text{(XI-26)}
$$

where E^n is taken as the reference state. Factoring, letting $k_p[E]_t = V_{max}$, and rearranging, the velocity equation becomes:

$$\frac{v}{V_{\max}} = \frac{[S]}{K_S \frac{\left(1 + \frac{[H^+]}{K_{e_1}} + \frac{K_{e_2}}{[H^+]}\right)}{\left(1 + \frac{\gamma[H^+]}{K_{es_1}} + \frac{\delta K_{es_2}}{[H^+]}\right)} + [S] \frac{\left(1 + \frac{[H^+]}{K_{es_1}} + \frac{K_{es_2}}{[H^+]}\right)}{\left(1 + \frac{\gamma[H^+]}{K_{es_1}} + \frac{\delta K_{es_2}}{[H^+]}\right)}} \tag{XI-27}$$

The velocity is a complicated function of pH but if the two successive pK values of the enzyme are well separated, the system can be analyzed as two hyperbolic mixed-type or partial segments.

A Diprotic System Where the Successive pK Values of the Enzyme are Closer Than 3.5 pH Units

In the preceding section we examined systems in which the two successive acid dissociation constants of the enzyme were separated by at least 3.5 pH units. This separation allowed us to ignore one of the terms of the slope and $1/v$-axis intercept factors in a given pH range and thereby determine the pK values in a relatively straight-forward manner. For all practical purposes, we could use the same plotting methods even if the two pK values were only 2 pH units apart. To be sure, the estimated pK values would be in error (by about 0.2 pH units). However, the object of the study is to suggest possible functional groups that might be involved at the active site, and an error of ± 0.5 pH unit is not likely to introduce any additional uncertainty.

If the pK values are separated by less than 2 pH units, then an alternate procedure (described below) must be used to determine pK_{e_1}, pK_{e_2}, and so on. This procedure is a general one and can be applied to any diprotic system where the active enzyme form is $E^n S$. For example, consider System A1 where now p$K_{es_1} = 6.0$ and p$K_{es_2} = 7.0$. The plot of observed $V_{\max}$ versus pH is shown in Figure XI-12. (The kinetic value plotted is called "observed $V_{\max}$" instead of $V_{\max_{app}}$ to differentiate between the peak value, called "apparent $V_{\max}$," and the theoretical maximal velocity, indicated as $V_{\max}$. In the systems described earlier, these latter two parameters were essentially identical.) Figure XI-12 also shows the fractional concentrations of the three enzyme species present at saturating [S]. The observed $V_{\max}$ is proportional to the fraction of "E"S present as E^nS. Note that the pH values that yield 1/2 apparent V_{max} are not pK_{es_1} and pK_{es_2}. The optimum pH still occurs at 1/2 $(pK_{es_1}+pK_{es_2})$. The observed V_{max} is given by:

$$\text{observed } V_{max} = \frac{V_{max}}{\left(1+\dfrac{[H^+]}{K_{es_1}}+\dfrac{K_{es_2}}{[H^+]}\right)} = \frac{V_{max}[H^+]K_{es_1}}{K_{es_1}[H^+]+[H^+]^2+K_{es_1}K_{es_2}} \tag{XI-28}$$

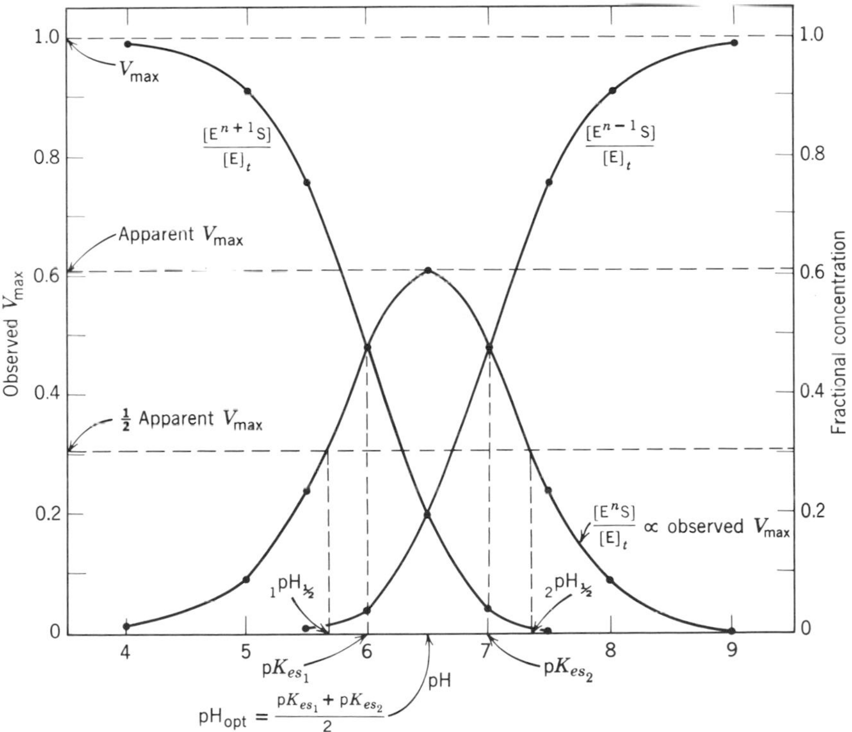

Fig. XI-12. Distribution of ionic forms of "E"S as a function of pH for a diprotic system in which $pK_{es_1}=6.0$ and $pK_{es_2}=7.0$. If only E^nS is catalytically active, the plot of observed V_{max} versus pH will be the same as that for $[E^nS]/[E]_t$. The pH optimum will occur at the average of pK_{es_1} and pK_{es_2}, but the pH values at half observed V_{max} will not coincide with the pK_{es} values.

When $[H^+]=[H^+]_{opt}$:

$$\text{apparent } V_{max} = \frac{V_{max}}{\left(1+\frac{[H^+]_{opt}}{K_{es_1}}+\frac{K_{es_2}}{[H^+]_{opt}}\right)} = \frac{V_{max}[H^+]_{opt}K_{es_1}}{K_{es_1}[H^+]_{opt}+[H^+]_{opt}^2+K_{es_1}K_{es_2}} \tag{XI-29}$$

Substituting,

$$[H^+]_{opt}=\sqrt{K_{es_1}K_{es_2}} \tag{XI-30}$$

$$\text{apparent } V_{max} = \frac{V_{max}\sqrt{K_{es_1}K_{es_2}}\,K_{es_1}}{K_{es_1}\sqrt{K_{es_1}K_{es_2}}+K_{es_1}K_{es_2}+K_{es_1}K_{es_2}} = \frac{V_{max}}{1+2\sqrt{\frac{K_{es_2}}{K_{es_1}}}} \tag{XI-31}$$

$$\therefore \quad \tfrac{1}{2}\text{ apparent } V_{max} = \frac{V_{max}}{2+4\sqrt{\frac{K_{es_2}}{K_{es_1}}}} \tag{XI-32}$$

One-half apparent V_{max} can also be expressed as (from equations XI-29 and XI-30):

$$\tfrac{1}{2}\text{ apparent } V_{max} = \frac{V_{max}[H^+]_{1/2}K_{es_1}}{K_{es_1}[H^+]_{1/2}+[H^+]_{1/2}^2+[H^+]_{opt}^2} \tag{XI-33}$$

Setting equation XI-33 equal to equation XI-32, canceling and rearranging:

$$[H^+]_{1/2}^2-(K_{es_1}+4[H^+]_{opt})[H^+]_{1/2}+[H^+]_{opt}^2=0 \tag{XI-34}$$

Equation XI-34 has two real roots, ${}_1[H^+]_{1/2}$ and ${}_2[H^+]_{1/2}$, which correspond to the $[H^+]$ at 1/2 apparent V_{max} on each side of the curve. The sum of the two roots of a general quadratic equation can be written as minus the ratio of the coefficient of X to the coefficient of X^2. For equation XI-34:

$$\boxed{{}_1[H^+]_{1/2}+{}_2[H^+]_{1/2}=K_{es_1}+4[H^+]_{opt}} \tag{XI-35}$$

Since ${}_1[H^+]_{1/2}$, ${}_2[H^+]_{1/2}$, and $[H^+]_{opt}$ can be obtained from the observed $pH_{1/2}$ and pH_{opt} values, K_{es_1} can be calculated. With K_{es_1} known, K_{es_2} can be calculated from equation XI-30 and V_{max} calculated from equation XI-32. In the same way, K_{e_1}, K_{e_2}, and V_{max}/K_S can be determined from a plot of observed V_{max}/observed K_S versus pH.

Displacement of pK Values Under Nonrapid Equilibrium Conditions

So far, we have considered only relatively simple rapid equilibrium unireactant systems in which the rate-limiting step is the conversion of ES to E+P. If the system is not at equilibrium, then the experimentally determined pK values may not correspond to the true pK values. Two examples are described below.

Consider the diprotic unireactant system shown below in which the catalytic step is fast compared to the dissociation of "E"S. For simplicity, we will assume that the vertical reactions involving H^+ addition or dissociation are at equilibrium and that the rate constants for the addition and dissociation of S are the same for all forms of "E" (i.e., $\alpha = 1$, $\beta = 1$; therefore, $K_{es_1} = K_{e_1}$ and $K_{es_2} = K_{e_2}$). Only E^nS is assumed to yield product.

$$
\begin{array}{ccccccccc}
E^{n+1} & + & S & \underset{k_{-1}}{\overset{k_1}{\rightleftharpoons}} & E^{n+1}S & & E^{n+1} & & \\
K_{e_1}\downarrow\uparrow & & & & \uparrow\downarrow K_{es_1}=K_{e_1} & & \uparrow\downarrow K_{e_1} & & \\
H^+ & & & & H^+ & & H^+ & & \\
+ & & & & + & & + & & \\
E^n & + & S & \underset{k_{-1}}{\overset{k_1}{\rightleftharpoons}} & E^nS & \underset{k_{-2}}{\overset{k_2}{\rightleftharpoons}} & E^n & + & P \\
K_{e_2}\downarrow\uparrow & & & & \uparrow\downarrow K_{es_2}=K_{e_2} & & \uparrow\downarrow K_{e_2} & & \\
H^+ & & & & H^+ & & H^+ & & \\
+ & & & & + & & & & \\
E^{n-1} & + & S & \underset{k_{-1}}{\overset{k_1}{\rightleftharpoons}} & E^{n-1}S & & E^{n-1} & &
\end{array}
$$

The velocity equation can be derived using the method of Cha where X represents the rapid equilibrium segment composed of the three forms of "E" and Y represents the rapid equilibrium segment composed of the three forms of "E"S.

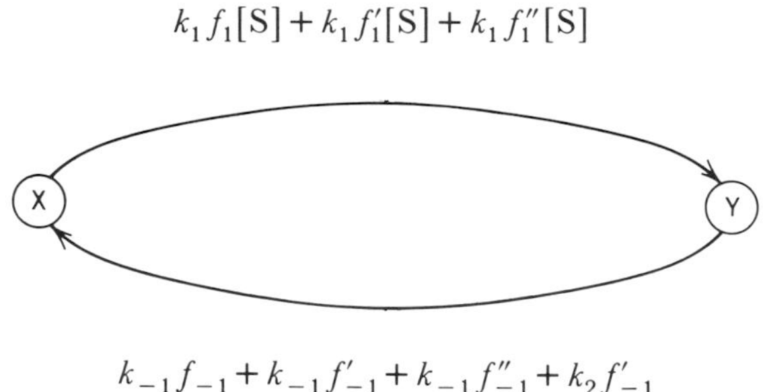

where f_1, f_1', etc. are fractional concentrations:

$$f_1 = \frac{[E^{n+1}]}{[X]} = \frac{\dfrac{[H^+]}{K_{e_1}}}{1 + \dfrac{[H^+]}{K_{e_1}} + \dfrac{K_{e_2}}{[H^+]}} = \frac{\dfrac{[H^+]}{K_{e_1}}}{f_E}$$

$$f_1' = \frac{[E^n]}{[X]} = \frac{1}{f_E}, \qquad f_1'' = \frac{[E^{n-1}]}{[X]} = \frac{\dfrac{K_{e_2}}{[H^+]}}{f_E}$$

The E^n is taken as the reference state. A similar set of distribution equations can be written for segment Y. Since $K_{es_1} = K_{e_1}$ and $K_{es_2} = K_{e_2}$, the denominator of both sets of distribution equations are identical. The King-Altman solutions yield:

$$\frac{[X]}{[E]_t} = \frac{k_{-1}(f_{-1} + f'_{-1} + f''_{-1}) + k_2 f'_{-1}}{[X] + [Y]} = \frac{k_{-1} + \dfrac{k_2}{f_E}}{[X] + [Y]}$$

$$\frac{[Y]}{[E]_t} = \frac{k_1[S](f_1 + f_1' + f_1'')}{[X] + [Y]} = \frac{k_1[S]}{[X] + [Y]}$$

Since $f_{-1} + f'_{-1} + f''_{-1} = 1$, and $f_1 + f_1' + f_1'' = 1$, the numerators simplify. The velocity is given by:

$$v = k_2[\mathrm{E''S}] = k_2 f_1'[\mathrm{Y}] \qquad \text{and} \qquad \frac{v}{[\mathrm{E}]_t} = \frac{k_2 f_1'[\mathrm{Y}]}{[\mathrm{X}]+[\mathrm{Y}]}$$

$$\therefore \quad \frac{v}{[\mathrm{E}]_t} = \frac{\dfrac{k_1 k_2[\mathrm{S}]}{f_\mathrm{E}}}{\left(k_{-1} + \dfrac{k_2}{f_\mathrm{E}}\right) + k_1[\mathrm{S}]} = \frac{k_1 k_2[\mathrm{S}]}{(k_{-1} f_\mathrm{E} + k_2) + k_1 f_\mathrm{E}[\mathrm{S}]} \qquad \text{(XI-36)}$$

Note that only part of the constant term is multiplied by f_E, the Michaelis pH function. Equation XI-36 can be written:

$$\boxed{\frac{v}{V_{max}} = \frac{[\mathrm{S}]}{\dfrac{k_2 + k_{-1}\left(1 + \dfrac{[\mathrm{H}^+]}{K_{e_1}} + \dfrac{K_{e_2}}{[\mathrm{H}^+]}\right)}{k_1} + [\mathrm{S}]\left(1 + \dfrac{[\mathrm{H}^+]}{K_{e_1}} + \dfrac{K_{e_2}}{[\mathrm{H}^+]}\right)}} \qquad \text{(XI-37)}$$

If $k_2 \ll k_{-1}$, the equation reduces to that for a simple rapid equilibrium diprotic system. Let us assume that $\mathrm{p}K_{e_1} = 4$ ($K_{e_1} = 10^{-4}$), $\mathrm{p}K_{e_2} = 7$ ($K_{e_2} = 10^{-7}$), $k_1 = 10$, $k_{-1} = 1$, $k_2 = 100$, $[\mathrm{E}]_t = 1$, and $V_{max} = 100$. $V_{max_{app}}$ and $K_{m_{app}}$ are given by:

$$V_{max_{app}} = \frac{V_{max}}{\left(1 + \dfrac{[\mathrm{H}^+]}{K_{e_1}} + \dfrac{K_{e_2}}{[\mathrm{H}^+]}\right)}, \qquad K_{m_{app}} = \frac{k_2 + k_{-1}\left(1 + \dfrac{[\mathrm{H}^+]}{K_{e_1}} + \dfrac{K_{e_2}}{[\mathrm{H}^+]}\right)}{\left(1 + \dfrac{[\mathrm{H}^+]}{K_{e_1}} + \dfrac{K_{e_2}}{[\mathrm{H}^+]}\right)}$$

The linear or log plot of $V_{max_{app}}$ versus pH will peak at pH 5.5 very close to the true V_{max}. The pH values at $1/2\,V_{max}$ yield $\mathrm{p}K_{e_1}$ and $\mathrm{p}K_{e_2}$. The plot of $V_{max_{app}}/K_{m_{app}}$ versus pH peaks close to 1.0, but the pH values at 0.5 do not correspond to pK_{e_1} and pK_{e_2}.

$$\frac{V_{\max_{app}}}{K_{m_{app}}} = \frac{V_{\max}}{k_2 + k_{-1}\left(1 + \dfrac{[\mathrm{H}^+]}{K_{e_1}} + \dfrac{K_{e_2}}{[\mathrm{H}^+]}\right)} = \frac{100}{100 + \left(1 + \dfrac{[\mathrm{H}^+]}{K_{e_1}} + \dfrac{K_{e_2}}{[\mathrm{H}^+]}\right)}$$

To obtain a value 0.5, the parenthetical term must equal 100. On the low pH side, this occurs at $[\mathrm{H}^+]/K_{e_1} = 99$, or at about pH 2. On the high pH side, $K_{e_1}/[\mathrm{H}^+] = 99$ at about pH 9. Thus the apparent pK_e values are displaced from the true values by 2 pH units. If $k_2 = 10k_{-1}$, the displacement would be about one pH unit. The same displacement phenomenon would be seen in systems where the true pK values are not separated by 3.5 pH units so that equations XI-30 and XI-35 must be used to calculate the values.

Effect of the Ionization of EP

In all the systems described so far, we have assumed that "E"S yields "E" + P directly. A more realistic reaction sequence would include the step "E"S $\rightleftharpoons$ "E"P as shown below.

$$\begin{array}{ccccccccccc}
\mathrm{E}^{n+1} & & & & \mathrm{E}^{n+1}\mathrm{S} & & \mathrm{E}^{n+1}\mathrm{P} & & \mathrm{E}^{n+1} & & \\
K_{e_1}\updownarrow & & & & \updownarrow K_{es_1} & & \updownarrow K_{ep_1} & & \updownarrow K_{e_1} & & \\
\mathrm{H}^+ & & & & \mathrm{H}^+ & & \mathrm{H}^+ & & \mathrm{H}^+ & & \\
+ & & & & + & & + & & + & & \\
\mathrm{E}^n & + & \mathrm{S} & \underset{k_{-1}}{\overset{k_1}{\rightleftharpoons}} & \mathrm{E}^n\mathrm{S} & \underset{k_{-2}}{\overset{k_2}{\rightleftharpoons}} & \mathrm{E}^n\mathrm{P} & \underset{k_{-3}}{\overset{k_3}{\rightleftharpoons}} & \mathrm{E}^n & + & \mathrm{P} \\
K_{e_2}\updownarrow & & & & \updownarrow K_{es_2} & & \updownarrow K_{ep_2} & & \updownarrow K_{e_2} & & \\
\mathrm{H}^+ & & & & \mathrm{H}^+ & & \mathrm{H}^+ & & \mathrm{H}^+ & & \\
+ & & & & + & & + & & + & & \\
\mathrm{E}^{n-1} & & & & \mathrm{E}^{n-1}\mathrm{S} & & \mathrm{E}^{n-1}\mathrm{P} & & \mathrm{E}^{n-1} & &
\end{array}$$

For simplicity we will assume that only E^n binds S but once $\mathrm{E}^n\mathrm{S}$ forms it can ionize to $\mathrm{E}^{n+1}\mathrm{S}$ or $\mathrm{E}^{n-1}\mathrm{S}$. The velocity equation can be written directly from the $[H^+]$-independent equation (equation IX-18):

$$\frac{v}{[E]_t} = \frac{k_1 k_2 k_3 [S]}{(k_{-1}k_{-2} + k_{-1}k_3 + k_2 k_3)\left(1 + \frac{[H^+]}{K_{e_1}} + \frac{K_{e_2}}{[H^+]}\right) + k_1\left[(k_{-2} + k_3)\left(1 + \frac{[H^+]}{K_{es_1}} + \frac{K_{es_2}}{[H^+]}\right) + k_2\left(1 + \frac{[H^+]}{K_{ep_1}} + \frac{K_{ep_2}}{[H^+]}\right)\right][S]}$$

or

$$\boxed{\frac{v}{[E]_t} = \frac{k_1 k_2 k_3 [S]}{(k_{-1}k_{-2} + k_{-1}k_3 + k_2 k_3) f_E + k_1[(k_{-2} + k_3) f_{ES} + k_2 f_{EP}][S]}} \tag{XI-38}$$

The constant term, representing free E, is multiplied by f_E, a factor describing the effect of varying pH on the fraction of "E" available for combination with S. That part of the denominator [S] term representing "E"S is multiplied by f_{ES}, while that part representing "E"P is multiplied by f_{EP}. The kinetic constants are:

$$V_{max_{app}} = \frac{k_2 k_3 [E]_t}{(k_{-2} + k_3) f_{ES} + k_2 f_{EP}} \tag{XI-39}$$

$$K_{m_{app}} = \frac{(k_{-1}k_{-2} + k_{-1}k_3 + k_2 k_3) f_E}{k_1[(k_{-2} + k_3) f_{ES} + k_2 f_{EP}]} \tag{XI-40}$$

$$\frac{V_{max_{app}}}{K_{m_{app}}} = \frac{1}{slope_{1/S}} = \frac{k_1 k_2 k_3 [E]_t}{(k_{-1}k_{-2} + k_{-1}k_3 + k_2 k_3) f_E} \tag{XI-41}$$

The $V_{max_{app}}/K_{m_{app}}$ term is a simple function of f_E, and the usual plots will yield the true pK_{e_1} and pK_{e_2} values. The $V_{max_{app}}$ now depends on two Michaelis pH functions, that is, on the pK values of "E"S and "E"P. The pK values obtained from plots of $V_{max_{app}}$ versus pH will depend on the relative values of k_2, k_{-2}, and k_{-3} and on the difference between the pK_{es} and pK_{ep} values. If k_2 is very much larger than k_{-2} or k_3 so that $k_2 f_{EP}$ is always larger than $(k_{-2} + k_3) f_{ES}$, then the rate limiting step will always be the breakdown of "E"P, and $V_{\max_{app}}$ will depend only on f_{EP}. The usual plots will yield pK_{ep_1} and pK_{ep_2}. If k_2 is very small compared to k_{-2} or k_3, so that $(k_{-2}+k_3)f_{ES}$ is always larger than $k_2 f_{EP}$, then the rate limiting step will always be the conversion of "E"S to "E"P. The usual plots will yield pK_{es_1} and pK_{es_2}. If the pK_{es} and pK_{ep} values are identical, the usual $V_{\max_{app}}$ plots yield the true $pK_{es}=pK_{ep}$ values regardless of the relative magnitudes of the rate constants. If the pK_{es} and pK_{ep} values are not the same and k_2 is neither very much larger nor very much smaller than $k_{-2}+k_3$, the apparent pK values obtained from $V_{\max_{app}}$ versus pH plots will be displaced from the true values. Figure XI-13 shows a plot of $V_{\max_{app}}/[E]_t$ versus pH for a system in which $K_{es_1}=10^{-3}$, $K_{es_2}=10^{-10}$, $K_{ep_1}=10^{-6}$, and $K_{ep_2}=10^{-9}$; that is, one

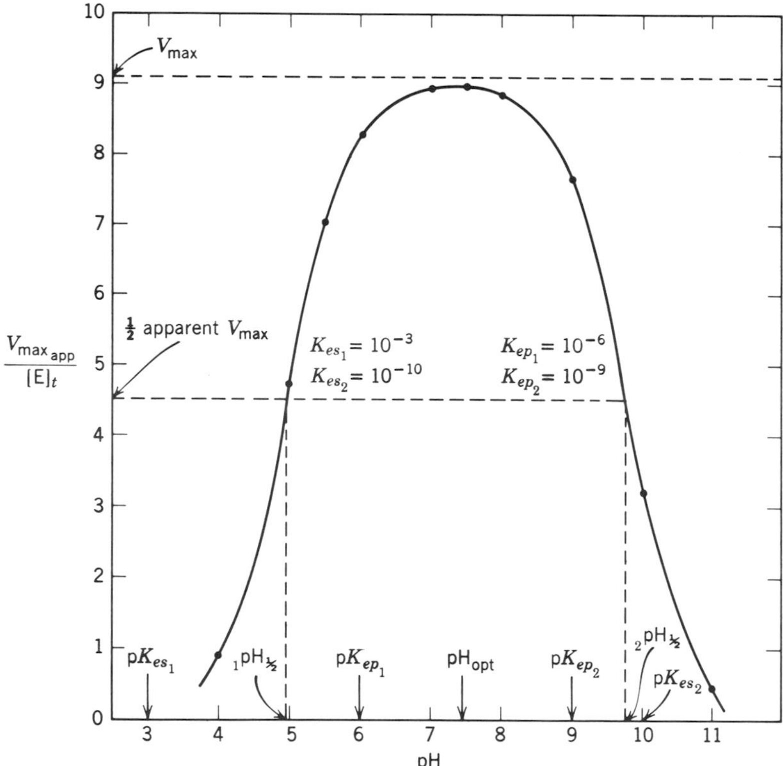

Fig. XI-13. The $V_{\max_{app}}$ versus pH plot for a diprotic system in which one prototropic group on E becomes a much weaker acid and one prototropic group becomes a stronger acid upon conversion of "E"S to "E"P. As a result, the pH values at $0.5\,V_{\max_{app}}$ do not coincide with either pK_{es} or pK_{ep} values; $k_2=10$, $k_{-2}=1$, $k_3=100$.

prototropic group on the enzyme becomes a much weaker acid and one prototropic group becomes a stronger acid as a result of the conversion of "E"S to "E"P. (In practice, just $V_{max_{app}}$ can be plotted.) The rate constants are assumed to be $k_2 = 10$, $k_{-2} = 1$, and $k_3 = 100$. In spite of the fact that $k_3 > k_2$, the conversion of "E"P to "E" + P is equally or more rate-limiting at some pH values. For example, at pH 4, >90% of the "E"S is present as the active E^nS form but only 1% of the "E"P is present as the active E^nP. The pK values of "E"S and "E"P are separated sufficiently to yield an apparent V_{max} very close to the theoretical V_{max}. Yet the apparent pK's do not correspond to either the pK_{es} or pK_{ep} values. The apparent pK values are displaced because the changing pH causes a change in the degrees to which the k_2 and k_3 steps are rate-limiting.

If all three forms of "E" bind S with the same affinity, the velocity equation would be:

$$\frac{v}{[E]_t} = \frac{k_1 k_2 k_3 [S]}{[(k_{-1}k_{-2} + k_{-1}k_3)f_{ES} + k_2 k_3] + k_1[(k_{-2} + k_3)f_{ES} + k_2 f_{EP}][S]} \tag{XI-42}$$

Now the $V_{max_{app}}/K_{m_{app}}$ versus pH plots will also show a displacement in apparent pK values if $k_2 k_3$ is significantly larger than $(k_{-1}k_{-2} + k_{-1}k_3)$.

Limitations of pH Studies

The main objective of pH studies is to identify the protropic groups at the active site of the enzyme. But as we have seen, the apparent pK values may or may not coincide with the true pK's even for the relatively simple systems described above. The equations presented in this chapter assume either rapid equilibrium conditions or steady-state conditions for the reactions of S (in which all forms of "E" are assumed to bind S with the same affinity). If these assumptions were not made, the velocity equations would be exceedingly complex. The assumptions may not hold for many enzymes. In fact, for some enzymes, the reactions involving the addition and dissociation of S or P and the interconversion of "E"S and "E"P may be as rapid as the proton transfer reactions. A full steady-state treatment of the unireactant, diprotic system yields an equation for $v/[E]_t$ containing 15 numerator terms, some with the substrate concentration to the second and third powers and the H^+ ion concentration to the second, third, and fourth powers. The denominator contains 117 such terms (Ottolenghi, 1971). For some combinations of rate constants, the reciprocal plot would theoretically be nonlinear.

The fact that most enzymes catalyze reactions involving two or more substrates (and that most biological molecules are weak acids or bases) further clouds the interpretation of any detailed pH study. Thus the pK's determined from the linear or log plots described earlier can only be considered as *tentative* values which may *suggest* the identities of the prototropic groups. Whenever possible, additional information should be sought (e.g., the ΔH_{ion} from studies of the effect of temperature or effects of site-specific reagents).

Choice of Buffers

Particular buffers are usually chosen because they have a pK_a in the region of the desired pH. This is good practice, but it is not always necessary. If only extremely small amounts of H^+ are produced or utilized during the assay period, then it may be possible to maintain the pH with a compound or mixture of compounds that have very little buffer capacity. For example, mixtures of $H_3PO_4 + KH_2PO_4$, $KH_2PO_4 + K_2HPO_4$, and $K_2HPO_4 + K_3PO_4$ will cover a pH range of 2 to 12. To prepare the solutions, we need only mix equimolar solutions of H_3PO_4 and K_3PO_4 to the desired pH. To be on the safe side, the pH of the reaction mixture should be measured at the start and at the end of the assay.

Dicarboxylic acids are frequently used in preparing buffers. If the two pK_a values are separated by two pH units or less, the acid can be used for buffers at p$K_{a_1} - 0.5$ pH units to p$K_{a_2} + 0.5$ pH units. The buffer capacity midway between pK_{a_1} and pK_{a_2} will be as high as at the pK_a values. For example, succinic acid can be used to prepare buffers from pH 3.7 to pH 6.1. Benzene pentacarboxylic acid can be used from pH 2 to pH 7 if ionic strength and metal ion chelation are not a problem. Mixtures of two or more compounds can be used to prepare high buffer capacity solutions over a wide pH range. For example, Tris + maleic acid can be used for pH 5.5 to 8.6.

Ionic Strength Effects

It is generally good practice to use at least two different kinds of buffers at any given pH to detect indirect effects of the buffer components. For example, an enzyme may be more stable or more active in citrate buffer at pH 5 than in acetate buffer at the same pH. The difference may stem from the fact that citrate chelates contaminating metal ions that inactivate the enzyme or inhibit the reaction. Alternately, the enzyme may prefer the higher ionic strength of the citrate buffer compared to the acetate buffer. Figure XI-14 shows an activity curve that might result when different buffers are used for different pH regions. If buffer A is used at pH 3 and 4, buffer B at pH 5, buffer C at pH 6 and 7, and buffer D at pH 8 and 9, we

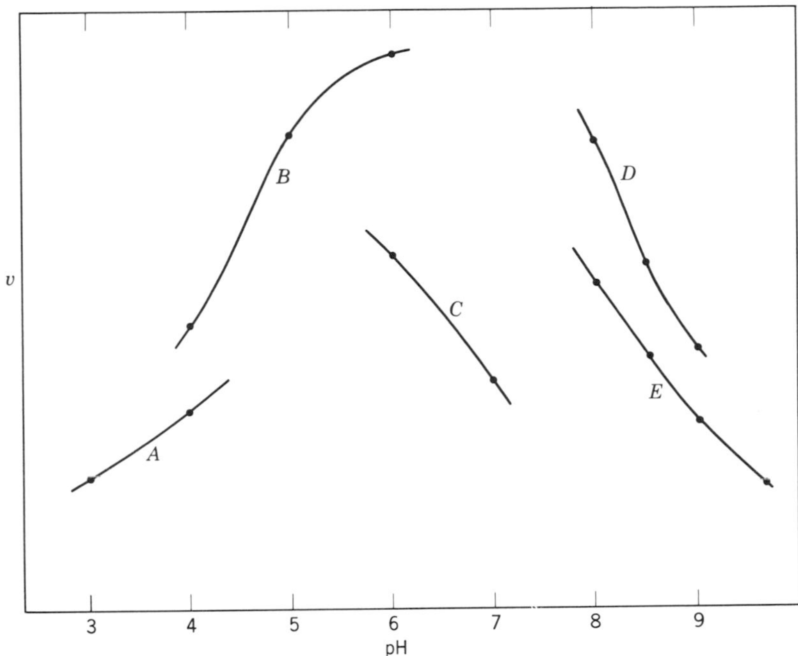

Fig. XI-14. The v versus pH plots that might be obtained with different buffers for different pH ranges.

might conclude that the enzyme displays two pH optima. By overlapping the buffers, we see that there is really only one pH optimum at about pH 6 and that buffer C is quite inhibitory. The inhibition may be a result of several factors. If a metal ion is needed as an activator, buffer C might chelate the metal. Alternately, one of the components of buffer C might resemble one of the substrates and, thereby, act as a competitive inhibitor. Differences in ionic strength must also be considered. The fact that two buffers of the same concentration and pH can have markedly different ionic strengths is often ignored. The ionic strength of a solution, $\Gamma/2$ (or μ or ω), is given by:

$$\boxed{\frac{\Gamma}{2} = \frac{1}{2}\Sigma M_i Z_i^2} \qquad \text{(XI-43)}$$

where M_i = the molarity of the ion

Z_i = the charge on the ion (regardless of sign)

To compensate for ionic strength differences in a pH study, the best procedure is to determine which buffer has the highest ionic strength and then make all the others up to the same value with a neutral noninhibitory salt such as NaCl or KCl. The ionic strength of a solution of KCl or NaCl (i.e., a 1-1 salt) is equivalent to its molarity. After adjusting the ionic strengths of all buffers, their pH's should be remeasured.

When calculating ionic strengths with equation XI-43, Z_i represents the net charge on the ion. Thus the zwitterion form of a neutral amino acid is assumed to be uncharged. For a glycine buffer at pH 9, only the concentration of the $H_2N—CH_2—COO^-$ form and its counterion (e.g., K^+) are used in calculating $\Gamma/2$. When calculating the ionic strength of a histidine buffer at pH 7, only the concentrations of the histidine$^+$ form (protonated imidazole) and its counterion (e.g., Cl^-) are taken into account. The charged COO^- is assumed to cancel the charged NH_3^+.

B. EFFECT OF TEMPERATURE

Temperature Effects on Enzyme Stability

Most chemical reactions proceed at a faster velocity as the temperature, T, is raised. An increase in T imparts more kinetic energy to the reactant molecules resulting in more productive collisions per unit time. Enzyme catalyzed reactions behave similarly, up to a point. Enzymes are complex protein molecules. Their catalytic activity results from a precise, highly ordered tertiary structure which juxtaposes specific amino acid R groups in such a way as to form the stereospecific substrate binding sites and the catalytic center. The tertiary structure of an enzyme is maintained primarily by a large number of weak noncovalent bonds. In practical terms, an enzyme molecule is a very delicate and fragile structure. If the molecule absorbs too much energy, the tertiary structure will disrupt and the enzyme will be denatured, that is, lose catalytic activity. Thus as the temperature increases, the expected increase in v resulting from increased E + S collisions is offset by the increasing rate of denaturation. Consequently, a plot of v versus T usually shows a peak, sometimes referred to as the "optimum temperature." Figure XI-15*a* shows the formation of P as a function of time at different temperatures. We can see that the "optimum temperature" depends on the assay time chosen. If t_2 is chosen, we could conclude that the "optimum temperature" is 40°C (Fig. XI-15*b*). However, if t_1 is chosen as the assay time, the apparent velocity at 50°C would be greater than that at 40°C. The true "optimum" temperature for an assay is the maximum temperature at which the enzyme exhibits a constant activity over a time

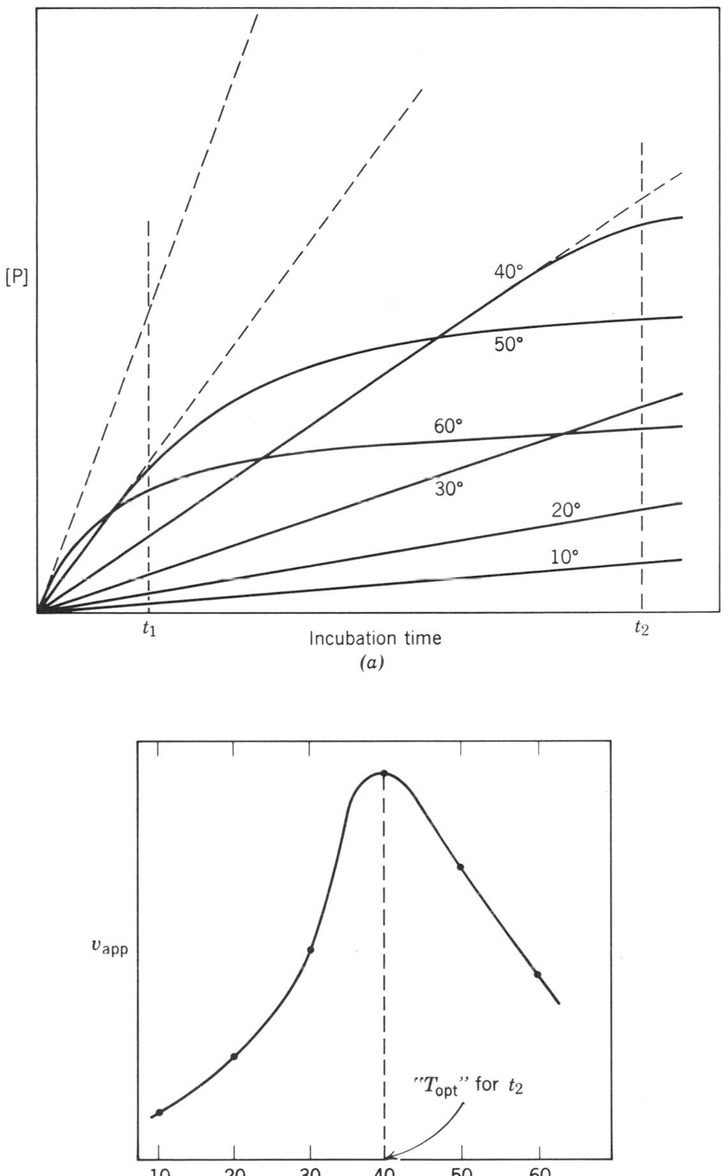

Fig. XI-15. Effect of incubation temperature on product formation. (*a*) [P] versus time. The broken lines show the true initial rate of P appearance at 50° and 60°. (*b*) v calculated from [P] at time $= t_2$.

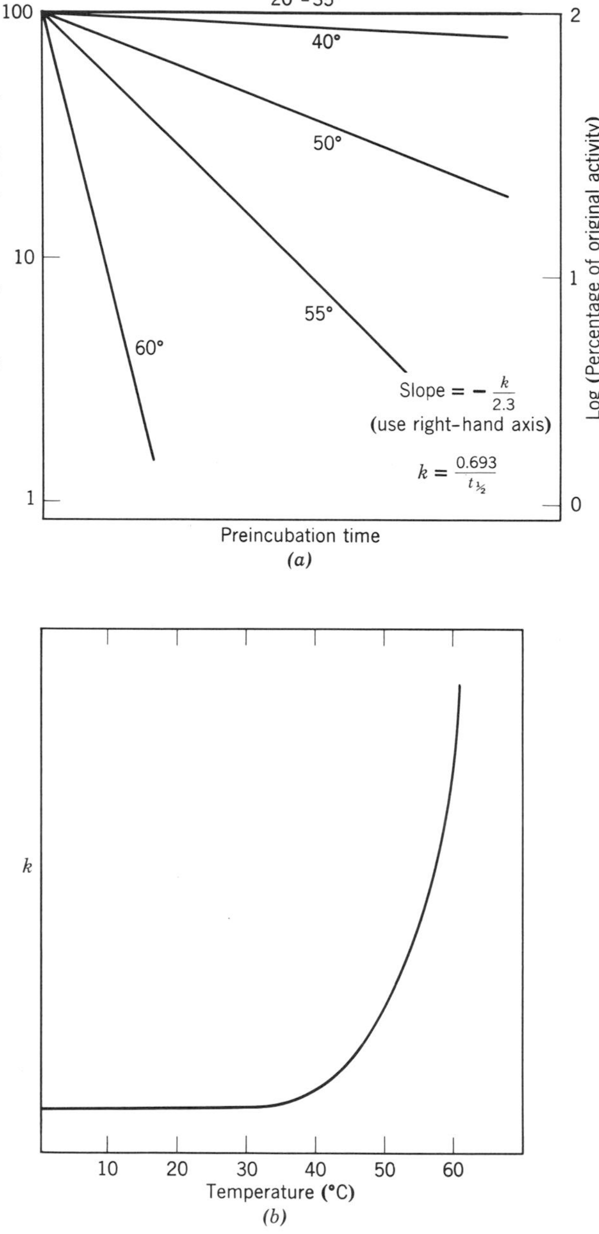

Fig. XI-16. Enzyme inactivation. (*a*) Log percentage of original activity versus preincubation time at different temperatures. Activity is measured at, for example, 35°C. (*b*) First-order rate constant for enzyme inactivation versus temperature.

period at least as long as the assay time. This can easily be established by preincubating the enzyme at different temperatures for one or two times the desired assay time and then measuring the activity at a temperature low enough to cause no denaturation. Figure XI-16 shows the results of such an experiment. For this particular enzyme, an assay temperature of 35°C would be optimum. The temperature stability of an enzyme depends on a number of factors including the pH and ionic strength of the medium and the presence or absence of ligands. Substrates frequently protect against temperature denaturation. Low molecular weight enzymes composed of single polypeptide chains and possessing disulfide bonds are usually more heat stable than high molecular weight, oligomeric enzymes. In general, an enzyme will be more heat stable in crude cell-free preparations containing a high concentration of other proteins (provided no proteases are present).

Identifying Prototropic Groups from ΔH_{ion}

The standard free energy change of a reaction is given by:

$$\boxed{\Delta G^0 = -2.3RT\log K_{\text{eq}}} \qquad \text{(XI-44)}$$

and

$$\boxed{\Delta G^0 = \Delta H^0 - T\Delta S^0} \qquad \text{(XI-45)}$$

where ΔH^0 is the standard enthalpy (heat content) change, ΔS^0 is the standard entropy change, and R is the gas constant (1.98 cal $\times$ mole^{-1} $\times$ degree^{-1}). Combining equations XI-44 and XI-45 we obtain:

$$\boxed{-\log K_{\text{eq}} = \frac{\Delta H^0}{2.3R}\frac{1}{T} - \frac{\Delta S^0}{2.3R}} \qquad \text{(XI-46)}$$

The effect of changing temperature on K_{eq} is given by the Van't Hoff equation:

$$\boxed{\frac{d\ln K_{\text{eq}}}{dt} = \frac{\Delta H^0}{RT^2}} \qquad \text{(XI-47)}$$

or

$$\frac{d \ln K_{eq}}{d\left(\frac{1}{T}\right)} = \frac{-\Delta H^0}{R} \tag{XI-48}$$

When the Van't Hoff equation is integrated between the limits of K_{eq_1} and K_{eq_2} at T_1 and T_2 we obtain:

$$\log \frac{K_{eq_2}}{K_{eq_1}} = \frac{\Delta H^0}{2.3R}\left(\frac{T_2 - T_1}{T_2 T_1}\right) = pK_{eq_1} - pK_{eq_2} \tag{XI-49}$$

Experimentally, ΔH^0 is obtained from equation XI-49 by measuring the K_{eq} at two or more different temperatures, or from equation XI-46 by plotting $-\log K_{eq}$ (i.e., pK_{eq}) versus $1/T$ (Fig. XI-17). The change in pK_{eq} with temperature can be used in conjunction with pH studies to identify the prototropic groups responsible for enzyme activity. Plots of $V_{max_{app}}$ versus pH (or $V_{max_{app}}/K_{m_{app}}$ versus pH when the substrate ionization does not change appreciably with temperature) are constructed for several temperatures. The observed pK_{es} and pK_e values are then replotted against $1/T$ as shown in Figure XI-17. The standard enthalpy of ionization, ΔH^0_{ion}, obtained from the slope of the plot can then be compared to tabulated values (Table XI-1). Since the pK_a values of the buffer components also change with temperature, the pH of each buffer should be measured at each temperature.

Effect of Temperature on K_m and K_i

Equations XI-44 to XI-49 can be used to determine the thermodynamic functions for any reaction that can be described by an equilibrium constant. Thus for rapid equilibrium systems where $K_m = K_S$, a plot of $-\log K_S$ versus $1/T$ gives ΔH^0 for the reaction $ES \rightleftharpoons E + S$. The ΔG^0 for the reaction can be calculated from equation XI-44. With ΔG^0 and ΔH^0 known, ΔS^0 can be calculated from equation XI-45. When K_m is not equivalent to K_S, the calculated thermodynamic values are only apparent ones. The "ΔH^0," for example, will actually be a complex function of the several rate constants comprising K_m and the ΔH^0 values of the corresponding steps. An additional complication is introduced if the substrate ionizes and only one ionic form is active. Temperature will affect the K_a value of the substrate as well as the K_a values of the groups of the active site. If the active substrate species is known and the ΔH_{ion} of the substrate is known, then the concentration of the active species can be calculated.

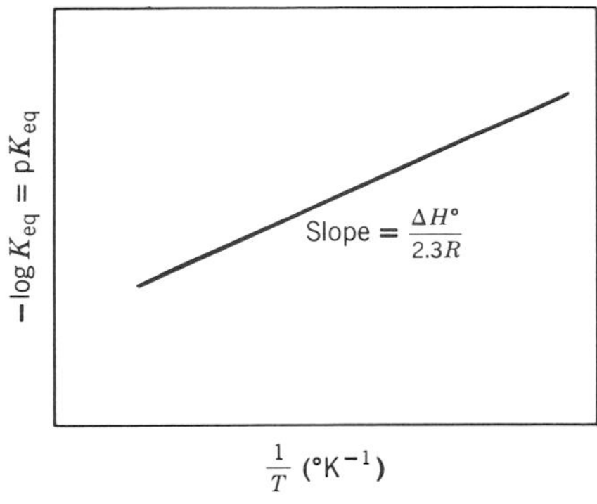

Fig. XI-17. $\Delta H°$ for a reaction can be determined by measuring the K_{eq} at several different temperatures and plotting pK_{eq} versus $1/T$.

The Collision Theory and the Arrhenius Equation—Energy of Activation

The velocity of any homogeneous chemical reaction depends on the frequency of collisions between reactant molecules. The collision frequency is influenced by the concentrations of reactant molecules and how fast they move about (i.e., their kinetic energy). Thus we can write for the reaction between E and S:

$$\text{collision frequency} = \text{constant } [\text{E}][\text{S}]$$

The collision frequency does not equal the reaction velocity because only a small proportion of the collisions occur with sufficient energy to promote the reaction. This minimum energy required for a fruitful reaction is called the *energy of activation*, E_a. The Maxwell-Boltzman law indicates that the fraction of molecules in a system which possesses at least an amount of energy E_a at a given temperature, T, is approximately proportional to the term $e^{-E_a/RT}$. Thus the frequency of *fruitful collisions* at a given temperature is:

$$\text{fruitful collision frequency} = v = \text{constant } [\text{E}][\text{S}]e^{-E_a/RT} \qquad \text{(XI-50)}$$

The velocity is also given by:

$$v = k[\text{E}][\text{S}]$$

Thus the rate constant, k, equals some constant times $e^{-E_a/RT}$, or more specifically:

$$\boxed{k = PZe^{-E_a/RT}} \qquad \text{(XI-51)}$$

where P = the probability of the reaction occurring when the reactant molecules collide; P depends on a number of factors including the orientation of the colliding molecules

Z = the collision frequency: the total number of collisions per unit time per unit concentration (for a bimolecular reaction)

$e^{-E_a/RT}$ = the fraction of the total reactant molecules with an energy of E_a or greater

k = the specific reaction rate constant (e.g., k_1 for the reaction $E+S \rightarrow ES$, or k_p for the reaction $ES \rightarrow E+P$)

The Z is a kinetic energy factor and thus varies with the absolute temperature. However, a small change in T has a relatively small effect on Z. For example, a ΔT of 10° will increase Z by 10/273 or 3.7%. In contrast, a 10° increase in T will double the $e^{-E_a/RT}$ term for an E_a of 12,000 cal/mole.

The relationship expressed in equation XI-51 was formulated empirically by Arrhenius in 1889, before the collision theory was expounded. The relationship is usually written as the Arrhenius equation:

$$\boxed{k = Ae^{-E_a/RT}} \qquad \text{(XI-52)}$$

Or, in linear form:

$$\boxed{\log k = -\frac{E_a}{2.3R}\frac{1}{T} + \log A} \qquad \text{(XI-53)}$$

The integrated form of the Arrhenius equation is:

$$\boxed{\log \frac{k_2}{k_1} = \frac{E_a}{2.3R}\left(\frac{T_2 - T_1}{T_2 T_1}\right)} \qquad \text{(XI-54)}$$

where k_2 and k_1 are the specific reaction rate constants at T_2 and T_1, respectively. The equations have the same form as the Van't Hoff equations except now we are dealing with rate constants instead of equilibrium constants. The E_a is determined experimentally by plotting $\log k$ versus $1/T$ (Fig. XI-18, Curve A). Since the slope of the plot depends on the ratio of k_2/k_1, the absolute values of the rate constants are not needed. Apparent values, or any parameter that is proportional to the rate constants, can be plotted. The effect of temperature on enzyme catalyzed reactions is

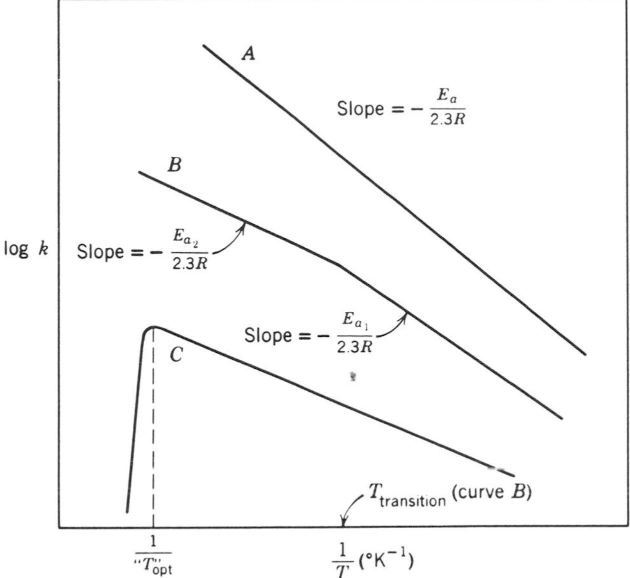

Fig. XI-18. The activation energy, E_a, for a reaction can be determined by measuring the reaction rate constant at different temperatures and plotting log k versus $1/T$. For enzyme-catalyzed reactions, log $V_{max}/[E]_t$ or just log V_{max} can be plotted. (A) The usual plot. (B) Sometimes the plot will show a change in slope if at some temperature a different step becomes rate-limiting. (C) A sudden drop in the plot indicates enzyme inactivation.

frequently expressed in terms of a temperature coefficient, Q_{10}, which is the factor by which the rate constant is increased by raising the temperature 10°C. From equation XI-54:

$$\boxed{E_a = \frac{2.3RT_1T_2\log Q_{10}}{10}} \qquad \text{(XI-55)}$$

The T^2_{aver} can be used for T_1T_2. A Q_{10} of 2 is equivalent to an E_a of about 12,000 cal/mole in the region of 10° to 40°C.

In a simple rapid equilibrium system, $V_{max}/[E]_t = k_p$, a first-order rate constant. Thus a plot of log $V_{max}/[E]_t$ versus $1/T$ yields E_a for the catalytic step. In practice, just log V_{max} can be plotted, since the V_{max} of a given preparation is proportional to k_p. Keep in mind that log V_{max} versus $1/T$ must be plotted on a linear scale. The slope cannot be read off directly from a plot of V_{max} versus $1/T$ on a semilog scale. Because the K_m varies with T, it cannot be assumed that a given concentration of substrate will be saturating at all temperatures. Ideally, V_{max} should be determined from a reciprocal plot at each temperature. For most enzyme-catalyzed reactions, V_{max} depends on several rate constants, each of which may be affected differently by changing temperature. As a result, the E_a calculated from the Arrhenius plot will be an apparent or "average" value. The Arrhenius plot itself may be nonlinear if different steps become rate-limiting at different temperatures. In some cases, the plot may show a sharp change in slope at some temperature ("transition temperature") where the dependency of V_{max} changes from one rate-limiting step to another (Fig. XI-18, Curve *B*). Fox and co-workers (1972) have observed sharp changes in the slopes of Arrhenius plots for β-glycoside transport by *E. coli*. The transition temperature could be correlated with the fatty acid composition of the cell membranes and, presumably, the transition results from a sharp change in the "fluidity" of the membrane. A sudden drop in the Arrhenius plot at low $1/T$ (high T) indicates protein denaturation (Fig. XI-18, Curve *C*).

Eyring Transition State Theory—Absolute Reaction Rates

To explain the need for a minimum energy for a fruitful reaction, Eyring, in 1935, proposed the transition state theory. This theory states that reactant molecules must overcome an energy barrier and pass through an activated complex before proceeding on to the product of the reaction. Reactant molecules that attain only a fraction of the required activation energy simply fall back to the ground state. The activated state is viewed as an unstable, "halfway," transient phase in which bonds and orientations are distorted. Once the reactants attain the transition state, they are committed to form the product of the reaction at a rate independent of temperature and the nature of the reactants; that is, they slide down the other side of the energy barrier to the next ground state. The "energy diagram" for the reaction of E+S to yield ES is shown in Figure XI-19. The $(E\cdot\cdot S)^{\ddagger}$ represents the transient activated complex. All molecules of $(E\cdot\cdot S)^{\ddagger}$ derived from E+S are committed to go on to ES. The same activated complex is involved in the reverse reaction of ES dissociating to E+S. All molecules of ES that attain sufficient energy to become $(E\cdot\cdot S)^{\ddagger}$ are committed to dissociate to E+S. There seems to be a contradiction here: *all* molecules of $(E\cdot\cdot S)^{\ddagger}$ cannot simultaneously yield ES and also E+S. The contradiction can be resolved if we distinguish between $(E\cdot\cdot S)^{\ddagger}$ formed from E+S and $(E\cdot\cdot S)^{\ddagger}$ formed from ES. This is shown in Figure XI-20. To form $(E\cdot\cdot S)^{\ddagger}$ from E+S, a molecule of E and a molecule of S must *approach* each other with sufficient energy. When the two molecules come close enough to form

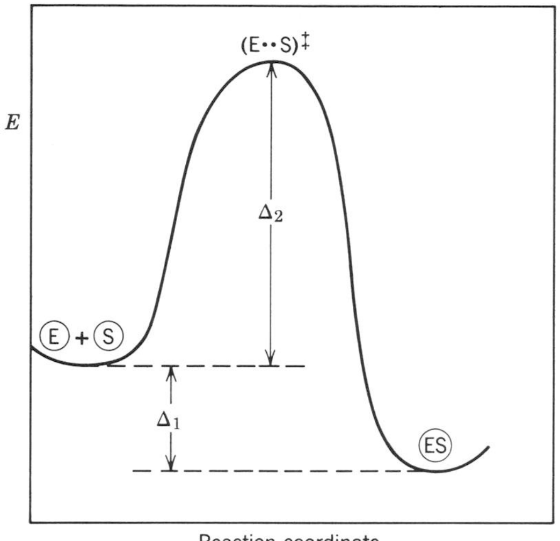

Fig. XI-19. Energy profile for the reaction of E+S to yield ES. $(E \cdot \cdot S)^{\ddagger}$ represents the activated transition state. If ΔH is plotted, $\Delta_1 = \Delta H^{\circ}$ for the reaction while $\Delta_2 = \Delta H^{\ddagger}$.

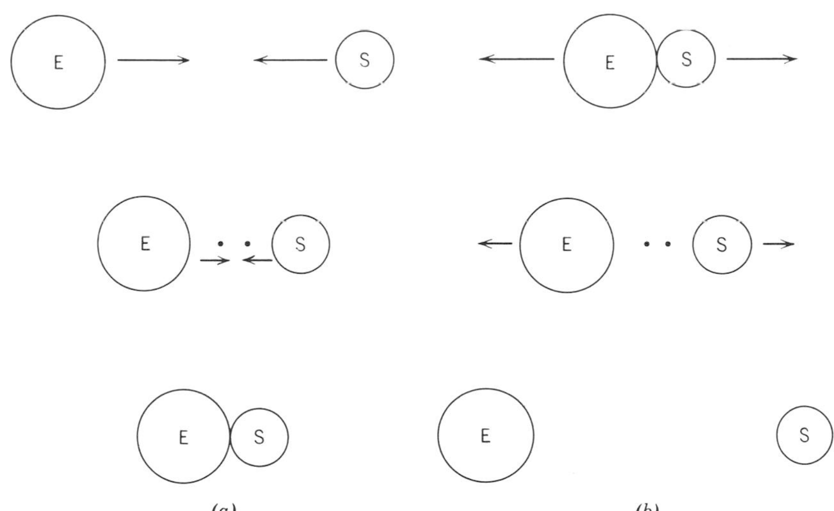

Fig. XI-20. Formation of the activated transition state from E+S and from ES. (*a*) All molecules of $(E \cdot \cdot S)^{\ddagger}$ derived from E+S must go on to form ES. (*b*) All molecules of $(E \cdot \cdot S)^{\ddagger}$ derived from ES must go on to dissociate to E+S.

$(E\cdot\cdot S)^{\ddagger}$ their "momentum" carries them on to ES (Fig. XI-20*a*). Similarly, to form $(E\cdot\cdot S)^{\ddagger}$ from ES, E and S must have sufficient energy to pull apart from each other. When sufficient energy is available to reach $(E\cdot\cdot S)^{\ddagger}$, the "momentum" of the molecules causes them to continue to pull apart yielding E+S (Fig. XI-20*b*). A similar energy diagram can be drawn for the catalytic step in which ES becomes EP. Only the molecules of ES which attain sufficient energy to attain the transition state $ES^{\ddagger}\cdot\cdot EP^{\ddagger}$ go on to become EP. $ES^{\ddagger}\cdot\cdot EP^{\ddagger}$ represents the "halfway state" or, more appropriately, the point of no return in which the bonds of ES have been distorted to such an extent that EP must form. $ES^{\ddagger}\cdot\cdot EP^{\ddagger}$ formed from EP must go on to yield ES.

The energy difference Δ_1 shown in Figure XI-19 can be described by K_S, ΔG^0, ΔH^0 and ΔS^0 for the overall reaction $E+S \rightleftharpoons ES$. If we assume that an equilibrium exists between reactant molecules climbing to the top of the energy barrier and those falling back to the initial ground state, then the energy difference Δ_2 can be described by an analogous set of thermodynamic quantities, $K^{\ddagger}$, $\Delta G^{\ddagger}$, $\Delta H^{\ddagger}$, $\Delta S^{\ddagger}$, as well as by an E_a and k_1. The $K^{\ddagger}$ is a pseudo equilibrium constant defined as:

$$K^{\ddagger} = \frac{[E\cdot\cdot S]^{\ddagger}}{[E][S]} \qquad \therefore \quad [E\cdot\cdot S]^{\ddagger} = K^{\ddagger}[E][S]$$

$[E\cdot\cdot S]^{\ddagger}$ is actually in a steady-state. Thus $K^{\ddagger}$ is a dynamic equilibrium constant analogous to a K_m rather than to a K_S. The rate of ES formation will be equal to the concentration of the activated complex multiplied by ν, the rate at which the complex falls down the energy barrier to the new ground state. ν is also the vibrational frequency of the E–S bond since all molecules of $E\cdot\cdot S^{\ddagger}$ formed from E+S go on to yield ES.

$$\text{rate} = \nu[E\cdot\cdot S]^{\ddagger} = \nu K^{\ddagger}[E][S] \qquad \text{(XI-56)}$$

At the transition state, the energy of bond vibration equals the potential energy of the bond; E_{vib} and E_{pot} are given by:

$$\boxed{E_{\text{vib}} = h\nu} \qquad \text{(XI-57)}$$

$$\boxed{E_{\text{pot}} = k_B T} \qquad \text{(XI-58)}$$

where h is Planck's constant (6.624×10^{-27} erg sec), and k_B is the Boltzmann constant (1.38×10^{-16} erg deg^{-1}, the gas constant per molecule). When $E_{vib}=E_{pot}$:

$$\nu=\frac{k_B T}{h}$$

$$\therefore \quad \boxed{\text{rate}=\frac{k_B T}{h}K^{\ddagger}[\text{E}][\text{S}]} \qquad \text{(XI-59)}$$

The rate of ES formation is also given by:

$$\text{rate}=k_1[\text{E}][\text{S}] \qquad \text{(XI-60)}$$

$$\therefore \quad \boxed{k_1=\frac{k_B T}{h}K^{\ddagger}} \quad \text{or} \quad \boxed{k_1=\frac{k_B T}{h}e^{-\Delta G^{\ddagger}/RT}} \qquad \text{(XI-61)}$$

or

$$\boxed{\Delta G^{\ddagger}=-2.3RT\log\frac{k_1 h}{k_B T}} \qquad \text{(XI-62)}$$

The $\Delta G^{\ddagger}$ can also be defined as:

$$\boxed{\Delta G^{\ddagger}=\Delta H^{\ddagger}-T\Delta S^{\ddagger}} \qquad \text{(XI-63)}$$

Substituting for $\Delta G^{\ddagger}$ from equation XI-62 we can obtain a linear equation:

$$\boxed{\log\frac{k_1}{T}=-\frac{\Delta H^{\ddagger}}{2.3R}\frac{1}{T}+\log\frac{k_B}{h}+\frac{\Delta S^{\ddagger}}{2.3R}} \qquad \text{(XI-64)}$$

Thus if k_1 is measured at different temperatures, a plot of $\log k_1/T$ versus $1/T$ will allow $\Delta H^{\ddagger}$ to be determined from the slope (Fig. XI-21). As noted earlier a plot of $\log k_1$ versus $1/T$ gives E_a, not $\Delta H^{\ddagger}$. The two quantities are related:

$$\boxed{E_a=\Delta H^{\ddagger}+RT} \qquad \text{(XI-65)}$$

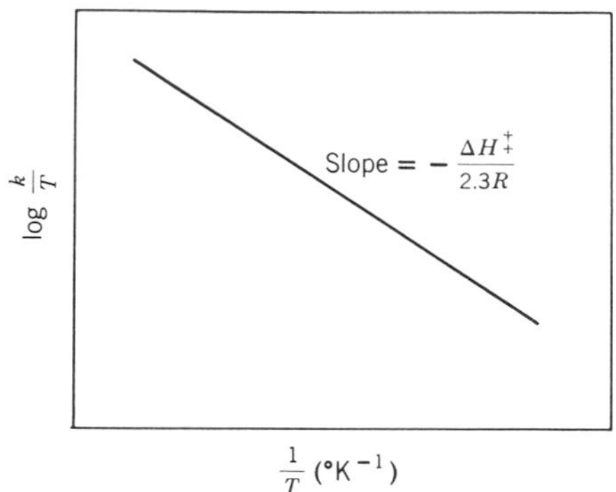

Fig. XI-21. The $\Delta H^\ddagger$ can be obtained by measuring the rate constant for the reaction $E+S \xrightarrow{k_1} ES$ at several different temperatures and plotting log k_1/T versus $1/T$. A plot of log k versus $1/T$ gives E_a. For a nonhomogeneous preparation, the apparent first-order rate constant, $k=k_1[E]$, can be plotted. To determine k_1 or k, a rapid reaction technique must be employed.

With $\Delta H^\ddagger$ determined from a plot of equation XI-64 and $\Delta G^\ddagger$ determined from equation XI-61, $\Delta S^\ddagger$ can be calculated from equation XI-63. The latter two calculations require the absolute value of k_1 while E_a and $\Delta H^\ddagger$ can be obtained by measuring any parameter that is proportional to k. Thus for a nonhomogeneous preparation, the apparent first-order rate constant, $k=k_1[E]$, can be plotted. Equation XI-64 can be used for any process that can be described by a rate constant. Thus a plot of log V_{max}/T versus $1/T$ will determine $\Delta H^\ddagger$ for the step(s) affecting V_{max}. Figure XI-22a shows an energy diagram for the overall reaction: $E+S \rightleftharpoons (E\cdot\cdot S)^\ddagger \rightleftharpoons ES \rightleftharpoons (ES\cdot\cdot EP)^\ddagger \rightleftharpoons EP \rightleftharpoons (E\cdot\cdot P)^\ddagger \rightleftharpoons E+P$. Figure XI-22$b$ shows the same energy diagram with the Δ values for the reverse reaction. (Read Fig. XI-22b from right to left). Any thermodynamic value (ΔH, ΔG, ΔS) can be plotted as the ordinate. The steps are described below.

Δ_1: $E+S \underset{}{\overset{K_{eq}}{\rightleftharpoons}} E+P$ (or $S \overset{K_{eq}}{\rightleftharpoons} P$). The Δ_1 represents the overall energy difference between the unactivated product and unactivated substrate. This energy difference is constant regardless of the reaction mechanism or whether the reaction is enzyme-catalyzed or not. The overall reaction is characterized by K_{eq}, ΔH^0, ΔG^0, and ΔS^0.

Δ_2: $E+S \underset{K_S}{\overset{K_{eq}}{\rightleftharpoons}} ES$. The Δ_2 represents the difference in energy levels between the ES complex and free E + free S. The equilibrium constant, K_{eq}, for the reaction as written $=1/K_S$. In rapid equilibrium systems $K_S=K_m$; K_S

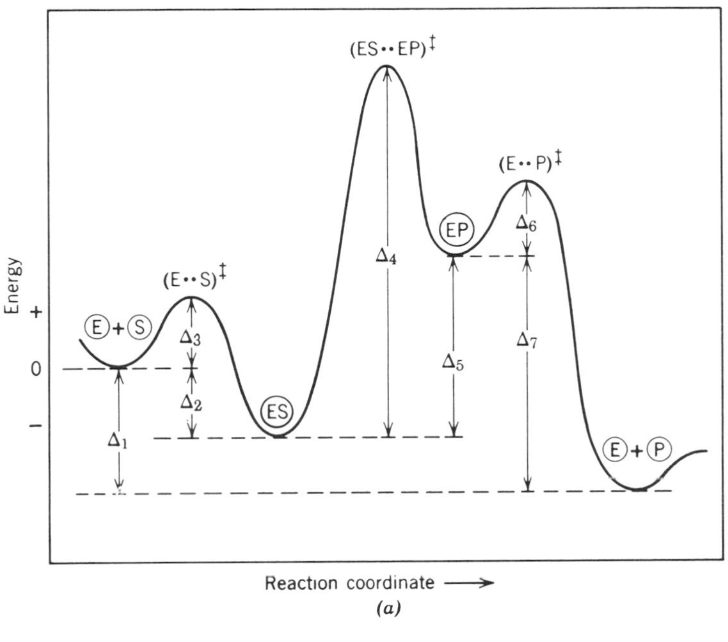

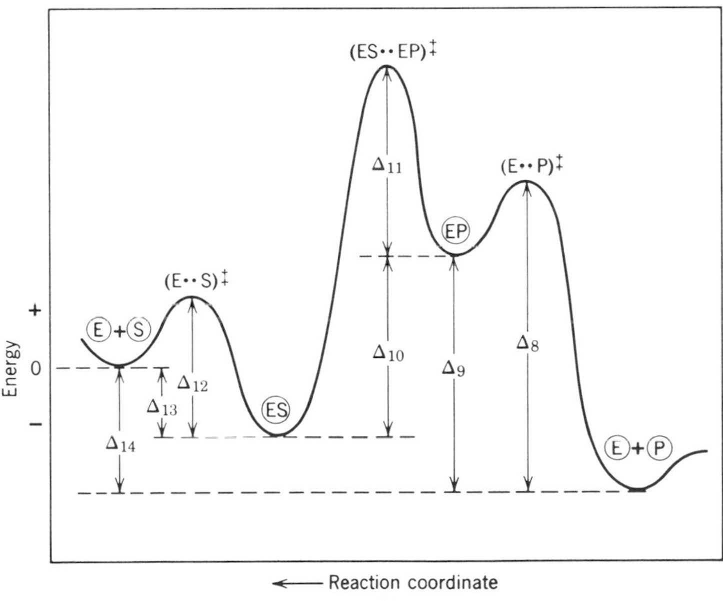

Fig. XI-22. (*a*) Energy profile for the overall reaction E+S ⇌ ES ⇌ EP ⇌ E+P. (*b*) Energy profile for the overall reaction E+P ⇌ EP ⇌ ES ⇌ E+S.

is usually in the range of 10^{-6} to 10^{-2} M so that ΔG^0 is in the range of -2700 to -8200 cal/mole.

Δ_3: $E+S \xrightarrow{k_1} (E\cdot\cdot S)^{\ddagger}$. The Δ_3 represents the energy required to form the activated $(E\cdot\cdot S)^{\ddagger}$ complex. The thermodynamic values associated with this step (E_a, $\Delta G^{\ddagger}$, $\Delta H^{\ddagger}$, and $\Delta S^{\ddagger}$) are obtained by determining the effect of temperature on k_1. The rate of ES formation is extremely rapid; k_1 values range from about 10^6 to 10^9 $M^{-1}\text{sec}^{-1}$. Consequently, to determine k_1, stopped flow or some other rapid reaction technique must be used.

Δ_4: $ES \xrightarrow{k_p} (ES\cdot\cdot EP)^{\ddagger}$. This step represents the bond breaking and/or bond making step. If this catalytic step is rate-limiting, a plot of log V_{max} versus $1/T$ gives E_a for the step. In the absence of the enzyme the E_a for the reaction would be considerably higher. A plot of log V_{max}/T versus $1/T$ gives $-\Delta H^{\ddagger}/R$ as the slope. From $\Delta H^{\ddagger}$ (which usually is in the range of 6000 to 15,000 cal/mole), the other thermodynamic values of the step can be calculated.

Δ_5: $ES \underset{}{\overset{K_{eq}}{\rightleftharpoons}} EP$. The Δ_5 represents the energy difference between the two central complexes. Values of Δ_5 cannot be measured directly but can be calculated from the other Δ values.

Δ_6: $EP \xrightarrow{k_3} (E\cdot\cdot P)^{\ddagger}$. The Δ_6 represents the energy required to raise EP to a transition state for dissociation. The thermodynamic values for this step are determined by measuring the effect of temperature on the first-order rate constant for the dissociation of EP. The value of the constant is usually about 10^1 to 10^4 sec^{-1}. Rapid reaction techniques are required to make the measurements.

Δ_7: $EP \underset{K_P}{\overset{K_{eq}}{\rightleftharpoons}} E+P$. The Δ_7 represents the energy difference between EP and free E + free P; $K_{eq}=[E][P]/[EP]=K_P$. In a rapid equilibrium unireactant system $K_P=K_{ip}=K_{m_P}$.

An analogous series of Δ values exist for the reverse reaction (Fig. XI-22*b*). There are several obvious relationships between the thermodynamic parameters shown in Figure XI-22*a* and Figure XI-22*b*.

$$\Delta_1=-\Delta_{14} \qquad \Delta_6+\Delta_7=\Delta_8$$

$$\Delta_2+\Delta_3=\Delta_{12} \qquad \Delta_7=-\Delta_9$$

$$\Delta_2=-\Delta_{13} \qquad \Delta_1=\Delta_2+\Delta_5+\Delta_7$$

$$\Delta_4=\Delta_{10}+\Delta_{11} \qquad \Delta_{14}=\Delta_{13}+\Delta_{10}+\Delta_9$$

$$\Delta_5=-\Delta_{10}$$

The numerical values of Δ_1, Δ_2, and Δ_7 will be negative, while the numerical values of Δ_3, Δ_4, Δ_5, and Δ_6 will be positive for the scheme shown in Figure XI-22*a*. In the reverse scheme shown in Figure XI-22*b*, Δ_8, Δ_9, Δ_{11}, Δ_{12}, Δ_{13}, and Δ_{14} will be positive, while Δ_{10} will be negative.

Thermodynamics of Enzyme Inactivation

The effect of temperature on the rate of enzyme denaturation can be described by the usual thermodynamic values. In general, the E_a for enzyme catalysis falls within the range of 5000 to 15,000 cal/mole, while E_a for inactivation is generally much higher (50,000–150,000 cal/mole). At low temperatures, the rate of denaturation is quite slow compared to the catalytic rate. But since $E_{a_{\text{denat}}} > E_{a_{\text{cat}}}$, the rate of denaturation increases faster than the rate of catalysis as the temperature is increased. Enzyme denaturation results from an unfolding of a precise tertiary structure; that is, the polypeptide chains become more random. Consequently, it is not surprising that $\Delta S^{\ddagger}_{\text{denat}}$ values are relatively high and positive (40 to 500 Kcal × mole^{-1} × degree^{-1}). The $\Delta H^{\ddagger}_{\text{denat}}$ values are generally 40,000 to 200,000 cal/mole, while $\Delta G^{\ddagger}_{\text{denat}}$ run 20,000 to 50,000 cal/mole at 25°C. The $E_{a_{\text{denat}}}$ and $\Delta H^{\ddagger}_{\text{denat}}$ can be determined from plots of log k_{denat} *versus* $1/T$ and $\log(k_{\text{denat}}/T)$ *versus* $1/T$, respectively. The k_{denat} values are conveniently obtained from the slopes of log percent remaining activity *versus* time plots (Figure XI-16a). $\Delta G^{\ddagger}_{\text{denat}}$ and $\Delta S^{\ddagger}_{\text{denat}}$ can be calculated without using a homogeneous preparation because k_{denat} is a first-order rate constant (it does not have units of concentration).

A knowledge of the effects of temperature on each step of an enzyme-catalyzed reaction would shed a good deal of light on the molecular mechanisms involved. Unfortunately, a complete study is not yet practical for most multireactant enzymes. To date, the most detailed studies have been on hydrolytic enzymes and other pseudo monoreactant enzymes.

REFERENCES

General References

Dixon, M. and Webb, E. C., *Enzymes*, 2nd ed., Academic Press, 1964, Ch. 4, pp. 116–166.

Plowman, K. M., *Enzyme Kinetics*, McGraw-Hill, 1972, Ch. 6.

Segel, I. H., *Biochemical Calculations*, John Wiley & Sons, 1968, Chs. I, II, III.

Whitaker, J. R., *Principles of Enzymology for the Food Sciences*, Marcel Dekker, 1972, Chs. 10, 11.

Williams, V. R. and Williams, H. B., *Basic Physical Chemistry for the Life Sciences*, Freeman, 1967, Ch. 6.

pH Effects

Alberty, R. A. and Bloomfield, V., *J. Biol. Chem.* **238**, 2804 (1963).

Chao, J. and Graves, D. J., *Biochem. Biophys. Res. Commun.* **40**, 1398 (1970).

Davis, F. W. J. and Lees, H., *Can. J. Microbiol.* **19**, 135 (1973).

Dixon, H. B. F., *Biochem. J.* **131**, 149 (1973).

Dyson, J. E. D., D'Orazio, R. E. and Hanson, W. H., *Arch. Biochem. Biophys.* **154**, 623 (1973).

Ottolenghi, P., *Biochem. J.* **123**, 445 (1971).

pH and Ionic Strength Calculations

Finlayson, J. S., *Basic Biochemical Calculations*, Addison-Wesley, 1969, Ch. 5.

Segel, I. H., *Biochemical Calculations*, 2nd ed., John Wiley & Sons (1976), Ch. 1.

Buffers

Gomori, G., "Preparation of Buffers for Use in Enzyme Studies," in *Methods in Enzymology*, Vol. 1, S. P. Colowick and N. O. Kaplan, eds., Academic Press, 1955, p. 138.

Good, N. E. and Izawa, S., "Hydrogen Ion Buffers," in *Methods in Enzymology*, Vol. 24, A. San Pietro, ed., Academic Press, 1972, p. 53.

Good, N. E., Winget, G. D., Winter, W., Connolly, T. N., Izawa, S. and Singh, R. M. M., *Biochemistry* **5**, 467 (1966).

Lewis, J. C., *Anal. Biochem.* **14**, 495 (1966).

Mallette, M. F., *J. Bacteriol.* **94**, 283 (1967).

Temperature Effects

Low, P. S., Bada, J. L. and Somero, G. N., *Proc. Natl. Acad. Sci. U.S.* **70**, 430 (1973).

Paule, M., *Biochemistry* **10**, 4509 (1971).

Wilson, J. E., *Arch. Biochem. Biophys.* **147**, 471 (1971).

Anomalies in Arrhenius Plots

Fox, C. F., "Membrane Assembly," in *Membrane Molecular Biology*, C. F. Fox and A. Keith, eds., Sinauer, 1972, Ch. 12.

Talsky, G., *Angew. Chem. Internatl.*, (ed. in English) **10**, 548 (1971).

APPENDIX

LEAST SQUARES METHOD

Experimental points seldom fall exactly on a straight line when the data are plotted in reciprocal form. If the errors are random, then the line of best fit can be calculated by the method of least squares. The method is illustrated in Table A-1. The experimental points and the calculated line are shown in Figure A-1.

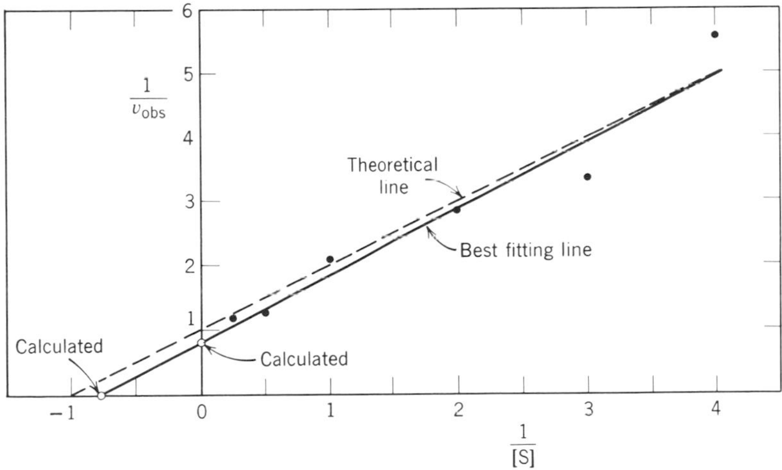

Fig. A-1. Plot of $1/v$ versus $1/[S]$ where the experimental points (dark circles) do not fall exactly on a line. The best fitting line was calculated by the method of least squares. The method gives the best $1/v$-axis intercept and the best slope, from which the best $1/[S]$-axis intercept can be calculated. The line gives $V_{max_{app}}=1.22$ (actual = 1.00) and $K_{m_{app}}=1.29$ (actual = 1.00).

Table A-1 Calculation of the Best Fitting Line by the Method of Least Squares

[S]	v_{true}	v_{obs}	$x=1/[S]$	$y=1/v_{obs}$	x^2	y^2	xy
0.250	0.200	0.18	4	5.56	16	30.91	22.24
0.333	0.250	0.30	3	3.33	9	11.09	10.00
0.50	0.333	0.35	2	2.86	4	8.18	5.72
1.00	0.500	0.48	1	2.08	1	4.33	2.08
2.00	0.667	0.80	0.5	1.25	0.25	1.56	0.63
4.00	0.800	0.82	0.25	1.22	0.0625	1.49	0.31
$n=6$			$\Sigma x=10.75$	$\Sigma y=16.30$	$\Sigma x^2=30.31$	$\Sigma y^2=57.56$	$\Sigma xy=40.98$
			$\bar{x}=1.79$	$\bar{y}=2.72$			
			$\bar{x}^2=3.20$	$\bar{y}^2=7.40$			
			$n\bar{x}^2=19.20$	$n\bar{y}^2=44.40$		$n\bar{x}\bar{y}=29.21$	

Line: $y=mx+b$ where $y=1/v_{obs}$, $x=1/[S]$, $m=slope$, $b=1/v$-axis intercept $=1/V_{max_{app}}$.

$n=$ number of points, $\bar{x}=$ average value of x, $\bar{y}=$ average value of y.

$$\Sigma X^2=\Sigma x^2-n\bar{x}^2=30.31-19.20=11.11$$

$$\Sigma Y^2=\Sigma y^2-n\bar{y}^2=57.56-44.40=13.16$$

$$\Sigma XY=\Sigma xy-n\bar{x}\bar{y}=40.98-29.21=11.77$$

$$Slope=m=\frac{\Sigma XY}{\Sigma X^2}=\frac{11.77}{11.12}=1.06$$

$$1/V_{max_{app}}=b=\bar{y}-m\bar{x}=2.72-1.06(1.79)=0.82$$

$$1/[S]\text{-axis intercept}=-\frac{1}{K_{m_{app}}}=\frac{0.82}{1.06}=-0.773$$

REFERENCES

Lark, P. D. and Craven, B. R., *The Handling of Chemical Data*, Pergamon Press, 1969.

Schefler, W. C., *Statistics for the Biological Sciences*, Addison-Wesley, 1969.

Bishop, O. N., *Statistics for Biology*, Houghton-Mifflin, 1966.

INDEX